HANDBOOK OF WESTERN PALEARCTIC BIRDS
Volume I

HANDBOOK OF WESTERN PALEARCTIC BIRDS

Volume I
Passerines: Larks to *Phylloscopus* Warblers

Hadoram Shirihai and Lars Svensson

HELM
LONDON · OXFORD · NEW YORK · NEW DELHI · SYDNEY

Photographs by
Daniele Occhiato, Markus Varesvuo, Amir Ben Dov, Hanne & Jens Eriksen,
Vincent Legrand, Carlos Gonzalez Bocos, René Pop, Hugh Harrop,
David Monticelli, Mathias Schäf, Aurélien Audevard
and numerous others

Maps by Magnus Ullman

Editorial advice and assistance by
Nik Borrow, José Luis Copete, Guy Kirwan, René Pop and Nigel Redman

Identification consultants
David Bigas, Javier Blasco-Zumeta, Simon S. Christiansen, José Luis Copete,
Aron Edman, Marcel Gil, Alexander Hellquist, Magnus Hellström, Steve Howell and Peter Pyle

HELM
Bloomsbury Publishing Plc
50 Bedford Square, London, WC1B 3DP, UK

BLOOMSBURY, HELM and the Helm logo are trademarks of Bloomsbury Publishing Plc

First published in the United Kingdom in 2018

Copyright © Hadoram Shirihai and Lars Svensson, 2018
Photographs © named photographers (see credits list on p. 639), 2018

Hadoram Shirihai and Lars Svensson have asserted their right under the Copyright, Designs and Patents Act, 1988,
to be identified as Author of this work

Every effort has been made to trace copyright holders of materials featured in this book.
The publisher would be pleased to hear of any omissions or errors so they can be corrected.

All rights reserved. No part of this publication may be reproduced or transmitted in any form or by any means, electronic or mechanical,
including photocopying, recording, or any information storage or retrieval system, without prior permission in writing from the publishers

A catalogue record for this book is available from the British Library

Library of Congress Cataloguing-in-Publication data has been applied for

ISBN: HB: 978-1-4729-3757-5
ePub: 978-1-4729-6057-3   ePDF: 978-1-4729-3758-2
Printed in China by RR Donnelley

2 4 6 8 10 9 7 5 3 1

Publisher: Jim Martin
Design by Julie Dando, Fluke Art

To find out more about our authors and books visit www.bloomsbury.com and sign up for our newsletters

CONTENTS

Acknowledgements ... 7

Introduction ... 8
    Layout and scope of the book ... 8
    Species taxonomy ... 9
    Subspecies ... 10
    Sequence ... 11
    Nomenclature ... 11
    Photographs ... 12
    Maps ... 13
    Species accounts ... 14
    Glossary and abbreviations ... 17
    Selected gazetteer ... 19

An approach to moult and ageing birds in the field ... 20

General references ... 25

List of passerine families: traditional and new order ... 27

A brief presentation of passerine families ... 28

Species accounts
    Larks ... 31
    Swallows and martins ... 103
    Pipits and wagtails ... 132
    Bulbuls ... 198
    Waxwings and Hypcolius ... 202
    Dippers ... 207
    Wrens ... 210
    Accentors ... 213
    Robins and nightingales ... 227
    Redstarts ... 253
    Rock thrushes and chats ... 272
    Wheatears ... 297
    Thrushes ... 362
    Bush and reed warblers ... 405
    Sylvia warblers ... 490
    Phylloscopus warblers ... 564
    Crests ... 611

Vagrants to the region ... 620

Checklist of the birds of the Western Palearctic – passerines ... 634

Photographic credits ... 639

Indexes ... 642

ACKNOWLEDGEMENTS

Anyone writing a detailed regional handbook is bound to be in debt to a very large number of helpful and knowledgeable people, and this book is no exception. It is impossible to produce a work like this without relying on the help and support of numerous persons, museums, institutions and libraries.

When one of us (HS) first came up with the idea of this book it was meant to be a somewhat more popular photographic guide, and early on photographers were asked to send in their best photographs. It started in the film era, but has subsequently moved into the digital world. This means that we have tested the patience of contributing photographers more than usual, and we thank all those who have stayed with the project despite its long gestation prior to publication. Although we have benefited from the cooperation of some 1200 photographers in Europe, Africa and Asia, of which about 750 are represented in the two passerines volumes, this is too many to be listed here (but all are credited in the list of photographers in each volume). Still, we must single out some who have either contributed significantly in numbers or quality, or who have made special efforts to fill gaps by travelling to remote parts of the treated region; hence our special thanks go to Abdulrahman Al-Siran, Rafael Armada, Aurélien Audevard, Amir Ben Dov, Carlos Gonzalez Bocos, Hanne & Jens Eriksen, Hugh Harrop, Lior Kislev, David Monticelli, Jyrki Normaja, Daniele Occhiato, René Pop, Mathias Schäf, Norman Deans van Swelm, Markus Varesvuo and Edwin Winkel. Many other photographers contributed commendably, and we would like to also mention Mike Barth, Arnoud B. van den Berg, Colin Bradshaw, Kris De Rouck, Axel Halley, Vincent Legrand, Alison McArthur, Werner Müller, Arie Ouwerkerk, Mike Pope, George Reszeter, Huw Roberts, Ran Schols, Ulf Ståhle, Gary Thoburn, Alejandro Torés and Matthieu Vaslin.

We have also benefited greatly from the use of a number of excellent internet collections of photographs, and their owners and staff have kindly cooperated with us to facilitate contacts with photographers and by publishing reports of progress and needs of the handbook project over the years. Those internet sites which we used the most are mentioned below under 'Photographs', but we would like to acknowledge here the significant help we received from Morten Bentzon Hansen (Netfugl), Dave Gosney and Dominic Mitchell (Birdguides), Emin Yoğurtcuoğlu (Trakus), Tommy P. Pedersen (UAE Birding), Tim Loseby and other members of the Oriental Bird Club (Oriental Bird Images), Askar Isabekov (Birds, Kazakhstan), Lior Kislev (Tatzpit) and David Bismuth (Ornithomedia). Dolly Bhardwaj, Prasad Ganpule and Abhishek Gulshan not only provided invaluable photographs of Indian birds, but also assisted with contacts with Indian photographers; we are very grateful to all these people.

There are a number of key persons whose contributions are of such importance for the successful completion of the book that they deserve warm thanks: Jim Martin (publisher), Guy Kirwan and Nigel Redman (text editing), Julie Dando (design), Magnus Ullman (maps), René Pop (photographic editor), José Luis Copete, Magnus Hellström and Alexander Hellquist (photographic consultants), and Nik Borrow (photographic assistance). For taxonomic input of various kinds we acknowledge valuable contributions by Per Alström, Nigel Collar, Joel Cracraft, Pierre-André Crochet, Edward Dickinson, Alan Knox, Mary LeCroy, Urban Olsson, Robert Prŷs-Jones and Frank Steinheimer. Alison Harding and Robert Prŷs-Jones helped on numerous occasions with tracing old references. Natalia Delvina and Alexander Hellquist helped with translation of some key Russian texts.

We are indebted to staff in the following museums, who have been cooperative, welcoming and helpful over the many years we have been preparing the manuscript, with often biannual visits to Tring, New York, Paris and Copenhagen. For covering travel expenses for one of many trips by LS to Tring, support received from the Synthesys project (financed by the European Community) is gratefully acknowledged. Three of several trips to New York by LS and one by HS were similarly kindly supported by the Chapman Collection Study Grant. Our particular thanks go to staff in visited museums: Robert Prŷs-Jones, Mark Adams, Hein van Grouw and Alison Harding (NHM, Tring), Joel Cracraft, Paul Sweet, Thomas J. Trombone and Peter Capainolo (AMNH, New York), Per Ericson, Göran Frisk and Ulf Johansson (NRM, Stockholm), Jon Fjeldså and Jan Bolding Kristensen (ZMUC, Copenhagen), Claire & Jean-François Voisin and Marie Portas (MNHN, Paris), Michael Brooke (UMCZ, Cambridge), Sylke Frahnert, Pascal Eckhoff and Jürgen Fiebig (ZMB, Berlin), Renate van den Elzen, Till Töpfer, Kathrin Schidelko and Darius Stiels (ZFMK, Bonn), Anita Gamauf and Hans-Martin Berg (NMW, Vienna), Steven van der Mije and Kees Roselaar (NBC, Leiden), Pavel Tomkovich, Eugeny Koblik and Sergey Grigoriy (ZMMU, Moscow), Andrey Gavrilov (ZIA, Almaty) and Daniel Berkowic and Roi Dor (TAUZM, Tel Aviv). We are also grateful to those museums which we had no opportunity to visit but who kindly arranged with loans of skins or sampling for DNA, and we thank Nicola Baccetti (ISPRA, Bologna), Georges Lenglet (IRSNB, Brussels), Jan T. Lifjeld (ZMUO, Oslo), Jochen Martens (JGUM, Mainz), Tony Parker (WML, Liverpool) and Kristof Zyskowski (YPM, New Haven).

We owe thanks to José Luis Copete, Nigel Cleere, Roy Hargreaves, Beth Holmes, Stephanie Peault, John Black and Claire Voisin for helping us to take some measurements, mainly in NHM, Tring. Guy Kirwan, Bo Petersson and Clive Walton provided other useful measurements.

For valuable help with double-checking identification, ageing and sexing of the selected photographs we are indebted to two independent groups of experts: (1) Magnus Hellström, Simon S. Christiansen and Aron Edman, and (2) José Luis Copete, David Bigas, Marcel Gil and Javier Blasco-Zumeta. Peter Pyle and Steve Howell kindly offered their expert advice on all American species as to captioned age and sex. Gabriel Gargallo helped with advice on moult and ageing and sexing of some species, while Yosef Kiat was consulted selectively regarding birds in Israel, and both are thanked for providing useful help. The input of all these helpful people has greatly improved the end result, but we should point out that we have in a very few instances opted to disregard their recommendations in favour of our convictions based on own research, and any lingering mistakes in how the photographs have been captioned are the full responsibility of the two authors.

Magnus Hellström contributed greatly to the section 'An approach to moult and ageing birds in the field'.

José Luis Copete and Marcel Haas kindly helped to compile a list of accidentals, generally with fewer records than ten, to be included in the passerine volumes.

We thank Norbert Teufelbauer, coordinator of the forthcoming Austrian breeding bird atlas, for helpfully checking several range maps concerning Austria.

Others who helped with various requests for information, or who provided other useful help, include: Mohammed Amezian, Raffael Ayé, Rosario Balestrieri, Mikhail Banik, Peter H. Barthel, Carl-Axel Bauer, Per-Göran Bentz, Anders Blomdahl, Leo Boon, Mattia Brambilla, Vegard Bunes, Ulla Chrisman, Alan Dean, Gerald Driessens, Gadzhibek Dzhamirzoev, Annika Forsten, Johan Fromholtz, Steve Gantlett, the late Martin Garner, Hein van Grouw, Morten Bentzon Hansen, Roy & Moira Hargreaves, Martin Helin, Niklas Holmström, Eugeny Koblik, Markus Lagerqvist, Antero Lindholm, Pekka Nikander, Barbara Oberholzer, David Parkin, the late David Pearson, Pamela Rasmussen, Yaroslav Redkin, Magnus Robb, the Scharsach family, Kathrin Schidelko, Maurizio Sighele, Brian Small, David Stanton, Darius Stiels and Menekse Suphi.

Finally, but most importantly, we thank our loving and supporting families for patience and encouragement throughout the years of preparation. Hadoram is grateful to María San Román, to his late daughter Eden and to all of his family, Lars to Lena Rahoult for endless support.

Hadoram Shirihai & Lars Svensson

INTRODUCTION

Layout and scope of the book

The original idea for this handbook was developed by HS in the late 1990s. Around 2000, LS was invited to be co-author. From the start, the aim for the project was to focus on identification and taxonomy, and to make it the most complete and profusely illustrated photographic guide to Western Palearctic birds. With the development of modern camera equipment, increasing numbers of birdwatchers travelling to distant parts of the region, and 'amateurs' making a significant photographic contribution through new digital camera equipment with image stabilisers and auto-focus, the time was ripe for producing the first comprehensive and complete photographic handbook to the birds of our part of the world.

The text aims to give a modern summary of advice on identification, ageing, sexing and moult, and to provide an up-to-date statement of the geographical variation within the treated range, describing in some detail all distinct subspecies. It also reflects the latest changes in taxonomy at lower levels and offers notes on these, and on unresolved problems. Whereas we have chosen to keep a traditional sequence, we still provide a summary of recent new insights in higher-level taxonomy. Each species account includes brief summaries of vocalisation and distribution, too. Maps produced especially for this book show the current summer and, often, winter ranges of nearly all treated species *and* subspecies.

The region covered is the Western Palearctic (Europe, Asia Minor, the Middle East and North Africa), but in this book we also treat the entire country of Iran (all, or the south-eastern part, are usually excluded) and all of Arabia (of which the southern parts are often omitted). Moreover, also covered are the Cape Verde Islands, the Azores, the Canary Islands, Madeira, Iceland (but not Greenland), Jan Mayen, Svalbard and Franz Josef Land. The limit in the south across the African continent follows the southern national or political borders of Western Sahara, Algeria, Libya and Egypt. We also include the disputed territory between Egypt and Sudan usually referred to as the Hala'ib Triangle (Gebel Elba). The eastern limit runs north from the Arabian Sea, excluding Socotra, along the eastern border of Iran through the Caspian Sea, then north along the Ural River and the 'ridge' of the Ural Mountains to the Kara Sea. A total of 408 species (plus seven extralimital ones afforded brief mention under species which they resemble) and 948 taxa are treated in the main section of the first two volumes covering the passerines (a further two volumes treating the non-passerines will follow). This first volume covers larks to warblers and crests, whereas Volume 2 will treat flycatchers to buntings and icterids.

Any species found within the covered range at least ten times is treated in the main section, whereas an additional 85 passerine species (34 of which are covered in volume 1) found on between one and nine occasions (in a few cases a little more) are given a brief presentation in an appendix. The grand total of passerine species

Map showing the treated region.

treated therefore becomes 500, and the number of taxa 1033. The limit of ten records must not be seen as firm. New records of rare species are made all the time, and it is impossible to be fully up-to-date. Our deadline for inclusion was the end of 2016. For a few of these vagrants the known number of records already exceeds nine, but either the species was a more frequent visitor in the past or it has become more regular only in the last few years.

We have included Iran in the Western Palearctic because we feel that its western parts have more in common with the Western than the Eastern Palearctic bird fauna, and since we only deal with whole nations or political territories rather than selected parts, eastern Iran is included as a bonus. The greater part of Iran belongs to the Palearctic fauna, and only for the south-eastern corner could a case be made for referring that to the Oriental region. The inclusion of southern Arabia seemed a logical change because, with the possible exception of the coastal corner of Yemen and Dhofar, the entire Arabian avifauna is more Palearctic than Afrotropical, and birdwatchers rarely visit only certain parts of these countries—they want to see every bird within the political borders once there.

It should be added that by expanding (or, to a much lesser extent—in southern Sahara—reducing) the conventional definition of the area known as 'the Western Palearctic' from a concept of natural and artificial borders to purely political ones, where they can be followed, we seek not to take issue with adherents of the former concept. Our decision is based purely on a pragmatic and birdwatcher-friendly approach.

During our work it became evident that with so many excellent photographs of nearly all possible plumages gathered in one handbook, photographs which often enable you to study feather wear and patterns in the finest detail, it would be a good time to expand and update the information offered on ageing and sexing. Similarly, we decided to expand the treatment of subspecies. Modern birdwatchers are knowledgeable and no longer interested only in species; they often seek to establish the subspecies, age and sex of birds, and undoubtedly the value of an individual record increases significantly if these further qualifications can be made.

Work was initially divided rather equally between the two authors, each writing first drafts of about half of the species. In the final stages HS did most of the work searching for and selecting photographs, drafting captions and checking the authenticity and data of the images, while LS undertook a final editing and review of the entire text including taking final decisions on geographical variation (which subspecies to be included), and on any contentious taxonomic issues, and also wrote the introductory and end matter. After the two authors were both content with the image selection and the detailed identification of the birds portrayed, all images with drafted captions were subject to a final check by two informal identification panels formed solely for this book (see Acknowledgements). Almost the entire text was also checked by Guy Kirwan.

Species taxonomy

Species concepts have been much discussed in recent years. The species is generally accepted as the basic unit in the natural hierarchy, yet its definition is still, to a certain extent, a matter of argument. As someone once put it, 'a species is what a majority of informed taxonomists agree about'. Just as family, genus and subspecies categorisations are somewhat diffuse and subjective in their definitions, so too is the species.

Today, several concepts exist side by side, the Biological Species Concept (BSC) which lays emphasis on reproductive isolation, the Phylogenetic Species Concept (PSC) which is defined as diagnosable populations that share a common ancestry, and the General Lineage Concept of Species (GLC) or the very similar Evolutionary Species Concept (ESC), which focus on the species as a lineage of directly related populations (and therefore focus more on the time aspect of evolution). There exist further concepts but generally speaking these are variations of the above-mentioned main concepts. In frustration over this unresolved situation leading to varying numbers of recognised bird species to protect, and in particular over the far-reaching and inconsistent influence of genetics in taxonomy (see also below), BirdLife International formed its own species concept based on five weighted criteria and noteworthy for excluding genetic evidence entirely (Tobias *et al.* 2010).

No concept can yet be said to be preferred by a majority of ornithologists, and in reality the main concepts, although theoretically very different, differ less than many believe in practical outcome. But there are differences. We basically favour the BSC,

While attempting to apply a consistent species concept, the authors still at times ran into difficult borderline cases, and not all readers will agree with every decision made. A few examples will illustrate the dilemmas facing us. Arabian Lark (top left: Israel, April) and Dunn's Lark (top right: Western Sahara, January) have been split, whereas reasons for keeping Arabian Accentor (bottom left: Yemen, October) separate from Radde's Accentor (bottom right: Turkey, June) were thought to be insufficient. (Top left, H. Shirihai: top right, C. N. G. Bocos; bottom left U Ståhle; bottom right, S. Tigrel)

noting at the same time that its application has been influenced in recent years by the other concepts. For instance, if hybridisation with other species was previously seen as a barrier to species status under BSC, it can now be acceptable where limited, and if the resulting offspring are either less fit or do not alter the two involved species more than marginally and in a stable way. That allopatric populations, with a wide sea or a high mountain range between them, do not meet means that we cannot test their reproductive capacity. Still, this can generally be overcome by using good sense and observing and comparing them with similarly different populations that do live in contact. The problems of diagnosability in the PSC are often on the same level, so neither of these two concepts can claim to be 'superior'.

We have tried to find a sensible balance between, on the one hand, a conservative approach that favours stability and awaits solid proof from different independent sources before proposed splits or other changes are adopted, and on the other an ambition to mirror the latest developments in taxonomy. While it is good to be up-to-date, nobody wants to have to retreat after adopting changes too hastily. Reversals only cause confusion, and undermine the authority of the changer, so a cautious approach and good compromise are always sought.

When estimating whether a proposed new species split is warranted we have applied the following criteria: (1) it should be morphologically distinct in all individuals of at least one sex or age, and differences should not be minute; (2) it should live in biological segregation; (3) its vocalisations or behaviour ought to differ at least in one aspect; and (4) it is a further reassurance if it differs genetically. The first two requirements are compulsory, whereas the other two criteria are important but supplementary; they would strengthen a case but are not absolutely necessary. We are reluctant to accept new species that fulfil criteria 2–4 but not number 1. Similarly, for sympatric or parapatric populations we prefer to wait if points 1, 3 and 4 are fulfilled but not number 2. 'Minute differences', insufficient for separating two taxa at species level, are those which require advanced mathematical formulae to be proven, such as principal components analysis (PCA) or variance analysis (ANOVA); differences should be obvious without these methods. Subtle average differences are not enough. 'Biological segregation' is to be understood, as explained above, as a state where any hybridisation between two species is marginal (localised and involving relatively fewer pairs than would be expected if no barriers existed) and not known to increase in extent.

The Basalt Wheatear *Oenanthe warriae* is the most recently described new passerine species in the treated region. Until the 1980s, this isolated and neglected 'black wheatear' of the Basalt Desert between Jordan and Syria was mistakenly identified as Strickland's Wheatear *O. opistholeuca*, and later as a 'black morph' of Mourning Wheatear. Due to many threats in its very restricted range it now occurs in very small numbers in Syria, with unclear prospects of survival. (A. Ben Dov)

We considered at one stage to show for a few borderline cases, like the Siberian Chiffchaff *tristis* or the Eastern Black-eared Wheatear *melanoleuca*, the species epithet (the second word in a species' scientific name) within rounded brackets, thus e.g. '*Phylloscopus* (*collybita*) *tristis*', to signal that it is an incipient species on its (long!) way to becoming an undisputed separate species. Eventually, however, we decided not to do so. Any such practice, including the more formal one to recognise so-called allospecies (or even 'phylospecies') and show the species epithet of these within square brackets, involves subjective choices and a certain muddling of species taxonomy and nomenclature, at least to the layman. Consequently, we elected to make choices in each case and instead comment on the difficulties and the subjective nature of some of these decisions at the end of each relevant species account. The important thing is to describe all distinct taxa. The best taxonomic treatment is still, to a certain degree, a matter of taste and preference.

To conclude, we have tried to follow certain criteria and in general be conservative and cautious rather than bold in our choices. Someone may note that for the genus *Oenanthe* we have accepted or proposed more splits than previous authors, but this is just the result of consistently trying to apply the above-mentioned species criteria.

Subspecies

We have made an effort to independently assess all described subspecies that appear in modern ornithological literature within the range covered by the book. This has forced us to spend much time in museum collections, the only place where series of closely related but differently named subspecies can be compared side by side. Between the two of us we have spent many, many months in museums in order to write this handbook. One important object of this laborious work has been to decide whether a subspecies is sufficiently distinct to be recognised and treated as separate from other subspecies of the same species. We have, by and large, tried to apply the so-called 75% rule (e.g. Amadon 1949), meaning that at least three-quarters of a sample of individuals selected at random (in some cases of one sex or age) of a subspecies must differ diagnosably from other described subspecies within the examined species. The specimens selected for assessment should as far as possible be from similar age categories and seasons to allow for the most comparable plumages to be fairly assessed. The main limiting factor for this part of our work has been the scarcity of comparable material of some taxa in collections.

We have long felt the need to focus on *distinct* subspecies and to dismiss the many subtle or even questionable subspecies that were perhaps based on too short series or on skewed or inadequate material, or which in reality represented only minor tendencies within a population, far from the minimum 75%. Natural variation within any one population has often been underestimated when new subspecies have been described. In our opinion, it is better to concentrate on clear differences, and to account for slight clinal tendencies or deviations from the most typical under a single subspecies heading, rather than to create a confusing mosaic of subtle or dubious but formally named subspecies. Most of the final assessment of subspecies taxonomy was done by LS.

Compared to other handbooks and checklists, we accept about 15% fewer subspecies. Those deemed to be too similar to a neighbouring race to be upheld have been lumped with that race and treated as a synonym.

Much geographical variation within continentally distributed species is smooth and gradual, this variation usually being termed clinal. There are taxonomists who tend to see this as a problem and who prefer to treat the entire such range as one subspecies. It is true that clines create circumscription difficulties. However, we prefer to judge every case by its own merits; if the ends of a cline are very different, a minimum of two named subspecies seems reasonable, one at each end and with a diffuse centre of intergrading characters. If there are hints of stepwise change along the cline, these suggest subspecific borders and lead to recognising additional subspecies along the cline.

We want to stress that the definition of a subspecies is based on morphology. If size, structure, colours or patterns differ with sufficient consistency (in this work, as stated, by at least 75% of the examined specimens of at least one sex or age) within a geographically defined part of the range of a species, then it is a valid and distinct subspecies *irrespective* of whether an examination of its mitochondrial DNA or another genetic marker reveals no clear genetic difference from neighbouring subspecies. Geneticists are sometimes so impressed by their new tool that they forget these prerequisites and distinctions. If we are to re-define subspecies by their genetic properties, all existing subspecies need to be discarded and all work on geographical variation redone in laboratories. But in such a case, scientists need first to agree on exactly which genetic definition all variation is to be judged against. Since this is unlikely ever to happen, we had better stick to the morphology-based subspecies taxonomy we already have!

It is best to add that our approach to subspecies taxonomy is quantitative rather than qualitative. If a described taxon does not in our view reach the required level of distinctness, we place it in synonymy under a senior name. But that does not mean that we claim to know better than others regarding the recognition of a valid taxon, only that it is not in our view distinct enough according to our requirements. If others prefer to describe every tiny variation under a formal name it is up to them.

Subspecies deemed by us to be borderline cases as to whether they are warranted or not, and subspecies only differing very subtly from other valid subspecies, have been marked with an open circle in front of their names. For subspecies that are either for some reason questionable, or for which we failed to find sufficient or indeed

& Christidis 2014) and ornithological societies (e.g. British Ornithologists' Union 2009) have already applied dramatically different family sequences compared to the one worked out by Voous. Although differing in details, these new orders share several main features and are no doubt taxonomically more correct, but they will probably be further refined in years to come, clearly demonstrated by the fact that the passerine sequence suggested by Joel Cracraft for the fourth edition of *The Howard and Moore Complete Checklist of the Birds of the World* (Dickinson & Christidis 2014) differs in several aspects from what was thought by some to be definitive only about ten years ago. Considering this prevailing relative turbulence, and the advantage of using a familiar order which makes it easier for the reader to find a family or genus, we have decided to employ a sequence that is still mainly that of Voous, and which is similar to the one applied in the widely used *Collins Bird Guide* (Svensson *et al.* 2009), which treats nearly the same region. The family sequence used here can be compared to the new order in Dickinson & Christidis (2014) in a table on p. 27. With this background, it was also decided not to give family introductions in the main part of the book, since some of the long-established families have either been split or lumped, and some unexpected close relationships have now been revealed and *vice versa*. For an insight into modern passerine family taxonomy we refer the reader to Cracraft *in* Dickinson & Christidis (2014), and to the brief overview on pp. 28–29.

Nomenclature

Scientific names

The scientific names in this handbook mostly follow those adopted by Svensson *et al.* (2009) and Dickinson & Christidis (2014), with a few deviations due to different taxonomic choices made by the authors. In the end, the precise nomenclature chosen is that preferred by the authors.

A scientific species name consists of two words in Latin or in Latinised form, the first written with an initial capital letter signifying the genus (a group of closely related species), and the second being the species epithet ('species name' in everyday conversation, although a species name formally is both these words). To stand out better, scientific names below family level are recommended to be written in a different font, usually by using *italics*. A typical example would thus be *Parus major* for Great Tit. The word *Parus* ('tit') is the generic name and the species epithet *major* ('large') makes it clear that this is the Great Tit.

In the late 19th century the subspecies concept evolved and was subsequently commonly adopted, with eventually a third word being added to the scientific name to indicate the exact geographic origin of a bird based on its morphology. The Great Tits breeding in Britain have thus been named *Parus major newtoni*, based mainly on their subtly darker back and yellowish-white rather than pure white wing-bars and tertial-edges compared to the populations living in continental Europe, which are called *Parus major major*. For the first described subspecies within a species, *major* in this case, the subspecies name is created by simply doubling the species epithet, which happens only when more than one subspecies of the same species is described. Such first-described subspecies are often referred to as 'nominate' (or 'nominotypical'), and are sometimes mistaken as being 'more important'; however, the term only indicates that these taxa happen to be first described and have (by chance) the oldest names within the respective species. (To avoid such misunderstandings we have refrained from using the term.) If there is no conceivable geographical variation within a species it is said to be monotypic (being true for *c.* 20% of passerine species) and its scientific name is limited to two words.

We have included the name of the original author and the year of first publication of each approved scientific name, often useful to make a scientific name unambiguous, and which becomes important if any changes of scientific names need to be carried out. The family name of each author is written out with one traditional exception: the abbreviation 'L.' stands for the Swedish naturalist Carl von Linné, father of the binominal name system. (In the English-speaking world he is generally known by his birth name, Carl Linnaeus; however, he was ennobled in 1757 and changed his name thereafter[1].) Scientific nomenclature officially starts on 1 January 1758, the

Museum collections are invaluable for the taxonomist, since examining a collection of bird skins (as bird study specimens are known by scientists) brought together from various parts of a species' range is the only way geographical variation can be properly studied. Here, as an example, four Asian and Arabian subspecies of Desert Lark *Ammomanes deserti* are compared, though many more have been described and are valid. These are just a few of the skins held in the Natural History Museum, Tring, England; the total is more than 700,000. (L. Svensson/Natural History Museum)

any material, a question mark has been placed in front of their names. Subspecies known or suspected to be now extinct carry a dagger mark in front of the name.

At the end of each subspecies entry a selection of synonyms has been listed alphabetically within brackets; these are subspecies names that appear in the literature or on some specimen labels in museums but which are either junior synonyms or are invalid for one reason or another. They are shown as a service to the reader and to make it clear that they have not simply been overlooked by us. We have as a rule only included the more recent and commonly seen synonyms, but a few older or more rarely used ones have also been included if considered helpful to the reader. Anyone seeking a more complete list is referred to Sharpe (1874–99), Hartert (1903–38), Peters *et al.* (1931–87) or Vaurie (1959, 1965).

Sequence

Within the passerine order, the sequence of families follows Voous (1977). The sequence within families and genera also generally follows Voous except where more recent insights have enabled revision. The Voous list was largely based on the so-called Peters Checklist (1931–87) and was the basis for influential handbooks like *The Birds of the Western Palearctic* ('*BWP*', Cramp *et al.* 1977–94) and is also largely used in *Handbook of the Birds of the World* ('*HBW*', del Hoyo *et al.* 1992–2012). Ornithology experienced a long period of relative taxonomic stability as long as it was based mainly on morphology and behaviour.

With the arrival of genetics as a new tool in systematics, starting with publications by Sibley & Ahlquist (1990) and Sibley & Monroe (1990), then followed by a stream of important contributions in the following two decades and more, it has become clear that the 'Voous order' does not correctly reflect the true relationship and evolutionary history of passerine families (or of non-passerines). A few regional faunas (e.g. Grimmett *et al.* 1998), world checklists (e.g. Dickinson 2003, Dickinson

---

[1] Since Linné's name still appeared as 'Linnæi' (genitive form of Linnaeus) on the title page of the important tenth edition of the *Systema Naturae* (1758), it is customary to refer only to this form of his name in nomenclature. However, several species names were first published in the twelfth edition (1766), and in this his name was written as 'a Linné' (Latinised form of von Linné). Formally, the Code issued by the International Commission on Zoological Nomenclature permits his later name to be used for such species. By abbreviating his name to 'L.' such potential confusion is avoided.

year in which the tenth edition of Linné's *Systema Naturae* was published. No earlier published names are permitted. If the author's name and year of publication appear within brackets it signifies that the species in question was originally described in a different genus. For instance it is *Alauda arvensis* L., 1758, for the Skylark, but *Lullula arborea* (L., 1758) for the Woodlark, since Linné originally described the latter as '*Alauda arborea*', thus in another genus than is the custom today.

Abbreviated first names, initials, are given only for authors like Brehm, Gmelin and Lichtenstein, where there existed more than one person with the same surname who each named several taxa. For people like Ernst Hartert or Otto Kleinschmidt, any namesake created only one or two names in all, so only the family name is given for the more famous person of the two.

Scientific names are governed by the rules laid out in the *International Code of Zoological Nomenclature* (ICZN, 4th edn. 1999, published by the International Commission on Zoological Nomenclature and available on the internet). Several amendments were published separately in 2012. The rules are logical and generally clear but require a wider knowledge of taxonomy, clear thinking and much practice to be properly understood. And even then you will often need advice from more experienced taxonomists or for particularly difficult questions a ruling by a commissioner linked to the ICZN.

Of the many rules in the Code only one needs to be commented on here, illustrating why the year of publication of a taxon name is sometimes important. The oldest available valid name has priority and should be used. This means that if an author decides to merge ('lump') two very similar subspecies (because they are not sufficiently distinct), as has happened in many places in this book, knowledge of which of the two merged subspecies was described and published first is essential, since the new 'larger' taxon must carry the older name.

English names

Choosing English bird names is a contentious subject, and although we are probably slowly approaching more international—or at least national or regional—agreement we are some way from achieving that, and it still seems impossible to please everybody. We have had the ambition to follow the attempts by the International Ornithological Congress (IOC) to create an international standard for English bird names (Gill & Wright 2006, with subsequent updates, also available online as Gill & Donsker 2016). In the end, we did not follow it 100%, but came very close. In a few instances we felt that selected modifiers in that work did not take into account recent taxonomic changes, and for some name choices we were sensitive to known criticism within Britain that several traditional and well-liked names (Rose-coloured Starling, Lapland Bunting) would be lost if the IOC standard was followed without exception. We also feel that for a small group of familiar birds, well known also to the layman, the modifier if needed should be 'Common' rather than 'Eurasian' or 'European' (e.g. Skylark, House Martin, Magpie, Raven, Starling and Bullfinch). Thus, in a few cases we followed our own conviction, always the privilege of independent authors.

A special problem in English names relates to whether to hyphenate compound words, write them as two or write them as one word. As one of us (LS) has been involved editorially in the fourth edition of *The Howard & Moore Complete Checklist of the Birds of the World*, it felt natural to adopt similar criteria for group names as those used in that checklist. In brief we avoid trying to create group names by the use of hyphens outside genera, and even within genera we avoid change to long-established names unless instructive gains are very obvious. Scientific names serve as indicators of relationship, therefore English ones do not need to attempt to do the same. If they do it is fine, but one should not introduce numerous new names in an effort to mirror the latest phylogenies published. It is better to have clear, unambiguous, brief, understandable and locally well-known names as far as possible. Since English indexes are arranged after the main (often last) word in a compound name, it is important to have this word unhyphenated. It is far more likely that you will find a sought-after lark under Lark than under Sparrow-Lark.

On the other hand we note that birdwatchers, patiently accepting and adopting the many changes of taxonomy and scientific names being imposed upon them at an ever-increasing rate, are surprisingly conservative and resistant to change when it comes to vernacular or English names. A little more flexibility and open-mindedness would not hurt if internationally accepted English names are eventually to be achieved.

Photographs
The search for images

One of the main features and strengths of this book is the comprehensive coverage of all important plumages with photographs taken in the field. It has been a delight to see the extremely high quality of many of the submitted photographs, and also to note the wealth of photographs now available for rarely photographed or even rarely seen species. The large 'team' of contributing photographers is to be warmly congratulated and thanked for its fine achievement.

The aim has been to include images of both sexes if they differ, young and older ages in as many stages as can clearly be distinguished, and often also the seasonal plumage stages, which usually means in spring and autumn. On top of this is geographical variation; we have strived to include at least one photograph, and often several, of each distinctive subspecies, or at least show examples of the variation if it is more subtle. There are on average more than ten photographs per species, with a maximum of 49 for Yellow Wagtail.

Despite being rather widespread and a well-known species, the Calandra Lark has rarely been photographed in fresh autumn plumage. In order to close such gaps, the project's photographers often targeted specific missing plumages. Israel, December. (H. Shirihai)

To achieve all this, early on the publisher enrolled René Pop to act as photographic editor for the project. Later his task was taken over by HS, who has been responsible for the major part of the picture research. Further, the publisher set up a homepage especially for the project, devised and managed by then-project editor Jim Martin, where photographers could conveniently find out what was still missing, and to which site they could transfer their images for submission. In the later years of production the authors wanted to ensure that the remaining gaps were filled, and more photographs were brought into the project to raise the level of quality overall. They therefore enrolled three additional photographic researchers, José Luis Copete, biologist and freelance photographic editor for *Handbook of the Birds of the World*, Magnus Hellström, warden at Ottenby Bird Observatory, Sweden, and former chairman of the Swedish Rarities Committee, and Alexander Hellquist, current member of the Swedish Rarities Committee; all three also helped with important targeted searches of websites and journals to track down missing plumages. As mentioned above, all selected images and drafted captions were then carefully checked by two independent panels of experts (see Acknowledgements). In specific cases, additional experts were sometimes consulted to ensure the highest possible accuracy.

The following are some of the important internet collections of bird photographs which we have made much use of and of which the owners and staff kindly cooperated with us when searching for missing plumages, thus providing an important source for achieving complete photographic coverage: www.artportalen.se; www.birdguides.com; www.birdphoto.fi; www.birds.kz; www.birdsofsaudiarabia.com; www.birds-online.ru; www.kuwaitbirds.org; www.nature-shetland.co.uk; www.netfugl.dk; www.oiseaux.net; www.orientalbirdimages.org; www.ornithomedia.com; www.ouessant-digiscoping.fr; www.rbcu.ru; www.tarsiger.com; www.tatzpit.com; www.trakus.org; www.uaebirding.com.

As mentioned above, HS put in years of work checking out many tens of thousands of photographs, corresponding with photographers in many countries and searching websites to find crucial plumages and to check the identifications of many already submitted images, as well as drafting all of the captions. Without the combination of all these dedicated efforts this handbook would not have become as complete as it now is.

INTRODUCTION

Particular effort was made to illustrate the rarest species and subspecies, such as the south-western Arabian population of African Pipit *Anthus cinnamomeus*, the poorly known and hardly ever photographed race *eximus* (Saudi Arabia, June), and the white-headed Central Asian race *leucocephala* of Yellow-headed Wagtail *Motacilla flava* (Mongolia, June). (Left: J. Babbington; right: H. Shirihai)

For the passerines, three photographers have between them contributed almost 25% of the total number of photographs, Daniele Occhiato, Markus Varesvuo and Hadoram Shirihai. Significant contributions have also been made by Hanne & Jens Eriksen and Amir Ben Dov. See also the List of photographers (p. 639).

Captions

Rather than keeping the captions as brief as possible and 'letting the images speak for themselves' we have opted for longer captions with selected identification advice and key points to note. From experience we know that many readers look at the images and read the captions first, long before they start penetrating the main text. By making the captions slightly longer and more interesting, we hope to bridge the difference between information conveyed by picture and word, and to activate and facilitate the way a photograph can be 'read'.

Much effort has been made to check not only identifications and ageing, but also to ensure that dates and localities for the photographs are correct. In a modern photographic handbook of birds it is no longer adequate to give only the species and the photographer's name for each picture. With a greater interest in taxonomy and subspecies, it is also necessary to give the location and time of year to help the reader independently evaluate whether a documented bird is correctly assigned. Therefore as a minimum we provide, whenever possible, country or smaller region and the month or season in the caption to every photograph.

It is still possible that the odd mistake has slipped through, and we will be happy to receive notification of any such claim via the publishers for possible future correction.

Maps

All of the maps have been commissioned specially for this book. We wanted clear and fairly large maps with up-to-date information about breeding range, winter range (if within the region covered) and main migration routes. We also wanted to show, for the first time for the majority of species, the ranges of the subspecies. The border between two fairly distinct subspecies has been indicated by an unbroken line, whereas a border between two very similar subspecies, or two that are clinally connected with intermediates over a wide hybrid zone, has been indicated by a dashed line. The borders are, for very natural reasons, approximate and do not claim to be precise, but they should give a fair overview of the geographical variation of a species.

The maps have been compiled by Magnus Ullman, well travelled in the entire region as a tour leader and an experienced bird cartographer. He has had access to practically every known important published checklist or bird atlas when preparing the maps. Boundaries between subspecies have been largely worked out by LS based both on the literature and on museum collections.

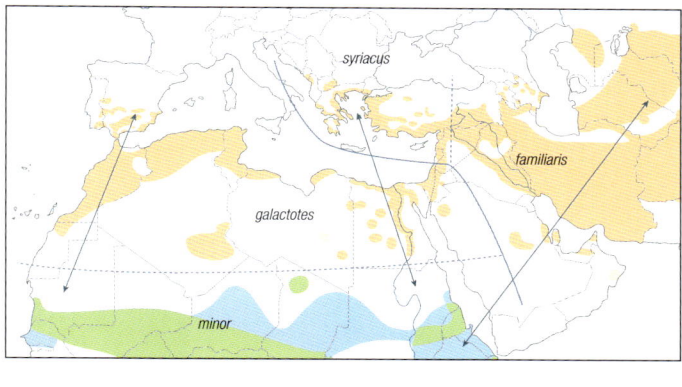

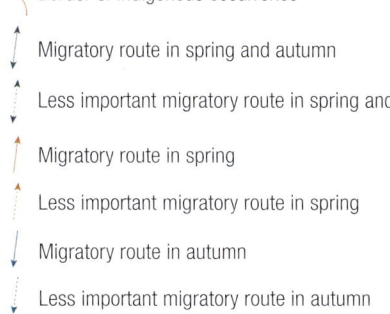

Example of range map, here showing that of Rufous-tailed Scrub Robin *Cercotrichas galactotes*. Colours and symbols are explained in the legend.

Species accounts

Names

Each species account is headed by a section giving the various names of the species. The principles for the selection of English and scientific names when more than one alternative exists has already been mentioned under 'Nomenclature'. A few alternative English or American names are given when these seemed called for. A short selection of foreign language names ends the heading.

Short introductions

Each species account opens with a brief outline of the species, a quick sketch to give a first presentation of the 'personality' of the bird, of what to expect from a first encounter and where to see it. These summaries are not meant to be complete or 'consistently' written. Therefore, do not expect to invariably find data on characteristic habits, favoured habitats or abundance here, to mention a few examples, although such information is frequently included. The larger type size (on tinted background) emphasises the preamble nature of these paragraphs.

Identification

Features helpful for identification of each species in the field are given under 'Identification'. Field characters thought to be more important than others *appear in italics*. Ageing and sexing are only briefly treated under this heading (but more thoroughly under a separate heading). If there is no significant geographical variation, the text is relevant for the entire area where the bird can be seen, but for species with more obvious geographical variation the text under 'Identification' generally relates to the most widely distributed subspecies in Europe. Whenever in doubt consult the information given under 'Geographical variation and range' further down in the species account.

The characters listed as important for identification 'in the field' are much about size, structure, colours and plumage details, and undeniably these often help in identifying a bird. But how a bird moves and behaves, and the 'personality' it conveys by this to the trained eye (often referred to as 'jizz') is almost equally important, only more difficult to describe in words and requiring more talent and training to use. Wherever such 'jizz' characters are obvious and useful, they are also included in the text, generally at the end.

Vocalisations

Of the various vocalisations of a bird, song is described first, followed by characteristic calls. By 'song' we mean not only what we conventionally think of as a pleasing and melodic repeated strophe, but any territorial or mate-seeking emittance of sound including the modest and irregular chuckles of a Fieldfare or feeble and 'pensive' notes by a White Wagtail. While the whole song, preceded and followed by a pause of varying length, can alternatively be called a *strophe*, a part of a song is referred to as a *phrase*, *syllable* or *note*.

Calls include contact calls, flight calls, alarm calls, anxiety calls and begging calls of young, but any calls are described only if they are characteristic enough to serve as aids for identification. Minor calls and minor call variations exist in many species, if not in most, but these variations have generally not been included unless they have some kind of significance. Voice descriptions are based largely on our own experience and on LS's personal recordings, but also on several published recordings (including those made available via free websites), and of unpublished recordings made available by courtesy of the recordists or by the National Sound Archive, British Library, London.

Anyone who has taken the trouble to memorise the songs and calls of birds knows how much easier field identification becomes. Waders, pipits and warblers, often so similar in appearance, suddenly become manageable due to their different calls and songs. True, just as when using 'jizz' as a tool in birding, sounds require a certain amount of gift to be successfully applied. But we believe that people often underestimate their own ability to learn and recognise bird vocalisations. All too often you hear resigned declarations that 'calls are just as variable as they are perceived and rendered by each observer'. This is simply not true. Bird sounds are rather more consistent than they are variable, and you too can probably learn to recognise bird sounds and describe in a meaningful way what you hear to others.

Rendering complex bird voices using letters and words can seem a blunt tool, but it is a start and often far better than what some sceptics tell you. Listen again and again to an unfamiliar call and write down your own rendering, and compare the call to other bird sounds (or other sounds than those given by birds) you are already familiar with. Little by little you will improve your skills and eventually be able to confidently sort out the entire chorus in a forest during a walk in May.

All sound transcriptions appear in *italics*, and **bold face** denotes emphasised (louder, stressed) syllables. An attempt has been made to render the pace at which syllables are uttered, *vivivi* being very quick, *vi-vi-vi* somewhat slower, *vi vi vi* still less rapid, *vi, vi, vi* yet again even slower, and *vi ... vi ... vi* more like individual calls given in a slow, hesitant series. The choice of vowels is meant to indicate roughly the pitch, from low to high *a, u, o, e, ü, ee, i*, thus *a* (as in 'after') is meant to be lower than *u* (as in 'use'), which is subtly lower than *o* (as in 'over'), the others in rising pitch being *e* (like in 'pet'), *ü* (like in German 'über'), *ee* (like in 'steel') and *i* (as in 'ring'). The inclusion of the German 'y' (ü, 'umlaut u') rather than the English letter is because the English y is pronounced in a variety of ways, e.g. like 'ju' (at the beginning of a word), often more like a diphthong ('ai') or, at the end of a word, in the same way as 'i' and therefore not well suited for call renderings. The way the consonants have been selected can help convey whether a call starts or ends with a hard, abrupt sound or if it is softer; *kek* or *tac* are sharper and harder sounds than *gep* or *bipp*.

Similar species

Reading the 'Identification' sections is a useful start when trying to put a name to an unfamiliar bird. The sections called 'Similar species' are then useful as a final check that you have arrived at the correct conclusion. Read through them with an open mind. Bird identification is not always simple, and only the knowledgeable realise that it is all too easy to fall victim to a delusion. Better to double-check the crucial characters once too often than not often enough, and compare them critically with what you have seen.

Ageing & sexing

This book is aimed mainly at ordinary birdwatchers who observe free-flying birds in nature and try to identify them with the help of eyes and ears, albeit usually aided by the use of binoculars, telescopes and perhaps cameras with powerful lenses. But it should prove useful also to ringers and museum workers with the bird in the hand. The sections on ageing and sexing are detailed enough to solve most problems facing a ringer or museum researcher. We have as a rule only left out certain specialised characters such as colour of gape, presence of tongue spots, brood patch, cloacal protuberance and growth bars on remiges or tail-feathers, all characters better reserved for specialist ringers' guides. Still, with optical equipment and cameras becoming ever better, and birders being increasingly knowledgeable (and demanding), we have strived to make the sections on ageing and sexing as complete as possible for both ordinary birdwatching and 'advanced birding' in the field.

Each section first provides a brief overview of what can be identified in the field as regards ageing and sexing. Then follows a summary of the moult strategy of the species, or, if moult differs geographically, the indicated subspecies. Often, knowledge of moult is essential for correct ageing, and sometimes also sexing. If you are not yet fully familiar with what moult is and how it is performed by birds we recommend that you consult specialist literature, since here is not the proper place to deal with it. A brief treatment is offered by Svensson (1992), a somewhat more detailed one in Ginn & Melville (1983), and quite comprehensive accounts are found for example in Stresemann & Stresemann (1966), Jenni & Winkler (1994) and Howell (2010). However, some comments on how knowledge of moult and feather wear can be useful for interested birdwatchers might be appropriate, and therefore the basics are covered in a separate chapter, 'An approach to moult and ageing birds in the field', which follows directly after the Introduction (p. 20).

After moult has been outlined in each species account, there follows a guide to ageing and sexing usually arranged in the two broad seasons of 'Spring' and 'Autumn' (though sometimes 'Summer' or other seasons or combinations of seasons are used). Spring here corresponds to plumages encountered in early spring to late summer (or mid autumn depending on timing of moult), whereas autumn refers to plumages worn in late summer or mid autumn to late winter or early spring. This may sound a bit loose, but since breeding, migration and moult all vary in details depending on species or population, the diffuse definitions are deliberate and unavoidable.

Juvenile plumage is treated separately, provided it differs from later plumages, and entries on this can appear either under Spring (rarely) or Autumn (often), if not even under a separate heading 'Summer', which is sometimes used. We have attempted to find practical and useful solutions that correspond to the plumage development of a certain species rather than adhere very strictly to a fixed formula of headings.

Biometrics

One of the reasons for the long gestation time of this book was the decision, wherever possible, to examine museum material (or in some cases live birds) rather than citing existing literature. We went back to the birds themselves as often as we could in our efforts to provide an independent view. Still, for some of the species or subspecies we rely on the thorough and careful work done mainly by C. S. Roselaar for *BWP*, and a few other fully referenced sources. Finally, for some species we commissioned the competent help of Nigel Cleere, José Luis Copete and Roy Hargreaves to take measurements, and a few others who helped take some measurements are credited in the Acknowledgements. The work is based on an extensive examination of museum specimens (mainly in Tring, New York, Stockholm, Copenhagen and Paris, but also to some extent in Berlin, Bonn, Leiden, Vienna, Tel Aviv, Almaty, St Petersburg and Moscow), and targeted fieldwork (in, for example, Armenia, the Caucasus, Israel, Kazakhstan, Siberia, Mallorca, the Pyrenees, Morocco, Tunisia and Turkey). Nearly every currently listed subspecies in major existing handbooks and checklists has been examined, measured (mainly by LS) and compared with related subspecies or populations in order to assess their distinctness and validity. In total, according to a rough estimate, *c.* 700,000 measurements of more than 40,000 specimens have been taken for the passerines alone.

Why so many measurements? Well, this book is meant to serve not only field ornithologists but also ringers and museum workers, and measurements and details of wing formulae are often indispensable tools when separating closely related and very similar taxa. Detailed measurements are required for a fair assessment of geographical variation and for establishing whether a described taxon really differs from those already known. And undoubtedly some basic size measurements also enable the successful identification of a bird in the field. Total length has been obtained from rather long series' of skins, calculating an average size span only after discarding all extremes and oddly prepared specimens. Only specimens prepared in such a way that the bill is pointing forward have been used.

The remaining measurements have been taken following the standard set out in Svensson (1992) with one exception: for measuring bill length to feathering we measure the exposed culmen, which means that we measure from the tip of the bill to where the culmen of the bill *disappears* under the forehead feathering (provided this is undamaged), thus not to the base of the forehead feathers. This is because pushing the calipers to these feather bases, frequently damaged on museum specimens and from time to time also on trapped live birds, gives on average more biased and variable values. In a few species with dense, hair-like nasal feathering covering the basal part of the culmen (e.g. redpolls, shrikes and some corvids, to name a few) these feathers have been discounted and the measurement taken to the border between the true forehead feathers and the nasal feathers.

Bill to feathering is the length from tip of bill to where culmen disappears under the true forehead feathers—thus visible culmen length. (L. Svensson)

The two most difficult measurements to obtain accurately, in our experience, are tail length and bill to skull. Added to this is tarsus length on skins if the feet are not well prepared. Tail length is easier to measure on live birds than on skins, because the latter can have glue or stained blood hidden under the tail-coverts, which will catch the end of the ruler before this reaches the true base of the tail-feathers, causing the measurement to be read misleadingly short. And the tail can be fixed in an unnatural way to the body by the taxidermist and cause either high or low values. Only by using a pair of compasses (rather than a thin ruler inserted between the tail and undertail-coverts) when such a skin is encountered can this problem usually be overcome. If tail length is important for identification and you arrive at a surprising value it is a good idea to re-do the measurement once or twice with various tools and, better still, to let someone else measure it independently.

Total length is taken with a live bird in a natural relaxed position and only slightly stretched, bill pointing forward, the tip touching the zero-stop of a ruler. Measure to the tip of the longest tail-feather. When measuring skins it is important to only select specimens with a similar posture. (H. Shirihai)

Tail length is conveniently measured by inserting under the tail a thin ruler with the scale starting from the outer edge of one end and pushing it gently between the long tail-feathers (rectrices) and the longest undertail-coverts until it reaches the root of the central tail-feathers. Measure to the tip of the longest tail-feather. (L. Svensson)

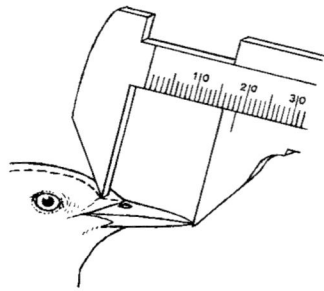

Bill to skull is best measured using sliding calipers of a sufficiently thick material (*c.* 3 mm). The outer caliper (the one pushed against the joint between skull bone and culmen of bill) should have a pointed profile. Push this caliper along the culmen until it reaches the start of the skull bone, then close the calipers until the inner one touches the tip of the bill. (L. Svensson)

The measurement of bill to skull shares the same basic limitation experienced with tail length: you cannot actually see the point to which you measure. There is sometimes an unevenness either on the foremost part of the skull bone or on the innermost part of the bill, and this creates a problem when locating (by feel only!) the groove or angle where the bill meets the skull. Always double-check any measurement troubled with such problems. In cases where we do not provide bill length to skull, only bill to feathering, this is because for these species in our experience the angle between bill and skull is too difficult to locate with accuracy for bill length to skull to be reliably applied.

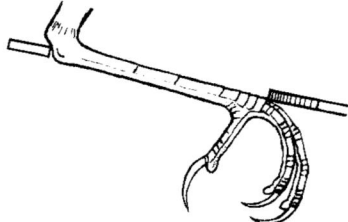

Technique for measuring the tarsus length on a skin. (L. Svensson)

Remember that the tarsus is measured from the groove or depression on the rear side of the ankle joint ('reversed knee') between the tarsus and the tibia, thus one should not include the lower knob of the tibia. However, it is more difficult to decide to where one should measure at the lower end of the tarsus. The rule-of-thumb to 'measure to the lower edge of the last complete scale on the front side before the toes diverge' is sometimes difficult to follow since you will encounter all kinds of intermediates between complete and divided scales at the base of the toes. The object should always be to obtain measurements that correspond to those given in the literature and ideally that are as similar as possible regardless of whether a skin or a live bird is measured. On a live bird it is natural to fold back the toes completely and measure to the lower end of the 'tarsus bone'. The toes of a dried skin are stiff so you need to figure out where the same point would be without being able to move the toes. We recommend that all complete scales are included and preferably one extra scale (towards the toes) rather than counting one too few if the scales are ambiguous.

The selection of measurements in this book varies somewhat between species reflecting their perceived usefulness for identification purposes. For all species wing and tail lengths are given with mean values (*m*) and sample sizes (*n*) in brackets. Measurements are divided between the two sexes if mean values differ in excess of 1–2%, or if sufficient specimens were available for examination. An average ratio between tail and wing lengths expressed as a percentage is also included for most taxa since this is often useful when assessing structure and for taxonomic considerations. Measurements of bill (sometimes several) and tarsus follow for nearly all species.

The section on biometrics is usually concluded with a description of the wing formula, i.e. the relative length of some key flight-feathers of the wing, usually expressed as the distance of the tip of each feather in relation to the wingtip (wt). One exception is the outermost short primary, which is measured in relation to the tip of the longest primary-covert (for a detailed explanation of the study and description of the wing formula, see for example Svensson 1992). Wing formula details have been limited to the basics thought to be particularly useful for identification.

Note that we number primaries from outside and inwards. Just as we do not walk backwards if we can avoid it we don't intend to describe the wing formula of a bird starting with p10, then p9 and so on backwards to p1. In moult studies the primaries are conveniently counted from inside and outwards as this is the same order in which they are replaced in 99.9% of the passerine species, but what works well for describing moult (just one specialised branch of ornithology) is not practical for dealing with the wing formula, where the interesting feathers are generally the outer ones. Just as it is possible for most people to cope with miles and kilometres at the same time, or euros and pounds, counting the primaries on a bird in two ways depending on circumstances presents no real obstacle.

We have included the distance between the two outermost primaries ('p1 < p2') for nearly all species except for those with a vestigial p1 (e.g. pipits, finches and buntings). The usefulness of this measurement for identification and for making taxonomic assessments was first noticed by HS in the 1980s. One advantage is that both these feathers, p1 and p2, are protected from excessive feather wear due to their sheltered position on the closed wing, allowing quite precise measurements to be taken even on heavily worn summer adults.

On the other hand, we do not include the so-called primary length (p3 length to base) since in our experience it is not unquestionably easier to take in a consistent way than the wing length according to the maximum method. But it is a fact that it ruffles the wing-feathers more, which is potentially harmful to the flight ability of handled live birds. In other words, we have independently arrived at partly similar conclusions as Gosler *et al.* (1995).

We nearly always provide both measurements and wing formula details for *each subspecies*, a novelty in this book. The current trend is to focus not just on species but on distinct subspecies as well—often the new species of tomorrow. Here we offer, possibly for the first time, fairly full coverage of all distinct subspecies of a large region.

Finally, it should be noted that total length is measured to the nearest half-centimetre, and feather-related measurements (wing and tail length, wing formula details) are measured to the nearest half-millimetre, and for these, whole-number values are not given with a decimal (thus we write '15' rather than '15.0'). For measurements of bare parts (bill, tarsus, claws, etc.), these are taken to the nearest decimal, i.e. tenth of a millimetre, and the decimal is invariably given also for even values (e.g. '6.0' rather than just '6'). These differences reflect in our experience a realistic level of accuracy that can be achieved and be repeated by others.

Geographical variation & range

This section opens with a summary of geographical variation within the species. If there is no geographical variation and no subspecies currently recognised, the species is said to be monotypic.

Under species with two or more recognised subspecies, each subspecies receives a separate entry. The entry starts with the scientific name plus author and year, followed by a brief summary of where it breeds and winters. Data on distribution is mainly based on available modern literature, but some information has also been taken from our extensive examination of museum specimens. A definition then follows of how a particular subspecies differs from others of the same species. We nearly invariably present separate biometrics and wing formula details for each subspecies. Where appropriate, comments are offered where our opinion differs from established handbooks and checklists. If our deviations are more complex to explain in the subspecies accounts, we have instead given them under 'Taxonomic notes' (see below). Each subspecies paragraph ends with a list of synonyms, mainly junior names which refer to the same population but which cannot be used under the present taxonomy and according to the rules of nomenclature. Since our research has led to a number of often cited subspecies being found insufficiently distinct and hence placed in synonymy, it is important to consult these lists to avoid the impression that they have been overlooked.

Our coverage of subspecies is complete only for the treated range, the Western Palearctic. For some species of particular interest we have also treated some (not necessarily all) extralimital subspecies to offer a better overview, or because occurrence of some of these within the treated range might be seen as reasonably likely in the future.

Taxonomic notes

Bird taxonomy has entered an era of great activity and exciting new results. This is largely due to the introduction of genetic analyses and new theoretical models. Some of the results and proposed changes appear to be solidly supported and are already followed by most authors, list editors and organisations. Others are more controversial and it seems a sound approach to let them rest a little, to be further discussed and tested independently by others before the proposals lead to changes in taxonomy. For studies relying only on mitochondrial DNA data, it is particularly desirable that the results are corroborated either by nuclear DNA or another 'independent' genetic method, or by morphology, vocalisations, behaviour, etc., before changes are implemented. Under 'Taxonomic notes' we give a few brief hints about ongoing work on the taxonomic scene and what currently is seen by us or others as a controversial question, or what remains to be discovered. We also hint at possible future splits of very distinct subspecies.

References

To save space we have restricted references under the species accounts to those in journals and a very few rarely seen books or sound recordings. We have omitted titles of papers, and only give the names of the journals (often abbreviated; see p. 18), volume or issue number, and page numbers. The inclusion of a reference does not mean that we necessarily agree with all of its conclusions (although this is frequently the case) but they are meant as a service to the reader to conveniently acquire a complete picture when studying a certain species. Handbooks, monographs and sound recordings have been used extensively throughout the preparation of this work and many would be cited under almost every species account if we were to include them there. Instead they are given in a comprehensive list of main references on pp. 25–26. Consulted atlases and checklists are also listed there.

Glossary and abbreviations

We have restricted the use of technical abbreviations in normal text to frequently utilised age categories, the four cardinal points in connection with geographical names (e.g. 'N Britain') and to numbering of certain wing-feathers or tail-feathers (e.g. 'p2' meaning 'second primary from the outside'). Abbreviations are listed below, and at the same time much of the terminology used is explained. In sections dealing with biometrics and wing formulae we have allowed for more abbreviations in order to gain space, since in these sections the same terms are repeated a multitude of times. They should still be easily understood given some practice. Some less common abbreviations are explained where they first appear. Note that we differ from several other books in focusing on the age of birds, not on plumages, since the latter vary considerably and often require specialist knowledge of moult.

Age and sex

ad	adult (mature bird in definitive plumage)
f.gr.	full grown(s) (fledged bird of unknown age)
imm	immature (young bird of unspecified age but still not adult)
juv	juvenile (young fledged bird in its first set of feathers)
1stS	first-summer bird (often = 2ndCY spring, a bird at age of about one year)
1stW	first-winter bird (strictly from start of replacement of juv plumage, often only partly, until when, in late winter or early spring, breeding/summer plumage starts to be attained; if exact plumage state cannot be ascertained, the term is used to refer to birds in their first autumn, winter and early spring)
1stCY	first calendar year (from hatching to 31 Dec of the same year)
1stY	first year of life, roughly speaking from hatching to first-summer
2ndCY	second calendar year (from 1 Jan the year after hatching to 31 Dec of the same year)
2ndS	second-summer bird (often = 3rd CY spring, at the age of about two years)
2ndW	second-winter bird (a bird which has attained winter plumage in its 2ndCY, either through moult or feather wear, often at the age of c. 14–16 months, and until the following spring)
♂, ♂♂	male, males
♀, ♀♀	female, females

Biometrics and wing formula

B	bill length (to skull unless otherwise stated; see opposite page)
B(f)	bill length to feathering (measured to where culmen is covered by real feathers, i.e. 'exposed culmen'; nasal hair or bristles not counted)
BD	bill depth (at feathering unless otherwise stated)
BD(n)	bill depth at inner (proximal) edge of nostril openings; when distal edge is meant this is stated as 'BD(n dist)'
BW	bill width (at feathering unless otherwise stated)
BW(g)	bill width at gape ('corner of mouth', i.e. widest place where bill is hard)
emarg.	emargination of outer web of a primary
HC	hind claw length
L	length (total length from tip of bill to tip of tail, head/bill pointing forward but neck only gently outstretched)
m	mean value (often elsewhere abbreviated as '±')
MC	median (front) claw length
n	number (= sample size)
p, pp	primary, primaries (long outer wing-feathers forming the wingtip; abbreviation only appears combined with feather numbers, e.g. 'p3', 'pp3–5')
p1	first (outermost) primary
p10	tenth (innermost) primary
pc	primary-coverts (in wing formula descriptions referring to 'tip of longest')
r, rr	tail-feather(s) (= rectrix, -ices) (abbreviation only appears combined with feather numbers, e.g. 'r6', 'rr4–6')
s, ss	secondary(-ries) (somewhat shorter inner wing-feathers constituting the 'arm'; abbreviation only appears combined with feather numbers, e.g. 's1', 'ss3–6')
s1	first (outermost) secondary
T	tail length (from base of central tail-feathers to tip of longest feather)
TF	tail fork (distance between longest and shortest tail-feathers)
TG	tail graduation (the same definition as tail fork, only that the tail is rounded)
Ts	tarsus length (explained on opposite page)
W	wing length (distance from wing-bend, i.e. carpal joint, to tip of longest primary, wing-feathers flattened and straightened sideways against the ruler)
WS	wingspan (distance from wingtip to wingtip, wings gently stretched straight out)
wt	wingtip
>	more than, larger than
<	less than, smaller than

General

C	central (in connection with geographical name)
c.	*circa* (Lat.), approximately
cf.	*confer* (Lat.), compare with, see also
cm	centimetre
E	east, eastern (in connection with geographical name)
edn	edition
e.g.	*exempli gratia* (Lat.), for example
et al.	*et alii* (Lat.), and others
etc.	*et cetera* (Lat.), and so on
i.e.	*id est* (Lat.), that is to say
in litt.	*in litteris* (Lat.), in correspondence, written information
in prep.	in preparation, a reference underway but not yet published
Is	Islands
Jan, Feb	January, February, etc.
m	metre
mm	millimetre
Mts	mountains
N	north, northern (in connection with geographical name)

p., pp.	page, pages
pers. comm.	personal communication, verbally
S	south, southern (in connection with geographical name)
sec.	second(s)
s.l.	*sensu lato* (Lat.), in the broad sense
s.s.	*sensu stricto* (Lat.), in the narrow sense
ssp.	subspecies (geographically defined morphological variation within a species; the same thing as race, and these two words are used alternatively without signifying any difference of meaning)
syn.	synonym(s)
var.	*varietatis* (Lat.), variety, variation (usually = colour morph, in connection with scientific name)
vs	*versus* (Lat.), against; as opposed to
W	west, western (in connection with geographical name)
○	open circle in front of taxon name denotes a subtle subspecies that differs only in a very minor way from neighbouring subspecies (symbol appearing only under 'Geographical variation & range')
?	denotes a questionable subspecies, or one for which we have been unable to examine any or adequate material

Museums

AMNH	American Museum of Natural History, New York, USA
ISPRA	Istituto Superiore per la Protezione e la Ricerca Ambientale, Bologna, Italy
IRSNB	Royal Belgian Institute of Natural Sciences, Brussels, Belgium
JGUM	Johannes Gutenberg Universität, Mainz, Germany
MNHN	Muséum National d'Histoire Naturelle, Paris, France
NBC	Naturalis Biodiversity Centre, Leiden, the Netherlands
NHM	Natural History Museum, Tring, UK
NMW	Naturhistorisches Museum, Vienna, Austria
NRM	Naturhistoriska Riksmuseet, Stockholm, Sweden
TAUZM	Tel Aviv University Zoological Museum, Tel Aviv, Israel
UMCZ	University of Cambridge, Dept. of Zoology, Cambridge, UK
USNM	National Museum of Natural History, Smithsonian Institution, Washington, USA
WML	World Museum, Liverpool, UK
YPM	Yale Peabody University Museum, New Haven, USA
ZFMK	Zoologisches Forschungsmuseum Alexander Koenig, Bonn, Germany
ZIA	Zoological Institute, Almaty, Kazakhstan
ZISP	Zoological Institute, St Petersburg, Russia
ZMB	Museum für Naturkunde, Berlin, Germany
ZMMU	Zoological Museum, Moscow University, Russia
ZMUC	Zoologisk Museum, Copenhagen, Denmark
ZMUO	Zoologisk Museum, Oslo, Norway

Handbooks, journals and other references

The following are the most commonly used abbreviations of reference names.

Acta Orn.	Acta Ornithologica
Amer. Mus. Novit.	American Museum Novitates
BB	British Birds
BBOC	Bulletin of the British Ornithologists' Club
Biol. J. Linn. Soc.	Biological Journal of the Linnean Society
Bonn. zool. Beitr.	Bonner zoologische Beiträge
Bonn. zool. Monogr.	Bonner zoologische Monographs
Bull. AMNH	Bulletin of the American Museum of Natural History
Bull. OBC	Bulletin of the Oriental Bird Club (name changed to BirdingASIA in 2004)
Bull. OSME	Bulletin of the Ornithological Society of the Middle East
BW	Birding World
BWP	The Birds of the Western Palearctic
DB	Dutch Birding
HBW	Handbook of the Birds of the World
H&M 4	The Howard & Moore Complete Checklist of the Birds of the World, 4th edn
J. Avian Biol.	Journal of Avian Biology
J. Bombay N. H. S.	Journal of the Bombay Natural History Society
J. Evol. Biol.	Journal of Evolutionary Biology
J. f. Orn.	Journal für Ornithologie
J. of Orn.	Journal of Ornithology
Linn. Soc. Zool. J.	Zoological Journal of the Linnean Society
Mol. Phyl. & Evol.	Molecular Phylogenetics and Evolution
Notatki Orn.	Notatki Ornitologiczne
Novit. Zool.	Novitates Zoologicae
Orn. Beob.	Ornithologische Beobachter
Orn. Fenn.	Ornis Fennica
Orn. Jahrb.	Ornithologische Jahrbuch
Orn. Monatsber.	Ornithologische Monatsberichten
Orn. Vestnik	Ornitologicheskii Vestnik (*Messager ornithologique*)
PLoS ONE	Public Library of Science (open-access journal)
Ring. & Migr.	Ringing and Migration
Riv. ital. Orn.	Rivista italiana di Ornitologia
Ross. Orn. Zhurn.	Rosskii Ornitologicheskii Zhurnal (*Russian Journal of Ornithology*)
Sandgr.	Sandgrouse
Vår Fågelv.	Vår Fågelvärld
Zool. Med. Leiden	Zoologische Mededelingen, Leiden
Zool. Verh. Leiden	Zoologische Verhandelingen, Leiden
Zool. Zhurn.	Zooligicheskii Zhurnal
Zoosyst. Rossica	Zoosystematica Rossica

Selected Gazetteer

Arabia	Sometimes loosely used for the Arabian Peninsula, by and large south of the political borders of Jordan and Iraq.
Balearics	Group of islands in W Mediterranean Sea, including Mallorca, Menorca, Ibiza and adjacent smaller islands.
Balkans	Imprecise region in SE Europe usually comprising Croatia, Bosnia and Herzegovina, Albania, Greece, Macedonia, Montenegro, Bulgaria and E Romania. It takes its name from the mountain range with the same name.
Baltic States	The three countries of Estonia, Latvia and Lithuania.
Britain	See *Great Britain*.
Central Asia	(sometimes referred to as 'Russian Turkestan') includes the countries east of the Caspian Sea but does not include Iran, Pakistan, Tibet or other parts of China, nor Mongolia. The area thus includes Kazakhstan, Uzbekistan, Turkmenistan, Kyrgyzstan, Tajikistan and Afghanistan.
Cyclades	Group of Greek islands in the S Aegean Sea north of Crete.
Eurasia	The vast landmass formed by Europe and Asia.
Fenno-Scandia	Part of N Europe including Sweden, Norway, Finland and westernmost Russia (Murmansk Oblast, Karelia), thus excluding Denmark and Iceland.
Great Britain	England, Wales and Scotland, but not including Northern Ireland, Isle of Man or the Channel Islands. Also known simply as 'Britain'.
Gulf States	The Arab states around the Persian Gulf: Iraq, Kuwait, Bahrain, United Arab Emirates, Qatar, Saudi Arabia and Oman. Generally, only areas near the Gulf are included.
Hala'ib Triangle	Also known as Gebel Elba. Disputed territory on the border between Egypt and Sudan, on the Red Sea coast.
Iberia	The SW European peninsula comprising mainland Spain and Portugal.
Kalmykia	The plains NW of the Caspian Sea.
Kirghiz Steppe	Traditional name for the mainly vast grassy plains in N Kazakhstan, from east of the Volga River eastward to Zaisan Lake (thus not within Kyrgyzstan as the name might imply). Frequently now called the Kazakh Steppe, though not in HWPB.
Kurdistan	The region inhabited mainly by Kurdish people which encompasses parts of E and SE Turkey, N Syria, N Iraq and NW Iran.
Levant	A rather imprecise term here applied to the land on the eastern shores of the Mediterranean Sea. Commonly, Cyprus, a corner of S Turkey, Syria, Lebanon, Israel, the Palestine territory, Jordan and the Sinai Peninsula are included.
Macaronesia	Group of islands in the N Atlantic including Azores, Madeira, Canaries and Cape Verde.
Maghreb	A region of NW Africa traditionally encompassing the Atlas Mts and the coastal plains of Morocco, Algeria, Tunisia and Libya. It now comprises the whole of these four countries plus Western Sahara and Mauritania, but excluding Egypt (cf. North Africa).
Middle East	The same as Levant but with the addition of the whole of Turkey and Egypt, plus Iraq, Iran and the entire Arabian Peninsula. (Sometimes called the 'Near East'.)
Nordic Countries	Comprises NW Europe and includes Iceland, Norway, Denmark, Sweden and Finland.
North Africa	Here defined as the African countries mainly north of the Sahara, thus Morocco, Algeria, Tunisia, Libya and Egypt.
North Sea	Part of the N Atlantic Ocean located between Norway, Denmark, the British Isles, Germany, the Netherlands, Belgium and France.
Palearctic Region	('Arctic region of the Old World') One of eight ecological regions on Earth, covering Europe, N Africa, N Arabia and N Asia north of the Himalayas (but see elsewhere for specific definition of the Western Palearctic region for this handbook).
Sahel	Arid savanna region along the southern border of the Sahara Desert, Africa.
Scandinavia	The three countries of Denmark, Norway and Sweden (but not including Finland).
Siberia	Defined here in the broadest sense, thus reaching from the Ural Mts eastward to the Bering Strait and along the Pacific coast south to Ussuriland and also including Sakhalin, in the south to the borders of Kazakhstan, Mongolia and China. The eastern part of Siberia is sometimes known as the 'Russian Far East', but is here included in 'Siberia'. The Russian Far East is sometimes restricted to the extreme SE part of Siberia, i.e. Amurland and Ussuriland.
Transbaikalia	Area south and south-east of Lake Baikal in SC Siberia, including regions often referred to as Dauria and Chita.
Transcaspia	Area east of the Caspian Sea, mainly W Kazakhstan, W Uzbekistan and Turkmenistan.
Transcaucasia	Area south of the Caucasus, mainly extreme E Turkey, Georgia, Armenia and Azerbaijan.
Turkestan	See 'Central Asia'.
Tyrrhenian Islands	Corsica, Sardinia and smaller adjacent islands along the Italian coast.

AN APPROACH TO MOULT AND AGEING BIRDS IN THE FIELD

Many birds can be aged even in the field, using knowledge of plumage variation. One needs to take into account both the normal plumage colours and patterns, and seasonal variation due to moult, wear and bleaching. Knowing the age of a bird often increases the value of an observation. A brief outline of moult in passerines is therefore offered. Moult, the replacement of an old feather and growth of a new one in its place, is essential for maintaining flight ability, heat and rain insulation, and sometimes serves to attain plumage patterns important for successful social interaction or camouflage. The example below of ageing Mistle Thrush in autumn based on moult, feather shape and feather wear explains well what to look for.

The adult Mistle Thrush (top left and right) has all greater coverts uniformly edged ochre-buff, while the tail-feathers are rather broad with rounded tips not yet showing any wear at the edges. The 1stW (bottom left and right) shows a moult contrast in the greater coverts between inner moulted feathers edged ochre-buff similar to the adult appearance, and the retained juvenile outer coverts being edged clearly more whitish. The tail-feathers of the young bird are slightly narrower, and tips are more pointed and already show some abrasion. Note also that juvenile feathers are of subtly inferior quality, have a more 'matt' surface and less dense barbules. (L. Svensson/ NRM)

Timing

The timing of the moult depends both on migrations and the breeding cycle. As is easily understood, moult involves production of numerous feathers and costs energy, and this is why it is performed either after breeding on the breeding grounds or in the winter quarters, thus not during the similarly energy-intensive migrations. Therefore, there are two main moulting periods—in *late summer* and during *winter*.

Terminology

The summer moult is often conducted close to the breeding grounds and generally completed before the start of the autumn migration (exceptions are few and insignificant). In adult birds that have just finished breeding, this moult is called the *post-nuptial moult* (literally 'after breeding moult'), while in young and newly fledged birds it is called the *post-juvenile moult*. In the post-juvenile moult the young birds replace all or most of their juvenile feathers that were grown in the nest. In a complete moult of remiges, primaries are nearly always replaced descendently (from inside out) and secondaries ascendently (from outside in), feathers being grown more or less symmetrically in both wings.

The winter moult is conducted sometime between the autumn and spring migrations, generally on the wintering grounds. It is called the *pre-nuptial moult* ('before breeding moult') regardless of the age of the bird. This is the moult in which many species acquire their breeding plumage. Most tropical migrants perform such a winter moult, while most short-distance migrants do not.

Keeping the above in mind, we now need to add the *extent* of the moult in order to describe it properly:

Complete moult includes the whole plumage and results in a uniform one-generation plumage.

Partial moult includes a part of the plumage, resulting in an often visible contrast between different feather generations. In some species the partial moult may be restricted to some head and body feathers, while other species moult more extensively and include all contour-feathers of the body, a variable number of wing-coverts, tertials, tail-feathers and sometimes even a few secondaries or primaries, but never the whole plumage.

With the above basic knowledge of passerine moult, it is convenient to combine the timing and extent of the moult in order to simplify the terminology. We follow Svensson (1992) using the following simple terms (upper case serves to signal complete, lower case partial):

Summer complete moult (**SC**) is a complete moult during late summer.

Summer partial moult (**sp**) is a partial moult during late summer.

Winter complete moult (**WC**) is a complete moult during winter.

Winter partial moult (**wp**) is a partial moult during winter.

Further, apart from the general rules outlined above and which hold true for the vast majority of species, it should be noted that a few species show a moult pattern that is more complicated, and in order to facilitate the description of these we use the following three additional terms or categories:

Arrested moult starts like a complete moult, involving the moult of some remiges, but is arrested (usually because of migration) before it is completed, and is not resumed later on. Hence, the rest of the unmoulted remiges will not be moulted until in the next complete moult.

Suspended moult starts like a complete moult, involving the moult of remiges, but is suspended (usually because of migration) before it is completed, but unlike in arrested moult it is subsequently resumed from the point of suspension, after arrival at winter or summer grounds. Thus, in the end all remiges will be renewed.

Partial moult of remiges is an energy-saving strategy to moult only some remiges, often the outer being most important for flight ability, while the rest are left unmoulted. This, for instance, is found in immatures of some species of warblers and shrikes wintering in the tropics.

The assessment of moult

Since adult and young birds often differ in the extent of the moult (complete vs partial), this provides a tool for separating these two categories. One could say that ageing a bird is much about assessing the moult history of the individual. Presence or absence of *moult limits* tells us whether the last performed moult was partial or complete, and combined with knowledge of both the normal plumage and the moult pattern of that very species, the age can frequently be derived with good certainty.

Finding a moult limit may be anything from straightforward to 'close to impossible'. In very many species, different feather generations normally differ from each other in colour, pattern, shape, structure and length. Our chances of finding such limits increase significantly if we know the (average) extent of moult used by the species we are looking at. Moult limits that are situated *within*, for example, the greater coverts, median coverts or tertials are generally much easier to find than those appearing *between* different feathers groups.

Note that in some groups of birds *both* adults and first-summers have a partial winter moult and therefore return in spring with a moult limit. It is thus essential to learn in which groups the method works in spring. Differences between the two age categories may still exist, but they are much more limited and often impossible to be certain of. Similarly, in some groups, juveniles perform a complete post-juvenile moult in the same season when the adults moult completely. These groups of birds accordingly cannot be aged using the moult limit method.

Juvenile feathers are grown rapidly and are inevitably of a slightly less endurable quality. This becomes especially important to know when examining spring birds with some retained juvenile feathers (often primary-coverts and remiges, sometimes also tail-feathers), since wear and bleaching tend to affect juvenile feathers more than feathers of later generations. This is helpful to know when trying to differentiate between worn adult feathers and worn juvenile feathers.

A warning is also called for: quite often '*false moult limits*' can appear, potentially leading to wrong conclusions about the age. Such false moult limits can occur when some inner (or outer) greater coverts naturally have a different colour or pattern from the rest, which therefore may look like a moult limit. Another common reason is when some feathers have been better preserved from wear and bleaching due to their more sheltered position in the wing, rendering them the appearance of being of a more recently moulted feather generation. Starvation or aberrations when feathers are grown might leave odd feathers shorter than their full length, again potentially appearing as a moult limit. Accidentally lost and replaced feathers may simulate a moult limit, so it is best to check both wings, since moult is nearly always symmetrical. The only way to avoid falling into these many traps is practice, and adopting a cautious approach in general.

The different main moult strategies

Below, comprehensive descriptions of the main moult strategies shown by Palearctic passerines are given. However, one should keep in mind that there are numerous exceptions from the rules, and individual variation may occur which is not described.

Strategy 1

Adult: SC – Young: sp (no winter moult in either age)

This moult strategy is found in, for example, some chats and most or all thrushes, tits, corvids and finches. The summer complete moult of adults results in a uniform one-generation plumage. The summer partial moult of young birds involves only a part of the plumage and results in two generations of feathers, with some contrast in wear or pattern between them. Since none of these groups conduct any moult during the winter, the same plumages are present also in spring, but by then are more worn.

In this adult Olive-backed Pipit (Korea, Oct) the impression of a moult contrast in the greater coverts is false and created by naturally differently patterned inner coverts having olive-buff (less pale) tips than the outer, but which still are of the same generation as the rest of the wing. (R. Newlin)

Blackbird, ad (above: Sweden, Oct) and 1stW (below: Sweden, Mar), is a typical example of a species having moult strategy 1. On the young bird note the contrastingly more worn and bleached brown juv outer greater coverts, primary-coverts, remiges and alula, unlike the evenly-feathered fresh and blacker wing of the old bird (with primaries still growing). (Above: M. Varesvuo; below: M. Lofgren)

Plumage development of ♂ **Common Redstart**, with 1stW (above: Kuwait, Sep) and 1stS (below: Finland, Jun). During autumn, the well-known summer plumage is partly concealed by pale tips to the body-feathers (above). These tips will in time wear off and in spring reveal the colourful breeding plumage (below). Notice how the unmoulted juv remiges and coverts become more worn and brown in spring than in autumn. (Above: J. Tenovuo; below: M. Varesvuo)

White Wagtail in autumn, with ad (above: Israel, Oct) and 1stW (below: Italy, Nov), a species moulting according to strategy 2. The adult wing is uniform and fresh after its complete summer moult. In 1stW, after a partial summer moult, a clear moult contrast is evident where the juv remiges, primary-coverts and outer greater coverts are worn and bleached brownish in contrast to the inner greater and median coverts, moulted and now fresh and adult-like. (Above: A. Ben Dov; below: D. Occhiato)

White Wagtail in spring, with comparison of ad and 1stS (both Finland, May). After the pre-nuptial moult in winter, the outer greater coverts of ad (above), retained from summer but due to their better quality still not too worn, are edged whitish-grey, and ageing is further supported by comparatively fresh ad-type primary-coverts. As can be seen, 1stS (below) also has superficially similar moult contrasts in the wing, but the outer greater coverts, remiges and primary-coverts are retained juv, being more worn and bleached browner, with reduced and off-white fringes. (Above: J. Normaja; below: M. Varesvuo)

Strategy 2

Adult: SC, wp – Young: sp, wp

This moult strategy is found in, for example, pipits, wagtails, *Sylvia* warblers and buntings. The summer complete moult of adults results in a uniform one-generation autumn plumage. The summer partial moult of young birds involves only a part of the plumage and results in two generations of feathers, with some contrast in wear or pattern between them. In winter, both ages have a partial moult, which means that all birds will show a moult contrast during spring, regardless of age. When ageing such birds during spring, the following should be noted: (i) In some young birds the winter partial moult is *more extensive* than the summer partial. For those species both ages will show a single moult contrast in the wing, and ageing must be based on an assessment of the feathers that were not moulted during winter (often the outer greater coverts, all primary-coverts and remiges). One needs to decide whether they are of adult type (moderately worn) as in 3rdCY birds or older, or whether they are of juvenile type (more worn and bleached) as in 2ndCY birds. – (ii) In some young birds, the winter partial moult is slightly *less extensive* than the summer partial. In such cases a 2ndCY bird will sometimes show *two* moult contrasts (with three feather generations involved), while a 3rdCY bird or older always will show only one moult contrast (two feather generations present). – (iii) In some species the winter partial moult is restricted to body-feathers alone and does not affect the wing at all.

AN APPROACH TO MOULT AND AGEING BIRDS IN THE FIELD

Three feather generations (juv, post-juv and pre-nuptial) are present in this White Wagtail (Sweden, April), which give two different moult contrasts in the greater coverts. Such pattern proves a bird to be 1stS. There are four juv outer greater coverts retained, the fifth is post-juv (differing only subtly) and the inner five are pre-nuptial (moulted in winter). (Ottenby Bird Observatory)

Strategy 3

Adult: SC – Young: SC

This moult strategy is applied by larks, starlings, some sparrows and a few other odd species. Both adult and young birds have a summer complete moult. Once this moult is completed, the ages are inseparable by plumage.

A fresh and recently moulted ♂ Black Lark in winter (above, Kazakhstan) showing much black in the wings and on the underparts, but extensive pale-fringed areas on the head and mantle compared to a ♂ in spring, showing its unmistakable almost completely black plumage (below, Kazakhstan). The transformation between these two stages is due to wear of differently coloured feather tips, not to moult. (Above: R. Chittenden; below: N. Blake)

Both these spring ♀ Rüppell's Warblers (Israel, March) show fresh greater coverts and tertials acquired in the pre-nuptial moult in winter, but an adult (above) has clearly fresher primaries and primary-coverts, as opposed to the browner and more worn juvenile ones of the 1stS bird (below). (H. Shirihai)

Strategy 4

Adult: sp, WC – Young: sp, WC

This moult strategy involves, for example, some swallows and many warblers wintering in the tropics, particularly members of the genera *Acrocephalus*, *Iduna* and *Hippolais*. Following the partial summer moult, all birds regardless of age category show moult contrasts. However, the ages are usually easily separated during the autumn since the remiges of adult birds are eight to ten months old and by now worn, while the same feathers in the young birds were grown in the nest only one to three months ago and are still fresh. During the winter, all birds have a complete moult after which the ages cannot be separated. It should also be said that quite a few species show prolonged or differentiated periods of moult during winter, which may result in moult contrasts that are of little or no use for ageing during spring.

Sand Martin with suspended moult. Above (Oman, September) note the rather large number of remiges already moulted before reaching Africa. Below a trapped bird in spring (Israel, March) with traces of suspended moult, the two innermost primaries being older than the rest. This individual probably started its moult of primaries in late summer and then suspended it prior to migration. After autumn migration it resumed from the point of suspension, moulting the rest of the primaries doing its pre-nuptial moult, resulting in a pattern of old inner and new outer primaries in spring. (Above: H. & J. Eriksen; below: A. Edman & S. S. Christiansen)

Upcher's Warbler during autumn migration with worn ad (above: Ethiopia, September) and fresh juv/1stW (centre: United Arab Emirates: September), and a fresh spring bird (below; Bahrain, May) after complete pre-nuptial moult, when ageing is no longer possible. (Above: H. Shirihai; centre: H. Roberts; below: A. Drummond-Hill)

Strategy 5

Arrested and suspended moults, partial moult of remiges

A general description of arrested and suspended moult can be found above under Terminology. It may be worth noting that both arrested and suspended moults are conducted in the same sequence as a regular complete moult, only being interrupted before completed (arrested moult) or introducing a lengthy pause halfway before completion (suspended moult), whereas the partial moult of remiges often starts about halfway out in the wing leaving the inner primaries unmoulted, and if any secondaries are moulted usually some inner ones are kept. Single species with either of these strategies are found in several diverse genera, good examples being Woodchat Shrike, northern populations of Isabelline Shrike, warblers such as several of the genera *Locustella* and *Sylvia*, and Great Reed Warbler. Barred Warbler and Ortolan Bunting are noteworthy for, as a rule, moulting their primaries in summer and secondaries in winter.

This 1stS ♂ Rüppell's Warbler (Turkey, May) has returned with several inner retained juvenile primaries being brown and some darker moulted outer ones, thus an example of partial moult of remiges. Most of the greater coverts were replaced sometime in early autumn, while the tertials were moulted more recently in a pre-nuptial winter moult. (D. Occhiato)

GENERAL REFERENCES

The books, checklists, atlases and sound recordings below have been consulted throughout the preparation of this handbook and have been an indispensable source of knowledge and reference. Only references directly referable to the content of this volume have been included.

Ali, S. & Ripley, S. D. (1987) *Compact Handbook of the Birds of India and Pakistan*. 2nd edn. Oxford University Press, Oxford.

Alström, P., Colston, P. & Lewington, I. (1991) *A Field Guide to the Rare Birds of Britain and Europe*. HarperCollins, London.

Alström, P. & Mild, K. (2003) *Pipits & Wagtails of Europe, Asia and North America*. Christopher Helm, London.

Alström, P. & Mild, K. (in prep.) *Larks of Europe, Asia and North America*. Christopher Helm, London.

Amadon, D. (1949) The seventy-five per cent rule for subspecies. *Condor*, 51: 250–258.

Andrews, I. J. (1995) *The Birds of the Hashemite Kingdom of Jordan*. Privately published, Musselburgh.

Ash, J. & Atkins, J. (2009) *Birds of Ethiopia and Eritrea – An atlas of distribution*. Christopher Helm, London.

Ayé, R., Schweizer, M. & Roth, T. (2012) *Birds of Central Asia*. Christopher Helm, London.

Balmer, D. E., Gillings, S., Caffrey, B. J., Swann, R. L., Downie, I. S. & Fuller, R. J. (2013) *Bird Atlas 2007–2011; The breeding and wintering birds of Britain and Ireland*. BTO, Thetford.

Baumgart, W. (2003) *Birds of Syria*. 2nd edn. OSME, Sandy.

Beaman, M. (1994) *Palearctic Birds*. Harrier Publications, Stonyhurst.

Beaman, M. & Madge, S. (1998) *The Handbook of Bird Identification for Europe and the Western Palearctic*. Christopher Helm, London.

Beolens, B., Watkins, M. & Grayson, M. (2014) *The Eponym Dictionary of Birds*. Bloomsbury, London.

Bergmann, H.-H., Helb, H.-H. & Bauman, S. (2010) *Die Stimmen der Vögel Europas*. DVD. Aula, Wiebelsheim.

Bønløkke, J., Madsen, J. J., Thorup, K., Pedersen, K. T., Bjerrum, M. & Rahbek, C. (2006) *Dansk Traekfugleatlas*. Rhodos, Humlebæk.

Borrow, N. & Demey, R. (2001) *Birds of Western Africa*. Christopher Helm, London.

Brazil, M. (2009) *Birds of East Asia*. Christopher Helm, London.

Brewer, D. (2001) *Wrens, Dippers and Thrashers*. Christopher Helm, London.

Bundy, G. (1976) *The Birds of Libya*. BOU, London

Chappuis, C. (1987) *Migrateur et hivernants*. 2 cassettes. Grand Couronne.

Chappuis, C. (2000) *African Bird Sounds – 1. North-West Africa, Canaria and Cap-Verde Islands*. 4 CDs. SEOF/NSA, London.

Clarke, T. (2006) *Birds of the Atlantic Islands*. Christopher Helm, London.

Clement, P. & Hathway, R. (2000) *Thrushes*. Christopher Helm, London.

Clements, J. F. (2000) *Birds of the World. A Checklist*. 5th edn. Pica Press, Robertsbridge.

Cramp, S. & Perrins, C. M. (eds.) (1988–92) *The Birds of the Western Palearctic*. Vols. 5–6. Oxford University Press, Oxford.

Cramp, S., Simmons, K. E. L., Perrins, C. M. & Snow, D. W. (eds.) (2006) *BWPi*, version 2.0. [Includes film footage and sounds.] Oxford University Press, Oxford.

Dementiev, G. P. & Gladkov, N. A. (eds.) (1953–54) *Ptitsy Sovietskogo Soyuza*. Vols. 5–6. [In Russian.] (English translation, 1968, *Birds of the Soviet Union*. Jerusalem.)

Dickinson, E. C. (ed.) (2003) *The Howard & Moore Complete Checklist of the Birds of the World*. 3rd edn. Christopher Helm, London.

Dickinson, E. C. & Christidis, L. (eds.) (2014) *The Howard & Moore Complete Checklist of the Birds of the World*. 4th edn. Vol. 2. Aves Press, Eastbourne.

Dickinson, E. C., Overstreet, L. K., Dowsett, R. J. & Bruce, M. D. (2011) *Priority! The Dating of Scientific Names in Ornithology*. Aves Press, Northampton.

Dunn, J. L., Blom, E. A. T., Alderfer, J. K., Watson, G. E., Lehman, P. E. & O'Neill, J. P. (2006) *National Geographic Field Guide to the Birds of North America*. 5th edn. National Geographic, Washington.

Equipa Atlas (2008) *Atlas das aves nidificantes em Portugal (1999–2005)*. ICN & B. Assírio & Alvim, Lisbon.

Eriksen, H. & Eriksen, J. (2010) *Common Birds of Oman – An Identification Guide*. 2nd edn. Al Roya Publishing, Muscat.

Eriksen, J. & Reginald, V. (2013) *Oman Bird List*. 7th edn. Center for Environmental Studies and Research, Sultan Qaboos University, Muscat.

Estrada, J., Pedrocchi, V., Brotons, L. & Herrando, S. (eds.) (2004) *Atlas dels ocells nidificants de Catalunya 1999–2002*. Lynx Edicions, Barcelona.

Flint, P. R. & Stewart, P. F. (1992) *The Birds of Cyprus*. 2nd edn. BOU, Tring.

Fransson, T. & Hall-Karlsson, S. (2008) *Swedish Bird Ringing Atlas*. Vol. 3. Passerines. Naturhistoriska Riksmuseet & Sveriges Ornitologiska Förening, Stockholm.

Fry, C. H., Keith, S. & Urban, E. K. (eds.) (1992–2004) *The Birds of Africa*. Vols. 4–7. Academic Press, London.

Gallagher, M. D. & Woodcock, M. W. (1980) *The Birds of Oman*. Quartet Books, London.

Gavrilov, E. & Gavrilov, A. (2005) *The Birds of Kazakhstan*. Tethys, Almaty.

Gill, F. & Donsker, D. (eds.) (2016) *IOC World Bird List*. Version 6.2. Doi: 10.14344/IOC.ML.6.2.

Gill, F. & Wright, M. (2006) *Birds of the World: Recommended English Names*. Princeton University Press, Princeton.

Ginn, H. B. & Melville, D. S. (1983) *Moult in Birds*. BTO guide no. 19. Tring.

Glutz, U. N., Bauer, K. & Bezzel, E. (eds.) (1966–1998) *Handbuch der Vögel Mitteleuropas*. Vols. 1–14. Aula-Verlag, Wiesbaden.

Goodman, S. M. & Meininger, P. L. (eds.) (1989) *The Birds of Egypt*. Oxford University Press, Oxford.

Gorman, G. (1996) *The Birds of Hungary*. Christopher Helm, London.

Gosler, A. G., Greenwood, J. J. D., Baker, J. K. & King, J. R. (1995) A comparison of wing length and primary length as size measures for small passerines. *Ring. & Migr.*, 16: 65–78.

Grimmett, R., Inskipp, C. & Inskipp, T. (1998) *Birds of the Indian Subcontinent*. Christopher Helm, London.

Grimmett, R., Inskipp, C. & Inskipp, T. (2011) *Birds of the Indian Subcontinent*. Helm Field Guide. Christopher Helm, London.

Grimmett, R., Roberts, T. & Inskipp, T. (2008) *Birds of Pakistan*. Christopher Helm, London.

Gulledge, J. (ed.) (1983) *A Field Guide to Bird Songs of Eastern and Central North America*. 2nd edn. 2 cassettes. Cornell University Press/Houghton Mifflin, Boston.

Hagemeijer, W. J. M. & Blair, M. J. (1997) *The EBCC Atlas of European Breeding Birds*. T. & A.D. Poyser, London.

Handrinos, G. & Akriotis, T. (1997) *The Birds of Greece*. Christopher Helm, London.

Harrison, C. (1982) *An Atlas of the Birds of the Western Palaearctic*. Collins, London.

Hartert, E. (1903–21) *Die Vögel der Paläarktischen Fauna*. Vols. 1–2. Friedländer & Sohn, Berlin.

Hartert, E. (1921–22, 1932–38) *Die Vögel der Paläarktischen Fauna*. Vols. 3–4. ('Ergänzungsband', vol. 4, ed. by F. Steinbacher). Friedländer & Sohn, Berlin.

Hollom, P. A. D., Porter, R. F., Christensen, S. & Willis, I. (1988) *Birds of the Middle East and North Africa*. T. & A.D. Poyser, Calton.

Howell, S. N. G. (2010) *Molt in North American Birds*. Houghton Mifflin, Boston.

Howell, S. N. G., Lewington, I. & Russell, W. (2014) *Rare Birds of North America*. Princeton University Press, Princeton.

del Hoyo, J., Elliott, A. & Christie, D. A. (eds.) (2004–12) *Handbook of the Birds of the World*. Vols. 9–16. Lynx Edicions, Barcelona.

International Commission on Zoological Nomenclature (ICZN) (1999) *International Code of Zoological Nomenclature*. International Trust for Zoological Nomenclature, London. (www.iczn.org)

Isenmann, P. & Moali, A. (2000) *Birds of Algeria*. SEOF, Paris.

Isenmann, P., Gaultier, T., El Hili, A. Azafzaf, H., Dlensi, H. & Smart, M. (2005) *Birds of Tunisia*. SEOF, Paris.

Jännes, H. (2002) *Calls of Eastern Vagrants*. CD. Early Bird, Helsinki.

Jännes, H. (2002) *Bird Sounds of Goa & South India*. CD. Early Bird, Helsinki.

Jenni, L. & Winkler, R. (1994) *Moult and Ageing of European Passerines*. Academic Press, London.

Jennings, M. C. (2010) *Atlas of the Breeding Birds of Arabia*. Fauna of Arabia, Frankfurt & Riyadh.

Jobling, J. A. (2010) *The Helm Dictionary of Scientific Bird Names*. Christopher Helm, London.

Kennerley, P. & Pearson, D. (2010) *Reed and Bush Warblers*. Christopher Helm, London.

Kirwan, G. M., Boyla, K. A., Castell, P., Demirci, B., Özen, M., Welch, H. & Marlow, T. (2008) *The Birds of Turkey*. Christopher Helm, London.

Knox, A. G. & Parkin, D. T. (2010) *The Status of Birds in Britain & Ireland*. Christopher Helm, London.

Kren, J. (2000) *Birds of the Czech Republic*. Christopher Helm, London.

Larsson, L., Ekström, G., Larsson, E. & Gandemo, M. (2008) *Birds of the World*. 2nd edn. CD. Lynx Edicions, Barcelona.

Leibak, E., Lilleleht, V. & Veromann, H. (1994) *Birds of Estonia. Status, Distribution and Numbers*. Estonian Academy Publishers, Tallinn.

Lindell, L., Wirdheim, A. & Zetterström, D. (ed.) (2002) *Sveriges fåglar*. 3rd edn. [Official check-list of Swedish birds.] Vår Fågelv. Suppl. 32. SOF, Stockholm.

Marti, R. & del Moral, J. C. (eds.) (2003) *Atlas de las Aves Reproductoras de España*. SEO/BirdLife, Madrid.

Mild, K. (1987) *Soviet Bird Songs*. 2 cassettes. Privately published, Stockholm.

Mild, K. (1990) *Bird Songs of Israel and the Middle East*. 2 cassettes. Privately published, Stockholm.

Mitchell, D. & Young, S. (1997) *Photographic Handbook of the Rare Birds*. New Holland, London.

Palmer, S. & Boswall, J. (1981) *A Field Guide to the Birds Songs of Britain & Europe*. 16 cassettes. SR Phonogram, Stockholm.

Panov, E. N. (2005) *Wheatears of Palearctic: Ecology, Behaviour and Evolution of the Genus Oenanthe*. Pensoft, Sofia.

Peters, J. L., Blake, E. R., Greenway, J. C., Howell, T. R., Lowery, G. H., Mayr, E., Monroe, B. L. Jr, Rand, A. L. & Traylor, M. A. Jr (1960–86) *Check-list of the Birds of the World*. Vols. 9–15. Harvard University Press, Cambridge, MA.

Phillips, A. R. (1991) *The Known Birds of North and Middle America*. Part 2. Denver Museum, Denver.

Porter, R. F. & Aspinall, S. (2010) *Birds of the Middle East*. 2nd edn. Christopher Helm, London.

Pyle, P. (1997) *Identification Guide to North American Birds*. Vol. 1. Slate Creek Press, Bolinas, CA.

Pyle, P., DeSante, D. F. Boekelheide, R. J. & Henderson, R. P. (1987) *Identification Guide to North American Passerines*. Slate Creek Press, Bolinas, CA.

Rasmussen, P. C. & Anderton, J. C. (2005, 2012) *Birds of South Asia. The Ripley Guide*. Vols. 1–2. 1st and 2nd edn. Smithsonian Institution, Washington & Lynx Edicions, Barcelona.

Redman, N., Stevenson, T. & Fanshawe, J. (2011) *Birds of the Horn of Africa*. 2nd edn. Christopher Helm, London.

Ridgway, R. (1912) *Color Standards and Color Nomenclature*. American Museum of Natural History, New York.

Roberts. T. J. (1992) *The Birds of Pakistan*. Vol. 2: Passeriformes. Oxford University Press, Oxford.

Robson, C. (2011) *A Field Guide to the Birds of South-East Asia*. 2nd edn. New Holland, London.

Roché, J. C. (1990) *All the bird songs of Britain and Europe*. 4 CDs. Sittelle, Mens.

Roché, J. C. & Chevereau, J. (eds.) (1998) *A sound guide to the Birds of North-West Africa*. CD. Sittelle, Mens.

Roché, J. C. & Chevereau, J. (eds.) (2002) *Birds sounds of Europe and North-west Africa*. 10 CDs. Wildsounds, Salthouse.

Rogacheva, H. (1992) *The Birds of Central Siberia*. Privately published, Husum.

Roselaar, C. S. (1995) *Songbirds of Turkey: an atlas of biodiversity of Turkish passerine birds*. GMB, Haarlem & Pica Press, Robertsbridge.

Roselaar, C. S. & Shirihai, H. (in prep.) *Geographical Variation and Distribution of Palearctic Birds*. Vol. 1: Passeriformes. Christopher Helm, London.

Ryabtsev, V. K. (2001) *Ptitsy Urala*. UrGu, Yekaterineburg.

Schubert, M. (1979) *Stimmen der Vögel Zentralasiens + Mongolei*. 2 LPs. Eterna, Berlin.

Schubert, M. (1984) *Stimmen der Vögel. VII. Vogelstimmen Südosteuropas (2)*. LP. Eterna, Berlin.

Schulze, A. (2003) *Die Vogelstimmen Europas, Nordafrikas und Vorderasiens*. 17 CDs or 2 mp3s. Edition Ample, Germering.

Sharpe, R. B. (1874–99) *Catalogue of the Birds in the British Museum*. Vols. 1–28. British Museum, London.

Shirihai, H. (1996) *Birds of Israel*. Academic Press, London.

Shirihai, H., Christie, D. A. & Harris, A. (1996) *Macmillan Birder's Guide to European and Middle Eastern Birds*. Macmillan, London.

Shirihai, H., Gargallo, G. & Helbig, A. J. (2001) *Sylvia Warblers*. Christopher Helm, London.

Sibley, C. G. & Ahlquist, J. (1990) *Phylogeny and Classification of Birds*. Yale University Press, New Haven, CT.

Sibley, C. G. & Monroe, B. L. Jr (1990) *Distribution and Taxonomy of Birds of the World*. Yale University Press, New Haven, CT.

Sibley, D. (2014) *The Sibley Guide to Birds*. 2nd edn. Knopf, New York.

Smithe, F. B. (1975, 1981) *Naturalist's Color Guide*. American Museum of Natural History, New York.

Snow, D. W. & Perrins, C. W. (eds.) (1998) *The Birds of the Western Palearctic*. Concise edition. 2 vols. Oxford University Press, Oxford.

Stresemann, E. & Portenko, L. A. (1960–98) *Atlas der Verbreitung palaearktischen Vögel*. Vols. 1–17. Akademie-Verlag, Berlin.

Stresemann, E. & Stresemann, V. (1966) Die Mauser der Vögel. *Journal für Ornithologie*, 107. Sonderheft. Berlin. 448pp.

Strömberg, M. (1994) *Moroccan Bird Songs and Calls*. Cassette. Privately published, Sweden.

Svensson, L. (1984) *Soviet Birds*. Cassette. Privately published, Stockholm.

Svensson, L. (1992) *Identification Guide to European Passerines*. 4th edn. Privately published, Stockholm.

Svensson, L., Mullarney, K. & Zetterström, D. (2009) *Collins Bird Guide*. 2nd edn. HarperCollins, London.

Svensson, L., Zetterström, D. & Andersson, B. (1990) *Fågelsång i Sverige*. CD and booklet. Mono Music, Stockholm.

Thévenot, M., Vernon, R. & Bergier, P. (2003) *The Birds of Morocco*. BOU Checklist No. 20. BOU, Tring.

Thorup, K. (2004) *Bird Study*, 51: 228–238.

Ticehurst, C. B. (1938) *A Systematic Review of the Genus Phylloscopus*. British Museum, London.

Tomiałojć, L. & Stawarczyk, T. (2003) *Awifauna Polski*. 2 vols. PTPP, Wrocław.

Turner, A. & Rose, C. (1989) *Swallows and Martins of the World*. Christopher Helm, London.

Ueda, H. (1998) *283 Wild Bird Songs of Japan*. Yama-kei, Tokyo.

Urquhart, E. (2002) *Stonechats*. Christopher Helm, London.

Vaurie, C. (1959) *Birds of the Palearctic Fauna. Passeriformes*. H. F. & G. Witherby, London.

Veprintsev, B. N. & Leonovich, V. (1982–86) *Birds of the Soviet Union: A Sound Guide*. 7 LPs. Melodia, Moscow.

Veprintsev, B. N. & Veprintseva, O. (2007) *Voices of the Birds of Russia*. mp3. Phonoteca, Moscow. (In Russian.)

Vinicombe, K., Harris, A. & Tucker, L. (2014) *The Helm Guide to Bird Identification. An In-depth Look at Confusion Species*. Christopher Helm, London.

Vinicombe, K. & Cottridge, D. M. (1996) *Rare Birds in Britain and Ireland*. HarperCollins, London.

Voous, K. H. (1977) *List of Recent Holartic Bird Species*. BOU, London.

Williamson, K. (1967) *Identification for Ringers 2. Phylloscopus*. 2nd edn. BTO, Oxford.

Williamson, K. (1968a) *Identification for Ringers 1. Cettia, Locustella, Acrocephalus and Hippolais*. 3rd edn. BTO, Oxford.

Williamson, K. (1968b) *Identification for Ringers 3. Sylvia*. 2nd edn. BTO, Oxford.

Witherby, H. F., Jourdain, F. C. R, Ticehurst, N. F. & Tucker, B. W. (1938–41) *The Handbook of British Birds*. H. F. & G. Witherby, London.

Wolters, H. E. (1979) *Die Vogelarten der Erde*. Paul Parey, Hamburg.

Zimmerman, D. A., Turner, D. A. & Pearson, D. J. (1996) *The Birds of Kenya and Northern Tanzania*. Christopher Helm, London.

Zink, G. (1973–85) *Der Zug europäischer Singvögel*. Vols. 1–4. Vögelzug-Verlag, Möggingen.

LIST OF PASSERINE FAMILIES: TRADITIONAL AND NEW ORDER

The order of families in this handbook is the traditional one found in Cramp et al. (1977–94), The Birds of the Western Palearctic (which used Voous 1977 as a basis), and with little variation in most field guides (e.g. Svensson et al. 2009). The choice to stick with this order is made solely to facilitate use; what is familiar to most readers is thought to be user-friendly. However, recent research using molecular biology clearly shows the traditional order to be in many instances wrong, largely based as it was on morphological similarity, which sometimes cannot separate homologous characters from those acquired through convergence or by chance. It is clear that soon we will all have to learn and use a new sequence of families. Interested parties are referred to vol. 2 of the recent fourth edition of The Howard and Moore Complete Checklist of the Birds of the World (Dickinson & Christidis 2014), where Joel Cracraft presents and explains the latest taxonomic developments. It should be stated that we are still a long way from full agreement over best taxonomy and sequence. The Howard and Moore checklist is just one of several options adopting a new order.

The table below offers a simple comparison between the traditional and new (sensu Howard and Moore, 4th edn) orders among the passerines. The number of families involved for the species in the main section of this handbook will increase under the new order by a quarter, from 34 to 44. As an example, the large family Sylviidae will be divided into several smaller families. Some of the changes are noteworthy, such as the fact that larks are closely related to cisticolas and prinias, the Bearded Reedling has proven not to be a timalid or parrotbill but closely related to the larks, further that swallows are surprisingly inserted among the various new warbler families, and that pipits and wagtails end up among sparrows and finches. Chats, redstarts and wheatears are not small thrushes, but part of the large flycatcher family. There is much new to digest and learn.

Traditional order
Sensu Voous / BWP

ALAUDIDAE – Larks
HIRUNDINIDAE – Swallows
MOTACILLIDAE – Pipits and wagtails
PYCNONOTIDAE – Bulbuls
BOMBYCILLIDAE – Waxwings and Hypocolius
CINCLIDAE – Dippers
TROGLODYTIDAE – Wrens
PRUNELLIDAE – Accentors
TURDIDAE – Nightingales, chats and thrushes
SYLVIIDAE – Warblers, crests and kinglets
MUSCICAPIDAE – Flycatchers
MONARCHIDAE – Paradise flycatchers
TIMALIIDAE – Reedling, babblers and parrotbills
AEGITHALIDAE – Long-tailed tits
PARIDAE – Tits
SITTIDAE – Nuthatches
TICHODROMIDAE – Wallcreeper
CERTHIIDAE – Treecreepers
REMIZIDAE – Penduline tits
NECTARINIIDAE – Sunbirds
ZOSTEROPIDAE – White-eyes
ORIOLIDAE – Orioles
LANIIDAE – Bush-shrikes and shrikes
CORVIDAE – Corvids
STURNIDAE – Starlings and mynas
PASSERIDAE – Sparrows and allies
PLOCEIDAE – Weavers
ESTRILDIDAE – Waxbills and allies
VIREONIDAE – Vireos
FRINGILLIDAE – Finches
PARULIDAE – New World warblers
THRAUPIDAE – Tanagers
EMBERIZIDAE – Buntings, New World sparrows and cardinals
ICTERIDAE – Bobolink and icterids

New order
Sensu Cracraft / H&M 4

VIREONIDAE – Vireos
ORIOLIDAE – Orioles
MALACONOTIDAE – Bush-shrikes
LANIIDAE – Shrikes
CORVIDAE – Corvids
MONARCHIDAE – Monarchs and paradise flycatchers
NECTARINIIDAE – Sunbirds
PRUNELLIDAE – Accentors
PLOCEIDAE – Weavers
ESTRILDIDAE – Waxbills, munias and allies
PASSERIDAE – Sparrows, snowfinches and allies
MOTACILLIDAE – Pipits and wagtails
FRINGILLIDAE – Finches
CALCARIIDAE – Longspurs and allies
EMBERIZIDAE – Buntings
PASSERELLIDAE – New World sparrows and allies
PARULIDAE – New World warblers
ICTERIDAE – Icterids
CARDINALIDAE – Cardinals, tanagers and allies
PARIDAE – Tits
REMIZIDAE – Penduline tits
ALAUDIDAE – Larks
PANURIDAE – Reedlings
CISTICOLIDAE – *Cisticola* warblers
LOCUSTELLIDAE – *Locustella* warblers
ACROCEPHALIDAE – Reed warblers
HIRUNDINIDAE – Swallows
PYCNONOTIDAE – Bulbuls
PHYLLOSCOPIDAE – Leaf warblers
SCOTOCERCIDAE – Bush warblers and allies
AEGITHALIDAE – Long-tailed tits
SYLVIIDAE – *Sylvia* warblers and allies
ZOSTEROPIDAE – White-eyes
LEIOTHRICHIDAE – Babblers and laughingthrushes
REGULIDAE – Crests and kinglets
BOMBYCILLIDAE – Waxwings
HYPOCOLIIDAE – Hypocolius
CERTHIIDAE – Wallcreeper and treecreepers
SITTIDAE – Nuthatches
TROGLODYTIDAE – Wrens
STURNIDAE – Starlings
CINCLIDAE – Dippers
MUSCICAPIDAE – Chats and flycatchers
TURDIDAE – Thrushes

A BRIEF PRESENTATION OF PASSERINE FAMILIES

As explained in the Introduction, the sequence adopted in this handbook is the traditional one, among the passerines starting with larks and ending with buntings and New World icterids. This is thought to be user-friendly and facilitate finding a sought species quickly. Taxonomy is in a dynamic phase with many new insights from molecular methods, and future lists and books will undoubtedly adopt a very different order. However, new findings surface continuously resulting in recommended further changes of the taxonomy, and interpretations of results vary somewhat between taxonomists; we are still not seeing the end of these changes. This state of uncertainty is another reason to stick to a familiar sequence for the time being.

Passerines comprise over half of the global number of bird species. They are traditionally kept together in one large order, Passeriformes, but this might change in the future. They are often divided into suboscines and oscines, the latter including the vast majority, and all the families listed below. Suboscines only concern a few vagrants to the treated region (tyrant flycatchers, and one pitta species).

While the main section of the book contains accounts of the more regularly occurring species arranged in one continuous sequence without division into family chapters, it is thought helpful to give brief family descriptions separately. The families below are those of the traditional order treated in this volume, but where appropriate comments on new thinking and alternative arrangement are offered.

Alaudidae – Larks

Small terrestrial birds usually encountered in open habitats. Mainly brownish plumage and rather strong bills suitable for feeding heavily on seeds, but insects are prominent part of food, too. Some species have an erectable short or obvious pointed crest formed by crown-feathers. More technically, larks are defined as a group by the detailed structure of tarsus and syrinx. Hind claw rather straight and often long. Sexes alike or very similar in plumage in majority of species, but males often somewhat larger; some species have obvious sexual dimorphism. Accomplished singers, prolonged or continuously repeated song often delivered on wing, at times from great height. Nest on ground in low vegetation. Rather hardy birds, but majority except in the south are migrants. 26 species breed or are regularly found in the Western Palearctic, and one more has been found as a vagrant. – In traditional classification usually placed first in sequence, but genetic evidence suggests that the larks are more closely related to the Bearded Reedling, and belong to a group of warbler-like birds including among others cisticolas, prinias, leaf warblers, members of the genus *Sylvia* but also swallows and long-tailed tits.

Hirundinidae – Swallows and martins

Small specialised flight-feeders catching flying insects with open mouth. Small flattened bill but wide gape, wings long and pointed, tail often forked, outer tail-feathers sometimes elongated to thin 'streamers'. Tarsus short and rather weak. Often dark above (glossy blue-black or brown) and mainly white below. Sexes alike or very similar, both as to plumage and size, but, rather unusually for passerines, females average subtly larger than males in some species. Unremarkable song a rather subdued twitter or chatter. Flight fast with sweeping glides but can be fluttering and acrobatic when required. Build nests, often with mud, on vertical cliff walls, house walls under eaves, on roof beams, under harbour bridges, etc. Many are highly migratory, wintering in tropical areas. Twelve species are regular within the region, a further five are vagrants. – Swallows have traditionally been placed second in order after the larks and before the pipits and wagtails, or rarely first of all Western Palearctic passerines, but DNA shows them rather unexpectedly to sit embedded in a superfamily consisting of various warbler lineages (e.g., *Locustella, Acrocephalus, Phylloscopus*) and bulbuls.

Motacillidae – Pipits and wagtails

Small, slim and delicately built birds with long tail that is often rhythmically wagged when feeding. Good fliers, flight path undulating. Largely terrestrial habits. Often walk with jerky head movements. Bill thin and pointed indicating insectivorous feeding habits. Pipits are discreetly plumaged in brown, buff and white, generally finely streaked, whereas wagtails are more strikingly plumaged, either in black-grey-white or in yellow, olive, brown and white. Sexes alike in pipits, differing rather clearly in wagtails. In general, pipits are accomplished singers, often performed in typical song-flight, while wagtails have the simplest song of all passerines that sing. Nest on ground in tussock (most species), or in holes in stone walls or houses (some wagtails). Most species are diurnal migrants. 20 species are treated in the main section, two among the vagrants. – The traditional position of this family, closely after larks and swallows, has been severely shaken by evidence from molecular methods showing it to be related to weavers (Ploceidae), sparrows (Passeridae) and finches (Fringillidae). It is obvious that bill shape, tail length and general habits and appearance are unreliable clues to correct relationships.

Pycnonotidae – Bulbuls

Medium-sized birds with usually long and somewhat rounded tail. Short neck and rounded head with short but strong bill. Mainly dark-plumaged, often with some white, yellow or red detail. Sexes alike. Wings rather short and rounded with resulting rather slow flight on fluttering wingbeats; flight path direct. Often seen in small parties, vocal and noisy. Song simple but loud. Nest cup of grass and fine twigs in tree or bush, sometimes on building. Sedentary. Four species occur within the treated region. – Although general appearance might indicate relationship with dippers or waxwings—the traditional taxonomic view—or with starlings, several recent genetic studies put bulbuls in the large and, based on external characters and habits, seemingly heterogeneous group comprising, e.g., different lineages of warblers, long-tailed tits, and swallows.

Bombycillidae – Waxwings and Hypocolius

Medium-sized birds with rather plump bodies, short necks and short (waxwings) or long (Hypocolius) tails. Bill short but stout and strong. Feet strong, too. Waxwings resemble starlings in shape, having similarly direct flight, broad-based but pointed wings and broad necks but immediately differ in mainly rufous-grey colour in combination with prominent pointed crest. Strong fliers, often occurring in parties outside breeding season, feeding fearlessly on berries even in gardens and towns. Apart from brownish or rufous-grey general colour have red and/or yellow ornaments. Song a high-pitched metallic trilling. Nest in conifer in dense taiga. Hypocolius differ markedly apart from the long tail in its greyish-buff, black and white plumage, the much flatter, shorter crest and its clear sexual dimorphism. It is also restricted to desert-like habitats. – Two species regular, a third is a vagrant to the region. In modern taxonomy the Hypocolius is generally afforded a separate family, Hypocoliidae.

Cinclidae – Dippers

Medium-sized, compact, short-tailed and short-necked birds, highly adapted to feed below the surface of water in rivers and mountain streams. Adaptations include some solid skeletal bones, nostrils that can be covered by a membrane during dives, advanced accommodation ability of the eyes, ear-openings covered by a skin flap, a very dense plumage and enlarged oil-gland for plumage care. Jumps in the water, even in midwinter, and walks against the current by holding tail high, which helps press the bird down against the bottom. Flight direct with quickly whirring wingbeats. Plumage black, brown and white. Sexes alike. Song squeaky, throaty and scratchy, but still carrying over the noise of a stream. Resident or moderately migratory. Breeds in steep riverbanks, on the walls of a river mill, under a stone bridge, etc. Sometimes double-brooded with first clutch on or near wintering site, the second further north. One species occurs within the region.

Troglodytidae – Wrens

Small or tiny rotund, short-necked birds, most with short folded tail often held erect. Bill fine and pointed, fairly long and slightly down-curved. Active, moves around a lot, mainly in cover. Dwell in scrub and forest thickets, usually close to ground but can take song post higher up in tree. Usually brown with darker barring or streaking, slightly paler below. Sexes alike. Very loud song for its size, delivered at a fast pace. Domed twig nest in dense undergrowth. Mainly sedentary, although some northern breeders move south for the winter. One species within the region.

Prunellidae – Accentors

Small sparrow-like rather compact birds, but compared to sparrows have thinner and more pointed bills. Mainly terrestrial feeders but also frequently seen in low, thick vegetation. Often moves energetically with constant short hops in crouched position with bent legs when feeding. Rather skulking habits. Most species have brown, grey and white plumage, upperparts usually dark-streaked, and some have rusty-buff or orange-tinged underparts. Sexes alike, or nearly so. Song rather discreet, a brief, warbling strophe with clear, whistling voice. Nest low in a dense bush. Short-range migrants or residents. Five species occur within the treated region.

Turdidae – Nightingales, chats and thrushes

In traditional taxonomy this family is seen as large, encompassing both real thrushes and smaller chats, nightingales, redstarts, wheatears, etc. (see comment below). Small to medium-sized birds with usually strong feet and rather erect stance adapted to terrestrial feeding. Food includes earthworms, snails, other invertebrates, berries and seeds. Feeding activity of true thrushes involves either walking a few steps, standing still to glean the ground for movements of worms or insects, or using bill to overturn leaves and debris on ground in order to expose food items. Bill of medium length, often rather thin and pointed but more stout in some. Flight in larger species often strong and undulating over longer distances, more straight and fluttering over short. Adult plumage variable, but that of juveniles nearly always spotted. Sexual dimorphism absent in some species, present in others. True thrushes (e.g. *Turdus*) are found in woods, parks and gardens, also in bushy areas and near water where there is scrub or scattered trees. The smaller chats are found in a variety of habitats, both in dense scrub or woods and in open areas like heaths, pasture land, deserts and barren mountainsides. Most species are accomplished singers with a loud voice and characteristic song. Nest is an open cup hidden in low vegetation, in a bush or a tree. Most species are migratory, but southern breeders can be sedentary. Mainly nocturnal movements, but some diurnal migration also by the true thrushes. 64 species are regular within the covered region, whereas another seven are vagrants only. – The modern view based on several extensive molecular studies is that the chats, robins, redstarts, nightingales, wheatears, rock thrushes, etc., are not part of Turdidae but are members of the Old World flycatcher family Muscicapidae, separate from the true thrushes.

Sylviidae – Warblers, crests and kinglets

Small or tiny insect-eating birds with fine bills and usually thin legs, mainly living in the canopy of trees, but some occur more often in scrub or in reeds or sedges near water. Quick movements in vegetation when feeding, and often skulking habits. Some species eat berries and nectar, as well as insects and spiders. Extensive radiation of species, but many within subgroups are very similar. Plumage variable within such a large group, but underparts nearly always paler than upperparts, and many have either a light supercilium, one or two light wing-bars, a dark cap or other characteristic pattern. Sexes often alike, but in some subgroups male and female differ. Highly evolved singers. Nest on ground in vegetation or low in bush or reeds, less often higher up in trees. Many are nocturnal long-distance migrants requiring insect food year-round. 76 species breed or occur fairly regularly within the covered region, and another four have been noted as vagrants. – This (in traditional taxonomy) quite large family has been thoroughly reformed by recent molecular research. Sylviidae in the old sense is polyphyletic with groups such as long-tailed tits, larks, swallows, bulbuls and timalids nested between traditional warbler genera or subgroups. Rather than fitting all these groups under the family label Sylviidae ('warblers'), the logical solution has been to split the large family up into several smaller families. In the strict sense a monophyletic Sylviidae family comprises only the *Sylvia* warblers, fulvettas and parrotbills (latter two extralimital in this context). It should be noted, finally, that the true relationship of crests and kinglets is still unresolved.

SINGING BUSH LARK
Mirafra cantillans Blyth, 1845

Fr. – Alouette chanteuse; Ger. – Buschlerche
Sp. – Alondra cantarina; Swe. – Sångbusklärka

Of the *c.* 25–30 species of *Mirafra* larks (a mainly Afrotropical genus), this is the sole representative in the covered region. Most often found in dry grassy plains, it is mainly resident in SW Saudi Arabia, south and east to Aden (Yemen) and SW Oman, but also occurs in the Indian subcontinent and across much of the Sahel. Like other *Mirafra* larks, often difficult to observe due to secretive habits, staying hidden in cover on the ground. Often has a rather peculiar, fluttering escape flight.

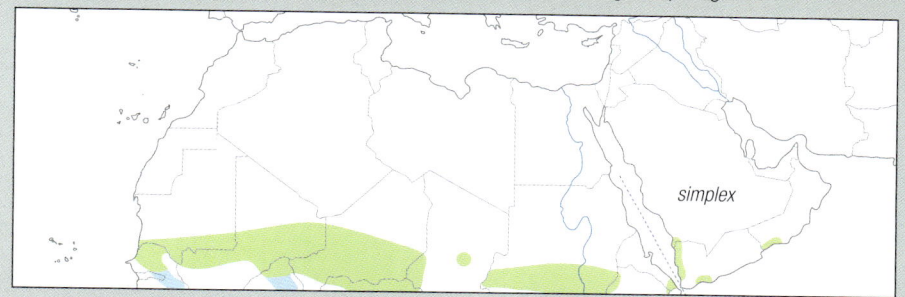

M. c. simplex, Oman, Apr: in song-flight on typical square-shaped wings; note face and breast patterns, and narrow short tail. (H. & J. Eriksen)

M. c. simplex, Oman, May: the only *Mirafra* lark in the treated region. Note pale fringes to upperparts and wings, rather strong face pattern and streaky breast-side patches. (H. & J. Eriksen)

M. c. simplex, Oman, May: often perches on bushes. With wear, pale fringes to wing-feathers much reduced; note rufous wing panel, most obvious if wing drooped. (H. & J. Eriksen)

IDENTIFICATION Streaky, medium-sized, stout-billed lark, with short, narrow tail (whitish outermost feathers visible on landing). No or very slight crest. *If flushed, escape flight typically high but short on broad, rounded, fluttering wings, then dives on stiff lowered wings with brief fluttering breaks* (semi-hovers), sometimes aborting, flying further and fluttering again, before finally diving into cover. On ground, note fairly *broad-based, yellowish-horn bill with rather curved, slightly darker culmen*. Small feet, short hind claw. Upperparts warm brown heavily streaked dark, with especially well-marked *scallop-like fringes to scapulars* and, to lesser extent, crown and mantle. *Flight-feathers rufous-tinged*, especially on outer upperwing. *Clear-cut pale fringes and tips to wing-coverts and long tertials* (often cloaking primaries). Head pattern consists of reasonably broad and *long, pale supercilium, lores and eye-ring*, dark brown-streaked ear-coverts and pale throat; *dark eye-stripe behind eye, rear edge to ear-coverts and, especially, moustachial stripe*; and *large dark eye*. Underparts rather pale buffish-white with *fine dark short streaks, and small blotches on breast and upper flanks* (can produce dark upper-breast-side patch). In very close view, note large nostrils. Usually solitary or in pairs, and often unobtrusive, flushing only when almost underfoot.

M. c. simplex, Oman, Feb: stout bill (horn-coloured with rather curved, darker culmen), narrow tail and short wings. When fresh, broad pale tips to median coverts form wing-bar. Presumed 1stY still in post-juv moult (H. & J. Eriksen)

Singing Bush Lark *M. c. simplex* (left: Oman, Jan) and Oriental Skylark *Alauda gulgula* (right: United Arab Emirates, Nov): if the thicker-based and more conical bill of Singing Bush Lark is not seen well, there is risk of confusion with Oriental Skylark, though usually the less scaly upperparts, and more densely and finely streaked breast of latter are appreciable, while voice is diagnostic. The Singing Bush Lark here is in fresh plumage after complete moult (post-juv or post-nuptial). All of upperparts and wing-coverts are crisply pale-fringed, and unlike similar juv feathers (cf. juv in image below), fringes are somewhat narrower and less buffish. (Mattias Ullman)

VOCALISATIONS Song a series of usually accelerating, slightly varied phrases, each falling in pitch, lasting 3–7 seconds and can recall sections of song of Water Pipit or Chaffinch more than a lark species. Sometimes utters more prolonged phrases, a mix of warbling, scratchy chirps, whistles and buzzes, occasionally including mimicry, and sometimes given continuously, e.g. *swee-swee, tir-wit-tir-wit, che, che, che, che, che...*, or (E Africa) *tsip-tsip-tsip chirrrrip tchew tchew tchew tchew tchew tchweep tchweep tchweep tsi-tsi-tsi-tsi tew tew-tew-tew-twi chwi chwi...* Given from ground bush or in circling bat-like display-flight with jerky wingbeats (for up to 40 minutes). – Commonest call a quiet *proop-proop*.

SIMILAR SPECIES Song and display-flight rather unique in SW Arabian range, where it is the only bush lark. Combination of heavily-fringed upperparts, comparatively large bill and narrowly-streaked underparts also relatively distinctive. – Small Skylark, which regularly reaches Arabia in winter, also has streaky breast and short primary projection, but overall shape and small bill, less scaly upperparts, buffy outer tail-feathers and more densely streaked chest (in Pectoral Sandpiper fashion) are useful separating features.

AGEING & SEXING Ageing impossible following complete post-nuptial and post-juv moults. Sexes alike. – Moults. Few moult data available for Arabian race *simplex*: three in Dec were fresh, two Mar–April moderately worn, while of four from Aug–Sep, two were still unmoulted and heavily worn or had just started, and two were in advanced moult with old outer primaries. – **SPRING Ad** In late breeding season, heavily-bleached birds overall darker above and less clearly patterned, with reduced upperparts fringes; underparts streaking bolder and slightly concentrated at sides. **1stS** Inseparable from ad. – **SUMMER–AUTUMN Ad** As spring but broader, fresher and warmer buff fringes, and scaly pattern enhanced. **1stW** Inseparable from ad. **Juv** Like ad but has softer body-feathering and is overall paler (more sandy-buff), with broad pale fringes above and more extensive breast spotting.

BIOMETRICS (*simplex*) **L** 13–14.5 cm; **W** ♂ 75–82 mm (n 12, m 79.2), ♀ 74–78.5 mm (n 7, m 76.0); **T** ♂ 47–55 mm (n 12, m 51.1), ♀ 46–54 mm (n 7, m 49.5); **T/W** m 64.9; **B** 12.3–14.9 mm (n 20, m 13.9); **BD** 6.0–6.8 mm (n 19, m 6.4); **Ts** 20.2–23.0 mm (n 20, m 21.4). **Wing formula: p1** > pc 2–7 mm, < p2 35–43.5 mm; **p2** < wt 1–4 mm; **pp3–5** about equal and longest; **p6** < wt 1–3.5 mm; **p7** < wt 7–11 mm; **p10** < wt 13–18 mm (once 11); **s1** < wt 13.5–19 mm (once 12). Emarg. pp3–6, although on p6 often only faint.

GEOGRAPHICAL VARIATION & RANGE Moderate to small variation with only one subspecies occurring within the treated range. Three extralimital subspecies generally recognised (but not treated here). Mainly resident.

M. c. simplex, juv, Yemen, Jan: broad scales on crown and mantle; wing still entirely juv. (H. & J. Eriksen)

M. c. simplex (Heuglin, 1868) (extreme SW Saudi Arabia, Yemen, SW Oman). Treated above. Rather intermediate between African and Indian populations. Bill longer and subtly more slender than in Indian birds, thinner than in African populations (which have strongest bills). Breast markings faint and restricted, on average even more subdued than in ssp. *cantillans* of India, clearly fainter than in African birds (which have strongest breast markings).

TAXONOMIC NOTE Often considered conspecific with extralimital Australasian Horsfield's Bush Lark *M. javanica* (of SE Asia and China to Australia), but behaviour, especially flight and vocalisations, differ rather clearly.

BLACK-CROWNED SPARROW LARK
Eremopterix nigriceps (Gould, 1839)

Fr. – Moinelette à front blanc; Ger. – Weißstirnlerche
Sp. – Terrera negrita; Swe. – Svartkronad finklärka

The most widespread of its genus, this dainty lark of Saharo-Sindian deserts only periodically reaches S Israel and S Morocco slightly north of its main range further south. Most of its congeners can be found in sub-Saharan African grasslands and savannas. Gregarious, often occurring in small parties. Rather confiding, permitting close approach, and escape flights often short.

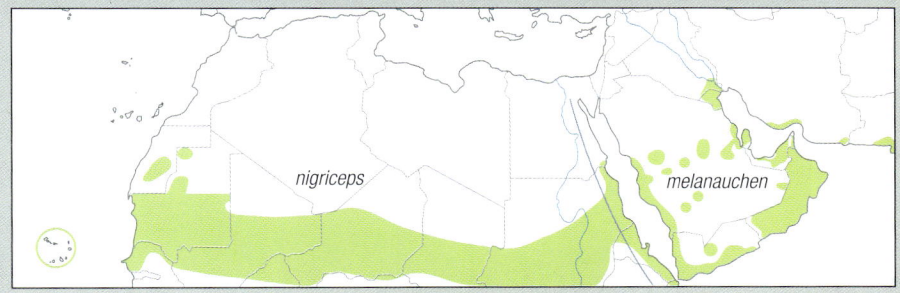

in flight, pale greyish-pink or horn-coloured legs, and *mainly dark tail with sandy-brown central feathers and sandy/whitish outermost pair*. Walks or quickly runs across open ground, often crouching low on stopping; also adopts upright stance, sometimes in exposed situations. In flight appears short-tailed with fast fluttering wingbeats and undulating motion, and escape-flight often long. Frequently gregarious.

VOCALISATIONS Song a melancholy 2–4-note phrase, e.g. *se-weee chu... se-weee chu... se-weee chu...*, rhythmically repeated. Commences with lower-pitched twittering notes and ends in a drawn-out whistle. Song appears to vary individually and geographically, e.g. in S Arabia a more complicated *chuteet teu teeu, te-wee te-wee*yup, and *zree tulitit weeh tisweeeeee*. Song-flight consists of fluttering ascent, undulating flight and staggered descent on slightly raised wings. — Commonest call a quiet *jip* or louder, raw, sparrow-like *chirp*, rather like Richard's Pipit (though more strident or buzzing). There is also a dry, low twitter, *rrrp*, and a short whistle.

SIMILAR SPECIES Ad ♂ unmistakable in Western Palearctic due to diagnostic combination of small size, piebald head, mainly black underparts, underwing-coverts and tail pattern. — ♀/young might be confused with *Bar-tailed Lark*, chiefly due to size and superficially similar jizz on ground. However, note proportionately heavier conical bill (smaller in Bar-tailed), dark tail with paler central and outer feathers (lacking black subterminal tail-band), as well as face pattern (see above), underwing colour, shorter primary projection (longer in Bar-tailed with darker tips) and calls separate them, even at moderate distances. — *Dunn's Lark* may also need to be eliminated (all three species often mix), but Black-crowned Sparrow Lark is smaller, with a dark underwing, heavy conical bill (greyish rather than pinkish), more uniform head pattern (lacking diagnostic pattern of latter), overall duller and less warm plumage (with dark median coverts) and shorter, blunter wings. Due to its gregarious nature, ad ♂♂ are usually present, making identification considerably more automatic. Congeners all occur south of the Sahara or in India. — Also, cf. *Sand Lark* for differences from that species.

AGEING & SEXING Ages usually alike after post-juv moult. Apparently only some ♂♂ can be aged by plumage, but ageing largely impossible as apparently most or all juv undertake complete moult. Sexes readily separable after post-juv moult. — Moults. Post-nuptial and post-juv complete (reports of partial post-juv moult unconfirmed), mostly Jul–Aug, but

E. n. nigriceps, ♂, Cape Verde Is, Oct: small size, piebald head and extensive black underparts render it unmistakable. Bold plumage suggests ad. Large white cheek patches, extending almost to bill base and continuing in broad collar, characteristic of this race. (E. Winkel)

IDENTIFICATION The smallest lark in the region: *tiny and rather dumpy, with a thick conical pale bill* (decurved, appearing triangular in profile and mostly horn-coloured or greyish), short rounded wings, and rather short legs, giving it a highly compact look. ♂ has striking piebald head, always with *bold white forecrown and pale cheek patches and white nape collar*, with *black continuing across entire underparts*. In contrast, upperparts almost plain, dull grey-brown indistinctly streaked dusky, except rather dark median-covert bar and ill-defined but obvious pale fringes to tertials (pale edges as broad as dark centres, giving striped look from behind). Nondescript ♀ very different, lacking black of ♂ and *recalls a small finch* or Dunn's Lark: pale rufous-sandy or cinnamon-brown above with streaked crown, and buff-white below with *indistinctly brown-streaked or mottled lower throat and breast*, with *darker median covert-bar* and, to some extent, centres of tertials and scapulars also darker (especially with wear). Face pattern rather weak but expression characteristic, with *pale eye-ring encircling small dark eye* which appears to protrude from contour of head, while 'open' appearance enhanced by *pale lores* and deep-based bill. Supercilium usually faint and often virtually absent with wear, but *pale and streaky neck-side surround behind ear-coverts* quite marked. In all plumages, *broad black* (less intense in ♀) *axillaries and underwing-coverts usually visible*

E. n. nigriceps, ♂, Western Sahara, Apr: ♂♂ of continental populations of *nigriceps* tend to have extensive white forecrown (extending beyond eye level) and large white cheeks, often reaching submoustachial region. (R. Armada)

E. n. nigriceps, ♂, Morocco, Mar: some ♂♂ browner, rather than black below and on head (cf. image on p. 33, bottom). (D. Monticelli)

E. n. nigriceps, ♂, Western Sahara, Apr: song-flight. Note black underwing-coverts, a feature of both sexes. (R. Armada)

considerably earlier in southern populations. – **SPRING Ad** See Identification for sex differences. **1stS** Much like ad, but some ♂♂ (probably referable to this age class) retain some atypical plumage, with black areas browner and smaller, and often variably admixed whitish to pale buff-brown on head and body (some even recall other African sparrow larks!). – **SUMMER–AUTUMN Ad** Much as in spring, but very fresh ♂ has some whitish tips to black parts, and upperparts of both sexes plainer and warmer (dark feather centres partially concealed). **1stW** Much like ad though young ♂ more frequently attains partial ad plumage, a few being almost intermediate between the sexes, or in extreme cases are much like ♀ but underparts partially black. **Juv** Close to ad ♀ but has softer plumage tinged more warm buff and extensively pale-edged above with better-demarcated fringes to tertials and wing-coverts, and crown faintly barred; chest faintly mottled (some obviously so).

BIOMETRICS (*nigriceps*) **L** 11–12.5 cm; **W** ♂ 76–83.5 mm (*n* 15, *m* 78.5), ♀ 72–77.5 mm (*n* 14, *m* 75.2); **T** ♂ 41–49.5 mm (*n* 16, *m* 45.8), ♀ 41–47 mm (*n* 14, *m* 43.9); **T/W** *m* 58.4; **B** 10.5–13.3 mm (*n* 29, *m* 12.3); **B**

E. n. nigriceps, ♀, Western Sahara, Apr: deep-based, conical bill and lack of obvious head pattern. (R. Armada)

E. n. nigriceps, 1stY ♂, Western Sahara, Apr: many young ♂♂ attain partial ad plumage following complete post-juv moult, but appearance highly variable. (R. Armada)

E. n. melanauchen, ♂, possibly ad, Oman, Nov: relatively extensive black on head, especially around bill, and smaller white forecrown patch, are features of this race. (D. Occhiato)

E. n. melanauchen, variation in ♀♀, Oman (left- and right-hand images: Nov; centre: Dec): all taken in Arabia and almost at same season, but note amount of variation. In this posture, the only character shared by all three is pale conical bill. Central image portrays typical ♀: open-faced with pale eye-ring encircling small dark eye, pale lores and neck-sides, irregular dark-streaked breast but plain upperparts with darker median-coverts bar. The plainest bird (left) could easily be misidentified as Bar-tailed Lark, while streaky example (right) would be a real challenge to separate from Afrotropical Chestnut-headed Sparrow-lark *E. signatus* (see vagrants section). (Left: D. Occhiato; centre: H. & J. Eriksen; right: M. Schäf)

8.8–11.2 mm (*n* 29, *m* 9.9); **BD** 6.1–7.3 mm (*n* 29, *m* 6.7); **Ts** 14.3–17.3 mm (*n* 30, *m* 16.3). **Wing formula:** p1 > pc 3 mm, to 2 mm <; **p2** < wt 0–3 mm; **pp3–4** usually equal and longest; **p5** < wt 0.5–3 mm; **p6** < wt 5.5–8 mm; **p7** < wt 9.5–15 mm; **p10** < wt 16–20.5 mm; **s1** < wt 16–22.5 mm. Emarg. pp3–5 (and often slightly on p6).

GEOGRAPHICAL VARIATION & RANGE Within the covered region only two distinct races occur, with differences most pronounced in ad ♂, but note the extent of individual variation (partially age-related and due to different amount of feather wear). ♀♀ are much more difficult to separate. A third extralimital subspecies (*affinis*) occurs in India. Resident or nomadic, making irregular irruptions or short-range movements in non-breeding season.

E. n. nigriceps (Gould, 1839) (Cape Verde Is, Western Sahara to Sudan, east to the Nile; south to Senegal, Mali, Chad). Treated above. The smallest subspecies. Some variation in ♂ plumage, but white forecrown patch usually rather large, and black lores narrow. Black of lower neck-sides does not extend onto upper mantle or nape, leaving very broad white collar connected to white cheeks. – Birds in Africa sometimes separated ('*albifrons*') on account of subtly larger size, paler overall plumage, fractionally shorter bill, and in ♂ plumage often larger white forecrown patch. However, all of these characters variable and differences subtle and far from consistent, showing much overlap, hence African breeders best included here. (Syn. *albifrons*; *modesta*.)

E. n. melanauchen (Cabanis, 1851) (E Sudan to Somalia, Arabia, S Iraq, SE Iran). Has more extensively black tracts in ♂, with only small white forehead patch, often not wider than bill (but some have a somewhat larger patch). Extensive black from lower neck-sides to upper mantle, hind-neck and, partially, onto nape, with white collar narrow, weak and, usually, disconnected from white cheeks. Both sexes relatively darker above than preceding two races. Subtly larger, longer-legged and longer-tailed: **W** ♂ 78–84 mm (*n* 11, *m* 80.4); **T** ♂ 47–50 mm (*n* 11, *m* 48.5); **T/W** *m* 60.4; **Ts** 16.5–17.9 mm (*n* 11, *m* 17.1). (Syn. *sincipitalis*.)

REFERENCES Morgan, J. H. & Palfery, J. (1986) *Sandgr.*, 8: 58–73.

E. n. melanauchen, ♀, Oman, Nov: in close views, nondescript ♀ shows darker centres to tertials, scapulars and wing-coverts, sometimes with hint of median-coverts bar. (A. Below)

DUNN'S LARK
Eremalauda dunni (Shelley, 1904)

Fr. – Alouette de Dunn; Ger. – Einödlerche
Sp. – Alondra de Dunn; Swe. – Saharaökenlärka

The Saharan counterpart of Arabian Lark, these two were previously treated as subspecies of the same species, but here the African population is given separate species status because of rather substantial differences (see Taxonomic notes). Resident or partially nomadic in the S Sahara from Sudan west to the Atlantic. It is apparently only an irregular breeder within the covered region, although recent influxes have been reported in Western Sahara and S Morocco. It occurs in similar habitats as Arabian Lark, open sandy or stony arid plains with scattered bushes. It is still a little-studied bird.

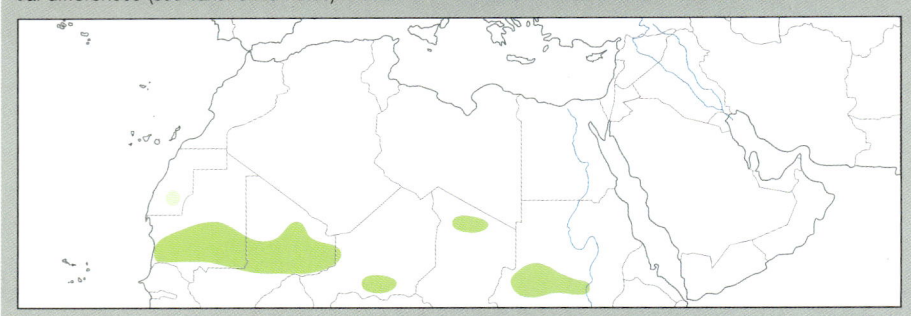

Western Sahara, Apr: moderately worn bird with very subdued face pattern. Some birds, like this, look distinctly dainty with a smaller and thinner bill, not unlike Bar-tailed Lark, but still have diagnostic finely-streaked crown and upperparts, and different tail pattern. Dark primary tips just visible beyond tertials. (R. Armada)

Western Sahara, Jan: rather fresh plumage in winter, when upperparts streaking still rather well developed. Note vague face pattern, weaker bill and unmarked breast compared to Arabian Lark. Also, upperparts streaked more reddish-brown. (C. N. G. Bocos)

IDENTIFICATION A small, pale and plain desert-dwelling African lark. Compared to Arabian Lark, Dunn's is *smaller-billed and smaller-bodied, plainer-headed* and *has less obvious streaking above*, increasing to some extent the risk of confusion with Bar-tailed Lark (which also favours sandy deserts) When seen close it is certainly streaked above, but *streaks rufous* rather than cold grey-brown and therefore less contrasting on the pinkish sandy-buff ground. Dunn's smaller size also invites greater risk of confusion with ♀ Black-crowned Sparrow Lark (see under Similar species). Beside its smaller size and more delicate build, Dunn's Lark differs from Arabian in its distinctly shorter, more *Calandrella*-like, conical bill shape, usually with a steeper-curved culmen towards its tip (more like Greater Short-toed Lark). As hinted above, also differs from Arabian Lark in its *more reddish-brown* (rather than grey-brown) *streaking on crown and hindneck*, more rufous centres to rest of upperparts, the *lack* (or near lack) *of the black moustachial stripe* of Arabian (though in some a thin or broken stripe is visible in close views), and its overall more obscure face pattern. The upperparts streaking tends to bleach to more plain appearance by early spring (later and less bleached in Arabian). Furthermore, Dunn's Lark also has fewer rufous (not brown-black) streaks on chest, and only rather indistinct grey-black tips to long primaries (more primaries visible with more extensive black tips in Arabian). The tail is relatively shorter than in Arabian Lark but shares the same pattern, and behaviour seems similar to Arabian.

VOCALISATIONS No comparative study undertaken, but presumably has similar vocalisations to better-studied Arabian Lark.

SIMILAR SPECIES Most similar to allopatric *Arabian Lark* but is smaller and differs in subtler and more pink-rufous streaking above (can look nearly unpatterned) and has less prominent or no obvious dark moustachial stripe. – For separation from *Bar-tailed* and *Desert Larks*, check for Dunn's rufous-streaked crown and upperparts (which can be obscured by feather wear, and can be somewhat less obvious in some birds, especially if distant), but in general, Bar-tailed and Desert Larks are unstreaked above, Dunn's not. Dunn's also has diagnostic bold black outer parts of its tail visible when spread (unlike in *Ammomanes*), a differently shaped, chunkier pinkish bill, and generally whiter underparts. Fine rufous breast-streaking differs notably from most Bar-tailed and Desert Larks (diffuser and blotchier in latter), but some cases of superficial overlap occur. Despite rather bland face pattern, Dunn's Lark still differs from Bar-tailed and Desert Larks by broad whitish eye surround, often narrow, angled dusky mark in front to below eye that sometimes continues

Western Sahara, Apr: some are quite strong-billed, but bill never as bulbous (instead more conical and *Calandrella*-like in shape) and less pinkish (more yellowish-flesh) than that of Arabian Lark. Also weaker face pattern, with only slight, broken moustachial stripe and almost no dark in front of eye. (R. Armada)

Western Sahara, Apr: some approach Arabian Lark in appearance, but overall head pattern weak compared to that species, including fine 'pencil-line' moustachial stripe. Upperparts streaking very thin, and bill rather narrow. (P. Adriaens)

Western Sahara, Apr: in flight, note characteristic short tail with paler central feathers, the rest being blackish with buffish-white outer edges, and dark-tipped outer primaries. (F. López)

as an ill-defined and thin eye-stripe posteriorly, and whitish-washed lores and rear supercilium. – The smaller size and weaker plumage markings mean that Dunn's Lark can resemble ♀ *Black-crowned Sparrow Lark* (streaky upperparts and quite similar tail pattern), but this confusion risk is dealt with under latter. – Especially in distant views, a *Calandrella* lark could appear superficially similar to Dunn's Lark, as these also have streaked upperparts (especially Lesser Short-toed Lark) and even a slight pale eye-ring (chiefly seen in Greater Short-toed Lark), but many features separate them.

AGEING & SEXING Ages alike after post-juv moult. Sexes alike in plumage, but size useful (see Biometrics). Seasonal variation insignificant. – Moults. As yet unstudied, but presumably similar to Arabian Lark. – **SPRING Ad** Head pattern to some extent less striking with wear, underparts noticeably whiter, and streaking often reduced or absent; rufous feather-centres to upperparts can be reduced. **1stS** Inseparable from ad. – **SUMMER–AUTUMN Ad** Overall warmer than in spring, including buff breast. **1stW** Identical to ad.

Juv Soft, fluffy body-feathering and overall gingery plumage, with numerous white tips to crown- and mantle-feathers.

BIOMETRICS L 13–14 cm; **W** ♂ 79–90 mm (n 25, m 83.0), ♀ 74–85 mm (n 16, m 78.1); **T** ♂ 46.5–57.5 mm (n 25, m 51.8), ♀ (42) 45–53 mm (n 15, m 48.2); **T/W** m 62.2; **B** ♂ 13.3–15.4 mm (n 25, m 14.5), ♀ 12.5–15.2 mm (n 16, m 13.8); **BD** 5.7–7.3 mm (n 39, m 6.7); **Ts** ♂ 19–22 mm (n 24, m 20.0), ♀ 17.0–21.5 mm (n 14, m 19.5); **HC** 3.8–7.9 mm (n 28, m 6.0). **Wing formula: p1** < pc 0–5.5 mm, or > to 2 mm; **p2** < wt 0–2.5 mm; **pp3–4** about equal and longest (rarely pp2–5); **p5** < wt 0–1.5 mm; **p6** < wt 2.5–7.5 mm; **p7** < wt 11–17 mm; **p10** < wt 18–28 mm; **s1** < wt 20–27 mm. Emarg. pp3–6.

GEOGRAPHICAL VARIATION & RANGE Monotypic. – Sub-Saharan Sahel belt from Western Sahara to Sudan, possibly regularly north to southernmost Morocco; all populations resident or nomadic, at times making irregular irruptions outside normal range. (Syn. *pallidior*.)

TAXONOMIC NOTES Dunn's Lark and Arabian Lark have often been viewed as conspecific in the past. Considering the marked size difference (no overlap in wing length), the small but consistent plumage differences, their allopatric ranges divided by the Red Sea, and the substantial genetic distance (> 7%. U. Olsson *in litt.*), it seems appropriate to afford them separate species status. – The generic treatment of this species (and of Arabian Lark) has varied over the years, they being sometimes included in *Ammomanes* or afforded a separate genus *Calendula* or *Pyrrhulauda*.

REFERENCES Alström, P. *et al.* (2013) *Mol. Phyl. & Evol.*, 69: 1043–1056. – Copete, J. L. *et al.* (2008) *Alula*, 14: 88–93. – Heard, C. D. R. & Kirwan, G. M. (1997) *Sandgrouse*, 19: 8–11. – Lees, A. C. & Moores, R. D. (2006) *BB*, 99: 482–484. – Mild, K. (1990) *Bird Songs of Israel and the Middle East*. Audiotapes and booklet. Stockholm. – Round, P. D. & Walsh, T. A. (1981) *Sandgrouse*, 3: 78–83. – Shirihai, H. (1994) *DB*, 16: 1–9. – Shirihai, H., Mullarney, K. & Grant, P. (1990) *BW*, 3: 15–21.

Western Sahara, Apr: viewed from below against blue sky, square-ended blackish tail contrasts strongly with rest of plumage. (R. Armada)

Western Sahara, Apr: in rather worn plumage, at certain angles can show very obscure upperparts streaking, inviting confusion with Bar-tailed or Desert Larks, but note shape and colour of bill, dusky mark in front of eye, and tail pattern. Arabian Lark never bleaches as plain above as this, especially so early in year. (R. Armada)

ARABIAN LARK
Eremalauda eremodites (Meinertzhagen, 1923)

Fr. – Alouette d'Arabie; Ger. – Arabienlerche
Sp. – Alondra árabe; Swe. – Streckad ökenlärka

Resident or partially nomadic in Arabia and S Middle East. It regularly breeds in Jordan and more sporadically in Israel and N Sinai, where small-scale influxes have been noted, and it has recently been seen in C Syria. Often treated as an eastern subspecies of its close relative in the Sahara, Dunn's Lark, but here afforded species status (see Taxonomic note under Dunn's Lark). It favours open sandy or stony arid plains with scattered acacias, always interspersed with patchy grasses and bushes, where it often occurs with Bar-tailed, Hoopoe, Temminck's and Thick-billed Larks; it tends to avoid the rocky mountain-slopes preferred by Desert Lark, despite some overlap in lower hills. Its sudden, irruptive winter movements are typical, following which it may remain to nest in new areas, which are equally rapidly evacuated post-breeding.

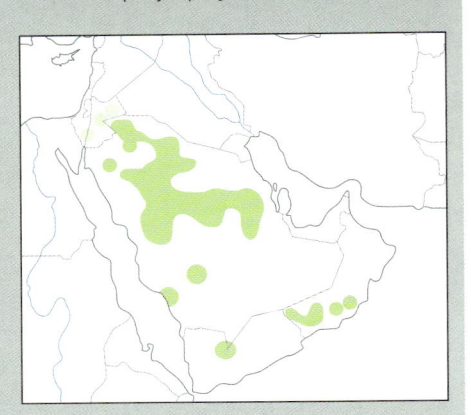

IDENTIFICATION Size roughly intermediate between Desert and Bar-tailed Larks, but larger-headed with *heavy pinkish bill* (convex culmen, bill-tip 'swollen'), *dark brown-streaked crown and* (less prominently) *streaked upperparts*, grading from sandy to pink-isabelline (pinkish-cinnamon or rufous tone to wings, particularly tertials, greater coverts and bases of flight-feathers). Nape and neck-sides can appear to be subtly greyer-tinged. Underparts mostly cream-white with variable *buff wash and narrow streaks on breast* (especially sides). Complex face pattern variable, but note *narrow dark moustachial stripe*, latter often connected to *narrow, angled dusky stripe below broad whitish eye-surround* that continues as eye-stripe behind eye. Whitish lores, rear supercilium and *ill-defined lower and rear ear-covert patches* often obvious, creating 'staring' look. Long pinkish-cinnamon *tertials almost entirely cloak flight-feathers*. In flight rather short *tail has striking bold black outer parts contrasting with pale red-brown central feathers* on upperside (overhead, whole tail appears contrastingly black, short and square). Closely related Dunn's Lark in Africa differs rather clearly from Arabian Lark in being more evenly streaked pink-rufous above and in having weaker or lacking dark markings on head-sides. Tail subtly longer in Dunn's but similarly patterned, and behaviour seems similar to Arabian Lark. Often very tame, permitting close approach and rarely flies from observer. Shuffling walk, when carriage horizontal, but can appear more upright, and runs across flat ground. Flight rather jerky, and wing shape recalls *Mirafra* lark, appearing round. Tail often fanned immediately prior to landing, revealing diagnostic pattern.

VOCALISATIONS Song delivered mostly from ground, a sustained, fast series of rather explosive and rambling, scratchy warbles and creaky twittering notes, with occasional churrs or wheezy notes mixed in, e.g. *twit che-wit tcherr wit-wee-ree-churr-re-ree che*. Aerial song, given while fluttering in one spot up to 20 m above ground, often comprises well-spaced shorter phrases, rather like song of Greater Short-toed Lark, hard warbles with low-pitched melodious whistles and rattling notes, but phrases sometimes longer and more elaborate, and occasionally ends in drawn-out *cheee*. – Contact call on ground a subdued *prt*, audible only c. 10 m. Flight call either a monosyllabic single *wazz* or liquid *pscheep* (not unlike Richard's Pipit), and a hard, purring, upwards-inflected *drrruee* (somewhat like Bar-tailed or Crested Larks), sometimes followed by a rather soft phrase of 3–5 notes. Also occasionally heard is a Tawny Pipit-like *chilp* or doubled *tchup-tchup* recalling Greater Short-toed Lark.

Israel, Mar: highly distinctive compared to Dunn's Lark, with rather large head, heavy-based pinkish bill with deep bulbous tip, and finely but distinctly brown-streaked crown and upperparts. Often has greyer-tinged nape and neck-sides, with narrow streaks extending onto breast-sides. Look for diagnostic narrow but unbroken dark moustachial stripe, broad whitish eye-surround, and ill-defined lower ear-coverts patch. (T. Muukkonen)

Kuwait, Feb: typical individual, with dark brown-streaked crown and upperparts, and long ginger tertials almost fully cloaking dark wingtips. Note also the bulbous bill. (G. Brown)

Israel, Apr: well-marked bird, which unlike Dunn's Lark has narrow streaks on neck-sides, extending to breast-sides, narrow but unbroken dark moustachial stripe, broad whitish eye-surround and supercilium, and darker eye-stripe. More buff- or rufous-tinged Greater Short-toed Larks could be confused with Arabian, although bill colour and shape, and tail pattern always separate them. (R. Mizrachi)

HC 4.0–7.5 mm (*n* 29 *m* 6.5). **Wing formula: p1** < pc 0–5.5 mm, or > to 2 mm; **p2** < wt 0–2.5 mm; **pp3–4** about equal and longest (rarely pp2–5); **p5** < wt 0–1.5 mm; **p6** < wt 2.5–7.5 mm; **p7** < wt 11–17 mm; **p10** < wt 18–28 mm; **s1** < wt 20–27 mm. Emarg. pp3–6.

GEOGRAPHICAL VARIATION & RANGE Monotypic. – Arabia, Jordan, irregularly to Israel and perhaps further north in Middle East; resident or nomadic, at times making irregular irruptions outside normal range. (Syn. *kinneari*.)

TAXONOMIC NOTE See Dunn's Lark.

REFERENCES Alström, P. *et al.* (2013) *Mol. Phyl. & Evol.*, 69: 1043–1056. – Copete, J. L. *et al.* (2008) *Alula*, 14: 88–93. – Heard, C. D. R. & Kirwan, G. M. (1997) *Sandgrouse*, 19: 8–11. – Lees, A. C. & Moores, R. D. (2006) *BB*, 99: 482–484. – Mild, K. (1990) *Bird Songs of Israel and the Middle East*. Audiotapes and booklet. Stockholm. – Shirihai, H. (1994) *DB*, 16: 1–9.

Israel, Apr: a rather pale example, with less heavily-streaked upperparts, and quite weak face markings. Very difficult or impossible to distinguish from Dunn's Lark, although the decidedly larger and heavier bill, if judged correctly, could support identification of tricky individuals outside either species' main range. Coloration almost blends into stony desert habitat, making bird difficult to locate, even when just 5 m away and singing. (H. Shirihai)

SIMILAR SPECIES Needs to be separated above all from *Bar-tailed* and *Desert Larks*, which are often present in same area. Especially note the relatively massive (swollen and blunt-ended), almost all-pinkish to straw-coloured bill of Arabian Lark, the (variably) streaked upperparts with usually noticeably-fine dark streaking on crown and more diffuse and faint streaking on mantle (both essentially plain in the others), more striking head pattern (fine dark moustachial stripe and eye-stripe, and invariably broad pale eye-ring noticeable), and quite different uppertail pattern with whole outertail black contrasting with pale brown centre when tail is spread. Also note variable breast-side streaking and strong pinkish-cinnamon tone to upperparts and wings (chiefly when fresh). Structural features are also useful: Arabian Lark has an obviously larger head, distinctly broader and more rounded wings, and no (or almost no) primary projection. Arabian recalls Bar-tailed in general behaviour but is less constantly active. While Arabian and Bar-tailed Larks favour open sandy habitats, Desert Lark occurs in rockier, often mountainous, landscapes, although all three could overlap. – Especially in distant views, Arabian can be confused with a *Calandrella* lark, as these also have streaked upperparts and a slight pale eye-ring, but many features separate them, notably the rather plain rufous-tinged tertials and all-black tail-sides with only narrow pale edge on outermost, lacking obvious dark tertial-centres and white tail-sides of *Greater Short-toed Lark*. – See *Black-crowned Sparrow Lark* for potential confusion with ♀ of that species. – Being widely allopatric, mainly sedentary Arabian and *Dunn's Larks* are hardly candidates for confusion. See under latter for differences.

AGEING & SEXING Ages alike after post-juv moult. Sexes alike in plumage, but size useful (see Biometrics). Seasonal variation insignificant. – Moults. Post-nuptial and post-juv (both complete), mostly Jun–Aug. – **SPRING Ad** Head pattern to some extent less striking with wear, underparts noticeably whiter, and streaking often reduced or absent; dark feather-centres to upperparts can be reduced. **1stS** Inseparable from ad. – **SUMMER–AUTUMN Ad** Overall warmer, including buff breast. **1stW** Identical to ad. **Juv** Soft, fluffy body-feathering and overall gingery plumage, with numerous white tips to crown- and mantle-feathers.

BIOMETRICS L 14–15 cm; **W** ♂ 93–100.5 mm (*n* 21, *m* 96.8), ♀ 84.5–90.5 mm (*n* 11, *m* 87.4); **T** ♂ 54–60 mm (*n* 21, *m* 57.3), ♀ 49–54 mm (*n* 11, *m* 51.6); **T/W** *m* 59.1; **B** ♂ 15.0–17.1 mm (*n* 21, *m* 16.1), ♀ 13.4–15.6 mm (*n* 10, *m* 14.6); **BD** 6–7 mm (*n* 31, *m* 6.7); **Ts** ♂ 20.3–23.6 mm (*n* 21, *m* 21.9), ♀ 20.1–22.3 mm (*n* 11, *m* 21.4);

Israel, May: the first-ever nest to be found, during a massive invasion in late 1980s. Note how upperparts streaking remains obvious even as late as May. (H. Shirihai)

Israel, Apr: in flight note striking black panels on outertail contrasting with pale red-brown central and whitish outer tail-feathers. (Y. Perlman)

BAR-TAILED LARK
Ammomanes cinctura (Gould, 1839)

Fr. – Ammomane élégante; Ger. – Sandlerche
Sp. – Terrera colinegra; Swe. – Sandökenlärka

Smaller than Desert Lark, from which it tends to be segregated by its favouring more open, flatter and sandier desert environments. It often strikes you as a daintier and more energetic bird than Desert Lark, constantly rushing ahead over the sand with only brief pauses. Essentially a Saharo-Sindian species, inhabiting true deserts to semi-deserts from the Cape Verde Is to the Middle East, S Afghanistan and S Pakistan. Wandering birds have reached north to Cyprus and Turkey in spring.

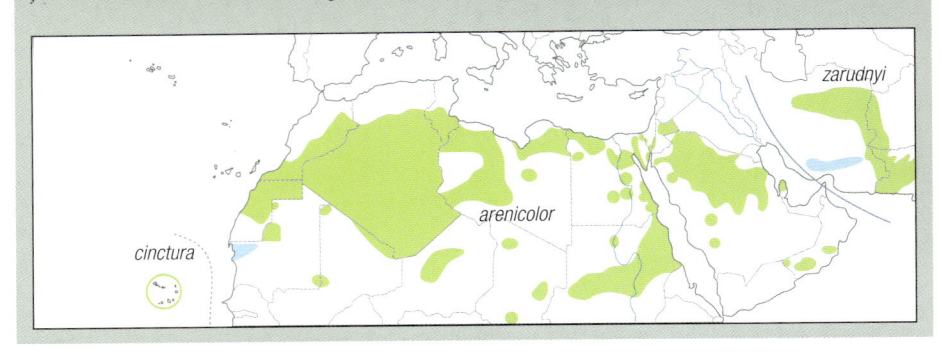

A. c. arenicolor, Egypt, Sep: unusual image depicting diagnostic features in flight, especially blackish terminal tail-band. Note also how rufous the remiges can look. (M. Heiß)

A. c. cinctura, Cape Verde Is, Dec: generally warmer/rustier than those in Africa and Middle East. Mottled and quite strong breast-streaking might indicate young bird, but reliable ageing impossible following post-juv moult. (H. & J. Aalto)

A. c. arenicolor, Jordan, Apr: in eastern Jordan, some are very pale, even in the dark rocky Basalt Desert (where the local race *annae* of the Desert Lark is the darkest). Often, as here, has prominent pale 'spectacles' around the small dark eyes. In the hottest part of the day, wings frequently held a bit open to help keep cool. (H. Shirihai)

IDENTIFICATION A rather small, neatly proportioned lark, with a *small rounded head, miniature bill*, compact shape and essentially *unstreaked pale grey-buff upperparts*, with obvious *orange tone to upperwings and tail base*. Bill thin but rather short, appearing stubby, and usually sandy-pink, with culmen and tip often darker. Diagnostic is the *deep dark rufous-brown tail with almost solid black terminal band* (contrast and width of band vary). Virtually unpatterned head (although crown of ♂ often finely streaked darker, visible at closest range) and body except *faint pale supercilium*, tips to wing-coverts and fringes to tertials, and primaries variably tipped dark to black (often quite contrasting). Underparts pale buff-white with buff tones strongest on slightly mottled chest and palest on chin, throat and undertail-coverts. Eye small and dark. Stance frequently upright, emphasizing long-legged appearance, and typically performs short fast runs with abrupt brief pauses (seldom still for long) across flat ground, heightening its compact appearance. Rarely permits close approach, and escape-flight is long, fast and low on jerky wingbeats (or runs off at some speed).

VOCALISATIONS Song is very distinctive, comprising mechanically and slowly repeated squeaky notes like creaking of iron gate in wind (or old-fashioned water pump!), each squeaky note preceded by one or two subdued short notes, e.g. (*te-eh*) *chu-weest* repeated 5–10 times during series of circular, steep undulating high flights (the whistle given in rhythm with dips); also sings from ground or bush (a slightly more complex low creaky warble). – Commonest calls a short, nasal, clicking *chep*, a little like Linnet, given singly. Also a thin, rolling *cherr* and a soft piping *peeyu*.

SIMILAR SPECIES Most obvious confusion risk is with *Desert Lark*, but Bar-tailed is generally smaller, with a more rounded body and head, shorter and slenderer bill, freer gait (possibly due to proportionately slightly longer, thinner legs), faster flight and, especially, more restless running behaviour. Bar-tailed also has a comparatively short primary projection, only *c.* 1/4 to 1/3 of tertial length, with 2–3 contrastingly blackish primary tips visible (Desert's primary projection is *c.* 1/2 tertial length, usually with four less-contrasting tips visible). These structural features, and the dark transverse terminal band on the tail, are diagnostic of Bar-tailed Lark, and should immediately distinguish it from Desert Lark and *Dunn's Lark*. Furthermore, Bar-tailed usually also has almost unmarked lores (dusky grey stripe through central lores in Desert) and, on most, a more pronounced pale crescent behind ear-coverts (much less so in Desert), but lacks Desert's extensive breast streaking, which is often quite obvious if ill-defined. – A perhaps unexpected confusion risk occurs in summer with juvenile *Temminck's Lark*, which has plain head and breast/neck pattern before post-juv moult, and very similar colours to Bar-tailed, including blackish primary tips visible beyond tertials. Note heavier bill of Temminck's (almost same size as in Desert Lark), faintly white-spotted upperparts, and, in worn plumage, hint of rufous blotches and all-blackish outer tail with white sides (not just black-tipped rufous feathers as in Bar-tailed). – For distinction from ♀/juv *Black-crowned Sparrow Lark*, see that species.

AGEING & SEXING Ages alike after post-juv moult. Sexes very similar in plumage, but subtle difference possibly useful (see below), as is sometimes size (see Biometrics). Seasonal variation moderate. – Moults. Post-nuptial and post-juv moults complete, mostly mid Jun–Sep, but earlier in some southern populations. No pre-nuptial moult. – **SPRING Ad** Sexes similar, but ♂ nearly always finely streaked darker brown-grey on crown, whereas ♀ is either uniform pinkish-sandy, or has only very diffuse darker mottling. (This sexual difference is rather obvious in examined specimens but requires confirmation with more material.) By late spring, or prior to moult, heavily worn with some dark feather-bases exposed, face pattern even less marked, orange tone to upperwing and tail, and distinctness of dark primary-tips and tail-band much reduced. **1stS** As ad. – **SUMMER–AUTUMN Ad** As spring, with some individual variation; a few being distinctly duller/greyer above or more rufous-toned. **1stW** Those with slightly more mottled and stronger breast-streaking are apparently young, but no safe ageing criteria known. **Juv** Soft fluffy body-feathering, less extensive and less sharply-defined black tips to primaries and tail-feathers, with overall buffier plumage and indistinct pale feather tips create sparse mottling on back and breast.

BIOMETRICS (*arenicolor*) **L** 13.5–14.5 cm; **W** ♂ 91.5–99 mm (n 23, m 94.8), ♀ 85–92 mm (n 12, m 89.4); **T** ♂ 50–59 mm (n 23, m 55.0), ♀ 50–54.5 mm (n 12, m 52.0); **T/W** m 58.0; **B** ♂ 12.4–14.5 mm (n 23, m 13.5), ♀ 12.0–13.4 mm (n 12, m 12.8); **BD** 4.3–5.4 mm (n 34, m 5.0); **Ts** 19.7–22.3 mm (n 34, m 21.1); **HC** 4–6 mm. **Wing formula: p1** > pc 0–7 mm (rarely 0.5 mm <); **p2** < wt 0.5–4 mm, =5 or 5/6; **pp3–4** longest (rarely either of p3 or p4 0.5–1 mm <); **p5** < wt 1–3.5 mm; **p6** < wt 6.5–9.5 mm; **p7** < wt 12.5–17 mm; **p10** < wt 23–29 mm; **s1** < wt 24–31 mm. Emarg. pp3–5 (sometimes faintly also on p6).

Bar-tailed *A. c. arenicolor* (left: Morocco, Dec) and Desert Larks *A. d. payni* (right: Morocco, Feb): three main differences (although not always as obvious as here): (i) Bar-tailed has rather short primary projection, with 2–3 contrastingly blackish primary tips visible (Desert has usually four less contrasting tips); (ii) diagnostic rufous tail with black terminal band (Desert has rusty-brown base with ill-defined but broad darker central and distal thirds), although contrast and width of band vary; (iii) shorter/narrower bill with lower base pinkish-horn (Desert has longer/stronger bill, with pale base mainly yellowish). (Left: G. Ekström; right: R. Schols)

A. c. arenicolor, Western Sahara, Apr: normally small bill not always obvious; this bird has a relatively long, pointed bill. Unmarked upperparts and blackish terminal tail-band just visible. (R. Armada)

A. c. arenicolor, Morocco, Apr: a rather atypical longer/thicker-billed bird with (misleadingly) slightly darker culmen; such variation must be taken into account. Birds with warm rusty tones apparently frequent in western populations, but here probably the result of soil discoloration. (T. Grüner)

GEOGRAPHICAL VARIATION & RANGE Three races generally recognised, geographically isolated and rather clearly different. Birds in the south-west are darker and warmer brown, those in the centre pale and sand-coloured, those in the north-east slightly darker again and much greyer, and also longer-winged. Mainly resident, but short-range movements by continental populations noted in winter.

A. c. cinctura (Gould, 1839) (Cape Verde Is). Has deeper rufous-cinnamon upperparts and breast than *arenicolor*, with black on tips to tail-feathers and primaries deeper and more distinctly contrasting. Also, upperwing more uniformly dark cinnamon or pinkish-brown. About same body-size but has shorter wings and tail than the other races, especially compared to *zarudnyi*. **W** ♂ 87–95 mm (n 14, m 91.3), ♀ 85–93 mm (n 12, m 88.9); **T** ♂ 49–57 mm (n 14, m 52.8), ♀ 45–55.5 mm (n 12, m 50.2); **T/W** m 56.8; **B** 11.2–14.0 mm (n 25, m 13.0); **BD** 4.7–5.4 mm (n 25, m 5.1); **Ts** 20.0–22.3 mm (n 24, m 21.3). **Wing formula: p10** < wt 20–26 mm; **s1** < wt 21–27.5 mm.

A. c. arenicolor (Sundevall, 1850) (N Africa, Middle East, much of Arabia). Described above under Identification and Biometrics. Compared to *cinctura* less rufous-tinged and paler, more sand-coloured, and black tail-band narrower and more diffusely marked. Compared to *zarudnyi*, clearly more pinkish-buff, less greyish. Has about same body-size as *zarudnyi* but is shorter-winged. – There is a deep divide genetically between *arenicolor* from Morocco and *arenicolor* from Arabia (Alström *et al.* 2013), thus with more study we might see former population described as separate subspecies based on hitherto overlooked morphological or other differences. Ssp. *arenicolor* was originally described from 'lower Egypt or Arabia petraea'. (Syn. *kinneari*; *pallens*.)

A. c. zarudnyi Hartert, 1902 (C & E Iran, Afghanistan, Pakistan). Distinctly duller and greyer than other forms, with drab brown-grey upperparts, and underparts more buff-grey (less white). Has even broader black tips to tail-feathers than *cinctura*, covering about half tail. **W** ♂ 95.5–103 mm (n 13, m 100.3), ♀ 91–95.5 mm (n 12, m 93.7); **T** ♂ 54–64 mm (n 13, m 60.3), ♀ 52–60 mm (n 12, m 55.3); **T/W** 59.7; **B** ♂ 11.5–13.6 mm (n 13, m 12.8), ♀ 11.7–13.0 mm (n 12, m 12.3); **BD** 4.6–5.5 mm (n 24, m 5.1); **Ts** 20.3–22.7 mm (n 25, m 21.5).

REFERENCES Alström, P. *et al.* (2013) *Mol. Phyl. & Evol.*, 69: 1043–1056. – Shirihai, H. (1994) *DB*, 16: 1–9. – Shirihai, H., Mullarney, K. & Grant, P. (1990) *BW*, 3: 15–21.

A. c. arenicolor, Morocco, Feb: misleading dark culmen and tip, but bill still small and delicate, with sandy-pink base to lower mandible (not yellowish as often true of Desert). Almost unmarked lores (no dusky stripe like Desert Lark). Some birds, like this, show weak moustachial stripe. Distinctive short black primary extension beyond tertial tips and black tail-tip. (R. Schols)

A. c. arenicolor, Israel, Mar: sand-coloured plumage permits perfect blending in desert surround. Often makes sudden freeze and then run, almost recalling a desert lizard motion. Thus, watch in the right habitat for the suspected movements. Secondly, beware that a few birds may have better-developed moustachial stripe, and especially breast streaking, which can invite confusion with Arabian Lark. (H. Shirihai)

DESERT LARK
Ammomanes deserti (M. H. C. Lichtenstein, 1823)

Fr. – Ammomane isabelline; Ger. – Steinlerche
Sp. – Terrera sahariana; Swe. – Stenökenlärka

Widespread in both true deserts and semi-deserts, Desert Lark is one of the few species able to survive in the driest rocky and virtually vegetation-free habitats. It is resident across the Sahara, and in Arabia north through the Levant to SE Turkey, thence east to NW India. Rather than inhabiting the flattest parts at the bottom of valleys, this species is often found on slightly elevated hillsides and ridges with more pebbles, boulders and scree than sand, readily also near steeper rocks, ravines and even buildings.

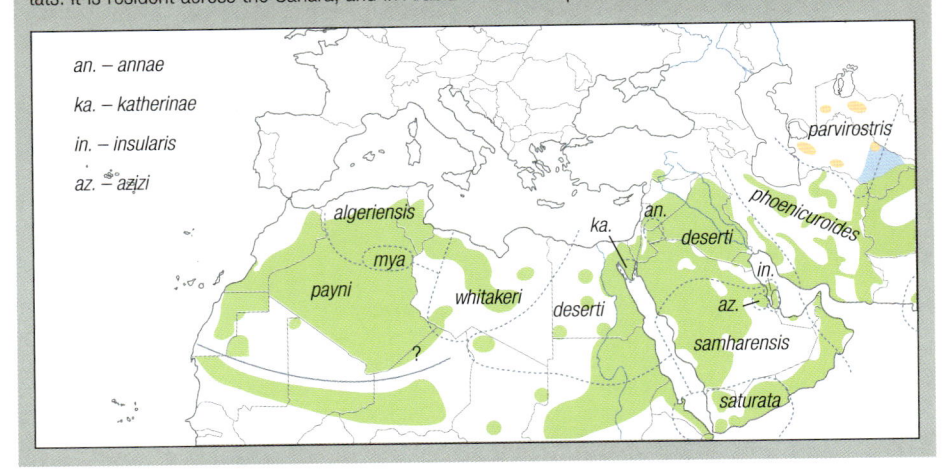

A. d. payni, Morocco, Apr: an extreme cinnamon-tinged individual (presumably due to soil discoloration or reflection from ground). (T. Grüner)

IDENTIFICATION Medium-sized, virtually unpatterned lark with a fairly long pointed bill, rather large head and quite long wings. Plumage subject to wide tonal variation according to habitat, but always essentially *plain above*, with at most very weak crown-streaking and very faint dark brown lateral throat-stripe and moustachial stripe, but pale fore-supercilium and narrow eye-ring, *dark loral streak*, and *diffuse streaking on throat and breast* often noticeable. However, note two key characters across entire range compared to Bar-tailed Lark: (i) *longer and thicker bill, at times bulbous-looking*, with *much of upper mandible and tip dark*, and basal lower mandible always pale horn-yellow, if anything with an orange tinge invariably lacking in Bar-tailed; and (ii) *darker central tail-feathers and distal 1/3 of outer rectrices, which grade into dull rusty-brown base of tail*. Upperparts dull greyish brown-isabelline (but some forms are sandier, others more rufous-buff or much duskier), with ill-defined paler tips and fringes to wing-coverts and tertials.

Underparts generally pale cream to buff, often with warmer elements. Eye dark and rather small. Usually walks unhurriedly, but occasionally runs. Compared to Bar-tailed Lark, less energetic, stops and pecks at ground for long periods, or perches prominently for some time. Often tame, permitting very close approach, and escape-flight short and high, on regular wingbeats.

VOCALISATIONS Song melodious and far-carrying, a mechanical-sounding phrase of 3–6 rather low-pitched syllables (repeated 6–15 times), well spaced, usually with characteristic rolling 'r' in voice, e.g. **wree**-chrre-*ay*... *choee*... **wree**-chrre-*ay*... *choee*... etc., or *chuwe-trutru*... *chu-***weeü***... *chuwe-trutru*... *chu-***weeü***... (some variations recalling song of Long-billed Pipit), usually given in short, undulating song-flight, but uttered partially also from ground. – Calls somewhat variable but three types emerge as most common, (i) a low-pitched, resounding, monosyllabic straight call without rolling 'r', *chip*, slightly recalling a single emphatic call from Greenfinch, or more drawn-out, *chuu*; (ii) a slightly drawn-out, rolling, metallic *chrreh*, which slightly recalls Alpine Chough; and (iii) a variation related to the second call but more extended and clearly undulating in pitch, *chrriiayee*.

SIMILAR SPECIES Generally requires careful separation from *Bar-tailed Lark*. Remember that Desert Lark inhabits broken, rocky and, above all, sloping habitats, not flat sandy deserts like Bar-tailed, and its jizz is quite different, with a longer primary projection, proportionately longer tail, and a longer, less stubby bill than Bar-tailed. Compared to Bar-tailed, the pattern of the spread tail is important, but some have more conspicuous dark, and perhaps more rufous ground colour, which to the inexperienced might suggest Bar-tailed. However, latter has dark coloration more confined to tail-tip, usually forming a clear-cut subterminal band, which does not grade into the ground colour of the tail base as in Desert. Wing pattern differs from Bar-tailed in lacking latter's distinctly darker primary tips and slightly more pronounced darker-centred median coverts, but tertials usually obviously patterned, breast often quite heavily streaked and white underparts always sullied grey to deeper buff, though differences not always clear. In combination, the long pointed bill with yellowish lower base, tail pattern, overall shape and generally less restless, more methodical behaviour, as well as habitat choice, are key to separation from Bar-tailed. – Distinguished from *Dunn's Lark* by plain crown and mantle, lacking any clear streaking (though may have diffuse mottling in worn plumage), and more pointed, less deep-based bill.

AGEING & SEXING Ages alike after post-juv moult. Sexes alike. – Moults. Post-nuptial and post-juv moults (both complete) mostly in Jun–Sep. Seasonal variation insignificant. – **SPRING Ad** Mantle may appear slightly mottled when heavily abraded and bleached, face pattern often appears less distinct, and entire plumage is overall greyer and duller, with rufous in tail, tertials and remiges duller and more restricted. **1stS** As ad. – **AUTUMN Ad** Warmer toned than in spring. **1stW** As ad. **Juv** Much as ad but has soft, fluffy body-feathering, with fewer streaks on breast.

BIOMETRICS (*deserti*) **L** 15–16 cm; **W** ♂ 96–106 mm (n 16, m 102.1), ♀ 92–100 mm (n 23, m 95.5); **T** ♂ 59–71 mm (n 16, m 65.5), ♀ 58–67 mm (n 23, m 60.9); **T/W** m 64.0; **B** ♂ 16.4–19.5 mm (n 16, m 17.9), ♀ 14.6–17.6 mm (n 22, m 16.2); **BD** 5.2–6.9 mm (n 41, m 6.2); **Ts** 20.0–23.7 mm (n 39, m 21.8). **Wing formula:** **p1** > pc 4–14 mm (once 17), < **p2** 33–48 mm; **p2** < wt

A. d. payni, Morocco, Jan: this race is tinged soft cinnamon-grey above, wearing more greyish. It has broad rufous fringes to remiges, a bright chestnut rump and tail-base, and is warm pink-buff below. (K. De Rouck)

A. d. payni, S. Morocco, Jan: contrasting cinnamon-tinged underparts, characteristic of this race, and relatively large appearance with longish bill, especially notable in montane populations—like this individual. Unlike Bar-tailed Lark, has strongly bicoloured bill (with orange-yellow lower mandible, not pinkish) but more diffuse tail pattern and lacks contrasting dark wingtip. (M. Schäf)

A. d. algeriensis, Tunisia, Apr: a quite distinctly pale race, with upperparts tinged pink-cinnamon, and rather whitish underparts. Note yellowish base to bill. (D. Occhiato)

Possibly *A. d. algeriensis*, Tunisia, Nov: short/delicate bill and pink-cinnamon tone to upperparts suggestive of this race. Unusually, bill appears very thin, but lower mandible is orange-yellow (not pinkish like Bar-tailed Lark). (A. Torés Sánchez)

4.5–10.5 mm, =6–7; **pp3–5** about equal and longest (but often either of p3 and p5, or both, subtly shorter; **p6** < wt 1–6.5 mm; **p7** < wt 6.5–13 mm; **p10** < wt 16–25 mm; **s1** < wt 18–27 mm. Emarg. pp3–7, but usually only faint on p7.

GEOGRAPHICAL VARIATION & RANGE Complex but predominantly clinal and often subtle geographical variation, and debatable whether merely tonal (and very limited size) differences invariably warrant naming, as these can represent local and quite restricted responses to soil, sunlight and moisture conditions. About 20 races described from the covered region alone (e.g. 18 in Peters 1960, 20 in Dickinson 2003), but some of these undoubtedly based on short series or very local variations, while we focus on the 13 most distinctive which differ in both colour and size, and which are sufficiently consistent over a reasonably large area. It is quite likely that examination of larger and more comprehensive material may yield results that could argue against the continued recognition even of some of these. It should be noted also that specimens in collections are unavailable from extensive areas. As to claims that *geyri* breeds in SE Algeria (e.g. H&M 4), this has not been possible to confirm, and it has been left untreated as it is extralimital. — It has been shown recently (Alström et al. 2013) that two samples from Pakistan (*phoenicuroides*) together with one from Israel (*deserti*) were quite well separated genetically from birds from Morocco (*payni*) and Saudi Arabia (near Taif, thus probably *samharensis*) and Jordan (*annae*), possibly indicating that the species should be split. However, until better studied and supplemented by morphological and other differences there is no immediate need to alter existing species taxonomy. — As usual, races below are arranged broadly from west to east. All populations are basically resident, or make only local movements in winter.

A. d. payni Hartert, 1924 (Morocco, S Algeria south of Tagemaït, east to Ahaggar and Tassili n'Ajjer). Predominantly medium greyish-brown with slight cinnamon-grey hue above, and tertials and secondary-coverts duskier brown contrasting with broad rufous fringes to wing-feathers. Chestnut rump, uppertail-coverts and tail-base often quite contrasting (but rufous tinge notable also on darker and greyer back). Also, compared to *algeriensis* warmer pink-buff below, with throat and breast whitish-cream heavily streaked dark grey, and belly warm buff. **W** ♂ 101–110 mm (*n* 19, *m* 105.2), ♀ 95–102.5 mm (*n* 16, *m* 99.6); **T** ♂ 64–74 mm (*n* 19, *m* 69.1), ♀ 60–70 mm (*n* 15, *m* 65.8); **B** ♂ 16.2–19.0 mm (*n* 18, *m* 17.8), ♀ 16.1–17.8 mm (*n* 16, *m* 16.8); **BD** 5.3–6.7 mm (*n* 38, *m* 6.1); **Ts** 20.3–23.9 mm (*n* 34, *m* 22.2). — Birds in Ahaggar and Tassili n'Ajjer often referred to separate ssp. '*janeti*' or to any of *algeriensis*, *mya* or *whitakeri*, but comparison of long series of *payni* and birds from this area (incl. type of *janeti*) revealed no consistent difference except a very slight clinal tendency to become less grey on back, hence included here. (Syn. *janeti*.)

A. d. algeriensis Sharpe, 1890 (N Algeria incl. Atlas Mts, northern edge of Algerian Sahara, C & S Tunisia, NW Libya). Quite distinctly paler and averages smaller than preceding race, being rather pinkish-tinged and sandy, with upperparts tinged pink-cinnamon, including most tertials and secondary-coverts, which have vinous-rufous cast, offering only slight contrast with rufous fringes to remiges. Throat whitish-cream and breast cream or whitish-buff with ill-defined, smaller, subdued brownish streaks on throat- and breast-sides, less prominent on average than in *payni* and can appear unstreaked. Lower breast and belly pale cinnamon-buff. **W** ♂ 97.5–106 mm (*n* 23, *m* 101.7), ♀ 90–99 mm (*n* 18, *m* 94.3); **T** ♂ 62.5–70 mm (*n* 23, *m* 66.1), ♀ 56.5–65 mm (*n* 17, *m* 61.8); **T/W** *m* 65.2; **B** ♂ 15.5–19.1 mm (*n* 22, *m* 17.0), ♀ 15.0–17.1 mm (*n* 17, *m* 15.8). **BD** 5.0–6.3 mm (*n* 40, *m* 5.7); **Ts** 20.2–23.5 mm (*n* 39, *m* 21.6). **Wing formula: p1** > pc 6–12 mm, < p2 37–47.5 mm; **p2** < wt 6.5–9 mm, =6–7; pp3–5 about equal and longest (p3 and p5 sometimes < wt 0.5–2 mm); **p6** < wt 1–4.5 mm; **p7** < wt 7.5–12 mm; **p10** < wt 17–23 mm; **s1** < wt 19–23 mm. Emarg. pp3–7 (less prominently on p7). (Syn. *lusitanica*.)

A. d. deserti, Israel, Mar: this race is usually dull or drab grey-brown above, but some variation, and this bird is tinged more sandy-buff. Streaking on breast often pronounced, as here. (O. Plantema)

A. d. deserti, Syria, May: a more sandy-buff or cinnamon-pink individual, less drab grey-brown above, with only slight contrast against rufous rump and tail-base. (A. Audevard)

○ **A. d. mya** Hartert, 1912 (C Algerian Sahara from El Goléa south to Tagemaït plateau). Differs subtly but apparently consistently in being a trifle larger and slightly paler than *algeriensis* with a heavier bill and longer tarsi. **W** ♂ 106–111 mm (*n* 15, *m* 108.1), ♀ 97–104 mm (*n* 14, *m* 100.3); **T** ♂ 70–75.5 mm (*n* 15, *m* 72.6), ♀ 64–71 mm (*n* 14, *m* 67.2); **T/W** 67.2; **B** ♂ 18.0–20.5 mm (*n* 15, *m* 19.1), ♀ 16.7–19.0 mm (*n* 14, *m* 17.7); **BD** 6.0–7.2 mm (*n* 33, *m* 6.6); **Ts** 21.3–25.4 mm (*n* 33, *m* 23.5). **Wing formula: p1** > pc 8–15 mm; **p10** < wt 18–26 mm; **s1** < wt 20–27 mm. (Syn. *intermedia*.)

A. d. whitakeri Hartert, 1911 (Jabal as Sawda in WC Libya, probably south-west to Fezzan in SW Libya). A very dark race, much darker than *algeriensis* and *deserti*, between which two it is sandwiched. Most similar to *payni* but slightly darker (in this feature between *samharensis* and *saturata*) and greyer overall, less rufous-tinged. Extent of dusky brown centres and rufous on wing intermediate between *algeriensis* and *payni*. Underparts dusky buff-brown with prominent dark grey-brown streaking on throat- and breast-sides, streaking far more extensive than in *payni*, and chest and belly much deeper buff or buff-grey. Long-winged, but bill relatively short: **W** ♂ 103–112 mm (*n* 19, *m* 106.5), ♀ 95–103 mm (*n* 17, *m* 98.5); **B** ♂ 17.3–19.9 mm (*n* 10, *m* 18.4), ♀ 16.7–17.6 mm (*n* 7, *m* 17.1).

A. d. deserti (M. H. C. Lichtenstein, 1823) (E Libya south to Tibesti in Chad; Egypt except locally in Sinai, N Sudan, Arabia north of *c.* 27°N, Levant, Iraq, SE Turkey, W Iran; mainly resident, though northernmost Asian populations may move to lower altitudes in winter). Usually fairly pale cinnamon or sand-coloured above, somewhat contrasting with rufous-cinnamon rump/uppertail-coverts and tail-base, and has hint of dusky brown centres to tertials and secondary-coverts; throat very lightly mottled or flecked darker. Differs from rather similar *samharensis* south of it in Arabian Peninsula in being clearly paler. Relatively small (see Biometrics above). – Apparently intergrades with neighbouring races where they meet. Note that there is a certain individual or local variation, some birds being slightly darker and greyer brown above. When series were compared it proved impossible to recognise *isabellina* as a consistently paler and sandier race than *deserti*. Instead, both slightly paler and sandier birds on the one hand, and somewhat darker and more grey-brown ones on the other, seem to coexist over large areas in NE Africa and Middle East. – Breeders in E Iraq and SW Iran ('*cheesmani*'; type examined) are subtly paler and more brown-tinged above, and form link from typical *phoenicuroides* to *deserti*. Differences not consistent and very slight, and not grounds for separation. – Those described from SE Turkey ('*coxi*'; type examined) fall well within variation of *deserti*; they tend to be on average a little larger, but scant material available for evaluation: **W** ♂ 102–108 mm (*n* 7, *m* 104.9), ♀ 95–100 mm (*n* 3, *m* 97.5); **B** ♂ 16–18 mm, ♀ 16.5 mm (NHM, BWP). (Syn. *arabs*; *coxi*; *fratercula*; *isabellina*; *mirei*.)

A. d. katharinae Zedlitz, 1912 (S Sinai). Darker and greyer with more prominent dark blotches or streaks on lower throat than typical *deserti*. Appears to have a fairly limited range (type locality Mt Sinai) and might require further study. Limited material examined, but those available consistently differ from *deserti*. **W** ♂ 99–104 mm (*n* 14, *m* 101.5), ♀ 94–102 mm (*n* 7, *m* 98.1); **T** ♂ 62–68 mm (*n* 15, *m* 65.0), ♀ 60–68 mm (*n* 7, *m* 63.6); **T/W** 64.3; **B** 15.2–17.4 mm (*n* 22, *m* 16.6); **BD** 5.0–6.3 mm (*n* 19, *m* 5.9); **Ts** 20.0–23.4 mm (*n* 22, *m* 21.9).

○ **A. d. annae** Meinertzhagen, 1923 (black lava deserts of Jordan and probably Syria). Has dark brown-grey upperparts and dusky grey-brown throat- and breast-sides and upper flanks, with limited area of cream-buff and dark streaks on chin to upper breast (some more heavily streaked on chest), and dark cinnamon-pink lower belly to undertail-coverts; very limited or no rufous-cinnamon on rump,

A. d. annae, Jordan, Dec: note incredible adaption to local rock coloration in the black lava deserts of Jordan (slate-grey upperparts and very warm underparts). This race is very similar to other named populations in similar rocky habitats, especially *saturata* of Yemen and W & C Arabia. (R. Pop)

uppertail-coverts and tail-base, or on fringes of tertials, secondary-coverts and remiges. However, some individual variation represented by slightly paler birds, and race less distinctive than usually suggested in the literature; may be a synonym of *saturata*. **W** ♂ 103–108 mm (*n* 6, *m* 104.8), ♀ 98–101.5 mm (*n* 4, *m* 100.5); **B** ♂ 16.1–18.6 mm (*n* 6, *m* 17.7), ♀ 15.0–16.4 mm (*n* 4, *m* 15.7); **BD** 6.1–7.4 mm (*n* 10, *m* 6.6).

A. d. saturata Ogilvie-Grant, 1900 (local on dark lava fields in Yemen and W Arabia). Surprisingly similar to widely allopatric *annae* or only fractionally more heavily or extensively streaked on breast, and some are slightly paler above (identical *saturata* include birds from Wadi Hamah, Jeddah, Saudi Arabia). The only real difference between these populations is in size, *annae* being slightly larger, but Desert Larks vary locally in size even within established races. **W** ♂ 95–105.5 (once 110) mm (*n* 12, *m* 102.6), ♀ 92.5–101 mm (*n* 14, *m* 97.7); **T** ♂ 61–71.5 mm (*n* 12, *m* 66.8), ♀ 59–67 mm (*n* 14, *m* 62.9); **B** ♂ 16.8–19.4 mm (*n* 12, *m* 17.8), ♀ 15.5–18.2 mm (*n* 14, *m* 16.9); **BD** 5.9–7.0 mm (*n* 11, *m* 6.4). – Strong correlation evident between darkest birds and black lava patches of S & WC Arabian plate, from Yemen (*saturata*) north to Jordan (*annae*). Should these two

A. d. deserti, Israel, Mar: in many areas it is impossible to label single birds to subspecies due to similarities between them, and because of both local and individual variation, often in response to soil coloration, some birds, like this one, are distinctly plainer. (D. Occhiato)

races prove to be better lumped, *saturata* would have priority. However, some *saturata* are paler and less dusky below, thus approaching *samharensis*; indeed, the type of *saturata* is closer to *samharensis* than to the darkest *saturata*. Probably *samharensis*, *saturata* and *annae* form a single variable 'ecological race' confined to igneous and dark pre-Cambrian rocks of Arabian plate, but more work is required before a taxonomic change can be recommended.

○ **A. d. samharensis** Shelley, 1902 (Red Sea Province of Sudan, N Eritrea, WC & SW Saudi Arabia and Yemen away from dark-soiled lava plains; probably more widespread also in E & C Arabia). Intermediate between *deserti* and *saturata*, being slightly darker and purer grey above, with warmer and streakier underparts than former, and quite distinctly paler with less slate-grey upperparts and noticeably paler underparts than latter. Very limited rufous-cinnamon on rump/uppertail-coverts and tail-base, and on fringes of tertials, secondary-coverts and remiges. Bill relatively long and strong. **W** ♂ 97–106.5 mm (n 11, m 101.8), ♀ 92.5–102 mm (n 12, m 97.3); **T** ♂ 61–71 mm (n 11, m 65.2), ♀ 60.5–68 mm (n 11, m 63.4); **B** ♂ 16.1–19.5 mm (n 11, m 17.7), ♀ 14.9–17.9 mm (n 13, m 16.6); **BD** 5.5–7.0 mm (n 21, m 6.2). Arabian birds slightly darker than E African birds including type of *samharensis* (Eritrea), thus approaching *saturata*. Intergrades with *saturata* and apparently *deserti* further north, and in Oman with *phoenicuroides*. – Note that birds referable to this subspecies occur on all sides of dark lava belts, where the darker *saturata* breeds. Note also that birds labelled '*hijazensis*' (type locality Hadda, Wadi Fatima, W Saudi Arabia) are in some collections very dark and actually closer to *saturata*, but since the type itself (NHM) is closer to *samharensis* it is here listed as a junior synonym of this and not of *saturata*. (Syn. *hijazensis*.)

○ **A. d. insularis** Ripley, 1951 (Bahrain). Differs from *deserti* in being colder and paler buff below, duskier buff-white and less warm rufous-buff, and from *samharensis* in being generally paler. It is rare in collections but those examined in NHM had same size as *deserti*. A very subtle and local race.

○ **A. d. azizi** Ticehurst & Cheesman, 1924 (known only from Al Hofuf, EC Saudi Arabia). Has very pale, sandy-cream or pale cinnamon upperparts and faintly pinkish buff-white underparts, appearing near-white below. Examined material small but recognition seems warranted (though it is less distinctly pale than often described and illustrated), and some *deserti* are equally pale or at least very similar. **W** ♂ 96.5–105 mm (n 5, m 100.5), ♀ 94–98 mm (n 7, m 96.1); **T** ♂ 64.5–66 mm (n 4, m 65.6), ♀ 60–63 mm (n 5, m 62.0); **B** ♂ 16.7–18.9 mm (n 5, m 17.8), ♀ 16.5–17.5 mm (n 7, m 17.1); **BD** 6.4–7.2 mm (n 9, m 6.8). (AMNH, NHM.)

A. d. phoenicuroides (Blyth, 1853) (N Oman, E United Arab Emirates, much of C & S Iran, Afghanistan, Baluchistan, Pakistan, S Kashmir, NW Punjab south to Sind). Somewhat darker and greyer than *deserti*, and paler and usually subtly greyer than *samharensis*. Comparatively short bill. Slightly variable within its fairly large range, but differences very subtle and far from clear. Very similar to *parvirostris* with same rather dark and greyish upperparts, and strongly flecked lower throat, differing only in that breast and belly are more washed pink-cinnamon, less yellowish-white. **W** ♂ 98–111 mm (n 20, m 103.1), ♀ 93–104 mm (n 10, m 98.3); **T** ♂ 62–74 mm (n 20, m 67.6), ♀ 62–70 mm (n 10, m 64.3); **B** ♂ 15.1–16.8 mm (n 20, m 16.2), ♀ 14.1–16.8 mm (n 10, m 15.2); **BD** 5.5–6.8 mm (n 29, m 6.2). – Birds in Iran ('*iranica*') tend to be slightly darker brown-grey on breast and have perhaps on average slightly thinner bill; still, much overlap and individual variation, and hardly grounds for separation. – Those from S & E Zagros, Iran ('*darica*'), are similar to typical *phoenicuroides* but on average slightly paler, less pure grey above and paler below, still only a subtle and far from consistent difference; size similar, too: **W** ♂ 101–113 mm (n 15, m 107.7), ♀ 93–106 mm (n 9, m 100.1); **B** ♂ 16.0–18.0 mm (n 16, m 17.5), ♀ 14.5–17.7 mm (n 11, m 16.3). – Breeders from E Iran, Afghanistan and adjacent areas sometimes slightly paler, warmer and less streaked on throat, and such birds seem the basis for the description of '*orientalis*' (Zarudny & Loudon, 1904); however, such pale unstreaked birds are collected from areas with normal *phoenicuroides*, thus seems to be local or individual variation. – Birds from the Muscat area, Oman, separated on dubious grounds as '*taimuri*' (in type description compared only with the much darker *samharensis* and *saturata*); appear subtly smaller but plumage colours variable and within normal variation of *phoenicuroides*. Type of '*taimuri*' not examined, but several specimens from Muscat match those from Baluchistan, montane N Oman and United Arab Emirates, thus inseparable from *phoenicuroides*. Size difference seems too slight to warrant separation. (Syn. *cheesmani*; *darica*; *iranica*; *orientalis*; *taimuri*.)

○ **A. d. parvirostris** Hartert, 1890 (Transcaspia in Turkmenistan, W Uzbekistan, NE Iran; in winter short-range movements, sometimes to lower elevations). Moderately large and strong-billed, differing from neighbouring very similar *phoenicuroides* by being subtly paler and more yellow-tinged below, and having a trifle stronger bill. **W** ♂ 98–112 mm (n 27, m 104.0), ♀ 93–104.5 mm (n 18, m 99.8); **T** ♂ 63–74 mm (n 27, m 67.6), ♀ 62–69 mm (n 18, m 64.8); **T/W** m 64.9; **B** ♂ 15.0–17.4 mm (n 26, m 16.1), ♀ 14.3–16.6 mm (n 18, m 15.3); **BD** 5.5–7.0 mm (n 44, m 6.2); **Ts** 20.3–24.2 mm (n 44, m 22.3).

REFERENCES Alström, P. *et al.* (2013) *Mol. Phyl. & Evol.*, 69: 1043–1056. – Shirihai, H. (1994) *DB*, 16: 1–9. – Shirihai, H., Mullarney, K. & Grant, P. (1990) *BW*, 3: 15–21. – Vaurie, C. (1951) *Bull. AMNH*, 97: 431–526.

A. d. annae, Jordan, Mar: not all individuals in the basalt shield of E Jordan are very dark, but they are always highly distinctive, and sometimes heavily streaked below (most contrasting on paler individuals). (H. Shirihai)

A. d. phoenicuroides, Iran, Jan: rather small bill, typically rather uniform medium grey above, with dull and less contrasting rufous elements, while breast and throat-sides are usually extensively streaked. (E. Winkel)

(GREATER) HOOPOE LARK
Alaemon alaudipes (Desfontaines, 1789)

Fr. – Sirli du désert; Ger. – Wüstenläuferlerche
Sp. – Alondra ibis; Swe. – Härfågellärka

Arguably one of the most special birds within the covered region, this unusual lark is perhaps best known for its fantastic display-flights and far-carrying fluting song (which, to some ears, recalls the British national anthem) that are evocative of otherwise silent desert dunes. Principally sedentary, Hoopoe Lark mainly inhabits sandy deserts and semi-deserts from the Cape Verde Is to Arabia, the Near East, and the NW Indian subcontinent, and has wandered north to S Turkey.

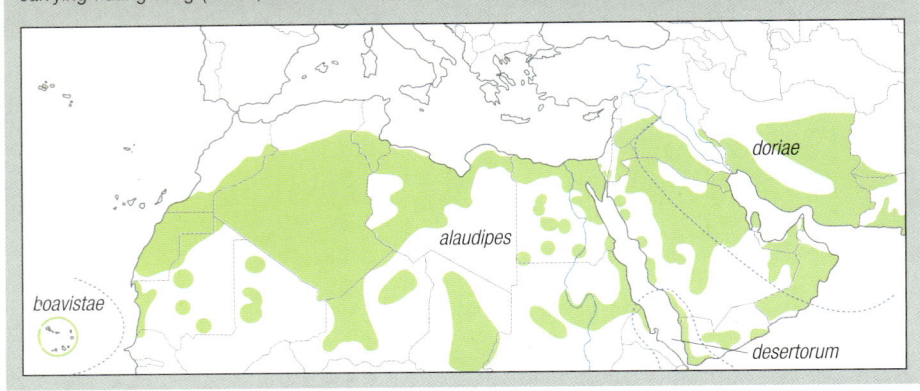

A. a. alaudipes, ♂, Morocco, Jan: a typical individual. Uniquely among larks, has very long bill, decurved distally (often held above horizontal). Fresh midwinter plumage when black markings of breast and face still partially obscured by whitish tips; warm buff-brown upperparts and contrastingly grey head are often features of *alaudipes*, especially ♂♂, as is tendency for broader moustachial stripe. (M. Schäf)

IDENTIFICATION Strikingly *large, long-legged lark, with very long, slightly decurved bill* (often held above horizontal), thus unique among larks, suggesting a small Cream-coloured Courser or in flight even a tiny Hoopoe. Plumage highly camouflaged, essentially uniform sandy-grey above and white below. *White eye-ring and supercilium, and narrow black lores, eye-stripe and moustachial stripe*, framing whitish-grey cheeks, while crown is slightly streaked. Thin lateral throat-stripe usually develops into *distinctly streaked and blotched black breast*. On folded wing, three dark bands (formed by dark bases to median and greater coverts, and outer primaries) contrast with three pale areas, the lesser coverts, the fringes and, especially, whiter tips to the greater coverts, and bases to the secondaries just below the dark-centred tertials. However, some, even adults, are plainer. *Tail long, mainly black with grey-brown centre and white-edged outermost feathers. Stunning pied appearance in flight, with double black and white bands*: black wingtip continuing as secondary bar (leaving broad white trailing edge) and along primary-coverts, while fore coverts are greyish or sand-coloured. Black panels (and white edges) to fanned tail afford further contrast. From below, wings strikingly white, with narrow dark secondary bar and primary tips. Walks and runs on ground with extremely confident appearance. Prefers to run rather than fly, escaping in a zigzag: if you take your eye from the bird it is gone! Flies reluctantly but loosely and freely on slightly jerky wingbeats, while escape-flights are rather long and low, often ending in a short glide. Spectacular song-flight starting either from a bush top or in low horizontal flight just above ground level, culminating in short steep climb (1–5 m) with spread tail, followed by vertical descent on folded wings before alighting on ground.

VOCALISATIONS Voice consists of remarkable fluted, piped and melancholy notes, with a marked ventriloquial quality, starting tentatively, accelerating and then slowing again, e.g. *voy voy… vüüü(cha) vüüü(cha) swe-swe-swe-swe-swe sisisi… svee, sveeh*, with the *vüüü* note piercingly thin and most distinctive. In Middle East, but apparently not in NW Africa, a peculiarly Black Woodpecker-like *kri-kri-kri-kri-…* is often inserted early in the song or uttered between strophes. – Call a thin tremulous *zrruee* or *dschrep* (upward-inflected).

SIMILAR SPECIES Adult unmistakable (especially in flight), although at distance might be confused, initially, with a large pipit or even a courser. Juvenile also immediately identifiable in flight, but on ground could suggest rufous form of Dupont's Lark, but latter is smaller and always has far stronger head and upperparts streaking, numerous spots and streaks on throat and breast, and a much shorter tail, among other structural and plumage differences.

AGEING & SEXING Ages (after post-juv moult) usually not separable. Sexes largely alike, but average difference exists in throat- and breast-streaking. Size is useful criterion in southern population (see Biometrics). – Moults. Post-nuptial moult (complete) and post-juv (complete or sometimes partial), mostly Jul–Aug, but slightly earlier in the south. Ageing by moult limits in young applies only to a few birds. Two colour types over much of range—one more cinnamon-tawny above and the other greyer—often considered morphs, but much individual variation and many intermediates, and all fresh birds usually warmer in colour. – **SPRING Ad** ♂ is larger and has longer bill and longer legs on average. Also, lower throat/upper breast has on average bolder and denser dark streaking. Although there is overlap, this difference can be useful when breeding pair seen together. **Ad** ♀ is shorter-billed and shorter-legged than ♂, with weaker head and breast markings, but much overlap in plumage. In both sexes, with wear, breast perhaps more heavily blotched, facial marks more pronounced, head and neck to mantle bleach greyer, and underparts become purer white; in heavily worn birds dark bases of greater and median coverts and tertials bleach duller and are less contrasting. **1stS** Much

A. a. alaudipes, ♂, Western Sahara, Jan: irrespective of age/sex, some are predominantly sandy rufous-buff above. (C. N. G. Bocos)

as ad, but some retain very worn juv wing-feathers. Poorly marked ♂♂ which approach ♀♀ are perhaps young birds, but much variation. – SUMMER–AUTUMN **Ad** Head and breast markings appear weaker, being partially concealed by pale fringes and tips. **1stW** Generally like ad, but late-hatched 1stW often retain some tail-feathers and coverts, and are reliably aged during first year of life, but others are inseparable from ad. **Juv** Shorter and less decurved bill. Upperparts and upperwing-coverts sandier and lightly spotted. Most lack heavy breast pattern, and facial marks absent or reduced (supercilium less distinct), while underparts are dirty white with pale buff breast and upper flanks.

BIOMETRICS (*alaudipes*; North Africa) **L** 20.5–22.5 cm; **W** ♂ 120–139 mm (n 23, m 127.8), ♀ 108–117 mm (n 15, m 113.8); **T** ♂ 83–95 mm (n 23, m 87.9), ♀ 74–84 mm (n 15, m 77.9); **T/W** m 68.8; **B** ♂ 27.0–35.2 mm (n 23, m 31.4), ♀ 26.3–30.0 mm (n 15, m 28.2); **BD** 4.7–6.4 mm (n 39, m 5.6); **Ts** 29–36 mm (n 40, m 32.4). **Wing formula: p1** > pc 4–9.5 mm, < p2 49–62 mm; **p2** < wt 5–11 mm; **pp3–4** about equal and longest; **p5** < wt 0.5–3 mm; **p6** < wt 4–6 mm; **p7** < wt 16–23 mm; **p10** < wt 31–39 mm; **s1** < wt 30–40 mm. Emarg. pp3–6.

A. a. alaudipes, Morocco, Feb: two images showing extreme variation in *alaudipes*. Left bird probably ♂ given heavy facial and breast markings, and relatively long bill, whereas right bird probably ♀ with reduced markings and shorter bill. These two also demonstrate colour variation irrespective of sex, with grey (left) and rufous morphs. Neither sexual differences nor colour morphs are well differentiated, with numerous intermediates. (Left: R. Schols; right: D. Occhiato)

A. a. alaudipes, ♂, Morocco, Mar: here demonstrating the species' remarkable song-flight with short vertical climb up to c. 5 m, tail spread during climb and upon 'falling down'. In flight striking pattern to dorsal wing surface, with black wingtip continuing as secondary bar (variable in width) and bordered in front and behind by broad white panels, and with grey-buff forewing. The underwing pattern is similar but the coverts are whitish-cream. Note also largely black tail with white edges to outer tail-feathers while central ones are contrastingly buffish-brown. (M. Southcott)

GEOGRAPHICAL VARIATION & RANGE Predominately clinal variation with many intermediates, while effects of wear and bleaching, and individual variation, often render racial separation unreliable. Furthermore, the existence of two colour types, which are not geographically segregated, must be taken into account. Combination of plumage tone and size useful, but separation never clear-cut and we suggest that a detailed review would reveal one or two names to be synonyms. All populations are resident.

A. a. boavistae Hartert, 1917 (Cape Verde Is). Relatively small and dark, being greyish-brown tinged rufous on crown to back, former being diffusely streaked dark. Flight-feathers and greater coverts somewhat darker brown on average than in *alaudipes*. Breast heavily streaked on cinnamon-grey ground. Gingery tones dorsally more frequent and somewhat deeper than in other populations. Some individual variation, but colour even of more greyish birds usually distinctly rufous-buff (however, with wear many become equally greyish above). **W** ♂ 124–128 mm (n 7, m 126.0), ♀ 109.5–118 mm (n 12, m 114.5); **T** ♂ 80–88 mm (n 7,

A. a. alaudipes, Oman, Nov: variable secondary bar can be very narrow, broken or virtually absent, leaving huge white area. (A. Blomdahl)

m 85.7), ♀ 72–82 mm (*n* 12, *m* 76.8); **T/W** *m* 67.7; **B** ♂ 28.0–34.3 mm (*n* 7, *m* 30.4), ♀ 25.7–29.0 mm (*n* 12, *m* 27.4); **BD** 4.9–6.7 mm (*n* 28, *m* 5.7); **Ts** 30.8–35.6 mm (*n* 28, *m* 33.6). **Wing formula: p2** < wt 3.5–8 mm; **s1** < wt 28–40 mm.

A. a. alaudipes (Desfontaines, 1789) (Mauritania, Niger, Morocco east to W Arabia, Levant). Very similar to *boavistae*, especially breeders in North Africa, but although sandy or cinnamon-buff, subtly duller and greyer, with breast streaks generally larger, especially in Middle East. – Those from Levant (e.g. Syria, Palestine & Jordan) tend to be slightly larger than those in N Africa (for which see Biometrics above), and they have a peculiarly Black Woodpecker-like *kri-kri-kri-kri-kri* call inserted between songs, which is seemingly never given by N African populations. However, much overlap in morphology, and available material scant, hence best kept in *alaudipes*. (Syn. *meridionalis*; *omdurmanensis*.)

A. a. doriae (Salvadori, 1868) (Iraq & E Arabia to Sind). Similar to *alaudipes*, but ♂♂ have bolder dark blotching on lower throat/upper breast, and streaking also more dense and concentrated on average. Further, pink-cinnamon wash

A. a. boavistae, probably ♂, Cape Verde Is, Mar: a somewhat strong-billed form. (C. Batty)

A. a. doriae, Iran, Jan: distinctly large and greyer race. Species' typical face pattern of white eye-ring and supercilium, narrow black lores, eye-stripe and moustachial, framing whitish-grey cheeks, while thin black lateral throat-stripe usually develops into distinctly streaked and blotched breast. (A. Ouwerkerk)

above on average strongly reduced, being at least as grey as *desertorum*. Very large, especially those from Iran eastward. **L** 20–25 cm; **W** ♂ 128–143 mm (*n* 20, *m* 137.3), ♀ 116–130 mm (*n* 13, *m* 123.7); **T** ♂ 82–102 mm (*n* 20, *m* 95.7), ♀ 81–93 mm (*n* 13, *m* 86.1); **T/W** *m* 69.6; **B** ♂ 30.7–35.9 mm (*n* 21, *m* 33.3), ♀ 25.2–32.3 mm (*n* 14, *m* 29.7); **Ts** 34.1–37.2 mm (*n* 13, *m* 35.2), ♀ 29.7–33.9 mm (*n* 13, *m* 31.8). **Wing formula: p1** < **p2** 51.5–71 mm. Birds from Iraq, Kuwait, Bahrain, etc., tend to be slightly smaller, closer to *alaudipes*. (Syn. *pallida*.)

○ *A. a. desertorum* (Stanley, 1814) (southern shores of Red Sea, coastal Yemen & N Somalia). Resembles *doriae* but is even greyer above (still brown), and lesser uppertail-coverts are grey instead of whitish-buff. Breast-streaking perhaps on average slightly bolder. Rather long-tailed, proportionately. **W** ♂ 124–132 mm (*n* 14, *m* 127.1), ♀ 108–118 mm (*n* 11, *m* 112.0); **T** ♂ 83–92 mm (*n* 14, *m* 88.1), ♀ 74–85 mm (*n* 11, *m* 79.5); **T/W** *m* 70.0; **B** ♂ 29.6–33.5 mm (*n* 14, *m* 32.0), ♀ (once 23) 26.0–29.7 mm (*n* 11, *m* 27.4); **BD** 5.0–6.8 mm (*n* 25, *m* 6.0); **Ts** 29–36 mm (*n* 25, *m* 32.7).

REFERENCES VAURIE, C. (1951) *Bull. AMNH*, 97: 431–526.

A. a. alaudipes, juv, Morocco, Mar: readily separated from ad by short bill and presence of some exposed dark feather-centres above. (D. Monticelli)

DUPONT'S LARK
Chersophilus duponti (Vieillot, 1824)

Fr. – Alouette de Dupont; Ger. – Dupontlerche
Sp. – Alondra ricoti; Swe. – Dupontlärka

For a lark, this is the odd one out. Most larks in our region are obvious, making their presence known by virtue of their incessant, loud songs and easily-observed song-flights over fields and meadows. However, Dupont's Lark—confined to arid plains with grasses and low scrub in Iberia and N Africa—is a contender for the title 'Europe's most elusive bird'. Singing its discreet song in the pitch dark, or in daylight at quite high altitude, the bird dives silently like a stone to the ground and then disappears in the grass. To get to know this species requires luck, or days of hard work—or both.

C. d. duponti, Spain, May: front-on, dense breast-streaking ending abruptly recalls Pectoral Sandpiper *Calidris melanotos*, while whitish eye-ring and supercilium are other good field marks. (C. M. Martin)

C. d. duponti, Spain, May: note peculiar bill shape, long and slightly decurved, with deep base. Also rather small head, long neck and characteristic upright stance. If not too worn, dark rear ear-coverts patch and subterminal marks to scapulars and greater coverts are important distinctions too. (J. Daubignard)

IDENTIFICATION As evident from the above, mainly noticed and identified by its characteristic song. When (finally!) seen close on the ground, it is a little smaller and slimmer than Crested Lark, but with more *upright stance*, has a rather *small head* and *long neck*, and *long and slightly downcurved bill*. Legs and tail are comparatively short on the other hand. The *throat and breast are rather boldly streaked*, but rest of underparts unstreaked. *Head rather dark*, but *unbroken pale eye-ring* and short supercilium behind eye stand out. There is an isolated pale loral stripe. Legs are pinkish-brown. It moves warily in brief bursts of jerky, 'nervous' movements, stopping all the time to watch intently, bill often held high. Very difficult to flush, it runs cleverly for cover in the vegetation. That said, it sometimes can be seen without a struggle, perched atop low scrub. In song-flight high up, mainly towards the end of night and at dawn, but often also during morning hours, the *tail is mainly kept folded* like Calandra Lark. – There is some geographical variation, birds in inland N Africa between C Morocco and Egypt being more reddish-tinged and having a slightly longer bill.

VOCALISATIONS Song is a brief verse with a ringing, slightly strained pitch, monotonously repeated without much alteration for minutes. The strophe can be transcribed as *wu-tlee-tre-weeüüih*, thus often consists of 5–6 syllables, generally with second and last stressed, the last rising in pitch. Shorter versions can be heard: *tlee-weeüüih*. The song is mostly delivered in flight, either in the dark of night or in the morning (not solely at dawn, as often claimed), in both cases usually from considerable height (but strangely it often sounds as if coming from the ground!), sometimes also in lower flight or from perch near the ground. Birds in NE Libya are reported to sing slightly but clearly differently, *dii-dii-drii, drrrrr*, thus with two fine introductory notes and a trill at the end (Hartert 1923). Wing-clapping between or inserted in songs have been reported (Hazevoet 1989). – The call, which is rarely heard, sounds like part of the song, *wu-tlee* or a ringing *pu-chee* (Smith *in* BWP). Also a shrill *tsiii* when alarmed and running away on the ground (Hartert 1923).

SIMILAR SPECIES Within its range, given only brief views (which are the norm!) the streaked breast and general size can recall *Skylark*, but note the long and slightly downcurved bill and lack of a pale trailing edge to the wing. The head pattern recalls those of *Crested Lark* and *Thekla Lark*, but these two have a crest (rarely missing in heavily moulting birds), and they both lack the white tail-sides of Dupont's, while Thekla has a much shorter and straighter bill. – There is the remote risk of confusion with *Lesser Short-toed Lark*, which also has a streaked breast, but its smaller size and short bill should be evident. The song can at first be confused with that of *Bar-tailed Lark* (which rarely occurs in same habitats) due to similar fine voice and ringing pitch, but differs in details once heard well. – *Tawny Pipit* and *Hoopoe Lark* share its habits, and both can resemble Dupont's Lark (e.g., juv Tawny Pipit has a streaked breast). Note the upright stance, short tail and dark crown with hint of pale median crown-stripe on Dupont's (van den Berg 1984).

C. d. duponti, Spain, Apr: a rather dark race with cold brown tinge and fewer rufous elements, but appearance varies depending on light (cf. images above). (M. Saarinen)

C. d. duponti, juv, Morocco, Apr: note prominent pale tips and fringes to upperparts and wing-coverts, as well as shorter and pinker bill. (M. Buckland)

C. d. duponti, Spain, Sep: following the late summer moult shows attractive neat whitish fringes on upperparts, superficially reminiscent of juv. The scaly pattern of the fresh plumage will be reduced or lost due to abrasion, and already from late autumn or early winter the pattern above changes to become more streaky. Very difficult bird to see or even flush (especially in non-breeding season when not singing). (H. Shirihai)

AGEING & SEXING Ages and sexes (after post-juv moult) alike as to plumage, but see Biometrics for marked sexual size dimorphism. Insignificant seasonal differences. – Moults. Complete moult of both ad and juv in late summer (mainly Jul–Sep), though moult of juv somewhat more protracted. – SPRING **Ad** In late breeding season heavily-bleached birds slightly darker above and less clearly patterned, with reduced pale upperparts fringes; underparts streaking a little bolder. **1stS** Inseparable from ad. – SUMMER–AUTUMN **Ad** Upperparts streaked dark, no prominent pale tips or edges. **1stW** Inseparable from ad. **Juv** Upperparts and wing-coverts with prominent pale, greyish- or buffish-white tips and edges. Tail-feathers rather narrow and pointed.

BIOMETRICS (*duponti*) **L** 16–18 cm; **W** ♂ 99–106 mm (*n* 26, *m* 102.1), ♀ 88–97 mm (*n* 19, *m* 92.0); **T** ♂ 59–68 mm (*n* 26, *m* 63.1), ♀ 51–60 mm (*n* 19, *m* 55.4); **T/W** ♂ *m* 61.8, ♀ *m* 60.2; **B** ♂ 21.3–24.3 mm (*n* 23, *m* 22.7), ♀ 19.0–23.8 mm (*n* 18, *m* 21.3); **B**(f) ♂ 18.2–22.4 mm (*n* 24, *m* 20.1), ♀ 16.0–21.7 mm (*n* 18, *m* 18.5); **Ts** ♂ 22.6–25.2 mm (*n* 24, *m* 23.9), ♀ 22.2–23.7 mm (*n* 19, *m* 22.9); **HC** 7.3–12.5 mm (*n* 42, *m* 9.8). **Wing formula: p1** minute; **p2** < wt 0–2.5 mm; **pp3–4**(5) about equal and longest; **p5** < wt 0–3 mm; **p6** < wt 3–7 mm; **p7** < wt (once 8) 10–16 mm; **p10** < wt 20–29 mm; **s1** < wt 22–31 mm. Emarg. pp3–6 (slightly less prominently on p6). – In a comprehensive study of live birds, sexed molecularly and trapped in Spain and Morocco, mainly in Ebro Valley, NE Spain (Vögeli *et al.* 2007), wing length separated 97.5% of ad, overlap affected only region 96–97 mm: **W** ♂ 96–108 mm (*n* 286, *m* 102.4), ♀ 87–97 mm (*n* 28, *m* 91.9). A more elaborate discriminant function formula combining wing and bill/skull length gave even better separation.

GEOGRAPHICAL VARIATION & RANGE Two rather distinct subspecies, but intermediates have been reported from a transitional zone in S Algeria and S Tunisia, and some specimens in collections are less distinct. All populations resident.

C. d. duponti (Vieillot, 1824) (Iberia, N Morocco, Atlas Mts of Algeria, C Tunisia; possibly local N Tunisia but requires confirmation). Dark brown with subdued rufous tones, streaks being cold brown to blackish (though juv slightly warmer brown). In particular, lateral crown, mantle, scapulars and tertial centres are darker brown compared to *margaritae*. Bill length also indicates race, being consistently shorter than in *margaritae*, but note that there is some overlap between the two, and in particular that ♀♀ have shorter bills often making a ♂ *duponti* appear similar to a ♀ *margaritae* in this respect.

C. d. margaritae (König, 1888) (N Africa from Atlas Mts in C & S Morocco, Algeria in zone between southern slopes of Atlas Mts and Sahara, S Tunisia, Libya, NW Egypt. More obviously rufous-tinged and somewhat paler, and bill is on average longer, whereas tail and hind claw are proportionately shorter. Crown being rufous-brown looks paler than in *duponti*, and median crown-stripe often less contrasting. **W** ♂ 95–105 mm (*n* 19, *m* 101.4), ♀ 88–92 mm (*n* 7, *m* 90.4); **T** ♂ 53–63 mm (*n* 19, *m* 59.2), ♀ 52–57 mm (*n* 7, *m* 54.6); **T/W** *m* 59.0; **B** ♂ 23.7–29.0 mm (*n* 18, *m* 25.7), ♀ 22.7–23.5 mm (*n* 6, *m* 23.2); **B**(f) ♂ 20.4–25.2 mm (*n* 18, *m* 22.8), ♀ 18.9–20.2 mm (*n* 6, *m* 19.9); **Ts** ♂ 23.0–26.3 mm (*n* 19, *m* 24.5), ♀ 21.8–24.5 mm (*n* 7, *m* 23.0); **HC** 5.5–9.3 mm (*n* 24, *m* 7.4).

REFERENCES VAN DEN BERG, A. (1984) *DB*, 6: 102–105. – GARCÍA, J. T. *et al.* (2008) *Mol. Phyl. & Evol.*, 46: 237–251. – HARTERT, E. (1923) *Novit. Zool.*, 30: 1–32. – HAZEVOET, C. J. (1989) *BBOC*, 109: 181–183. – VÖGELI, M. *et al.* (2007) *Ardeola*, 54: 69–79.

C. d. margaritae, S Morocco, Apr: note characteristic paler and more rufous-tinged streaking in this race. That bill averages longer than in ssp. *duponti* (much overlap, though) is hardly visible on this bird. (T. Grüner)

C. d. margaritae, Egypt, Apr: note long, powerful bill, dense streaking of breast and neck, and rufous-tinged upperwing characteristic of this race. Note that streaking of head and breast in *margaritae* does not always appear obviously rufous. (P. Kaestner)

THICK-BILLED LARK
Ramphocoris clotbey (Bonaparte, 1850)

Fr. – Alouette de Clotbey; Ger. – Knackerlerche
Sp. – Calandria picogorda; Swe. – Tjocknäbbad lärka

A strikingly-marked lark with an eye-catchingly heavy bill, but is often mysteriously elusive within its desert range. Resident or semi-nomadic at the northern edge of the Sahara in a belt running from Western Sahara through E Morocco to N Libya. It has not been found breeding in Egypt, but has done so in CS Syria and regularly in E Jordan south to NW Saudi Arabia. Has bred irregularly in S Israel, and is a vagrant to Oman.

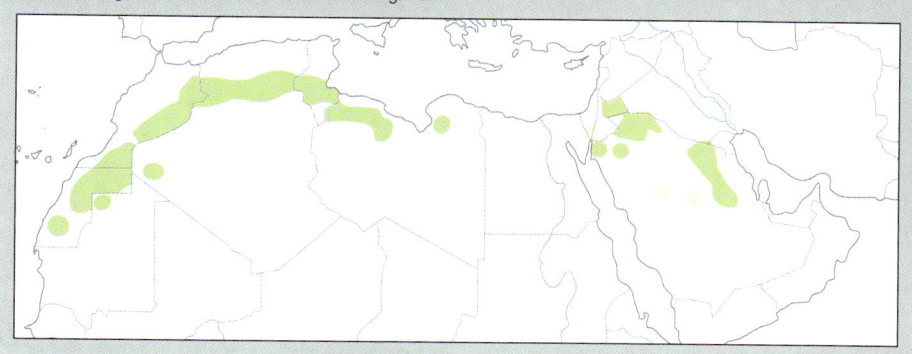

♀ summer, Morocco, Mar: limited dark below, with relatively narrow dark feather-centres usually a good indication of ♀ in spring. (D. Monticelli)

♂ summer, Israel, Mar: with wear black face pattern and black streaks below can be especially bold and solid, while whitish lores, eye-crescents, chin and (especially) cheek patches are reduced. (A. Ben Dov)

IDENTIFICATION A bulky lark with a *remarkably swollen pale bill* and large-headed appearance, upright stance, and relatively long-winged, short-tailed flight silhouette. Often appears on the scene as if from nowhere, but diagnostic wing pattern and calls make it easily identified under most conditions. Plumage surprisingly cryptic in the arid, rocky terrain it favours. Rather plain, drab grey-brown upperparts, sharply contrasting with *black-splashed face, bold whitish spots* at base of each mandible, *whitish lower eye-crescent*, chin and cheek patches, which together with *bold black spotting from upper breast to belly* (the spots coalescing in centre) give it a diagnostic appearance. Eye-ring greyish-white. Upperwing-coverts and tertials slightly rufous with paler fringes. Both surfaces of flight-feathers blackish (darkest in ♂) *with very broad white trailing edge* obvious in flight. Near-black subterminal mark on all but central tail-feathers. Central tail-feathers rufous-brown. Flight powerful, very much recalling *Melanocorypha* larks. Escape-flight occasionally long and low, ending abruptly, the bird often immediately becoming invisible, while often heard in high flight across desert but remains unseen against bright blue sky. Song-flight, given at some height, involves slow parachuting or zigzagging descent. Moves with slightly restless short walks or longer runs and hops. Usually occurs in small flocks.

VOCALISATIONS Song a quiet drawn-out and upwards-inflected whistle, now and then with a few short rapid twittering notes added afterwards, e.g. *tweeeiip qule-trrrtk-tk… tweeeiip…* – Calls include a low, somewhat plaintive ('querying') *coo(ee)* and *oo-ep* in flight, and a more rolling, frizzled, high-pitched and rising *zrrree-ih* or *zrrreh*. There is also a more clicking *chop*, somewhat recalling Desert Lark. Some birds are remarkably silent, even in flocks.

SIMILAR SPECIES All ages unmistakable, even poorly patterned fresh ♀ and juv, given clear view of the exceptional bill. Given general structural similarities to some *Melanocorypha*, confusion with any of these is possible when distant, particularly if only the dark underwing is seen, but problems should not persist once the bill and plumage details become apparent. Even in flight from below, when the large bill may be much less apparent, identification is still relatively unproblematic given the species' rather unique underwing pattern (among larks), consisting of grey-brown coverts, a striking dark wedge extending across virtually all of the primaries and secondary bases to the junction with the body, and broad

Presumed ad ♀ or 1stY ♂ (front), and probable ad ♂, Morocco, Dec: sexual differences sometimes already obvious by midwinter, but individual variation leads to much overlap. (K. De Rouck)

Tunisia, Mar: bold head and underparts patterns, and size of bill, make species unmistakable. However, some not easy to sex and can appear intermediate: less solid back areas with more broken-up face pattern and narrower/sparser streaking below suggest ♀ (cf. centre-left image of p. 51, of ♂ at same season). (D. Occhiato)

Israel, Mar: short-tailed and large-headed silhouette, and broad white trailing edge. (A. Ben Dov)

white trailing edge petering out on the inner primaries. This pattern is repeated on the upperwing. Compared to *Calandra Lark* the white trailing edge is much broader, almost twice as broad, and compared to *White-winged Lark* the white edge is narrower but more uniformly broad. Note also the lack of any rufous on the upper forewing.

AGEING & SEXING Ages usually inseparable after post-juv moult. Sexes often separable, especially in spring/summer; size differences helpful (see Biometrics). Strong seasonal and individual variation, mainly due to wear and bleaching, affecting overall amount of exposed black on face and underparts. – Moults. Complete post-nuptial and post-juv moult mainly in Jul–Aug, after which ageing becomes unreliable or impossible. – **SPRING Ad** ♂ With wear, black on face and underparts becomes very extensive and almost solid, sharply demarcated from white facial marks. Head to mantle wears greyer. Bill is pale greyish with yellowish tinge near cutting edges and base of lower mandible, has darker tip and often faint bluish tone to outer parts. **Ad** ♀ Usually slightly smaller than ♂, with more limited (and browner) black and more ill-defined pale marks on face, bill less bluish-grey (often yellower grey), and underparts less heavily blotched, more streaked (hardly ever acquires large black areas). **1stS** Some less-advanced ♂♂, which approach ad ♀ in amount of black, are probably 1stS. – **SUMMER– AUTUMN Ad** After complete moult, extensive pale fringes obscure most dark areas. Generally streaky below, but fore face and central breast to abdomen still heavily marked black (especially dense and broad in ♂♂, although sexing unreliable at this season due to individual variation). **1stW** Much like same-sex ad, or more ♀-like. **Juv** Soft, fluffy and overall sandier plumage with only faint facial and underparts markings (head almost uniform), and often has slightly weaker bill.

BIOMETRICS L 17–18 cm; **W** ♂ 125–134 mm (*n* 26, *m* 129.6), ♀ 119–125 mm (*n* 12, *m* 122.3); **T** ♂ 57–64 mm (*n* 10, *m* 59.4), ♀ 53–58 mm (*n* 12, *m* 56.0); **T/W** *m* 45.8; **B** 17.8–22.0 mm (*n* 25, *m* 19.7). – For live birds (Perlman & Kiat (2011): **W** ♂ 128–133 mm (*n* 7, *m* 130.6), ♀ 124–128 mm (*n* 9, *m* 125.7); **T** ♂ 60–65 mm (*n* 7,

Winter, Morocco, Jan: two birds showing midwinter plumage variation, when dark feathers are partially concealed by pale tips, but differences in amount of black suggest left bird is ♂ and right ♀. Following complete moult in late summer by both ad and juv there are no certain ageing criteria. (M. Schäf)

LARKS

m 62.4), ♀ 57–66 mm (n 9, m 60.8); **B**(f) ♂ 17.7–19.4 mm (n 7, m 18.7), ♀ 16.6–19.4 mm (n 9, m 18.1); **Ts** 22.8–27.9 mm (n 16, m 24.0). – **Wing formula: p1** minute in ad, slightly longer and more rounded in juv; **p2** < wt 1–3 mm; **p3** longest; **p4** < wt 1–3 mm; **p5** < wt 8.5–13 mm; **p6** < wt 19–25 mm; **p10** < wt 43–53 mm; **s1** < wt 46–55 mm. Emarg. pp3–5.

GEOGRAPHICAL VARIATION & RANGE Monotypic. – Breeds at northern edge of Sahara desert, from Western Sahara, extreme NW Mauritania and E Morocco across NC Algeria to N Libya; apparently absent in Egypt but breeds irregularly in S Israel, SC Syria, E Jordan and NW Saudi Arabia; resident or nomadic in winter, generally making shorter movements only.

REFERENCES Perlman, Y. & Kiat, Y. (2011) *DB*, 33: 1–10.

Presumed 1stW ♂, Tunisia, Sep: bill still lacks bluish-grey colour and dark tip of ad, but head already relatively heavily marked. Note ongoing moult of primaries. (A. B. van den Berg)

Presumed 1stW ♀, Tunisia, Aug: following completion of moult, ages and sexes are the same, but combination of pinkish straw-coloured bill base (less grey) and weak dark markings on head and face suggest young ♀. (G. Conca)

Juv, Israel, Mar: overall sandy, finch-like appearance (quite unlike ad); bill pinkish without dark tip, and still relatively small. Note broad pale eye-ring and wing and tail patterns. (C. Batty)

CALANDRA LARK
Melanocorypha calandra (L., 1766)

Fr. – Alouette calandre; Ger. – Kalanderlerche
Sp. – Calandria común; Swe. – Kalanderlärka

If Skylark is the most abundant and well-known lark in northern Europe, Calandra Lark shares this title with Crested Lark in the Mediterranean region. Calandra Lark is common on grassy plains at moderate elevations from Morocco to Kazakhstan and Central Asia, frequently occurring in natural habitats but being particularly abundant in cultivated fields. Mostly resident but gathers locally into flocks in winter, and some are migrants. Etymologically, 'calandra' apparently means simply 'lark' (Gr.), thus the bird's English name in fact translates as the 'lark lark'.

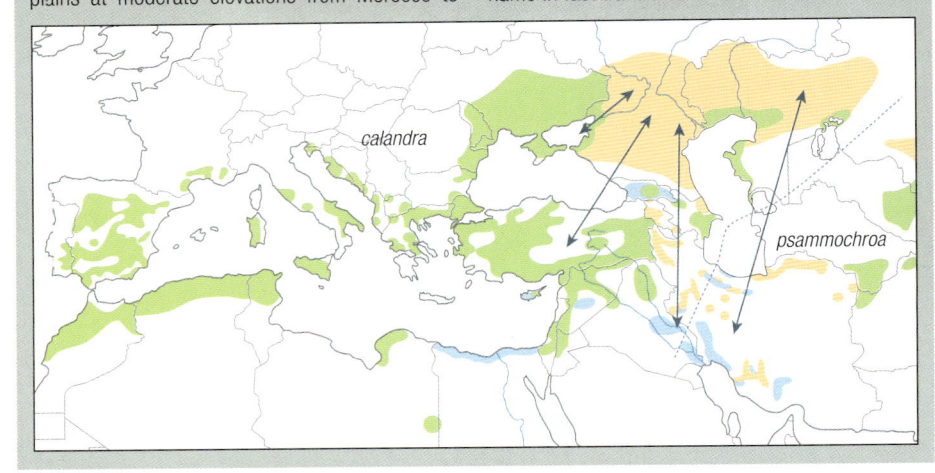

M. c. calandra, Tunisia, Mar: invariably has black breast-side patches, often becoming more solid (less streaky) with wear. Note large bill with straw-coloured base, and face pattern. (D. Occhiato)

M. c. calandra, Spain, Feb: unlike any other lark in range, note prominent (in this case massive) dark patches on breast-sides, hefty bill and thick neck. (B. Baston)

IDENTIFICATION A compact bird with short tail, thick neck and strong bill. Almost as large as a small thrush, *its size* is the first thing to strike the observer, and is enhanced when the bird is flushed, since it has *long and dark wings*, which adds to its large impression. That said, a crouching bird does not appear that big, and you will need to look for finer details to eliminate even Greater Short-toed Lark, both having similar colours and *dark patches on the upper breast-sides*. The *bill is hefty* on Calandra, appearing almost oversized for its head (never so in Short-toed), and in flight the *rear of the wing is broadly* and distinctly *edged white* (much stronger contrast than in Skylark). Note that on some freshly moulted autumn birds, the dark breast patches are largely concealed by pale feather-tips. Often seen in characteristic song-flight at moderate height: long *wings with black-looking undersides, slow-motion wingbeats* with wings held straight (making the bird look even bigger than it is, almost like a small hovering raptor) and *folded tail* (still with white sides visible). *Flight* in territorial chases usually *fluttering* and at times semi-hovering, revealing *rather broad, blunt-tipped wings* (cf. White-winged Lark).

VOCALISATIONS Song superficially similar to the well-known song of Skylark, an 'endless' pouring of chirruping notes and fast-paced twittering, but differs in, relatively speaking, slightly slower tempo and hinted pauses (Skylark has none of those), but easiest on inclusion of long, dry rolling notes akin to the call. Will include mimicry in its song, too, notably of waders (but so too will Skylark). – Call a dry, fast rolling or 'frizzling' *shrrreep*, often drawn out and repeated in series when the birds chase each other and tumble down in the grass to 'fight it out'; shorter and more Skylark-like notes are sometimes heard.

SIMILAR SPECIES Very similar to *Bimaculated Lark*, both in appearance and song, but distinguished in song-flight by the blackish-looking underwings (Bimaculated rather pale brown-grey), at times slow motion-like wingbeats (Bimaculated constantly fluttering), and folded and longer-looking tail (Bimaculated spread and short-looking). When flushed, note

M. c. calandra (left: Israel, Mar; right: Spain, May) in flight: from below underwing often appears uniformly dark, even blackish, but forewing is paler; prominent white trailing edge to wing and lack of broad white tail-tip. Broad white trailing edge also visible from above, emphasised by blackish secondaries. (Left: Y. Perlman; right: S. Hage)

dark wing with white trailing edge, and white tail-sides and white 'corners' (Bimaculated has paler wing without white trailing edge, and tail has broad white tip, not sides). On the ground, Bimaculated has a more striking facial pattern, with whiter supercilium and darker lores. Other suggested differences, like bill size and size of breast-side patches, are less useful due to much overlap. – A distant flying *Thick-billed Lark* will also show a prominent white trailing edge to wing but this is even broader and more prominent in Thick-billed, and usually the heavier bill, slightly shorter tail and overall more front-heavy look will be evident. – ♀ *Black Lark* has similarly dark wings but lacks white trailing edge. (Further differences from ♀ Black Lark given under that species.) – *Greater Short-toed Lark* is much smaller and slimmer, has finer bill and longer tail, and is usually slightly paler overall.

AGEING & SEXING Ages and sexes (after post-juv moult) alike in plumage, but see Biometrics for rather marked sexual size dimorphism. Insignificant seasonal differences. – Moults. Complete moult of both ad and juv in late summer (mainly Jul–Sep), though moult of juv somewhat more protracted. – SPRING Ad In late breeding season heavily-bleached birds slightly darker above and less clearly patterned, with reduced pale upperparts fringes; underparts streaking a little bolder. **1stS** Inseparable from ad. – SUMMER–AUTUMN Ad Upperparts streaked dark, no prominent pale tips or edges. 1stW Inseparable from ad. **Juv** Upperparts and wing-coverts have prominent pale tips and edges, greyish- or buffish-white.

BIOMETRICS (*calandra*) **L** 17–20 cm; **W** ♂ 125–139 mm (n 60, m 131.8), ♀ 111–126 mm (n 40, m 118.6); **T** ♂ 60–72 mm (n 22, m 64.8), ♀ 53–59 mm (n 17, m 56.4); **T/W** m 48.9; **B** ♂ 17.2–20.7 mm (n 60, m 19.2), ♀ 15.0–18.3 mm (n 38, m 16.9); **BD** ♂ 8.7–11.0 mm (n 39, m 9.9), ♀ 7.8–9.0 mm (n 28, m 8.5); **Ts** ♂ 27.0–28.5 mm (n 18, m 27.7), ♀ 24.0–27.4 mm (n 12, m 26.0). **Wing formula: p1** minute; **pp2–3**(4) usually equal and longest (or p2 0.5–3 mm <); **p4** < wt 0.5–4 mm (rarely = wt); **p5** < wt 8–15 mm; **p6** < wt 19–28 mm; **p7** < wt 27–37 mm; **p10** < wt 40–51 mm; **s1** < wt 42–55 mm. Emarg. pp3–5 (sometimes less distinct on p5).

GEOGRAPHICAL VARIATION & RANGE Slight variation, and partially obscured by a degree of individual variation. Differences in wear and bleaching (and even soil-staining) add to the difficulties of identifying single birds outside breeding season.

M. c. calandra (L., 1766) (Eurasia east to Ural steppes, Turkey, Caucasus, Transcaucasia, extreme SW & NW Iran, Levant, N Africa; western populations mainly sedentary, eastern partly or wholly migratory wintering on north coast of Egypt, Levant, Iraq, Kuwait, S & E Iran). Rather dark, mantle boldly streaked, streaks cold brown to blackish, with some variation in darkness of plumage and boldness of streaking over much of range, and some of this variation has been taken as grounds for description of further subspecies. Birds in the Levant (N Israel) have been separated as '*hebraica*' on account of bolder streaking above and larger breast patches,

M. c. calandra, Bulgaria, Feb: in early spring (and often well into summer) still shows some streaks below breast-side patches, which are often narrower (though shape varies with posture, e.g. narrower if crouching). In such circumstances, might be confused with Greater Short-toed Lark, though note solid black line across breast, pale patch on fore ear-coverts below eye, and large yellowish base to lower mandible. (M. Varesvuo)

M. c. calandra, Bulgaria, May: from behind (especially with poorly-marked individuals and when breast-side patches mostly invisible)—as here—could be misidentified as Skylark, especially in mixed flocks. The overall bulkiness, larger bill and pale patch on lower ear-coverts become important in such circumstances. (B. Nikolov)

M. c. calandra, Ukraine, Apr: when singing or displaying on ground, often raises tail, and raises crown-feathers like Skylark. (M. Vaslin)

M. c. calandra, SE Turkey, May: birds in SE Turkey represent intergrades between two races, being *calandra* but inclining towards *psammochroa*. Note pale and slightly less saturated upperparts, the feathers broadly fringed sandy-grey when worn, with dark centres small and inconspicuous; breast and flanks pale buff (although often discoloured by reddish soil), with breast markings smaller on average. Amount of streaking on breast (left bird) varies individually, with degree of feather wear and, to some degree, sex. Bill shorter and more pointed in these birds, especially that on right, but note diagnostic white outer tail-feather edges (not tips)—an important distinction in areas of sympatry with Bimaculated Lark. (D. Occhiato)

M. c. calandra, Israel, Dec: when still very fresh, breast-side patches can be largely concealed, and instead whole breast is strongly streaked, causing such birds to be overlooked among winter flocks of Skylarks, with which it often mingles. (H. Shirihai)

M. c. calandra, juv, Bulgaria, Jun: prior to complete post-juv moult, dark spots on breast and whitish scales above are distinctive. (R. Wilson)

but these alleged differences could not be confirmed by us. Birds from near Gaza, S Levant, were described as race '*gaza*', being more rufous-tinged, but available series short, and soil-staining can perhaps not be excluded, so best included in *calandra* until better substantiated. Birds in extreme SE Turkey ('*hollomi*') said to be slightly paler and less cinnamon-tinged than typical *calandra*, but this is part of a cline towards *psammochroa*, and we follow Roselaar (1995) in lumping it with *calandra*. (Syn. *dathei*; *gaza*; *hebraica*; *hollomi*.)

○ **M. c. psammochroa** Hartert, 1904 (much of Iran, Central Asia; winters SE Middle East, SW Asia). Slightly paler on average, and perhaps tinged more pinkish-brown in fresh plumage. Streaks somewhat narrower and less dark. A little larger on average than *calandra* and subtly more long-tailed. A subtle race. **W** ♂ 130–141 mm (n 13, m 134.3), ♀ 116–127 mm (n 12, m 122.1); **T** ♂ 66–73 mm (n 13, m 68.4), ♀ 53–67 mm (n 12, m 60.3); **T/W** m 50.2; **B** ♂ 18.1–20.6 mm (n 12, m 19.4), ♀ 16.2–19.7 mm (n 12, m 17.6). (Syn. *raddei*.)

TAXONOMIC NOTES As evident from the above, the situation in the Levant is not fully understood. Taxa '*hollomi*', '*hebraica*' and '*gaza*' are morphologically close to *calandra*, while '*dathei*' (S Turkey from Gaziantep eastwards) and especially '*raddei*' (originally collected in W Iran, but similar birds found in Syria and Jordan) to some degree approach *psammochroa*. Thus, the situation in the Levant can be described as 'unstable variation', representing part of the clinal integration of *calandra* into *psammochroa*. The above lumping of many subspecies should be regarded as tentative until further material is available, but is already based on fairly long series from most regions.

REFERENCES Hartert, E. (1923) *Novit. Zool.*, 30: 1–32. – Lindroos, T. & Tenovuo, O. (2000) *Alula*, 6: 17–24.

BIMACULATED LARK
Melanocorypha bimaculata (Ménétries, 1832)

Fr. – Alouette monticole
Ger. – Bergkalanderlerche
Sp. – Calandria bimaculada
Swe. – Asiatisk kalanderlärka

The Bimaculated Lark is an Asian species, similar and related to Calandra Lark. In areas where both occur, Bimaculated Lark prefers barren and elevated ground, whereas Calandra Lark thrives on more lush and lower-altitude plains. Due to the relative harshness of its breeding grounds, Bimaculated Lark is mainly a migrant, wintering in NE Africa, Arabia and NW India, but also locally in Central Asia where the climate is favourable.

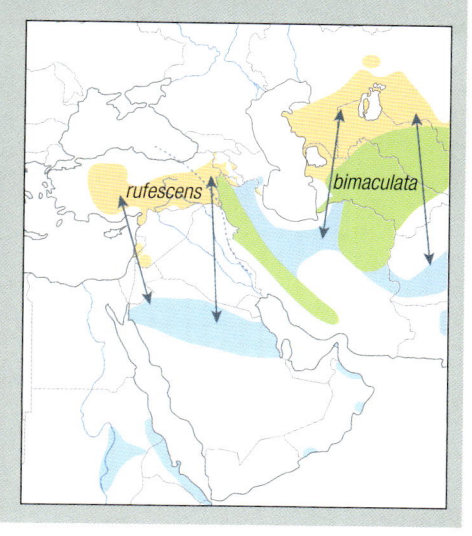

differs much more than the song from that of Calandra Lark, and is more similar to Greater Short-toed Lark, often a dry double twitter, *trip-trip*, but many variations including a more Calandra Lark-like *tshrree*.

SIMILAR SPECIES Very similar to *Calandra Lark*, which see. (See also above under Identification and Vocalisations.)

AGEING & SEXING Ages and sexes (after post-juv moult) alike as to plumage, but see Biometrics for sexual size dimorphism. Insignificant seasonal differences. – Moults. Complete moult of both ad and juv in late summer (mainly Jul–Sep), though moult of juv somewhat more protracted. – SPRING **Ad** In late breeding season heavily-bleached birds slightly darker above and less clearly patterned, with reduced pale upperparts fringes; underparts streaking slightly bolder. **1stS** Inseparable from ad. – SUMMER–AUTUMN **Ad** Upperparts streaked dark, no prominent pale tips or fringes. **1stW** Inseparable from ad. **Juv** Upperparts and wing-coverts with prominent pale, greyish- or buffish-white tips and edges.

BIOMETRICS (*bimaculata*) **L** 16–17.5 cm; **W** ♂ 115–130 mm (*n* 66, *m* 122), ♀ 109–120 mm (*n* 34, *m* 113.7); **T** ♂ 50–61 mm (*n* 29, *m* 54), ♀ 46–56 mm (*n* 17, *m* 51); **T/W** *m* 45.3; **B** ♂ 17.8–22.5 mm (*n* 66, *m* 19.5), ♀ 17.3–21.0 mm (*n* 27, *m* 19.2); **BD** 8.4–10.2 mm (*n* 23, *m* 9.3); **Ts**

Presumed *M. b. rufescens*, Israel, Mar: bold facial pattern with clear-cut and long supercilium, and rufous cheeks with bold white foremost patch, as well as dark breast-side patches. When perched, diagnostic white tail-tip often invisible or partly concealed by central tail-feathers. Locality suggests ssp. *rufescens*, even though plumage is more grey-brown than rufous-tinged. (H. Shirihai)

M. b. bimaculata, Kazakhstan, May: compared to Calandra Lark, generally has more contrasting head pattern, with whiter supercilium and darker loral stripe and cheek pattern (but much overlap). (C. Bradshaw)

IDENTIFICATION Readily recognised as either Bimaculated or Calandra Lark based on *compact build* with *short tail* and *heavy bill*, *dark breast-side patches* and pale supercilium. Compared to Calandra Lark, slightly smaller and proportionately shorter-tailed, has more *contrasting head pattern* with whiter supercilium and darker loral stripe and cheek pattern. When perched, it is also often possible to see that the *tail is white-tipped* (except the central pair of tail-feathers), but lacks white sides, a character which is of course more easily ascertained when the bird takes off. In song-flight, note the *spread tail* (again, white-tipped!) and the moderately dark, brown-grey underwing without paler trailing edge (Calandra has a blackish-looking underwing with prominent white trailing edge). Song-flight is quickly fluttering and lightly undulating, with 75% spread tail (Calandra has slow motion-flight on outstretched wings, tail closed; often 'standing still' in the air for long periods).

VOCALISATIONS Song very similar to that of Calandra Lark, an 'endless' pouring of various twitters at fast pace, but differs on, relatively speaking, slightly deeper pitch and by 'harder voice', often also appearing a shade more monotonous with fewer imitations inserted. In practice, however, often inseparable from that of Calandra. – Call, surprisingly,

Bimaculated (left: Turkey, May) and Calandra Larks (right: Bulgaria, Feb): very similar when perched, but white-tipped tail of Bimaculated usually visible (white tail-sides in Calandra), though most easily seen on take-off. Strongly contrasting head pattern in Bimaculated, but some overlap (and here white feathers of Bimaculated sullied reddish by soil). Markedly rufous ear-coverts also characteristic of many Bimaculated. Bill size and breast-side patches vary individually in both, latter partly due to freshness of plumage and posture. On ground, Bimaculated tends to show shorter tail, projecting less beyond wingtip, with on average shorter primary projection and often noticeably short space between two outermost visible primaries. (Left: D. Occhiato; right: M. Varesvuo)

24.5–28.0 mm (*n* 23, *m* 26.0). **Wing formula: p1** minute; **pp2–3** usually equal and longest (or p2 0.5–3 mm <); **p4** < wt 2.5–5 mm; **p5** < wt 11–16 mm; **p6** < wt 20–25 mm; **p10** < wt 40–46 mm; **s1** < wt 42–47.5 mm. Emarg. pp3–4 (rarely a hint also on p5).

GEOGRAPHICAL VARIATION & RANGE Slight and clinal, mainly involving colour shades. Single birds generally impossible to identify due to individual variability, wear, etc.
○ ***M. b. rufescens*** C. L. Brehm, 1855 (S & C Turkey, Levant, Iraq; winters in southern parts of range or moves to NE Africa, Arabia, SE Middle East). Tinged a fraction more rufous- or cinnamon-brown in fresh plumage than *bimaculata* (though beware of soil aberrations!). Dark streaks on lower mantle and back perhaps slightly heavier. Subtly larger than *bimaculata*, but large overlap. Supercilium reportedly a fraction less conspicuous on average but difficult to assess on skins. A very subtle form. **W** ♂ 120–129 mm (*n* 16, *m* 123.6), ♀ 112–119 mm (*n* 14, *m* 115.1); **T** ♂ 48–58 mm (*n* 16, *m* 53.5), ♀ 48–54 mm (*n* 14, *m* 51.2); **B** ♂ 18.8–22.2 mm (*n* 16, *m* 20.5), ♀ 18.1–21.5 mm (*n* 14, *m* 19.4); **BD** 8.0–10.5 mm (*n* 29, *m* 9.1).
○ ***M. b. bimaculata*** (Ménétries, 1832) (NE Turkey, Caucasus, Transcaucasia, Iran, Central Asia, W China; winters in southern parts of range or moves to C & S Iran, Pakistan and NW India). Described above. In general the plumage is rather cold brown. Whitish supercilium prominent. No differences detected between birds of NE Turkey or Caucasus and those of Central Asia ('*torquata*') when comparing long series. (Syn. *torquata*.)

REFERENCES Lindroos, T. & Tenovuo, O. (2000) *Alula*, 6: 17–24.

Bimaculated Lark in flight (left: India, Feb; right: Turkey, May): more uniform brown-grey underwing (cf. Calandra Lark) and lack of clear white trailing edge, but white tail-tip. (Left: K. Malling Olsen; right: D. Occhiato)

Presumed *M. b. rufescens*, Israel, Mar: compact with short tail and heavy bill. Note individual variation in size and shape of dark breast-side patches, but also due to posture. Not so strongly rufous-tinged individual (especially cheeks duller), still matching best the very subtly different *rufescens* breeding in Levant. (H. Shirihai)

M. b. bimaculata/*rufescens*, Israel, Mar: some still fresh autumn birds can be rather patternless. Those with largely concealed dark mark on breast and less bold facial pattern can resemble Greater Short-toed Lark, though general size, bill size, key plumage characteristics and voice will readily separate the two. Separation from Calandra Lark best achieved by combination of wing and tail pattern (in flight), but as shown here, strong supercilium and rusty-tinged ear-coverts are also useful clues. (L. Kislev)

M. b. bimaculata/*rufescens*, Oman, Nov: still in rather fresh plumage on wintering grounds in S Arabia. (M. Varesvuo)

BLACK LARK
Melanocorypha yeltoniensis (J. R. Forster, 1768)

Fr. – Alouette nègre; Ger. – Mohrenlerche
Sp. – Calandria negra; Swe. – Svartlärka

Like the White-winged Lark, Black Lark is a steppe specialist that is confined to the vast so-called Kirghiz plains (which, oddly, are in northern Kazakhstan, not in Kyrgyzstan). Quite abundant locally, in particular on the flattest parts of natural steppe, but does not avoid moderate cultivation, and is often found where sparse bushes invade the steppe. The ♂ is a real character, appearing considerably larger than the ♀ (partly an illusion), and courting her with a frenzy that can at times be described as less than well-mannered. Resident or, at most, a short-distance migrant.

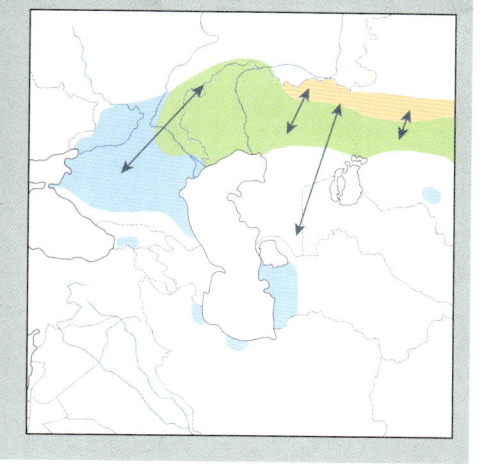

(wings lifted high, wingbeats slow, still good speed), utters squeaky sounds and flaps wings above back a few times, then climbs to 75–100 m with energetic wingbeats while either singing loudly, more subdued or in silence, before circling for some minutes, alternating between bursts of song and seemingly silent spells in which only subdued notes are uttered. Most intense song often delivered during descent in parachuting flight.

VOCALISATIONS Song, often from ground but frequently given in spectacular song-flight (see above), is superficially similar to Skylark, a prolonged twittering and chirruping, sometimes with mimicry interwoven. Often possible to identify by its higher pitch and feebler notes, by the inclusion of passages of squeaky, guttural or 'frizzling' notes somewhat recalling the sounds produced by a colony of breeding Rose-coloured Starlings, and of slowly-repeated, softer calls. The song is only delivered at full strength in certain parts of the song-flight, mainly in descent; at others, the bird is near silent, only giving 'hesitant', subdued 'miaowing' or 'pleading' notes. – Call a fizzing trill, somewhat variable, but in essence recalling the 'Rose-coloured Starling-like' parts of the song.

♂, Kazakhstan, Jun: in late spring/summer, pale fringes to fresh autumn plumage lost, and ♂ is more or less all black. (E. Yoğurtcuoğlu)

♂, Britain, Jun: some ♂♂ retain many whitish tips well into summer, making plumage look untidy. (N. Blake)

IDENTIFICATION Both sexes are large larks, the ♀, however, being somewhat smaller than the ♂. The ♂ can only be confused with a Starling, which it actually can resemble at times, being nearly *all-black* with a *short tail* and rather *thickset body*, and similar direct flight. The bill is strong (considerably shorter than in Starling) and bone-white, often with a bluish sheen, and the legs are blackish (Starling: pinkish). In fresh winter plumage, and into late spring or even early summer, the black plumage of ♂♂ is partly obscured by buff-white feather-tips, remaining longest on mantle, scapulars and tail-coverts. The ♀ is more nondescript, a brown-grey bird with darker mantle and wings, streaked dark above, and variably streaked or blotched dark on breast. Larger scapulars are quite dark-centred with paler fringes, creating a *row of dark spots on the folded wing*. On some, the dark breast-streaking forms larger blotches at the sides, vaguely recalling the pattern of Calandra Lark, and the *dark underwing* can further add to this similarity. Note, however, the *lack of any paler trailing edge* to the wing, and the generally *more greyish than pinkish legs*. (For further differences, see below.) – In courtship on ground, ♂ cocks tail, droops wings and fluffs up head- and neck-feathers. In song-flight, which varies according to wind force and situation, the ♂ often circles low in remarkable butterfly flight

♀ Kazakhstan, Jun: rather untidy spotty and streaky plumage, with characteristic densely-streaked breast-band and broadly pale-fringed upperparts, as well as large pale horn-coloured bill, mainly greyish rather than pink legs and prominent pale mask around eye. (E. Yoğurtcuoğlu)

♀, **Kazakhstan, May**: surprisingly Calandra-like, with large breast-side patches and pinkish legs, but has uniformly pale horn-coloured bill, broad white outer edge to one visible outer secondary, and more dark-streaked rump and uppertail-coverts. Also, compared to Calandra, face pattern rather subdued, especially lack of contrast in supercilium and lores. (A. B. van den Berg)

♂, **Kazakhstan, Jun**: almost starling-like impression in flight, black with horn-coloured bill, though jizz is very characteristic. (T. Lindroos)

SIMILAR SPECIES The similarity between the ♂ and a *Starling* is covered under Identification; the resemblance of ♀ to *Calandra Lark* was mentioned too, but further differences should be elucidated, in particular because ♀ Black can have dull pinkish legs instead of grey, albeit rarely. ♀ Black has a rather uniformly pale (grey, straw- or horn-coloured) bill, whereas Calandra usually has at least a hint of a contrasting darker culmen (a few exceptions noted). On Calandra, the underwing is rather uniformly dark (if anything, the coverts slightly paler grey than the flight-feathers), whereas on ♀ Black, the underwing-coverts are contrastingly blacker than the remiges (as on Crag Martin), thus the opposite of Calandra. On ♀ Black, unless heavily worn, longest tertials and outer secondaries have broad, sharply demarcated whitish outer fringes, whereas on Calandra these are diffuse and tinged rufous-buff (and therefore not noticeable). ♀ Black, unless heavily worn, tends to have diffuse broad pale subterminal bars inside tips of tertials, lacking on Calandra. Head pattern on ♀ Black on average more evenly brown-grey, appearing rather diffuse and plain-looking, whereas Calandra is slightly more contrasting and distinctly patterned. – *Bimaculated Lark*, which lacks white trailing edge to wing like Black Lark, has somewhat heavier bill, much less dark underwing, has pink-brown, not greyish, legs. It also has a white-tipped tail (♀ Black: virtually all-dark tail, outer edges only narrowly white).

♂, **Kazakhstan, Feb**: late winter ♂♂ still heavily pale-fringed, which makes them unmistakable, but extensive individual variation. (A. Isabekov)

♂, **Kazakhstan, Nov**: like other larks, all age classes undertake complete moult in late summer, thus after moult ageing becomes impossible, but extensive black plumage makes sexing straightforward here. (R. Chittenden)

AGEING & SEXING Ages (after post-juv moult) alike. Sexes usually readily separable. – Moults. Complete moult of both ad and juv in summer (mainly Jun–Aug). Seasonal changes due to abrasion of feather-tips rather obvious. – **SPRING F.gr.** ♂ Mainly black, all feathers tipped cream or buff-white in fresh plumage, very narrowly on centre of underparts, neck and flight-feathers, extensively on mantle, back, scapulars, flanks and tail-coverts, moderately over rest. General impression more black than buffish from midwinter or early spring. **F.gr.** ♀ Brown-grey above, off-white below, upperparts streaked, and breast variably streaked or blotched dark, wings and tail dark grey or blackish. As ♂, plumage pale-tipped when fresh, becoming darker with abrasion. **1stS** Inseparable from ad. – **SUMMER–AUTUMN Ad** ♂ Plumage black, no or only few traces of white feather-tips. **Ad** ♀ Upperparts streaked dark, no prominent pale tips or edges. Breast variably blotched dark grey. **1stW** After completion of moult generally inseparable from ad. **Juv** Upperparts and wing-coverts have prominent pale, greyish- or buffish-white tips and edges. Breast without larger dark blotches.

BIOMETRICS **L** 17.5–20 cm; **W** ♂ 130–144 mm (*n* 30, *m* 135.5), ♀ 116–127 mm (*n* 22, *m* 120.8); **T** ♂ 66–78 mm (*n* 24, *m* 71.2), ♀ 58–68 mm (*n* 22, *m* 62.5); **T/W** *m* 52.2; **B** ♂ 18.2–20.2 mm (*n* 23, *m* 18.8), ♀ 15.3–17.8 mm (*n* 21, *m* 16.9); **BD** ♂ 8.7–9.5 mm (*n* 12, *m* 9.1), ♀ 8.1–8.9 mm (*n* 6, *m* 8.4); **Ts** 22.1–26.4 mm (*n* 18, *m* 24.4). **Wing formula: p1** minute; **pp2–3** (4) about equal and longest; **p4** < wt 0–3 mm; **p5** < wt 5–12 mm; **p6** < wt 17–23 mm; **p10** < wt 41–51 mm; **s1** < wt 43–53 mm. Emarg. pp3–5 (rarely less distinct on p5).

GEOGRAPHICAL VARIATION & RANGE Monotypic. – Russia from middle Volga eastward to Irtysh basin, steppes in Kazakhstan to Zaisan in the east; mostly short-range movements in winter to SW & S Central Asia reaching to Sea of Azov.

REFERENCES Croft, K. (2003) *BW*, 16: 238–243. – Lindroos, T. & Tenovuo, O. (2000) *Alula*, 6: 170–177.

Presumed 1stY ♂, **Kazakhstan, Nov**: much paler and less heavily marked than bird on p. 60 (bottom), but streaking below extends to belly, suggesting ♂, probably young ('blackest' ♀ not known to develop black feathers below breast). (R. Chittenden)

♀♀, **Kazakhstan, Nov**: when wing folded, can resemble ♀ White-winged Lark, but lacks any rufous on wing-bend, has dark subterminal marks on tertials and no dark centres on these; breast spotted, rather than streaked, with no rufous on hindneck as often shown by White-winged Lark. Ageing impossible due to complete moult, but lack of black spots on rather clear white belly-feathers indicative of ♀♀. (R. Chittenden)

Juv, **Kazakhstan, May**: dark, profusely scaled buff-white, with whitish supercilium behind eye, and white throat. May breed sympatrically with White-winged Lark, but juveniles are readily separated, with Black Lark lacking rufous hindneck of White-winged. (A. van den Berg)

(GREATER) SHORT-TOED LARK
Calandrella brachydactyla (Leisler, 1814)

Fr. – Alouette calandrelle; Ger. – Kurzzehenlerche
Sp. – Terrera común; Swe. – Korttålärka

One of the most widespread and familiar small larks of southern parts of the Western Palearctic, a clear knowledge of its identification assists in identifying several other rather similar species. The Short-toed Lark is a summer visitor to plains and cultivation of the Mediterranean east to the Middle East, Mongolia and W China, and winters mostly in the Sahel and S Asia. There is also a mainly resident population in S Arabia. In many parts of NW Europe it is a regular autumn rarity, but in N Africa and the Middle East it forms spectacular dense flocks that migrate by day low over the desert in typical undulating flight and are conspicuous due to their continuous chirruping calls.

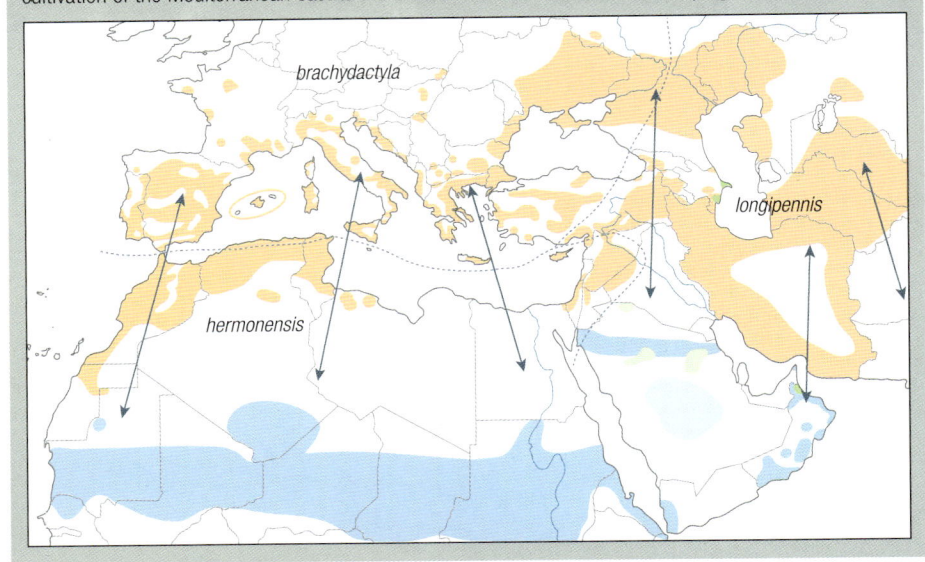

C. b. brachydactyla, Bulgaria, May: identified to race by locality and season. Tertials cloak primary tips, and thus lacks primary projection of Lesser Short-toed Lark (unless longest feather is missing or not fully grown). Characteristic face and wing markings. Rather dull (with less russet) plumage suggests ♀, but individual variation makes it usually impossible to sex birds with certainty. (M. Schäf)

less heavy than larger larks, with rapid take-off, fast undulating progress and abrupt dives. Flocks often maintain tight formation, both in flight and on ground (where can be very unobtrusive but also alert). Habitually shuffles on ground, but often runs or hops. Slow drifting (at about same height) and non-stop fluttering song-flight in tight circles, with strophes delivered synchronously with wingbeats in each undulation.

VOCALISATIONS Song a high-pitched, musical trill involving several brief phrases (repeated, but each segment can be completely different), often commencing falteringly on slightly lower pitch and accelerates and ends on slightly higher pitch with twittering or chirruping notes in very fast succession and seemingly 'in a mess', or opening with a few strong warbling sounds (almost Whitethroat-like) and ends with the fast twitter. Variable and melodic with brief pauses between strophes, 1–3 sec. long, pauses usually slightly longer than song strophes but varies depending on 'mood' and wind force. Longer songs with drawn-out phrases often contain more mimicry, making it much harder to separate from Lesser Short-toed. Beware occasional resemblance to (or mimicry of) *Melanocorypha* larks. – Commonest calls a short *tewp* or a dry chirruping *drit* or *drrr-t-t*, or a harder trill, or a few clipped, sparrow-like or Skylark-like notes (but less rippling than latter), e.g. *tchirrup* or *trriep-diu*, and longer *tue-*

pale lower neck-sides; also *whitish eye-surround* and (variable) patch on lower ear-coverts, creating *rather strong head pattern*. Upperparts prominently streaked dull black-brown (with paler nape and slightly gingery rump and uppertail-coverts), but most conspicuous are *boldly dark-centred median coverts* and double pale wing-bars. Lacks extensive breast-streaking (sometimes just a few blotches or indistinct streaks at sides), and often has *brown-black upper-breast-side patches* on warmer sandy-buff breast, while rest of underparts are whitish. In flight, look for *dark tail* with pale central feathers and *bold white outermost pair*. Bill mostly pale horn-coloured, with darker culmen and tip. Flight much

C. b. brachydactyla, Italy, Apr: identified to race mainly by locality and season, but also warm tone above and on breast. Prominent rusty tinge on crown is feature of some ♂♂, and well-marked black breast-side patches also indicate ♂, but much individual variation and overlap between sexes. (M. Schäf)

IDENTIFICATION Markedly smaller than Skylark, with a shorter cardueline-like bill, square-shaped head (no crest, except a hint when excited), and characteristic *long pointed wing with long tertials* and rather long tail. Coloration of the upperparts and breast, as well as bill size, somewhat variable, and colours vary seasonally, with age and sex, and due to soil staining. Western and southern birds warmer sandy-buff above (often with a stronger red-brown crown), but eastern and northern populations duller grey-ochre. Structure, however, is constant, as usually are face pattern and calls, which are useful for identification. Look for rather *distinctive buff-white supercilium*, emphasized by *dark eye-stripe* which usually appears to surround darker cheeks, and

C. b. brachydactyla, Spain, Apr: well-marked individual, showing black breast-side patches and classic face and wing patterns, including distinctive whitish eye-surround and supercilium, and dark median-coverts wing-bar; long tertials very nearly cloak primaries. Race inferred by locality. As both ad and juv moult completely by late summer, ageing generally impossible (like virtually all larks in the covered region). (C. N. G. Bocos)

LARKS

C. b. brachydactyla / longipennis, C Turkey, May: at this season, some can perhaps be sexed, with ♀ (right) generally duller with less rufous in more streaky crown and duller buff breast-sides, but substantial individual variation and overlap. Face pattern, conical, cardueline-like bill, dark upper breast-side patches, and long tertials almost reaching wingtip identify species. Racial identification in C/E Turkey difficult due to intergradation of *brachydactyla* into *longipennis*. (D. Occhiato)

teu-trettip-trp with tonality of a House Martin; at other times, e.g. when flocks migrate, a more liquid note quite similar to Tawny Pipit, *tshilp*. Usually lacks purring and rattling drawn-out flight calls of Lesser Short-toed, but the odd bird, or the odd call, can still be misleadingly similar!

SIMILAR SPECIES See quite similar Hume's Short-toed Lark as to separation from that. — Most significant confusion risk within the treated region is with Lesser Short-toed Lark (broad geographical overlap), which is dealt with under that species. Other species of larks occurring in the same area are much more easily separated, at least by experienced observers.

AGEING & SEXING Ages alike after post-juv moult. Sexes generally alike, but size differences in some populations perhaps useful in extreme cases (see Biometrics). — Moults. Both juv and ad moult completely, mostly Jun–mid Aug; apparently only limited partial body moult in late winter and only in some populations. — SPRING **Ad** Dark breast-side patches most obvious at this season (often with sharp blotches and streaks at sides); also more obvious individual variation in colour and amount of streaking on forehead and crown, most being greyish-buff and evenly and heavily streaked, others hardly streaked, and sometimes has brighter rufous-cinnamon crown (extremes apparently ♂♂). **1stS** Indistinguishable from ad. — SUMMER–AUTUMN **Ad** Broadly fringed above, with generally more sandy-buff or cinnamon hues, and dark feather centres partially concealed. Super-cilium is generally less well defined. Warm breast-band enhanced but dark breast-side patches less distinct, though mottling or mostly ill-defined streaky blotches characteristic at this season. **1stW** Some retain juv breast markings (more streaky blotches), even beyond Sep, otherwise inseparable from ad (both are evenly fresh). **Juv** Upperparts spotted/scalloped pale and white; breast pale buff with irregular dark brown spots and ill-defined blotches which may coalesce into slight gorget, while dark shoulder patch indistinct if present.

BIOMETRICS (*brachydactyla*) **L** 14–15.5 cm. **W** ♂ 92–101 mm (*n* 30, *m* 96.4), ♀ 86–96 mm (*n* 31, *m* 90.9); **T** ♂ 55–64 mm (*n* 19, *m* 59.2), ♀ 51–60 mm (*n* 22, *m* 55.3); **T/W** *m* 60.5; **B** 11.2–14.5 mm (*n* 59, *m* 13.2); **BD** 5.1–6.4 mm (*n* 41, *m* 5.5); **Ts** 19.0–22.0 mm (*n* 37, *m* 20.5). Wing formula: **p1** minute; **pp2–3** equal and longest; **p4** < wt 0.5–3.5 mm (rarely = wt); **p5** < wt 7.5–12 mm; **p6** < wt 14–19 mm; **p7** < wt 18.5–25 mm; **p10** < wt 27.5–36 mm; **s1** < wt 29–38 mm; **tert. tip** < wt 2–13 mm, =4/6 (rarely =4 or 6). Emarg. pp 3–4 (sometimes faintly also on p5).

GEOGRAPHICAL VARIATION & RANGE Mainly slight and predominantly clinal variation (upperparts and ear-coverts greyer and less rufous towards east, and supercilium and underparts whiter, but in Asia this trend is reversed, with upperparts browner again). Variation in the treated region is represented by the races listed below. Observers should bear in mind the substantial extent of individual variation and that all birds in fresh plumage (irrespective of race) are

Israel, Mar: race impossible to determine. Long-billed birds like this can almost recall Tawny Pipit. Breast-side patches sometimes less obvious when birds stretch upwards, and due to individual variation they can be reduced or lacking in spring/summer. (O. Plantema)

more similar. Also beware of the tendency for ♂♂ to appear brighter overall, and the normal occurrence of migratory birds of one subspecies within the breeding range of another. Add to this sometimes strong effect of staining or discolorations by local soil type, 'soil aberrations', and other environmental factors, which in our opinion, at least in part, were the reason for several unwarranted subspecies (see Taxonomic notes). Here, accordingly, only the more distinct ones are kept, based on careful examination of specimens in several large collections. Races differ mainly by plumage, and are most reliably identified when examining sufficiently long series of specimens and when comparing related taxa. Races in the covered region hardly differ in size or structure (substantial overlap and insignificant average differences) with the possible exceptions that *hermonensis* has on average slightly shorter legs and less pointed wings. Tail length in relation to wing length is subtly shorter in *longipennis*, but the difference is very slight.

C. b. brachydactyla (Leisler, 1814) (S Europe, W Turkey, apparently also locally north coast of Africa, at least in some years; winters in N Africa between Mauritania/Somalia and Chad/Sudan). Rufous-brown above (duller cinnamon when worn), heavily streaked dull blackish, sometimes with slightly rustier crown and uppertail-coverts (but some European populations, predominantly in the east, e.g. '*hungarica*', tend to be subtly duller and greyer). When series are compared with series of *longipennis* it is subtly paler and

Netherlands, May: outside normal range, as with this obvious vagrant, racial separation usually unsafe, but extreme rufous-tinged birds are usually well-coloured ♂♂ of *hermonensis*, particularly N African populations. (N. D. van Swelm)

warmer brown above, but some individuals of the two often inseparable. Similarly, it is usually subtly but clearly darker and colder brown than *hermonensis*, but again odd birds of both are inseparable. Grades seamlessly into *longipennis* in Ukraine and E Turkey. (Syn. *hungarica*.)

C. b. hermonensis Tristram, 1865 (NW Africa in Morocco to Tunisia, NW Libya, Malta, Levant; largely resident in Africa or sub-Saharan movements to N Mali, S Algeria east to Sudan). Averages paler and more reddish above than *brachydactyla* due to both paler ground colour and less blackish and less broad streaks, being more constantly saturated pinkish-cinnamon or rufous (sandy-isabelline when worn) above, pronounced on crown and uppertail-coverts, and streaks generally narrower and browner. There is often a hint of a saddle effect (more than in other races) in that streaked back stands out darker between paler and slightly less streaked rump and nape/crown. Underparts usually (but to variable extent) tinged rufous-cinnamon, strongest on upper breast and flanks. Originally described from Mt Hermon, Israel, but breeders in NW Africa at least from Morocco to S Tunisia ('*rubiginosa*'), possibly also N Libya, do not differ (four syntypes and series of *hermonensis* compared to long series from NW Africa in NHM), being similarly pale rufous or cinnamon-brown above. **W** ♂ 88–101 mm (n 28, m 95.7), ♀ 87–95 mm (n 16, m 90.6); **T** ♂ 54.5–62 mm (n 28, m 58.7), ♀ 51–59 mm (n 16, m 54.1); **T/W** m 60.7; **B** 12.0–14.7 mm (n 46, m 13.3); **BD** 5.1–6.3 mm (n 44, m 5.7); **Ts** 18.2–21.7 mm (n 46, m 20.1); **HC** 6.0–10.5 mm (n 26, m 7.2). **Wing formula: p5** < wt 6.5–9.5 mm; **p6** < wt 13.5–17 mm. (Syn. *rubiginosa*; *woltersi*.)

○ **C. b. longipennis** (Eversmann, 1848) (E Ukraine, Caucasus, Transcaucasia, E Turkey, Iran, Russian steppes east through Central Asia to N Mongolia and W Transbaikalia; winters mainly in India and S Central Asia, possibly also in Arabia and E Africa). Subtly darker brown and more heavily streaked black than *brachydactyla*, strongly streaked also on crown, less cinnamon-brown or sandy-toned as that often appears. Rufous never prominent on crown (but can be present to some extent). Still, single birds come close and can be impossible to identify away from breeding grounds. Same wing length as *brachydactyla* (despite scientific name *longipennis* referring to 'long wings'). — Specimens collected in winter in N Africa labelled '*longipennis*' are often misidentified darker examples of *brachydactyla*, perhaps stemming from E Europe or W Turkey, but a few (e.g. recorded in Tunisia, Biskra in Algeria and south to Chad) seem genuine enough and could represent overlooked normal African wintering

Kuwait, Mar: probably *longipennis* by range and overall rather greyish and narrowly-streaked upperparts. Prominent dark median-coverts bar. The slightly drooped wings have created an apparent (misleading) primary projection. This bird has unusually large breast-side patches. (A. Halley)

India, Apr: range and overall fairly dull upperparts suggest *longipennis*. Note individual variation in breast markings. Large migrant flocks may pause to drink in the desert, here with one Bimaculated Lark (left of centre) for size comparison. (A. Deomurari)

C. b. longipennis, Kazakhstan, Sep: race assigned by locality. Rather strong breast-streaking in fresh autumn plumage, thus confusion with Lesser Short-toed Lark more likely, making it essential to check face pattern, wing markings and relative lack of primary projection. (R. Pop)

Kuwait, Sep: probably *longipennis* by locality, but racial identification away from breeding areas uncertain, and in autumn differences less evident. Note few blotches or streaks on breast-sides, long pointed wing with long tertials, long tail, and dark-centred median coverts. (J. Tenovuo)

United Arab Emirates, Nov: race unknown. A tricky individual, displaying a primary projection (due to still-growing longest tertial) and predominately streaked breast-sides, both of which could suggest Lesser Short-toed Lark. However, shape, especially of head and bill, and face and wing patterns clearly identify the species. (I. Boustead)

Juv, Israel, May: prior to complete post-juv moult, entire upperparts and upperwings scaled darker and boldly tipped whitish, forming strong marbled pattern above. Face pattern recalls that of ad. (L. Kislev)

grounds of this race. **W** ♂ 90–102 mm (*n* 28, *m* 95.9), ♀ 87.5–95.5 mm (*n* 18, *m* 89.8); **T** ♂ 53–61.5 mm (*n* 27, *m* 57.6), ♀ 51.5–58 mm (*n* 18, *m* 54.4); **T/W** *m* 60.4; **B** 11.6–14.7 mm (*n* 45, *m* 13.3); **BD** 5.2–6.7 mm (*n* 45, *m* 5.9); **Ts** 18.7–22.2 mm (*n* 43, *m* 20.3). – Birds from Caspian region, N & W Iran and Transcaucasia ('*artemisiana*') have on average *c.* 3.5 mm shorter wing and very slightly less wide bill with marginally finer tip (though width differs only by some 0.2 mm or less) but are otherwise identical; said to be paler and greyer and to have longer bill, differences which we found impossible to corroborate, and these populations are best included in *longipennis*. (Syn. *artemisiana*; *orientalis*.)

TAXONOMIC NOTES Slight variation in colour has provided basis for other, less distinctive and perhaps debatable, races in addition to those discussed above. We fail to recognise '*hungarica*' (for Hungarian breeders), '*orientalis*' (Russian Altai) or '*woltersi*' (NW Syria & S Turkey) as the claimed differences appear too slight to warrant recognition and are best considered to represent insignificant local variation or parts of general clines. Even when, as here, adopting somewhat wider (and hence slightly more 'flexible') races, the ranges and borders between races are difficult to draw. These problems become most noticeable among birds from S Turkey, Middle East and the Levant. It follows that ranges of subspecies given here are tentative, and we would suggest that a re-evaluation of the taxonomy of this species is called for, especially of the controversial populations mentioned above, and of migration habits and winter grounds. – Formerly often considered conspecific with sub-Saharan Red-capped Lark *C. cinerea*, but there is now wide agreement to treat them separately. Still, together with Rufous-capped and Blanford's Short-toed Larks they form a so-called superspecies, meaning that they are closely related. – Recently the race *dukhunensis* was shown to be sister to Hume's Short-toed Lark *C. acutirostris* and not closely related to the Greater Short-toed Lark taxa sampled (Alström *et al.* 2013; Stervander *et al.* in press). The first study sampled only cyt b mtDNA, whereas the more recent one used also nuclear data; it seems therefore best to treat *dukhunensis* as a separate species, and a brief account is offered here.

Extralimital: **MONGOLIAN SHORT-TOED LARK** *C. dukhunensis* (Sykes, 1832) (montane steppe of Tibetan plateau, and through E Xinjiang and S & E Mongolia to E Transbaikalia and W Manchuria; winters in S Asia, but single birds with *dukhunensis*-like characters noted on migration as far west as Israel). Somewhat resembles eastern populations of *longipennis*, but upperparts rather consistently darker ochre-brown with broader and darker streaks on mantle, back and scapulars; uppertail-coverts almost unstreaked. Streaking of often rufous-tinged crown narrower, and forecrown often but not invariably unstreaked and usually duller greyish-brown. Pale buff-white or pinkish-buff supercilium from at least above eye and behind. Pale buff eye-surround and lores. Ear-coverts brown with blackish upper edge. Most characteristic is coloration below, with breast (mostly sides) and upper flanks distinctly washed rufous-buff surrounding the rather well-defined dark breast-side patches. (Underparts somewhat recall Greater Short-toed *hermonensis*, but *C. dukhunensis* differs by its darker and colder brown upperparts with bolder blacker streaking.) Also averages subtly longer-winged and thicker-billed. Bill orange-straw with hint of dark tip and culmen (recalling Hume's Short-toed Lark), legs brown. **W** ♂ 96–103 mm (*n* 14, *m* 100.4), ♀ 91–97 mm (*n* 13, *m* 94.2); **T** ♂ 58–62 mm (*n* 14, *m* 59.5), ♀ 50–56 mm (*n* 13, *m* 53.2); **T/W** *m* 58.1; **B** 12.4–14.8 mm (*n* 27, *m* 13.5); **BD** 5.6–7.0 mm (*n* 24, *m* 6.3); **Ts** 19.1–22 mm (*n* 27, *m* 20.8). Wing formula identical to that found in *C. b. brachydactyla* and *b. longipennis*. (Syn. *puii*.)

REFERENCES ALSTRÖM, P. *et al.* (2013) *Mol. Phyl. & Evol.*, 69: 1043–1056. – ALSTRÖM, P., MILD, K. & ZETTERSTRÖM, D. (1991) *BW*, 4: 422–427. – LANSDOWN, P. (1999) *BB*, 92: 308–312. – STERVANDER, M. *et al.* (2016) *Mol. Phyl. & Evol.*, 102: 233–245.

Mongolian Short-toed Lark *C. dukhunensis*, E Mongolia, May: overall dusky impression and noticeably slender, pointed bill of this extralimital close relative of Hume's Short-toed Lark; the bill is also rather dusky with even darker tip. Characteristic pale greyish tinge to underparts, but washed warm buff on breast-sides below dark patches. Unlike Hume's Short-toed Lark has well-developed whitish eye-surround and unmarked lores. Note also rufous crown with rather subdued dark streaking. These birds were a breeding pair, showing some sexual differences, the ♀ (right) being generally duller with less rufous on crown and duller buff breast-sides. (H. Shirihai)

RUFOUS-CAPPED LARK
Calandrella eremica (Reichenow & Peters, 1932)

Fr. – Alouette de Peters; Ger. – Jemenlerche
Sp. – Terrera capirrufa; Swe. – Brunkronad lärka

For a long time, all Short-toed Lark forms in Europe, Asia and Africa were united into one polytypic and widespread species, *Calandrella cinerea*. Recently, however, there is increasing agreement that the complex is better treated as several closely related species, one of which is the here presented Rufous-capped Lark, with one SW Arabian race, and another being the extralimital Blanford's Short-toed Lark, with two similar-looking races in Eritrea, Ethiopia and N Somalia (for latter, see Taxonomic notes).

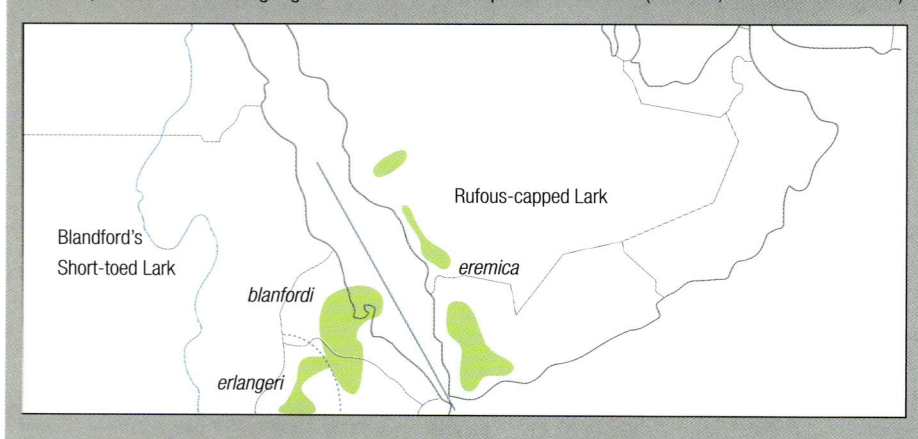

C. e. eremica, Yemen, Jan: rather dainty appearance, with small, thin bill, finely black-streaked chestnut crown and usually rather obvious black breast-side patches with only limited rufous hue or dark streaking below them. The chestnut crown, but also often the lead-grey forehead and neck-sides, are characteristic of ♂♂. (H. & J. Eriksen).

IDENTIFICATION (*eremica*) Closely recalls Greater Short-toed Lark ('Short-toed' henceforth) in structure and behaviour (just slightly smaller and more delicate), being a small to medium-sized lark with a tapering body, long wings and tail, and *long tertials concealing wingtip*. Compared with Short-toed Lark, centre of *crown noticeably rufous-brown and less streaked*, while forecrown is duller grey-brown (and can be quite dark when worn). *Nape rather pale brown distinctly streaked black*, giving it a different look than Short-toed. It also has *more extensive black breast-side patches* than the average Short-toed Lark, with *dark eye-stripe emphasizing broader whitish supercilium*. Mantle and back medium dark grey-brown diffusely streaked dark (not very different from Short-toed Lark ssp. *brachydactyla*). Wings identical to Short-toed Lark, only shorter with even *less primary projection*. *Tail pattern*, however, identical to that of African Red-capped Lark *C. cinerea* (extralimital; not treated) *with very limited whitish-buff fringes* restricted to outer webs (and extreme tips) of outer two tail-feathers (r6 lacks Short-toed's deep pale wedge on inner web). Compared to Red-capped Lark, *eremica* Rufous-capped has a paler, slightly more streaky and less deeply-saturated rufous cap, has black (not chestnut) breast-side patches and lacks the extensive chestnut rump and uppertail-coverts of Red-capped. May form large flocks in non-breeding season. Pecks rapidly on ground, running occasionally or walking rapidly, and may raise crest when excited or to cool crown in hot conditions. Flight sometimes appears somewhat lighter, more pipit-like, probably due to the smaller size.

VOCALISATIONS (*eremica*) Only a single recording studied by us, but song apparently recalls that of Short-toed Lark (perhaps with even less inclusion of mimicry), and pattern of display-flight also similar. – Flight-call recalls Short-toed, *chrit*, but short single calls may be followed by a short twitter, *chrit, pit-wit-pit*, or twitter given alone, and also utters a Corn Bunting-like *ptk* in flight.

SIMILAR SPECIES In Arabia most likely to be confused with migrant *Short-toed Lark*, but has noticeably more rufous on crown in all plumages (with clearly-defined blackish streaks on sides and, mainly, at rear), more distinctly and contrastingly streaked nape and, diagnostically, reduced pale edges to tail (most useful in the hand). See Biometrics, especially wing shape, which readily separates the two. Some Short-toed can have a markedly rufous crown and, equally, dull *eremica* Rufous-capped are very close to most Short-toed; in fact, it is unknown whether dull birds can be reliably identified using crown coloration alone. Rufous-capped also tends to have a greyish-mottled (even darkish) loral line (paler and cleaner in Short-toed, albeit with overlap), and a somewhat less obvious dark median coverts bar with clearcut pale edges, while the black breast-side patches tend to be more horizontal and can extend onto the upper flanks or central breast, but again overlap in both species renders these features of limited use. Overall smaller and less strongly built with, on average, a narrower darker horn-grey culmen (generally paler in Short-toed), but these differences are not diagnostic. – Both *Red-capped Lark* and African race *daaroodensis* of Rufous-capped Lark differ from Arabian *eremica* in having strong chestnut tinge on breast-sides, and Red-capped has deep chestnut breast-side patches, not black, whereas *erlangeri* Blanford's Short-toed (C Ethiopia) has these patches both black and chestnut. Further, these forms have much more striking head pattern with darker eye-stripe, more prominent pale supercilium, blackish forecrown and bright chestnut rump/uppertail-coverts. – Separated from *Lesser Short-toed* by same features as for Short-toed.

AGEING & SEXING (*eremica*) Ageing impossible after complete post-juv moult. Sexes alike. – Moults. Post-nuptial and post-juv moults complete. Pre-nuptial moult unknown. Few moult data from Arabia: four from Feb quite fresh or only slightly/moderately worn, one from Apr quite worn;

Comparison between Rufous-capped Lark *eremica* (Yemen, Jan) and (Greater) Short-toed Lark, probably *hermonensis* (Israel, Mar): although Rufous-capped has wing formula and tail pattern of *C. cinerea* group, it is generally closer to Short-toed in plumage. They can be very similar, true even for bright birds (the rusty crowns and strong face and breast patterns suggest ♂♂ in both cases). Rufous-capped's rather characteristic face pattern somewhat recalls Hume's Short-toed Lark in blacker lores and whiter supercilium behind eye; rather well-developed black lateral crown-stripe also often characteristic. Unlike Short-toed, note slender bill that often has darker culmen and tip, rather like Hume's Short-toed. Still, the similarity between the two is striking. (Left: H. & J. Eriksen; right: R. Pop).

C. e. eremica, Yemen, Oct: dark lores, emphasising whitish supercilium above and behind eye, and black lateral-crown stripe bordering supercilium and rufous cap. Mantle and back medium dark grey-brown diffusely streaked dark. (U. Ståhle)

C. e. eremica, Yemen, Oct: sexes alike, but birds with less rufous on crown and weaker face pattern are probably ♀♀. Note slender bill with darker culmen and tip, recalling Hume's Short-toed Lark. (U. Ståhle)

C. e. eremica, juv, Yemen, Oct: prior to complete post-juv moult, entire upperparts and upperwings are tipped whitish. (U. Ståhle)

five specimens from Jun very worn; two fresh full-grown juvs 26 Apr and 5 May. – **SPRING Ad** Red cap enhanced with feather wear, and dark streaks on rear crown almost lost in some. **1stS** Indistinguishable from ad. – **SUMMER–AUTUMN Ad** Freshest bird from Arabia is from Feb. Rufous cap less deep, and dark streaks on rear and sides of crown pronounced. **1stW** Apparently inseparable from ad. **Juv** Scaly pattern to upperparts and upperwings created by most feathers having a blackish centre and paler tip. Breast indistinctly spotted/blotched, and belly buffish.

BIOMETRICS (*eremica*) **L** 12.5–13.5 cm; **W** ♂ 84–89 mm (*n* 8, *m* 86.0), ♀ 77–80.5 mm (*n* 2, *m* 78.8); **T** ♂ 50–58 mm (*n* 8, *m* 54.1), ♀ 42 mm (*n* 2, *m* 42.0); **T/W** *m* 62.3; **B** 12.1–13.9 mm (*n* 11, *m* 13.0); **BD** 5.0–5.5 mm (*n* 10, *m* 5.2); **Ts** 16.9–18.3 mm (*n* 11, *m* 17.5). **Wing formula:** **p1** minute; **p2** < wt 0–2 mm; **pp3–4** (pp2–5) about equal and longest; **p5** < wt 0–2 mm; **p6** < wt 3–11 mm; **p7** < wt 14–16.5 mm; **p10** < wt 21–24 mm; **s1** < wt 19–24 mm. Emarg. pp3–5 (exceptionally very slightly also on p6).

GEOGRAPHICAL VARIATION & RANGE One race occurs within the treated range, another is extralimital (*daaroodensis*; not treated).

C. e. eremica (Reichenow & Peters, 1932) (SW Saudi Arabia, Yemen; resident or makes mainly short movements to lower altitudes in winter). Treated above. Slightly more greyish-buff, less sandy, and darker than similar NE African race (*daaroodensis*); also, has on average a slightly deeper saturated rufous cap. (Syn. *philbyi*.)

TAXONOMIC NOTES Red-capped Lark *C. cinerea* has traditionally been considered a single species following split from Greater Short-toed Lark *C. brachydactyla*, but occasionally two (e.g. Dickinson 2003, who separated the forms in Horn of Africa & SW Arabia as Blanford's Short-toed Lark, following Hall & Moreau 1970) or even three (e.g. Wolters 1982, Sibley & Monroe 1990, del Hoyo *et al.* 2004, who further separated 'Erlanger's Lark *C. erlangeri*' as a monotypic species). We adopt a different line still, based on own research on morphology and recent genetic research by Alström *et al.* (2013) and Stervander *et al.* (2016) meaning that the four taxa appearing in Arabia and Horn of Africa are treated as two species, Rufous-capped Lark (treated above) and Blanford's Short-toed Lark *C. blanfordi* with the two extralimital races *blanfordi* and *erlangeri*. – Arabian *eremica* and Somali *daaroodensis* are intermediate between Greater Short-toed Lark on the one hand, and Red-capped Lark and Blanford's Short-toed Lark on the other: they possess only black breast-side patches, not entirely or mostly red patches (as in Red-capped) or mixture of red and black (Blanford's), and lack the extensive and deep rufous-red crown and rich chestnut rump/uppertail-coverts of these forms, but share with Red-capped the limited pale edges to outer tail-feathers and rounded wing shape, which also separate *eremica* Rufous-capped Lark from Greater Short-toed Lark. Arabian *eremica* (together with Somali *daaroodensis*) is quite distinctive compared to *erlangeri* (Ethiopian highlands) and *blanfordi* (E Eritrea & NE Ethiopia, latter incl. *asmaraensis*, probably a synonym, from C & S Eritrea); *blanfordi* is closer to *erlangeri* in having substantial rufous on breast-sides and extensive chestnut rump and uppertail-coverts.

REFERENCES Alström, P. *et al.* (2013) *Mol. Phyl. & Evol.*, 69: 1043–1056. – Stervander, M. *et al.* (2016) *Mol. Phyl. & Evol.*, 102: 233–245.

Blanford's Short-toed Lark *C. blanfordi erlangeri*, C Ethiopia, Sep: endemic to Ethiopian Highlands. Blanford's Short-toed Lark has not been recorded in the treated region, but it performs both latitudinal and seasonal movements. Compared to Rufous-capped, it is larger with richer russet plumage, larger black-and-rufous breast-side patches, deeper buffish flanks, and heavily dark-streaked and browner upperparts, with narrow buffish-white outer edges to outer rectrices. (H. Shirihai)

HUME'S SHORT-TOED LARK
Calandrella acutirostris Hume, 1873

Fr. – Alouette de Hume; Ger. – Tibetlerche
Sp. – Terrera de Hume; Swe. – Höglandslärka

This close relative of Greater Short-toed Lark breeds at high altitudes from E Iran (local and rare) to NW India, and in S Central Asia. At least some populations are migratory, moving south or to slightly lower altitudes in winter. There is one record in Israel in February 1986. Exceptional views or preferably in-hand examination will probably be needed to prove further records.

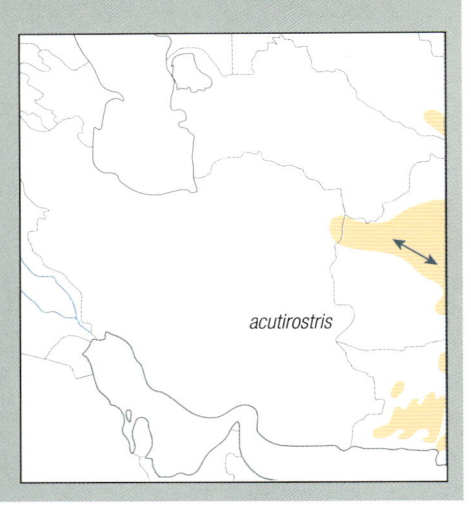

acutirostris

IDENTIFICATION Generally very similar to Greater Short-toed Lark ('Short-toed' henceforth) but, despite some overlap and much individual variation in both, Hume's is *overall greyer*, has a *blunter wing*, more *unstreaked crown* and a *distinct dark eye-stripe*. Crown either virtually *unstreaked* or more narrowly streaked than in Short-toed, and *ear-coverts are plainer, lores more prominently dark*, while the *white supercilium usually is bolder above and behind eye and very ill-defined in front* (in Short-toed, equally prominent throughout). The *whitish eye-surround is pale and indistinct, appearing broken by the black loral line* (in Short-toed, the surround is both broad and more complete). *Bill proportionately slightly longer, more slender and pointed* than in Short-toed, and *more obviously bi-coloured*, with a *pale yellowish-orange base and blackish culmen and tip*. In general, the upperparts tend to be less heavily streaked and greyer, but rump and uppertail-coverts slightly more rufous than in Short-toed. Belly and vent whitish (less buff-tinged than in Short-toed), whereas chest typically is rather uniform sandy-grey or dull brown-grey with no or reduced streaking compared to Short-toed, thus has on average smaller dark breast-side patches and a virtually unstreaked chest. Tail patterns are similar, but the white outer edge is noticeably narrower in Hume's (confined to the outer web and a narrow wedge inside shaft on inner web of the outermost feather). Wing structure differs from Short-toed in that the fifth outermost primary (p5) is long, almost reaching wingtip, giving the wing a blunter shape (p5 much shorter in Short-toed), and linked to this is the fact that three long primaries have emarginated outer webs (generally only two in Short-toed).

VOCALISATIONS Song (from ground or in broad-circling song-flight with alternating rapid wingbeats and dipping glides similar to Short-toed Lark) mellow and more variable than Short-toed, e.g. *trew-tew-telee-lieu*, often climaxing in a series of chipping notes before parachuting to earth; often exhibits considerable variation in song composition, and may include mimicry. – Calls include a sharp *trree* or more rolling *tiyrr* and disyllabic *tre-lit*, very different from Short-toed.

SIMILAR SPECIES Main confusion risk is with *Short-toed Lark*, of which the main separating characters are discussed above. Due to marked variation and much overlap in most plumage features, as well as in bill length and colour, in both species, it is vital to base identification on as many characters as possible. Calls are very important too. Identification of a vagrant Hume's Short-toed will require exceptionally detailed documentation. – Note also that the extralimital Mongolian Short-toed Lark *C. dukhunensis* (briefly treated

C. a. acutirostris, **India, Jun**: very similar to Greater Short-toed Lark, and often first detected by call. Unlike that species, has narrower and more pointed yellowish bill with contrasting dark culmen and extensively dark tip. Whitish supercilium clearer above and behind eye, and dark eye-stripe often tends to appear bolder and blacker on lores, becoming paler and more diffuse behind eye (in both respects unlike in Greater Short-toed), but much variation. (J. Kuriakose)

C. a. acutirostris, **India, Jul**: narrow, pointed yellowish bill with dark culmen and tip; dark eye-stripe often appears bolder and blacker on lores; supercilium usually bolder and whiter above and behind eye. (C. Francis)

C. a. tibetana, **China, Jun**: often has rather slim, elongated and long-tailed appearance; bold whitish supercilium from eye back, and contrasting bill coloration. This extralimital race breeds on the Tibetan Plateau and differs only subtly in being on average more greyish, less warm brown. (W. Müller)

Hume's Short-toed Lark *C. a. acutirostris* (left: India, Feb) and (Greater) Short-toed Lark *C. b. brachydactyla* (right: Israel, Mar): Hume's has similar structure (including very short or no primary projection) and plumage to Greater Short-toed, but usually has more slender and pointed bill that is more obviously bicoloured than in Greater Short-toed, with yellowish-orange base and blackish culmen and tip. Note that Hume's here rather atypically has clear whitish supercilium also in front of eye, but still a classic black and better-defined loral stripe, and also triangular-shaped dark centres to the scapulars and clear white underparts, all important differences for Hume's compared to Greater. Race impossible to decide outside the breeding area/season. (Left: G. Ekström; right: A. Ben Dov)

under Short-toed Lark, Taxonomic notes) approaches Hume's Short-toed in some respects.

AGEING & SEXING Ages after post-juv moult alike. Sexes alike. – Moults. Post-nuptial and post-juv moults complete and performed prior to any migration (often mainly Aug–Sep). If present, pre-nuptial body moult probably insignificant. – **SPRING Ad** Dark breast-side patches (some blotches and streaks), dark upperparts streaking and rufous wash to crown and uppertail-coverts all prominent at this season. Also, supercilium above and behind eye more obvious. Heavily worn and bleached from late spring. **1stS** Inseparable from ad. – **SUMMER–AUTUMN Ad** Broadly fringed above, with generally more sandy-buff tones and less obvious dark feather centres; supercilium generally less well defined and dark breast-side patches less distinct, partially or entirely overlain by fresh buff fringes. 1stW Indistinguishable from ad. **Juv** Spotted pale above and on most upperwing-coverts and tertials, giving lightly scalloped appearance; washed pale buff below with rather indistinct and irregular dark spots on upper breast.

BIOMETRICS (*acutirostris*) **L** 14–15 cm; **W** ♂ 90–98 mm (n 16, m 93.5), ♀ 85–91 mm (n 13, m 88.5); **T** ♂ 58–68 mm (n 16, m 63.3), ♀ 54–63 mm (n 13, m 59.5); **T/W** m 67.3; **B** ♂ 14.0–16.3 mm (n 15, m 14.7), ♀ 13.3–15.0 mm (n 12, m 14.1); **B**(f) ♂ 10.2–12.8 mm (n 15, m 11.4), ♀ 9.9–12.0 mm (n 11, m 10.8); **BD** 5.0–6.1 mm (n 26, m 5.4); **Ts** 19.7–22.0 mm (n 29, m 20.9); **HC** 4.8–9.7 mm (n 22, m 7.3). **Wing formula: p1** minute; **p2** often < wt 0.5–1.5 mm; **p3–4** usually equal and longest; **p5** < wt 0.5–3 mm; **p6** < wt 9–12 mm; **p7** < wt 14–18 mm; **p10** < wt 20–27 mm; **s1** < wt 23.5–28 mm; **tert. tip** = wt, or =5 (or =5/6). Emarg. pp3–5 (sometimes faintly also on p6).

GEOGRAPHICAL VARIATION & RANGE Two fairly distinct races of which only one occurs within the treated area, the other is extralimital (*tibetana*, Tibetan plateau). There is some intergradation where the two meet. – See Taxonomic notes under Greater Short-toed Lark, where closely related Mongolian Short-toed Lark *C. dukhunensis* is briefly described.

C. a. acutirostris Hume, 1873 (extreme E Iran, SE Central Asia east to N & W Xinjiang, NW China, Afghanistan, NW India, much of Kashmir, Pamir; in winter to lower altitudes or other moderately long movements). Described above. Tends to be more brown-tinged, less greyish than *tibetana*, and has slightly shorter wing but longer bill on average.

REFERENCES Shirihai, H. & Alström, P. (1990) *BB*, 83: 262–272. – Vaurie, C. (1951) *Bull. AMNH*, 97: 431–526.

C. a. acutirostris, India, Jun: besides the diagnostic bill colour, this bird also shows the characteristic orange-buff hue below the breast-side patches, and the tendency for the species to show triangular-shaped dark centres to the scapulars, as well as cold greyish upperparts and very clean whitish underparts. (S. Castelino)

C. a. acustirostris, India, Sep: here in fresh plumage when the breast-side patches are partially concealed, but instead some ill-defined breast and flank streaking often developed. Pay attention to the classic combination present here, with clear bicoloured and largely orange bill, and dark eye-stripe being strongest in front of eye whereas supercilium is whiter above and behind. Further, note the cold stony-grey dorsal colour and purer white underparts. (R. Balar)

LESSER SHORT-TOED LARK
Calandrella rufescens (Vieillot, 1819)

Fr. – Alouette pispolette; Ger. – Stummellerche
Sp. – Terrera marismeña; Swe. – Dvärglärka

An incredible songster of arid plains, usually known well only by keen birdwatchers. The taxonomy of this and the (possibly distinct) 'Asian Short-toed Lark' (see Taxonomic notes) offers a great challenge due to wide range and intricate variation.

Lesser Short-toed Lark breeds in the Canaries, Iberia and N Africa, and from Turkey and the Middle East to Lake Balkhash, with northern and eastern populations moving south and west in winter, primarily to the Middle East. For breeding typically prefers semi-deserts, dry salt pans and arid plains, but is slightly more catholic in winter.

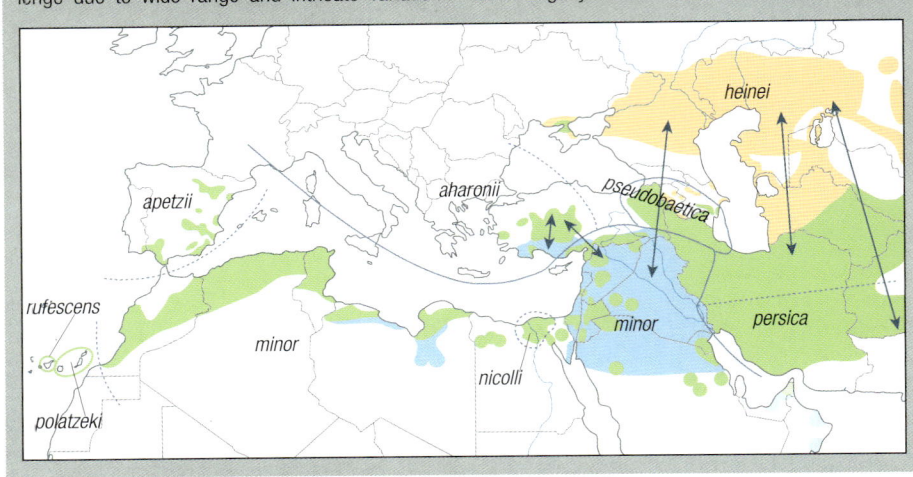

C. r. apetzii, Spain, Dec: small rounded head, small-looking bill, and heavy streaking even on neck-sides, lower throat and head-sides. This individual is rather heavily streaked, which is especially obvious in fresh plumage. (C. Bradshaw)

IDENTIFICATION Similar to Short-toed Lark (for clarity hereafter usually called Greater Short-toed Lark), although generally rather *rounder-headed and smaller-billed*. Western Lesser Short-toed Larks are marginally smaller and more compact than Greater Short-toed, but this is more or less reversed in the east. In all populations, primaries extend beyond tertials (diagnostic *primary projection* with typically three primary tips visible on folded wing). Also constant is *streaked breast*, with dense, fine streaks from lower neck-sides, over breast, sometimes reaching upper belly or upper flanks (where streaks diffuser and buff-brown). Note that when seen in profile, fine breast-streaking may merge into apparent large blotch inviting confusion with Greater Short-toed Lark. Close views usually reveal *more evenly and finely streaked head* which, compared to Greater Short-toed, has an *obscure pale supercilium* and only vague darkish eye-stripe, with distinctly flecked ear-coverts and fine crown-streaking, producing Linnet-like facial impression, augmented by the smaller, stubby bill, which is deeper-based and appears more curved on both mandibles. Head-on, often has an obviously pale forehead where the pale buff supercilia meet. Compared to Greater Short-toed Lark in S Europe, ground colour sandier (more greyish or browner in Middle East) with better-defined black-brown feather-centres to upperparts, but dark-centred median coverts usually less obvious. The short, deep-based bill is predominantly horn-coloured with a darker tip and

C. r. polatzeki, Fuerteventura, Canaries, Jan: compared to Greater Short-toed, head and entire breast extensively streaked, with less obvious (or almost no) pale supercilium and darkish eye-stripe. Note rounded head and small, mainly yellowish bill. (R. van de Weijde)

Lesser Short-toed *C. r. minor* (left: Jordan, Mar) and (Greater) Short-toed Larks *C. b. brachydactyla/longipennis* (right: Israel, Mar): with experience, even at first glance often easily distinguished by jizz (and calls), especially by former's more compact impression (rounder head and smaller bill), with more contrastingly streaked crown and rear cheeks. Note that dark patches on breast-sides of Greater Short-toed not always obvious. These two birds are different enough, but many are not and require careful check. It is often helpful to know that Greater Short-toed in autumn are streakier on the breast. Also, always check primary projection and face pattern. (H. Shirihai)

C. r. minor, Morocco, Feb: characteristically pale and sandy-tinged, generally less heavily streaked above and on cream-buff breast; white on outer tail-feathers rather extensive. (I. Yúfera)

culmen, and legs are bright ochre. General flight and behaviour much as Greater Short-toed but relatively rounder wings and shorter tail sometimes evident in flight. Compared to Greater Short-toed, has a rather distinct low-level, circling song-flight which is often performed with slow 'rowing' and high wingbeats (recalling song-flight of some finches) and with semiopen tail which is closed, then opened, then closed, etc. The low song-flight lacks obvious undulations with brief glides on closed wings. Note, however, that a high-level, more Greater Short-toed-like song-flight with bursts of wingbeats and brief glides occurs too. Song also given on ground or (unlike Greater Short-toed) atop a bush.

VOCALISATIONS Low-altitude song usually quite unlike Greater Short-toed Lark, being jaunty, quicker and richer, with more varied phrases. Often commenced in ascent (on high, slow, yet energetic wingbeats) with a series of well-spaced thin or drawn-out, emphatic whistles, *pssii-eh… pssii-eh… pssii-eh…*, etc., suddenly becoming a frenzy with much mimicry, dry churrs, rattles, piping, babbling and characteristic frizzling notes. Can also resemble Crested Lark or Calandra Lark, and superficially even Skylark (Lesser Short-toed may mimic all of the above-mentioned species). Low-level song also characterised by absence of pauses, thus no short strophes or obvious patterns are formed. High-level song much more similar to Greater Short-toed Lark, often in strophes opening with a few faltering notes difficult to distinguish from latter. Best separated by longer strophes, faster pace, 'more notes', and inclusion of a dry call note, *drrrrt*, or similar trills. — Commonest call characteristic: a purring rattle and drawn-out *drrrrd*, usually slightly falling or slowing at end, *drrrrt-t-t*. (On Fuerteventura, ssp. *polatzeki*, call noted to ascend at end, *drrre-itt*.) This rather loud, dry rippled voice can recall Sand Martin, is often repeated, and is more protracted and sharper than Greater Short-toed. There is also a surprisingly sparrow-like *chett*.

SIMILAR SPECIES For separation from often quite similar *Greater Short-toed Lark*, note the following. Distinctly and often wholly streaked breast (if present in Greater Short-toed, markings restricted to sides and appear as uneven streaks and blotches), but must be seen head-on for safe evaluation. Lacks dark upper-breast-side patches of most (but not all) Greater Short-toed. Primary projection of *c.* 3 exposed tips beyond tertials may require prolonged close views to confirm, as tertials and primaries often concolorous (primaries almost entirely cloaked by tertials in Greater Short-toed, but check for worn, broken, missing or growing tertials). Slightly more compact, with marginally shorter tail and slightly stubbier bill (relatively slender, pointed and more conical in Greater Short-toed), and more rounded head. Evenly and finely streaked head, with less obvious pale supercilium and darkish eye-stripe indistinct or absent, but often has distinct lateral throat-stripe (lacking or very narrow in Greater Short-toed), while some Greater Short-toed have an unstreaked rufous crown. Many populations possess more sharply demarcated and narrower dark feather-centres to upperparts (in Greater Short-toed, streaks bolder and often slightly more diffuse). Less distinct dark bar on median coverts, and outer web of central alula narrowly and diffusely fringed (much broader and clearly-defined pale fringe in Greater Short-toed, the pale area often extending over much of outer web). Usually differentiated by vocalisations, which enable identification of flying birds (although a superficially similar flight-call is occasionally given by Greater Short-toed). However, plumage features and biometrics of both vary geographically, forcing observers to closely scrutinise birds in some regions. Bill differences not always apparent, especially if comparing some eastern races of Lesser (in which bill is larger and more conical) and a relatively small-billed Greater Short-toed. Furthermore, a Lesser with extensive breast-streaking, especially on sides, can appear to have a vague dark breast-side mark (somewhat reminiscent of Short-toed) or, conversely, some pale birds may possess much paler streaking confined to the breast, appearing almost non-existent at distance; Greater Short-toed, on the other hand, frequently has continuous narrow streaking across upper breast (albeit concentrated on sides). Streaking often rather diffuse in some Turkish Lesser Short-toed and further east. — Both *Alauda* species, especially *Oriental Skylark*, may offer a pitfall. However, Lesser Short-toed is smaller, especially compared to *Skylark* (which furthermore has a whitish trailing edge to wing), and has a shorter stubbier bill, indistinct crest, and usually paler and less heavily streaked plumage (although breast-streaking can be rather similar, especially in Oriental Skylark); hind claw of skylarks diagnostically longer. Experienced observer can easily eliminate skylarks by vocalisations. Cf. also Sand Lark and Dunn's Lark.

AGEING & SEXING Ages alike after post-juv moult. Sexes alike. — Moults. Post-nuptial and post-juv moults both complete, mostly Jul–Sep, but in Canaries and Mediterranean

Kuwait, Mar: features best fit *pseudobaetica*, or perhaps intermediate between that subspecies and *heinei* or *persica*. Especially note the fewer and finer streaks on chest than typical of *minor*. (A. Halley)

C. r. aharonii, C Turkey, Apr: distinctive, being large and plump, with relatively large (deeper-based) and yellow-tinged bill. Colder and greyer streaking above, and typically has shorter, narrower and sparser breast-streaking. (A. Karataş)

C. r. aharonii, C Turkey, May: heavy, almost bulbous yellow-tinged bill with restricted dark on culmen, and general greyish tinge above. (R. Armada)

Presumed *C. r. heinei*, Azerbaijan, Apr: generally rather similar to C Turkish *aharonii*, with large bill, but differs in slightly darker and warmer brown plumage, and more prominent streaking. (K. Gauger)

Presumed *C. r. heinei*, Kazakhstan, May: note fairly pale, grey-tinged brown plumage, with some quite prominent streaking, and large bill, all suggesting *heinei*. (C. Bradshaw)

C. r. pseudobaetica, E Turkey, May: proportionately weaker-billed, warmer and less pure grey above than *persica* and especially *aharonii*, with darker and broader streaks above; breast-streaking usually quite extensive but fine. (D. Monticelli)

C. r. pseudobaetica, E Turkey, Jun: common around Lake Van, and compared to other races in Near East is quite typically darker or browner (warmer), and heavily streaked. From *minor* in Levant, slightly darker and more greyish-tinged, but smaller and smaller-billed than *aharonii*. (P. Alberti)

region earlier, May–Aug. Pre-nuptial moult absent. – **SPRING Ad** Dark streaks and face pattern obvious; upperparts and ear-coverts more sandy-grey and underparts almost white (western races). Only moderately worn in early spring, but abrasion and bleaching obvious from May. **1stS** Indistinguishable from ad. – **SUMMER–AUTUMN Ad** Broadly fringed above and has generally warmer sandy-buff to rufous-cinnamon pigments, and dark streaking less contrasting or to some extent concealed (e.g. on head). **1stW** Inseparable from ad. **Juv** Upperparts and most upperwing-coverts and tertials spotted/scaled paler and subterminally bordered black; breast pale buff with less regular dark brown streaks and blotches.

BIOMETRICS (*apetzii*) L 13.5–15 cm; **W** ♂ 87.5–94 mm (n 15, m 90.6), ♀ 80.5–86 mm (n 16, m 83.4); **T** ♂ 51–57 mm (n 15, m 53.9), ♀ 46.5–52 mm (n 16, m 48.9); **T/W** m 59.1; **B** ♂ 11.0–13.2 mm (n 15, m 12.0), ♀ 10.7–12.6 mm (n 16, m 11.5); **BD** 4.5–5.6 mm (n 30, m 4.9); **Ts** 18.9–21.8 mm (n 31, m 20.2); **HC** 5.5–11.7 mm (n 26, m 8.1). **Wing formula: p1** minute; **p2** 0–2 mm < wt; **pp2–4** (or pp3–4) about equal and longest; **p4** < wt 0–1.5 mm; **p5** < wt 3.5–6 mm; **p6** < wt 9–14 mm; **p7** < wt 12.5–20 mm; **p10** < wt 19.5–29 mm; **s1** < wt 23–30 mm; **tert. tip** < wt 11–20 mm, =6 or 6/7 (52%), =7 or 7/8 (31%), or =8 (17%). Emarg. pp3–5, but frequently fainter on p5.

GEOGRAPHICAL VARIATION & RANGE About 15 races have been recognised, including those often assigned to the *cheleensis* group ('Asian Short-toed Lark'). Whether the latter merits species status is unresolved: claimed sympatry in breeding ranges of *leucophaea* and *heinei* led to suggestions that *cheleensis* be considered a separate species. It has also been proposed that birds breeding in C Turkey, C & E Iraq, Iran and Afghanistan belong to the *cheleensis* group, but this lacks foundation (see below under Taxonomic notes). We favour a more cautious approach, maintaining a single species. Even so, the most recent recovered phylogeny of Alaudidae (Alström *et al.* 2013) indicated that Sand Lark *C. raytal* is sister to *rufescens*, not *cheleensis*, which was basal to both these. If corroborated by future studies, this provides an argument to either split the *rufescens* group from the *cheleensis* group, or to include not only *cheleensis* but also *raytal* in a large polytypic *rufescens*. – We tentatively arrange the taxa in three groups, but these should not be regarded as clear-cut taxonomic entities since several characters overlap, and since some forms are rather distinctive within their groups, e.g. *rufescens* and *leucophaea* (see below). – Variation predominantly clinal and often insignificant. Groups and races differ in coloration and, to a lesser extent, in size, although eastern birds are larger or longer-winged, or both. Due to marked individual and to some extent seasonal variation, effects of soil and other environmental factors, racial assignment is rarely possible without biometrics and access to specimens. Seven races occur in the treated region, but we also deal briefly with two potential vagrants, *leucophaea* and *cheleensis*. The following is based on extensive field and museum studies of almost the entire complex. We have also benefited from input by P. Alström & K. Mild.

WESTERN POPULATIONS
(Canaries, North Africa & S Middle East)
Relatively small (L 13–14.5 cm) and short-winged, also relatively short-billed and short-tailed (T/W ratio *c*. 55–61), especially compared to Asian subspecies, although *minor* somewhat intermediate, approaching *heinei*. Coloration and breast- and upperparts-streaking vary, but apart from *rufescens*, which is quite dark, essentially paler and thus superficially closer to *heinei* and *cheleensis*, but distinctly smaller with smaller bills. Wing usually more rounded, and extent of white on outer pair of tail-feathers reduced, but *minor* intermediate or approaches *heinei*.

†? **C. r. rufescens** (Vieillot, 1819) (Tenerife; now apparently extinct). Heavily streaked above and on chest, slightly smaller, and decidedly darker and more rufous than neighbouring *polatzeki*. Darker rufous-tinged and generally clearly more streaked than *minor*. Whether the rufous colour is the effect of soil-staining or not is currently unknown, but juv already have this colour, and we prefer to accept the race until contradictory evidence becomes available. One specimen from Tenerife was paler like *polatzeki* (ZMB 2000/8485), but in this case a straggler from other parts of the Canaries can perhaps not be excluded. **W** ♂ 87.5–93 mm (n 16, m 89.8),

C. r. persica, Iran, Jan: paler above than *pseudobaetica* and *heinei*, with typical buffish-cinnamon fringes to dark feather-centres, fine streaking above, rather pale brown tertials, whitish below with very faint buffish wash, and relatively little and narrow dark breast-streaking. Differs from *minor* in being larger and larger-billed, paler and colder above, lacking rufous tinge, with thinner streaking. Longer primary projection, longer tail and stronger bill differentiate this species from Indian Sand Lark in Iran. Note how bill shape changes with angle and how a small crest can be partially raised. (M. Hornman)

♀ 81.5–91 mm (*n* 8, *m* 84.9); **T** ♂ 49–56.5 mm (*n* 16, *m* 53.7), ♀ 46–54 mm (*n* 8, *m* 50.3); **T/W** *m* 59.6; **B** 10.3–12.4 mm (*n* 24, *m* 11.6); **BD** 4.7–5.9 mm (*n* 23, *m* 5.3); **Ts** 17.6–20.0 mm (*n* 23, *m* 18.8); **HC** 5–8 mm (*n* 22, *m* 6.8). **Wing formula: p5** < wt 3–6 mm; **p6** < wt 8.5–13 mm; **p7** < wt 13–19.5 mm; **p10** < wt 23–29 mm; **s1** < wt 24–31 mm; **tert. tip** < wt 11–21 mm, =6/7 or 7 (common) or =7/8 or 8 (rare). Emarg. pp3–5. (Syn. *canariensis*.)

C. r. polatzeki Hartert, 1904 (Canaries except Tenerife; resident). Has same size and amount of streaking above and on chest as *rufescens* (though streaks average slightly narrower) but differs clearly in being less rufous, more cinnamon-brown. It also has subtly heavier bill, but much overlap. Darker brown and generally clearly more streaked than *minor*. **W** ♂ 85–92 mm (*n* 15, *m* 88.8), ♀ 81.5–88 mm (*n* 6, *m* 84.0); **T** ♂ 50–56 mm (*n* 15, *m* 52.8), ♀ 46–53 mm (*n* 6, *m* 49.6); **T/W** *m* 59.3; **B** 10.7–12.7 mm (*n* 30, *m* 11.7); **BD** 4.7–6.3 mm (*n* 29, *m* 5.4); **Ts** 17.2–20.0 mm (*n* 30, *m* 18.9); **HC** 5.5–8.1 mm (*n* 30, *m* 6.8). Wing formula very similar to *rufescens*. (Syn. *distincta*.)

C. r. apetzii (A. E. Brehm, 1857) (S & E Iberia; resident or makes short-range movements in winter). Duller brown (less reddish-tinged) than *rufescens* with much heavier streaking on greyish-brown upperparts and yellowish cream-buff breast; streaking generally better developed than in other subspecies, often invading throat and upper flanks. Has more distinct lateral throat-stripe and pronounced ear-coverts streaking. Pale areas on rr5–6 reduced and typically sullied buffish, dark shaft on r6 almost reaches tip, and on r5 dark outer web often reaches beyond base of pale wedge, while undertail-coverts more extensively streaked than in other races; darker-centred median coverts bar, and dark centres to tertials and central tail-feathers more pronounced. Biometrics given separately above. (Syn. *baetica*.)

C. r. minor (Cabanis, 1851) (North Africa east to Levant, SE Turkey, Arabia; resident or makes short-range movements in winter). Paler and sandier than previous races, with a stronger cinnamon tone and less heavily streaked above and on cream-buff breast. Rump and uppertail-coverts often tinged rufous-cinnamon. In general, white on outer two pairs of tail-feathers rather extensive, but variable, on r5 usually >50% of outer web white, often all. Perhaps more than other races, *minor* tends to vary strongly according to local soil conditions. Call perhaps not as dry as in Europe and on Canaries, more liquid and more similar to Short-toed Lark. **W** ♂ 87–96 mm (*n* 20, *m* 90.9), ♀ 81–94 mm (*n* 16, *m* 86.5); **T** ♂ 50–59 mm (*n* 20, *m* 54.0), ♀ 48–59 mm (*n* 16, *m* 52.0); **T/W** *m* 59.8; **B** 10.0–13.2 mm (*n* 38, *m* 11.6); **BD** 4.6–5.9 mm (*n* 37, *m* 5.3); **Ts** 18.7–21.3 mm (*n* 36, *m* 20.0); **HC** 5.0–10.5 mm (*n* 35, *m* 7.3). **Wing formula: p6** < wt 9–15 mm; **p7** < wt 14–21.5 mm; **p10** < wt 22.5–30.5 mm; **s1** < wt 24–32 mm; **tert. tip** < wt 10–22 mm, = 6/7 or 7 (when much worn =7/8 or 8). Emarg. pp3–5 (sometimes a little fainter on p5).

○ **C. r. nicolli** Hartert, 1909 (Nile Delta; resident). Poorly differentiated from *minor*, being slightly more heavily streaked and marginally darker. It is warm brownish, with bolder streaks on crown, mantle and scapulars, and buffier on chest and flanks, which have blotchier streaking; also averages smaller and more delicately built. **W** ♂ 85–90.5 mm (*n* 13, *m* 87.8), ♀ 81–84 mm (*n* 5, *m* 82.9); **T** ♂ 49–55 mm (*n* 12, *m* 52.6), ♀ 45–51 mm (*n* 5, *m* 48.0); **B** 11.2–12.7 mm (*n* 18, *m* 11.9); **BD** 4.6–5.8 mm (*n* 18, *m* 5.3); **Ts** 19.2–20.7 mm (*n* 18, *m* 20.0).

NEAR EASTERN AND N MIDDLE EASTERN POPULATIONS
(SE Europe, Near East, Turkey to Kazakhstan)
In several aspects intermediate between western and eastern populations, e.g. T/W ratio c. 57–64, thus has proportionately slightly longer tail than western birds. However, more importantly birds belonging to this group are quite markedly larger (L 15–17 cm) than western birds and have larger bill; also, duller and greyer, rather profusely and broadly streaked above, and typically have shorter/narrower and diffuser breast-streaks that often extend onto flanks. Size much as eastern populations, but latter have somewhat paler remiges and rectrices, and significantly more extensive white on outermost pair of tail-feathers. Wing shape intermediate, but typically emargination on p5 indistinct or even lacking (usually obvious in eastern populations). General coloration and streaking close to eastern populations.

C. r. aharonii Hartert, 1910 (C Turkey; most make short-range movements south-east in winter to N Levant, some possibly reaching N Africa). Considerably larger than *minor*, with obviously deeper-based and more bulbous or conical bill, and is proportionately longer-tailed and paler above, being noticeably paler and greyer brown in comparison to usually rufous-tinged *minor*. Upperparts boldly streaked dark, more contrasting than in *minor* (more like *heinei*). Differs from quite similar *heinei* in being paler, especially above, and from *persica* in being larger and greyer, not so warm brown. When fresh, pure white underparts washed pale buffish on breast (paler than *minor*, breast having narrower and fewer streaks). Amount of white on rr5–6 relatively extensive: on r5, 75–100% of width of outer web white in 94% (but lacks white tip); dark on inner web of r6 well marked, reaching 3–9 mm from tip. **W** ♂ 96.5–104 mm (*n* 18 *m* 99.3), ♀ 91.5–98.5 mm (*n* 7, *m* 94.4); **T** ♂ 60–69 mm (*n* 18, *m* 64.4), ♀ 57–66 mm (*n* 7, *m* 61.3); **T/W** *m* 64.8; **B** ♂ 12.4–14.0 mm (*n* 17, *m* 13.3), ♀ 12.5–14.2 mm (*n* 7, *m* 13.3); **BD** ♂ 6–7 mm (*n* 18, *m* 6.5), ♀ 5.8–6.7 mm (*n* 7, *m* 6.1); **Ts** 19.4–22.6 mm (*n* 23, *m* 21.0). **Wing formula: p7** < wt 17–21 mm; **p10** < wt 27–33 mm; **s1** < wt 29–35 mm; **tert. tip** =6/8 (when heavily worn in summer =8/10), < wt 18–29.5 mm. Emarg. pp3–5 (though some less prominently on p5). (Syn. *niethammeri*.)

C. r. heinei (Homeyer, 1873) (Ukrainian shores of Black Sea east to Central Asia including Kirghiz steppe, Transcaspia, Caucasus, N Iran; short-range or medium-range movements south in winter, some reaching Middle East and Egypt). Resembles *aharonii* but is darker above and on breast, has on average slightly weaker (though similarly often slightly yellow-tinged) bill, warmer pink-buff wash and sharper streaking on breast, on average less white on r6 and outer web of r5 (prominent and well-marked dark wedges on inner web of r6, almost reaching tip), and is proportionately shorter-tailed. Generally darker, more heavily streaked and subtly finer-billed than *persica*. For differences from *pseudobaetica* see that race. Differs from *minor* in being larger and proportionately longer-tailed, with darker brownish-grey, not so rufous or ochre-tinged brown upperparts with slightly darker and broader streaks. Also has on average bolder and blacker streaking on breast compared to *minor*, and has less rufous-toned supercilium, ear-coverts, breast, flanks, fringes to wing- and central tail-feathers. **W** ♂ 96–107 mm (*n* 24, *m* 101.4), ♀ 91–101 mm (*n* 16, *m* 95.3); **T** ♂ 58–67 mm (*n* 24, *m* 63.3), ♀ 55–62 mm (*n* 16, *m* 59.5); **T/W** *m* 62.5; **B** ♂ 11.1–15.0 mm (*n* 17, *m* 13.2), ♀ 10.9–13.8 mm (*n* 16, *m* 12.7); **B**(f) ♂ 8.5–13.0 mm (*n* 14, *m* 10.9), ♀ 8.7–10.3 mm (*n* 6, *m* 9.6); **BD** 5.3–6.3 mm (*n* 35, *m* 5.8); **Ts** 19.0–21.9 mm (*n* 24, *m* 20.8); **HC** 6.5–11.3 mm (*n* 24, *m* 8.7). **Wing formula: p5** < wt 4–9 mm; **p6** < wt 11–18 mm; **p7** < wt 16–22 mm; **s1** < wt 29–35 mm. Emarg. pp3–4 (sometimes also on p5).

○ **C. r. pseudobaetica** Stegmann, 1932 (E Turkey,

C. r. heinei, juv, Kazakhstan, Jul: all races share pale scaling and subterminal black markings above in this plumage; note facial pattern. (H. & J. Eriksen)

C. r. cheleensis ('Asian Short-toed Lark'), Mongolia, Jun: extremely similar to eastern populations of the Western Palearctic, and populations as far east as Kazakhstan, e.g. *heinei*, and like this fairly sizeable and large-billed. Nearly always has narrow, neat black streaks on cream-coloured breast, coalescing to form characteristic dark patches below neck-sides. Rather prominent whitish supercilium above and behind eye. Relatively more white in tail, rather paler bases and extensive pale fringes to remiges, rectrices, tertials and secondary-coverts, which contribute to general pale appearance. (H. Shirihai)

N Iraq, Transcaucasia, south-western shores of Caspian Sea; apparently also Turkish Black Sea coast and Bulgaria; winters Middle East, N Arabia). Very similar to *heinei* but is on average slightly smaller and when series compared subtly warmer brown and a little paler; black streaks on upperparts a trifle less bold. Palest birds very close to darkest extremes of *aharonii* but differ in somewhat smaller general size, less bulbous and somewhat shorter bill, and in stronger streaking on more ochre-brown, less pinkish-grey or pale cinnamon-grey ('stone grey') crown and back. **W** ♂ 94–101.5 mm (*n* 14, *m* 96.9), ♀ 88–92 mm (*n* 6, *m* 90.5); **T** ♂ 56–63 mm (*n* 14, *m* 60.0), ♀ 55–58 mm (*n* 6, *m* 56.6); **T/W** *m* 61.8; **B** ♂ 12.1–13.2 mm (*n* 14, *m* 12.8), ♀ 11.8–12.9 mm (*n* 6, *m* 12.5); **BD** 5.6–6.4 mm (*n* 16, *m* 6.1). **Wing formula:** Emarg. pp3–5 (although usually more faintly on p5).

C. r. persica (Sharpe, 1890) (SE Iraq east through S Iran to S Afghanistan; winters in S Iran, Gulf States, Arabia). More buffish-cinnamon and paler above than *pseudobaetica* and *heinei*, with rather pale brown tertials lacking extensive dark centres. Distinct but fine streaking above, whitish below with very faint buffish wash, and relatively few and narrow dark streaks on breast. Heavy-billed. White in tail reduced and more like *heinei*, or even more limited. Its pale and finer-streaked head, back and breast are intermediate between *heinei* and *leucophaea*; also rather close to *aharonii* but clearly more buffish or sandy, less stone-grey above, with slightly narrower, paler and fewer dark streaks on breast. Differs from *minor* in being noticeably larger and larger-billed, paler and a little colder brown above, lacking rufous tones, and has thinner streaks on crown; also whiter below with narrower and fewer breast-streaks (especially in centre). **W** ♂ 93.5–104 mm (*n* 20, *m* 98.7), ♀ 91–97 mm (*n* 10, *m* 93.5); **T** ♂ 55–66 mm (*n* 20, *m* 60.3), ♀ 53–62 mm (*n* 10, *m* 56.7); **T/W** *m* 61; **B** ♂ 12.6–15.2 mm (*n* 20, *m* 13.8), ♀ 12.2–14.1 mm (*n* 10, *m* 13.2); **BD** 5.6–6.8 mm (*n* 23, *m* 6.3); **Ts** 18.5–21.5 mm (*n* 22, *m* 20.5). **Wing formula: p2** < wt 0–2.5 mm; **p3** (pp2–4) longest; **p4** < wt 0–2 mm; **p5** < wt 5–10 mm. Emarg. pp3–4 (sometimes faintly also on p5). (Syn. *seistanica*.)

EASTERN POPULATIONS

(Extralimital: Central & East Asia), 'ASIAN SHORT-TOED LARK'
Three races east of the Caspian Sea: *leucophaea*, *cheleensis* (including '*beicki*' and '*obscura*') and *kukunoorensis* (including '*tangutica*', '*seebohmi*', '*tuvinica*' and '*stegmanni*'). No confirmed records within the treated region. Generally, extremely similar to previous group and separation without examination of tail pattern and wing formula often impossible. Fairly large (L 15–17 cm) with large bill, and in overall proportions well differentiated from western populations (T/W ratio *c.* 63–70; thus only some overlap with previous group, and none with western group). Generally paler (chiefly ash-grey, with some buff or rusty tones in *cheleensis*), finely streaked above, with narrow, neat black streaks on cream-coloured breast, and more white in tail, as well as more rounded wing (emarg. of p5 as a rule distinct, although in *cheleensis* rather faint). Further characterised by paler remiges, tail-feathers, tertials and secondary-coverts, which have comparatively pale centres and rather whitish tips/fringes.

C. r. leucophaea Severtsov, 1873 (Kazakhstan, Uzbekistan & Turkmenistan, although precise range poorly known after apparent decline and requires confirmation; wintering habits poorly known, may be nomadic or make short movements). Slightly smaller than *heinei* with a proportionately longer tail and finer bill, and is much paler overall with finer breast- and upperparts-streaking, in paleness closer to *persica* but lacking nearly any cinnamon tinge above, being more grey-brown. Plumage often 'rich and fluffy', almost like an Arctic Redpoll compared to a Common Redpoll. Ear-coverts paler and less streaked than in *heinei*, again similar to *persica*. Also compared to *heinei* tends to have paler culmen and tip, and overall more yellowish bill, but some overlap. Best clue is pattern of r6, *leucophaea* having more white on inner web: the dark wedge is much narrower than the white portion and reaches no nearer tip than *c.* 15–40 mm (2–14 in 99% of *heinei* and *persica*; once 17 mm), and outer part of dark wedge often pale and ill-defined (invariably dark and well marked in *heinei* and *persica*). On average, *leucophaea* has more distinctly demarcated whitish tips to secondaries (*c.* 1 mm-wide tips), but overlap. Wing formula appears less useful than claimed in some Russian literature, and it is better to use a combination of tail length, pattern of white on tail, relatively small bill and pale and greyish-tinged plumage. Generally, *leucophaea* is quite distinctly paler and finer-streaked than extralimital *cheleensis* or *kukunoorensis*, but shares with these extensive white on outer tail-feathers. **W** ♂ 96–101 mm (*n* 15, *m* 97.7), ♀ 90–95 mm (*n* 12, *m* 92.1); **T** ♂ 60–69 mm (*n* 15, *m* 66.1), ♀ 57–64.5 mm (*n* 12, *m* 61.2); **T/W** *m* 67.1; **B** 10.3–13.2 mm (*n* 31, *m* 12.0); **BD** 4.8–5.8 mm (*n* 32, *m* 5.2). **Wing formula: p2** < wt 0–2.5 mm; **pp3–4** about equal and longest; **p5** < wt 2.5–6 mm; **p6** < wt 9.5–14 mm; **p7** < wt 16.5–20 mm. Emarg. pp3–5 (rarely a hint also on p6). Notch on p2 (ill-defined!) 27.5–31 mm (in *heinei*: 24–28.5 mm).

C. r. cheleensis (Swinhoe, 1871) (C & E Mongolia, Transbaikalia, Manchuria, NE China; wintering habits poorly known, but makes medium-range movements to S Asia, and could straggle into covered range in winter). Rather similar to *heinei*, *persica* and *pseudobaetica* but is essentially darker and browner above in direct comparison, often slightly rufous-tinged brown on sides of chest and on upperparts (often rufous-toned uppertail-coverts) and has more extensive white wedges on inner webs of r6, similar to *leucophaea*. Fine streaks on chest often coalesce into larger spot on sides, almost as in Greater Short-toed. Much darker and warmer brown than *leucophaea* with more prominent streaking above and on chest. Rather large size variation indicating the possibility that more than one subspecies is involved. **W** ♂ 87–98 mm (*n* 12, *m* 93.1), ♀ 84–94.5 mm (*n* 12, *m* 89.8); **T** ♂ 58–70 mm (*n* 11, *m* 63.5), ♀ 58–67 mm (*n* 12, *m* 62.0); **T/W** *m* 68.8; **B** 11.0–14.2 mm (*n* 24, *m* 12.1); **BD** 4.8–6.3 mm (*n* 24, *m* 5.3). (Syn. *beicki*; *obscura*.)

TAXONOMIC NOTES Relationships within the *rufescens* complex are as yet unresolved, and more research is needed in the Middle East and, especially, Central Asia, where several taxa critical to our understanding of the group breed. – Roselaar (1995) postulated that breeders in high plateaux of C Turkey, together with *persica* (SE Iraq to S Afghanistan), belong to the *cheleensis* group. However, *persica* and *aharonii* morphologically clearly cluster with *heinei*, and all share the tail pattern of *minor*. Therefore it seems wiser to retain all of these forms under the umbrella of one species, at least for now. – The breeding range of *leucophaea* reportedly overlaps with *heinei* between the former Aral Sea and Lake Balkhash (e.g. Stepanyan 1967), although we have not seen any breeding-season specimens of *leucophaea* from within the breeding range of *heinei*. Furthermore, recent searches in Kazakhstan failed to locate *leucophaea* within the core of their claimed area of sympatry, finding only *heinei* there. Such absence could be the result of a sharp decline (perhaps the result of habitat destruction), or nomadic habits typical of the species. Thus, sympatry between *rufescens* and *cheleensis* is unproven, offering a significant stumbling block to the theory that the *cheleensis* group is a separate species. However, new evidence has become available recently (Alström et al. 2013) indicating that *rufescens* and *cheleensis* might be best split after all. This study also recovered *Calandrella* as non-monophyletic, with the recommendation to move *rufescens*, *cheleensis* and *raytal* to a separate genus, *Alaudala*. We prefer to await more evidence before that recommendation is followed. – Roselaar (1995) claimed that the population in C Turkey should be named *niethammeri* Kumerloeve, 1963. However, both Alström & Mild (in prep.) and we have examined museum specimens, including type material (AMNH), and breeders in C Anatolia, and independently reached the conclusion that *niethammeri* is a junior synonym of *aharonii*.

REFERENCES ALSTRÖM, P. et al. (2013) *Mol. Phyl. & Evol.*, 69: 1043–1056. – ALSTRÖM, P., MILD, K. & ZETTERSTRÖM, D. (1991) *BW*, 4: 422–427. – LANSDOWN, P. (1999) *BB*, 92: 308–312. – STEPANYAN, L. S. (1967) *Acta Orn.*, 10: 97–107. – VAURIE, C. (1951) *Bull. AMNH*, 97: 431–526.

(INDIAN) SAND LARK
Calandrella raytal (Blyth, 1845)

Fr. – Alouette des sables; Ger. – Uferlerche
Sp. – Terrera raytal; Swe. – Sandlärka

Found in SE Iran, this close relative of Lesser Short-toed Lark is one of the few species covered here that arguably belongs to a different faunal region, the Oriental rather than the Palearctic, but included because all of Iran is treated. However, it is a borderline case, and some would argue that the entire Iranian fauna is essentially Palearctic. Sand Lark inhabits dry salt flats, riverbanks and other areas close to water, and its range also extends further east, through Pakistan to Assam and Bangladesh.

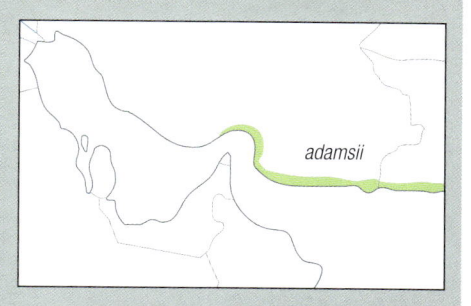

IDENTIFICATION Slightly *smaller than Greater Short-toed or Lesser Short-toed Larks*, with a proportionately *longer, slenderer and less deep-based bill*, a stocky body, *considerably shorter tail*, rather rounded wings and a *short but obvious primary projection*. Upperparts cold sandy-grey with diffuse streaking (strongest on crown and, to lesser extent, scapulars) and rather uniform wings (median coverts being less dark), but slightly paler rump and uppertail-coverts. On the whole, *upperparts are both greyer and plainer* than in congeners. *Whitish supercilium relatively prominent*, chiefly above and behind eye, with a broad whitish eye-surround, often broken by the dark loral mark, creating a pale crescent below eye, often with an ill-defined pale ear-coverts patch. Underparts whitish or slightly creamy-white with *very fine, sparse breast-streaking*. Uppertail mostly blackish (or dark brown) contrasting with greyer central feathers, and outer edges have limited white (r6 has white outer web and narrow, evenly broad white inside shaft) lacking broader white wedges on outertail of congeners in the area. Behaviour similar to Lesser Short-toed Lark, running in zigzagging spurts across open ground. Sometimes raises crown feathers in a short crest. Flight rather jerky and fluttering, but this in a way typical for all *Calandrella* or other closely related species.

C. r. adamsi, Iran, Apr: at first glance bears resemblance to several *Calandrella* larks in the region, but is especially similar to Lesser Short-toed Lark due to breast-streaking and similar primary projection. Note relatively thin breast-streaking (mostly confined to shafts), grey-tinged upperparts, and largely dark uppertail. As with other larks, take into account individual variation in bill length and shape; in particular the right bird has an unusually long and heavy bill. (C. N. G. Bocos)

VOCALISATIONS Song mostly recalls that of Lesser Short-toed Lark, a varied series of short, rapidly-repeated warbling notes, *treet-truit-it-it-du-du-du-du-di-di-di-dit* (thus usually longer and more complex than Greater Short-toed). Song-flight apparently involves a series of swoops before descending steeply to the ground. – Calls involve a guttural *prrr… prrr*, or *trrrit*, apparently similar to congeners (Roberts 1992, Grimmett *et al*. 1998).

SIMILAR SPECIES Slightly smaller size and different structure, including thinner bill and especially stockier appearance (emphasized by shorter tail and wings), very useful in separating this species from *Greater* and *Lesser Short-toed Larks*. Sand Lark lacks any hint of dark breast-side patches, and its facial pattern differs from both, being more like *Hume's Short-toed Lark*. The short primary projection helps separate it from both Short-toed and Hume's Short-toed Larks, but not always from Lesser Short-toed Lark (though projection often shorter than in Lesser Short-toed with only 2–3 visible primary tips, rather than 3–4 as in Lesser Short-toed). Streaking on breast less marked than latter, and on upperparts is generally less heavy than either of the other species. – Given the relatively compact build, bears some resemblance to ♀ *Black-crowned Sparrow Lark* but streaked upperparts of present species, as well as different

C. r. adamsi, Iran, Jan: rather pale greyish and finely streaked upperparts; also note reduced, thin breast-streaking, small bill, and rather short tail and primary projection. The squat shape with rounded head is very apparent on this bird, as well as the plain pattern of head and wing. (A. Ouwerkerk)

Sand Lark (ssp. *adamsi*; Iran, Jan) and Lesser Short-toed Lark (ssp. *aharonii / pseudobaetica*; Israel, Nov): the first clue in recognising Sand Lark in the field is its stocky or compact jizz, with quite short tail and proportionately thinner (less deep-based) bill. Secondly, Sand Lark usually appears slightly colder sandy-grey above with diffuse streaking (strongest on crown). Thirdly, some slight differences as to facial pattern may be useful: relatively prominent pale supercilium, chiefly above and behind eye, with a broad whitish eye-surround, often broken by thin dark loral mark, creating a pale crescent below eye, and often with an ill-defined pale submoustachial patch. As a whole, the facial pattern is weaker than in Lesser Short-toed, with obscured (almost lacking) dark lateral throat-stripe. (Left: A. Ouwerkerk; right: L. Kislev)

C. r. adamsi, Gujarat, India, May: this typically long-billed race reaches east to W India, but irrespective of any geographical variation some can show slightly stronger face pattern and breast-streaking, especially when worn. However, compared to Lesser Short-toed Lark visibly smaller and more compact, with shorter tail and longer bill and rather greyish-tinged upperparts with reduced streaking. (P. Ganpule)

C. r. adamsi, NW India, Jan: a rather pale and less streaky individual (often the case with birds in fresher plumage), in typical posture running fast across open ground. Very small, slender bill, and pale supercilium is usually most pronounced above and behind eye. (N. Srinivasa Murthy)

bill colour and shape, primary projection, and head and wing patterns offer readily-appreciated separating features.

AGEING & SEXING Ages alike after post-juv moult. Sexes alike. — Moults. Post-nuptial and post-juv moults complete. Not known if there is a partial pre-nuptial body moult. — SPRING **Ad** Duller and less sandy-grey at this season; wear and bleaching noticeable from May. **1stS** Indistinguishable from ad. — SUMMER–AUTUMN **Ad** Broadly fringed above and is slightly more sandy-grey and sandy-buff, and dark streaks less contrasting. **1stW** Inseparable from ad. **Juv** Spotted/scaled pale and bordered dark above and on wings. Breast has irregular dark brown streaking and blotches on buffish ground colour.

BIOMETRICS (*adamsi*) **L** 12–13.5 cm; **W** ♂ 80–89 mm (n 20, m 84.9), ♀ 78–84 mm (n 14, m 81.2); **T** ♂ 40–51 mm (n 20, m 46.8), ♀ 42–48 mm (n 14, m 43.7); **T/W** m 54.6; **B** 12.0–14.5 mm (n 34, m 13.1); **B**(f) 10.0–11.4 mm (n 7, m 10.8); **Ts** 18.3–19.3 mm (n 7, m 18.7); **HC** 4–7 mm (n 6, m 6.2). **Wing formula: p1** minute; **pp2–4** about equal and longest; **p5** < wt 2–5 mm; **p6** < wt 7–12 mm; **p7** < wt 12–18 mm; **tert. tip** < wt 11.5–16.5 mm, =6/7 or =7 (rarely =6 or 7/8). Emarg. pp3–5 (exceptionally faintly on p6). (NHM, MNHN; partly from Vaurie 1951.)

GEOGRAPHICAL VARIATION & RANGE Three races generally accepted, two of them extralimital: *krishnakumarsinhji* is restricted to its type locality in WC India, whereas *raytal* occurs across much of rest of India east to Burma, and the longer- and finer-billed *adamsi* occupies the rest of the range (including SE Iran). All populations mainly sedentary.

C. r. adamsi (Hume, 1871) (SE Iran, SE Afghanistan, S Pakistan, NW India; resident). Described above.

TAXONOMIC NOTES New evidence has become available recently (Alström et al. 2013) indicating that *Calandrella* is polyphyletic. These authors recommended moving *rufescens*, *cheleensis* and *raytal* to a separate genus, *Alaudala*. We agree, but like these authors prefer to await more evidence before their recommendation is followed.

REFERENCES Alström, P. et al. (2013) *Mol. Phyl. & Evol.*, 69: 1043–1056. – Vaurie, C. (1951) *Bull.AMNH*, 97: 431–526. – Wallace, D. I. M. (1973) *BB*, 66: 382.

CRESTED LARK
Galerida cristata (L., 1758)

Fr. – Cochevis huppé; Ger. – Haubenlerche
Sp. – Cogujada común; Swe. – Tofslärka

Crested Lark is a widespread and common species in dry, warm, open habitats and sandy soils in C and S Europe, and also in large parts of N Africa and Middle East. This is the lark you often flush in vineyards, in sandy agricultural fields, shunting yards, dock areas with silos, near villages in semi-deserts, etc., either running away on the ground looking warily over its shoulder or taking off with fluttering, broad wings and desolate, piping calls. Mainly a resident species wherever it occurs.

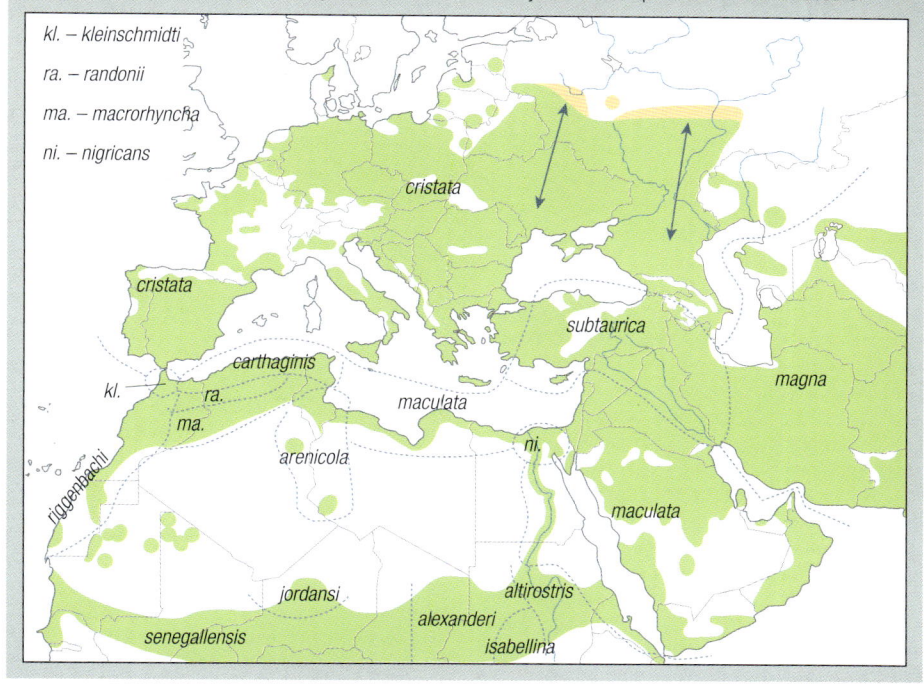

G. c. arenicola, or intergrade with *carthaginis*, C Tunisia, May: note strong, long bill with convex culmen and almost concave lower mandible outline, rather greyish-brown general colour with finely but distinctly streaked whitish breast. Photographed in Al Hammah between Gabès and Douz. (R. Pop)

IDENTIFICATION A grey-brown lark of medium-large size which first of all needs to be separated from Skylark. To do so, note *absence of paler trailing edge to the wing*, a spiky erect *crest*, short *tail without white sides* (but can be pale brown with rusty tinge), broad wings, and a rather *long, pointed and downcurved bill*. Breast variably streaked dark (in desert-dwelling subspecies less prominently). Head pattern rather typical of genus *Galerida* with *dark moustachial stripe and lateral throat-stripe joining at their lower ends*, dark lores, dark eye-stripe behind eye, and pale eye-surround enhanced by darkish border below. In good light often possible to see warm, russet tinge to underwing (though this can be faint or nearly absent), and sometimes faintly on uppertail-coverts. Crest can be folded back, but never so far as to become invisible. For separation from Thekla's Lark, see below under Similar species.

VOCALISATIONS Song of two types, either a short variant from ground or low song post, a simple combination of various piping notes with desolate ring, or much more elaborate and pleasing, given in prolonged song-flight at great height (fluttering wingbeats, tail partially spread), an 'endless' and entertaining mixture of desolate piping notes (same type of notes as in short song, and calls), clever mimicry, and 'standard lark sounds' of trills and twitters. The pace of the long song is moderate, the tone musical. Best recognised by the interwoven call-like piping notes. The short song variant is inseparable from, or very similar to, the normal multisyllabic call, only is repeated for long periods from song post. – Calls include multisyllabic call, a short series of piping, drawn-out, desolate notes (as in the short song), often 3–5 in combination, *tree-lee-puuh* or *wuu-tee-te-wuu*. In low flight often gives a strident, whimpering *brshew*, and more cheerful and melodic *dwuuee*, frequently repeated. The latter two calls are related and can grade into each other.

SIMILAR SPECIES Has a longer and more pointed crest than *Skylark* (which can erect a blunt crest when agitated), and crest is invariably visible. The tail of Crested is shorter and lacks white sides, while the wings are slightly broader and lack the pale trailing edge of Skylark. – Most similar to *Thekla's Lark*, and separation from it requires care and knowledge, some birds presenting a real identification challenge. To the trained ear, vocalisation is a useful clue, Thekla's sounding more soft and melodious, Crested more piping with desolate ring. There is some overlap in calls, but generally these differ subtly. Further, Thekla's has a slightly shorter and more bluntly pointed bill, with lower edge of lower mandible slightly convex in outline so that the tip does not appear as downcurved as in Crested. However, a few Thekla's have a straight outline to the lower mandible, so care is needed and all features need to be considered (bill of Crested longer, more pointed, with lower mandible straight or even subtly concave in outline, giving a downcurved, 'mean' expression). On average, Thekla's has more distinct, dense and extensive breast-streaking, the streaks being shorter and broader, not so thin and pointed. Thekla's also has darker and greyer plumage (though see Geographical variation), a slightly greyer, less rufous-tinged underwing (though this can be hard to judge), and slightly more rufous-tinged uppertail with better contrast against greyish lower back/rump (often useful in Europe, less so in Africa where many Crested have quite rufous uppertail). Behaviour including song-flight is similar, although Thekla's tends to perch in trees more frequently.

AGEING & SEXING Ages alike after post-juv moult. Sexes alike as to plumage, but size helpful. – Moults. Complete moult of both ad and juv in summer and early autumn (mainly

G. c. cristata, Greece, Apr: versus Skylark, note longer, spiky, erect crest, usually longer bill and different breast-streaking and face pattern. Rather dark dull brown above with slight olive-grey tinge, and extensively streaked breast are features of this race. (R. Pop)

G. c. cristata, Spain (left: Mar; right: Apr): with Thekla's Lark shares distinctive pattern of prominent dark lateral throat and moustachial stripes, which are absent or much less distinct in Skylark. Breast-streaking prominent but often rather blurred. Lower edge of usually long bill straight or very nearly so (slightly convex in shorter-billed Thekla's). Depending mainly on light, impression varies from commonly rather greyish to sometimes warmer brown. (Left: A. Domínguez; right: C. N. G. Bocos)

Jun–Sep, sometimes into Oct). – **SPRING Ad** In late breeding season heavily-bleached birds become slightly darker above and less clearly patterned, with reduced pale upperparts fringes. Underparts streaking a little bolder. **1stS** Inseparable from ad. – **SUMMER–AUTUMN Ad** Upperparts streaked dark, no prominent pale tips or edges. Breast streaked dark grey. **1stW** After moult inseparable from ad. **Juv** Upperparts and wing-coverts have prominent and rather distinct pale buffish-white tips and edges. Breast blotched diffusely dark with a hint of paler tips.

BIOMETRICS (*cristata*) **L** 17–18 cm; **W** ♂ 105–114 mm (n 47, m 108.8), ♀ 97–105 mm (n 35, m 101.5); **T** ♂ 60–68 mm (n 47, m 64.7), ♀ 57–64 mm (n 35, m 60.3); **T/W** m 59.4; **B** ♂ 17.6–21.3 mm (n 47, m 19.3), ♀ 16.7–20.7 mm (n 35, m 18.6); **BD** 5.3–6.5 mm (n 78, m 5.9); **Ts** 22.5–26.6 mm (n 83, m 24.9); **HC** 8–15 mm (once 17.5) (n 69, m 11.1). **Wing formula: p1** short, < pc 2–7 mm, < p2 51.5–67 mm; **p2** < wt 0–4 mm; **pp3–4**(5) longest; **p5** < wt 0.5–1.5 mm (rarely = wt); **p6** < wt 4.5–8 mm; **p7** < wt 15–21 mm; **p10** < wt 23.5–30 mm; **s1** < wt 24–31 mm. Emarg. pp3–6 (sometimes a little less distinct on p6).

GEOGRAPHICAL VARIATION & RANGE Complex with no fewer than c. 45 subspecies named within treated range, 24 of them listed in *HBW* and (apparently) 22 in *BWP*. We apply stricter rules for a minimum of distinctness and acknowledge only 11 plus several more extralimital. Variation in Europe is slight and clinal, whereas populations in Africa and Central Asia differ more markedly, in general being paler, warmer-coloured and longer-billed. Local, largely insignificant, variations occur within described subspecies, at times leading to proposals of separate races. Also, a fair amount of individual variation both in general darkness and prominence of streaking makes separation of subtle subspecies less convincing. Several published subspecies in our opinion questionable and represent clinal intergradations into neighbouring ones (or even in some cases colour morphs or mere local 'soil aberrations'!), rather than separable populations that merit formal naming. In summary, here only the more distinct ones are kept, based on several collections (mainly NHM, AMNH, MNHN, NRM, ZMUC, NBC, ZFMK). Geographical variation in N Africa and partly in W Asia appears to be still incompletely studied and mapped. Taxa below broadly arranged after 'coastal and lowland' or 'desert and mountain' distribution, but numerous cases of intermediates exist between them. – See also comments on possible division into two species under Taxonomic notes. – All populations mainly sedentary, with restricted movements only by those breeding in the extreme north and north-east.

'COASTAL AND LOWLAND' FORMS
G. c. cristata (L., 1758) (Europe including W Russia, W Turkey, NW Kazakhstan). Treated above. Rather dark, greyish-brown plumage with slight ochre or cinnamon cast throughout including belly if not bleached or stained (upperparts can appear to have slight grey tinge), breast well streaked, but streaks usually (though not invariably) rather ill-defined, in particular at lower end. Bill usually rather stout and darkish in outer part. – Birds in Spain ('*pallida*') tend to be a shade paler and subtly colder grey-brown above, and to have more off-white, less brown-tinged belly, but differences very slight and not constant. Birds in S & SE Europe, including Italy ('*neumanni*'), and Dalmatia and Bulgaria ('*meridionalis*') are sometimes subtly or markedly darker (suggesting dark morph), but majority identical to *cristata* or at least insufficiently different to warrant separation when series are compared. Birds on Crimea ('*moltschanowi*'), in lower Volga region ('*tenuirostris*') and in the Caucasus ('*caucasica*') appear identical to *cristata* when long series compared. Grades into *subtaurica* in W Turkey somewhere in E Aegean region or W Anatolia, and also in Transcaucasia. (Syn. *apuliae*; *balcanica*; *caucasica*; *gallica*; *ioniae*; *madaraszi*; *magdae*; *meridionalis*; *moltschanowi*; *mühlei*; *neumanni*; *pallida*; *tenuirostris*.)

○ **G. c. kleinschmidti** Erlanger, 1899 (coastal extreme NW Morocco in Tangier region). Somewhat darker than geographically closest population of *cristata*, which is in S Iberia ('*pallida*'). Typically has quite blackish bold streaking on crown and crest, and bold streaking on back and breast. Subtly darker than *carthaginis* in NE Morocco and further east, and differs clearly from *riggenbachi* to south in being darker with less rufous-tinged underparts, although they clearly intergrade at c. 33°N. Somewhat smaller than *cristata*: **L** 15.5–17 cm; **W** ♂ 105–110 mm (n 13, m 106.6), ♀ 96–105 mm (n 11, m 99.6); **T** ♂ 60–68 mm (n 13, m 61.9), ♀ 54–62 mm (n 11, m 57.5); **T/W** m 57.9; **B** 18.3–21.1 mm (n 24, m 19.9); **Ts** 23.3–27.5 mm (n 23, m 25.6). – It should be noted that almost as dark and heavily-streaked subpopulations or individual birds (then a morph?) occur also within *cristata*, notably in Italy ('*neumanni*') and in N Balkans and west of Black Sea ('*meridionalis*').

G. c. riggenbachi Hartert, 1902 (Western Sahara, W Morocco except in extreme north). Similarly dark as *kleinschmidti* (or only a trifle paler on average) but larger and clearly richer cinnamon-brown or even tinged rufous, especially below. Streaking above variable, sometimes slightly stronger than in *cristata*. Usually markedly long-billed, but bill length also quite variable for one subspecies, not only related to sex but probably due to secondary intergradation of shorter-billed coastal populations with neighbouring large-billed montane forms. **W** ♂ 106–115 mm (n 14, m 109.2), ♀ 96–109 mm (n 14, m 102.8); **T/W** m 60.5; **B** 18.5–26.0 mm (n 54, m 21.6).

○ **G. c. carthaginis** Kleinschmidt & Hilgert, 1905 (coastal NE Morocco, N Algeria & Tunisia south to c. 35°N). Differs from *kleinschmidti* in being slightly larger and a little paler with narrower and often more brown streaks on breast, and from *riggenbachi* by less rufous-tinged underparts and subtly finer streaking. Paler and sandier brown above with more subdued streaking than *cristata* and has more distinct breast-streaking and marginally whiter belly. Somewhat longer bill than *cristata*. – Grades into much paler *arenicola*

G. c. macrorhyncha, SE Morocco, Mar: in NW Africa, birds of montane interior and deserts paler, less streaked and longer-billed than coastal forms (and separate species status suggested). Note on this bird from Zagora, Atlas Mts in SE Morocco, heavy, long bill, which is almost decurved; plumage above pale sandy-brown (with cinnamon-pink tinge), and with very subdued rufous-tinged streaking. (D. Barnes)

G. c. maculata, Israel, Sep: typically pale form found in many areas from Levant eastwards. Usually light sandy-brown with olive-grey tinge above (never cinnamon). Breast-streaking often blotchy. (H. Shirihai)

G. c. maculata, Kuwait, Feb: slightly warm cinnamon-brown tinge to plumage, especially on breast and flanks; breast-streaking quite bold and blackish. Note spiky crest and straight lower mandible profile. (A. Halley)

G. c. maculata, Oman, Feb: easily separated from Skylark in flight by absence of clear paler trailing edge to wing, and by shorter tail without white sides. (M. Varesvuo)

of S Tunisia roughly between 35° and 34°N, and intermediate population has been named 'gafsae'. Differs from *maculata* along N African coast from Libya eastward by slightly larger size, less distinct streaking above, and in particular the lower back can be almost uniform greyish-cinnamon. **W** ♂ 103–113 mm (*n* 15, *m* 109.0), ♀ 100–107 mm (*n* 11, *m* 103.5); **T/W** *m* 58.2; **B** 18.5–23.6 mm (*n* 26, *m* 20.9). **Wing formula:** p1 < pc 1.5–7 mm (*m* 3.5). (Syn. *gafsae* [part]; *macrorhyncha*)

G. c. maculata C. L. Brehm, 1858 (N Libya except extreme west; N Egypt except Nile Delta; Nile Valley in the north from south of the delta to Aswan, Red Sea coast south to Eritrea, Arabia except extreme north, also Sinai, Israel, Jordan, S Syria). Slightly smaller and darker and more prominently streaked above than *cristata* and *carthaginis*, and breast-streaking is a little bolder and very dark. Often tinged rufous-cinnamon on breast and flanks, and also on feather-edges of wings and slightly on back. Rather similar to *subtaurica* with which it intergrades in extreme S Turkey, W Iraq and N Syria, but generally differs in being a little darker and more rufous-tinged, less greyish. **L** 16–17.5 cm; **W** ♂ 101–110 mm (*n* 18, *m* 104.5), ♀ 96–101 mm (*n* 10, *m* 98.0); **T/W** *m* 58.6; **B** ♂ 18.5–21.5 mm (*n* 18, *m* 19.7), ♀ 17.5–20.4 mm (*n* 10, *m* 18.6). – Several subspecies have been described from this large area, but we are unable to uphold races such as *brachyura*, *festae* or *cinnamomina* when series compared and types examined. Birds described from Wadi Natrun, N Egypt, as '*caroli*' cannot be separated from series of *maculata*, but peculiarly seem to bleach heavily in early spring, becoming completely different in appearance; merit further study. (Syn. *brachyura*; *caroli*; *cinnamomina*; *eritreae*; *festae*; *halfae*; *imami*; *moeritica*; *tardinata*; *thomsi*; *zion*.) – Note that *altirostris* is an extralimital cinnamon-buff subspecies (see below) with almost unstreaked upperparts, originally described from Dongola, N Sudan. This name can thus not be applied to any West Palearctic race of Crested Lark.

G. c. nigricans C. L. Brehm, 1855 (Nile Delta). The darkest of all subspecies, slightly darker even than *kleinschmidti*, dark olive-grey above with prominent dark streaks; crown often very dark in worn plumage; heavily streaked also on breast, and partly on upper flanks. Slightly smaller and longer-tailed than *cristata*. Grades into *maculata* near Cairo. **L** 17–18 cm; **W** ♂ 101.5–108 mm (*n* 14, *m* 105.1), ♀ 94.5–107 mm (*n* 12, *m* 100.9); **T/W** *m* 60.5; **B** 18.2–21.0 mm (*n* 26, *m* 19.4); **BD** 5.4–6.5 mm (*n* 25, *m* 6.0). (Syn. *deltae*.)

'DESERT AND MOUNTAIN' FORMS

G. c. randonii Loche, 1860 (Haut Plateaux of N Algeria east to about Djelfa, west to interior NE Morocco). Differs from *carthaginis*, with which it apparently intergrades, by slightly larger size, longer wings and much longer and stronger-based bill, though some variation. Birds in NE Morocco appear particularly large. Undertail-coverts sometimes streaked brown. Compared to *macrorhyncha*, larger and subtly darker, with darker and slightly more obvious streaks on breast and upperparts. **L** 18–20 cm; **W** ♂ 113–119 mm

G. c. maculata, N Oman, Nov: some appear larger and bulkier, with very strong bills and elongated crests, and plumage can vary somewhat in darkness. (Left: A. Blomdahl; right: D. Occhiato)

(n 12, m 116.1), ♀ 107–111 mm (n 10, m 108.2); **T** ♂ 62–72.5 mm (n 12 m 68.2), ♀ 61–66 mm (n 10, m 63.4); **T/W** m 58.6; **B** 22.5–26.7 mm (n 22, m 24.4); **BD** 6.4–8.3 mm (n 21, m 7.0); **Ts** 26.2–28.7 mm (n 21, m 27.4). – See Taxonomic notes.

G. c. macrorhyncha Tristram, 1859 (interior of northernmost Algeria, S of Atlas, apparently also E & SE Morocco and extreme SW Tunisia). A fairly pale sandy-brown race with pale rufous tinge, especially to wing. Long-billed. It resembles *randonii*, but is slightly smaller with less heavy (thinner) bill, and is paler and more rufous-tinged. Undertail-coverts usually unmarked. **L** 17–18 cm; **W** ♂ 105–115 mm (n 20, m 111.4), ♀ 99–110 mm (n 20, m 104.0); **T/W** m 59.5; **B** 19.5–25.5 mm (n 49, m 22.1); **BD** 5.7–7.0 mm (n 9; m 6.3); **Ts** 24.5–26.9 mm (n 9, m 25.4). – Distinction from *arenicola* in S Tunisia less well understood. See Taxonomic notes.

G. c. arenicola Tristram, 1859 (deserts of NE & extreme E Algeria, S Tunisia, W Libya). Resembles *macrorhyncha*, thus pale, but is somewhat more grey-tinged ('colder') brown above with less dark streaking on mantle/back. Underparts whitish, breast moderately but rather distinctly spotted on cream-white ground, flanks largely unmarked. Rather long bill. Grades into subtly darker *carthaginis* in the north, between c. 35° and 34°N (intermediates have been named 'gafsae', but not recognised here). There is also apparent intergradation into subtly larger and stronger-billed *macrorhyncha* in SW Tunisia and adjacent Algeria, and this area is populated by intermediate birds between three races. – Some birds in S & SE Tunisia more saturated rufous above and below, difficult to fit into subspecific pattern here presented. **L** 16.5–18.5 cm; **W** ♂ 107–116 mm (n 17, m 110.7), ♀ 101–105 mm (n 12, m 103.3); **B** 18.5–23.8 mm (n 30, m 21.3). **Wing formula: p1** < pc 1.5–7 mm, rarely 1 mm > pc (n 19, m 3.9). (Syn. *gafsae* [part]; *helenae* of Ajjer in EC Algeria.)

○ **G. c. subtaurica** Kollibay, 1912 (Cyprus, C & E Turkey, SW Transcaucasia, N Syria, Iraq, Kuwait, extreme W & SW Iran). Resembles *cristata* but is a trifle paler with more distinct dark streaking on breast and upperparts, not so diffusely patterned. Grades into *cristata* in W Turkey and W & S Transcaucasia, and into the slightly smaller and more cinnamon-tinged *maculata* in N Levant and Iraq. **W** ♂ 106–112 mm (n 11, m 108.5), ♀ 99–114 mm (n 15, m 103.4); **T** ♂ 60–68 mm (n 11, m 64.5), ♀ 58–67.5 mm (n 15, m 61.6). **B** and **Ts** of same size as *cristata*. (Syn. *ankarae*; *cypriaca*; *weigoldi*.) – Admittedly there are some minor variations within this race. However, we have compared as long series as possible and failed to corroborate the claimed differences between other proposed races when applying a 75% rule. A very complex pattern of subtle variation has been described for Turkey with four or even five races involved (BWP, Roselaar 1995), but we find it difficult to confirm the presence of more than two races from the material available to us.

G. c. magna Hume, 1871 (Iran and much of Central Asia). Somewhat paler and 'sandier' brown above than *cristata*, and only weakly streaked, especially so on lower back and rump. Underparts are paler brownish-white, especially on belly, than *cristata*, and streaks on breast are distinct but rather small. Bill and wings approximately of same length as in *cristata*, thus slightly longer than in *maculata*. Grades into *subtaurica* in E Transcaucasia and E Iraq. **W** ♂ 106–116 mm (n 10, m 109.8), ♀ 99–108 mm (n 10, m 103.3); **B** ♂ 20.2–22.5 mm (n 10, m 21.1), ♀ 19.2–21.2 mm (n 10, m 20.2). (Syn. *alaschanica*; *iwanowi*; *retrusa*; *submagna*; *vamberyi*.)

Extralimital races: **G. c. jordansi** Niethammer, 1955 (Aïr Mts, N Niger but apparently not in extreme S Algeria). Pale, warm ochre or rufous-brown (almost reddish) and only very lightly streaked or spotted.

G. c. alexanderi Neumann, 1908 (E Niger, Chad, W Sudan but apparently not reaching extreme S Libya). Pale and less reddish, more cinnamon-tinged.

G. c. altirostris C. L. Brehm, 1855 (Nile Valley in N Sudan, but probably not occurring as far north as extreme S Egypt). Pale sandy-brown or cinnamon-coloured and much paler streaked above (at times almost entirely unstreaked on lower back) and has much finer streaks on breast than *maculata* to north of it.

TAXONOMIC NOTES Recently it has been demonstrated that the largest N African forms living in more arid habitats away from the coastal zone, which have the largest bills (*randonii* and *macrorhyncha*), have somewhat different DNA (by c. 2%) and probably represent an incipient or cryptic second Crested Lark species in the area, 'ATLAS LARK' *G. macrorhyncha* (Guillaumet et al. 2005, 2010). The interrelation between this *macrorhyncha* clade and Crested Lark populations east of it (*arenicola*; mainly in E Algeria and S Tunisia) requires resolution before the situation can be evaluated more definitively. Studies in the contact zone between long-billed inland forms and shorter-billed forms living in more vegetated areas in N Morocco and Algeria (*carthaginis*) are also desirable. – In a recent comprehensive multilocus phylogeny of Alaudidae (Alström et al. 2013), three samples of ssp. *randonii* of the proposed 'Atlas Lark' showed this to be sister to *G. malabarica* (India, extralimital and not treated), placed together in a different subclade than the remaining races sampled of Crested Lark. If this most surprising result correctly represents relationships it is another strong reason to split Crested Lark. However, considering the high degree of improbability for such a relationship it is desirable that further research include more Crested Lark taxa and additional specimens.

REFERENCES ALSTRÖM, P. et al. (2013) *Mol. Phyl. & Evol.*, 69: 1043–1056. – GUILLAUMET, A., CROCHET, P.-A. & GODELLE, B. (2005) *Molecular Ecology*, 14: 3809–3821. – GUILLAUMET, A. et al. (2010) *Ecography*, 33: 961–970. – VAURIE, C. (1954) *Amer. Mus. Novit.*, 1672: 6–7.

G. c. maculata, juv, Oman, Feb: prior to first complete moult has prominent pale buffish-white tips and edges above and on wing-feathers. Also less typical shape to still-growing bill. (C. van Rijswijk)

G. c. nigricans, Nile Delta, N Egypt, Mar: the darkest race. (V. Legrand)

G. c. magna, N Iran, Jan: a pale and large race. Upperparts pale sandy-grey, and dark streaking moderately prominent, underparts white and cleaner with dark breast-streaking narrow and rather restricted in extent. Note also relatively long, decurved and slender bill. (E. Winkel)

THEKLA'S LARK
Galerida theklae A. E. Brehm, 1857

Fr. – Cochevis de Thékla; Ger. – Theklalerche
Sp. – Cogujada montesina; Swe. – Lagerlärka

Very similar to Crested Lark, but at least in Europe favours higher elevations, more arid, rockier, pristine habitats, and areas with tall garrigue, trees and orchards. It is a fraction smaller and more compact on average than Crested Lark, but this is usually not as helpful as the somewhat shorter bill, its slightly mellower and more melodious calls, and its habitat preferences. A resident or short-distance migrant. Often referred to as 'Thekla Lark', but Brehm named it in honour after his deceased daughter Thekla, so it really should be Thekla's Lark.

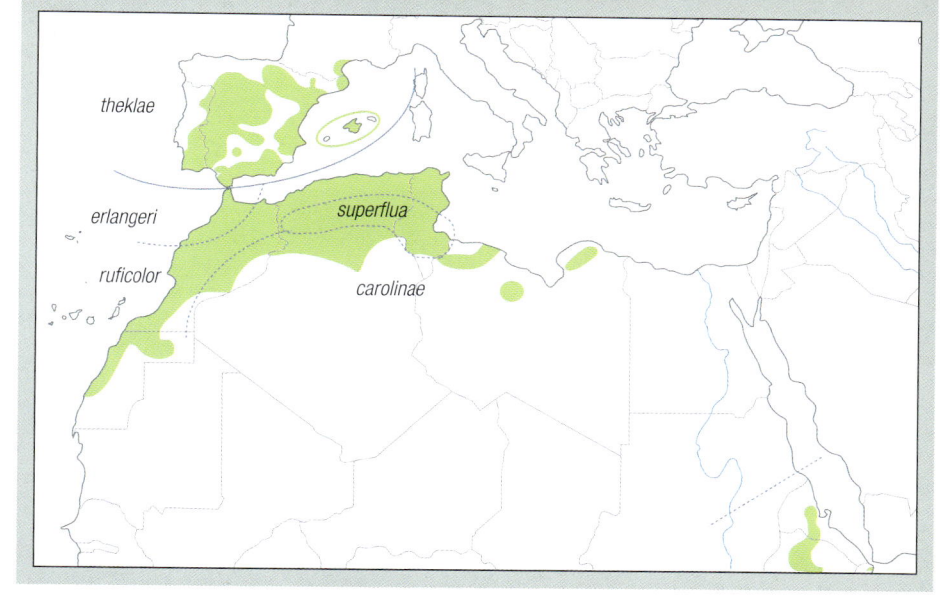

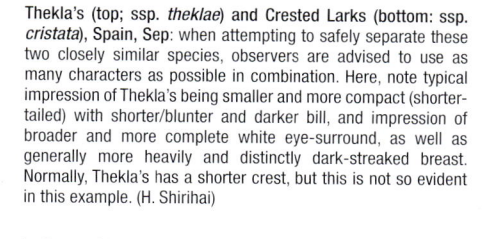

Thekla's (top; ssp. *theklae*) and Crested Larks (bottom: ssp. *cristata*), Spain, Sep: when attempting to safely separate these two closely similar species, observers are advised to use as many characters as possible in combination. Here, note typical impression of Thekla's being smaller and more compact (shorter-tailed) with shorter/blunter and darker bill, and impression of broader and more complete white eye-surround, as well as generally more heavily and distinctly dark-streaked breast. Normally, Thekla's has a shorter crest, but this is not so evident in this example. (H. Shirihai)

IDENTIFICATION Closely related to Crested Lark, and often quite difficult to separate from it. Slightly smaller and more compact, with proportionately *slightly shorter tail*, often giving same short-tailed impression as Woodlark in flight. Also useful is slightly *shorter and more bluntly-pointed bill*, with the *lower mandible more convex in outline near the tip* (Crested almost straight, or even concave, giving its bill a more acutely pointed, downcurved and 'angry' look; a few have same or at least very similar bill shape though). On average, Thekla's has a *slightly darker bill* (Crested habitually appears pale-billed, Thekla's not), but this requires much practice and will not separate every bird. The crest is on average a little shorter and more 'complete' when fanned (Crested longer and more 'spiky'), and the outer tail-feathers are on average more rusty-tinged (though latter frequently difficult to assess, or overlaps with Crested). *Breast* generally more heavily and *distinctly streaked dark* than on Crested (though, as in Crested, in desert-dwelling African subspecies streaking is finer and less prominent). Underwing rather plain grey-white, generally lacking any clear rufous tones. Conversely, uppertail-coverts on average more rufous-tinged than in Crested, in Europe and coastal areas of NW Africa, with better contrast to rather greyish lower back/rump. (In rest of Africa less obvious or no difference between Thekla's and Crested in this respect due to paler and browner lower back/rump in both.)

G. t. theklae, Spain, May: compared to Crested Lark, note rather small and blunt bill with straight or slightly convex lower outline, and on average shorter and denser crest. The Iberian race *theklae* is fairly dark with mainly greyish-brown tones, and has a prominently streaked breast. (C. N. G. Bocos)

G. t. ruficolor, W Morocco, Jan: similar to European *theklae*, but averages subtly paler and more rufous-tinged. Outer tail-feathers on average more rusty-tinged than in Crested Lark. (C. N. G. Bocos)

Thekla's (left: *superflua*, S Tunisia, May; centre, probably *erlangeri*, NW Morocco, Apr) **and Crested Larks** (right: presumed *arenicola*, Tunisia, Feb): in many areas in NW Africa, Crested usually readily eliminated on long and almost decurved, 'mean-looking' bill and long, spiky crest, very different from left-hand Thekla's, which has shorter bill with convex outline to lower mandible, quite bold breast-streaking, and rather short crest. However, more difficult Thekla's (centre) occur, especially in NW Africa, with somewhat longer and straighter bill, and paler, less boldly streaked plumage, meaning identification must rely also on habitat and voice. The Crested Lark belongs to a rather dark and long-billed race, *arenicola*, slightly larger than *cristata*. (Left: R. Armada; centre: R. Pop; right: A. Fossé)

Morocco, Apr: race unknown. Slightly more compact than very similar Crested Lark, with shorter and blunter bill (lower mandible more convex). Breast often more heavily and densely streaked than Crested (though, as with other features, geographical variation can obscure differences). (A. B. van den Berg)

G. t. erlangeri, or intergrade between this and *ruficolor*, NW Morocco, Feb: compared to Crested Lark, bill shorter with slightly more convex lower mandible, and plumage overall darker. This bird has a quite confusingly long-crested appearance. (D. Occhiato)

VOCALISATIONS Song and calls similar to Crested, but consistently more mellow and musical in tone, less piping with desolate ring. Short song type from ground or song post (often in trees or bush-tops, unlike in Crested) a pleasing combination of soft, 'yodelling' notes, on average more melodious than corresponding song of Crested, but at times quite similar. Long song, generally in song-flight at some height (fluttering wingbeats, tail partly spread), very similar to that of Crested but usually differs in the interwoven more melodious calls. Sometimes long song differs in being shorter than in Crested and rather scratchy and metallic, at times recalling song of Stonechat. – Calls very similar to those of Crested, like the commonly heard short series of yodelling, melodious notes, often 3–6 in combination, *du-delli-**dew**-dil-dee*, but can often be separated with practice by its softer and more melodious tone. Other calls of Crested have their counterparts in Thekla's, too, only the tone is, again, a little softer and more pleasing. Often heard is the up-slurred *druuee*, miaowing but a little discordant in tone.

SIMILAR SPECIES Differences from very similar *Crested Lark* are discussed above under Identification and Vocalisations. Behaviour, including song-flight, similar, although Crested only exceptionally perches in trees.

AGEING & SEXING Ages (after post-juv moult) alike.

Sexes alike in plumage but often separable on size. – **Moults.** Complete moult of both ad and juv in summer (mainly Jun–Sep). – **SPRING Ad** In late breeding season heavily-bleached birds slightly darker above and less clearly patterned, with reduced pale upperparts fringes. Underparts streaking a little bolder. **1stS** Inseparable from ad. – **SUMMER–AUTUMN Ad** Upperparts streaked dark, no prominent pale tips or fringes. Breast streaked dark grey. **1stW** After moult inseparable from ad. **Juv** Upperparts and wing-coverts have prominent pale buffish-white tips and edges. Breast blotched diffusely dark with a hint of paler tips.

BIOMETRICS (*theklae*) **L** 15.5–16.5 cm; **W** ♂ 99–109 mm (*n* 34, *m* 103.7; once 97, but possibly mis-sexed), ♀ 94–105.5 mm (*n* 33, *m* 99.1); **T** ♂ 54–63.5 mm (*n* 33, *m* 58.4), ♀ 51–62 mm (*n* 33, *m* 56.6); **T/W** *m* 57.5; **B** 15.2–18.0 mm (*n* 28, *m* 16.7); **BD** 5.0–6.5 mm (*n* 28, *m* 5.8); **Ts** 23.0–26.1 mm (*n* 27, *m* 24.5). **Wing formula: p1** > pc 3 mm, to 2 mm < (rarely > 4.5 or < 3 mm); **p2** < wt 0.5–4.5 mm; **pp3–5** about equal and longest (though p5 sometimes to 2 mm <); **p6** < wt 0.5–6 mm; **p7** < wt 10–18 mm; p8 < wt 14.5–23 mm; **p10** < wt 18–28 mm; **s1** < wt 19–27 mm. Emarg. pp3–6.

GEOGRAPHICAL VARIATION & RANGE Moderate variation, mainly affecting plumage colours, and mostly clinal. Tail tends to be proportionately subtly shorter in the north. Plumage differences in Europe and NW Africa generally small and characters variable, probably partly caused by soil staining. Many birds cannot be identified to subspecies. For some general comments on lark variation, see under *G. cristata*. Three extralimital subspecies exist, not treated here. – All populations mainly sedentary.

G. t. theklae A. E. Brehm, 1857 (Iberia, Balearics, S France). Treated above. Rather dark, greyish-brown plumage, breast prominently streaked, streaks well marked. (Syn. *polatzeki*.)

○ *G. t. erlangeri* Hartert, 1904 (NW Morocco south to about Casablanca). Slightly darker on average than *theklae*, with a little heavier streaking. Grades into *ruficolor* in south and east. **W** ♂ 99–107 mm (*n* 17, *m* 103.4), ♀ 94–103 mm (*n* 13, *m* 97.9); **B** 15.4–17.4 mm (*n* 30, *m* 16.5).

○ *G. t. ruficolor* Whitaker, 1898 (C & SW Morocco south of preceding, Western Sahara, coastal area of Algeria and Tunisia). Rather broadly streaked above like preceding two, but is on average slightly paler overall and more rufous- or pinkish-brown on upperwing and back, and breast streaking is often slightly finer. A trifle larger. Grades into *erlangeri* in the north. Perhaps on average slightly paler and more rufous-tinged in SW Morocco and Western Sahara ('*theresae*'), but subtle, clinal and hardly grounds for separation.

G. t. ruficolor, N Tunisia, May: compared to Crested Lark, note shorter crest (although differences not always as clear-cut as here) and short bill. (R. Pop)

G. t. ruficolor, SW Morocco, Jan: somewhat variable but plumage usually tinged rufous or cinnamon-pink, and broadly streaked. (M. Schäf)

W ♂ 98–111 mm (*n* 25, *m* 105.2), ♀ 94–105 mm (*n* 15, *m* 98.7); **B** ♂ 15–18 mm (*n* 25, *m* 16.8), ♀ 14.3–17.3 mm (*n* 15, *m* 15.9). **Wing formula: p1** > pc 0–3 mm, or < pc 0–1.5 mm (*n* 23). (Syn. *aguirrei*; *harterti*; *theresae*.)

G. t. superflua Hartert, 1897 (Algerian high plateaux, S Tunisia). Paler and more pinkish-buff or cinnamon-buff than *ruficolor*, and streaks are finer. Wings often noticeably rufous-tinged. Cheeks often fairly pale. A certain degree of individual variation in plumage colour in this race, some being paler and more rufous-tinged than others. Fairly large and long-tailed but proportionately short-billed. **W** ♂ 102–109 mm (*n* 18, *m* 105.7), ♀ 96–101.5 mm (*n* 10, *m* 99.4); **T/W** *m* 60.6; **B** ♂ 15.3–18.7 mm (*n* 17, *m* 17.2), ♀ 14.6–17.4 mm (*n* 11, *m* 15.9). (Syn. *hilgerti*.)

G. t. carolinae Erlanger, 1897 (SE Morocco, N Algerian Sahara east to N Libya). Subtly smaller than *superflua*, and plumage is even paler and more rufous and cinnamon-buff. The palest birds (in E Algeria and W Tunisia) are sometimes called '*deichleri*', but judging from collecting localities this appears to be a pale morph rather than a separate subspecies. **W** ♂ 99–107 mm (*n* 14, *m* 104.4), ♀ 96–101 mm (*n* 11, *m* 98.4); **T/W** *m* 60.0; **B** 15.6–18.6 mm (*n* 26, *m* 17.1). (Syn. *cyrenaicae*; *deichleri*.)

Probably *G. t. carolinae*, Morocco, Mar: very long and here unusually erect crest. Note how pale this bird is compared to that in the previous image, indicating adaption to desert habitat. (D. Monticelli)

WOODLARK
Lullula arborea (L., 1758)

Fr. – Alouette lulu; Ger. – Heidelerche
Sp. – Alondra totovía; Swe. – Trädlärka

A less well-known lark to the layman, partly due to its arboreal habits. The only representative of its genus. Often sings in flight by night, high above its breeding forest, when unsurprisingly it escapes general attention. In daytime, it is rather unobtrusive. Has a lovely song for those who notice it. Summer visitor, returning mainly in March–April, wintering in S Europe or N Africa. Favoured habitats include open pine forest on rocky or sandy soils, clear-felled areas, deciduous woods with glades and meadows, and heaths with scattered copses.

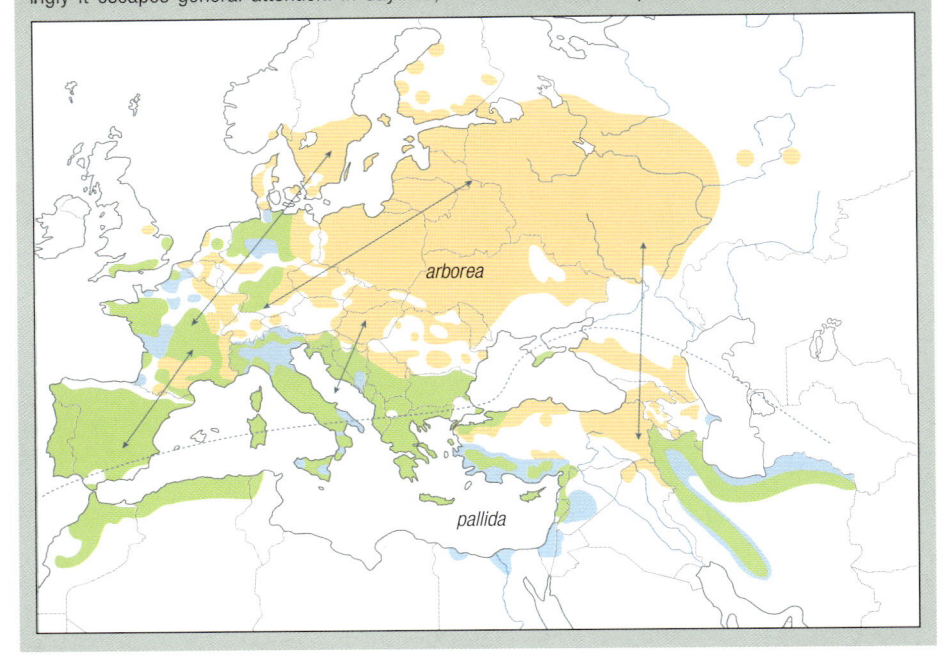

L. a. arborea, Sweden, Oct: immediately recognised in flight by broad, rounded wings and short tail. Due to often slow and deliberate flight some key patterns often appreciated easily also in flight, namely the 'pale-dark-pale' pattern at wing-bend and the striking white tips to tail. (M. Varesvuo)

IDENTIFICATION A medium-sized lark with rounded wings and short tail. In flight, *deep undulations and short tail* are striking. The shortness of the tail is optically enhanced by its *white tip*, which 'disappears' against a bright sky. When perched, on the ground, on a telephone wire or in a treetop, this brown-and-white bird with dark-streaked buffish breast has a characteristic *'pale-dark-pale' pattern at wing-bend* (on primary-coverts and alula), *long and obvious pale supercilia*, which meet at nape in an angle, and rufous-*brown cheeks outlined by a whitish bar* across neck-sides. Crown quite boldly streaked dark, the streaks often forming broad, dark 'stripes'. When agitated (song, anxiety), can raise a hint of a short crest in tit-fashion. Compared to Skylark, note rather *uniform and reddish-brown cheeks* in contrast to more grey-brown rest of head.

L. a. arborea, Netherlands, Mar: typical posture when perched. Plain rufous-tinged cheeks, prominent pale supercilium and diagnostic 'pale-dark-pale' pattern at wing-bend. (P. Cools)

VOCALISATIONS Song is uniquely melodious and lovely, delivered both under the morning sun and in pitch-dark night, at its best in a prolonged, 'wandering' song-flight at some height, and therefore carrying far. Various themes are presented in accelerating pace, falling slightly in pitch, and increasing in loudness, separated by brief halts, e.g. *lee,, lee, lee, lee lee lee-lee-lee-leeleelala… eh-la, eh-la eh-la-ela-eluelululu… vivivivivi… luh luh lu lu-lu-lu-lulululu…*, etc. Most themes are slow and simple, but some are fast trills or more elaborate. The song sounds poetic and contemplating with its absence of the usual frenzy and 'horror vacui' of most other larks. – Call is a soft yodel or musical whistle, multisyllabic, e.g. *tlueet-tlueet-tlueet* or *düdluu-ee*.

SIMILAR SPECIES Due to its appearance in wooded areas, and avoidance of vast fields, rarely liable to be confused with other larks in summer. The short tail and 'pale-dark-pale' pattern on the wing-bend should eliminate *Tree Pipit*, which shares its habitats, or the odd *Skylark* appearing on a neighbouring forest allotment. On migration might mix with other larks of roughly the same size, like *Lesser Short-toed Lark*, which has similarly streaked breast, but note Woodlark's fine bill, striking head pattern, wing-bend pattern, and white-tipped tail.

AGEING & SEXING Ages (after post-juv moult) alike. Sexes alike as to plumage. Little seasonal change. – Moults. Complete moult of both ad and juv in summer (mainly Jun–Aug). – SPRING **Ad** In late breeding season heavily-bleached birds overall darker above and less clearly patterned, with reduced pale upperparts fringes. Underparts streaking a little bolder. **1stS** Inseparable from ad. – SUMMER–AUTUMN **Ad** Upperparts streaked dark, no prominent pale tips or edges. Breast streaked dark grey. **Juv** Upperparts and wing-coverts have prominent pale buffish-white tips and edges. Breast blotched diffusely dark with a hint of paler tips.

BIOMETRICS (*arborea*) **L** 14.5–15 cm; **W** ♂ 95–101 mm (n 28, m 98.0), ♀ 91–96 mm (n 13, m 93.7); **T** ♂ 50–55.5 mm (n 28, m 52.9), ♀ 48–52.5 mm (n 13, m 50.7); **T/W** m 54.0; **B** 13.2–14.8 mm (once 15.3) (n 27, m 14.1); **BD** 4.1–5.0 mm (n 25, m 4.6); **Ts** 19.5–22.6 mm (n 27, m 21.1); **HC** 9.5–15.0 (once 17.0) mm (n 27, m 14.1). **Wing formula: p1** < pc 1–7 mm, < p2 50–58 mm; **p2** < wt 1–4 mm (rarely = wt); **pp3–4**(5) about equal and longest; **p5** < wt 0.5–2 mm (rarely = wt); **p6** < wt 5–10 mm; **p7** < wt 13.5–19 mm; **p8** < wt 17–23 mm; **p10** < wt 23–28 mm; **s1** < wt 25–30 mm. Emarg. pp3–6 (sometimes a little less distinct on p6).

GEOGRAPHICAL VARIATION & RANGE Very slight, and

lone birds often impossible to attribute to subspecies due to individual variation.

L. a. arborea (L., 1758) (Europe, except in south & extreme south-east; breeders in Fenno-Scandia and Russia winter mainly in S & W Europe, rarely to N Africa, more southerly breeders sedentary or make short-range movements). Treated above. Birds of Corsica and several other Mediterranean islands often referred to *pallida* but majority inseparable from *arborea*. (Syn. *familiaris*; *wagneri*.)

○ **L. a. pallida** Zarudny, 1902 (S Spain, S Italy including Sicily, extreme SE Europe, N Africa, Asia Minor, Levant, Crimea, Caucasus; birds of Sardinia intermediate; short-range movements or sedentary). Very slightly and only on average paler grey above, while breast is less tinged ochre, more cream-buff or off-white, making breast-streaking somewhat more obvious on some. A very poorly differentiated race. **W** ♂ 95–102 mm (n 16, m 97.9), ♀ 91–97 mm (n 15, m 93.5); **T** ♂ 49–56 mm (n 16, m 52.9), ♀ 46–55 mm (n 15, m 50.0); **T/W** m 53.8; **B** 13.2–15.4 mm (n 32, m 14.1). (Syn. *flavescens*; *harterti*; *wettsteini*.)

L. a. arborea, Netherlands, Apr: when flank feathers or greater coverts cover characteristic pattern of wing-bend, becomes more similar to Skylark, but still has prominent pale supercilium and short tail with white tips on outer few feathers. (P. Cools)

L. a. pallida, Spain, Apr: long whitish supercilia, which meet at nape, and plain rufous-toned cheeks bordered by pale neck-side band. Also plump body, small head and short tail. This southern race is subtly larger and paler than *arborea*. (A. Torés Sánchez)

L. a. pallida, Spain, Sep: in advanced post-nuptial moult. The long whitish supercilia typically meet at the nape, and the diagnostic bold white patterns of the primary-coverts and alula are even better exposed than usual. (H. Shirihai)

L. a. pallida, Israel, Dec: typical impression in winter when fresh. Note paler-tinged, more cream-buff breast, making breast-streaking somewhat more obvious, which is one of the reasons for the recognition of this very poorly differentiated race. Note also how the crown-feathers are often raised a bit to form a short crest. (A. Ben Dov)

L. a. pallida, juv, Turkey, Jun: like in other larks, this short-lived plumage is distinctive due to prominent buffish-white tips and fringes above. Unlike superficially similar juv Skylark note the diagnostic 'pale-dark-pale' pattern at wing-bend and the long whitish supercilia that meet at the nape, as well as the usually clearly much broader black subterminal centres to greater coverts. Also typical of juv Woodlark are broader white tips than buff edges to the primaries. (H. Shirihai)

ORIENTAL SKYLARK
Alauda gulgula Franklin, 1831

Alternative name: Small Skylark

Fr. – Alouette gulgule; Ger. – Orientfeldlerche
Sp. – Alondra oriental; Swe. – Mindre sånglärka

A compact version of the familiar and widespread Skylark, this mainly E Palearctic lark has only recently proven to be regular in the covered region. A resident or partial migrant in SW Siberia south to E Iran and east to the Philippines, small numbers are now known to reach E Arabia and Israel, even Egypt, in winter. Often noticed when flushed from a field by its characteristic buzzing call. Rather wary and difficult to approach closely in open terrain, while keeping well hidden when cover present.

inconspicua

Israel, Mar: in worn spring plumage when overall paler and greyer, being less tinged buff or sandy as when fresh (see next image below). Note very short primary projection, proportionately slightly longer bill, lack of white outer tail-feathers (best evaluated on take-off or landing, latter when typically hovers close above ground and fans tail), and finer breast-streaking, all of which distinguish it from Skylark. (H. Shirihai)

IDENTIFICATION Similar to Skylark but *overall more squat with a notably shorter and narrower tail*, sandier or buffish plumage, *slightly longer bill*, and shorter, broader-based and *rounder wings* (when folded, *lacking Skylark's obvious primary projection*). Plumage generally like Skylark but has rusty ear-coverts and wing panel (formed by rufous-tinged outer webs to many flight-feathers), more *narrowly streaked* (rather than blotchy) *breast with a buffier ground colour* and a *narrower sandy* (instead of whitish) *trailing edge to wings*. Also *sandy* (not white) *outer tail-feathers* and overall buff or sandy underparts distinguish it from Skylark, while flight behaviour and calls usually are conclusive. Quicker wingbeats than Skylark, song-flight often swaying around in the sky, rarely staying in one spot for a while like Skylark.

VOCALISATIONS Song basically like Skylark but more feeble and includes buzzing notes, and is often distinguished by more mechanical repetition of rather hard and unmusical notes for long periods, frequently sticking to rather monotonous repetition of dry, harsh series like *pri-pri-pri-rivi-rivi-rivi-rivi-tri-tri-tri-pre-pre-pre-*...; unlikely to be heard singing within the covered region. – Common and distinctive flight calls are a harsh, buzzing *pzeebz*, a *baz-baz*, or *bzrü* and *baz-terrr*, some of these being somewhat reminiscent of Richard's Pipit, others recalling Sand Martin or Reed Bunting. Also a soft *pyup*. Other calls include a somewhat Barn Sallow-like *plip* and several Skylark-like notes, *prit-üt*, *drreeü*, etc.

SIMILAR SPECIES The principal confusion species is Skylark, and in late autumn and winter (the season when all records in the treated region, away from Iran, have occurred), Oriental Skylark appears noticeably smaller, more compact and more neatly streaked on breast than more robust Skylark. The latter is longer-winged (with 2–3 widely-spaced primary tips always visible; projection virtually non-existent in Oriental, but beware heavily abraded tertials), longer-tailed and boldly streaked on breast (forming more clearly-defined pectoral band), with a clearly broader white trailing edge (in Oriental Skylark very narrow and sandy, and usually difficult to see). Skylark's whiter underparts and outer tail-feathers are obvious in fresh plumage, and supercilium is generally narrower (sandier and broader over lores in Oriental). In flight, Oriental has short, rounded wings which produce a slower progress and more flapping action, and together with the shorter tail create a more compact silhouette. Unlike Skylark, often briefly hovers just before landing (as if investigating the spot), a habit also employed by Richard's Pipit in tall-grass areas. Also frequently more reluctant to fly from observer, keeping tight to ground or under bush, and when flushed rises more horizontally with slower flaps and buzzing calls. – Inexperienced observers could confuse Oriental Skylark with Short-toed Lark and especially Lesser Short-toed Lark, but attention to the more prominent breast-streaking, noticeably longer and narrower bill, slightly larger size without the long primary projection of Lesser Short-toed or the blackish breast-side patches of many Short-toed, their different facial and scapular patterns, and altogether different flight action and calls, will avoid such pitfalls. – Woodlark, though approaching Oriental Skylark in size and structure, is readily eliminated by its pale-dark-pale wing-bend pattern, bolder white tail-corners, and moderate primary projection (intermediate between Oriental and Skylarks). Also note better-developed crest, whiter supercilium conspicuous above and behind eye, even reaching rear crown, and rather diffuse fringes to tertials and greater coverts (well defined in Oriental). They also differ in facial pattern, as well as in scapular and breast patterns, and Woodlark has a characteristic jerky, strongly undulating flight and melodic 'yodelling' *toolooeet* call.

AGEING & SEXING Ages following post-juv moult alike. Sexes similar. – Moults. Complete post-nuptial moult and post-juv moult in late summer. No pre-nuptial moult. – **SPRING Ad** Especially by late spring can be heavily worn and bleached, with much duller ground colour, primary projection slightly longer due to tertial wear, and (usually small

Israel, Dec: some show an extremely long crest when excited; typical rusty-tinged ear-coverts and wings, and narrowly (rather than boldly) streaked breast on buffier ground colour than in Skylark, which colour faintly continues onto belly. (A. Ben Dov)

and often hidden) crest often appears longer at this season. Both trailing edge of wing and outer tail-feathers often whiter but former never as prominent as in Skylark. **1stS** As ad. — SUMMER–AUTUMN **Ad** Fresh plumage is tinged warmer/sandier. Broadly pale-fringed above. **1stW** As ad. **Juv** Soft, fluffy body-feathering remarkably spotted and speckled above, with more drop-shaped streaks on breast than in later plumages.

BIOMETRICS (*inconspicua*) **L** 14–16 cm; **W** ♂ 94–102 mm (n 16, m 98.3), ♀ 85–96 mm (n 12, m 91.5); **T** 50–58 mm (n 22, m 52.9); **T/W** m 55.5; **B** 15.3–17.3 mm (n 28, m 16.2). **Wing formula: p1** > pc 9.5–12 mm; **p2** < wt 0.5–1 mm; **pp3–4** (rarely 2–5) about equal and longest; **p5** < wt 0.5–5 mm (m 1.3) (in Skylark: 5–9.5 mm, m 7.0). (Largely after *BWP*; for live birds in Eilat, Israel, see Shirihai 1986a, 1986b.)

GEOGRAPHICAL VARIATION & RANGE Moderate to slight variation, but several subspecies separated. Of these only one, *inconspicua*, thought to occur within covered region. Most races are distinctly darker and often smaller than *inconspicua*, which breeds in Turkmenistan, Iran and S Kazakhstan, and apparently largely winters in India but also west to Arabia. However, those regularly recorded in Israel appear to have more heavily marked upperparts than typical *inconspicua*, with rusty tone to ear-coverts and outer fringes to most flight-feathers more pronounced (somewhat approaching *lhamarum* of Pamirs, Kashmir & W Himalayas), thus exact race of stragglers to Israel and other parts of Middle East remains unresolved.

N India, Jan: these two images show the diagnostic lack of white trailing edge to wing (unlike Skylark). In low flight, Oriental Skylark may keep the legs uncovered, ready to hover before landing. (H. Taavetti)

A. g. inconspicua Severtsov, 1873 (E Iran, W Turkmenistan and much of Central Asia, south-east to NW India; winters S Levant, Egypt, Arabia east to W India). Described above. A comparatively pale form.

REFERENCES SHIRIHAI, H. (1986a) *BB*, 79: 186–197. – SHIRIHAI, H. (1986b) *Sandgrouse*, 7: 47–54.

Oriental Skylark (top left, Israel, Nov; right, Oman, Jan) and Skylark (bottom left, Netherlands, Oct; right, Israel, Nov) in fresh plumage: compared to Skylark, Oriental is more squat and short-tailed, although bill is relatively stronger, and plumage more uniformly tinged sandy or buffish. Also lacks Skylark's obvious primary projection, the primaries only just reaching beyond tertials in some. Breast-streaking in Oriental generally narrower, less dense and has less clear-cut lower border. However, some Skylarks (bottom right) are more narrowly streaked and can appear closer to Oriental, whereas not all Oriental are evenly buff below (top left). Diagnostic flight calls are always helpful. (Top left: L. Boon/Cursorius; top right: H. & J. Eriksen; bottom left: N. D. van Swelm; bottom right: L. Kislev)

(COMMON) SKYLARK
Alauda arvensis L., 1758

Alternative name: Eurasian Skylark

Fr. – Alouette des champs; Ger. – Feldlerche
Sp. – Alondra común; Swe. – Sånglärka

The most widespread and numerous lark in the region, representing 'The lark' to most people, hence 'Common' is preferred as a modifier over 'Eurasian'. Sings its 'endless' song, which forms an integral part of the European summer, from high in the sky throughout spring and summer. A migrant in the north, resident in the south and west, in winter it often forms large flocks on stubble fields, pastures and seashores. Found everywhere in Europe where there is open land, absent only from Iceland. The European population has been estimated at *c.* 17 million pairs, but is probably steadily declining due to habitat changes and modern agricultural methods.

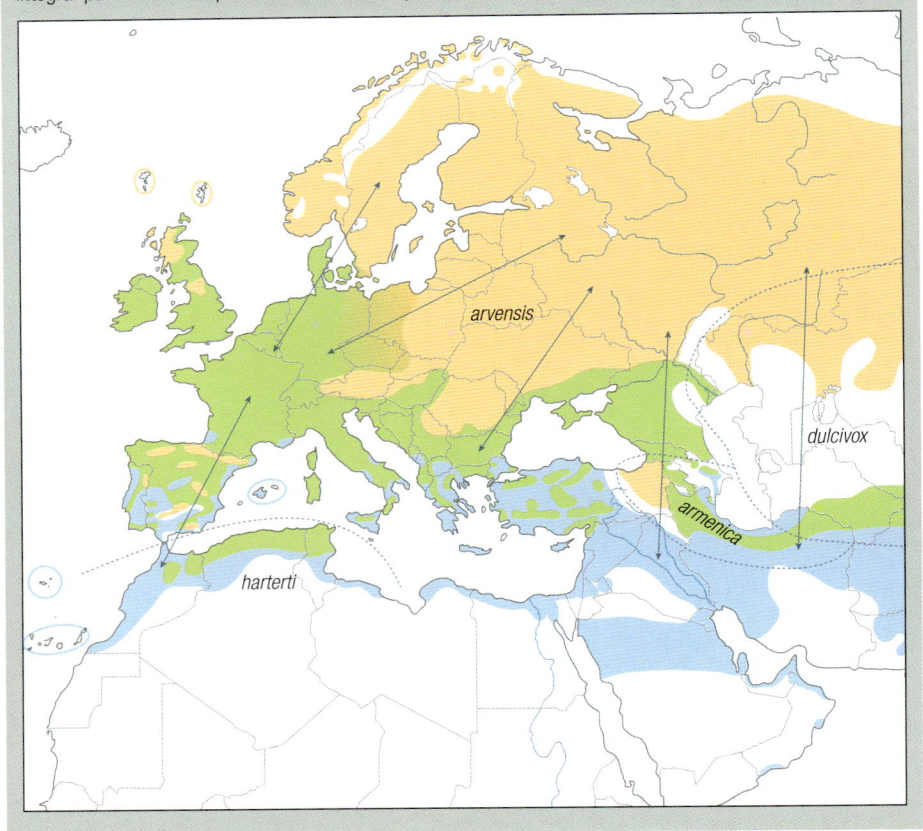

IDENTIFICATION Medium-sized brown-and-white lark with *streaked breast* on cream-buff ground. Has a *small crest*, which can be half-raised (in song or courtship) or, more commonly, be folded to disappear entirely. In flight, note *pale trailing edge to the wing* (not clean white like in Calandra Lark, but still contrasting) and *white sides* on often-spread tail. When perched, on the ground or a fence-post, the tail appears reasonably long, but it is 'cloaked' in flight by the broad innerwing. Also, notice the unstreaked white belly (though flanks are cream-tinged and have a few thin streaks). Neck-sides slightly paler than ear-coverts and breast-sides, often creating a hint of paler semi-collar. Often recognised by its song-flight, starting to sing already early in the ascent, climbing with fluttering wings and half-spread tail to considerable height (50–100 m), remaining in pretty much *the same spot for long periods*, all the time with *fluttering wingbeats* (never slow-motion flight like Calandra Lark), finally coming down in stages, still singing, and then dropping silently the last 15 m like a stone to the ground. When flushed, like most larks takes off with rather 'explosive' wing-beats and a few subdued calls, flying off rather low with fluttering wingbeats mixed with brief glides on lowered wings.

VOCALISATIONS Song is well known to many, but difficult to describe other than as an uninterrupted stream of twittering, rolling and chirruping notes, delivered at a fast pace, now and then relieved by mimicry (often of Wood and Green Sandpipers) and some piping sounds. Can sing without the briefest pause for 15 minutes or more! – Call rather variable, generally fairly dry but 'full' rolling sounds, *prreet, prrlüh, prreeh-e*, etc. In anxiety a more finely piping note, *peeh*.

SIMILAR SPECIES Of other medium-sized larks with a streaked breast, distinguished from *Woodlark* by longer tail with white sides, pale trailing edge to the wing, more grey-brown (less rufous) and slightly streaked cheeks, and different vocalisations. – When raising its crest might be mistaken for a *Crested Lark*, but note shorter, more rounded crest and the same characters as separate it from Woodlark. – With only brief views can resemble a *Lesser Short-toed Lark*, but is larger, has somewhat longer tail and slimmer shape, and bill is thinner and longer (though note that bill shape varies within vast range of Lesser Short-toed, some having a slightly longer bill). – The real challenge is to separate Skylark from *Oriental Skylark*. Latter is smaller with somewhat shorter tail and faint cream-buff tinge to lower breast and belly (Skylark has white belly contrasting with buff breast). Oriental best told by short primary projection and 'buzzing' flight call; supporting characters are sandy-buff (not white) tail-sides, no or only very slightly paler trailing edge to wing, and on average more rusty tinge to edges of folded primaries. – Might even resemble a *Richard's Pipit* to the inexperienced, but that species has much longer legs, longer bill and tail, moves differently, etc.

AGEING & SEXING Ages alike after post-juv moult. Sexes alike in plumage. – Moults. Complete moult of both ad and juv in summer (mainly Jun–Aug). No pre-nuptial moult. – SPRING **Ad** In late breeding season heavily-bleached birds slightly darker above and less clearly patterned, with reduced pale upperparts fringes; underparts streaking a little bolder. **1stS** Inseparable from ad. – SUMMER–AUTUMN **Ad** Upperparts streaked dark, no prominent pale tips or edges. Breast streaked dark grey. **1stW** After moult inseparable from ad. **Juv** Upperparts and wing-coverts have prominent pale buffish-white tips and edges. Breast blotched diffusely dark with a hint of paler tips.

BIOMETRICS (*arvensis*) **L** 16.5–18 cm; **W** ♂ 106–119 mm (*n* 60, *m* 113.9), ♀ 99–109 mm (*n* 50, *m* 103.5); **T** ♂ 65–77 mm (*n* 60, *m* 70.5), ♀ 59–69 mm (*n* 50, *m* 64.3); **T/W** *m* 62.0; **B** 11.9–16.0 mm (*n* 58, *m* 14.1); **Ts** 21.4–25.7 mm (*n* 59, *m* 23.5); **HC** 11–19 mm (*m* 14.3). **Wing formula:** $p1$ minute; $pp2$–4 about equal and longest; $p5$ < wt 5–9 mm; $p6$ < wt ♂ 17–21 mm, ♀ 14–16.5 mm; $p7$ < wt ♂ 24–29 mm, ♀ 20–24 mm; $p10$ <

A. a. arvensis, Italy, Apr: in song-flight (note hint of pale trailing edge to wing and obvious white tail-sides), starts to sing early during ascent, climbing on fluttering wings and half-spread tail to considerable height. (D. Occhiato)

wt ♂ 36–40 mm, ♀ 30–36 mm; **s1** < wt ♂ 38–44 mm, ♀ 32–37 mm. Emarg. pp3–5.

GEOGRAPHICAL VARIATION & RANGE Slight and clinal. Several subspecies described, some of which appear quite subtle (and some even questionable). Individual variation, plumage condition, and even soil-staining of the plumage, make it difficult to assign many birds to subspecies. Only comparatively distinct ones, in our view meaningful to uphold, are listed below.

A. a. arvensis L., 1758 (most of Europe, Russia, SW Siberia, W & C Turkey, possibly W Caucasus; southern populations resident, northern short-range migrants to W & C Europe). Treated above. Birds in S & SE Europe, Turkey and Caucasus ('*cantarella*') reportedly on average slightly paler and more ochre-brown above, with fractionally bolder breast-streaking, on average slightly shorter bill, and secondaries tinged on average more rufous, but much overlap in all these respects. Reports of subtly differing populations within Great Britain could not be confirmed. (Syn. *cantarella*; *divergens*; *guillelmi*; *scotica*; *sierrae*; *tertialis*; *theresae*.)

○ **A. a. harterti** Whitaker, 1904 (NW Africa; resident).

A. a. arvensis, Italy, May: a worn breeding bird with unstreaked white belly contrasting with ochre-tinged and densely streaked breast and flanks. Small rounded crest, which can be erected when agitated. (D. Occhiato)

Individual variation in *A. a. arvensis* (left: Sweden, Jul; right, Poland, Jun): though a rather nondescript brown-and-buff bird, there is a certain degree of variation in plumage and posture. Note, apart from pointed but fairly strong bill, white tail-sides and rather indistinct head pattern, how amount of breast-streaking varies with posture, and how drooping wings (right) can create a longer-tailed impression. (Left: J. Tufvesson; right: A. Kant)

Very similar to *arvensis*, or perhaps slightly paler above, but often with finer-streaked upperparts and breast. Birds in west tend to have stronger rufous-buff tinge on breast. **W** ♂ 107–117 mm (n 10, m 113.5), ♀ 101–108 mm (n 13, m 108.8); **B** 12.0–16.2 mm (n 27, m 13.9).

A. a. dulcivox Hume, 1872 (steppes of Volga, S Urals, N & W Kazakhstan, C Siberia, Transcaspia, NE Iran; winters SE Europe, Middle East, SW Asia). Also fairly similar to *arvensis*, but slightly larger and somewhat paler, with finer streaking above. **W** ♂ 110–120 mm (n 20, m 116.1), ♀ 102–110 mm (n 14, m 106.4); **B** ♂ 13.5–17.3 mm (n 20, m 15.1), ♀ 13.3–15.6 mm (n 12, m 14.3); **HC** 9–16 mm; **p5** < wt 4.5–9 mm. (Syn. *cinerascens*; *cinerea*; *schach*. – Extralimital race *kiborti* of E Siberia and Transbaikalia is very similar, and perhaps a synonym.)

○ **A. a. armenica** Bogdanov, 1879 (E Turkey, S Caspian, much of Iran; mainly resident). Somewhat larger and very slightly paler than *arvensis*, being on average paler grey above, and breast is less tinged ochre, more cream-buff. Generally colder brown (greyer) than *dulcivox*, but some are close and difficult to separate. **W** ♂ 113–122 mm (n 10,

A. a. arvensis, juv, Greece, Jun: boldly pale-scaled upperparts typical of young larks. Juvenile Woodlark eliminated by largely plain brown alula and wing edge. (M. Schäf)

A. a. arvensis, Spain, Feb: rather indistinct supercilium that fades shortly behind eye, typical bill shape and rather uniform boldly streaked upperparts. (E. Winkel)

m 117.3), ♀ 102–109 mm (*n* 10, *m* 106.0); **B** ♂ 13.8–16.3 mm (*n* 10, *m* 15.4), ♀ 13.0–15.0 mm (*n* 10, *m* 14.0). (Sometimes labelled '*intermedia*' in collections, but this ext-alimital race occurs in Russian Far East. Others are inadvertently labelled '*cantarella*'. Syn. *subtilis*.)

REFERENCES IVANOV, A. (1928) *Annuaire du Musée Zoologique de l'Acad. des Sciences de l'URSS*, 6: 279–287.

Israel, Feb: in flight note typical shape and flock formation, as well as whitish trailing edge to wing and white tail-sides. (H. Shirihai)

Presumed *A. a. armenica*, E Turkey, Jun: this race is subtly larger and very slightly paler than *arvensis*, while breast has on average thinner streaking. (H. & J. Eriksen)

A. a. dulcivox / kiborti, Mongolia, May: subtly larger and slightly paler, with on average generally thinner breast-streaking. (H. Shirihai)

Israel. Nov: rather pale-looking Skylarks on migration and winter in the Levant. Birds outside their breeding ranges are generally impossible to assign to race. Variation includes birds with quite thin breast-streaking (right) that might invite confusion with Oriental Skylark. (Left: L. Kislev; right: A. Ben Dov)

WHITE-WINGED LARK
Alauda leucoptera Pallas, 1811

Fr. – Alouette leucoptère; Ger. – Weissflügellerche
Sp. – Calandria aliblanca; Swe. – Vitvingad lärka

This is one of the specialised steppe-dwelling larks which requires 'endless' open grassy fields for comfort. Compared to Black Lark, to which it until recently was believed to be closely related, it is more often found on slightly elevated or undulating areas, not on the flattest plains. Nowhere abundant, some searching or luck will be needed to locate it, especially as favoured breeding sites may alter between years. But, when you find the first bird, there may soon be many more, since it often forms small colonies. A short-distant migrant wintering in SW Central Asia, Transcaspia and SE Russia.

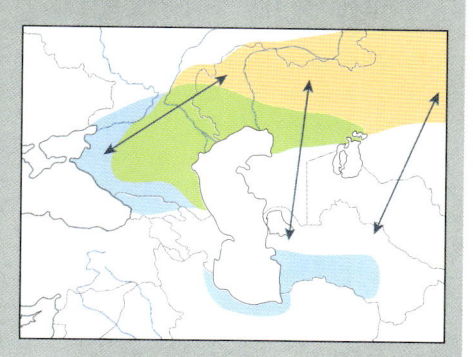

IDENTIFICATION A fairly *large* lark with striking wing pattern and some rufous in plumage. When flushed, immediately apparent is tricoloured wing with *broad white trailing edge*, blackish centre and brown or rufous forepart. In song-flight, in which wings are held rather outstretched and tail folded, the *wings look long and narrow* since the broad white rear edges 'disappear' against the sky, rendering it an almost wader-like or pratincole-like appearance. Normal flight, when chasing each other low over ground with *backswept, pointed wings*, like a mixture of thrush and martin, on *quick agile wingbeats* at high speed (Calandra Lark: more rounded wings and fluttery wingbeats). Both sexes have a *rufous wing-bend* and often some *rufous tinge on uppertail*, but only the ♂ has an *unstreaked rufous crown*. No black breast-side patches. Bill is strong, but not huge like in some other lark species.

VOCALISATIONS Song is basically similar to Skylark, a prolonged twittering and chirruping with lots of variation. It can often be recognised by the inclusion of brief pauses and faltering, accelerating notes in Short-toed Lark fashion. The voice is somewhat harder and drier than that of Skylark, too. The inclusion of mimicry, and of piping, miaowing notes, is

♂ **Kazakhstan, May**: head extensively rufous, lores and eye-ring pale, pale area extending as prominent white supercilium behind eye. Upperparts densely streaked dark, contrasting with rufous wing-bend (lesser coverts) and rump. Unstreaked uniformly rufous cap and rear ear-coverts, and limited streaking below are features of 'classic' ♂♂. (M. Koshkin)

♂ **winter–summer, Finland, Apr**: best character of ♂ is uniform rufous crown, ♀ being streaked dark and much less rufous. Both sexes have rufous wing-bend. White inner primaries just visible. (S. Tuomela)

♀ **Kazakhstan, Jun**: note lack of obvious rufous on cheeks, crown and rump/uppertail-coverts, but instead strongly developed dark breast-streaking. Note rufous on upper scapulars and on primary-coverts and wing-bend. The white secondary panel on the folded wing is less contrasting but still there. (E. Yoğurtcuoğlu)

♂, **Kazakhstan, Jul**: some ♂♂ have dark-streaked crown but note on this bird still predominantly rufous colour typical of ♂, and rufous also evident on parts of cheeks and wings. Note also quite white underparts and almost entirely unstreaked breast, further supporting sexing. (A. Forsten)

also frequently heard. Unlike Skylark during ascent of song flight usually delivers song in sections separated by silent pauses. — Call resembles Skylark (somewhat reminiscent of Bimaculated Lark or Short-toed Lark, too), a dry rolling double twitter, *drre-drre*.

SIMILAR SPECIES Needs to be separated from *Calandra Lark*, which can occur in the same fields, and which is also a large lark with white trailing edge to the wing. Note that Calandra has much darker underwing overall (White-winged dark only on outer parts), lacks any rufous in plumage and usually flies with fluttering wingbeats of blunter wings. — In easternmost part of range there is the remote risk of confusion with a vagrant extralimital *Mongolian Lark* (*M. mongolica*; not treated), but this is a slightly larger bird, and has black patches on breast-sides, a buff-white central crown-patch and a neat pale supercilium which runs around nape.

AGEING & SEXING Ages (after post-juv moult) alike in plumage. Insignificant seasonal differences. Sexes often separable on plumage, but some ad ♀ and 1stS ♂ can be rather similar. — Moults. Complete moult of both ad and juv in late summer and early autumn (mainly Aug–Oct, sometimes Nov). — **SPRING Ad** In late breeding season

Presumed ♂, Kazakhstan, Dec: a more ♀-like ♂ with some streaking on rufous crown and on breast, and separation from more rufous ♀ not always easy. Note large white portion visible in wing, and bold flank-streaking, making identification easy. Ageing impossible once post-juv moult concluded. (A. Isabekov)

♂ summer, Kazakhstan, Jun: note tricoloured upperwing, with very broad white trailing edge, blackish centre and brown or rufous forewing. Wings typically narrow and long, an impression enhanced when white rear disappears against bright sky (e.g. from below). (Left T. Lindroos; right: J. Normaja)

Azerbaijan, Feb: in mostly fresh winter plumage with streaked breast. (M. Heiß)

♀ (left) and ♂, Kazakhstan, Jan: typically hardy birds, in winter often has to cope with cold and snowy conditions. The ♀ at left has smaller white portion in wing, bolder dark streaking on breast and flanks, and less rufous, if any, on head. (A. Salemgareyev)

♀, presumed 1stW, Kazakhstan, Aug: a rather dull ♀ in summer, presumably young. When flanks and scapulars feathering sometimes, as here, conceals the diagnostic rufous wing-bend, such birds could easily be overlooked as ♀ Black Lark, Skylark or Lesser Short-toed Lark. (V. Fedorenko)

Kazakhstan, Nov: mixed flock of White-winged and Black Larks (three birds in centre and a fourth near front) showing variation in wing pattern. (R. Chittenden)

heavily-bleached birds slightly darker above and less clearly patterned, with reduced pale upperparts fringes. Underparts streaking a little bolder. **Ad ♂** Whole crown unstreaked pale rufous. Breast often nearly unstreaked, off-white with a little rufous at sides. **Ad ♀** Crown grey-brown streaked dark, sometimes with a little rufous partly concealed. Breast usually dusky with much streaking (and only a little rufous admixed in some), though some possess fewer streaks and approach ♂ pattern. **1stS** Inseparable from ad. – **SUMMER– AUTUMN Ad** Upperparts streaked dark, no prominent pale tips or fringes. **1stW** Inseparable from ad. **Juv** Upperparts and wing-coverts have prominent pale, greyish- or buffish-white tips and fringes.

BIOMETRICS L 16.5–18 cm; **W** ♂ 116–128 mm (n 28, m 122.3), ♀ 108–117 mm (n 17, m 113.1); **T** ♂ 63–70 mm (n 28, m 66.8), ♀ 58–65.5 mm (n 17, m 62.4); **T/W** m 54.8; **B** ♂ 13.9–16.6 mm (n 27, m 15.2), ♀ 13.2–15.3 mm (n 17, m 14.0); **BD** 5.6–6.8 mm (n 31, m 6.3); **Ts** 23.0–25.7 mm (n 13, m 24.5). **Wing formula:** p1 minute; pp2–3 usually equal and longest; p4 < wt 4–7 mm; p5 < wt 14–18 mm; p6 < wt 21–29 mm; p10 < wt 43–55 mm; s1 < wt 45–56 mm. Emarg. pp3–4 (often faintly suggested also on p5).

GEOGRAPHICAL VARIATION & RANGE Monotypic. – S Russia north of Caucasus, Kalmykia and eastward, SW Siberia, steppes in Kazakhstan; short-range movements to SW & S Central Asia in winter.

TAXONOMIC NOTE Until recently traditionally referred to genus *Melanocorypha*, but a recent phylogenetic study (Alström *et al.* 2013) showed it to be basal to Common and Oriental Skylarks, not closely related to Calandra Lark or the other relatives of that, and it is hence best moved to *Alauda*.

Juv, Kazakhstan, Jun: upperparts and wing-coverts heavily scaled, with rufous tinge to rear supercilium, hindneck and tertials. (J.-M. Breider)

REFERENCES Alström, P. *et al.*. (2013) *Mol. Phyl. & Evol.*, 69: 1043–1056. – Lindroos, T. & Tenovuo, O. (2000) *Alula*, 6: 170–177. – Mild, K. (1987) *Soviet Bird Songs*. Stockholm. – Veprintsev, B. N. & Leonovich, V. (1986) *Birds of the Soviet Union: A Sound Guide*. Larks, pipits. Moscow.

RASO LARK
Alauda razae (Alexander, 1898)

Fr. – Alouette de Razo; Ger. – Rasolerche
Sp. – Alondra de Razo; Swe. – Rasolärka

Confined to the tiny island of Raso, in the Cape Verde Is, with a population of just 98 individuals in 2003, which may now have risen to *c.* 150, this species is adapted to the extreme aridity of its desolate volcanic island home, where it is usually found in close proximity to small clumps of vegetation. The population fluctuates in response to climatic conditions, and its nests are almost certainly predated by a near-endemic gecko. Apart from the tiny total number of birds there is also the problem of uneven sexual ratio with twice as many ♂♂ as ♀♀.

Cape Verde Is, Mar: compared to image below left, note small bill, which probably reflects sexual differences (probably ♀), but it could be an immature with a weaker bill. (H. Shirihai)

with almost whitish-grey base to lower mandible; long sturdy legs and long strong toes and claws. Very short primary projection; in flight wing looks short and round, and *lacks obvious pale trailing edge of Skylark*. Flight and gait mostly suggest Skylark; may form flocks of up to 25 in non-breeding season.

VOCALISATIONS Similar to Skylark but song apparently less varied and consequently more repetitive, given from perch on ground but also in display-flight high in the sky (but not to such extreme heights as Skylark). Birds also seen to rise steeply up higher and higher, and then dive back to ground.

SIMILAR SPECIES Somewhat intermediate between *Skylark* and *Lesser Short-toed Lark*. More flapping flight than the former and lacks any trace of white trailing edge to wing, but confusion with other species extremely unlikely, given that Raso Lark is restricted to a single island, and the only other breeding larks in the Cape Verde Is (Black-crowned Sparrow Lark, Bar-tailed and Hoopoe Larks) are all quite different and none has yet been recorded on Raso.

AGEING & SEXING Ages after post-juv moult alike. Sexes alike. – Moults. Post-nuptial and post-juv moults both complete, generally late Apr–Sep (but breeding season erratic and governed by irregular rains). No pre-nuptial moult. – SPRING **Ad** Both sexes prior to moult worn and bleached, upperparts streaking more marked or blotchy, and ground colour of underparts whiter. ♀ has noticeably shorter bill

Cape Verde Is, Feb: endemic to Raso Island, and unique in appearance, especially considering its long, thick and slightly decurved bill. A rather worn and bleached individual. (R. Pop)

IDENTIFICATION Appears rather like a long- and *thick-billed Skylark* (especially the ♂), and *bill also typically slightly decurved*; otherwise a compact, *short-crested* and streaky lark. Predominantly dull brown-grey above, with paler grey to warmer buff or off-white fringes to dark-centred crown, mantle, scapulars, tertials and most upperwing-coverts. There is an *obvious dark median-covert bar*, and fresh birds often show a *scaly mantle*. Streaks shorter and finer on usually paler hindcrown and neck. Face pattern comprises pale *cream-white eye-ring*, lores and indistinct supercilium that continues around buffy, streaked ear-coverts creating *prominent whitish neck-side crescents*; dark loral, eye, moustachial and lateral throat-stripes usually narrow and indistinct, but there is often a dark border to the lower part of the white eye-ring. Mostly white or cream underparts with buffish wash to breast and flanks, and *neat black streaks, broadest on breast-sides*. Tail dull black, contrasting with pale rump and prominent white outer feathers. Bill dark horn,

Cape Verde Is, Mar: when still rather fresh, shows hint of scaly fringes, especially on mantle and scapulars. Note also rather Skylark-like shape. (H. Shirihai)

(apparently due to different feeding habits; ♀♀ dig much less for food). **1stS** As ad. – SUMMER–AUTUMN **Ad** See above. **1stW** As fresh ad. **Juv** Resembles ad but overall rather buffier, with more black-spotted upperparts, better-defined pale fringes to wing-coverts and flight-feathers, and much less defined breast-streaking.

BIOMETRICS **L** 12–13 cm; **W** ♂ 83–89 mm (*n* 13, *m* 85.4), ♀ 76–80 mm (*n* 9, *m* 78.6); **T** ♂ 45–53 mm (*n* 10, *m* 48.1), ♀ 42–45 mm (*n* 10, *m* 43.8); **B** ♂ 17.1–18.7 mm (*n* 10, *m* 18.0), ♀ 13.8–15.6 mm (*n* 10, *m* 14.8). **Wing formula: p1** minute; **p2** < wt 1–2 mm; **pp3–4** longest; **p5** < wt 0.5–2 mm; **p6** < wt 4–7.5 mm; **p10** < wt 18–25 mm; **tert. tip** ~ p4. Emarg. pp3–5.

GEOGRAPHICAL VARIATION & RANGE Monotypic. – Raso, Cape Verde Is; resident.

TAXONOMIC NOTE Recent genetic data indicate a close relationship with Skylark.

REFERENCES ALSTRÖM, P. *et al.* (2013) *Mol. Phyl. & Evol.*, 69: 1043–1056. – ALSTRÖM, P., MILD, K. & ZETTERSTRÖM, B. (1991) *BW*, 4: 422–427. – DONALD, P. F. & BROOKE, M. DE L. (2006) *BB*, 99: 420–430. – HAZEVOET, C. J. (1989) *BBOC*, 109: 82–87.

Cape Verde Is, Feb: diffuse head pattern but neat dark streaks on breast (rather well developed on this individual). (R. Pop)

Left: Cape Verde Is, Oct: note neat dark streak through fore part of white eye-ring, and white outer tail-feathers. Characteristic are pale cutting edges and base to lower mandible, and the strong feet. Primary projection very short. (E. Winkel)

Below: Juv, Cape Verde Is, Mar: young bird in post-juv moult, still showing many soft, fluffy body-feathers and scaly pale fringes with dark subterminal marks on upperparts. (A. B. van den Berg)

HORNED LARK
Eremophila alpestris (L., 1758)

Alternative name: Shore Lark

Fr. – Alouette haussecol; Ger. – Ohrenlerche
Sp. – Alondra cornuda; Swe. – Berglärka

With its 'Viking-like horns' and characteristic head pattern, this most attractive lark is widespread across the Holarctic, in our region being represented by two distinct populations, western/northern *flava* (of the so-called *alpestris* group) and the 'Levant Horned Lark' (or *penicillata* group). The presence of the slightly smaller Temminck's Lark in the Middle East and N Africa can make identification problematic in some areas. Horned Larks are usually attracted to moderately high open mountain-slopes or ridges in alpine or arid habitats.

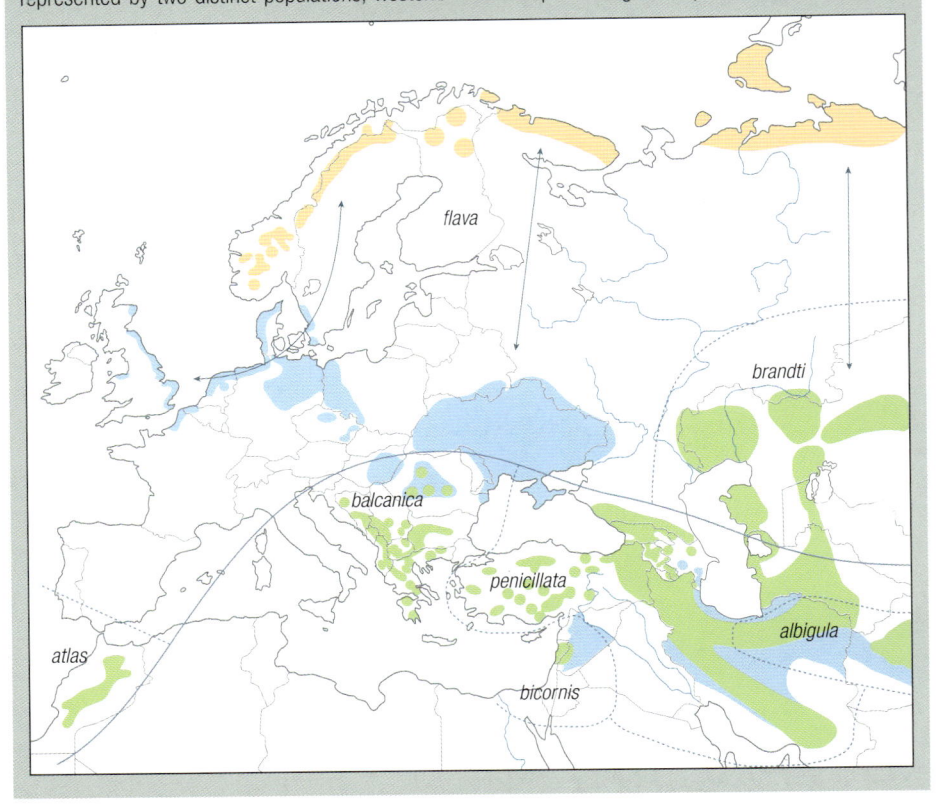

E. a. flava, ♀, Finland, May: worn plumage. Narrower black head patches than ♂, shorter horns (here somewhat worn off), narrower breast patch, and faintly streaked dull pinkish-tinged nape. Mantle more obviously streaked dark. (J. Peltomäki)

VOCALISATIONS Song is a rather shortish squeaky jingle, involving a high-pitched, repeated strophe, stuttering and accelerating slightly after faltering start (in Corn Bunting fashion), perhaps most recalling Short-toed Lark but higher-pitched, e.g. *zih ze zi-zi zreh-zri-zeeo* (sometimes also sounding like distant Black Redstart or Radde's Accentor). – Calls include typically thin, piping and squeaky notes, on even pitch, often a soft mellow *tsee* or thin, high-pitched *eeht* or *zieh*, or 2–3 in rapid succession, *tseee-se-se* or *zee-zee-lee*, somewhat like a drawn-out Meadow Pipit but more piercing; often more melodic, *eeh-dü* or *eeh-deedü*. Also has a harsher *prsh* or *tsrr*. Another call is a short *chew*, a little like Lapland Bunting.

SIMILAR SPECIES Only likely to be confused with Temminck's Lark, but no or very slight overlap in range, and the two are usually clearly segregated by habitat. – Spangled juv must be separated from juv Skylark (easy if seen well).

AGEING & SEXING (*flava*) Ages after post-juv moult largely alike. Sexes separable mainly in spring/summer. – Moults. Post-nuptial and post-juv moults both complete and rather rapidly completed, mostly late Jul–mid Sep. No pre-nuptial moult. – SPRING Ad ♀ resembles ♂ but has duller face, on average less extensive dark forecrown patch with smaller black centres and broader pale tips in fresh plumage, shorter horns, narrower breast patch, faintly streaked hindcrown and nape (plain in ♂), more obviously streaked and

IDENTIFICATION Medium-sized lark with a *short bill, longish narrow tail* and *essentially pink-brown, slightly mottled upperparts* and white underparts tinged dusky pink on breast-sides and flanks. *Black forecrown patch* with *narrow horns above eyes* (not always obvious), narrow *black loral line reaching broad cheek patch* drooping at rear, and which do not meet (except in 'Levant Horned Lark') *bold triangular black gorget on upper breast* with broad, square-shaped central extension onto lower throat, while *face and throat are washed pale yellow, cream or whitish* – together a pattern that is most striking head-on. Largely *black tail has pink-brown central feathers, and narrow white outer edges*, but no obvious wing pattern in flight. Juvenile differs markedly in being dark brown above, from crown to rump, *densely spotted yellow-buff* or whitish. Underparts whitish with dusky brown-grey breast-band and dark ear-coverts, both diffusely spotted off-white. Flight rather buoyant and pipit-like. Wing-beats rapid, and wings close almost completely with each stroke. Escape-flight usually fairly short but occasionally high and long. Song-flight begins with a high fluttering ascent, followed by undulating rising and falling, songs delivered in falling glide with stretched wings and slightly spread long-looking tail, before swiftly descending. Sings less from ground. Walks or hops swiftly on ground when feeding, but when approached close runs mouse-like, only briefly pausing. Perches on flat or raised ground, rocks and buildings but not vegetation.

E. a. flava, ♂, Finland, Jun: solid and extensive black head patches, well-developed horns (here laid flat), lack of streaks on cinnamon-tinged nape, and indistinctly streaked greyer upperparts typical of adult ♂. Note that *flava* is bright yellow on head, has pinkish-tinged nape, and warm brown wash to upperparts and wing-coverts. (J. Peltomäki)

E. a. flava, Netherlands, Oct: sexing in fresh autumn plumage difficult due to individual variation, and often impossible. However, some ♂♂ distinctive, like here, by solid black mask and pectoral spot, as well as by less streaked upperparts. (S. Schilperoort)

E. a. flava, Netherlands, Nov: following post-juv moult sexing very difficult, but dark facial markings still extensively concealed suggest ♀. (N. D. van Swelm)

E. a. flava, Denmark, Oct: tail pattern always distinctive, meaning that even if facial markings are not seen, identification is straightforward (this bird probably ♂ given relatively strong face pattern and moderately streaked mantle). (K. Malling Olsen)

duller upperparts, and is smaller (see Biometrics). With heavy wear (mainly breeders in summer), head pattern of both sexes less bold, horns partially broken and face pattern heavily obscured, even somewhat scruffy by summer. **1stS** Birds with reduced and less bold black on crown and face, very little pinkish on nape, and more prominently streaked above are probably always 1stY. Others are more difficult to age, being quite similar to ad. – AUTUMN **Ad** Horns of ♂ shorter or absent, as in all ♀♀, and both sexes have poorly demarcated face and breast marks due to buff-grey tips to fresh feathers. **1stW** Similar to ad, but face and breast marks even more concealed than in fresh ad (difference most obvious in 1stW ♀; 1stW ♂ is more like ad ♀). Still moulting birds with some juv flight-feathers not yet shed recognised as such unmoulted juv feathers are quite fresh and broadly pale-fringed. Some must be left unaged after completion of moult due to overlap in characters. **Juv** Soft, fluffy body-feathering and boldly pale-spotted dark brown upperparts, sharply defined buff-white fringes and tips (with subterminal dark brown marks) to wing-coverts and tertials, and underparts and throat/sides of neck whitish (sometimes faintly washed yellow) with dusky brown-grey gorget and dark ear-coverts, both with diffuse paler mottling; lacks horns. Yellow on head/face starts to develop in Aug on chin/throat, simultaneously with first black feathers growing on breast.

BIOMETRICS (*flava*) **L** 16–18 cm; **W** ♂ 109–115 mm (*n* 24, *m* 111.8), ♀ 100–106 mm (*n* 10, *m* 102.9); **T** ♂ 67–77 mm (*n* 25, *m* 71.3), ♀ 61–67 mm (*n* 10, *m* 63.3); **T/W** *m* 63; **B** 12.3–14.9 mm (*n* 35, *m* 13.7); **BD** 4.8–5.7 mm (*n* 35, *m* 5.3); **Ts** 20.0–23.4 mm (*n* 33, *m* 21.8); **HC** 7–13 mm (*n* 28, *m* 9.7). **Wing formula:** p1 vestigial; pp2–4 about equal and longest (p2 sometimes < wt 0.5–1.5 mm, p4 commonly < wt 0.5–3 mm); p5 < wt 7–11 mm; p6 < wt 16.5–22 mm; p7 < wt 24–32 mm; p10 < wt 36–47.5 mm; s1 < wt 38–49 mm. Emarg. pp3–5 (on p5 not quite as distinct in many).

GEOGRAPHICAL VARIATION & RANGE Marked and complex variation, often arranged in subspecies groups differing in amount of yellow on face and details of black pattern on head and chest. Pale races lack yellow but have more extensive black on head and breast. Variation rather distinct between groups, but largely clinal within them.

ALPESTRIS GROUP
(N Eurasia, including NW Europe & NW Africa, and the Americas)
Diagnostically, dark cheek mark does not reach and merge with breast patch, and pale facial areas generally yellow in Europe but cream-white further east. Upperparts, except in *brandti*, slightly to markedly streaked, especially when worn, and more distinct in ♀♀. Distinctly smaller than next group; note relatively shorter hind claw as given for *flava* compared to *penicillata*, but much overlap due to extensive variation within all races. More than ten races described, mostly from Asia and North America, but in the treated region only the following three occur regularly, while a fourth extralimital has straggled to W Europe.

E. a. flava (J. F. Gmelin, 1789) (N Europe and much of Siberia north of *c.* 55°N; winters coastal North Sea, W Baltic, NC Europe, S Russia, Kazakhstan). Described above. ♂ has pale facial areas bright yellow, crown tinged yellowish-olive, and nape, neck-sides, lesser and median coverts pinkish-brown or chestnut-brown; mantle and scapulars dull olive-grey. Black mark on forecrown comparatively narrow. No or only very restricted black above nostrils on forehead. See Biometrics. (Syn. *euroa*.)

○ *E. a. atlas* (Whitaker, 1898) (Atlas Mts in Morocco; short-range altitudinal movements in winter). Differs only subtly from *flava*, with horns of breeding ♂ very marginally longer, and tends to have slightly broader black mark on cheeks, more black on forecrown, rest of upperparts warmer brown, a greenish-yellow tone to crown and hindneck, and less heavily streaked mantle and scapulars. ♀ somewhat less heavily streaked above. Measurements virtually

E. a. atlas, ♂, Morocco, Feb: birds in Atlas Mts have rather long horns and bill, somewhat broader black cheek marks, and upperparts are rather plain. Breeding plumage develops early. (D. Occhiato)

E. a. atlas, ♂, Morocco, Jan: still winter, thus black face markings retain paler tips, obscuring the pattern, especially the bar on forecrown. (M. Schäf)

E. a. atlas, ♀, Morocco, Mar: noticeably duller head colours than ♂, much shorter horns, and narrower and less neat breast and cheek patches, as well as faintly streaked crown to nape (plain in ♂). (R. van Rossum)

E. a. brandti, ♂, Mongolia, May: a member of the *alpestris* group, the black cheek marks do not connect with the black breast patch, but gap is smaller than in *flava*, with cream-white face, not yellow. (H. Shirihai)

identical to *flava*, except slightly longer and stronger bill. **W** ♂ 106–118 mm (*n* 17, *m* 111.6), ♀ 101–109 mm (*n* 10, *m* 104.9); **T** ♂ 70–79 mm (*n* 17, *m* 74.7), ♀ 65–72 mm (*n* 10, *m* 68.2); **T/W** *m* 66.2; **B** 14.2–17.5 mm (*n* 26, *m* 15.7); **BD** 5.0–6.3 mm (*n* 25, *m* 5.7); **Ts** 20.5–23.4 mm (*n* 26, *m* 22.1).

E. a. brandti (Dresser, 1874) (lower Volga, Kirghiz Steppe east through Kazakhstan, Omsk and Novosibirsk region and Kuznetsk Basin in SC Siberia, Altai, Tuva, Sayan, N Mongolia, W Manchuria, south to Tien Shan range; winters Central Asia, Mongolia, N China, apparently wintering regularly in extreme east of treated region). Clearly paler and slightly plainer than *flava*, with a cream-white face and throat (at most a very faint trace of yellow on forecrown and chin). Same size as *flava* but with proportionately longer tail, slightly shorter legs and blunter wing. Black marks on forecrown, lores and cheek patch somewhat broader, and black reaches higher on neck-sides; typically, crown to upper mantle and lesser coverts pale vinous-pink, becoming pale brown-grey with pink-buff hue on lower mantle, scapulars and tertials, with ill-defined and rather narrow dark brown centres. **W** ♂ 104–118 mm (*n* 24, *m* 111.3), ♀ 98–107 mm (*n* 14, *m* 104.0); **T** ♂ 70–84.5 mm (*n* 24, *m* 77.7), ♀ 63.5–73 mm (*n* 14, *m* 68.4); **T/W** *m* 68.3; **B** ♂ 12.3–16.8 mm (once 17.2; *n* 24, *m* 15.1), ♀ 12.6–15.6 mm (*n* 14, *m* 14.0); **BD** 4.9–5.9 mm (*n* 37, *m* 5.5); **Ts** 19.1–22.5 mm (*n* 38, *m* 20.9); **HC** 7.0–14.0 mm (*n* 24, *m* 10.1). **Wing formula: p5** < wt 4–10.5 mm; **p6** < wt 14–21 mm; **p7** < wt 21–29 mm; **p10** < wt 34–42 mm; **s1** < wt 36–45 mm. – Birds from Altai ('*altaica*') claimed to be subtly smaller and darker with more pronounced streaking above, but complete overlap of characters in series compared by us. (Syn. *altaica*; *hachlowi*; *montana*; *parvexi*.)

Extralimital: **E. a. alpestris** (L., 1758) (E Canada, Newfoundland; vagrant to W Greenland, recently proven vagrant to Iceland, Britain, Denmark). Similar to *flava*, but has broader black mark on forecrown, bolder blackish streaking on back and richer pinkish-chestnut on wing-bend, lesser and median coverts, nape and uppertail-coverts, and sides of breast and flanks have slightly more pink-chestnut tinge. At least in some, pale area behind dark cheek patch narrower and more diffuse. Bill averages longer and often has more pointed tip. (A claimed difference in shape of r6 could not be confirmed.) **W** ♂ 110–118 mm (*n* 15, *m* 114.0), ♀ 103–109 mm (*n* 14, *m* 105.5); **T** ♂ 64–74 mm (*n* 15, *m* 70.1), ♀ 58–66.5 mm (*n* 14, *m* 62.6); **T/W** *m* 60.5; **B** ♂ 13.9–17.9 mm (*n* 14, *m* 15.3), ♀ 13.0–15.4 mm (*n* 14, *m* 14.7); **BD** 5.0–6.1 mm (*n* 28, *m* 5.3).

E. a. brandti, ♀, Mongolia, May: pale parts of head typically very pale, cream-white; this race is also proportionately longer-tailed. (H. Shirihai)

E. a. penicillata, ♂, Georgia, Jun: very limited whitish throat surrounded by very broad black pattern; variation (which is not fully understood) includes birds, like this, with distinct rufous crown. (R. Pop)

E. a. penicillata, ♂, Turkey, May: like all races in the *penicillata* group, has broad black cheek patches merging with extensive black breast, isolating small pale cream-white throat, and very long horns. Note that some can have yellowish tinge to pale facial area. (D. Occhiato)

PENICILLATA GROUP
('LEVANT HORNED LARK': Middle East to Central Asia)
Differs from the *alpestris* group in larger size and in that broad black cheek patch of ♂ is broadly connected to extensive black breast patch (but note some variation in size of latter), isolating small pale throat; black horns on crown-sides long, and usually has a black band at base of forehead, above nostrils and base of culmen. Rest of plumage generally very similar to previous group, but has paler/greyer upperparts and lesser coverts, and tends to have whiter underparts, almost white throat, upper forehead and supercilium. In ♀♀, however, black connection on lower neck-sides often narrow and indistinct, breast patch narrower and mottled, and black on forecrown often broken or ill-defined; crown, hindneck, mantle and scapulars sandier-buff. When fresh rather plainer, being narrowly streaked or virtually unstreaked above but in heavily worn plumage, most noticeably in ♀♀, rather heavily blotched, though dark marks diffuse. Note often considerably longer hind claw in *penicillata* (cf. race *flava*), but overlap. Vocalisations, jizz and behaviour not known to differ substantially from *alpestris* group. None of the races in this group can generally be identified other than to group level when dealing with birds outside breeding range.

E. a. balcanica (Reichenow, 1895) (Balkans, Romania, Bulgaria; resident or short-range altitudinal movements in winter) differs from *penicillata* in being slightly smaller on average and having yellower facial areas in fresh plumage (but not as rich yellow as *flava*), though bleaches whiter and is virtually indistinguishable from *penicillata* in worn plumage. Typically, brownish crown and hindneck of fresh ♂ often tinged yellowish (producing olive effect), and lower mantle/back and scapulars are ash-grey (greyer than in either *penicillata* or *bicornis*) and narrowly (but clearly) streaked. Black connection between ear-coverts and chest-band often narrowing to 'waistline', narrower than in *penicillata*, but variable, and one type (ZMB 31522) has very broad black connection. Scant material available for examination in major collections. **W** ♂ 117.5–119.5 mm (*n* 4, *m* 118.8); **T** ♂ 77–80 mm (*n* 4, *m* 78.1); **T/W** *m* 65.4; **B** 15.5–17.3 mm (*n* 5, *m* 16.3); **BD** 5.4–6.5 mm (*n* 5, *m* 5.9); **Ts** 23.0–24.8 mm (*n* 5, *m* 24.2); **HC** 12.0–15.2 mm (*n* 5, *m* 13.3). One ♀: **W** 110 mm, **T** 70 mm, **Ts** 23.0 mm. – **W** ♂ 111–120 mm (*n* 25, *m* 115.9), ♀ 106–112 mm (*n* 5, *m* 109.4) (*BWP*).

E. a. penicillata (Gould, 1838) (Turkey, Caucasus, Transcaucasia, N & W Iran; resident or short-range altitudinal movements in winter). Compared to *balcanica* has pale areas whiter, only indistinctly washed pale yellow when fresh (distinctly paler than *flava*, but not dissimilar to *brandti*),

E. a. penicillata, ♀, Turkey, May: in ♀♀ of *penicillata* group, black bridge between cheek and breast patches on lower neck-sides often slightly broken, and breast patch is narrower. Horns much shorter and forecrown band often ill-defined. (G. Reszeter)

E. a. penicillata, juv, Georgia, Jun: boldly pale-spotted upperparts and breast, with white fringes and tips (and dark subterminal marks) to upperwing-coverts and tertials (similar in all juveniles of the species). (R. Pop)

Approaching *E. a. albigula*, ♂ (left) and ♀, NE Iran, Jun: note vinaceous-pink hindneck, sharply demarcated from rest of pale upperparts; black bridge between cheek and breast patch on lower neck-side is broken in ♀. (E. Winkel)

bleaching to yellowish-white or white in spring. ♂ has dull grey-buff mantle and scapulars (with very faint yellowish or pale olive tinge), both usually narrowly streaked (especially when worn), but some fresh birds almost unstreaked above like *bicornis*. Note that there is a surprising amount of variation in amount of black on lower throat and chest in ♂♂, some having very extensive black chest patch reaching wing-bend, others a much more restricted collar more like *flava*. **W** ♂ 113.5–124 mm (*n* 30, *m* 119.6), ♀ 104–115 mm (*n* 12, *m* 111.3); **T** ♂ 75–86 mm (*n* 30, *m* 81.0), ♀ 70–81 mm (*n* 12, *m* 74.0); **T/W** *m* 67.4; **B** 14.5–18.2 mm (*n* 34, *m* 16.1); **BD** 5.1–6.3 mm (*n* 34, *m* 5.7); **Ts** 21.8–24.5 mm (*n* 30, *m* 23.2); **HC** 7.7–16.0 mm (*n* 24, *m* 11.4). **Wing formula:** Has similar rather blunt wing as *brandti*. – Birds in W & C Turkey tend to be subtly smaller and have less distinct streaking on back ('*kumerloevei*'), but much individual variation and overlap, and mainly effect of clinal variation with influence from *bicornis* in the south and *balcanica* in the west. No other differences could be detected when types and series (42 *penicillata* with 19 *kumerloevei*) were compared. (Syn. *kumerloevei*.)

    E. a. bicornis (C. L. Brehm, 1842) (Lebanon, Syria, N Israel; resident). Distinctly smaller than *penicillata* and, in fresh ad ♂ plumage, almost pale greyish vinous-buff

Probably *E. a. penicillata / albigula*, ♂, Iran, Jan: note black face markings have some paler tips, while horns are shorter too. (E. Winkel)

E. a. bicornis, ♂, N Israel, May: small, rather plain and greyish-brown above, while pale areas on head and neck are cream-white (very faintly tinged yellow at most) in this race. (A. Ben Dov)

E. a. bicornis, ♀, N Israel, Jun: much duller with less solid dark face marks than ♂, and shorter horns (perhaps abraded during nest attendance). (A. Ben Dov)

dorsally, lacking contrast between hindcrown and rest of upperparts, unlike previous two races, and face and throat usually whitish (sometimes pale yellowish when very fresh, but soon bleaching to white). Upperparts of ♂ rather faintly streaked, whereas upperparts of ♀ pale sandy grey-brown with more streaks. Bill generally slender. **W** ♂ 110–118 mm (*n* 12, *m* 113.7), ♀ 105–108 mm (*n* 11, *m* 106.6); **T** ♂ 72–82 mm (*n* 12, *m* 77.0), ♀ 66–74 mm (*n* 11, *m* 69.4); **B** ♂ 15.2–18.0 mm (*n* 12, *m* 16.7), ♀ 14–18 mm (*n* 11, *m* 15.8). (Syn. *aharonii*.)

    ○ **E. a. albigula** (Bonaparte, 1850) (NE Iran through Afghanistan, Pamirs to NW Himalayas; resident or short-range altitudinal movements in winter). Close to *penicillata* and *bicornis* but throat white (even in fresh plumage), and rest of underparts subtly whiter, less buff-tinged on average. ♂ has vinous-pink hindcrown to upper mantle slightly more clearly demarcated from rest of upperparts, which are dull sandy-grey to slightly brownish and very weakly streaked (except when worn). Upperparts of ♀ similar but less pinkish on nape and almost buffy olive-brown on back with more blotchy dark streaks, becoming very heavily streaked when worn. Black of cheeks and lower neck in ♂ connects broadly with breast patch (but only very narrowly or broken in ♀). Similar to *brandti*, with which it reportedly intergrades, but black of cheeks and lower neck usually broader and connected (disconnected in *brandti*). **W** ♂ 113–124 mm (*n* 18, *m* 118.4), ♀ 102–116 mm (*n* 17, *m* 109.8); **T** ♂ 75–87 mm (*n* 18, *m* 80.6), ♀ 68–77 mm (*n* 17, *m* 72.8); **T/W** *m* 67.2; **B** 13.5–17.5 mm (*n* 35, *m* 15.5); **BD** 5.0–6.4 mm (*n* 25, *m* 5.6); **Ts** 20.8–24.0 mm (*n* 34, *m* 22.6); **HC** 9.0–13.2 mm (*n* 23, *m* 11.6). (Syn. *diluta*; *oreodrama*; *pallida*.)

TAXONOMIC NOTES Recent genetic data rather surprisingly indicate a close relationship between *Eremophila* and Short-toed Lark and its allies (Alström et al. 2013). – A comprehensive phylogeny based on multilocus genetic markers (Drovetski et al. 2014) found that *Eremophila* (including *E. bilopha*) requires taxonomic change. The S Asian race *elewesi* of the Horned Lark (extralimital, not treated) was basal to the entire tree in which *bilopha* was nested. A minimum change if *bilopha* is to be retained as a separate species is to split *elewesi*. These authors suggested more far-reaching changes, but we prefer to await further data before a drastic split of Horned Lark is made.

REFERENCES Alström, P. et al. (2013) *Mol. Phyl. & Evol.*, 69: 1043–1056. – Drovetski, S. V. et al. (2014) *PLoS ONE*, 9(1): e87570. – Kirwan, G. M. (2006) *Sandgrouse*, 28: 10–21. – Small, B. (2002) *BW*, 15: 111–120. – Vaurie, C. (1951) *Bull. AMNH*, 97: 431–526.

TEMMINCK'S LARK
Eremophila bilopha (Temminck, 1823)

Alternative name: Temminck's Horned Lark

Fr. – Alouette bilophe; Ger. – Saharaohrenlerche
Sp. – Alondra sahariana; Swe. – Ökenberglärka

Mainly sedentary and nomadic in desert-like lowlands of N Africa, parts of Arabia, the Levant, and Iraq. Studies in Israel showed Temminck's Lark to be mainly a summer visitor, with migrant flocks in spring in the Rift Valley, and an erratic breeder in response to rainfall. Compared to Horned Lark it is much more associated with vast arid lowlands than mountain ridges. It has wandered south to Yemen and, more exceptionally, to The Gambia.

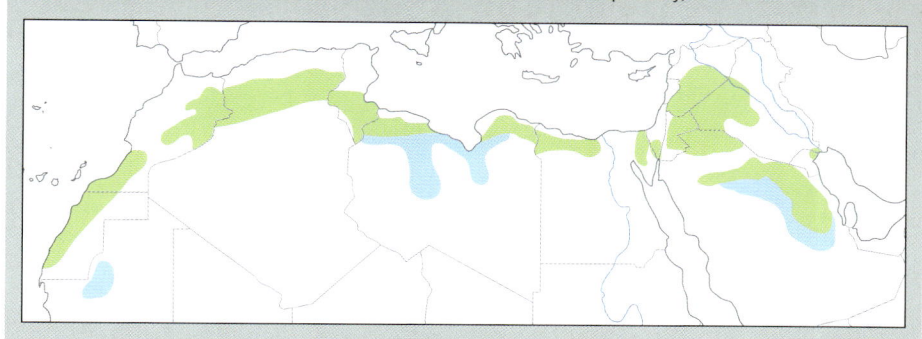

♂, Morocco, Jan: black markings still not fully immaculate. One of the most attractive larks, and usually encountered in unspoilt desert. (M. Schäf)

♂, Morocco, Mar: unlike larger Horned Lark, has pink-tinged pale brown (and almost unmarked) entire upperparts. White face and black breast patch that never connects to black cheek patch also diagnostic (the only form of Horned Lark with this pattern is Asian *brandti*, which is entirely allopatric). Sexed by relatively long horns (an extreme example), broader and uniform jet-black head markings, and warmer and more rufous above. (G. Conca)

IDENTIFICATION A rather small lark with basic plumage pattern as Horned Lark, but has *pink-isabelline upperparts*, appearing generally unmarked except at close range, with paler fringes especially on wing-coverts and tertials evident when fresh. Ground colour of *face white*, with striking *black band on forecrown above eye*, black forehead, lores and cheeks forming *mask-like patch*, and has *conspicuous black band on chest*, which is *triangular* as in Horned Lark, *but upper edge tapers to a point*. Tiny, narrow and pointed black horns on crown-sides often erected rather conspicuously. Rest of underparts essentially white, with faint sandy-pink flush to breast and flanks. *Mainly black tail has pale rusty-isabelline central feathers and white outermost feathers* (when folded looks mostly rusty-isabelline from above and largely black from below). *Primary projection notably shorter and wing shorter and more rounded than Horned Lark* (emphasizing more compact appearance). Behaviour recalls latter, but due to smaller size and very different habitat (more open sandy areas, rather than rocky montane environments of Horned Lark), flight, gait and jizz more like Bar-tailed Lark, with long rapid runs and hops. Song delivered from ground or in rather brief song-flight with fluttering climb to mid height, followed by steep descent, somewhat similar to Horned Lark.

VOCALISATIONS Song recalls Horned Lark with repetition of short phrases, but sounds finer with high-pitched metallic whistles. Pace more even and less obviously accelerated, but utters explosive (*tzzirrritt*) calls and occasionally some cracked notes. Full song more complex and phrases often trisyllabic with fine warbling Robin-like sounds admixed with a few Bar-tailed Lark-like elements (also indistinct twittering or chirruping sounds), e.g. *psi-psi-psi-crchhhh*, … *pite-pite-zeee-tzzirrritt-tzi-ti-ti*, … *petsche-yoo-etch-yoo*… – Flight calls varied, rather metallic, quite loud and tinkling, e.g. *chizz-chizz*, *tzip*, *cee-u* or *see-yoo*, or *swiie-teei-oo*, and a softer nasal *tzeu*.

SIMILAR SPECIES Generally like *Horned Lark*, although they overlap only to a very limited degree outside the breeding season, in the Levant and NW Africa, and are usually well separated by altitude and habitat. Temminck's Lark is overall smaller, longer-tailed and rustier or sandier above with pinkish-sand to warm rufous upperparts, lacking obvious black centres to feathers, and has a shorter primary projection. Additionally, it has a white face and the black breast patch never meets that on the cheeks (like Horned Lark in NW Africa, which fortunately has a yellowish face –

Right: ♂, Jordan, Mar: there is some variation in length of horns. Thinner horns and less solid head markings might indicate younger ♂. (H. Shirihai)

never shown by Temminck's). 'Levant Horned Larks' (ssp. *penicillata* and *bicornis* Horned Lark) also have a white face, but the black bands meet and they have a different-shaped breast patch. Juv Temminck's Lark is much brighter and more uniform rusty-isabelline than equivalent plumage of Horned Lark, and is more likely to require separation from one of the *Ammomanes* larks. – Told from juv *Bar-tailed Lark* by proportionately elongated body, longer tail, thinner and darker bill, and more rusty-orange and clearer pale spotting above, more obvious pale supercilium and diagnostic tail pattern (as in ad).

AGEING & SEXING Ages after post-juv moult largely alike. Sexes similar. – Moults. Post-nuptial and post-juv (complete), mostly Jun–Sep. No pre-nuptial moult. – **SPRING Ad ♂** Upperparts slightly warmer and more rufous, and black on head and breast broader and better defined than ♀ or perhaps 1stY ♂. **Ad ♀** Black areas on head and breast usually smaller and duller (browner, admixed pale brown-grey), and horns shorter. However, much overlap and many cannot be sexed on plumage alone. Wing- and horn-length useful in the hand (see Biometrics). **1stS** Like ad, some perhaps less neatly patterned, albeit with much overlap. All age groups near end of breeding season (especially ♀♀ and 1stY) have facial marks more scruffy and horns partially worn off, whilst white between cheeks and breast patch narrower, sometimes almost absent. – **AUTUMN Ad** As spring but

♀, Israel, Mar: in spring–summer, most ♀♀ readily separated by washed-out face markings and shorter horns (sometimes entirely absent). The more rufous and sandy-tinged and almost unmarked upperparts identify the species from Horned Lark. (D. Barnes)

Ad ♂, Morocco, Dec: fresh winter plumage with solid black lores and most of cheeks. (K. De Rouck)

1stW ♂, Morocco, Sep: mottled (tipped whitish) dark face and breast markings. (G. Conca)

black forehead and facial areas narrower, and pale fringes to wing-coverts and tertials broader and warmer. **1stW** At least some have less-advanced face and breast markings (often partly mottled), and shorter horns, but much overlap especially with fresh ad ♀. **Juv** Lacks bold head pattern of ad and has faintly white-spotted upperparts on sandy-rufous ground, and warm buff breast (but wear soon to whitish).

BIOMETRICS L 14.5–16 cm; **W** ♂ 95.5–103 mm (*n* 17, *m* 99.1), ♀ 87–95.5 mm (*n* 13, *m* 92.5); **T** ♂ 58–71 mm (*n* 17, *m* 67.0), ♀ 55.5–63 mm (*n* 13, *m* 59.4); **T/W** *m* 66.2; **B** ♂ 13.4–16.2 mm (*n* 17, *m* 15.2), ♀ 13.0–15.3 mm (*n* 13, *m* 14.6); **BD** 4.9–6.4 mm (*n* 30, *m* 5.5); **Ts** 18.2–21.5 mm (*n* 30, *m* 19.9); **HC** 5.5–8.2 mm (*n* 12, *m* 7.0). **Horn-length:** ♂ usually > 20 (16–27) mm, ♀ usually < 15 (10–19) mm. **Wing formula: p1** minute, < wt 35–40 mm; **pp2–4** about equal and longest (sometimes either p2 or p4, or both, 0.5–1 mm <); **p5** < wt 3–6 mm; **p6** < wt 6–14 mm; **p7** > wt 14–22 mm; **p10** < wt 27–34 mm; **s1** < wt 25–35.5 mm. Emarg. pp3–5.

GEOGRAPHICAL VARIATION & RANGE Monotypic. – N Africa, N Arabia, S Middle East north to S Syria; resident or only minor dispersal. – Very slight variation noted, and only in size: those in Mauritania (extralimital) are particularly small and have been named *elegans*, but this proposed subspecies is rarely, if nowadays ever, recognised. (Syn. *elegans*.)

TAXONOMIC NOTE Formerly considered a race of Horned Lark, but consistently differs in ecology and plumage, especially that of juv, and absence of pronounced sexual dimorphism. Claimed hybridisation from Syria between the two requires confirmation.

Juv, Morocco, May: oddly, differs rather markedly from that of Horned Lark, being virtually unspotted, rusty-buff and off-white. (A. Faustino)

BROWN-THROATED MARTIN
Riparia paludicola (Vieillot, 1817)

Alternative name: Plain Martin
Fr. – Hirondelle paludicole
Ger. – Braunkehl-Uferschwalbe
Sp. – Avión paludícola africano
Swe. – Brunstrupig backsvala

This small martin frequents a wide variety of open habitats but is always closely tied to water and is much less gregarious than Sand Martin. Widespread across sub-Saharan Africa, and a small population is resident in W Morocco (where it breeds in winter). Mostly a partial migrant. Related to Chinese Martin of E Central Asia and South Asia.

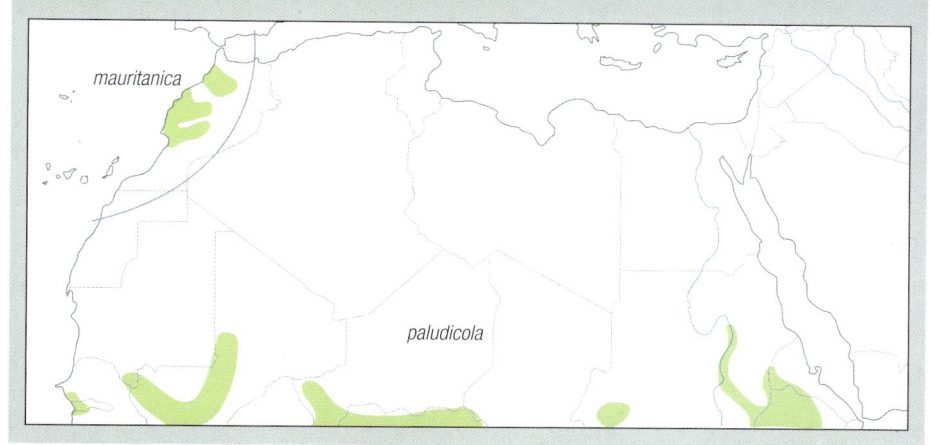

R. p. mauritanica, Morocco, Dec: some birds are smoothly patterned buff-brown on throat, upper breast and flanks. (M. Putze)

R. p. mauritanica, Morocco, Mar: NW African race is rather uniform dull fuscous-brown above. Note characteristically darker (cream-brown) chin and throat, and greyish-sullied face-sides; dark eye-patch makes bird appear large-eyed. Moult limit in greater coverts at this season suggests 1stY. (A. B. van den Berg)

IDENTIFICATION One of the smallest West Palearctic hirundines, usually visibly smaller and *distinctly daintier* (with shorter and blunter wings) than Sand Martin. Diagnostically lacks Sand Martin's pectoral breast-band, but instead has *both throat and upper breast dusky grey-brown*. Predominantly dull fuscous-brown to mouse-brown above (rump/back can appear slightly paler), by and large not very different from Sand Martin. Belly whitish but lower flanks often sullied brown-grey. Tail contrastingly dark (no whitish spots). Some (especially in distant views) may appear duskier overall, even lacking the paler ventral area. Dark eye-surround is usually well developed, but in darkest birds may be less contrasting. Underwing has slightly darker, dusky-brown wing-coverts offering some contrast with underparts (though generally less than in Sand Martin). Flight and behaviour strongly recall familiar Sand Martin, but given close comparative views can sometimes be picked out by its weak, fluttering flight (coupled with tiny size). – For in-hand identification, note that tarsi are completely unfeathered unlike in similarly-sized Pale Martin.

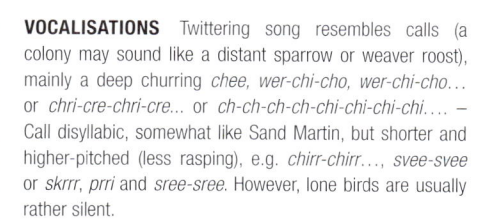

VOCALISATIONS Twittering song resembles calls (a colony may sound like a distant sparrow or weaver roost), mainly a deep churring *chee, wer-chi-cho, wer-chi-cho…* or *chri-cre-chri-cre…* or *ch-ch-ch-ch-chi-chi-chi-chi…*. – Call disyllabic, somewhat like Sand Martin, but shorter and higher-pitched (less rasping), e.g. *chirr-chirr…*, *svee-svee* or *skrrr*, *prri* and *sree-sree*. However, lone birds are usually rather silent.

SIMILAR SPECIES Only significant confusion risks are with *Sand Martin* and recently split *Pale Martin* (see below), from which it differs most appreciably in its strikingly different underparts pattern (has darkish face, chin, throat apart from chest, and thus lacks a contrasting dark breast-band but has otherwise white underparts). Beware that Egyptian race *shelleyi* of Sand Martin is confusingly small and closer in size to Brown-throated, and may show a narrower or more diffuse breast-band than *riparia*. Further, *shelleyi* when fresh in spring sometimes shows dark mottling on chin and part of (mostly lateral) throat which could give wrong impression of darker frontal areas on fast flypast. – *Pale Martin* is rather small and similarly plainer brown-grey above. Note that some natural variation in darkness of throat and chest in Brown-throated Martin occurs, and that palest birds may invite confusion with congeners. The main issue, however, is that all these features can be difficult to confirm, making very good, prolonged views essential, especially if dealing with a potential vagrant. – Could theoretically be confused with closely related *Chinese Martin*, but this is restricted to Asia and will therefore hardly be a common problem. Chinese Martin is even smaller and a little paler grey-brown with contrastingly paler drab rump, but is otherwise very similar, and some could be inseparable unless trapped or photographed. – *Rock* and *Crag Martins* (which see), both of which

R. p. mauritanica, Morocco, Apr: note lack of Sand Martin's breast-band, but rather duskier frontal areas. Paler vent noticeable, contrasting especially with darker throat, breast and underwing-coverts. (F. Trabalon)

may overlap with present species, should appear obviously larger, more powerful and agile fliers, and they possess white tail spots among other plumage differences.

AGEING & SEXING (*mauritanica*) Ages and sexes alike. – Moults. Complete post-nuptial moult in (Mar) Apr–Jul (Sep). Post-juv moult reportedly complete, but available material suggests only partial. When ad finishes moult in summer, first-years are still fresh, and no proof of when moult starts found. Pre-nuptial moult perhaps partial or absent. – **ALL YEAR Ad** Variably duller and plainer with wear; when very fresh indistinctly fringed greyish, most noticeably on tertials (lacks broad rufous fringes of juv). Note that when ad moults completely, usually May–Jul, body-feathers and wing-coverts may have paler fringes, thus appearing almost as worn juvenile plumage. Also worth noting that unmoulted ad primaries can be either fresh and dark or heavily abraded and brown, variation presumably caused by different breeding habits and habitats. – **1stY** Readily separated from ad in first few months as mostly retains juv plumage (with extensive pale rufous or ochre-buff fringes) but as fringes wear off (and some juv feathers perhaps moulted) very much as ad. – **Juv** Like ad but browner; wing-coverts, tertials, scapulars and rump have obvious pale rufous-buff fringes. Looser underparts feathering buffier and has less obvious darker throat and breast than ad.

BIOMETRICS (*mauritanica*) Sexes very nearly the same size. **L** 10.5–11.5 cm; **W** ♂ 97–106 mm (*n* 19, *m* 102.0), ♀ 96–109 mm (*n* 22, *m* 101.5); **T** 40–47 mm (*n* 45, *m* 42.9); **T/W** *m* 42.9; **TF** 0–6 mm (*n* 41, *m* 3.3); **UTC < tail-tip** 8–17 mm (*n* 41, *m* 10.4; usually < 12 mm; UTC =tip of longest undertail-coverts; **B** 7.8–9.8 mm (*n* 36, *m* 3.8); **Ts** 9.3–11.5 mm (*n* 34, *m* 10.4). **Wing formula: p1** minute; **p2** longest (rarely pp2–3); **p3** < wt 0–3 mm; **p4** < wt 5.5–10 mm; **p5** < wt 12–19 mm; **p6** < wt 19–26.5 mm; **p10** < wt 41–49 mm; **s1** < wt 46–54 mm. No emarg. pp.

GEOGRAPHICAL VARIATION & RANGE Two races may occur within the covered region, one of which is a regular breeder, the other is a possible visitor or vagrant.

○ ***R. p. mauritanica*** (Meade-Waldo, 1901) (Morocco; resident). Treated above. Rather small and dark, uniform fuscous-brown above; chin and throat greyer brown and rest of underparts washed whitish-cream, with pale brownish-grey body-sides.

Extralimital: ***R. p. paludicola*** (Vieillot, 1817) (Sahel from Senegal to Sudan, Eritrea and Ethiopia, north along the Nile to Wadi Halfa, Sudan; S Tanzania to South Africa; resident but may straggle to S Egypt). Very similar to *mauritanica* but averages subtly smaller and has often (though not invariably) slightly paler, more greyish-brown upperparts. **W** 94–107.5 mm (*n* 12, *m* 100.6); **T** 41–51 mm (*n* 12, *m* 44.8); **T/W** *m* 44.5. (Syn. *minor*, *paludibula*).

TAXONOMIC NOTE See under Chinese Martin *R. chinensis* for reasons to treat this as a separate species.

R. p. mauritanica, Morocco, Jul: small, compact and dull and brownish in flight; against the sky, often the most obvious pattern is the contrast between brownish throat/breast and paler belly, darker underwing-coverts and black eye. (T. Muukkonen)

R. p. mauritanica, Morocco, Dec: compared to Sand Martin has daintier/slender appearance. Rather paler underparts, with even throat showing only some small brown flecks (so not all are clearly 'brown-throated'). Apparent moult limits in wing suggest 1stY. (M. Varesvuo)

R. p. paludicola, Ethiopia, Sep: this sub-Saharan race is slightly paler than NW African *mauritanica*. Some appear to have almost uniform greyish throat and chest. (H. Shirihai)

R. p. paludicola, Ethiopia, Sep: almost uniform dark throat, chest and upper flanks. Note tiny size compared to larger Sand Martin, although difficult to appreciate in the field, especially given that in NE Africa and Middle East the small race *shelleyi* of Sand Martin, and very rarely Pale Martin, also occur. (H. Shirihai)

CHINESE MARTIN
Riparia chinensis (J. E. Gray, 1830)

Alternative name: Grey-throated Martin

Fr. – Hirondelle chinoise; Ger. – Graukehl-Uferschwalbe
Sp. – Avión paludícola oriental; Swe. – Kinesisk backsvala

A little known Asian counterpart of the African Brown-throated Martin, previously usually treated as a subspecies of this. Breeds in E Central Asia and widely in East Asia, mainly in China, wintering in India and southern Asia. Found in open habitats often near water.

India, Feb: perhaps due to shade and bright background the contrast below is exaggerated, and this impression is not uncommon in the field. The brownish chin/throat immediately separates *chinensis* from Pale Martin and can sometimes appear even darker than this. (S. K. Sen)

IDENTIFICATION The smallest West Palearctic hirundine of all, on average smaller even than ssp. *shelleyi* Sand Martin or ssp. *indica* Pale Martin, and plumage very similar to both requiring careful observation of plumage features and shape. (Several specimens misidentified even in museum collections.) Upperparts mouse-grey or medium grey-brown, but *rump and uppertail-coverts contrastingly paler brown-grey*. Importantly, *chin to upper breast usually uniformly brown-grey* (but slightly paler than upperparts), *contrasting with rest of whitish underparts*. There is a certain amount of individual variation, including birds with slightly paler throat, but in general the rather uniformly brownish-grey throat and chest is the best character to eliminate Pale Martin (which invariably has a whitish throat with at most only a faint pale brownish tinge on chin). Separating pale-throated birds from Pale Martin requires careful check of sides of neck, where Chinese Martin has *whole side of head and side of neck diffusely brown-grey* but Pale Martin whitish side of neck surrounding brown-grey cheek patch. Flight-feathers quite dark grey-brown. The *dark eye-surround is usually well developed* and could extend to the side of forehead. Underwing with slightly *darker, dusky-brown wing-coverts in contrast to rest of underparts* (though generally not as dark as in Sand Martin). *Tail fork very shallow, or tail nearly square* unlike in Sand Martin (but similar to many Brown-throated Martins). Due to tiny size, flight is fluttering and light, and rest of behaviour recalls Sand and Brown-throated Martins.

India, Nov: this small and rather pale E Asian martin has wandered to the Middle East. Chin to breast varies individually and with feather wear from typically pale drab-brown to partly cream or (rarely) almost whitish. Although here seemingly rather pale-throated note wide extension of brown below eye towards lateral throat-sides (wider brown than width of eye). Both ages can apparently moult completely (in late summer) making ageing as a rule impossible. (A. Gulshan)

VOCALISATIONS Little known but presumed to be similar to that of Brown-throated Martin. Call in one recording from India an almost sparrow-like fine chirping *tzrih*.

Cambodia, Feb: characteristically, chin to breast is lightly tinged brownish, merging gradually into rest of whitish underparts, with only limited brownish-grey on upper flanks. (J. Eaton)

Taiwan, Jan: note overall very compact appearance, and from above paler (pale grey-brown) upperparts and wing-coverts, with the rump paler still. Whole side of head brown-grey, dark 'sunglasses' extensive, reaching forehead; note the vague pale throat, which varies within the species. (J. Ting)

SIMILAR SPECIES Differs from *Brown-throated Martin* in being obviously smaller and somewhat paler, and often having a somewhat contrasting paler rump. Otherwise shares with this the shaded throat and breast in contrast with whitish belly (thus no contrasting breast-band), but compared to Brown-throated Martin usually has more restricted brownish-grey on flanks. – Another clear confusion risk is with *Pale Martin*, of which the smallest race, *indica*, is similarly small and similarly plain brown-grey above; see above under identification. Still, safe separation of some Pale Martins from Chinese Martins with maximum pale throat probably often impossible unless handled. – Beware that Egyptian race *shelleyi* of Sand Martin is confusingly small and closer in size to Chinese, and that *shelleyi* may have hitherto undetected breeding populations also in Middle East, thus not restricted to Egypt. However, all Sand Martins have a largely whitish throat in contrast to brown breast-band, hence separation should be straight-forward. Those *shelleyi* which have brown-spotted chin and upper throat-sides still have mainly white lower throat.

AGEING & SEXING Ages and sexes alike. – Moults. Complete post-nuptial moult in late summer (mainly late Jul–Sep). Extent of post-juv moult not closely studied, but available material suggests partial. When ad finishes moult in summer, first-years are still fresh. Pre-nuptial moult perhaps partial but more likely usually absent. – **ALL YEAR Ad** In spring becomes variably duller and plainer with wear. When very fresh in autumn indistinctly fringed greyish, most noticeably on tertials (lacks broad rufous fringes of juv)

India, Mar: diagnostic completely brownish throat and upper breast, separating it from Sand and Pale Martins; separation from African Brown-throated Martin remains particularly challenging (see main text). Note shallow tail fork. (N. Srinivasa Murthy)

Taiwan, Oct: unlike Pale Martin, has brown-grey sides to head and neck, with usually (as here) no white semicollar from throat to neck-sides (only rarely a hint). Also, the dark face mask ('sunglasses') tends to be larger, often even reaching bill and onto forehead. Nevertheless, variation and often-brief field views render this feature of limited use. (J. Wang)

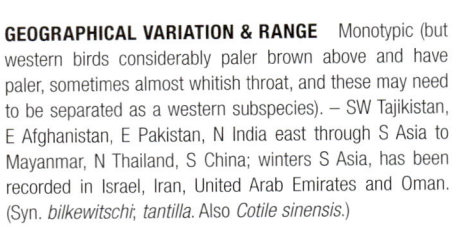

India, Oct: regardless of season or feather wear, many Chinese Martins can have, or appear to have, rather pale throat with hint of paler neck-side and darker upper breast, hence strongly resembling Pale Martin, and some must be left unidentified unless trapped or well photographed, especially when encountered away from breeding range. Clearly, more research is needed into plumage variation within these two Asian martins. Chinese Martin seems to have on average more extensive and diffusely outlined dark head-sides, and lacks (or has less developed) pale neck-sides ('semicollar'), but even these subtle differences will at times be ambiguous. (A. Banerjee)

GEOGRAPHICAL VARIATION & RANGE Monotypic (but western birds considerably paler brown above and have paler, sometimes almost whitish throat, and these may need to be separated as a western subspecies). – SW Tajikistan, E Afghanistan, E Pakistan, N India east through S Asia to Mayanmar, N Thailand, S China; winters S Asia, has been recorded in Israel, Iran, United Arab Emirates and Oman. (Syn. *bilkewitschi*, *tantilla*. Also *Cotile sinensis*.)

TAXONOMIC NOTES The taxon *chinensis* has previously usually been considered a subspecies of Brown-throated Martin, but see Rasmussen & Anderton (2012) and Ayë et al. (2012). Recent unpublished research (Schweizer, *in litt.*) shows *chinensis* to be distinctly different genetically, and considering its small size and paler colours, and the very wide gap in distribution between this and nearest Brown-throated Martin, it seems a well-supported move to render *chinensis* species status.

Juv, India, May: note buffish fringing above and on frontal areas, and obscure facial pattern, otherwise similar to ad. (N. Srinivasa Murthy)

and more broadly on rump and uppertail-coverts. – **1stY** Readily separated from ad in first few months on extensive pale rufous or ochre-buff fringes to wing-coverts, scapulars, tertials, rump and uppertail-coverts, but as fringes wear off (and some juv feathers perhaps moulted) it becomes very similar to ad. – **Juv** Like ad but subtly browner. Wing-coverts, tertials, scapulars, rump and uppertail-coverts have obvious pale rufous-buff fringes. Looser underparts feathering buffier, and often has less obvious darker throat and breast than ad.

BIOMETRICS Sexes of virtually the same size. **L** 10–11 cm; **W** 85–97 mm (*n* 61, *m* 91.3); **T** 37–45 mm (*n* 60, *m* 40.7); **T/W** *m* 44.5; **TF** 0–7.5 mm (*n* 59, *m* 3.9); **UTC < tail-tip** 4.5–14 mm (*n* 53, *m* 8.9; UTC =tip of longest under-tail-coverts); **B** 7.0–9.5 mm (*n* 48, *m* 8.4); **Ts** 9.3–11.0 mm (*n* 35, *m* 10.2). **Wing formula: p1** minute; **p2** longest (rarely pp2–3); **p3** < wt 0–6.5 mm; **p4** < wt 4–11 mm; **p5** < wt 10–17 mm; **p6** < wt 16–23 mm; **p10** < wt 34–43 mm; **s1** < wt 39–48.5 mm. No emarg. pp.

Chinese Martin (left: India, Oct) and Pale Martin, presumed ssp. *diluta* (right: United Arab Emirates, Feb): two very similar species, with subtle differences. Dark cheek patch in Chinese is more extensive below the eye (at least as wide as the eye diameter, often much more), while in Pale it is narrower (less than eye width). Chinese tends to have whole throat and upper breast sullied brownish, whereas throat in Pale is usually whiter creating at least a faint impression of darker breast-band (although these images also show how slight the difference can be). Finally, Chinese habitually shows a characteristically larger dark mask ('sunglasses'), often reaching to forehead. (Left: K. Dasgupta; right: D. Clark)

SWALLOWS AND MARTINS

SAND MARTIN
Riparia riparia (L., 1758)

Alternative names: Collared Sand Martin, Common Sand Martin, Bank Swallow

Fr. – Hirondelle de rivage; Ger. – Uferschwalbe
Sp. – Avión zapador; Swe. – Backsvala

Widespread across the entire Holarctic, the Sand Martin is a summer visitor to Eurasia, where it is one of the first harbingers of spring. It nests in dug-out holes in vertical sandy riverbanks, sand quarries and the like, and congregates in huge flocks at pre-migration and migratory roosts. Sand Martin winters from the Sahel to southern Africa, and in SE Asia. In common with many other recent works, we regard Pale Martin of Central Asia as a separate species.

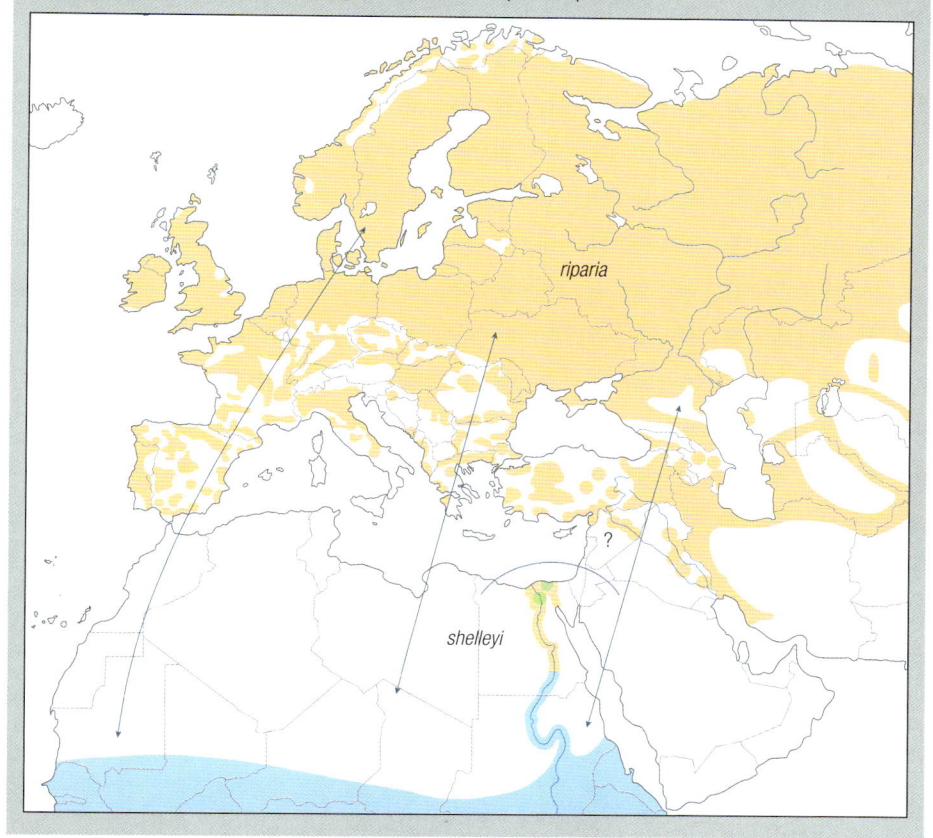

R. r. riparia, Finland, May: note compact shape, especially shortish tail; also, white half-collar behind sharply demarcated dark ear-covers, and well-defined dark breast-band. (J. Peltomäki)

IDENTIFICATION A small hirundine, significantly smaller and less bulky than Barn Swallow, slightly smaller also than House Martin, with a somewhat less forked (more square-ended) tail. Almost *uniform darkish brown upperparts and breast-band*, contrasting with *otherwise white underparts* (although often some brown flank smudges); underwing dusky-brown with axillaries and wing-coverts slightly darker than rest. At close range slight grey (when fresh) or sandy tone (worn) occasionally detectable on forehead, rump, wing-coverts and tertials, and face slightly duskier, especially around eye. Chin and throat typically clean white in adults, but juveniles frequently either faintly washed buff or even speckled brown on chin. *Cheek patches* invariably *rather dark brown, well demarcated from whitish throat-sides and neck*. Note large geographical variation in size. There is some individual variation in prominence, darkness and pattern of breast-band, a few having a hint of a divide at centre (though never completely divided), and some have rather narrow yet largely well-marked band. Flight typical of hirundines but rather weaker and more flickering than, for instance, Barn Swallow: involves more fluttering with sudden short jinking changes in direction and less gliding, but may fly quite slowly when hunting over water. Very restless colonies often suddenly break into panic in response to perceived danger. On passage forms large roosts with Barn Swallow, mainly in vast reedbeds.

VOCALISATIONS Generally quieter than most hirundines: the song is a harsh twittering which is merely a repetition of the commonest call, but perhaps harsher, faster and louder, with characteristic tempo: *tshe-tete-tshe…* – Main call a dry, soft rasping note, a grating *trrrsh*, *tschr* or *tchrrrp*; often uttered in long series, e.g. *trrrsh, trre-trre-trre-rrerrerre…*; in flight some squeaking or high-pitched *twizt*-like notes, especially in alarm.

SIMILAR SPECIES Unmistakable given a clear view. Most significant confusion risks are marginally smaller *Brown-throated Martin* (which see). – There is less risk to confuse Sand Martin with larger *Crag* and *Rock Martins*, none of which has an obvious breast-band contrasting with an otherwise white lower breast and belly. With experience, flight quite different from either of the two latter species, being less powerful and steady. – With recent split of *Pale Martin*, a new identification challenge has arisen; see that species for details. – Differs significantly from *House Martin* and *Barn Swallow*, both of which breed sympatric with Sand Martin in Europe and elsewhere. – Banded Martin (*Neophedina cincta*; extralimital and only a rare vagrant to SW Arabia and extreme

R. r. riparia, Netherlands, Jun: this highly colonial breeder often nests in vertical riverbanks. (K. Mauer)

R. r. riparia, Netherlands, Jul: breeding ad (right) and its chicks at nest entrance approaching fledging (buffy throat and forehead, also rather diffuse collar behind dark ear-coverts and breast-band). Breast-band of ad invisible at this angle. As the species undergoes complete prenuptial moult, ageing becomes impossible thereafter. (N. D. van Swelm)

R. r. riparia, Netherlands, Jun: triangular wings, slightly forked, short tail, and dark underwing with coverts darkest are characteristic. Also well-defined dark breast-band across otherwise white underparts. (K. Mauer)

R. r. riparia, Norway, Jun: in flight from above—dark feather-centres often create mottled effect. Note distinct white half-collar behind dark ear-coverts (M. Varesvuo)

S Egypt; see vagrants section, p. 623) recalls Sand Martin, but s distinctly larger, with a white fore supercilium (visible at close range), much broader breast-band, and much more sluggish, less agile flight. – Chinese Martin is smaller, paler brown above and lacks a clear dark breast-band. (See this species for details.)

AGEING & SEXING Possible to age only prior to post-juv moult (often Aug–Oct). Sexes alike. – Moults. Post-nuptial moult variable as to timing and extent, usually performed in winter quarters, rarely started on breeding grounds, i.e. involving some remiges, but often suspended to be concluded in winter quarters (some finishing in Dec–mid Feb). Post-juv partial and limited, involving renewal of some body-feathers, secondary-coverts and tertials, and no or very few remiges and tail-feathers (if any flight-feathers, more commonly in southern populations); moult in winter quarters more extensive than that in breeding area, involving resuming and completing post-breeding moults (no real pre-nuptial moult appears to exist, merely prolonged post-nuptial and post-juv moult, which both take place mostly in winter quarters but may start in late summer). – **SPRING Ad** Evenly fresh (pale silvery fringes to upperparts and wing-coverts give paler/greyer appearance). **1stS** Inseparable from ad. – **AUTUMN** Ageing possible based on feather wear and plumage pattern. **Ad** Usually retains all or virtually all old/worn feathers before commencing autumn migration; a few moult prior to or during migration, including some body-feathers, a limited number of small inner wing-coverts, tertials and, rarely, a few remiges and tail-feathers (creating strong moult limits). Forehead and crown concolorous dark brown. **1stW** Differs in being uniformly fresh; scapulars, tertials, wing-coverts and rump feathers have rufous-buff to greyish-white fringes. Forehead tipped buff-white or pale grey, contrastingly paler than all-dark crown. **Juv** Like ad but dark areas less uniform, narrowly (sometimes broadly) fringed buff to dull white, often with buff-tinged and sometimes finely brown-speckled chin and upper throat, and buff fringes to central parts of breast-band.

BIOMETRICS (*riparia*) Sexes of very nearly same size (if anything, ♀ a fraction larger). **L** 12–13 cm; **W** ♂ 101.5–114 mm (*n* 25, *m* 108.4), ♀ 101–113.5 mm (*n* 23, *m* 108.9);

Variation in breast-band in European *R. r. riparia*, Finland, May: from solid, dark and wide breast-band (left) to extreme end of variation (right), a bird with narrow and diffuse band. Also note some variation in face pattern, especially in prominence of dark eye-surround, latter partly an effect of light and shadow. (A. Juvonen)

T 46–59 mm (*n* 85, *m* 52.5); **T/W** *m* 48.2; **TF** 5.5–12.5 mm (*n* 88, *m* 9.0; only 3.6% < 7 mm); **UTC < tail-tip** 7.5–22 mm (*n* 56, *m* 12.7; UTC =tip of longest undertail-coverts); **B** 7.5–10.9 mm (*n* 81, *m* 9.3); **BD** 2.2–3.2 mm (*n* 72, *m* 2.7); **Ts** 9.4–11.5 mm (*n* 49, *m* 10.3). **Wing formula: p1** minute; **p2** usually longest (rarely < wt 0.5–1 mm), or pp2–3 equal; **p3** < wt 0–3.5 mm; **p4** < wt 5–12.5 mm; **p5** < wt 13–20.5 mm; **p6** < wt 20–27 mm; **p10** < wt 43–57 mm; **s1** < wt 51.5–65 mm. No emarg. (only subtly narrower webs on pp3–4).

GEOGRAPHICAL VARIATION & RANGE Although there is very little variation within the covered region over a vast area in the north (*riparia*), there is rather marked variation in the south involving at least one distinctive race (see below). In far NE Asia there is at least one more extralimital race (*ijimae*, see below). Recently, three more extralimital subspecies were proposed (Evtikhova & Redkin 2012), although these, if proven valid, may be more relevant in a different context using a much finer subspecies concept than adopted here.

R. r. riparia (L., 1758) (Europe, Siberia east to Kolyma, Baikal and Altai, south to Mediterranean Sea, Turkey, Middle East, Kazakhstan; North America; has bred locally NW Africa; winters mainly in sub-Saharan Sahel zone and in E Africa, from Senegal to Horn of Africa and south to Mozambique). Treated above. Rather large, and relatively dark brown above, with a variably broad but complete dark breast-band. Tarsi unfeathered or with a few, short vestigal feathers near hind toe in some, only rarely more extensive and covering about one-third. – Population in North America tends to be slightly smaller but overlap substantial and deemed best kept within *riparia*. – Breeders in Iran and adjacent areas of S Central Asia ('*innominata*') tend to be subtly paler and warmer tawny brown above, but much overlap and far from consistent, thus best included here. – Populations in SC Siberia perhaps averaging subtly darker and larger, but again much overlap and scant material available for proper assessment. (Syn. *dolgushini*; *innominata*; *kolymensis*.)

R. r. shelleyi (Sharpe, 1885) (Egypt, possibly also somewhere in Levant or W Middle East; winters Sudan, N Ethiopia). Distinctly smaller than *riparia*. Very subtly paler and warmer brown above with a similarly variable breast-band, as to width and prominence, paler birds hence approaching Pale Martin in some respects. Further, usually has lower rear tarsi feathered with a small tuft near hind toe, or with feathers on up to c. 1/3 length of tarsi (just as is commonly case in Pale Martin). Like *riparia* it often has paler brown forehead making dark eye-surround stand out more when seen head-on. Also, *shelleyi* may show some tiny dark spots on chin or upper throat when fresh in early spring as first noted in migrants at Eilat, S Israel (so-called '*eilata*'; see below). Egyptian population: **L** 11–12 cm; **W** 89–99 mm (*n* 37, *m* 94.3); **T** 40–51 mm (*n* 37, *m* 45.2); **T/W** *m* 47.9; **TF** 2–8.5 mm (*n* 37, *m* 6.3); **UTC < tail-tip** 5–17 mm (*n* 37, *m* 9.9); **B** 7.6–9.7 mm (*n* 36, *m* 8.7); **BD** 2.1–3.1 mm (*n* 33, *m* 2.5); **Ts** 9.0–10.9 mm (*n* 36, *m* 9.8). **Wing formula: p10** < wt 39–50 mm; **s1** < wt 44–54 mm. (Syn. *eilata*, see below; *littoralis*.)

Extralimital: ○ *R. r. ijimae* (Lönnberg, 1908) (SE Transbaikalia, S Yakutia east to Amur, Sakhalin, Kurils, Hokkaido; winters SE Asia). Subtly smaller and darker than *riparia*, but much overlap in both respects. Tends to have quite dark and extensive breast-band, most extensive at sides (but much overlap with *riparia* in this respect too). Frequently pale-tipped feathers on lower back, rump and shorter uppertail-coverts, often also in ad in spring. **W** 96–110 mm (*n* 58, *m* 103.4); **T** 45–57 mm (*n* 58, *m* 50.9); **T/W** *m* 49.2; **TF** 5.5–14 mm (*n* 58, *m* 8.7); **UTC < tail-tip** 8.5–17 mm (once 25; *n* 54, *m* 12.4); **B** 7.5–10.6 mm (*n* 43, *m* 8.8); **Ts** 9.3–11.3 mm (*n* 36, *m* 10.3). **Wing formula: p5** < wt 11.5–18 mm; **p6** < wt 19–26.5 mm; **p10** < wt 45–54.5 mm; **s1** < wt 50.5–60 mm. (Syn. *stoetzneriana*; *taczanowskii*.)

TAXONOMIC NOTES Considering its distinctly smaller size, often partly-feathered rear tarsi and somewhat paler plumage, *shelleyi* is a very distinct subspecies. Speculations that it could constitute a cryptic separate species received no support from a preliminary examination of its genetic properties (U. Olsson unpubl.; M. Schweizer unpubl.). Several samples of its mtDNA cluster close to those of *riparia*. – Migrants passing Eilat, S Israel, were described as a separate race '*eilata*' (Shirihai & Colston, 1992) based on similar small size to *shelleyi*, and like this usually slightly paler brown or greyish-brown upperparts and breast-band, but differing in more frequently having finely brown-spotted chin and sides of throat, a character otherwise only found in juvenile Sand Martins. Although this little-known form passes S Israel in fair numbers every year in early spring (peak mid Mar–early Apr), nothing concrete is known of its breeding range (though two recently fledged juv, apparently of this form, recorded in N Israel Apr 2012), nor of its true taxonomic status. Genetically, it is very close to both *riparia* and *shelleyi* (M. Schweizer unpubl.), and with very nearly the same size and plumage as *shelleyi* it is probably best treated as a northern population of that subspecies with unknown breeding range. However, these small migrants merit further study and could yet prove to be a subtly different race of *riparia*. As a basis for future research some biometrics are presented. **W** 90–99 mm (*n* 24, *m* 94.0); **T** 43–50 mm (*n* 22, *m* 46.3); **T/W** *m* 49.1; **TF** 4.5–9.5 mm (*n* 11, *m* 6.5); **B** 8.4–11.5 mm (*n* 14, *m* 9.9); **Ts** 9.1–10.5 mm (*n* 14, *m* 9.8). Often partly-feathered tarsi like *diluta* and *shelleyi*. – Loskot (2006) ventured to review Zarudny's race '*innominata*' described from Central Asia (type locality Tashkent) without personally examining the type series, always a risky method in taxonomy, and concluded that the type series included both Sand and Pale Martins (something which was not entirely convincingly proven). To clarify, a lectotype was selected from Seistan, SW Afghanistan, and the range was defined as 'mainly in Iran'. The selected type, an ad ♀, had W 99 mm and TF 6 mm, both measurements indicating *diluta* rather than *riparia*. We have examined a series of presumed breeders from Luristan, W Iran, including at least two birds from mid May kept in AMNH, and also several birds from May and Aug in Iran at NHM, and find these by and large inseparable from *r. riparia*.

REFERENCES Evtikhova, A. N. & Redkin, Y. A. (2012) *Russ. J. Orn.*, 21: 2845–2872. – Kirwan, G. M. & Grieve, A. (2013) *Sandgrouse*, 35: 114–125. – Loskot, V. M. (2006) *Zool. Med. Leiden*, 80(13): 213–223. – Pavlova, A. et al. (2008) *Mol. Phyl. & Evol.*, 48: 61–73. – Shirihai, H. & Colston, P. (1992) *BBOC*, 112: 129–132.

R. r. riparia, juv, England, Jul: extensive buff fringing above, and narrow buff-white collar behind ear-coverts. With wear fringes becomes less neat and greyer. Some juv and fresh ad have (highly variable) dark mottling on throat. (S. J. M. Gantlett)

Presumed *R. r. shelleyi*, Azraq, NE Jordan, Mar: this taxon is tiny, averaging about 15% smaller than *riparia*, and is even smaller than Pale Martin. Note mottled chin and lateral throat, paler grey rump, and breast-band being diffuse in centre, as well as Pale Martin-like oval-shaped 'sunglasses'. That such birds, once named '*eilata*', have been observed in NE Jordan in early/mid Mar suggests they breed in the Middle East. More research required to confirm that they really are the same taxon as Egyptian *shelleyi*. (H. Shirihai)

PALE MARTIN
Riparia diluta (Sharpe & Wyatt, 1893)

Alternative name: Pale Sand Martin

Fr. – Hirondelle pâle; Ger. – Fahluferschwalbe
Sp. – Avión pálido; Swe. – Blek backsvala

This Central Asian and Siberian martin was for many years considered a species in the Russian literature but only recently has it become better known among Western ornithologists, leading to its more widespread acceptance as a separate species. There is now proof that it even breeds in the same colonies as Sand Martin without any sign of interbreeding, as well as in separate colonies in different habitats, offering ample evidence that the two forms are biological species. Pale Martin breeds from E Kazakhstan to Mongolia, and from the lower Ural River east to S Ussuriland. It has wandered at least to Arabia, whereas claimed records from the Levant are still under consideration.

R. d. diluta, Mongolia, May: a tricky bird with complete breast-band, like Sand Martin, but still has diffuse, washed-out ear-coverts and seemingly shallow tail fork and long undertail-coverts. Such birds are difficult to separate in the field without good photographs to evaluate all characteristics. (J. Normaja)

R. d. diluta, W Mongolia, Jun: compared to Sand Martin, breast-band usually paler and more diffuse (chiefly in centre), while pale half-collar is ill-defined. More subdued border to rather pale brown ear-coverts patch makes this appear blurred, while oval-shaped dark eye-surround ('sunglasses') is also typical. Upperparts paler and drabber. Much variation and overlap with Sand Martin make it necessary to use as many characters as possible. Both age classes undergo complete pre-nuptial moult, making ageing impossible from late winter onwards. (H. Shirihai)

IDENTIFICATION Very similar to Sand Martin, being almost a 'cryptic species'. However, it is smaller and paler than most Eurasian populations of Sand Martin, though subtly larger than the Egyptian race *shelleyi* of that, and larger than Brown-throated Martin in Asia. It shares the characteristic breast-band of Sand Martin, but this is *clearly paler and less well-defined*, often looking 'diluted' or washed out, with the effect that it looks slightly greyer brown; a few have a slightly more prominent breast-band, so better to concentrate on *paler brown and more diffusely marked cheek-patch and crown* making dark eye-surround contrast much stronger ('sun-glasses' effect). Pale Martin is further characterised by being on average *paler and drabber, sometimes quite distinctly tinged cream-brown above* (oddly enough often making it appear slightly more greyish than Sand Martin in field encounters), and—visible at close range only—has chin nearly always washed buff (Sand Martin white, but could be stained). Forehead variably plain brown like rest of crown or, more commonly, somewhat or clearly paler brown. Rump is subtly paler brown than rest of upperparts. Also *differs in calls* (see below). Jizz and behaviour much as Sand Martin, but *tail on average less deeply forked* (a certain slight overlap, especially in young birds, still a useful clue), and white undertail-coverts reach nearer to tail-tip than in *riparia*. Flight similar but on average more light and fluttering, more like in Brown-throated (though often difficult to assess with certainty).

VOCALISATIONS At colonies gives a short, grating twittering song with unclear pattern, not that different from song of House Martin. – Main call is a hard chirping *dret* or *brrit*, subtly but clearly different from Sand Martin in being less dry rasping and more hard chirping, more akin to House Martin calls by comparison.

SIMILAR SPECIES Main confusion risk is with Sand Martin which has almost identical plumage, including the general brown and white colour and dark breast-band below a white throat. Slightly weaker breast-band and paler, cold cream-brown or greyish wash to upperwing-coverts and back of *diluta* often difficult to appreciate, mainly because these require suitable light conditions and viewing angle to appreciate, and ideally, in the case of vagrants, comparative views. Beware that fresh Sand Martins have pale fringes, making the upperparts appear overall paler, and in brief views could seem very similar to Pale Martin. Perched Pale Martins show rather clearly paler cheeks (Light Cinnamon-Drab) with only eye-surround dark, creating the typical 'sun-glassed' appearance and differing rather clearly from Sand Martin's evenly dark brown and sharply outlined cheek patches including (only marginally to moderately darker) eye-surround. There is possibly some slight overlap in the prominence of the dark breast-band, although any bird with a deep brown and complete and well outlined band is a *riparia* Sand Martin, and a bird with a washed out pale and more diffusely defined and greyish-tinged band is Pale Martin. On perched birds in close views useful to note colour of chin: tinged buff-brown in nearly all Pale Martins but white in nearly all Sand Martins (while fine brown spotting on white throat would indicate juvenile Sand Martin or rarely older *shelleyi* of the same species). With experience, the calls of the two should also prove diagnostic. Pale Martin averages smaller than most Eurasian races of Sand Martin, but is very slightly larger than *shelleyi* of Sand Martin, making size in certain regions only of limited use for identification. – For in hand identification of tricky birds it has been suggested that presence or not and amount of small feathers on rear edge and inner side of the tarsus is of significant importance (Zarudny 1916, Goroshko 1993, Loskot & Dickinson 2001, Loskot 2006, Schweizer & Aye 2007), with Pale Martin having tarsi partially or almost completely feathered, whereas Sand Martin has tarsi mainly bare, any feathering restricted to a small tuft above the base of the hind toe, rarely with some feathering also on lower

R. d. diluta, Mongolia, May: paler upperbody (greyish with slight sandy suffusion), often contrasting with darker upperwing, but variation and overlap with Sand Martin (and effects of lighting) render such differences of limited use. (J. Normaja)

Pale Martin *R. d. diluta* (right bird in left image and top-left bird in right image) and Sand Martin *R. r. ijimae* (left bird in left image and bottom-right bird in right image), NW Mongolia (Jun): seen side-by-side, especially when comparing *diluta* to a relatively dark race of *riparia* with often somewhat broader breast-band, is an eye-opener—especially note former's paler upperparts, ill-defined breast-band, borders to ear-covers, and more obvious 'sunglasses' effect. (H. Shirihai).

part of tarsus (never on its upper half). However, we found much more overlap than previously stated and would recommend caution to use this character on single birds. Clearly, more research is needed on live birds (tarsal feathering of skins being difficult to assess accurately) before the criterion can be confidently applied. Claims that the outermost minute primary is paler on average in Pale than in Sand Martin were difficult to confirm and use, and there seems to be substantial overlap. – Separation of Pale Martin from Brown-throated Martin is usually less problematic, as the latter lacks a contrasting breast-band and has a dusky brown throat (and breast), but beware variation in Brown-throated (in particular in extralimital *chinensis*) including birds with subtly paler throat than chest (see images). Separation of Pale Martin from all other martins in the region is as treated under Sand Martin.

AGEING & SEXING Possible to age at least in summer and autumn. Sexes alike. – Moults. Post-nuptial moult variable as to timing and extent, often started in Jun, at times completed on breeding grounds, but often suspended to be concluded in winter quarters. Post-juv partial and limited, involving renewal of some body-feathers, secondary-coverts and tertials and no or very few remiges and tail-feathers (if any flight-feathers more commonly in southern populations). – AUTUMN Ageing possible based on feather wear and plumage pattern. **Ad** Usually commences autumn migration without any renewal of feathers (or moults only just a few), retained plumage being old/worn. No prominently paler edges to greater coverts, scapulars or tertials. Forehead and crown uniformly dark brown. **1stW** Differs in being uniformly fresh; scapulars, tertials, wing-coverts and rump feathers have rufous-buff to greyish-white fringes. Forehead tipped buff-white or pale grey, contrastingly paler than all-dark crown. **Juv** Like ad but dark areas less uniform, fringed extensively narrow buff to dull white, often with buff-tinged throat and buff fringes to central parts of breast-band.

BIOMETRICS (*diluta*) As sometimes found among hirundines, sexes have same size (or ♀♀ are even a fraction larger than ♂♂). **L** 11–12 cm; **W** 94–107 mm (n 52, m 102.5); **T** 42–53 mm (n 51, m 48.2); **T/W** m 47.0; **TF** 3–9 mm (n 52, m 6.4); **UTC** < tail-tip 5.5–14 mm (n 31, m 9.1, only 9% > 11.0; UTC =tip of longest undertail-coverts); **B** 7.0–10.6 mm (n 48, m 9.4); **BD** 2.1–2.9 mm (n 48, m 2.5); **Ts** 9.1–11.5 mm (n 34, m 10.3). **Wing formula: p1** minute; **p2** longest; **p3** sometimes equal to p2, but more commonly < wt 0.5–3.5 mm; **p4** < wt 5.5–10 mm; **p5** < wt 13.5–18.5 mm; **p6** < wt 19–27 mm; **p10** < wt 44–55 mm; **s1** < wt 51–58.5 mm. No emarg.

GEOGRAPHICAL VARIATION & RANGE Slight variation. Some authors recognise up to seven races within Pale Martin, but differences between some of these appear very slight and far from consistent, and seem to require confirmation on larger material. Birds in Tibet and SW China (*tibetana*) are very slightly larger than *diluta* and subtly darker above on average and often have throat, flanks and belly suffused grey-brown, hence breast-band poorly marked; breeders in S & C China (*fohkienensis*) are subtly smaller than *diluta*,

Presumed *R. d. diluta*, United Arab Emirates, Dec: washed-out breast-band often darkest at sides, while pale half-collar behind ear-coverts can also be less obvious. Upperparts look paler, even greyish cream-brown, compared to Sand Martin; on at least some birds, the characteristic pale buffish forehead as shown by most Pale Martins is well developed. (T. Pedersen).

Presumed *R. d. diluta*, United Arab Emirates, Jan: showing individual variation in breast-band that is essentially washed-out, especially in centre but which can be more solid as shown by upper bird. Active pre-nuptial moult can also influence plumage variation, including shape of breast-band. Lowest bird perfectly matches *diluta* photographed on breeding grounds in W Mongolia on p. 110. (N. Moran).

have slightly shallower tail fork and on average a little darker crown and cheeks (but still have washed-out, diffuse breast-band and do not differ in other plumage colours contrary to statements elsewhere); and breeders in N India and N Pakistan (*indica*) are clearly smaller than *fohkienensis*, hence these three are recognised here as separate races apart from *diluta*, but others named seem extremely similar or are inseparable in collections. Since large parts of the material examined by us refer to birds collected on passage or in winter a more comprehensive study based solely on breeding birds could come to a different conclusion.

R. d. diluta (Sharpe & Wyatt, 1893) (SC Siberia, Altai, Sayan, Transbaikalia, N Mongolia, much of Kazakhstan and through Central Asia east to NW India; winters E Kazakhstan, India, S China, rarely E Arabia). Treated above. It is a rather small race with pale brown-grey upperparts and usually pale and washed-out breast band. (Syn. *gavrilovi*; *plumipes*; *transbaykalica*.)

Extralimital races: **R. d. tibetana** Stegmann, 1925 (Tibet east to NE Mongolia, SW China, but details poorly known; short-range movements to southern part of range and to India and Indo-China). The largest race. Rarely has lower throat and whole breast sullied pale grey-brown, and since rest of throat and sometimes chin and flanks can be slightly sullied, too, dark breast band at times indistinct. Those with darkest throat appear like large Chinese Martins *R. chinensis*. **W** 103–111.5 mm (*n* 21, *m* 107.1); **T** 47–60 mm (*n* 21, *m* 51.1); **T/W** *m* 47.8; **TF** 2–8 mm (*n* 21, *m* 5.4); **UTC < tail-tip** 3–14 mm (*n* 20, *m* 9.8; UTC =tip of longest undertail-coverts); **B** 8.2–9.9 mm (*n* 15, *m* 9.1). **Wing formula: p5** < wt 11–19 mm; **p6** < wt 16.5–26 mm; **p10** < wt 46–52 mm; **s1** < wt 51–58 mm.

R. d. indica Ticehurst, 1916 (N Pakistan, NW India; resident or short-range movements in winter). Very similar to *diluta* but consistently smaller, and tail fork is markedly shallower. **L** 10.5–11.5 cm; **W** 88–99 mm (*n* 38, *m* 94.1); **T** 33.5–48 mm (*n* 41, *m* 42.2); **T/W** *m* 44.7; **TF** 2–7 mm (*n* 42, *m* 3.6); **UTC < tail-tip** 4.5–12 mm (*n* 30, *m* 7.0; UTC =tip of longest undertail-coverts); **B** 7.5–9.7 mm (*n* 32, *m* 8.8). **Wing formula: p5** < wt 9.5–15.5 mm; **p6** < wt 17–24 mm; **p10** < wt 39–46.5 mm; **s1** < wt 41.5–52 mm.

? R. d. fohkienensis (La Touche, 1908) (C & S China; resident or short-range movements in winter). An interesting race requiring further study (see under Taxonomic notes). Intermediate between *diluta* and *indica* in size, having by and large the same plumage colours as these two with same diffusely outlined often diluted breast band (although crown and cheeks average darker than in both, thus *fohkienensis* approaches Sand Martin in these respects). It shares with *indica* a quite shallow tail fork, slightly shallower than in *diluta*. **W** 90–105 mm (*n* 40, *m* 98.2); **T** 36–47 mm (*n* 40, *m* 42.7); **T/W** *m* 43.5; **TF** 1–6.5 mm (*n* 39, *m* 3.2); **UTC < tail-tip** 4–11.5 mm (*n* 35, *m* 6.9; UTC =tip of longest undertail-coverts); **B** 7.8–10.0 mm (*n* 38, *m* 8.9). **Wing formula: p10** < wt 40–49.5 mm; **s1** < wt 44–56 mm.

Variation in Pale Martin (left: presumed ssp. *diluta*, United Arab Emirates, Dec; right: United Arab Emirates, Feb): individual variation in breast-band prominence and paleness of throat and neck-side half-collar in Pale Martin sometimes creates severe identification challenges, in particular when encountering Chinese Martins with paler throat and hint of pale half-collar. The right-hand bird is truly intermediate and we do not claim to know for sure that it is a Pale Martin, only that on balance it looks best for this. (Left: M. Almazrouei; right: M. Barth)

TAXONOMIC NOTES Gavrilov & Savchenko (1991) and Goroshko (1993) treated Sand Martin and Pale Martin as separate species based on morphological and vocal differences and their sympatric occurrence, the two showing a tendency to segregate according to habitat in areas of sympatry. We have been able to confirm in Kazakhstan that Pale Martin sometimes breeds in the same colonies as Sand Martin without mixing. Hence, treating Pale Martin as a separate species seems uncontroversial. – The racial variation within Pale Martin requires further study. Zarudny (1916) separated birds from upper Syrdar'ya, Fergana, Tashkent and adjacent areas as '*plumipes*' on account of more extensive tarsus feathering and paler and more greyish upperparts and more washed-out breast band, differences which we (like Loskot & Dickinson 2001) are unable to confirm. Loskot (2001) described a new race ('*gavrilovi*') from Siberia and adjacent parts of N Central Asia based on slightly larger size than *diluta*. However, he based this on a comparison with ten *diluta* which clearly were too small to be representative; his averages for the claimed new Siberian race exceed only fractionally what we got from 31 *diluta* from E & SE Kazakhstan south to N Afghanistan and Baluchistan. Such small differences (c. 2 mm for wing length) are only part of normal north–south clines in migratory birds, and other separating characters were not mentioned. – The subspecies *fohkienensis* is worthy of further study since it has a mixture of characters of Sand and Pale Martins, but has traditionally been referred to the Pale Martin. Genetically it has proven to be highly different from both (M. Schweizer unpubl.), and it actually might represent an undescribed Asian cryptic *Riparia* species. – Claims of occurrence in Levant, Egypt and Red Sea coast of Arabia lacks foundation and are most probably due to misidentifications. – See also notes under Sand Martin.

REFERENCES Gavrilov, E. I. & Savchenko, A. P. (1991) *Bull. Mosk. O-va Ispyt. Pri.*, 96 (4): 34–44. – Goroshko, O. A. (1993) *Russ. J. Orn.*, 2 (3): 303–323. – Kirwan, G. M. & Grieve, A. (2013) *Sandgrouse*, 35: 114–125. – Loskot, V. M. (2001) *Zoosyst. Rossica*, 9: 461–462. – Loskot, V. M. (2006) *Zool. Med. Leiden*, 80-5 (13): 213–223. – Loskot, V. M. & Dickinson, E. C. (2001) *Zool. Verh. Leiden*, 335 (15): 167–173. – Pavlova, A. et al. (2008) *Mol. Phyl. & Evol.*, 48: 61–73. – Schweizer, M. & Aye, R. (2007) *Alula*, 13: 152–158. – Shirihai, H. & Colston, P. (1987) *BBOC*, 112: 129–132. – Zarudny, N A. (1916) *Orn. Vestnik*, 7: 25–38.

Pale Martin, *diluta* (left: United Arab Emirates), small-sized Sand Martin, presumed *shelleyi* (centre: Jordan, Mar), and Chinese Martin (right: India, Dec): separation of these three, especially if appearing as vagrants away from breeding areas, is complicated by variation and some degree of overlap in plumage and size. Typically, Pale Martin has pale and blurred ear-coverts patch and rather diffuse breast-band, especially at centre. Compared to Pale Martin, Sand Martin *shelleyi* usually has better-defined breast-band and darker cheeks, but some are intermediates. As a further complication, some Chinese Martins have paler throats and impression of slightly darker breast-band, but unlike the other two species tends to have wider brownish cheeks extending to side of throat, limiting the pale throat area (if any). Some birds may have to be left unidentified. (Left: D. Clark; centre: H. Shirihai; right: K. Dasgupta)

ROCK MARTIN
Ptyonoprogne fuligula (M. H. C. Lichtenstein, 1842)

Alternative name: Pale Crag Martin

Fr. – Hirondelle isabelline; Ger. – Steinschwalbe
Sp. – Avión isabel meridional; Swe. – Afrikansk klippsvala

Mainly resident in montane regions, where it prefers gorges and ravines in deserts, the rather nondescript Rock Martin occurs from Morocco to Egypt, through the S Levant, Arabia and Iran, reaching to Pakistan. In these dry regions it replaces Crag Martin, indeed the two are completely segregated by range and habitat when breeding, with only some mixing on migration and in winter.

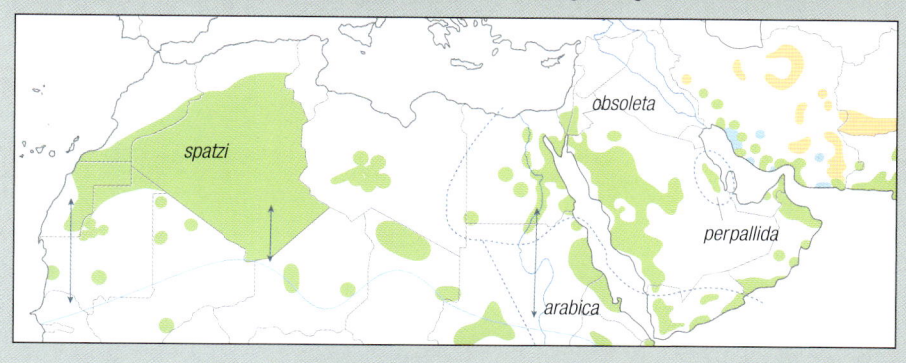

cheep-cheep-churr, or a quick *chirrup* uttered in different situations (sometimes with the song).

SIMILAR SPECIES Most likely to be confused with *Crag Martin* but Rock is somewhat smaller (appreciable in comparative views or with experience) and a little paler below, the body generally whitish-grey with a pink-buff tinge, and pale, unmarked chin and upper throat (somewhat darker upper throat contrasts with paler lower throat/breast in Crag), while flanks and especially undertail-coverts are only subtly darker than breast and belly, unlike in Crag which has much darker vent and undertail-coverts. On average, pale fringes to undertail-coverts better developed. Underwing-coverts generally less obviously darker than flight-feathers (compared to Crag), sometimes appearing almost equally pale, except darker axillaries and leading edge of inner wing, which form dark carpal area, darker than rest of underwing (but in some lights can appear rather pale; see below). Also generally paler brown above (though difficult to judge in strong sunlight, and extremes come very close), with better-developed pale fringes, notably to tertials and inner greater coverts, with a paler inner wing and darker primaries, while rump and uppertail-coverts are also often slightly paler with rather clear pale tips (Crag is essentially uniform cold grey or brownish-grey above, perhaps slightly sandier when fresh). Both have whitish spots on inner webs of all but central and outermost rectrices (visible only on spread tail and most obvious from below). Beware that in strong sunlight, against

H. f. obsoleta, Israel, Mar: these images perfectly capture the jizz and coloration of this arid-desert hirundine, which is often very tame and may even follow people to locate disturbed aerial insects. (M. Varesvuo)

IDENTIFICATION *Small to medium-sized, short-tailed hirundine with washed-out colours that is most likely to be encountered in desert environments.* Somewhat smaller than Crag Martin to which it bears a strong resemblance in flight, behaviour and plumage. Generally, rather *paler grey-brown above*, paler and more *cream-white and buff below* with a *paler, unstreaked throat* and *less contrastingly dark underwing-coverts*. White spots in spread tail (as in Crag Martin) and undertail-coverts also rather pale. The main issue when attempting an identification is to reliably evaluate the colour pattern of a bird that is constantly changing in appearance with direct and reflected light, angle and background, as it flies to and fro. Not always easy!

VOCALISATIONS Song is a muffled, slightly raucous twitter, e.g. *tche-pte-tche-tec-t-t-che-che-chrrrrrr*, ending emphatically, but other vocalisations unobtrusive and easily overlooked. – Main contact calls are a dry *trrt*, like other related hirundines, and a slightly nasal *vick*; also high-pitched *twee*, a repeated rapid *chir chir chir…* or

H. f. obsoleta, Israel, Mar: collecting mud for its nest; the pale creamy grey-brown upperparts almost merge into the background. (A. Ouwerkerk)

H. f. obsoleta, Israel, Mar: overall pale plumage, darkest areas being limited to carpal and around eyes, important in separation from Crag Martin. Only when tail is spread are white spots revealed, but variation and overlap in their size and shape make this feature of no use to separate the two species, although they tend to reach closer to the tips in Rock. (M. Varesvuo)

Rock (left: *obsoleta*, Oct, Israel) and Crag Martins (right: Oct, Israel): Rock is generally Crag Martin-like but with slightly shorter tail. Given good views, in Levant usually readily distinguished by overall paler grey/cream coloration, with more uniform underparts—look for unmarked chin and throat, and dark on underwing restricted to carpal (whereas Crag has broader and more contrasting underwing-coverts, axillaries and trailing edge). Also less contrasting dark undertail-coverts in Rock compared to Crag. (H. Shirihai)

P. f. obsoleta, juv, S Egypt, May: note pale fringes above. (J. S. Hansen)

P. f. arabica, Oman, Nov: rather fragmented range in NE Africa and SW Arabia, reaching extreme SE Egypt in Gebel Elba; unlike *obsoleta* overall browner and less clean below, being suffused buffish-brown and to some extent marbled darker. (M. Schäf)

a blue sky, Rock's underwing-coverts can appear solidly darker than the translucent remiges, strongly recalling the pattern of Crag Martin, making close views vital for correct assessment of this feature. Also note that some races of Rock in N Africa and SW Arabia are darker, and closer to Crag, further complicating the identification issue in these areas. — Compare also the somewhat smaller *Sand Martin* and *Pale Martin*.

AGEING & SEXING Ageing often possible (with experience and in close views or handled birds) by feather age and wear. Sexes alike. — Moults. Post-nuptial moult generally complete during and after breeding (earlier in N Africa, later in Levant and Middle East), and post-juv moult mainly partial (both moults slow and protracted). Pre-nuptial moult perhaps absent or very limited, but little studied. — SPRING **Ad** Bleaches purer grey above and paler or whiter below. **1stS** Similar to ad, but in the hand aged by retained juv primaries, primary-coverts and some secondary-coverts, chiefly greater coverts, but telling these reliably from later feather generations often extremely difficult. — AUTUMN **Ad** Evenly feathered. **1stW** As ad but retains most juv secondary-coverts and tertials, also all remiges, primary-coverts and (most or all) tail-feathers, which are softer-textured and browner. In early autumn easily aged by broad, often rather well-defined buffy fringes to tertials and greater coverts, but these wear rapidly, and by Nov are hardly visible, if at all. Beware that moult limits within greater coverts are difficult to detect or lacking, usually because no greater coverts are moulted, or difference between worn outer and new inner greater coverts is slight; also difference between retained juv and new ad primary-coverts is not sharp. **Juv** As ad, but less pure grey above (fringed buff, especially on mantle, scapulars, rump and uppertail-coverts). Wing-coverts and tertials fringed buffish-cream or rufous. Often deeper buff below.

BIOMETRICS (*obsoleta*) **L** 12.5–14 cm; sexes are same size: **W** 113–125 mm (n 32, m 119.6); **T** 47–53 mm (n 32, m 50.5); **TF** 2–6 mm (n 25, m 4.1); **T/W** m 42.2; **B** 9.9–11.9 mm (n 14, m 10.6); **BD** 2.5–3.1 mm (n 12, m 2.8); **Ts** 9.5–11.0 mm (n 12, m 10.4). **Wing formula:** **p1** minute; **p2** usually longest (rarely < wt 0.5–1 mm); **p3** < wt 0.5–2.5 mm (sometimes =p2); **p4** < wt 8–13 mm; **p5** < wt 17–24 mm; **p10** < wt 57–64 mm; **s1** < wt 62–71 mm. No emarg. (only subtly narrower webs on pp2–3). **White 'window' on r4** (length on inner web): 6–13 mm.

GEOGRAPHICAL VARIATION & RANGE On the whole only slight variation. Still, several races described and often arranged in two groups, northern *obsoleta* and southern *fuligula*. These two groups sometimes given species status, but not here due to obvious similarities and lack of compelling arguments. Four races in the covered region, all referred to *obsoleta* group. Variation mostly involves plumage tones, which are clinal, and much overlap in measurements renders these of little use in subspecies designation. Two extralimital subspecies described, but not treated here.

○ ***P. f. spatzi*** (Geyr von Schweppenburg, 1916) (NW & WC Africa from Morocco to SW Libya, south to Chad; mainly resident). Medium brown-grey above, clearly darker than *obsoleta*, and underparts notably suffused dirty pink-buff when fresh; very close to *arabica* and therefore a very marginal race. **W** 118–132 mm (n 19, m 121.7); **T** 48–54 mm (n 19, m 51.2); **TF** 0–6 mm (n 19, m 3.7); **T/W** m 42.1; **B** 10.1–12.3 mm (n 17, m 11.3). — Birds referred to '*presaharica*' (NC Algeria to S Morocco, N Mauritania) are very similar to *spatzi* and only differ very subtly and on average in being a shade paler and more sandy-brown above; many are inseparable. (Syn. *presaharica*.)

P. f. obsoleta (Cabanis, 1851) (Egypt, S Israel, much of Arabia except south-west & north-east; Iran, Pakistan; mainly resident, but some movements noted). Described above.

SWALLOWS AND MARTINS

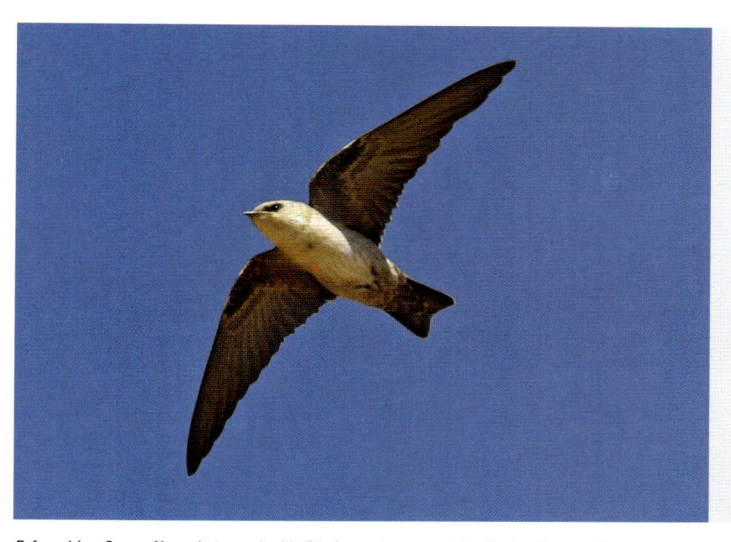

P. f. arabica, Oman, Nov: photographed in Dhofar and compared to *obsoleta* is overall brownish above and typically suffused buffish-brown below, where often faintly streaked darker, especially on lower belly; tail averages slightly shorter and more deeply forked. Pale throat prevents confusion with Crag Martin, and darkest part of underwing is concentrated on carpal region, while undertail-coverts are not so contrastingly dark. However, such differences constantly change with light and angle. (M. Schäf)

H. f. obsoleta / arabica, NE Oman, Nov: all of Oman (except Dhofar) is assumed to be occupied by *obsoleta*, but most Omani birds are clearly browner overall, being warmer below, to some extent approaching *arabica*. (T. Lindroos)

Rather intermediate between palest and darkest races in this group, slightly smaller and paler overall than *spatzi*. – Birds of C & E Iran and into Pakistan ('*pallida*') average (in spite of name) subtly darker above and are sometimes marginally more brown-tinged overall, but much variation, overlap and subtle differences, hence best included here. (Syn. *pallida*.)

P. f. arabica (Reichenow, 1905) (Yemen, Sudanese Red Sea north to Gebel Elba, N Horn of Africa, Socotra; resident). Rather brown above and suffused deeper buff-grey below than other Middle Eastern races. Perhaps a fraction shorter-tailed with slightly deeper tail fork than *obsoleta*, but differences certainly minute. Bill slightly shorter but thicker. **W** 112.5–124.5 mm (*n* 19, *m* 118.7); **T** 44–51.5 mm (*n* 19, *m* 48.5); **TF** 2–6 mm (*n* 25, *m* 4.1); **T/W** *m* 42.2; **B** 9.9–11.9 mm (*n* 14, *m* 10.6); **BD** 2.5–3.1 mm (*n* 12, *m* 2.8); **Ts** 9.5–11.0 mm (*n* 12, *m* 10.4). **White 'window' on r4** (length on inner web): 5–10.5 mm.

P. f. perpallida (Vaurie, 1951) (Al Hufuf region, NE Arabia; possibly S Iraq; resident or makes short-range movements in winter). Palest of all subspecies, with plain greyish-cream upperparts (very little brown), and has obviously darker wings, while underparts, especially throat and upper breast, are almost white (faint pink-buff suffusion only); dusky under-wing-coverts more contrasting. Scant material examined. **W** 116–123 mm (*n* 9, *m* 120.4); **T** 48–52 mm (*n* 9, *m* 50.2); **TF** 2–4.5 mm (*n* 9, *m* 3.6); **T/W** *m* 42.1; **B** 9.2–11.9 mm (*n* 8, *m* 10.3); **BD** 2.5–3.6 mm (*n* 8, *m* 3.1). **White 'window' on r4** (length on inner web): 10–13 mm.

TAXONOMIC NOTES Sometimes considered conspecific with Crag Martin *P. rupestris*, but not followed here. – Rock Martin *P. fuligula* may comprise two species (as adopted in Dickinson & Christidis 2014): the paler *P. obsoleta* in N Africa & Asia, and the darker *P. fuligula* in Afrotropics. Others prefer to recognise three groups: *obsoleta* (N Africa & Asia), which is pale and comparatively small; *fusciventris* (W, C & E Africa) which is much darker and distinctly smaller than the *obsoleta* group; and *fuligula* (S Africa, north to Angola & W Mozambique) which is somewhat paler and larger than the previous group. Many races have been described in both the *obsoleta* and *fusciventris* groups, but some are probably just junior synonyms.

H. f. obsoleta / arabica, NE Oman, Apr: some Omani birds have darker/browner upperparts. (H. & J. Eriksen)

CRAG MARTIN
Ptyonoprogne rupestris (Scopoli, 1769)

Fr. – Hirondelle de rochers; Ger. – Felsenschwalbe
Sp. – Avión roquero; Swe. – Klippsvala

A typical inhabitant of cliff faces in montane regions, the Crag Martin is mostly resident or a partial or altitudinal migrant, although northernmost populations are migratory. Widespread across alpine regions from Iberia to Greece, and N Africa to the Middle East, thence east to China and the Himalayas, wintering south as far as the Nile Valley, Ethiopia and India. Often spotted when scanning the steepest rock faces for raptors, or in or around road tunnels in craggy mountain regions.

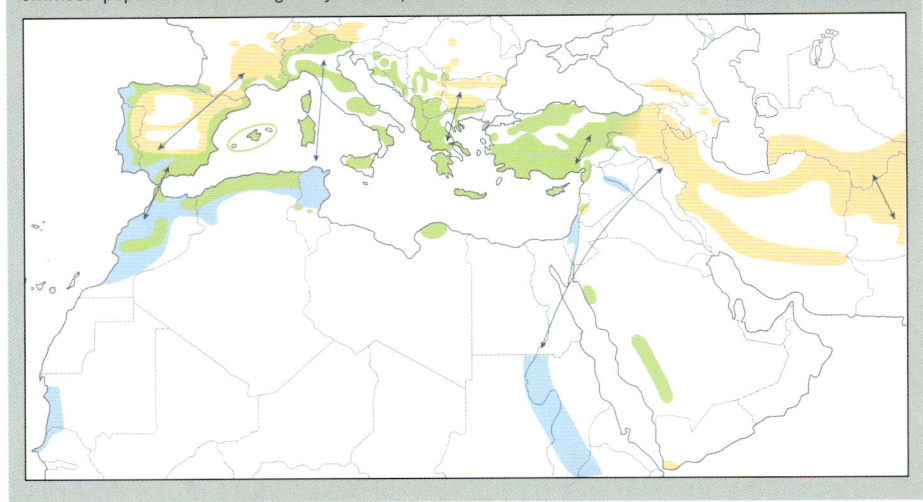

Spain, Jul: compact with broad, pointed wings, short and only very slightly forked tail. A typical individual showing dark foreanc innerwing-coverts, and streaky grey-brown chin and throat-sides. Growing inner primaries in summer suggest ad during post-nuptial moult. (T. Lindroos)

Mallorca, Balearic Islands, Spain, Jan: compared to Rock Martin, generally darker and browner above. Note (just visible) diagnostic streaky grey-brown chin and throat-sides. (J. Bazán Hiraldo)

Greece, Apr: again, note blotchy grey-brown chin and throat. (B. Baston)

IDENTIFICATION Characteristically robust and rather *compact hirundine with broad, yet pointed wings, short, almost square tail*, and mainly *greyish-brown* plumage with a duskier face. Medium grey-brown *underwing with strongly contrasting dull blackish coverts* (greater coverts slightly less dark); also characteristic is the *greyish-white chin/throat with tiny grey-brown streaks and spots* extending to sides of face (visible at very close range and often forming a bar on lower cheeks). Pale buff (wearing to whitish) breast, pale brown-grey flanks and belly, and *dark brown-grey* (can appear almost blackish) *undertail-coverts* (with pale tips visible only at close range). Also when close, often shows *white subterminal spots to underside of all but central and outermost tail-feathers* (which are readily visible from below when tail is spread). Flies with frequent steady glides, but capable of surprising agility, twisting and turning following a few rapid wingbeats, and sometimes flies quite high like a small swift, but more usually found patrolling a favoured cliff face.

VOCALISATIONS Rather quiet and unobtrusive, can be silent for long periods. – Song a rapid throaty twittering and chirruping without any clear pattern, a mixture of various calls and scratchy notes. – Calls recall House Martin but are generally more quiet and more often uttered singly or just a few notes at a time, a disyllabic or chirpy *zrit* or *pri-pit*, or single very short and subdued *trp* or *drt* (almost like distant Little Bustard); also a sharp, short and very high *tzi*; further, a cheerful Linnet-like rising *poeeh* or *sfooee*, and *chwit* or *chwitit*; during aerial chases, a rapid *chu-chu-chu-chu-chu*, similar to song variant of Redpoll.

SIMILAR SPECIES Over most of the treated region, identification is easy, given combination of brownish plumage, compact silhouette with short tail, and contrastingly dark underwing-coverts, as well as whitish tail-spots. However, identification complicated by presence of *Rock Martin* (which see) in Middle East and North Africa, where the two are often found in close proximity during the non-breeding season.

AGEING & SEXING Ageing possible by feather wear and moult patterns if seen close or handled. Sexes alike. – Moults. Post-nuptial moult in late summer–early autumn (often end Jul–Sep) complete but rather protracted, may continue almost to midwinter (and often suspended halfway through). Post-juv moult mainly partial, often starting late. Pre-nuptial moult thought to be very limited or even absent. – **SPRING Ad** With wear generally duller above, and underparts paler with more contrasting dark throat, face, flanks and underwing-coverts. **1stS** Retained juv primaries, primary-coverts, usually all greater coverts and tertials, and perhaps some other wing-coverts, all more heavily bleached and duller than ad. If not heavily worn, greater coverts still typically fringed paler. – **AUTUMN Ad** Evenly fresh but some older feathers diagnostic (i.e. moult occasionally suspended, with flight-feathers often still moulting in Oct–Nov, and freshly moulted primaries in winter, unlike 1stY). **1stW** Retains some juv wing-coverts, including all greater coverts, tertials, wing- and tail-feathers (softer textured and duller). Greater coverts and tertials have better-defined rufous-buff fringes giving a scaly look. At least until mid autumn base of lower mandible yellowish (darker in ad). Primaries and primary-coverts slightly worn in autumn, moderately so in winter and distinctly worn from late spring. **Juv** As ad but warmer and slightly paler due to extensive rufous fringes on upperparts. Throat paler and only indistinctly mottled. Tail-spots average smaller.

BIOMETRICS L 14–15 cm. A rare case where a ♀ passerine has on average longer wing than the ♂: **W** ♂ 124–135 mm (n 18, m 130.6), ♀ 126.5–140 mm (n 12, m 131.8); **T** 51–58 mm (n 40, m 54.6); **TF** 1–7 mm (n 39, m 3.5); **T/W** m 41.6; **B** 10.4–11.9 mm (n 18, m 11.3); **BD** 2.7–3.3 mm (n 16, m 3); **BW** 6.5–8.3 mm (n 17, m 7.4); **Ts** 10.5–12.0 mm (n 13, m 11.4). **Wing formula: p1** minute; **p2** usually longest (rarely < wt 0.5–1 mm); **p3** < wt 0.5–3 mm (sometimes =p2); **p4** < wt 8–13.5 mm; **p5** < wt

SWALLOWS AND MARTINS

18–24 mm; **p10** < wt 63–72 mm; **s1** < wt 70–78 mm. No emarg. (only subtly narrower webs on pp2–3).

GEOGRAPHICAL VARIATION & RANGE Monotypic. – S Europe, NW Africa, Asia Minor, Central Asia east to NW Mongolia, China; northern populations make shorter movements in winter to Mediterranean basin, Middle East, S Asia. – There is a tendency for birds in Asia to be slightly more saturated brownish below, with more restricted whitish throat/upper breast, but difference is still slight, many are identical in appearance, and this vague difference is insufficient for formal separation.

Italy, Jun: diagnostic streaks/spots on chin and throat not always easy to detect, but versus Rock Martin note blackish-brown fore primary- and median coverts, and undertail-coverts. (D. Occhiato)

Spain, Sep: some birds can have limited streaks or spots on chin and throat, or appear to lack such marks (more often when worn towards end of breeding), when plumage can also be bleached paler. Versus Rock Martin, note chunkier appearance, duskier overall impression, browner underparts including undertail-coverts. (H. Shirihai)

Italy, Dec: another rather tricky individual, with superficial impression of distinctly darker carpal patches and overall pale appearance. However, note just-visible diagnostic streaks/spots on sides of throat, browner belly, and even darker undertail-coverts. (M. Viganò)

Spain, Feb: against the light, flight-feathers contrast spectacularly with blackish-looking wing-coverts; also comparatively broad-based short wings, 'fat' body and broad tail base, which separate it from any other hirundines in region. (T. Kolaas)

Italy, Jun: diagnostic streaks/spots on chin and throat not always easy to detect, but versus Rock Martin note blackish-brown fore primary- and median coverts, and undertail-coverts. (D. Occhiato)

117

BARN SWALLOW
Hirundo rustica L., 1758

Fr. – Hirondelle rustique; Ger. – Rauchschwalbe
Sp. – Golondrina común; Swe. – Ladusvala

Among swallows and martins, to the layman this is probably the best-known species, although the House Martin regularly lives closer to man. When a 'swallow', so often a symbol of summer and warmth, is to be depicted in a book or paper, the Barn Swallow—with its long tail-streamers—is generally chosen. The Barn Swallow shows a preference for nesting—not surprisingly—inside barns and stables, whenever there is a hole or broken window to allow access, but almost as common a nest site is a beam or hole under bridges or rafts in marinas and harbours. Often seen over ponds and watercourses, and associates with grazing cattle. The majority of the European population migrates to winter in S Africa, with a minority staying in N Africa and S Spain.

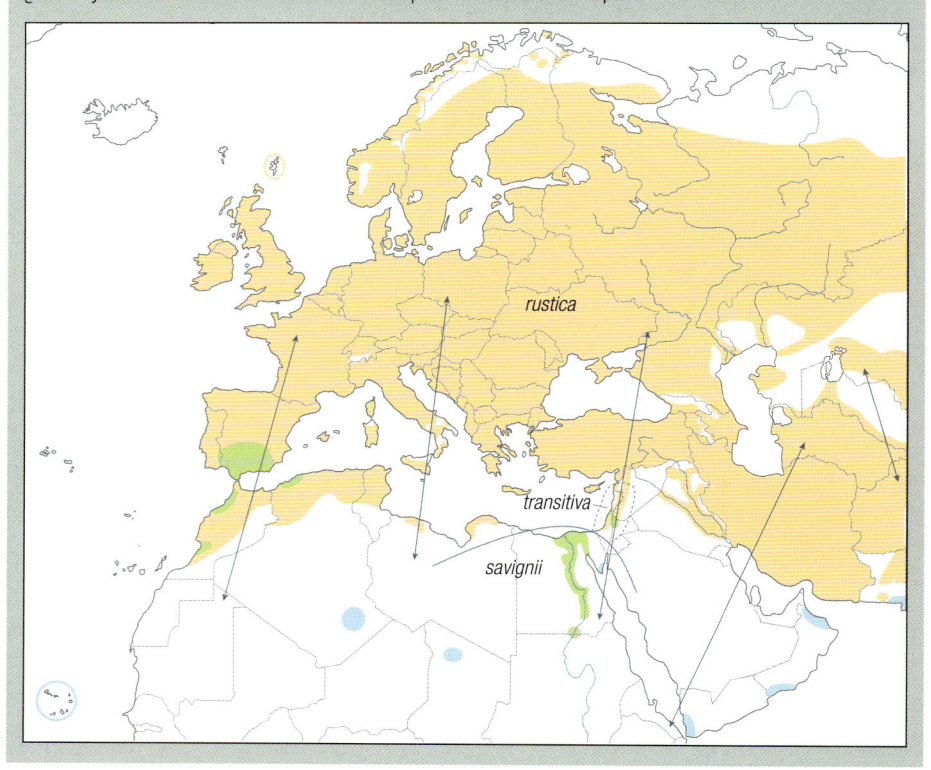

H. r. rustica, ♂, Finland, Jul: long thin tail-streamers; combination of metallic dark blue above and on breast-band, red forehead and throat patch, and snowy-white underparts make identification to subspecific level possible. Undergoes complete moult in winter quarters making ageing impossible thereafter. (M. Varesvuo)

IDENTIFICATION Readily told by *long, thin tail-streamers* on a *deeply forked tail*, pointed wings, *metallic dark blue upperparts* lacking any paler pattern (except tiny white spots on many outer tail-feathers, visible at close range when the tail is spread), *dark throat* in contrast to paler underparts, and fast, *agile flight with 'clipping' wingbeats* but very little fluttering and gliding on outstretched wings (as in House Martin or Red-rumped Swallow). When seen perched in good light the throat and forehead are dark rufous-red, the throat edged below by a glossy blue breast-band, but at a distance the throat appears just 'dark', the red being surprisingly difficult to see. Rest of underparts vary somewhat with geography, but in most of Europe they are white, sometimes with a faint buff or pink flush. In Egypt and Middle East the underparts darken to a saturated rufous. Young birds *lack the long tail-streamers*, having only a hint of them, and the *red on throat and forehead is much paler*; to separate a distantly flying bird from a House Martin, note above all the different mode of flight. Calls are useful, too.

VOCALISATIONS Song fairly loud and very cheerful, a fast twittering now and then interrupted by a drawn-out croaking or 'choking' sound, which can become a dry rattle. The only other species with a similar song structure is the Siskin. Delivered both from perch and in flight. – Calls loudly and frequently around its nest-site, be it a farm or a marina. In 'good mood' the conversation between preening birds perched on a roof-beam in a stable can recall Budgerigars' *Melopsittacus undulatus* cosy twitter. Normal contact call, mostly on wing, is a sharp but cheerful *vit!*, often repeated a few times. Alarm caused by a cat or Sparrowhawk is given by even louder disyllabic *see-flit!*

SIMILAR SPECIES Quite a characteristic bird that should not cause many identification problems. The only other swallow with pointed tail-streamers, the *Red-rumped Swallow*, has pale throat, a pale (rufous-buff) rump patch, a narrow rufous neck collar and all-black undertail-coverts. Its flight is also more like that of House Martin, with frequent glides on outstretched wings. – *House Martin* has a pure white rump patch, only moderately forked tail and by and large a white throat.

AGEING & SEXING Ages differ in 1stCY. Sexes usually differ slightly (after first complete moult), but a few are alike or at least very similar. Almost no seasonal variation. – Moults. Complete moult of both ad and juv protracted (to provide unhampered flight ability), for majority of European birds performed in winter quarters (Dec–Mar, sometimes starting earlier), but minority start with a few inner primaries while still in Europe, these apparently invariably being ad. For breeders in N Africa most start primary moult in summer. – **SPRING** ♂ Tail-streamers on average longer and thinner, and tail-fork thereby deeper, a difference often useful when breeding pair seen together. (See Biometrics.) On average more metallic blue gloss on upperparts and breast-band, but much overlap, and majority identical. ♀ Plumage as in ♂ (on average less metallic blue gloss of little practical use), but tend to have less uniform and deep rufous throat. Tail-streamers on average a little shorter and broader, and tail-fork thereby shallower. – **AUTUMN Ad** ♂ Much as ♂ in spring, with long tail-streamers. Forehead and throat usually deep rufous-red, but there is some individual variation, some birds presenting rather patchy mixture of old bleached feathers and new darker-coloured ones. Upperparts dark metallic blue with slight purplish tinge, some having rather worn and brownish feathers admixed. In some, blue breast-band narrower or almost broken, and one or both tail-streamers worn or broken, and such birds with maximum number of bleached and dull feathers resemble 1stCY at first glance. **Ad** ♀ Much as ad ♂, but tail-streamers (as in spring) on average shorter and slightly broader (though some overlap). **Juv/1stCY** Tail-streamers short and broad (on average a little longer and more pointed in ♂♂). Forehead and throat pale, light rufous or pinkish-buff. Breast-band sooty-grey tinged brownish, with no or only insignificant gloss. Upperparts mixture of blue and sooty-grey with limited metallic gloss, and tinge of blue more greenish than purplish. Early moulters may be difficult to separate from ad from late autumn onwards.

BIOMETRICS (*rustica*) **L** 15–21 cm (of which tail-streamers in ad c. 4–8); **W** ad ♂ 119–132 mm (n 35, m 126.5), ad ♀ 120–129.5 mm (n 13, m 124.6); **T** ad ♂ 91–130 mm (n 35, m 108.9), ad ♀ 81–106 mm (n 13, m 90.3), **1stCY** 60–72 mm (n 23, m 66.7); **TF** ad ♂ 48–85 mm (n 35, m 65.3), ad ♀ 34–60 mm (n 13, m 45.4), **1stCY** 16–30 mm (n 23, m 23.2; ♀ < 26); **tail-streamers** (width measured 15 mm from tip in ad, at tip of r5 in **1stCY**)

SWALLOWS AND MARTINS

H. r. rustica, ♂, Estonia, May: tail-streamers very thin and long. Note whitish (only slightly buff-tinged) underparts and underwing-coverts, and red throat-patch. (M. Varesvuo)

H. r. rustica, presumed ♀, Latvia, Jun: tail-streamers appear rather short and slightly broader than in previous images, thus probably ♀. Beside diagnostic red facial pattern and attractive glossy blue upperparts, note the neat pattern of white spots on outer tail-feathers. (M. Varesvuo)

H. r. rustica, ♂ (left: Denmark, May) and ♀ (right: Italy, Jun): small percentage of birds have some pinkish-buff hue below, even quite similar to paler *transitiva*. Such birds are most frequent in S Europe. (Left: P. Schans Christensen; right: D. Occhiato)

ad ♂ 1.3–2.8 mm (*n* 35, *m* 1.9), ad ♀ 1.8–3.1 mm (*n* 13, *m* 2.5), **1stCY** 3.2–5 mm (*n* 24, *m* 4.1); **B** 10.3–12.4 mm (*n* 37, *m* 11.6); **Ts** 10.2–11.5 mm (*n* 22, *m* 10.9). **Wing formula: p1** minute; **p2** longest; **p3** < wt 0.5–6 mm (rarely = wt); **p4** < wt 7–13 mm; **p5** < wt 12–22.5 mm; **p6** < wt 23–32 mm; **p10** < wt 56.5–66 mm; **s1** < wt 63–74 mm. No emarg. to outer webs of primaries.

GEOGRAPHICAL VARIATION & RANGE Moderate variation, the most obvious being amount of reddish colour on underparts. Largely clinal variation, and individual birds frequently impossible to name subspecifically. At least two extralimital races occur in E Asia, not covered here.

H. r. rustica L., 1758 (Europe, W Siberia, Caspian region, Turkey, Iraq; winters mainly in transequatorial Africa, commonly south to South Africa, but eastern population in S Asia). Treated above. Underparts (lower breast to undertail-coverts) white to creamy- or pinkish-white, but in south (e.g. Caucasus, E Turkey) some variation and intergradation evident into *transitiva*, with more saturated rufous-buff underparts. Birds closely matching *transitiva* found in many European countries in summer, making racial identification of such dark-breasted birds problematic. (Syn. *afghanica*; *ambigua*.)

H. r. transitiva (Hartert, 1910) (coastal Levant, from Syria to Gaza; some populations resident, others short-range migrants to Red Sea region but birds matching this plumage collected as far south as Kenya and Uganda). Has invariably reddish-buff underparts, never whitish, but darkness is somewhat variable. Interestingly, in Israel two distinct populations breed, summer visitors and residents, with the latter mostly found in NE valleys. These residents are the deepest red below, with some extremes closely approaching *savignii* (see below). On the other hand, some ad *transitiva*

H. r. rustica, ♀ feeding juv, Finland, Jul: tail-streamers shortest in juv (hardly longer than wingtips); dark plumage more sooty-grey, tinged brownish, while forehead and throat-patches are pale rufous or pinkish-buff. (M. Varesvuo)

H. r. rustica, ad, Ethiopia, Sep: much plumage variation among moulting ad in autumn; note partially broken breast-band, but rufous face still solid and intense. (H. Shirihai)

are paler, more pinkish-red and impossible to separate from warmer-toned *rustica*. Whether these paler birds are always part of the migratory population is not known. The least migratory populations perform complete post-nuptial moult, with many ad ending moult of remiges as early as (Sep) Oct–Nov, migratory moult Jan–early Mar. All measurements a fraction smaller than *rustica*. **W** ad ♂ 118–128 mm (n 13, m 124.5), ad ♀ 116–122 mm (n 7, m 119.4); **1stCY** 114–123 mm (n 9, m 119.8); **T** ad ♂ 70–113 mm (n 13, m 98.1), ad ♀ 69–95 mm (n 7, m 82.8), **1stCY** 61–68 mm (n 8, m 64.4); **B** 9.5–12.0 mm (n 19, m 11); **Ts** 9.5–11.5 mm (n 20, m 10.7). **Wing formula: p10** < wt 54–65 mm; **s1** < wt 60–69 mm.

H. r. savignii Stephens, 1817 (Egypt; resident). Rather distinct with deep reddish-brown underparts, approaching colour of throat. Slightly smaller than *rustica*, with shorter and broader tail-streamers. **W** ♂ 117–124 mm (n 15, m 119.6), ♀ 114–121 mm (n 13, m 117.0); **T** ♂ 90–105 mm (n 15,

H. r. rustica, juv, (left: Spain, Sep; right: Italy, Jun): pale forehead and light brownish- to pinkish-buff throat bordered by dark breast-band, and pale underparts and underwing-coverts with light brownish-buff suffusion. Tail-streamers short and broad. Upperparts are a mix of blue and sooty-grey with limited metallic gloss. (Left: M. Varesvuo; right: D. Occhiato)

H. r. rustica, juv, Italy, Jun: extensive variation in amount of red and blue in juv plumage. Even prior to moult in autumn, young can resemble winter-plumaged ad (in particular left and right birds). However, they have diffuse, mottled breast-band, and paler rufous face. (D. Occhiato)

H. r. transitiva, ♂, Israel, Apr: Levantine race characteristically has rich (albeit variable) pinkish-buff underparts, with extreme examples even deeper and darker red. (P. Hytönen)

m 96.8), ♀ 74–92 mm (n 13, m 80.6); **TF** ♂ 46–63 mm (n 15, m 52.3), ♀ 29–46 mm (n 13, m 35.8); **tail-streamers** (see above) ♂ 1.7–3.0 mm (n 15, m 2.3), ♀ 2.4–4.0 mm (n 13, m 3.2).

Extralimital races: **H. r. tytleri** Jerdon, 1864 (SC Siberia, N Mongolia; possible migrant through Iran and Arabia to E Africa, but main wintering area most likely in S Asia). Has narrow black breast-band at centre, intermediate in this respect between *rustica* and *erythrogaster*, and deep reddish underparts like *savignii*. Apparently distinct sexual dimorphism by size and tail-streamers, much more so than in *rustica*. **W** ♂ 115–125 mm (n 11, m 121.4), ♀ 106–120 mm (n 8, m 113.4); **T** ♂ 92–120 mm (n 11, m 108.3), ♀ 66–89 mm (n 8, m 77.7); **TF** ♂ 45–76 mm (n 11, m 63.6), ♀ 26–45 mm (n 7, m 34.6); **tail-streamers** (see above) ♂ 1.8–2.6 mm (n 11, m 2.1), ♀ 2.3–3.2 mm (n 7, m 2.7).

H. r. erythrogaster Boddaert, 1783 (North America; straggler to Azores). Has broken or narrow and washed-out bluish-black breast-band, generally forming two dark patches either side of breast with diffuse bridge in centre. Underparts variably tinged rufous-buff, strongest on vent and lower flanks. Slightly smaller than *rustica*. **W** ♂ 118–128 mm (n 12, m 121.3), ♀ 116–122 mm (n 10, m 117.0); **T** ♂ 82–99 mm (n 12, m 91.5), ♀ 71–83 mm (n 10, m 77.4); **TF** ♂ 40–61 mm (n 12, m 48.7), ♀ 26–36 mm (n 10, m 32.6); **tail-streamers** (see above) ♂ 1.8–2.9 mm (n 12, m 2.3), ♀ 2.5–3.5 mm (n 10, m 3.0). (Syn. *insularis*.)

TAXONOMIC NOTES The question of species status for American Barn Swallow *erythrogaster* was raised by Zink et al. (1995) and later by Sheldon et al. (2005) based on genetic research. The difference in mtDNA between *rustica* and *erythrogaster* is 1.7%. We feel it is best to keep the latter form within *H. rustica* until more data are available including also all taxa in E Asia. Morphologically, *tytleri* together with E Asian *gutturalis* Scopoli, 1786, and *saturata* Ridgway, 1883, offer a natural link between European and American taxa. – Whether ssp. *transitiva* winters south of Egypt is a matter of opinion. Birds with appropriate features have been recorded south to equatorial Africa, whereas virtually no records are known from Zimbabwe or South Africa. This pattern of occurrence would support the view that *transitiva* winters south of Egypt, but until there are recoveries of ringed birds it is always possible to regard these as unusually strongly pigmented *rustica*.

REFERENCES Dor, R. *et al.* (2010) *Mol. Phyl. & Evol.*, 56: 410–418. – Jiguet, F. & Zucca, M. (2005) *BW*, 18: 475–478. – Sheldon, F. H. *et al.* (2005) *Mol. Phyl. & Evol.*, 35: 254–270. – Zink, R. M. *et al.* (1995) *Condor*, 97: 639–649.

H. r. transitiva, ♀, Israel, Apr: sexed by less immaculate colours below and relatively short tail-streamers. (L. Kislev)

H. r. transitiva, ad, ♂ (left) and ♀ during complete post-nuptial moult, Israel, Sep: distinctive even in autumn. Clear distinction between summer visitor and resident populations, with the latter nesting earlier in spring and many ad completing post-nuptial moult as early as Sep/Oct. (A. Ben Dov)

SWALLOWS AND MARTINS

H. r. transitiva, ad, Israel, Apr: typical gathering on electric wire. (L. Kislev)

H. r. savignii, ad, presumed ♂, Egypt, Sep: almost evenly deep reddish underparts and throat; note long tail-streamers typical of ♂, although they seem untypically broad. (E. Winkel)

H. r. savignii, ad ♀, Egypt, Mar: ♀ tends to have reddish underparts visibly paler than throat (but some overlap between sexes); note shorter and relatively broad tail-streamers. (I. Weiß)

H. r. savignii, ad, Egypt, Jan: deep reddish colour also extends to underwing-coverts. Tail-streamers (one missing, the other short but narrow) do not allow safe sexing. (J. J. Nurmi)

H. r. savignii, juv, Egypt, Mar: even young birds have obvious reddish underparts, but tail-streamers are short (hardly reaching beyond wingtips), and upperparts more sooty-grey or brownish than blue. (I. Weiß)

H. r. tytleri, ♂ (left) and ♀, Mongolia, Jun: this race characteristically has narrow or broken breast-band concentrated on sides, and deep reddish underparts like darkest *transitiva* or even *savignii*. (H. Shirihai)

H. r. erythrogaster, imm and ad (left: Costa Rica, Dec) and 1stY (right: Venezuela, Dec): non-breeding plumages of N American race, in which a vagrant to W Europe is most likely to appear. Note warm reddish underparts and broken and thin dark breast-band. (H. Shirihai)

SWALLOWS AND MARTINS

WIRE-TAILED SWALLOW
Hirundo smithii Leach, 1818

Fr. – Hirondelle à longs brins; Ger. – Rotkappenschwalbe
Sp. – Golondrina colilarga; Swe. – Trådstjärtsvala

This relative of the Barn Swallow has a wide distribution both in tropical Africa and S Asia, being a short-range migrant in the north and largely resident in the south. Usually inhabits open country and clearings near water in lowlands and hills, and from montane grassland to savanna landscapes to an altitude of 2750 m. Has straggled several times to Oman in winter, and equally often to United Arab Emirates in winter and early spring. (A record in March 1995 in Egypt has recently been rejected.)

H. s. filifera, ad-like, United Arab Emirates, Apr: here with partially grown tail-streamers. Note that at certain angles the dark mark on the breast-sides can be partially hidden between neck and wing-bend. Moult and ageing has not been much-studied by us, but this bird is either an ad or a young bird that has extensively moulted completely in winter. (M. Almazrouei)

H. s. filifera, ad ♂, India, Aug: unlikely to be confused if combination of glossy blue-black upperparts and head-sides, rufous-chestnut crown, pure white underparts with blue-black mark on rear flanks, and diagnostic very long, wire-like tail-streamers, are seen. The latter also confirm the sex. (J. Parekh)

IDENTIFICATION Generally unmistakable, showing a combination of *glossy blue-black upperparts and side of head, rufous-chestnut crown, pure white underparts* with a *blue-black rear flanks mark* (from sides reaching to legs), and at least in adult ♂♂ diagnostic *very long wire-like and flexible tail-streamers*. Tail-streamers of ♀ short or half length. In juv, or birds missing the tail-streamers, the tail is by and large square-ended. Some birds have small dark pectoral patches at sides of breast, in extreme cases more extensive almost forming hint of breast band. Underwing-coverts and vent/undertail-coverts white.

VOCALISATIONS Has a rather twittering song (typical of hirundines). – Calls a variety of short unobtrusive notes, *twit* or *chit*, etc.

SIMILAR SPECIES Compared to the potential regional confusion species, Barn Swallow and Ethiopian Swallow, Wire-tailed Swallow has, just as its common name implies, very long and thin tail-streamers, particularly striking in adult ♂♂. ♀♀ and juveniles may be more easily confused with the other species, but should still cause few problems given the wholly chestnut crown and nape, and clean white throat and underparts.

AGEING & SEXING Ages differ in autumn but apparently not in spring. Sexes differ slightly in most (after 1st complete moult), but a few are alike or at least very similar. Almost no seasonal variation. – Moults. Little studied, but apparently both ad and 1stY moult completely before next breeding season. – **SPRING Ad** Shiny blue upperparts, including ear-coverts, chestnut crown and nape, and very clean white underparts, from throat to undertail-coverts, and extensive white on the undertail, as well as very long tail-streamers. Sexes similar, but ♀ has shorter tail-streamers. **1stY** Much like ad after post-juv moult. **Juv** Overall duller than ad, with a paler crown and considerably shorter tail-streamers.

BIOMETRICS (*filifera*) **L** 11–13 cm (exclusive of tail-streamers); **W** ♂ 112–125 mm (*n* 16, *m* 118.4), ♀ 107.5–121 mm (*n* 14, *m* 114.2); **T** (exclusive of streamers) ♂ 26–41 mm (*n* 16, *m* 34.5), ♀ 30–42 mm (*n* 14, *m* 35.5); **L tail-streamers** (measured from tip of r5) ad ♂ 92–157 mm (once 42 mm; *n* 16, *m* 115.8), ad ♀ 18–52 mm (*n* 14, *m* 29.7), **1stCY** 0–33 mm (*n* 7, *m* 9.3); **width tail-streamers** (measured 15 mm from tip in ad, within 10 mm from tip in 1stCY) 0.5–1.0 mm (*n* 30, *m* 0.8); **B** 9.0–12.4 mm (*n* 25, *m* 10.8); **Ts** 9.5–13.6 mm (*n* 25, *m* 11.1). **Wing formula: p1** minute; **p2**(3) longest; **p3** < wt 1.5–5.5 mm (rarely = wt); **p4** < wt 8–16 mm;

H. s. filifera, ad-like, India, Dec a pair showing dramatic difference in tail-streamer length, being much longer in ♂ (on left), and the deeper chestnut crown of this sex. (C. Robson)

p5 < wt 16–25.5 mm; **p6** < wt 23–33.5 mm; **p10** < wt 52–65 mm; **s1** < wt 57–72 mm. No emarg. of outer webs of primaries.

GEOGRAPHICAL VARIATION & RANGE Polytypic, two subspecies recognised, one Asian and one African, both extralimital. Vagrancy to Arabia seems to involve only ssp. *filifera*.

H. s. filifera L., 1758 (S Uzbekistan, S Tajikistan, Afghanistan through the Indian subcontinent east to S Vietnam; winters in southern parts of range). Treated above. Differs from ssp. *smithii* (Afrotropics) mainly through its darker chestnut (not light rufous) cap. Ad also has on average longer tail-streamers. (Syn. *bobrinskoii*.)

H. s. smithi Leach, 1818 (much of trans Saharan Africa from Gambia to Sudan and Eritrea, south to N South Africa; mainly resident. Similar to *filifera* but cap is lighter rufous and tail-streamers of ♂ are shorter.

H. s. filifera, juv, **United Arab Emirates, Feb**: duller than ad, with paler crown and considerably shorter tail-streamers. (H. Roberts)

H. s. smithii, ad-like, **Botswana, Aug**: very similar to Asian race, but chestnut crown averages paler, and ad has shorter tail-streamers. (L. Dürselen)

H. s. smithii, juv, **Tanzania, Apr**: even at a young age, rufous-chestnut crown and dark marks on rear flanks and breast-sides easily enable identification. (H. Shirihai)

RED-RUMPED SWALLOW
Cecropis daurica (Laxmann, 1769)

Fr. – Hirondelle rousseline; Ger. – Rötelschwalbe
Sp. – Golondrina dáurica; Swe. – Rostgumpsvala

In Europe, this charming swallow is confined to the Mediterranean region, often occurring in mountains or in rocky areas near the sea, but it is also widely distributed in Africa, Middle East and Asia. It builds an igloo-shaped nest of mud, fixed snake-proof to the roof of a cave, recess, road tunnel or under a stone bridge. Usually a rather scarce bird, rarely seen in great numbers. Summer visitor to Europe which migrates to W & NE Africa to winter.

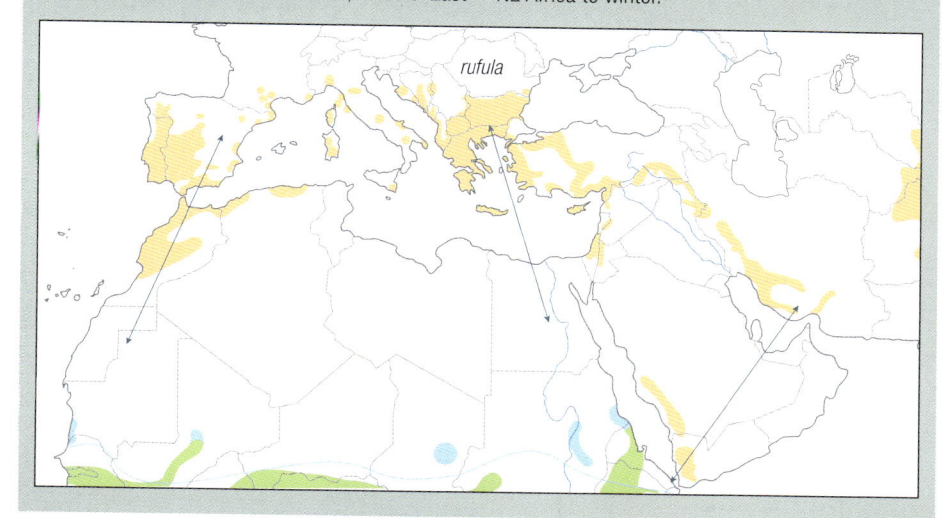

C. d. rufula, Israel (top: Mar; bottom: Sep): variation in amount of streaking on underparts, with top bird showing thin streaking, while bottom bird is unstreaked. Orange on head-sides and collar visible from most angles. Black undertail and undertail-coverts squarely delimited from pale body. (A. Ben Dov)

C. d. rufula, Israel, Mar: fresh plumage after complete winter moult. Rufous-orange collar and rump make species unmistakable in the covered region. (H. Shirihai)

IDENTIFICATION Whereas shape is very similar to that of Barn Swallow, with *pointed wings* and *long, pointed tail-streamers*, flight is a little reminiscent of House Martin with frequent *glides on outstretched wings*, not as fast with active wingbeats as Barn Swallow. As its name implies has a large *rufous-buff* or *rusty-red rump patch*, which can appear bicoloured in that the rear part often is paler, whitish, whereas the foremost is more rusty. Due to this, the rump can look whiter at a distance than it is (and it *is* whiter on juv). The *cheeks are rusty-red* too, and this colour encircles the blue-black crown and joins narrowly at the back of the neck (the rufous collar is not as easy to see in the field as the rump patch). A useful field mark when flying birds are seen from below is the *jet-black undertail-coverts*, which seem to separate tail squarely from body. And compared to Barn Swallow it lacks any white 'windows' at the base of the tail, which is *solidly dark*. Throat and chest are pale like rest of underparts, being *light pinkish-buff* with very discreet, narrow streaking (visible only close-up). Mantle and inner upperwing-coverts glossy blue-black but rest of upperwing dull black (Barn Swallow has some bluish gloss also on coverts and flight-feathers).

VOCALISATIONS Song has some resemblance to Barn Swallow, containing some strained and croaking sounds, but the pitch is lower, the pace slower and the strophes more disrupted and hard (can recall Dipper in general structure and tone). Delivered both from perch and in flight. – Contact call a cheerful, rather subdued or soft, somewhat nasal *tveyk*, which can vaguely recall the common call of Tree Sparrow. A territorial call by the ♂ has been rendered as a miaowing sound. Alarm is a repeated short *ki* or a *kii-e* with falling pitch. Another call is a quarrelling *krr*.

SIMILAR SPECIES Should usually be straightforward to identify within treated range (but beware of several similar-looking species in Africa and Asia). The only other W Palearctic swallow with pointed tail-streamers, the *Barn Swallow*, has dark throat, all-dark upperparts and white or rufous (not black) undertail-coverts. – *House Martin* has white undertail-coverts, black cheeks and nape, a pure white rump patch (beware that juv Red-rumped may appear white-rumped), only moderately forked tail and by and large all-white underparts.

AGEING & SEXING Ages differ in 1stCY. Sexes differ usually very slightly (after first moult), but a few are alike. – Moults. Complete moult of both ad and juv for majority of European birds apparently entirely in winter quarters (Oct–Jan), but exceptionally innermost primary moulted in late Sep while still near breeding area. No post-nuptial moult of body-feathers in summer, at least not in Europe. – **SPRING** ♂ Tail-streamers on average longer, and tail-fork thereby deeper, a difference sometimes useful when breeding pair seen together (but much overlap; see Biometrics). ♀ Plumage as in ♂. Tail-streamers on average a little shorter, and tail-fork therefore shallower. – **AUTUMN** Ad ♂ Much as ♂ in spring, with long tail-streamers. Upperparts (except rump and back of neck) glossy blue-black. Rump rather saturated rufous-buff. **Ad** ♀ As ad ♂, but tail-streamers (as in spring) on average shorter. **1stCY** Tail-streamers somewhat shorter and broader than in ad. Upperparts, especially crown, have no or little gloss. Tertials and inner greater coverts tipped rusty buff-white. Flight-feathers dull brown-grey.

BIOMETRICS (*rufula*) **L** 17–19 cm (of which tail-streamers *c.* 4–5); **W** ♂ 119–129 mm (*n* 23, *m* 123.4), ♀ 114–127 mm (*n* 18, *m* 120.0); **T** ad ♂ 94–112 mm (*n* 22, *m* 101.3), ad ♀ 80–100 mm (*n* 18, *m* 92.0); **TF** ad ♂ 51–69 mm (*n* 23, *m* 57.7), ad ♀ 38–56 mm (*n* 18, *m* 48.2); **width tail-streamers** (measured 15 mm from tip) ad ♂ 2.6–3.8 mm (*n* 23, *m* 3.1), ad ♀ 2.8–4.7 mm (*n* 18, *m* 3.5); **B** 9.3–11.6 mm (*n* 18, *m* 10.5); **Ts** 12.5–14.0 mm (*n* 17, *m* 13.1). **Wing formula: p1** vestigial; **pp2–3** equal and longest, or p3 < wt 1–4 mm; **p4** < wt 6–10 mm; **p5** < wt 14–19 mm; **p6** < wt 22–27 mm; **p10** < wt 55–62 mm; **s1** < wt 62–70 mm. No emarg. of outer webs of primaries.

GEOGRAPHICAL VARIATION & RANGE Marked and complex variation, with at least nine distinct subspecies described. Only one occurs regularly within treated range, but two more Asian taxa, treated below, have straggled to Norway, and one of them (unknown which) to Scotland.

C. d. rufula (Temminck, 1835) (Europe, NW Africa, W & S Turkey, Middle East, W, S & N Iran, parts of Central Asia including much of Tien Shan and Afghanistan; winters mainly in sub-Saharan Africa north of forest zone, also SW Asia). Treated above. Streaking on pale ochre-buff underparts always very light, streaks being very thin, often concentrated on breast. Rufous neck-collar narrow but complete in virtually all birds. Tail-streamers long.

C. d. rufula, Israel, Mar: rufous collar and buffish underparts, contrasting with blue cap and upperparts, and blackish wings. Fresh, following complete winter moult. (D. Occhiato)

C. d. rufula, Finland, May: narrow body-streaking, complete orange collar and bicoloured rump identify this bird as *rufula*, rather than one of the Asian races, which have also wandered to N Europe. Some (like this bird) have contrastingly whiter uppertail-coverts, especially with wear, producing white patch near tail-base. (H. Taavetti)

C. d. rufula, Israel, Apr: fresh plumage, after complete winter moult. Note fine streaks below. (R. Pop)

C. d. rufula, Egypt, May: some even show almost pectoral-like breast-streaking in spring, mirroring juv plumage, but as species undertakes complete winter moult it is impossible to relate this feature to age. (E. Winkel)

EXTRALIMITAL RACES THAT HAVE STRAGGLED TO EUROPE:
C. d. daurica (Laxmann, 1769) (S Siberia and NE Kazakhstan east through N Mongolia to Russian Far East including W Amur, south to NE China; winters S & SE Asia). Differs from *rufula* in being somewhat larger, and having more distinct streaking below, ground colour of underparts being less evenly ochre-buff but more patchily rusty-tinged off-white. Has incomplete rufous collar across nape (dark blue 'bridge' at centre of nape). Rump all rufous, very finely streaked (usually invisible unless handled), lacking whitish rear of most *rufula*. **L** 19–20.5 cm; **W** ♂ 127–136 mm (*n* 10, *m* 131.5), ♀ 123–129 mm (*n* 9, *m* 126.2); **T** ad ♂ 96–116 mm (*n* 10, *m* 107.3), ad ♀ 80–106 mm (*n* 9, *m* 96.7); **TF** ad ♂ 50–71 mm (*n* 10, *m* 60.5), ad ♀ 48–61 mm (*n* 8, *m* 53.9). **Wing formula: p10** < wt 55–62 mm; **s1** < wt 63–70 mm. (Syn. *gephyra*; *tibetana*.)

C. d. japonica (Temminck & Schlegel, 1845) (E Amur, Japan, Korea, E & S China; winters SE Asia, N Australia). Resembles *daurica* but is smaller and has even stronger black streaking on off-white underparts unevenly tinged rufous-buff. Uniformly rufous rump has fine streaking, usually more distinctly so than in *daurica*, often visible in good field views. Upperparts on average subtly darker than in *daurica*, and sides of head often more densely streaked dark. Some birds on migration and winter grounds difficult to safely separate from *daurica*. **W** ♂ 116–126 mm (*n* 14, *m* 121.5), ♀ 113–121 mm (*n* 13, *m* 117.1); **T** ad ♂ 89–107 mm (*n* 12, *m* 100.3), ad ♀ 79–99 mm (*n* 13, *m* 90.4); **TF** ad ♂ 50–60 mm (*n* 12, *m* 56.5), ad ♀ 35–55 mm (*n* 12, *m* 3.5).

REFERENCES THORNE, R. & THORNE, S. (2014) *BB*, 107: 405–412. – TVEIT, B. O. (2011) *BW*, 24: 327–341. – WILSON, J. D. et al. (2006) *Ring. & Migr.*, 23: 57–61.

C. d. rufula, Israel, Sep: fresh juv, with denser streaks on breast, tail-streamers shorter, upperparts duller blue, and some coverts tipped buff-white. Yellow-white corner of gape, typical of recently fledged juv, still obvious. (A. Blomdahl)

C. d. daurica, Khabarovsk, Russia, Jun: two eastern races have been recorded in N Europe; they differ from *rufula* in having much stronger underparts streaking and incomplete rufous collar (though dark blue 'bridge' is often broken and/or very narrow), but mostly by all-rufous rump. (N. Loginov)

C. d. japonica / daurica, Scotland, Jun: note partial break in rufous collar, uniform orange-rufous rump and well-developed streaking below. Biometrics are often required to separate *japonica* from *daurica*. (I. Fulton)

STREAK-THROATED SWALLOW
Petrochelidon fluvicola (Blyth, 1855)

Alternative name: Indian Cliff Swallow

Fr. – Hirondelle fluviatile
Ger. – Braunscheitelschwalbe
Sp. – Golondrina india; Swe. – Indisk svala

Seems now an almost annual vagrant to the region's south-eastern corner (United Arab Emirates, Kuwait and Oman), but also once to Egypt; possibly previously overlooked. Breeds mainly in the W & C Indian Subcontinent. Western limit is in tropical Jalalabad area (E Afghanistan) and the N Indus valley (Pakistan), fairly close to the treated region. Usually frequents open hilly country with crags and cliffs, but sometimes associated with human habitation.

Oman, Dec: short, shallow-forked and streamer-less tail; glossy blue upperparts, contrasting with lightly mottled brownish rump and rufous-orange cap make species unmistakable in the treated region. (H. & J. Eriksen)

Oman, Dec: a distinctive swallow with densely streaked throat (tinged dusky) becoming broader but irregular lines on breast, which is best distinction from other hirundines; also note rufous crown. (H. & J. Eriksen)

India, Mar: small and compact, with a short, streamer-less tail; upperparts glossy blue, contrasting with chestnut crown to nape; many show blotchy white streaks on mantle. A breeder, but as moult and ageing of this species has not been studied closely by us, we do not attempt to age any of the birds shown here. (S. Singhal)

IDENTIFICATION A distinctive swallow, and as its name suggests, the *streaked throat* is best distinction from several similar more compact martins and swallows. Characteristically *small and compact*, with a *short, shallowly forked tail* lacking tail-streamers. *Upperparts glossy blue*, contrasting with the lightly *streaked chestnut crown and nape* and brownish wings, uppertail and rump. Sides of face and entire *underparts buff-white with narrow but extensive black-brown streaking* (heaviest on face and upper breast). Undertail-coverts vary from plain white, through white with brown shafts to some having extensive brown markings on longest undertail-coverts. Many birds show *blotchy white streaks or patches on mantle* (exposed feather bases). Much due to its small size has a weak and *fluttering flight, appearing slower* than most other swallows. Highly gregarious.

VOCALISATIONS Song a series of deep-voiced churring notes lacking clear pattern. – Call a very dry, crackling, short, abrupt, slightly rolled *brrrt, brrrt* ... , given in short quick irregular series. Most similar perhaps to Sand Martin, although not quite as dry and rasping. There is also a similar but flatter, more trilled, rude-sounding *bur'r'r't*.

SIMILAR SPECIES Combination of chestnut cap, blue-glossed upperparts, heavy black-streaked throat, breast and upper flanks on dirty-white ground and lack of tail-streamers should prevent confusion with other swallows.

AGEING & SEXING **Ad** Sexes similar. **Juv** Predominantly brown with buff fringing: crown browner than in ad, and mantle and wings brown-toned with buff fringes (most obvious on scapulars and tertials).

BIOMETRICS Sexes of nearly same size. **L** 10–11.5 cm; **W** ♂ 87.5–97 mm (n 12, m 91.0), ♀ 86–94.5 mm (n 12, m 90.1); **T** 33–44 mm (n 23, m 38.4); **T/W** m 42.1; **B** 6.9–9.9 mm (n 24, m 8.2); **Ts** 10.0–12.4 mm (n 19, m 11.0). **Wing formula: p1** minute; **p2** longest or = p3; **p3** < wt 0–2 mm; **p4** < wt 3.5–8 mm; **p5** < wt 8–15 mm; **p6** < wt 15–22 mm; **p10** < wt 38–45 mm; **s1** < wt 44–49 mm.

GEOGRAPHICAL VARIATION & RANGE Monotypic. – Afghanistan, C Pakistan, India; winters S India.

India, Nov: streaked throat and breast offers best distinction from other hirundines; underwing rather dark, contrasting with white body. (S. Sen)

United Arab Emirates, Jan: at certain angles and when fresh, streaked throat can appear like dusky bib; there is some slight variation in degree and pattern of streaks on breast. (S. Lloyd)

United Arab Emirates, Feb: weak, fluttering flight, appearing slower than most swallows, partially attributable to compact structure. Note streaked throat and breast, and contrasting chestnut crown. Blotchy white streaks on mantle can form a patch. (H. Roberts)

(AMERICAN) CLIFF SWALLOW
Petrochelidon pyrrhonota (Vieillot, 1817)

Fr. – Hirondelle à front blanc; Ger. – Fahlstirnschwalbe
Sp. – Golondrina risquera; Swe. – Stensvala

This widespread American swallow has straggled to the W Palearctic, notably several times to Britain, but also to Ireland, France, Sweden, Iceland and Tenerife in the Canary Islands. To a European it might at first appear like a Red-rumped Swallow that has lost its tail-streamers. Breeds in dense colonies. Migrates to South America in late summer to spend the winter.

P. p. pyrrhonota, ad/2CY, Sweden, Nov: in post-breeding moult. Short-tailed swallow, in ad plumage, comprising white forehead, deep rufous-red cheeks to neck, and large rufous-buff rump. (I. Kruys)

IDENTIFICATION Rather a *compact* swallow, roughly the size of a House Martin, with *square, broad tail* (becoming rounded when spread) and broad but pointed wings. Flight is not particularly fast, and it often soars and glides rather leisurely on outstretched wings. *Upperparts largely brownish-black with bluish gloss*, relieved by a *large rufous-buff rump patch*, paler collar (both greyish and rufous) and some *irregular whitish streaks on mantle and back*. When seen well there are some paler brown feather bases visible above, and wing-feathers including tertials are dark brown. *Forehead* has a buffish-white or *white patch*, which is quite noticeable on adults but darker and more subdued in juveniles. *Chin, cheeks and sides of neck are deep rufous-red* (again duller and more brownish in juveniles), and on *centre of lower throat* there is a *blackish patch*. Upper and undertail-coverts are dark brown edged whitish. Sexes alike. Juvenile less glossy blue above, more dull and dark grey, and rufous and black on cheeks and throat is duller, more brown-grey.

VOCALISATIONS Calls include a nasal, sparrow-like *chiev* and a low *chrr*.

SIMILAR SPECIES First, a juvenile Red-rumped Swallow (or a hypothetical adult that has lost its tail-streamers) needs to be eliminated. Note on Cliff Swallow the square tail, whitish streaking on mantle/back, and the lack of black undertail-coverts. Also, the dark cap reaches just below the eye (well above eye in Red-rumped). – There is an outside chance that a *hybrid* House Martin × either of Barn or Red-rumped Swallow might resemble a Cliff Swallow, but this possibility should be eliminated by carefully going through the characters listed under Identification. – The real pitfall is offered by a close American relative, the Cave Swallow (*P. fulva*; not yet recorded in W Palearctic; Hough 2000). This has same size, jizz and flight, and shares most of the plumage characters of Cliff Swallow, but differs in having a dark rufous forehead patch (which is therefore less striking at a distance), a more dark-capped look (the cap of Cliff merges more into sides of neck), paler and purer orange throat and neck-sides, and lacks a blackish patch on centre of throat.

AGEING & SEXING Ages differ in 1stCY. Sexes alike (although ♂ said to have on average larger blackish throat patch than ♀; Pyle 1997). – Moults. Complete moult of both ad and juv in winter quarters (flight-feathers replaced Nov–Mar), but post-nuptial moult starts already in summer or during autumn migration with body-feathers. – **SPRING** Sexes and ages alike. – **AUTUMN Ad** Much bluish gloss on crown and mantle. Forehead patch buffish-white and well marked. Whitish streaks on mantle/back fairly obvious. Cheeks and neck-sides deep rufous-red. Tertials thinly edged white. **1stCY** Very little bluish gloss above and virtually none on crown. Forehead patch sullied brown, less distinct. Pale streaks on mantle/back diffuse. Cheeks and neck-sides dull brown-grey with little or no rufous. Tertials edged rufous-buff.

P. p. pyrrhonota, ad/2CY, California, USA, autumn: worn plumage in autumn; with wear, rufous and buff on forehead, collar and rump can become duller or washed-out. (S. N. G. Howell)

BIOMETRICS (*pyrrhonota*) Sexes of nearly same size. **L** 12.5–14 cm; **W** 101–119 mm (n 24, m 111.5); **T** 47–54 mm (n 24, m 50.1); **T/W** m 45.0; **B** 9.4–11.3 mm (n 24, m 10.4); **Ts** 12.2–13.5 mm (n 15, m 12.7). **Wing formula: p1** minute; **p2** longest or = p3; **p3** < wt 0–2 mm; **p4** < wt 5–9 mm; **p5** < wt 11.5–17 mm; **p10** < wt 46–53 mm; **s1** < wt 52–60 mm.

GEOGRAPHICAL VARIATION & RANGE Moderate and mostly clinal, with several subspecies described, but only *pyrrhonota* thought to have occurred in W Palearctic. The others breed in W & SW North America and Middle America.

P. p. pyrrhonota (Vieillot, 1817) (Alaska, Canada, USA east of the Rockies except in the south and south-west; winters S South America). Treated above. Slightly larger than other races. Forehead patch pale, whitish or pale buffish (slightly or much darker and more rufous in some other races). Underparts often slightly dusky greyish-white, but variable. (Syn. *lunifrons*.)

REFERENCES Hough, J. (2000) *BW*, 13: 368–374.

P. p. pyrrhonota, 1stW, France, Oct: broad- and round-winged structure, and distinctive plumage, especially rufous-red cheeks, buff collar and large rufous-buff rump, make this rare visitor almost unmistakable. Aged by wear to unmoulted primaries, secondaries and tail-feathers, pattern and moult limit in tertials, lack of white on forehead, and by extension and moderate intensity of metallic blue hood. (A. Audevard)

P. p. pyrrhonota, Alaska USA, Jun: note diagnostic rufous collar and throat, and contrasting buffish-white forehead against bluish crown. (R. Pop)

(COMMON) HOUSE MARTIN
Delichon urbicum (L., 1758)

Fr. – Hirondelle de fenêtre; Ger. – Mehlschwalbe
Sp. – Avión común; Swe. – Hussvala

One of the best-liked wild birds of all! A welcome sign of spring when it returns from Africa, confiding and living close to man, often returning to the same house, where it builds its cup-shaped mud nest under roof eaves, lively and spirited, dashing to and fro in elegant flight. Charming, its cheerful twitter is ready compensation for the droppings on the windowsill. When the House Martins have left again for Africa in late August and through September, it becomes rather empty and quiet around the house.

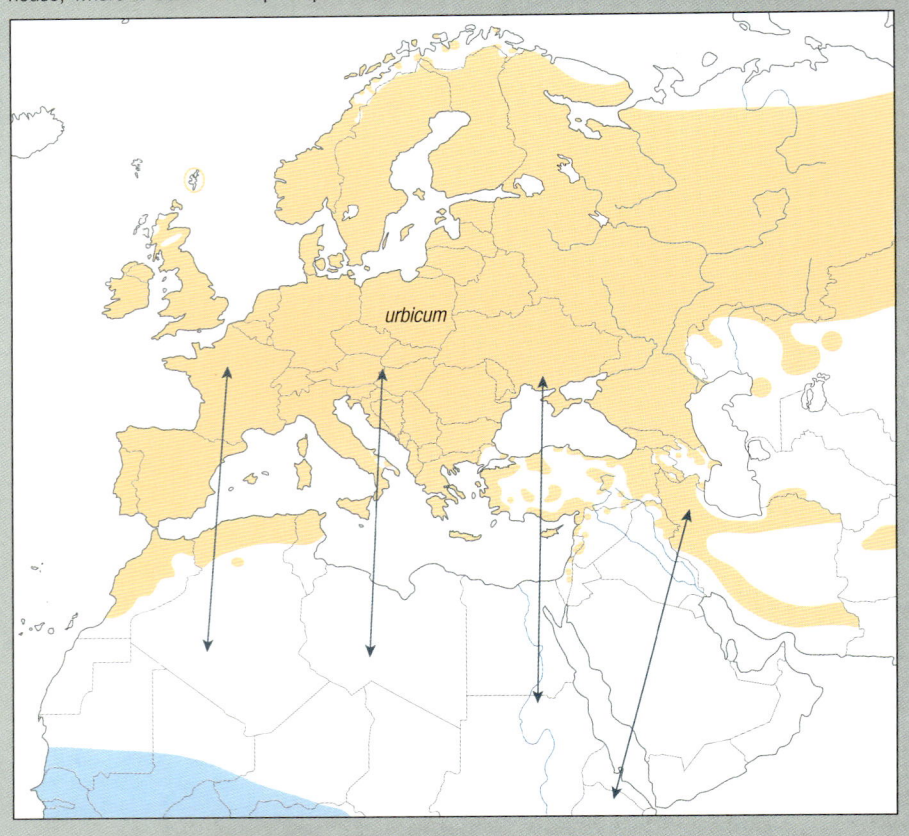

IDENTIFICATION Dressed up in black-and-white evening dress all year: *upperparts black* with bluish metallic gloss except for *large pure white rump patch*; all-white underparts, including throat and undertail-coverts (although throat and breast on average more sullied dirty brown-white in ♀). The *black tail* is rather short and narrow and *moderately forked*. Wings pointed but comparatively broad. In flight told by frequent spells of quickly fluttering wingbeats relieved by short glides on outstretched wings. Tarsi and even toes are neatly feathered white in grouse fashion. Sexes similar. Juvenile similar to adult but has less metallic gloss above and throat is dusky white, not pure white.

VOCALISATIONS Quite noisy when many pairs breed together; then there seems to be constant reasons for arguments! The song sounds cheerful, a modest but fast and energetic, dry twittering, lacking structure (to the human ear!). Mainly given when perched, but brief bursts can be delivered on wing. – Call a dry rolling or chirping *prrit*, repeated and varied in details. When alarmed (cat, Sparrowhawk *Accipiter nisus*) it utters a repeated emphatic and more drawn-out *chierr*.

SIMILAR SPECIES Should be straightforward to identify once mode of flight and white rump patch are noted. A distant juvenile Red-rumped Swallow could theoretically appear to have a white rump patch and no tail-streamers, but once a House Martin is seen close and for long enough it should be obvious that it has all-dark crown and back of neck, white undertail-coverts, and somewhat different jizz and flight mode. – Extralimital *Asian House Martin* (see vagrants section, p. 624) is very similar but fractionally smaller and more short-tailed and compact, has a more fluttering and 'light' flight and contrasting blackish underwing-coverts. White rump patch averages smaller than in Asian race *lagopodum* of House Martin but it has similarly shallowly forked tail as that. It may be prudent to point out that many of these characters would fit at least a few extremes of recently fledged Common House Martins. These can rarely have quite dark-looking underwings due to incomplete growth of white coverts, making it vital to secure the fullest possible description. At least ad have a narrow black rim below bill on chin and under gape.

AGEING & SEXING Ages differ in 1stCY. Sexes very similar, often inseparable. – Moults. Complete moult of both ad and juv for majority of European birds in winter quarters (Nov–Apr), but minority start with a few inner primaries and some tertials while still in Europe. Towards the end of the protracted moult of flight-feathers in late winter, most replace an unknown number of body-feathers, apparently a substantial part or even all. – **SPRING** ♂ Underparts, especially chin and throat, and rump pure white (latter rarely has very fine dark shaft-streaks). Blue gloss on uppertail-coverts strong. ♀ Either similarly white on underparts and rump as ♂, or, more commonly, recognised by having chin, throat and in particular upper breast tinged brown-grey, rarely also rump (often with a few brown blotches and dark shaft-streaks) and undertail-coverts. Blue gloss on uppertail-coverts usually very subdued, confined to narrow fringe, and gloss weak. – **AUTUMN Ad** Tertials either faintly tipped white or entirely black. Bill all black. Crown has some bluish gloss even when

D. u. urbicum, England, Jun: all-white underparts and rump, contrasting with metallic blue upperparts and black mask are unmistakable. All birds undergo complete moult in winter making ageing impossible in spring. (I. Dickey)

D. u. urbicum, England, May: wings characteristically swept back; underwing-coverts often paler/whiter than remiges, but degree of contrast changes with light and angle. (B. Liggins)

D. u. urbicum, Denmark, Jun: contrasting dark cap and obvious tail fork. ♀ (like this) typically sullied brownish on chin, throat and sometimes breast. (P. Schans Christensen)

SWALLOWS AND MARTINS

D. u. urbicum, ad, Estonia, Jun: contrastingly brown wings can appear more worn than they really are. Possible ♂ given very pure white chin and throat (but some variation and overlap between sexes). (M. Varesvuo)

D. u. urbicum, juv, England, Sep: brownish upperparts (very little gloss), and broadly white-tipped tertials. (S. Oakes)

D. u. urbicum, juv, England, Jul: brownish nape and limited blue gloss above. White rump also diagnostic for juv, but when very fresh it is variably sullied brown-grey. (D. Tipling)

D. u. urbicum, juv, England, Aug: chin, throat and upper breast, as well as body-sides, tinged brown-grey, while sides of undertail-coverts are mottled. (S. Round)

D. u. lagopodum, Thailand, Apr: distinctive by having far larger white rump that even covers larger uppertail-coverts; tail-fork relatively shallower than ssp. *urbicum*. (A. Jearwattanakanok)

worn. At least ♂ has chin and throat pure white (see Spring as to sexing). **1stCY** Tertials broadly tipped white. Base of lower mandible yellowish. Crown greyish-black without gloss. Chin, throat and upper breast (at least sides of breast), often also rump, sullied brown-grey.

BIOMETRICS (*urbicum*) **L** 14–15 cm. Sexes similar size, but ♀ on average fractionally larger (sic!). **W** 107.5–117 mm (*n* 38, *m* 113.7); **T** 60–66 mm (*n* 38, *m* 63.3); **TF** 15–23 mm (*n* 38, *m* 18.6); **B** 8.6–10.2 mm (*n* 26, *m* 9.2), **B**(f) 6.8–8.1 mm (*n* 20, *m* 7.3); **Ts** 10.5–12.3 mm (*n* 32, *m* 11.3). Wing formula: $p1$ minute; $p2$ longest (or $pp2$–3 longest); $p3$ < wt 0.5–3 mm (or = wt); $p4$ < wt 7–11 mm; $p5$ < wt 15.5–21 mm; $p6$ < wt 23–30 mm; $p10$ < wt 54–59 mm; $s1$ < wt 59–64 mm. No emarg. of outer webs of primaries.

GEOGRAPHICAL VARIATION & RANGE Moderate and clinal, with two distinct subspecies described. Only one occurs within the treated range, the other as far as known only in E Asia, west to the Altai.

D. u. urbicum (L., 1758) (Europe, N Africa, Turkey, Middle East, W Central Asia). Treated above. Southern birds tend to be subtly smaller ('*meridionale*' described from Algeria and Morocco) but this part of cline lacks marked steps, and hence best included here. (Syn. *fenestrarum*; *meridionale*; *vogti*.)

Extralimital: *D. u. lagopodum* (Pallas, 1811) (Yenisei basin, Altai, NW Mongolia, NW China). Differs in being slightly smaller and more compact in shape, having much larger white rump patch extending to uppertail-coverts and somewhat further up on back, and also in dusky brown-grey underwing, often looking all-dark in the field. Tail-fork shallower than in *urbicum*. Differs from Asian House Martin (p. 624) in larger white rump patch extending to lower back and diagnostically to uppertail-coverts, and perhaps on average slightly paler underwing-coverts (lacking contrastingly darker coverts) and subtly more forked tail (but not a substantial difference). (Syn. *alexandrovi*; *whiteleyi*.)

TAXONOMIC NOTES Closely related Asian House Martin *D. dasypus* was formerly treated as conspecific with Common House Martin, but the two overlap in Transbaikalia and south to NW India, apparently without interbreeding. They differ in several morphological traits and are hence better treated as separate species (see Vagrants, p. 624). – Note that ssp. *lagopodum* of Common House Martin is suspected to locally overlap with *urbicum* in NW China and possibly in some areas of NW Mongolia and Russian Altai, but this needs to be confirmed and better studied; it would have obvious taxonomic consequences if established.

REFERENCES Hill, L. A. (1992) *Ring. & Migr.*, 13: 113–116.

D. u. lagopodum, Buryatia, Russia, Jun: from above (left) note huge white rump, almost 2–3 times wider than exposed tail base (unlike *urbicum* in which it is 1–1.5 times); from below note dusky brown-grey underwing. Rather compact shape, with relatively shallower tail-fork. (M. Hellström)

RICHARD'S PIPIT
Anthus richardi Vieillot, 1818

Fr. – Pipit de Richard; Ger. – Spornpieper
Sp. – Bisbita de Richard; Swe. – Större piplärka

One of the most regular Siberian visitors to NW Europe in autumn; experienced observers will locate it by virtue of its unique flight-calls. Richard's Pipit breeds from C Siberia eastwards, and mainly winters in S Asia, but it frequently wanders west to Europe and N Africa in autumn, with small numbers recently discovered wintering in Spain, Italy, Israel, Turkey and Arabia. It breeds on drier bogs and fields and other open grassy land in Siberia, Mongolia and N China.

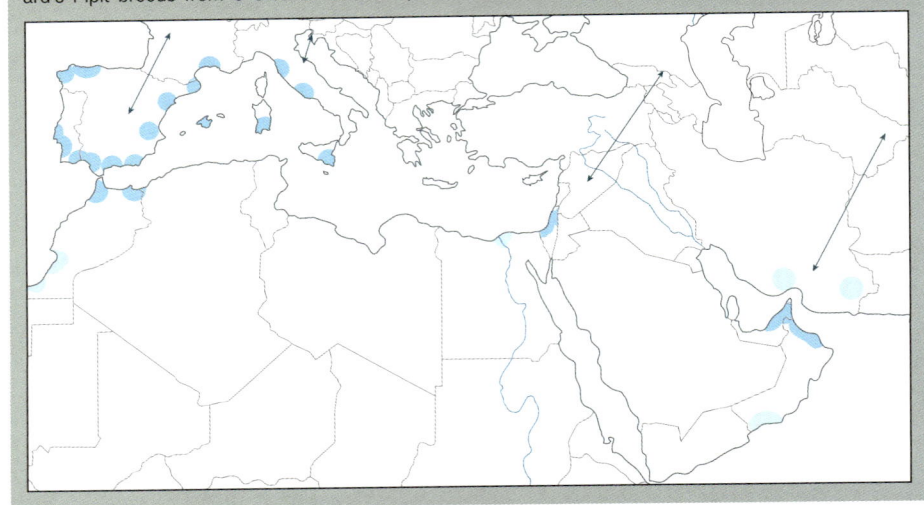

1stS, Netherlands, Mar: some retain many juv wing and scapular feathers (heavily worn and bleached); unmarked pale lores. Can appear rather compact, especially when not alert. (T. Beeke)

Sweden, May: heaviest-built pipit, with somewhat thrush-like bill (typically often held slightly upwards), long tail and very long legs. With feather wear, crown to mantle streaking becomes stronger black-brown. Some can have narrow dark grey lores. Exposed wing-feathers all appear ad-like, though it could equally be 1stY that has moulted extensively in winter. (D Erterius)

Mongolia, Jun: especially when alert, appears typically erect, but long legs can be hidden by grass. Note rather strong breast-streaking and face markings. Despite some detectable moult limits in wing (e.g. outer greater coverts unmoulted), ageing uncertain. (H. Shirihai)

IDENTIFICATION The largest and *heaviest-built pipit in the treated region, heavily streaked dark brown and buff above, with a somewhat thrush-like bill* (lower mandible often slightly angled upwards from gonys, and a rather strongly downcurved upper mandible near tip), long tail, and *very long legs and hind claws*. Size and plumage vaguely recall Skylark (a well-publicised identification pitfall for the unwary), but the main identification issues are with other large pipits. When fresh or moderately worn, crown to mantle buffish dun-brown streaked black-brown, becoming less streaked on lower back and rump, and dark brown tail has buff-brown central feathers and broadly buff-white edged outermost. *Broad, unmarked pale cream-buff lores, eye-ring and long, broad supercilium*, buff-brown ear-coverts with dusky eye-stripe behind eye and narrow brown-black moustachial stripe and pale lower cheeks; also has pale surround to head-sides and *long brown-black lateral throat-stripe that usually terminates in black streaks on upper breast-sides*. Tends to have dark lateral crown-stripes (though rather diffuse and not invariably present). Underparts mostly creamy whitish but *variably buffish on breast and flanks, with sparse black-brown streaks over rest of breast*. Wings generally brown, with conspicuous buff to almost whitish fringes to coverts and tertials, but darker-centred median coverts and pale wing-bars usually less obvious. Bill dark horn-brown, paler ochre-grey on cutting edges and base of lower mandible, and legs and feet dull yellow-flesh. Flight powerful, with long, shallow undulations, bursts of flapping wingbeats and, *on landing, may first hover well above ground* (especially in tall grass). *Long legs often noticeable on landing or when flushed.* Gait typically a strutting walk, but also hops and leaps. *Carriage often erect, with stretched neck* and exposed tibia, especially when alert, but more horizontal when relaxed or walking. Wary but may permit close approach. Less gregarious than small pipits.

VOCALISATIONS Prolonged high, circling and undulating song-flight involves a monotonous and rapidly repeated series of (4–12) buzzy *zeevu* or *tschivü* notes, e.g. *tschivü-tschivü-tschivü-tschivü-tschivü*, but is unlikely to be heard in covered region. – Most common flight-call, especially when flushed, is a drawn-out *shreep* or *pschreep* (somewhat House Sparrow-like), often initially at its loudest and with a strong *r*-sound; less frequently a soft *chup*, even rarer a *chirp* (often given twice and quite similar to Tawny Pipit); also a faintly downslurred *pshee* or atypical *cheep*, some of which may provoke confusion with Blyth's Pipit.

SIMILAR SPECIES Richard's is bulkier than *Tawny Pipit*, with more upright posture, proportionately heavier bill, longer and stouter legs, longer tail and disproportionately long hind claws. Plumage differences usually rather pronounced, with most Tawny Pipits having much paler sandier-brown or even yellowish-grey upperparts, dark lores and very little underparts streaking. Richard's usually has much richer brown upperparts with heavier streaking, especially on mantle and scapulars. Beware heavily worn or (rarely) fresh 'plainer-backed' Richard's which may show paler upperparts. A further complication is that young Tawny (including autumn migrants) can retain many juvenile feathers quite late and appear confusingly heavily streaked. Richard's almost always lacks a darker loral stripe, which together with pale area around eye contrasts with prominent eye-stripe behind eye (reverse on Tawny), but at certain angles Richard's may show hint of dark area, while loral stripe on Tawny occasionally faint. Also note Richard's more prominent supercilium, indistinct dark median-covert bar (sometimes concolorous with surrounding feathers), much broader and diffuser edges to secondary-coverts and tertials, deep and contrasting gingery-buff breast/flanks with better-developed streaking, lateral throat-stripe and patch, and conspicuous white in tail, whereas Tawny has a prominent dark median-covert bar, narrower and better-defined edges to secondary-coverts and tertials (although these features are invalid for birds of either species with any retained juvenile feathers), more creamy/buff underparts with more scattered, irregular streaking and reduced lateral throat-stripe and patch, and buffish-sullied pale area of tail. Richard's has much more powerful and undulating flight, and diagnostic calls; also, unlike Tawny,

PIPITS AND WAGTAILS

Oman, Nov: often hovers with dangling legs before landing. Note typical face and breast patterns. The very long hind claws cannot be appreciated at this angle. (M. Varesvuo)

Presumed ad, Oman, Nov: here in rear view, this large pipit can appear almost thrush-like, especially due to upright posture and strong bill. Note also typical streaky upperparts, and pale cream-buff lores with only limited narrow grey loral stripe. Appears to have uniformly ad wing-feathers. (M. Varesvuo)

often but not always hovers prior to landing. – Differences from extremely similar but in W Palearctic much rarer *Blyth's Pipit* discussed in detail under that species. Main differences compared with Blyth's are vocalisations, larger size including longer tail, tarsi, hind claws and bill, thicker bill-base, post-juvenile median coverts pattern with pointed dark centres, longer and narrower, more even-width white wedge on inner web of penultimate tail-feather (shorter and wider-angled wedge in Blyth's) and slightly more prominent pale supercilium. – Differences from *Paddyfield Pipit (A. rufulus,* extralimital and not treated) are small and often difficult to appreciate in the field when faced with a single bird, but the short, hard *chep* call of Paddyfield is distinctive when heard, and size is slightly smaller, tail shorter, loral stripe darker and upperparts less prominently streaked.

AGEING & SEXING Ageing possible with close inspection of moult and feather pattern or wear in wing, more practical in autumn than in spring. Sexes alike in plumage (but size appears useful). – Moults. Post-nuptial moult complete and post-juv moult partial, both being extended, sometimes suspended, usually starting before migration and lasting into winter (Jul–Jan, rarely later). Pre-nuptial (partial), mostly Feb–Mar. Post-nuptial moult extremely variable: some undergo complete pre-migration moult, but many suspend (some or most of body- and flight- and tail-feathers), resuming in winter quarters. Post-juv moult including most or all of head, body and lesser coverts, usually virtually all median coverts, some or all tertials and alula, and inner or virtually all greater coverts, and often r1, but extent highly variable, and typically pauses in moult stop-over in autumn–winter. Unclear to what extent pre-nuptial moult marks end of extended autumn moult or new complementary moult, with many individuals reported moulting some or most body-feathers and none to all lesser, median and greater coverts and tertials, and perhaps s6, r1 and alula, but presumably such extensive late autumn moult more frequent in 1stY. –

SPRING Experience essential to age late winter and spring birds, as both ad and 1stY show strong moult contrast in wing. **Ad** Wing- and tail-feathers uniformly abraded, lacking any moult contrast. With wear, buff fringes to upperparts and wing-feathers become narrower and paler, and dark centres more pronounced and blotchy, underparts much paler. **1stS** Very similar to ad but some perhaps distinguishable if any juv secondary-coverts retained (with noticeably sharper dull brown centres and heavily reduced whitish fringes); juv primaries and primary-coverts more worn (contrasting with surrounding new feathers), and primary-coverts slightly more pointed with more distinct, whiter and more clear-cut pale tips. Also, if retained, central alula has whiter, more clear-cut tips and fringes. – **AUTUMN Ad** Readily separated from 1stW by being evenly fresh while those that suspend moult show far stronger moult contrast in wing. **1stW** Usually readily aged by variable number of juv outer median and greater coverts, and often some tertials (with narrow, whiter fringes/tips sharply delimited from dark centres), contrasting with fresher inner ad-type feathers in same tracts (which have more diffuse and warmer-coloured fringes). Post-juv moult often suspended extensively, making such birds intermediate between ad and juv, e.g. some have juv scapulars, secondary-coverts and tertials, and ad breast pattern. 1stW that have renewed all wing-feathers, but no remiges or primary-coverts, are less easily aged, though primary-coverts subtly weaker textured and already slightly worn, especially when compared with renewed secondary-coverts and tertials; less deep brown iris than ad sometimes useful in autumn, even into early winter. **Juv** Pale, rounded fringes to feathers of upperparts create scalloped effect; underparts paler than ad, with less buff and broader streaks extending further on flanks; wing pattern conspicuous owing to off-white fringes and sharply delimited black centres to tertials, median and greater coverts.

BIOMETRICS L 18–19.5 cm; **W** ♂ 93–103 mm (*n* 42, *m* 97.6), ♀ 88–96 mm (*n* 28, *m* 92.6); **T** ♂ 71–82 mm (*n* 42, *m* 75.8), ♀ 70–79 mm (*n* 28, *m* 73.4); **T/W** *m* 78.3; **B** 16.5–19.6 mm (once 15.2 in juv ♀; *n* 76, *m* 17.9); **B**(f) 11.6–15 mm (*n* 72, *m* 13.3); **BD** 4.3–5.4 mm (*n* 63, *m* 4.9); **BW** 4.4–5.9 mm (*n* 63, *m* 5.2); **Ts** 28.0–33.0 mm (*n* 77, *m* 30.4); **HC** 13.0–22.0 mm (once 24.0) (*n* 74 *m* 16.0); **HC/W** 13.4–25.8 (*n* 74, *m* 16.8). **Wing formula: p1** minute; **pp2–4** about equal and longest (or either p2 or p4, or both, 0.5–2 mm <); **p5** < wt 1–4 mm; **p6** < wt 9–13.5 mm; **p7** <

Presumed ad, Oman, Nov: when very fresh, upperparts can appear almost plain; breast-streaking is sometimes reduced, and less contrasting due to extensive warm buff wash on breast and flanks. Moult limit within median coverts (both generations ad-type) could indicate ad with suspended moult. (D. Occhiato)

1stW, Egypt, Oct: 'classic example' in plumage pattern and coloration, especially strong breast-streaking, warm buff flanks, thin and diffuse grey loral line. Strong bill and proportionately very long legs. Contrasting juv greater coverts confirm age. (D. Occhiato)

Richard's (left: United Arab Emirates, Oct) and Tawny Pipits (right: Israel, Oct), both 1stW: irrespective of age and season, separation usually straightforward. Separation of an unusually pale/grey Richard's (exaggerated here by strong light) from a streakier young autumn Tawny can require care. Note Richard's stronger build (including bill and legs), and browner and rustier plumage. Almost lacks Tawny's usually contrasting dark loral line, though in this Tawny it is unusually narrow. Richard's has much longer hind claw. Both have many juv wing-feathers. (Left: H. Roberts; right: A. Ben Dov)

1stW, Israel, Nov: face markings (subject to individual variation) include pale cream-buff lores (loral stripe at most narrow and dark grey), eye-ring, eye-surround and long, broad supercilium; also variable dusky eye-stripe behind eye and narrow moustachial stripe; lateral throat-stripe terminates in blotchy streaks on upper breast. Some juv wing-coverts confirm age. (L. Kislev)

Israel, presumed ad, Nov: seemingly in active moult during autumn migration. (L. Kislev)

Oman, Nov: this bird has rather poorly-marked face, with narrow breast-streaking (cf. previous images of better-marked individuals). (M. Varesvuo)

wt 15–19 mm; **p10** < wt 23–29 mm; **s1** < wt 23.5–31 mm. Emarg. pp3–5. **White wedge r5** (inner web) 26–54 mm (n 75, m 42.8; very few < 35 mm), invariably narrow, often with almost parallel sides and no tendency to widen at edge of feather. Many have slightly diffuse delimitation, dark portion of inner web being slightly paler basally. Outer web of r5 usually extensively dark along shaft to near tip (only outer edge pale), but some have slightly more pale at tips, and extremes overlap with Blyth's in this respect.

GEOGRAPHICAL VARIATION & RANGE Monotypic. – Siberia from Irtysh and upper Ob east to upper Lena and Sea of Okhotsk, south to E Kazakhstan, Tibet, China; winters mainly in India and SE Asia, but regularly also locally in S Iberia, S Italy, N Africa, Levant, S Turkey, Arabia, probably also Gulf States. – Variation very slight once *A. richardi* is split as a separate species. Often considered polytypic, but Alström & Mild (2003) treated *richardi* as monotypic, and from material examined by us this seems the most sensible policy. Most material examined is from western half of *richardi* range, which part is the probable source of all birds occurring within the treated region. – Birds from Amur, Ussuriland and NE & EC China, wintering in SE Asia ('*sinensis*') sometimes separated on account of smaller size, claimed darker upperparts and more saturated underparts, but size differences slight and of clinal character, and colours very similar or the same, hence not followed here. Biometrics of birds in E Siberia and Far East: **W** ♂ 90.5–99 mm (n 17, m 94.8), ♀ 85–96.5 mm (n 18, m 90.3); **T** ♂ 67–81 mm (n 17, m 74.8), ♀ 66–77 mm (n 18, m 71.6); **T/W** m 79.1; **B** 15.9–19.5 mm (n 35, m 18.0); **BD** 4.5–5.4 mm (n 24, m 5.0); **Ts** 28.8–32.5 mm (n 35, m 30.7); **HC** 12.7–22.0 mm (n 34 m 16.4); (Syn. *centralasiae*; *dauricus*; *sinensis*; *ussurensis*.)

TAXONOMIC NOTE *A. richardi* was long treated as a race of a widespread and polytypic species *A. novaeseelandiae* of Africa, Asia and Australasia, but more modern treatments tend to recognise as many as five different species, with *A. rufulus* breeding in S Asia, *A. australis* and *A. novaeseelandiae* in Australasia, and *A. cinnamomeus* in Africa and SW Arabia. This may be a sensible taxonomy, or prove to be too many, but we refrain from passing a judgement without deeper knowledge, and since only *richardi* is of concern to the present work.

REFERENCES GRANT, P. J. (1972) *BB*, 65: 287–290. – LARSSON, H. (2008) *Roadrunner*, 16 (3): 28–33. – LÓPEZ-VELASCO, D. *et al.* (2012) *DB*, 34: 11–19. – SCHMIDT, C. (1993) *Limicola*, 7: 178–190. – VINICOMBE, K. (2006) *Birdwatch*, 172 (Oct): 24–26.

AFRICAN PIPIT
Anthus cinnamomeus Rüppell, 1840

Alternative name: Grassland Pipit

Fr. – Pipit africain; Ger. – Zimtpieper
Sp. – Bisbita africano; Swe. – Afrikansk piplärka

A large pipit of S & E Africa also found in the extreme SW Saudi Arabian and Yemeni highlands. African Pipit, a close relative of Richard's Pipit, is apparently non-migratory within the treated region, where it inhabits elevated plateaus with low grass, the typical habitat over most of its African range.

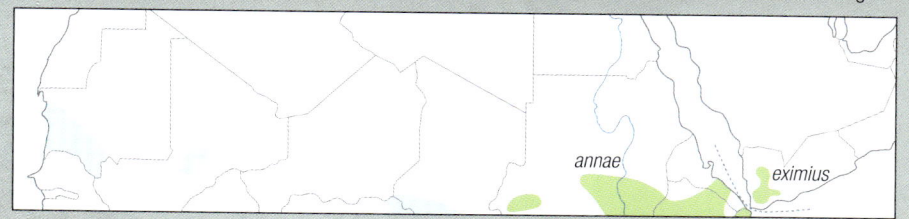

IDENTIFICATION A large, rather bright brown, streaked pipit, most resembling Richard's Pipit, but size, jizz and action approach Tawny and Long-billed Pipits. Face and wing patterns are very similar to Richard's but in several respects also resemble the other two. Upperparts greyish or dull buff-brown with *conspicuous dark streaks on crown and mantle, long, broad supercilium* (whiter behind eye), pale buff-brown ear-coverts with dusky eye-stripe, *short brown-black moustachial stripe*, and *long brown-black lateral throat-stripe usually terminates in black streaks that spread over entire breast*. Below mostly whitish-buff, but sometimes deeper *buffish on unstreaked flanks*. Whitish-buff fringes to secondary-coverts and tertials (covert centres triangular with pointed tips, but bases not sharply delimited, thus recalling Richard's, but obvious dark median-covert bar and paler wing-bars approach Tawny). Whitish-cream outer fringes to two outermost tail-feathers striking on take-off. Hind claws moderately long. Bare parts mainly as Richard's but has *bright orange-flesh base to lower mandible*, and legs are the same colour. Often adopts rather more horizontal posture than Richard's, and on ground gait quite strongly recalls a wagtail, while flight is generally less powerful than Richard's.

VOCALISATIONS (Based on notes from Kenya and Tanzania; not known whether a difference exists between these breeders and Arabian ones.) Song a repeated *trrlitt-trrlitt-trrlitt...* or *tree-tree-tree...*; given continually from perch or in flight-display, in flight often more protracted. – Most common flight-calls are a repeated *trit*, or *chip* or *trip*, somewhat recalling Tawny Pipit.

SIMILAR SPECIES In SW Arabia most likely to be confused with migrant *Tawny Pipit* (regular) but less so with *Richard's Pipit* (probably only a vagrant), and apparently to some extent overlaps with resident *Long-billed Pipit*. Combination of Richard's-like streakier crown and mantle/scapulars, open-faced appearance (dark loral patch usually weak or absent), obvious lateral throat-stripe, extensive breast-streaking, deeper buffy flanks, and wing pattern, should eliminate both Tawny and Long-billed. Furthermore, African's legs and hind claws are proportionately longer (hind claws intermediate between Tawny and Richard's), and bill base and legs more orange. Beware that African approaches Tawny and Long-billed in size, sometimes has a dark loral patch, and that Arabian birds appear slightly greyer and paler. Also note that separation in juvenile plumages is extremely difficult, with Tawny and African Pipits being virtually identical and perhaps only separable in the hand. Long-billed is further separated by its longer and more pointed bill, plainer upperparts, more saturated and evenly greyish underparts, weaker facial pattern, and more diffuse pattern to the secondary-coverts and tertials. African Pipit is probably only safely separated from Richard's by smaller size, shorter hind claws and (sometimes) the clear dark loral line. All these forms should be best distinguishable by calls.

AGEING & SEXING Birds in SW Arabia poorly studied, but African birds generally possess same seasonal variation and moult patterns as Richard's, but being primarily a tropical resident moult is less seasonally fixed. Generally, ad-like imm plumages distinguished by retained juv feathers. – Moults. Post-nuptial moult complete, post-juv partial. Pre-nuptial moult partial. Only 11 specimens in NHM adequate for analysis (from SW Arabia, Somalia): two moderately fresh (Feb, Apr); seven moderately to heavily worn (Jun–Sep), one of which (23 Sep) nearing end of complete post-nuptial moult; two juv (Sep) one of which partially moulted.

BIOMETRICS (*eximius*) **L** 16–17 cm; **W** 82.5–89 mm (n 10, m 86.1); **T** 59–66 mm (n 10, m 63.3); **T/W** m 73.5; **B** 15.8–17.4 mm (n 9, m 16.7); **BD** 4.2–4.9 mm (n 10, m 4.6); **Ts** 23.9–26.6 mm (n 10, m 25.6); **HC** 10.0–11.5 mm (n 9, m 10.9). **Wing formula: p1** minute; **p2** < wt 0–2 mm; **pp3–4** (pp2–4) longest; **p5** < wt 0.5–3 mm; **p6** < wt 8–12 mm; **p7** < wt 13–17 mm; **p10** < wt 20–25 mm; **s1** < wt 22–26.5 mm. Emarg. pp3–5(6). **Pale wedge r5** (inner web): 20–39 mm.

GEOGRAPHICAL VARIATION & RANGE Widespread in sub-Saharan Africa with a population in SW Arabia, usually with moderate variation, but more marked differences in S & SC Africa might indicate separate species status. About a dozen races of African Pipit have been described but only one occurs within the covered region. One extralimital race might straggle to the treated area and is therefore briefly mentioned below. All or most populations thought to be resident.

○ **A. c. eximius** Clancey, 1986 (Yemen, SW Arabia). Described above. Very similar to *annae* and differs only subtly in being slightly richer coloured, not quite as pale and greyish-tinged; slightly smaller and shorter-legged on average.

Extralimital: **A. c. annae** Meinertzhagen, 1921 (Eritrea, Djibouti and Somalia to Kenya and NE Tanzania). Differs from *eximius* in being slightly larger with longer legs, and being on average a little paler and more greyish brown-buff, not as saturated. Still, only a subtle difference. **W** 83–98 mm (n 15, m 91.3); **T** 61–75 mm (n 15, m 67.1); **T/W** m 73.5; **B** 15.0–18.2 mm (n 14, m 16.4); **BD** 3.9–5.0 mm (n 12, m 4.6); **Ts** 24.5–29.0 mm (n 15, m 27.3); **HC** 9.0–13.0 mm (n 15, m 11.2).

TAXONOMIC NOTE The *A. cinnamomeus* group was long treated as a subspecies group within *A. novaeseelandiae* of Africa, Asia and Australasia, but more modern treatments tend to recognise 4–6 distinct groups or species.

A. c. eximius, Saudi Arabia, Jun: told from Richard's and Blyth's Pipits by on average slightly darker loral line (though variable and depends on light), smaller size and daintier jizz, and from Richard's on shorter hind claw, as is just visible here. (J. Babbington)

A. c. eximius, presumed 1stY, Saudi Arabia, Jun: erect posture, long legs and general plumage pattern recall Richard's Pipit, especially as loral line appears fairly pale here due to light and viewing angle. However, it is daintier and smaller than Richard's, and has shorter hind claw, hence similarity with Blyth's (possible vagrant) becomes real challenge and requires flight calls or handling to sort out. (J. Babbington)

BLYTH'S PIPIT
Anthus godlewskii (Taczanowski, 1876)

Fr. – Pipit de Godlewski; Ger. – Steppenpieper
Sp. – Bisbita estepario; Swe. – Mongolpiplärka

Only recently have observers become aware that this Asian species is a rare wanderer to Europe, mainly to Britain and Fenno-Scandia, although at least one 19th-century specimen collected in Europe lay unnoticed in a museum collection for many years. Blyth's Pipit breeds on dry grassy slopes and forest glades in foothills and mountains from S Siberia to NE China, and winters in the Indian subcontinent, where it associates with Richard's Pipits in grassy areas.

Presumed 1stS, Mongolia, Jun: rather worn plumage. Compared to Richard's, note rather short pointed bill and shorter legs. Also, adult-type median coverts with more square-shaped and sharply demarcated centres that produce better-marked wing-bar. Ageing unsure, though apparent presence of three generations of greater coverts suggest 1stY. (H. Shirihai)

IDENTIFICATION Large, rather buffy-tinged, streaked pipit that most closely recalls Richard's Pipit. Vagrants *best separated by voice and pattern of adult-type median and greater coverts*, while useful too are *slightly smaller size and slighter build* (with gait closer to Tawny than Richard's Pipit, and *less upright carriage* than latter), less bold head pattern, *paler and buffier-cinnamon edges to dark-streaked feathers of crown, nape and mantle* creating more contrasting pattern, *stronger wing pattern and more evenly saturated underparts*. If some central median coverts have been moulted to adult pattern these usually offer good help to separate Richard's from Blyth's, since Blyth's Pipit acquires more Tawny Pipit-like pattern after first moult, having *more square, even and distinct border between blackish base and quite pale buff tips* (only a small dark 'tooth' at the shaft), whereas Richard's has *more pointed dark bases and less contrastingly pale tips/edges* which are less distinctly demarcated. (Note that both innermost and outermost few median coverts have more similar and therefore inconclusive pattern in both species.) If bird preens and spreads tail, pattern of white on outer tail-feathers useful clue: Blyth's has *shorter and more obtuse (wide-angled) white wedge with stronger delimitation* (never diffuse borders on inner web as in some Richard's) *on tip of inner web of fifth tail-feather*, whereas Richard's typically has *long and narrow (almost parallel-sided) white wedge on inner web* of fifth tail-feather. A very few (< 1%) are more similar as to shape of white wedge, but the character does not overlap exactly (as stated elsewhere). Many features should be regarded as supportive, rather than diagnostic, and identification should always rely upon a combination of features. In most cases, it is vital to age a bird first, for which good views and knowledge of Richard's Pipit are required. For ageing and to determine the wing-covert pattern, telescope views or photographs are essential. However, it is now apparent that, with experience, identification of Blyth's Pipit on plumage and structure is rather more straightforward than some past literature suggested. Flight is generally as Richard's but appears decidedly more 'small-pipit-like', and flight-calls usually clearly differ. Note that Blyth's briefly hovers when landing in grass but not on bare ground, just like Richard's Pipit.

VOCALISATIONS Song (extremely unlikely to be heard within the treated region) is more complex than the monotonous song of Richard's and is usually readily recognised by the mixture of a characteristic phrase repeated a few times, *zee-viii-dyay, zee-viii-dyay, zee-viii-dyay* (can vaguely recall Red-throated Pipit song), and a hammering steady series of buzzing, abrupt notes *bzre bzre bzre bzre...*, etc. – Most common flight-calls a *pshee(o)* or *pscheeu*, slightly down-bent with emphasis on central part; although audibly rough, much less explosive and softer than Richard's, and often sounds closer to Yellow Wagtail. Also a dry clicking *chep* (or *tchip* or *chup*), often doubled, which might recall a distant Redpoll. There is also a more buzzing call, *bzre*, rather like the repeated note from the song.

SIMILAR SPECIES Although tends to be slightly more uniformly greyish-buff above than most Richard's Pipits, while size, jizz and actions approach Tawny, Blyth's is extremely similar to the former in plumage (and thus differs quite markedly from Tawny). Extreme care is required in separating Blyth's and Richard's, and the following are the most critical features of Blyth's: (i) marginally smaller overall (particularly the head and bill) with a more slender body and proportionately shorter legs and tail (affording a more horizontal, Tawny-like carriage); (ii) shorter hind claws, roughly equal in length to hind toes (hind claws much longer than hind toes in Richard's, but extremes overlap slightly);

Blyth's (left) and Richard's Pipits (right), Mongolia, Jun: former has smaller, more spiky-shaped bill, rounder head with clearer crown streaking, stronger-patterned mantle and denser breast-streaking. Also better-marked (more Tawny Pipit-like) pale wing-bar, with newly moulted adult-type median coverts and visibly shorter hind claw. Ageing difficult: the older greater coverts of the Blyth's indicate either pcst-nuptial ad or juv, while the Richard's seems to have only ad-type wing-feathers. (H. Shirihai)

(iii) rather shorter and more evenly pointed bill, lacking markedly curved culmen near tip of Richard's; (iv) on average slightly paler and often slightly more buff-cinnamon upperparts, which are usually more contrastingly streaked than Richard's on crown to mantle; (v) generally more evenly buff below (usually lacks Richard's more evident contrast between breast/flanks and belly) but breast-streaking averages denser. Best three characters: (vi) the pattern of adult-type coverts (on adults or first-years) with the median (especially) and greater covert-centres in Blyth's more sharply set off and more square-shaped (whereas on Richard's, dark centres are triangular and laterally diffuse), and the greater coverts have sandier tips, producing more distinct pale wing-bars than on Richard's (which has buff-brown/warmer tips lacking such strong contrast); (vii) the white wedge on inner web of penultimate tail-feather is usually short and 'broad' (wide-angled), often widening further at the edge of the feather, an impression enhanced by the tendency to have extensive pale outer part of outer web (outer web of r5 in Richard's usually dark except extreme tip); very rarely birds occur with a tail pattern more similar to Richard's (still, no overlap when attention is paid to fine details), and the wedge is invariably well delimited

1stW, Finland, Nov: smaller and daintier than Richard's. From latter by thinner/more pointed bill (lacking curved culmen of Richard's); also shorter legs, shorter hind claw and densely streaked breast, while underparts are more evenly buff (lacking whiter belly of Richard's). No ad-type median coverts visible, so this feature cannot be used. (J. Peltomäki)

Blyth's (left: Germany, Oct) and Richard's Pipits (right: Germany, Oct): structural characters again useful. Tail and legs of Blyth's proportionately shorter, but crown also more evenly streaked and supercilium comparatively short (compared to long supercilium and dark crown-sides of Richard's). In addition to underparts pattern and rather short hind claw, the newly moulted ad-type median coverts of Blyth's have more square-shaped centres (triangular and more diffuse in Richard's). Both birds are 1stW, with juv greater coverts. (Left: S. Pfützke; right: A. Halley)

1stW, France, Oct: typical thin/pointed bill, shorter legs, heavy breast-streaking (with small delicate spots forming patch on sides), and rather short supercilium (poorly demarcated behind eye). This bird has less typical whiter belly (more like Richard's). Most wing-feathers are juv. (A. Audevard)

United Arab Emirates, Feb: compared to Richard's, note compact appearance, especially the rounder head, and shorter legs and tail. Like Richard's, has unmarked pale lores, but supercilium shorter, with stronger upperparts streaking. All median coverts are ad-type. (H. Roberts)

1stW, Sweden, Dec: heavily worn; birds often wear greyer but are rather heavily streaked above. Rather evenly buffish underparts and denser breast-streaking are typical of Blyth's. Most of the greater coverts are juv. (M. Nord)

(at times very diffusely outlined in Richard's); and (viii) Blyth's more contrasting, solidly warm brown ear-coverts, but less prominent supercilium (chiefly confined to just above and behind eye), and the evenly streaked crown, lacking hint of darker lateral crown-stripes as often found in Richard's. Regarding the secondary-coverts, note that evaluating their pattern requires great care, as this character is not valid for juvenile coverts, and beware variation due to extended moult season, wear and bleaching. Furthermore, when assessing adult-type feathers it is important to concentrate on the central ones, as the innermost and outermost in Richard's are quite frequently more Blyth's-like in pattern. – While theoretically also rather similar to *Tawny Pipit*, Blyth's should be easily distinguished in being warmer and browner and more heavily streaked above than Tawny, and deeper buff below with better-developed underparts streaking (although this feature is probably of little use with young Tawny), narrower and more clear-cut tertial fringes, and, importantly, unmarked pale lores (Tawny has a dark loral line).

AGEING & SEXING Ages differ, but ageing requires close attention to details and moult contrast, and is mainly possible in autumn. Sexes alike in plumage (but see Biometrics). – Moults. Post-nuptial moult complete and post-juv partial, mostly pre-migratory, but protracted and much variation. Post-juv moult generally in two phases: pre-migratory phase includes most or all of head, body and lesser coverts, but usually the rest of wing-feathers are retained (some birds may moult a few median coverts and ninth greater covert, and a very few all lesser and median coverts and 8–9th greater coverts); resumes moult after migration, continuing in late autumn and winter, when most birds renew all secondary-coverts (presumably complementary), tertials and r1, and some or most body-feathers (some may retain a few outer, rarely most, juv greater coverts, and sometimes the longest tertials). Renewal in winter quarters essentially partial and presumably mostly complementary; perhaps only a few undertake true pre-nuptial moult (chiefly Mar). – **SPRING Ad** Only ad-type (worn or fresh) wing- and tail-feathers discernible. With wear more greyish-brown above, and fringes narrower/paler enhancing dark blotchy centres. Also underparts become whiter, and wing-bars reduced. **1stS** Very similar to ad but, in the hand, has retained outer juv median and greater coverts (sharper dull brown centres and heavily-reduced whitish fringes). All juv primaries and primary-coverts retained and are rather more worn and bleached browner (also, retained primary-coverts have clearer-cut narrow whitish fringes). Beware that both ages can show contrasting moult limits (following winter renewal of some wing-coverts, tertials and tail-feathers, as well as a few or most body-feathers). – **AUTUMN Ad** Usually undergoes complete moult before migration, but not always entirely completed before onset of migration, meaning that many migrants have only partially moulted body-feathers, and a few unmoulted secondary-coverts, tertials and r1, which will be shed in winter (unlike in Richard's and Tawny, suspended remiges moult unknown in this species). **1stW** Much like fresh ad but retains some juv secondary-coverts, tertials and scapulars, thus shows clear moult limits and has characteristic pattern to juv secondary-coverts as in Richard's, but unlike latter retains more secondary-coverts. Unfortunately, very few 1stW will have acquired ad-like central median and greater coverts (useful for specific identification) by early autumn. Birds that have replaced all secondary-coverts can sometimes still be aged (in the hand and on sharp photographs at least) by retained primary-coverts, central alula (duller with clearer-cut whitish fringes) and primaries. However, some attain intermediate plumage between ad and juv and are then more easily aged, showing some contrasting retained juv coverts or tertials. **Juv** A bird in complete juv plumage is unlikely to be seen in the covered region, but see 1stW; much as juv Richard's.

BIOMETRICS **L** 16–17.5 cm; **W** ♂ 89–99 mm (*n* 28, *m* 93.8), ♀ 83–94 mm (*n* 20, *m* 89.4); **T** ♂ 63–74 mm (*n* 24, *m* 68.4), ♀ 62–70 mm (*n* 20, *m* 66.2); **T/W** *m* 73.4; **B** 14.9–17.6 mm (*n* 57, *m* 16.4); **BD** 4.1–5.0 mm (*n* 49, *m* 4.4); **BW** 4.0–5.3 mm (*n* 39, *m* 4.6); **Ts** 23.5–28.0 mm (once 29.3) (*n* 53, *m* 26.3); **HC** 9.0–13.4 mm (once 14.3) (*n* 53, *m* 11.4); **HC/W** 9.7–14.3 (once 15.4) (*n* 53, *m* 12.5). **Wing formula: p1** minute; **pp2–4** about equal and longest (p2 sometimes 0.5–2 mm <); **p5** < wt 0.5–2.5 mm; **p6** < wt 8–12 mm; **p7** < wt 13–17 mm; **p10** < wt 22–27 mm; **s1** < wt 22–30 mm. Emarg. pp3–5. **White wedge r5** (inner web) 13–34 mm (*n* 45, *m* 24.5), usually rather widely angled, often widening further near edge of feather; very rarely more oval-shaped, when recalls shape in Richard's with extreme short wedge. Invariably well-marked (sometimes more diffuse outline and contrast in Richard's). Usually (> 90% of examined specimens) extensive whitish tip to outer web of r5, leaving only narrow dark shaft-streak over about outer third of feather, and often more (cf. Richard's).

GEOGRAPHICAL VARIATION & RANGE Monotypic. – Tuva, Mongolia, Transbaikalia and adjacent SE Siberia, N China; winters India, SE Asia. (Syn. *striolatus*.)

REFERENCES Alström, P. (1988) *BW*, 1: 268–272. – Alström, P. & Mild, K. (1997) *Limicola*, 11: 97–117. – van den Berg, A., van Ree, L. & Roselaar, C. S. (1993) *DB*, 15: 198–206. – Bradshaw, C. (1994) *BB*, 87: 136–142. – Larsson, H. (2008) *Roadrunner*, 16 (3): 28–33. – López-Velasco, D. et al. (2011) *DB*, 34: 11–19. – Maasen, E. & van den Berg, A. (2003) *DB*, 25: 44–48. – Schmidt, C. (1993) *Limicola*, 7: 178–190. – Small, B. (1997) *DB*, 19: 189–190. – Vinicombe, K. (2007) *Birdwatch*, 184 (Oct): 32–35.

Blyth's (left) and Tawny Pipits (right), India, Jan: this pitfall is not always considered, but there is scope for confusion, especially with distant or silent birds. Three key characters are especially useful, namely Blyth's plainer lores (lacking Tawny's clear dark eye-stripe), better-streaked mantle (some 1stY Tawny have exposed dark centres), and more evenly buff-washed underparts. The Blyth's is probably ad, but it is difficult to eliminate 1stY with advanced pre-nuptial moult; the Tawny is 1stW with juv greater coverts. (S. Sridhar)

TAWNY PIPIT
Anthus campestris (L., 1758)

Fr. – Pipit rousseline; Ger. – Brachpieper
Sp. – Bisbita campestre; Swe. – Fältpiplärka

Much the commonest and most widespread of the large pipits in the treated region, Tawny Pipit occurs across continental Europe to W China, and south to N Africa and the Middle East, where it is a summer visitor to dry, grassy, treeless areas, including sandy seashores, heaths and sandy-soiled clear-fellings in forests. It winters widely in the Sahel and NE Africa and Arabia, as well as east to India. On migration, expect to encounter the species in lowlands and coastal areas, often in the same habitats used by migrant Greater Short-toed Larks and wheatears.

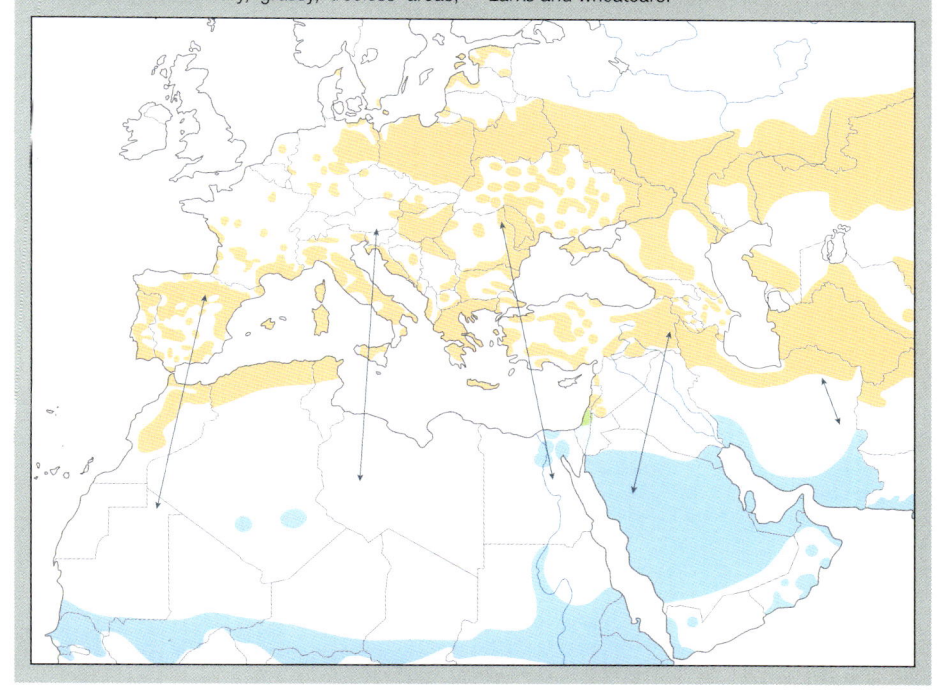

Ad, Israel, Mar: note pleasant yellowish-buff tinge to face and upper breast, often a feature of breeding ad. Some eastern Tawny are plainer and greyer above. In parts of Middle East, there is also a risk of confusion with the very similar Long-billed Pipit. Only ad-like wing feathers visible here. (H. Shirihai)

IDENTIFICATION A mid-sized to large, *slim pipit* with something of the *appearance of a pale wagtail*. Over much of the treated region should be readily distinguished from the main confusion species, the much rarer Richard's and Blyth's Pipits. Overall *rather pale, indistinctly-streaked plumage* with more pronounced facial pattern, especially *diagnostic dark loral line* and *distinct dark median-covert bar*, and vocalisations and behaviour also offer strong clues. Usually shows *prominent long pale supercilium* and *hint of pale submoustachial stripe*, while dark *lateral throat-stripe is only thin* and poorly marked in many. Juvenile or first-winter Tawny present a greater confusion risk (than adult) with either of the other two species. All plumages have *whitish-cream outer edges to two outermost tail-feathers* (striking on take-off). Bill dark brown with flesh-coloured base to lower mandible, and legs mostly bright yellow-flesh; *hind claws relatively short*. Gait may quite strongly recall a wagtail, with furtive runs and darts; wags tail frequently (but not vigorously or constantly), tail and rear body in one piece much like a Common Sandpiper, and has rather upright carriage, though usually much less erect, more horizontal than Richard's. Flight also generally less powerful than Richard's, more wagtail-like (with marked undulations and slender, longer-tailed silhouette), while escape-flight, often quite long and direct, usually ends on open ground (and never hovers before landing in grass). Song-flight comprises a fluttering ascent, reaching a considerable height, followed by a parachuting descent (sometimes makes slightly longer undulating flights). Sometimes forms small flocks on passage and in winter.

VOCALISATIONS Song poorly developed, a monotonous rather slow repetition of 2–3 rising syllables with a ringing tone, e.g. *tsirliih… tsirliih… tsirliih…* or *cherlee… cherlee… cherlee…*, often drawn-out at end and repeated 7–12 times, mainly given in descent or in each undulation, as well as from perch or on ground. Some variation noted, e.g. in Balkans *sr'r'rüh*, trembling and falling in pitch. – Varied flight-calls, of two main types: a *rather quiet tchriip* (often likened to a House Sparrow) and a slurred *chirrup* or *chirlup* often doubled (somewhat recalling Richard's but less rasping) or mixed, e.g. *chilp-chrp-chi-chup*, sometimes sounding like a distant Short-toed Lark; also a loud Yellow Wagtail-like *tshilp* or softer *chilrp*; beware that a single short, blunt *chip* or *chup* can be misleadingly similar to Blyth's Pipit's *chep* call, but is harsher and lacks nasal quality of latter.

SIMILAR SPECIES Usually markedly less heavily built than *Richard's Pipit*, with shorter/slimmer bill, tail and especially shorter legs (to some degree size and structure closer to *Blyth's Pipit*). Compared to Richard's and Blyth's Pipits, usually readily identified by following characters: long, better-defined pale supercilium emphasized by sharper (often almost black) loral line and broader/diffuse eye-stripe behind eye (note that loral line can be less obvious due to angle of light or wear); rather plain, almost unstreaked upperparts, varying geographically from pale sandy-brown to dull greyish; bolder pale wing-bars and contrasting dark-centred median coverts; adult has virtually unstreaked breast or (thin) streaking more restricted to breast-sides (none on flanks),

Italy, May: compared to Richard's and Blyth's Pipits, usually readily identified by rather plainer, almost unstreaked upperparts, long and better-defined supercilium, emphasised by sharper dark loral line; relatively short hind claw separates Tawny from Richard's. Ageing in spring, however, often difficult without close examination of primary-coverts and outer greater coverts. (D. Occhiato)

Tawny Pipits, variation in spring, Israel, Mar: left bird has Richard's Pipit-like impression, especially deep buff body-sides and overall jizz; the other is plainer and greyer, more like Long-billed Pipit, but note proportionately shorter and slenderer bill and tail, and relatively stronger head and upperwing patterns. Ageing in spring is very difficult, even in the hand, but left bird perhaps ad (only ad-type wing-feathers visible), though advanced 1stY difficult to eliminate; right bird presumably 1stS (in extensive moult) due to worn, seemingly juv shorter tertials, but moult limit in greater coverts difficult to relate to age. (A. Ben Dov)

although this is subject to quite some variation; moustachial stripe and lateral throat-stripe thin but often quite prominent (though variable), while rest of underparts typically whiter, slightly tinged buffish at sides; and the relatively short hind claws separate Tawny from Richard's (note that extremes still come close) but not from Blyth's. – In winter, local *African Pipit* could be a pitfall, since this resembles a first-winter Tawny. However, first-winter Tawny on their African winter grounds generally attains a more adult-like plumage with reduced breast-streaking. Also, African Pipit generally differs on subtly longer tarsus, hind claws and on long, narrow white wedge to inner web of r5. – In Middle East, a further confusion risk is the slightly larger and more robust but otherwise extremely similar *Long-billed Pipit*, although this has a far more restricted distribution and is sedentary in the covered region. They overlap only to limited degree during migration or in the breeding season, when Tawny might be found in areas used by Long-billed. Long-billed has generally less prominent dark median-coverts bar with less distinct pale-tipped feathers, and has more subdued or lacks dark moustachial stripe.

AGEING & SEXING Ageing possible with close check of moult, feather pattern and wear in wing, easier in autumn, when some 1stW have retained a number of juv body-feathers. Sexes alike (though size sometimes useful; see Biometrics). – Moults. Post-nuptial moult complete and post-juv partial, mostly pre-migration, though extent and timing of moult vary greatly for both, and ad, especially in northern populations, suspend (sometimes many remiges), resuming moult in winter (completed Dec). Post-juv moult includes most or all of head and body, some to all lesser and median coverts, a few inner greater coverts, one to all tertials and r1 (rarely, all secondary-coverts and tertials are renewed); some, presumably from late broods, often retain many juv body-feathers and all wing-feathers until Oct; resumes post-juv moult in winter, including part or all of head- and body-feathers, some to all lesser and median coverts, a few inner to all greater coverts, 1–3 tertials and r1 (a few or all tail-feathers and alula). However, at least some 1stW retain all juv secondary-coverts, tertials and a few body-feathers until midwinter. Pre-nuptial moult partial in winter quarters (chiefly Feb–Mar) usually includes few or most (sometimes all) body-feathers, and usually at least some wing-coverts, tertials and tail-feathers, but is often difficult to separate from late post-breeding or post-juv moult. – **SPRING** As a rule ageing is very difficult. **Ad** Only ad-type (old or fresh) wing- and tail-feathers discernible (cf. 1stS); especially those which moulted extensively in winter appear very neat in spring. **1stS** Extremely similar to ad but perhaps distinguishable (with experience and on sharp photographs and in the hand) if has retained some juv secondary-coverts (with dull brown centres and heavily-reduced whitish fringes); retained juv primaries and primary-coverts decidedly more worn and browner (primary-coverts also narrower with hardly any, very narrow pale fringes; ad has ill-defined and slightly broader fringes); also, if retained, central alula has whiter, more clear-cut tips and fringes. – **AUTUMN Ad** Aged by having only ad-type body-, wing- and tail-feathers; those suspending wing- and tail-feather moult show contrasting limits with heavily-worn outer wings (unlike uniform 1stW). **1stW** Readily aged by retained juv secondary-coverts (clearer-cut pointed dark centres and narrower whitish-buff fringes), including usually all or most median and greater coverts, tertials and even body-feathers (creating moult limits with freshly-moulted ad-like secondary-coverts, which have more rounded and less sharply-defined centres); many moult very few wing- and body-feathers, and to some extent are intermediate between ad and juv, or most like juv (including prominent lateral throat-stripe and breast-streaking); also all primaries and primary-coverts juv (weaker textured and already slightly worn). At this season, those with mostly juv wing-feathers or extensive wing-coverts moult easily separated from ad with very worn remiges and coverts. Rare 1stW that have renewed all secondary-coverts less easily separated from uniformly fresh ad, but note slight contrast between remiges and primary-coverts on the one hand and surrounding fresher coverts on the other. Iris less deep brown than in ad. **Juv** Differs from ad in having much scalier upperparts, more obvious dark brown lateral throat-stripe, and some short but partly bold streaks on breast reaching flanks.

BIOMETRICS L 16–17.5 cm; W ♂ 85–98.5 mm (n 27, m 91.8), ♀ 83–94 mm (n 15, m 88.8); T ♂ 63–76 mm

Oman, Mar: strong moult limits in wing (and ongoing moult in remiges), with many heavily worn wing-feathers that are presumably juv. However, it could be an ad with suspended moult, which is very frequent and sometimes striking in this species. Note diagnostic sharp dark loral line. (M. Varesvuo)

Tawny Pipits, variation in 1stW, Denmark (left, Aug), and Spain (right, Sep): left bird has many juv feathers, including some scapulars and tertials, and most wing-coverts (clearer-cut dark centres and narrower whitish-buff fringes); breast-streaking blotchy. Right bird appears to be from early brood having already moulted most body-feathers and many greater coverts, but left others unmoulted; many unmoulted juv median coverts now heavily abraded. Juv/1stW more likely to be confused with Richard's or Blyth's Pipits. (Left: P. Dam; right: H. Shirihai)

(n 27, m 69.6), ♀ 62–70 mm (n 15, m 66.5); **T/W** m 75.4; **B** 15.0–18.9 mm (n 46, m 17.2); **B**(f) 12.3–14.2 mm (n 24, m 13.3); **BD** 4.0–5.1 mm (n 16, m 4.6); **BW** 3.8–5.3 mm (n 15, m 4.7); **Ts** 23.5–27.4 mm (n 44, m 25.5); **HC** 7.9–11.9 mm (n 44, m 9.2). **Wing formula: p1** minute; **pp2–4** about equal and longest (sometimes p4 0.5–1 mm <); **p5** < wt 1–3.5 mm; **p6** < wt 9–15 mm; **p7** < wt 15–20 mm; **p10** < wt 22.5–30 mm; **s1** < wt 23–31.5 mm. Emarg. pp3–5. **White wedge r5** (inner web) 18–41 mm (n 44, m 32.4), narrow (exceptionally diffuse or lacking).

GEOGRAPHICAL VARIATION & RANGE Monotypic. – Europe except the north, NW Africa, Turkey, N Middle East, Central Asia, SW Siberia, W Mongolia; winters Sahel belt in N Africa, Arabia and India. – Often considered polytypic, with up to four subspecies recognised (westernmost *campestris*, as well as *boehmii*, *kastschenkoi* and *griseus*), but we fail to discern any consistent differences, and like Alström & Mild (2003) we treat *A. campestris* as monotypic. Still, eastern birds between E Iran and through Central Asia to E Kazakhstan, usually referred to *griseus*, seem rather consistently smaller, noticeably duller and overall greyer and plainer above (some lack any sandy-ochre or buff pigments of most western populations); they also tend to have whiter underparts and supercilium, and a pale greyish, rather than buff, suffusion to the flanks. This noted, some are inseparable from European birds, and more material should be examined before this race is acknowledged. (Syn. *boehmii*; *griseus*; *kastschenkoi*.)

REFERENCES Schmidt, C. (1993) *Limicola*, 7: 178–190. – Vinicombe, K. (2007) *Birdwatch*, 172 (Oct): 24–26.

Israel, Oct: possibly ad due to lack of any obvious juv wing-feathers and less streaky plumage. However, 1stW following extensive post-juv moult of wing-coverts cannot be eliminated. (L. Kislev)

1stW, Portugal, Sep: after still very limited post-juv moult with mostly scaly juv plumage remaining on upperparts, and with sharply pointed dark centres of mainly juv wing-coverts. Note quite obvious dark facial markings, and throat and breast streaking, too. Typical dark lores also well developed. (A. Gonçalves)

LONG-BILLED PIPIT
Anthus similis (Jerdon, 1840)

Fr. – Pipit à long bec; Ger. – Langschnabelpieper
Sp. – Bisbita piquilargo; Swe. – Långnäbbad piplärka

Patchily distributed from the Levant through Arabia and E Africa south to the Cape and in W Africa, as well as in S Iran east to the Indian subcontinent, mainly in dry, rocky hills with sparse vegetation. In montane Israel and surrounding areas (where it overlaps with Tawny Pipit), Long-billed Pipit is usually not found above 700 m. Some birds descend from higher altitudes in winter, and it is occasionally recorded at this season alongside Richard's and Tawny Pipits. However, further east in Iran and Pakistan most are apparently migrants.

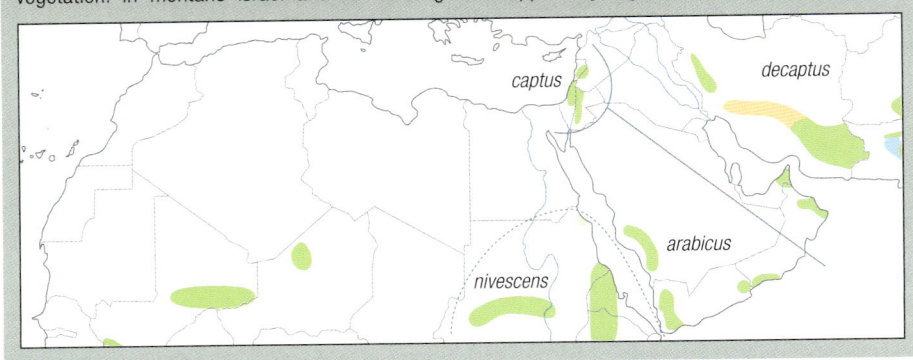

IDENTIFICATION In the treated region might be confused with the quite similar Tawny Pipit. Generally a subtly-patterned large pipit, lacking prominent markings on the predominantly dun-coloured plumage, although combination of generally strong build, weak facial pattern apart from *long supercilium*, long bill and voice permits correct identification, even by those lacking previous experience. Note in comparison to Tawny generally *slightly darker and browner above* and *less prominent median-coverts bar* (bases of coverts less blackish, pale tips warmer brown and less contrasting), plus *poorly marked or no dark moustachial stripe*. Of considerable value for identification is that Long-billed is *almost exclusively encountered on grassy montane slopes* with exposed boulders (on which it frequently perches rather more horizontally). Furthermore, identification is usually not difficult, even of the least distinctive race in the treated region, *captus* of the Levant, while separation of *arabicus* in S Arabia, *nivescens* which probably reaches extreme SE Egypt, or *decaptus* of Iran and Pakistan, from Tawny Pipit is even more straightforward. All plumages share pale yellow-brown legs, and dark culmen/bill tip and yellowish-pink base to lower mandible. Creeps among rocks and bushes (often entering cover far ahead of approaching observer, from where it can be difficult to flush), and tail movements generally less pronounced than in Tawny, involving slight fans or upward motions (never wags). Rather shy or secretive, not easy to approach closely.

VOCALISATIONS Song rather simple, typically involving two far-carrying deliberate phrases separated by a pause, the first rising and second falling, *tir-ee...tiu* and *chu-weet...chee...tseeueet* or in full song 3–4 phrases, e.g. *siu...chürr...sui...chivü...srrüi...siu...* or *shreet-shuwit-shree* (first and last notes harsh, the second purer, louder and disyllabic), given from an exposed rock or in slow fluttering song-flight. – Commonest flight-calls *dyeep* or *chupp* similar to Desert Lark (which can breed in same areas) and often repeated, e.g. *chip... chüp... chüp... chip... chüp... chüp-chüp-chüp*.

SIMILAR SPECIES Size, shape and jizz generally recall *Tawny Pipit*, though averages larger and bulkier (not as sleek or *Motacilla*-like as Tawny), appearing heavy-chested with a visibly longer bill (stouter with a slight droop due to the marginally more curved culmen; Tawny has a thinner, more pointed bill). Proportionately somewhat shorter-legged, and tail rather longer and fuller (obvious when flushed). Generally appears to have rather uniform grey-brown upperparts, but eastern populations of Tawny might appear to have more greyish upperparts, and these migrate through range of Long-billed. Closed tail of latter always darker, contrasting

A. s. captus, presumed 1stW, Israel, Dec: often has cold, bland plumage. Long-billed and long-tailed, but comparatively short-legged. Inner three greater coverts often longer than rest, but in this case they represent post-juv renewal; also blotchier breast-streaking suggests 1stW. (L. Kislev)

A. s. captus, Israel, Jan: rather long and slightly curved bill, along with characteristic upward stance. Levantine race usually pale and rather unpatterned. The dull stony grey-brown and diffusely streaked upperparts often contrast with darker wings and relatively long tail. Streaking on breast often profuse but thin, and flanks and lower belly often tinged warm buff. Rather well-defined and narrow whitish supercilium often tends to narrow in front of eye; warmer russet ear-coverts typical. Ageing uncertain, but rather heavy wear to tertials and inner greater coverts, and pointed rectrices, indicate 1stW. (H. Shirihai)

PIPITS AND WAGTAILS

Long-billed (left: *arabicus*, Yemen, Jan) and Tawny Pipits (right: Oman, Nov): former has bulkier, and proportionately longer tail; bill averages longer, with more curved culmen. Head pattern weaker in Long-billed, with narrower and less contrasting supercilium, and rather poorly-developed moustachial and lateral throat-stripes, than Tawny. Underparts deeper buff-grey (lacks contrasting whiter belly of Tawny), and breast-streaking more even. The Tawny is apparently ad, by evenly rather fresh wing-feathers; Long-billed best left un-aged as impossible to evaluate age-related feather patterns. (Left: H. & J. Eriksen; right: A. Below)

more with rest of plumage, but has less striking pattern when spread, since pale outer edges/tips of two outermost tail-feathers are less extensive and less pure white (being more buff). Head pattern generally poorly marked, with prominent dark eye but weaker and narrower supercilium, albeit quite long (Tawny has an obvious broad supercilium emphasized by dark crown-sides and eye-stripe), and narrower loral line than Tawny. Moustachial stripe indistinct and shorter, lateral throat-stripe reduced, paler and diffuse (both at times well marked on Tawny), and submoustachial region poorly defined in comparison. Underparts usually show minimal contrast with upperparts, as former, including undertail-coverts, saturated buff-grey (lacking contrastingly whiter underparts with two-toned appearance to flanks and belly of Tawny). Long-billed has diffuse breast-streaking, lacking, or virtually so, on adult Tawny (but can be rather heavy in juvenile and first-year Tawny, albeit paler and less heavy than same plumages of Long-billed). Generally, upperwing more uniform with upperparts. Also, has slightly different pattern to adult-type median coverts: on Long-billed, dark centres project as a long 'tooth' usually to the feather tip, and centres more diffuse, especially on inner webs, than in Tawny (on which blackish centre is better defined, highly contrasting and rarely projects towards tip). Greater coverts and tertial pattern similar, being duller, with more diffuse feather-centres than in Tawny. Juveniles of Long-billed and Tawny Pipits may occur together (e.g. in Israel), even before acquiring first-winter feathering, but can still be separated (see below). – Separation from *Richard's Pipit* fairly straightforward on less prominent dark streaking above and on breast, shorter legs and considerably shorter and well-curved hind claws, finer pointed bill, buffish brown-tinged pale sides (rather than white) to tail, vocalisations, etc. – Cf. also *African Pipit* (in SW Arabia), which may overlap with the more streak-backed race of Long-billed (*arabicus*), although the two are still readily differentiated.

AGEING & SEXING Ageing possible by close inspection of moult, feather pattern and wear in wing (easier in autumn). Sexes alike by plumage (though see Biometrics). – Moults. Varying geographically, depending on timing of breeding, but all have complete post-nuptial and partial post-juv moults, mostly mid Jul–mid Sep, sometimes into Nov. Partial post-juv moult usually includes most or all head- and body-feathers and lesser coverts, all median coverts, most or all greater coverts, tertials and alula, and often r1 (or a few more tail-feathers). Suspended moult in autumn among ad probably infrequent or absent, at least in most populations. Partial pre-nuptial moult absent in most populations within the treated region, or limited to some body-feathers, but perhaps usually no or only few wing-coverts, sometimes all tertials and several tail-feathers (such moult mostly involves 1stY). – **SPRING Ad** Uniformly rather fresh wing- and tail-feathers. With wear duller greyish-brown above, sometimes with dark blotchy feather centres. Also paler below, and wing and face markings reduced. **1stS** Perhaps separable in hand or in sharp photographs by some retained juv secondary-coverts. All primaries and primary-coverts juv and more worn, bleached browner, with narrow pale fringes to primary-coverts. Unlike ad, a few 1stS show three generations of greater coverts (a few heavily worn juv outer, less strongly worn ad-type post-juv central, and fresher pre-nuptial inner). 1stS with no moult limits among secondary-coverts extremely difficult or impossible to age. – **AUTUMN Ad** By Sep/Oct evenly fresh. **1stW** Look for moult limits with some juv outer secondary-coverts, tertials and even some remaining juv body-feathers. 1stW having renewed entire wing except remiges and primary-coverts (rather frequent) less easily aged, but primary-coverts subtly weaker textured and already slightly worn (especially compared to renewed

Variation in *A. s arabicus*, Oman (left, Nov) and Yemen (right, Jan): this race is often streakier and warmer-toned, albeit some are much paler and greyer. Like other races, variation also age- and feather wear-related: blotchier breast-streaking on right-hand bird is often seen in 1stY birds. Generally (as shown by ad on left), upperparts often appear duller, tinged stony grey-brown, with less contrasting blackish-centred median-coverts bar. Both show species' characteristic long-billed appearance, with slightly curved culmen. Left bird appears to have uniformly ad wing-feathers. (Left: D. Occhiato; right H. & J. Eriksen)

secondary-coverts and tertials). Also, retained primary-coverts and (occasionally) central alula have clearer-cut whitish fringes/tips. **Juv** As juv Tawny but has relatively weaker-patterned upperparts, lacking scaly pattern of Tawny; also (unlike Tawny) has much weaker lateral throat-stripe and fewer, sparsely scattered, small triangular brown spots on breast-sides, producing pattern less different from ad than in juv Tawny. Underparts rather uniform pale cinnamon (underparts of most Tawny Pipits generally whitish, especially belly).

BIOMETRICS (*captus*) **L** 17–18.5 cm; **W** ♂ 90–98 mm (*n* 18, *m* 95.1), ♀ 86–94.5 mm (*n* 9, *m* 89.9); **T** ♂ 71–82 mm (*n* 17, *m* 75.9), ♀ 70–77 mm (*n* 9, *m* 73.1); **T/W** *m* 80.4; **B** 18.3–22.0 mm (*n* 28, *m* 20.1); **BD** 4.2–5.2 mm (*n* 12, *m* 4.6); **BW** 4.3–5.5 mm (*n* 12, *m* 5.0); **Ts** 23.6–27.8 mm (*n* 26, *m* 25.9); **HC** 7.3–10.9 mm (*n* 26, *m* 9.4); **HC/W** 7.4–12.0 (*n* 27, *m* 10.0); **pale wedge r5** (inner web) 8–22 mm. **Wing formula: p1** minute; **pp2–5** about equal and longest, but p2 and p5 often < wt 0.5–1.5 mm; **p6** < wt 2.5–5 mm; **p7** < wt 10–14 mm; **p10** < wt 18–20.5 mm; **s1** < wt 18–22 mm. Emarg. pp3–5.

GEOGRAPHICAL VARIATION & RANGE Rather marked variation and several subspecies separated, many of which are extralimital and not treated here. All populations resident within the covered region, although some may descend from higher altitudes in winter. (Measurements for *captus* and *nivescens* based on material in AMNH, NBC, NHM, NHMW and TAUZM, for *decaptus* and *arabicus* on data from Alström & Mild 2003.)

A. s. captus Hartert, 1905 (Levant). Described above, a rather small race with proportionately long slender bill, rather pale, plain grey upperparts and very faint and narrow streaking of underparts. Generally speaking, *captus* is close in size to *arabicus* but its colours are nearer *decaptus*. See above for biometrics.

A. s. decaptus Meinertzhagen, 1920 (Iran, N Oman east to NW India). Larger, slightly more long-tailed and long-billed than *captus*, with rather dull upperparts. Also, deeper ochre-buff below with better-defined breast-streaking. **W** ♂ 99–109.5 mm (*n* 14, *m* 103.3), ♀ 94.5–101 mm (*n* 11, *m* 97.5); **T** ♂ 79.5–99 mm (*n* 14, *m* 86.6), ♀ 79–85 mm (*n* 9, *m* 82.7); **B** ♂ 19.4–22.8 mm (*n* 13, *m* 20.7), ♀ 19.5–21.0 mm (*n* 11, *m* 20.3).

A. s. arabicus Hartert, 1917 (S Arabia). Rather similar to *captus* in size but browner and streakier above, more heavily and blotchier streaked on breast, with warmer brown underparts and fringes to outer tail-feathers, broader/blacker centres and more rusty-buff fringes to wing-coverts and tertials. **W** ♂ 92.5–101.5 mm (*n* 11, *m* 97.7), ♀ 89–93.5 mm (*n* 10, *m* 91.6); **T** ♂ 72.5–79.5 mm (*n* 11, *m* 75.7), ♀ 69.5–77 mm (*n* 9, *m* 73.2); **B** ♂ 18.7–20.3 mm (*n* 11, *m* 19.7), ♀ 18.2–20.0 mm (*n* 10, *m* 19.0).

A. s. nivescens Reichenow, 1905 (NE Sudan to N Somalia and perhaps SE Egypt, where the species has only recently been discovered). Differs from *arabicus* (although they apparently intergrade in Yemen) by paler and sandier upperparts (less dusky grey-brown), and has buffier underparts with slightly whiter belly and narrower, sharper breast-streaking. Relatively small: **W** ♂ 86–98 mm (*n* 7, *m* 93.8), ♀ 84–93.5 mm (*n* 12, *m* 88.3); **T** ♂ 65–75.5 mm (*n* 7, *m* 71.8), ♀ 62–71 mm (*n* 12, *m* 66.8); **T/W** *m* 76.9; **B** 17.2–19.9 mm (*n* 12, *m* 18.8).

REFERENCES Laird, W. & Gencz, A. (1993) *BB*, 86: 6–15.

A. s. decaptus, presumed ad, United Arab Emirates, Dec: a larger, longer-tailed and longer-billed race, also tends to have deeper ochre-buff underparts, and well-marked breast-streaking. Note ad-type median coverts, with dark centres projecting like 'teeth' towards tip, and more diffuse centres, especially on inner webs. Evenly fresh wing suggests ad. (I. Boustead)

A. s. decaptus, Iran, Jan: rather dull-coloured race, with birds in drier habitats (and worn plumage) tending to have greyer/paler plumage. Ageing uncertain without closer inspection of wing. (A. Ouwerkerk)

BERTHELOT'S PIPIT
Anthus berthelotii Bolle, 1862

Fr. – Pipit de Berthelot; Ger. – Kanarenpieper
Sp. – Bisbita caminero; Swe. – Kanariepiplärka

This endemic to the Macaronesian islands in the NE Atlantic prefers dry, open, rocky or sandy country on seashores, heaths and mountain slopes with much grass and scattered bushes. It is found in the Canaries, where it has been recorded to quite high altitudes (often 2500 m, even to 3000 m), and in the Madeira group and the Selvagens, where it is usually the only pipit, and should therefore pose few identification problems. Due to the open habitat it favours, it is often easily spotted once you are within its range.

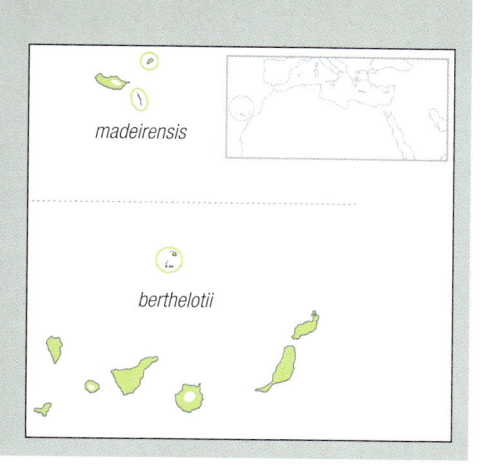

A. b. berthelotii, Canaries, Dec: strongly marked dark streaking on breast, much thinner on flanks. When fresh, whitish supercilium, small dark loral stripe, lateral throat-stripe and moustachial stripe, and pale submoustachial region are all pronounced. (J. Vakkala)

A. b. berthelotii, Canaries, May: the smallest pipit in the covered region, endemic to NE Atlantic islands, where often found in volcanic habitats. Often rather tame. Here a quite worn bird; some have much reduced dusky loral-stripe and narrower dark streaking on breast and flanks. Age uncertain without handling. (H. Shirihai)

IDENTIFICATION A rather small and strongly-streaked pipit. *Slightly smaller than Meadow Pipit*, but structure and actions recall Rock and Water Pipits, with several structural and especially plumage similarities to large pipits, particularly Tawny Pipit. *Predominantly brownish-grey upperparts* (impression more grey than brown) rather noticeably streaked darker (but nape, lower back and rump almost unmarked), and *wings rather strongly patterned* (distinctive whitish median-covert bar with contrasting black centres, but less distinct pale buff-white fringes to greater coverts and tertials). *Head pattern obvious*, with well-developed *cream-white supercilium* (though sometimes diffuse) and lower-ear-coverts patch, and small *dusky loral patch* (thus appears to have long dark eye-stripe), while whitish eye-ring is rather pronounced. *Obvious black-brown lateral throat-stripe and moustachial stripe, and pale submoustachial region rather well developed*. Otherwise has *rather clean whitish underparts* with variable pale pink or buff suffusion to neck-sides, breast and flanks, and *pronounced black-brown streaking on breast and upper flanks*. Black-brown tail in flight shows distinctive whitish edges to outermost feathers. *Bill slender* (but not really long) with dark horn-coloured culmen/tip and pale yellowish (even reddish-tinged) base to lower mandible, while legs (short and set well back) are yellow-brown. On ground generally recalls Water Pipit, with noticeable head movement when walking and runs swiftly in quick forward dashes (with brief stops), hopping on boulders. Wags tail frequently, often rather vigorously and constantly. Typically clambers over small plants when feeding. Flight not markedly bounding, unlike many pipits, and often extremely tame and prefers to run rather than escape by flight. Solitary, or forms small groups in winter.

VOCALISATIONS Song most similar to Tawny Pipit (given in undulating song-flight), a plaintive simple disyllabic note, delivered rather energetically and higher-pitched than Tawny Pipit, *tzirle* or *tsivirr*, repeated 4–7 times, and sometimes even acquires a prinia-like rattling and rolling quality, *tzir'r'le*. – Calls essentially recall Tawny Pipit and Yellow Wagtail; commonest is a low-pitched *tsri(e)* or *tchelee* and *pschirlip*. Also a short *chup* and a soft rolling *kree*.

SIMILAR SPECIES The only pipit in its range, thus confusion unlikely except with vagrants, especially Meadow, Tree, Water and Tawny Pipits, but these should be readily eliminated by distinctive structural, size and behavioural differences, as well as by plumage characteristics and calls.

AGEING & SEXING Ageing best achieved by close examination of moult, feather pattern and wear in wing, but never

A. b. berthelotii, Canaries, Dec: much individual variation in streaking on underparts; here it is extensive and bold on breast, but becomes more diffuse and paler on flanks. Median coverts apparently juv, and together with heavily worn rest of wing suggest 1stW. (M. Høegh Post)

easy, and perhaps only correctly achieved with obvious examples. Sexes similar. – Moults. Post-nuptial moult complete and post-juv moult partial, latter including all head, body, lesser and median coverts, variable number (none to all) of greater coverts and tertials, and often r1, mainly Jun–Aug. No pre-nuptial moult. – SPRING **Ad** As autumn, but especially when heavily worn much plainer and greyer above. **1stS** As ad but at least sometimes distinguished in hand (or sharp photographs) by moult limits between replaced and some retained juv secondary-coverts. All primaries and primary-coverts juv and more heavily worn. Due to wear and bleaching, differences from ad obscured, especially on those lacking juv secondary-coverts. – AUTUMN **Ad** By Sep/Oct evenly fresh: look for presence of only ad-type greater coverts with more smoothly demarcated dark centres (and no sign of moult limits in wing); typically, ad also tends to be more diffusely streaked on breast and body-sides, while the upperparts are greyer and less distinctly streaked (see photo). **1stW** Aged by unmoulted juv tertials and outer greater coverts (slightly looser structure, clearer-cut dark centres and paler tips than adjacent more recently moulted ad-type feathers). Also retains many tail-feathers (but central rectrices often renewed), primaries and primary-coverts, which are subtly weaker textured and more worn than ad (but experience required to determine such differences). Some moult more extensively and are therefore less easily separated. **Juv** Resembles ad but more intensely streaked above and below (and more spotted on breast). Upperparts have greater number of more contrasting dark centres, and wing-coverts and tertials have sharper dark centres (pattern nearer juv Tawny Pipit).

BIOMETRICS (*berthelotii*) **L** 13.5–14.5 cm; **W** ♂ 75–80 mm (n 13, m 77.3), ♀ 73–77 mm (n 6, m 74.8); **T** ♂ 59–64 mm (n 13, m 61.2), ♀ 58–61.5 mm (n 6, m 59.6); **T/W** m 79.3; **white wedge on r5** 20–35 mm (n 15, m 26.4); **B** 14.0–15.7 mm (n 19, m 14.8); **B**(f) 10.5–12.3 mm (n 19, m 11.4); **BD** 3.2–3.8 mm (n 18, m 3.5); **Ts** 21.4–23.4 mm (n 19, m 22.6); **HC** 6.7–9.5 mm (n 19, m 8.4). **Wing formula: p1** minute; **p2** < wt 0.5–3 mm; **pp3–4**(5) longest; **p5** < wt 0–1 mm; **p6** < wt 6–8 mm; **p7** < wt 11–13 mm; **p10** < wt 16–21 mm; **s1** < wt 17–21 mm. Emarg. pp3–5.

A. b. berthelotii, Canaries, presumed 1stY, Feb: moderately worn. Streaked dark brownish-grey upperparts, rather strongly patterned wings, with distinctive whitish median-coverts bar and contrasting black centres. Moult contrast among tertials and median coverts, worn, pointed and bleached brownish primary-coverts, and patchy moult among mantle and scapulars suggest 1stY. (R. Armada)

GEOGRAPHICAL VARIATION & RANGE Slight variation only. Two very similar subspecies recognised.

A. b. berthelotii Bolle, 1862 (Canaries; resident). Treated above. Plumage colours and patterns on average the same as in the following race, but *berthelotii* has shorter bill and a fraction shorter hind claw.

○ *A. madeirensis* Hartert, 1905 (Madeira; resident). Very similar to *berthelotii*, and plumage apparently the same, but *madeirensis* differs subtly in its longer and slightly stronger bill and fractionally longer hind claw and shorter white wedge on r5. **W** ♂ 74–80 mm (n 11, m 77.5), ♀ 72–78 mm (n 6, m 75.3); **T** ♂ 55–63 mm (n 11, m 59.6), ♀ 54–62.5 mm (n 6, m 59.1); **T/W** m 77.4; **white wedge on r5** 13–30 mm (n 17, m 23.9); **B** 15.5–17.4 mm (n 16, m 16.7); **B**(f) 12.0–13.6 mm (n 17, m 13.0); **BD** 3.4–4.2 mm (n 17, m 3.9); **Ts** 21.0–23.5 mm (n 17, m 22.6); **HC** 7.0–10.5 mm (n 18, m 8.9).

TAXONOMIC NOTES Long treated as being closely related to smaller pipits, especially Rock and Water Pipits, but modern treatments tend to recognise it as being closer to the larger pipits, especially Tawny Pipit, despite it being the smallest pipit of the treated region. – Although Alström & Mild (2003) considered *A. berthelotii* to be monotypic, we feel that the longer and slightly stouter bill of *madeirensis* constitutes reason enough to recognise the latter.

A. b. madeirensis, Madeira, Nov: possibly fresh ad, with no sign of any juv feathers; only ad-type greater coverts with more smoothly demarcated dark centres (juv coverts have clearer-cut dark centres). Extensive and bold streaking on breast is more diffuse on upper flanks. When still fresh, upperparts tinged greyer and less distinctly streaked. (H. Nussbaumer)

A. b. madeirensis, Madeira, Oct: another fresh bird, with breast heavily blotched; apparently ad, by evenly fresh, ad-like wing-coverts. Longer bill is a feature of Madeiran birds. (R. Pop)

OLIVE-BACKED PIPIT
Anthus hodgsoni Richmond, 1907

Fr. – Pipit à dos olive; Ger. – Waldpieper
Sp. – Bisbita de Hodgson; Swe. – Sibirisk piplärka

Olive-backed Pipit is a close relative of Tree Pipit, occurring in NE Russia and the Siberian taiga. In the western part of its range the two species overlap, but further east only Olive-backed breeds. Olive-backed Pipit seems to prefer slightly more closed forests and higher altitudes than Tree Pipit, but there is overlap as to both. Its song sounds like a mixture of Tree Pipit and an accentor—or even a bunting or *Phylloscopus* warbler. It is usually delivered from an elevated perch in a treetop. It winters in SE Asia but straggles rather frequently to many W European countries.

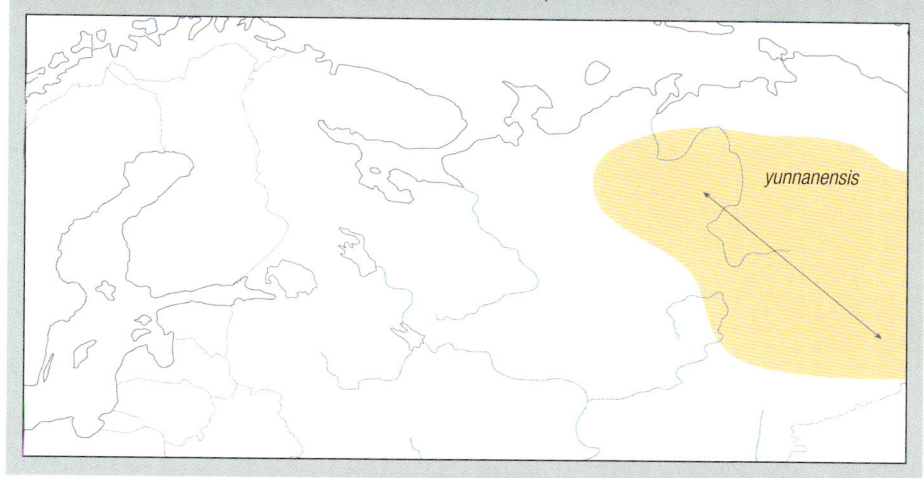

of Red-throated Pipit song. It does *not* slow down now and then with repeated whistling notes *see*va *see*va *see*va..., like Tree Pipit. Other species that can come to mind when the song is heard, at least distantly, are as unrelated as Chestnut Bunting and Pallas's Leaf Warbler. – Call very similar to that of Tree Pipit, a slightly hoarse, high-pitched drawn-out *speez*. Perhaps on average slightly thinner and less 'buzzing' than Tree Pipit, more akin to Red-throated Pipit, but the calls of all pipits vary a little individually. Whether the call of Olive-backed can be separated with confidence from that of Tree Pipit in W Europe is therefore debatable, to say the least. Alarm is a rather distinctive short, shrill, slightly hoarse *zrit* (thinner and sharper than in Tree Pipit), steadily repeated.

SIMILAR SPECIES Differs from *Tree Pipit* by combination of: (i) stronger contrasting head pattern, with more prominent bicoloured supercilium (ochre-buff in front of eye, the rest white; normally only a hint of this in Tree Pipit), black margin above the supercilium, and a white and a black spot behind and slightly below the rear end of the supercilium; (ii) weaker streaking on more greenish back (the normally subdued streaking can in certain angles and lights still look fairly obvious); and (iii) in autumn more uniform greenish, less contrasting tertials, lacking a whitish edge; and (iv) breast-streaking usually somewhat bolder than in Tree Pipit (but some overlap).

AGEING & SEXING (*yunnanensis*) Ages very similar after post-juv moult and generally not separable. Sexes alike. – Moults. Complete post-nuptial moult of ad in Aug–Sep prior to autumn migration. Simultaneous partial post-juv moult does not involve flight-feathers or tertials, and only rarely some inner greater coverts. Juv tertials have same pattern as ad, and keep in good condition through autumn, making ageing more difficult than in Tree Pipit. Both age groups undergo a partial pre-nuptial moult in winter quarters. – **SPRING** Ad and 1stS alike in the field (but see under Tree Pipit for possible in-hand characters). – **SUMMER** **Ad** Plumage worn. Due to wear, streaking of upperparts rather prominent. Breast heavily blotched black. **Juv** Plumage fresh. Upperparts somewhat more brown-tinged, less greenish than ad. Streaks of underparts more diffuse and long, not such well-marked blotches as ad. – **AUTUMN** **Ad/1stW** These two generally alike unless 1stW separable by (slight!) contrast between moulted and unmoulted wing-coverts.

BIOMETRICS (*yunnanensis*) **L** 14–15.5 cm; **W** ♂ 80–89 mm (*n* 22, *m* 85.3), ♀ 81–85.5 mm (*n* 16, *m* 83.5); **T** ♂ 59.5–65 mm (*n* 22, *m* 61.8), ♀ 57–63 mm (*n* 16, *m* 59.7); **T/W** *m* 71.9; **B** 13.7–15.5 mm (*n* 38, *m* 14.6);

a *black upper margin*. At rear the *supercilium appears to be 'broken'*, a white patch being separated/surrounded by some black on virtually all birds (but beware of the odd Tree Pipit with a head pattern that approaches this). Tertials in autumn invariably fresh and greenish in both adults and first-years (cf. Tree Pipit). Legs pale pink, *hind claws short and curved* like Tree Pipit.

VOCALISATIONS Song best compared with Tree Pipit but is a little weaker and higher-pitched, and although it is quite varied it is in a sense more monotonous with fewer changes in pitch, pace or structure, hence can recall Black-throated Accentor and Dunnock. The inclusion of dry trills or rattles and some high, thin notes can also remind you

A. h. yunnanensis, Oman, Jan: typical greyish-green coloration and subdued darker streaking above; head pattern distinctive. Ageing uncertain without handling: juv inner greater coverts often more olive-tipped than rest, giving superficial impression of moult limits. (H. & J. Eriksen)

IDENTIFICATION A typical small pipit in shape and plumage, and must be separated primarily from Tree Pipit. *Upperparts greyish-green* with only subdued darker streaking on mantle and back (usually clearly less streaked than Tree Pipit), and *unmarked rump*. Crown and nape streaked dark. Underparts white with buff tinge on sides of throat, across breast and on sides, often becoming more olive-grey on lower flanks. Birds with strongest colours can have throat (especially sides) and upper breast tinged orange-buff, almost recalling ♀ Red-throated Pipit. *Boldly streaked* or spotted *black on breast*, more thinly on upper belly and flanks. Head pattern distinctive with *prominent supercilium*, largely white but in most birds *tinged obviously ochraceous-buff in front of eye*. Prominence of supercilium in majority enhanced by

Autumn Olive-backed (left: England, Oct) and Tree Pipits (right: Ethiopia, Sep): Olive-backed has weaker streaking on subtly greener mantle and bicoloured supercilium, enhanced by black upper border, rear ear-coverts spots, and more evenly greenish, less contrasting tertials, lacking whitish edge of Tree at this season; streaking below usually bolder. Uniform wing-feathers of Olive-backed suggest an ad, but as the species has very similar ad and juv feathers this can only be confirmed in the hand. In contrast, ageing of Tree in autumn is easier, with this bird being a 1stW (most wing-coverts juv). (Left: R. Nason; right: H. Shirihai)

South Korea, Sep: characteristic head pattern with bicoloured supercilium, whitish behind eye, tinged ochraceous in front, white-and-black rear ear-coverts patches, and black lateral crown-sides. Ageing and subspecific determination uncertain. (A. Audevard)

China, Oct: highly characteristic, bold head pattern and silky white underparts with bolder streaking (cf. Tree Pipit). Ageing uncertain as is subspecific determination without handling. (I. Fisher)

BD 3.6–4.4 mm (*n* 28, *m* 4.1); **BW** 3.9–4.7 mm (*n* 28, *m* 4.2); **Ts** 19.5–21.7 mm (*n* 38, *m* 20.8); **HC** 6.5–8.5 mm (*n* 37, *m* 7.5). **Wing formula: p1** minute; **pp2–4** about equal and longest; **p5** < wt 0.5–3 mm; **p6** < wt 9–14 mm; **p10** < wt 20–26 mm; **s1** < wt 23–27 mm; **tert. tips** < wt 0–8.5 mm.

GEOGRAPHICAL VARIATION & RANGE Two subspecies generally recognised differing mainly in upperparts streaking and wing shape.

A. h. yunnanensis Uchida & Kuroda, 1916 (E Russia, Siberia, N Mongolia; winters in S & SE Asia). Treated above. Upperparts streaking generally moderate, especially in fresh plumage. Streaking strongest on crown (but rather dull olive-grey ground makes streaking appear still rather moderate). The only race known to have occurred within the treated range. (Syn. *inopinatus*.)

Extralimital: **A. h. hodgsoni** Richmond, 1907 (Himalaya, C China, Japan; winters in SE Asia). Has more prominently streaked crown, mantle, back and scapulars (though slightly variable, and difference not always obvious); bold black streaks on crown often stand out due to rather pale brownish-olive ground. More strongly streaked also on breast (sometimes very heavily blotched, and reaching upper belly

A. h. yunnanensis, 1stW, United Arab Emirates, Oct: some are washed deep buff below (an extreme example pictured). Ageing uncertain. (M. Barth)

and flanks). Slightly more blunt wing, apparently mainly due to longer secondaries. **W** ♂ 83–88 mm (*n* 10, *m* 85.2), ♀ 79–85 mm (*n* 10, *m* 82.1); **T** ♂ 57–66.5 mm (*n* 10, *m* 61.4), ♀ 55–65 mm (*n* 11, *m* 61.0); **T/W** *m* 73.0; **HC** 7.5–9.0 mm (*n* 13, *m* 7.8). **Wing formula: p5** < wt 0–2 mm, **p6** < wt 6.5–11.5 mm, **p10** < wt 18–23 mm; **s1** < wt 19–23.5 mm. (Syn. *berezowskii*, *maculatus*.)

A. h. yunnanensis, Mongolia, May: in spring/summer when worn, Olive-backed is to some extent less strikingly coloured, chiefly because it becomes greyer/duller and somewhat streakier above, with less obvious head markings. Ageing uncertain. (H. Shirihai)

PIPITS AND WAGTAILS

TREE PIPIT
Anthus trivialis (L., 1758)

Fr. – Pipit des arbres; Ger. – Baumpieper
Sp. – Bisbita arbóreo; Swe. – Trädpiplärka

Tree Pipit, as its name suggest, occurs in woodlands up to the tree-line, but like other pipits also in many types of open habitat. It would be easily overlooked if it were not for its loud and persistently-delivered song. It is a summer visitor to open wooded areas (from lowland to mountains, the latter preferred in southern parts of the range), heaths with copses of trees, more open meadows at high altitude in the south or in bogs and tundra in the far north provided there are a few taller trees. It breeds across Eurasia to eastern Russia and Mongolia, south to N Turkey, N Iran and the Himalayas, wintering in sub-Saharan Africa and the Indian subcontinent.

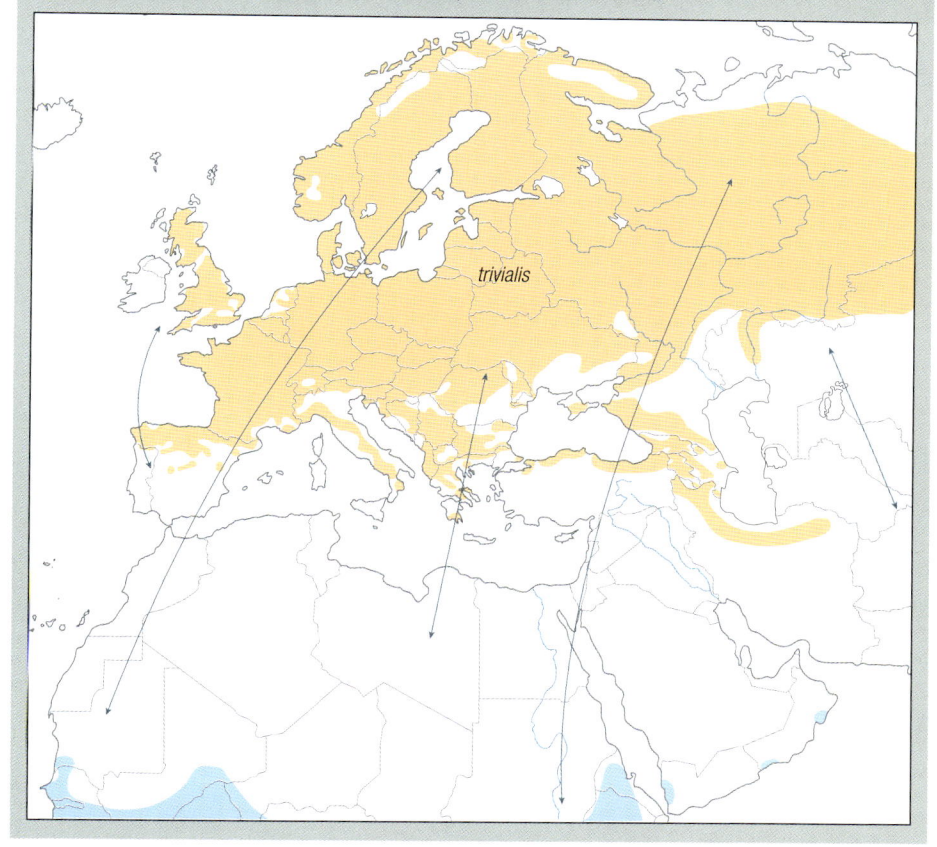

trivialis

started from treetop ascending some 5–10 m in fluttering flight, then descending into trees at end, parachuting on stiff wings.

VOCALISATIONS Full song comprises a loud series of trills and whistling notes repeated with a varied tempo and rising and falling pitch, often including highly typical retarded passages with more and more drawn-out notes followed again by very fast series, e.g. zit-zit-zit-zit cha-cha-cha-cha sürrrrrrrrr *siiiii*-a tvet-tvet-tvet-tvet siva *siiva siiiva*, *siiihva* cha-cha-cha. Full song usually delivered only now and then and in song-flight, while a shorter or interrupted song version is given repeatedly from treetop between outbursts of full song. – Flight call a straight, slightly hoarse, high-pitched *pzeeh* or *spihz*, quite like Olive-backed Pipit though by comparison subtly lower-pitched and harsher. Alarm call a short, metallic, penetrating, *sütt*, persistently repeated.

SIMILAR SPECIES Greatest confusion risk is with Meadow Pipit, although their song, calls and behaviour are quite different. A silent Tree Pipit feeding on the ground is distinctive by its more shuffling and crouching gait, and deliberate slow walk, pumping (rather than twitching or wagging) the folded tail and typically flying up into the branches when alarmed. Walks branches in trees agilely, again often pumping the tail. Look for greater contrast between buffish and heavier-streaked breast and whiter belly to undertail-coverts, with noticeably more delicate streaking on flanks. Also differs by its slightly heavier bill, bolder head markings (e.g. longer, more obvious pale supercilium, submoustachial stripe and pale ear-coverts spot), less distinct dark markings on mantle, almost unstreaked rump, and shorter, curved hind claws. – For separation from Olive-backed Pipit see that.

AGEING & SEXING Ageing often difficult, and experience needed, even in the hand, using moult, feather pattern, wear in wing and iris colour (easier in autumn). Sexes alike in plumage. – Moults. Complete post-nuptial moult of ad in late Jul–Sep prior to autumn migration. Partial post-juv moult simultaneously, in northern populations including head and body, usually some lesser but only occasionally some median coverts, rarely 1–3 (4) inner greater coverts, and exceptionally 1–3 tertials and r1, but replacement more extensive in southern populations, including more secondary-coverts and tertials, and more frequently r1. Partial pre-nuptial moult (Jan–Mar) of both age categories involves some head- and body-feathers, none to all (usually some) lesser coverts, some (often all) median coverts, none to all greater coverts (usually 3–5 inner), usually all (rarely no) tertials, and usually r1 (rarely all tail-feathers). – **SPRING** Ageing very difficult and

IDENTIFICATION A small, olive-brown, streaky pipit, readily separated by voice, jizz and behaviour from several congeners. In W Palearctic main confusion risk is with Meadow Pipit, versus which Tree Pipit is slightly more robust with a marginally *stronger bill* and *more contrast between snow-white belly and buffish breast/flanks*, and between *well-spaced bolder streaks on breast and much finer streaks on flanks*. (Only exceptionally has slightly bolder streaks also on flanks; check hind claws, rest of plumage and structure on all controversial birds.) That upperparts markings are on average subtly more diffuse (less black and distinct) on Tree Pipit is usually difficult to discern on a single bird. Face pattern of Tree on the other hand is usually clearly better marked, with creamy-yellow *supercilium especially pronounced in front of and above eye* (but sometimes almost entirely obscured), *yellow-white eye-ring broken* (unbroken in Meadow) by dusky loral stripe and eye-stripe, and usually has *pale spot on rear ear-coverts*. On average, creamy-yellow submoustachial stripe is broader and cleaner, and dark lateral throat-stripe and patch somewhat more pronounced. Wing pattern quite well marked with double buff-white wing-bars (especially on median coverts) and well-defined paler tertial fringes. Clear whitish corners visible on spread tail. *Often associated with trees*, and has several typical behavioural characteristics (see below). Song-flight

A. t. trivialis, Finland, Apr: obvious contrast between warmer, heavier-streaked breast and whiter belly, with noticeably more delicate flanks-streaking. As to ageing, moult limits among wing-coverts in spring difficult to evaluate without handling. (M. Varesvuo)

— 149 —

Tree Pipit (left) and Meadow Pipit (right), Finland, Apr: usually readily separated by voice and behaviour, but silent Meadow can be eliminated (with practice) by Tree's more robust bill (with pinkish rather than orange-yellow lower mandible), more contrast between whiter belly and buffish breast/flanks, and much weaker flanks-streaking. Face pattern (especially creamy-yellow supercilium) usually better marked, whitish eye-ring is faintly broken (complete in Meadow) by dusky loral-stripe, while dark lateral throat-stripe somewhat more pronounced, and pale spot on rear ear-coverts is feature of many Tree (absent in Meadow). Ageing of both is uncertain. (M. Varesvuo)

often impossible. **Ad** Only ad-type (worn or fresh) wing- and tail-feathers discernible. With wear buffish hue below paler, and underparts more concolorous; also duller/greyer above with streakier appearance. **1stS** Exceptionally separable in the hand by some retained and very worn juv secondary-coverts; all primaries and primary-coverts juv and more worn and bleached browner than ad (primary-coverts also slightly more pointed and looser textured at tips), but difference in wear and pattern slight, and intermediate-looking birds frequent. Note that both age classes show moult contrast, with at least some new median and greater coverts and tertials, and perhaps tail-feathers, usually central ones, at this season. Iris exceptionally still slightly more dull brown (less warm rufous-brown) than ad. – AUTUMN **Ad** By Aug–Oct evenly fresh (retained unmoulted and heavily bleached wing-feathers rare but diagnostic). **1stW** Often shows clear moult limits with variable number of juv secondary-coverts and tertials (both juv median and greater coverts shorter, subtly looser textured and more worn at tips with slightly paler fringes than adjacent fresh ad-like coverts, while retained median coverts have more triangular clearer-cut dark centres with a more pronounced 'tooth' or pointed tip). Also, retained tail-feathers, primaries and primary-coverts more worn and weaker textured than ad-type feathers, and primary-coverts slightly more pointed and more frayed at tips. 1stW that do not renew any secondary-coverts or which lack obvious moult contrast are more difficult to age. Iris grey-brown (rufous-brown in ad). **Juv** Heavy dark upperparts streaking to rump and uppertail-coverts, otherwise as fresh ad. On average subtly more brown-tinged, less olive above.

BIOMETRICS (*trivialis*) **L** 15–16 cm; **W** ♂ 86–97 mm (*n* 45, *m* 90.2), ♀ 83–91 mm (*n* 18, *m* 87.0); **T** ♂ 59.5–67 mm (*n* 45, *m* 62.6), ♀ 58–64 mm (*n* 17, *m* 61.2); **T/W** *m* 69.6; **B** 13.2–15.7 mm (*n* 37, *m* 14.4); **BD** 3.6–4.4 mm (*n* 22, *m* 4.0); **BW** 4.1–4.9 mm (*n* 21, *m* 4.4); **Ts** 20.0–22.1 mm (*n* 36, *m* 21.0); **HC** 6.3–8.9 mm (*n* 35, *m* 7.6). **Wing formula: p1** minute; **pp2–4** about equal and longest (sometimes p4 0.5–1.5 mm <); **p5** < wt 2.5–6.5 mm; **p6** < wt 10–15 mm; **p7** < wt 14.5–20.5 mm; **p10** < wt 22–31 mm, **s1** < wt 23–31 mm. Emarg. pp3–5. **White wedge r5** (inner web): 1–10 mm (very rarely diffuse or lacking).

GEOGRAPHICAL VARIATION & RANGE Slight and clinal variation with eastern and south-eastern populations become progressively heavier streaked and perhaps subtly paler or greyer. At same time, bill becomes subtly heavier and wing more rounded in the south-east. Only two subspecies warranted based primarily on structure and boldness of streaking, supplemented by wing length and bill size. A few more subspecies suggested but these appear too subtle and variable to be warranted.

A. t. trivialis (L., 1758) (Europe, much of Siberia, N Turkey, Caucasus, Transcaucasia, N Iran, N Kazakhstan, N Mongolia, possibly also S Mongolia and N China; winters

A. t. trivialis, Finland, Apr: characteristic song-flight descent. (R. van der Weijde)

A. t. trivialis, presumed 1stS, Italy, Apr: in spring/summer can wear slightly greyer. Without handling, ageing often uncertain (as both ages moult partially in winter), but such strong moult contrast in wing, and apparently heavily-abraded juv median coverts, strongly suggest 1stS. (D. Occhiato)

PIPITS AND WAGTAILS

A. t. trivialis, Finland, Apr: note rather plain uppertail-coverts, much finer flanks-streaking, broken whitish eye-ring and well-developed dark lateral throat-stripe and neck-side patch. Ageing uncertain. (D. Occhiato)

A. t. trivialis, Israel, Mar: a rather pale spring bird, still showing clear contrast between buffish and heavily-streaked breast and whiter belly to undertail-coverts, with noticeably thinner streaking on flanks. Note also rather short, curved hind claws. Despite moult limit in greater coverts it is unsure if the outer older feathers are ad or retained juv. (H. Shirihai)

A. t. trivialis, 1stW, United Arab Emirates, Oct: typical jizz when feeding on open ground; note characteristic upperparts streaking, broken white eye-ring and thinner flanks-streaking. Later in autumn, increased feather wear may enhance age differences, with contrasting worn juv median coverts. (T. Pedersen)

A. t. trivialis, 1stW, Spain, Sep: stout jizz, including bill (and note pinkish lower mandible), with whiter belly and much finer flanks-streaking, as well as quite well-developed face pattern (creamy-yellow supercilium, broken whitish eye-ring and dark lateral throat-stripe and patch). Moult limits with juv median coverts; remiges and primary-coverts heavily worn. (A. Torés Sánchez)

A. t. trivialis, ad, Denmark, Sep: evenly fresh after complete post-nuptial moult; plumage also neat, and tinged warmer and greenish (less greyish than when worn in spring or summer). (B. Tureby)

mainly Afrotropics and India with some also in Mediterranean basin east to Gulf States). Described above. Biometrics above refer mainly to Swedish birds, whereas some southern populations in Europe and Asia have slightly blunter wing (p5 < wt 2–3.5 mm) and bolder breast streaking, forming a cline to following subspecies. (Syn. *arboreus*; *differens*; *salomonseni*; *schlueteri*; *sibiricus*.)

Extralimital: ○ **A. t. harringtoni** Witherby, 1917 (NW Himalaya, Tien Shan north to about Chu River in SE Kazakhstan, extreme NW China; winters mainly India and adjacent areas). Differs only very subtly by slightly bolder black streaking on crown, mantle and breast (but much overlap, and not difficult to find less boldly streaked birds, as it is to also find boldly streaked breeders in N Europe and Siberia), and by slightly stronger bill. Wing a little blunter on average than in European *trivialis*, but similar to populations in Central Asia and Iran. Only separable when series are compared; individual birds frequently impossible to assign to subspecies. **W** ♂ 83–91 mm (*n* 17, *m* 88.4), ♀ 84–88 mm (*n* 10, *m* 85.9); **T** 59–66 mm (*n* 28, *m* 63.0); **T/W** *m* 72.1; **B** 14.0–16.0 mm (*n* 27, *m* 14.9); **BD** 4.1–5.3 mm (*n* 27, *m* 4.6); **BW** 4.5–5.5 mm (*n* 27, *m* 5.0). **Wing formula: p5** < wt 0–3.5 mm; **p6** < wt 8–11.5 mm; **p7** < wt 13.5–17 mm; **p10** < wt 22–26.5 mm, **s1** < wt 23–28 mm. (Syn. *burzii*.)

REFERENCES KARLSSON, L., PERSSON, K. & WALINDER, G. (1993) *Ornis Svecica*, 3: 69–80.

PECHORA PIPIT
Anthus gustavi Swinhoe, 1863

Fr. – Pipit de la Petchora; Ger. – Petschorapieper
Sp. – Bisbita del Pechora; Swe. – Tundrapiplärka

Among European birdwatchers this is a much sought-after rarity, which sometimes straggles in autumn to seashores and islands of NW Europe (particularly Britain and Ireland, but has also reached, e.g., Fenno-Scandia and France), mainly in late September and October. Pechora Pipit breeds in brushy tundra and the remote taiga zone, from the Pechora River to the Bering Strait, and locally in NE China, moving south-east and south to winter in Indonesia and the Philippines. Birds reaching W Europe, generally thought to be mainly young and inexperienced birds, seem to sometimes make 180° mistake in their first navigation.

Presumed *A. g. gustavi*, Scotland, Oct: diagnostic primary projection, with two exposed tips beyond tertials, plus characteristic bold streaking above, with black-and-white tramlines on mantle/scapulars. (H. Harrop)

IDENTIFICATION (*gustavi*) Recalls first-winter Red-throated Pipit, but averages subtly smaller and shorter-tailed, and unlike that (and other pipits) has *diagnostic short but noticeable primary projection* (2–3 exposed tips beyond longest tertials). Characteristic is the bolder streaking on crown and upperparts, with black-and-white stripes appearing as marked *buff-white 'tramlines' on mantle and scapulars*. Upperparts and especially head and neck are generally rather warm rufous-brown (Red-throated: colder grey-brown). Rump narrowly but still distinctly streaked, and wing-bars more prominent and purer white than in Red-throated. Whitish underparts show only vestigial dark lateral throat-stripe but well-developed lateral patches where throat meets breast. Streaking on breast and flanks quite prominent, and Pechora tends to have buffish breast and whiter belly (in Red-throated these are uniformly buffish or whitish). *Pale eye-ring incomplete, broken by short dark lores and eye-stripe* (Red-throated: eye-ring complete, dark stripes reduced), though head is relatively poorly marked, with only vague pale buff supercilium. Spread tail shows warm-toned buff-white (thus less pure white) wedges to outer feathers. *Rather long, fine bill* (slightly longer than in Red-throated; dark horn with pink base). Legs pink-flesh with hind claws usually quite longish. On ground, stance rather upright, but almost infamously skulks in long grass and is extremely difficult to flush, often walking exceptionally close to observers but in deep cover. Flight less buoyant than Red-throated, perhaps more recalling Meadow Pipit with erratic bursts of wingbeats. Flight-call (given only occasionally) diagnostic. May hover briefly prior to landing. Has peculiar prolonged (can last several minutes) song-flight at *c.* 50 m height, somewhat like Skylark, drifting around with quivering wingbeats, dropping slightly with somewhat raised quivering wings and raised folded tail during each song strophe.

VOCALISATIONS Song very odd for a pipit, a rather subdued jingling and 'mechanical' series of sharp notes, quickly repeated and sometimes comprising short, hard, dry trills, usually accelerating a little and often ending with a few extended and emphatic, harsh notes, *zezezezeze-chi-chi-chi-chi-zer-zer-zer-zerrrrr-chreeh-chreeh-chreeh*. The 'electric' quality of some of the notes and the rhythm can almost recall a displaying Siberian snipe species, e.g. Pin-tailed *Gallinago stenura*. Song variation noted to be rather prominent. – Commonest call a chipping or emphatic whistling note with almost electrical or snapping overtone, a monosyllabic *dzepp*, *tsip* or *tsi-tsip*, often steadily repeated, e.g. *tsip, tsip, tsip,...* or *tsi-tsi-tsi-tsip* with a somewhat explosive quality (may recall flight-call of Grey Wagtail), but usually silent when flushed. Other variants are softer and can recall common call of Chaffinch, *hweit*.

SIMILAR SPECIES Most pertinent confusion species is juvenile/first-winter *Red-throated Pipit* (both are vagrants in countries where serious risk of confusion exists), as both are sharply and heavily streaked on the crown, mantle and rump (no other pipit in the covered region has strong rump streaking). Pechora, however, has a suite of characters that in combination should secure identification (and eliminate other potential confusion risks such as *Meadow* and *Tree Pipits*), described in detail under Identification. Observers should take into account that Red-throated could have the longest tertial still partially growing, broken or missing, and could misleadingly appear to have a primary projection outside tertials (making it important to check both wings, count the number of feathers and ensure their tips are equally spaced), and note that some Red-throated may show a vague loral mark. Individual variation in both species (especially with wear) means that some appear intermediate in pattern and tone of upperparts, wing-bars and underparts pigmentation. Note also that dark-streaked longer undertail-coverts are far more common in Red-throated, whereas Pechora usually has unmarked ochre-buff undertail-coverts, only rarely with a thin dark shaft-streak on the longest ones; a bold streak eliminates Pechora. Finally, Pechora has narrow, rather long pale wedges on penultimate tail-feathers, while in Red-throated (and Meadow and Tree Pipits) these wedges are shorter and more triangular.

AGEING & SEXING Ageing usually difficult but sometimes possible in the hand by close inspection of moult and feather wear in wing. Sexes alike. – Moults. Complete

Presumed *A. g. gustavi*, Germany, Oct: rufous tone to head, indistinct (almost lacking) supercilium, pale eye-ring broken by dark lores, but conspicuous white wing-bars, and heavy neck-side patches and breast- and flanks-streaking. Ageing uncertain. (G. Schuler)

PIPITS AND WAGTAILS

Presumed *A. g. gustavi*, Scotland, Oct: note long and rather straight hind claws, rufous tone to supercilium and ear-coverts, and pale eye-ring broken by dark lores; white wing-bars highly contrasting, as are large neck-side patches and streaking on snowy-white underparts. (H. Harrop)

Presumed *A. g. gustavi*, England, Nov: obscure supercilium and brown-tinged, finely streaked head and nape. Bill rather strong but pointed, often with much pink at base. Note also clear primary projection outside the longest tertial. Mantle characteristically boldly striped off-white and black. Ageing uncertain. (P. Morris)

post-nuptial moult and partial post-juv moult in late summer (Jul–early Sep). Post-juv moult rather restricted and does not involve any median or greater coverts. Partial pre-nuptial moult (Feb–Mar) for both age groups involves renewal of some body-feathers, at least some inner greater coverts and tertials, and sometimes central pair of tail-feathers. – **SPRING Ad** Wing- and tail-feathers less worn than 1stS. With wear ground colour duller/paler above, mantle-stripes and wing-bars less obvious or (if heavily abraded) almost lost, and buffish tinge to underparts much fainter. **1stS** Perhaps separable by more striking moult contrast with more heavily-worn retained juv secondary-coverts, primaries and primary-coverts. (Note that ad also shows discreet moult limits among secondary-coverts, tertials and sometimes tail-feathers in spring.) – **AUTUMN Ad** Only ad (evenly fresh) wing- and tail-feathers discernible. **1stW** Post-juv moult involves mainly head- and body-feathers, but juv secondary-coverts, remiges, primary-coverts, tertials and tail-feathers are all retained, thus no moult limits in wings or tail. Careful in-hand examination, however, can reveal primaries, primary-coverts and tail-feathers to be more worn and weaker textured than in ad. **Juv** Much like ad but has very indistinct supercilium, better-streaked crown and ear-coverts, and slightly blacker streaking on underparts and rest of upperparts. Sometimes weaker, softer plumage on nape and vent discernible.

BIOMETRICS (*gustavi*) **L** 14–16 cm; **W** ♂ 81–88 mm (*n* 24, *m* 84.9), ♀ 79–87 mm (*n* 14, *m* 82.1); **T** ♂ 52–58 mm (*n* 24, *m* 55.3), ♀ 52–58 mm (*n* 14, *m* 53.9); **T/W** *m* 65.3; **wedge r5** 15–36 mm (once 2 mm; *n* 30, *m* 26.1); **B** 15.0–17.0 mm (*n* 37, *m* 16.1); **B**(f) 11.0–13.5 mm (*n* 28, *m* 12.1); **BD** 3.8–4.7 mm (*n* 26, *m* 4.3); **Ts** 22.5–25.2 mm (*n* 32, *m* 23.7); **HC** 8.3–12.3 mm (*n* 35, *m* 10.5). Wing formula: **p1** minute, **pp2–4** equal and longest (or p2 and/or p4 0.5–1 mm <); **p5** < wt 3–7 mm; **p6** < wt 9–15 mm; **p7** < wt 12.5–19 mm; **p10** < wt 22–28 mm; **s1** < wt 24–30 mm; **tert. tip** < wt 8–16 mm (once 19) (cf. 0–6 mm in Red-throated Pipit), =pp5/8. Emarg. pp3–5 (but usually not as distinct on p5).

GEOGRAPHICAL VARIATION & RANGE Very slight over much of range, with only one population in southeast differing more markedly in size, general coloration and streaking, and apparently in song. Further races described but appear too subtle and variable to be warranted. Apparently all vagrants in Europe are of the western race *gustavi*.

A. g. gustavi Swinhoe, 1863 (N Russia at least west to Pechora, possibly to Mesen, and east to Bering Sea, in much of Siberia south to *c*. 63°N, in far east to N Kamchatka; winters SE Asia). Described above. (Syn. *commandorensis*; *stejnegeri*.)

Extralimital: **A. g. menzbieri** Shulpin, 1928 (NE China, extreme SE Siberia in Ussuriland, S Amur; winters SE Asia, including Philippines). Slightly smaller than *gustavi* and usually obviously less rufous-tinged on head and neck, tends to be more greyish-green above. Also, dark streaking of upperparts often bolder making it appear darker; in particular, forecrown and back of neck heavier streaked than in *gustavi* and more uniform with hindcrown and mantle (often slightly paler brown in *gustavi*). Legs appear rather short (but scant material measured by us). Despite these differences frequently difficult to identify away from breeding range due to variation and some overlap of characters. Following available recordings and sonograms (e.g. in Alström & Mild 2003), the song seems to contain several fast, drawn-out buzzing rattles, *zrrrrrrrrrr*, and be more varied with some buzzing rattles being deeper-pitched and almost frog-like, *torrrrr*, compared to the mainly fast repetition of the same sharp note as in *gustavi*, *zezezezezezeze…*, and if this proves to be a consistent difference it could have implications for taxonomy (see below). **W** ♂ 77–82 mm (*n* 17, *m* 79.9), ♀ 74–79 mm (*n* 9) (partly after Payevski, in Bub et al. 1981). – **W** 74–82 mm (*n* 12, *m* 78.8); **T** 49–54 mm (*n* 12, *m* 52.9); **T/W** *m* 65.9; **B** 14.7–15.9 mm (*n* 12, *m* 15.4); **B**(f) 10.8–12.3 mm (*n* 12, *m* 11.8); **Ts** 20.5–24.3 mm (*n* 12, *m* 22.5). Wing formula: **p5** < wt 2.5–5 mm; **p6** < wt 9–12 mm; **p7** < wt 13–16 mm; **p10** < wt 20–24.5 mm; **s1** < wt 22–26 mm (NHM, AMNH, ZMB, YIO).

TAXONOMIC NOTE Based on claims of distinct song (which require confirmation) and fairly marked morphological differences, eastern race *menzbieri* has been suggested to be a separate species (Wolters 1979, Leonovich et al. 1997). A more comprehensive study, covering vocalisations over larger parts of the two ranges, and extensive genetic sampling, seems desirable.

REFERENCES LEONOVICH, V. V., DEMINA, G. V. & VEPRINTSEVA, O. D. (1997) *Byul. Mosk. O-va. Ispitatelei Prirodi Otd. Biol.*, 102: 12–22.

Presumed *A. g. gustavi*, China, May: little seasonal variation between autumn and spring. Note highly obscure dark lateral throat-stripe, but well-developed patch on neck-side. Also typically, entire sides of nape tinged warm olive-buff and very finely streaked (no other pipit has same pattern). (B. van den Boogaard)

MEADOW PIPIT
Anthus pratensis (L., 1758)

Fr. – Pipit farlouse; Ger. – Wiesenpieper
Sp. – Bisbita pratense; Swe. – Ängspiplärka

For most users of this handbook, the Meadow Pipit is the epitome of a pipit due to its being extremely widespread and common, breeding in most open grassy areas on tundra, fells and forest bogs from SE Greenland and Iceland east to NW Siberia. Mainly a short-distance or partial migrant, Meadow Pipits move south and west in winter (chiefly to open damp and grassy fields, seashores or by lakes), usually evacuating Fenno-Scandia but being found over much of W Europe south to NW Africa, the Mediterranean fringe, and even to Iran and Pakistan.

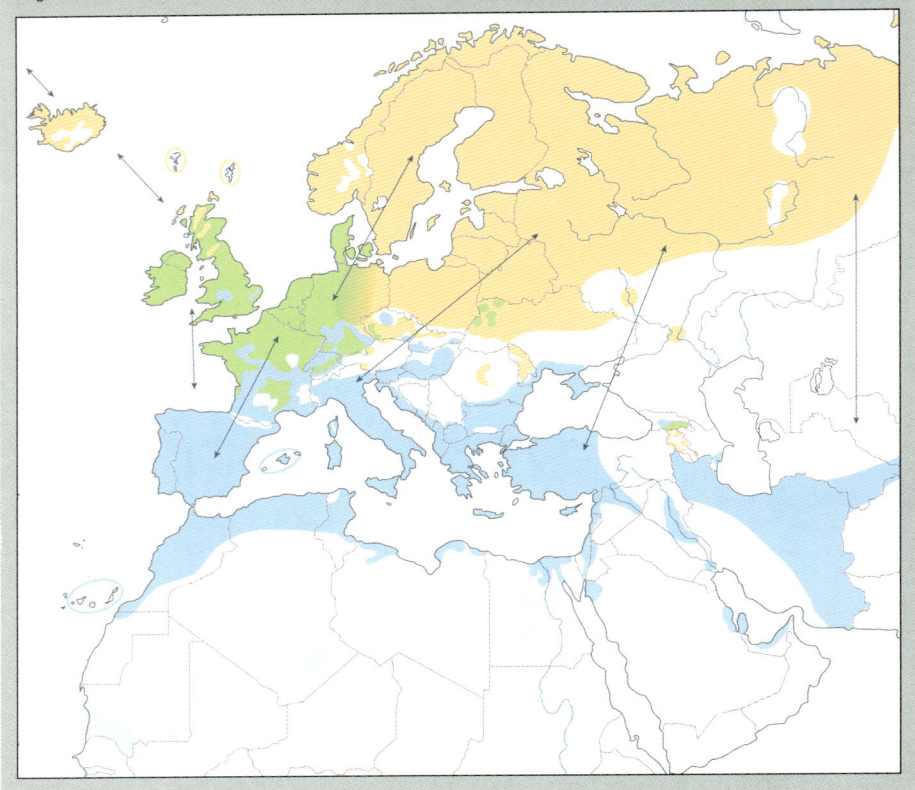

IDENTIFICATION Rather sleek but dumpy-looking pipit, without any striking or otherwise diagnostic field characters. Typically green-olive to duller greyish olive-brown above, and mantle/scapulars extensively (but not boldly) *streaked dark*, with a couple of ill-defined narrow buff-white 'tramlines', but lower back and rump by and large unstreaked. Entire *underparts rather concolorous* pale buffish when fresh (much whiter when worn) and *evenly streaked on breast and flanks*, forming long coarse lines over entire flanks. Bill slender and short, usually with a noticeable *pale brownish-orange lower base*. Face pattern rather indistinct with poorly marked yellow-white supercilium, narrow but *obvious and unbroken pale eye-ring*, plain ear-coverts and indistinct dark eye-stripe. Lateral throat-stripe and patch moderate or small, but submoustachial stripe rather pronounced. Pale wing-bars, formed by buff-white margins to dark-centred median and greater coverts, and fringes to tertials (buff elements bleach away in late spring–summer). In flight shows whitish outermost tail-feathers. Legs pale brown with rather long but not strikingly curved hind claws. Gait rather shuffling, occasionally walking or running more upright, and has distinctive creeping walk on ground with indistinct tail twitching and wagging. Flight rather jerky and fluttering, being rather hesitant, and escape-flight often high and erratic (accompanied by reasonably characteristic calls), ending with bird plummeting back to open ground. Song-flight has fluttering ascent, and gliding descent with tail often raised and legs slightly dangling. Gregarious, especially on migration and in winter.

VOCALISATIONS In song-flight gives an accelerating, then decelerating, sequence of repeated thin, piercing monosyllabic notes, *zi zi zi zi zi zi zi zi zi zi zü-zü-zü-zü-zü- zü-zü-zü-svisvisvisvisvi tüü tüü tüü tüü tii-svia*, the motif changing 3–4 times during the strophe but constant in structure (not unlike Rock Pipit, but initial notes higher and less jangling, and strophe often ends with a slight flourish); perched singer reels off intro after intro, *zi zi zi zi zi*, without concluding full song. – Commonest call given in flight and when flushed is a sibilant, thin, hurried *ist-ist-ist* (most like Water Pipit), with shorter and less emphatic variations *ist* and *ist üst*. Alarm a tremulous double note, *sitt-itt*. When breeding, a shorter, slightly metallic *chip*, with anxious quality, and some variations to this.

SIMILAR SPECIES Over most of its range only likely to be confused with *Tree Pipit*, as these are the most frequently encountered members of *Anthus*. For separation from Tree Pipit as well as from *Red-throated* (in non-red-throated plumages), *Pechora*, *Berthelot's* and *Rock Pipits*, see those species. In much of Western Palearctic the commonest pipit and easily identified with experience and knowledge of the diagnostic call. Silent birds may require more careful separation from other species. In addition to vocalisations, it is important to concentrate on the rather less bold upperparts streaking and unstreaked lower back/rump, the evenly marked underparts streaking and pigmentation, and rather featureless facial pattern. Note that undertail-coverts invariably lack dark central streaks (present in most Red-throated).

AGEING & SEXING Ageing in autumn possible with close inspection of moult, feather pattern/wear in wing, and iris coloration; in spring variation in wear usually prevents safe ageing. Sexes alike. – Moults. Complete post-nuptial moult and partial post-juv moult in late summer (Jul–Sep). Post-juv renewal including head- and body-feathers, lesser and a variable number of other secondary-coverts (generally just a few inner, or none), tertials and tail-feathers (often all tertials and r1; much greater retention in northern populations). Partial pre-nuptial moult mostly Jan–Mar. – **SPRING** Safe ageing rarely possible, requiring much experience and only possible with typical birds. **Ad** Only ad-type wing- and tail-feathers discernible. In worn plumage upperparts usually more grey-brown and rather more heavily dark-streaked, and underparts more whitish than in fresh autumn plumage. **1stS** Extremely similar to ad; some perhaps separable in hand by retained juv secondary-coverts, and rarely shows diagnostic three generations of median and/ or greater coverts. All primaries and primary-coverts juv and often more worn/bleached. Beware that all ages usually show moult contrast among secondary-coverts and tertials at this season (due to pre-nuptial renewal of some secondary-coverts and tertials, perhaps tail-feathers and also most body-feathers). Furthermore, with bleaching, differences in

Norway, Jun: during breeding season, upperparts wear greyer brown with heavier dark streaking and acquires whiter underparts. Bold flanks-streaking and unbroken pale eye-ring eliminate Tree Pipit. Despite moult limit in greater coverts, ageing in spring/summer is usually unsafe. (M. Varesvuo)

coloration and extent of wear in median and greater coverts very difficult to detect. – AUTUMN **Ad** From Sep–Oct evenly fresh (suspended and heavily-bleached wing-feathers rare but diagnostic). **1stW** Often shows moult limits, with some juv secondary-coverts, tertials and tail-feathers (contrasting with newly-moulted coverts which are marginally longer and fresher). Juv median coverts also have diffuse, more triangular dark centres with more pronounced 'tooth' and whiter fringes, and greater coverts have less well-defined centres with indistinct 'indented' tip on outer web compared to ad (important when ageing birds with unmoulted secondary-coverts). Juv primary-coverts narrower and more pointed, and primary-coverts and primaries more worn and weaker textured than in ad; also juv tail-feathers usually slightly more worn and pointed, and iris grey-brown (rufous-brown in ad). **Juv** Compared to fresh ad, dark streaking on upperparts and crown bolder, entire underparts tinged paler yellow-buff and more faintly streaked, while face pattern tends to appear more blurred. Sometimes weaker, softer plumage on nape and vent discernible.

BIOMETRICS L 13.5–15 cm; **W** ♂ 78.5–84 mm (*n* 22, *m* 81.5), ♀ 75–83 mm (*n* 17, *m* 78.2); **T** ♂ 55–62 mm (*n* 22, *m* 58.9), ♀ 53–62 mm (*n* 17, *m* 56.7); **T/W** *m* 72.4; **B** 13.0–15.0 mm (*n* 34, *m* 14.1); **BD** 3.0–3.7 mm (*n* 18, *m* 3.4); **BW** 3.3–4.2 mm (*n* 19, *m* 3.9); **Ts** 19.8–22.3 mm (*n* 35, *m* 20.7); **HC** 9.0–12.4 mm (*n* 34, *m* 10.6). **Wing formula**: $p1$ minute; $pp2$–5 usually about equal and longest (though $p5$ often < wt 0.5–1 mm); $p6$ < 8–11.5 mm; $p7$ < 13–16.5 mm; $p10$ < wt 18–23 mm; $s1$ < wt 18–23.5 mm. Emarg. $pp3$–5. **White wedge** $r5$ (inner web) 2–20 mm (exceptionally lacking).

GEOGRAPHICAL VARIATION & RANGE Monotypic. – W & N Europe east to W Siberia, south to France, N Italy, Hungary, N Ukraine; winters C & S Europe, N Africa, Middle East, N Arabia, Iran and S Central Asia. – Very slight and clinal variation with birds from E Ireland & W Scotland, or even throughout British Isles, Iceland & Faeroes, described as being slightly darker and more brown-tinged above and deeper buff below ('*whistleri*'), and birds from W Ireland as being slightly darker still and more rufous-tinged ('*theresae*'), but neither of these appear sufficiently distinct to warrant separation. (Syn. *theresae*; *whistleri*.)

Netherlands, Apr: often first identified by its diagnostic flight-call, but silent birds on ground by featureless face pattern (poorly-marked pale supercilium, though unbroken pale eye-ring often prominent), less bold upperparts streaking and unstreaked back/rump; note also evenly-pigmented underparts, dark-streaked breast/flanks, and rather short bill with pale brownish-orange lower base. (R. Pop)

Presumed 1stW, Italy, Feb: indistinct facial pattern and unstreaked lower back to rump/uppertail-coverts are important characteristics. By late winter/early spring, can wear greyer with much-reduced olive tones to upperparts, while ill-defined dark-streaked mantle/scapulars are enhanced. Moult contrast in greater coverts (left image) at this season strongly suggests 1stY. (D. Occhiato)

Ad, Netherlands, Oct: quite prominent whitish eye-ring and brownish-orange lower base to bill. Ageing impossible, as apparent moult contrast in greater coverts and tertials cannot be confirmed at this angle. (R. Pop)

Presumed 1stW, Germany, Oct: when fresh, characteristically has greenish hue to plumage. In general, streaking on sides of throat and breast becomes neither weaker nor narrower on flanks, only more diffuse. As many wing-coverts are not fully visible, assessment of age not definitive, although some of the greater coverts and tertials are apparently juv. (M. Schäf)

Juv, Netherlands, Jun: soft, fluffy body-feathers, bold dark streaking above and entire underparts tinged yellow-buff. Note long, slightly curved hind claws. (N. D. van Swelm)

RED-THROATED PIPIT
Anthus cervinus (Pallas, 1811)

Fr. – Pipit à gorge rousse; Ger. – Rotkehlpieper
Sp. – Bisbita gorjirrojo; Swe. – Rödstrupig piplärka

In summer, the attractive Red-throated Pipit occurs in wet tundra and on open mountain slopes with watercourses and willow stands from northernmost Fenno-Scandia east to Kamchatka, with smaller numbers in W Alaska. It winters in open agricultural and other grassy areas in N & C Africa, Middle East and across S Asia. Its migration routes are well defined, most noticeably through the E Mediterranean and Levant, but almost entirely missing in extreme W Europe, where it is only an uncommon visitor.

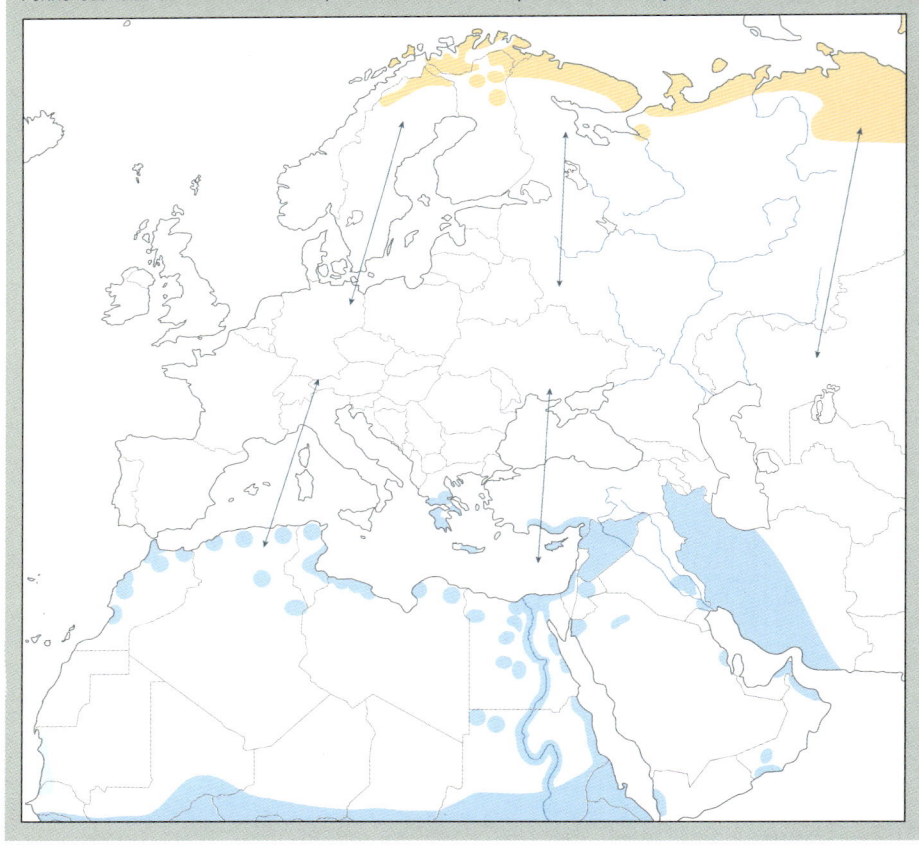

Ad, presumed ♀, Norway, Jun: only ad-like wing-coverts visible and remiges not heavily worn (another image of same bird shows similar primary-coverts). Red stops short of breast, or just reaches it, best fitting older ♀ with relatively well-developed red face, but extensive overlap means sexing unsure. (H. Harrop)

IDENTIFICATION A small streaky pipit, in size and proportions close to Meadow Pipit, thus slightly shorter-tailed than Tree Pipit. Adult plumages have *pinkish-rufous or brownish-orange face and throat, often extending to upper breast* (and sometimes to supercilium and forehead). Young (and a very few adult ♀♀) lack reddish face and breast. All plumages have brownish-toned and *broadly-streaked upperparts* (more rufous-tinged, less olive-grey, than Tree and Meadow Pipits), *including lower back, rump and uppertail-coverts*, mantle with a few *dark and pale 'tramlines'*; evenly buffy underparts *spotted boldly on breast and sides*, forming 2–3 almost unbroken black stripes on flanks, and *bold lateral throat-stripe and patch* (chiefly in first-winter birds, and sometimes lacking in very colourful ♂♂); *whitish wing-bars and tertial fringes quite conspicuous*. In non-rufous-throated birds face pattern still distinctive, especially the pale supercilium, eye-ring and lores. Bill rather short (and thinner than Tree Pipit) with yellowish lower base, legs usually pale reddish-brown, and hind claws rather long and moderately curved. Has

Presumed ad ♂, Israel, Mar: seemingly only ad wing-feathers (worn and fresh) visible, while extensive red (with virtually no streaks on central breast or traces of lateral throat-stripe/patch) strongly suggest ad ♂. (D. Occhiato)

♂, presumed ad, Greece, Apr: clear blackish streaks on rump and uppertail-coverts. Ageing uncertain, given that both age classes show moult limits in wing following partial pre-nuptial moult. Amount of red on chest indicative of ♂ (either ad or well-advanced 1stS). (R. Pop)

PIPITS AND WAGTAILS

diagnostic drawn-out, very high-pitched flight-call. Gait on open ground like Meadow Pipit, and flight mode probably the same or very similar. May forms large flocks on migration and in winter. Song-flight well developed, being higher than Meadow Pipit and often sustained longer before parachuting descent. — For in-hand identification or extreme close views (in particular of non-adults), note that 65% have dark central streak on buff-tinged longest undertail-coverts (invariably unstreaked and whitish undertail-coverts in Meadow Pipit).

VOCALISATIONS Song comprises repeated rhythmic multisyllabic themes often with last note stressed, sudden changes to fine, drawn-out high-pitched notes (similar to characteristic flight-call) and Redpoll-like dry buzzing trills, e.g. *svü-svü-svü-**svüh**, svü-svü-svü-**svüh**, svü-svü-svü-**svüh**, psiü psiiiü psiiiiüh sürrrrrrrr wi-wi-wi-wi tsvü-tsvü-tsvü*. — Call is very characteristic by its fine, 'penetrating' tone and almost straight pitch, only slightly downslurred, *pssiiih*, rather drawn-out and when close usually with audible slight harsh quality. Sometimes you hear from a migrant flock or wintering party shorter variations, often doubled with second note at lower pitch, a little like Meadow Pipit, *psih-pseh*. Alarm a short *chüp* or *chu*, rather like Ortolan Bunting's, and this is often heard from wintering birds, too.

SIMILAR SPECIES Typical adult unmistakable by rufous-pink or orange-buff throat (at times extending to forehead, supercilium and upper breast), but all first-years and rarely some poorly-developed adult ♀♀ lack reddish colours and are hence prone to confusion with *Meadow Pipit*. Note however that Meadow has an unstreaked rump, less boldly streaked upperparts, often with a greenish rather than brownish tone, more diffuse wing-bars, less contrasting and less blackish underparts streaking, plainer face pattern with less well-marked supercilium, and—very importantly—a distinctly different call. — Red-throated Pipits lacking reddish could theoretically also be confused with *Tree Pipit*, but latter has unstreaked rear upperparts, less boldly streaked mantle/scapulars, narrower flanks-streaking and different face pattern, gait, behaviour and call. — *Pechora Pipit* poses most significant confusion risk with non red-throated plumages, but note on Red-throated the lack of any significant primary projection (primaries completely cloaked by tertials on folded wing, or only shows hint of single primary tip), the somewhat paler lores with lack of any prominent dark eye-stripe in front of eye, the less bold black upperparts streaking, less prominently whitish median-coverts bar, proportionately longer tail and different call.

Presumed 1stS ♂, Israel, Mar: strong moult contrast among greater coverts perhaps indicative of 1stS. Heavily-streaked breast, lack of dark lateral throat-stripe and red reaching to upper breast are features of 1stS ♂. (D. Occhiato)

Presumed 1stS ♀, Greece, Apr: heavily abraded wing-feathers (as early as Apr), except fresh longest tertial, suggest 1stS. Combination of age and heavily-streaked underparts, including much of breast and well-developed lateral throat-stripe, with dull pinkish-orange restricted mostly to throat, suggest young ♀, but some young ♂♂ can match this plumage at this season. (R. Pop)

AGEING & SEXING Ageing often possible with close inspection of moult, feather pattern and wear in wing. Sexes differ mainly in ad breeding plumages, but extensive overlap and we recommend labelling only classic individuals of known age (wing length also useful—see Biometrics). — Moults. Complete post-nuptial and partial post-juv moults before autumn migration. Post-juv renewal includes only head- and body-feathers, some lesser coverts and only occasionally a few median coverts, one or two inner greater coverts or the odd tertial. Partial pre-nuptial winter moult (mostly Jan–Mar) rather variable, involves many head- and body-feathers, most or all lesser and median coverts, inner (3–6) greater coverts, all tertials, and perhaps r6 (some, especially 1stS, retain most secondary-coverts and tertials). — SPRING Due to age-related variation in amount of red, best to age birds before attempting sexing: ad has only ad-type wing- and tail-feathers; 1stS often has many retained juv secondary-coverts and all primaries and primary-coverts of juv type, which are distinctly more worn/bleached (primary-coverts also narrower, weaker textured and browner). However, note that *all* ages usually show moult contrasts in spring among secondary-coverts; ageing safest in the hand. **Ad** ♂ Red on throat brighter and usually extends to supercilium and upper breast, where dark streaks and lateral throat-stripe/patch narrower or lacking, but some closely approach ♀ or are intermediate. **Ad** ♀ Throat colour less saturated, tending

Ad, presumed ♂, Israel, Nov: evenly-fresh ad wing-coverts and extensive red face readily age this bird as ad; some red elements on lower breast, and lack of dark lateral throat-stripe and patch, suggest ♂. (A. Ben Dov)

Ad ♂, Israel, Nov: combination of evenly fresh wing after post-nuptial moult and extensive red on head and breast in late autumn readily age this bird as ad ♂. (A. Ben Dov)

towards pinkish-orange (or paler), with coloration restricted to chin and throat. Underparts on average better-streaked, including breast and lateral throat-stripe/patch well developed. **1stS** Like ad but has less-advanced reddish face and throat, and often ♂ has better-streaked breast and rather obvious lateral throat patch (largely ad ♀-like), while ♀ usually dullest with little or no pink on throat and pattern almost like 1stW. — **AUTUMN** Ad from Sep/Oct evenly fresh and usually has at least some rufous/pinkish-orange on throat; 1stW has all tail-feathers and most wing-feathers juv (a variable number of lesser coverts, occasionally a few inner median and greater coverts, and more rarely tertials, are renewed), and usually lacks any pinkish-orange on throat. **Ad ♂** Compared with spring plumage generally less pinkish-rufous, more brownish-orange on face, throat and breast, and has bold black spots on breast and quite pronounced lateral throat-stripe/patch. **Ad ♀** As ad ♂, amount of reddish on throat/face slightly reduced and more brownish-orange; also, more heavily streaked below than in spring. A very few have so reduced brown-orange on throat as to resemble 1stW with warmest buff throat, but note completely fresh plumage lacking any moult contrast. **1stW** Throat entirely whitish-buff or cream, but some 1stW ♂♂ rarely show vestigial pinkish-buff feathering. Has boldly streaked upperparts, heavy breast- and flanks-streaking, and a prominent wedge-shaped blackish lateral throat-stripe/patch. Generally like non-red-throated ♀ but usually overall more heavily/boldly marked. As 1stW occasionally show a little pinkish-orange, and a few ad ♀♀ hardly any or even none, other features must be assessed before a bird with little or no pinkish-orange can be safely aged in autumn: juv primary-coverts (more pointed) and primaries (less fresh), while retained secondary-coverts and tertials less fresh, weaker textured and have whiter tips/fringes than ad, and juv median coverts also have more clear-cut triangular dark centres with pronounced 'tooth' (ad: more evenly broad and buffish tips, dark centres without 'tooth' at centre). Juv tail-feathers usually more worn and pointed, and iris grey-brown (rufous-brown in ad). The few that show moult limits among secondary-coverts are more easily aged as 1stW. **Juv** Generally as 1stW but has soft, fluffy body-feathering (notably on vent and nape) and sharp buff lines on mantle-sides; entire underparts suffused yellow-ochre, and throat, breast and flanks all heavily streaked, although streaks often appear narrower, especially on flanks.

BIOMETRICS L 14–15.5 cm; **W** ♂ 83–92 mm (*n* 27, *m* 87.6), ♀ 82–87 mm (*n* 22, *m* 84.2); **T** ♂ 56–64 mm (*n* 26, *m* 60.0), ♀ 55–59 mm (*n* 22, *m* 57.0); **T/W** *m* 68.1; **B** 12.9–15.4 mm (*n* 47, *m* 13.9); **BD** 3.2–4.0 mm (*n* 33, *m* 3.5); **BW** 3.4–4.3 mm (*n* 37, *m* 3.9); **Ts** 20.0–22.2 mm (*n* 47, *m* 21.3); **HC** 8.6–12.7 mm (*n* 44, *m* 10.5). **Wing formula: p1** minute; **pp2–4** often equal and longest (p4 sometimes < wt 0.5–1 mm); **p5** < wt 1–3 mm; **p6** < wt 10–13 mm; **p7** < wt 16–18.5 mm; **p10** < wt 23–28 mm; **s1** < wt 25–29 mm. Emarg. pp3–5. **White wedge r5** (inner web) 2–14 mm (very rarely lacking). **Tert. tip** = wt (or sometimes < wt 0.5–10 mm).

GEOGRAPHICAL VARIATION & RANGE Monotypic. — Extreme N (Arctic) Europe east through N Russia to NE Siberia, south to Kamchatka in the east; winters N & C Africa, Nile Valley, Red Sea coast, Middle East, locally in Arabia and Gulf States, and commonly in SE Asia (Burma eastwards). — Very slight and clinal variation has been described, with birds in east tending to be smaller and less streaked on throat and breast in breeding plumage, whereas western birds ('*rufogularis*') should be larger and more streaked. These differences appear very slight and inconsistent and do not hold when applying 75% rule. (Syn. *rufogularis*.)

Egypt, Dec: typical midwinter appearance. However, since ageing is unsure, sexing is also impossible. Reddish feathers on face suggest an ad, while heavily-streaked breast, well-developed lateral throat-stripe and patch, and dull pinkish-orange restricted largely to throat suggest ♀. (E. Winkel)

Presumed ad, Israel, Oct: wing-coverts apparently ad-type, but seemingly two generations of tertials confusing. Heavily-streaked breast, well-developed lateral throat-stripe and patch, and pinkish-brown throat in Oct, suggest ad ♀. (A. Ben Dov)

1stW, Egypt, Oct: overall more heavily streaked than Meadow Pipit, with more striking face pattern. Note juv secondary-coverts and tertials, with juv median coverts having typical triangular, dark centres, pronounced 'teeth' and whiter edges. Sexing 1stW by plumage impossible. (D. Occhiato)

1stW (late autumn to winter), left: Egypt, Oct; right: Israel, Dec: unlike Meadow Pipit, rump and uppertail-coverts heavily streaked, upperparts more boldly streaked, with clearer dark and pale lines on mantle (lacking Meadow's greenish hue), bolder white wing-bars, and more contrasting, blacker underparts streaking. Face pattern usually more striking, especially pale supercilium, eye-ring and lores, while dark wedge-shaped lateral throat patch larger and more solid. (Left: P. Alberti; right: M. Varesvuo)

ROCK PIPIT
Anthus petrosus (Montagu, 1798)

Fr. – Pipit maritime; Ger. – Strandpieper
Sp. – Bisbita costero; Swe. – Skärpiplärka

Chunkier and duskier than other small pipits and most likely to be encountered alone or in pairs, typically on rocky islets, seaweed-covered rocks or tideline debris and, in winter, also found on estuaries and rarely inland at freshwater lakes. Breeding range restricted to rocky seacoasts in the Faeroes, Britain, NW France, Scandinavia, Finland and extreme NW Russia; it is mostly sedentary in the centre and the south of its range, but Fenno-Scandian birds disperse southwards, and the species is considerably more widespread and catholic in habitat choice in winter, although it is still restricted to seacoasts, from the British Isles and Denmark south to Gibraltar and Morocco.

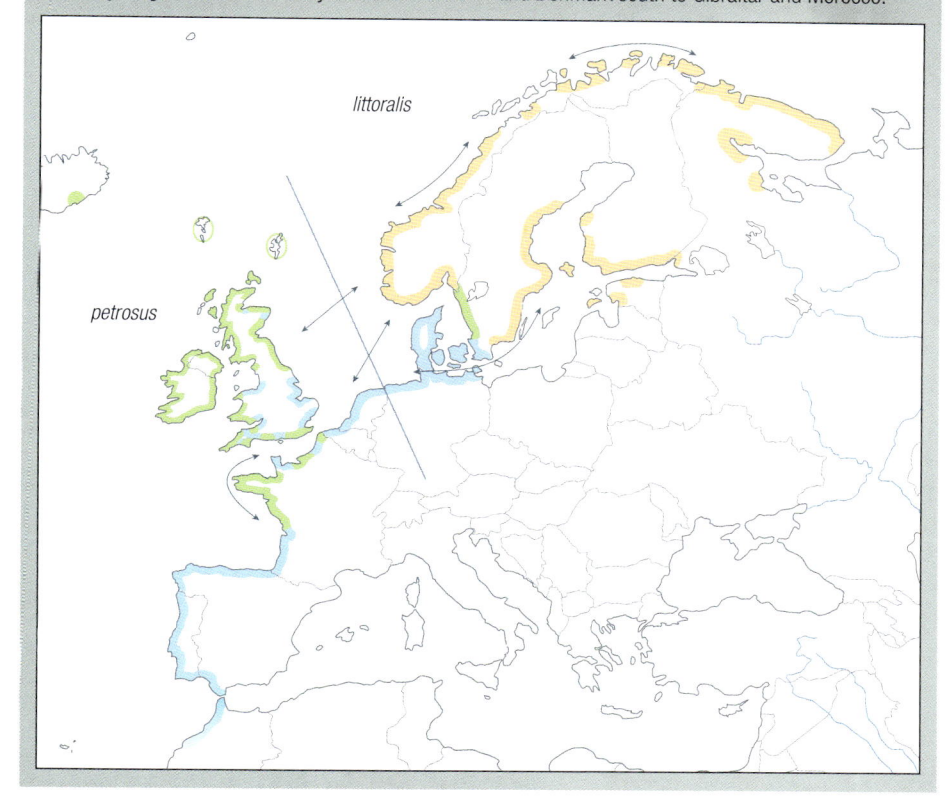

Presumed *A. p. petrosus*, France, Apr: some birds can appear paler, like winter *littoralis*; while locality and date are seemingly indicative of *petrosus*, the apparently extensively renewed greater coverts seem better for *littoralis*. Without closer views of wing-coverts, ageing also uncertain. (M. Vaslin)

IDENTIFICATION Note that Rock Pipit breeds only *on rocky seashores* and is very rarely seen far inland even in the non-breeding season. *Rather dark and bulky, with a longish, strong bill.* Shares with Water Pipit (with which previously regarded conspecific) *dark reddish-brown to blackish legs* (but note some variation, rarely medium pink-brown). In breeding plumage, British and W European birds (*petrosus*) *almost uniform dark olive-grey to olive-brown above*, being only vaguely streaked darker, and rump and uppertail-coverts unmarked olive-grey. Underparts typically *rather broadly and profusely streaked brownish on deep buff lower throat, breast and flanks*, leaving almost clean white chin and upper throat. Lateral throat-stripe rather narrow but patch is fairly broad. All plumages have rather *obscure pale wing-bars* and tertial fringes, and a generally *short faint supercilium*, as well as a narrow whitish and broken eye-ring. Note diagnostic *greyish wedges to outer webs of outer tail-feathers* (except whitish tips). Observers should bear in mind that the rather different Baltic form (*littoralis*) to some extent approaches Water Pipit. Gait is less creeping and typically hops, walks and runs among rocks or seaweed and on short grass. Flight more confident and direct than Meadow Pipit with distinctive, bulkier silhouette. Song-flight much like Meadow Pipit, usually given on rigid wings in typical ascent, before descending to rock or cliff edge, with the parachuting descent sometimes appearing slower.

VOCALISATIONS Song generally very similar to Water Pipit (and also somewhat recalls Meadow Pipit), but is louder and harder, with individual notes less fine, more hoarse and compound, comprising repetition of shrill notes with 3–4 motifs, e.g. *zrü-zrü-zrü-zrü-zrü-zre-zre-zre-zre-zre-zre-zre-sui-sui-sui-sui-zri-zri-zri*, often ending with soft rattling *chrrrrr*. – Flight- and alarm calls *viisst* or *weesp* and other variants, apparently identical to Water Pipit, or at least very similar. (On average perhaps subtly longer, higher-pitched and more rising than call of Water Pipit; Fijen 2014.) Both Rock and Water Pipit calls differ from Meadow Pipit in being sharper, more metallic and shriller; mainly given singly, but if in twos or sometimes tripled then slower than in Meadow Pipit. Also gives high-pitched insect-like *tzi* notes (often in series) and harsh Serin-like notes.

SIMILAR SPECIES Confusion with *Meadow Pipit* rather frequent, but latter slightly smaller and shorter-billed (with a pale lower base to bill, rather than this being evenly dark as in most Rock), has paler, pinkish legs (though some Rock have pale flesh-brown legs), distinctly paler, more olive

A. p. petrosus, Ireland, Apr: race based on locality and date, and even in spring dusky plumage with dense blotchy-streaked breast and flanks. Supercilium short and indistinct. Ageing at this angle difficult and best left unattempted. (J. Lowen)

A. p. petrosus, 1stS, England, Apr: race based on locality and date. Migrant *littoralis* could lack grey on head and have streaky underparts at this season. Strong moult limits in wing (as early as Apr) indicative of 1stS. (M. Lawrence)

upperparts with better-defined streaking, less extensive but blacker underparts streaking (not brownish and diffuse as in Rock), whiter outer tail-feathers and different vocalisations. Meadow also has a streakier crown, paler lores, more distinct moustachial stripe and better-marked supercilium, imparting a less dusky facial impression than Rock Pipit. However, Baltic race *littoralis* in worn breeding plumage is paler with more streaks above and sharper streaking below, and can appear superficially closer to Meadow, but sometimes has pale pinkish-buff tinge on breast and greyish head/mantle in spring/summer. – Separating non-breeding Rock from *Water Pipit* often requires care in areas of overlap, as at this season they may share certain seashores, estuaries and adjacent habitats, although it should be remembered that only Water Pipit is likely to be found far inland. In summer, however, there is no overlap in range, and Rock Pipit of the race *petrosus* has more heavily-streaked underparts, lacking any trace of pinkish. Of the different populations of Rock Pipit, *littoralis* is the one most liable to resemble Water Pipit, mainly in breeding plumage (when they are unlikely to meet) and only in extreme birds with a much greyer head/mantle than normal, subtly pink-buff breast and reduced underparts streaking. However, Rock is still more evenly olive-grey (tinged greenish) above (Water: often subtly purer grey head, greyish-brown mantle and back, and brown-tinged rump/uppertail-coverts), with boldly but diffusely-streaked flanks, and lateral throat-stripe/patch all more evident (unlike Water). In winter, problems can occur due to individual variation, light conditions, etc. Rock has usually more evenly warm olive-tinged brown-grey upperparts with at least some green tones (Water: browner, duller and more mottled mantle/scapulars, often contrasting with greyer head and even warmer brown rump and uppertail-coverts); face more uniform with more pronounced whitish eye-ring but almost no supercilium (quite unlike majority of Water, though a few are similar); underparts distinctly warmer tinged yellowish-grey with more extensive and diffuser streaks that coalesce (Water: whiter ground colour and sharper/sparser streaks); and has darker outer tail (see above; beware that strong backlight can make greyish outertail appear whitish). Water Pipit also shows on average more yellow on bill-base than Rock (Andersson 2012), latter being duller and more orange, though some overlap.

AGEING & SEXING Ageing usually possible by close examination of moult, feather pattern and wear in wing, but less easy in spring than autumn. Sexes largely alike in plumage (but see Biometrics); *littoralis* has sometimes separate breeding plumage and slight tendency to show sexual

A. p. petrosus, Ireland, Sep: locality and date, but also dusky (olive grey-brown) appearance and densely streaked underparts, best fit *petrosus*. Ageing of some birds without handling or just from single image often impossible. (C. Bradshaw)

A. p. petrosus, 1stW, England, Oct: dark above with heavily-streaked underparts, most like *petrosus*. 1stW by moult limits with new inner greater covert. (N. Hallam)

A. p. petrosus / littoralis, 1stW, England, Dec: heavily and densely streaked underparts usually indicative of *petrosus*, but some *littoralis* are identical. 1stW by moult limits with two new inner greater coverts. (M. Goodey)

A. p. petrosus, juv, England, Jun: browner plumage with obvious black feather-centres above, while lateral throat and underparts streaking stronger. Also note yellow gape flanges, typical of recently fledged young. (M. Rose)

dimorphism, but much overlap in both respects (see below and Geographical variation & Range). – Moults. Complete post-nuptial and partial post-juv moults in late summer (mainly Aug–early Oct). Post-juv moult (generally Jul–Sep) involves head- and body-feathers, some to all median coverts, one to a few inner greater coverts, one to all tertials and—rarely—r1, but second broods, and northern populations of *littoralis*, generally do not renew any median and greater coverts or tertials (only some lesser coverts). Partial pre-nuptial moult also variable (in northern populations of *littoralis* may include some head- and body-feathers, median and greater coverts, tertials and often r1, but is rather limited or absent in southern *petrosus* populations). – SPRING **Ad** ♂ tends to have fewer streaks on breast and flanks, where slightly more pinkish-tinged buff-white, than ad ♀, and to have darker legs, but complete overlap, and sexing by plumage probably only feasible when studying breeding pairs (if even then); aged as ad by having only ad-type wing- and tail-feathers. **1stS** Might be aged in the hand by some retained juv secondary-coverts forming moult limits with coverts renewed during post-juv and pre-nuptial moults (and a few 1stS diagnostically show three feather generations; two generations in wing, however, common in both age classes, but mainly in southern *petrosus* populations); also, all primaries, primary-coverts and tail-feathers are juv and more worn and bleached browner. – AUTUMN **Ad** By Sep evenly fresh with olive tone to upperparts, and streaking and buffish ground colour to underparts increase, becoming duskier, while in some bill acquires slight yellowish base to lower mandible; also, especially in *littoralis*, supercilium and wing-bars more obvious. Median and greater coverts tipped more olive-brown, less dirty white, and tips to greater coverts (save innermost three) mainly confined to outer webs and comparatively well marked against dark centres. **1stW** May show clear moult limits, with several moulted and fresh inner secondary-coverts and tertials; juv median coverts have diffuse, more triangular dark centres with more pronounced 'tooth' and slightly more pale buffish fringes, while greater coverts have less dark and less well-defined centres with more diffuse buff-tinged (less olive-brown) tips and outer edges (unlike ad). Also, retained primaries and primary-coverts often more worn and subtly weaker textured than in ad. Tail-feathers usually retained (slightly less fresh and more pointed), but occasionally renews central pair. **Juv** Much like winter ad but has browner upperparts with black feather centres and pale fringes more obvious, and cleaner underparts, with lateral throat-stripe/patch and underparts streaking more clearly defined.

A. p. littoralis, ad, Finland, May: not all attain grey head and almost unstreaked pink-buff underparts in summer (some remain in winter-like plumage while others are even more colourful); aged as ad by relatively fresh wing-feathers. (M. Varesvuo)

Summer plumages of *A. p. littoralis* (left: Sweden, Apr) and Water Pipit (right: Switzerland, Apr): can be extremely difficult to separate, especially *littoralis* with greyer head and pinker underparts with relatively little streaking, or Water with less pure grey head and better-streaked underparts. Nevertheless, *littoralis* is usually more evenly olive-grey above (clearer contrast between greyer head and browner rest of upperparts in Water), tends to have better-streaked flanks and lateral throat patch than Water, and central belly usually lacks pure white area of Water. Both birds have completed pre-nuptial moult, but degree of feather wear and moult contrast suggest that left bird is 1stS, and the other an ad. (Left: M. Nord; right: R. Aeschlimann)

A. p. littoralis, 1stS, Finland, May: some attain only 'partial summer plumage' with still heavy breast- and flanks-streaking, virtually no pink-buff below, and grey on head and upperparts darker and less pure. This duller plumage is typical of younger birds (note moult contrast, with many old and heavily-abraded juv wing-feathers). (M. Varesvuo)

BIOMETRICS (*littoralis*) **L** 16–17 cm; **W** ♂ 88–94 mm (n 17, m 90.6), ♀ 81–92 mm (n 12, m 85.4); **T** ♂ 61–69 mm (n 17, m 65.4), ♀ 55–66 mm (n 12, m 60.5); **T/W** m 71.6; **B** 15.2–17.8 mm (n 30, m 16.8); **B**(f) 11.3–15.0 mm (n 29, m 13.1); **BD** 3.6–4.2 mm (n 28, m 3.9); **Ts** 21.5–24.0 mm (n 30, m 22.6); **HC** 7.5–11.5 mm (n 30, m 9.9). **Wing formula: p1** minute; **pp2–5** equal and longest (or pp3–4 longest and p2 < wt 0.5–3 mm, p5 < wt 0.5–2 mm); **p6** < wt 9–12 mm); **p7** < wt 15–18 mm; **p10** < wt 22–27 mm; **s1** < wt 23.5–28 mm. Emarg. pp3–5.

GEOGRAPHICAL VARIATION & RANGE Moderate clinal variation in that northern birds are slightly paler and more distinctly but sparsely streaked, at least in summer plumage. Two rather similar races recognised.

A. p. petrosus (Montagu, 1798) (Britain, NW France; resident or makes shorter movements). Poorly differentiated seasonal plumage variation, thus is at all seasons darker, more unstreaked and more dusky-olive above than *littoralis*, with an indistinct supercilium, dull buff-brown wing-bars and tertial fringes, and is deeper saturated below with copious streaking covering virtually entire underparts. Very similar to non-breeding *littoralis*. **W** ♂ 92–98.5 mm (n 30, m 95.0), ♀ 85.5–92.5 mm (n 17, m 89.1); **T** ♂ 62–70.5 mm (n 30, m 67.5), ♀ 59–66.5 mm (n 16, m 62.9); **T/W** m 71; **B** ♂ 17.2–19.2 mm (n 32, m 18.0), ♀ 16.5–17.9 mm (n 21, m 17.3) (Alström & Mild 2003). – Birds on Faeroes ('*kleinschmidti*') tend to be slightly larger on average and tinged a shade more yellowish-olive, but differences very slight and inconsistent, hence best included in *petrosus*. Similarly, birds of Outer Hebrides ('*meinertzhageni*') same size as *petrosus* but tend to be on average subtly darker brown above and paler cream-buff below, but again differences are very slight and we follow Svensson (1992) by including them in *petrosus*. (Syn. *hesperianus*; *kleinschmidti*; *meinertzhageni*; *obscura*; *ponens*.)

A. p. littoralis C. L. Brehm, 1823 (Fenno-Scandia, NW Russia; winters coastal W Europe). To some degree intermediate between *petrosus* and Water Pipit, closer to former in non-breeding plumage and approaches latter when breeding. At latter season acquires greyer crown and nape, and paler underparts with rosy-buff flush on chest, and streaks typically restricted to upper flanks; lateral throat-stripe still present, and tends to have better-developed pale supercilium than Water Pipit. In non-breeding plumage perhaps separated from *petrosus* by paler and purer olive mantle and scapulars, and perhaps more pronounced dark feather-centres. At all seasons tends to have pale portion of outer tail-feathers greyish-white with purer white fringes.

TAXONOMIC NOTES Until comparatively recently, Water, Rock and Buff-bellied Pipits were regarded as diagnosable taxa within a single species, but since the mid 1980s they have more commonly been treated as separate species based on morphology, vocalisations and genetic data, and on the claimed sympatry between Water Pipit (ssp. '*blakistoni*', here treated as synonymous with *coutellii*) and Buff-bellied Pipit (ssp. *japonicus*) in the Baikal region (Nazarenko 1978). Most recently, the evidence of mtDNA suggests that Buff-bellied Pipit is more closely related to Meadow Pipit than to either of the other two species with which it was long considered conspecific.

REFERENCES Andersson, R. (2012) *Ornis Svecica*, 22: 33–38. – Fijen, T. P. M. (2014) *DB*, 36: 87–95. – Königstedt, D. & Müller, H. E. J. (1983) *Beiträge zur Vogelkunde*, 29: 77–88. – Nazarenko, A. A. (1978) *Zool. Zhurn.*, 57: 1743–1745.

Presumed *A. p. littoralis*, 1stW, Denmark, Jan: race assigned by locality. In winter plumage, there is no consistent single feature to separate the races. Nevertheless, olive-buff hue on head and breast, and overall pale coloration better fit *littoralis*. Pale reddish-brown legs (not all Rock have dusky-brown legs). 1stW with many juv wing-feathers. (J. Dam)

Presumed *A. p. littoralis*, probably 1stW, Norway, Sep: race assigned by locality. Similar in overall appearance to *A. p. petrosus*. Ageing difficult, but abraded whitish tips to inner greater coverts and tertials suggest 1stY, and very pointed rectrices support this. (A. Clarke)

Winter plumages of *A. p. littoralis* (left: Norway, Sep) and Water Pipit (right: Italy, Nov): taking variation into account, some Rock and Water Pipits are very similar, as can be ascertained here; compared to Water, Rock is usually tinged more olive-grey above (thus often tinged greenish), not as brown on mantle/scapulars, and has warmer, more extensive and diffuse flanks-streaking that coalesces (sharper/sparser streaks on whiter underparts in Water). Rock usually has more pronounced whitish eye-ring in plainer face, whereas Water usually has more prominent pale supercilium (but rarely both are very similar as can be seen here). Both are 1stW with many juv wing-feathers. (Left: E. Bergersen; right: D. Occhiato)

WATER PIPIT
Anthus spinoletta (L., 1758)

Fr. – Pipit spioncelle; Ger. – Bergpieper
Sp. – Bisbita alpino; Swe. – Vattenpiplärka

Water Pipit is an alpine bird which usually breeds above the tree-line on wet and moist mountain slopes. Listen there in summer for its shrill and repetitive but rhythmic song. Spotting the bird is less easy, it being small and rather nondescript on the wide alpine stage. It breeds in WC & S Europe, Turkey, the Caucasus and N Iran, and widely in Central Asia to N & W China, Mongolia and Lake Baikal. The species is largely migratory, wintering near water or at lower elevations, at which season it is much more widespread and occurs north to S Scandinavia—thus some birds possibly perform a northward autumn migration—and south to NE Africa, Arabia and east to NW India and SE Asia.

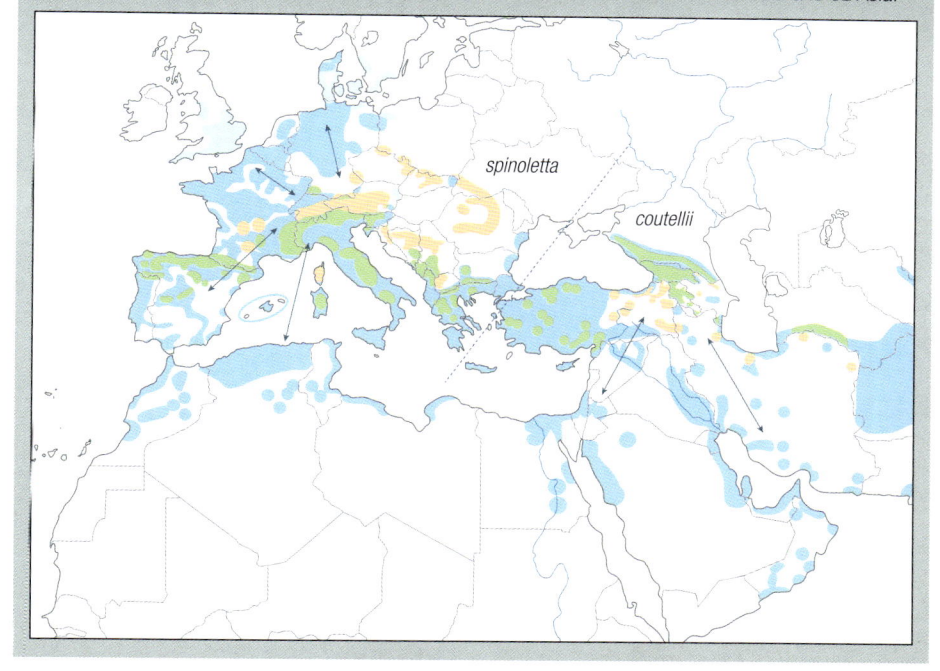

A. s. spinoletta, presumed ad, Italy, Jun: summer plumage already heavily bleached, with pale buff-pink wash concentrated on breast, and no streaking. Wing, however, very uniform, relatively fresh and evenly fringed greyish suggesting ad (juv wing-feathers should show extensive wear and less colour). (D. Occhiato)

IDENTIFICATION A small pipit that differs from all others in the region, except Rock Pipit, in typically having *blackish or very dark brown legs*. Calls also by and large identical to those of Rock Pipit, thus careful separation using plumage features is essential. In breeding plumage has *medium brownish-grey or grey head* with *whitish or cream supercilium* usually quite well developed (note that odd birds have less distinct supercilium), grey of head often subtly contrasting with brownish olive-grey mantle (but can look rather uniform in some birds), both faintly streaked darker. Rump and uppertail-coverts usually tinged brownish, not grey (but rarely bleached with more greyish appearance). Underparts dull white with variable *pale buff-pink wash on throat to mid belly*, diagnostically *unstreaked or streaks confined to upper flanks, and fairly narrow and distinct*. In winter, *crown slightly greyer*, and creamy *underparts extensively and rather obviously streaked* dark brown, more distinctly (less diffusely) on average than in Rock (albeit streaking relatively sparse). Wings have pale tips to median coverts and fringes to greater coverts, forming double wing-bars, and pale grey-buff tertial fringes. *Tail has extensive white edges* to outer web of outermost feather, white wedge on inner web and white tip to adjacent feather. On average bill more bicoloured with *more yellow on base of bill* than in Rock Pipit (Andersson 2012). More gregarious in winter and on passage than Rock Pipit. Song-flight much like Rock Pipit.

VOCALISATIONS Song close to Rock Pipit, but has more rhythmic motifs and thin, drawn-out notes, e.g. *zru zrü zrü zrü-zrü-zrü-zrü-zü-zü-züzü-züzüzüzüzü sviririrr-sviririrr-sviririrr süüü-süüü psiieh-psiieh-psiieh*. – Usually most flight-calls are indistinguishable from those of Rock (if anything perhaps on average subtly shorter, lower-pitched and flatter than call of Rock Pipit; Fijen 2014), but both are readily separable from Meadow Pipit in being less strident and delivered slower than latter, though typically rather explosive, harsh and more drawn-out with a faintly shrill quality, e.g. *vüisst* or *wheest* and *peezp* or *peez-piz*. In 'conversation' upon alighting, calls are more like Meadow Pipit, *viss, vüss, vüss-vuss* and similar.

SIMILAR SPECIES Most likely to be confused with Rock Pipit, in winter or on migration (summer-plumaged adults should pose few problems and the two are unlikely to come into contact at this season); see Rock Pipit for main differences. – See *Buff-bellied Pipit* for comparison with the present species. – Confusion also possible with *Meadow Pipit* in winter, but latter is slightly smaller and shorter-billed, usually with paler pinkish or reddish/yellowish-brown legs (but some Water Pipits have less clearly dark, even almost pale flesh-coloured legs), a distinctly yellowish-green tinge to upperparts with better-defined streaking, and rump/uppertail-coverts less plain. Meadow Pipit is also typically suffused buffish-yellow below with blacker flanks-streaking (not almost whitish with few but broader streaks like Water Pipit), paler lores, more distinct moustachial stripe and better-marked lateral throat-stripe/patch, but supercilium much less distinct. With experience they are also readily separable by vocalisations.

AGEING & SEXING Ageing and sexing largely as in Meadow and Rock Pipits. – Moults. Complete post-nuptial and partial post-juv moult in late summer (Jul–Sep). Post-juv renewal includes all head- and body-feathers, most to all lesser coverts, usually no or few (rarely all) median coverts, mostly none or just 1–2 inner greater coverts, and occasionally 1–2 (rarely all) tertials. Partial pre-nuptial moult in late winter (mainly Feb–Mar), involving most or even all head- and body-feathers, usually some (varying from none to

A. s. spinoletta, France, Mar: in advanced moult to summer plumage, with greyish head (quite pronounced whitish supercilium) contrasting with brownish olive-grey mantle, faintly streaked darker; pale buff-pink wash on breast to mid-belly, with streaking confined to upper flanks. Ageing uncertain. (A. Audevard)

A. s. spinoletta, presumed 1stS, Netherlands, Mar: in summer plumage, some are browner on back, approaching *coutellii*. Ageing uncertain, but note strong moult contrast between fresh long two tertials and distinctly worn (possibly juv) short innermost, suggesting 1stY. (N. D. van Swelm)

A. s. spinoletta, presumed ad, Spain, Sep: some individual variation in winter plumage, but identification is straightforward here. Note dull grey-brown upperparts, with browner rump, blotchy streaking on breast but more distinct on flanks, well-developed cream-white supercilium and dusky legs. Ageing difficult, but probably ad given evenly-fresh ad-type wing-feathers. (G. Thoburn)

most) median coverts, a few (0–7, usually 1–3) inner greater coverts, all tertials, perhaps r1 and very rarely inner secondaries, but moult generally most extensive in 1stS, though some of this age still renew very few feathers or none. — **SPRING** Ageing very difficult, often impossible due to slight differences and some variation in wear and pattern between ad and 1stS. **Ad** Sexes similar, but ♀ may have less grey on head than ♂. Note that both sexes can have underparts largely unstreaked and tinged pinkish. Prior to moult in summer, wing-bars and tertial fringes less obvious, and pink flush to underparts and supercilium much reduced. **1stS** Very similar to ad, and usually not separable unless there are three generations of coverts, thus with a few heavily-worn retained juv feathers (very rare). — **AUTUMN Ad** By Sep evenly fresh with grey tone to head and upperparts, pink on underparts lost, and all races browner above and more streaked below. When fresh, inner median and greater coverts usually warmer than outer feathers, misleadingly suggesting moult limits, potentially leading to incorrect ageing. **1stW** May show clear moult limits on inner wing, with some inner median and greater coverts and tertials renewed; tail-feathers retained (but central pair occasionally renewed), primaries and primary-coverts more worn and weaker textured than ad; juv greater and median coverts have tips to dark centres more triangular with pronounced point (like Meadow Pipit), but differences extremely hard to detect. **Juv** Like ad winter but crown, nape and rest of upperparts more obviously streaked, with shorter and more diffuse breast-streaking and more obvious and triangular dark centres to median coverts (and, to lesser extent, greater coverts).

BIOMETRICS (*spinoletta*) **L** 15.5–16.5 cm; **W** ♂ 90–95 mm (*n* 15, *m* 92.2), ♀ 82–92 mm (*n* 13, *m* 86.3); **T** ♂ 64–73 mm (*n* 15, *m* 67.5), ♀ 57.5–66 mm (*n* 13, *m* 62.7); **T/W** *m* 73.2; **B** ♂ 15.3–17.7 mm (*n* 15, *m* 16.6), ♀ 15.0–16.9 mm (*n* 12, *m* 16.1); **BD** 3.6–4.3 mm (*n* 19, *m* 4.0); **Ts** 22.5–25.5 mm (*n* 27, *m* 23.5); **HC** 8.5–12.7 mm (*n* 22, *m* 10.7). **Wing formula: p1** minute; **pp2–5** about equal and longest (sometimes p2 < wt 0.5–2 mm, p5 < wt 0.5–1.5 mm); **p6** < wt 8–12 mm; **p10** < wt 21–26 mm; **s1** < wt 21–27 mm. Emarg. pp3–5.

A. s. spinoletta, 1stW, Italy, Oct: 'typical' dull winter plumage, showing dark legs and streaked pattern. Aged by juv median and greater coverts, and tertials, with brown centres, white tips and already abraded fringes, especially to tertials. (D. Occhiato)

A. s. spinoletta, 1stW, Netherlands, Dec: a rather atypical, heavily streaked *spinoletta*, which could be mistaken for Rock Pipit, but note distinctly streaked flanks on whitish ground, good supercilium, much orange-yellow on base of bill and brownish back. (N. D. van Swelm)

A. s. spinoletta, Italy, Jan: some Water can resemble *littoralis* Rock Pipit, as demonstrated by this bird. Nevertheless, streaking on central flanks distinct, and just visible white in outer rectrix supports the identification (although worn *littoralis* often has whitish outer webs of t6, too). Limited yellow on bil base (superficially suggesting Rock) is due to mud covering tip. Greater coverts and tertials are juv indicating 1stW. (D. Occhiato)

A. s. coutellii, 1stS, Turkey, Apr: underparts are usually predominantly pinkish-buff by Apr. Compared to *spinoletta* (in same stage), pink averages deeper and more extensive. With further abrasion, pink becomes diluted and concentrated on chest, like *spinoletta*. Sharp moult limits in wing and juv median coverts identify bird as 1stS. (Z. Kurnuç)

PIPITS AND WAGTAILS

GEOGRAPHICAL VARIATION & RANGE Fairly moderate variation despite wide and partially disjunct range. We recognise only two races since *coutellii* and *blakistoni* are inseparable by plumage when longer series are compared, and since they only differ very slightly in some measurements with extensive overlap. See Alström & Mild (2003) for a different view.

A. s. spinoletta (L., 1758) (C & S Europe east to Greece and Bulgaria; short-range movements in winter to suitable coasts, estuaries and rivers, north to Sweden and Denmark, south to Morocco, east to Greece and W Black Sea). Described above. Typically has rather uniform upperparts with subdued darker streaking on crown and mantle, and in summer plumage a fairly pale pink lower throat to breast, slightly contrasting with whitish belly.

A. s. coutellii Audouin, 1826 (Turkey, Caucasus, N Iran, mountains in Central Asia from Tajikistan to E Kazakhstan and W Mongolia, Altai, Sayan; winters at or close to sea level in NE Africa, Arabia and SW Asia, south to Nile Valley, Oman, Iran, Pakistan, NW India, China). Very similar to *spinoletta* but generally has subtly paler and warmer grey-brown ground colour to its upperparts, with usually bolder and blacker streaking (so general impression can be even darker), though streaking somewhat variable, some being moderately streaked. In breeding plumage underparts are richer and more uniformly rufous-pink with ochre hue reaching belly/vent (*spinoletta* has whitish belly/vent) and tends to have more flanks-streaking. Sometimes has paler/purer grey forehead/forecrown and ear-coverts compared to *spinoletta*. Also has on average broader, whiter supercilium. Non-breeders have upperparts still decidedly paler brown with buffier tinge, slightly more pronounced dark olive-brown streaking, and greyer wash to crown. Underparts have, on average, narrower, more restricted streaking on richer buff flanks. Call often like short, 'listless' Tree Pipit, *beez*, but can be more emphatic and upslurred *weeist*. **W** ♂ 86–93 mm (n 14, m 89.2), ♀ 80–88 mm (n 12, m 84.6); **T** ♂ 62–69 mm (n 14, m 65.5), ♀ 56–65 mm (n 12, m 61.3); **T/W** m 73.0; **B** 15.0–17.5 mm (n 25, m 16.4); **Ts** 21.3–25.0 mm (n 24, m 23.8). – Breeders in Central Asia often separated ('*blakistoni*') on account of subtly larger size and on average paler colours and less prominent streaking, also subtly shorter bill and tarsus, but differences very slight and show much overlap, not amounting to 75% rule requirement, hence included here. White tip of r5 averages slightly larger in east, but much overlap. Biometry for extralimital eastern birds ('*blakistoni*'): **W** ♂ 87–95 mm (n 16, m 90.7), ♀ 81.5–88 mm (n 10, m 84.4); **T** ♂ 62.5–73 mm (n 16, m 67.8), ♀ 61–67.5 mm (n 10, m 63.5); **T/W** m 74.9; **B** 15.3–17.5 mm (n 23, m 16.1); **Ts** 20.8–23.9 mm (n 26, m 22.6). (Syn. *blakistoni*; *caucasicus*.)

A. s. coutellii, presumed 1stS, Israel, Mar: in summer plumage tends to have purer grey head than *spinoletta*, paler sandy olive-brown and more pronounced dark streaking over rest of upperparts. Ageing difficult, but strong moult contrast suggests 1stS. (B. Nikolov)

TAXONOMIC NOTE Regarding relationships with Buff-bellied and Rock Pipits, see note under Rock Pipit.

REFERENCES Andersson, R. (2012) *Ornis Svecica*, 22: 33–38. – Fijen, T. P. M. (2014) *DB*, 36: 87–95. – Koning, F. J. (1982) *Limosa*, 55: 155–160. – Königstedt, D. & Müller, H. E. J. (1983) *Beiträge zur Vogelkunde*, 29: 77–88. – Nazarenko, A. A. (1978) *Zool. Zhurn.*, 57: 1743–1745.

A. s. coutellii, presumed 1stS, Oman, Mar: especially during Dec to Mar, one can expect to encounter moulting birds on wintering grounds, usually with rather 'scruffy' appearance to head, breast and mantle. Strong moult limits in wing. (H. & J. Eriksen)

A. s. coutellii, 1stW, Oman, Nov: some are duller overall, with more olive pigments above. Compared to superficially similar Meadow Pipit (if plainer rump/uppertail-coverts not seen), the upperparts have diffuse streaking, with poorly-indicated whitish eye-ring, but better-developed pale supercilium; also note blotchier streaking below, and darker legs. (H. & J. Eriksen)

Variation in *A. s. coutellii*, Israel, Dec: much individual variation, and plumage overlap with *spinoletta* (e.g. right-hand bird). Some variation is age- and possibly sex-related, influenced by timing of pre-nuptial moult. Both are 1stW, but left bird has moulted to ad-type wing-coverts, with greyer head, browner and more diffuse streaking on breast, and hardly-indicated lateral throat-stripe. Right bird has olive-grey and slightly more uniform upperparts, blacker-centred and sharper streaks on breast, and better-developed lateral throat-stripe. (A. Ben Dov)

A. s. coutellii, 1stW, Israel, Nov: young *coutellii* can have some buff or rufous-tinged pigmentation on mantle and scapulars, clearly indicating race involved. Aged as 1stW by number of juv wing-feathers contrasting with post-juv inner greater coverts. (A. Ben Dov)

BUFF-BELLIED PIPIT
Anthus rubescens (Tunstall, 1771)

Alternative name: American Pipit

Fr. – Pipit d'Amérique; Ger. – Pazifikpieper
Sp. – Bisbita norteamericano; Swe. – Hedpiplärka

Buff-bellied Pipit occupies a wide range, breeding in tundra to alpine meadows across E Siberia, and, on the opposite side of the Bering Sea, across North America. Winters mainly in Japan and China, and in North America south to Middle America, being more catholic at this season, using a variety of open, grassy and waterside habitats. In the covered region, *rubescens* has been recorded as a vagrant in NW Europe and small numbers of *japonicus* winter in Israel, Egypt and, perhaps, E Arabia, especially in the United Arab Emirates, with vagrants recorded west as far as Sweden.

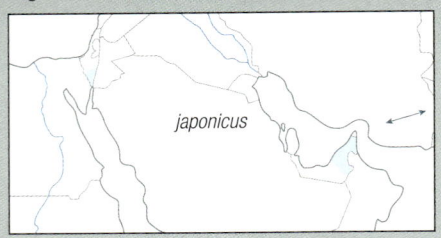

japonicus

IDENTIFICATION Very like Water Pipit, but smaller and considerably *slighter with a Meadow Pipit 'feel'*. Compared to Water and Rock Pipits has diagnostic *complete white eye-ring* (not broken by dark loral stripe) and usually *pale flesh-coloured legs* in race *japonicus* (but dark brown to black in *rubescens*). Other important characters are the rather *plainer greyish olive-coloured upperparts* (although streaking and coloration vary geographically and with wear), and better-defined, whiter wing-bars. In non-breeding plumage (most likely to be seen in the treated region), rather *heavily streaked below* on pale yellowish-buff ground, and *blackish lateral throat-stripe and*, especially, *throat patch, well developed*. In spring/summer slightly greyer above with an orange or pink-buff wash below, and (unlike most Water Pipits) clearly streaked on body-sides. In all plumages appears to have quite *well-defined whitish supercilium* and rather plain (but often very faintly flecked) lores. Pale area of outer tail-feathers rather pure white. Short, narrow-based bill has rather obvious pale lower base. On ground appears rather dainty, with erratic movements more reminiscent of Meadow than Water Pipit.

VOCALISATIONS Song (unlikely to be heard within covered region) a series of rapidly-repeated notes, rather similar to Water Pipit. – Calls useful for identification, one or often two high, short *zeep*, *tsipp* or *tsiit* notes, lacking obvious shrill quality of Water or Rock Pipits, generally more like those of Meadow Pipit, albeit higher, thinner, more drawn-out, and more liquid. On take-off gives rapid series of *tsi-tsip* or *tsiip-iip-iip* notes, much higher and more nervous than Meadow.

SIMILAR SPECIES In size, structure and calls, Buff-bellied recalls *Meadow Pipit* and could easily be dismissed as such. In particular, a warm-coloured and less well-marked Meadow is very like non-breeding Buff-bellied of race *japonicus*, which has rather similar pale brown legs. With experience, *japonicus* should be identified by (among many small differences) its paler lores, more complete pale eye-ring and plainer and greyer upperparts. – Principal confusion risk is with *Water Pipit* in autumn/winter plumage, but note that Buff-bellied is smaller than Water, and proportionately slighter with a finer, shorter bill. Legs of race *japonicus* are ochre- to reddish-brown, darker than pinkish legs of Meadow but paler than most Water; race *rubescens*, however, usually has dark brown or blackish legs. Unlike Water, Buff-bellied has complete and conspicuous white eye-ring and plainer lores, and a more prominent deep yellow base to lower mandible; also dark moustachial stripe rather distinct (almost entirely lacking in Water), much more distinct lateral throat-stripe (at first faint but broadens into obvious blackish patch or wedge on lower throat-sides) and pronounced creamy submoustachial stripe. Upperparts uniform dark olive-brown (non-breeding) or greyish olive-brown (breeding), unstreaked or with very indistinct dark feather centres, lacking Water's pale (grey or buff-brown) ground colour (with warmer rump and greyer nape) and rather pronounced dark mottling. In non-breeding plumage, underparts have much heavier spots or streaking (chiefly *japonicus*; *rubescens* has less extensive and narrower streaking), with longer blackish-brown streaks on flanks on typically whitish to pale buff (chiefly in *japonicus*) or much richer buff (most *rubescens*) ground. In breeding plumage, underparts mostly pinkish-buff or slightly orange-tinged, still strongly streaked/spotted (Water: pinkish or pale vinaceous underparts with much-reduced markings). Also, compared to Water, has obvious double whitish wing-bars and whiter area on outer tail-feathers. – In W Europe confusion also possible with *Rock Pipit* which is relatively darkish, plain-backed and has rather heavily-marked underparts. Buff-bellied, however, is clearly smaller and daintier than Rock with a smaller bill; also usually has plainer upperparts (Rock: tends to show at least some dark feather centres), more obvious white wing-bars and supercilium (Rock: wing-bars usually more diffuse, dull cream-buff, and supercilium more subdued), an unbroken white eye-ring, and distinctly whitish outer tail-feathers (Rock: conspicuous loral stripe disrupts eye-ring, and outer tail-feathers creamy or pale brown). Differences in leg colour as between Buff-bellied and Water, but note that nominate Buff-bellied has darker legs approaching Rock in W Europe. Non-breeding Buff-bellied differs further in having predominantly whitish-buff (*japonicus*) or deeper buff (*rubescens*) underparts, respectively, streaked bolder and denser or rather heavily but narrowly, and lacking Rock's warm yellowish-olive/cream underparts with broad, heavy (but mainly diffuse) streaks. In the covered region the two species are unlikely to come into contact in breeding plumage.

AGEING & SEXING (*rubescens*) Age distinctions largely as in Meadow, Water and Rock Pipits. Sexes alike. – Moults. Complete post-nuptial moult, and partial post-juv and pre-nuptial moults, similar in extent and timing to Water Pipit. In North America, post-juv renewal includes none to all median coverts, 0–4 inner greater coverts, and sometimes 1–2 tertials, but no tail-feathers. Pre-nuptial moult in late winter includes 0–4 inner greater coverts, usually 1–3 tertials and often one or both r1; ssp. *japonicus* mostly similar but few birds studied. – **SPRING Ad ♂** in breeding plumage tends to have sparser underparts streaking than ♀; aged by having only ad (worn or fresh) wing-feathers, but moult contrast among secondary-coverts usually discernible and thus hardly differs from 1stS. **1stS** Extremely similar but more worn, especially juv primaries, primary-coverts and tail-feathers; retained (heavily worn) and renewed juv secondary-coverts create more obvious moult limits. – **AUTUMN Ad** Evenly fresh; see above for details of non-breeding plumages. **1stW** As (non-breeding) ad, but moult limits with variable number of juv secondary-coverts and tertials; retained primaries, primary-coverts and tail-feathers more worn and weaker textured than ad. Tips of dark centres to juv median coverts more triangular with a sharper point close to tip and along shaft, and tips whiter, while fringes to greater coverts less clear-cut and lack indentation at tip of outer web of ad, but such differences extremely hard to detect in the field, and intermediates occur. **Juv** Distinctive on breeding grounds by being more heavily streaked above but has less distinct streaking on flanks than fresh autumn birds. Unlikely to be recorded in the covered region.

BIOMETRICS (*rubescens*) **L** 14–16 cm; **W** ♂ 84–89 mm (n 12, m 85.5), ♀ 78–82 mm (n 12, m 80.3); **T** ♂ 60–65 mm (n 12, m 62.3), ♀ 56–63 mm (n 12, m 59.4);

A. r. rubescens, 1stW, France, Sep: records in W Europe refer to North American race, which is plainer and paler above, and richer buff below with less extensive and narrower and diffuser streaking. Legs often dark but can be paler, as here. Aged by many retained juv wing-feathers. (A. Audevard)

PIPITS AND WAGTAILS

A. r. rubescens, 1stW, Sweden, Jan: some could easily pass for Rock Pipit; note almost unstreaked upperparts (appearing dark/warm-toned due to light), distinctly narrow-streaked warmer buff underparts and plain pale lores. Legs are usually rather dark (as here). All or almost all wing-feathers are juv. (C. Halkjær)

A. r. rubescens, presumed 1stS, Canada, Jun: summer plumage recalls Water Pipit, with grey upperparts, pinkish pigmentation and reduced streaking below. This plumage never recorded in the treated region, but a late-staying spring bird could develop partial summer plumage. Strong moult contrast in wing suggests 1stS. (K. Pedersen)

T/W m 73.5; **B** 14.2–16.8 mm (n 24, m 15.2); **BD** 3.4–4.4 mm (n 24, m 3.9); **BW** 4.2–4.8 mm (n 24, m 4.5); **Ts** 17.4–23.2 mm (n 24, m 21.9); **HC** 9.5–12.0 mm (n 23, m 10.5). **Wing formula: p1** minute; **p2** < wt 0–1.5 mm; **pp3–4** (pp2–5) equal and longest; **p5** < wt 0–2 mm; **p6** < wt 10.5–13 mm; **p7** < wt 15–18 mm; **p10** < wt 20–23 mm; **s1** < wt 22–26 mm. Emarg. pp3–5.

GEOGRAPHICAL VARIATION & RANGE Rather marked variation within wide range. Three subspecies can be recognised, of which two have been recorded in the treated region.

A. r. rubescens (Tunstall, 1771) (N North America mainly in Canada and Alaska; resident or makes short-range movements; vagrant to NW Europe). In autumn/winter appears plainer and paler than *japonicus*; also richer buff below with less extensive and narrower streaking. In breeding plumage wears greyer and duller above than *japonicus* and underparts on average more extensive and deeper orange-buff with relatively fewer and finer streaks. Tarsus dark brown or blackish, not unlike Water Pipit. (Syn. *harmsi*; *pacificus*.)

A. r. japonicus Temminck & Schlegel, 1847 (NE Siberia west to Baikal region, Burjat, in the east south to Amur,

Buff-bellied Pipits (both left images: top left *A. r. rubescens*, Shetland, Scotland, Sep; bottom left *A. r. japonicus*, Oman, Jan) versus Rock Pipit (top right: presumed *A. p. littoralis*, Italy, Dec) and Water Pipit (bottom right: *A. s. spinoletta*, Italy, Nov) in late autumn/winter: both *rubescens* and *japonicus* share an unbroken white eye-ring and plain lores, rather heavily streaked underparts, underparts being variably tinged pale yellowish-buff, large blackish lateral throat-patch, dark-toned and plain (*rubescens*) to weakly streaked (*japonicus*) greyish-olive upperparts and contrasting whitish wing-bars. Note that *japonicus* unlike the other three taxa tend to have paler legs (though exceptions known). All birds seemingly have many juv feathers in wing and pointed and already worn tips to some tail-feathers, suggesting young age. (Top left: H. Harrop; bottom left: H. & J. Eriksen; top & bottom right: D. Occhiato)

— 167 —

A. r. japonicus, presumed ad, Israel, Dec: Middle Eastern records relate to this darker (NE Asian) race, which in winter plumage has dark olive-brown upperparts, striking white wing-bars when fresh, heavy blackish lateral throat-stripe and streaking on breast and flanks. Ageing difficult, but still rather fresh uniform greater coverts best for ad. (A. Ben Dov)

A. r. japonicus, presumed ad, Israel, Dec: very rarely *japonicus* can have black legs, like Rock and Water Pipits and some *rubescens* Buff-bellied. This bird also has an incomplete white eye-ring (but this could be due to abrasion). Note also the unusual lack of prominent yellow base to lower mandible. However, distinct lateral throat-stripe, greyish olive-brown and streaked upperparts, and quite heavy streaking of underparts all favour *japonicus*. Only ad-like wing feathers seem to be visible. (H. Shirihai)

Sakhalin; winters S China, Korea, Japan; regular visitor as far west as Middle East). Particularly distinctive in non-breeding plumage with darker olive-brown upperparts, which usually appear unstreaked, with on average darker feather centres than *rubescens* (dark centres largely concealed but visible in close view). Underparts sometimes paler with more extensive whiter belly, but in fresh spring plumage often richly tinged ochre-buff on throat, breast and flanks with only slightly paler belly. Blackish lateral throat-stripe and patch bolder and better developed, continuing as much heavier and darker streaking, even over entire breast (often forming hint of dark gorget) and flanks. In breeding plumage tends to be less grey dorsally and more strongly streaked below than *rubescens*. Legs usually pale (brown or pink) but sometimes darker (very rarely as blackish as in Water Pipit). Averages slightly larger, but much overlap: **W** ♂ 84–92 mm (*n* 22, *m* 88.4), ♀ 80–90 mm (*n* 12, *m* 85.3); **T** ♂ 57–69 mm (*n* 23, *m* 62.7), ♀ 59–67 mm (*n* 12, *m* 61.9); **T/W** *m* 71.3; **B** ♂ 13.7–17.2 mm (*n* 23, *m* 15.6), ♀ 13.8–16.7 mm (*n* 11, *m* 15.2); **Ts** 20.1–23.6 mm (*n* 30, *m* 22.4); **HC** 8.6–13.1 mm (*n* 30, *m* 10.4). (Alström & Mild 2003; ZFMK.)

TAXONOMIC NOTE Formerly considered a race of Water Pipit, but eastern range of race *coutellii* of Water Pipit ('*blakistoni*') and *japonicus* of Buff-bellied Pipit breed in sympatry near Lake Baikal, apparently without hybridisation (Nazarenko 1978). However, it should be noted that the two do not breed side by side but are ecologically segregated, thus paralleling situation between Hume's Lesser Whitethroat *Sylvia curruca althaea* and Kazakh Lesser Whitethroat *S. c. halimodendri* in Central Asia where desert plains meet mountains. Depending on taxonomic preferences such sympatry can be interpreted either to support a split or as having no relevance.

REFERENCES French, P. (2006) *BW*, 19: 499–515. – Knox, A. (1988) *BB*, 81: 206–211. – Nazarenko, A. A. (1978) *Zool. Zhurn.*, 57: 1743–1745.

A. r. japonicus, United Arab Emirates, Mar: moulting to summer plumage on wintering grounds. Age uncertain. (N. Moran)

A. r. japonicus, 1stS, Korea, Apr: in summer, upperparts more uniformly grey, underparts mostly pinkish-buff, and still to some degree streaked; also usually has conspicuous white eye-ring. Most of wing-feathers apparently juv and strongly abraded, but has contrasting new inner greater covert. (K. Hyun-Tae)

FOREST WAGTAIL
Dendronanthus indicus (J. F. Gmelin, 1789)

Alternative name: Tree Wagtail

Fr. – Bergeronette de forêt; Ger. – Baumstelze
Sp. – Lavandera forestal; Swe. – Trädärla

Forest Wagtail breeds in NE Asia along edges of forests near water, shallow rivers, in clearings in damp forests or plantations, from sea-level to c. 2000 m. Migratory or partly migratory; wintering south to S China, India and SE Asia including in Indonesia and the Philippines. In recent years found to be an almost annual vagrant and winter visitor to the United Arab Emirates, with records also in Kuwait and Oman.

IDENTIFICATION An unmistakable strikingly patterned mainly olive-brown and black-and-white wagtail with short slender tail. Unique in having *double broad black necklace bands*, one across upper breast, often with broad line extending down centre of breast to the lower, less well-marked and often incomplete at centre. Complex wing markings with black-based wing-coverts broadly tipped white (tinged yellow when fresh), forming *prominent double wingbars* on folded wing (also very evident in flight), and a *white secondary panel* and a small *white basal primary patch*. Also distinctive is the *white narrow supercilium* and the three-coloured tail with obvious white outer feathers, blackish inner feathers, and an olive-brown central pair. Otherwise upperparts greyish olive-brown, contrasting with white underparts. Bill long and largely pink, *legs pale pinkish-buff*. *Sways body and tail from side to side* (not wagging it up and down as other wagtails), walking easily along horizontal branches, well-camouflaged against foliage. Often seen feeding in shaded moist patches of open forest floor, and when flushed may alight quickly into canopy.

VOCALISATIONS Song delivered from perch in a tree somewhat resembles slow song of Great Tit, a repeated squeaky or scratchy disyllabic note, first syllable high-pitched, second lower, *zlic-zheee zlic-zheee zlic-zheee* ... (Alström & Mild 2003). – Main call is a metallic, abrupt Chaffinch-like *pink* or *pink-pink*. Occasionally a very quiet *tsip* while foraging.

SIMILAR SPECIES Can hardly be confused with any other species within the covered range once diagnostic double necklace band and characteristic wing pattern seen.

AGEING & SEXING Ageing usually not possible once post-juv and post-nuptial moults concluded. Sexes alike. – Moults. Complete post-nuptial moult after breeding in summer (often Jul–Aug), and partial post-juv at the same time or earlier, latter including usually all wing-coverts and tertials. Partial pre-nuptial moult in late winter (Feb–Apr), variable in extent, and e.g. no to all tertials renewed; r1 generally moulted, too. – **SUMMER Ad** First wing-feathers heavily worn, then these in active moult. **1stW** Plumage as ad once post-juv moult finished, and generally inseparable. Only if some juv feathers retained with resulting slight difference in wear can a few be aged. **Juv** Slightly browner above, with less complete necklace-bands than in ad, but post-juv moult starts very early and fledged birds very soon look like ad. Still, told by their fresh plumage, while ad in summer are very worn or in active wing moult.

BIOMETRICS L 15–17 cm; **W** ♂ 76–83 mm (*n* 15, *m* 79.5), ♀ 73.5–86 mm (*n* 12, *m* 80.3); **T** ♂ 63–71 mm (*n* 15, *m* 68.2), ♀ 61–70 mm (*n* 17, *m* 65.2); **T/W** *m* 84.6; **B** 15.5–18.1 mm (*n* 34, *m* 16.7); **BD** 3.3–4.3 mm (*n* 34, *m* 3.8); **Ts** 20.1–22.4 mm (*n* 34, *m* 21.2); **HC** 4–6 mm. Wing formula: p1 minute; pp2–4 equal and longest; p5 < wt 3–6 mm; p6 < wt 10–13 mm; p7 < wt 14–17.5 mm; p10 < wt 21–26 mm; s1 < wt 22–28 mm. Emarg. pp3–5 (on p5 sometimes less clearly).

GEOGRAPHICAL VARIATION & RANGE Monotypic. – SE Siberia, Manchuria, Korea, NE China; winters SE Asia.

Malaysia, Jan: annual visitor to E Arabia, including United Arab Emirates. Unmistakable due to its strikingly unique breast and wing patterns. Ageing impossible (as with all images shown here). (C. Wai Mun)

United Arab Emirates, Jan: despite rather uniform head pattern, the double blackish necklaces and complex banded wing pattern make this wagtail wholly unique in the treated region. (M. Barth)

Oman, Nov (left) and United Arab Emirates, Oct: the double blackish necklaces can already be detected in side view (left), but are best viewed head-on (right). In some birds, dark bands are more solid and blacker, and from certain angles appear almost connected below the 'bib patch' (left), but in others the bands are less tidy, appearing browner and broken (right). (Left: E. Yoğurtcuoğlu; right: M. Barth)

YELLOW WAGTAIL
Motacilla flava L., 1758

Fr. – Bergeronnette printanière; Ger. – Schafstelze
Sp. – Lavandera boyera; Swe. – Gulärla

An unobtrusive, charming bird of pastures and damp meadows, often associating with grazing cattle. The ♂ is as yellow as the buttercup flowers growing where it breeds, whereas the ♀ is more discreet. While the ♂ may be handsome to look at, it has possibly the most modest song of all 'songbirds'. In late summer and early autumn it migrates for the winter to tropical Africa, where it occurs in similar grassy or wet habitats as in Europe, but now with buffaloes and rhinos.

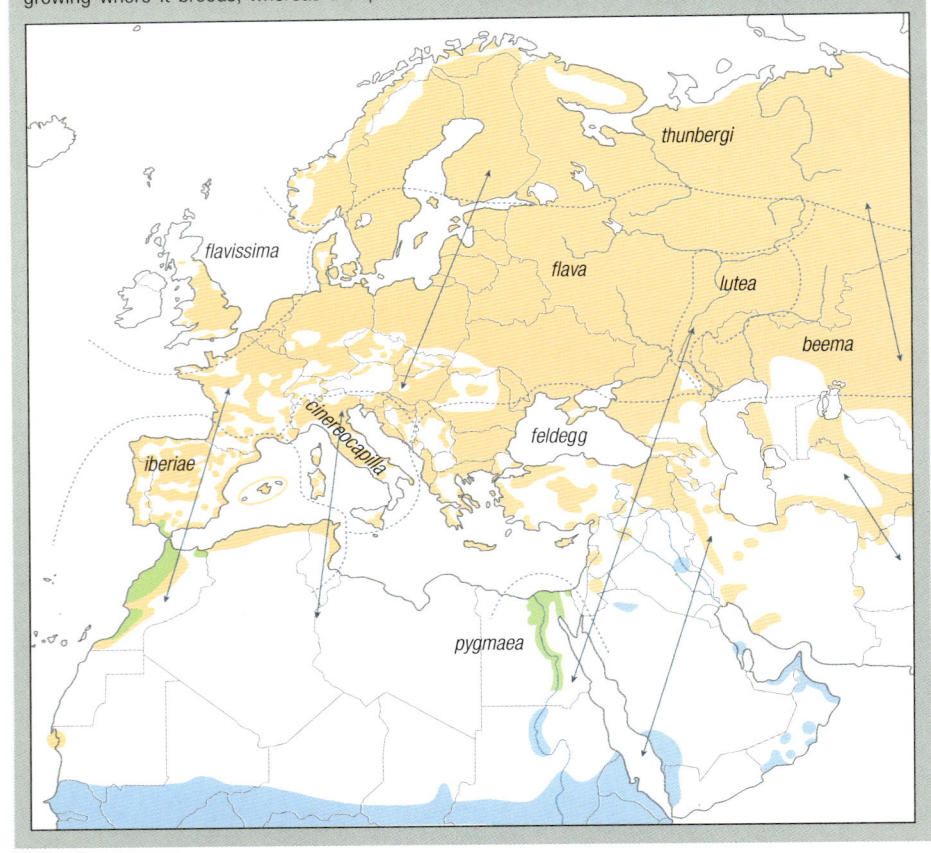

M. f. flava, ♀, presumed 1stS, Latvia, Jun: race assigned by locality and presence of *flava* ♂♂ nearby, but plumage also matches race *thunbergi*. This bird has relatively weak supercilium. Although most wing-feathers are apparently juv, due to extreme wear in wing, ageing at this season often difficult. (M. Varesvuo)

IDENTIFICATION Wagtails are all slim-shaped and long-tailed birds, walking with nodding head-movements and—as the name implies—tail-waggings, and the Yellow Wagtail is no exception, although its movements are generally less vigorous than those of its relatives. The *bill is thin, pointed and dark*, and the *dark tail has white sides*. In spring, the ♂ is bright *yellow below*, whereas the ♀ has yellowish belly but a pale yellow or off-white breast. Both sexes have dull *greenish or olive-brown mantle/back*. Wings have *double pale wing-bars* (generally yellowish) of variable prominence, and the *tertials are dark with yellowish-white edges*, which can be quite obvious when fresh. In autumn, the ♂ attains a somewhat more ♀-like or immature-like plumage. There is extensive geographical variation affecting the head pattern of ♂♂, and these variations are detailed below (see Geographical variation & Range), but some main features are outlined here. In most of Europe (*flava*), the *crown and nape are bluish-grey*, while the cheeks and ear-coverts are plainer grey, at times a fraction darker than crown, and the lores are clearly darker grey. The supercilium is *prominent and pure white*, and there is usually a small whitish mark below the eye. *Centre of throat is bright yellow*, often separated from the grey cheeks by a narrow white submoustachial stripe. In N Europe (*thunbergi*) the crown is fractionally darker grey, usually with dark feather-centres, the *lores and ear-coverts are near-black* (dark slate-grey) lacking a white mark below eye, and the supercilium is missing (> 90%) or is very short and incomplete; yellow breast frequently with some dark blotches at sides or forming a complete gorget (more often than in *flava*). ♂♂ in Britain and on adjacent coasts from S Norway to NW France (*flavissima*) have olive-green crown (sometimes paler, with yellow forecrown) and ear-coverts, and *yellow supercilium*. Those in Iberia, Italy, NW Africa and Egypt resemble *flava* but have white throat without any yellow; see below for details. Breeding ♂♂ of Balkans, SE Europe and Turkey (*feldegg*) differ in their *glossy black cap* and lack of any white on head; the black cap often reaches far down on nape and upper mantle, and is *sharply defined* from greenish rest of mantle. – Juveniles of all races are entirely brown(-grey) and whitish-buff, lacking any obvious yellow or green (though flanks and vent can be quite warm yellowish-buff), and are told by their *broad dark lateral crown-stripes* and *dark patches on sides of throat and upper breast*.

M. f. flava, ♂, presumed ad, Italy, Apr: typical example, with bluish-grey head, but darker ear-coverts and especially lores. Supercilium prominent and pure white, and at least centre of throat is bright yellow. Entire wing-feathers seemingly ad. (D. Occhiato)

M. f. flava, ♂, presumed 1stS, Germany, Apr: variation includes birds with almost uniform dark head-sides (i.e. ear-coverts uniform with crown). Exposed basal remiges and outer greater coverts could be juv feathers (heavily bleached browner). (M. Schäf)

PIPITS AND WAGTAILS

M. f. flava, ad ♂, Israel, Sep: even though some yellowish-green tips obscure grey of plumage, many ad ♂♂ can be identified to race as they appear similar to summer plumage. Dusky necklace on breast typical of winter plumages; entire wing is in fresh post-nuptial plumage. (A. Ben Dov)

1stW, Italy, Sep: on autumn migration, many 1stW appear ♀-like, with variable yellow wash below (throat and breast least effected); also has clear whitish/cream supercilium and streak below eye. Cannot safely be labelled as to race or sex. Entire wing comprises juv feathers. Brown smudging on breast also typical of this age. (D Occhiato)

VOCALISATIONS The song is extremely simple, delivered from atop a higher plant on a meadow, or from a fence-wire or low post, commonly comprising two similar, sometimes three or four (and occasionally just one) strident or scraping notes, *tsreep-tsriiip*, the last note often slightly more emphatic, drawn-out and higher-pitched. Song phrases of four notes often introduce some little irregularity in rhythm or pitch, and a variation of the two-note song has a rather softer last note, *tsreep tsweeo*. The notes closely resemble the alarm call, and, incidentally, the flight-call of Citrine Wagtail. Upon arrival at breeding site sometimes has a low prolonged twittering sub-song, much like other wagtails. – Call in much

Presumed *thunbergi*, ad ♂, Israel, Sep: at first glance, can appear superficially like ad ♀ *M. f. feldegg*. However, combination of being ad (secondary-coverts have diffuse fringes, and much yellow below) and some visible grey feathers on crown strongly indicative of ad ♂ *M. f. thunbergi*. Just one of countless challenges in attempting to age, sex and identify members of this complex. (L. Kislev)

of N & W Europe is a soft, fine *psit* (alternatively rendered *psie*) and a slightly fuller *pseeoo* (or *sleeoo*), and intermediates between these two. There is some geographical variation: birds in Balkans and parts of E Europe (*feldegg*) utter a more grating, frothy *tsrree(eh)* or *tseerrp*, thus with audible rolling 'r' in and often slightly downwards-inflected; similar calls, although perhaps somewhat softer, may be heard from SW & S European breeders (*iberiae, cinereocapilla*). Birds of Ukraine and from north and east of Caspian Sea (*beema, lutea*) are variable, some having grating or scraping calls, others softer calls like *flava*. In alarm, most if not all races seem to use a strident *tsreep*, similar to the notes building up the song.

M. f. thunbergi, ♂, Russia, May: usually has slate-grey crown (darker than *flava*), with blackish ear-coverts, but lacks white supercilium and mark below eye. Little white on chin-sides, rest of chin and throat yellow, while breast-sides have dark greenish blotches. Wing concealed by flanks (preventing definite ageing), though very brownish and faded upper tertials suggest 1stS. (P. Parkhaev)

M. f. thunbergi, 1stS ♀, Finland, May: race assigned by locality, but plumage also matches *flava*; most *thunbergi* have whitish supercilium (as here) and are inseparable from same plumage of *flava*. Strong moult limits in wing, with heavily-bleached browner juv feathers makes this bird 1stS. (A. Below)

M. f. flavissima, 1stS ♂, England, May: crown, lores, eye-stripe (even most of ear-coverts) and rest of upperparts essentially greenish-olive, forming variable contrast with yellower forecrown, and prominent bright yellowish supercilium and underparts. A few have very little yellow on forehead. Note relatively thin wing-bars (cf. *M. f. lutea*). Outer greater coverts are juv feathers (bleached browner), contrasting with both post-juv and pre-nuptial newer coverts. (R. Chittenden)

SIMILAR SPECIES Main confusion risk is with juvenile and first-winter *Citrine Wagtail*. Juveniles of both Yellow and Citrine Wagtails lack any clear yellow or green element in the plumage, and have similar dark lateral crown-stripes and dark breast patches, and they differ mainly in (i) wing formula, (ii) width of pale wing-bars and (iii) pattern of longest uppertail-coverts. Fifth primary (counted from outside) typically 3.5–7 mm < wingtip (exceptionally just 2.5 mm), and second primary usually equal to third and fourth (Citrine: fifth primary 1–2.5 mm, rarely 3 mm, < wingtip, and second often a little shorter than third and fourth). Wing-bars in Yellow are 1.5–2.5 mm wide, very rarely 3 mm (Citrine: 3–4 mm). Longest uppertail-coverts are grey, brownish-grey or olive-grey with diffusely darker grey or blackish centres (Citrine: largely all-black or slate-coloured coverts with only narrow paler margins when fresh; exceptions very rare). First-winters can be similarly difficult to separate, since a minority of Yellow can appear to lack any yellow below in field encounters (nearly all have a little pale lemon on centre of belly, on vent and undertail-coverts, difficult to see unless handled), and upperparts can appear brownish-grey without any obvious olive cast. Note that Yellow has dark lores; the forehead concolorous with crown down to bill; a bill that is not all-black but has paler pinkish or greyish base to lower mandible; and lacks a pale surround to the ear-coverts (Citrine: pale lores, often slightly paler forehead, virtually always all-dark bill, and ear-coverts surrounded by diffuse pale band). For reliable identification use all of the above characters in combination, since odd birds can show 'wrong' or intermediate appearance for one of them, but hardly for all. — Separation from immature *Grey Wagtail* generally straightforward given immature Yellow's shorter tail, whitish or pale yellow undertail-coverts and dark legs (Grey: long tail, bright yellow undertail-coverts, and pinkish legs).

AGEING & SEXING (*flava*) Ages separable in summer and autumn. Ageing in spring without trapping virtually impossible, and even with handled birds difficult. Sexes separable in spring and summer, for ad sometimes also in autumn (generally only extremes, and preferably when first aged). — Moults. Complete post-nuptial moult of ad in (Jun) Jul–Sep prior to autumn migration (on average earlier in S part of range, and may even start in spring in N Africa, where breeding season is early). Partial post-juv moult in Jul–Sep does not involve flight-feathers. Northern birds replace fewer wing-coverts than breeders in W & S Europe. Both age groups undergo a partial pre-nuptial moult in winter quarters, to some extent in Nov–Dec but mainly Jan–Apr, which apparently involves body-, head-, all tail-feathers, most tertials and a variable number of secondary-coverts. Both ad and 1stS have a moult contrast in greater coverts in spring. — SPRING Ageing difficult even when handled. Ad tends to have greyer and glossier primaries and primary-coverts, both with more fresh tips in late spring and summer, 1stS browner and more worn. If greater coverts not too worn, ad has as a rume smoother and more ill-defined paler edges, 1stS whitish and more distinct. ♂ Throat and breast vivid yellow. Whole crown and nape bluish-grey (sometimes with a little green admixed), ear-coverts similar or slightly darker grey. Supercilium pure white, narrow but distinct. Mantle and back (greyish-) green without brown element. ♀ Throat and breast a washed-out yellowish-white, breast sometimes with pale brownish cast, and often with some dark blotches on upper breast. Crown and nape brownish-grey, often with some greenish admixed, ear-coverts dull brownish-grey. Supercilium off-white, slightly tinged buff or pale yellowish, usually less distinctly marked than on ♂. Mantle and back olive-grey with some brown element. — SUMMER **Ad** As in spring. **Juv** Crown dark brown with blackish sides. Upperparts brown-tinged dark grey, rather blotchy, very little green element. Wing-bars (tips of median and greater coverts) buff-white or cream-white.

M. f. flavissima, ♀, presumed 1stS, England, May: race assigned by locality and presence of ♂♂ nearby, but also tendency for yellow-tinged supercilium. Many inseparable from *flava*. Apparent strong moult contrast in wing indicates age. (G. Catley)

M. f. flavissima, ad ♂, England, Sep: like other races, ad ♂♂ in autumn attain (to varying degree) more ♀/immature-like plumage; here crown, ear-coverts and rest of upperparts essentially greenish-olive, contrasting with bright yellow supercilium. Note ad-type greater coverts with diffuse yellowish-buff fringes. (G. Thoburn)

1stW, England, Sep: by range could be *M. f. flavissima*, but inseparable from other races, including *flava*. Note juv-type greater coverts with contrasting and sharply demarcated whitish tips; remnants of brown patch on upper breast, also typical of juv. (G. Thoburn)

1stW, England, Sep: especially 1stW ♀ *flavissima* can be much whiter below, with limited buff or yellow. Median coverts, inner greater coverts and tertials have been moulted (even more extensive replacement is frequent in southern and western Yellow Wagtail populations). (G. Thoburn)

M. f. flavissima, juv, England, Jul: in all races, juv entirely brown-grey and whitish, lacking any obvious yellow or green, and has characteristic dark upper-breast patch, with dark lateral throat- and crown-stripes. Also fluffy juv body-feathers. (N. Blake)

PIPITS AND WAGTAILS

M. f. iberiae, ♂, Spain, May: diagnostic pure white throat and grey head; ear-coverts sometimes even darker; white supercilium narrower than in *flava*, frequently being ill-defined or even lacking in front of eye. This race tends to lack or almost lack white sub-ocular streak typical of *flava* (here a minimal spot). Ageing impossible. (C. N. G. Bocos)

M. f. iberiae, ♀, Spain, Aug: in extensive post-nuptial moult, but has rather pure white throat, rather distinctly separated from yellowish breast. ♀ can be inseparable from other races, especially *flava* and *cinereocapilla*. Too heavily worn and bleached to assess age, though such extremely worn primaries should be juv (i.e. 2CY in early post-breeding moult). (Q. Marcelo)

Underparts buffish off-white; upper breast has 'gorget' of dark brown blotches or solid dark patch. – AUTUMN **Ad ♂** Differs from ♂ in spring by much-reduced yellow on throat and upper breast, and on brown tinge to head, breast and mantle. Often some yellow streaks on sides of throat. Lower breast and belly rather saturated yellow, breast generally with some dark blotches. Crown and nape with blue-grey entirely or partially concealed by brownish tips. Greater coverts, median coverts and tertials freshly moulted, edged (tertials) or tipped (secondary-coverts) dull buffish- or yellowish-olive, pale edges or tips grading smoothly into darker feather-centres. **Ad ♀** As ad ♂ but even less yellow below, usually with none on throat and breast. (Still, some sexual overlap in amount of yellow could exist.) Crown and nape brownish-grey (no blue-grey concealed). Greater and median coverts and tertials freshly moulted, pattern as in ad ♂. **1stW** Much as ad ♀, or even paler and less yellow below (= ♀♀). Unmoulted tertials and greater coverts slightly worn and have whitish (rather than buffish or yellowish) edges or tips, these whitish edges/tips being rather sharply defined from dark feather-centres (not smoothly fading into darker centres as

1stW, Spain, Aug: by range could be *iberiae*, but inseparable from other races, including *flava*. Especially autumn *iberiae* can appear to have narrower white supercilium, more uniform ear-coverts, and 'extended' white throat. Note juv-type greater coverts with contrasting whitish tips. (C. N. G. Bocos)

M. f. cinereocapilla, ad ♂, Italy, Apr: unlike similar *thunbergi*, entire throat white, a feature shared with *iberiae*, but present race lacks white supercilium; note lead-grey on nape reaches further onto upper mantle. Aged by evenly rather fresh wing-feathers. (D. Occhiato)

♀, presumed 1stS, Italy, Apr: *M. f. cinereocapilla* by locality and presence of ♂♂ nearby, but plumage also typical of this race, with no whitish supercilium, head tinged lead-grey, and white throat distinctly separated from yellowish breast. However, many ♀♀ inseparable from *iberiae*, *feldegg* and even *thunbergi*. Exposed basal remiges and outer greater coverts appear to be juv (bleached browner), thus probably 1stS. (D. Occhiato)

M. f. feldegg, ♂, Greece, Apr: note sharply delimited black head, but some have hindneck and mantle greener. Ageing without closer inspection of wing impossible. (M. Schäf)

M. f. feldegg, ♂, presumed 1stS, Israel, Mar: sometimes black of head extends diffusely onto upper mantle, while whiter chin-sides are also variable. Exposed remiges are apparently juv (heavily worn and bleached brown). (D. Occhiato)

M. f. feldegg, ♀, presumed ad, Greece, Apr: much individual variation among ♀♀, here a 'classic example' with dark/blackish-grey cap, while many show variable pale supercilium. Many wing-feathers apparently ad. (M. Schäf)

M. f. feldegg, ad ♀, Kuwait, Feb: some ♀♀ strongly mirror ♂♂, with all-black but less glossed cap, upperparts still greyish-tinged (instead of greenish-yellow) and underparts off-white, with yellow tinge on vent. Aged by only ad-type coverts and 'advanced' plumage. (R. Al-Hajji)

M. f. feldegg, ♀, Syria, May: a few ♀♀ develop quite substantial pale supercilium. Ageing impossible in this case. (A. Audevard)

in ad). Any bird with contrast between yellowish-olive inner greater coverts (moulted) and outer ones tipped off-white (unmoulted juv feathers) is 1stW. **1stW ♂** Those 1stW with strong yellow on lower breast and belly and long wing (see below) are nearly always ♂♂. **1stW ♀** Those 1stW with very little yellow on underparts and short wing (see below) are nearly always ♀♀, although beware late-hatched ♂♂ with very little yellow that can occur as late as mid Sep. Not all 1stW can be sexed!

BIOMETRICS (*flava*) **L** 15–16.5 cm; **W** ♂ 78–86 mm (*n* 67, *m* 82.2), ♀ 74–81 mm (*n* 26, *m* 78.1); **T** ♂ 67–76.5 mm (*n* 67, *m* 71.1), ♀ 64–74 mm (*n* 26, *m* 68.3); **T/W** *m* 86.8; **B** 14.3–16.9 mm (*n* 74, *m* 15.6); **BD** 3.3–4.3 mm (*n* 38, *m* 3.9); **BW** 3.7–4.6 mm (*n* 22, *m* 4.1); **Ts** 21.7–25.2 mm (once 26.2; *n* 75, *m* 23.5); **HC** 7.0–12.6 mm (*n* 68, *m* 9.3). **Wing formula: p1** minute; **pp2–4** roughly equal and longest (p4 sometimes 1–2 mm <); **p5** < wt 3.5–6.5 mm (very rarely 2.5–3 mm <); **p6** < wt 10–15 mm; **p10** < wt 22–29 mm; **s1** < wt 24–30.5 mm. Emarg. pp3–5, although on p5 sometimes less prominently.

GEOGRAPHICAL VARIATION & RANGE Marked and complex variation. Numerous subspecies described, most of which are connected and interbreed freely, producing birds with intermediate characters. Some races repeat closely the patterns of other races despite these not being neighbours. Variation mainly manifested in head pattern of ♂♂, and the descriptions below refer to ♂ plumage unless otherwise stated. Subspecies arranged from west to east and starting with northerly taxa, then this pattern repeated for southerly taxa.

M. f. flava L., 1758. 'Blue-headed Wagtail' (S Scandinavia, much of C Europe; winters widely in trans-Saharan savannas, rarely on N African coasts). Treated above (see Identification, Ageing & Sexing and Biometrics). – Note that *flava* intergrades with all surrounding subspecies (*thunbergi, flavissima, iberiae, cinereocapilla, feldegg* and *beema*) and produces birds of intermediate appearance in these zones of contact. Birds with less than classic *flava* plumage should preferably not be labelled as such.

M. f. thunbergi Billberg, 1828. 'Grey-headed Wagtail' (N Fenno-Scandia, N Russia, N Siberia east to Kolyma; winters in trans-Saharan Africa, possibly also in SW Asia east to at least India). Differs in that ♂♂ have darker grey forecrown (tendency to have darker grey feather-centres) and much darker, slate-grey or blackish ear-coverts lacking white mark below the eye. No sharp contrast between grey nape and green mantle (as often found in *feldegg*). Also, in general there is no white supercilium, but a minority (possibly the result of interbreeding with *flava*) have 'remnants' of white above or

M. f. feldegg, ad ♂, Israel, Sep: black of head in fresh plumage variably obscured by yellowish-green tips. These, plus ad-type secondary-coverts, readily identify bird as ad. (L. Kislev)

immediately behind or (rarely) in front of eye. Chin and throat either entirely yellow, or with a little white on chin and sides of throat as in *flava*. Many have some dark greenish-grey or blackish blotches on breast-sides, on some forming a gorget across breast (somewhat less common in *flava* and exceedingly rare in *feldegg*). ♀♀ sometimes possible to recognise by lack of whitish supercilium, and a rather dark head, but most are inseparable from ♀ *flava*. Biometrics very similar to those of *flava*, the following being the only differences of any importance: **W** ♂ 80–87 mm (once 78; *n* 65, *m* 82.6), ♀ 76–82.5 mm (*n* 34, *m* 79.4); **T** ♂ 67–78 mm (*n* 65, *m* 72.3), ♀ 67–75 mm (*n* 34, *m* 69.7); **T/W** *m* 87.6; **HC** 7.8–11.2 mm (*n* 60, *m* 9.5). Bill, tarsus and wing formula similar to *flava*. (Syn. *alakulensis*; *plexa*.) – Very similar to extralimital *macronyx* of C & E Mongolia, NE China and S Russian Far East, and often inseparable on morphology, but separated geographically and apparently clearly genetically; see below. Also, the call is sharp and strident in *macronyx*, not as soft as usually in *thunbergi*. For further differences see below.

M. f. flavissima (Blyth, 1834) 'British Yellow Wagtail' (Britain, Ireland, sparsely and locally on coasts of S Norway, Low Countries, N France; winters Senegambia, Liberia to Ivory Coast). Has crown and nape greenish and concolorous with mantle; ear-coverts, too, are greenish and apparently never

1stW, presumed ♀, Italy, Sep: has *M. f. feldegg*-like head pattern, with almost uniform (vaguely dark) head (at this season, some birds can show short, diffuse supercilium above and behind eye). Typical pattern of juv secondary-coverts; almost lack of yellow below (only faintest tinge on vent) suggest ♀. (D. Occhiato)

M. f. pygmaea, ♂, Egypt, May: superficially most similar to larger *cinereocapilla*, with no supercilium and mostly white throat. This race has characteristic greenish-olive mantle merging into hindneck. Due to strong feather wear in wing, ageing without handling impossible. (K. Haataja)

M. f. beema, 1stS ♂, Kuwait, Mar: resembles *flava* but has paler grey-blue head, more prominent white supercilium and sub-ocular line; also, white on chin and throat-sides better developed. Race *beema* is similar to hybrid *flava* × *flavissima* (so-called 'Channel Wagtail', which are frequent in Holland, Belgium and N France). Exposed remiges and outer greater coverts are juv (heavily worn and bleached browner). (M. Pope)

M. f. beema, 1stS ♂, Kazakhstan, Jun: in some extremely pale birds (rare), chin and throat very white, merging with white of sub-ocular line. Strong moult limits in wing age it as 1stS. (J. Normaja)

become predominantly yellowish. Supercilium prominent and vividly yellow. Some ♂♂ have a yellowish forehead, and very rarely worn summer ♂♂ can attain quite pale and yellowish crown, and have slightly paler green ear-coverts than normal, making them very similar to *lutea*. (Cf. that race.) Double wing-bars usually moderately broad and whitish, but that on median coverts can be yellowish. ♀♀ have on average more frequently yellow-tinged supercilium than in previous two races, but many are inseparable. **L** 15.5–17 cm; **W** ♂ 80–88 mm (*n* 20, *m* 83.1), ♀ 77–84 mm (*n* 16, *m* 79.3); **T** ♂ 68–78 mm (*n* 20, *m* 71.4), ♀ 64–75 mm (*n* 16, *m* 69.4); **T/W** *m* 86.6. **Wing formula: p5** < wt 4–5.5 mm, **p6** < wt 10.5–13 mm, **s1** < wt 25–29 mm. (Syn. *rayi*.) – Hybridises with *flava* outside British Isles; hybrids sometimes known as 'perconfuscus' ('Channel Wagtail'), some of which show remarkable resemblance to ssp. *beema* (cf. below), and may even be inseparable. Important to note full set of characters and compare with classic *beema*. See Dubois (2007).

M. f. iberiae Hartert, 1921. 'Iberian Wagtail' (Extreme SW France, Iberia, Balearics, NW Africa; winters mainly W Africa, less commonly east to Chad). Differs from *flava* by having white throat without any yellow, and on average less bluish tinged grey crown and nape. Ear-coverts and lores are darker lead-grey without any white mark below eye. White supercilium narrower than in *flava* (in particular in front of eye) and is frequently incomplete in foremost part. There is apparently some variation as to length and prominence of supercilium in extreme south-west of range, since van den Berg (2011) reported occurrence of apparent breeders in SW Morocco with very short or no supercilium, and some probably inseparable from Ashy-headed Wagtail *cinereocapilla*. Tail on average proportionately slightly longer than in *flava*. ♀♀ variable, some have rather pure white throat with distinct border to yellow-tinged breast, but others closely resemble *flava* and *cinereocapilla*. Call usually a harsh *tsreep* with audible 'r', different from softer *psiie* of *flava* and other preceding races. **L** 15–16.5 cm; **W** ♂ 78–84 mm (*n* 20, *m* 80.4), ♀ 74–79 mm (*n* 13, *m* 76.7); **T** ♂ 69–74.5 mm (*n* 20, *m* 71.5), ♀ 65–70 mm (*n* 13, *m* 67.4); **T/W** *m* 88.5. **Wing formula: p5** < wt 3–4.5 mm; **p6** < wt 9–11.5 mm; **p10** < wt 22–24.5 mm; **s1** < wt 25–29 mm. (Syn. *fasciata*.)

M. f. cinereocapilla Savi, 1831. 'Ashy-headed Wagtail' (SE France including Corsica, Italy, Slovenia, W Croatia; winters Mali, Nigeria, Chad and adjacent areas). Resembles *thunbergi* except that entire throat is white like in *iberiae* and *pygmaea*, and that on many the lead-grey on nape runs further back onto upper mantle. Also, very slightly smaller than *thunbergi* with proportionately a little longer tail. Lacks

M. f. beema, ♂, presumed ad, Russia, May: some ♂♂ attain more *flava*-like head pattern, but supercilium usually longer, with paler ear-coverts and more white on chin; also sub-ocular line better marked than most *flava*. Apparently evenly ad-type wing-feathers suggest ad. (Y. Belousov)

Presumed *M. f. beema*, ♀, Russia, May: probably *beema* based on location and pale grey tinge to head, merging into pale green upperparts, and well-developed whitish supercilium and sub-ocular line. (Y. Belousov)

white supercilium or has, like in *thunbergi*, a little white above or just behind eye (rarely also a trace above lores). Call as *iberiae*, i.e. generally a rather harsh *tsreep* or *psrreeh*. **W** ♂ 78–85 mm (*n* 19, *m* 81.8), ♀ 76–83 mm (*n* 14, *m* 78.8); **T** ♂ 68–77 mm (*n* 19, *m* 72.5), ♀ 65–74 mm (*n* 14, *m* 68.7); **T/W** *m* 88.1. **Wing formula**: p5 < wt 2–4 mm; **p6** < wt 8.5–11.5 mm; **p10** < wt 20–25 mm; **s1** < wt 21.5–27 mm. – Intergrades with *flava* in SE France and Switzerland, and along coast of S France to Spanish border with *iberiae*. Birds with a little yellow on lower throat, or with more complete white supercilium, could be hybrids or merely represent natural variation within *cinereocapilla*; the best labelling of such birds depends on assessment of all characters.

M. f. feldegg Michahelles, 1830. 'BLACK-HEADED WAGTAIL' (SE Europe, Turkey, Caucasus, Near East to Caspian Sea, N Iran, S Kazakhstan; winters Sudan and E Africa, Sahel, south to Rwanda, Congo, former Zaïre). Generally quite distinct with black cap and no white supercilium, the black crown being glossy unless heavily worn. Black cap often reaches far down at rear, onto upper mantle, and is distinctly marked against the bright greenish rest of mantle. (Birds with diffuse border are probably hybrids with neighbouring paler-headed race, but could be poorly developed 1stS.) Black of head in fresh

M. f. beema, ♀, Kazakhstan, May: often inseparable from *flava*, though in this bird whitish supercilium and sub-ocular line are rather prominent. Ageing impossible. (A. Audevard)

1stW, N Kazakhstan, Aug: somewhat *beema*-like (but also *flava*-like) head pattern with moderately prominent pale supercilium and pale sub-ocular line. Rather greyish upperparts. (Y. Belousov)

1stW, Israel, Oct: *M. f. beema*-like head pattern with prominent supercilium, which can be broader, especially in front of eye, and can even reach sides of nape; pale sub-ocular line well developed here. Except lesser and median coverts, most of wing apparently juv. (A. Ben Dov)

M. f. lutea, 1stS ♂, Kazakhstan, May: similar to *flavissima* but tends to have more evenly yellow head, which with wear becomes more uniform (lacking clearly darker ear-coverts), while overall yellow coloration of breeding ♂ is deeper than *flavissima*; yellow wing-bars usually distinctly broader. Strongly abraded outer greater coverts are juv. (M. Vaslin)

PIPITS AND WAGTAILS

M. f. lutea, 1stS ♂, Israel, Apr: spring birds like this, with slightly darker ear-coverts and crown, recall *flavissima*, but note deep yellow forecrown, largely pale yellowish lores, hardly any trace of eye-stripe, faint greenish wash to ear-coverts, and broad deep lemon wing-bars. Strongly abraded about five outer greater coverts are juv. (A. Ben Dov)

M. f. lutea, ♀, Kazakhstan, Apr: many ♀♀ are inseparable from other races, but some are highly distinctive, with strong yellow on supercilium, lower cheeks and underparts, and broad yellow wing-bars. Since both ad and 1stS have a partial winter moult both can show a moult limit in greater coverts in spring; hence this bird is difficult to age. (A. Vilyayev)

M. f. lutea, ad ♂, Bahrain, Sep: in fresh plumage virtually identical to *flavissima*, but note more yellow on forehead and less marked supercilium, which is best developed behind eye. Entire wing fresh ad. (A. Drummond-Hill)

M. f. leucocephala, 1stS ♂, Kuwait, Apr: this race is only a vagrant to the treated region (mostly in Middle East). Mainly white head, with faint pale grey nape and ear-coverts. Moult limit, with juv remiges and outer greater coverts, identify this bird as 1stS. (P. Fågel)

plumage may have some (yellowish-) green tips admixed (never any blue-grey, which is indicative of *thunbergi*). Underparts including throat deep yellow. Normally no white on chin or sides of throat, these being deep yellow. Lacks dark blotches on upper breast (which are common in *thunbergi*). Birds E of Caspian Sea ('*melanogrisea*') tend to have less extensive and less glossy black cap, and often a little white on chin and sides of throat, although much overlap, and characters form a stepless cline, hence not recognised as a separate race. ♀ varies extensively, many being distinctive with blackish cap, greyish-tinged upperparts and off-white underparts (faintly yellow only on belly, vent and undertail-coverts), others having more yellow below, greenish tinge on grey-brown back, and a partial pale supercilium. Throat nearly always whitish as in *cinereocapilla*. No whitish mark below eye. Lores rather plain, uniformly dark (black or slate-grey) or fairly pale grey-brown, lacking dark narrow streak under pale supercilium as in most races. Some 1stW birds, especially ♀♀, can lack nearly all yellow in plumage (only faint yellow on vent, difficult to see in the field) and look very brown-grey and white. Call invariably a harsh *tsreep* or *tzree(eh)* with audible '*r*'. **W** ♂ 81–89 mm (once 79; *n* 17, *m* 83.8), ♀ 75–84 mm (*n* 15, *m* 80.0); **T** ♂ 68–74 mm (*n* 17, *m* 71.5), ♀ 67–74 mm (*n* 15, *m* 69.6); **T/W** *m* 85.7; **B** ♂ 15.7–18.0 mm (*n* 17, *m* 16.8), ♀ 15.0–16.5 mm (*n* 14, *m* 15.8); **Ts** 22.0–25.1 mm (*n* 23, *m* 23.7). **Wing formula: p5** < wt 3–5 mm; **p6** < wt 9–13 mm; **p10** < wt 22–26 mm;

s1 < wt 23.5–28 mm. (Syn. *aralensis*; *melanocephala*; *melanogrisea*.) – Hybridises with *flava*, *cinereocapilla*, *beema* and *lutea* in zones of contact, and resulting mixed populations can be of striking appearance, sometimes leading to formal names, although they are better regarded as too instable and heterogeneous to be recognised as subspecies, also since exclusive breeding ranges are lacking or at least unknown: '*superciliaris*' has a black cap and white supercilium; '*xanthophrys*' is similar but has a yellow supercilium; '*dombrowskii*' resembles *flava* but has darker grey or blackish ear-coverts without a white mark under eye, and a subtly darker grey crown and nape.

  M. f. pygmaea (A. E. Brehm, 1854) 'EGYPTIAN YELLOW WAGTAIL' (Egypt; mainly resident or short-range movements south along Nile). Similar to *cinereocapilla*, with white throat (or yellow wash on lower throat) and generally no supercilium (sometimes a little white behind eye), but is decidedly smaller. Grey of crown and nape has some green admixed when fresh, and upperparts are darker and duller olive-grey in general than in *cinereocapilla*. **L** 14.5–15 cm; **W** ♂ 73–79 mm (*n* 20, *m* 75.4), ♀ 70–74 mm (*n* 12, *m* 71.7); **T** ♂ 61–69 mm (*n* 20, *m* 65.1), ♀ 58–64 mm (*n* 12, *m* 61.0); **T/W** *m* 85.9; **Ts** 20.5–23.3 mm (*n* 29, *m* 22.5); **HC** 7.5–10.0 mm (*n* 24, *m* 8.8). **Wing formula: p5** < wt 1–3.5 mm; **p6** < wt 7–11 mm; **p10** < wt 17.5–23 mm; **s1** < wt 19–25 mm.

  ○ *M. f. beema* (Sykes, 1832) 'SYKES'S WAGTAIL' (SE

M. f. leucocephala, ♀, Mongolia, Jun: in general, a much duller version of ♂, with greyish head and very faint yellow below. (H. Shirihai)

Presumed *M. f. tschutschensis*, 1stS ♂, South Korea, Apr (racial identity by location and appearance): very similar to *M. f. flava* (and has wandered to W Europe), but has on average less bluish, purer grey crown/nape in ad ♂, and nearly always lacks (or has just small) white patch below eye in quite dark mask (which is consequently more solid). 1stS by retained juv remiges and outer greater coverts. (C. Soonkyoo)

***M. f. macronyx*, ♂, Mongolia, Jun**: extremely similar to *M. f. thunbergi* and perhaps only separable by DNA and call (here identity confirmed by breeding locality). Ageing unsure as only small amount of wing surface visible. (H. Shirihai)

Russia, Kirghiz Steppe in Kazakhstan, S Ural area, SW Siberia; winters S & E Africa, Arabia, Pakistan and India). Resembles *flava* but has less blue in grey of crown and nape, is pale ash-grey or medium grey with no or only slight lead-blue tinge, on average paler than *flava*, but some come close. In fresh plumage grey of nape often has some green tips. Ear-coverts similarly pale ash-grey (*flava*: dark slate-grey), and the white mark under eye is usually much larger than in *flava*. Supercilium as a rule broader and more prominent, in particular in front of eye (where often narrow in *flava*), and usually reaches further back, to sides of nape. On average slightly more white on chin and sides of throat (a few have white also on centre of uppermost throat), but others are similarly yellow on chin and centre of throat as *flava*. Typical ♀ is a little paler grey on head and paler green on mantle, and has more prominent supercilium. Several inseparable from *flava*, though. Rarely, 1stW can lack *any* yellow on underparts (Shirihai & Gellert 1987), although hardly as frequently as claimed by Alström & Mild (2003; 'in a quarter to a third of all'), but perhaps in 10%. Note that many which in the field appear to lack any yellow below actually have faint yellow on centre of belly and vent, and partly on undertail-coverts (never so in 1stY *M. citreola*). Call variable, either (usually) the softer *flava* type, or (rarely) the harsher *feldegg* type. Also heard is a more metallic *plit*. **L** 14.5–16 cm; **W** ♂ 78.5–86 mm (*n* 38, *m* 81.7), ♀ 74–82 mm (*n* 15, *m* 77.9); **T** ♂ 67–74 mm (*n* 20, *m* 70.1), ♀ 64–72 mm (*n* 15, *m* 67.6). – Although rather variable and close to *flava*, we see no reason to lump *beema* with *flava*; well over 75% of ♂♂ can be separated in collections. A different problem is presented by the occurrence of similar or near-identical birds in NW Europe caused by hybridisation between *flava* and *flavissima* (cf. latter).

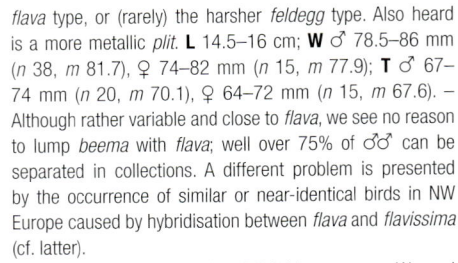

M. f. lutea (S. G. Gmelin, 1774) 'Yellow-headed Wagtail' (allegedly lower Volga and middle Ural region, possibly locally NW Kirghiz Steppe, but details poorly known; cf. Alström & Mild 2003; winters S & E Africa and India). Surprisingly similar to Yellow Wagtail *flavissima* (considering distributional gap of > 2000 km). Generally though, breeding ♂ *lutea* has a near all-yellow head with mainly yellowish and only very slightly green-tinged ear-coverts, pale lores (yellow, or with a light grey 'smudge'), and only a pale green tinge on hindcrown and nape. Wing-coverts, especially median coverts, are often broadly tipped yellow, more prominently so than in *flavissima*, which usually has narrower and more yellowish-white or white tips. A bird with completely deep yellow forehead and forecrown, largely pale yellowish lores, washed-out (diffuse) greenish-yellow eye-stripe and sides of neck (yellow encircling faint greenish wash on ear-coverts), and broad deep lemon wing-bars, can only be a *lutea*, but not all are this characteristic. In the palest *flavissima* ♂♂ (generally worn and abraded summer birds), the forecrown can become pale yellow (still with faint olive hue), but at least the ear-coverts and loral stripe seem to remain slightly darker greenish than in *lutea*. Reversely, very rarely what appears to be breeding *lutea* ♂♂ (judging from localities of three specimens in NHM) can copy the appearance of normal *flavissima*, with all-greenish crown and forehead, darker olive-grey lores and eye-stripe, and a contrasting long supercilium to base of bill. (Luckily, this is more a problem for birdwatchers in the lower Volga region than for British birders.) The only slight difference, and then on direct comparison, seems to be an often deeper yellow in most *lutea*. – Although we cannot find support in museum collections for the sometimes-voiced claim that *lutea* and *flavissima* ♂♂ cannot be separated (e.g. Alström & Mild 2003), we advocate great caution when identifying an extralimital *lutea*; most ♂♂ in spring differ from *flavissima*, but some birds are in practice inseparable. ♀♀ are inseparable from *flavissima*, and closely resemble many *beema* and *flava*, too, when full range of variation is considered. **W** ♂ 80–85.5 mm (*n* 22, *m* 82.3), ♀ 77–83 mm (*n* 10, *m* 79.6); **T** ♂ 67–75 mm (*n* 22, *m* 70.9), ♀ 66–72 mm (*n* 10, *m* 68.6). (Syn. *campestris*.)

Extralimital Asian subspecies, normally not occurring within the treated region (though a very few records of accidental occurrence are known). At least the three last-mentioned tend to be very subtly longer-legged and longer-billed, and have on average longer and less curved hind claw (though much overlap) and proportionately longer tail, while wingtip is slightly blunter.

M. f. leucocephala (Przevalski, 1887) 'White-headed Yellow Wagtail' (SE Altai, NW Mongolia, NW China, but details poorly known, and frequently impossible to locate when searching for it within claimed breeding range; winters SE Africa & SW Asia). The most typical ♂♂ have a largely white head with only a very pale grey tinge on hindcrown and nape in some, and pale grey ear-coverts. There is quite some variation in this race, and some ♂♂ are darker, having pale grey crown to nape and ear-coverts, with a prominent supercilium, approaching the appearance of palest *beema*. (It could even be that *leucocephala* is better viewed as a colour morph of easterly *beema*.) Green upperparts slightly paler and tinged more yellowish than in *flava* or *beema*. Beware of partially leucistic, or aberrant birds of other races looking like this race.

***M. f. taivana*, ♂, presumed ad, South Korea, May**: all olive-green upperparts and crown, pure yellow prominent supercilium and very dark slate mask (subtly tinged green); underparts (including throat) are all yellow, without any dark blotches on upper breast. Seemingly uniform ad-type wing feathers suggest ad. (A. Audevard)

Presumed *M. f. taivana*, possible 1stW ♂, South Korea, Dec: well-developed supercilium tinged yellow, and quite dark lores, ear-coverts and upperparts with olive tinge, all fit this race. Nevertheless, due to extensive variation and overlap with other races, many birds in winter plumage away from breeding range are difficult to safely identify to subspecies. Much of wing is still juv, and this together with yellow underparts and well-marked head pattern infer sex and age. (Park Jong-Gil)

M. f. 'superciliaris', 1stS ♂, Israel, Mar: apparently a stable hybrid population involving *M. f. feldegg*. Note black cap and white supercilium, but this often varies. Exposed basal part of remiges and outer greater coverts are juv (heavily bleached browner). (A. Ouwerkerk)

M. f. 'xanthophrys', ♂, presumed ad, Israel, Mar: hybrid population involving *feldegg* with black cap but yellow (not white) supercilium. Apparently evenly ad-type wing-feathers suggest ad. (D. Occhiato)

M. f. tschutschensis J. F. Gmelin, 1789 'Eastern Yellow Wagtail' (Sayan and Tuva in SC Siberia, Baikal region, Yakutia, upper Lena east to Chukotskiy Peninsula and Kamchatka, also W & N Alaska; possibly this or intergrades with *beema* in Zaysan Basin, Altai and W Mongolia; winters SE Asia; a very few stragglers to W Europe confirmed by DNA). Very similar to Blue-headed Wagtail (*flava*) but has on average slightly longer and straighter hind-claw (though much overlap), crown/nape in ad ♂ on average less bluish, more pure grey, ear-coverts variably dark but often quite dark (even blackish) with no or only hint of white below eye. White supercilium often subtly narrower than in Blue-headed. 1stW ♀ frequently lacks visible yellow in plumage, being very grey and white. **W** ♂ 78–84.5 mm (*n* 20, *m* 81.0), ♀ 71.5–81 (*n* 12, *m* 76.8); **T** ♂ 66–75 mm (*n* 20, *m* 70.2), ♀ 63–73 mm (*n* 12, *m* 66.8); **T/W** *m* 86.8; **B** 13.9–16.9 mm (*n* 31, *m* 15.7); **Ts** 21.5–26.1 mm (*n* 32, *m* 23.8); **HC** 9.2–12.4 mm (*n* 31, *m* 10.9). (Syn. *angarensis*; *simillima*; *zaissanensis*.)

M. f. macronyx (Stresemann, 1920) 'Manchurian Wagtail' (C & E Mongolia, S Transbaikalia and extreme SE Siberia, Manchuria and NE China; winters SE Asia; not known to have straggled to Europe). Extremely similar to Grey-headed Wagtail (*thunbergi*) and perhaps only separable by DNA and more rasping, sharp call. Some might be separated using combination of call, slightly paler grey forehead and crown, and broader and yellower wing-bars (Alström & Mild 2003). Does not seem to have dark-blotched breast as regularly as *thunbergi*. As in *tschutschensis*, young birds, especially ♀♀, can be very grey and white, showing no yellow in normal field views. **W** ♂ 77–85 (*n* 12, *m* 80.5), ♀ 76.5–84 mm (*n* 12, *m* 78.8); **T** ♂ 67.5–74 mm (*n* 12, *m* 71.0), ♀ 63–73.5 mm (*n* 12, *m* 68.8); **T/W** *m* 86.8; **B** ♂ 13.7–16.6 mm (*n* 12, *m* 15.9); **Ts** ♂ 22.5–25.6 mm (*n* 12, *m* 24.3); **HC** ♂ 9.2–13.9 mm (*n* 12, *m* 11.3).

M. f. taivana (Swinhoe, 1863) 'Green-headed Wagtail' (Russian Far East in lower Amur, Sakhalin, and NW Hokkaido, Japan, possibly also adjacent areas but imperfectly known; winters SE Asia). Ad ♂ has olive grey-green upperparts and crown relieved by bold lemon-yellow supercilium, whereas broad eye-stripe and ear-coverts are very dark, slate or greenish-black. Whole throat and rest of underparts vivid yellow. Ad ♀ quite similar to ♂, being almost as yellow beneath, only slightly duller and often with grey-brown tinge on breast; head pattern slightly less neat. 1stW quite often of grey-and-white type with little or no yellow visible as described under preceding two taxa, but sometimes separable by their broader, more prominent whitish supercilium and slightly darker and more uniform ear-coverts. **W** ♂ 77–86 (*n* 23, *m* 81.9), ♀ 75.5–81.5 (*n* 11, *m* 78.4); **T** ♂ 68–77 mm (*n* 23, *m* 72.7), ♀ 66–76 mm (*n* 11, *m* 70.4); **B** 14.9–17.2 mm (*n* 26, *m* 16.2); **Ts** 23.6–26.5 mm (once 22.1; *n* 27, *m* 24.8); **HC** 9.4–13.3 mm (*n* 26, *m* 11.1).

TAXONOMIC NOTES Preliminary genetic analyses (Ödeen & Alström 2001, Alström & Ödeen 2002, Voelker 2002) show at least two different racial groups, with a division in Asia, but morphological characters and distribution do not support a simple split into two species based on the genetic results, at least not on present knowledge. Examination of mitochondrial and nuclear DNA showed partly contradicting results. For instance, the mitochondrial data showed three E Asian Yellow Wagtail taxa to be closer to nominate Citrine Wagtail *M. citreola* (and to some Pied Wagtail taxa *M. alba*, as well as to Grey Wagtail *M. cinerea*) than to European taxa of Yellow Wagtail, but this was not supported by nuclear data or vocalisations (apart from morphology), which both placed *citreola* outside the *flava* complex. More research seems desirable before the taxonomy is revised. – Ssp. '*zaissanensis*' of Zaysan Basin and upper Irtysh River described as 'a poorly differentiated race, intermediate between *beema* and *angarensis*' (Vaurie 1959), the latter extralimital and close to *thunbergi*, but since these two are quite different

M. f. 'perconfuscus', ♂, presumed 1stS, England, Apr: so-called 'Channel Wagtail', apparently a stable hybrid population between *flavissima* and *flava* outside British Isles, closely resembles *beema*. Primaries, primary-coverts and lesser coverts still juv. (M. Lawrence)

Presumed *M. f. 'dombrowskii'*, ♂, Poland, May: another stable hybrid population, in E Europe, perhaps involving *M. f. feldegg* and *flava*; '*dombrowskii*' is variable, and many ♂♂ have darker ear-coverts and prominent white mark below eye. (P. Alberti)

'*zaissanensis*' is better regarded as a zone of intergradation. Here, it is provisionally included in *tschutschensis* following Alström & Mild (2003).

REFERENCES Alström, P. & Ödeen, A. (2002) In: Alström, P. Ph.D. thesis, Uppsala Univ. – van den Berg, A. B. (2011) *DB*, 33: 117–121. – Bot, S., Groenendijk, D. & van Oosten, H. H. (2014) *DB*, 36: 295–311. – Dubois, P. (2007) *BW*, 20: 104–112. – Ödeen, A. & Alström, P. (2001) In: Ödeen, A. Ph.D. thesis, Uppsala Univ. – Shirihai, H. & Gellert, M. (1987) *Proc. 4th Intern. Bird Ident. Meet., Eilat 1986*. Eilat. – Voelker, G. (2002) *Condor*, 104: 725–739. – Waern, M. (2006) *Roadrunner*, 14(2): 36–38.

CITRINE WAGTAIL
Motacilla citreola Pallas, 1776

Fr. – Bergeronnette citrine; Ger. – Zitronenstelze
Sp. – Lavandera cetrina; Swe. – Citronärla

This strikingly beautiful wagtail breeds on wet meadows or taiga bogs, at lakesides and along rivers primarily in Asia, but its distribution stretches also far west into E Europe. There it has expanded its range westwards in recent decades and now seemingly is about to colonise Finland, the Baltic states, Poland and other east European countries. It winters in the Middle East, E Arabia and S Asia and is particularly abundant in N India and Burma.

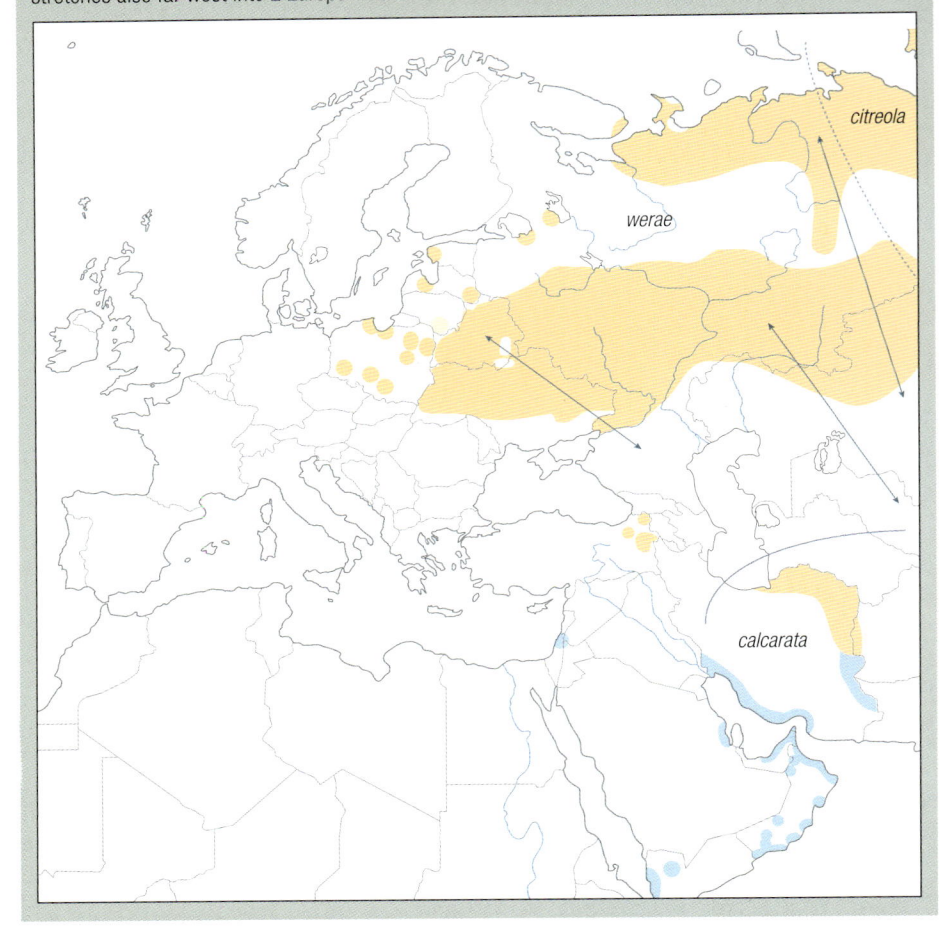

M. c. citreola, ♀, presumed 1stS, Oman, Mar: unlike Yellow Wagtail (even *flavissima*) has greyer upperparts, whiter/broader wing-bars, and pale yellow supercilium continuing around darker ear-coverts. Rather dark grey upperparts match *citreola*. Ageing difficult, though probably 1stS, as apparently three generations of wing-feathers, with juv remiges, post-juv greater coverts, and innermost greater coverts (probably median coverts, too) are pre-nuptial. (M. Varesvuo)

IDENTIFICATION Of similar size and shape as the other European wagtails with a tail-length intermediate between White and Yellow Wagtails. Adult ♂ generally straightforward with *all-yellow head and underparts*, *grey back* (*citreola*) with a *black band across lower nape/upper mantle*, and *two broad white wing-bars*, which in fresh plumage may appear to merge into one large white patch. In the southern mountain form (*calcarata*) the entire mantle and back are black. Undertail-coverts pale yellow or whitish. ♀♀ share with the *citreola* ♂ the pure *grey back* (lacking olive tinges) and the two broad *white wing-bars*. However, the crown and nape are greyish, as are the ear-coverts, leaving a prominent *yellow supercilium* (and often yellow forehead), which *continues down behind the ear-coverts and joins the yellow throat*. The breast is yellow, in some the belly too (though paler), whereas others have a whitish belly. Undertail-coverts are invariably white. Tertials in all plumages blackish with prominent whitish edges. In autumn, the adult ♂ attains a ♀-like plumage. First-winter birds lack all yellow in plumage until late autumn or early winter, are (brownish-) grey above and off-white below with double broad white wing-bars, appearing like mixture of first-winter White and Yellow Wagtails. Important to note *call* (see below), fairly pale *lores lacking dark central streak*, *pale stripe encircling darker ear-coverts* in practically all birds, and *all-dark bill*. Juvenile

M. c. citreola, ♂, presumed ad, Oman, Mar: vivid yellow head and underparts, with black band on lower nape/upper mantle, and extensive white wing-bars. Blotchy black on crown varies individually. Grey upperparts, deep yellow underparts and broad black nuchal band identify it as *citreola*. Presumed ad, as bases to primaries and outer, unmoulted, greater coverts do not appear sufficiently worn for juv. (M. Varesvuo)

M. c. citreola, ad ♂ (left) and 1stS ♀ (right), E Mongolia, May–Jun: further east, *citreola* has more white in wing, forming solid panel (especially in ♂), thereby resembling *calcarata*. ♂ has only ad wing-feathers discernible, while 1stY ♀ is aged by juv remiges and outer greater coverts. (H. Shirihai)

PIPITS AND WAGTAILS

M. c. werae, ♂, Poland, May: averages paler than *citreola*, sometimes almost ash-grey above, with less deep yellow head and body (not all birds are as pale as this). Black nuchal band either fairly prominent (as here) or thin and even nearly missing. (H. Harrop)

M. c. werae, 1stS ♀, Turkey, Jun: pale grey and yellow coloration is feature of ♀ *werae*; grey on flanks average slightly paler and more restricted. Note broad supercilium, surrounding darker ear-coverts, and hint of dark lateral crown-stripe. Juv remiges and outer greater coverts confirm age as 1stY; blotchy necklace is another vestige of immaturity. (R. Debruyne)

is entirely dark brown (-grey) and off-white with broad dark lateral crown-stripes and a dark patch on upper breast, in other words extremely similar to juvenile Yellow, but told on call and width of wing-bars.

VOCALISATIONS Song is very simple and resembles most song of White Wagtail, one or a few modest notes in combination with at least second-long pauses between stanzas, delivered from top of higher plant or low tree at meadow or bog. There is some variation; some phrases are more like Yellow's 1–3 strident notes, *tzreep*, but unlike in Yellow these are often alternated with more soft notes reminiscent of White Wagtail, e.g. *tshilp* or *tsle-tsle*. At times gives a low prolonged and fast twittering sub-song like in other wagtails. – Calls from perched birds in the breeding area both a rather soft Yellow Wagtail-like *psiie* (*flava* type call) and a sharp and strident, straight *tzreep*. In flight only the last call type is used, also by migrants, and is repeated identically several times while the bird is within hearing distance. Yellow Wagtails of S and SE Europe, and of Asia, have a similarly sounding strident call with 'r' in, but in our experience the one of Citrine is straighter, more emphatic and harder, and is repeated in a more mechanical fashion. Alarm is the same as the flight call. From wintering birds also, apart from the strident straight *tzreep*, a more Yellow Wagtail-like, soft disyllabic *tsi-lih*.

SIMILAR SPECIES The summer ♂ can be confused with ♂ of *Yellow-headed Wagtail* (*M. flava lutea*) and with the most yellow-headed ♂♂ of British *Yellow Wagtails* (*M. flava flavissima*), but note on Citrine pure grey lower back and rump, usually blackish uppertail-coverts, and (in most) black nape band, and white (not yellowish) wing-bars. Immature ♂♂ and ♀♀ more similar to British Yellow Wagtail but again greyer, less olive-tinged back/rump and whiter wing-bars best clues along with call. On Yellow, the supercilium does not continue down around darker ear-coverts, and there is a dark loral stripe (lores slightly paler in Citrine). Juveniles of both Yellow and Citrine Wagtails lack any clear yellow or green element in the plumage, and have similar dark lateral crown-stripes and dark breast patches, and they differ mainly in (i) wing formula, (ii) width of light wing-bars, and (iii) pattern of longest uppertail-coverts. (For details, see under Yellow Wagtail.) First-winters can be equally difficult to separate, since some Yellow seem to lack any yellow below. Note that on Citrine, lores are rather pale; forehead is slightly paler than crown, and forepart of supercilium often slightly tinged buff; bill is all-black; there is a pale surround to the ear-coverts. (Yellow has a dark loral stripe, forehead concolorous with crown, whole supercilium uniformly buff-white and well defined in forepart, bill with slightly paler grey or pinkish base

M. c. citreola, ad ♂, Oman, Nov: by late autumn, some winter ♂♂ have brighter yellow parts and purer grey upperparts, being more similar to summer plumage, but have variable olive-grey wash to crown, nape and usually on ear-coverts, enhancing broad yellow supercilium. This plumage varies strongly individually. Aged by ad wing-feathers. (P. Dubois)

to lower mandible, and no pale band encircling dark ear-coverts.) Young Citrine has a faint brownish cast on mantle, which in certain lights can appear olive-brown, but at least lower back and rump are invariably pure grey (olive-tinged brownish in Yellow). For reliable identification use all these characters in combination, since odd birds can show 'wrong' or intermediate appearance for one of them, but hardly for all. – Note that there is a superficial resemblance between first-winters of Citrine and *White Wagtail*, but the latter is easily separated on its dark grey breast patch. – Separation from immature *Grey Wagtail* generally straight-forward on immature Citrine's shorter tail, whitish undertail-coverts and slightly longer and black legs (Grey: long tail, bright yellow undertail-coverts, and pinkish shorter legs).

AGEING & SEXING (*citreola*) Ages separable in summer and autumn, rarely also in spring. Sexes separable in spring and summer, ad often also in autumn (generally only extremes, and preferably when first aged). – Moults.

Presumed *M. c. werae*, 1stS ♀, Greece, Apr: some ♀♀ have pale breast and belly, but yellow on face remains bright. 1stS by distinctive moult limits, with heavily-bleached browner juv remiges and outer greater coverts. (M. Schäf)

M. c. citreola, Kazakhstan, Aug, showing variation in early autumn ad plumages: following complete post-nuptial moult, ad acquires duller plumage, and sexual dimorphism often unclear. However, left bird presumably ad ♂ by darker elements in crown and ear-coverts, and generally brighter yellow areas, as well as broader and purer white in wing; the other is presumably ♀ by paler crown and ear-coverts, with diluted yellow in face, and reduced white in wing. Both aged by ad wing-feathers. (R. Pop)

Complete post-nuptial moult of ad in Jul–Aug (Sep) prior to autumn migration. Partial post-juv moult in Jun–Sep does not involve flight-feathers. Both age groups undergo a partial pre-nuptial moult in winter quarters, to some extent in Oct–Jan but mainly in Feb–Apr, which apparently involves a part of body, head, a variable amount of secondary-coverts and tertials (none to all). Many 1stY renew tail-feathers as well (Alström & Mild 2003). Ageing in spring without trapping the bird virtually impossible, and even with handled birds difficult. Both ad and 2ndY frequently have a moult contrast in greater coverts. – **SPRING** ♂ Head and underparts vividly yellow (except undertail-coverts being more whitish). Rarely a little green-grey or black on hindcrown, nape and ear-coverts (more often, but not only, in 1stS). A black band across lower nape and upper mantle, of variable width and prominence, usually rather broad. White wing-bars extensive in fresh plumage, sometimes partly merging to large whitish patch. ♀ Much of crown and ear-coverts olive-grey, but broad supercilium and band down behind ear-coverts yellow, joining yellow throat/upper breast. Forehead (sometimes forecrown, too) and diffusely centre of ear-coverts dusky-yellow. Lower breast and belly variable, often paler

M. c. citreola, ad, presumed ♀, Oman, Nov: following partial autumn moult into winter plumage, with yellow only on face. Despite heavily-worn tertials, remiges and primary-coverts fresh and greyish (ad-like), indicating that it is probably an ad after partial/suspended moult. (D. Occhiato)

M. c. citreola, 1stW, presumed ♂, Oman, Nov: 1stW moulting into winter plumage, still with juv secondary-coverts, and renewing tertials. Greyer upperparts and brighter yellow areas indicate ♂. Clear yellow surround to dusky ear-coverts, and blotchy gorget on breast. (D. Occhiato)

M. c. citreola, 1stW, Oman, Nov: first yellow coloration usually attained from late autumn, starting on head and to some extent below. Note juv remiges, primary-coverts and outer greater coverts, as well as longest tertial. (D. Occhiato)

yellow, sometimes off-white. Flanks often greyish. Undertail-coverts white. Never a complete black band across lower nape/upper mantle. Rarely attains advanced plumage with more yellow on head and some blackish patches at sides of upper mantle; to separate such from less advanced 1stS ♂, try ageing: heavy wear of tertials and brownish primaries indicate 1stS ♂, whereas a fresher plumage would signify an ad ♀. A few may be inseparable, though. – **SUMMER** **Ad** As in spring. **Juv** Crown dark brown with blackish sides. Upperparts brown-tinged grey. Wing-bars (tips of median and greater coverts) broad and prominent, buff-white or cream-white, soon bleaching to white. Underparts buffish off-white; across upper breast a solid dark gorget or patch. – **AUTUMN** **Ad** ♂ Attains a plumage which is quite similar to ♀ in spring, thus olive-grey on much of crown, nape and on ear-coverts. Supercilium and throat clear yellow, breast and sometimes belly pale yellow or yellowish-white. Greater and median coverts and tertials freshly moulted, edged (tertials) or tipped (secondary-coverts) greyish-white, pale edges or tips fading smoothly into darker feather-centres. Across breast a gorget of grey spots. Mantle is not so pure grey as in spring, often being slightly tinged olive-brown (thus closer to Yellow Wagtail). **Ad** ♀ Very similar to ad ♂, and majority inseparable. Extremes can be sexed: those ad with very pale yellow, and practically no yellow tinge on forehead are ♀♀, those with stronger yellow, especially on head, and some yellow on forehead, are ♂♂. Secondary-coverts and tertials freshly moulted, pattern as in ad ♂. **1stW** Lacks any yellow

PIPITS AND WAGTAILS

Aberrant bird or hybrid *M. citreola* × *M. flava*, ♂, Oman, Nov: extensive olive-grey on crown but otherwise mostly bright yellow head, lacking Citrine Wagtail's characteristic yellow supercilium; upperparts also tinged slightly olive-brown (less pure grey), and has Yellow Wagtail-like wing-bars. If not just an aberrant bird, possibly a hybrid between the two discussed species. (D. Occhiato)

M. c. citreola / werae, 1stW, Scotland, Sep: transition between juv and winter plumages, characteristically lacking yellow, at least until mid/late autumn. Essentially grey above and off-white below, with prominent double white wing-bars, easily overlooked as young White Wagtail; unlike Yellow Wagtail, note pale surround to ear-coverts and dark lateral crown-stripe. (H. Harrop)

M. c. citreola / werae (left: Kazakhstan, Sep; right: Israel, Sep), variation in 1stW: on left a distinctive bird, with well-developed head pattern and wing-bars; however, right-hand bird has rather obscure face pattern, with only faintest indication of dark lateral crown-stripe, and white wing-bars much narrower. (Left: R. Pop; right: H. Shirihai)

M. c. calcarata, 1stS ♂, India, Apr: all-black mantle and back, huge white wing panel, and even more saturated yellow plumage make this race highly distinctive. Exposed basal secondaries and outer greater coverts are juv (bleached browner). (N. Devasar)

(first yellow attained in late autumn or early winter, starting on supercilium and ear-coverts). Double wing-bars broad and prominent, white or whitish (can have faint grey-buff hue when fresh), tips sharply defined from blackish centres (not smoothly fading into darker centres as in ad). Sexes alike.

BIOMETRICS (*citreola*) **L** 16–18 cm; **W** ♂ 79–92.5 mm (*n* 37, *m* 86.4), ♀ 76–86 mm (*n* 18, *m* 81.3); **T** ♂ 67–86.5 mm (*n* 24, *m* 79.0), ♀ 65–81 mm (*n* 19, *m* 73.5); **T/W** *m* 90.9; **B** 15.3–18.2 mm (*n* 39, *m* 16.8); **BD**(f) 3.3–4.2 mm (*n* 15, *m* 3.6); **Ts** 23.0–26.3 mm (*n* 23, *m* 24.7); **HC** 9.0–12.8 mm (*n* 15, *m* 11.4). **Wing formula:** p1 minute; pp2–4 roughly equal and longest (but p2 often 0.5–1.5 mm <); p5 < wt 0.5–2.5 mm (very rarely 3–4 mm <); p6 < wt 7.5–12 mm; p10 < wt 19.5–26 mm; s1 < wt 22–29 mm. Emarg. pp3–5.

GEOGRAPHICAL VARIATION & RANGE Rather limited variation, but one subspecies very distinct. All taxa connected clinally or at least with extensive hybridisation.

○ **M. c. werae** (Buturlin, 1907) (E Europe, Ukraine, S Russia, E Turkey, NW Kazakhstan, SW Siberia, NW China; winters SW & S Asia). Differs from *citreola* only very slightly and on average in being a fraction smaller and having a proportionately slightly shorter tail, and in that ad in summer are sometimes slightly paler, less saturated yellow on head

M. c. calcarata, ad ♀, India, Jul: many ♀♀ of this race separable by darker grey (often blotchy) mantle and back. An extreme example in respect of the little amount of yellow at this season. Aged by relatively fresh and blacker ad remiges and primary-coverts. (C. Francis)

M. c. calcarata, 1stW, India, Dec: note strong contrast between darker/purer grey above and off-white below, with broad white stripe around dark ear-coverts, and broad wing-bars. (B. van den Boogaard)

and underparts, and paler grey on back. The black band across lower nape in ♂ is on average narrower (but much overlap), in some birds hardly extending onto upper mantle, and extreme birds can lack any black or just have a narrow grey-black line. The grey on flanks are on average slightly paler and more restricted. Summer ♀ differs often in same way from *citreola* as ♂, but even fainter, and few can be separated. **W** ♂ 81–90 mm (*n* 15, *m* 85.5); **T** ♂ 72–82 mm (*n* 15, *m* 76.0); **T/W** *m* 88.3; **B** ♂ 15.6–18.0 mm (*n* 15, *m* 16.9); **Ts** 22.8–26.4 mm (*n* 15, *m* 24.9).

M. c. citreola Pallas, 1776 (NW and C Siberia, E Kazakhstan, Mongolia, China; winters mainly in India). Treated above (see Ageing & Sexing and Biometrics). Grades into very similar *werae* over large area, and in practice numerous birds are unassignable to a particular race out of these two.

M. c. calcarata Hodgson, 1836. 'Mountain Citrine Wagtail' (S Turkmenistan, Kopet Dag, Afghanistan, Tien Shan, Tibet, Himalaya; mainly in mountains, but some in deserts; winters SW & S Asia). Quite distinctive in that ad ♂ has mantle and back completely black, and rump is black or dark grey, or a mixture. Also, the yellow colour of underparts tends to be even more saturated than in *citreola*. Size similar to *citreola*, but bill and legs are a trifle longer, while the wing and tail are slightly shorter. Undertail-coverts pale yellow (usually yellowish-white or white in other races). 1stS ♂ resembles ♀, but have generally clearly more black feathers admixed on mantle and back, and are deeper yellow on head, with unbroken yellow forehead (♀ usually grey-green on centre of forehead reaching bill; odd ad ♀ with pale forehead has this washed-out yellowish-white). A few 1stS ♂♂ are more developed and approach ad ♂ appearance but with greyer rump and some dark patches on otherwise yellow crown and nape. ♀♀ are very similar to ♀♀ of other taxa, and usually inseparable. Yellowish undertail-coverts and dark smudges on mantle and back might indicate *calcarata*, but variation still incompletely known. 1stS ♀ are (surprisingly) often less yellow below than in other races, pale birds being proportionately more common in *calcarata*. Call *tsrre*, perhaps subtly softer than *citreola*, and in anxiety a more Yellow Wagtail-like *tsrlee-e*. **W** ♂ 80–90 mm (*n* 20, *m* 85.7), ♀ 77–85 mm (*n* 12, *m* 79.8); **T** ♂ 72–80 mm (*n* 20, *m* 76.0), ♀ 69–73 mm (*n* 12, *m* 71.0); **T/W** *m* 88.8; **B** 16.6–19.0 mm (*n* 32, *m* 17.5); **Ts** 24.5–27.8 mm (*n* 32, *m* 26.3). **Wing formula: p5** < wt 0.5–1.5 mm; **p6** < wt 6–10 mm; **p10** < wt 18–25 mm; **s1** < wt 19–26.5 mm. (Syn. *citreolides*; *weigoldi*).

TAXONOMIC NOTES The distinctive, in ♂ plumage black-backed, *calcarata* is a candidate for being recognised as a separate species. However, it is known to intergrade with the other two races where they meet, and knowledge of the extent of this, and of the full range of variation, is in our opinion still insufficient for a confident decision. As always, it is desirable to complement data on morphology and genetics with studies of vocalisation, behaviour, etc. – In Ödeen & Alström (2001), *calcarata* was shown to be widely separated from *citreola* based on mtDNA (these two being non-monophyletic), but sister taxon based on nuclear DNA. – We agree with Alström & Mild (2003) that *werae* is a questionable race, being very close morphologically to *citreola*. However we have kept it as a subtle but valid taxon on account of rather markedly smaller size, shorter tail and consistent average colour differences, although admittedly these are small.

REFERENCES Ödeen, A. & Alström, P. (2001) In: Ödeen, A. PhD thesis, Uppsala Univ.

Presumed *M. c. calcarata*, 1stY ♀, Afghanistan, Mar: note darkish upperparts, strong contrast between renewed tertials and juv remiges, and three generations of greater coverts (outermost are juv). (F. Joisten)

Juv, Russia, Jul: juv of all Citrine races characterised by clear blackish lateral crown-stripes and solid dark gorget or breast patch; white wing-bars broad and prominent; underparts buffish off-white. (J. Peltomäki)

GREY WAGTAIL
Motacilla cinerea, Tunstall, 1771

Fr. – Bergeronnette des ruisseaux; Ger. – Gebirgsstelze
Sp. – Lavandera cascadeña; Swe. – Forsärla

Of the wagtails, this is perhaps the least known, despite being widespread over much of Europe. With the longest tail of all, which it often wags intensely, deep yellow vent, and shortish pink legs it looks a bit 'different'. It is a specialised insect-feeder along stony streams and tiny, shallow water-courses with exposed stones and boulders, both in woods and in alpine habitats. Another oddity for a wagtail is its habit of perching high in trees. In S Europe follows rivers into towns and villages. Northern breeders move south-west for the winter, southern birds are mostly resident.

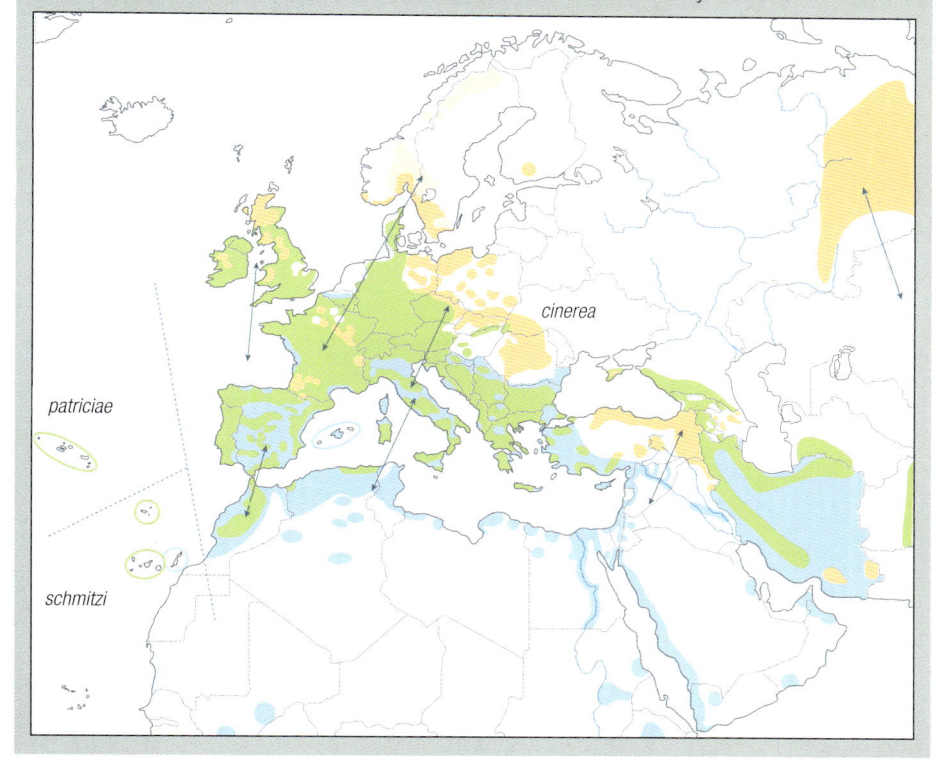

IDENTIFICATION The proportionately *very long tail* and rather short, *pinkish or pale brown-grey legs* give this wagtail a distinctive look. Adding to its distinctiveness is its habit of *vigorously wagging its tail and its entire rear body*, more or less in Common Sandpiper fashion. Characteristic flight with long, deep undulations and long tail striking. In all plumages notice the *deep yellow vent and undertail-coverts*, and *yellowish-green rump*. Grey upperparts contrast with blackish wings. *Tertials are black with prominent white edges* in fresh plumage. *Narrow supercilium white* (or whitish). No double wing-bars on the folded wing (like in Yellow Wagtail), but instead *a single broad white central wing-bar* is visible in flight, mainly from below against the sky, but also often from above. Adult ♂ summer has a *black throat bib* separated from grey ear-coverts by a *prominent white submoustachial stripe*, and below the black bib the *breast is* just as *deep yellow* as the vent, whereas belly is paler yellow and flanks are even paler yellow or whitish. Although the black bib is distinctive enough when seen front-on, it can be surprisingly difficult to see, or unimpressive, from other angles. The ♀ sometimes has a hint of the black bib, and rarely it is rather more pronounced (inviting confusion with less than perfect ♂♂), but normally it is totally absent. Immatures and autumn birds of both sexes have a buff-brown wash over breast, and any yellow generally restricted to vent (although ad ♂ often has a yellow tinge on breast and belly). The adult ♂ loses its black bib in autumn, partly or completely, and the throat usually becomes tinged rusty-buff.

VOCALISATIONS Song is simple and monotonous, a fast series of repeated, usually identical, sharp notes, *ziss-ziss-ziss-ziss-ziss*, designed to be audible above the streaming water, delivered with little variation. (The only risk of confusion would occur in Central Asia in alpine habitats, where White-capped Bunting *Emberiza stewarti*, extralimital and not treated, actually has a quite similar-sounding song.) Many ♂♂ alternate between a faster and a slower variant of this strophe. Less commonly the strophe suddenly changes pace and pitch, *ziss-ziss-ziss-ziss chu chu* (in Ortolan fashion, though no confusion risk). Like apparently all wagtails, also has a prolonged twittering and warbling sub-song, but in our experience this is much less often used than, e.g., in White Wagtail. – Flight-call resembles sharper variants of White Wagtail call, *tzit-tzit*, recognised by shrill sharpness and that both notes are identical, and identically stressed (slightly different in both respects in White). Alarm is a repeated disyllabic *sü-eeht*, often given from elevated perch near nest.

SIMILAR SPECIES There is little risk of confusion once the combination of very long tail (black with white sides), bright yellow vent/undertail-coverts and yellowish-green rump is seen. The paler than black legs and lack of two narrow wing-bars clinch the identification. Once the call is learnt the species can be easily separated in flight from White and Yellow Wagtails even without the single broad, white central wing-bar being seen.

AGEING & SEXING (*cinerea*) Ages separable in summer, and sometimes in autumn if seen well or trapped. Sexes separable in spring and summer, ad rarely also in autumn (generally only extremes, and preferably if first aged). – Moults. Complete post-nuptial moult of ad in late Jun–Sep prior to autumn migration. Partial post-juv moult in May–Sep does not involve flight-feathers or primary-coverts, but often some inner greater coverts and tertials. A few tail-feathers also often moulted. Both age groups undergo a partial pre-nuptial moult in winter quarters mainly in Jan–Mar, which apparently involves part of body, head, a few inner greater coverts, tertials (none to all) and a few tail-feathers (often r1 and r6, sometimes whole tail). – **SPRING** Ageing very difficult and frequently impossible if not three feather generations present among greater coverts, these being 1stS with some retained juv coverts. ♂ Throat black, or black with some limited white admixed at upper centre; black bib sometimes reaches upper breast. Narrow distinct white supercilium and prominent white submoustachial stripe. Lores blackish. Breast clear yellow, belly similar or slightly paler, flanks pale yellow (or whitish), vent and undertail-coverts saturated yellow. Rump yellowish-green, with sides inclining to purer yellow. A few, presumably 1stS, have less-developed black bib with rather much white admixed, and these are very difficult to separate from advanced ad ♀♀ in the field. ♀ lacks black throat bib of ad ♂, or has only grey mottling on lower throat (very rarely advanced ♀♀ can have quite extensive dark grey or black on throat, especially on lower parts, inviting confusion with some 1stS ♂♂). Submoustachial stripe less white and prominent than in ♂. Supercilium slightly tinged cream or buff (all white in ♂). Lores greyish. Breast and belly variable, often paler yellow than in ♂, sometimes near-white, and breast may have slight buffish tinge. Flanks whitish. – **SUMMER Ad** Worn. Plumages as in spring. **Juv** Fresh. Appears like a 'washed-out' and slightly brownish-tinged ♀. Supercilium short, buffish and indistinct, breast and belly tinged rusty-buff, grey of upperparts

M. c. cinerea, ♂, presumed ad, Georgia, Jun: very long tail (vigorously wagged) and rather short, pinkish legs. Summer ♂ typically has bold black bib, just reaching upper breast, bordered by white submoustachial stripe. White supercilium varies (much reduced here). Deep yellow below, deepest on vent, with yellowish-green rump. Ageing difficult, but solid black throat suggests ad. (R. Pop)

M. c. cinerea, presumed 1stS ♂, Israel, Mar: despite clear moult limits in wing, difficult to confirm age of older feathers, and hence bird's age. Sexing therefore also requires caution, but being possibly a young bird and already having a rather solid black bib in March indicates ♂. (H. Shirihai)

M. c. cinerea, ♀, presumed 1stS, Kazakhstan, Jun: remiges and most secondary-coverts seemingly juv (heavily bleached browner) suggest 1stS, which seldom has black bib. Less pure grey upperparts, narrower supercilium, greyish lores and duller yellow breast also indicate young ♀♀. (M. Vaslin)

with brownish hue. Generally some pale yellow on vent, but some are so light as to make the yellow difficult to discern in the field. — AUTUMN **Ad** Plumage resembles ♀ in spring, black on throat of ♂ reduced or absent, yellow of breast and belly more subdued, breast and supercilium with buff tinge. Grey of crown, nape and mantle lightly tinged olive-brown. All flight-feathers, tertials and wing-coverts blackish and of same age (though beware that innermost greater coverts are slightly greyer also in ad). Sexes very similar and often inseparable. Birds with deepest yellow on breast and belly to undertail-coverts with contrast between buff-white throat (sometimes even tinged rufous) and yellow-tinged breast, and with most prominent pale supercilium and submoustachial stripe, should be ♂♂. **1stW** Very similar to ad, and not always separable. Flight-feathers slightly more brownish-grey (more blackish in ad), and in some a slight contrast discernible between blacker moulted secondary-coverts and tertials, and greyer unmoulted juv feathers. Sexes alike or very similar.

BIOMETRICS (*cinerea*) L 17–19.5 cm; **W** ♂ 81–87 mm (n 23, m 83.2), ♀ 79–85.5 mm (n 12, m 82.1); **T** ♂ 83–103 mm (n 23, m 90.9), ♀ 81–97 mm (n 12, m 89.3); **T/W**

M. c. cinerea, ♀, E Kazakhstan, May: rather extensive partial black bib. Ageing without handling often unsafe in spring, especially as it is difficult to infer age of older wing-feathers (these moult limits can indicate either ad or 1stS). (J. Normaja)

M. c. patriciae, São Miguel, Azores, May–Jun: this breeding pair (feeding same juv) demonstrates how throat can be similar between sexes. Both are too heavily worn to age using moult pattern, but relatively uniformly-feathered wing and still reasonably fresh primary-coverts of left-hand bird suggest dark-throated ad ♀, which characteristically has white central throat. Presumed ♂ (on right) has poorly-developed dark throat with white admixed; extremely worn, it is probably 1stS. (J. Normaja)

PIPITS AND WAGTAILS

M. c. cinerea, 1stY ♂, Oman, Feb: some are less difficult to age, like this bird, which has clear moult limits between heavily worn and browner juv remiges, primary-coverts and outer greater coverts, contrasting with several fresh secondary-coverts and tertials. Black bib still mainly concealed. (H. & J. Eriksen)

M. c. patriciae, ad, Azores, Oct: racial ID by location only. Ageing and sexing frequently difficult, but here the even, very fresh wing, with still growing post-nuptial remiges, conclusively gives the age, but since this bird is already in winter plumage sexing is impossible. Species told by slim jizz with long, slender tail, and invariably deep yellow vent. (H. Shirihai)

m 109; **B** 15.2–16.9 mm (*n* 32, *m* 16); **Ts** 18.0–20.7 mm (*n* 35, *m* 19.7); **HC** 5.0–7.0 mm (*n* 15, *m* 6.1). **Wing formula: p1** minute; **pp2–4** roughly equal and longest (p4 sometimes 1 mm <); **p5** < wt 5–7 mm; **p6** < wt 11.5–18.5 mm; **p10** < wt 26–30 mm; **s1** < wt 28–33 mm. Emarg. pp3–5.

GEOGRAPHICAL VARIATION & RANGE Surprisingly limited variation despite wide range with large gaps, e.g. absent over much of Russia, N Kazakhstan, Tibet and W China. Populations on Atlantic islands a little darker and shorter-tailed.

M. c. cinerea Tunstall, 1771 (much of Europe, NW Africa, Caucasus, Turkey, N & SW Iran, Urals and Siberia; winters widely from W Europe and W Africa through E Africa to SE Asia). Treated above (see Ageing & Sexing and Biometrics). A fair amount of individual variation as to tail length, tail pattern and yellow saturation occurs across entire range, making it difficult to uphold several claimed races. Tail pattern usually has all-white r6, white r5 with partially dark outer web or narrowly dark along shaft, r4 sometimes same or with slightly more extensive dark portions. (Syn. *caspica*; *melanope*; *robusta*.)

M. c. schmitzi Tschusi, 1900 (Madeira, Canary Is; resident). Has slightly narrower and shorter supercilium, narrower submoustachial stripe, and on average a shorter wing and longer tail. Yellow of underparts in some birds is even stronger than in *cinerea*. Tail pattern as *cinerea* or with slightly more dark on r4. Originally described from Madeira, but birds on Canaries ('*canariensis*') very similar and included here, although tending somewhat towards *cinerea*. **W** ♂ 80–86 mm (*n* 15, *m* 82.6), ♀ 80–84 mm (*n* 10, *m* 81.2); **T** 89–100 mm (*n* 25, *m* 93.3); **T/W** *m* 113.8; **B** 15.2–17.4 mm (*n* 24, *m* 15.8); **Ts** 18.5–21.1 mm (*n* 25, *m* 20.0). (Syn. *canariensis*.)

M. c. patriciae Vaurie, 1957 (Azores; resident). Resembles *schmitzi* but differs by on average slightly longer bill and tarsus, slightly shorter tail and less white on outertail (r4 with much black on inner web and r5 with narrow black inner edge). **W** ♂ 79.5–88 mm (*n* 19, *m* 83.5), ♀ 79–84.5 mm (*n* 13, *m* 81.4); **T** 85–95 mm (*n* 32, *m* 89.5); **T/W** *m* 108.4; **B** 16.1–17.7 mm (*n* 32, *m* 16.9); **Ts** 19.1–21.7 mm (*n* 32, *m* 20.6). **Wing formula: p5** < wt 3–5 mm; **p6** < wt 10–13 mm; **p10** < wt 23–27.5 mm; **s1** < wt 24–27.5 mm.

TAXONOMIC NOTE We agree with Alström & Mild (2003) that variability within *cinerea* makes separation of further subspecies than listed above in Asia and Macaronesia problematic.

REFERENCES Cf. under *M. flava*.

M. c. cinerea, 1stW, Italy, Nov: rather dull individual, although still distinctive by disproportionately long tail, bright yellow vent and undertail-coverts, and yellowish-green rump. As autumn progresses, contrast between juv remiges, primary-coverts and outer greater coverts, and fresh tertials and inner secondary-coverts becomes more obvious. (D. Occhiato)

M. c. cinerea, juv, Switzerland, Jun: like ♀, with narrow and reduced supercilium, whitish-cream underparts tinged dirty rusty-buff, usually much duller yellow on vent, and grey of upperparts less pure. (R. Kunz)

WHITE / PIED WAGTAIL
Motacilla alba L., 1758

Fr. – Bergeronnette grise; Ger. – Bachstelze
Sp. – Lavandera blanca; Swe. – Sädesärla

One of the loveliest companions to man, delicately built, landing on the farmyard or garden lawn, wagging its long tail and walking tirelessly around, or cheering you up with its unobtrusive calls from the top of your roof. However, it not only lives in close vicinity to man, but is also a bird of remote taiga bogs and lakesides. Most birds migrate for the winter to the Mediterranean region, Middle East and N & E Africa. Returns early in spring, usually in Feb and Mar, though in N Europe often not until early Apr.

IDENTIFICATION Typical slim, *long-tailed* wagtail, walking energetically with *nodding head* or rushing for insects on *quick, thin legs* over short grass, roads, roofs, etc. *Wags tail* vigorously just after landing or after a short rush. In flight notice pronounced undulations and long tail. Colours only *black, grey and white* (olive and yellowish tinges in autumn subordinate), but detailed pattern quite variable within wide range in Palearctic, particularly in ♂♂. Adult ♂ in N & Continental Europe (*alba*) has grey back but *black hindcrown and nape with sharp border to grey mantle*. Forecrown and sides of head and neck pure white, *beady eye* thus encircled by white. Throat and breast black. Flanks pale ash-grey. Adult ♀ similar but has a somewhat less smart head pattern, with usually less black on crown, and nape is largely grey (no sharp division with black on crown); some ♀♀ are quite ♂-like with black crown and nape and fair contrast to grey mantle, while others lack any black on crown and have restricted white on forehead (rarely none). White parts of head often less pure, and white forecrown often finely streaked or spotted. Adult ♂ in Britain & Ireland (*yarrellii*) has black mantle and back, is darker grey on flanks, and ♀ is dark grey above with variable number of blackish patches, or is uniformly blackish-grey. Birds in Morocco (*subpersonata*) have different head pattern with a black band across lower cheeks (merging black of nape with black of throat) and narrow black eye-stripes and moustachial stripes. There is a white patch isolated on the side of the neck. Birds in N Iran (and further east; *personata*) are easily told by grey back and largely black head, neck and breast with a white mask. – In autumn, both sexes of all races attain *whitish chin and throat*, leaving only a *blackish breast patch*. Ad ♂ typically has unspotted extensive white forehead and forecrown, while ad ♀ has either greyish forehead or limited white with some dark spotting; note that amount of visible black in winter plumage is variable, a few ♂♂ having much of it concealed by grey, while a few ♀♀ have rather a lot visible. First-winters resemble ad ♀ but are on average less black and white and more tinged yellowish-olive on sides of head. Juveniles resemble first-winters but have hint of darker lateral crown-stripes, more diffuse head pattern and a grey (not black) patch on upper breast which connects with grey submoustachial stripes.

VOCALISATIONS Song of two types (as in all wagtails, but most obvious in this species), the normal song being very simple, a slow repetition (every 1.5 sec. or slightly shorter pauses) of a modest note or combination of 2–3 notes, *cher-li... chilp... chewee... chü-lee...*, etc. It thus differs from Yellow Wagtail by its varied nature and absence of sharp,

M. a. alba, ♂, **Estonia, May** solid black nape, distinctly demarcated from pure white forecrown and head-sides; chin and throat uniformly black, usually lacking visible white. (M. Varesvuo)

M. a. alba, ad ♂, **Italy, Mar**: note ad's grey-fringed primary-coverts, and ♂'s head and bib pattern. A few ad ♂ *alba* develop quite substantial white in wing-coverts, but panel still not as large or solid as more eastern races. Upperparts medium grey or marginally paler. (D. Occhiato)

PIPITS AND WAGTAILS

M. a. alba, 1stS ♂, Finland, May: sexing problematic without accurate ageing. That this bird is 1stS (juv remiges, primary-coverts and outer greater coverts), yet very much ad-like, safely identifies it as 1stY ♂. (M. Varesvuo)

M. a. alba, ad ♀, Finland, May: post-nuptial outer greater coverts are ad, being less worn and edged whitish-grey; ageing further supported by fresh ad-type primary-coverts. Being ad, dark spots on forehead, grey feathering on otherwise blackish crown and dark smudge behind eye indicate ♀. General pattern recalls 1stS ♂. (J. Normaja)

M. a. alba, 1stS ♀, Finland, Apr: poorly-developed head pattern with mostly grey crown sufficient to identify this bird at this season as 1stS ♀. In addition, outermost greater coverts, primary-coverts, remiges and alula are brownish-bleached juv feathers. (J. Normaja)

M. a. alba, ad ♂, Israel, Oct: freshly moulted, attains white chin/throat, otherwise as summer plumage. Note pattern of ad-type greater coverts with broad, smooth ash-grey edges. (A. Ben Dov)

M. a. alba, ad ♂, Italy, Dec: rather easily aged due to overall ad-type wing-feathers, with extensive grey-white fringes and pale tips to remiges. Sex by extensive white forecrown and much black on hindcrown and nape. (D. Occhiato)

straight, buzzing notes, but can be quite like some variants of Citrine song (see latter). The other song type is a prolonged and fast, irregular twittering, containing the common call and several other elements; often used in territorial fights, or when chasing a Sparrowhawk *Accipiter nisus* or Cuckoo *Cuculus canorus*, or mobbing a cat, but sometimes heard also from newly-arrived feeding birds in early spring ('joie de vivre'!). – The normal call, uttered both by perched birds and in flight, is a disyllabic, cheerful, 'bouncing' *che-lit* (or *tsli-vit*, *chi-chik*, etc.), second syllable being slightly more stressed. At times it sounds more complex, almost trisyllabic, especially in excitement. Alarm given with repeated normal calls, which may run together and turn into twittering song variant when bird takes off in great anxiety. Juveniles have similar calls, only somewhat more 'metallic' or sharp.

SIMILAR SPECIES Post-juveniles can hardly be confused with other species within the treated range (though must be separated from some closely related extralimital species in Asia and Africa; see Alström & Mild 2003). Separation from juvenile Grey Wagtail generally straightforward given juvenile White's pure white undertail-coverts, grey breast-patch and longer and black legs (Grey has yellow undertail-coverts, no dark breast patch and shorter pinkish or pale brown-grey legs). – Juvenile Citrine Wagtail is more similar but calls differ, and is usually eliminated by White's slightly less prominent light wing-bars, more diffuse pale supercilium (often lacking in front of eye) and slightly more olive-tinged grey upperparts (Citrine has slightly broader wing-bars, somewhat better-marked supercilium and lightly brown-tinged grey back. First-winter Citrine is quite similar to first-winter White but lacks a black breast patch (at most has a few dark blotches at sides).

AGEING & SEXING (*alba*) Ages separable in summer and autumn. Sexes generally separable in spring and summer, for ad often also in autumn (usually only extremes, and preferably when first aged). – Moults. Complete post-nuptial moult of ad usually in Jul–Sep prior to autumn migration. Partial post-juv moult in Jun–Sep highly variable, but does not involve flight-feathers or primary-coverts. A variable number of greater coverts moulted, often about inner half, but none or all occur also. Both age groups undergo a partial pre-nuptial moult in winter quarters, mainly in Jan–Mar, which involves body and head, a variable number of median and inner greater coverts, tertials and tail-feathers (none to all). Ageing in spring usually requires close observation or trapping, 1stS often having more brownish and heavily worn primaries but some are quite similar. Both ad and 2ndCY frequently have a moult contrast in greater coverts. – **SPRING** Ageing frequently impossible unless three generations of greater coverts present, these being 1stS with a few retained juv coverts. Outer 4–5 greater coverts differ from inner in having only narrow pale outer edges, in ad edges usually being more pure grey-white, in 1stS more brownish-white, but difficult intermediates occur. ♂ Sexing frequently more problematic than understood due to age-related and individual variation. Typical ♂ plumage: hindcrown and nape black with distinct demarcation against grey mantle. Much pure white on forecrown, with neat border to black. Sides of head rather pure white. Chin and throat uniformly black, usually lacking visible white. See Biometrics. Very rarely, even ad ♂ can have traces of white on chin and grey admixed in the black on nape, approaching very closely to advanced ♀♀. ♀ Crown variable, but often has some blackish, grading smoothly into greyer nape. A few (presumably 1stS) have no black on crown at all. White on forehead restricted and less pure white, often with dark spots near border to solid dark. A few (presumably 1stS) have very little white on forehead at all, and this is tinged olive. Sides of head white with some greyish and olive tones. Rarely, quite ♂-like ♀♀ appear with nearly all-black crown and nape, and well-marked border to grey mantle. Still, they can often (but not always) be identified by a few grey feathers admixed on black nape, and a few white spots on chin and upper throat. Forecrown and forehead often have tiny black spots too. White on sides of head usually less pure white than in ♂. – **SUMMER Ad** Worn. As in spring. **Juv** Fresh. Crown dull brownish-grey with blackish sides. Upperparts olive-tinged brown-grey. Underparts off-white; across upper breast a solid dark patch. – **AUTUMN Ad** ♂ As in spring but freshly moulted, and attains white chin and throat, leaving only black patch on breast. Also, black on nape and hindcrown largely replaced by grey. Greater and median coverts, and tertials, freshly moulted, edged (tertials) or tipped (secondary-coverts) greyish-white, pale edges or tips grading smoothly into darker feather-centres. No moult contrasts in wing. Mantle less pure grey than in spring, being often slightly tinged olive. Forehead and face largely white (very restricted amount of yellowish-green tinge in some). **Ad ♀** Similar to ad ♂, but has even less black on crown (sometimes none), and none on nape. On average less white on forehead than in ♂. **1stW** Very similar to ad of each sex, but on average slightly less black and white on head, and more tinged yellowish-olive. Most trapped birds can be aged by slight moult contrast in the wing, unmoulted juv greater coverts or tertials having slightly paler brownish-grey ground colour but whiter and more distinctly demarcated tips concentrated on outer webs (ad has blackish ground colour with greyer and broader tips being slightly more diffusely set off).

BIOMETRICS (*alba*) **L** 17–18.5 cm; **W** ♂ 86–94.5 mm (*n* 58, *m* 89.8), ♀ 81–88 mm (*n* 36, *m* 84.9); **T** ♂ 82–92 mm (*n* 58, *m* 87.1), ♀ 80–88 mm (*n* 34, *m* 83.9); **T/W** *m* 97.7; **B** ♂ 15.3–17.3 mm (*n* 22, *m* 16.1), ♀ 14.7–

M. a. alba, ad ♀, **Spain, Dec**: highly variable, but many fresh ad ♀♀ have black on crown heavily obscured, yet possess rather large breast-patch. Ad undergoes complete post-nuptial moult and entire plumage is fresh in autumn. Some 1stY also moult all greater coverts, making ageing unsafe. (R. Cediel-Algovia)

M. a. alba, 1stW ♂, **Italy, Nov**: being 1stW (remiges, primary- and outer greater coverts are juv, heavily worn and bleached brownish), the rather extensive black on crown readily identifies this bird as ♂. Yellowish hue to face also typical of juv and 1stW (ad being on average purer white). (D. Occhiato)

M. a. alba, 1stW ♀, **Sweden, Nov**: note moult limits in wing, with unmoulted juv outer greater coverts and exposed remiges. Not all 1stW are readily sexed by resembling respective ad, including this one. At first glance, extensive black on breast suggest 1stW ♂, but virtually all-grey crown indicates 1stW ♀. (L. Olsson)

M. a. alba, 1stW, presumably ♀, **Italy, Nov**: mostly grey crown and broken/ill-defined black breast-band, which together with unmoulted juv outer greater coverts and exposed remiges, strongly suggest 1stW ♀. (D. Occhiato)

M. a. alba, juv, **Finland, Jul**: resembles 1stW, but has fluffy body-feathers, and hardly any head pattern, further being mostly grey (not black) on breast, though most have more solid dark patch across upper breast. (J. Peltomäki)

PIPITS AND WAGTAILS

M. a. dukhunensis, ad ♂, moulting from winter to summer, Iran, Jan: note large whitish wing panel (an extreme example), which automatically confirms the bird's age as ad. The size and solid white panel matches wing pattern of *M. a. personata*. (E. Winkel)

M. a. dukhunensis, 1stS ♂, Azerbaijan, May: pale grey upperparts and broad white edges to wing-coverts and tertials, which merge to form large wing panel. Partially exposed remiges and outer coverts are juv, worn and already bleached browner; wing panel less solid than older ♂ *dukhunensis*. (M. Heiß)

M. a. dukhunensis, 1stW ♂, Armenia, Oct: wing-coverts already extensively fringed whitish-grey, forming quite extensive pale panel. This 1stW demonstrates that some young ♂♂ have extremely little black on crown. (V. Ananian)

M. a. dukhunensis, 1stW ♀, Kazakhstan, Sep: young ♀♀ often develop slightly broader and whiter tips and edges to wing-coverts (median post-juv, and greater juv in this case) than equivalent plumage of *alba*; upperparts also relatively pale grey. Identifying young, especially ♀, *dukhunensis* outside their normal range or in areas where *alba* also occurs is impossible. (Y. Belousov)

16.0 mm (*n* 22, *m* 15.4); **Ts** ♂ 22.0–23.7 mm (*n* 22, *m* 22.9), ♀ 21.0–23.0 mm (*n* 22, *m* 22.1); **HC** 5.5–7.5 mm (*n* 18, *m* 6.5). **Wing formula: p1** minute; **pp2–4** roughly equal and longest; **p5** < wt 3–4 mm; **p6** < wt 12.5–15 mm; **p10** < wt 26–30.5 mm; **s1** < wt 28–33 mm. Emarg. pp3–5. (Length of white on inner web of r5 measured to tip 40–67 mm (*n* 44), thus complete overlap with *yarrellii*, including shape of white wedge, hence is of very limited value for racial determination.)

GEOGRAPHICAL VARIATION & RANGE Marked and complex variation with several subspecies described of which five occur regularly within the treated range. There is some intergradation where races meet, but apparently not quite as frequently as in Yellow Wagtail. For extralimital subspecies see Alström & Mild (2003).

M. a. yarrellii Gould, 1837. 'PIED WAGTAIL' (Britain, Ireland; some sedentary, others winter SW Europe to W Africa). Differs from *alba* in spring by much darker upperparts (hindcrown to uppertail-coverts of ad ♂ jet-black, almost glossy, rarely with some dark grey admixed on back; ad ♀, too, is blackish above, or dark grey with some blackish parts). Odd birds of both sexes do not develop maximum darkness. Differences from *alba* in upperparts coloration generally prevail in autumn, but a few birds moult to a somewhat paler shade, or have partly grey feathers mixed in the

M. a. yarrellii, 1stS ♂, Netherlands, Apr: diagnostic jet-black upperparts, but being 1stS (heavily worn and bleached brownish juv remiges just discernible) some dark grey feathers persist among the black. (R. Pop)

M. a. yarrellii, ad ♀, England, Jul: ♀ darker above than *alba*, being blackish-grey, including rump and uppertail-coverts. Differs from summer-plumaged ♂, including 1stS ♂, by combination of being ad (blacker remiges and primary-coverts) and by contrast between blacker crown, nape and mantle, paler/greyer rest of upperparts, and less sharply defined white face. (N. Blake)

M. a. yarrellii, ♀, presumably ad, Netherlands, Mar: partially exposed black on upperparts and dusky flanks; only ad-type wing-feathers visible (though short tertial has not been renewed in pre-nuptial moult). (N. D. van Swelm)

M. a. yarrellii, ad ♂, Ireland, Sep: freshly and evenly moulted, and similar to summer plumage, but has white throat and more restricted black on chest. (M. Grimes)

M. a. yarrellii, ad ♀, France, Feb: even in autumn, still has more blackish elements above than *alba*. Paler crown and nape than winter ad ♂, but best differentiated from similar 1stW ♂ by being ad (rather fresh, blacker remiges and primary-coverts; no moult limits or juv feathers visible in wing-coverts). (A. Audevard)

M. a. yarrellii, 1stW ♂, Ireland, Sep: even in this young ♂, the upperparts (and flanks) are darker grey than any *alba*. Being 1stW (note juv longest tertial), safely differentiated from quite similar ad ♀, while black cap and white forehead separate it from 1stW ♀. (C. Bradshaw)

M. a. yarrellii, 1stW ♀, England, Sep: separation of young ♀ *yarrellii* from *alba* can be problematic. Nevertheless, in 'neutral' light, the former is usually somewhat darker grey above (in particular on rump and uppertail-coverts) and usually has, as here, some blackish blotches on mantle. Limited black on crown and breast is indicative of 1stW ♀. Still no moult of wing-coverts. (N. Bowman)

PIPITS AND WAGTAILS

M. a. subpersonata, ♂♂, presumed 1stS, Morocco, Mar (left) and Jan: truly unique and very attractive taxon with its white neck-side patch and complex head pattern comprising white mask, white patch below eye and thin black loral and moustachial stripes. Both birds seem to be 1stS by a number of retained juv outer greater coverts (brownish-tinged, contrasting strongly to moulted inner), while lack of grey on crown indicates ♂; some mottling on forecrown is characteristic of 1stY ♂, too. Note that the right bird still in Jan has white bib of winter plumage. (Left: D. Occhiato; right: A. Ouwerkerk)

M. a. subpersonata, ad ♀, Spain, Mar: some ad ♀♀ more easily sexed by having dark grey crown and nape, as shown by this vagrant to S Spain. (S. Peregrina)

M. a. subpersonata, ad ♂, Morocco, Sep: freshly and evenly moulted, similar to summer plumage, but has white throat and more extensive white neck-side patch. (P. Öberg)

blackish upperparts. Only 1stW (in particular ♀♀) can sometimes be so relatively pale grey above as to cause problems, but on direct comparison are still somewhat darker grey than *alba*. As a rule, whole rump is black or dark grey in *yarrellii*, but paler grey in *alba*; however, odd *yarrellii* have less dark rump, and some *alba* have at least lower rump and uppertail-coverts quite dark grey, inviting confusion on this character alone. Diffuse dark flecking or spotting on otherwise grey upperparts is a strong (probably conclusive) indication of *yarrellii*. Flanks and sides of breast in *yarrellii* are darker olive-grey, not as comparatively pale ash-grey as in *alba* (although quite a few *alba* have somewhat darker grey flanks than average, and can be similar to average *yarrellii*). 1stW ♂ can resemble ad ♀, and the two can be difficult to separate in the field, but 1stW ♂ generally has large jet-black sharply defined irregular patches on otherwise dark grey mantle and back, whereas dark portions of ad ♀ usually are less deep black and more diffuse. Note that both ad ♀ and 1stW ♀ have much visible black on crown and nape, also in autumn (like ♂ *alba*, but unlike ♀ of that race). Amount of white on forehead is on average larger than in *alba* when comparing same ages and sexes. Finally, presence of some dark spots or smudges on belly seems to be a good indicator of *yarrellii* (Adriens et al. 2010), since

M. a. subpersonata, 1stW ♂, Morocco, Aug: most wing-feathers new (only outermost greater coverts are juv); unlike 1stW female, more black feathers on crown visible. Black loral line, greyish-black ear-coverts, and clear white patch below eye bordered by thin black line below, make even young plumage unmistakable. (I. Yúfera)

M. a. subpersonata, 1stW ♀, Morocco, Oct: like 1stW ♂ but crown purer grey. (H. Overduin)

M. a. subpersonata, juv, Morocco, Jun: juv also rather distinctive, with clear whitish supercilium and quite pronounced dark breast patch. (A. B. van den Berg)

38% in a sample of 69 showed it, whereas no *alba* (n 136) did. **W** ♂ 86–94 mm (n 34, m 89.7), ♀ 84–91 mm (n 34, m 86.8); **T** ♂ 80–91 mm (n 34, m 86), ♀ 78–88 mm (n 34, m 83.2); **T/W** m 95.9. (Length of white on inner web of r5 37–68 mm (n 51), thus nearly complete overlap with *alba*, including shape of white wedge, hence is of very limited value for racial determination.)

M. a. alba L., 1758. 'WHITE WAGTAIL' (Iceland, N & Continental Europe, W Siberia north of following race, Turkey; winters W & S Europe, N & sub-Saharan Africa, Middle East). Treated above in main text (see Ageing & Sexing and Biometrics); recognised by its medium grey mantle and back, its usually comparatively pale grey flanks and by having black on nape and throat-sides separated by white. No black eye-stripe. This race covers quite a large area without any significant morphological variation. Only in easternmost Russia and W Siberia south to Altai and Kirghiz Steppe, and in E Turkey, does it tend to have broader white edges and tips to greater coverts, but also to median coverts, and thus grade into *dukhunensis*. – Especially young ♀♀ can sometimes be surprisingly difficult to separate from corresponding plumage of *yarrellii* (see above).

○ **M. a. dukhunensis** Sykes, 1832 (C & E Caucasus, SE Russia north to lower Volga, SW Siberia, W Central Asia; winters NE Africa, Arabia, SW Asia). Slightly paler grey on back and has broader white edges to wing-coverts and tertials so that in fresh plumage, in particular in ♂♂, these merge into a large whitish wing patch. In Caucasus, at least, these differences seem to be consistent (easily > 75%), whereas other populations are more variable and grade into *alba*. Biometrics same as *alba*. (Syn. *persica*; see also Taxonomic notes.)

M. a. subpersonata Meade-Waldo, 1901. 'MOROCCAN WAGTAIL' (Morocco; resident). Resembles *alba* but is quite distinctive in that black throat merges with black nape-sides, leaving only a white patch below the black necklace either side of neck, and a broad white mask, broken by a black eye-stripe. There is also a thin black moustachial stripe. Sexes similar, ♀ alone has slightly less distinct head pattern but generally differs like in *alba* by greyer hindcrown and nape with less distinct border between black and grey. The *che-lit* call is said to be 'slower' than in European races (Robb & van den Berg 2010). Subtly larger than *alba*. **W** ♂ 88–95.5 mm (n 15, m 91.6), ♀ 83–88.5 mm (n 6, m 86.3); **T** ♂ 85–94 mm (n 15, m 90.0), ♀ 79–89 mm (n 6, m 86.8); **T/W** m 98.9; **B** 15.2–18.0 mm (n 21, m 16.6); **Ts** 20.9–24.7 mm (n 21, m 23.3).

M. a. personata Gould, 1861. 'MASKED WAGTAIL' (N & E Iran, east across Central Asia, north to Altai and W Baikal, east to W China; winters SW & S Asia). Similar to *subpersonata* but even more striking in ad ♂ summer, which

M. a. personata, ad ♂, Israel, Mar: extensive black hood, leaving just a broad white mask including forehead. Also large white panel on wing-coverts and inner remiges, concealing black bases/centres in ad ♂. (A. Ben Dov)

M. a. personata, ad ♂, Iran, Jan: evenly moulted (post-nuptial), and similar to summer plumage, but still has white throat. (E. Winkel)

has entire head, neck, upper mantle and breast uniformly black relieved only by neat white mask and forehead. Lower mantle and back medium grey, sometimes 'stained' black. Tertials very broadly edged white. Ad has extensive white on secondary-coverts, forming large unbroken patch on closed wing. ♀ has less extensively black head, nape and upper mantle, these parts having a varying amount of grey. Only 1stW ♀ has visible dark centres to greater coverts, creating a wing pattern more similar to *alba*. Trifle larger than *alba*. **L** 17.5–19 cm; **W** ♂ 90–98 mm (*n* 18, *m* 94.1), ♀ 85–91 mm (*n* 11, *m* 88.0); **T** ♂ 86–101 mm (*n* 18, *m* 92.7), ♀ 82–95 mm (*n* 11, *m* 88.0); **T/W** *m* 99.0; **B** 14.3–17.5 mm (*n* 24, *m* 16.2); **Ts** 21.9–25.4 mm (*n* 26, *m* 23.1).

TAXONOMIC NOTES Various genetic studies (Ödeen & Alström 2001, Alström & Ödeen 2002, Voelker (2002) have demonstrated that all taxa of *M. alba* are closely related and therefore in our opinion best kept as one species, at least on present knowledge. Differences in DNA are small, indicating fairly recent evolution of taxa. – Ssp. *personata* afforded species status in some Russian literature based on sympatry with *M. a. dukhunensis* (without interbreeding), e.g. in parts of Kazakhstan and Russia, but elsewhere there is extensive intergradation with *dukhunensis* (locally resulting in description of separate taxon, '*persica*'), and separation of *personata* is not followed here. – There have been a few claims in Europe, including in Britain, of extralimital *M. a. leucopsis* Gould, 1838 (Amur, Ussuriland, SE China, Korea), 'WHITE-FACED WAGTAIL' (or 'AMUR WAGTAIL'), but some (or all?) of these could be aberrant Pied Wagtails *M. a. yarrellii*. One should be aware of the existence among wagtails of hybrids between races, or aberrant birds, copying closely the pattern of other races, some of which may be rather unlikely rarities far from their usual ranges. A cautious and restrictive approach when dealing with wagtail races is called for more than with most other groups of birds, as indeed is the official policy adopted with much more likely claimed vagrants of *beema*.

REFERENCES ADDINALL, S. G. (2010) *BB*, 103: 260–267. – ADRIENS, P., BOSMAN, D. & ELST, J. (2010) *DB*, 32: 229–250. – ALSTRÖM, P. & ÖDEEN, A. (2002) In: Alström, P. Ph.D. thesis, Uppsala Univ. – ÖDEEN, A. & ALSTRÖM, P. (2001) In: Ödeen, A. Ph.D. thesis, Uppsala Univ. – ROBB, M. S. & VAN DEN BERG, A. B. (2010) *DB*, 32: 251–253. – ROWLANDS, A. (2010) *BB*, 103: 268–275. – VOELKER, G. (2002) *Condor*, 104: 725–739.

M. a. personata, Oman, Nov: solid white wing panel and mask make this taxon distinctive at any season. The slightly less clearly patterned head, with dark grey crown and nape best fit ♀; greyish-tinged bases and rounded tips to primary-coverts, plus solid white wing panel suggest ad (winter ad, both sexes, develops variable white throat and narrow submoustachial stripe). However, remiges are not fresh, and longest tertial apparently juv, suggesting this bird is a 1stW ♂. (D. Occhiato)

M. a. personata, 1stW ♂, Oman, Oct: moult limits, with mostly white inner greater coverts new and ad-like, contrasting with darker-based juv outer coverts, forming incomplete panel. Being 1stW, amount of black on foreparts identify this bird as young ♂, rather than ad ♀. (D. Occhiato)

M. a. personata, presumed ♀, Kazakhstan, Sep: dark neck-sides, rather pronounced white mask and wing-coverts; grey crown suggests a ♀. Pale yellowish tinge to bill base could suggest young bird, but entire wing appears evenly fresh and ad-like. Thus, both sex and age are best considered unconfirmed. (R. Pop)

M. a. personata, 1stW ♀, Kazakhstan, Oct: the dullest plumage, here with mostly juv greater coverts, and no white wing panel. Still has whitish mask, bordered by dark ear-coverts. (Y. Belousov)

AFRICAN PIED WAGTAIL
Motacilla aguimp Temminck, 1820

Fr. – Bergeronnette pie; Ger. – Witwenstelze
Sp. – Lavandera africana; Swe. – Brokärla

Widespread in sub-Saharan Africa, African Pied Wagtail barely reaches the covered region in extreme S Egypt, in the Nile Valley around Lake Nasser north to Aswan, where it breeds on small islets in this man-made lake. Like several other wagtails it is quite at home in man-altered habitats, and nearness to water seems a basic requirement at least when breeding. Its smart plumage pattern makes it easily spotted.

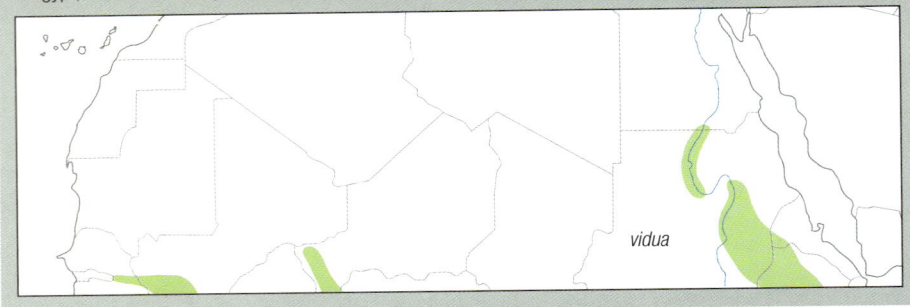

M. a. vidua, ad ♂, Kwazulu-Natal, South Africa, Nov: strikingly patterned black-and-white plumage, with rather linear-patterned face, especially long, broad white supercilium, while throat and neck-side patches appear almost connected; broad black crescent-shaped breast-band, and bold white wing-panel. Aged by evenly ad wing-feathers; sexed as ♂ by black areas being jet- to matt-black. (H. Shirihai)

IDENTIFICATION Overall larger and plumper than White/Pied Wagtail (obvious even in flight) and strikingly patterned, being predominantly black and white, with different head, breast and wing patterns compared to the more familiar species. *Crown and rest of upperparts almost entirely jet-black* with rather linear-patterned face: *long, broad white supercilium* and white throat, prominent but variable *white neck-side patch* which often almost appears to connect the latter two, and *broad black crescent-shaped breast-band* dividing white throat from white underparts, all usually immediately obvious. *Large white panel covering most of folded wing*, and is even more obvious in flight, when large white bases to flight-feathers are visible, also from below. Tail edged broadly white. ♀/young similar in general pattern but black parts of ♂ replaced by blackish-grey or grey-brown. At all ages, bill and legs black. Gait and behaviour as White Wagtail.

VOCALISATIONS Song is either simpler and rather monotonous, or richer and prolonged, latter recalling Atlantic Canary in rhythm and phrasing, containing clear whistled notes that are slightly varied and repeated in short series. Each phrase separated by a variably long pause, e.g. *weet-weet, wip-wip-wip, weet-wee-wee…*, or *tchuu-tchuu… tchuu-tchuu-tchee… tchuu-tchuu-tchick… tchuu-tchuu-tcherewe… tchu-tchu wee… tchu-tchu-wee… tchuu-tchuu-tchu-tche…* (mainly after Alström & Mild 2003 and recordings on XenoCanto website). – Commonest call a sharp, shrill and metallic *zchip* or *tzchik*, thus monosyllabic and clearly different from White/Pied. In alarm the call may be quickly repeated in short series.

SIMILAR SPECIES Velvet-black and white plumage may recall ♂ Pied Wagtail (*M. alba yarrellii*; British race of White Wagtail), but they are very unlikely to come into contact, and African Pied differs from all races of White Wagtail in: (i) black of crown/forehead reaches bill; (ii) bold white supercilium; (iii) all-black lores; (iv) ear-coverts reach hindneck (despite variable white neck-side patch); (v) distinct breast-band enclosing large broad white throat and upper-breast patch; and (vi) broad white blaze on wing obvious both at rest and in flight. ♀/1stW, and even juvenile, might be confused with White Wagtail but easily distinguished by large area of white on flight-feathers and wing-coverts, and by head and breast patterns.

AGEING & SEXING Ageing possible with close inspection of moult pattern and feather wear in wing. Sexes often similar. – Moults. Complete post-nuptial moult of ad apparently in summer (e.g. Jun–Aug in S Egypt). Partial post-juv moult, probably at same time but depends on date of hatching, includes all or most head- and body-feathers and lesser coverts, usually all median coverts, a variable number of greater coverts (usually inner 4–7), 1–3 tertials, and most or all tail-feathers (♀♀ tend to retain more juv feathers than ♂♂). Partial pre-nuptial moult in early spring (probably mostly Feb–Mar) includes head- and body-feathers, lesser coverts, tertials and a variable number of median and greater coverts and tail-feathers. – **SPRING Ad** In breeding plumage, both sexes very similar and often inseparable in the field. Best criterion is on average blacker mantle/back of ad ♂, subtly oilier grey-black or slate colour of ad ♀. Sometimes possible to discern that crown and uppertail-coverts are blacker than mantle/back in ♀ (upperparts more uniformly black in ♂), but some ♀♀ are just as black as ♂♂. Also, ♀ has on average less extensive white in wing-coverts and remiges, and sometimes narrower black breast-band (but much overlap). Both sexes aged by having only ad (fresh or worn) wing- and tail-feathers. **1stS** Like ad, but usually retains some juv secondary-coverts, contrasting with coverts renewed in post-juv and pre-nuptial moults (ad also shows moult contrast at this season). Also, all juv primaries, primary-coverts and sometimes alula and variable number of tail-feathers relatively more worn and bleached, sometimes being contrastingly browner. Renewal in 1stY often less extensive, especially in ♀♀, which usually retain more feathers from non-breeding plumage, including on head and upperparts or some tertials; rump and uppertail-coverts often slightly greyer/duller than ad. Wings also show less white on primary bases and, on closed wing, secondary panel smaller and less well defined than in ad, or already worn off completely. – **AUTUMN** Winter plumage. **Ad** Evenly

M. a. vidua, 1stS ♀, Egypt, May: aged by juv outer wing-feathers, otherwise this bird has replaced most body-feathers and wing-coverts during pre-nuptial moult. Sexing can be problematic, but greyish crown and mantle indicate ♀. (D. Occhiato)

M. a. vidua, 1stW ♂, Botswana, Nov: some young ♂♂ have relatively limited or only partial first pre-nuptial moult. This bird has moulted head and breast, which have dark areas jet-black. Note mainly brownish upperparts and greyish rump. Not safely sexed unless aged first, in this case as 1stS (exposed remiges juv and very worn, contrasting with new tertials). (H. Shirihai)

PIPITS AND WAGTAILS

M. a. vidua, 1stY ♂, Egypt, Jan: ongoing pre-nuptial moult, with new blackish feathers growing on mantle, and growing tertials, otherwise rest of wing-feathers juv. (D. Monticelli)

M. a. vidua, 1stW ♂, Kenya, Jan: unlike ad, 1stW has greyish crown, nape and rest of upperparts. Identified to species by complex white and dark patterns, which generally mirror those of ad. Sexing not easy, but relatively large white patch on basal primaries, broader breast-band, and some black feathers on forecrown indicate 1stW ♂. (H. van Diek)

M. a. vidua, 1stY ♀, Egypt, Sep: mid-grey crown and mantle (instead of black to greyish-black) automatically age this bird as 1stY, while very tiny white patch on basal primaries confirm that it is ♀, supported by narrower and almost broken neck-side connection to breast-band. (V. & S. Ashby)

fresh with upperparts in ♂ now greyer, while upperparts of ♀ mostly dull brownish olive-grey. **1stW** Close to ad ♀ in appearance but may show clear moult limits with retained juv outer secondary-coverts and tertials, but some renew these completely and are closer to ad. Retained tail-feathers (some), primaries and primary-coverts more worn and weaker textured than ad. More dark feather-bases in wing exposed, reducing extent of white panel, and juv remiges have considerably less white (e.g. length of dark pattern on outer web of s3 32–35 mm in 1stY, but 25–30 mm in ad; Alström & Mild 2003). Some retain juv feathers on crown, mantle, scapulars and lesser coverts, which are paler, browner and more worn, contrasting with dark greyish fresh feathers. A few ♀♀ also easily aged by lack of black on head and upperparts. Sexes often inseparable, but ♀ may have more greyish feathers. **Juv** Dark plumage mostly dark grey-brown, with head pattern less distinct, supercilium narrower, breast-band mottled and wing markings less bold.

BIOMETRICS (*vidua*) **L** 18.5–20 cm; **W** ♂ 88–101 mm (n 25, m 94.3), ♀ 85–97 mm (n 23, m 90.7); **T** ♂ 87–103 mm (n 25, m 93.3), ♀ 83–95 mm (n 23, m 90.3); **T/W** m 99.3; **B** 16.0–21.1 mm (n 47, m 18.1); **BD** 3.5–4.9 mm (n 40, m 4.2); **Ts** 22.0–27.0 mm (n 47, m 24.8). **Wing formula: p1** minute; **p2** < wt 0.5–3 mm; **pp3–4** about equal and longest (sometimes pp2–5); **p5** < wt 0–2.5 mm; **p6** < wt 8–12 mm; **p7** < wt 13–18 mm; **p10** < wt 20–29 mm; **s1** < wt 22–30 mm. Emarg. pp3–5.

GEOGRAPHICAL VARIATION & RANGE Rather well-marked difference between two existing subspecies, one in South Africa being extralimital and not treated here.

M. a. vidua Sundevall, 1850 (S Egypt and south throughout Afrotropics except extreme south; resident or makes shorter movements in non-breeding season). Described above and the only race in the covered region.

M. a. vidua, juv, Egypt, Jul: essentially dark grey-brown above, with obscure head pattern, comprising only faint supercilium (mostly behind eye), and washed-out dusky breast-band; white wing panel less bold and more restricted, but still obvious. (R. Armada)

WHITE-EARED BULBUL
Pycnonotus leucotis (Gould, 1836)

Fr. – Bulbul à oreillons blancs; Ger. – Weißohrbülbül
Sp. – Bulbul orejiblanco; Swe. – Vitkindad bulbyl

The White-eared Bulbul is largely a Middle Eastern bird. It is resident in lowlands, including gardens and cultivation, from S & E Iraq and parts of Arabia, where it is currently rapidly expanding, east to W India. There are records from Syria and Jordan, and recently from SE Turkey, suggesting either further expansion or that it was previously overlooked there, and escapes have been noted in Israel and Jordan.

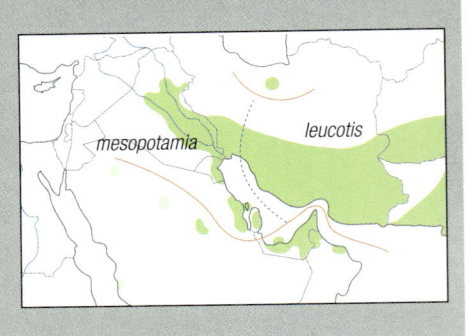

P. l. mesopotamia, NE Jordan, Apr: square-headed with bold white ear-coverts patch and tip to tail, and bright orange-yellow undertail-coverts. Yellow eye-ring apparently a feature of this race. (H. Shirihai)

P. l. mesopotamia, Syria, May: moult limits in greater coverts and tail point to 1stY, but ageing generally impractical as vast majority of both age classes have moulted completely by late autumn. (K. De Rouck)

P. l. mesopotamia / leucotis, Oman, Nov: large white patch on head-sides and rather warmer underparts are features of *mesopotamia*, but dark grey eye-ring is better fit for *leucotis* (unclear if eye-ring coloration is geographically governed, or if it merely varies with age/sex and season, or all of these). (D. Occhiato)

IDENTIFICATION A bulky-bodied bulbul with head and throat black, a *bold white, fan-shaped patch behind and below the eye*, covering also ear-coverts and reaching onto neck-sides (recalling Great Tit pattern), dull grey-brown upperparts, with darker flight-feathers. It has a *blackish-brown tail with white terminal spots* which are rather obvious in flight. Crown-feathers at least in easternmost populations slightly elongated and can be raised in short, stubby crest. Underparts rather nondescript, the pale brownish or dull cream-coloured breast becoming pale grey-buff to grey-white on belly, with *rich yellow vent* (more orange-yellow than lemon) being the only noticeable feature. Eye dark red-brown, enclosed by narrow dark (grey or brown-tinged, rarely paler and yellowish) eye-ring. Bill and legs black. Typically feeds in bushes and low trees, where easily followed due to its typically sluggish behaviour. Closely associated with human settlements.

VOCALISATIONS Song typical of the genus, being quite similar to White-spectacled Bulbul: a full, melodious jingle three or four syllables long, frequently repeated, and suggesting a speeded-up Golden Oriole, *oo-toodle-oo*, rendered as 'take me with you' or 'tea for two'; may include bubbling, chuckling or warbling sounds on a rising and falling scale, also a protracted series of *chip-chop* phrases. – Utters a sharp *pit* in alarm, frequently repeated, *pit-pit pit-pit* or *pit-a pit-a*.

SIMILAR SPECIES Most closely resembles *White-spectacled Bulbul*, but both are unlikely to be confused, given the present species' striking white cheeks, contrasting with the black head and throat, and white tail-tip, although both species have yellow undertail-coverts. – Overlaps in SE Asia with *Red-whiskered* (*P. jocosus*; extralimital, not treated) and *Red-vented Bulbuls*, both of which have much more prominent crests, red undertail-coverts and quite different head patterns, although the former also has white ear-coverts, but has a red spot behind the eye and white continues onto the throat. Red-vented has an all-dark head, this coloration also extending across much of the underparts.

AGEING & SEXING Ages (following post-juv moult) are alike, and there is very little seasonal variation. Sexes alike in plumage, but size is helpful (see Biometrics). – Moults. Both ad and juv moult completely, mostly Jul–Oct, but timing in juv more variable, depending on hatching date. – SPRING **Ad** Overall paler (head becomes black-brown) due to extensive wear in late spring and summer. **1stS** Inseparable from ad. – AUTUMN **Ad** Plumage rather warmer and darker when very fresh. **1stW** Inseparable when last juv feathers shed, but some retain juv remiges and coverts until Nov (probably even later), and if so reliably aged (also worn juv feathers never as strongly bleached as unmoulted ad feathers). **Juv** Soft, fluffy body-feathering and head browner than ad, with white cheeks less clear-cut, and duller vent. – Note: whether presence of yellow eye-ring is merely geographically related (see below), or also varies with age/sex and seasonally, or perhaps according to all these factors, is as yet unknown.

BIOMETRICS (*mesopotamia*) **L** 19–20 cm; **W** ♂ 85–94.5 mm (n 12, m 90.4), ♀ 84–92 mm (n 12, m 87.8); **T** ♂ 80–90 mm (n 12, m 85.4), ♀ 76–89.5 mm (n 12, m 82.9); **T/W** m 94.4; **B** 16.1–18.9 mm (n 24, m 17.6); **B**(f) 12.7–13.5 mm (n 8, m 13.1); **BD** 5.7–7.1 mm (n 24, m 6.3); **Ts** 19.3–23.7 mm (n 24, m 21.8). **Wing formula: p1** > pc 14–19.5 mm; **p2** < wt 12–17 mm, =9/10 or 10 (common) or <10 (less frequent); **p3** < wt 2–4 mm; **pp4–6** about equal and longest (rarely p6 < wt 0.5–1.5 mm); **p7** < wt 2–4 mm; **p8** < wt 5–9 mm; **p10** < wt 11–14 mm; **s1** < wt 13–15 mm. Emarg. pp3–7 (and usually faintly on p8).

GEOGRAPHICAL VARIATION & RANGE Only slight variation throughout its range. All populations are sedentary. Man-assisted spread outside feral range (brown line) indicated on the map.

P. l. mesopotamia Ticehurst, 1918 (C Iraq, Euphrates & Tigris Valleys, NE Arabia, SW Iran; local Jordan). Has restricted black on head, leaving very large white patch, and lacks crest; eye-ring at least in some birds yellow-orange (but see note above), more drab grey on upperparts and somewhat more saturated below. (Syn. *dactylus*.)

P. l. leucotis (Gould, 1836) (S Iran to NW India). Similar to *mesopotamia* but has hint of a crest on hindcrown, a prominent narrow grey or black eye-ring and tends to have more extensive black on head corresponding with on average smaller white cheeks, a less deep-based bill, a whiter belly, and shorter wings and tail. **W** 84–93 mm (n 15, m 88.2); **T** 78–88 mm (n 15, m 82.3); **T/W** m 93.3; **B** 15.3–18.1 mm (n 15, m 16.8); **B**(f) 12.3–14.7 mm (n 8, m 13.6); **BD** 5.8–6.7 mm (n 15, m 6.1); **BW** 5.9–6.8 mm (n 15, m 6.4); **Ts** 19.6–22.7 mm (n 14, m 21.1). – **W** ♂ 81–91 mm, ♀ 80–88 mm; **T** ♂ 66–84 mm, ♀ 73–80 mm; India (Ali & Ripley 1987). – Intergrades with *mesopotamia* in SW Iran and possibly to some degree in United Arab Emirates and Oman.

TAXONOMIC NOTE *P. leucotis* is increasingly frequently considered, as here, a separate species from *P. leucogenys* (Himalayan Bulbul; not treated) of the Himalayan foothills, although many authorities still treat them conspecifically (as they hybridise over much of N Pakistan) under the name 'White-cheeked Bulbul'. Forms a superspecies with Red-vented Bulbul, with which it occasionally hybridises, and White-spectacled Bulbul, Common Bulbul and several other S Asian species.

P. l. leucotis, Kuwait, Jan: dark eye-ring, less extensive white cheeks and paler underparts, suggesting *leucotis*. Yellow undertail-coverts shared with White-spectacled Bulbul, but easily separated by striking white cheeks. (D. Monticelli)

WHITE-SPECTACLED BULBUL
Pycnonotus xanthopygos (Hemprich & Ehrenberg, 1833)

Alternative name: Yellow-vented Bulbul

Fr. – Bulbul d'Arabie; Ger. – Gelbsteißbülbül
Sp. – Bulbul árabe; Swe. – Levantbulbyl

Most European birders first encounter this species in Israel, where it is often the first herald of dawn. Widespread and generally a common resident wherever there are bushes and low trees, often in gardens, orchards and around habitation, from S Turkey, over much of the Levant south and east through coastal and lowland Arabia. Recently recorded in Iraq, where its status is unclear.

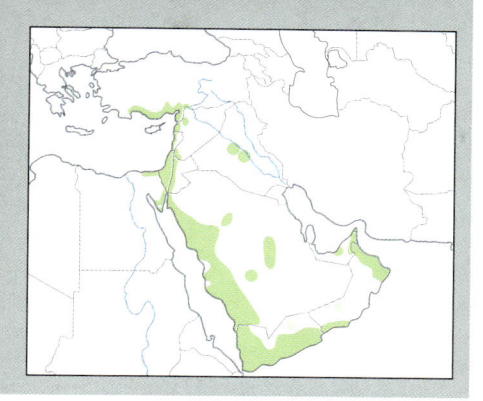

Israel, Mar: in the Levant, the species is one of the best-adapted passerines to human cultivations and habitations, most numerous in palm groves and gardens. Generally unmistakable given its shape, black hood, whitish-grey spectacles, pale-tipped black tail, and, the most obvious field mark, the rich yellow vent. (H. Shirihai)

IDENTIFICATION Long-tailed and bulky-bodied bulbul with well-demarcated *black hood*, a conspicuous narrow *grey-white eye-ring*, dull brown-grey upperparts but clearly *blackish tail* (often with very small, ill-defined white terminal spots), and generally rather nondescript dusky underparts, the *rich yellow vent* being the only obvious feature. Flight direct, though slightly wavering, on rounded wings with energetic, flapping wingbeats. Like other bulbuls, commonly forms small gatherings which can be quite noisy.

VOCALISATIONS Song a loud, fluty bubbling with a jerky rhythm, usually comprising short phrases of 2–8 uneven syllables, e.g. *chick chillewee, chuwü… chook-whee-too chellewee…*, or *twur-tu-twee-teeroo* (structure somewhat like Golden Oriole), and repetitive, fast series, *buli-buli-buli…* (hence the name bulbul). – Calls include a harsh burring *wreck* (alarm) and a grating, repeated *chaar-chaar-chaar-…* (irritation), a nuthatch-like *pwitch*, and often, a sharper, persistent series of harsh notes, *kutchi-kutchi-kutchi…* (a characteristic element of the 'dawn chorus' in its range). Reported to occasionally mimic other birds, but we have never noticed this.

SIMILAR SPECIES Unmistakable within its range if diagnostic yellow vent seen, though recently discovered in Iraq, where *White-eared Bulbul* shares this feature, but they are easily separated by other traits. – If yellow vent not seen, confusion possible with *Common Bulbul* which has an overall paler and browner head and lacks the striking whitish eye-ring. However, both are sedentary and have separate breeding ranges (though they come close near the Suez Canal). Juveniles would pose a tough identification challenge, and their separation would probably necessitate use of structural features (also applicable to adults), namely the slightly longer narrower bill (with more sharply pointed tip), slightly longer crown-feathers forming a slight crest when erect (admittedly only in adults), and marginally more square-cut tail of White-spectacled.

AGEING & SEXING Ages alike after post-juv moult in summer. Sexes alike in plumage (but see Biometrics), and very little seasonal variation. – Moults. Both ad and juv moult completely, mainly Jun–Sep, but timing, especially in juv, variable depending on hatching date. – **SPRING Ad** Rather duller and less richly saturated compared to fresh autumn plumage. When hood is heavily worn it becomes slightly blacker (with lower throat/breast browner), thus more sharply demarcated from brownish-grey upperparts and cream-white underparts. **1stS** Inseparable from ad; no moult limits found in examined birds (both age groups) from Sep–Apr. – **AUTUMN Ad** When very fresh appears rather warmer and darker. **1stW** Once moult completed, inseparable from ad, but some may retain juv remiges and coverts until late autumn or even winter, and these can be reliably aged. Ad can still be moulting in early autumn (latest ad seen in moult on 30 Oct; NHM), but contrast between older and new feathers far stronger. At least in early stages, iris more greyish-tinged (dark or warm brown in ad). **Juv** Soft, fluffy body-feathering and overall slightly paler than ad, with duller head and loosely-feathered vent.

Israel, Mar: both age classes undergo complete post-breeding moult, but completion especially in juv varies, depending on hatching date, with some evidence that ad moult can also be delayed. Slightly greyer/fresher inner secondaries and greater coverts suggests moult is completed in two waves, rather than two generations, but impossible to be sure, especially in spring. (Y. Perlman)

Oman, Nov: unlikely to be mistaken in range, given square-shaped head, short wings, relatively long tail, bill shape, black hood and yellow undertail-coverts. Though treated as monotypic, birds in S Arabia, at least in Oman, appear overall warmer/browner, more saturated below (less greyish) with less pronounced whitish spectacles. In autumn ageing is impossible, and birds that moult early can be somewhat worn by late autumn. (D. Occhiato)

Oman, Nov: this bird has either unmoulted (very worn and bleached) or still-growing remiges and rectrices, and moult limits among primary-coverts, suggesting a youngster. Furthermore, the exposed left tail-feather has a very obscure pale brown-buff (more juv-like) tip. (M. Schäf)

BIOMETRICS L 19–20.5 cm; **W** ♂ 96.5–102 mm (n 10, m 98.9), ♀ 88–97 mm (n 12, m 92.3); **T** ♂ 87–97 mm (n 10, m 92.3), ♀ 84–93 mm (n 12, m 88.7); **T/W** m 94.8; **B** 19.5–21.6 mm (n 15, m 20.5); **B**(f) 14.7–17.3 mm (n 21, m 16.3); **BD** 6.2–7.2 mm (n 18, m 6.7); **BW** 6.0–7.1 mm (n 20, m 6.6); **Ts** 20.3–23.5 mm (n 22, m 22.0). **Wing formula: p1** > pc 15–22 mm, < p2 22–27 mm; **p2** < wt 12–19 mm, = ss, or < ss; **p3** < wt 3–4.5 mm; **pp4–6** about equal and longest (though sometimes either or both of p4 and p6 0.5–2 mm <); **p7** < wt 1–2.5 mm; **p8** < wt 4–5.5 mm; **p10** < wt 9–15 mm; **s1** < wt 10–15.5 mm. Emarg. pp2–7 (sometimes faintly also on p8).

GEOGRAPHICAL VARIATION & RANGE Monotypic. – S Turkey, Levant, E Sinai, Red Sea coast in W & S Arabia, Oman, local S Iran; possibly local Iraq; resident.

TAXONOMIC NOTE Here treated as monotypic, though birds in Arabia tend to be considerably paler in spring than those in the north. Formerly considered conspecific with Common Bulbul, and some Afrotropical races of latter occasionally still treated within *xanthopygos*.

COMMON BULBUL
Pycnonotus barbatus (Desfontaines, 1789)

Fr. – Bulbul des jardins; Ger. – Graubülbül
Sp. – Bulbul naranjero; Swe. – Trädgårdsbulbyl

A noisy and gregarious grey and off-white bird, often showing well and frequently. One of the most widespread and taxonomically challenging African birds; some races closely recall White-spectacled Bulbul (but these fortunately do not occur in the covered region). In the W Palearctic the species is represented by two races, one breeding in NW Africa from Morocco to N Tunisia, the other in the Nile Valley.

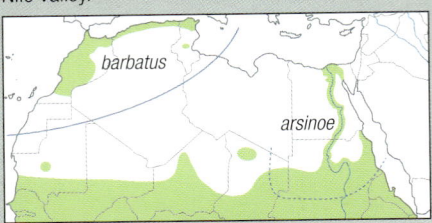

P. b. arsinoe, Egypt, Mar: a smaller subspecies, with deeper and more extensive black hood. (J. S. Hansen)

BIOMETRICS (*barbatus*) **L** 20–21 cm; **W** ♂ 100–107 mm (n 17, m 103.4), ♀ 93–100 mm (n 18, m 96.8); **T** ♂ 87–95 mm (n 16, m 90.2), ♀ 83–89 mm (n 14, m 85.6); **T/W** m 92.0; **B** 18.5–21.9 mm (n 30, m 19.9); **B**(f) 14.5–17.5 mm (n 14, m 16.0); **Ts** 20.9–23.5 mm (n 14, m 22.7). **Wing formula: p1** > pc 18–22 mm; **p2** < wt 14–20 mm, < ss (rarely = ss or 10/ss); **p3** < wt 2–5 mm; **p4** < wt 0–1.5 mm; **p5** longest (sometimes pp4–6 equal); **p6** < wt 0–1.5 mm; **p7** < wt 1–4 mm; **p8** < wt 3–8.5 mm; **p10** < wt 10–14 mm; **s1** < wt 14–17 mm. Emarg. pp3–7 (sometimes faintly on p8, too).

GEOGRAPHICAL VARIATION & RANGE Slight variation in W Palearctic; two races recognised. Several extralimital races described in Africa, not treated here (but see Taxonomic note). All populations mainly sedentary.
P. b. barbatus (Desfontaines, 1789) (Morocco, N Algeria and N Tunisia). Described above.
P. b. arsinoe (M. H. C. Lichtenstein, 1823) (Egypt in Nile Valley, south to C Sudan and E Chad). Distinctly smaller, face more extensively and deeper black, with a paler (greyer) rump, small silvery-white patch behind ear-coverts (variable, but usually better developed than in *barbatus*) and undertail-coverts even more rarely have yellow at sides; also, breast darker or more evenly washed (less blotchy). **W** ♂ 91–97 mm (n 21, m 93.6), ♀ 86–91 mm (n 12, m 88.7); **T** ♂ 82–89 mm (n 11, m 85.5), ♀ 75–83 mm (n 8, m 79.2). (BWP.)

TAXONOMIC NOTE Common Bulbul comprises four groups, sometimes treated at species level, though some forms hybridise locally. Only one group occurs within the treated region: (i) the *barbatus* group in N, W & C Africa (white undertail-coverts and uniform mantle and breast); (ii) *somaliensis* in Ethiopia & N Somalia (white undertail-coverts, white spot behind ear-coverts, and scaly mantle and breast); (iii) *dodsoni* in E Somalia, SE Ethiopia & E Kenya (scaly mantle and breast, white ear-coverts spot and yellow undertail-coverts); and (iv) *tricolor* of C & S Afrotropics (similar to *barbatus* group, but has yellow undertail-coverts).

P. b. barbatus, Morocco, Feb: the least-colourful bulbul in the covered region, with diffuse blackish-brown hood, which contrasts with uniform grey-brown upperparts, and cream-washed grey-brown underparts. Evenly-feathered wing (following moult) prevents ageing. (D. Occhiato)

IDENTIFICATION Medium-sized rather nondescript bulbul with a high crown, an (ill-defined) *fuscous-brown or blackish-brown hood* and even darker lores and face, with uniform grey-brown mantle to tail. Underparts cream or dirty white, washed brown-grey with some slightly *darker mottling on breast*, and mostly *whitish vent and undertail-coverts*. Eye dark brown, and sturdy bill and legs dull black. Behaviour and noisy habits closely recall White-spectacled Bulbul.

VOCALISATIONS Song much like White-spectacled Bulbul, e.g. a slow series of somewhat abrupt, clipped notes, *chick chillewi, chuwü…*, some syllables deep and fluty in Blackbird fashion; also a rapid *pil-pil, pele-pele-li-li-li* noted. Does not seem to contain any mimicry of other birds. – Calls, too, recall White-spectacled Bulbul, e.g. *cheep* or *tchup* and a low *cheedle cheedle cheedlelit*. Gives persistent *tschirr* when nervous or alarmed.

SIMILAR SPECIES Most closely resembles *White-spectacled Bulbul*, but immediately separated by lack of yellow vent, whitish eye-ring or contrasting blackish head. Common Bulbul is a more uniformly brownish-grey and off-white bird lacking any striking features.

AGEING & SEXING Ages and sexes alike (but see Biometrics), and almost no seasonal variation. – Moults. Few moult data from the treated region, but both ad and juv generally moult completely after breeding. Timing, especially in juv, rather variable (in W Palearctic mostly Jul–Nov). – **SPRING Ad** With extensive wear, underparts become progressively whiter and upperparts and hood browner. **1stS** Inseparable from ad; reports of 1stY in NW Africa with apparently juv remiges and primary-coverts unconfirmed. – **AUTUMN Ad** When very fresh whole plumage warmer tinged, including greyish-cream underparts, which can show tiny patches of yellow on vent-sides. Iris warm or dark brown. **1stW** Largely as ad, but at least some late-hatched birds delay replacing (or retain?) a few juv wing- and tail-feathers, and can then be aged (ad can still be moulting in early autumn but contrast between older and new feathers then far stronger). At least in early stages iris greyer (variable but usually discernible). **Juv** Soft, fluffy body-feathering but otherwise as ad and not always separable in the field.

P. b. barbatus, Morocco, Feb: in some birds, hood more fuscous-brown, face appears darker and there can be some mottling on breast. Ageing impossible. (D. Occhiato)

P. b. barbatus, ad, Spain, Sep: among the very first self-established settlers in Europe (and the continent's first breeding species of the genus), at Tarifa, Strait of Gibraltar. A breeder (seen to feed two chicks) in advanced post-nuptial moult. (H. Shirihai)

P. b. arsinoe, Egypt, Apr: darker and more evenly brown-washed breast is a feature of this race. (D. Occhiato)

RED-VENTED BULBUL
Pycnonotus cafer (L., 1766)

Fr. – Bulbul à ventre rouge; Ger. – Rußbülbül
Sp. – Bulbul cafre; Swe. – Rödgumpad bulbyl

A S Asian bulbul, which is resident in lowlands from Pakistan east to SW China and south to Sri Lanka. Successfully introduced in parts of E Arabia, and on several SW Pacific islands (formerly also in parts of Australia and New Zealand). It is obviously an adaptable bird, and it is now rather common, e.g. in the United Arab Emirates.

P. c. haemorrhousus, presumed ad, Sri Lanka, Dec: an attractive bulbul, with the most prominent features being the blackish hood, contrasting with broadly whitish-scalloped nape, mantle and lesser wing-coverts, whitish wing-bars, and diagnostic deep red undertail-coverts (here only just visible). Strongly contrasting and heavily worn still unmoulted feathers may suggest ad during post-nuptial moult. (H. Shirihai)

IDENTIFICATION A medium-sized bulbul with a reasonably long, boldly white-tipped tail and sooty-brown or blackish plumage. *Head to upper breast particularly dark*, hindcrown having somewhat elongated feathers forming a *blunt crest*, while *mantle and back are greyer, broadly scalloped dusky white*, and wings browner with extensive grey-white fringes and tips forming *two wing-bars*. Largely sooty-brown tail and *off-white rump and uppertail-coverts*. Rest of underparts dusky grey with coal-black chest finely scalloped off-white, and diagnostic *carmine-red undertail-coverts*. Noisy, like other bulbuls, and usually in pairs, or small parties outside breeding season, when it may form larger concentrations in fruit-bearing bushes and low trees.

VOCALISATIONS Song rich and loud, much as in White-eared Bulbul, oft-repeated and usually comprises about three short phrases with emphasis on last or penultimate note, e.g. *prit-prit-preeht* or *pe, tche-pi-pepte, pepte-pe* or *wit-tut-weee-ho*, some variants aptly rendered 'be-care-ful' or 'be quick-quick'; also simple, gruff *pri-pri-ruu*. – In alarm typically gives a repeated sharp grating *peep-peep-peep*; also a simple *pri* and *pri-pru*.

SIMILAR SPECIES Difficult to misidentify given a good view. Separation from *White-eared Bulbul* covered under that species. Main confusion risk is Red-whiskered Bulbul (*P. jocosus*, extralimital and not treated; has escaped in several Gulf States, but not well established), which has red spot behind eye, white throat and ear-coverts with narrow black lower border to cheeks, slightly graduated tail, tall-crested appearance, and no scalloping on underparts, which are overall much paler and whiter. Uniform brown upperparts lack wing-bars or scalloping.

AGEING & SEXING Ages alike in plumage after post-juv moult. Sexes alike (but see Biometrics). Very little seasonal variation. – Moults. Both post-nuptial and post-juv moults complete. Scant moult data (32 specimens of *intermedius* with known collection date examined, NHM, AMNH): 24 (ad/1stY) from 31 Oct–2 Aug showed no moult limits, and birds evenly fresh or moderately so until Feb, moderately to heavily worn from Mar, and usually heavily worn 18 May– 2 Aug; two still unmoulted juv 6 Jun and 14 Sep; five (19 Oct–7 Nov) finishing moult had growing outer primaries and inner/central secondaries, one definitely ad, as it had one very worn secondary. One bird in Himalaya on 31 Dec in final stage of primary moult. – SPRING **Ad** Especially when heavily worn, overall paler than in autumn with strongly reduced scaly pattern. **1stS** Inseparable from ad. – AUTUMN **Ad** When very fresh, appears darker, with characteristic pale fringes to upperparts and underparts more prominent. **1stW** At least *intermedius* seems to undergo complete post-juv moult, thus inseparable from ad. **Juv** Soft, fluffy feathering, paler than ad, with duller head and vent and virtually no scaly marks.

BIOMETRICS (*intermedius*) **L** 20–22 cm; **W** ♂ 101–108 mm (n 15, m 104.9), ♀ 93–104 mm (n 9, m 97.5); **T** ♂ 91–102 mm (n 15, m 97.8), ♀ 89.5–100 mm (n 9, m 92.7); **T/W** m 94.3; **B** 18.0–22.3 mm (n 23, m 20.5); **B**(f) 15.4–17.5 mm (n 23, m 16.4); **BD** 6.3–7.7 mm (n 25, m 6.8). **Wing formula: p1** > pc 16–27 mm; **p2** < wt 14–22 mm, < 10 (rarely =10); **p3** < wt 3–7 mm; **pp4–6** about equal and longest (rarely pp4–5 or p5 longest); **p7** < wt 2–5 mm; **p8** < wt 4.5–11 mm; **p10** < wt 12–22 mm; **s1** < wt 12.5–21 mm. Emarg. pp3–7 (and usually faint also on p8).

GEOGRAPHICAL VARIATION & RANGE Up to nine races recognised, but not known which of these found in the covered region, apparently solely after introductions. Local in Kuwait, United Arab Emirates, N Oman and elsewhere in the Gulf region; rarely recorded from S Oman. Three races briefly described below, one or more of which could be involved in most of the introductions. All are resident.

P. c. cafer (L., 1766) (C & S India). Described above. Black or sooty-brown mainly restricted to head and throat.

P. c. haemorrhousus (J. F. Gmelin, 1789) (Sri Lanka). Very similar to *cafer*, only averaging slightly darker and browner overall, in particular below.

P. c. intermedius Blyth, 1846 (Himalayas, from Pakistan to NW Uttar Pradesh). Biometrics given above. Differs from preceding by having more extensive black, this not being largely confined to head and throat but extending down both on breast and mantle.

TAXONOMIC NOTE Hybridisation with White-eared and other bulbuls frequent (Roberts 1992, Grimmett *et al.* 1998).

Presumed ssp. *cafer* / *haemorrhousus*, Bahrain, Dec: note bold white tail-tip, typical of the species. As can be seen here, the vast majority of birds recorded in the Middle East best match plumages found in races *cafer* and *haemorrhousus*, of S India and and Sri Lanka respectively. Ageing impossible. (G. Lobley)

P. c. haemorrhousus, Sri Lanka, Dec: there is considerable individual variation, unrelated to age, sex or geographical variation, and especially in the extent of the whitish scalloping around the nape, mantle and breast, and in the degree of black extension on throat to upper breast. Note the short crest, often more evident when the bird is excited. Ageing impossible. (H. Shirihai)

(BOHEMIAN) WAXWING
Bombycilla garrulus (L., 1758)

Fr. – Jaseur boréal; Ger. – Seidenschwanz
Sp. – Ampelis europeo; Swe. – Sidensvans

During large irruptions, this exotic-looking silky-feathered passerine is a major attraction for birders in countries bordering the North Sea, as the Bohemian Waxwing seems to epitomise the far north. It is widespread across the N Holarctic, breeding in lichen-rich, mature conifer forests in damp, mossy terrain from Fenno-Scandia to Kamchatka, in the east south to Amurland, and occurring also in NW North America. It irregularly ranges south in the non-breeding season as far as continental Europe, Transcaucasia, Central Asia and the southern USA, with some evidence of roughly a ten-year cycle to such movements.

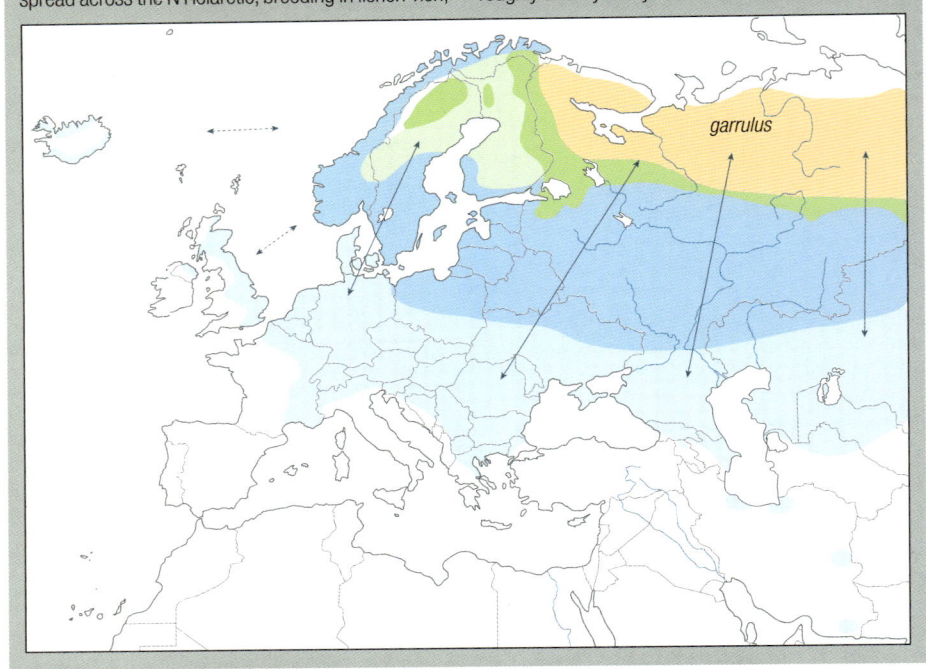

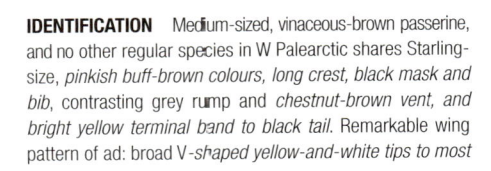

Ad ♀, Finland, Oct: aged by yellow outer and white inner webs to primary tips, and sexed by short waxy red appendages (some missing) and rather narrow terminal tail-band; rather narrow white inner webs to primary tips and shorter crest are other characteristics of ♀, while the bib definition is difficult to judge in this posture and angle. (M. Varesvuo)

IDENTIFICATION Medium-sized, vinaceous-brown passerine, and no other regular species in W Palearctic shares Starling-size, *pinkish buff-brown colours, long crest, black mask and bib*, contrasting grey rump and *chestnut-brown vent, and bright yellow terminal band to black tail*. Remarkable wing pattern of ad: broad V-shaped yellow-and-white tips to most primaries, and *primary-coverts and secondaries with obvious white bars at their tips and variable waxy red appendages* (requires closer views to see). Young lack the V-shaped yellow-and-white primary-tips, have only a straight yellowish or off-white mark on each primary tip. Young, especially young ♀♀, are slightly duller with much-reduced secondary appendages. Flight silhouette and flight mode recall that of Starling, employs fast direct flight with long undulations on triangular wings and short tail. Social behaviour also recalls Starling, as flocks are dense, and progress is fast and restless. On perched birds, overall body shape can change markedly from sleek to very plump depending on temperature, 'mood' and activity. In winter attracted to berry-laden shrubs and trees (mature, orange-red rowanberries favourite in N Europe), even in gardens (in breeding season more insectivorous, sometimes making aerial flycatching sorties). Rather shy and wary, perching atop taller tree in neighbourhood to check out situation before coming down to lower bushes. Essentially arboreal, rarely venturing to the ground to feed. Usually in small parties in winter, but sometimes in larger aggregations involving several hundreds.

VOCALISATIONS Song is mainly a variant of the high-pitched trilled call, being slow and halting, and mixed with hard, raucous sounds, *sirrr sirrrrr chark-chark chi-chark sirrr sirrrrr...* etc. – Commonest call a pleasant, weak but quite high-pitched sibilant *sirrrrr*, like a small silver bell, audible close to and given when perched or in flight, and often in

Ad ♂, Netherlands, Nov: bright yellow outer and broad white inner webs on primary tips, and six long waxy red appendages, rather broad terminal tail-band, while well-defined throat patch and elongated crest (exceptionally long in this case) are characteristic of ad ♂. (A. Ouwerkerk)

Ad ♂, Finland, Nov: aged by white on inner webs of primary tips; very broad terminal tail-band and fairly distinct lower border of bib suggest ♂. However, this is a less typical ♂, since the six waxy red appendages appear intermediate in length and the white inner webs to the primary tips are narrower and taper midway, almost suggesting a ♀. (M. Varesvuo)

WAXWINGS AND HYPOCOLIUS

1stY ♂, Switzerland, Jan: juv remiges with yellow/white primary tips to outer webs (but not on inner webs); sexed as ♂ by the four or five rather long waxy red appendages and rather extensive yellow tail-tip, while lower throat border is rather sharply defined. (H. Nussbaumer)

1stY ♂, Finland, Jan: juv remiges with yellow/white primary tips to outer webs only; sexed as ♂ by rather extensive and bright yellow tail-tip. The waxy red appendages are unusually short and few on this bird. (M. Varesvuo)

pealing chorus from flock prior to flight; wingbeats make characteristic low rattling sound.

SIMILAR SPECIES Generally unmistakable. *Cedar Waxwing* (*B. cedrorum*; extralimital, treated among vagrants) of North America, a potential escape and very rare vagrant to the covered region, is very similar, but marginally smaller with a white vent and pale yellow belly, and the flight-feathers lack 'ornaments' other than the waxy red appendages on the secondaries. – At some distance, especially in flight, might be confused with *Starling* given structural similarities, flight action and silhouette, but has a shorter bill, fuller head and subtly bulkier body, and less rapid wingbeats in the generally faster and more graceful flight (trilling flight-call distinctive, too).

AGEING & SEXING (*garrulus*) Ageing possible during 1stY. Sexes rather similar but separable (with very few exceptions) in the field when seen close or studied on photographs. No obvious seasonal variation. – Moults. Post-nuptial moult complete (mainly Aug–Nov), though some may suspend (or arrest?) feather renewal, even including some remiges, resuming in winter quarters. Post-juv moult partial, mostly confined to head- and body-feathers and possibly some secondary-coverts, but if so probably only lesser coverts. Probably no separate pre-nuptial moult in ad or 1stY (any moult being rather complementary to post-nuptial). – **ALL YEAR Ad** Aged by lack of moult limits, and any unmoulted/suspended feathers are distinctly more worn; tips of primaries have bright yellow fringes to outer webs and white on inner webs forming characteristic V (but many ad ♀♀ have part of V on inner web rather narrow or even broken, or restricted to pp5–8, but pale margin of outer web never terminates abruptly at shaft, as in juv primaries). **Ad ♂** Throat patch generally better defined, especially at lower border (but there can be a limited zone of greyish outside the black, or the lower border is slightly uneven or patchy). Tends to have brighter yellow fringes to primaries, and broader white tips to inner webs of primaries. More and longer waxy red appendages on secondary tips (6–8 in number; longest measuring 5–9.5 mm). Broader yellow terminal tail-band (on r1 5–9 mm; on r5 6.5–11 mm; both measured parallel with shaft and irrespective of on which web). **Ad ♀** Less well defined and sootier grey throat patch with more diffuse lower border (patch being black mainly on chin). Narrower (sometimes broken) white fringes to inner webs of primaries. Fewer

1stY, ♀ (left) and ♂, Finland, Jan: an instructive image depicting sexual dimorphism; both easily aged by yellow/white tips to outer webs of juv primaries, lacking white edges to inner webs. Left bird ♀ by having only two (short) appendages, white fringes to outer webs of primaries (no yellow), and very narrow and pale yellow tail-tip. In contrast, the young ♂ has four appendages, bright yellow fringes to outer webs of primaries, and very broad and bright yellow tail-tip. However, many young are difficult to sex. (M. Varesvuo)

1stW ♀, Scotland, Nov: as well as only white fringes to outer webs of juv primaries, all other features also point to this age/sex. (H. Harrop)

— 203 —

and shorter red appendages on secondary tips (5–8; longest 3–7 mm). Narrower yellow tail-band (on r1 4–6.5 mm; on r5 6–8.5 mm). **1stY** Juv flight-feathers have straight yellowish or white markings confined to outer webs of primary tips (ending at shaft), and hence lack the V-shaped marks of ad. (Very rarely, 1stW ♂ has some white on extreme tip of inner web of a few primaries, but not enough to form V-shaped marks.) Correctly-aged birds can be sexed as follows. **1stY** ♂ More and longer waxy red tips to secondaries (4–8; longest 2.5–6 mm) and rather more extensive yellow tips to tail-feathers (on r1 5.5–9 mm, on r5 6.5–10 mm). **1stY** ♀ Fewer and shorter waxy red tips to secondaries (0–5 [6]; longest if any 1.5–4.5 mm) and less extensive yellow tips to tail-feathers (on r1 2–5.5 mm, on r5 3–6.5 mm). In both sexes, head and throat pattern as in ad, or intensity of throat patch often slightly less advanced. **Juv** Resembles 1stY but essentially much drabber and less immaculate, with much shorter crest, and has diffuse white supercilium, diffusely brown-streaked throat and underparts, and duller vent.

BIOMETRICS (*garrulus*) **L** 18–20 cm; sexes nearly same size: **W** ♂ 112–125 mm (n 61, m 117.0), ♀ 111–121 mm (n 55, m 116.3); **T** ♂ 57–65.5 mm (n 38, m 61.9), ♀ 57–64 mm (n 40, m 60.1); **T/W** m 52.2; **B** 14.3–17.9 mm (n 44, m 16.0); **B**(f) 9.6–12.0 mm (n 44, m 10.8); **BD** 5.2–6.2 mm (n 43, m 5.7); **Ts** 19.0–21.2 mm (n 45, m 20.3). **Wing formula: p1** minute; **pp2–3** longest; **p4** < wt 3–6 mm; **p5** < wt 11–16 mm; **p6** < wt 19.5–24 mm; **p10** < wt 40–48 mm; **s1** < wt 43.5–53 mm. Emarg. pp3–4.

GEOGRAPHICAL VARIATION & RANGE Slight variation, and probably only two races warranted, one of which (*pallidiceps* Reichenow, 1908) is confined to North America. Only ssp. *garrulus* breeds in the entire Palearctic. Easterly populations ('*centralasiae*') in C & E Siberia sometimes separated based on very slightly paler plumage, but we follow Stepanyan (1960), who found differences too slight, and existence of vast overlap.

 B. g. garrulus (L., 1758) (Fenno-Scandia east to E Siberia; mainly short-range movements in winter, sometimes nomadic and making irregular irruptions mainly south and south-west, reaching W & S Europe). Described above.

Sweden, Nov: large flocks form during irruptions. The birds almost blend into their background. (R. Pop)

Ad ♂, Finland, Mar: an exceptional image demonstrating that ageing and sexing are possible even in flight. Aged by white on inner webs to primary tips, while seven waxy red appendages are visible on secondary tips, and broad terminal tail-band and well-defined black border to the lower throat indicate ♂. (M. Varesvuo)

1stY ♀, Finland, Mar: a young bird with yellow only on outer webs of primaries (not extending to tips) and narrow terminal tail-band; the latter, plus the apparently whitish tips to flight-feathers, suggest ♀. (M. Varesvuo)

Finland, Oct: in flight has starling-like jizz and tight flock formation, with ad birds easily aged by being in final stage of outer primary moult. (M. Varesvuo)

(GREY) HYPOCOLIUS
Hypocolius ampelinus Bonaparte, 1850

Fr. – Hypocolius gris; Ger. – Seidenwürger
Sp. – Hipocolio; Swe. – Hypokolius

Shares bill shape and soft feathering of boreal-living waxwings, but this grey shrike-like bird is essentially Middle Eastern, where it is one of the most sought-after species by birders. With access to its main breeding grounds in the Tigris and Euphrates basin of Iraq currently largely impossible, most of our knowledge of this remarkable bird comes from its wintering grounds in E Arabia, where it favours bushy scrub, date palms and gardens. An occasional winterer in Oman, Yemen and NE Africa, Hypocolius is a vagrant to Turkey, and recently about semi-annual in Israel, usually occurring in late winter and spring.

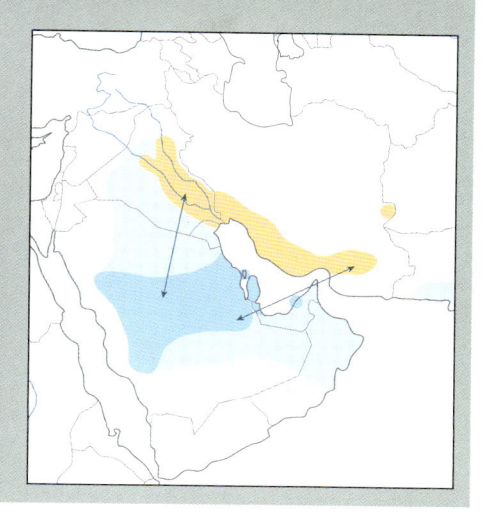

1stY ♂, Kuwait, Mar: compared to ad ♂, has overall ochre-tinged or buffish-grey plumage with white primary tips tipped grey-brown, forming white subterminal area; note moult contrast with renewed smallest and central tertials. (A. Halley)

Ad ♂, Kuwait, Mar: in ♂ plumage is somewhat grey shrike-like; jet-black bases to remiges, primary-coverts, mask and broad black tail-band, while white primary tips (unusual in a passerine) are extensive and pure. Soft waxwing-like feathering is notable, with mostly light greyish-blue plumage of ♂. (A. Halley)

IDENTIFICATION Almost as large as a Great Grey Shrike, and shares *slim body, short wings and long tail* of that species but has a rather *angular head*, and *lacks hooked bill*. Behaviour recalls waxwings and *Turdoides* babblers. Upperparts blue-grey (♂) to dull greyish cream-brown (♀) with a slight isabelline-buff tinge to underparts. ♂ has black mask extending to, and meeting on, nape and in narrow frontal band above bill, but ♀/immature less strikingly marked. Pattern of primaries provides essential age and sex criterion, being most striking in adult ♂, broadly tipped pure white (a very unusual feature since it makes feathers less durable), while adult ♀ has subterminal black primary marks, which together with exposed pale tips form *neat black-and-white pattern*. Tail is broadly tipped dark on both surfaces. Both wing and tail pattern useful in flight identification. At all ages bill is dark (black or horn), and legs yellow-flesh. Typically occurs in small groups, occasionally acting very tame but unobtrusive. May fly to some height on rather rapid, whirring wingbeats. Sometimes cocks tail on landing but often perches upright, though also observed to adopt more horizontal postures.

VOCALISATIONS Complex. Does not appear to have a regular song, but apparently territorial sounds include a loud, repeated *kirrrkirrrkirrr...*, and possibly also some of the miaowing calls. – Calls many and varied, including a mellow, fluid, bee-eater-like *tre-tur-tur*, short, quick and frequent miaowing or whistling sounds (like distant Common Buzzard *Buteo buteo*), e.g. *peeu, mee(ow), quee* or *meee*. At times long series of such notes, *wheew a-wheeew whee-di-du whee-du di-di-du du-du-du*, or *quei-yerr, quei-yerr*, and a descending *whee-oo* given when perched (resembles ♂ Wigeon *Anas penelope*); also a fairly low-pitched, harsh monosyllablic *chirr* or drawn-out disyllabic *trirr* or *quireerr*, and a higher *chirp* in scold. Flocks can be noisy, with calls audible some distance, but singles in winter mostly silent.

SIMILAR SPECIES Resembles at first glance a slim *Great Grey Shrike*, but similarities end there. Indeed, given the difficulties in placing this remarkable species taxonomically, it is perhaps unsurprising that, given a good view, even the relatively nondescript non-adult plumages are unlikely to be confused with any other bird in the covered region.

AGEING & SEXING Ages and sexes distinctive. – Moults. Complete post-nuptial moult and partial post-juv moult, both mostly Aug–Sep (sometimes into Oct). Post-juv renewal involves most of head, body, lesser, median, a few to most greater coverts, some or all tertials, and sometimes r1. Pre-nuptial moult probably absent, but few data available.
– **ALL YEAR Ad ♂** White primary tips extensive and pure, *c.* 25.5–29 mm on p3 (along outer web); primary-coverts black (a few with very thin white fringes); black tail-band always broader and more contrasting than in ♀, at shaft *c.* 25–38 mm on r1, 10–32.5 mm on r6. With wear upperparts less bluish and underparts dirty white with pale grey tone to sides. **Ad ♀** Shorter tail and easily separated by lack of black head markings; primaries have distinct blackish subterminal bands (*c.* 5–8 mm wide on inner web of p3) contrasting with dull white tips (2–3.5 mm), in flight appearing as a well-demarcated bar near edge of wing; primary-coverts have ill-defined blackish tips. Black tips on rr1–5 ill-defined and narrower than on ♂ (*c.* 8–15 mm on r1 along shaft), and virtually lacking on r6. With wear upperparts duller and underparts paler buff or dirty white. **Imm** Aged by juv primaries, primary-coverts, (often) all except central tail-feathers, and variable number of outer greater coverts, showing

1stY ♂, Oman, Mar: especially when fresh, black mask less intense and narrower; the primary-coverts are just visible, revealing the diagnostic age discriminator of black bases with extensive pale fringes and tips. (H. & J. Eriksen)

Ad ♀, Iran, Jan: easily aged and sexed by lacking ♂'s black head markings, primaries have distinct blackish subterminal bands and white tips, black tail-band narrow, while much of plumage is dull greyish cream-brown and isabelline-buff. (H. van Diek)

1stY ♀, Kuwait, Apr: virtually uniform drab, including primary-coverts and remiges, which at most show slightly ill-defined duskier tips, and pale buff fringes, i.e. lacking ad ♀'s subterminal black bars and distinct white fringes to tips, and broad white tips with partial grey-brown wash of 1stY ♂; tail uniform buff-brown, sometimes slightly darker brown subterminally. (R. Bonser)

characteristic juv pattern of the respective sexes. **1stY ♂** Much as ad ♂ but dull or buffish-grey; crown more grey-buff and contrasts less with mantle, and black mask sometimes less intense and narrower; white primary tips smaller and partially tipped grey-brown (along outer web on tip of p3 c. 25–29 mm), forming white subterminal area (resembles ad ♀ but lacks blackish subterminal bands and narrow dull white tips), and secondaries brownish-grey, with less extensive blackish bases; tail-bar narrower and duller (at shaft c. 24–35.5 mm on r1, 10–30 mm on r6); juv tertials and greater coverts tinged brownish and fringed buff-white on tips and base of outer webs; primary-coverts black with variable but usually extensive greyish-white fringes and quite broad ill-defined tips. **1stY ♀** Virtually uniform drab flight-feathers, but sometimes has slightly ill-defined duskier tips, and inner primaries and secondaries faintly fringed pale buff (lacking ad ♀'s subterminal black bars and distinct white fringe to tips; no broad white tips with partial grey-brown wash as in 1stY ♂). Primary-coverts uniform drab-grey (no black on tips like in ad ♀). Tail uniform buff-brown, sometimes slightly darker brown subterminally (some black on tip in ad ♀ and 1stY ♂). Juv tertials and greater coverts as 1stS ♂ but faintly marked. **Juv** Soft, fluffy body-feathering is entirely pale sandy-brown, with wing and tail markings like 1stY.

BIOMETRICS L 22–23 cm; **W** ♂ 95–108 mm (*n* 18, *m* 100.8), ♀ 95–101 mm (*n* 16 *m* 99.3); **T** ♂ 100–115 mm (*n* 19, *m* 108.3), ♀ 96–108 mm (*n* 15, *m* 102.3); **T/W** *m* 105.5; **TG** 7–21 mm (*n* 35, *m* 11.8); **B** 17.0–20.2 mm (*n* 35, *m* 18.6); **B**(f) 12.4–15.2 mm (*n* 36, *m* 13.7); **BD** 5.8–7.4 mm (*n* 31, *m* 6.5); **Ts** 23.0–26.0 mm (*n* 37, *m* 24.2). **Wing formula:** p1 < pc 3–7.5 mm, < p2 48.5–61 mm; **p2** < wt 0.5–7 mm (sometimes = p3 and longest), =3 or 3/5 (rarely =3 or 5, =5/6); **pp3–4** about equal and longest (sometimes p4 < wt 0.5–2 mm); **p5** < wt 1.5–5 mm; **p6** < wt 5.5–10.5 mm; **p7** < wt 8–13 mm; **p10** < wt 18.5–23.5 mm; **s1** < wt 21–28 mm. Emarg. pp3–4.

GEOGRAPHICAL VARIATION & RANGE Monotypic. – Iraq, S Iran, Arabia; resident, nomadic or makes only shorter winter movements, occurring to S Afghanistan, Pakistan and NW India. (Syn. *orientalis*.)

1stY ♀, Oman, Mar: Even with a head-on view of the most nondescript plumage, jizz, size, uniform coloration and long tail with dark tip readily identify the species. In other images of the same individual, the lack of white primary-tips and the renewed central tail-feathers confirm age. (H. & J. Eriksen)

United Arab Emirates, Nov: typical formation and height in flight, usually on rather rapidly whirring wings; long tail striking. Several ad ♂♂ (with broad black bases to remiges, mask and black tail-band, and extensive white primary tips) and ad ♀♀ (plain faces). (H. & J. Eriksen)

(WHITE-THROATED) DIPPER
Cinclus cinclus (L., 1758)

Fr. – Cincle plongeur; Ger. – Wasseramsel
Sp. – Mirlo acuático europeo; Swe. – Strömstare

This peculiar, amphibious passerine can be found at shallow streams and narrower rivers with boulders and pebbly islets in hill country from the British Isles and Fenno-Scandia to Siberia and Mongolia, south to NW Africa, W Asia and the N Indian subcontinent. It is capable of diving and fishing for insects and larvae, also from ice-edge in winter. Often seen flying low over water in direct flight with whirring wingbeats. Resident or short-range migrant, latter with peculiar habit of sometimes breeding first once on or near winter grounds, then a second time much further north.

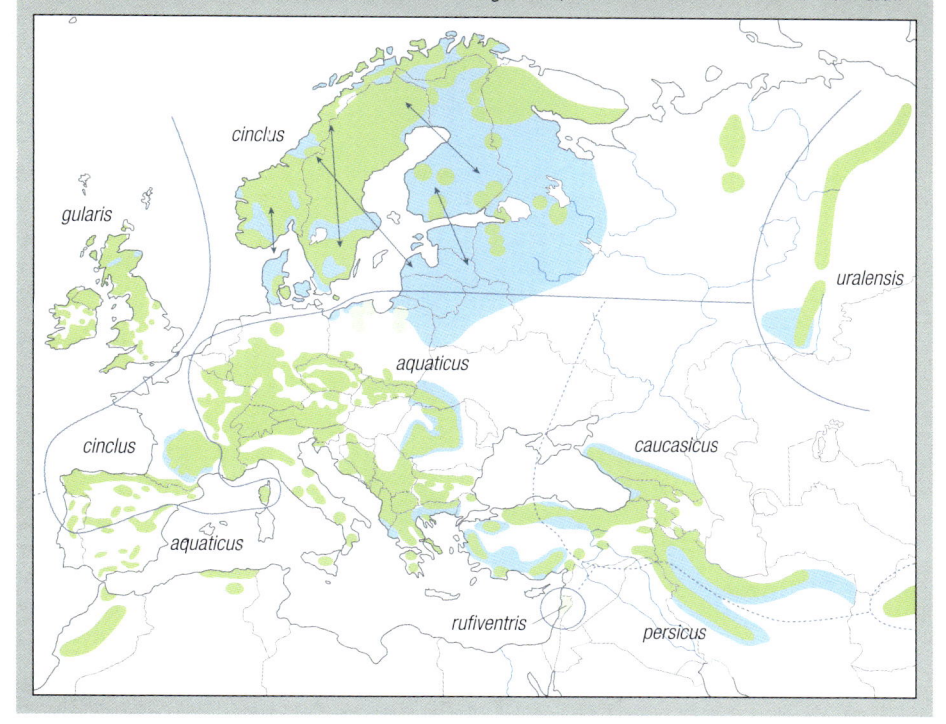

C. c. cinclus, Finland, Jan: showing the species' typical habitat of clean-water streams and rivers. (M. Varesvuo)

VOCALISATIONS Song a peculiar mixture of harsh, squeaky and throaty notes and brief rippling trills or short softer phrases, delivered in a steady stream but with little structure; frequently contains a few repeated notes or phrases in pairs, a bit like song of Bullfinch or Common Crossbill. Song heard from edge of stream or river islet, on a sunny day at times even in midwinter from edge of ice. – Common call is a sharp, penetrating *zrits!*, well audible above rushing waters.

SIMILAR SPECIES Plumage and habits make the species usually unmistakable within treated region, where it is the only species of dipper. – In a brief and poor view, a ♂ *Ring Ouzel* could theoretically be a pitfall for the beginner being also black with white breast patch, but apart from different habits and habitats note larger size, long tail, more elongated shape, smaller and crescent-shaped white breast patch, and yellow bill of the thrush.

IDENTIFICATION A small thrush-sized bird, but appears more compact and bulky given *rounded body*, accentuated by *neck-less appearance*, *short wings* and *half-cocked, short tail*. Rotund and pot-bellied, its shape recalls more a huge wren, with *overall dark plumage*, comprising *dark brown head*, *slate-coloured upperparts*, wings and tail, *black-brown rear underparts* (with variable chestnut on breast), and *stunning white bib from chin down to breast*. Bare parts largely dark. On ground, *stands broad-legged*, walks, runs and, occasionally, hops, but most obvious are *body-bobbing and curtsy movements*, with *downward tail-flicks* and *blinks of white eyelids*. Flight usually direct and rapid, low over water, on whirring wings, swinging around bends in river. Dives or walks partially submerged in water, even 'swimming' upstream with flicks of wings propelling the bird downward and forward, or walks on bottom.

AGEING & SEXING (*cinclus*) Ageing difficult but sometimes possible in the hand, using moult, feather pattern and wear in wing, and iris colour (easier in autumn). Sexes alike in plumage but differ in size. Seasonal variation slight. – Moults. Complete post-nuptial moult at end of summer, mostly Aug–Sep (early Oct). Partial post-juv moult at the same time. Post-juv renewal usually involves head- and body-feathers, lesser and median coverts, and sometimes a few inner greater coverts. Presence of a pre-nuptial moult unconfirmed, and probably absent. – **SPRING** Ageing difficult. **Ad** In summer, wear has reduced or removed any pale fringes, and birds become slightly paler brown on head and a little greyer elsewhere. **1stS** Very similar to ad, and differences less clear compared to autumn, as many pale greyish fringes to belly-feathers and juv greater coverts, tertials and primary-coverts are now worn off. Even handled birds difficult to age due to similarity of older and newer greater coverts, but sometimes primaries are more worn than in ad, primary-coverts also being narrower or more pointed with remnants of pale tips on inner ones, and iris colour differs by being duller brown rather than deep sepia-brown. (Ageing thus possible at least of some birds and with practice.) – **AUTUMN Ad** Much as in spring. Differs from 1stW in lacking, or having very indistinct and narrow, pale tips to tertials, primary-coverts and greater coverts, and primary-coverts being broadly rounded with neat tips. Underparts more uniformly dark. Iris often deep sepia-brown. **1stW** Resembles ad but primary-coverts (especially inner), greater coverts (especially unmoulted outer ones) and tertials narrowly tipped pure white or greyish-white, pale edges being on average somewhat broader or more prominent than in fresh ad. (Note that juv and ad-type greater coverts are neatly grey-edged on outer webs, providing no or only very subtle moult limits.) All remiges and primary-

C. c. cinclus, ad, Finland, Jan: the N European race is typically dark above with limited rufous below. Note evenly feathered and not heavily worn wing of ad. (M. Varesvuo)

C. c. cinclus, Finland, Jan: typical flight posture, with the fluttery wingbeats blurring the tips. (M. Varesvuo)

C. c. cinclus, 1stY, Finland, Feb: N European birds can have slightly wider and brighter rufous below than shown here (limited to a hint on lower chest). Note juv greater coverts (with typical white tips), while juv remiges and large alula are already worn and bleached brown. Seeing Dippers on snow is not uncommon during N European winter. (M. Varesvuo)

coverts juv being more worn and a little weaker textured, and primary-coverts slightly narrower and rather pointed, tips slightly frayed. Iris olive-grey or dull grey-brown (rather than deep sepia-brown). Lower breast and belly generally paler and less uniform, feathers more or less prominently tipped greyish-white (often appearing as a grey-white zone just below white chest), but these pale tips soon wear off. **Juv** Soft fluffy body-feathering and entire upperparts dark slate, with black fringes producing obvious dark-mottled effect. Underparts mostly white to dull grey near tail, heavily spotted and closely barred dusky and black-brown on rear flanks, with less marked pale scaling on undertail-coverts, and secondaries fringed white.

BIOMETRICS (*cinclus*) **L** 17.5–19.5 cm; **W** ♂ 95–100 mm (*n* 24, *m* 97.1), ♀ 85.5–94 mm (*n* 26, *m* 89.4); **T** ♂ 51–59 mm (*n* 24, *m* 54.3), ♀ 46–53 mm (*n* 25, *m* 49.3); **T/W** *m* 55.6; **B** 19.3–22.3 mm (*n* 47, *m* 20.8); **B**(f) 13.8–17.5 mm (*n* 46, *m* 15.6); **BD** 4.1–5.7 mm (*n* 45, *m* 4.9); **Ts** 26.5–30.2 mm (*n* 24, *m* 28.2). – Two independent studies of live birds from Sweden and Norway gave following wing-lengths for sexing (mm): ♂ > 95 (*n* 248), ♀ < 97 (*n* 383) (Vuorinen 1991); ♂ > 94 (*n* 266), ♀ < 94 (*n* 358) (Andersson & Wester 1971). **Wing formula:** **p1** > pc 0–4 mm; **pp2–4** about equal and longest (p2 and p4 sometimes slightly shorter); **p5** < wt 1–5 mm; **p6** < wt 6–11 mm; **p7** < wt 9–16 mm; **p10** < wt 20–29 mm; **s1** < wt 22–30 mm. Emarg. pp3–5 (p5 usually slightly or much less prominently).

GEOGRAPHICAL VARIATION & RANGE Marked and complex variation, with peculiarly disjunct distribution of one race (*cinclus*), considerable differences in plumage colour within W & C Europe, but only small and clinal differences elsewhere. Taxa in N & NE Iberia and C Europe rather variable in colour. Seven races recognised within treated region. Another three extralimital races occur in Asia (accordingly not treated).

○ ***C. c. gularis*** (Latham, 1801) (Britain, Ireland; mainly resident). Usually has very obvious and rather bright chestnut area on breast below white bib, sometimes even extending to fore flanks and mid belly, though some are duller and darker with more limited chestnut. Upperparts, too, vary, some being almost uniform sooty black, others (more commonly) having variably broad greyish centres with black subterminal fringes. Head typically brownish. Juv darker above than juv *cinclus*, with more heavily marked underparts including rufous-buff undertail-coverts. Very similar to *aquaticus* with same underparts, differing only subtly in that mantle/back is a little darker and sootier black, less greyish. **W** ♂ 87.5–98 mm (*n* 18, *m* 93.6), ♀ 83–91 mm (*n* 15, *m* 87.5); **T** ♂ 48–57 mm (*n* 18, *m* 52.5), ♀ 44–54 mm (*n* 15, *m* 48.5); **T/W** *m* 55.8; **B** 19.1–23.2 mm (*n* 33, *m* 21.4); **B**(f) 14.6–18.4 mm (*n* 27, *m* 16.2); **BD** 4.7–6.0 mm (*n* 24, *m* 5.2); **Ts** 26.0–31.0 mm (*n* 31, *m* 28.9). – Hartert (1910) separated birds of Ireland and W Scotland ('*hibernicus*') on account of being darker above, but total overlap when series compared, and very far from meeting 75% rule. (Syn. *britannicus*; *hibernicus*.)

○ ***C. c. cinclus*** (L., 1758) (Fenno-Scandia, NC Russia west of Pechora, C & S France, Pyrenees, NW & C Iberia, Corsica; northern populations winter in NW Europe, Baltic basin or EC Europe, others are mainly resident). Described above, typically being dark slate-grey or blackish above, with no or only very slight rufous tinge below. The range is peculiarly disjunct in that chestnut-bellied populations in W & C Europe divide more dark-bellied ones in Fenno-Scandia from those in C & S France, Pyrenees and NC Iberia. Corsica is also inhabited by dark-bellied birds. All dark-bellied populations of S Europe have been described as separate subspecies, but as they are either identical to *cinclus* or differ only very marginally morphologically and in a minority of birds, we prefer (like Vaurie 1959, *BWP*, Dickinson & Christidis 2014) to include them here, although some of them may have evolved independently. (Syn. *atroventer*; *pyrenaicus*; *sapsworthi*.)

C. c. gularis, England, Oct: the rather distinctive British race has extensive chestnut on lower chest and upper belly, extending to flanks and mid belly. Ageing impossible. (S. Round)

C. c. aquaticus, Switzerland, Mar: dark slate-grey upperparts contrasting with duller brown head, very extensive and deep rufous-chestnut lower breast and belly, black flanks and rear belly/vent. Ageing and moult pattern difficult to evaluate given that the apparent browner (older?) smallest tertial, alula, primary-coverts and primaries, contrasting with fresher rest of wing, does not fit the species' known moult strategy. (G. Schuler)

C. c. aquaticus Bechstein, 1797 (C & SE Europe including E Belgium, Germany, Poland, E France, C & S Spain, Italy including Sicily, Austria, Balkans, NW Africa, W Turkey; mainly resident but some are nomadic or make short-range movements in winter). Slate-grey and not as dark or blackish on back and vent as *cinclus*, and head is paler brown. Lower breast and belly deep rufous-chestnut, with flanks and rear belly/vent blackish-grey. Very similar to *gularis* with virtually same underparts but differs by having on average subtly paler and more greyish mantle/back, less blackish, and slightly paler brown head/nape, but a few extremes are probably inseparable. Clinally slightly darker in E Europe and Balkans than in west of range. **L** 16.5–17.5 cm; **W** ♂ 91–98 mm (*n* 14, *m* 94.5), ♀ 84–87.5 mm (*n* 11, *m* 86); **T** ♂ 48–55 mm (*n* 14, *m* 50.6), ♀ 45–51 mm (*n* 11, *m* 46.6); **T/W** *m* 53.8; **B** 18.5–22.0 mm (*n* 24, *m* 20.5); **B(f)** 14.2–17.0 mm (*n* 24, *m* 15.5); **BD** 4.1–5.0 mm (*n* 45, *m* 4.9); **Ts** 26.5–30.2 mm (*n* 24, *m* 28.2). **Wing formula: p1** > or < pc 0–3 mm. – Birds in NW Africa ('*minor*') appear by and large indistinguishable from *aquaticus* (except perhaps by marginally longer bill in some). (Syn. *alpinus* and three more alpine taxa with very limited range and subtle differences proposed by von Burg 1924; *medius*; *minor*; *montanus*; *orientalis*.)

C. c. caucasicus Madarász, 1903 (Turkey except western part, NE Iraq, Caucasus, Transcaucasia, N Iran through Elburz Mts; resident or short-range movements in winter). Has rather uniform dusky tawny-brown underparts, lacking more clear division between warm brown lower breast/upper belly and greyer lower belly (lower belly/vent being only marginally darker). Rather characteristically has crown deep brown extending onto mantle, lacking more blackish or dark slate-grey mantle and back of *aquaticus* and *cinclus*. Bill on average slightly thinner in dorsal view, sides being more compressed than in other races. Seems to grade into *aquaticus* in NC Turkey. Birds in NE Turkey have even been referred to *cinclus*, but not followed here. **L** 17–18 cm; **W** ♂ 92–101 mm (*n* 16, *m* 96.2), ♀ 83–94 mm (*n* 13, *m* 88.3); **T** ♂ 45–57 mm (*n* 16, *m* 52.4), ♀ 44–53 mm (*n* 13, *m* 47.5); **T/W** *m* 54.2; **B** 19.3–23.0 mm (*n* 28, *m* 21.3); **Ts** 26.4–31.5 mm (*n* 26, *m* 28.7). **Wing formula: p1** > pc 0–8 mm (rarely 0.5–2 mm < pc); **p5** < wt 0–3 mm; **p6** < wt 5–8.5 mm; **p7** < wt 9.5–12 mm; **p10** < wt 15–23 mm; **s1** < wt 17–26 mm. – Birds formerly living on Cyprus (extinct) and in S & SW Turkey ('*olympicus*') sometimes separated but appear poorly defined and better included here (*contra* Kirwan *et al.* 2008, who referred them to *aquaticus*); the few specimens examined by us had slightly rufous-tinged rather than grey mantle, just like *caucasicus*. (Syn. *amphitryon*; *olympicus*.)

†? ***C. c. rufiventris*** Tristram, 1884 (Lebanon, apparently now extinct). Close to *caucasicus* and *persicus* but is smaller and darker overall than latter, and slightly paler and warmer brown than former. It has dull rufous-brown lower breast to mid belly and quite uniform warm tawny-brown crown to back. Only lower back to uppertail-coverts are darker and more slate-grey. Equally narrow bill as *caucasicus* and *persicus*. Scant material available (*n* 4); a subtle race that requires further study as to be recognised or not. **W** 84–93 mm (*m* 89.3), **T** 45–48 mm (*m* 46.4), **B** 21.3–22.3 mm (*m* 21.8), **Ts** 27.2–29.2 mm (*m* 27.9).

C. c. persicus Witherby, 1906 (SW Iran, including Zagros Mts; mainly resident). Resembles *caucasicus* but is clearly paler above and below, and subtly larger. Entire upperparts brown, paler and warmer on crown to mantle, successively duller and darker on back to uppertail-coverts. Lacks any slate-grey colour. Underparts (below white bib) pale but bright chestnut, only darkening slightly on vent. **L** 18–19 cm; **W** ♂ 98–106 mm (*n* 10, *m* 102.0), ♀ 91.5–94 mm (*n* 4, *m* 92.9); **T** ♂ 51–60 mm (*n* 10, *m* 55.6), ♀ 46–53 mm (*n* 4, *m* 51.0); **T/W** *m* 54.6; **B** 20.8–23.4 mm (*n* 14, *m* 22.0); **Ts** 27.3–32.0 mm (*n* 14, *m* 29.7).

? ***C. c. uralensis*** Serebrovski, 1927 (Ural Mts; short-range movements mainly towards south in winter). Said to be very near *cinclus* but subtly paler above with chocolate-brown lower breast and darker brown belly, although 'not as rufous as *aquaticus*' (Vaurie 1959). No material seen and should perhaps be reassessed and be compared again with full variation of *aquaticus*.

TAXONOMIC NOTE Variability, overlap and claimed poor support from genetic traits of *cinclus* and *aquaticus* in N & C Spain led Campos *et al.* (2010) to question the wisdom of upholding current taxonomy, but such problems are commonplace and not reason to abandon subspecific division. Furthermore, genetic properties have generally very little bearing on definition of morphologically described geographical subspecies.

REFERENCES Andersson, S. & Wester, S. (1971) *Orn. Scand.*, 2: 75–79. – Campos, F. *et al.* (2010) *Ring. & Migr.*, 25: 3–6. – Vuorinen, J. (1991) *Cinclus Scand.*, 4: 65–67.

Ad, Netherlands, Dec: racial identification outside breeding season and of migrants often tricky, but this appears to be N Eurasian *C. c. cinclus*. Dark upperparts and no rufous below. However, it is impossible to certainly exclude the possibility of an extreme dull *C. c. aquaticus*. Aged by evenly feathered and not heavily worn wing-feathers. (R. Jansen)

C. c. caucasicus, Georgia, Jun: this race typically has deep brown crown extending onto mantle, and rather uniform dusky tawny-brown underparts. (R. Pop)

C. c. aquaticus, juv, Italy, Aug: note fluffy body-feathers, scaly upperparts and mostly white underparts, with extensively white-fringed wing-feathers. (F. Ballanti)

(EURASIAN) WREN
Troglodytes troglodytes (L., 1758)

Fr. – Troglodyte mignon; Ger. – Zaunkönig
Sp. – Chochín común; Swe. – Gärdsmyg

This small bird more than makes up for its tiny size with its loud high-pitched voice, which is a familiar sound of woodlands and gardens, in lowlands to high mountains, from Iceland to Russia and south to NW Africa, east to the Himalayas (where it reaches 5500 m), Kamchatka, and many E Pacific islands including Japan and Taiwan. (North American populations are now commonly separated as two different species, Winter Wren *T. hiemalis* and Pacific Wren *T. pacificus*.) Northerly Eurasian Wren populations winter to the south, but it is a hardy bird that often attempts wintering even in snow-covered habitats.

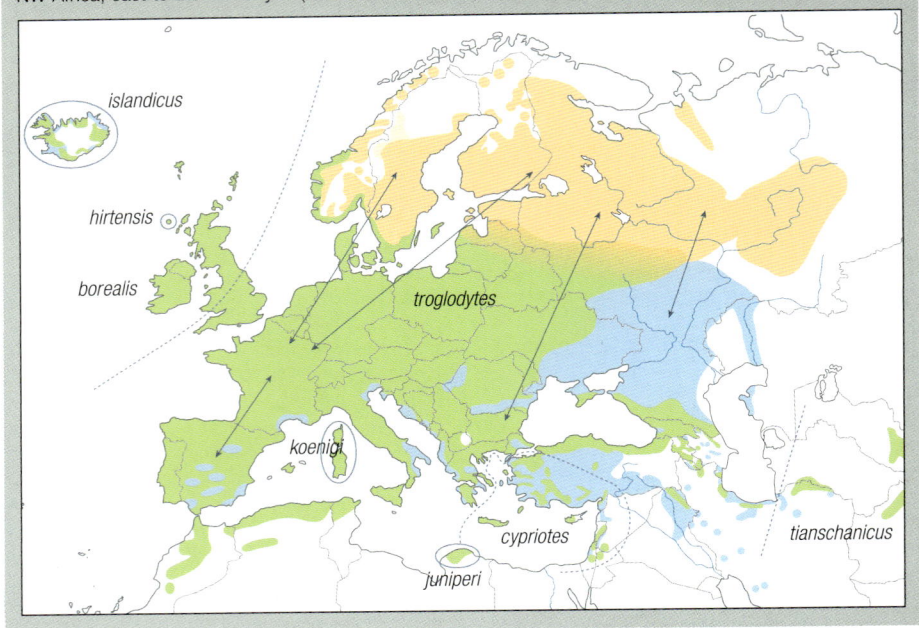

T. t. troglodytes, Italy, Dec: rather long-legged with short rounded wings, short tail often cocked sharply. Apparently has two generations of wing-feathers, which together with contrastingly rufous remiges, could suggest 1stY, but ageing uncertain. (D. Occhiato)

IDENTIFICATION Among the smallest Western Palearctic birds, relatively compact and round-bodied, quite long-legged with short rounded wings. *Habitually restless, cocking short tail and keeping on or near ground*. Usually skulks quietly and is heard before seen. *Russet-brown above* (brighter and more rufous on rump/uppertail-coverts) and *paler buff-brown to off-white below*, with even paler chin and throat. Rather conspicuously *thin bill, pale buff supercilium and barred wings and flanks*. Bill dark horn with yellower lower mandible and base, eyes blackish, and legs usually pale brown. Flight short and direct on rapid, whirring wings. Rather mouse-like, creeping, hopping and climbing through low vegetation. Sings mostly from cover. Not social, except at roosts in winter.

VOCALISATIONS Song amazingly loud for so tiny a bird, a brief and explosive trill of metallic ringing notes, pulsating almost as if bird is trying to sing as loudly as possible (with gape fully open and entire body vibrating), e.g. *zü-zü-zü-zü- si-zirrrrrr svi-svi-svi siyu-zerrrrr sivi*. Some notes in the song quite high-pitched and may be difficult to hear by elderly people. – Calls include a hard clicking *zeck* and a churr or hard rattle when nervous, *zer'r'r'r* (the two sometimes combined). The churring call is harder and more rattling than the reasonably similar but more slurred call of Red-breasted Flycatcher (sometimes in same habitat in summer).

SIMILAR SPECIES Unmistakable within the treated region, as all other wrens occur solely in the New World. Given a very brief view, could be confused with a *Dunnock* (often in same habitat), but latter is larger, more elongated and longer-tailed and, in addition to obvious plumage differences (when seen well), has quite different behaviour.

AGEING & SEXING Ageing possible only on close inspection of moult pattern and feather wear in wing. Sexes similar with no seasonal variation. – Moults. Complete post-nuptial moult and partial post-juv moult in late summer (mostly Jul–Oct). Post-juv renewal involves head- and body-feathers, all lesser and median coverts, and variable number of greater coverts (usually 2–7), tertials and tail-feathers. Pre-nuptial moult absent or insignificant; if any, in Jan–Apr. – SPRING **Ad** Aged by lack of moult limits in greater coverts or between major feather tracts. Primaries and primary-coverts relatively less worn; whitish tips to wing-coverts often still present (see Autumn). Barring of remiges on slightly spread wing not orderly and continuous from feather to feather but 'disordered' (Taylor 2012); on average denser and more distinct barring (but some overlap). When heavily worn, generally slightly duller and browner, enhancing dark bars and whitish supercilium, throat and chest. **1stS** Identical to ad, but in the hand sometimes possible to age by moult limits, with contrastingly more worn and duller juv primaries, primary-coverts and some outer greater coverts. Barring of remiges on slightly spread wing orderly and continuous from feather to feather (Taylor 2012); on average fewer and less distinct bars on secondaries, poorly marked in particular on mid section of secondaries. Both ad and 1stS may undergo limited winter moult (involving all or part of head, throat and lesser coverts, only occasionally other body-feathers and even a few median coverts), but ageing criteria should remain valid. – AUTUMN **Ad** Evenly fresh. Greater coverts may have subtle greyish cast (still largely brown) and some, at least, have whitish tips. On average less continuous dark bars on flight-feathers (showing more irregularities from feather to feather), and wing-coverts lack moult steps. Number of pale (buff) spots on outer web of p4 9–12 (Hawthorn 1971, Ward & du Feu 2006). **1stW** Look for contrast between outer unmoulted (juv) rich rufous-brown greater coverts, and renewed, paler and fresher inner ones, which are medium brown or even bronze/greyish (also longer, and often show small white tips as in ad), thus forming abrupt change in colour pattern and length of these two blocks of coverts, though sometimes difficult to detect in birds that have moulted only few inner greater coverts. Check

T. t. troglodytes, Spain, May: typical stance when singing, delivering its amazingly far-carrying, metallic, ringing song. (C. N. G. Bocos)

T. t. troglodytes, Netherlands, Apr: tiny, mouse-like and acrobatic. Strongly barred wings and clear pale supercilium behind eye. Ageing impossible in this case. (O. Plantema)

also that new inner greater coverts are equally fresh, densely textured and of the same colour as median and lesser coverts. Occasionally replaces some tertials, thus showing similar moult contrast as in greater coverts. Dark barring of remiges, having been grown simultaneously, form continuous and orderly bars across wing. Number of pale (buff) spots on outer web of p4 7–9 (Hawthorn 1971, Ward & du Feu 2006). **Juv** Soft, fluffy body-feathering and pale yellow gape in first weeks after fledging; plumage generally warmer than ad but markings less obvious, secondary-coverts lack whitish tips, and undertail-coverts are more uniform.

BIOMETRICS (*troglodytes*) **L** 9–10 cm; **W** ♂ 47–52 mm (n 24, m 49.5), ♀ 44–51 mm (n 19, m 47.2); **T** ♂ 28–35 mm (n 24, m 31.4), ♀ 27–32 mm (n 19, m 29.4); **T/W** m 63.0; **B** ♂ 12.0–14.2 mm (n 24, m 13.4), ♀ 12.0–13.7 mm (n 14, m 13.0); **BD** 2.4–3.0 mm (n 32, m 2.8); **Ts** 15.2–18.2 mm (n 37, m 17.3). **Wing formula: p1** > pc 8–13 mm; **p2** < wt 3.5–7.5 mm, =8/10 (common), =8 or <10 (rare); **pp3–5**(6) about equal and longest (sometimes p3 < wt 0.5–1 mm); **p6** < wt 0.5–2.5 mm; **p7** < wt 1.5–3 mm; **p10** < wt 5–8 mm; **s1** < wt 5–9 mm. Emarg. pp3–6.

GEOGRAPHICAL VARIATION & RANGE The Eurasian Wren is one of the most fascinating species within the covered region for studying geographical variation and evolution. Up to 30 races have been recognised within Eurasia by most recent works (e.g. 28 in Dickinson & Christidis 2014). In our view, eight subspecies occur in the W Palearctic and are recognised here as distinct. Variation is largely clinal in continental areas, but some races, chiefly insular ones, are more strongly differentiated. Without an appreciation of the rather large individual variation of plumage colours and barring, and without assessing sufficiently long series, it is all too easy to recognise a mosaic of taxa which on a broader appraisal do not reach a 75% level of distinctness. Our assessment of W Palearctic *troglodytes* taxa, albeit based on examination of hundreds of specimens in several large museum collections, is still tentative and incomplete, mostly due to lack of complete series from some regions.

T. t. islandicus Hartert, 1907 (Iceland; short-range movements in winter to coastal areas). Averages the largest form and is intermediate in general saturation, darkness and extent of barring above between the darkest and palest races. Unlike *borealis* has paler underparts than *troglodytes*, with less buff and strongly reduced bars on throat to chest (or with very little barring even down to upper belly), thus bars are concentrated on flanks, lower belly and ventral area. Sexes very nearly the same size (♀, interestingly, a fraction larger except bill). **L** 11–12 cm; **W** 54–61.5 mm (n 28, m 58.0); **T** 33–42 mm (♀ 28, m 37.7); **B** 13.7–17.0 mm

T. t. troglodytes, individual variation, Italy, Nov: irrespective of age or season, some birds are much plainer, with limited barring above and below (right-hand bird). Ageing very difficult and requires experience, even with birds in the hand. (D. Occhiato)

T. t. troglodytes, Tring, England, Feb: *troglodytes* intergrades with *borealis* in northern England, but at least in Tring north of London, populations should be clearly *troglodytes*, showing chiefly rusty sepia-brown upperparts and fairly conspicuous and dense barring overall. Ageing difficult without a closer view. (H. Shirihai)

T. t. troglodytes, Azerbaijan, May: the Greater Caucasus population is genetically well differentiated, but we have failed to find sufficient morphological evidence to recognise these birds as different from *T. t. troglodytes*. (M. Heiß)

(n 27, m 15.2); **BD** 3.1–3.7 mm (n 23, m 3.3); **Ts** 18.8–21.3 mm (n 28, m 19.9). **Wing formula:** Much as in *troglodytes*, but **p1** < wt 10–15 mm.

T. t. borealis Fischer, 1861 (Faeroes, Outer Hebrides, Shetlands, Scotland, Ireland, England at least in north and west; resident). Warm rusty sepia-brown above, and fairly conspicuously and densely barred overall (though as in all races this varies individually). Slightly larger than *troglodytes* with marginally longer legs and stronger and longer bill. – Breeders from Shetlands ('*zetlandicus*') much the same size as *borealis* from Faeroes (*zetlandicus* averages very slightly smaller, but vast overlap) and similarly prominent barring above, but on average very subtly darker rusty sepia-brown upperparts (in particular on crown, nape and upper mantle), whereas underparts tend to be slightly less barred. Bill is on average a fraction longer; still, differences in all these respects are very slight (average difference in bill length c. 0.4 mm) and inconsistent, majority falling within variation of *borealis*, and we prefer to include them in latter subspecies. – Birds from Ireland, Inner Hebrides, mainland Scotland and England ('*indigenus*') are intermediate between *borealis* and *troglodytes*, but form part of a cline and are often rather indistinct. At least western and northern birds are best included in *borealis*. – Two additional insular Scottish races have been separated ('*hebridensis*' of Outer Hebrides, and '*fridariensis*' of Fair Isle, S Shetlands) but appear to fall well within overall natural variation of *borealis*, i.e. differ only very slightly and inconsistently in size or degree of darkness and saturation, and do not fulfil minimum requirement for accepted subspecies. – Measurements for Faeroes breeders: **L** 10–11 cm; **W** ♂ 50–55 mm (n 11, m 52.5), ♀ 46–52 mm (n 13, m 48.7); **T** ♂ 30–36 mm (n 10, m 33.2), ♀ 27.5–33 mm (n 13, m 30.4); **B** ♂ 13.9–16.1 mm (n 11, m 15.4), ♀ 13.6–15.8 mm (n 13, m 14.5); **BD** 3.0–3.2 mm (n 10, m 3.1); **Ts** 18.4–20.4 mm (n 11, m 19.3). (Syn. *fridariensis*; *hebridensis*; *indigenus*; *zetlandicus*.)

T. t. hirtensis Seebohm, 1884 (St Kilda I, Scotland; resident or makes only minor movements in winter). A distinctive subspecies in being paler brown (more greyish) and having a rather long strong bill. Large, nearly as large as *islandicus*, and same size as *borealis*. Pale rufous-brown with greyish tinge above, heavily barred on flanks, lower belly and ventral area, whereas throat to upper belly are largely unbarred pale brown-white with only tiny spots. **W** ♂ 52–55.5 mm (n 14, m 53.4), ♀ 49–52 mm (n 11, m 50.3); **T** ♂ 32–38 mm (n 14, m 34.0), ♀ 30–36 mm (n 11, m 33.3); **T/W** m 64.8; **B** ♂ 13.9–16.6 mm (n 13, m 14.7), ♀ 13.5–15.5 mm (n 10, m 14.4); **BD** 2.6–3.6 mm (n 20, m 3.2); **Ts** 18.0–20.4 mm (n 23, m 19.0).

T. t. troglodytes (L., 1758) (widespread in N & continental Europe south to at least NW Africa, mainland Italy, Sicily and Balkans; probably also S England; east to Russia, Crimea, Turkey, Caucasus, Transcaucasia, NW & W Iran, perhaps adjacent parts of Iraq; either resident or, in northern parts, a proportion of population makes short- or medium-range movements in winter reaching W & C Europe). Treated above. Small and generally bright rufous (but somewhat variable and sometimes more dull tawny-brown) with bars mainly confined to scapulars, wings, tail and flanks (reduced over rest of upperparts. Grey- or buff-white below with rufous-tinged flanks, vent and undertail-coverts; throat and breast largely unbarred dusky buff or pale rufous-tinged, whereas flanks and lower belly/vent are densely barred. – Birds in S Spain and Balearics ('*kabylorum*') claimed to differ from *troglodytes* by slightly greyer or colder brown tones and more extensive and prominent barring, but we find variation within both populations creating complete overlap in these subtle differences, and best included here. Breeders in NW Africa (again referred to '*kabylorum*') reportedly paler above and below than *troglodytes* but this only a slight tendency with much overlap, hence separation not warranted. Birds of Crimea, Caucasus and NW & W Iran ('*hyrcanus*', '*zagrossiensis*') usually separated on account of more greyish upperparts, paler throat and heavier barring but such differences impossible to confirm in long series. Only detectable subtle difference is an on average 1 mm longer bill, deemed insufficient for separation. Interestingly, the Caucasian population has been found to be genetically highly distinctive (Drovetski *et al.* 2004). (Syn. *bergensis*; *erwini*; *europaeus*; *hyrcanus*; *kabylorum*; *meinertzhageni*; *muelleri*; *occidentalis*; *parvulus*; *weigoldi*.)

○ **T. t. koenigi** Schiebel, 1910 (Corsica & Sardinia, with birds approaching this race in Portugal and NW Spain; resident). Very similar to *troglodytes*, and could perhaps be synonymised with it. Upperparts virtually identical to *troglodytes* (contrary to statements, e.g., by Vaurie 1959), differing only very subtly in being more rufous-tinged white below, less sullied greyish, and being on average more boldly barred on flanks than *troglodytes*. Bill very subtly longer, but much overlap. **W** ♂ 47.5–51.5 mm (*n* 12, *m* 49.1), ♀ 45–49 mm (*n* 14, *m* 46.8); **T** ♂ 31–34 mm (*n* 12, *m* 32.4), ♀ 28–34.5 mm (*n* 14, *m* 31.4); **T/W** *m* 66.0; **B** ♂ 13.8–14.8 mm (*n* 12, *m* 14.4), ♀ 12.5–14.7 mm (*n* 14, *m* 13.8).

T. t. juniperi Hartert 1922 (apparently isolated around Jebel Akhdar, in *pistachio* bushy habitat, coastal NW Cyrenaica, NE Libya; resident); is subtly paler rufous above than *troglodytes*, whereas underparts are more evenly greyish-tinged rufous, and it has weak, reduced barring both above and below. Rather small. Bill markedly curved and long. **W** ♂ 45.5–50 mm (*n* 13, *m* 47.7); **T** ♂ 31–36 mm (*n* 13, *m* 33.2); **T/W** *m* 69.7; **B** ♂ 15.3–16.0 mm (*n* 13, *m* 15.6); **BD** ♂ 2.5–3.2 mm (*n* 13, *m* 2.9); **Ts** ♂ 16.0–18.5 mm (*n* 13, *m* 17.0).

○ **T. t. cypriotes** (Bate, 1903) (Aegean Is, Crete, Rhodes, Lesbos, W & S Turkey, Cyprus, Levant south to Israel/Jordan; resident). Very similar to *troglodytes*, on average very slightly darker rufous above, and is more exten-

T. t. borealis, individual variation, Shetlands (left, Nov) and Faeroes (right: Jun): the darkest above of all NW Atlantic populations and conspicuously barred; on average also more rusty, with rather strong bill and legs. However, there is much individual variation, partially due to wear (left bird fresh and warmer in colour), but also some inter-island variation. (Left: H. Harrop; right: S. Skov)

T. t. borealis, juv, Shetlands, Aug: juv/fledglings of all races have fluffy body-feathers. Dark and heavily barred plumage typical of this race already evident. (H. Harrop)

T. t. hirtensis, St Kilda, Scotland, Jul: one of the largest and palest races, with rather greyish plumage, and throat and breast largely unbarred. Ageing impossible. (S. Fisher)

sively barred on flanks and belly, leaving throat and breast more contrastingly whitish (still with slight brown tinge). Bill on average slightly longer. Sexual size difference small. **W** ♂ 47–53.5 mm (*n* 23, *m* 49.8), ♀ 46–50 mm (*n* 8, *m* 47.4); **T** ♂ 29–35 mm (*n* 22, *m* 31.6), ♀ 29–33.5 mm (*n* 7, *m* 31.4); **T/W** *m* 64.3; **B** 12.5–15.9 mm (*n* 31, *m* 14.1); **BD** 2.6–3.2 mm (*n* 27, *m* 2.9); **Ts** 16.0–18.3 mm (*n* 29, *m* 16.9). Geographical ranges and differences between *cypriotes* and *troglodytes* unclear and need further investigation. Birds from Crete ('*stresemanni*') are best regarded as a subtle variation, being slightly less saturated (subtly greyer brown, less vividly rufous), dark barring slightly more extensive and bill slightly longer. These slight differences are frequently difficult to discern, hence best included here. (Syn. *seilerni*; *stresemanni*; *syriacus*.)

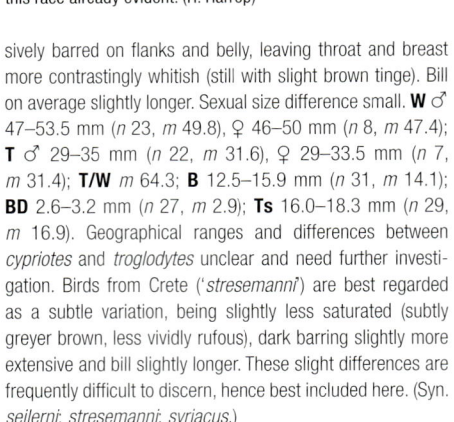

○ **T. t. tianschanicus** Sharpe, 1882 (NE Iran, S Transcaspia, east through Central Asian mountain ranges; resident or makes local movements in winter). Similar to *troglodytes* but on average slightly duller and more greyish-tinged pale brown with subdued barring. It is same size as *troglodytes* but has proportionately longer tail. – There is some slight variation in colour tone and prominence of barring within this race but hardly distinct and consistent enough to warrant separation of more subspecies. Sexes very nearly same size. **W** 46.5–55.5 mm (*n* 27, *m* 49.9); **T** 31–38 mm (*n* 27, *m* 33.6); **T/W** *m* 67.5; **B** 12.7–14.3 mm (*n* 24, *m* 13.6); **BD** 2.4–3.0 mm (*n* 25, *m* 2.6); **Ts** 16.0–18.0 mm (*n* 22, *m* 16.8). (Syn. *cineraceus*; *pallidus*; *subpallidus*; *tarimensis*.)

TAXONOMIC NOTE Although traditionally viewed as a single Holarctic species, recent genetic analysis and study of song differences indicate that three species are involved, one in Eurasia and two in the Nearctic (Drovetski *et al.* 2004, Toews & Irwin 2008). AOU has recently elevated the two Nearctic forms, *pacificus* ('Pacific Wren') and *hiemalis* ('Winter Wren'), to species level, separate from Eurasian Wren, a view followed here.

REFERENCES Drovetski, S. V. *et al.* (2004) *Proc. Roy. Soc. Lond.*, 271: 545–551. – Hawthorn, I. (1971) *Ringers' Bull.*, 3: 9–11. – McGowan, R., Clugston, D. & Forrester, R. (2004) *BW*, 17: 71–75. – Taylor, R. C. (2012) *Ring. & Migr.*, 27: 106–108. – Toews, D. P. L. & Irwin, D. E. (2008) *Molecular Ecolog.*, 17: 2691–2705. – Ward, R. M. & du Feu, C. (2006) *Ring. & Migr.*, 23: 62–64.

T. t. cypriotes, Cyprus, Oct: overall duller and less heavily barred plumage than *T. t. troglodytes*, rufous being mostly restricted to wings and tail. (H. Shirihai)

DUNNOCK
Prunella modularis (L., 1758)

Alternative name: Hedge Accentor

Fr. – Accenteur mouchet; Ger. – Heckenbraunelle
Sp. – Acentor común; Swe. – Järnsparv

Mainly restricted to the W Palearctic, the unassuming Dunnock could potentially be one of the most familiar garden and urban parkland birds, yet few outside the ranks of birdwatchers know it. In Fenno-Scandia, this is because the species is shy there, avoiding man. In W & C Europe, where it truly is a garden dweller, it is too insignificant to be much noticed. The Dunnock is mainly sedentary (or partially migratory), but northern birds are all migrants, wintering in SW Europe. Fond of open farm or pasture land with hedgerows or dense bushes, but also seashores with junipers, brambles and roses.

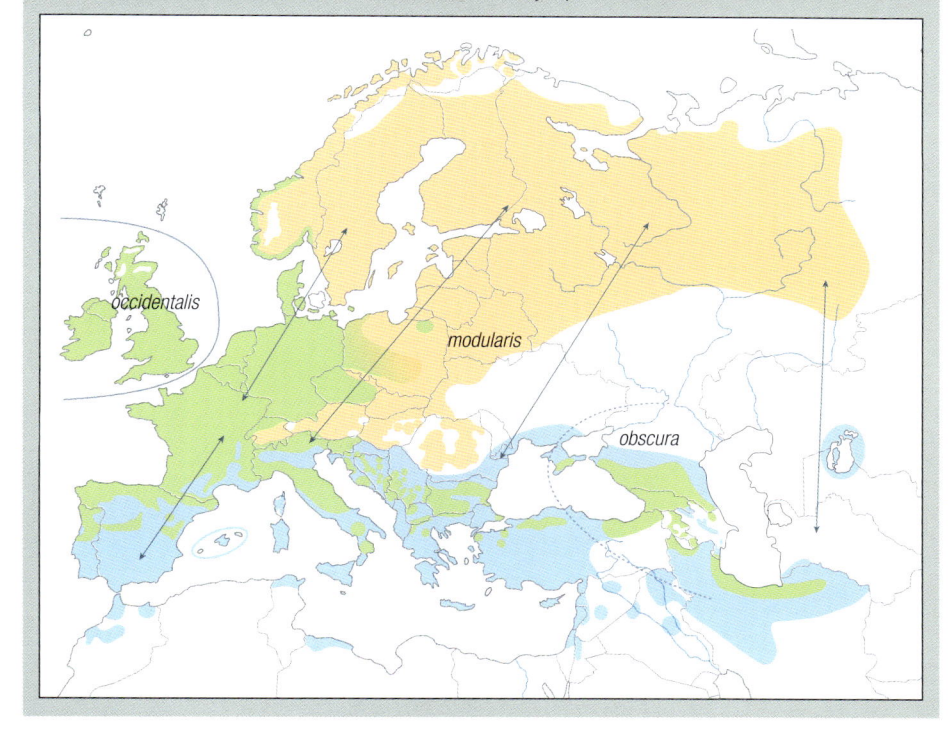

P. m. modularis, 1stW, presumed ♂, Italy, Dec: most wing-feathers juv (albeit with clear moult limits among greater coverts, three outermost visible ones being juv, the fourth replaced, fifth juv, and rest new), while duller iris and paler lower mandible age this bird as 1stW. Probably a ♂ by amount of grey on foreparts, but sexing 1stY often unsafe. (D. Occhiato)

IDENTIFICATION A typical accentor, shaped like a slim sparrow with a *spiky bill*. Three plumage characteristics make identification straightforward: *lead-grey head, with paler grey surround to browner, streaky ear-coverts* and plain or mottled brown-black crown; *profusely streaked brown-black mantle and scapulars* (rather in House Sparrow fashion, though darker and more rufous); and hint of *narrow whitish greater-covert bar*. The overall impression is of a rather uniformly dark bird. Most show a *rusty secondary panel and buff-brown body-sides streaked rufous-brown*; smoky-grey throat and breast grade to dusky white on central belly, with boldly streaked vent and undertail-coverts. Also unstreaked dull brown rump/uppertail-coverts (contrasting slightly with black-brown tail) and tiny whitish spots on tertials. The dark horn-coloured bill has pink-brown lower base, legs are pink to red-brown and *iris is red-brown* (adults). Habitually quite restless, seldom venturing far from cover. Characteristic mouse-like shuffle or creeps with short sparrow-like hops (as if legs fixed together), but also more active hopping with simultaneous wing-flicking, and sometimes jerks tail. Flight typically whirring, low but rather fast, with longish tail obvious, rather like a clumsy *Sylvia* warbler. Usually

P. m. modularis, ad ♂, Italy, Apr: in spring, foreparts purer plumbeous-grey; reddish iris no longer a useful ageing criterion, as many 1stS can develop such colour. However, the wing is evenly feathered and not heavily worn, greater coverts have relatively small and ill-defined pale tips, while primary-coverts are ad-like; sexed as ♂ by amount of grey below and on face. (D. Occhiato)

P. m. modularis, presumed ad ♀, Germany, Feb: evenly-feathered wing, including greater coverts (impression of obvious whitish tips to outer feathers but ill-defined tips on inner ones probably an artefact) indicate ad; primary-coverts also ad. Once aged, assumed to be ♀ by relatively small amount of grey on foreparts. Reddish (ad-like) iris, but most of bill still pale, the latter a winter feature. (S. Pfützke)

encountered singly or in pairs and often in same habitat as Linnet, Robin and Wren.

VOCALISATIONS Song (usually given from atop bush or low tree, and lasting *c.* 2 seconds) is a modest but familiar warble, clear and quite loud, somewhat suggestive of Wren but much quieter, less explosive, lower and more even-pitched, and more formless, notes running into each other without clear structure, e.g. *tütellititelletititütellütotelitelleti*; given year-round. — Commonest call an uninflected discordant *tiih* which becomes more insistent in alarm. On migration a fine rapid series on same pitch, a very thin *zizizizi* or *tihihihi*, almost mouse-like.

SIMILAR SPECIES Unlikely to be confused with any congener within treated range, given a reasonable view, as *Alpine Accentor* is much larger and more boldly patterned, and the Asian species all possess striking (usually dark or black) head markings, with distinctive supercilia and obviously much paler or brighter underparts. — *Japanese Accentor* (*P. rubida*; extralimital, not treated) is very similar, only differing in browner crown and lack of paler grey supercilium. — Given a brief view, might be confused with any other small, ground-hugging passerine, e.g. juvenile *Robin* or a small bunting *Emberiza*; indeed, the plumage pattern most closely recalls Black-faced Bunting!

AGEING & SEXING Ageing in usual encounters rarely feasible, but sometimes possible by close inspection of moult pattern and feather wear in wing, and iris coloration. Sexes largely alike with little seasonal variation. Generally, age-related and sexual differences complex and rather slight, requiring much expertise and practise with birds in the hand, and even then many will prove impossible or at least difficult to label due to variability and effect of bleaching and wear (Menzie & Malmhagen 2013). — Moults. Complete post-nuptial moult and partial post-juv moult in late summer (mostly Jul–Sep, sometimes Oct). Post-juv moult involves all head- and body-feathers, lesser coverts, usually all median coverts, usually none to few inner greater coverts (but some replace more and very rarely all), usually no tertials (but occasionally smallest feather moulted in early-fledged birds), and very occasionally some or even all tail-feathers renewed. Pre-nuptial moult absent. — **SPRING Ad ♂** Sexes usually not safely separable, although ad ♂ averages cleaner dull plumbeous-grey on head (and even purer grey by late spring). Crown often plain dark brown-grey, rarely streaked or mottled

P. m. modularis, 1stW, presumed ♀, Spain, Sep: aged primarily on dull olive-brown iris and pale base to lower mandible. Supporting evidence provided by rather bold buffish tips to greater coverts. Despite being ♀-like, sexing unsafe because of degree of variation in grey foreparts between the sexes, especially when plumage very fresh. (H. Shirihai)

P. m. modularis, juv, Italy, Jun: soft, fluffy body-feathers and juv greater coverts. Dull eye colour, pale base to bill, and yellowish gape-flanges still visible. (D. Occhiato)

P. m. occidentalis, presumed ad ♂, England, Mar: upperparts of the endemic British race only slightly darker and more olive-brown, with slightly heavier, broader and blacker streaking; overall less rufous than *P. m. modularis*, while lead-grey foreparts are darker still. Sepia- or amber-brown iris, black bill and evenly fresh wing-feathers, with extensive and purer grey on head and below, suggest ad ♂. However, many birds in spring are difficult to age and sex; iris and bill colours are age-, sex- and season-related, while in spring juv greater coverts can appear more ad-like. (C. Bovis)

blackish. Chin and upper throat lead-grey and concolorous with rest of throat/breast, with only hint of paler chin in a few. **Ad ♀** On average slightly paler, less clean grey on face, chin, throat and breast (fringed paler and more mottled dirty grey and brown), while breast often tinged olive-brown, not quite uniformly grey, belly more extensively white, but allow for some overlap with less classic ♂♂; bleaches paler and less pure grey in late spring/summer, face and breast retain faint olive fringes, with even stronger olive-brown tone to crown and ear-coverts. Grey-brown crown nearly always boldly streaked blackish, exceptionally plain like in ♂. Chin/upper throat commonly paler than rest of throat, more grey-white and mottled. **1stS** Like respective ad, but in the hand has on average relatively more worn and browner juv wing-feathers (fringes often frayed or uneven), though moult limits and differences in pattern of greater coverts difficult to detect. Some 1stS have iris more olive or duller brown. Duller imm ♂ approaches ad ♀, and imm ♀ shows on average least pure grey, and streakier crown and ear-coverts, but much overlap. — **AUTUMN Ad** Iris almost invariably reddish-brown and bill all dark. Plumage evenly fresh. Greater coverts have pale (whitish or ochre-whitish) tips, especially outer ones, with greyish-black along shaft and faintly olive-tinged brownish fringes, both rather ill-defined and blending into each other; coverts on average slightly broader and more rounded at tips than in juv; pale tips restricted to outer webs or only faintly invading inner webs on inner coverts (Drost 1951; cf. 1stW). Plumage generally much as in spring, but pale fringes and

ACCENTORS

tips broader, and plumage overall less obviously streaked. Grey of face and foreparts washed olive-brown. **1stW** Much as fresh ad, but iris dull brown or greyish olive-brown through Oct and often Nov, and bill has paler base to lower mandible (in both respects becoming more ad-like from late autumn). Primaries, primary-coverts and tail-feathers, and some outer or all greater coverts are juv (though in rare cases birds moult all), at times slightly less fresh, especially by late autumn. Also, retained juv greater coverts sometimes subtly narrower distally, with bolder and on average larger yellowish-buff tips (on inner and central coverts nearly equally well marked on inner webs), contrasting with better-defined and darker black field around shaft, while renewed inner greater coverts (if any) have less contrasting dark centres, and less sharply defined and on average smaller pale (more whitish) tips, at times absent on inner coverts (Drost 1951). Still, moult limits often very difficult to detect in autumn due to individual variation and frequent similarity of juv and post-juv coverts pattern. Young ♀ tends to be duller overall, more heavily mottled off-white to pale buffish on throat, breast more brownish/olive with more buff-white fringes, and belly more extensively off-white; young ♂ and ad ♀ intermediate in relation to purer grey ad ♂. **Juv** Soft, fluffy body-feathering, overall more ochre or buff, lacking any grey, and has whitish throat, more prominently striated underparts (also on breast and lower throat) with narrow dull white central belly, pale buff tips to greater coverts forming striking wing-bar, and pale supercilium.

P. m. occidentalis, presumed 1stS ♂, England, Feb: no other species has same combination of size, jizz and plumage, including strong but pointed bill, heavily streaked upperparts and some grey surrounding brownish ear-coverts. Ageing and sexing more complicated, in particular in spring when colour of iris and bill is generally no longer helpful. This could be a 1stS ♂ but since ageing is not conclusive, ad ♀ cannot be excluded. (H. Shirihai)

P. m. occidentalis, 1stS ♀, England, Apr: olive-tinged iris, seemingly all-juv wing-feathers and extensively streaked brown crown, indicate 1stS ♀. Generally, it is only possible to age and sex extremely well-marked examples, like this one. (S. Round)

P. m. obscura, presumed ad ♂, N Iran, May: subspecific designation by range, and overall duller and less russet plumage; this race also tends to have more ill-defined dark fuscous-brown streaking above, more limited and duller grey areas, browner underparts, but quite extensive rufous-brown blotching on flanks. Seemingly ad-like wing, with sufficient grey feathering on head, neck and breast to best fit ad ♂. (C. N. G. Bocos)

BIOMETRICS (*modularis*) **L** 14–15 cm; **W** ♂ 66–73 mm (*n* 24, *m* 70.4), ♀ 66–70.5 mm (*n* 14, *m* 68.5); **T** ♂ 54–64 mm (*n* 24, *m* 58.1), ♀ 54–62 mm (*n* 14, *m* 57.0); **T/W** *m* 82.8; **B** 12.6–15.2 mm (*n* 37, *m* 13.8); **BD** 3.8–4.5 mm (*n* 25, *m* 4.1); **Ts** 18.9–21.2 mm (*n* 37, *m* 19.9). **Wing formula: p1** > pc 2.5 mm, to 1 <; **p2** < wt 3.5–7.5 mm, =6 or 6/7 (94%) or =5/6 (6%); **pp3–5** about equal and longest (but p3 often subtly shorter); **p6** < wt 1.5–4 mm; **p7** < wt 7–9 mm; **p8** < wt 9–11 mm; **p10** < wt 12–16 mm; **s1** < wt 13.5–17 mm. Emarg. pp3–6.

GEOGRAPHICAL VARIATION & RANGE Generally only slight variation (and several described races best regarded as synonyms), involving coloration, wing shape and size, but much overlap in all of these traits. Colour variation within all populations fairly large. Here only the following three races are recognised.

P. m. occidentalis (Hartert, 1910) (British Isles; resident). Somewhat darker than *modularis* with a rather dark lead-grey face and breast, often swarthy-looking crown and nape and more heavily streaked back. Also, it has a somewhat more rounded wing, proportionately longer tail and subtly stronger feet and bill. Still, some are difficult to separate from *modularis*, which winter in same area. **W** ♂ 68–72 mm (*n* 14, *m* 70.8), ♀ 66–71.5 mm (*n* 18, *m* 68.4); **T** ♂ 59–65 mm (*n* 14, *m* 61.3), ♀ 55–62 mm (*n* 18, *m* 58.5); **T/W** *m* 86.0; **B** 13.5–15.5 mm (*n* 31, *m* 14.4); **BD** 4.1–4.9 mm (*n* 18, *m* 4.4); **Ts** 20.0–22.4 mm (*n* 29, *m* 21.3). **Wing formula: p1** > pc 3 mm, to 0.5 mm <; **p2** < wt 3.5–7 mm, =6/7 or 7 (75%), =6 (15%) or =7/8 (10%); **p6** < wt 0–2 mm; **p7** < wt 4–7 mm; **p8** < wt 7–9.5 mm; **p10** < wt 10–13.5 mm; **s1** < wt 12–15 mm. – A comparison of series in NHM of the present subspecies with proposed '*hebridium*' did not reveal any consistent differences. (Syn. *hebridium*; *hibernicus*; *interposita*.)

P. m. modularis (L., 1758) (much of Europe except British Isles, east to Urals, N Turkey; breeders of Fenno-Scandia and much of Russia winter W & C Europe, breeders in W & C Europe largely resident). Described above in main account and under Biometrics. (Syn. *arduennus*; *euxina*; *lusitanica*; *mabbotti*; *meinertzhageni*.)

○ *P. m. obscura* (Hablizl, 1783) (Crimea, Caucasus, Transcaucasia, perhaps E Turkey and N Iran; mainly short-range movements in winter). Subtly paler than *modularis* and slightly browner overall, on average with more ill-defined dark brown or fuscous (rather than black) streaking above and less pure grey face. Grey of breast admixed off-white and brownish (forming somewhat scaly pattern even when moderately worn), and flanks can show quite extensive rufous-brown blotching (but some overlap with *modularis*). Same size as *modularis* (despite statements otherwise). **W** ♂ 68–72 mm (*n* 15, *m* 70.3), ♀ 66–71 mm (*n* 14, *m* 68.3); **T** ♂ 55–65 mm (*n* 15, *m* 59.4), ♀ 53–62 mm (*n* 14, *m* 57); **T/W** 84.0; **B** 13.4–15.8 mm (*n* 28, *m* 14.5); **BD** 4.0–4.6 mm (*n* 10, *m* 4.3); **Ts** 20.3–22.3 mm (*n* 28, *m* 21.1). (Syn. *blanfordi*; *enigmatica*; *fuscata*; *orientalis*.)

REFERENCES MENZIE, S. (2014) *Ringing News*, 13: 4. – MENZIE, S. & MALMHAGEN, B. (2013) *Ringers' Bull.*, 28: 57–62.

SIBERIAN ACCENTOR
Prunella montanella (Pallas, 1776)

Fr. – Accenteur montanelle; Ger. – Bergbraunelle
Sp. – Acentor siberiano; Swe. – Sibirisk järnsparv

Only a marginal breeder in the W Palearctic, inhabiting overgrown thickets in the taiga of the Pechora River basin and N Urals, usually in willow and birch scrub, but also in conifer forest and tundra edge. From its W Palearctic outpost it occurs east to Anadyrland, mostly north of the Arctic Circle, more locally to S Siberia, and winters south to NE China and Korea. It occasionally wanders west to N & NW Europe in autumn (most records in Fenno-Scandia), exceptionally even reaching SE Europe and the Middle East.

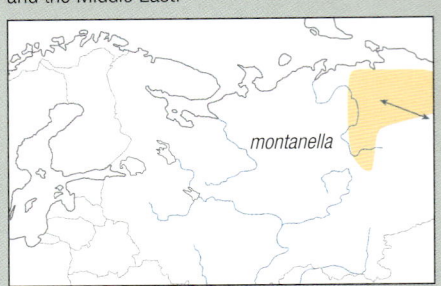

P. m. montanella, ad ♂, Russia, Sep: characteristic are the ochre or orange-tinged colours and boldly patterned head with invariably pale throat. No apparent moult limits in wing, reddish iris, and solid head markings indicate fresh ad ♂. (S. Cherenkov)

P. m. montanella, 1stW, Sweden, Oct: sex unclear, but possibly ♀ considering slightly duller colours and pattern than most. Ageing based on dark iris and slightly worn and pointed tail-feathers. Although there is probably a moult limit among greater coverts, this is difficult to establish without handling. Rather dull grey-brown upperparts lacking rufous also support ageing. (P. Lundgren)

IDENTIFICATION Dunnock-sized *long-tailed* accentor, with *long, flared orange-tinged ochre-buff supercilium* (broadest behind eye) and *similarly-coloured throat and breast*, *brownish-black crown and ear-coverts*, narrow greyish shawl, *warm reddish-brown upperparts with darker red-brown* (common) *or blackish* (less common variation) streaking, but greyer rump. Dark wings with broad rufous-brown fringes and tips, and longish brown tail. Some distinctly colder brown and heavily streaked darker above. Rest of underparts buffish with dark rufous-brown streaking, especially on rear flanks; usually has dark *blotches on breast*, but dark feather-centres often concealed when fresh. At close range, paler central crown usually evident, with dark lateral crown-stripes and blackish ear-coverts, with variable but distinct pale rear spot. When fairly fresh has *narrow whitish wing-bars*, and some have whiter abdomen, but at all ages bill black or horn-coloured, with ochre base to lower mandible, and legs and feet bright yellow-brown. Habits and jizz typical of accentors, but note proportionately long tail. Rather shy on breeding sites and keeps mostly in cover, although vagrants can be rather tame.

VOCALISATIONS Song Dunnock-like, thin, high-pitched and similarly uninterrupted, a brief outburst of sharp notes. Compared to Dunnock slightly less clear, more sharp and scratchy. Possibly also subtly more low-pitched, e.g. *tri-tie-tirz-te-tir-zi*. – Call (including in flight) a fine *ti-ti-ti*, or *se-se-se*, very Dunnock-like and doubtfully separable.

SIMILAR SPECIES Main confusion risk is with Black-

P. m. montanella, presumed 1stW ♂, Sweden, Feb: same individual from different angles; note rusty-buff throat and long supercilium (broader behind eye), darker lateral crown-stripes and dark ear-coverts (with distinct pale spots). Upperparts streaked, but rump greyer, while ochre-buff underparts have dark rufous-brown streaking; dark bill with paler lower mandible, and legs bright flesh-brown. In right-hand image, remiges and tail seem rather worn for Feb (presumed juv feathers), and iris lacks deep red tone of ad. Sexing not easy, but for young bird in late winter, plumage best fits ♂. (L. Olsson)

throated *Accentor*, especially in fresh non-breeding plumage, when latter can appear to lack a dark bib. Their separation is covered under the latter's account. – May also require separation from *Radde's Accentor*, which has dirty white supercilium, orange-tinged breast (especially in winter), a hint of a dark lateral throat-stripe, obvious pale submoustachial stripes (lacking in Siberian), generally paler upperparts with bolder blackish streaks, and tends to have a browner central crown. – Very similar also to extralimital *Brown Accentor* (*P. fulvescens*; not treated), but this has pinkish flush rather than ochre in buff underparts, is more greyish-brown above with more subdued streaking, and has slightly less dark crown even when worn. Lacks tendency of Siberian to have blackish feather-centres on breast partly showing, is invariably smoothly pink-buff below.

AGEING & SEXING Ageing possible by close inspection of moult, feather pattern and wear in wing, and using iris coloration. Sexes alike or very similar in plumage (but see Biometrics). Seasonal variation rather limited. – Moults. Complete post-nuptial moult and partial post-juv moult in late summer (mostly Jul–Sep). No pre-nuptial moult. – SPRING **Ad** Both sexes differ from 1stS in being evenly feathered and in having less worn primaries and primary-coverts. Independent of wear, age or sex, some have upperparts less orange-chestnut, and more heavily streaked blackish-brown and rufous. **Ad** ♂ With wear, black (actually very dark brown) of head deeper, supercilium may fade to pale buff, rounded darker markings on underparts, especially breast, more visible, and upperparts become duller rufous with stronger streaking. **Ad** ♀ Often inseparable from ♂ on plumage, but has slight tendency to be subtly duller in colours and pattern. Crown generally browner with sometimes less obvious darker lateral crown-stripes (but a few similar to ♂), cheeks browner and upperparts less rufous; also averages less bright or deep ochre-yellow below, and dark feather-centres on breast less obvious. **1stS** Very similar to ad, but juv primaries, primary-coverts and tail relatively more worn, sometimes forming subtle moult limits (see 1stW in Autumn for moult patterns). – AUTUMN **Ad** Differs from 1stW in being evenly fresh, and has dark sepia or reddish-brown iris. Plumage largely as breeding, but supercilium may appear brighter, and dark marks below less obvious. **1stW** Much like ad, though juv primaries, primary-coverts, tail and usually all/most greater coverts and tertials retained (subtly more worn and weaker textured), sometimes slightly contrasting with newly moulted ad-like feathers (any new greater coverts edged more greyish, tipped purer white, dark central streak less bold, often not reaching tip). Iris often dark grey-olive, but from Oct some develop warm brown colour. **Juv** Soft, fluffy body-feathering is less rufous on upperparts, and head pattern is overall duller, while underparts are dull cream-buff with large brown spots on breast and throat-sides; lacks grey patch on neck-sides of ad.

BIOMETRICS (*montanella*) **L** 13–14.5 cm; **W** ♂ 67–76.5 mm (*n* 36, *m* 72.4), ♀ 64–74 mm (*n* 22, *m* 70.3); **T** ♂ 58–70 mm (*n* 36, *m* 64.1), ♀ 55–66 mm (*n* 22, *m* 62.0); **T/W** *m* 88.5; **B** 12.0–15.1 mm (*n* 56, *m* 13.3); **BD** 3.8–4.7 mm (*n* 25, *m* 4.2); **Ts** 18.0–20.2 mm (*n* 29, *m* 19.1). Wing formula: p1 > pc 3 mm, to 3 mm <; **p2** < wt 3–4.5 mm, =6 or 6/7 (rarely =5/6); **pp3–5** about equal and longest (but p3 often subtly shorter); **p6** < wt 1.5–4 mm; **p7** < wt 7–10 mm; **p8** < wt 9–13.5 mm; **p10** < wt 12.5–17 mm; **s1** < wt 14–18.5 mm. Emarg. pp3–6.

GEOGRAPHICAL VARIATION & RANGE Two races, but extralimital *badia* from NE Siberia only weakly differentiated from *montanella* (see below).

P. m. montanella (Pallas, 1776) (extreme NE Europe, Pechora, N Urals, east to C Siberia, S Yakutia, Amur; winters C & E Asia). Described above.

Extralimital: ○ *P. m. badia* Portenko, 1929 (extreme NE Siberia, lower Lena, Anadyr, Chukotski; winters E Asia). Very similar to *montanella* but averages subtly darker on crown (some have near-uniform blackish crown and very dark cheeks), darker rufous mantle and scapulars, perhaps on average a little deeper ochre below when fresh. Size very similar but tail proportionately subtly shorter and bill a little longer. **W** ♂ 68.5–75 mm (*n* 10, *m* 72.3); **T** ♂ 59.5–65 mm (*n* 10, *m* 62.3); **T/W** *m* 86.2; **B** 12.7–15.0 mm (*n* 12, *m* 13.8).

TAXONOMIC NOTE For relationships within *Prunella*, see Black-throated Accentor.

P. m. badia, ad ♂, South Korea, Jan: reddish iris (just discernible) and solid black head markings make ageing straightforward. Aside of more boldly patterned blackish head, *badia* also characterised by deep chestnut streaks on upperparts, while deeper cinnamon-buff underparts are streaked more thinly and more rufous (less blackish) on flanks. (K. Hyun-Tae)

P. m. badia, 1stW, presumed ♂, China, Dec: ageing often difficult, but relatively worn rectrices and dark-looking eye indicate 1stW, supported by distinct black central streaks on greater coverts reaching tips (similar to age difference in Dunnock), and if so plumage fits better for ♂. Predominantly rufous mantle streaking is a feature of race *badia*. (M. Parker)

P. m. montanella, 1stW ♂, Turkey, Nov: ageing difficult (like all *Prunella*) as moult limits in wing often difficult to discern, and ad and juv greater coverts very similar. However, primaries, primary-coverts and rectrices are quite obviously juv (browner, more worn and latter pointed), the bill's base is pinkish and iris is apparently olive, indicating 1stW. Blacker head markings and richer upperparts suggest ♂. This bird's predominantly rufous mantle and flanks streaking approaches *P. m. badia*, but some *montanella* can show this amount of rufous. (S. Bekir)

Comparison between Siberian Accentor (top left: England, Oct), Black-throated Accentor (top right: Sweden, Oct), Radde's Accentor (bottom left: Israel, Dec) and Brown Accentor (bottom right: Kazakhstan, Oct): in 2016, there was an exceptional influx of Siberian Accentors, with at least 200 birds recorded in N & W Europe. Most birds were tame and allowed close views and a better understanding of the variation in autumn plumages, but also drew attention to how similar several of the Asian accentors can be, hence this comparison of four species, which at times can be most challenging to separate. — Separation of **Siberian Accentor** and **Black-throated Accentor** is usually unproblematic as long as latter shows anything of its dark bib. Rarely in autumn, however, some Black-throateds appear entirely pale-throated (at least in certain angles and lights), requiring careful observation. In fresh plumage, Black-throated can have such broad yellow-buff feather-tips on throat that it masks nearly every sign of the in summer black bib. To complicate matters further, some Siberians (as on the one shown here) can have some apparent dusky upper-throat markings. Although difficult to make out here, subtle differences might help, such as supercilium in Siberian tends to be be paler and whiter at rear (reverse in Black-throated) and on average more rufous tinge on mantle and flanks and more even orange-buff underparts (in Black-throated tendency to contrast between orange-buff breast and slightly more whitish belly). — Separation of **Radde's Accentor** (bottom left) from **Siberian Accentor** and pale-throated **Black-throated Accentor** can be especially problematic if (rarely) Radde's lacks the clear thin dark lateral throat-stripe, a stripe which as a rule defines an obvious pale submoustachial stripe. Still, note Radde's characteristic combination of bolder head pattern (with uniform or at least less streaky crown and ear-coverts, and with pale rear patch of ear-coverts restricted to few dots), but better-developed whitish supercilium that typically broadens behind eye, and subtly better-marked buffish-orange breast-band. The extralimital **Brown Accentor** (bottom right; not treated, but see under Siberian Accentor, similar species) combines many characteristics of all the other three species, but has an especially Radde's-like head pattern. Faced with a potential vagrant, check for extensive pinkish-buff flush on much of the rather unstreaked underparts. — Considering the overall rather dull plumages, all four are probably 1stW, with the first two probably ♀♀, being so relatively pale and lacking stronger contrasts. (Top left: D. Hutton; top right: J. Rosquist; bottom left: O. Sherer; bottom right: A. Isabekov)

BLACK-THROATED ACCENTOR
Prunella atrogularis (Brandt, 1843)

Fr. – Accenteur à gorge noire
Ger. – Schwarzkehlbraunelle
Sp. – Acentor gorginegro; Swe. – Svartstrupig järnsparv

One of the 'ultimate' vagrants to find in Europe, Black-throated Accentor is predominantly an E Palearctic species which breeds in high-altitude spruce forests and in lowland taiga east to Altai, Dzungaria and W China, and winters further south and at lower elevations, including in E Iran. It occasionally wanders west and south, having reached several European countries (most frequently in Fenno-Scandia), Israel, Kuwait and Oman.

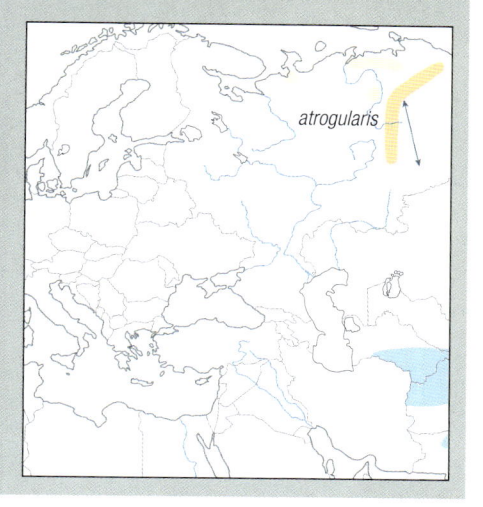

P. a. atrogularis, ♂, Urals, Jun: summer plumage develops during breeding season, with black bib merging into ear-coverts, and pale submoustachial stripe and supercilium wearing away to various degrees. In this extreme case, only rear supercilium is visible, making the bird appear 'black-headed', rather than 'black-throated'. (D. Monticelli)

IDENTIFICATION Slightly larger than Dunnock and Siberian Accentor, but structurally close to both. Most obvious are the *long whitish-buff supercilium and submoustachial stripe*, and *diagnostic black bib*. In fresh plumage the black *bib is partially or nearly wholly concealed by pale feather tips*. Rest of head looks black or blackish-brown, but when seen close note *central crown is visibly brown* (with narrow darker lateral crown-stripes), while *dark ear-coverts almost always have pale rear mark*, and, when fresh, *an area below eye is mottled paler*. In worn plumage, whitish-buff submoustachial stripe often narrower, shorter and sometimes totally absent due to feather abrasion, thus *bib and cheeks merge and face can appear all black*. Conspicuously *black-streaked, dull greyish-brown upperparts*, and boldly dark-centred wings with well-defined pale fringes, especially to tertials and greater coverts, and whitish tips to greater coverts, forming *narrow wing-bar*; rump almost plain olive-brown. Yellowish-ochre or rusty-buff breast and flanks rather clearly set off from whitish belly, *streaked blackish or brownish* (never rufous), *most heavily on rear flanks*. Small pale grey neck-sides (variable). Habits and jizz typical of accentors.

VOCALISATIONS Song typical of accentors, a brief, high-pitched outburst of uninterrupted notes, perhaps most similar to Radde's Accentor, with short strophes, few changes of pitch and at comparatively low volume, but still explosive and high-pitched, e.g. *tri-trzite-se-tri-trzite-zeri*. To the inexperienced can sound rather like fragment of song of Olive-backed Pipit. – Call is a thin *ti-ti-ti*, very similar to those of congeners; louder and more tremulous in alarm.

SIMILAR SPECIES Could be confused with Siberian Accentor, but Black-throated has (i) diagnostic black bib, obvious in breeding plumage but sometimes much less striking in autumn when fresh. Indeed, some young (chiefly first-winter ♀) Black-throated have black bib largely concealed by broad pale feather-tips (especially in lateral view), when following characters come into play: Black-throated has (ii) paler/whiter supercilium (particularly in front of eye, the reverse in Siberian) and centre of chin/throat (latter may show some dark mottling); (iii) in Black-throated, mantle is generally duller brown and heavily streaked with bolder black-brown marks (lacking Siberian's more rufous tinge, thinner and shorter streaking); (iv) central crown paler with some diffuse streaks (both these features obscure in Siberian); (v) Black-throated typically is bicoloured below, with an orange-buff breast and whiter belly (Siberian has much less contrasting underparts); and (vi) tends to have mottled lower ear-coverts and blacker, less rusty flanks-streaking. In the hand, a pale-throated bird can be confirmed by the diagnostic dark bases to the chin and throat feathers, which are usually more extensive and solid than in Siberian (or Radde's) Accentor. – From *Radde's Accentor* by (i) lack of thin lateral throat-stripe creating obvious pale submoustachial stripe; (ii) pale rear patch to ear-coverts (lacking or very obscure in Radde's, almost always distinctive on Black-throated, albeit somewhat variable); (iii) at least some dark mottling on throat, if not obviously dark (Radde's tends to have more uniform throat being whitish, slightly tinged cream-buff when fresh); and (iv) upperparts usually paler brown and more boldly dark-streaked than in Radde's.

AGEING & SEXING Ageing and sexing very difficult, and in spring often impossible. Size of some use for sexing (see Biometrics). – Moults. Complete post-nuptial moult of ad, and partial post-juv moult, in late summer (mostly Jul–Sep). Post-juv renewal involves all head- and body-feathers, lesser and median coverts, a variable number of inner greater coverts, and all to no tertials. Pre-nuptial moult limited or insignificant, and little known. – **SPRING Ad** Both sexes differ from 1stS in having on average less worn primaries and primary-coverts. **Ad ♂** With wear, black of head and throat deeper and more solid (and pale submoustachial stripe often worn-off, thus black bib merges with side of head), supercilium and underparts paler than in autumn, and belly whiter, with more dark spots on breast-sides, occasionally even on breast. Mantle and scapulars more prominently streaked black. **Ad ♀** Very similar to ad ♂, and many cannot be sexed on plumage, but average duller overall with less intensely black lores, ear-coverts and bib, and duller crown (though much overlap). **1stS** Juv primaries, primary-coverts, some outer greater coverts and tail relatively more worn and bleached duller than ad, often creating slight moult contrast (see 1stW); least-advanced imm ♀ usually duller and less contrastingly patterned than ad ♂, but differences from both ad ♀ and imm ♂ indistinct. Apparently some undergo partial winter moult (e.g. of tertials and even some body-feathers and inner wing-coverts) but scale unknown, and unclear if such a moult only involves 1stY or also ad. – **AUTUMN Ad** Ear-coverts and bib have pale tips and fringes, giving scaled appearance (stronger in ♀), and breast lacks or has reduced dark feather-centres; generally paler areas, including supercilium, more saturated buff. Differs from 1stW in being evenly fresh with dark brown or reddish-brown iris. **1stW** Much like ad, but dark throat feathering largely concealed by broad pale feather-fringes; it may be virtually impossible to see any dark feathering on some young ♀♀ when viewed from side. Juv remiges, tail, primary-coverts, and variable number of outer greater coverts more worn and weaker textured than ad feathers; iris dull olive-brown or greyish (at least in early autumn). **Juv** Duller above, and all of yellowish off-white underparts are extensively streaked; head pattern

P. a. huttoni, variation in summer ♂♂, Kazakhstan, May (left) and Jun: irrespective of racial variation, or age and sex, during summer the pale submoustachial stripe and supercilium wear off to various degrees, in some being lost completely prior to post-nuptial moult. Considering the rather solidly dark heads, both seem to be ♂. (A. Isabekov)

more diffuse with thinner ill-defined supercilium, lacking grey neck-sides (cf. juv Siberian, as they overlap in range).

BIOMETRICS (*atrogularis*) **L** 13–14 cm; **W** ♂ 74–76 mm (*n* 10, *m* 74.9), ♀ 70–73.5 mm (*n* 8, *m* 71.4); **T** ♂ 59–64 mm (*n* 10, *m* 61.5), ♀ 55–61.5 mm (*n* 8, *m* 58.5); **T/W** *m* 82; **B** 11.3–12.8 mm (*n* 18, *m* 12.1); **B**(f) 8.5–10.0 mm (*n* 18, *m* 9.3); **BD** 3.6–4.5 mm (*n* 17, *m* 4.1); **Ts** 17.7–20.1 mm (*n* 17, *m* 19.1). **Wing formula: p1** < pc 3, to 0.5 mm >; **p2** < wt 2–6 mm, =6 or 6/7 (60%) or =5/6 (40%); **pp3–5** about equal and longest; **p6** < wt 1.5–4.5 mm; **p7** < wt 6–10 mm; **p8** < wt 9–13 mm; **p10** < wt 14–18 mm; **s1** < wt 15–21 mm. Emarg. pp3–6.

GEOGRAPHICAL VARIATION & RANGE Two rather well-separated subspecies, both as to range and morphology, but differences between extremes still rather small, and perhaps not all birds can be labelled. Southern birds tend to be slightly larger, with larger black bib and more extensive and yellowish-ochre tinge below.

P. a. atrogularis (Brandt, 1844) (NE Russia and Urals; winters Afghanistan, N Pakistan, NW China). Described above. Typically small with small black bib and rather concentrated chest-band of orange-ochre contrasting with white belly. Between the black bib and orange chest there is almost invariably a hint of a whitish band. Many breeders in summer have no or only small pale submoustachial stripe resulting in impression of all-black face. Upperparts warm tawny with moderate dark rufous-brown streaking. Bill, tarsus and tail average somewhat shorter than in *huttoni*. Wingtip somewhat more pointed.

Extralimital: *P. a. huttoni* (Moore, 1854) (Altai, Zaysan area, W Mongolia, Dzungaria, E Kazakhstan, Tien Shan, east to W China; short-range movements in winter, mainly to Central Asia, Kashmir). Slightly larger on average, with upperparts generally more heavily streaked (streaks sometimes nearly black) and crown in worn adult ♂ a little darker brown. Bib averages larger and is not diffusely bordered whitish below (between black and ochre-buff). Cream-white submoustachial stripe of some birds broken or finishes well short of bill so that black bib merges largely with black ear-coverts, but variable in both races. Breast more extensively cinnamon-buff (with almost yellow-ochre tinge rather than orange), diffusely invading belly too, and flanks often more distinctly streaked. Also clearly longer legs, bill and blunter wing shape, with only slight overlap. **L** 14–15 cm; **W** ♂ 72–80 mm (*n* 19, *m* 75.7), ♀ 70–76.5 mm (*n* 10, *m* 73.8); **T** ♂ 60–69 mm (*n* 19, *m* 65.5), ♀ 60–68 mm (*n* 10, *m* 64.3); **T/W** *m* 86.5; **B** 12.5–14.7 mm (*n* 28, *m* 13.8); **B**(f) 9.2–11.4 mm (*n* 14, *m* 10.7); **BD** 4.1–4.7 mm (*n* 14, *m* 4.4); **Ts** 20.0–21.5 mm (*n* 29, *m* 20.6). **Wing formula: p1** > pc 3, to 1 mm <; **p2** < wt 3–5.5 mm, =6/7 (91%) or =6 (9%); **p6** < wt 0.5–3 mm; **p7** < wt 5–8 mm; **p8** < wt 8.5–11 mm; **p10** < wt 13–16 mm; **s1** < wt 14–17 mm. – Birds in north of range, in Mongolia, Altai, Dzungaria and Tien Shan (sometimes separated as '*menzbieri*' and '*lucens*'), tend towards *atrogularis* in being slightly smaller, having more orange tinge on chest with whiter belly, and slightly less strong streaking on upperparts, but a subtle and average difference that constitutes a natural clinal transition towards *atrogularis*. (Syn. *lucens*; *menzbieri*.)

TAXONOMIC NOTES Considering that *atrogularis* in the north has an allopatric range and that virtually all can be separated from *huttoni* using a combination of morphological criteria, it is perhaps worth considering whether it has reached a level of distinctness worthy of species status focusing on vocalisations. However, more field research is needed first. – Forms superspecies with Kozlov's Accentor *P. koslowi* (not treated) of Mongolia and neighbouring China, and these two form a species group together with Siberian Accentor and Radde's Accentor.

REFERENCES Drovetski, S. V. et al. (2013) *Ecology & Evolution*, 3: 1518–1528.

♂, presumed *P. a. atrogularis*, India, Nov: diagnostic black-and-buff head pattern, but in winter plumage bib smaller (confined to chin and uppermost throat) while pale submoustachial stripe well developed. Contrasting and boldly-patterned head indicate ♂; also ad-like in plumage, even having apparently rather fresh (and even) wing-feathers, but ageing still uncertain. Orange-buff tinge below seems to be confined to neck and breast, while belly is whitish typical of this race. (C. Francis)

P. a. atrogularis, 1stW ♂, Sweden, Oct: black areas of head, including bib, intense and suggest a ♂ (though there is much individual variation at this season); seemingly mostly juv wing-feathers. Racial identification based on rather diffuse streaking on comparatively pale upperparts and flanks. (J. Morin)

P. a. atrogularis, 1stW, Sweden, Oct: black bib largely concealed, crown paler, much brown on ear-coverts patch and pointed rectrices indicate a young bird, presumably ♀. In such plumage, especially with dark bib concealed, could be confused with Siberian Accentor. Black-throated has paler/whiter supercilium in front of eye (reverse in Siberian), duller and more heavily-streaked mantle (Siberian more rufous-tinged), central crown paler with diffuse streaking, lower ear-coverts distinctly mottled, and underparts more contrastingly patterned, with orange-buff breast, whiter belly and flanks streaking much less rusty. (T. Holmgren)

RADDE'S ACCENTOR
Prunella ocularis Radde, 1884

Fr. – Accenteur de Radde; Ger. – Felsenbraunelle
Sp. – Acentor de Radde; Swe. – Kaukasisk järnsparv

This discreet passerine of bushy alpine terraces and steep bouldery slopes at 2000–3500 m, at times just below but mostly above the tree-line, breeds only in C & E Turkey, Armenia, Georgia, Azerbaijan, parts of N Iran and in an isolated population in Yemen, where it is one of the most sought-after specialities. Radde's Accentor is generally a short-distance migrant or local resident, wintering at lower elevations in SE Iran and irregularly in N Israel, in Golan Heights. Vagrants have reached Jordan, Lebanon and Oman.

P. o. ocularis, presumed 1stS ♂, Armenia, Apr: with wear, variable number of dark streaks and mottling on breast frequent in all birds. As to age and sex, see image at bottom. (W. Müller)

P. o. ocularis, ♂, Turkey, Jun: long whitish supercilium, buff-white submoustachial stripe and throat, latter usually with few dark streaks at sides, and uniform dark brown ear-coverts and crown. Quite dark head markings and narrower and sparser streaking below indicate ♂. Apparently ad-type primary-coverts, all-black bill and red-brown iris could suggest ad, but ageing uncertain, especially as tips of tail-feathers are confusingly pointed and worn, and therefore perhaps juv. (S. Tigrel)

IDENTIFICATION A Dunnock-sized accentor, with *long off-white supercilium, whitish-buff throat, indistinct dark streaking on sides of throat* (most obvious on lower parts), and *dark brown* (in some lights blackish-looking) *ear-coverts and crown* (with slightly darker brown lateral crown-stripes especially obvious in all non-adult ♂ plumages, which have a paler brown and slightly streaked central crown). Upperparts pale rufous or tawny-brown with brownish-black streaking, tail mostly dark brown, and underparts whitish-cream with *ill-defined but broad buffish-orange* (mustard-yellow) *breast-band* and often some irregular, diffuse streaks on belly and flanks. Compared to several congeners, *pale rear spot on ear-coverts lacking or indistinct*, and *greyish shawl on neck-sides rather obscure or lacking*. Rump can appear to be slightly tinged olive-grey. Tertials and secondary-coverts pale-fringed and latter, when fairly fresh, have *narrow whitish wing-bars* (formed by whitish tips to greater coverts, with indistinct pale tips also to median coverts). Like other accentors, characteristically flicks wings and appears to 'shuffle' along ground, hops forward in abrupt moves. Usually in pairs, or small groups in winter, typically on the ground, hopping or running between patches of cover, but slightly more arboreal in winter, and can be difficult to observe (occasionally remaining concealed in one place for long periods).

VOCALISATIONS Song crystal-clear, given from low perch, a brief, rapid strophe in which several notes are repeated a few times, often also containing a few short, trembling trills, e.g. *se, se, titititerrr-siwe te te* **teswee pe-sewee terrrsi** (strophes vary in length, structure and tempo). Up to ten songs given per minute. Sometimes, at distance or in windy conditions, recalls song of Levant Black Redstart *Phoenicurus ochruros semirufus* (which partly breeds in same areas). – Call a Dunnock-like, high *ti-ti-ti*.

SIMILAR SPECIES Potential confusion with *Siberian* and *Black-throated Accentors* discussed under those species.

AGEING & SEXING (*ocularis*) Ageing rarely possible, and if so only by close inspection of moult pattern in wing and iris colour. Sexes virtually alike (at all seasons, sexing more reliable of birds of known age, and using wing length—see Biometrics), while seasonal variation limited to effects of feather wear and bleaching. – Moults. Complete post-nuptial moult in late summer (generally Jul–Sep). Partial post-juv moult at same time. Post-juv renewal involves all head- and body-feathers, lesser and median coverts, and variable number of inner greater coverts and tertials. Pre-nuptial moult absent. – **SPRING Ad** Both sexes differ from 1stS in having on average less worn primaries and primary-coverts. **Ad ♂** With wear, supercilium and throat-sides sometimes

P. o. ocularis, presumed 1stS ♂, Armenia, Apr: same bird as in top right image. Apparently juv wing-feathers, most notably primary-coverts and tertials, suggest 1stS; quite dark head markings indicate ♂. Some have grey-tinged upperparts like this bird. (W. Müller)

even whiter than in autumn, and dark brown head darkens to more blackish appearance (though still dark dull brown when handled); dark streaks above denser (pale fringes worn). **Ad ♀** Very similar to ad ♂ but on average has less solidly dark, more brown crown streaked darker, and ear-coverts sometimes less solidly dark too; breast usually slightly paler orange than in ♂, and throat and belly paler still; especially with wear becomes scruffier below with exposed dark streaks and mottling. **1stS** Juv feathers relatively more worn than ad, at times forming rather clear moult limits: primaries and primary-coverts abraded browner with frayed edges and tips, and latter also narrower and more pointed; some/most greater coverts also juv (heavily abraded) with pattern as described for 1stW, but wear obscures differences in pattern; iris still useful for ageing in some least-advanced birds (see Autumn). Plumage overall duller, with imm ♂ approaching ad ♀, and imm ♀ dullest (has browner/streakier head and paler central crown), but much overlap. — **AUTUMN Ad** Smoothly and broadly fringed above, breast deeper/purer orange, especially in ♂, dark brown of head generally duller and underparts streaking mostly concealed; head of ♀ subtly paler brown than ♂. Differs from 1stW in being evenly fresh, and by having reddish sepia-brown iris. **1stW** Like ad, but

P. o. ocularis, ♀, Kuwait, Mar: overall plumage, with extensive flanks streaking, browner head and clear blackish lateral crown-stripes indicate ♀. Greater coverts, single alula and primary-coverts are seemingly juv (pointed and bleached brown), and worn primaries, paler bill base and dull iris all suggest 1stY, but difficult to be sure. (P. Fågel)

P. o. ocularis, presumed 1stW ♂, Turkey, Dec: primary-coverts, and possibly at least one of outer greater coverts are apparently juv, while duller iris, broad pale base to bill and spotty ear-coverts suggest a young bird, while relatively darker head markings indicate a ♂. Note how streaks on body-sides are more diffuse when plumage still fresh. (H. Meşe)

P. o. ocularis, Turkey, Jan: unlike Siberian Accentor, Radde's typically has dirty white supercilium (never distinctly orange-buff), but breast obviously orange-tinged (especially in winter), with some scattered dark streaks on sides of throat (lacking in Siberian). From bib-less immature winter Black-throated note whiter supercilium, plainer brown central crown and more uniform ochre underparts with rather unstreaked breast of Radde's. (A. Atahan)

P. o. fagani, 1stW ♂, Yemen, Oct: aside from geography, the most consistent difference from *ocularis* is smaller size, though note shorter primary projection, stronger lateral throat-stripe and patch, duller underparts with more extensive and even streaking, more olive-grey upperparts, and subtly paler brown head and slightly stronger bill. Moult limits among tertials and greater coverts, and pointed tail-feathers of 1stY, with rather intense dark head markings of ♂. (U. Ståhle)

ACCENTORS

P. o. fagani, ♂ (left) and ♀, Yemen, Oct: head pattern overall darker and more intense, with more rusty-buff on breast in ♂, and underparts are more extensively streaked blacker on breast and flanks, but overlap exists between sexes, and variation in relation to age is poorly understood. (U. Stähle)

juv primaries, primary-coverts and tail-feathers, and many or all greater coverts somewhat less fresh and weaker textured. Retained greater coverts show narrower, looser and slightly more ill-defined buffish-rufous fringes to outer webs, and whitish tips broken by dark centres, which have more pointed ends reaching edge of coverts and are duller and browner with less clearly-defined borders (in ad centres darker and sharply demarcated but more rounded at tip and usually do not reach edges, thus not breaking pale tips, which are more smoothly rounded, and fringes to outer webs wider, better defined and more greyish-olive). Primary-covert fringes narrower and more buffish-rufous (in ad, fringes are greyer, broader and more ill-defined). Iris olive grey-brown. Young ♀ dullest with browner/streakier head (young ♂ and ad ♀ intermediate compared to ad ♂), but sexing even less easy than in spring, and impossible if birds not aged first. **Juv** Brownish crown, streaked darker, underparts, especially breast and flanks, more heavily streaked, underparts paler, inclining to buff-white, supercilium less distinct and throat less clean.

BIOMETRICS (*ocularis*) **L** 14.5–15 cm; **W** ♂ 75–78.5 mm (n 12, m 76.8), ♀ 71–77 mm (n 10, m 73.5); **T** ♂ 61–70 mm (n 12, m 65.6), ♀ 60–66 mm (n 10, m 63.3); **T/W** m 85.8; **B** 13.4–15.1 mm (n 22, m 14.4); **B**(f) 10.0–12.0 mm (n 22, m 11.1); **BD** 3.9–5.2 mm (n 21, m 4.5); **Ts** 20.0–22.3 mm (n 21, m 21.2). **Wing formula: p1** > pc 0–4.5 mm, < p2 34–39 mm; **p2** < wt 3–6 mm, =6/7 or 7; **pp3–5**(6) about equal and longest; **p6** < wt 0.5–2 mm; **p7** < wt 3.5–6.5 mm; **p8** < wt 7–10.5 mm; **p10** < wt 11–15 mm; **s1** < wt 13–16.5 mm. Emarg. pp3–6.

GEOGRAPHICAL VARIATION & RANGE Only slight variation, but range disjunct, and isolated population in Yemen regularly separated as 'Yemen Accentor'. However, both genetically (Drovetski *et al.* 2013) and morphologically the Yemen population is very close to Radde's Accentor and is best treated subspecifically.

P. o. ocularis (Radde, 1884) (SC & E Turkey, Transcaucasia, N, W & SW Iran; resident or makes only short movements to lower altitudes). Treated above. The slightly brighter subspecies of the two, although difference slight. Averages less streaked below, crown is slightly darker, particularly in ♂, and breast more saturated mustard-yellow.

P. o. fagani (Ogilvie-Grant, 1913) 'Yemen Accentor' (Yemen; resident, might occasionally move to lower altitudes, but no proof of this). Similar to *ocularis* but differs in being somewhat darker and duller brown on similarly streaked mantle and back but paler on crown, this being more streaked and less black-looking even in ♂. White supercilium narrower and throat on average more streaked, whereas breast is slightly less orange-tinged. Belly and flanks usually more streaked than in *ocularis*. Blackish bill appears from specimens to have more extensive yellow-pink on base of lower mandible. Wing shorter and more rounded, bill a fraction stronger. **W** 67–73.5 mm (n 13, m 69.8); **T** 55–67 mm (n 13, m 60.4) **T/W** m 86.4; **B** 12.5–15.9 mm (n 13, m 14.7); **B**(f) 10.4–12.4 mm (n 13, m 11.6); **BD** 4.3–5.0 mm (n 13, m 4.7); **Ts** 19.4–21.8 mm (n 13, m 21.0). **Wing formula: p1** > pc 2–5 mm, < p2 27–34 mm; **p2** < wt 4.5–9 mm, =7/8, 8 or 8/9; **pp3–6** about equal and longest; **p7** < wt 2–3.5 mm; **p8** < wt 5–7.5 mm; **p10** < wt 9–10.5 mm; **s1** < wt 10–11 mm. Emarg. pp3–7 (on p7 sometimes more faint).

TAXONOMIC NOTE The two subspecies are widely allopatric and gene flow between them is probably non-existent. They have therefore usually been regarded as different species, albeit closely related and forming a superspecies. Here they are treated as races of the same species based on modest genetic distance (Drovetski *et al.* 2013) and morphological differences that are small and on a level consistent with treatment of other disjunct species (e.g. *Cyanistes teneriffae*, *Cyanopica cyanus* and *Pica pica*).

REFERENCES Drovetski, S. V. *et al.* (2013) *Ecology and Evolution*, 3: 1518–1528.

P. o. fagani, juv, Yemen, Oct: head tinged greyer. The degree of individual variation in underparts streaking is highly remarkable. (H. & J. Eriksen)

ALPINE ACCENTOR
Prunella collaris (Scopoli, 1769)

Fr. – Accenteur a pin; Ger. – Alpenbraunelle
Sp. – Acentor alpino; Swe. – Alpjärnsparv

Alpine Accentor typically breeds in some of the highest and remotest places on Earth, up to 4000 m in the covered region, where it occurs from Morocco and Iberia discontinuously to the N Middle East, Caucasus and Transcaspia, thence to Japan and Taiwan, reaching 5500 m in the Himalayas. Winters mainly in the same areas but descends to lower elevations, sometimes reaching near sea level and south to Israel and Jordan. Often seen in small parties feeding on ground among scree and boulders. Perhaps most frequently encountered on skiing holidays in winter near top stations.

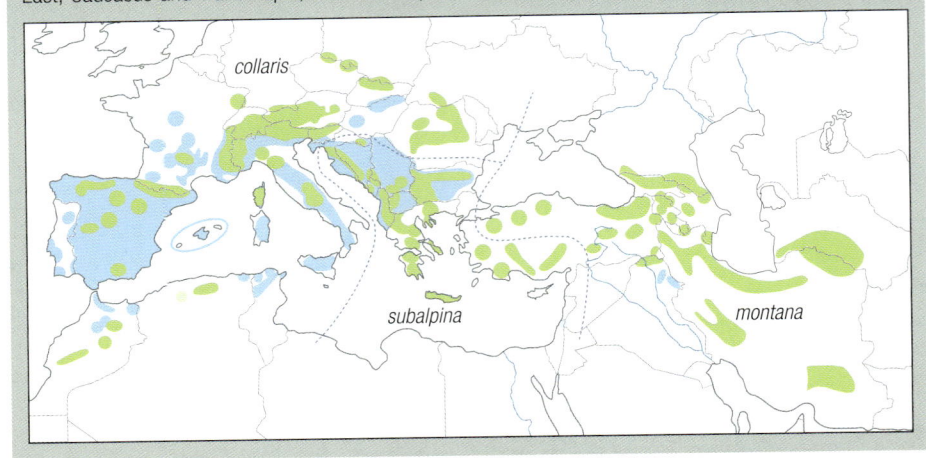

IDENTIFICATION Chunky, grey, brown and rufous accentor of montane regions, mainly dark-looking and usually encountered in barren boulder areas or on cliffs with sparse or low plants. Combination of greyish head and foreparts, pale lower mandible, *dark wing panel*, two thin *whitish wing-bars* and bold chestnut flanks ease identification. *Head to breast ash-grey*, tinged bluish or brown, and in close views may show *black-and-white speckling on chin and throat*. Flanks and breast-sides diffusely but extensively streaked chestnut, and *whitish undertail-coverts have black-brown centres*. Greyish-brown above with *pale mantle lines and rufous-tinged scapulars*; lower back and rump grey, but uppertail-coverts tinged rufous, streaked dark and slightly scaled white. Black median coverts with small buff-white tips, mainly *black greater coverts and primary-coverts* (often forming clearly visible dark wing panel) tipped white, tips forming dotted wing-bar ('string of pearls'); black tertials with rufous fringes and white tips. *Tail dark brown*, with *all but central feathers having broad white tips. Base of lower mandible bright yellow*. Sexes very similar, rarely if at all separable in the field. Walks, interspersed with short runs and hops, on long near-straight legs, and carriage typically more erect than most accentors, affording starling-like silhouette. Rarely perches on plants but often sings from atop a rock. Unobtrusive when breeding but more conspicuous and forms small, quite noisy flocks in winter, with far-carrying flight-calls. Song-flight short—upward flutter, brief hover and downward glide—recalling a lark.

VOCALISATIONS Song is a continuous low-pitched warble or twitter with creaking or harshly trilling or rolling notes frequently interwoven (quality of song therefore somewhat resembles Skylark, especially at distance, but is drier and lower-pitched, perhaps more like Linnet). The song is often rather monotonous due to mechanical repetition of basic phrase, e.g. *chürr-dire-jer-de-prrr teje chürr-dire-jer-de-prrr teje...*, etc.; sometimes the song is more complex and extended. – Commonest call when flushed is a short, husky, trilling *churr*, often repeated a few times, *churr-churr-churr*. Often heard is a more toneless *che-che*, somewhat recalling a distant Redpoll. These two calls can be combined and uttered alternately.

SIMILAR SPECIES Unmistakable in close views, differing from *Dunnock* in different habitat and generally in altitude, in larger size, greyer plumage, markedly patterned wings, in particular with dark greater coverts panel, and chestnut flanks. Less distinctive at long range, when some of these features may be difficult to see, but even then the rather strong, 'free', undulating flight should readily distinguish it from congeners and most other species found at similar altitudes.

AGEING & SEXING (*collaris*) Ageing possible if seen close or handled using feather pattern and shape, moult limits in wing or feather wear, and iris colour. Sexes mostly alike in plumage (but see Biometrics for slight difference in size). Seasonal variation rather limited. – Moults. Complete post-nuptial moult and partial post-juv moult in late summer or early autumn (mostly Jul–Oct, rarely Nov). Post-juv renewal involves all head- and body-feathers, lesser and median coverts, a variable number of inner greater coverts but no tertials. Pre-nuptial moult apparently absent or insignificant/partial. – **SPRING Ad** Primaries, primary-coverts, all secondary-coverts and tail-feathers evenly fresh and rather dark-centred with neat edges. Ad ♀ tends to have duller and less contrasting plumage than ad ♂, especially paler and less extensive chestnut below, but much overlap and variation. Both sexes by summer noticeably duller and greyer above, barring on chin and throat reduced, chestnut body-sides bleach to cinnamon-buff, rufous fringes to flight-feathers largely abraded, and white wing-bars and tail markings much reduced. **1stS** Sexual differences more subdued and therefore difficult to use. Juv primaries, primary-coverts, some or all greater coverts and tail-feathers relatively more worn with abraded/frayed fringes (see 1stW for moult pattern); coloration/pattern of retained greater

P. c. collaris, Switzerland, Jun: bulky-bodied, and has characteristic combination of grey head and chest, chestnut-blotched flanks and contrasting blackish wing panel. Pure grey foreparts and deep chestnut flanks suggest ♂, although sexing (and indeed ageing) in this species normally impossible. (P. Cools)

P. c. collaris, presumed ad, Netherlands, Apr: distinctive yellow base to bill and complex wing pattern unlike any other species in the covered region. Uniformly ad wing-feathers, with white tips to greater coverts restricted to outer webs, primary-coverts boldly tipped white, and warm brown iris indicate ad. With wear, pale throat patch becomes broader and whiter, with well-developed black-and-white flecks, which together with bright and extensive chestnut on flanks and purer grey fringes to remiges and foreparts could suggest ♂. (B. Veen)

coverts, alula and primary-coverts as in autumn, but due to bleaching differences less obvious. Apparently, some of both age classes undergo partial winter moult which could produce misleading moult contrast in wing. – AUTUMN **Ad** Evenly fresh. In the hand, iris dark sepia-brown. Pure white tips to greater coverts better marked and pointed basally along shaft, lacking any buff and blurred edges. Note distinctive white fringes to tips of outer webs of alula and primary-coverts, also mainly lacking any buff edges. Tail-feathers, primaries and primary-coverts generally more smoothly and widely rounded, darker-centred and strongly textured. **1stW** Much as ad, but iris dull greyish olive-brown. Primaries, primary-coverts, tail-feathers and many or all greater coverts juv with more worn fringes/tips. Retained juv greater coverts show broader whitish tips (less sharply pointed basally than in ad), and tips less pure white, often with brownish or buff subterminal area. Length of spot measured along shaft of third outermost juv greater covert 3–4.5 mm, in ad *c.* 2–3 mm, but inner greater coverts often replaced in 1stW to ad type. Often has tiny pale spot at tips of other retained feathers, and alula and primary-coverts more brownish with ill-defined buffish fringes to tips of outer webs. Also, tends to have grey of head and breast tinged more brownish. **Juv** Soft fluffy body-feathering and is duller, less ashy-grey above, and more uniform below, with buff-grey unmarked throat. Rest of underparts browner, with buff or pale rufous flanks, narrowly streaked darker, lacking contrast of ad, while both wing-bars are buffier. Cream-buff undertail-coverts very faintly streaked.

P. c. collaris, ad, Spain, Sep: note dotted white greater-coverts wing-bar (being ad, white tips restricted to outer webs, and grey bases do not reach tips along outer webs). Note diagnostic chevron pattern on undertail-coverts. Ageing further confirmed by still active moult in secondaries (end of post-nuptial). Likely ♀ by general dull coloration, with buffish-brown hue on cheeks, chest and upperparts, and ill-defined throat pattern, still sexing of this species usually impossible. (H. Shirihai)

P. c. collaris, presumed ad, Italy, Nov: whole wing fresh seems to indicate ad, and this further supported by outer web of primary-coverts being edged boldly white, and tail-feathers being rather broad at tips. Sexing always difficult but head and neck have limited pure grey and flanks subdued chestnut perhaps suggesting ♀. (D. Occhiato)

P. c. collaris, 1stW, Spain, Sep: sometimes seen in small winter flocks at ski stations. Aged by juv greater coverts and primary-coverts (latter with ill-defined white edges and worn tips), and already worn remiges. Extensive pure grey head and neck, well-developed throat mottling and deep chestnut flank-blotching infer ♂, but safe sexing impossible. (H. Shirihai)

BIOMETRICS (*collaris*) **L** 16.5–18 cm; **W** ♂ 96–111 mm (*n* 56, *m* 103.4), ♀ 96–106 mm (*n* 38, *m* 100.3); **T** ♂ 58–70 mm (*n* 31, *m* 64.5), ♀ 57–68 mm (*n* 24, *m* 62.1); **T/W** *m* 61.9; **B** 15.5–18.0 mm (*n* 27, *m* 16.8); **BD** 4.6–6.1 mm (*n* 22, *m* 16.9); **Ts** 23.2–26.5 mm (*n* 28, *m* 25.0). **Wing formula: p1** < pc 2–7 mm, < p2 56–68.5 mm; **p2** < wt 0.5–3 mm, =5 or 5/6 (=6); **pp3–4** longest; **p5** < wt 0.5–3.5 mm; **p6** < wt 7–10 mm; **p7** < wt 13.5–19 mm; **p10** < wt 23.5–30 mm; **s1** < wt 26–33.5 mm. Emarg. pp3–6.

GEOGRAPHICAL VARIATION & RANGE Three races breed within the treated region, which exhibit only slight differences. About five more extralimital subspecies in Asia (not treated). A few more subspecies have been named but they all seem to be junior synonyms with no or insignificant differences from previously described taxa.

P. c. collaris, 1stY, Italy, Feb: rather dull, with grey parts less pure, ear-coverts tinged brownish, and flanks streakier and subtly paler chestnut; almost no speckling on chin and throat. This bird is particularly dull due to its being young, and likely could be ♀. Juv remiges (primary tips already heavily worn and lack pale tips) and primary-coverts worn and pointed (with indistinct pale fringes); note moult contrast among tertials. (D. Occhiato)

P. c. collaris, juv, Germany, Jun: before or during post-juv moult, note streaky underparts and still rather downy head. (T. Grüner)

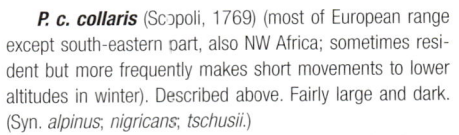

P. c. subalpina, C Turkey, May: a rather subtle race, and identification relies mainly on locality. However, weak dark streaks on paler grey ground colour above and rather pale chestnut on flanks are features of *subalpina*. Ageing or sexing not possible. (M. Ünlu)

P. c. collaris (Scopoli, 1769) (most of European range except south-eastern part, also NW Africa; sometimes resident but more frequently makes short movements to lower altitudes in winter). Described above. Fairly large and dark. (Syn. *alpinus*; *nigricans*; *tschusii*.)

○ **P. c. subalpina** (C. L. Brehm, 1831) (SE Europe including Balkans and Bulgaria, east to S Turkey; mainly resident or makes shorter movements in winter). Slightly paler with less rufous on flanks, purer grey above and on upper breast (but subtle and many are very similar), while ground colour of chin/throat appears a little whiter. Whitish wing-bars slightly broader and more prominent (white tips *c.* 1 mm longer at all ages than in *collaris*, still odd birds overlap). Somewhat smaller than *collaris*. **W** ♂ 97–106 mm (*n* 15, *m* 101.5), ♀ 94–101 mm (once 90; *n* 12, *m* 97.0); **T** ♂ 60–72 mm (*n* 15, *m* 64.6), ♀ 56–65 mm (*n* 12, *m* 61.2); **T/W** *m* 63.1; **B** 15.0–18.5 mm (*n* 27, *m* 17.4); **BD** 4.7–6.0 mm (*n* 25, *m* 5.3); **Ts** 22.9–26.3 mm (*n* 27, *m* 25.0). (Syn. *reiseri*.)

P. c. montana (Hablizl, 1783) (N & E Turkey, Caucasus, Transcaucasia, Iran, SW Turkmenistan; resident or moves to lower altitudes in winter). Subtly smaller and paler than *collaris*, with subtly thinner bill and blunter wingtip. Upperparts slightly more buff-brown and lightly streaked (streaking less black, more diffuse and dirty grey-brown). Chin/throat more boldly barred and blotched than in the other races so that bib looks darker with less white visible. Grey of breast not so pure, more tinged ochre. Less pure greyish than same-sized *subalpina* and is more diffusely streaked on back. **W** ♂ 97–109 mm (*n* 12, *m* 102.2), ♀ 91.5–104 mm (*n* 12, *m* 96.1); **T** ♂ 61–71.5 mm (*n* 12, *m* 66.9), ♀ 57–72 mm (*n* 12, *m* 63.6); **T/W** *m* 65.8; **B** 15.4–18.2 mm (*n* 23, *m* 17.1); **BD** 4.2–5.7 mm (*n* 19, *m* 4.9); **Ts** 22.5–26.1 mm (*n* 24, *m* 24.1). (Syn. *caucasicus*.)

Presumed *P. c. montana*, Israel, Dec: subspecific identification away from breeding grounds, and where more than one race could occur, difficult. Nevertheless, here combination of very thin bill, and rather pale and thinly streaked upperparts indicate this race. Age uncertain, but predominantly greyish-edged remiges and primary-coverts, the latter boldly tipped white subterminally, and greater coverts with well-defined white tips, suggest ad. Being possibly ad but overall rather dull, with less pure grey upperparts may suggest ♀. (L. Kislev)

P. c. montana, presumed ad, NE Turkey, May: typically hopping with both feet together. Race inferred by locality, and by apparently more limited streaking above on paler grey ground; also has thinner bill. Pureness of grey areas suggests ad ♂, but ageing uncertain, making sexing even less possible. (T. Tozsin)

P. c. montana, 1stW, NE Turkey, Nov: race inferred by locality, but extensive chestnut-brown on flanks, weakly streaked dull brown upperparts, and very thin bill typical of this race. Combination of young age (note worn and pointed juv primary-coverts, worn juv greater coverts and remiges), extensive pure grey head and neck, and deep chestnut flank-blotching could suggest ♂. (H. Kahraman)

RUFOUS-TAILED SCRUB ROBIN
Cercotrichas galactotes (Temminck, 1820)

Alternative names: Rufous Bush Robin, Rufous Bushchat, Rufous Scrub Robin

Fr. – Agrobate roux; Ger. – Heckensänger
Sp. – Alzacola rojizo; Swe. – Trädnäktergal

A summer visitor to the W Palearctic, this so-called scrub robin—a relative of nightingales and chats, which all belong to the flycatcher family—prefers sunny, dry, open country with groves of trees and scattered bushes and scrub. It is a relatively common but late-arriving breeder in S Iberia, and has a near-continuous distribution in N Africa, from S Morocco east to Egypt, the Levant and parts of Arabia, through Transcaucasia and over much of Central Asia. The species winters across sub-Saharan Africa.

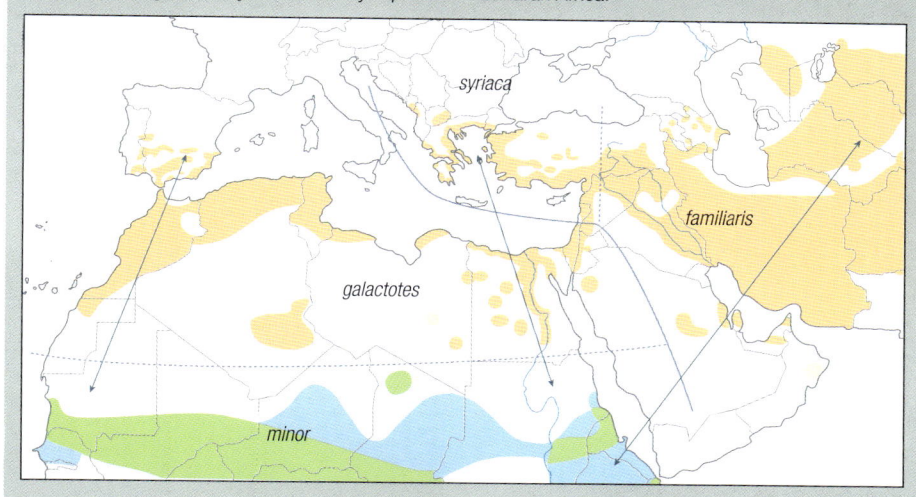

C. g. galactotes, Tunisia, May: relatively broader white tips to tail also visible from below. As ad and 1stY both moult completely in autumn/winter, ageing uncertain in spring/summer. Differences in freshness between some feather tracts ('false moult limits') and between individuals will occur related to moult. Highly terrestrial behaviour can augment individual variation and mislead assessment of feather wear. (R. Pop)

C. g. galactotes, Spain, Jun: may spread wings when hunting ants on ground. Relatively narrower black subterminal tail-band but broader white tips in this race, but much individual variation as to this character. (S. Fletcher)

IDENTIFICATION Medium-sized chat being only slightly smaller than Nightingale, but considerably more easily observed due to its habit of *running on open ground between bushes*, its *long tail fanned and cocked strikingly upright*. Essentially *plain russet-buff to cream-brown upperparts*, richest on rump and even *brighter rufous on tail*, which especially when fanned shows *diagnostic black subterminal band and white tips to all but central feathers*. Obvious *pale supercilium and eye-ring, dark eye-stripe*, ill-defined wing-bars (that on greater coverts more obvious) and paler-fringed dark brown tertials. Underparts mostly washed whitish-cream, with whiter throat and pale pinkish-brown breast and flanks. Relatively deep-based bill mostly horn-brown, with dull flesh-coloured base to lower mandible, and the rather long legs are pale pinkish-grey or flesh-coloured. Flight is chat-like, but with large-warbler impression. Hops on ground, with tail-jerks, sometimes cocks tail right up over back, then closes and slowly lowers it, and although often skulking it is not particularly shy. Wings frequently held open and drooped with slow forward flicks.

VOCALISATIONS Song rich and varied, a thrush-like series of high-pitched, clear melancholy and ringing notes, and pulsating, high-tension chirps, given either from hidden or elevated and visible perch, or sometimes in butterfly-like, parachuting display-flight, recalling both fragments of full song of Robin and subdued 'bedtime' song of Song Thrush, e.g. *tor, pip-prerep, tirip… pepli, pyeperp-te-ruip… pshe-ter tite-petret, trete-showeshowi…*, somewhat rising and falling in pitch and strength with intervals. — Commonest calls a hard tongue-clicking *chak*, sometimes doubled to *teck teck* or *chack chack*, and a soft short, straight whistle, *üh*, a bit like Nightingale but softer; also gives an insect-like buzzing, *bzzzzz* in anxiety.

SIMILAR SPECIES Unmistakable within the covered region, especially if its long-legged structure, long tail, and its wing and tail movements are observed, in particular its tail-cocking motion revealing to full effect the chestnut-red tail with black subterminal marks and white tips. Tail pattern, if seen, is diagnostic and separates from all other W Palearctic species.

AGEING & SEXING (*galactotes*) Ageing usually straight-forward on close views and if able to check moult pattern in wing (but see 1stS). Sexes alike and virtually no seasonal variation. — Moults. Complete post-nuptial moult of ad, and partial post-juv moult, usually immediately after migration in autumn or in early winter (but exact timing rather variable between populations, depending on breeding and migration periods and areas), often concluded before end of Dec (or rarely extended until midwinter), but occasionally starts on breeding grounds and is then suspended. Partial post-juv renewal in winter includes head- and body-feathers, some or all tertials, central, all or no tail-feathers, and many or all wing-coverts; apparently some juv also moult a few or several remiges in winter, but few data available. — SPRING **Ad** Evenly fresh. **1stS** Like ad, but juv primaries, primary-coverts and tail-feathers worn, fringes/tips frayed. In particular, primary-coverts more pointed and abraded; usually retains at least some juv secondary-coverts contrasting with coverts renewed in winter. However, frequency of complete juv moult in winter unknown, and probably a few are indistinguishable from ad. — AUTUMN **Ad** Until mid or late autumn, usually retains all or virtually all primaries, primary-coverts, greater and median coverts, which bleach to browner and lack pale fringes of juv feathers. Occasionally moults prior to or during migration, including some body-feathers and a limited number of small inner wing-coverts, tertials and, rarely, a few remiges and tail-feathers (then strong moult limits usually visible). **1stW** Differs from ad in wholly fresh feathering (post-juv moult before migration limited and rare, and never creates moult contrast as in ad), or evenly worn or frayed pale tips to coverts and flight-feathers. Greater coverts tipped and fringed paler and more obviously buffish-white. Juv outer tail-feathers have less extensive and not so sharply outlined

C. g. galactotes, Spain, Jun: compared to eastern races, *galactotes* has less pronounced pale wing panel and dark face markings (moustachial stripe can be virtually lacking, and eye-stripe is narrower and fainter). It has also more sandy-buff and russet (less brownish-grey) tones above, and cleaner and whiter underparts, including breast. (C. N. G. Bocos)

C. g. galactotes, Israel, Apr: long, cocked and contrastingly rufous tail (richest on rump and uppertail-coverts). Fairly rusty tones above suggest *galactotes*. (A. Ouwerkerk)

C. g. galactotes / syriaca, NE Jordan, Apr: the range of *galactotes* ends somewhere in the Levant, where it is replaced by the eastern races. This bird illustrates the difficulties in assigning birds to race in this region: it could be *galactotes* by sandy-rufous upperparts, and little contrast between mantle/back and rump. Conversely, it has rather strong pale wing panel, and face pattern approaches *syriaca* while tail pattern is intermediate or closer to latter. (H. Shirihai)

black subterminal bars, with less pure white tips. **Juv** Much like ad but upperparts paler and sandier, with faintly speckled or mottled throat, breast and fore flanks, and much less conspicuous tail markings (in Israel, occasionally has only very faint dark subterminal bars to tail-feathers).

BIOMETRICS (*galactotes*) **L** 16.5–17.5 cm; **W** ♂ 83–93.5 mm (*n* 15, *m* 88.9), ♀ 83–89 mm (*n* 10, *m* 86.4); **T** ♂ 69–75 mm (*n* 15, *m* 70.7), ♀ 65–77 mm (*n* 10, *m* 70.4); **T/W** *m* 80.4; **TG** 5–10 mm; **white on r6** 10–17 mm; **B** ♂ 19.0–21.6 mm (*n* 15, *m* 20.1), ♀ 18.7–20.5 mm (*n* 10, *m* 19.7); **BD** 4.3–5.3 mm (*n* 24, *m* 4.8); **Ts** 25.1–28.8 mm (*n* 24, *m* 27.1). **Wing formula: p1** > pc 3–5 mm (ad) or 4–8 mm (juv); **p2** < wt 2–6 mm, =5 or 5/6 (48%), =6 or 6/7 (48%) or =4/5 (4%); **pp3–4**(5) longest; **p5** < wt 0–4 mm; **p6** < wt 2–8 mm; **p7** < wt 6.5–12 mm; **p10** < wt 15–22 mm; **s1** < wt 17.5–24 mm. Emarg. pp3–5 (rarely faintly also on p6).

GEOGRAPHICAL VARIATION & RANGE Rather prominent and largely clinal variation, with western (*galactotes*, *minor*) and eastern races (*syriaca familiaris*) quite dissimilar and possibly constituting different species, though division in two groups not straightforward and best kept as one species until further studied.

C. g. galactotes, juv, Spain, Aug: like ad, but close views may reveal faintly mottled breast. Like 1stW, differs from ad by wholly fresh feathering. (J. Knorr Alonso)

C. g. galactotes (Temminck, 1820) (C & S Iberia, N Africa, Sinai, Israel, Jordan, apparently S Syria; winters Sahel zone of sub-Saharan W Africa east at least to Niger). Described above; the most reddish-tinged subspecies with broader white tips but narrower black subterminal marks on tail-feathers.

C. g. syriaca (Hemprich & Ehrenberg, 1833) (S Balkans, S & C Turkey, N Levant south to W Syria and Lebanon; winters Somalia, Horn of Africa). Much more greyish-brown, less russet than *galactotes*, and has on average smaller white tips but broader black subterminal marks on tail-feathers. Complete moult unlike in *galactotes* in late winter, Jan–Mar (early Apr). First-winters usually moult completely in winter, but some replace only a variable number of inner primaries. **L** 15.5–16.5 cm; **W** ♂ 84.5–92 mm (*n* 12, *m* 88.6), ♀ 83–89 mm (*n* 10, *m* 86.4); **T** ♂ 62–71 mm (*n* 13, *m* 70.7), ♀ 65–77 mm (*n* 10, *m* 70.4); **T/W** *m* 76; **TG** 4–9 mm; **white on r6** 8.5–16 mm; **B** 17.1–21.0 mm (*n* 22, *m* 18.9); **BD** 4.2–5.3 mm (*n* 20, *m* 4.6); **Ts** 23.9–26.5 mm (*n* 22, *m* 25.3).

C. g. familiaris (Ménétries, 1832) (SE Turkey, Iraq, Transcaucasia, Iran east to E Kazakhstan; winters NE Africa including Kenya). Similar to *syriaca* but has upperparts on

C. g. syriaca, Turkey, May: close in coloration to *C. g. familiaris*, but some are browner (less greyish or drab); face pattern also approaches (or is trifle weaker than) latter, with especially dark lateral crown, moustachial and eye-stripes, and pale wing panel quite well developed. (D. Monticelli)

Variation in S Turkey, with typical *syriaca* (left) and bird approaching *galactotes*, May: Levant and especially SE Turkey an area of possible intergradation between three races, but in Turkey typical *syriaca* mostly occurs west from about Gaziantep (SE Anatolia). Conversely, those east of this approach *familiaris*, although many are inseparable from *galactotes*, or even intermediate between all three races, as shown by right-hand bird (from Birecik, SE Turkey). Latter has rather strong face markings like *familiaris*, but colour above tends towards *galactotes*. (D. Occhiato)

C. g. familiaris, Kuwait, Apr: very broad black subterminal tail-band (with relatively narrow white tips), dark face makings obvious, and rather extensive pale wing panel, while drab upperparts clearly contrast with rufous tail. However, much individual variation, while in autumn, especially young *familiaris* tend to be less distinctive or inseparable, while *syriaca* is somewhat intermediate. (K. De Rouck)

C. g. familiaris, United Arab Emirates, Mar: typical face markings of this race, and greyish wash to breast. This bird, however, has intermediate tail pattern with almost equally wide black subterminal bands as white tips. (P. Arras)

average paler and even greyer, rufous of rump and uppertail slightly paler, and underparts whiter, creating a stronger contrast between greyish back and rufous rump. Tail pattern similar to *syriaca*. Crown, mantle and wings only occasionally show any rufous tones. There is usually more contrast in the head pattern with a whiter eye-ring and supercilium, dark lateral crown-stripes and darker eye-stripe. Breast and flanks are slightly greyer. Wing-coverts and tertials have slightly more obvious pale fringes even in ad plumage. Wing formula differs subtly. Apparently winter moult is complete for both ad and 1stW. **L** 15–16.5 cm; **W** ♂ 84–90.5 mm (n 14, m 86.5), ♀ 82.5–86 mm (n 11 m 84.8); **T** ♂ 62–68.5 mm (n 14, m 65.0), ♀ 62–68 mm (n 11, m 63.5); **T/W** m 64.9; **white on r6** 6–15 mm; **B** 17.0–20.0 mm (n 21, m 18.9); **BD** 4.2–4.9 mm (n 15, m 4.5); **Ts** 24.2–26.5 mm (n 17, m 25.4). **Wing formula: p1** > pc 0–4 mm or 0.5–1 mm <, < p2 43–47 mm; **p2** < wt 1.5–4.5 mm, =5 or 5/6 (common) or =4/5 or 6 (rare); **pp3–4** longest; **p5** < wt 1–3 mm; **p6** < wt 4–6.5 mm; **p7** < wt 7.5–11 mm; **p10** < wt 13.5–20 mm; **s1** < wt 15.5–22.5 mm. Emarg. pp3–5 (rarely faintly also on p6).

Extralimital: ***C. g. minor*** (Cabanis, 1851) (sub-Saharan Africa north to Sudan; resident, apparently unrecorded in the covered region). Has upperparts tinged pink-brown rather than rufous-brown as in *galactotes*; also distinctly smaller than *galactotes*, with 10% shorter wing and bill.

TAXONOMIC NOTE Due to rather well-marked differences in morphology it is possible that Rufous-tailed Scrub Robin is better split into two species. The two groups of taxa (treated as one species here) are largely allopatric, show rather obvious plumage differences, and apparently overlap in N Levant, but the relationships and interactions between them in this region are as yet largely unstudied.

C. g. familiaris, ad, Kuwait, Aug: race based on range and colour; note whiter underparts and dirty breast, typical of this form. In early autumn, prior to moult, ageing rather straightforward, here an ad with heavily worn and bleached wing-feathers, lacking broad white tips to worn primaries and tail. (M. Pope)

C. g. familiaris, 1stW, Kuwait, Aug: race based on range and, compared to *galactotes* (and even *syriaca*), colder and greyer upperparts, with rufous of rump and tail slightly duller, but still highly contrasting. Dark facial markings and pale wing panel less obvious than in ad/spring birds. Aged as 1stW by fresh plumage in early autumn. (R. Al-Hajji)

BLACK SCRUB ROBIN
Cercotrichas podobe (Statius Müller, 1776)

Alternative name: Black Bush Robin

Fr. – Agrobate podobé; Ger. – Rußheckensänger
Sp. – Alzacola negro; Swe. – Svart trädnäktergal

A sub-Saharan chat of dry acacia savannas which in recent years has been a regular summer visitor to extreme S Israel, although most records are of ♂♂, and the species has not yet been found breeding. There are also older records from Algeria, and recent sightings in Egypt and Jordan. Elsewhere, it occurs in vegetated lowlands from Mauritania east to the Horn of Africa, and from coastal Saudi Arabia south to Yemen, but is currently spreading north and east in Arabia.

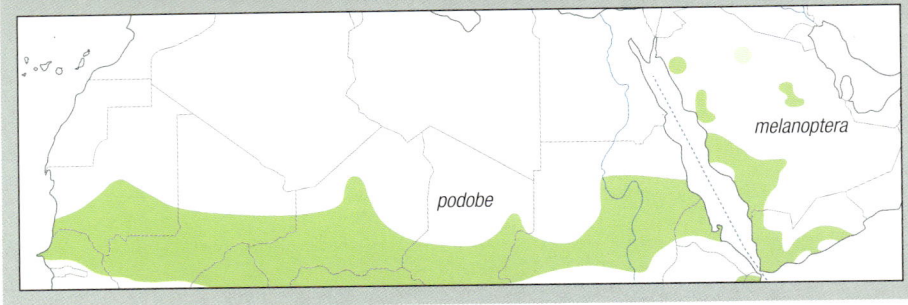

SIMILAR SPECIES Given habits, habitat and range, and combination of all-black plumage and long tail with white tips, this species should prove unmistakable, even given a relatively poor view.

AGEING & SEXING Ages generally inseparable once juv plumage replaced (but little studied). Sexes virtually alike in plumage (but see Biometrics), and no seasonal variation. – Moults. Few data, especially from north of range, e.g. Arabia, but based on specimens (NHM), and birds trapped in Israel, both post-nuptial and post-juv moults are complete (albeit some, probably 1stY, retain some remiges and coverts), generally Jun–Oct. – **ALL YEAR Ad** ♂ tends to be glossier and blacker (fewer brown tones) than ♀, but much overlap and majority cannot be sexed on plumage; best sexed by size, with ♂ considerably larger (see Biometrics). Amount of rufous in wings subject to considerable individual variation, but is, on average, more pronounced in ♀ (and 1stY with new remiges). ♂♂ tend to have more white in tail and on undertail-coverts, but much overlap. **1stY** Indistinguishable from ad once moult completed, but a few retain some juv secondaries and greater coverts, and others (apparently rarely) retain most remiges, tail-feathers, primary-coverts and greater coverts. However, both age classes, even in early spring, can appear extensively abraded on primaries, primary-coverts and greater coverts, thus misleadingly

C. p. melanoptera, ad ♂, Yemen, Jan: lead-black (and glossy) plumage typical of ♂, while relatively fresh plumage and orange-tinged sepia-brown iris indicate ad. (H. & J. Eriksen)

C. p. melanoptera, 1stY ♂, Israel, Mar: regular spring visitor to Israel, mostly 1stY ♂♂, which match birds from coastal W Arabia in having some rufous-cinnamon in remiges, mainly in 1stY. Note blacker, glossier tones of ♂, while moult limits in inner wing (new tertials, secondaries and inner greater coverts) indicate 1stY. (H. Shirihai)

IDENTIFICATION A sprightly, bush-haunting chat, with size and structure somewhat recalling more familiar Rufous-tailed Scrub Robin. However, it has longer tail and slightly longer legs, and, most importantly is *entirely lead or sooty black*, except *white tips to undertail-coverts and to graduated outermost four tail-feathers*, which are typically conspicuous when the tail is cocked (frequent). Also, may have rufous-buff underwing-coverts, and *rufous inner webs to primaries and secondaries*, which are sometimes visible on folded wing or in flight. Flight, gait, and behaviour all bear strong resemblance to Rufous-tailed Scrub Robin, but perhaps spends even more time on the ground than that species.

VOCALISATIONS Song variable, a rather strong and deep-voiced, melodious warble, rather *Sylvia*-like, most like Arabian Warbler of same habitat, consisting of fairly short phrases, with sweet, fluted notes and some more scratchy ones woven in. At times more agitated when strophes become longer. Imitation of other bird calls sometimes appears to be involved, but true mimicry unknown. – Call a hoarse but soft *chech*, though in Israel noted mostly to be silent unless singing.

C. p. melanoptera, 1stY ♂, Israel, Mar: typical ant-hunting behaviour, revealing rufous-cinnamon inner web to remiges. (Y. Perlman)

seeming to exhibit moult contrast. Iris usually dark olive-brown or dull brown (rather than sepia/reddish-brown of ad). **Juv** Resembles ad but has soft, fluffy body-feathering and is duller, more sooty-brown, with pale tips to undertail-coverts narrow or indistinct; less white on tail-tip.

BIOMETRICS (*melanoptera*) **L** 19–22 cm; **W** ♂ 87–98 mm (*n* 12, *m* 91.0), ♀ 83–95 mm (*n* 12, *m* 87.3); **T** ♂ 100–110 mm (*n* 12, *m* 106.3), ♀ 87–110 mm (*n* 12, *m* 99.9); **T/W** *m* 116.2; **TG** 26–32 mm; **white on r6** 12–17 mm; **B** 17.2–19.6 mm (*n* 25, *m* 18.3); **BD** 4.1–5.7 mm (*n* 25, *m* 4.9); **Ts** 25.5–29.4 mm (*n* 25, *m* 27.5). **Wing formula: p1** > pc 11–18 mm, < p2 27–35.5 mm; **p2** < wt 8–14 mm, =8 or 8/9 (52%), =7 or 7/8 (43%), or = 9 (5%); **p3** < wt 0.5–2 mm; **pp4–5** about equal and longest (p5 rarely to 1 <); **p6** < wt 1–2.5 mm; **p7** < wt 4–7.5 mm; **p10** < wt 11–16 mm; **s1** < wt 14–17.5 mm. Emarg. pp3–6.

GEOGRAPHICAL VARIATION & RANGE Moderate variation with two races described.

C. p. melanoptera (Hemprich & Ehrenberg, 1833) (W & S Arabia; resident or makes shorter movements in winter). Differs from *podobe* in tendency to have no (especially in ♂♂) or reduced rufous-cinnamon inner webs to flight-feathers, but some, especially coastal birds (mostly ♀♀ and apparently 1stY with renewed remiges, also juv prior to moult), can have equal amount of rufous to African birds. Said to average smaller than *podobe* but this could not be confirmed by material examined by us. Has subtly more rounded wing.

C. p. podobe (Statius Müller, 1776) (sub-Saharan Africa, possibly marginally entering southern edge of treated region; resident). Has extensive rufous-cinnamon inner webs to flight-feathers in all plumages. Wing on average subtly more pointed than in *melanoptera*. **W** ♂ 85.5–99 mm (*n* 23, *m* 90.3), ♀ 83–92 mm (*n* 19, *m* 87.4); **T** ♂ 97–116 mm (*n* 23, *m* 107.7), ♀ 97–111 mm (*n* 19, *m* 102.6); **T/W** *m* 118.6; **B** ♂ 17.3–19.9 mm (*n* 22, *m* 18.6), ♀ 16.3–19.1 mm (*n* 19, *m* 17.7); **BD** 3.9–5.4 mm (*n* 39, *m* 4.6); **Ts** 25.0–30.7 mm (*n* 42, *m* 27.5). **Wing formula:** p2 < wt 5–10 mm, =7 or 7/8 (52%), =8 or 8/9 (33%) or = 6/7 (15%).

C. p. melanoptera, presumed ♀, Israel, Mar: often seen feeding below Acacia trees. Duller black upperparts (with slight greyish tone) lacking gloss are features of ♀, and here a dull young ♂ can almost certainly be excluded. Rather dull olive-brown iris suggests 1stS, but evenly feathered wing at this date could indicate either age class. (R. Mizrachi)

C. p. melanoptera, presumed 1stS, Israel, Mar: long tail often cocked and fanned, revealing diagnostic bold white tips. Some birds, mostly 1stY, may retain some juv remiges, as here, the unmoulted primaries contrasting with fresh secondaries and tertials. (T. Krumenacker)

C. p. melanoptera, ♂, Israel, Mar: uniquely long-tailed chat, especially appreciable when seen in low escape flight, revealing striking tail- and wing-patterns. Amount of rufous in wing varies slightly with age, sex and geography. The glossy lead-black plumage is typical of ♂. (D. Gochfeld)

C. p. melanoptera, juv, Saudi Arabia, Aug: soft, fluffy body-feathers duller, with pale tips to undertail-coverts narrow or indistinct; all wing-feathers still juv, and yellow gape flanges visible. (D. Alhashimi)

(EUROPEAN) ROBIN
Erithacus rubecula (L., 1758)

Alternative name: Eurasian Robin

Fr. – Rougegorge familier; Ger. – Rotkehlchen
Sp. – Petirrojo europeo; Swe. – Rödhake

To many the typical small garden bird, much loved for its red-breasted appearance and varied song, which can be enjoyed almost year-round. In spite of its small size and cute look, it is a surprisingly aggressive bird, which can fiercely attack other Robins or intruders into its territory. Robins occur in woodlands and gardens from sea level to mid elevations, from Fenno-Scandia in the north and the British Isles in the west south to the NE Atlantic islands, NW Africa, and east to Iran and the Caucasus.

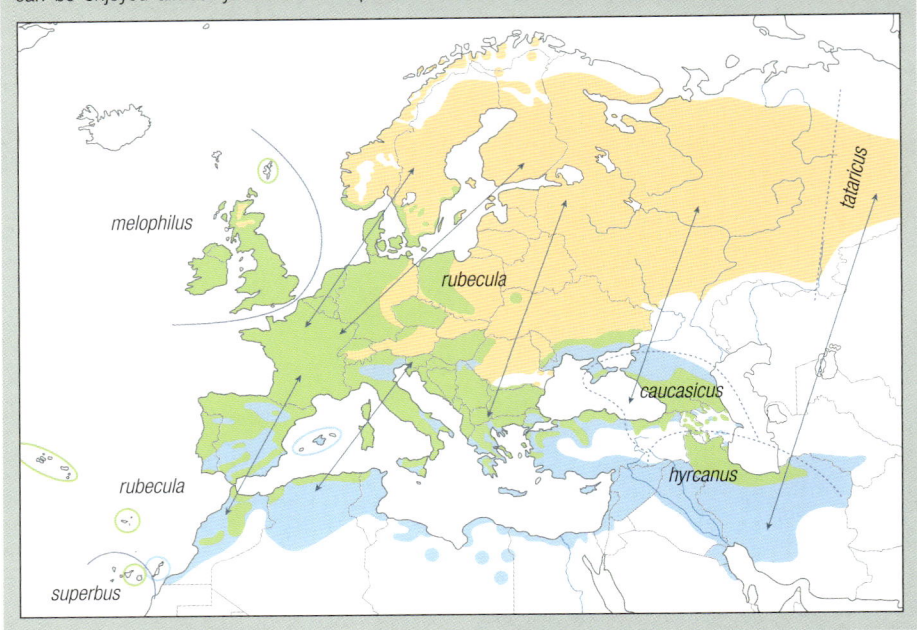

E. r. melophilus, England, Feb: the British race differs only subtly from *rubecula* by a little darker and browner upperparts and warmer brown uppertail-coverts. Difficult to age, but pedagogic: no moult contrast visible (perhaps partly due to the inner greater coverts being concealed by the scapulars). The primary-coverts, alula and outer greater coverts look very fresh and ad-like, while their tips are rather deep rusty, which is often seen in ad. Thus, apparently an ad with unusually broad (and therefore juv-like) rusty tips to greater coverts, but without handling best left unaged. (H. Shirihai)

E. r. melophilus, England, Feb: orange-rufous on face and bib, with grey surround, but apart from this easily spotted by its distinctive song, given also in winter. When not heavily worn, orange-buff tips to greater coverts often visible. (H. Shirihai)

IDENTIFICATION A rather bold, so-called chat with conspicuous *orange-brown face, throat and upper breast*, delimited from *olive-brown upperparts* and whitish underparts by *a pale grey or bluish-grey band on forecrown, sides of head, neck and on lower breast*. Rump variably washed grey-brown and often contrasts slightly with *warmer brown uppertail-coverts and tail*, while *flanks are suffused warm buff-olive*. Sexes alike. Characteristic behaviour: makes both short and long hops on ground in quick succession, then stops in upright posture. Freely uses natural and artificial low perches, but also sits high up in trees and skulks in cover for prolonged periods. In excitement may bob and cock tail. Stance otherwise normally rather upright. Frequently flicks wings (which are often held slightly drooped) and tail. Flight usually low and flitting.

VOCALISATIONS Song variable (both in tempo and capacity): a liquid warble, each motif usually lasting 2–4 seconds, often commencing with a few high, drawn-out notes, then dropping in pitch before the accelerated finale of shrill notes, trembling and excited squeaky (even bubbling) sounds. Typically changes in strength from pianissimo to forte and back in intricate, irregular pattern. Each strophe is varied, there is very little repetition, but the song is easily recognised on characteristic quivering quality (like 'trembling Aspen leaves') and clear 'glass chimes tone'. Excited ♂♂ deliver ever-changing motifs with only brief pauses for 1–2 minutes duration. Both sexes sing year-round, though ♀ less frequently and with shorter strophes. – Commonest call a scolding hard, short *tic*, which if nervous or prior to roost and 'reveille' is repeated into fast clicking series, *tic-ic-ic-ic* ('winding up an old clock'). Single such calls can be confused with normal call of Hawfinch if more modest voice and less 'explosive' quality of Robin is not evident. Alarm an extremely thin, sharp, ventriloquial *tsiih*; also utters a thin but slightly hoarse *tsi* or *tsri*, sometimes heard from migrating birds (at night).

SIMILAR SPECIES Unmistakable under reasonable observation conditions but can resemble other small chats (e.g. any of the *nightingales* or *Red-flanked Bluetail*) in a brief glance. In both instances, face pattern, tail pattern and behaviour provide easy means of separating adult Robin from any potential confusion species. – Juvenile Robin is quite similar to juvenile *Siberian Rubythroat* and requires careful separation. See under that species (Similar species) for differences. – To the beginner any bird with some red on the throat or breast could be a potential confusion species; in the case of uncertainty you are advised to first of all check out carefully any of *Redstart, Linnet, Redpoll, Bullfinch* and *Red-breasted Flycatcher*.

AGEING & SEXING Ages usually separable in summer and autumn, frequently also in spring. Sexes alike in plumage and almost in size. – Moults. Complete post-nuptial moult of ad in late Jul–Sep prior to autumn migration. Partial post-juv moult generally includes replacement of all head, body, lesser and median coverts, occasionally some tertials, and a variable number of greater coverts (often innermost 1–3 coverts). Apparently no moult in late winter. Ageing in spring usually requires close observation or trapping. – **SPRING Ad** From late spring often subtly greyer/duller above and paler below. (Claimed difference between sexes in that '♀ tends to have slightly paler orange areas than ♂' could not be confirmed when checking extensive material from throughout Europe. Just as many ♂♂ as ♀♀ were slightly paler than average, indicating that this variation is individual or age-related rather than sexual.) **1stS** Much as ad but has relatively more worn primaries, primary-coverts and tail-feathers (relatively fresher and more rounded in ad), and duller brown iris (warmer and more rufous in ad). Also, moult limits among greater coverts usually detectable (but in both age classes pale tips or wing-bar on greater coverts normally worn off or virtually so). – AUTUMN **Ad** Often has wing-bar formed by buff tips to evenly-fresh greater coverts (usually more evident on outer coverts but can be absent altogether; unlike young birds, buff tips narrower and wedge-shaped closer to shaft). On average, tips of tail-feathers slightly broader and more obtuse, nearly entirely lacking 'spiky' extensions at shafts, in particular on r1, though some overlap and requires practice to be used as supporting character. **1stW** Juv plumage replaced from Aug in migratory populations, by late Sep in sedentary ones, thereafter much like ad, but has usually more obvious buff wing-bar. Best aged by contrast between unmoulted outer greater coverts (subtly looser textured, a fraction less olive-tinged but browner, with broader yellowish tips on much of distal outer web, sometimes also extending to inner) and fresh inner ones (more ad-like, and usually lacking pale tips), though birds that moult very few or most greater coverts have less obvious moult limits. Sometimes buff tips also to tertials. Primary-coverts slightly more pointed in shape and a little more frayed, not so neatly edged as ad. Tail-feathers on average slightly narrower/more pointed with rather obvious 'spiky' extreme tips including on r1 (though some overlap with ad). Iris dark grey (brown in ad). **Juv** Lacks uniform upperparts and orange chest. Brownish above with copious pale buff spotting on crown, mantle and scapulars, and mainly buffish below, with duskier crescent-like mottling and pale buff spots on lower face, breast and flanks. Wings have buff-spotted lesser and median coverts and very obvious wing-bar on greater coverts.

BIOMETRICS (*rubecula*) **L** 13–14 cm; **W** ♂ 70–76 mm (*n* 33, *m* 72.6), ♀ 68.5–74 mm (*n* 25, *m* 71.3); **T** ♂ 53.5–60 mm (*n* 32, *m* 57.4), ♀ 54–60 mm (*n* 25, *m* 56.5); **T/W** *m* 79.1; **B** 12.8–15.1 mm (*n* 58, *m* 13.7); **BD** 3.0–4.0 mm (*n* 30, *m* 3.5); **Ts** 23.0–26.3 mm (*n* 58, *m* 24.8). Wing formula: p1 > pc 7–13 mm, < p2 19–25.5 mm; p2 < wt 9–13 mm, =7/8 or 8 (58%), =8/9 or 9 (37%) or <9 (5%); p3 < wt 1.5–3 mm; pp4–5 about equal and longest; p6 < 0.5–2 mm; p7 < 5.5–7.5 mm; p10 < 10–14 mm; s1 < 13–16 mm. Emarg. pp3–6.

GEOGRAPHICAL VARIATION & RANGE Eight races in Western Palearctic, but differences clinal and much affected by wear, bleaching and individual variation. Much overlap in measurements, and only with long series from breeding grounds is it sometimes possible to determine racial differences, but note quite distinctive birds from around Black/Caspian Seas, and *superbus* in the Canaries. See also Taxonomic notes.

E. r. rubecula, ad, Italy, Dec: typically adopts relaxed posture with fluffed up body feathers creating rounded shape. Aged by relatively fresh and more round-tipped primary-coverts, remiges and greater coverts, all wing-coverts having an olive tinge. (D. Occhiato)

E. r. rubecula, 1stS, Finland, Apr: singing its high-pitched and clear, 'tremulous' song from perch in coniferous forest. Much of the wing is retained worn juv feathers except a few inner greater coverts, which are fresher and subtly more olive-tinged than outer ones. (M. Varesvuo)

E. r. rubecula, ad (left: Spain, Sep) and 1stW (right: Italy, Dec): both ad and 1stW in autumn can have rufous-buff tips to some or all greater coverts, but when seen well ad recognised on having all wing subtly olive-tinged and rufous-buff tips to greater coverts changing size and shape only smoothly and stepwise, lacking marked steps, while 1stW has browner wing-feathers except for any moulted inner greater coverts, which are more olive-tinged like in ad. Note also better defined buff tips to inner tertials in 1stW bird. (Left: H. Shirihai; right: D. Occhiato)

E. r. rubecula, Switzerland, Jan: when seen from in front shows hint of divide in red bib at centre of belly, white protruding slightly into orange-rufous chest. Impossible to age at this angle of view. (H. Shirihai)

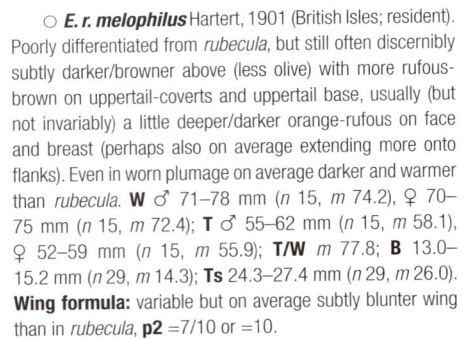

E. r. superbus, ad, Tenerife, Canary Is, Oct: rather distinctive, with better-defined facial bib, which is also deeper brick-red and smaller with better definition. This race has also hint of white eye-ring (best visible behind eye), and often extensive ash-grey forecrown to neck-sides and flanks. Wing evenly fresh typical of ad. (H. Shirihai)

○ **E. r. melophilus** Hartert, 1901 (British Isles; resident). Poorly differentiated from *rubecula*, but still often discernibly subtly darker/browner above (less olive) with more rufous-brown on uppertail-coverts and uppertail base, usually (but not invariably) a little deeper/darker orange-rufous on face and breast (perhaps also on average extending more onto flanks). Even in worn plumage on average darker and warmer than *rubecula*. **W** ♂ 71–78 mm (*n* 15, *m* 74.2), ♀ 70–75 mm (*n* 15, *m* 72.4); **T** ♂ 55–62 mm (*n* 15, *m* 58.1), ♀ 52–59 mm (*n* 15, *m* 55.9); **T/W** *m* 77.8; **B** 13.0–15.2 mm (*n* 29, *m* 14.3); **Ts** 24.3–27.4 mm (*n* 29, *m* 26.0). **Wing formula:** variable but on average subtly blunter wing than in *rubecula*, **p2** =7/10 or =10.

E. r. rubecula (L., 1758) (Fenno-Scandia, Continental Europe, Turkey except in north-east, NW Africa, W Canaries, Madeira, Azores; northern and eastern populations winter W & S Europe and NW Africa, southern and Atlantic are resident). Treated above under Identification, etc. Entire upperparts rather uniform olive-brown, lacking contrastingly rufous uppertail-coverts and uppertail (but can have very slight tendency towards warmer tail base than rest of upperparts). – Birds in C & E Algerian Atlas and in Tunisia described as a separate subspecies ('*witherbyi*'), but claimed differences tiny and far from consistent. Birds in Madeira and Azores tend to be subtly darker reddish-orange on breast than typical *rubecula*, though not as dark red as *superbus*; this is variable, though, and majority are inseparable from *rubecula*, hence included therein. (Syn. *armoricanus*; *hispaniae*; *maior*; *microrhynchos*; *sardus*; *witherbyi*.)

○ **E. r. caucasicus** Buturlin, 1907 (Crimea, NE Turkey, Caucasus & Transcaucasia; probably short-range winter movements reaching E Mediterranean and Middle East). Less distinctive than *hyrcanus* (below) but still just discernibly different from *rubecula*: upperparts subtly more warm brown, less olive-tinged, and many have moderately or markedly pronounced rufous-orange on uppertail-coverts and on basal half of tail, as well as on fringes to flight-feathers. Size and wing length as *rubecula*, but tail proportionately subtly shorter and bill slightly longer. – Birds breeding in Crimea ('*valens*') said to be paler and less olive above than *rubecula*, and more rusty on upper tail-base, but considering that there is a fair amount of variation within *caucasicus* it seems best to include it here; we were unable to separate *valens* from a long series of *caucasicus*. **W** ♂ 71–76 mm (*n* 15, *m* 73.0), ♀ 68.5–75 mm (*n* 12, *m* 71.2); **T** ♂ 53–60 mm (*n* 15, *m* 56.2), ♀ 52–59.5 mm (*n* 12, *m* 55.8); **T/W** *m* 77.6; **B** 14.3–16.8 mm (*n* 28, *m* 15.6); **Ts** 23.9–26.5 mm (*n* 27, *m* 25.1). **Wing formula:** p2 usually =8 or 8/9. (Syn. *valens*.)

E. r. hyrcanus Blanford, 1874 (S Azerbaijan & N Iran; wintering grounds poorly known but assumed to include Middle East and possibly N Arabia). Resembles *caucasicus* but is even darker with deeper rufous-red bib (almost as dark as *superbus*) and more obvious rufous-orange on uppertail-coverts and basal half of tail, as well as on fringes of flight-feathers. Still, some come close to *caucasicus* and cannot be separated without comparing series of both. Long-billed and long-legged. **W** ♂ 71–76 mm (*n* 15, *m* 73.5), ♀ 69–75 mm (*n* 12, *m* 71.3); **T** ♂ 53–62 mm (*n* 15, *m* 56.9), ♀ 53–59.5 mm (*n* 12, *m* 56.0); **T/W** *m* 77.9; **B** 14.5–17.0 mm (*n* 30, *m* 16.0); **Ts** 24.2–27.1 mm (*n* 29, *m* 25.8). **Wing formula:** p2 usually =8, 8/9 or 9.

? **E. r. tataricus** Grote, 1928 (S Urals, W Siberia; winter grounds not known but might include SW Central Asia and Middle East). Said to be paler/greyer above, with duller orange face and whiter underparts, otherwise as *rubecula*. Still, such birds occur within *rubecula* too, and *tataricus* should perhaps be synonymised with it. Limited material examined, and those seen did not differ from *rubecula*.

E. r. superbus Koenig, 1889. 'TENERIFE ROBIN' (Tenerife, Gran Canaria; resident). Differs consistently in morphology, having markedly darker red breast and face, and red bib averages slightly smaller and better defined versus purer white belly/vent. Has hint of pale eye-ring and broader ash-grey band from forecrown to flanks. Brown of upperparts slightly darker than in *rubecula* and tends to have more pure grey, less olive tinge. Vocalisations differ too, the song being shorter, simpler, can appear slightly lower-pitched and more thrush-like, and contains more trills and mimicry of other birds. Wing more rounded (making tail appear long), and size rather small. Bill subtly longer than in *rubecula*. – Breeders on Gran Canaria have been proposed to constitute a different taxon ('*marionae*') based on rather clearly different mtDNA, though no significant morphological or other differences presented (Dietzen et al. 2003), or established by us. **W** ♂ 68–73 mm (*n* 12, *m* 70.3), ♀ 66.5–74 mm (*n* 12, *m* 69.0); **T** ♂ 55–60 mm (*n* 12, *m* 57.6), ♀ 54–61 mm (*n* 12, *m* 56.9); **T/W** *m* 82.2; **B** 13.8–15.3 mm (*n* 24, *m* 14.5); **Ts** 23.2–25.9 mm (*n* 24, *m* 25.0). **Wing formula:** p2 =9/10 or 10 (50%) or =10/ss or ss (50%). (Syn. *marionae*.)

TAXONOMIC NOTES Separate species status has been proposed for the C Canarian race *superbus*, and there seems a good case for this considering the rather marked differences in plumage and song, and difference of 2.9% in the cytochrome b gene in mtDNA (Dietzen et al. 2003). Also, Bergmann & Schottler (2001) reported lower reaction to playback of *rubecula* song than to local *superbus* song. However, this might need more consideration before being implemented. – The fact that birds on Gran Canaria have been shown to be older and more basal in a phylogenetic tree than those of Tenerife, differing by as much as *c.* 3.7% in cyt b mtDNA (Dietzen et al. 2003), is interesting, but until other differences have been found we recommend recognition of just one subspecies for these two islands. It serves to remind us that genetic distance does not necessarily indicate any evolutionary or taxonomic difference other than genetic change itself if conditions have been equal over time on the two islands and no morphological differences have evolved.

REFERENCES BERGMANN, H.-H. & SCHOTTLER, B. H. (2001) *DB*, 23: 140–146. – DIETZEN, C., WITT, H.-H. & WINK, M. (2003) *Avian Sci.*, 3: 115–131. – STOCK, M. & BERGMANN, H.-H. (1988) *Zool. Jahrb. Physiol.*, 2: 197–212.

E. r. rubecula, juv, Switzerland, Jun: extensively spotted, mottled and streaked pale rusty-buff. The typical posture and rounded shape is already evident. (H. Shirihai)

WHITE-THROATED ROBIN
Irania gutturalis (Guérin, 1843)

Fr. – Iranie à gorge blanche; Ger. – Weisskehlsänger
Sp. – Petirrojo de Irán; Swe. – Vitstrupig näktergal

An attractive bird of warm, arid, usually high-altitude maquis slopes with scattered trees, often noticed by the fast, twittering or warbling song. Once spotted, the ♂ is a real beauty with its orange breast, pure white throat and supercilium, black head-sides and uniform lead-grey upperparts. Migrates in August–September to winter in E Africa, presumably following a route through E Arabia (as so few are seen in Israel). It returns to its breeding sites in Turkey, the Middle East and Central Asia mainly in April.

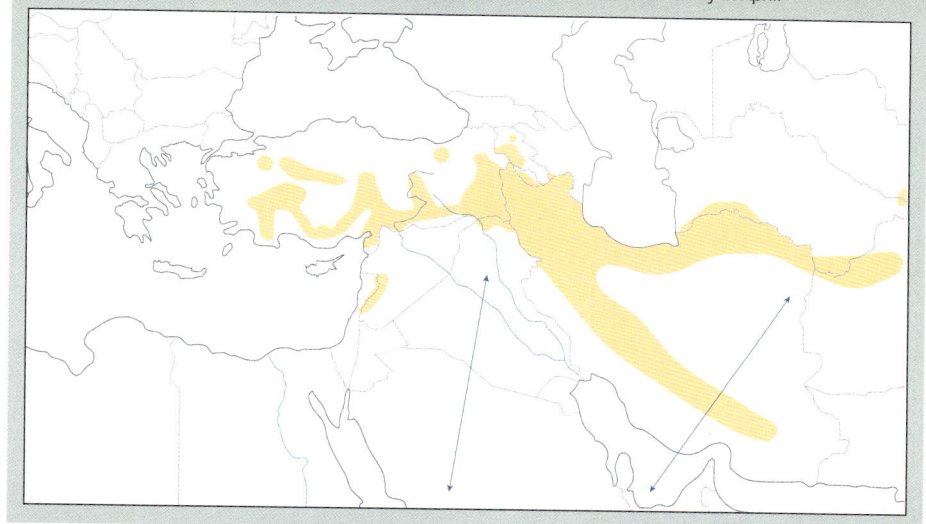

Ad ♂, Turkey, May: unmistakable in this plumage, and note especially all-dark tail with white undertail-coverts. Aged by lead-grey and less worn edges to primaries and primary-coverts. (D. Occhiato)

IDENTIFICATION A medium-sized, slim chat, rather like a Nightingale or Wheatear in shape and movements. Whether seen on the ground or in low vegetation it often *carries head and tail high*, and keeps its *wings slightly drooping* in thrush fashion. *Long-legged* and *strong-billed*. The rather *long tail is all dark*, black in ♂ and dark grey in ♀. The bill is black and the legs dark. The adult ♂ has an *orange or ochraceous-brown breast* (of varying saturation; some are paler ochraceous-buff), *black sides of head* and throat leaving a *narrow pure white central throat patch*, narrow white eye-ring (at times lacking) and *prominent white supercilium*, with attractive uniform *lead-grey upperparts*, save the black tail. The large *alula-feather of ♂ is blackish and contrasts strongly against paler rest of wing* except in some first-years. A few ♂♂ have a hint of a black border between white throat and orange breast (rarely a broader band, 3–10 mm wide). ♀ and first-winter ♂ have less clean upperparts, the grey being tinged olive-brown. Underparts are off-white with a pale ochraceous-buff or orange wash on flanks, the *breast being diffusely blotched or mottled grey*. There is a faint and short, *dirty white supercilium in front of the eye*. Rarely, ♀♀ practically without any warm buff or ochraceous visible below occur (Rijpma & Bakker 2006), having pale grey breast and flanks, and cream-white belly; underwing-coverts, however, seem to be invariably at least slightly orange-tinged. Juvenile is finely spotted like in most thrushes. Note size, general shape and the all-dark tail.

VOCALISATIONS Song is a very fast, prolonged scratchy warble that stays at roughly the same pitch throughout. Songs vary in length but are often 5–10 sec. Shorter strophes, well spaced, occur too when uninspired. Best recognised by mix of hard or creaky notes and whistles, and by its sheer speed and frenzy, which rules out most other songbirds, save perhaps some ecstatically singing *Sylvia* warblers, or a sub-song from a *Lanius* shrike. Mimicry of other birds may be interwoven into the song but is not a prominent feature of this species. – Calls are a variety of whistling notes, e.g. *viiüht*, vaguely downslurred, like a squeaking gate, or short, abrupt *viht* or lower-pitched *vü*, all these mixed by same bird (sounding like many species, which might be 'intentional'). Alarm call is a low, creaky *cherrr*, most resembling similar calls of Nightingale or Thrush Nightingale.

SIMILAR SPECIES The ♂ looks superficially like a ♂ *Common Redstart*, although is bigger, longer-legged and has a white throat patch (Redstart ♂ has completely black throat) and of course a black tail (Redstart orange-red). – A closer similarity exists with the extralimital *Indian Blue Robin* (*Luscinia brunnea*; not treated), both having reddish breast, greyish upperparts, black sides of head and a white supercilium, but this species lacks the white throat, is deeper reddish below from chin to belly, and has a shorter tail and pinkish legs. – ♀ is superficially like *Red-flanked Bluetail*, greyish with orange-tinged flanks, but differs in larger size, longer legs and tail, stronger bill, etc. – ♀ *Siberian Ruby-throat* is more similar in size and shape, but this has no orange tinge on flanks (a warm brown wash at most), a much more distinct dark loral stripe, a pale supercilium and short pale submoustachial stripe. – Extralimital ♀ *Himalayan Rubythroat* (*L. pectoralis*; not treated) is rather similar to the ♀ but has a warmer brown tinge on wings, back and rump, and note white outer corners on dark tail. – Rare variant of ♀ lacking any orange on breast or flanks may be quite similar to *Blackstart*, but note larger size, narrow whitish throat patch outlined with brown-grey, and breast often diffusely barred grey and white (throat and breast of Blackstart uniform grey), wing has lead-grey edges (contrasting pale grey or whitish outer edges to tertials and flight-feathers in Blackstart when fresh), a little stronger and straighter bill (slightly finer with down-curved tip in Blackstart), longer legs, and a somewhat more pointed wing with only three primaries emarginated (four in more rounded wing of Blackstart).

AGEING & SEXING Ages separable during 1stY. Sexes differ clearly in 1stS after pre-breeding moult in winter. – Moults. Complex, complete (or largely complete but suspended) post-nuptial moult of ad usually in mid Jun–early Sep (rarely later), on or near breeding grounds; when suspended, 1–4 secondaries left unmoulted, presumably replaced in Africa but not known whether immediately upon arrival in late autumn or in normal pre-breeding moult in late winter/early spring. Partial post-juv moult in late summer does not involve flight-feathers, tertials, greater coverts or primary-coverts. Partial pre-breeding moult of ad in late winter or early spring involves tertials and partly head and body, possibly also 1–4 secondaries (see above). Partial pre-breeding moult of 1stY in late autumn or early winter variable in extent, usually involving head and body, tertials and outer 1–3 secondaries and all or at least some inner greater coverts, and none or up to eight primaries and primary-coverts (those moulting most extensively thus moult

1stS ♂, Turkey, May: similar to ad ♂, but note strong moult contrast with heavily worn and narrowly brownish-buff-fringed primaries and primary-coverts. (D. Occhiato)

nearly completely in first-winter). – **SPRING Ad ♂** Upperparts (crown to rump) pure lead-grey. Tail black. Supercilium white, extending behind eye. Sides of head black. Throat pure white (with or without black lower border to orange breast). Breast and flanks variably orange-red to paler cream-buff or ochraceous-tinged. Tertials, greater coverts and primary-coverts uniform grey, lacking whitish spots on tips. Large alula-feather dark slate-grey, contrasting with paler rest of wing. **Ad ♀** Upperparts grey with slight olive-brown tinge. Tail dark grey or blackish. Supercilium very short and dusky white, not extending behind eye. Sides of head brownish-grey. Throat off-white. Breast diffusely blotched or mottled grey, flanks tinged ochraceous-buff, breast sometimes has scattered ochraceous-buff or orange blotches admixed with grey (rarely, 'advanced' ♀♀ have much orange below). Tertials, greater coverts and primary-coverts uniform without whitish spots on tips. **1stS ♂** As ad ♂ but shows contrast in wing between moulted inner wing-feathers and secondary-coverts edged lead-grey (as in ad) and outer which are more brownish and worn. Primary-coverts nearly always retained juv-feathers, and sometimes many tertials and greater coverts retained, too, having whitish spots on tips. If large alula-feather is still unmoulted from juv plumage shows poor contrast with rest of wing. **1stS ♀** As ad ♀ but often shows contrast in wing between moulted inner wing-feathers and secondary-coverts edged grey (as in ad) and outer which are more brownish and worn (but less obvious contrast than in ♂!). When most or all greater coverts replaced still moult contrast with retained brownish and worn primary-coverts and primaries. Retained primary-coverts, tertials and greater coverts have whitish spots on tips. – **AUTUMN Ad ♂♀** As in spring. Plumage fresh, in many birds save a few inner secondaries, which are kept during autumn migration. No pale spots on tips of tertials, greater coverts or primary-coverts. **1stW** Plumage as ad ♀, but has pale spots on tips of tertials, greater coverts and primary-coverts. Often some subdued dark barring on breast. Sexes alike. A very few birds are paler below with no visible orange on flanks, and only faint buff tinge on lower flanks and vent, and these are presumed to be 1stW ♀♀ only. **Juv** Olive grey-brown upperparts with buffish spots on crown, nape, mantle and shoulders. Buff-tinged dusky white underparts with dark-tipped feathers, giving spotted appearance. Dark tail-feathers.

BIOMETRICS L 16.5–18.5 cm; **W** ♂ 91–100 mm (n 23, m 95.8), ♀ 89–98 mm (n 19, m 93.4); **T** ♂ 67–78 mm (n 23, m 72.8), ♀ 64–77 mm (n 19, m 71.2); **T/W** m 76.0; **B** 16.7–20.2 mm (n 35, m 18.5); **Ts** 24.4–27.0 mm (n 30, m 25.7). **Wing formula:** p1 > pc 0–6 mm (rarely < pc 1.5 mm), < p2 42.5–47.5 mm; **p2** < wt 5–8.5 mm, =6 or 5/6; **pp3–4** about equal and longest (p4 sometimes 1 mm <); **p5** < wt 0–1 mm; **p6** < wt 1.5–4 mm; **p7** < wt 11–14 mm; **p10** < wt 20–23 mm; **s1** < wt 24–27 mm. Emarg. pp3–5.

GEOGRAPHICAL VARIATION & RANGE Monotypic. – S Turkey, Iraq, W Iran and in much of Central Asia; local in Levant and foothills of W Tien Shan; winters Kenya and adjacent areas of E Africa.

REFERENCES Rijpma, U. & Bakker, A. (2006) DB, 28: 302–304.

Ad ♀, United Arab Emirates, Mar: black bill and eye (enhanced by whitish eye-ring), greyish upperparts and whitish underparts with orange-tinged flanks. Aged by relatively fresh wing-feathers, with greyish edges to primaries and primary-coverts. (I. Boustead)

1stS ♀, Azerbaijan, Jun: like ad ♀, but has browner and more worn edges and tips to juv primaries and primary-coverts. (K. Gauger)

Ad ♂, Kuwait, Aug: an early migrant that has already completed post-nuptial moult, thus very fresh. Overall similar to spring/summer ad ♂, but some fine whitish tips on orange chest, and black on head-sides is greyer. Underwing-coverts orange-buff, characteristic of the species (paler in ♀ and young). (M. Pope)

Ad ♀ Kuwait, Aug: distinctive white eye-ring and orange-buff flanks. The narrow whitish throat is typical. Aged by greyish-fringed primary-coverts and greater coverts (lacking whitish tips of young birds). (M. Pope)

1stW, Kuwait, Aug: diffuse barring on breast often shown by young, while diagnostic buff-orange on flanks here mostly hidden (creating vague resemblance with Blackstart). Aged by juv wing-coverts with white tips. Sexing of 1stW usually not possible, but ♂ tends to have greyer upperparts. (P. Fågel)

1stW, Ethiopia, Sep: typical posture, with wings lowered and tail slightly cocked; distinctive buff-white tips to juv primary-coverts, greater coverts and tertials. Perhaps 1stW ♀ by much-reduced buff on flanks, and browner upperparts, but sexing at this age unsafe. (A. Faustino)

THRUSH NIGHTINGALE
Luscinia luscinia (L., 1758)

Fr. – Rossignol progné; Ger. – Sprosser
Sp. – Ruiseñor ruso; Swe. – Näktergal

A more northerly and easterly distributed close relative of the famous Nightingale, replacing it east of a line running from S Denmark, N Germany, S Poland and through E Europe to the western shores of the Black Sea. Thrush Nightingale is noted for its louder and more 'mechanical' almost rattling song, though many still enjoy its performance. Others claim it is so loud as to break your sleep if you have one singing outside your bedroom window. In spring it returns in mid May. Like the Nightingale it winters in sub-Saharan Africa, but on average with more easterly distribution.

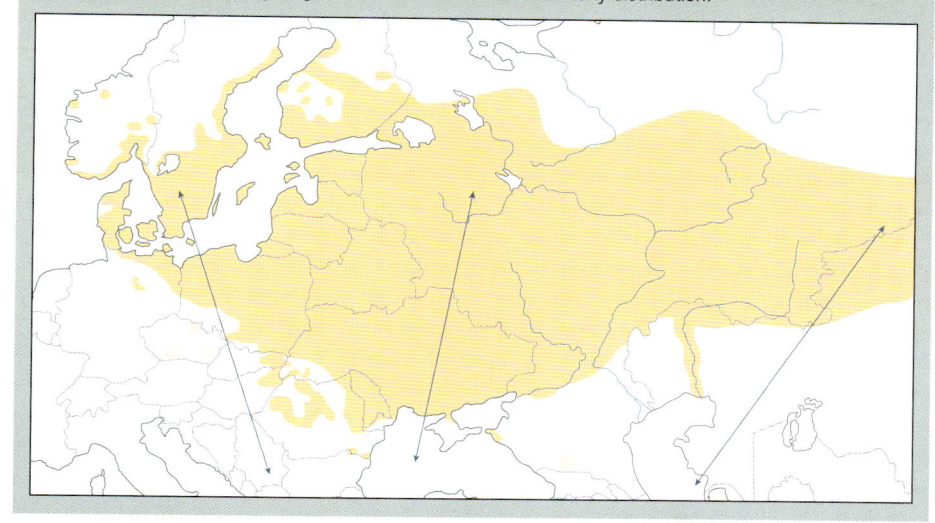

IDENTIFICATION Clearly smaller and slimmer than a *Turdus* thrush, but slightly larger than a wheatear. *Long-legged*, fairly long-tailed with a rather small head, short bill and long neck. It wags and twists its tail and bobs its rear body rather slowly and mechanically. The plumage is plain with *dark brown upperparts*, slightly more *rufous-tinged uppertail* and *dusky greyish underparts*, with slightly paler, off-white throat and centre of belly. When seen well (which happens rarely due to its skulking habits), note thin, diffuse *pale eye-ring*, *sides of throat diffusely blotched dark* (occasionally forming diffuse stripes either side of throat), vaguely vermiculated or *blotched breast*, flanks diffusely streaked darker, and at times also some dark marks visible on sides of undertail-coverts. Legs pink-brown, bill and eye dark. Best means of identification is the song (see below). Juvenile is very similar to juvenile of Nightingale, Robin and a few other close relatives in Asia. Differs from Robin in slightly larger size and longer tail, more rufous-tinged uppertail and colder brown, less rufous-tinged spotting of breast, from Nightingale (with difficulty) on less vividly russet uppertail.

VOCALISATIONS Song given from perch in cover (with lowered tail kept slightly fanned), many would say more loud than pleasing, recalling Great Reed Warbler in strength. The phrases are slightly longer, less fluty and mellow than those of Nightingale, a series of hard tongue-clicking, deep *chok* notes, dry rattling sounds and loud whistling notes run together in a varying mixture. A quite characteristic motif is a 'galloping' *chucko-choo chucko-choo chucko-choo...*, and this and series of deep *chok-chok-chok-chok-...* notes can carry more than a mile on calm nights! Each strophe often opened by a few slow, pensive whistles, *pew, pew, pew* (although these do not accelerate and increase in pitch like the typical crescendo uttered by Nightingale). Sometimes these whistles are upwards-inflected in Willow Warbler fashion, *hooeet*. Certain phrases by accomplished singers are a little softer and more pleasing, and can invite confusion with Nightingale, but a prolonged listen will reveal the identity, and on the whole the song is quite consistent over its range. Sings mainly at dusk, during the night and at dawn, but especially newly arrived or unpaired ♂♂ frequently sing in the bright morning, too. – Has no unique contact call, but apparently may use either of two alarm calls also for contact purposes. Alarm is a straight 'inhaling' whistle *iihp*, quite similar to the call of Collared Flycatcher. Also has a low, hard creaking *errrrr*. The two alarm calls are often used in combination.

SIMILAR SPECIES Easily confused with *Nightingale*. Note that Thrush Nightingale has a darker and less bright reddish uppertail (it is somewhat more rufous-tinged than rest of upperparts but not as bright reddish as in Nightingale). Also, Thrush Nightingale has a more patterned throat and upper breast, with a hint of darker grey lateral throat-stripes, and diffuse dark vermiculations or fine blotching on upper breast, and these markings may continue on flanks and sides of undertail-coverts (Nightingale has a more uniform breast and no or only very lightly suggested darker lateral throat-stripes). – For a more far-fetched similarity with some of the American *Catharus* thrushes, in particular with *Hermit Thrush*, see under Nightingale.

AGEING & SEXING Ages separable in summer and autumn, sometimes also in spring. Sexes alike. There seems to be complete overlap in measurements between the sexes (*contra* BWP and other sources). Dittberner & Dittberner (1989) claimed that ♂ is on average darker and more strongly patterned on breast, but a series of specimens from Sweden (NRM) gives no support to this, showing total overlap. – Moults. Complete post-nuptial moult of ad usually rapid in (Jun) Jul–Aug prior to autumn migration.

Israel, Apr: typical posture, showing diagnostic diffuse dark lateral throat-stripe, and mottling on breast that is often enhanced with wear in spring. Greater coverts apparently juv, being bleached and warm brown, but degree of wear to primaries difficult to judge, hence ageing unsure. (G. Shon)

Comparison in spring between Thrush Nightingale (left: Finland, May) and Common Nightingale (right: Netherlands, May): Thrush has diagnostic blotchy breast and often (mostly in spring/summer) flanks; Common has cleaner and whitish underparts. Common also tends to have diffuse greyish supercilium (usually lacking in Thrush) and better-marked whitish eye-ring, while Thrush has dark lateral throat-stripe (can be difficult to see). (Left: M. Varesvuo; right: R. Schols)

Thrush Nightingale (left) and Common Nightingale (right), Finland, May: darker uppertail (usually deeper rufous than rest of upperparts) in Thrush, i.e. usually not as bright reddish (almost orange) as Common, but often difficult to evaluate. Rest of upperparts of Thrush usually darker or greyer (less chestnut). (M. Varesvuo)

Partial post-juv moult in Jun–Aug does not involve flight-feathers but may include a few tertials and some inner greater coverts. Apparently no moult in winter. Ageing in spring usually requires close observation or trapping. — **SPRING Ad** Tertials and greater coverts uniformly brown, without paler spots at tips. Tips of tail-feathers quite fresh. Tips of long primaries reasonably fresh. **1stS** A few or several outer greater coverts generally have tiny buff spots remaining at tips (unless too worn). Tail-feathers sometimes a little worn at tips. Tips of long primaries at times heavily abraded, in which case such wear supports ageing. — **AUTUMN Ad** As in spring but freshly moulted, all tertials and greater coverts usually uniformly brown, though very rarely there can be a hint of paler ochre-buff thin edges around tips (still not in the shape of distinct spots). **1stW** Like ad but separated by tiny buff spots on tips of all or most tertials and greater coverts, sometimes also a hint of spots on some inner primary-coverts. Sometimes a few pale-tipped juv long uppertail-coverts retained. (May moult a few tertials and inner greater coverts to uniform ad type before autumn migration.) **Juv** Upperparts spotted ochre-buff, underparts densely scalloped dark brown.

BIOMETRICS Sexes are same size. **L** 15–17.5 cm; **W** 84–92.5 mm (n 31, m 88.8); **T** 60–70 mm (n 31,

1stW, Turkey, Sep: much plumage variation in autumn; this is a rather well-marked individual. Apparent moult limit in greater coverts, and some remnant buff spots on juv coverts and tertials (already very small) and relatively strong mottling below, indicate 1stW. (E. Yoğurtcuoğlu)

Thrush Nightingale (left: Turkey, Sep) and Common Nightingale (right: Spain, Aug): in autumn separation can be more problematic, requiring close and prolonged views to evaluate Thrush's dark lateral throat-stripe and mottling on breast. Note duller upperparts in Thrush (without rusty uppertail-coverts) and slightly dark-centred undertail-coverts (plainer and whiter in Common). Both are 1stW by juv primary-coverts, with Thrush having some juv (buff-tipped) greater coverts and tertials. (Left: A. M. Doğan; right: C. N. G. Bocos)

m 65.5); **T/W** *m* 73.9; **B** 14.4–17.4 mm (*n* 29, *m* 15.9); **Ts** 25.0–28.0 mm (*n* 29, *m* 26.2). **Wing formula: p1** minute and pointed, < pc 6–11 mm, < p2 50–56 mm; **p2** < wt 2–4.5 mm, =4 (sometimes =3/4 or 4/5); **p3** longest; **p4** < wt 3–5 mm; **p5** < wt 6.5–10 mm; **p6** < wt 9.5–13.5 mm; **p10** < wt 21–26 mm; **s1** < wt 25–29 mm. Emarg. p3.

GEOGRAPHICAL VARIATION & RANGE Monotypic. – Scandinavia, Finland, Germany and south to Romania, Ukraine, Russia east through N Kazakhstan to Altai; also local in W & C Caucasus; winters SE Africa.

REFERENCES Berger, W. (1967) *J. f. Orn.*, 108: 320–327. – Dittberner, H. & Dittberner, W. (1989) *Falke*, 36: 255–259, 314–317, 321.

Ad, Kuwait, Sep: moderately-marked dark lateral throat-stripe and rather shaded dark-marbled breast just visible, typical in autumn. Uppertail only slightly tinged rufous, less so than in Common Nightingale. No moult limits in wing and no tiny buff spots on tips of tertials or greater coverts, edges of these being evenly olive-brown, infer ad. (A. Audevard)

1stW, Scotland, Sep: Thrush Nightingale can lack dark lateral throat-stripe and breast mottling, and some may be inseparable from Common without biometrics. Ageing also difficult, especially as juv buff tips to wing-feathers can be small or lost by Sep/Oct. Here, approximately half of greater coverts seem moulted; quality of primaries, primary-coverts and tertials also suggest juv, while tips to tail-feathers are clearly pointed. (M. Breaks)

1stW, Scotland, Aug: long primary projection (with exposed primary extension usually longer than exposed tertials, unlike Common Nightingale), while spacing of primaries is more even, not gradually increasing towards tip as in Common). (M. Breaks)

Juv, Poland, Jul: densely scalloped and spotted dark brown on buff-white ground throughout. Already, rump is rufous-tinged. (A. Stankiewicz)

(COMMON) NIGHTINGALE
Luscinia megarhynchos C. L. Brehm, 1831

Fr. – Rossignol philomèle; Ger. – Nachtigall
Sp. – Ruiseñor común; Swe. – Sydnäktergal

Known far beyond the circles of birdwatchers for its pleasing song, this unobtrusive-looking bird with a golden throat has earned a place in poetry and common knowledge. The Nightingale, or 'Common Nightingale' to unambiguously separate it from its similar and close relative the Thrush Nightingale, winters in sub-Saharan Africa, returning in late Apr or May to W Europe. One evening you note that it is back again by its explosive but mellow, fluty song from some shady thicket or riparian forest. There it will stay and entertain you by night for a few weeks to follow.

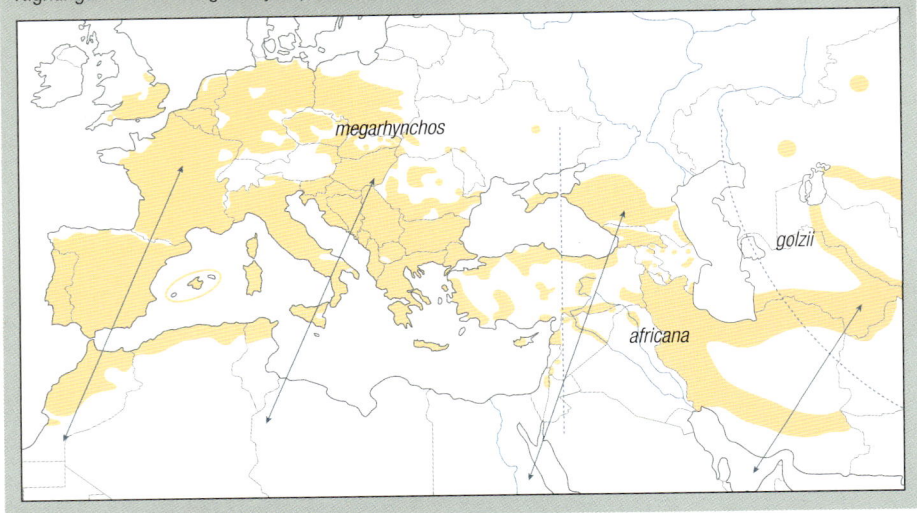

L. m. megarhynchos, presumed ad, Finland, May: pale eye-ring, rather large dark eye and vague greyish supercilium; off-white underparts washed pale brown on breast. Diagnostically, first primary longer than tips of primary-coverts, but in Thrush first primary tiny and often hidden. With difficult birds, check for prominent emargination of two longest primaries (just one in Thrush Nightingale). Ageing inconclusive, but seemingly evenly fresh, possibly ad wing. (M. Varesvuo)

IDENTIFICATION A small songbird related to the Robin with rather *plain, brown* and dusky white colours. The *rufous-red uppertail* is the most noteworthy distinction, whereas the rest of *upperparts are medium brown with a reddish tinge*, especially on the upperwing. Other characters are more subtle and require a good, close look, often not provided by this shy, skulking bird that usually keeps to cover and shady ground. There is a *narrow, diffuse, pale eye-ring* around the rather *large, dark eye*. The throat is off-white with pale brown sides (only rarely a hint of a darker stripe on each side). Rest of underparts are nondescript off-white with a pale greyish-buff or *greyish wash across upper breast and on flanks*. The breast is as a rule uniformly tinged brown-grey without blotches or vermiculations (although very rarely there can be a slight unevenness). *Undertail-coverts uniform whitish, tinged pale rufous-buff* (rarely more cream-white). Bill brown with a slightly paler base, the *rather long legs* being dull pink-brown. Sexes alike. Ages very similar, except that juveniles are characteristic enough, having upperparts spotted pale buff and feathers of underparts tipped dark. It wags and twists its tail and bobs its rear body in a rather slow and mechanical way. Best means of identification is the song (see below). Juvenile very similar to juvenile Thrush Nightingale but has somewhat more reddish uppertail like adult.

VOCALISATIONS Song given from perch in cover, in upright posture, loud and pleasing, an explosive outburst of fluty, whistling or trilling notes lasting 2–4 sec., each strophe separated by a brief pause of about same length or a little shorter. Like in Blackbird it constantly varies the motifs. Some strophes open with a most characteristic series of 'sucking', drawn-out whistles (only 'copied' by Moustached Warbler, although this occurs in different habitat and has weaker song), often rising faintly in pitch and delivered with growing strength like a crescendo, *loo loo luu lüü lee*, which metamorphoses into the normal outburst of fluty notes or trills. Sings mainly at dusk, during the night and at dawn, but newly arrived ♂♂ will habitually sing in full daylight, too. Nearly always easy to tell from Thrush Nightingale by much shorter and more varied, mellow strophes and by the inclusion of the above-mentioned crescendo. The very few which sound more borderline could actually be hybrids, or at least aberrant birds. – Lacks unique contact call, but apparently may use either of alarm calls for contact purposes. Alarm a straight (or rarely slightly upwards-inclined), short whistle *iihp* (or *eeiihp*), at times recalling a similar call of Chaffinch in SE Europe, but as a rule louder and more 'fierce'. Also a low, hard and dry purring or 'hollow' creaking *errr*. The two alarm calls are often used in combination.

SIMILAR SPECIES The most obvious risk of confusion is with *Thrush Nightingale*. Note that Common Nightingale has more vivid rufous-red uppertail, often more reddish-tinged upperwing, and slightly paler and brighter brown rest of upperparts. (Thrush Nightingale has darker red-brown and less bright uppertail, and whole upperparts slightly darker greyish-brown creating less of contrast with tail.) Further, Common Nightingale has more smoothly shaded, nearly un-patterned sides of throat and breast (Thrush virtually always slightly blotched or diffusely vermiculated sides of throat and breast), and uniform undertail-coverts tinged rufous-buff (Thrush Nightingale has colder cream-grey undertail-coverts with some dark blotches at sides). To this add a tendency in Common Nightingale to slowly pump its tail down and up and semi-fan it (Shirihai *et al.* 1996, Campbell 2011), while Thrush Nightingale holds its tail more still and folded. However, this difference might be valid only for eastern populations of Common Nightingale, which have a slightly longer tail on average than European birds. – A more far-fetched similarity is with some of the American *Catharus* thrushes, especially *Hermit Thrush*, which is also brown above with a rusty-red uppertail, but note larger size, distinctly black-spotted or black-streaked whitish throat and breast, and the characteristic underwing pattern of all *Catharus* species with broad white and dark bands. – In

L. m. megarhynchos, presumed 1stS, Tunisia, May: typical posture on landing, with tail partially fanned; note russet tones above, with bright rufous-red uppertail. Ageing, especially without handling, difficult in spring, but relatively worn primaries and primary-coverts, and pale tips to some greater coverts and tertials, suggest a 1stS. (R. Pop)

ROBINS AND NIGHTINGALES

L. m. megarhynchos, presumed 1stS, Spain, May: typical posture, with half-cocked tail. Especially in harsh morning light, russet tones above (especially on uppertail and uppertail-coverts) appear strong. Rather worn primary tips and primary-coverts suggest 1stS. (C. N. G. Bocos)

fleeting views can be confused with *Cetti's Warbler*, but note smaller size of latter and uniform rufous upperparts, lacking a contrastingly more reddish uppertail.

AGEING & SEXING (*megarhynchos*) Ages separable in summer and autumn, sometimes in spring. Sexes inseparable (except a few extremes on measurements when handled). – Moults. Complete post-nuptial moult of ad usually rapid, in Jul–Aug (rarely a little earlier or later) prior to autumn migration. Partial post-juv moult in Jun–Aug (Sep) does not involve flight-feathers but may include a few tertials and often some inner greater coverts, less often all greater coverts (Magnani 2004). Apparently no moult in winter. Ageing in spring usually requires close observation or trapping. – SPRING **Ad** Tertials and greater coverts uniform brown, without paler spots at tips. Tips to tail-feathers reasonably fresh. **1stS** A few or several outer greater coverts have tiny ochre-buff spots at tips (if not, wear makes moult contrast difficult to see). Tail-feathers often somewhat worn at tips. – AUTUMN **Ad** As in spring but freshly moulted, all tertials and greater coverts uniformly brown. **1stW** Like ad but separated by tiny ochre-buff spots on tips of all or two longer tertials, and all or outer greater coverts unmoulted. (Often moults innermost tertials and some inner greater coverts to uniform ad type before autumn migration.) **Juv** Upperparts spotted ochre-buff, underparts densely scalloped dark brown.

BIOMETRICS (*megarhynchos*) **L** 15–16.5 cm; **W** ♂ 81.5–93 mm (*n* 19, *m* 85.9), ♀ 80–88.5 mm (*n* 16, *m* 85.0); **T** ♂ 60–71 mm (*n* 19, *m* 66.2), ♀ 60–69 mm (*n* 16, *m* 64.7); **T/W** *m* 76.6; **B** 15.3–17.8 mm (*n* 39, *m* 16.5); **Ts** 25.2–27.5 mm (*n* 39, *m* 26.2). **Wing formula: p1** > pc 1–5 mm (rarely 0–2 mm <), < p2 36.5–46 mm;

p2 < wt 3–8 mm, =4/5 or 5 (rarely =5/6); **pp3–4** about equal and longest (sometimes p4 < wt 0.5–1.5 mm); **p5** < wt 4–7 mm; **p6** < wt 8–11 mm; **p10** < wt 17.5–22.5 mm; **s1** < wt 21.5–26.5 mm. Emarg. pp3–4.

GEOGRAPHICAL VARIATION & RANGE Moderate variation in Europe, NW Africa, Turkey and Middle East, more obvious in Central Asia (*golzii*). Differences concern plumage, size and proportions. Several subtle races proposed or hypothesized by Eck (1975) could not be validated when extensive material was examined.

L. m. megarhynchos C. L. Brehm, 1831 (W & S Europe, NW Africa, Turkey except in east; W Levant; winters sub-Saharan Africa from Senegal through Ghana, Nigeria and Cameroon, possibly to former N Zaïre). Treated above. Upperparts rather warm brown and cream-white below with brown-grey wash on breast and flanks. Tail generally < 69 mm, rarely up to 71 mm. (Syn. *baehrmanni*; *caligiformis*; *corsa*; *luscinioides*.)

L. m. africana (Fischer & Reichenow, 1884) (E Ukraine, Kalmykia, E Turkey, Syria, Iraq, Caucasus, Transcaucasia, W & N Iran, S Turkmenistan; winters E Africa, mainly Kenya). Differs from *megarhynchos* in being slightly duller (but not darker) brown-grey above, upperwing, mantle and back being slightly greyer, less rufous-tinged. Tail longer, invariably exceeding 65 mm. Sometimes a hint of a paler supercilium.

L. m. megarhynchos, presumed 1stW, Spain, Aug: similar to ad, especially if orange-buff tips to juv greater coverts and tertials are small or have started to wear off; if so, ageing must be based on assessment of primary tips and primary-coverts, which are here already worn; vestiges of buff tips to at least one greater covert, and large alula, as well as rather abraded tail-feathers, further suggest 1stW. (C. N. G. Bocos)

L. m. megarhynchos, juv, Italy, Jul: extensively spotted ochraceous-buff above, and densely scalloped dark below. (F. Ballanti)

L. m. africana, 1stS, United Arab Emirates, Apr: rather dull brown-grey above, less rufous-tinged than *megarhynchos*, with proportionately longer tail and more obvious paler supercilium. Hint of paler tips to wing-coverts and uppertail-coverts typical of eastern origin. Aged by moult limits in greater coverts. (T. Pedersen)

L. m. africana, 1stW, United Arab Emirates, Oct: especially in fresh plumage, to some degree recalls *golzii*, though compared to that buff fringing to wing-coverts, tertials and uppertail-coverts reduced. Characteristic dull brownish tone above, less pale sand-brown than *golzii*, and less rusty-toned than *megarhynchos*. 1stW by juv remiges, outer greater coverts and primary-coverts. (M. Barth)

L. m. africana / golzii, 1stS, United Arab Emirates, Mar: large, predominantly sand-brown plumage, and less obvious rufous tail than *golzii*, but lacks pale fringes to wing-feathers that is feature of latter. In S Iran *africana* and *golzii* intergrade, while early spring date also supports southern origin. Aged by juv outer greater coverts, with very narrow buffish tips (almost lost), and juv primary-coverts being worn and pointed. (I. Boustead)

L. m. golzii, 1stS, Kazakhstan, May: pale brown above and whitish below with characteristic hint of paler supercilium, and fine pale fringing of wing-coverts revealing this eastern race. Aged as 1stS by moult limit in greater coverts, with outer ones being shorter and white-tipped. Much of rest of wing is also juv feathers. (C. Bradshaw)

L. m. golzii, 1stW, Oman, Nov: in autumn, birds typically appear strikingly pale overall, adding to their subtly lager appearance. Fresh and pale-fringed 1stW are also distinctive during autumn migration, pale edges forming hint of pale wing panel and double wing-bars. Note also quite broad paler supercilium. Moult limits in greater coverts in this bird infer 1stW. (A. Audevard)

L 16–17.5 cm; **W** 84–93 mm (*n* 26, *m* 87.4); **T** 66–84 mm (*n* 26, *m* 74.5); **T/W** *m* 85.3; **B** 15.6–18.3 mm (*n* 25, *m* 17.0); **Ts** 25.2–28.3 mm (*n* 24, *m* 26.6). – Birds in S Iran can have hint of paler tips to wing-coverts and uppertail-coverts, and are intergrades between this race and *golzii*. (Syn. *taurida*.)

L. m. golzii Cabanis, 1873 (NE Iran through Central Asia to Mongolia; winters E & SE Africa). Differs from *africana* in being very slightly paler and sandier brown overall with a tendency to have pale-tipped and pale-edged greater and median coverts in fresh plumage, creating suggestion of double wing-bars. Also pale-tipped uppertail-coverts unless heavily abraded. On average slightly paler ear-coverts and lores, and tends to have hint of a diffuse, short, paler supercilium and pale-tipped longer lesser coverts. Roughly same size as *africana* (slightly larger than *megarhynchos*) but has longer and more pointed wings and clearly longer tail (> 73 mm). Moult strategy might differ, since a bird in Pamirs on 12 Oct (ZFMK) has just started complete moult, hence at least in this bird complete moult occurred after, rather than before, autumn migration. The song is on average slightly more 'noisy', less mellow, at times slightly reminiscent of the song of Thrush Nightingale, although should be possible to separate from that when listening to a longer sequence. The piping or whistling crescendo usually stays on one pitch, not slightly rising in pitch like *megarhynchos*. Also, there seems to be a higher incidence of dry rattling or purring series. **L** 17–18.5 cm; **W** ♂ 91–99 mm (*n* 20, *m* 93.9), ♀ 89–96 mm (*n* 12, *m* 92.3); **T** ♂ 77–86 mm (*n* 20, *m* 81.4), ♀ 74–82 mm (*n* 12, *m* 77.8); **T/W** *m* 86.8; **B** 15.9–19.0 mm (*n* 23, *m* 17.7); **Ts** 25.8–29.5 mm (*n* 23, *m* 27.7). **Wing formula:** **p1** < **p2** 44–51 mm; **p5** < wt 4.5–7 mm; **p6** < wt 9–12 mm; **p10** < wt 22–27 mm; **s1** < wt 24.5–31 mm. (Syn. *hafizi*.)

REFERENCES CAMPBELL, O. (2011) *BB*, 104: 40–41. – DICKINSON, E. C. (2008) *BBOC*, 128: 141–142. – ECK, S. (1975) *Beitr. Vogelk.*, 21: 21–30. – MAGNANI, A. (2004) *Ring. & Migr.*, 22: 59–60.

BLUETHROAT
Luscinia svecica (L., 1758)

Fr. – Gorgebleue (à miroir); Ger. – Blaukehlchen
Sp. – Ruiseñor pechiazul; Swe. – Blåhake

During a walk in summer through alpine or tundra willows, especially near a rippling mountain stream, in the light Arctic night, you are likely to hear a remarkable and strong song containing mimicry of other birds, sometimes even of reindeer bells. It is the 'Nordic Nightingale', the Bluethroat, that has enlivened your walk. N European breeders winter mainly in the Middle East, Pakistan and India, whereas W European birds head south or south-west to Spain and N Africa, or even cross the Sahara. Return migration occurs mainly in Apr and early May.

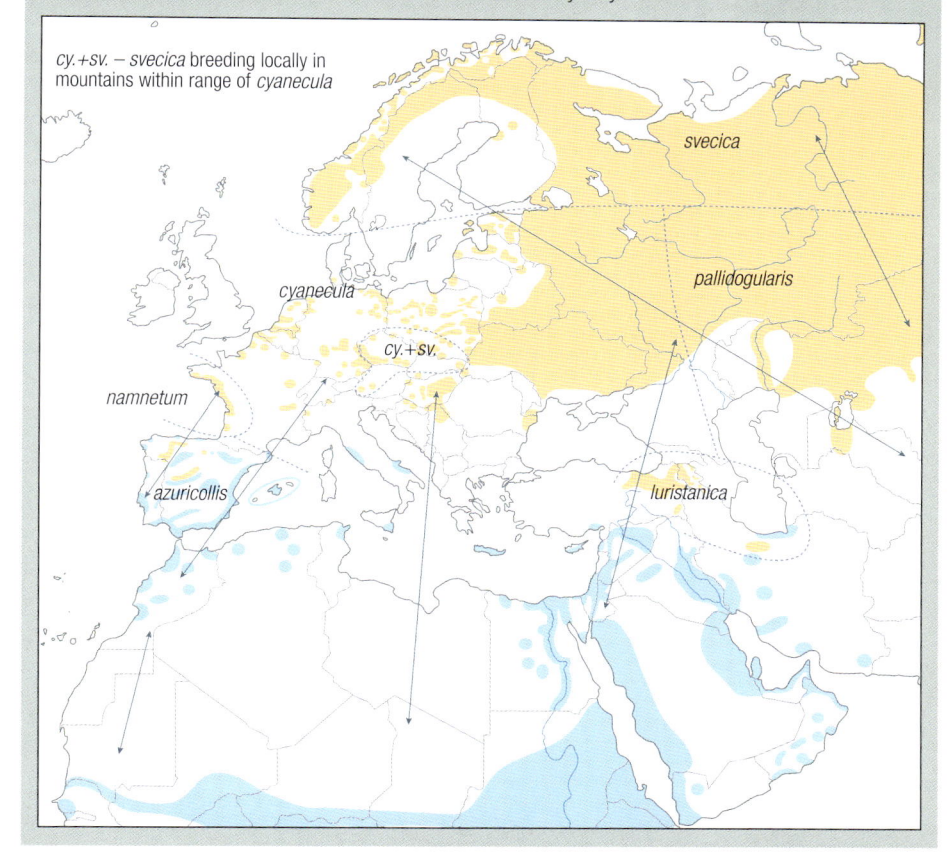

L. s. svecica, ♀, presumed ad, Norway, Jul: throat/breast pattern of ♀ highly variable, with those showing this amount of blue and rufous probably invariably older birds, but much individual variation and overlap, while colours become more exposed with wear. (B. Nikolov)

IDENTIFICATION A small 'chat' or robin with *long, thin legs* and rather *long* neck and *tail*, the latter often *cocked upwards*. When the neck is not stretched, and the body-feathers are fluffed up, the shape becomes more rounded, not that unlike a Robin, still with a somewhat longer and fuller tail. When feeding on ground, which it does frequently, movements are quick and 'nervous', and the tail is constantly cocked upwards and lowered slowly. In all plumages a *pale supercilium* is prominent, and the *outertail has rufous-red base*, most easily seen when the bird flies away from you. The adult ♂ in summer is easily recognised by *brilliant blue throat bib* (reaching onto the upper breast) with a black lower margin, and below this across the breast a variably distinct *narrow white and broader rufous band*. In the centre of the blue bib there is either a rufous or white spot, or the bib is all blue, depending on geographical origin (see below). A few birds of either sex even have an orange spot, and these are believed to be aberrant white-spotted (ssp. *cyanecula*), but that at least some are hybrids between this race and *svecica* can perhaps not be excluded. Autumn plumage of adult ♂ basically same but colours slightly less brilliant, and chin and upper throat become off-white, and sides of throat have black stripes like ♀. Adult ♀ variable, at least some lacking any blue on throat and breast, where replaced by off-white with prominent black stripes at sides of throat joining via a *broader transverse black band on upper breast*, others having blue patches on each side of throat and variable amount of blue at bottom of bib. Often has a *rufous band across breast* below the black. Less commonly, advanced adult ♀♀ have much blue across lower breast. 1stY ♀ invariably lacks any blue on throat, and has dark transverse breast-band often broken up into separate blotches or streaks. Juvenile recalls juvenile Robin with its spotted plumage but has much denser and darker streaking on head and upper body ('swarthy' look), and rufous-red uppertail-coverts and sides of tail.

VOCALISATIONS Song is a loud series of accelerating shrill or clear but hard notes which often increase in strength and culminate in some melodic figure or flourish, or turn into multisyllabic notes at lower pitch. Each series slightly different from previous, but some favourite themes used more often than others. Mimicry sometimes inserted in the endings, by some accomplished birds more and better than by others. Often some dry rasping sounds are inserted in the flourish recalling Black Redstart. At a distance, part of the song can also recall that of, e.g., Woodlark or Pied Flycatcher but is, when heard well, unmistakable. There seems to be some slight regional dialects (not surprising considering wide and partly disjunct range), but the species is readily recog-

L. s. svecica, ♂, Finland, Jun: diagnostic rufous bases to tail-sides (in all plumages). The rufous-red central throat spot is not exclusive to *svecica*; bib deep cobalt-blue. Due to already strong feather wear it is difficult to be sure of feather generation in wing, and this bird is better left un-aged. (M. Varesvuo)

L. s. svecica, 1stS ♀, Norway, Jul: especially 1stS ♀♀ tend to lack any obvious blue and rufous on breast, and instead have blackish streaks or blotches. Aged also by remnants of buffish tips to juv greater coverts. (V. Dell'Orto)

L. s. cyanecula, ♂, Netherlands, Apr: smaller, pure silk-white central throat spot and substantial rufous breast-band, leaving relatively narrow black band. Some birds, especially without handling, are difficult to age in spring. Apparent moult limits in greater coverts suggest 1stY, but condition and shape of primary-coverts more ad-like, therefore best left un-aged. (H. Gebuis)

nised all the same. – Calls are a whistling *whiit*, not unlike Black Redstart, or, as a variation, upwards-inflected *whuit*, like Redstart or Willow Warbler. Also a dry clicking, 'fat' *tlek* (also transcribed as *trak*), likened to the sound from a loose flag halyard flapping in the wind (Rosenberg 1953). The whistling and clinging calls are often combined and heard from migrants in cover, for instance in a reedbed or thicket by a ditch, *whiit tlek tlek whiit tlek*... A further call, sometimes heard in autumn, is a hoarse, cracked *bzrü* or *zerr*.

SIMILAR SPECIES Whereas adult ♂ summer should be fairly straightforward to identify, ♀ and autumn birds require some care. Important to establish combination of prominent pale supercilium and rufous bases to tail-sides. *Siberian Rubythroat* also has a light supercilium and largely similar size and shape, but lacks blackish stripes at sides of throat, and the dark marks on breast of ♀ and 1stW Bluethroat (at most has a brownish diffuse breast-band). Siberian Rubythroat also has an olive-grey tail without any rufous basally.

AGEING & SEXING (*svecica*) Ages separable in 1stW, frequently also in 1stS. Sexes differ after post-juv moult. – Moults. Complete post-nuptial moult of ad usually rapid, in mid Jul–Aug (early Sep) prior to autumn migration. Partial post-juv moult in Jul–Sep does not involve flight-feathers, and only rarely tertials, but includes all median and a few inner greater coverts. Southerly populations time their moult later, sometimes leaving flight-feathers until after autumn migration. Pre-nuptial moult in Feb–early Apr, mainly involves head and breast, and is more extensive in ♂♂ (apparently no moult at all in some ♀♀). – **SPRING Ad** ♂ Entire bib brilliant glossy blue, including chin and upper throat. No black stripes at sides of throat. All primary-coverts and greater coverts uniform brown-grey (very exceptionally has paler rufous-buff spots or thin edges to tips of some inner primary-coverts and outer greater coverts). Wing length helpful as to sex for some extremes. Tips of primaries on average fresher than in 1stS (but some overlap). **Ad** ♀ Throat pattern variable, some with some blue, others without any, but invariably has blackish stripes at sides of throat, black streaks or blotches on breast, and chin and centre of throat buffish-white. Commonly has narrow blue patches on each side of throat and often (but not invariably) some blue patches on upper breast. Primary-coverts and greater coverts like in ad ♂. Wing length helpful as to sex for some extremes. **1stS** Sexes differ largely as described for ad. Many 1stS ♂♂ have black or grey patches in the blue on sides of throat, but by and large have typical ♂ pattern of bib. Often retains some whitish or ochraceous-buff tips on a few outer greater coverts, and sometimes on inner primary-coverts. Tips of primaries on average slightly more worn than in ad (but some overlap). – **AUTUMN Ad** ♂ All primary-coverts and greater coverts uniformly brown, or some inner primary-coverts and outer greater coverts neatly edged pale brown (very exceptionally has paler spots on some inner primary-coverts and outer greater coverts, like in all 1stW). Wide blue band across breast, with black and rufous bands below (rufous band fairly broad and complete). Prominent blue patch on side of throat. Large red central spot on lower throat (with only limited buff-white admixed, although a few lowest feathers can have concealed pure white bases). **1stW** ♂ Resembles ad ♂ but separated by presence of rufous-buff rounded or wedge-shaped spots at tips of inner primary-coverts and outer greater coverts (though see above for rare exception of ad having such spots as well). Also, throat/breast pattern less neat, central spot on lower throat more buff-white than red, and blue on side of throat more restricted. **Ad** ♀ All primary-coverts and greater coverts uniform brown (for rare exception see ad ♂). Resembles poorly marked 1stW ♂, but separable using ageing criteria for ad. Usually some blue patches on breast, but blue can be entirely lacking. Those with most blue on breast may have tiny blue spots on sides of throat as well (but, less than in any ♂). No complete rufous band across lower breast, only

L. s. azuricollis, 1stS ♂, Spain, May: close to *cyanecula*, but usually has all-blue bib, with no white spot, narrower lower black border to blue area (though not on this bird!), and no pale line between rufous and black. Moult limit in greater coverts and juv primary-coverts indicative of 1stS. (C. N. G. Bocos)

L. s. cyanecula, ♂, presumed 1stS, Netherlands, Jun: variation includes some ♂♂ that lack or virtually lack throat spot (here just two white feather tips exposed). With progression of season and feather wear, ageing is more uncertain, although remiges, primary-coverts, alula and most if not even all greater coverts apparently juv; remnants of buff spots on central greater coverts support identification as 1stY. (P. de Knijff)

patches. **1stW ♀** Aged as 1stW on rufous-buff spots on tips of inner primary-coverts and outer greater coverts. Invariably lacks any blue in plumage. Very little rufous on breast, if any. Blackish band across breast often broken up into blotches or streaks. **Juv** Very different plumage: whole head, breast and upperparts to rump very dark, ochraceous-buff heavily streaked blackish. Belly paler, dusky yellowish buff-white. Uppertail-coverts rufous.

BIOMETRICS (*svecica*) **L** 13–14.5 cm; **W** ♂ 73–80 mm (*n* 50, *m* 76.0), ♀ 69–75 mm (*n* 30, *m* 72.4); **T** ♂ 51–58 mm (*n* 20, *m* 54.6), ♀ 48–55 mm (*n* 20, *m* 52.2); **T/W** *m* 71.9; **B** 14.0–16.7 mm (*n* 78, *m* 15.6); **Ts** 25.0–28.3 mm (*n* 33, *m* 26.7). Data from live birds at Gävle, EC Sweden (H. Ellegren *in litt.* 1989): **W** ♂ 73–82 mm (*n* 1880, *m* 77.5), ♀ 70–79 mm (*n* 1212, *m* 75.4). **Wing formula: p1** > pc 2.5–5.5 mm, < p2 27–35 mm; **p2** < wt 4–8 mm, =6 or 6/7 (=5/6); **pp3–5** about equal and longest (p5 sometimes 0.5–1 mm <); **p6** < wt 2.5–5 mm; **p7** < wt 5–9 mm; **p10** < wt 11–15 mm; **s1** < wt 13.5–19 mm. Emarg. pp3–5, sometimes slightly also on p6.

GEOGRAPHICAL VARIATION & RANGE Morphologically fairly moderate variation, but breeding ♂♂ differ in colour of throat/upper breast, and several subspecies have been described based on this. Some intergradation where subspecies meet, and not all migrants can be assigned to taxon. ♀♀ generally inseparable, except *namnetum* and possibly *magna* on size, so plumage descriptions below refer to ♂♂.

L. s. svecica (L., 1758) 'Red-spotted Bluethroat' (Fenno-Scandia, N Russia south to St. Petersburg, N Siberia; also, some fairly recently established isolated pockets in C European mountains, including in Czech Republic, apparently this race; winters mainly in Arabia, Central Asia and east to Pakistan, but some western birds at least partly in SW Europe and N Africa). Treated above. Bib deep cobalt-blue bordered sooty black below with rather large central rufous-red spot (*c.* 9×17 mm). Prominent rufous band across breast below the black. Has narrow pale (buff-white or off-white) stripe between the rufous band and the black lower edge of bib in most, but some have it broken, ill-defined and hardly visible. – Populations of S Norway and WC Sweden ('*gaetkei*') have been separated based on subtly longer average wing length, but difference tiny and much overlap (Svensson 1992, Hellgren *et al.* 2008). – Breeders in NE Ukraine, SC Russia east to Volga sometimes separated as '*volgae*', but these are variable and better seen as intergradation between *svecica*, *cyanecula* and *pallidogularis*. Blue colour of bib same as in *svecica* or slightly lighter (but still a trifle darker than in *pallidogularis*). Spot at centre small or mid-sized (from 4×12 to 6×15 mm are common), varying from all white, white with a few red stains, through rufous with white rim to all rufous-red. Pale band on breast between rufous and black is pure white and well defined in most (before too worn). Size as *svecica* or a trifle smaller. See also Hogner *et al.* (2013). – Several Asian races described but none examined by us differ noticeably from *svecica*. (Syn. *altaica*; *gaetkei*; *grotei*; *kaschgariensis*; *occidentalis*; *robusta*; *saturatior*; *tianchanica*; *volgae*; *weigoldi*.)

L. s. cyanecula (Meisner, 1804) 'White-spotted Bluethroat' (S, N & NE France to Netherlands, C & E Europe, W Russia south of St. Petersburg; winters mainly N & W Africa south to Senegal). Differs from *svecica* in having a pure silky white and smaller central throat-spot (often *c.* 7×10 mm, or a little smaller); sometimes the white spot is missing entirely (var. '*wolfi*', 5–8% in Germany and N France; cf. also ssp. *azuricollis* and *luristanica* below). White throat-spot in autumn often mixed with some rufous tips, making subspecific determination difficult, at least in the field; a few are

Quick key to racial variation in Bluethroat (spring/summer ♂♂): *L. s. svecica* (top left: Norway, May) has rather large central rufous-red throat-spot; *L. s. cyanecula* (top right: Spain, Mar) smaller and white central throat-spot (size varies), and blue colour often slightly purplish-tinged; *L. s. pallidogularis* (bottom left: Kazakhstan, May) as *svecica* but bib less deep blue, with rounder and slightly paler red spot, but also has relatively narrower rufous breast-band (and often very prominent black band); *L. s. luristanica* (bottom right: Armenia, May) has entire bib rather pale blue and very broad rufous breast-band (blackish band can be absent, broken or washed-out). Individual and geographic variation between these examples must be borne in mind, and beware that size and shape of throat-spot and breast-bands change with bird's posture. Black and rufous breast-bands are often admixed with white feathering. (Top left: D. Occhiato; top right: R. Armada; bottom left: J. Normaja; bottom right: V. Ananian)

1stS ♂, Italy, Feb: even in late winter/early spring, many young ♂♂ show partially developed throat/breast pattern, with a 'dirty' pale throat spot difficult to match with any race, even approaching most colourful ad ♀. This bird appears relatively *cyanecula*-like, but racial identification inconclusive. Age by some pale-tipped greater coverts. (D. Occhiato)

♂, Netherlands, Aug: in fresh plumage, colour of throat spot does not provide a safe racial distinction; here pale rufous spot could indicate *svecica*, but also *cyanecula* with buffish-rufous tips concealing white spot. Extensive blue on submoustachial region typical of fresh ad ♂, although some advanced 1stW can develop this amount of blue (probably most frequently in *svecica*). Without assessing wing-feathers, best left un-aged. (R. Debruyne)

Presumed *L. s. cyanecula*, ad ♂, Italy, Oct: a well-marked ♂ in autumn, tentatively identified to subspecies based on the clear white spot; also typical of *cyanecula* is the very broad rufous lower breast-band. (D. Occhiato)

1stW ♂, Scotland, Oct: pale rufous throat spot could suggest *L. s. svecica*, but inconclusive at this season. Throat/breast pattern of 1stW ♂ usually more heavily concealed by white tips. Unlike many (but not all) ad ♂♂, only limited blue on submoustachial region. Aged by juv greater coverts with buffish tips. (H. Harrop)

1stW ♀, England, Sep: lack of blue or rufous with only indistinct buff tones on breast, instead having numerous blackish blotches or streaks. Rufous sides to tail base just visible. Juv greater coverts confirm age. (I. H. Leach)

largely white-spotted even in autumn (notice pure white, glossy colour under rufous tips). Apparently a few can have orange central throat-spot (Ebels & van Duivendijk 2010), but all or some of these could also be the result of mixing with *svecica*. Note that a few *svecica* have large off-white spot with limited red in autumn, but the white is matt and slightly tinged cream. Blue colour of bib when worn (in summer) sometimes a trifle darker and more purplish-blue than in *svecica*. Rufous breast-band on average broader than in other races (though equally prominent in *luristanica*). ♀♀ inseparable from *svecica* and other races. Pale line between black and rufous on lower breast variable, but perhaps on average less visible than in *svecica*. **W** ♂ 73–81 mm (*n* 37, *m* 76.6), ♀ 71–77 mm (*n* 15, *m* 73.3); **T** ♂ 52–59 mm (*n* 36, *m* 54.9), ♀ 49–55 mm (*n* 15, *m* 52.2); **Ts** 23.9–27.2 mm (*n* 44, *m* 25.9). **Wing formula: p1** > pc 1–6.5 mm, < p2 29–35 mm; **p10** < wt 11.5–16.5 mm; **s1** < wt 15–19 mm. (Syn. *volgae*.)

L. s. namnetum Mayaud, 1934 (W France; winters probably mainly SW Europe, partly also NW Africa). Very similar plumage-wise to *cyanecula* with a white throat-spot, but is decidedly smaller and shorter-winged. A majority of birds can be identified using size alone. **L** 12.5–14 cm; **W** ♂ 67–72 mm (*n* 27, *m* 68.9), ♀ 63–70 mm (*n* 12, *m* 66.4); **T** ♂ 46–52.5 mm (*n* 27, *m* 49.6), ♀ 43.5–51 mm (*n* 12, *m* 47.5); **Ts** 23.0–26.6 mm (*n* 40, *m* 24.5). **Wing formula: p1** > pc 3–7 mm, < p2 25–30 mm; **p10** < wt 10–14 mm; **s1** < wt 12.5–16.5 mm.

○ **L. s. azuricollis** (Rafinesque, 1814) (NW & C Iberia; winters probably NW Africa). Resembles *cyanecula* but differs in more frequently having all-blue bib, lacking a white spot in >50%, by usually having a narrower black border at lower end of blue bib, and often lacking a pale line between rufous and black. The occurrence of spotless ♂♂ increases with age (Campos *et al.* 2011), averaging 42.7% among 2ndCY, 59.1% among older ♂♂. It is debateable whether this race fulfils the 75% rule requirement, but is still tentatively accepted here. Campos *et al.* (2011) also documented the rare occurrence among Spanish breeders of a partly or all-red throat-spot, indicating limited introgression of *svecica* genes. – Despite morphological resemblance to *luristanica*, these two did not seem closely related in a genetic study (Johnsen *et al.* 2006). Size similar to *cyanecula* (Campos *et al.* 2005). See also Hogner *et al.* (2013) for support to recognise this subspecies.

L. s. luristanica (Ripley, 1952) (E Turkey, Caucasus and E Transcaucasia, N Iran; winters NE Africa, Arabia). A little larger than *svecica*, being the largest race of all, with a stronger bill and a fractionally longer tarsus, with an all-

L. s. luristanica, ♂, presumed ad, Georgia, Jun: a rather large race that usually has an all-blue throat, although some show hint of white central spot. Blue rather bright, marginally paler than *svecica*. Rufous lower breast-band broad, black band narrow or, commonly, absent. Seemingly uniform ad wing-feathers. (J.-M. Breider)

Presumed *L. s. luristanica*, ♂, Turkey, May: although photographed on breeding grounds of *luristanica*, has rather prominent black band, which is often lacking in typical *luristanica*, making this bird almost *cyanecula*-like, without the white spot (but broader red band on lower breast as typical for *cyanecula*). Such birds are apparently frequent among Turkish *luristanica* and would be difficult to label on migration. Ageing unsafe at this angle. (A. Öztürk)

blue bib, or in a few a blue bib with a tiny white spot. Blue throat somewhat paler than in *cyanecula* (same colour as *pallidogularis*). Prominent breast-band deep rufous. Blackish band between blue and rufous absent or narrow and washed-out. Surprisingly similar to Iberian *azuricollis*, but usually differs in absence of black lower border to blue bib and larger size, as well as genetic differences (Johnsen *et al.* 2006). **L** 14.5–15.5 cm; **W** ♂ 79–84 mm (n 20, m 81.4), ♀ 74–80 mm (n 5, m 77.7); **T** ♂ 54–64 mm (n 20, m 58.7), ♀ 52–56 mm (n 5, m 54.2); **B** 15.0–18.1 mm (n 22, m 16.6); **Ts** 24.8–28.8 mm (n 24, m 27.2). **Wing formula: p1** > pc 4–8 mm, < p2 30.5–36 mm; **p2** =6/7 (=6 or =7, rarely =5/6); **p6** < wt 2.5–5.5 mm; **p7** < wt 6.5–9 mm; **p10** < wt 13.5–17 mm; **s1** < wt 16–20 mm. (Syn. *magna*.)

L. s. pallidogularis (Zarudny, 1897) (SE Russia, SW Siberia, plains and foothills of Kazakhstan, Transcaspia, Kyrgyzstan, Tien Shan; winters SW & S Asia). Differs from *svecica* in having slightly paler blue bib (same colour as *luristanica*). It has a rufous-red central throat-spot as in *svecica*, but the spot is sometimes larger and slightly paler rufous. The rufous breast-band is usually rather poorly developed,

L. s. pallidogularis, ad ♂, Kazakhstan, Jul: like other races, much individual variation. Blue throat bib often quite pale (almost 'silvery' blue); also rather large, round red throat spot, but narrow rufous lower breast-band, and black band often relatively well developed. Visible wing-feathers, including primary-coverts, are ad. (A. Isabekov)

Ad ♂, Kuwait, Dec: evidently one of the 'red-spotted' forms, but which? Only by measuring wing length of several birds can one speculate if they are the shorter-winged *pallidogularis* or the larger *svecica*. This dilemma is common in many parts of the Middle East. (A. Al-Sirhan)

♂, presumed ad, Israel, Mar: many wintering or migratory birds in the Levant are red-spotted, but the combination of having pale blue bib with diluted orange-buff spot and being rather short-winged (mean *c.* 74.5 mm) indicates eastern origin, possibly intergrades between *luristanica* and *pallidogularis*. Although being rather worn in Mar gives impression of 1stS, the lack of contrasts (in colour or wear) in wing imply ad, thus probably ad but ageing unsafe. (H. Shirihai)

broken or narrow and pale (but some have more extensive rufous). Usually a fairly prominent black band across breast below blue. **L** 13.5–15 cm; **W** ♂ 70–77 mm (n 20, m 74.1), ♀ 68–74 mm (n 12, m 70.8); **T** ♂ 52–62 mm (n 20, m 56.3), ♀ 49–55 mm (n 12, m 52.3); **B** 14.5–16.6 mm (n 28, m 15.5). **Wing formula: p1** > pc 3.5–9 mm, < p2 26.5–31.5 mm; **p2** < wt 5–9 mm, =7 or 7/8 (rarely ≤8); **p6** < wt 1–3 mm; **p7** < wt 4–6 mm; **p10** < wt 9–15 mm; **s1** < wt 11–17 mm. Emarg. pp3–6 (on p6 only rarely less prominently). (Syn. *discessa*; *kaschgarensis*; *kobdensis*; *przevalskii*; *volgae*.)

Extralimital: **L. s. abbotti** (Richmond, 1896) (N Kashmir, Ladakh, NW India; winters NW India). Dark with rather small usually red throat-spot (rarely white) and long and thin bill.

TAXONOMIC NOTES The Iberian population *azuricollis* (Rafinesque, 1814) is tentatively accepted as valid with support from, e.g., Johnsen *et al.* (2006) and Hogner *et al.* (2013), although still somewhat doubtful that it is sufficiently distinct from *cyanecula*. A more serious problem might exist with respect to its name. Since Rafinesque rarely deposited type specimens for his names, no type is known, and Rafinesque described it as 'recorded in Sicily and Spain', hardly an accurate range definition; the brief description does not clearly note that it usually has an all-blue bib, and the birds he described could theoretically be migrant *cyanecula*. Still, Mayaud (1958) referred this name to Iberian breeders. Designating a Spanish breeder as a neotype for the name *azuricollis* would appear sensible. – The subspecific name *luristanica* was introduced by Ripley in 1952 as replacement for *magna* Zarudny & Loudon, 1904, which was claimed to be preoccupied. Since this happened before 1961, the Code does not allow reinstatement of the original name after preoccupation is no longer an issue due to change of genus (from *Erithacus* to *Luscinia*).

REFERENCES Campos, F., Peris, S. J. & Lópes-Fidalgo, J. (2005) *Airo*, 15: 95–98. – Campos, F. *et al.* (2011) *Ardeola*, 58: 267–276. – Ebels, E. & van Duivendijk, N. (2010) *BW*, 23: 301–304. – Hogner, S. *et al.* (2013) *Behav. Ecol. Sociobiol.*, 67: 1205–1217. – Hellgren, O. *et al.* (2008) *Ecography*, 31: 95–103. – Johnsen, A. *et al.* (2006) *Mol. Ecol.*, 15: 4033–4047. – Mayaud, N. (1958) *Alauda*, 26: 290–301.

Juv, Russia, Jul: entire plumage brown and ochraceous-buff, heavily streaked and blotched, but tail as ad. (P. Parkhaev)

SIBERIAN RUBYTHROAT
Calliope calliope (Pallas, 1776)

Fr. – Calliope sibérienne; Ger. – Rubinkehlchen
Sp. – Ruiseñor calíope; Swe. – Rubinnäktergal

One of the 'taiga nightingales', this species is more often heard than seen due to its secretive habits, usually keeping well inside dense undergrowth. But when it is finally seen, the ♂ proves to be captivatingly attractive! And on a rare occasion a singing ♂ can sit exposed in the open. Arrives at its breeding grounds in the Siberian taiga in late May or June, raises its single brood and then leaves due east in order to circumvent the large Asian mountain ranges on its way south to winter in SE Asia, from E India eastwards. Hence it is quite rare in Central Asia. It breeds as far west as the Urals, and vagrants have been recorded in W Europe.

Ad ♂, Thailand, Feb: unmistakable if face and throat markings seen. Aged by lack of moult limits in wing, which was entirely replaced during post-nuptial moult; thick black border to red throat and broad grey lower breast-band typical of older ♂. (I. Waschkies)

IDENTIFICATION Compared to a Bluethroat, the Rubythroat is slightly larger and a trifle longer-tailed, but their habits and lifestyles are rather similar. Both spend much time on ground, hopping agilely on *long legs*, often with *raised tail*, or cocking it upwards repeatedly with quick twitches. The ♂ is easily recognised by its complete *ruby-red throat*, usually finely outlined black, and well-marked *pure white supercilium and submoustachial stripe*, both being enhanced by black on either or both sides. Upper breast is washed greyish, lower breast and *flanks are washed olive-brown* while centre of belly is whitish. Upperparts are uniform brown, although crown often is a shade darker and more reddish-tinged. Note that the *tail is all brown*, lacking black, white or any pattern. Young ♂♂ attain the attractive ♂ head pattern already in first autumn, although the chin and upper throat are initially often more whitish than red. Adult ♀ can either lack any red on throat or possess just a little, and have the black-and-white head pattern of ♂ subdued, brown-grey and off-white. However, some adult ♀♀ have more red on the throat, rarely nearly as much as ♂, and fairly well-developed white supercilium and submoustachial stripe (exceptionally even to become confusingly ♂-like). – Juvenile has buff-spotted brown upperparts with rufous-tinged uppertail-coverts, and diffusely streaked or spotted buff-white underparts, thus are very similar to juveniles of both nightingale species and Robin, from which they are not always separable in the field.

VOCALISATIONS Song is a varied warble with a characteristically hard voice, described as a rippling, unstructured ('chatty') strophe of 3–15 sec., delivered at a calm pace, and now and then containing fractions of mimicry and some hard, metallic notes. To give a rough description, it resembles song of Garden or Barred Warblers, or of Daurian Redstart (*Phoenicurus auroreus*; extralimital, treated under Vagrants), but is scratchier than the former two, with inserted whistling notes. It can also slightly recall song of Finsch's Wheatear by its strained voice and peculiar hard elements. Often recognised by a recurring characteristic phrase once learnt, '*chu-wedya-wedya-wah*'. – Call is a squeaking disyllabic *ii-lü* with the second syllable falling in pitch, a bit like Red-breasted Flycatcher. It also has a Fieldfare-like but softer *chak* call, and a Nightingale-like creaking *arrr*.

SIMILAR SPECIES Difficult to mistake an adult ♂ for anything else, but do make sure to check that the tail is all brown. ♀ might be confused with ♀ Bluethroat, but note all-brown tail, lack of black gorget or streaks on breast, and lack of dark crown-sides. – Rarely, ♀ Rubythroat with poorly developed supercilia and submoustachial stripes occur, and are possible to confuse with Red-flanked Bluetail, but the all-brown tail and longer pink-brown legs should usually prevent confusion. – Juvenile very similar to juvenile Robin, both having rufous uppertail-coverts and otherwise brown and buff-spotted plumage, but note slightly larger size, stronger legs and bill, longer tail, and crown and nape are somewhat darker brown with slightly narrower and more distinct buff-white streaks, not rounded buff spots, on paler brown ground.

Ageing ♂♂ in summer, Mongolia, Jun (left) and May (right): some ad ♂♂ (left) have limited or no black border to red bib and less clear grey band below it, whereas some advanced 1stS ♂♂ (right) have these features as well developed as ad ♂. Check moult first: ad has uniform wing-feathers, while 1stS has more brown edges to juv outer greater coverts, and primary-coverts are also juv. Ad has more worn wing, presumably due to active breeding in June. (H. Shirihai)

AGEING & SEXING Ages separable in summer and autumn, frequently also in spring. Sexes differ from 1stW. – Moults. Complete post-nuptial moult of ad usually rapid, in late Jul–Sep prior to autumn migration. Partial post-juv moult in Jul–Aug does not involve flight-feathers but may include a few tertials and some inner greater coverts. Apparently no pre-nuptial moult in winter. Ageing in spring usually requires close observation or trapping. – **SPRING Ad ♂** Pale ruby-red bib, narrowly outlined black (black often also runs across breast below red bib). Jet-black lores, and a black line from base of lower mandible at sides of throat. White supercilium bordered black below. Prominent white submoustachial stripe distinctly set off. Tertials and greater coverts uniformly brown, without paler spots at tips. Wing length sometimes helpful. **Ad ♀** Throat bib variable, off-white, pink or pale ruby-red, but those with most red (rarely closely approaching ♂ plumage) invariably lack black border below, and lores are dark grey or nearly black (though hardly solidly and broadly jet-black like typical ♂). Submoustachial stripe poorly marked, off-white or olive-tinged, fading in lower part. Tertials and greater coverts like ♂. Wing length sometimes helpful. **1stS** Not always separable, even in the hand, but typical birds should be possible to age; tertials and greater coverts with traces of pale spots at tips, or with indentations at tips where pale spots used to be, now abraded. Tail-feathers often rather worn at tips. Sexing as for ad. – **SUMMER Ad** Worn. Sexing as in spring. **Juv** Fresh. Brown upperparts with large buff spots, crown and nape darker brown with rather narrow and distinct buffish-white streaks. Rump and uppertail-coverts rufous. Ochre-tinged dusky white underparts with dark-tipped feathers, giving diffusely streaked appearance. – **AUTUMN Ad** Sexing as in spring. Freshly moulted, all tertials and greater coverts uniformly brown. Tail-feathers sometimes finely pale-tipped. Tail-feathers on average broader and more rounded at tips, and stay fresh longer. **1stW** Like ad (thus ♂♂ attain adult head pattern directly after post-juv moult) but separated by pale buff spots on tips of all or many tertials and greater coverts, most often on two shorter tertials and a few inner greater coverts. Tail-feathers on average somewhat narrower with more pointed outer parts, and tend to become abraded at tips rather quickly.

BIOMETRICS L 14.5–16 cm; **W ♂** 74.5–86 mm (n 53, m 79.4), ♀ 70.5–83.5 mm (n 46, m 75.9); **T ♂** 53.5–64.5 mm (n 25, m 59.5), ♀ 52–61 mm (n 18, m 55.8); **T/W** m 76.1; **B** 14.3–18.0 mm (n 97, m 16.0); **Ts ♂** 27.9–31.4 mm (n 23, m 29.4), ♀ 27.0–29.7 mm (n 18, m 28.4). Wing formula: **p1** > pc 6.5–13 mm, < p2 23–31.5 mm; **p2** < wt 6–9.5 mm, =7, 7/8 or 6/7; **pp3–4** about equal and longest (rarely pp3–5); **p5** < wt 0.5–2 mm; **p6** < wt 3–7 mm; **p7** < wt 7–11 mm; **p10** < wt 13–20 mm; **s1** < wt 16.5–23 mm. Emarg. pp3–5.

GEOGRAPHICAL VARIATION & RANGE Monotypic. – C Ural Mts, Siberian taiga east to Anadyr, Kamchatka, Sakhalin, south to N Mongolia, N Manchuria, N Japan; also C China; winters S & SE Asia, Philippines. – Wing-length increases slightly and clinally towards east, wing in W & C Siberia usually < 82 mm, whereas wing of ♂♂ in Far East usually 78–86 mm (breeders of Kamchatka and Kurils sometimes treated as 'camtschatkensis', of Anadyr as 'anadyrensis'), and eastern birds have on average slightly less rufous-tinged wing. However, these are subtle differences without clear borders and form a continuous cline, and all best included in calliope. There is also much variation in bib and breast pattern. Breeders on Sakhalin seem to be smaller than neighbouring populations, and have on average darker flanks (sullied brown-grey), and these could perhaps constitute a warranted race; still, scant material available, and we prefer to await further studies. NC Chinese breeders were described as separate subspecies ('beicki') by Meise (1937) as being more rounded-winged and longer-tailed. The slightly more rounded wing is an effect of the slightly shorter migration, and the longer tail just an effect of the rounded wing. There are no other differences, and substantial overlap with calliope is evident, hence best included within it. (Syn. anadyrensis; beicki; camtschatkensis; sachalinensis.)

TAXONOMIC NOTE A recent genetic-based phylogenetic study (Sangster et al. 2010) demonstrated that the genus Luscinia as generally treated was paraphyletic, with members appearing disconnected on different branches of the tree. To solve this Siberian Rubythroat, Siberian Blue Robin and a few other closely related chats were moved from Luscinia to separate genera. Siberian Rubythroat becomes Calliope calliope.

REFERENCES BARTHEL, P. (1996) Limicola, 10: 180–189. – LEADER, P. J. (2009) BB, 102: 482–493. – SANGSTER, G. et al. (2010) Mol. Phyl. & Evol., 57: 380–392.

♂, presumed 1stW, Germany, Oct: without handling, ageing uncertain. Most greater coverts seem to be juv having already lost their buffish tips or have just a trace of them, while two inner ones (darker) could represent post-juv renewal. Forepart pattern as ad ♂, but no black lower border to red bib and limited grey breast-band, further suggesting 1stW ♂. Tail-feathers appear narrow in their outerpart. (S. Pfützke)

1stS ♀, China, May: lacks red on throat of ♂, with subdued head pattern; some variation in this, and this bird has particularly plain head pattern. Aged by juv greater coverts with remnants of pale spots on tips. (R. Schols)

1stW ♀, England, Oct: fresh buffish tips to juv greater coverts and tertials, with rusty-tinged crown, dusky lores, and dirty whitish supercilium and submoustachial stripe. (I. Fisher)

♀, presumed ad, Thailand, Dec: lack of black lores and weak supercilium eliminate least-advanced 1stY ♂. Possible juv outer greater coverts create apparent moult limits, but no suggestion of buff tips, implying that impression is artificial and result of heavy wear. Shape and wear on primary-coverts intermediate between age classes, therefore best labelled cautiously. (C.-J. Svensson)

Presumed 1stS ♀, Mongolia, May: moult limits in wing (with juv outer greater covert) and only very little trace of reddish on chin suggest advanced 1stY ♀. 1stS ♂ with ♀-like forepart pattern, lacking a red bib, unknown. (J. Normaja)

RED-FLANKED BLUETAIL
Tarsiger cyanurus (Pallas, 1773)

Fr. – Rossignol à flancs rouge; Ger. – Blauschwanz
Sp. – Ruiseñor coliazul; Swe. – Blåstjärt

In order to find this enigmatic and beautiful bird in summer, you need to seek out ravines, hills or low fells in the remote, northern taiga, with mature coniferous forest and some windfalls, lichen and undergrowth. There, its simple song with a melancholy ring seems well suited for the habitat. Although now found as far west as N Finland, this is due to a fairly recent range expansion by this originally Siberian species, and in autumn it still migrates all the way to its traditional wintering grounds in SE Asia. In spring it does not reach its outposts in N Fenno-Scandia until late May or June.

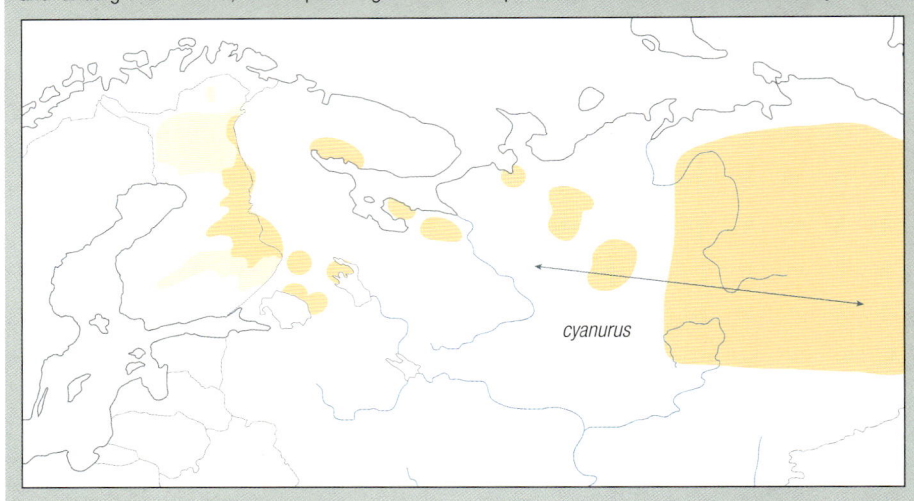

IDENTIFICATION About the size of a Robin, with similar quick movements, although it has a different habit of *constantly dipping its proportionately longer tail downwards*. Few small passerines have a *bluish-tinged tail and rump*, so seeing this narrows down the choices in our part of the world to this species and Siberian Blue Robin. However, the blue is not always easy to see; often the uppertail appears just 'dark'. Instead, concentrate on noting presence of *orange-red upper flanks* (poorly developed in a few, though) and the *neatly outlined off-white throat, surrounded by olive-grey* (bib more diffusely outlined in some). The *pale throat* (which has a very faint buff tinge when fresh) *is narrow*, being limited to the centre of the throat, and widens only at the bottom. Legs are rather short, thin and dark, often blackish. Adult ♂ in summer is a beautiful bird with largely *greyish-blue upperparts*, brilliantly blue wings and *sides of tail*, and sometimes has a little clear blue on sides of the crown as well. The rather *prominent supercilium is pure white* from above the dark lores to the eye or just behind it, but can be tinged pale blue over its rear part, at least when fresh. Note that there is variation in the amount of blue on upperwing of the ♂, which appears not to be age-related: a few have completely blue wings, but more commonly the wing is olive-brown with only limited blue near the wing-bend, or any intermediate appearance between these two extremes. Adult ♀ and young birds (up to one year old) are *olive-tinged brown-grey above* with *blue restricted to uppertail and rump*. Breast and sides of throat are olive-grey leaving a variably distinct *narrow off-white bib* as the only striking feature of this plumage. (Very rarely, adult ♀ can have some blue on lesser coverts and scapulars as well, and forepart of supercilium can show some off-white, but confusion with adult ♂ is prevented by the lack of blue on crown and mantle.) Autumn birds are all like ♀, with the exception that adult ♂ has some blue visible at wing-bend and sometimes on shoulders, and some off-white on fore supercilium. Due to the rather *uniform olive-grey sides of the head*, the thin *pale eye-ring* is rather prominent. Adult ♂ starts to show its blue upperparts (attained through plumage wear) in late Dec and Jan, and all are readily identified as ♂♂ soon after. Juvenile resembles juvenile Robin but has bluish-tinged uppertail.

VOCALISATIONS Best means of detecting this species is by its song, mainly heard during light Nordic summer nights, preferably between 1 and 4 am, delivered from as elevated perch as possible ('song post with a view'). The song is brief and loud, repeated without much variation, with short pauses between, a few clear notes with 'r' and a melancholy or desolate ring, little variation in pitch (though commonly third note is a little higher and last few notes fall subtly) and about seven syllables, third often slightly stressed, *itri-**chirr**-tre-tre-tru-trurr*. Last note sometimes more trilling, and voice of some birds a little clearer, but otherwise little variation in normal song of *cyanurus* (contra *BWP*, *HBW*). However, a completely different variation, probably aberrant, has been noted once in N Sweden: a monotonous repetition of desolate, piping notes, *teeu teeu teeu teeu...*, surprisingly similar to song of Willow Tit (Lind 2010)! – Calls are a straight whistling *viht*, recalling Black Redstart and audible at some distance, and a muffled, hard, slightly throaty *track*, often repeated quickly in twos or threes, *trk-trk*. Both calls are used as alarm near nest when repeated incessantly.

SIMILAR SPECIES The only other small species in treated range to have a bluish-tinged uppertail is *Siberian Blue Robin*, but apart from several minor differences this lacks orange on flanks and has stronger and paler legs (greyish-pink instead of blackish). – In case the exact colour of the tail has not been established, a ♀ or first-winter *White-throated Robin* could theoretically be a pitfall, it usually having orange flanks, too. However, apart from normally appearing in more open thickets and on the ground, this is a clearly larger and slimmer bird with stronger bill and longer legs and tail, the tail often being held raised above back. – *Siberian Rubythroat* shares breeding range and similar habitat with Red-flanked Bluetail, and first-winters and some ♀♀ can have a similar pale throat patch, but Siberian Rubythroat is larger and invariably lacks orange on flanks. – There is an extralimital twin species in the Himalaya and S China, '*Himalayan Bluetail*' (*T. rufilatus*, sometimes treated as a distinct subspecies of Red-flanked Bluetail; see Taxonomic notes), for which ♀ and immature ♂ require care and preferably handling to be safely separated; see Taxonomic notes.

AGEING & SEXING (*cyanurus*) Ages separable in 1stW. 1stS ♂ wears a ♀-like plumage, hence differs clearly from ad ♂, but unless singing rarely separable from ♀. (Existence of two ♂ colour morphs as per *BWP*, one being ♀-like, unknown to us.) As demonstrated by retraps of same birds in Japan (T. Shimizu, pers. comm., 2014), at least some ♂♂ develop paler blue plumage in first complete moult (still becoming obviously ♂-like), attaining darker and deeper blue colours only after following moult. In summary, sexes differ from 2ndW, rarely slightly from 1stW. – Moults. Complete post-nuptial moult of ad usually performed rapidly in late Jul–late Sep (thus unusually late for a long-distance migrant), apparently always prior to autumn migration

T. c. cyanurus, ad ♂, Finland, Jun: upperparts tinged cerulean-blue, but lower mantle/scapulars, neck-sides, wing bend and uppertail-coverts are brightest. White supercilium often bordered by bright blue lateral crown-stripe, while orange flanks add to this chat's uniqueness. (M. Varesvuo)

ROBINS AND NIGHTINGALES

T. c. cyanurus, ad ♂, Mongolia, Jun: compact with short neck and legs, but rather long tail. Patchy bright blue above with whitish supercilium, predominantly whitish underparts (in particular well-defined white throat patch) and fluffy orange flanks. (H. Shirihai)

T. c. cyanurus, ad ♂, China, Oct: in fresh autumn plumage, when bright blue is reduced, and underparts often slightly mottled. White supercilium often pale blue behind eye when fresh. (M. Parker)

(but one ♀ on Sakhalin 18 Aug still not started, others nearly finished 23 Sep). Partial post-juv moult in Jul–Sep does not involve flight-feathers, tertials, greater coverts or primary-coverts. Apparently no moult in late winter, abrasion being responsible for the transformation of ad ♂ to summer plumage. – **SPRING Ad ♂** Upperparts (crown to back) largely greyish-blue. Rump, uppertail-coverts, some lesser and median coverts bright blue. Supercilium (dull) blue behind and above eye, whitish in front of eye, reaching sides of forehead; abrasion will often turn some bluish parts whiter as season progresses. Wings variable (as far as known, unrelated to age), either (rarely) with all feathers (including primaries, primary-coverts, secondaries and scapulars) edged clear blue, or (more commonly) with all feathers edged olive-brown with ochraceous tinge except bluish lesser and median coverts, or a mixture of these two extremes, e.g. with some blue on inner greater coverts and scapulars (seemingly stepless variation rather than two morphs). **1stS ♂** Crown to back greyish-olive. As a rule, only a short and ill-defined olive-white supercilium (if any), but very rarely a near-complete ad type supercilium developed on otherwise ♀-like head. Rarely some faint blue on scapulars and some wing-coverts, but not a reliable sex character since some ad ♀♀ also can show this. Not possible to separate from ♀ unless by song. **♀** As 1stS ♂, with crown to back greyish-olive. Blue restricted to rump, uppertail-coverts and uppertail (exceptionally also blue on lesser coverts and some outer scapulars). Usually inseparable from 1stS ♂ (see above). – **AUTUMN Ad ♂** Blue of upperparts and pale supercilium partly (often largely) concealed by brown tips. Usually some blue visible, apart from rump and uppertail, on back, upper scapulars and lesser coverts, often also on sides of crown (above supercilium). White of fore supercilium often slightly visible. Some nearly as neat as in spring. Already in Jan, and perhaps earlier, full breeding colours start to emerge through abrasion on those with partly concealed blue colour in autumn. Shape of central tail-feathers variable, being either well rounded at tips or slightly more pointed, quite similar to 1stW with broadest tail-feathers. All greater coverts of same generation and subtly tinged olive-brown, lacking a colour contrast. **Ad ♀** Very similar to 1stW and only separable when trapped or seen very close. Generally, primary-coverts are rather uniform brown with olive tinge. Central tail-feathers often rather rounded at tips (but beware that many are more pointed, almost like in 1stW!). As in ad ♂, greater coverts uniform and subtly tinged olive-brown. **1stW** Very similar to ad ♀ and often inseparable. First, check greater coverts for any subtle moult contrast created by some inner moulted to ad type, differing from outer retained juv coverts being a trifle more rusty-tinged. Generally, primary-coverts are rather pointed, have rather frayed or 'loose' tips, and outer webs are rather pale brown and rufous-tinged. Central tail-feathers invariably acutely pointed (but beware that a few have slightly broader feathers approaching ad shape). Sexing of 1stW generally not possible, but with practice some extremes can be told, ♂♂ having deeper blue rump and uppertail and darker orange flanks than ♀♀. **Juv** Olive grey-brown upperparts with small buffish spots on crown, nape, mantle and shoulders. Ochre-tinged dusky white underparts with dark-tipped feathers, giving spotted appearance. Uppertail-coverts bluish.

BIOMETRICS (*cyanurus*) **L** 13–14 cm; **W** ♂ 75–83 mm (*n* 123, *m* 79.7), ♀ 73.5–80 mm (*n* 42, *m* 76.4); **T** ♂ 54–64 mm (*n* 122, *m* 59.3), ♀ 53–60 mm (*n* 42, *m* 56.8); **T/W** *m* 74.4; **B** 11.9–14.4 mm (*n* 165, *m* 13.1); **Ts** 20.4–23.5 mm (*n* 166, *m* 21.8); **HC** 5.2–7.2 mm (*n* 81, *m* 6.1). **Wing formula: p1** > pc 7–13 mm, < p2 23–31 mm; **p2** < wt 10–16 mm, =7 or 7/8 (71%), =8 or 8/9 (28%) or =6/7 (1%); **p3** < wt 1.5–3 mm; **pp4–5** longest; **p6** < wt 1–4 mm; **p7** < wt 8–12 mm; **p10** < wt 15–22 mm; **s1** < wt 16.5–23 mm. Emarg. pp3–6.

GEOGRAPHICAL VARIATION & RANGE Very slight variation, only noticeable in extreme south, where song differs clearly and morphology slightly. – The distinct form *rufilatus* (including *pallidior*) of Himalaya and S China is regarded by us as a separate species (see below).

T. c. cyanurus (Pallas, 1773) (entire taiga belt from Fenno-Scandia to Sea of Okhotsk, south to Altai, Sayan, N Mongolia, Manchuria; winters S & SE Asia, from Myanmar to SE China). Treated above. Typically short-legged with white fore supercilium and moderate amount of blue above in ad ♂, and slightly buff-tinged white bib in fresh ♀-like plumage. (Syn. *pacificus*; *ussurensis*.)

Extralimital: **T. c. albocoeruleus** Meise, 1937 (NC China, in N Gansu and NE Qinghai; mainly short-range movements to lower levels but some reach Myanmar and N Thailand). Differs mainly in having a markedly different song, falling in pitch like a common song variety of Redwing. Subtly larger than *cyanurus* and has markedly longer legs, intermediate in size between this and 'Himalayan Bluetail' *T. rufilatus*. Plumage very similar to *cyanurus* but has on average slightly less white and more pale blue on fore supercilium in fresh plumage. A limited sample (*n* 9) had the following biometrics: **W** 77–85 mm (*m* 81.1), **T** 55–68 mm (*m* 60.6), **B** 12.5–14.2 mm (*m* 13.5), **Ts** 23.1–25.0 mm (*m* 24.0). Has been suggested to constitute a cryptic separate (third) species, but more study required. Scant material available.

T. c. cyanurus, ad ♀, Finland, Jun: 1stS ♂ often inseparable from ♀. However, entire wing, including primary-coverts, are ad, meaning this bird can be automatically sexed as ♀. (M. Varesvuo)

TAXONOMIC NOTES Considering the generally clear and considerable morphological differences between *cyanurus* and *rufilatus*, the known genetic distance, their ranges being allopatric, and our inability to find true intermediates in collections (save the odd exceptions), we suggest that these two are better treated as two separate species. The few reports of alleged intermediates are more likely due to under-rated variation within *cyanurus* than to hybridisation. Song and calls also differ between the two (Martens & Eck 1995; own observations). Although *rufilatus* is extralimital and hardly a candidate for vagrancy to the treated region, we offer a description as a service to museum workers.

HIMALAYAN BLUETAIL, *T. r. rufilatus* (Hodgson, 1845) (Himalaya except in west, China except in north; winters lower altitudes, in Myanmar, Indochina). In ad ♂, much deeper and clearer ultramarine blue upperparts with black undertone; all cerulean-blue supercilium broadly joining on forehead (in *T. cyanurus*, supercilia have pure white forepart and are divided by dark forehead), a plumage which it retains in autumn–early winter (unlike ad ♂ *cyanurus*); wing-feathers usually entirely darkly blue-edged, some having limited brown admixed; sides of head and neck extensively blackish-blue; throat purer white, and white

T. c. cyanurus, 1stW, England, Oct: in autumn, young and ♀♀ dull olive-brown above, with brownish-tinged breast and throat-sides, whitish belly, variable buff-white streak on mid-throat (not always visible), and orange-buff flanks; all share blue tail of ad ♂. Aged by brown juv primary-coverts and greater coverts, the latter especially contrasting with moulted median coverts. (S. Elsom)

T. c. cyanurus, ad ♀, Netherlands, Feb: any blue in autumn/winter mostly confined to rump and uppertail, while orange flanks and thin but distinctive whitish eye-ring also obvious. White streak on mid-throat is highly distinctive. Age by uniform olive-brown primary-coverts and lack of moult contrast in greater coverts. Sex follows ageing as ad, supported by fairly moderate blue tinge on uppertail and rather subdued orange on flanks. (P. Cools)

T. c. cyanurus, presumed 1stW ♂, Germany, Oct: similar to ad ♀ (left), but since the tail-feathers are very pointed and seem rather narrow, primary-coverts are rather dull and a little worn at pointed tips (at least inner can be judged), and iris seems to be dark, it is most likely 1stW. Based on this, uppertail looks quite deep blue and upper flanks saturated rufous-ochre, indicating a ♂. (G. Schuler)

area narrower, being more extensively bordered bluish-black down to breast-sides; often 'swarthy' look due to sooty-black feathers admixed with off-white breast-feathers (much more so than in *T. cyanurus*). ♀-coloured are also generally separable, although some come close to *cyanurus*: ♀ *rufilatus* has deeper orange on flanks and nearly always narrower and cleaner off-white throat bib (*cyanurus*: broader, often tinged buff), being bordered darker olive-brown on sides and below. Also, upperwing and back often deeper rufous-brown than in *T. cyanurus*, which is more olive grey-brown. Wing, tail, bill and claws on average longer, and tarsus generally significantly longer (but see ssp. *albocoeruleus* above). **W** ♂ 80–89 mm (n 46, m 84.5), ♀ 77–85 mm (n 32, m 81.7); **T** ♂ 59–70 mm (n 46, m 65.1), ♀ 58–69 mm (n 32, m 62.6); **T/W** m 76.9; **B** 12.4–15.2 mm (n 79, m 13.9); **Ts** (once each 22.0 and 22.5) 23.0–26.9 mm (n 82, m 24.5); **HC** 5.8–8.0 mm (n 46, m 6.9). **Wing formula: p1** > pc 10–15 mm, < p2 22.5–29 mm; **p2** < wt 10.5–18 mm, =9 or 9/10 (52%), =8 or 8/9 (40%) or ≤ 10 (8%); **p3** < wt 2.5–4.5 mm; **pp4–5** longest (sometimes pp4–6); **p6** < wt 0–2.5 mm; **p7** < wt 6–11.5 mm; **p10** < wt 14–19 mm; **s1** < wt 16–21 mm. (Syn. *practicus*.)

T. c. cyanurus, juv, Russia, Jul: rather dark and spotted, but still has bluish uppertail. (P. Parkhaev)

T. r. pallidior (Baker, 1924) (Simla, Kashmir, Punjab; winters India). Subtly but consistently paler above (both sexes) and on average a fraction smaller: **W** 76–84 mm (m 80.3); **T** 57–65 mm (m 60.7); **Ts** 22.6–25.2 mm (m 23.7) (n 37).

REFERENCES HELLSTRÖM, M. & NOREVIK, G. (2013) *BB*, 106: 669–677. – LEADER, P. J. (2009) *BB*, 102: 482–493. – LIND, H. (2010) *Orn. Svecica*, 20: 54–56. – MARTENS, J. & ECK, S. (1995) *Bonn. Zool. Monogr.*, 38: 1–445.

EVERSMANN'S REDSTART
Phoenicurus erythronotus (Eversmann, 1841)

Fr. – Rougequeue à dos roux; Ger. – Sprosserrotschwanz
Sp. – Colirrojo de Eversmann; Swe. – Altajrödstjärt

This Central Asian redstart is a rather uncommon visitor to the covered region. Only small numbers winter in Iran and E Oman, and it is even rarer in Iraq and E Arabia, with vagrants recorded west to NW Saudi Arabia and Israel. The species inhabits sparse woodland with glades and is fond of forest edges with access to dry, grassy slopes with scree and scrub, often at some altitude but at times lower down. It is found from N Tajikistan and E Kazakhstan through the Altai to NW China, NW Mongolia and SW Siberia.

IDENTIFICATION The ♂ is an attractive medium-sized redstart, with a *grey-white forehead and prominent but thin supercilium, bluish ash-grey crown and nape*, black lores and ear-coverts reaching onto neck-sides and meeting predominantly black wing-coverts of the folded wing. Further, has *diagnostic elongated white shoulder panel* just below scapulars (when less worn also a pale inner secondary panel), and a small but conspicuous *bold white patch on primary-coverts*. Rest of upperparts and outertail, and underparts (from below black chin to breast and body-sides) *deep orange-chestnut*, with cream-white belly and vent. ♀ distinctly different (recalls ♀ Common Redstart), being grey-brown above and paler dull greyish-buff on breast and flanks, with subtly paler throat and belly, and appears rather nondescript with only *buff-white fringes to tertials and wing-coverts* (producing quite prominent wing-bars), and rufous tail. Visibly *larger and bulkier than Common and Black Redstarts*, also larger-headed, and tail proportionately shorter augmenting heavier appearance. Behaviour and jizz (erect posture) generally rather similar to latter two but, diagnostically, tail movements consist of distinct upward flicks without any quivering. Frequently perches on rocks and bushes, but does not avoid trees. Solitary or in pairs.

VOCALISATIONS Song is calm and pensive, a rather deep-voiced thrush-like warble or twitter with many metallic trilling sounds woven-in, just a few notes together at a time and each strophe spaced by 2–3 sec. pauses, e.g. *se sire-sree si-trrr … sesjetzirrirr-tsrriie … suuh-tsrrie tse-tse …* etc. – Call is a soft, slurred croaking or short grating *errr* and a very thin, fine whistling *eeh* or disyllabic *few-eet*. There is also a hoarse *chach-chach* in anxiety.

SIMILAR SPECIES Like many redstarts, adult and most immature ♂♂ unmistakable, but ♀ far less distinctive, generally recalling ♀ *Common* and *Black Redstarts*. However, ♀ Eversmann's is patently bulkier, the tail is not nervously quivered (but jerked up and down), and the bird is duller above with characteristic broader pale fringes to wing-coverts and tertials (creating clearer wing-bars and inner wing panel on edges of tertials and innermost secondaries—pattern not unlike fresh Spotted Flycatcher, but edges wear off during spring and summer). Also, tends to have buff lower back and upper rump (paler than rest of rump and uppertail-coverts), and a duller rufous tail, somewhat more mottled face and perhaps more conspicuous pale eye-ring. Lastly, ♀ Eversmann's lacks any orange-buff below or sharp contrast with whiter belly often shown by ♀ Common. Beware that some eastern races of Common and Black Redstarts could, rather confusingly, have better-marked wings, which must be taken into account.

AGEING & SEXING Ageing of ♂ easy by moult, feather wear and pattern in wing (♀ requires closer inspection, and some are difficult even in the hand). Sexes differ, and seasonal variation is pronounced in ♂ plumages. – Moults. Complete post-nuptial moult of ad in late summer after breeding, mostly Jul–Sep, and partial post-juv moult at the same time, usually including all head- and body-feathers, lesser and median coverts (and occasionally a few inner greater coverts). No pre-nuptial moult in winter. – **SPRING Ad** Both sexes aged by having only ad-type feathers, without moult limits of 1stS. **Ad ♂** From about Apr–May, white on forehead and crown-sides appears like bold supercilium. Remiges, tertials and coverts solidly blacker (though note that also in ad there is a very slight contrast between jet-black greater coverts and alula-feathers and subtly greyer brown primaries and tips to primary-coverts). White shoulder panel and primary-coverts patch more contrasting. Primary-coverts invariably have white bases covering more than half of visible length, often two-thirds, leaving only tips dark.

1stY ♀, Oman, Feb: bulkier than Common and Black Redstarts, in erect posture rather like a wheatear. Clear wing-bars and hint of innerwing panel, and lacks any orange-buff below or sharp contrast with whiter belly, which is often shown by either of the other redstart species. Aged by moult limit in greater coverts. (H. & J. Eriksen)

Ad ♀ Rather duller in summer than in autumn, with pale fringes to wing-feathers and wing-bars much reduced. **1stS** Young ♂, at least until early spring, may retain some paler fringes to wing-coverts and even upperparts. Primary-coverts all dark or with only limited amount of white basally diagnostic. (Those with some white at base of primary-coverts have generally at most half visible length white, rarely a little more, and often outer webs partly dark.) Both sexes often readily aged by worn and abraded primaries, primary-coverts and some outer greater coverts, forming moult limits, most pronounced in ♂. – **AUTUMN Ad** Both sexes aged by being evenly fresh, lacking moult limits of 1stW, with bolder-patterned wings in ad ♂; both white shoulder panel and especially primary-coverts patch are contrasting. **Ad ♂** Pale greyish-brown fringes to crown and nape impart slight scaly appearance. Pale fringes to wing-coverts and tertials evident. Black mask faintly speckled pale buff and bordered by vague pale grey supercilium. Reddish underparts tipped pale buff. **Ad ♀** Slightly warmer fawn grey-brown, sandier than in worn plumage. Characteristic pale fringes to tertials and wing-coverts most evident at this season. **1stW ♂** Much as fresh ad, but pale fringes broader, and black area further concealed; elongated white shoulder panel rather well

Ad ♂, Kazakhstan, Jun: unmistakable given diagnostic white shoulder panel, and conspicuous bold and very extensive white patch on primary-coverts (which also confirms age). (M. Vaslin)

1stS ♂, Mongolia, Jun: as ad ♂, both having grey crown, black mask, rufous throat, breast and mantle, and long white shoulder panel. Juv primary-coverts lack white patch, while juv remiges are bleached browner, forming strong moult contrast. (H. Shirihai)

1stY ♀, Oman, Feb: overall cold plumage, making double whitish wing-bars rather noticeable; also paler edges to bases of dark-tipped primary-coverts help identify birds before too worn. Aged by moult limit in greater coverts. (H. & J. Eriksen)

1stS ♀, Kazakhstan, May: pale wing-bars almost lost during summer. Still, appears larger and bulkier than Common and Black Redstarts, with proportionately shorter tail, as well as buff lower back and upper rump (paler than rest of rump and uppertail-coverts) are important for identification. White-tipped lesser coverts never occur in ♀ Common or Black. Moult limits among inner greater coverts indicate 1stS. (C. Bradshaw)

developed but white primary-coverts patch limited or lacking (see under Spring). Juv wing-feathers, primary-coverts, tail-feathers, tertials and all or most outer greater coverts retained, these being relatively weaker textured and rather more worn and browner. Retained greater coverts (nearly always all are retained) form moult contrast with renewed much blacker median coverts (and, if any, inner moulted greater coverts). Buff fringes of retained tertials and greater coverts broader, creating clearer edges and bars. **1stW ♀** Resembles fresh ad ♀ and best aged by moult limits as in 1stW ♂, but contrasts less sharp. **Juv** Rather like fresh ♀ but is brown with dull rusty-yellow spots above, cream with black mottling below.

BIOMETRICS L 14–15.5 cm; **W** ♂ 85–90.5 mm (n 24, m 87.1), ♀ 81–87 mm (n 15, m 83.9); **T** ♂ 66–73 mm (n 24, m 68.7), ♀ 64.5–70 mm (n 15, m 67.2); **T/W** m 79.3; **B** 13.5–15.9 mm (n 39, m 14.7); **Ts** 21.0–23.8 mm (n 39, m 22.8). **Wing formula: p1** > pc 8.5–14 mm, < p2 26.5–31 mm; **p2** < wt 9–16.5 mm, =7 or 7/8 (80%) or =8 or 8/9 (20%); **p3** < wt 0.5–2.5 mm; **pp4–5** about equal and longest (p5 sometimes to 1 mm <); **p6** < wt 1–3 mm; **p7** < wt 7–10 mm; **p8** < wt 11–13 mm; **p10** < wt 14–19 mm; **s1** < wt 15–22 mm. Emarg. pp3–6.

Ad ♂, Oman, Jan: unlike summer ad ♂, grey crown duller, blackish mask concentrated on lores, and rufous above and below limited due to broad pale fringes. Nevertheless, white-black-white pattern on wing always identify species (and also useful for ageing). (H. & J. Eriksen)

1stW ♂, Oman, Jan: compared to fresh ad ♂, pale fringes broader, further concealing black and rufous areas, prominence of long white shoulder panel varies, and white primary-coverts patch on juv feathers much reduced or as here lacking (dark tips broader, and any white edges on bases shorter and buff-tinged). Nevertheless, wing pattern, grey crown and dark mask are prominent, and sufficient for species identification. Most of greater coverts are juv, being more worn and browner. (H. & J. Eriksen)

GEOGRAPHICAL VARIATION & RANGE Monotypic. – Mountains with trees and bushes in Central Asia, from N Tajikistan and Kyrgyzstan north through Altai and Sayan into W Mongolia and NW China; winters Iran, S Central Asia, W Himalaya.

TAXONOMIC NOTE Formerly often regarded as polytypic, with *P. alaschanicus* (Przevalski's Redstart; extralimital and not treated) as a second race. However, most modern literature separate them as two different species, a view followed here.

1stW ♀, United Arab Emirates, Jan: generally dullest plumage, especially those with many juv wing-feathers and reduced pale wing-bars and innerwing panel. Nevertheless, despite six outer greater coverts being juv, pale wing-bars are still evident, and note bulky appearance and generally cold grey plumage. (N. Moran)

Juv–1stW, presumed ♀, Kazakhstan, Aug: even when very young, there is faint greater coverts wing-bar, while pale-edged and dark-tipped primary-coverts are also evident. (M. Valkenburg)

BLACK REDSTART
Phoenicurus ochruros (S. G. Gmelin, 1774)

Fr. – Rougequeue noir; Ger. – Hausrotschwanz
Sp. – Colirrojo tizón; Swe. – Svart rödstjärt

Originally a bird of steep cliffs and rocky mountains, this cousin of the Common Redstart has adapted to similar man-made habitats: factory buildings, church towers, castles and silos. Although widespread over much of Europe, and despite living so near man, few people other than birdwatchers know it by name. Northern birds move south to winter in S Europe and N Africa, while breeders in S & W Europe are resident. Returns early to breed, in late February to April.

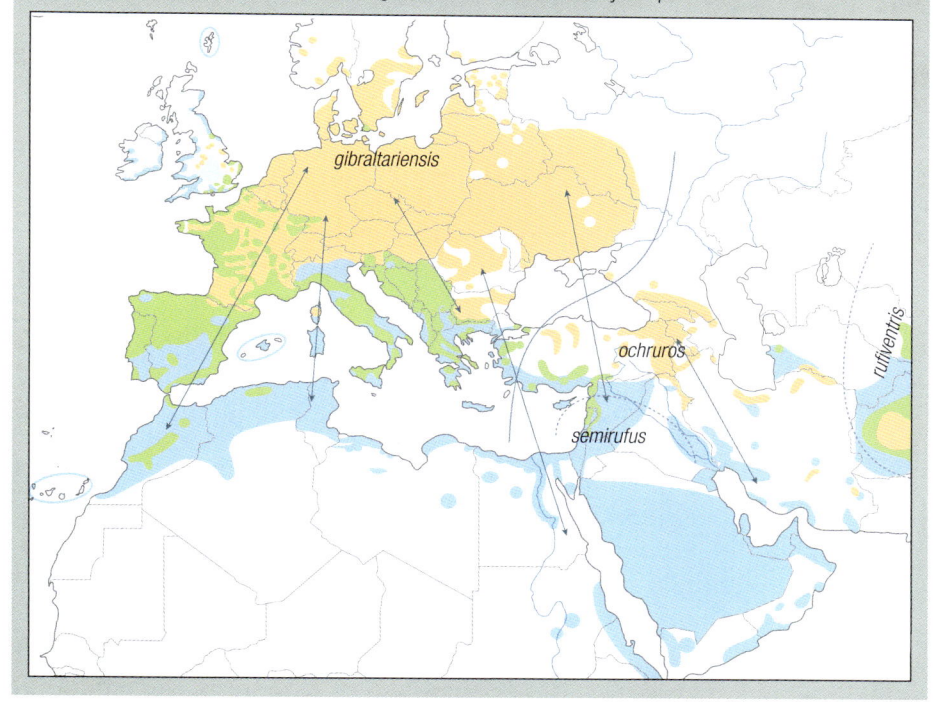

P. o. gibraltariensis, 1stS ♂, England, Mar: some 1stS ♂ '*paradoxus*' may have advanced plumage, with much black developed but a reduced white wing panel. However, ageing should take priority: note newly replaced inner greater coverts (of those that are visible), forming strong moult limit with brownish juv outer greater coverts, remiges, primary-coverts, alula and tertials. (J. Tymon)

IDENTIFICATION This is a small, slim and fairly long-tailed chat. Just like the similarly-shaped Common Redstart it *constantly quivers its orange-red tail* when perched. Quick, 'nervous' movements and dashing flight are typical of the redstarts. Complex geographical variation within its wide distribution does not concern European populations, where all adult ♂♂ are *sooty-black and grey* (darkest on face, throat and breast) with a prominent *white wing patch* (*gibraltariensis*). There is some slight variation as to rufous tones also on lower belly and lower flanks, most having none, a minority a little more. (Birds with substantial amount of rufous on much of belly and flanks, and irregularly mixing with blackish on breast, are more likely candidates to be ssp. *ochruros*, or intergrades with this, or hybrids between Redstart and Black Redstart.) Those in the Levant and Central Asia have *orange-red belly, sharply defined from black breast and throat*, and they generally *lack the white wing patch*. Breeders in Turkey and the Caucasus area are intermediate between these two types, in W Turkey generally a little closer to the European birds, but further east quite variable with some birds closely approaching Central Asian races in appearance. The ♀ and young ♂ in Europe are *sooty brown-grey on head and body* ('mouse grey'), somewhat darker and 'dirtier' than ♀ Redstart, lacking the orange or buffish tinge on the breast usually found in that species, and having an on average less pale throat. Just like ♂, ♀ and young have an *orange-red tail*, their most striking feature. Eastern ♀♀ have warmer, brown-tinged mouse-grey head and breast, slightly orange-tinged belly and are smoky brown above. They still differ from ♀ Redstart by their general darkness, although admittedly a few come close. The juvenile of the European race is very similar to the ♀; due to its general darkness it does not appear to have spotted plumage (unlike in most other chats and flycatchers), but when seen well there is some unevenness created by dark sub-terminal bars and slightly paler tips. Other clues to its age are yellow gape flanges and initially slightly shorter tail, food-begging behaviour, etc.

P. o. gibraltariensis, ad ♂, Switzerland, Jun: ♂ wears to solid black on face and chest, and tends to be cleaner grey above. Contrasting reddish posterior and white innerwing panel make this plumage unmistakable. In Jun, relatively less worn remiges and primary-coverts, the latter subtly fringed greyish, and overall neat plumage, indicative of ad. (H. Shirihai)

P. o. gibraltariensis, ad ♀ with begging juv, Sweden, Jun: being ad (evenly feathered and less worn, despite the late date), this bird can be sexed by plumage as ad ♀, rather than 1stS ♂ '*cairii*'. (M. Johansson)

VOCALISATIONS Song is habitually among the first you hear when day breaks (while it is still mainly dark), delivered from high song post, a slowly repeated rather brief phrase of loud, clear, metallic or whistling notes with a peculiar passage of low crackling sounds (like walking on 'gravel of glass'!) in the middle. There can be a second-long pause in the phrase, usually before the crackling sounds, and sometimes the components are delivered in a different order. Otherwise there is rather little variation in the song, neither geographically nor related to age. – Call is a straight, sharp whistling *vist*, stronger than the call of European Stonechat but otherwise similar. Alarm call is a low, hard, very short tongue-clicking *tk*, which may be combined with the whistling call, *vist tk tk*, or run together in fast series, *tk-tk-tk-tk*, rather like 'electricity sparks'.

SIMILAR SPECIES European adult ♂ is fairly straightforward to identify, being red-tailed and all dark with white wing patch. However, rufous-bellied birds in Middle East and Central Asia, lacking the white wing patch, invite confusion with several species: ♂ *Redstart* bears a vague resemblance, but note that the black bib on Redstart is confined to the throat, and does not extend to the breast. Note also the reversed pattern in that eastern Redstarts have white wing patch, whereas western not, while the opposite is true for

P. o. gibraltariensis, ♀-like 1stY, Feb, Spain: moderately developed paler secondary panel to otherwise uniform, mostly grey ♀-like plumage; latter, combined with young age (rather worn and brownish wing and possibly moulted post-juv inner greater coverts), identify it as either 1stS ♂ '*cairii*' or ♀. (C. N. G. Bocos)

P. o. gibraltariensis, ad ♂, Italy, Oct: evenly fresh, with black parts tipped grey, sometimes limited to face and throat; note ad-like, darker and fresher remiges and primary-coverts, with neat grey edges and small whitish primary tips. (D. Occhiato)

P. o. gibraltariensis, ad ♂, Spain, Sep: Iberian ad ♂♂ often extensively (but variably) blacker above and on wing-coverts, even when fresh. (G. Thoburn)

P. o. gibraltariensis, ad ♀, Scotland, Nov: when fresh often tinged uniform grey, with rather obvious paler innerwing panel, rufous posterior and thin whitish eye-ring. Evenly feathered, still fresh and grey-fringed wing-feathers and rounded tips to tail-feathers strongly indicate ad, hence automatically sexed as ♀. (H. Harrop)

P. o. gibraltariensis, 1stW ♂, Norway, Nov: 1stW '*paradoxus*' ♂ has variably advanced plumage. Note dark face, breast and newly moulted (post-juv) greater coverts, single secondary, and tertials, while juv primaries and primary-coverts are already contrastingly worn and bleached. Such very extensive post-juv moult is rare, at least in northern parts of range. (G. Gundersen)

REDSTARTS

P. o. gibraltariensis, 1stW ♂, Finland, Oct: this bird can be aged by juv remiges, primary-coverts and (at least outer half of) greater coverts (all rather worn and bleached). Identified as young ♂ by renewed tertials having darker centres with broad pure white edges (unlike similar ♀). (H. Taavetti)

P. o. gibraltariensis, juv, Lithuania, Jun: fluffy and slightly mottled juv feathers, but some almost lack scalloping and are even plainer, though plumage more uniform than juv Common Redstart. (G. Gražulevičius)

Black Redstart. – One needs also to consider the possibility that ♂ with general plumage pattern of an eastern race of Black Redstart is a hybrid Redstart × Black Redstart; these do occur. True Eastern Black Redstarts have an orange-red underwing and variably orange-tinged vent (hybrids: both underwing and vent are usually completely or partly whitish with limited orange-buff), and black bib reaches to upper breast below wing-bend in line roughly with median coverts (hybrids: black usually ends short of or at wing-bend, mainly restricted to throat). Pure white edges to long tertials is problematic for Eastern Black, but a few can have buffish edges when fresh, then looking pale enough. There is a tendency in Eastern Black to have some black on mantle/back even in grey-mantled birds, especially on sides of neck/upper mantle, but theoretically hybrids could show the same. – Extralimital Hodgson's Redstart (*P. hodgsoni*; not treated) is fairly similar, but although the black extends to upper breast it is not as extensive as in eastern Black Redstart. Further, it has a small white wing patch. – Another extralimital similar species is Blue-fronted Redstart (*P. frontalis*; not treated), but this has a bluish, metallic tinge on head and throat, and an inverted black 'T-pattern' on tail (black centre and end-band). – With a poor and distant view even a Güldenstädt's Redstart

Young eastern 'red-bellied' ♂ Black Redstart (left: probably *phoenicuroides*, England, Nov), hybrid *P. o. gibraltariensis* × *P. p. phoenicurus* (centre: Italy, Oct) and Common Redstart (right: *phoenicurus*, Finland, Sep): Black Redstart's lower border to bib is, in this case, only slightly lower down than in Common. On average, Common has more obvious whitish forecrown as here (still, in some much reduced, conversely sometimes better developed in *phoenicuroides* Black Redstart). Consider possibility of hybrids, especially if showing mixed characters—in the central image, note how similar it is to Eastern Black; its lack of whitish forecrown separates from most Common Redstarts. All three are 1stW ♂♂, judging by some juv wing-feathers, including primary-coverts. (Left: J. Gearty; centre: D. Occhiato; right: J. Vakkala)

P. o. ochruros, ad ♂, Georgia, Jun: orange restricted to lower belly and vent; further east this race is usually blacker above. (M. Lagerqvist)

Ad ♂, presumably red-bellied *gibraltariensis* (or intergrade *gibraltariensis* × *ochruros*), Italy, Feb: *ochruros* is rather variable and has some orange admixed in grey and white lower chest and belly. However, W European *gibraltariensis* very rarely shows much rufous on belly, almost like this bird, which appears to have a grey back and rather large wing patch, indicative of *gibraltariensis*, but not *ochruros*. Note ad primary-coverts. (D. Occhiato)

P. o. ochruros, ad ♂, Georgia, April: the highly variable race *ochruros* may include extensively rufous-bellied birds that are hardly separable from *semirufus*, or even from *phoenicuroides*, and especially so when the rufous reaches almost to the lower breast. Such ♂♂ can only be identified by range; this bird is from the Caucusus, where only *ochruros* is known to breed. (D. Tipling)

Presumed *P. o. ochruros*, ♀-like, Turkey, Jan: note reddish-tinged underparts, a tendency found in *ochruros*, but individual variation prevents definitive racial identification, especially as in winter *ochruros* and *gibraltariensis* overlap. Uniformly grey-edged greater and primary-coverts, and remiges, suggest ad, and if so can only be ♀. However, obviously brown centres to greater coverts and tertials could indicate 1stY with less worn juv feathers, in which case presumably 1stY '*cairii*'. (H. Yılmaz)

could be confused with one of the Asian Black Redstarts, sharing the same alpine habitats, but Güldenstädt's is much larger and plumper, has a whitish crown and nape, and a large white wing patch. – ♀ resembles several ♀♀ redstart species, and identification not always possible. Important to note absence of obvious pale wing-bar or wing patch (but can have a hint), and lack of blackish terminal band on tail (only central feathers are darker than rest). Darkish, 'mouse-grey' underparts including throat exclude *Redstart* and a few other alternatives.

AGEING & SEXING (*gibraltariensis*) Ages separable for ♂♂ during 1stY; ageing of ♀♀ sometimes possible, requires handling or very close observation. Sexes differ invariably from 2ndW, for a small proportion of advanced ♂♂ from 1stW (so-called '*paradoxus*' morph, the undeveloped ♀-like morph being called '*cairii*'). – Moults. Complete post-nuptial moult of ad usually in Jul–mid Sep (Oct), near breeding grounds. Partial post-juv moult in late summer does not involve flight-feathers, tertials, outer greater coverts or primary-coverts. Apparently no pre-breeding moult in winter. – **SPRING Ad** ♂ Forehead, throat and breast black (finely tipped grey before too worn). Belly and flanks greyish-white, usually with some blackish blotches. Large white wing patch formed by broad edges on outer webs of tertials, and narrow white edges of secondaries. All wing-feathers relatively fresh, glossy grey. **1stS** ♂ '*paradoxus*' Much as ad ♂ but more grey tips to black feathers remain, and white wing patch lacking or is much smaller (size depending on number of moulted tertials). Primaries, and often some outer greater coverts, more worn and brownish, creating contrast with greyish-edged inner greater coverts. ♀♀/**1stS** ♂ '*cairii*' Crown, throat and breast dark brownish-grey (without black), belly and flanks same or slightly paler. No white wing patch. Some ad ♀♀ identifiable by very fresh greyish wings and uniformly edged greater coverts, but this generally requires handling. – **AUTUMN Ad** ♂ As in spring, although black on head and body tipped grey; dark primaries neatly edged grey. **1stW** ♂ '*paradoxus*' Much as ♀♀/1stS ♂ in spring, although many of the black parts hidden by broad brownish-grey tips. ♀♀/**1stW** ♂ '*cairii*' Much as ♀♀/1stS ♂ in spring. Some ad ♀♀ can be identified by their uniformly edged greater coverts, whereas 1stW usually has contrast between the few moulted inner greater coverts, being edged grey, and outer more brownish. However, this generally requires handling. **Juv** Resembles ♀ but can sometimes be separated if seen close by yellow gape flanges and looser feathering on back, rump and tail-coverts.

BIOMETRICS (*gibraltariensis*) **L** 13.5–14.5 cm; **W** ♂ 82–88 mm (*n* 18, *m* 85.3), ♀ 82–85 mm (*n* 14, *m* 83.1); **T** ♂ 58–66 mm (*n* 17, *m* 61.3), ♀ 57.5–62 mm (*n* 14, *m* 59.6); **T/W** *m* 71.7; **B** 13.5–15.6 mm (*n* 29, *m* 14.6); **Ts** 22.0–23.6 mm (*n* 27, *m* 22.8). **Wing formula: p1** > pc 4.5–10.5 mm, < p2 31–35.5 mm; **p2** = wt 7–12 mm, =7 or 6/7 (~8); **pp3–5** about equal and longest (though p3 often 0.5–2 mm <, and p5 sometimes 0.5–1 mm <); **p6** = wt 2–5 mm; **p7** < wt 8–11 mm; **p10** < wt 16–20 mm; **s1** = wt 18–22 mm. Emarg. pp3–6.

♂♂, *P. o. semirufus* (left: Israel, Feb), *P. o. ochruros* (centre: Turkey, Jun) and *P. o. phoenicuroides* (right: Kazakhstan, Apr): note how *ochruros*, even in NC Turkey, recalls *semirufus*, rendering identification of latter impossible outside Levant. Many *ochruros* have extensive black mantle/scapulars, like most worn *semirufus*; black on breast in *semirufus* often reaches further back, with quite well-defined orange-rufous belly, but *ochruros* can be identical. *P. o. phoenicuroides* is more Common Redstart-like, with slightly smaller black bib, greyer upperparts and whitish forecrown, but subject to individual variation, feather wear and age-related variation. Differences in the innerwing panel age-related: the *semirufus* and *phoenicuroides* are advanced 1stS, the *ochruros* ad. (Left: T. Krumenacker; centre: S. Mutan; right: A. Isabekov)

P. o. semirufus, 1stS ♂, Israel, May: black usually reaches quite well down breast and is rather sharply separated from rich orange-rufous belly; note relatively high black forehead. Aged by juv remiges, primary-coverts and outer greater coverts. (A. Ben Dov)

P. o. semirufus, ♀, Israel, May: heavy wear early in summer implies 1stS, but ageing requires handling. Some ♀♀ quite distinctly tinged rusty-buff or chestnut below, and can be assigned to one of the eastern red-bellied forms. (A. Ben Dov)

GEOGRAPHICAL VARIATION & RANGE Marked and complex variation. At least five subspecies recognised falling in two groups. Western 'group' (only *gibraltariensis*) connected and intergrading where in contact with eastern group in SE Europe. Eastern taxa form a rather different-looking group, two of which races are in contact and intergrade, the third being apparently isolated. See Taxonomic notes.

WESTERN GROUP
P. o. gibraltariensis (J. F. Gmelin, 1789) (Europe, W Turkey; winters W & S Europe, N Africa, Middle East). Described above. Body of ad ♂ black, grey and white, usually with orange restricted to vent and undertail-coverts, but a few have faint orange on belly and lower flanks. Large white wing patch. Mantle and back variable, either ash-grey, black or a mixture. – Breeding ♂♂ of Iberian Peninsula ('*aterrimus*') tend to be on average blacker above than in rest of Europe, but much variation and overlap makes this race difficult to uphold following a 75% rule of distinctness. (Syn. *aterrimus*.)

EASTERN GROUP
P. o. ochruros (S. G. Gmelin, 1774) (C & E Turkey, Caucasus, NW Iran; short-range winter movements to Cyprus, Iraq, SW Iran). Rather variable, especially towards east. Shows signs of secondary intergradation in C Turkey with *gibraltariensis* (although poorly studied in detail) and in SE Turkey with *semirufus*. Ad ♂ differs from preceding by variable amount of rather prominent orange (mixed with grey and white) on belly and lower flanks, usually with patchy or diffuse border to grey and black. Birds approaching following races in general appearance, i.e. with more orange-red on belly, usually differ in having uneven or diffuse border between black breast and orange-red belly. White wing patch of ad ♂ usually narrow or small, sometimes washed greyish or missing, never large. ♀ as in *gibraltariensis*, but some have slight orange tinge in grey of belly. **W** ♂ 80–87 mm (n 15, m 83.4), ♀ 76–82 mm (n 15, m 79.3); **T** ♂ 54–66 mm (n 15, m 60.1), ♀ 53–60 mm (n 15, m 57.4). Bill, tarsus and wing formula as *gibraltariensis*.

P. o. semirufus (Hemprich & Ehrenberg, 1833) 'LEVANT BLACK REDSTART' (Syria, Lebanon, N Israel; short-range winter movements to S Iraq, Egypt, may reach Sudan). Differs clearly from preceding two races by having in ad ♂ plumage entire belly deep rufous-red, sharply defined from black upper breast/throat. Quite black on mantle and back. Crown and lower back dark ash-grey, forehead black. Forecrown palest grey, nearly white in some. Only a narrow pale wing patch, longest tertials in ad ♂ edged ash-grey when fresh, disappearing when worn. ♀ either like *ochruros* or with hint of ♂ pattern, with almost sooty-black breast and comparatively stronger orange tinge to the belly.

Presumed *P. o. semirufus*, ad ♂, Israel, Jan: black above often partially concealed by grey tips when fresh; aged by evenly feathered, ad-like blackish wing. Although some dark grey tips obscure border between black breast and chestnut belly, the latter at most reaches lower breast. Outside breeding range and season many *semirufus*, even ad ♂♂, cannot be distinguished from *ochruros* and *phoenicuroides* with certainty. (F. Heintzenberger)

Pale eastern Black Redstart ♀ (left: probably *P. o. phoenicuroides* or *semirufus*, Israel, Nov) and dull ♀ Common Redstart (right: *P. p. phoenicurus*, England, Sep): eastern ♀ Black Redstart often has pale/sandy-buff appearance and pronounced whitish eye-ring. Common has warmer underparts and whiter throat, which tend to be even duller in Black. Diagnostic calls and wing formula may be only clues to distinguish difficult examples. Both birds 1stW by juv primary-coverts. (Left: A. Ben Dov; right: I. Fisher)

P. o. phoenicuroides, ♂ (left) and ♀, Mongolia, Jun: worn summer plumage, showing ♂'s black bib reaching about mid-breast level, grey upperparts and large white forecrown; ♀ characteristically more brownish-buff (less grey) with warmer underparts, and despite worn plumage, pale eye-ring is evident. Ageing not possible when wing heavily worn. (H. Shirihai)

A rather small race. **L** 13–14.5 cm; **W** ♂ 76–81.5 mm (*n* 15, *m* 78.9); **T** ♂ 55–62 mm (*n* 12, *m* 58.5); **T/W** *m* 74.3; **B** ♂ 13.5–16.0 mm (*n* 12, *m* 14.9); **Ts** ♂ 21.7–24.0 mm (*n* 12, *m* 22.7). ♀ slightly smaller. **Wing formula: p1** > pc 5–8 mm (once 14), < p2 26–33 mm; **p7** < wt 6–9 mm; **p10** < wt 13.5–17 mm; **s1** < wt 15–19 mm.

Extralimital races: ○ ***P. o. phoenicuroides*** (F. Moore, 1854) 'RED-BELLIED BLACK REDSTART' (Kazakhstan, Altai, SW Siberia, NW Mongolia; winters S Central Asia, Pakistan, N India). Resembles *semirufus* but is larger. Black bib in ad ♂ averages slightly smaller than in *semirufus*, and ♂ has grey mantle and back with variable amount of black admixed in many (then patchy, looking 'untidy'; in very grey-mantled birds often still some blackish partly visible at sides of upper mantle), in fresh plumage with some brown tips. Those with most black above are often very similar to or inseparable from *rufiventris*. A little white on forecrown on many, becoming more visible in worn plumage, but some have uniform rather dark grey crown. Pale wing patch absent (or vestigial, and then not pure white). Imm ♂ basically ♀-coloured in 1stY, just as in European *gibraltariensis*. ♀ reddish-tinged mouse-grey below (paler and less grey than *gibraltariensis*), uniform smoky grey above with a brown hue. **L** 14–15.5 cm; **W** ♂ 81–87.5 mm (*n* 20, *m* 83.8); **T** ♂ 58–67 mm (*n* 20,

Presumed *P. o. phoenicuroides*, ad ♂, Oman, Jan: as *semirufus* and *ochruros*, but on average black border is on upper/central breast (much individual variation in all three races), which together with less deep red (more orange) rest of underparts, purer grey upperparts (less black), paler crown and whiter forecrown, afford more Common Redstart-like appearance. (H. & J. Eriksen)

m 61.6); **T/W** *m* 73.8; **B** ♂ 14.2–16.0 mm (*n* 14, *m* 15.0); **Ts** ♂ 21.8–23.7 mm (*n* 14, *m* 22.8), ♀ slightly smaller. **Wing formula: p1** > pc 6–10.5 mm, < p2 30.5–35.5 mm; **p7** < wt 6–11 mm; **p10** < wt 14.5–20 mm; **s1** < wt 17–21.5 mm. (Syn. *alexandrovi, turkestanicus*.)

P. o. rufiventris (Vieillot, 1818) (Turkmenistan east to Pamir, Himalaya, C China; winters S Central Asia, Pakistan, lower parts of Himalaya, N India). Very similar to *phoenicuroides*, and some are inseparable (e.g. no consistent difference in amount of black on breast), but is on average somewhat larger and longer-winged, and ad ♂ generally has much black on mantle (often nearly completely black mantle and back), and less white on forecrown (none or only a hint on feather-bases). Some even have crown largely black, almost concolorous with mantle (never so in *phoenicuroides*). However, apparently much overlap, and discrimination not always possible. Pale wing patch absent (or vestigial). ♀ similar to *phoenicuroides* and probably inseparable. **L** 14.5–16 cm; **W** ♂ 83–94 mm (*n* 25, *m* 88.0), ♀ 80–87.5 mm (*n* 12, *m* 83.3); **T** ♂ 59–67 mm (*n* 12, *m* 63.0); **T/W** *m* 71.5; **B** 14.4–16.0 mm (*n* 15, *m* 15.1); **Ts** 23.0–24.9 mm (*n* 15, *m* 24.0), ♀ somewhat smaller. **Wing formula: p1** > pc 3.5–9 mm, < p2 32–37.5 mm; otherwise similar to *phoenicuroides*.

Presumed *P. o. semirufus / phoenicuroides*, 1stS, ♀-like, Israel, Mar: an unusually pale ♀-type Black Redstart; chestnut tinge on underparts indicates one of the eastern 'red-bellied' forms; note well-developed whitish eye-ring. Lesser and median coverts look grey-edged, contrasting with abraded greater coverts, tertials, primaries, alula and primary-coverts, suggesting 1stY, thus 1stY '*cairii*'. (H. Shirihai)

TAXONOMIC NOTES The very obvious plumage differences in ♂♂ of European and Asian populations give rise to the possibility of a split into two species. Before this can be done the amount of contact between the two should be investigated. Although it is currently believed that most Turkish breeders can be referred to *ochruros*, not least because this subspecies is so loosely defined, details of range and contact are still scant. Strictly, due to its extensive morphological variability, and rather diffuse geographical delimitation, ssp. *ochruros* (much of Turkey except north-west, also Caucasus, Transcaucasia, N Iran) does not quite fulfil usual criteria for a sufficiently distinct race, appearing more as a wide zone of presumed secondary intergradation between European and Asian populations. However, it is maintained here for reasons of stability, as its suppression would require a new scientific name for the whole species.

REFERENCES ANDERSSON, R. (1991) *Ornis Svecica*, 1: 53–55. – NICOLAI, B., SCHMIDT, C. & SCHMIDT, F.-U. (1996) *Limicola*, 10: 1–41. – STEIJN, L. B. (2005) *DB*, 27: 171–194. – VINICOMBE, K. (2006) *Birdwatch*, 171 (Sep): 26–28.

P. o. phoenicuroides, ♂, presumed ad, Kazakhstan, Oct: paler grey upperparts, orange-tinged underparts, white-tipped black bib and paler supercilium, all characteristics of fresh *phoenicuroides*. Southern/eastern birds usually develop clearer sexual dimorphism when young, so ageing of autumn ♂ requires close check of moult. Grey-fringed primary-coverts indicative of ad, but other relevant plumage tracts cannot be seen. (A. Wassink)

P. o. phoenicuroides, 1stW ♂, United Arab Emirates, Jan: in general has Common Redstart 'feel', but border on breast lower, and head, breast and back more evenly dark than most Common Redstarts. Aged by juv primaries, primary-coverts and outer greater coverts. (I. Boustead)

Presumed *P. o. phoenicuroides* / *rufiventris*, ♀-like, Pakistan, Nov: eastern ♀-like birds often pale sandy-grey above and richer buff/rufous-brown below. Ageing without close inspection of wing moult impossible in this case, thus sexing unsafe too. (G. Mughal)

Eastern ♂ Black (left: *P. o. phoenicuroides*, Oman, Nov) and Common Redstarts (right: *P. p. phoenicurus*, ♂, Netherlands, Apr): differences such as dark border to breast, redder (less orange) belly and reduced whitish forecrown not always obvious, thus caution often necessary. Left bird is ad by evenly dark-centred and well-fringed grey coverts and remiges; right is 1stY by juv primary-coverts. (Left: B. Johansson; right: N. D. van Swelm)

P. o. rufiventris, ad ♂, Pakistan, May: rather variable race, overall like *phoenicuroides*, but more extensive black on mantle/scapulars and wing-coverts, and lacks white on forecrown and wing panel. Wing-feathers are ad. (G. Mughal)

(COMMON) REDSTART
Phoenicurus phoenicurus (L., 1758)

Fr. – Rougequeue (à front blanc);
Ger. – Gartenrotschwanz
Sp. – Colirrojo real; Swe. – Rödstjärt

Although the ♂ may be attractive and easy to recognise, Common Redstart is not always easy to get to know—unless you have it as a summer visitor near your home. In N Europe it is primarily a bird of uninhabited mature boreal pine forests but also lives rather unobtrusively in parks and larger gardens further south, and in mixed woods close to man without making itself much known. However, once noticed it can easily become a favourite due to its pretty colours and sweet song. It is one of the sub-Saharan migrants, moving by night in August and September to C Africa where it spends the winter. It returns in April and early May.

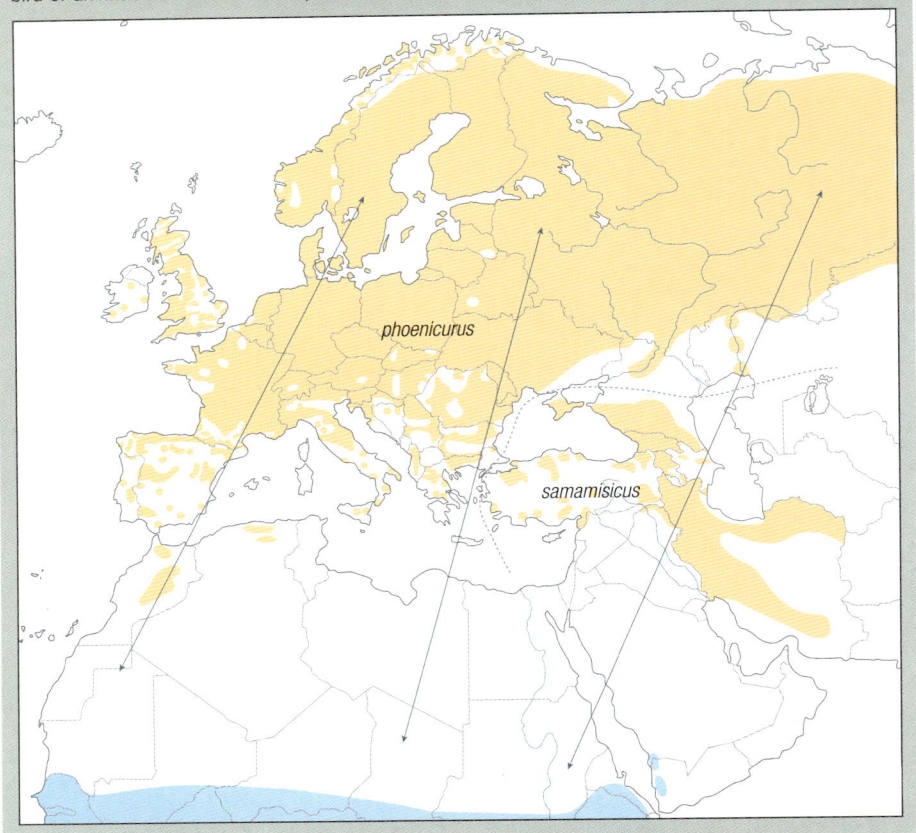

P. p. phoenicurus, 1stS ♂, Finland, Jun: over much of range, ♂ readily separated from other redstarts by relatively short black bib (reaching only to upper breast), white forecrown, grey crown and mantle, and orange underparts. Aged by juv remiges, primary-coverts and (most) greater coverts. (M. Varesvuo)

IDENTIFICATION A small, slim and fairly long-tailed chat, quite similar to Black Redstart but with a fractionally more pointed and relatively longer wing. Habits the same with characteristic frequent *quivering movements of its orange-red tail*. Stance rather upright when perched. Very active, changing perch all the time, flight fast and direct. The adult ♂ is an attractive bird with *orange-red breast and flanks, a black throat, white forehead, ash-grey crown to mantle, and orange-red tail*. The wing is uniformly brownish-grey, wing-coverts edged grey, but those in E Turkey and Middle East (*samamisicus*) have a white wing patch of variable size (prominent only in adult ♂). In autumn, much of the bright spring plumage is hidden by dull feather-tips, upperparts becoming brownish, underparts paler, but it is still possible to see some of the dark bib and orange breast, and generally a little white on the forehead. The first-winter ♂ is similar to adult ♂ in autumn but even more dull and ♀-like. The ♀ is *off-white below* with a *buffish-brown wash on breast and flanks*, or even lightly orange-tinged (but never darker greyish, which would indicate Black Redstart), while upperparts are rather plain grey-brown. Very rarely adult ♀ attains a plumage approaching ♂, with some blackish on throat and much orange on breast. The juvenile is similar to the ♀, but has a lightly spotted plumage created by pale feather-centres and darker subterminal tips above, and faintly scalloped dusky buff underparts.

VOCALISATIONS (*phoenicurus*) Song is discreet and difficult for the beginner to pick out in the chorus from a wood. But once the melancholy ring and peculiar rhythm and structure is learnt, the softly whistling phrase carries through the other voices. Very often the opening is the same (or one of a few variants), both from each bird and geographically, a clear whistling note followed by a quick repetition (2–4 times) of a straight or, more often, up-curled note with 'r' sound, *siih truee-truee-truee*. After this there is more variation as to the ending, but common to most is a sweet, Willow Warbler-like voice and a Treecreeper-like flourish. Some include fragments of mimicry in the ending, others harsher or more mechanical sounds. The whole, high-pitched phrase invariably has a desolate ring and lasts only *c.* 2–2.5 sec. It is one of the very first species you hear when day breaks. – Call is a distinct, upslurred whistling *hueet*, resembling the call of Willow Warbler to the point of confusion (but is on average slightly stronger, less sweet and feeble); can also call a slightly more drawn-out and liquid *hueest*, almost as Yellow-browed Warbler. Alarm call is a low, short clicking *tick*, which may be combined with the whistling call, *hueet tick-tick*, or run together in fast series, *tic-tic-tic-tic* (higher-

P. p. phoenicurus, ad ♂, Finland, Apr: aged by remiges, primary-coverts and greater coverts that are evenly grey-fringed; note narrow pale greyish lining to inner wing forming vague panel when still fresh (not rare in this subspecies). (M. Varesvuo)

P. p. phoenicurus, ♀, **Netherlands**, Apr: off-white chin to upper breast, ochraceous-brown wash to body-sides, with purer orange vent-sides and undertail-coverts, all typical. Edges of remiges and primary-coverts appear ad-like, being not heavily worn and greyish, but definitive ageing impossible. (P. van Rij)

P. p. phoenicurus, ♀, **Sweden**, May: with wear, becomes rather drab, buff tinge below very strongly reduced, but orange uppertail-coverts and tail, and whitish eye-ring remain prominent. Greater coverts and primary-coverts have ad-like pattern, but this bird cannot be aged with certainty at this late date and when wing heavily worn. (L. Waara)

pitched and not as dry as in Black Redstart, and much softer than the clicking call of Robin).

SIMILAR SPECIES Adult ♂ needs to be separated from eastern races of *Black Redstart*, which also have rufous-red lower breast and belly. Best character is the smaller black bib in Redstart, which never extends onto upper breast as in Black Redstart. Redstart also has more of a white mark on forecrown (only faintly suggested on some Black). Further, nape, mantle and scapulars are grey in Redstart but dark sooty-grey or black in Black Redstart, and rufous-red of underparts in Redstart somewhat paler and more orange, less dark and rufous. Although eastern Black Redstarts lack a white wing patch, whereas most Asian Common Redstarts have one, this does not rule out a displaced migrant of either species. Before the race *samamisicus* is reported from W Europe based on a white wing patch one should also consider the possibility of a hybrid between Redstart and Black Redstart, and a full description including an assessment of wing formula is vital. – For some more remote possibilities of misidentification, see Black Redstart. – ♀ resembles several ♀ redstart species, and identification often relies on geographic location and expectations. Important to note absence of pale wing-bar or wing patch,

P. p. phoenicurus, ad ♂, **Netherlands**, Sep: ageing of ♂♂ possible in the field, in this case an ad by lack of moult limits in wings; fresh ad ♂ very variable, but usually has more solid black face and purer white supercilium. (T. Luiten)

P. p. phoenicurus, 1stW ♂, **Kuwait**, Sep: most wing-feathers are juv, but at least two inner greater coverts fresh (contrastingly greyer fringe). 1stW ♂ highly variable, but usually has extensively white-tipped black face and incomplete and buffish-tinged supercilium. (J. Tenovuo)

P. p. phoenicurus, 1stW, presumed ♂, **Italy**, Sep: aged by post-juv renewal of inner two greater coverts, which are greyish-edged and longer than outer juv, rufous-edged, coverts. In general, juv and ad primary-coverts are very similar. Underparts blotched orange, with rather purer grey edges to inner greater coverts suggesting least advanced ♂, although no other features of ♂ visible in this image. (D. Occhiato)

and lack of blackish terminal band on tail (only central tail-feathers are a little darker than rest). Rather pale, buff-white underparts tinged ochre-brown or orange on breast and flanks exclude *Black Redstart* and a few other alternatives. – ♀ *Moussier's Redstart* is quite similar but clearly smaller with both tail and primary projection being shorter. See also under Black Redstart. – A more remote pitfall is the faint similarity between ♀ Redstart *samamisicus* and some first-year ♀♀ of *Daurian Redstart* (*P. auroreus*; extralimital, reaching west to Buryatia; not treated), having a white or at least a small pale primary patch. Daurian Redstart is subtly smaller, has a well-defined and pure white primary patch (*samamisicus* sometimes a small and at the most diffuse and pale primary patch) and seems to have a slightly narrower tail, a difference that is sometimes visible when perched birds quiver their tail. Also, the central tail-feathers in Daurian appear to be darker creating a more marked contrast in the tail.

AGEING & SEXING (*phoenicurus*) Ages separable for ♂ in 1stW, usually also in 1stS; ageing of ♀ sometimes possible in 1stW, but requires handling or very close observation. Sexes differ from 1stW onwards. – Moults. Complete post-nuptial moult of ad usually in Jul–early Sep, near breeding grounds. Partial post-juv moult in late summer does not involve flight-feathers, tertials, outer greater coverts or primary-coverts. Apparently no pre-breeding moult in winter. – **SPRING Ad ♂** Forehead, sides of head and throat black (fine pale tips wear off by late winter or early spring). Breast and flanks orange-red. Belly largely white, undertail-coverts buff. Forecrown white, centre of crown to mantle and scapulars lead-grey (brown tips lacking or insignificant). Primaries on average less abraded, glossier grey, less brownish. All greater coverts appear uniformly edged grey (but often tipped buff.brown). Primary-coverts rather rounder, broader and darker. **1stS ♂** Much as ad ♂ but black of throat admixed with white tips, and crown to mantle and scapulars brown-grey or grey with variable amount of brown tips. Primaries on average more worn and brownish. Either all greater coverts retained juv being both edged and tipped brown or a contrast is visible between outer brown greater coverts and greyish-edged inner. Primary-coverts on average slightly more pointed and abraded brown than in ad. **♀** Crown to back greyish-brown (without white on forecrown). Chin to breast off-white with buff-grey tinge, breast and flanks usually tinged ochraceous-brown or orange. Belly whitish, undertail-coverts cream-white. Rarely, ad ♀ can develop advanced ♂-like plumage, with strong orange-buff tinge on breast, and some blackish

P. p. phoenicurus, ad ♀, Switzerland, Sep: remiges and primary-coverts rather fresh, blacker-centred and greyer-fringed, indicating ad, and in such dull plumage can only be ♀. (R. Aeschlimann)

P. p. phoenicurus, 1stW ♀, England, Aug: aged by most wing-feathers being juv, with worn primary-coverts and rectrices, and reduced pale tips to primaries; dull plumage with hardly any grey or intense orange suggest ♀. (N. Brown)

P. p. samamisicus, ad ♂, Israel, Mar: unlike *phoenicurus*, ♂ has variable white wing patch; aged by wholly ad wing-feathers (no moult limits) and notably grey-fringed primary-coverts. (L. Kislev)

P. p. samamisicus, ad ♂, Kuwait, Mar: an extreme example with very large white wing patch; aged by uniform wing-feathers (no moult limits). (M. Pope)

P. p. samamisicus, ad ♂, Azerbaijan, May: more restricted white wing patch, perhaps due to intergradation with *phoenicurus*; apparently an unusually worn ad, at least no moult limits visible. (K. Gauger)

P. p. samamisicus, ♀, presumed ad, Israel, Mar: compared to image below, whitish inner wing panel is reduced, but its appearance varies, not only individually but also with age and feather wear, angle and light. Apparently evenly feathered and not heavily worn wing-feathers, and rounded, more greyish-fringed primary-coverts, suggesting ad. (H. Shirihai)

on throat (exceptionally whole throat), but they do not attain lead-grey upperparts or full white forecrown. Once aged as ad, confusion with imm ♂ is more easily avoided. Ageing difficult but possible with a few after experience using wear and shape of primary-coverts as in ♂, and by any retained more worn juv secondary-covert. – AUTUMN **Ad ♂** As in spring, although black on head and throat now finely tipped white, and grey of upperparts partly concealed by brown tips. White on forecrown partly concealed by dark tips, but usually visible. All greater coverts uniform, neatly edged grey. **1stW ♂** Much as ad ♂, although pale tips on throat broader and tinged brown. White on forecrown less developed, more concealed by dark tips or even lacking. Often some inner greater coverts moulted to ad type, edged greyer than more brownish outer. **♀** Much as ♀ in spring. Some 1stW ♀♀ can be aged in close observation, or when handled, by paler buff tips to some greater coverts, rather pointed tail-feathers with frayed tips, and primary-coverts with diffusely paler fringes (ad has more uniformly tawny-buff tips to greater coverts, more rounded and neat-edged tail-feathers, and more distinctly but narrowly pale-fringed primary-coverts). However, many are difficult and better left un-aged. **Juv** Roughly like ♀ but has lightly spotted plumage created by pale yellow-buff feather-centres and darker subterminal tips above, and scalloped yellowish-white underparts.

BIOMETRICS (*phoenicurus*) **L** 13–14.5 cm; **W** ♂ 76–86 mm (*n* 125, *m* 79.9), ♀ 73–83 mm (*n* 136, *m* 77.9); **T** 51–61 mm (*n* 44, *m* 56.9); **T/W** *m* 71.4; **B** 13.0–15.7 mm (*n* 38, *m* 14.5); **Ts** 20.2–22.7 mm (*n* 38, *m* 21.4). **Wing formula: p1** > pc 3–6 mm, < p2 31.5–38 mm; **p2** < wt 5.5–9.5 mm, =6 (48%), =6/7 (36%), =5/6 (15%), or =7 (1%); **pp3–4** about equal and longest; **p5** < wt 0.5–3 mm; **p6** < wt 6–9 mm; **p7** < wt 10–13 mm; **p10** < wt 18–20.5 mm; **s1** < wt 20–23 mm. Emarg. pp3–5; very rarely a slight hint of emarg. also on p6 near tip.

GEOGRAPHICAL VARIATION & RANGE Fairly well-marked variation. Only two subspecies recognised, apparently connected by fairly wide zone of intergradation but details poorly known, and specimens from contact zone rare in collections.

P. p. phoenicurus (L., 1758) (Europe, NW Africa, Siberia east to Baikal; winters across sub-Saharan Africa in savannas, with extension into former E Zaïre and Uganda, east to Ethiopia; also to lesser extent on NW African coast). Described above. Lacks large white wing patch, but can have buff-white or cream-white edges to long tertials and secondaries, which in fresh plumage can appear as narrow paler patch. Very rarely these edges can be even pure white but are invariably narrow, never widening as in following race. (Syn. *algeriensis*.)

P. p. samamisicus (Hablizl, 1783) 'EHRENBERG'S REDSTART' (Crimea, Kalmykia, Turkey, Transcaucasia, Middle East, Central Asia; winters in E sub-Saharan Africa, Arabia). Rather variable, but typical ad ♂ differs from preceding race in having variable amount of white on outer edges of tertials, secondaries and inner primaries, widest (*c*. 2.5 mm) on ss5–8, forming well-defined pure white patch on folded wing. Some have very large patch, others a more narrow one; differences partly due to individual variation, or to age and wear (1stS ♂ usually lacks patch entirely, but a few have large patch, depending on individual variation and extent of post-juv moult). Those *phoenicurus* in Europe with narrow pure white wing panel are fortunately rare, but birds with more prominent white in wing also documented (Small 2009); such birds in our opinion controversial and perhaps better regarded as either genuine *samamisicus*, intergrades between the two races, or hybrids (not necessarily first-generation) between Redstart and Black Redstart. 1stW ♂ *samamisicus* sometimes separated from Redstart by greyer, less brown upperparts and more extensive black or blackish on cheeks. ♀ usually not separable from *phoenicurus*, but

P. p. samamisicus, ♀, presumed 1stS, Israel, Mar: paler/purer grey upperparts and slight whitish innerwing panel often characterise this race. Hint of orange-tinged breast-band, especially at sides. Primary-coverts and most of greater coverts pointed and patterned as with worn juv coverts, but unclear if impression of greyish-fringed innermost greater coverts is true post-juv renewal. (H. Shirihai)

P. p. samamisicus, ad ♂, Kuwait, Oct: aged by lack of moult limits in wing; even fresh ♂ has extensive white wing patch, albeit usually less solid; like fresh ad ♂ *phoenicuroides*, black face usually rather solid with hint of pure white lower throat band and more prominent supercilium. (M. Pope)

P. p. samamisicus, 1stW ♂, Kuwait, Sep: some 1stW ♂♂ have clear hint of wing panel. Very obvious whitish eye-ring, contrastingly fresh inner greater covert (fringed lead-grey), while rest of greater coverts, primary-coverts and remiges are juv (contrasting with moulted median and lesser coverts); also rather dark cheeks and pure grey upperparts, indicating young ♂. (M. Pope)

note difference in call (see below), and a few have a large pale buff version of the white wing patch of ♂ (but note that ♀ *phoenicurus* in fresh plumage commonly has slightly pale-edged outermost secondaries and longest tertials too). Measurements and structure are very nearly the same as in *phoenicurus*. In post-juv moult, more inner greater coverts replaced on average than in *phoenicurus*, but variable. Song differs from *phoenicurus* by being rather harder and lower-pitched in tone, at times somewhat recalling song of Taiga Flycatcher. Call differs, too, in being a straight whistle, *iiht*, very like Collared Flycatcher or Thrush Nightingale. However, it should be noted that these vocal differences do not coincide with subspecies borders; breeders in Altai (*phoenicurus*) have the same call as *samamisicus*. – Breeders in W Turkey seem to be intergrades with *phoenicurus* or at least less typical, but incompletely studied, and scant material available in museums. – In N Iran and adjacent areas sometimes a dark morph occurs, '*incognita*', with broader black on forehead, and largely black mantle, scapulars and upper secondary-coverts. (Syn. *mesoleucus*.)

TAXONOMIC NOTE Considering the generally clear difference in ♂ plumage, the race *samamisicus* seems a likely candidate for elevation to species status. However, vocal differences do not coincide with racial borders, and more work is needed on DNA and morphological variation in the region closest to ssp. *phoenicurus*.

REFERENCES Small, B. J. (2009) *BB*, 102: 84–97.

P. p. samamisicus, ad ♀, Ethiopia, Nov: clear hint of wing panel, but can be much reduced or lacking in many ♀ *samamisicus*, while some fresh *phoenicuroides* can show trace of one. Greyish-fringed remiges, primary-coverts and greater coverts suggest ad, supported by broad, almost square-shaped primaries (especially inner ones). (O. Ellestrøm)

Juv, Turkey, Jun: typically prominently spotted plumage. Juv is paler and overall more spotted both above and below than juv Black Redstart. (F. Izler)

MOUSSIER'S REDSTART
Phoenicurus moussieri (Olphe-Galliard, 1852)

Fr. – Rougequeue de Moussier
Ger. – Diademrotschwanz
Sp. – Colirrojo diademado; Swe. – Diademrödstjärt

A resident of open forest, maquis and bare ground in the Atlas Mountains and foothills, Moussier's Redstart is one of the most special and distinctive endemics of NW Africa, where it occurs from S Morocco east to N Tunisia. It may disperse slightly in winter, when it also descends to near sea level, and has exceptionally wandered north to Malta, Iberia and even Wales.

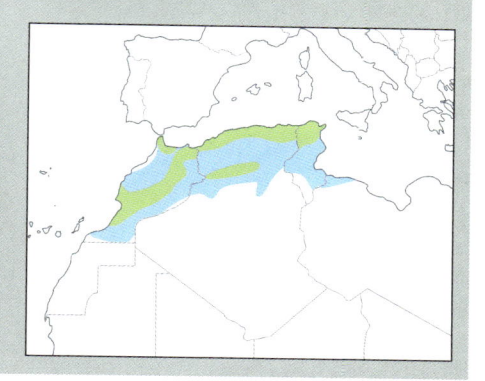

IDENTIFICATION A *small, compact redstart* with striking ♂ plumage, which has crown, ear-coverts, mantle and wings black, marked by a *bold white diadem on forehead continuing as a line above and behind ear-coverts, broadening and terminating on nape-sides*. Also has a large *white wing patch on the inner primaries and secondaries*, and a *bright orange-rufous rump and tail* with dark brown central feathers. Throat and underparts also orange-chestnut, only slightly paler on abdomen. In contrast, ♀ is brown-grey above, including head, with no or only ill-defined buff supercilium, and darker wings slightly fringed buffish on tertials and greater coverts (but wing rather plain brown when worn in spring). Breast, flanks, rump and tail pale orange-chestnut, centre of belly and vent being slightly paler. Throat and sides of head dusky grey-brown with some orange-chestnut tinge on throat. Eye-ring inconspicuous and pale buff. The *rather short-tailed and large-headed appearance* may recall a Stonechat rather than other redstarts, but it quivers its tail like several congeners. Usually keeps close to ground level, where not shy and is occasionally inquisitive.

VOCALISATIONS Song a thin, clear twitter, commencing with a short, simple warble, lasting 2–5 sec., often quite similar to Dunnock due to clear voice, staying much on the same pitch and 'unstructured', but at times revealed by insertion of hard and scratchy or squeaky elements recalling Black Redstart or Sand Martin. Can also be interrupted by explosive *Sylvia*-like notes (particularly recalling Dartford Warbler), and has then quite irregular composition. – The commonest calls are a thin whistle, *hiit*, a little like European Stonechat, and a soft, rasping, drawn-out *cherrrr* or *pscheeh*.

SIMILAR SPECIES Adult ♂ unmistakable, but ♀ and immature much less distinctive. Features to concentrate on are the small size, compact appearance (accentuated by relatively short tail) and warm rufous underparts (rather than buff-white). 1stW ♂ may show a slight hint of a white forehead and is therefore superficially similar to same-plumage Redstart, but this is larger, longer-tailed and has a mottled black throat, thus immediately eliminating Moussier's. On ♀ compared to ♀ Redstart, note more compact shape, smaller size and shorter, more rounded wing.

AGEING & SEXING Ageing of ♂ possible in the field; ♀ requires close inspection of moult pattern and feather wear in wing, and even then ageing not always possible. Sexes very nearly same size but differ strikingly in plumage (see above), with seasonal variation moderate in ♂ plumages. – Moults. Complete post-nuptial moult of ad in summer, generally Jun–Sep, and partial post-juv moult at same time, including all head- and body-feathers, lesser and median coverts, and a few inner (or sometimes all) greater coverts. Apparently no pre-nuptial moult. – SPRING Sexing usually straightforward (see Identification), but exceptionally ♀ can approach ♂ appearance by attaining small whitish wing patch and some white and black partly showing on crown and nape. **Ad** Aged by having only ad-type wing- and tail-feathers, and primary-coverts and primaries less worn than 1stS. **1stS** Similar to ad but has juv-type primary-coverts (being subtly more pointed and less neatly edged), primaries, tail-feathers and sometimes outer greater coverts (being browner and more heavily abraded), while some least-advanced ♂♂ further separated by being less neatly patterned. – AUTUMN **Ad** Both sexes evenly fresh, with ad ♂ developing pale buff-brown fringes to head, body, and wing-feathers, producing marked scaly appearance to dark parts of plumage and reducing contrasts. **1stW ♂** Most rather noticeably less black (more heavily fringed pale than fresh ad ♂), more brown-grey with reduced wing panel. Note also moult limits with some juv outer greater coverts and tertials; also juv primaries, primary-

♀, Morocco, Mar: rufous underparts, otherwise rather nondescript. Tertials (the little of them visible) appear new and grey-fringed, compared to already bleached remiges, suggesting moult limits, thus 1stY, but best left un-aged. (D. Monticelli)

Ad ♂, Morocco, Jan: striking bold white diadem from forehead to neck-sides (where broadens). Dark central feathers. No moult limits in wing, and quite dark primaries and primary-coverts equal an ad. (R. Schols)

1stS ♂, Tunisia, Mar: white panel in remiges, with otherwise glossy black upperparts and bright orange-rufous underparts and rump are features of ♂. Note brown juv wing-feathers, including primary-coverts and remiges. (D. Monticelli)

coverts and tail-feathers are weaker textured and more worn than ad. **1stW ♀** Best aged by moult limits as in 1stW ♂, but contrasts less marked. **Juv** Most like ♀ but has extensively spotted feathering, with slight rufous tinge above, scaly breast markings, and ♂ already displays distinct wing panel.

BIOMETRICS **L** 12–13 cm; **W** ♂ 63.5–69.5 mm (n 15, m 66.3), ♀ 63–67.5 mm (n 10, m 65.6); **T** ♂ 45–50 mm (n 15, m 47.8), ♀ 45–49 mm (n 10, m 47.5); **T/W** m 72.2; **B** 13.2–15.9 mm (n 25, m 14.2); **Ts** 21.3–25.0 mm (n 25, m 22.8). **Wing formula: p1** > pc 6.5–12.5 mm, < p2 20–25 mm; **p2** < wt 6–9.5 mm, =8–9 (rarely =7/8 or 9/10); **p3** < wt 0–1.5 mm; **pp4–5** about equal and longest (or pp3–5); **p6** < wt 2–5 mm; **p7** < wt 5–7.5 mm; **p10** < wt 8–11 mm; **s1** < wt 9–12 mm. Emarg. pp3–6.

GEOGRAPHICAL VARIATION & RANGE Monotypic. – Mountain slopes with trees and bushes in Atlas and adjacent areas of S Morocco, N Algeria and N Tunisia; largely resident or makes only shorter movements to lower levels or coastal areas in winter.

♀, **Tunisia, Dec**: compact, with short tail and warm rufous underparts. Round-tipped and fresh primary-coverts with grey edges suggest ad, and greyer upperparts and head an ad ♀. However, ageing uncertain as tertials appear very worn (juv), especially as species' moult is still poorly known. (P. Alberti)

1stW ♀, **Tunisia, Nov**: almost Stonechat-like appearance due to compact shape with short tail. Heavily worn and brown juv outer wing. (A. Torés Sánchez)

Juv ♀, **Morocco, Jul**: dark and extensively mottled or diffusely spotted plumage; ♀ by lack of distinct white wing panel. (C. Batty)

GÜLDENSTÄDT'S REDSTART
Phoenicurus erythrogastrus (Güldenstädt, 1775)

Fr. – Rougequeue de Güldenstädt
Ger. – Riesenrotschwanz
Sp. – Colirrojo de Güldenstädt; Swe. – Bergrödstjärt

This strikingly coloured giant redstart is one of the highest-breeding passerines of the region, being always found above the tree-line and often extending in summer to the summits of the tallest mountains in the Caucasus. Outside the treated range it occurs in the Himalayas, the Tien Shan system, Mongolia and SW Siberia. In winter descends to foothills, valleys and plains, often reaching 1500 m (rarely even 900 m), remaining in the main breeding range, although it has wandered south to Kuwait.

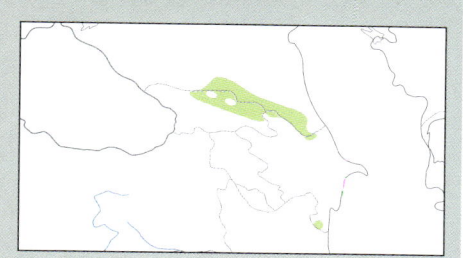

Ad ♂, Mongolia, Jun: diagnostic and stunning white wing flashes, which together with white cap and rufous tail make the species unmistakable. (H. Shirihai)

IDENTIFICATION Considerably larger than Common Redstart and distinctly chunkier, with a *proportionately larger head*, longer wings and shorter tail, and overall size is close to that of a small thrush. ♂ has striking *all-white crown and nape, and a huge white wing-patch across entire remiges*, prominent on a perched bird but even more obvious in flight. Also has black face to breast, back and wings, and dull orange-brown to deep chestnut belly, rump and tail, latter with indistinctly darker central feathers. ♀ is *like large version of most redstarts*, being rather nondescript compared to ♂, having head, back and scapulars greyish-brown isabelline, contrasting with reddish-chestnut lower rump to tail (slightly browner central feathers), a rather *obvious buff-white eye-ring* (emphasized by dusky-rufous cheeks), and lower face and breast dull greyish-fulvous, becoming warmer on flanks, while centre of belly is whiter. (See also Ageing & Sexing as to apparent existence of rufous-bellied morph.) Wings rather concolorous brown-grey with slightly paler sandy-buff edges to flight-feathers, especially tertials. Flight buoyant, the rather heavy body and long wings producing less fluttering wingbeats, and is overall somewhat more thrush-like than smaller congeners. Shy, but perches

♂ (left) and ♀, Pakistan, May: the largest redstart, and ♂ is unmistakable. ♂ aged by evenly ad wing-feathers, including black (instead of brownish) primary-coverts; age of ♀ uncertain without closer inspection of wing-coverts. (G. Mughal)

Ad ♂, Georgia, Apr: contrasting white crown and large wing panel, contrasting with black upperparts and bib, and deep rufous rest of underparts and tail. No moult limits in wing suggest ad. (W. Müller)

1stS ♂, Mongolia, Jun: as ad ♂ summer, but has mostly juv wing-feathers, including primary-coverts, with dark areas brownish (not black). White wing panel of juv ♂ is at most only slightly smaller than in ad. (H. Shirihai)

♀, **Georgia, May**: large and stocky, unlike other ♀ redstarts. Pale grey hue to brown above, no obvious pale inner wing panel, and less extensive rufous-orange on tail (even less visible when folded), but quite contrasting on rump/uppertail-coverts. Vent-sides and face often tinged buffish. With wear, pale tertial fringes lost, contributing to more uniform appearance. Extensive wear prevents certain ageing. (G. Darchiashvili)

Ad ♂, **India, Nov**: as ad ♂ summer, but black above and on breast tipped greyish, with grey wash to white crown and nape, and rufous-red below less deep. Pale tips to primary-coverts and no moult limits in wing indicate ad. (O. Pfister)

♀, **Kyrgyzstan, Apr (left), and N India, Nov (right)**: regardless of freshness of plumage variation is remarkable, from Black Redstart-like dark greyish plumage (left) to cream-buff or orange-tinged birds (right). If size not evaluated correctly, separation from ♀ Black Redstart can be challenging. Can retain fine pale edgings to coverts and tertials into spring. Note also some fulvous on face, and rufous on uppertail concentrated on sides. Broad whitish eye-ring often obvious (both), and sometimes hint of pale orange primary patch (right). (Left: M. Westerbjerg Andersen; right: D. Laishram)

boldly in open, on boulders and bushes, and most actions and behaviour are typical of the genus.

VOCALISATIONS Song somewhat like Blue Rock Thrush or, rather surprisingly, Pine Grosbeak, with several pleasant brief, hurried, fluty or whistling phrases of rather low-pitched, clear notes, large tonal steps, e.g. *chü-cha chee-trrrü che che-wee*. The song can be interspersed with varied chirping and wheezy sounds. Song given from ground, a boulder or in display-flight. – Calls include a straight whistling *heeht* and a harder *teck* in alarm, or a combination of these in Northern Wheatear fashion. In social interaction utters a dry rolling *drrrrt*.

SIMILAR SPECIES Adult ♂ is unmistakable, given a clear view, but ♀ is less distinctive. Nevertheless, compared to other redstarts in the covered region, the distinctly large size, stocky build (although admittedly it may prove difficult to always appreciate size correctly in distant views), less bright and more uniform chestnut tail, and overall coloration should quickly eliminate all potential confusion species.

AGEING & SEXING Ageing of ♂ easy (♀ considerably more difficult and requires close view or handling, and many still ambiguous) by moult, feather pattern and wear in wing. Sexes differ markedly. Slight seasonal variation in ♂ plumages. – Moults. Complete post-nuptial moult of ad in late summer, mostly Jul–Sep, and partial post-juv moult at same time. Post-juv moult includes all head, body, lesser and median coverts, but apparently as a rule no greater coverts (only very occasionally some inner greater coverts moulted). Apparently no pre-nuptial moult in winter. – **SPRING** ♂ White crown and nape, black throat and breast, extensive pure white wing patch. ♀ Grey-brown crown and nape, no white wing patch. Underparts variable (appearing to form two colour morphs), either rather plain ochre grey-brown or alternatively ghosting ♂ plumage, with dark grey or even blackish throat and breast, contrasting with rufous-tinged brown belly. **Ad** See above under Identification. **1stS** Like respective ad, but ♂ can—to some extent—retain greyer crown and nape. Best aged by distinctly more worn and browner primaries, primary-coverts and usually some outer greater coverts, producing clear-cut moult limits in ♂ but much less contrasting in ♀. – **AUTUMN Ad** Both sexes evenly fresh, but some suspend moult of some body-feathers and secondaries until arrival in winter quarters, forming diagnostic strong moult contrast between new and retained one-year-old feathers. **Ad** ♂ Plumage for a short period indistinctly washed (tipped) dusky, less deep black above and on throat/breast, with greyer crown and nape. All median and greater coverts and primary-coverts

REDSTARTS

1stY ♂, Kazakhstan, Jan: as ad ♂ winter, but crown washed more extensively greyish and less uniform rufous-chestnut below, while juv primary-coverts are browner and lack pale tips. (A. Isabekov)

♀, presumed ad, Georgia, Feb: much variation, with some ad/older (?) ♀♀ having deeper and more orange-brown underparts. Overall greyish-tinged plumage, including wing fringes, suggest ad, as does deep coloration below. Unclear if rufous-bellied ♀♀ represent dimorphism, rather than being age-related. Separation of such ♀ Güldenstädt's from red-bellied ♀/winter ♂ Black Redstart can be difficult, especially if size not appreciated, but note typical gingery tone to face. (G. Darchiashvili)

Ad ♀, Georgia, Feb: heavily-built redstart with cold grey upperparts; this ♀ has intermediate pigmentation below. Compared to red-bellied Black Redstart, ♀ Güldenstädt's lacks extensive rufous in tail and is colder grey above. (G. Darchiashvili)

Juv ♂ (front) and ♀, Mongolia, Jun: sexual dimorphism already clear at this age, especially large white wing patch of ♂. Note blackish greater coverts broadly tipped yellow-buff in ♂. (H. Shirihai)

uniformly black without paler tips (save very thin greyish tips to some feathers). **Ad ♀** Slightly warmer (less grey) than in worn plumage. – **1stW ♂** Much as fresh ad, but at least until late autumn has extensive greyish feather tips on otherwise white crown/nape, and is less uniform rufous-chestnut below. Juv remiges, tail-feathers, tertials, usually all greater coverts and primary-coverts retained (relatively weaker textured and somewhat browner than ad, not jet-black), best recognised by rather broad pale tips, forming hint of wing-bar and creating moult contrast with blacker post-juv median (and, if any, moulted inner greater coverts). **1stW ♀** Aged by moult limits and presence of pale-tipped wing-coverts as in ♂, but contrast much less sharp and many best left un-aged. **Juv** Resembles ♀ but has foreparts (chiefly lower throat to breast and flanks) clearly dark-scaled with buffy spots. Median and greater coverts tipped yellow-buff, forming clear wing-bars. Juv ♂ has extensive white wing panel and blacker coverts and remiges, lacking in juv ♀.

BIOMETRICS L 15–18 cm; **W** ♂ 100–107 mm (n 30, m 104.1), ♀ 96–101 mm (n 14, m 98.7); **T** ♂ 69–76 mm (n 30, m 72.4), ♀ 65–73 mm (n 14, m 69.8); **T/W** m 69.9; **B** 14.4–18.5 mm (n 43, m 16.0); **BD** 3.6–4.7 mm (n 26, m 4.0); **Ts** 23.4–27.1 mm (n 44, m 25.3); **white on ss** (visible length on folded wing) 23–32 mm. **Wing formula: p1** > pc 8.5–14 mm, < p2 34–42 mm; **p2** < wt 10–15 mm, =7/8 or 8 (95%) or =7 (5%); **p3** < wt 0.5–3.5 mm; **pp4–5** about equal and longest; **p6** < wt 1–3 mm; **p7** < wt 6.5–10 mm; **p8** < wt 12–15 mm; **p10** < wt 17–23 mm; **s1** < wt 19–26 mm. Emarg. pp3–6.

GEOGRAPHICAL VARIATION & RANGE Monotypic. – Caucasus, E Transcaucasia, N Iran, Tien Shan, Pamirs, Tarbagatai and east to SE Transbaikalia, Mongolia, N China; in winter descends to lower altitudes or makes shorter movements. – At least two subspecies often recognised, *erythrogastrus* in the Caucasus region and *grandis* in mountain ranges further east, where the latter is claimed to differ in being on average not quite as deeply black in ♂ plumage (in particular on mantle and scapulars), more matt sooty-black, and in having on average subtly smaller bill but slightly longer legs. However, black variable, and any difference is very subtle and inconsistent (about half look identical), bill in *grandis* averages only 0.3 mm thinner (and is of same length), and tarsus averages a mere 1.0 mm longer, with much overlap. The often-mentioned larger white wing patch in *grandis* is similarly of little use due to extensive overlap. It is therefore reasonable to merge these two taxa into a monotypic species. (Syn. *grandis, maximus; vigorsi*.)

LITTLE ROCK THRUSH
Monticola rufocinereus (Rüppell, 1837)

Fr. – Monticole rougequeue; Ger. – Schluchtenrötel
Sp. – Roquero chico; Swe. – Dvärgstentrast

A smaller version of Common Rock Thrush, confined to mountains in the S Arabian Peninsula, and to elevated parts of E Africa. It occupies similar rocky habitats as its European relatives, but is perhaps more associated with thickets and trees. Largely sedentary, but often moves to lower elevations or makes short migrations in winter. Breeding season on the Arabian Peninsula protracted, from February to August.

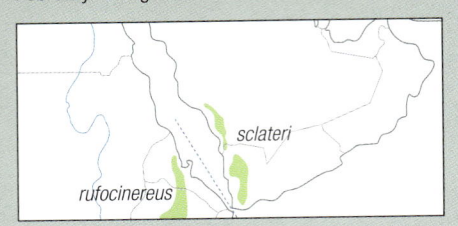

M. r. sclateri, ad ♂, Saudi Arabia, May: unmistakable due to lead-grey head and upper breast, contrasting with orange-red underparts; also rather stout and large-headed. Still not heavily worn, the primary-coverts are fresh and grey-fringed, indicating an ad. (G. Lobley)

M. r. sclateri, ♂, presumed ad, Yemen, May: plump with large head, while tail is short being orange-red with dark tip (including when seen from below). Cannot be reliably aged at this angle. (W. Müller)

M. r. rufocinereus, ad ♂, Ethiopia, Sep: NE African race is paler with browner (less grey) upperparts, but has slightly stronger face pattern (especially darker lores). Wing evenly fresh. (H. Shirihai)

IDENTIFICATION No larger than a wheatear but a little stouter and *larger-headed* with a *stronger bill*, which often appears *slightly downcurved*. Rather confiding, feeding fairly close to man. Sometimes when perched quivers tail in redstart fashion, especially in anxiety. *Head, throat and upper breast lead-grey*, crown down to back brownish-grey, whereas *rump, lower breast, belly and flanks are orange-red*. The *tail is orange-red with an inverted dark 'T'*, in other words has both a dark centre and terminal band. Sexes very similar, the ♀ is only slightly duller overall, with a marginally paler throat. Juvenile differs by having orange confined to rump and tail, while the rest of underparts are dull buffish off-white with dark vermiculations or blotches, and upperparts and wing-coverts are brown-grey with pale spots.

VOCALISATIONS The song resembles both Common Rock Thrush and, in particular, Blue Rock Thrush, brief strophes with a mix of warbling, clear notes and more scratchy sounds, often shuttling up and down the scale, and fading or falling slightly in pitch towards end. Song-flight not described. – Calls include a wheatear-like straight whistle, *vüht* or *tyyt*, and a dry rattling or rasping *trrt* (Clement & Hathway 2000).

SIMILAR SPECIES Due to its small size is only likely to be confused with an eastern race of *Black Redstart*, but these have no black end-band on the tail, and are slimmer with smaller head and bill, proportionately. – *Common Redstart* of race *phoenicurus* is eliminated by same characters, and additionally on having white forecrown and black forehead, lores and throat (although the black feathers are pale-tipped when fresh in winter).

AGEING & SEXING (*sclateri*) Ages differ very slightly in autumn, rarely also in spring. Sexes similar year-round, but sometimes separable when seen close, especially if pair seen together. – Moults. Complete post-nuptial moult of ad in late spring and summer, mainly May–Aug. Partial post-juv moult at about same time does not involve flight-feathers, tertials or primary-coverts. Usually no greater coverts moulted, but some birds replace a few inner. Apparently no pre-nuptial moult in winter. – **SPRING** ♂ Most of head and upper breast, including chin and throat, dark lead-grey. Crown to mantle rather pure grey. Lower breast and rest of underparts mainly saturated orange-red. ♀ Very similar to ♂, and many inseparable. Sometimes possible to recognise by slightly paler chin and throat, more brownish-tinged and less bluish head and mantle. Upper lores sometimes slightly paler grey. – **AUTUMN Ad** Much as in spring, but plumage now fresh. All greater coverts uniformly dark grey without obvious paler tips. Sexes differ less than in spring due to slightly brown tinge above also in ♂. **1stW** Much as ad, but colours on average a little duller, and greater coverts either uniform brown-grey with obvious buff-white tips, or with contrast due to some inner moulted greater coverts edged greyer and lacking prominent pale tips. A few moult all greater coverts to ad type. **Juv** Grey-brown, prominently spotted and vermiculated. Orange-red restricted to rump, uppertail-coverts and tail-sides. Yellow gape flanges often obvious.

BIOMETRICS (*sclateri*) **L** 15–16 cm; **W** ♂ 82.5–86.5 mm (*n* 12, *m* 84.3), ♀ 79.5–85 mm (*n* 7, *m* 81.9); **T** ♂ 55–60 mm (*n* 12, *m* 57.6), ♀ 52–59 mm (*n* 7, *m* 54.9); **T/W** *m* 67.9; **B** 17.5–21.1 mm (*n* 18, *m* 19.8); **BD** 4.7–6.0 mm (*n* 16, *m* 5.3); **Ts** 22.5–25.7 mm (*n* 19, *m* 24.5). **Wing formula:** p1 > pc 9–15 mm, < p2 24–29 mm; **p2** < wt 6.5–14 mm; =8 or <; **p3** < wt 0.5–3 mm; **pp4–5**(6) about equal and longest; **p6** < wt 0–4 mm; **p7** < wt 2.5–7 mm; **p8** < wt 6.5–10 mm; **p10** < wt 11.5–14.5 mm; **s1** < wt 11–15 mm. Emarg. pp3–7.

GEOGRAPHICAL VARIATION & RANGE Slight. Two races, one of which is extralimital.

○ **M. r. sclateri** Hartert, 1917 (SW Arabia; resident). Described above.

Extralimital: **M. r. rufocinereus** (Rüppell, 1837) (E Africa; resident). Very similar to *sclateri* but on average a little larger, and ♀ has sometimes slightly paler chin and throat, thus somewhat more pronounced sexual dimorphism (but available material not conclusive as to this).

M. r. rufocinereus, ad ♀, Ethiopia, Sep: ♀ resembles ♂, sometimes making separation difficult, but those of this NE African race are generally clearly browner above and on head and breast, and have whitish-mottled throat. Note again large-headed and short-tailed impression. (H. Shirihai)

(COMMON) ROCK THRUSH
Monticola saxatilis (L., 1766)

Alternative name: Rufous-tailed Rock Thrush

Fr. – Monticole de roche; Ger. – Steinrötel
Sp. – Roquero rojo; Swe. – Stentrast

Mainly a montane species, preferring rocky outcrops on steep alpine meadows, but also found at somewhat lower elevations in dry, open habitats with access to limestone cliffs. Since ♂ is a stunningly beautiful bird with strong colours it is sometimes noticed by non-birdwatchers, not least since it may come down to feed near farms and perch on walls or rooftops. A migrant, leaving Europe in early September, wintering mainly in sub-Saharan Africa. Vanguard ♂♂ return to their favoured European rocks in late February, but majority not until late March to mid April.

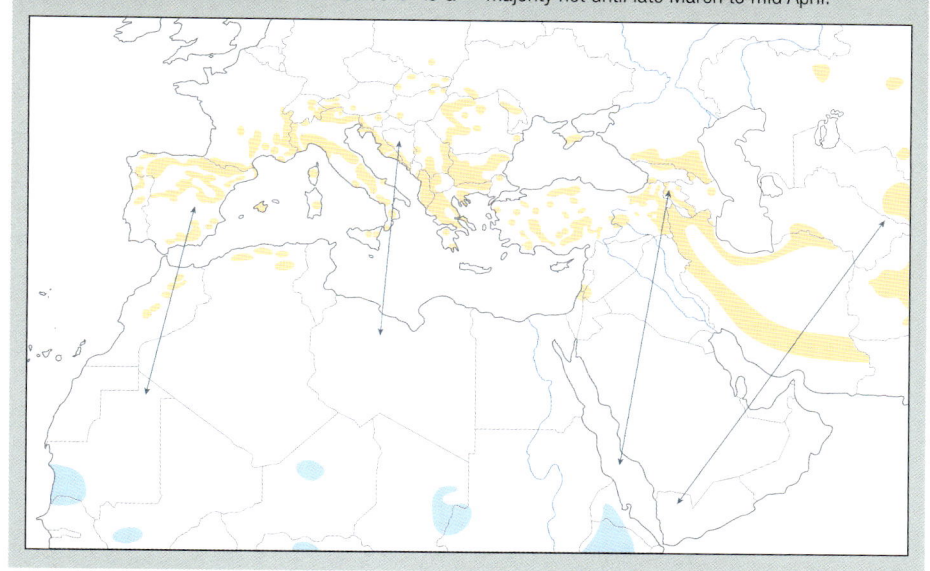

1stS ♂, Israel, Mar: on spring migration, when blue-grey mantle is typically partially brownish-tipped and rufous-orange underparts can have narrow white feather-tips. Mainly aged by restricted partial moult, with two generations of lesser and median coverts, excluding ad; note broad white tips to primary-coverts. (P. Alberti)

and greyer-looking head with a bluish-grey sheen visible through the vermiculations. Juvenile is like ♀, only with paler brown ground colour and more diffusely marked streaks and vermiculations.

VOCALISATIONS The song is a rather brief, pleasing strophe of soft clear whistling notes and trilling sounds which generally falls a little in pitch and has a certain desolate ring. Also, the rhythm varies in a melodious way. General voice recalls Black Wheatear, found in similar habitats in Iberia and NW Africa, but that species usually lacks the falling pitch and pleasing variation, is more monotonously warbling. The real challenge is to separate it from Blue Rock Thrush, and these two are at times so similar that you prefer to see the bird to be sure. Blue Rock Thrush sounds a little stronger, deeper, more tremulous (more 'r' sounds), even more melancholy, whereas Common Rock sings slightly softer, more reminiscent of sub-song of Song Thrush or Blackbird. Some birds show a preference for Woodlark-like repetition of same one or two notes for long spells. Practice makes perfect! In song-flight, strophes are drawn-out and often sound more warbling and very pleasing, and frequently include mimicry of other birds. – Calls include a wheatear-like straight or near-

blackish wings with an irregular *white patch on lower back*, sometimes partly extending onto scapulars. Rump is dark. The adult ♀ is duller-looking, basically brownish with dense, *dark streaking and vermiculations on head and underparts*; only the reddish tail stands out. Although some attain a faint bluish cast on head, there is no large white patch on back. In autumn, all birds are ♀-like, dark brown above and ochraceous- or orange-buff below, with dense vermiculations, dark below and pale above. A few ♂♂ can be picked out as they have a hint of pure white visible on back, but don't expect all to show it. Some adult ♂♂ have a trifle darker

Ad ♂, Turkey, May: unmistakable given greyish-blue head, nape and upper mantle, rufous-orange breast, belly and much of tail, and blackish wings with ragged white patch on lower mantle and back; long wings, strong bill, but rather short legs, and upright stance. Ageing (for once) rather straightforward: all wing-feathers are post-nuptial ad, not heavily worn, with little contrast, including still mostly blackish primary-coverts tipped whitish. (G. Thoburn)

IDENTIFICATION A compact, medium-small and rather *short-tailed* but *long-winged Monticola* thrush with *rufous-orange tail*. Sits rather upright, as a rule quite still. When alarmed bobs and quivers tail like a large redstart. Rather shy and wary, usually flies off early. Flight direct and fast. Performs song-flight with one peak, or undulations, and a concluding glide on outstretched wings. Bill rather long and strong. Adult ♂ in spring has beautifully *blue-grey head, nape and mantle, rufous-orange breast and belly*, and

1stS ♂, Turkey, May: white patch generally increases in size with feather wear. Despite wing-coverts being partially concealed by flanks feathers, greater coverts, tertials and remiges can be seen to be juv. ♂ in May with some dark subterminal bars to orange underparts and brownish hue to ear-coverts usually 1stY. (D. Occhiato)

1stS ♂, Israel, Apr: very similar to ad ♂ at this season, but some rather easily aged by more worn and brownish juv remiges, primary-coverts and greater coverts (the primary-coverts especially are narrow and lack bold pale tips), while blue areas are less neat, with more white tips. (L. Kislev)

Ad ♀, Georgia, Apr: much less colourful than ♂ in summer, with dense, dark scalloping over much of underparts and rather contrasting reddish-rufous tail. Aged by relatively dark remiges and coverts, while edges of greater coverts and especially upperparts are suffused greyish, with rather broad white tips to primary-coverts, i.e. unaffected by strong wear. (W. Müller)

straight whistle, *veet* or *uiht*, sometimes quickly doubled. A short clicking *tac*, or a hoarser *schack*, is given singly or in series when alarmed, the latter sound being harder than Fieldfare.

SIMILAR SPECIES Whereas the spring ♂ should not cause any problems within the treated area (♂ *Little Rock Thrush* of Yemen is much smaller, lacks white on back and has inverted black 'T' on red tail), ♀♀ and all autumn birds require some care. To eliminate *Blue Rock Thrush* and juvenile *Blackbird* note reddish tail, but remember that central feathers are darker brown and might cover much of the red in poor view of perched bird. Much larger size, longer wings and more direct, powerful flight should rule out a flying ♀ *redstart* even at longest range.

AGEING & SEXING Ages differ very slightly in autumn, rarely also in spring. Sexes separable from 1stS, sometimes from 1stW (latter requires close observation or handling). – Moults. Complete post-nuptial moult of ad in late summer, mainly Jul–Aug (Sep). Partial post-juv moult at about same time does not involve flight-feathers, tertials, primary-coverts or outer greater coverts. Partial pre-nuptial moult in winter (Dec–Feb) involves head and body and sometimes some inner lesser and median coverts. – **SPRING Ad ♂** Head (including throat), nape and mantle uniformly blue-grey. Back has irregularly outlined pure white patch. Whole breast and belly orange-red without dark vermiculations. Primaries blackish and reasonably fresh (but some are slightly more worn or brownish, difficult to separate from 1stS ♂). **1stS ♂** As ad ♂, and many inseparable, but sometimes possible to recognise by more worn and brownish primaries and primary-coverts, some remaining juv greater coverts with large pale spots, and on average less neat plumage with more dark and white tips on head and body (often on centre of belly). **♀** Crown to rump grey-brown streaked or vermiculated dark. (Very rarely a limited lead-grey tinge on crown and mantle.) If at all pale on back (only in heavily abraded ad), this is limited to feather-centres being tinged off-white or buff-white (never pure white). Throat off-white densely streaked and vermiculated dark. Breast and belly with rich orange-buff tinge, densely vermiculated. Ageing according to colour of primaries and primary-coverts more difficult than in ♂, but amount of wear and any remaining broadly pale-tipped greater coverts sometimes helpful. – **AUTUMN Ad ♂** As ♀ in spring, largely grey-brown above and orange-buff below

1stS ♀, Turkey, May: similar to ad ♀, but all greater and primary-coverts are juv, as are the remiges. (D. Occhiato)

Presumed ad ♀, Iran, Jun: in summer some (usually ad) ♀♀ can develop ♂-like whitish mantle patch and almost unbarred rufous-buff underparts, hence ageing should precede sexing. Relatively less worn and dark remiges and coverts, including primary-coverts. There is, however, moult limits with renewed innermost greater coverts and some median coverts, which makes both age and sex questionable. (E. Winkel)

Ad ♂, Spain, Sep: colourful plumage now concealed by ♀-like feather-tips, until late winter or spring when pale tips wear off gradually. However, some fresh ad ♂♂ sexed by blue cast or spots on head, nape and mantle, as well as greyish suffusion to fringes of wing-coverts and remiges, and more rufous bases on belly. Aged by being evenly fresh with dark/blackish centres to wing-feathers, and by greater coverts having faint blue-grey edges and whitish tips. (G. Thoburn)

ROCK THRUSHES AND CHATS

Ad ♀, Denmark, Oct: evenly fresh after post-nuptial moult, and no clear exposed lead-grey tinged feathers or the usual fresh ad ♂'s darker-hooded impression, but some rufous feathering on face allow inference on age and sex. (H. Brandt)

1stW, possibly ♂, Oman, Nov: 1stW plumages very similar to ♀-like fresh ad plumages, but unlike latter, wing-feathers slightly more worn with browner centres, while some grey-blue feather bases above strongly suggest a ♂. (A. Audevard)

with dark vermiculations, but sometimes recognised by blue cast or spots on head (also throat) and mantle, and by some pure white visible on back. Aged as ad by mint fresh and dark primaries and primary-coverts, and by greater coverts with faintly blue-grey edges and narrow whitish tips (sometimes slightly more extensive). **Ad ♀** As ad ♂, but never any bluish tinge on head or mantle, nor any white visible or hidden on back. Plumage fresh like in ad ♂. **1stW** As ad ♀, but primaries often slightly worn at tips from c. Aug, and rather more worn from late Sep. Sometimes a few outer greater coverts have extensive pale tips. (Many 1stW ♂♂ can be sexed in the hand by presence of some pure white feather-bases on back.) **Juv** Resembles ♀ but is somewhat more uniformly pale buff-brown above with pale spots and dark vermiculations. Yellow gape flanges often obvious. All secondary-coverts pale-tipped.

BIOMETRICS L 17.5–21 cm; **W** ♂ 116–132 mm (n 70, m 122.9), ♀ 113–127 mm (n 21, m 119.2); **T** ♂ 57–67 mm (n 31, m 61.3), ♀ 55–62 mm (n 12, m 59.0); **T/W** m 49.4; **B** 22.1–26.5 mm (n 27, m 24.4); **Ts** 25.3–29.6 mm (n 27, m 27.5). **Wing formula: p1** minute, < pc 7.5–16 mm, < p2 73–86.5 mm; **p2** < wt 0–3 mm, =3, 3/4, or 4; **pp3–4**

♀-like, Spain, Sep: wing is juv-like, with greater coverts already worn, and has extensively exposed rufous bases below, strongly suggesting a young ♂. However, it could also be an early-moulted ad ♀. (R. Martínez Fraga)

equal and longest (though rarely p4 0.5–1 mm <); **p5** < wt 9.5–15 mm; **p6** < wt 17.5–22 mm; **p10** < wt 37–45 mm; **s1** < wt 42–51 mm. Emarg. pp3–4.

GEOGRAPHICAL VARIATION & RANGE Monotypic. – S & C Europe and N Africa east to Mongolia and C China, north to Germany, Carpathian Mts, Ukraine, Caucasus, Kopet Dag, Altai, W Baikal, south to Zagros and S Iran; winters Africa, nearly entirely south of Sahara, in Sahel zone and E Africa, commonly in Eritrea.

1stW, possibly ♀, Israel, Sep: unlike ♀-like fresh ad plumage, the wing-feathers, including primary-coverts, are already slightly more worn (with browner centres), while overall paler plumage suggests ♀. (A. Ben Dov)

Juv, Iran, Jun: pale buff-brown with spotty and mottled appearance; secondary-coverts broadly tipped buffish-white; note fluffy body-feathers and yellow gape flanges. (E. Winkel)

BLUE ROCK THRUSH
Monticola solitarius (L., 1758)

Fr. – Monticole bleu; Ger. – Blaumerle
Sp. – Roquero solitario; Swe. – Blåtrast

This close relative of the Common Rock Thrush is more versatile in its preferred habitats. Its clear, fluty song can be heard both on rocky mountain-slopes, from ruins, in abandoned quarries and from cliffs by the sea. In winter some stay near their breeding territories but descend to lower levels, whereas others migrate to Africa. In March or April its song echoes again among the cliff faces.

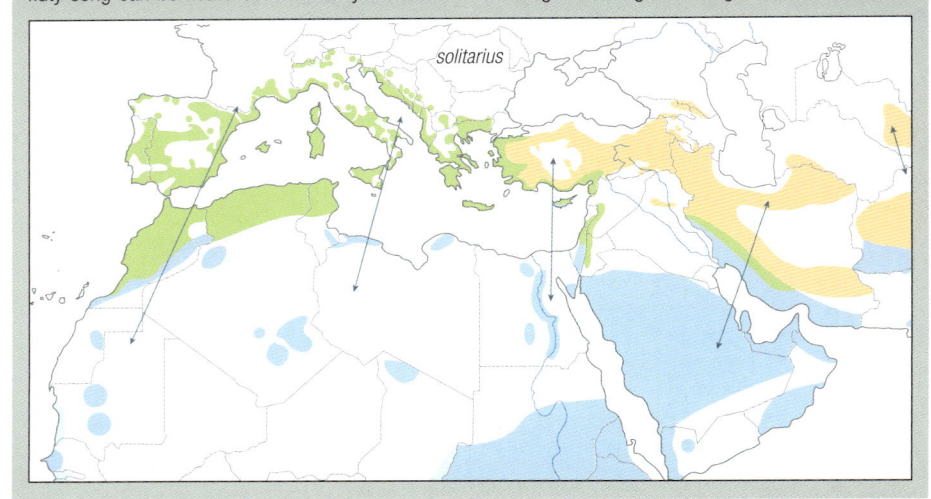

M. s. solitarius, 1stS ♂, Malta, Apr: some ♂♂ greyer, with less intense turquoise-blue tones. Aged by moult limits between renewed, bluish greater coverts and juv remiges and primary-coverts, which are very worn and pointed and bleached brownish. Usually fewer or no greater coverts are moulted in post-juv moult, and ageing is less straightforward than here. (N. Fenech)

IDENTIFICATION The size of a Song Thrush, but with somewhat more *elongated, slim shape* with proportionately longer tail, and thinner and *longer bill*. Due to the shape and all-dark colours may appear larger than its actual size, to which impression the frequently long-range observations and mountainous surroundings also contribute. Generally shy and wary, flying off early. Flight direct and fast. Performs song-flight with concluding glide on outstretched wings. Adult ♂ in spring has whole plumage beautifully *dark greyish-blue*, with *blackish wings, tail, bill and legs*. The adult ♀ is duller-looking, being entirely *brownish-grey* (in some with faint slate-grey tinge on sides of head and upperparts including uppertail) with dense, *dark streaking and vermiculations* on head and body. Cheeks and throat have a somewhat paler and warmer ochraceous-buff ground colour than rest of underparts, which are darker and greyer. In autumn, all birds are more or less ♀-like, dark brown-grey with dense dark and pale vermiculations, although the adult ♂ can generally be picked out by an obvious bluish cast on upperbody, in particular on rump/uppertail-coverts. First-winter birds are very similar to adults of either sex, but have on average slightly broader pale tips and edges to wing-coverts and tertials. Juvenile resembles ♀, only with paler brown ground colour and more diffusely marked streaks and vermiculations. Note yellow gape flanges.

VOCALISATIONS The song is a brief, pleasing strophe of loud and deep, clear whistling notes and much trilling or tremulous sounds, which generally fall in pitch, and has an obvious melancholy ring. Strophes are usually well spaced. The song is very similar to Common Rock Thrush, and it takes some practice to separate them (and it is still not always possible!), but the voice of Blue Rock is a little deeper and more monotonously fluty, with some shuttling back and forth, lacking the mimicry and higher-pitched warbling elements that you sometimes find in the song of Common Rock. In song-flight, like in Common Rock Thrush, strophes are drawn-out and may sound louder still. – Calls include a wheatear-like straight whistle, *viht*, or a disyllabic *üh-vih*. The whistle is sometimes quickly doubled or, more commonly, combined with hard clicking *chac*, thus *viht chac chac, viht chac viht chac-chac*, etc., much like Northern Wheatear, only stronger.

SIMILAR SPECIES The spring ♂ should not be difficult to identify with its entirely dark grey-blue plumage. ♀ and autumn birds look all dark and could at long range be confused with a ♀ or juv *Blackbird*, but note slimmer shape

M. s. solitarius, ad ♂, Jordan, Mar: aged by evenly-feathered wing, including grey-fringed and still whitish-tipped primary-coverts; bluish-grey plumage and lack of barring (except on undertail-coverts) sex the bird as ♂. (H. Shirihai)

M. s. solitarius, ad ♂, Greece, Apr: ♂'s blue and grey tones vary individually and change with light, here appearing turquoise or cerulean blue, in other lights more ultramarine. Ageing as image to the left. (R. Wielinga)

ROCK THRUSHES AND CHATS

M. s. solitarius, ad ♀, Malta, Feb: ♀ brownish, often washed pale greyish-blue on sides of head and upperparts, but best characterised by dark barring on underbody, while cheeks and throat are often spotty with warmer, more buffish wash. Aged by still fresh, white-tipped primaries and primary-coverts. (M. Sammut)

M. s. solitarius, 1stY ♀, Kuwait, Feb: irrespective of age, even young ♀♀ are often patchily tinged dull slate-grey on head-sides and upperparts. Aged by wing-feathers, including juv primary-coverts (very worn and pointed, with pale tips almost lost) and by moult limits with renewed innermost greater covert. (D. Occhiato)

M. s. solitarius, ♀, Turkey, Apr: characteristic barred underparts, spotty buffish cheeks and throat, and pale-streaked breast. Cannot be aged from this angle, when most of wing is invisible. (R. Debruyne)

M. s. solitarius, ad ♂, Malta, Nov: predominantly blue plumage with dark subterminal marks and numerous whitish tips (which may persist even into spring), as well as evenly feathered wing, with remiges, primary-coverts and greater coverts centred black and fringed bluish. (N. Fenech)

and longer bill. Told from ♀ or autumn *Common Rock Thrush* by rather long and blackish tail, and less ochraceous-tinged underparts.

AGEING & SEXING (*solitarius*) Ages differ very slightly in autumn, rarely also in spring. Sexes invariably separable in spring, sometimes also in autumn (ad ♂ often blue enough to be picked out). – Moults. Complete post-nuptial moult of ad in late summer, mainly mid Jul–Sep (Oct). Partial post-juv moult at about same time does not involve flight-feathers, tertials or primary-coverts. Usually no greater coverts are moulted, but some birds replace some inner. Apparently no pre-nuptial moult in winter. – **SPRING Ad** ♂ Whole plumage basically dark (greyish-) blue. Chin and throat bluish, tinged grey when fresh. Undertail-coverts bluish-grey tipped buffish-white and have dark subterminal marks. Primaries dark, glossy and fresh until breeding starts. Primary-coverts dark with no or only very narrow pale tips. **1stS** ♂ As ad ♂, and some inseparable, but often possible to age by slightly more worn and brown-grey, less blue-edged primaries, primary-coverts having broader pale tips, slight contrast between moulted inner and unmoulted outer greater coverts, and by whole plumage being on average less deep blue, more greyish-blue with a slight brown tinge, and heavier dark markings. ♀ Plumage largely brown-grey vermiculated dark, but ad often has some dull greyish-blue on sides of throat, parts of breast, flanks, mantle, back and uppertail (though never bluish and largely unbarred on chin, throat and breast). Chin and throat greyish buff-brown, streaked or barred dark when fresh. Undertail-coverts buff-grey (sometimes tinged rufous) tipped off-white with dark subterminal marks (never bluish). Ageing according to wear of primaries and number of pale tips on primary-coverts more difficult than in ♂, but sometimes possible; close views are necessary. – **AUTUMN Ad** ♂ Much as ad ♂ in spring, largely all blue, but fresh blue feathers now partly have pale tips and dark subterminal vermiculations. Primaries mint fresh and dark. Primary-coverts dark with only narrow whitish tips, if any. Greater coverts uniform, narrowly pale-tipped, edged blue. **1stW** ♂ Much as ad ♂, but blue on average a little duller and greyer, dark markings heavier. Primaries very slightly more brownish-grey and more worn from mid autumn. Primary-coverts same with rather broad whitish tips, often forming small white patch on folded wing. Greater coverts either uniform greyish with diffuse off-white tips, or show a contrast due to some inner moulted greater coverts edged bluish-grey and neatly and narrowly tipped white. A few moult all greater coverts to ad type. **Ad** ♀ Can appear similar to 1stW ♂, with some dull bluish tinge as in spring, but never any blue on chin, throat or under-tail-coverts, these being buff-brown. Plumage fresh like ad ♂. **1stW** ♀ As ad ♀, but invariably no blue below and no or very little above (if any, just lead-grey spots mixed with brown-grey general colour), and primaries often slightly worn at tips from about Sep. Unless seen close in good light, ♀♀ without much blue cast generally impossible to age. **Juv**

— 277 —

M. s. solitarius, 1stW ♂, Spain, Aug: some young ♂♂ show clear moult limits in wing, with contrastingly renewed inner greater coverts, and are more safely aged. Note that even juv wing-coverts in ♂ are bluish-edged. (A. M. Domínguez)

M. s. solitarius, 1stW ♂, Cyprus, Nov: fresh young ♂♂ can be extensively vermiculated below, especially on breast and belly, which can be rather pale. Aged by moult limits in wing, with renewed and contrasting inner greater coverts. (A. McArthur)

M. s. solitarius, ad ♀, Turkey, Oct: evenly fresh appearance (especially remiges and primary-coverts, which are greyish-fringed) and extensive slate-grey crown, nape and upperparts suggest ad ♀. Innermost greater covert greyer and newer, and paler than rest. (Z. Kurnuç)

Resembles 1stW ♀ but is somewhat more uniformly pale grey-brown above with more prominent pale spots. Yellow gape flanges often obvious.

BIOMETRICS (*solitarius*) **L** 20–24 cm; **W** ♂ 118–134 mm (*n* 57, *m* 124.7), ♀ 117–126 mm (*n* 26, *m* 122.2); **T** 74–94 mm (♀ < 88 mm; *n* 54, *m* 81.9); **T/W** *m* 65.9; **B** 26.0–32.4 mm (*n* 48, *m* 28.7); **Ts** 26.5–30.5 mm (*n* 23, *m* 28.7). **Wing formula: p1** < pc 0–8 mm (rarely 0–2.5 mm >), < p2 62–70 mm; **p2** < wt 5–9 mm, =5 or 4/6; **pp3–4** equal and longest (though rarely p4 0.5–2 mm <); **p5** < wt 4.5–8 mm; **p6** < wt 13–17.5 mm; **p7** < wt 19–26 mm; **p10** < wt 33–38 mm; **s1** < wt 38–42 mm. Emarg. pp3–5.

GEOGRAPHICAL VARIATION & RANGE Slight variation except very well-marked extralimital subspecies in E Asia. Only one race within treated region.

M. s. solitarius (L., 1758) (S Europe, N Africa, Turkey, Caucasus, Levant; short-range movements to lower altitudes or to winter grounds in coastal N Africa, Arabia). Described above. (Syn. *behnkei*, *longirostris*. We can find no consistent differences as to paleness, amount of barring, or biometrics between *longirostris* and *solitarius*.)

Extralimital: **M. s. pandoo** (Sykes, 1832) (E Afghanistan to C China; winters S & SE Asia but could theoretically straggle to E Iran). A little smaller and very slightly darker than *solitarius*. ♀ not as buff- or rufous-tinged below as *solitarius*, more dull greyish. **L** 19.5–22 cm; **W** ♂ 116.5–125 mm (*n* 12, *m* 120.5), ♀ 113–121 mm (*n* 12, *m* 117.8); **T** 75–83 mm; **B** 25.6–29.5 mm (*n* 24, *m* 27.6); **Ts** 26.6–28.6 mm (once 30.0; *n* 24, *m* 27.6). **Wing formula:** s1 < wt 34–40 mm.

M. s. madoci Chasen, 1940 (SE Asia; resident). Smallest with more rounded wing. Ad ♂ is blue like *solitarius*.

M. s. philippensis (Statius Müller, 1776) (E China; winters SE China, Philippines, Indochina). Differs clearly in that ad ♂ has deep rufous-red breast, belly and undertail-coverts, and 1stW ♂ has irregular rufous-red among blue and barred feathers. (Syn. *magnus*.)

TAXONOMIC NOTE Recently, it has been proposed that *philippensis* constitutes a separate species based on genetic analyses (Zuccon & Ericson 2010), which showed the Blue Rock Thrush complex to be deeply divided into two clades, each constituting a monophyletic group. Still, a conservative approach is preferred here, in particular since the concerned taxon is purely extralimital.

REFERENCES ZUCCON, D. & ERICSON, P. G. P. (2010) *Mol. Phyl. & Evol.*, 55: 901–910.

M. s. solitarius, 1stW ♀, Spain, Nov: browner wing-feathers are juv; tips of primaries and primary-coverts rather pointed with limited pale edges. This plumage is superficially similar to ♀-like plumages of Common Rock Thrush, but rufous tail of latter always prevents confusion. (A. M. Domínguez)

M. s. solitarius, juv ♂ with ad ♂, Greece, May: loose body-feathers and uniform pale grey-brown above, spotted and barred; sexed as ♂ by very bluish wing-feathers and almost no vermiculations on back. (D. Verdonck)

WHINCHAT
Saxicola rubetra (L., 1758)

Fr. – Tarier des prés; Ger. – Braunkehlchen
Sp. – Tarabilla norteña; Swe. – Buskskvätta

On a warm, quiet evening in the first half of May you suddenly hear a brief, explosive warble from an open lakeside and realise that the Whinchat has returned from tropical Africa, often to precisely the same territory it occupied the previous summer. The spring chorus is beginning to become complete. This is hardly a well-known bird among non-birdwatchers, but one well worth getting acquainted with. It is only a brief summer visitor; one August morning the birds are already gone, having started the long return journey to Africa the preceding night.

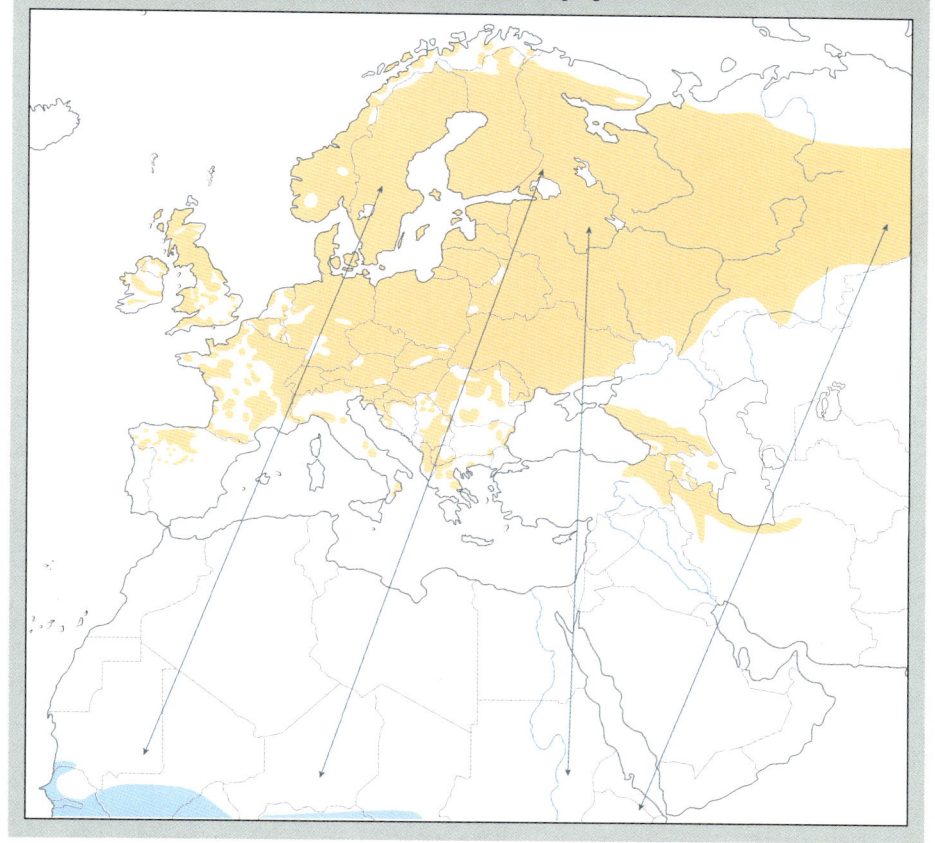

Ad ♂, **Germany**, May: in typical posture perched at top of a stalk. Sexing and ageing as image below, but variation includes ad ♂♂ with relatively smaller white primary-coverts patch. (M. Schäf)

IDENTIFICATION A small, rather compact, short-tailed chat. Sits upright atop a weed or fence wire, flicks wings and wags or twitches tail, flies off low in slightly jerky, hopping flight, alights on new perch after brief, quick ascent. Sits scanning from perch for some time. In all plumages basically a buffish- or *cinnamon-brown streaked* bird, with a warm, peachy or orange tinge on breast. *Small white patches at sides of base of tail* visible in flight but usually largely concealed when perched. Invariably has *rump and uppertail-coverts pale rufous-brown streaked dark*. Adult ♂ has a *white supercilium*, dark brown-black cheeks and a dark crown, some *white patches* or marks *on the wing*, *orange throat and breast* framed by white chin and white stripes along each side of throat. The ♀ and young ♂ are more nondescript buff-brown and streaked birds with cream-buff supercilium and less white in wing, if any. Important when identifying these plumages to note combination of whitish (or pale buff) base to tail-sides, pale, cream-white throat and entire pale brown rump being boldly streaked.

VOCALISATIONS A diurnal bird but has crepuscular habits, often singing at dusk and during early night, as well as at dawn. The song is loud and 'sudden', a brief outburst of clear, fluty notes mixed with rasping sounds and clever mimicry. Commonly, the strophe opens a little hesitantly, gathers speed and ends abruptly. A variant with peculiar resemblance to song of Corn Bunting, or elements of it, is practised by many. Song strophes are fairly well spaced. – Call a short, soft, low whistle, *yu*, when alarmed often combined with a dry, clicking *tec*, thus *yu tec-tec yu tec...* A rarely heard anxiety call has been described as a repeated *yup*, like the flight call of Chaffinch but more emphatic.

SIMILAR SPECIES Should only be possible to confuse with ♀ or first-winter *Eastern Stonechat* (*maurus* or other races), but note that Whinchat has: (i) dark spots or streaks over pale brown rump/uppertail-coverts; (ii) white base to outertail (visible in flight, but will not eliminate Caspian race of Eastern Stonechat *hemprichii*); (iii) more distinctly white- and black-spotted mantle/scapulars; (iv) somewhat more prominent buff-white supercilium; and (v) often some dark spots on breast. – ♀ *Fuerteventura Stonechat* is subtly smaller, with a proportionately longer bill and tail. Also, Whinchat has white tail-base and is brighter tawny- or rufous-brown above with more distinct black and buff spotting. – European *Stonechat* differs invariably by dark bib, black in ♂ and at least shaded grey in ♀.

Ad ♂, **Italy**, May: solid blackish-brown head-sides framed white and bright ochre-orange chest and flanks. Immaculate coloration plus broad white basal patch to reasonably fresh primary-coverts make this ♂ ad. (D. Occhiato)

1stS ♂, Malta, May: primary-coverts are juv, pointed, worn and bleached, with limited white bases. Browner head-sides, deep and purer pink-buff throat and breast also typical of 1stS ♂. Sharply-defined pure white in tail also indicates ♂. The species' diagnostic white bases to tail are largely concealed when perched. (N. Fenech)

♀, Poland, May: dull face pattern, including cream-white supercilium, few dark spots on duller orange-buff breast and very limited white in inner wing-coverts are ♀ features. Ageing of spring ♀♀ challenging, though quite worn and bleached wing already in May, rather pointed and worn primary-coverts, and limited white in median coverts are indicative of 1stS, while orange-buff tinge on centre of throat and dark brown iris might be better for ad. Best left unaged. (R. Messemaker)

Ad ♂, Germany, Oct: following post-nuptial moult attains more ♀-like plumage, but note very extensive white primary-coverts patch, blackish greater coverts (inner one has broad white edge) and broad white tips to some median coverts. Being ad and evenly fresh, blacker remiges still whitish-tipped. This bird is exceptionally well marked, and easy to age/sex, but many ad ♂♂ at this season are much less so. (A. Halley)

AGEING & SEXING Ages separable for ♂ in 1stS if seen well; more difficult in 1stW, but possible with a few in the field and nearly all when handled. Sexes differ invariably from 1stS, and for many autumn birds, too, if first reliably aged (generally requires handling or very close observation; ♀ often impossible to age unless trapped). – Moults. Complete post-nuptial moult of ad usually in Jul–Aug (Sep), near breeding grounds. Partial post-juv moult in late summer does not involve flight-feathers, tertials, primary-coverts or outer greater coverts. Some birds (hatched late or breeding far north) retain all juv greater coverts. Pre-breeding moult in late winter involves head, body, median and lesser coverts; rarely some tertials and inner greater coverts. Occasional birds returning in spring with a few inner primaries replaced are 1stS. – **SPRING Ad ♂** Lores and moustachial stripe dark brown-black to black. Ear-coverts dark brown or blackish. Supercilium, chin and lateral throat-stripes pure white and prominent. Vivid ochre-orange centre of throat and breast. Invariably sharp division between white and dark brown portions of tail-feathers. Never has dark spots on breast. One or, more often, several inner median coverts pure white, and often 1–2 pure white inner greater coverts, too. Blackish tips to primary-coverts small (half length or less of exposed feathers), invariably distinctly set off from pure white bases; a few have extensively or all-white inner primary-coverts, but more commonly these are darker than outer. All greater coverts and primaries reasonably fresh in early May (though some a little more abraded at tips), greater coverts with only slight (if any) contrast between black of inner and brown-black of outer (contrast due to partial pre-nuptial moult in winter). **1stS ♂** Differs from ad ♂ as follows: brownish tips to primary-coverts extensive (about half length or more of exposed feathers), sharply or diffusely set off from buff-white or white bases (limited overlap with ad ♂ pattern); beware that a few have inner primary-coverts all dark but outer white-based. Usually marked contrast between black of any moulted inner greater coverts and brown (and worn) outer (but contrast can be subtle, more like in ad). Primaries somewhat or heavily worn at tips, and on average more brownish (though some have darker and less worn primaries, more like in ad). Note that exceptionally some inner primaries are moulted in winter. Very rarely breast is spotted dark. **♀** Supercilium and chin buff-white. Ear-coverts brown, lores brown and buff-white (not largely black). Most (but not all) have some dark spots on breast, which is less orange more ochre-buff. Generally less clear-cut division between white (at times buffish-white, off-white) and dark brown portions of tail-feathers (only 10% have ♂-like pattern). Inner median and greater coverts can be extensively white-tipped but never all white. Primary-coverts variable, some have no or only limited white basally (1stS), others extensive white (up to 50% basally) on outer feathers (ad), with many intermediate patterns also occurring. Ageing usually difficult even in the hand and relies mainly on solid experience of difference in

1stS ♀, Greece, Apr: boldly streaked upperparts, short tail and long legs typical of the species. Dull head pattern and buffish underparts characteristic of ♀. Aged by already heavily abraded primaries and primary-coverts, supported by lack of any white in wing. (R. Pop)

ROCK THRUSHES AND CHATS

Ad ♂, Israel, Sep: in autumn many are difficult to age, making sexing also problematic. Blackish tips to primary-coverts, tertials and remiges ad-like; black bill also supports this. Blackish tips to primary-coverts small (less than half of exposed feathers) and clearly set off from pure white bases, matching ad ♂. (L. Kislev)

Presumed 1stW ♂, Netherlands, Sep: fresh-plumaged presumed 1stW, very similar to ad ♀, but has slightly paler base to lower mandible, and broad pale tips to primary-coverts might fit best with juv. Many greater and median coverts hidden by scapulars, preventing safe sexing, but bases to both primary-coverts and quite black outer tail-feathers appear pure and extensively white indicating ♂. (N. D. van Swelm)

wear and shape of primaries and primary-coverts, but also on iris colour, ad having brownish eye, 1stW grey. – **AUTUMN Ad ♂** When post-nuptial moult completed much closer to ♀ in spring, as white, black and orange are largely concealed or replaced, and breast may attain spots. Check primary-coverts pattern (see under Spring) for age, and tail-feather pattern for sex, as both are helpful (albeit not infallible). Entire plumage fresh and uniform, and no contrast among greater coverts. One or a few inner median coverts all white. Much white basally on inner 3–4 greater coverts. (A few might still be inseparable in the field from ad ♀.) **Ad ♀** Very similar to ad ♂, but check tail-feather pattern (see above), and inner secondary-coverts: never any all-white median coverts, and much less white basally on greater coverts. A few have all secondary-coverts brown-grey with pale tips and edges (thus effectively inseparable from 1stW ♀ on this character). Wing

Juv, Russia, Jul: longitudinal cream-white streaks above on rufous, dull brown and black ground; throat and breast densely streaked and spotted. (P. Parkhaev)

1stW ♀, Turkey, Oct: all plumages share extensive pale bases to outer tail-feathers, producing wheatear-like black T-shaped terminal-band. Limited white on inner wing-coverts and base of primary-coverts, plus narrow and pointed central rectrices, support ageing as 1stW. The blurred and buff-tinged division between dark and pale areas of tail shows it to be ♀. (I. Hangül)

length rather helpful when sexing handled birds. Sometimes outer primary-coverts have much white (50%), whereas inner are largely all dark. **1stW** Much as ad ♀, although some have contrast between inner moulted greater coverts being more blackish and contrasting slightly with unmoulted juv outer greater coverts. (Many birds retain all juv greater coverts, though; e.g. c. 40% of Swedish birds do not moult any.) Whitish wedge-shaped tips at shafts of brown-grey inner greater coverts (and median coverts) is good sign of 1stW. Some 1stW ♂♂ can be identified by extensive white bases to 1–3 inner greater coverts (and rarely median coverts), whereas 1stW ♀ invariably lacks any white bases to secondary-coverts. **Juv** Resembles ♀ but differs in having Treecreeper-like longitudinal cream-white streaking above on rufous, dull brown and black ground. Throat and breast densely streaked and spotted dark on buff ground.

BIOMETRICS L 12–14 cm; **W** ♂ 73–80 mm (n 50, m 76.5), ♀ 72–77 mm (n 32, m 74.1); **T** ♂ 42–48.5 mm (n 50, m 44.9), ♀ 41–47 mm (n 32, m 43.9); **T/W** m 59.1; **B** 12.5–14.8 mm (n 48, m 13.9); **Ts** 20.3–22.9 mm (n 48, m 21.7). **Wing formula:** p1 > pc 0–2 or < pc 0–2 mm, < p2 37–43.5 mm; p2 < wt 0.5–4 mm, =4/6; pp3–4 about equal and longest (though sometimes p4 0.5–1 mm <); p5 < wt 1.5–4 mm; p6 < wt 5–8.5 mm (7–10 mm when heavily worn); p7 < wt 10–14 mm; p10 < wt 17–22.5 mm; s1 < wt 19–25 mm. Emarg. pp3–5, although on p5 sometimes very slight.

GEOGRAPHICAL VARIATION & RANGE Monotypic. – Much of Europe east to W Siberia, Caucasus, Transcaucasia, NE Turkey; winters mainly sub-Saharan Africa, though some remain further north.

REFERENCES KARLSSON, L., PERSSON, K. & WALINDER, G. (1993) Vår Fuglefauna, Suppl. 1: 31–49.

FUERTEVENTURA STONECHAT
Saxicola dacotiae (Meade-Waldo, 1889)

Alternative name: Canary Islands Stonechat

Fr. – Tarier des Canaries; Ger. – Kanarenschmätzer
Sp. – Tarabilla canaria; Swe. – Kanariebuskskvätta

Anyone building up a long list of bird species seen in the W Palearctic, a common pastime among serious birders, is sooner or later bound to visit the volcanic island of Fuerteventura, one of the two eastern islands in the Canaries, to see this stonechat. At least 7000 pairs are estimated to exist of this Canary Islands endemic, confined solely to Fuerteventura. There it is a resident, often found in or near the dry, stony but sparsely vegetated little riverbeds on the slopes locally known as barrancos. It clearly prefers rather wooded slopes over more open and flat land.

♂, probably ad, Fuerteventura, Canary Is, Dec: many ♂♂ difficult to age, especially without close examination of greater coverts and tertials, but blackish head markings and long, well-defined whitish supercilium better fit ad. (A. Juvonen)

Ad ♂, Fuerteventura, Canary Is, Feb: blackish head with narrow white supercilium, rather dark upperparts, buffish-cream underparts, and rather small peachy breast patch. Together with lack of visible moult limits in wing and extensive white inner wing patch, these features indicate ad ♂. (T. Holmgren)

VOCALISATIONS The song is very similar to that of European Stonechat, a brief strophe consisting of a mixture of scratchy and warbling, clearer notes. Song strophes short and fairly well spaced. – Calls, too, resemble those of European Stonechat. A short clicking *tec* is uttered singly or 2–3 in combination. There is also a short, straight whistling *vist*, and these two call types are often combined like in European Stonechat.

SIMILAR SPECIES Needs to be separated from *European Stonechat* (ssp. *rubicola*) and *Whinchat*, both of which occur occasionally between autumn and early spring on the island. Note smaller size and subtly longer bill and tail of Fuerteventura Stonechat, and the invariably white chin and throat (*rubicola* has darkish throat even in ♀). Any white on uppertail-coverts, and lack of white on base of tail-feathers, will eliminate Whinchat. – There is a very remote risk that you could run into a vagrant *Eastern Stonechat* (*maurus*; not yet recorded on Fuerteventura), which in non-adult ♂ plumage would be rather similar. Note the streaked rump/uppertail-coverts on Fuerteventura Stonechat (*maurus* has a large unstreaked rufous-buff rump patch when fresh, unstreaked whitish when bleached), and the slightly longer bill and tail.

AGEING & SEXING (*dacotiae*) Ages differ slightly in autumn, and for ♂♂ often, for ♀♀ rarely, in spring, although

IDENTIFICATION Slightly *smaller* than a European Stonechat with a *thinner but longer bill*. Otherwise it resembles European Stonechat in general shape and behaviour. Sits upright when perched atop a bush or weed. Flicks wings and wags or twitches tail like other chats. Has proportionately a slightly *longer tail* than European Stonechat, and the *tail seems narrower*, too. The plumage is fairly similar to Eastern Stonechat (race *maurus*), but the ♂ differs clearly in having a *narrow white supercilium* and a *white chin and throat*. Further, *crown and sides of head are dark sepia-brown*, not black (though can appear blackish). There is a rather *narrow white semicollar*, which runs back over the neck-sides. A variably large (but usually small) *white wing patch* is visible on the folded wing. The breast has a rather small pale peachy patch (much paler than any Eastern Stonechat), the belly is off-white. *Rump tawny-brown streaked dark*, with some *narrow white* showing in worn plumage on the *streaked uppertail-coverts*. ♀ and young are more plain brown and off-white, with no or (in ad ♀) only a very small pale patch on the wing. Chin and *throat white*. Supercilium more dusky buff-white and less prominent than in ♂. Wing-feathers narrowly edged pale brown, creating a somewhat *paler wing* than in the ♂, and in European Stonechat.

♂, probably ad, Fuerteventura, Canary Is, Nov: deep ochraceous underparts, rather dark head-sides, and long pale supercilium are good indicators of ad ♂. These features, plus whitish chin and throat, make the species unmistakable. (G. Peña-Tejera)

1stY ♂, Fuerteventura, Canary Is, Feb: browner head markings, weaker and dirty buff-white supercilium, and poorly developed peach-coloured breast patch all indicate young ♂; remiges, primary-coverts and outermost greater coverts are juv (already very worn and heavily bleached), forming moult contrast with those that have been replaced. (G. Peña-Tejera)

some ad ♀/1stY ♂ are difficult. Sexes separable at least from 1stS, sometimes from 1stW (latter requires close observation or handling). – Moults. Complete post-nuptial moult of ad in spring and summer (mainly late Mar–Jul) due to early breeding. Partial post-juv moult in spring shortly after fledging (Mar–Jul) does not involve flight-feathers, tertials, primary-coverts or outer greater coverts (very rarely all but outermost). Apparently limited pre-nuptial moult; some winter birds have two generations of greater coverts. – LATE WINTER **Ad ♂** Head largely uniformly blackish-brown, with narrow (sometimes broken, rarely lacking) white supercilium. Chin and throat pure white. A white patch on side of neck forming semicollar, reaching far back. Upperparts brown streaked darker. Rump/uppertail-coverts paler brown streaked dark, tail-coverts with some narrow and irregular white in most birds. Underparts off-white tinged buffish-cream, breast with ochraceous or pale orange patch. Wings dark with white patch (usually narrow and small on folded wing, more obvious in flight) formed by innermost three greater coverts. **1stS ♂** As ad ♂ but head not as dark, having some brown streaking on crown and ear-coverts, or whole crown mainly brown (feathers with darker centres giving slightly spotted look). Ochre-coloured patch on breast paler or almost absent. White wing patch on average smaller. Flight-feathers more worn, browner and paler, with pale 'panel' on folded secondaries often more obvious than in ad. Often a visible contrast between some blacker moulted inner greater coverts and browner outer. **♀** Head brown, streaked. Ill-defined supercilium buffish-white. Chin and throat off-white, sometimes tinged cream. No white semicollar, but sometimes this area is pale brownish-white. Rest of underparts off-white with rather even buff tinge (those with slightly contrasting ochre-tinged breast generally ad). No or only a small white patch on wing (those with white patch are usually ad), often concealed when folded. If seen very close, sometimes possible to see contrast between inner moulted greater coverts and unmoulted outer in 1y, whereas ad has all greater coverts of same age. **Juv** Resembles ♀ but has fine dark vermiculations on breast and fine cream-white streaks over much of upperparts (longest on sides of crown and nape). This plumage replaced by 1stW already in spring or early summer. – LATE SUMMER **Ad ♂** As in breeding plumage, but black and white portions partly concealed by brownish-grey tips of freshly moulted plumage. **♀/1stW** As ♀ in spring, but plumage fresh, overall appearance slightly paler. Plumages very similar, and separation generally requires close observation or handling. 1stW ♂ usually has large white wing patch, ♀ not.

BIOMETRICS (*dacotiae*) Sexes of similar size. **L** 11–12.5 cm; **W** 59–65 mm (n 34, m 61.8); **T** 41–51 mm (n 34, m 45.5); **T/W** m 73.7; **B** 14.1–16.7 mm (n 34, m 15.5); **Ts** 20.5–23.2 mm (n 32, m 22.1). Wing formula: **p1** > pc 7–11 mm, < p2 20–24 mm; **p2** < wt 3–7 mm, =7/8 or 8 (65%), =6/7 or 7 (30%) or =8/9 (5%); **pp3–6** about equal and longest (either of p3 and p6, or both, sometimes 0.5–1 mm <); **p7** < wt 1.5–4 mm; **p8** < wt 4.5–6.5 mm; **p10** < wt 9–12 mm; **s1** < wt 11–13 mm. Emarg. pp3–6.

GEOGRAPHICAL VARIATION & RANGE Slight, and presumably now historical: the subtle difference there was is apparently now gone, since one of only two subspecies appears to have become extinct. The extinct race is known only from 12 specimens, all collected in 1913; it has never been relocated since.

S. d. dacotiae (Meade-Waldo, 1889) (Fuerteventura, E Canary Is; resident). Described above. Slightly smaller with more contrasting ♂ head and underparts pattern in spring plumage compared to ssp. *murielae*.

? *S. d. murielae* Bannerman, 1913 (Allegranza, Montaña Clara, NW of Lanzarote; now apparently extinct). Said to differ from *dacotiae* by more concolorous and brown (less blackish and contrasting) crown in fresh ad ♂ summer plumage. However, only one ad ♂ *murielae* from June available for comparison with very few June ♂♂ *dacotiae* in collections, so a chance difference cannot be excluded. Still, existing *murielae* sample of 12 (NHM, AMNH) differ in being subtly larger and proportionately longer-tailed, and ♂♂ seem to have tendency to show a diffuse white patch on either side of forehead (in *dacotiae*, narrow and distinct white supercilium), and hence *murielae* accepted until proven invalid. (Variation in freshly moulted ♂ plumage of *dacotiae*, in particular as to head pattern, urgently needs to be established.) Also, Illera (2008) found a genetic difference of 0.3% between samples of *murielae* and *dacotiae*, but the length of the sequence analysed was very short, and the result is deemed inconclusive. **L** 12–12.5 cm; **W** 62–65.5 mm (n 12, m 63.6); **T** 45–49 mm (n 12, m 47.3); **T/W** m 74.3; **B** 15.4–16.2 mm (n 12, m 15.8); **Ts** 21.8–23.5 mm (n 12, m 22.8).

TAXONOMIC NOTE Illera et al. (2008) showed that Fuerteventura Stonechat is sister to European Stonechat (*rubicola*/*hibernans*), these two being more closely related than *rubicola* is to *maurus*. The Canary Is have probably been colonised from NW Africa or SW Europe by *rubicola*/*hibernans* and developed a taxon with distinct morphology due to isolation.

REFERENCES Bannerman, D. A. (1913) *BBOC*, 33: 37–38. – Bannerman, D. A. (1914) *Ibis*, 56: 75–77. – Illera, J. C. & Atienza, J. C. (2002) *Ardeola*, 49: 273–281. – Illera, J. C. et al. (2008) *Mol. Phyl. & Evol.*, 48: 1145–1154. – Seoane, J. et al. (2010) *Ardeola*, 57: 387–405.

♂, Fuerteventura, Canary Is, Aug: blackish head-sides largely concealed by whitish tips, suggesting young ♂, but extensive peach-coloured underparts and long whitish supercilium better fit ad. (J. Anderson)

♀, Fuerteventura, Canary Is, Aug: ♀ rather plain-looking, especially when fresh, and extensively peach-tinged below, though more whitish throat is noticeable; lacks clear white supercilium. Cannot be aged. (G. Peña-Tejera)

1stY ♀, Fuerteventura, Canary Is, Dec: ♀ generally brownish above and off-white below, with streaky crown, no or only limited whitish inner wing patch, and diffuse pale supercilium; lacks white semi-collar of ♂, while underparts are mainly off-white in worn state. Aged by contrast between replaced inner greater coverts and juv outer ones, while primary-coverts are also juv. (A. Juvonen)

1stW ♀♀, Fuerteventura, Canary Is (left: May; right: Mar): during or at end of post-juv moult. Unlike ad ♀ at this season, young are fresh (♀ prior to completing post-nuptial moult still has some very worn, bleached feathers). Clear rufous-pink breast with some dark spotting; both birds have mainly juv wing-feathers, including primary-coverts, and bill (at least of right hand bird) has some paler elements, also characteristic of young. (Left: H. Shirihai; right: B. Baston)

(EUROPEAN) STONECHAT
Saxicola rubicola (L., 1766)

Alternative name: Common Stonechat

Fr. – Tarier pâtre; Ger. – Schwarzkehlchen
Sp. – Tarabilla común; Swe. – Svarthakad buskskvätta

Although nowhere abundant, still a widespread and frequently seen small chat in open country with scattered scrub, often found on heaths, moors, seashores, wasteland in agricultural areas, and in vineyards. In Britain often associated with gorse and heather. Commonly perches on a fence or atop a bush, scanning like a small shrike. European Stonechat is a member of a large group of closely related stonechats, which also inhabits Africa and Asia. Some European birds winter in more southerly and maritime climates, and some reach the deserts of the N Middle East and N African coast.

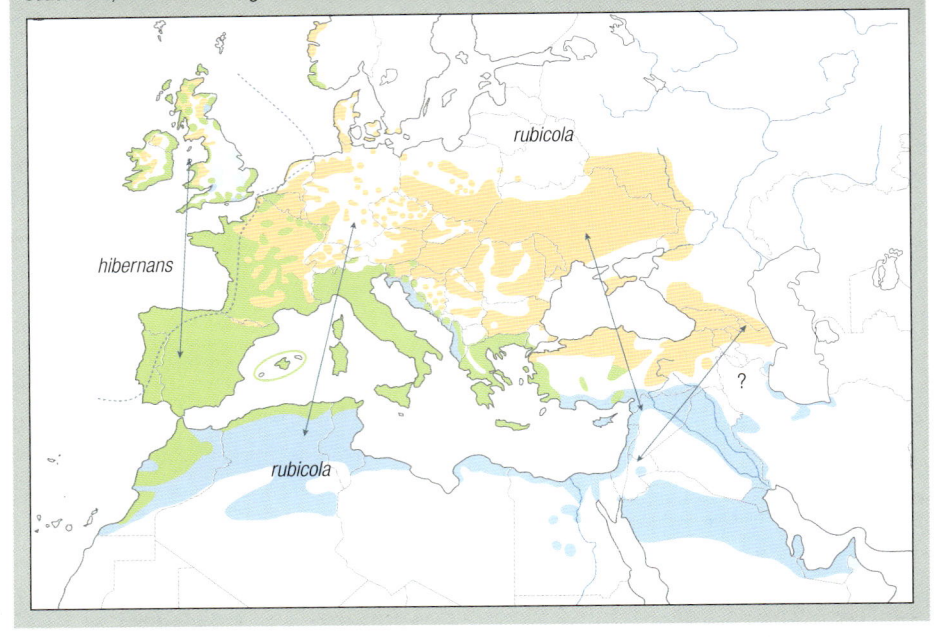

S. r. rubicola, ad ♂, Denmark, May: orange breast often strongly reduced, and white areas cleaner as spring progresses. Note bold dark streaks at both ends of white rump patch eliminating similar Eastern Stonechat. Aged by having only ad wing-feathers; this species' remiges tend to wear quickly, creating misleading impression of moult contrast. (H. F. Nielsen)

IDENTIFICATION Subtly smaller than Whinchat but of similar shape, differing in having slightly *shorter, more rounded wings* and can at times acquire a *more compact, neckless and dumpy impression*. Breeding ♂ has *all-black head* and *rufous-orange chest with white half-collar*, black-streaked dark brownish upperparts and a *narrow white patch on the innerwing-coverts* (in fresh plumage, patterns partially concealed and less striking). *Rump/uppertail-coverts* variable, usually fairly dark being *a mixture of black, brown and white streaks*, but may show narrow unstreaked white band on rump (buff when fresh). Pale-rumped ♂♂ still have black-streaked uppertail-coverts (although this sometimes requires close view to be seen). *Tail blackish-brown* (with narrow buff edges to outer rectrices). Axillaries and underwing-coverts medium dark grey with fairly broad pale fringes. ♀ is more nondescript: upperparts duller brown (still streaked), *head and throat mostly greyish-brown* (mottled paler when fresh), with less white in wings or in rump/uppertail-coverts, *no white half-collar*, and is more *uniform dull brownish orange-red below*. Even in fresh plumage *throat appears darker than breast* (or equally dark but greyer). Often perches on visible lookout. May hop on ground, but mostly feeds after sallies from perch, where very upright, despite plump body. Flies on rapid whirring wingbeats but is agile and capable of brief hovering. Always flicks wings and tail on landing, and often also when agitated or nervous. Perky, somewhat restless, but often inquisitive and, especially in breeding season, territorial and pugnacious, though less obtrusive in winter (when often still in pairs) and on passage. Song often given in a 'dancing' song-flight with vertical rises of >5 m, descending with tail and wings spread and legs conspicuously trailing, even with brief hovering.

VOCALISATIONS Song is a short twitter or scratchy warble, chiefly a series of high-pitched, jingling notes, rather monotonously repeated (lacking variation and fluty notes of Whinchat), recalling more Dunnock and especially Horned Lark, e.g. *zezri vrie-zrih sre-zi-zreh*. – Most characteristic call is a sharp, straight whistle *vist* and a single or doubled throaty click, *vist trak-trak, vist...*, the *trak-trak* sounding a little like two stones being knocked together, but resembles most alarm of Black Redstart. The *vist* call, even that rather like Black Redstart, can be very subtly upturned, but this is generally only perceptible when studying sonograms.

SIMILAR SPECIES Separating European Stonechat from *Eastern Stonechat* requires much care and experience, being problematic all year but especially in autumn. Beware that apart from colour of underwing-coverts in ♂♂, every character can be shared between European Stonechat (*rubicola*) and Eastern Stonechat, including characters usually associated with Eastern: large white neck patch, concentrated rufous breast patch, and large unmarked white rump patch (although at least some black streaks are present on uppertail-coverts, outside the unstreaked rump, which usually separates *rubicola* from Eastern). To this comes a tendency to have a few indistinct streaks also on lower flanks, never seen in Eastern (Hellström & Waern 2011; M. Hellström *in litt.*). Furthermore, classic *rubicola* is streaked and partly brown above even in worn summer plumage, generally not becoming blackish to the extent of *variegatus, hemprichii* and *maurus*. – Summer ♂ Eastern Stonechat of ssp. *variegatus* (which can occur together with European Stonechat on migration and on same mountains in summer, though usually separated by altitude) differs by (i) being a slightly larger bird and (ii) having, especially in worn spring–summer plumage, deeper reddish-chestnut breast patch with better contrast to whitish rest of underparts, (iii) on average larger white neck-side patch, and (iv) often more uniformly black upperparts with no or only

S. r. rubicola / hibernans, ♂, presumed ad, Spain, Mar: in S of range ad ♂ can acquire immaculate plumage by early spring, but strong wear in wing makes this bird impossible to age with certainty, still primary-coverts are ad-like. Locality (Valladolid) just within breeding range of *rubicola*, but extensive rufous underparts approach *hibernans*. It could be a migrant, but variation between *rubicola* and *hibernans* is such that many birds cannot be assigned to race. (A. Torés Sánchez)

limited brown elements. The white on base of outer tail-feathers in *variegatus* (often in the shape of a pointed wedge on inner webs of some outer tail-feathers) does not usually show on perched birds, and is often difficult to see in flight; still, conclusive if seen, since European birds never have white at base of tail-feathers. – Eastern Stonechats from Kazakhstan, Siberia and adjacent areas (*maurus, stejnegeri*) are rather frequent autumn stragglers to W Europe, have same size and broadly same plumage colours as European Stonechat, but separation in autumn can often rely on three characters, their paler throat, a hint of a pale supercilium and the pale and unstreaked rather large rump patch. Note however that a few *rubicola* European Stonechats also have a partly unstreaked pale rump patch (though there are invariably some dark streaks on longest uppertail-coverts), and some *rubicola* have a rather pale throat due to broader than usual pale tips in fresh plumage. Both sexes differ from European Stonechat by being on average paler and more ochre-tinged above in fresh autumn plumage, being not so dark brown as European breeders. Eastern ♂♂ have blackish underwing, not medium grey with paler fringes, and this can sometimes be seen when a bird alights or takes off. Finally, the orange breast patch of ♂ Eastern is usually more concentrated, not extending so far onto upper belly and flanks as in the average European Stonechat, and the white half-collar is usually more extensive. – Eastern Stonechats breeding north of the Caucasus west to E Ukraine and north to lower Volga region (ssp. *hemprichii*) look very much like those from Siberia but are slightly larger and readily separated by their extensive white in the tail in ♂♂, several outer tail-feathers having large white basal portions creating a Wheatear-like pattern, visible on flying or preening birds. – For the relatively inexperienced observer, Whinchat, especially ♀/young, might prove a stumbling block (♂♂ with their obvious pale supercilium and small white primary-coverts patch, among many other differences, should not be a significant confusion risk). However, even non-adult ♂ plumage of Whinchat has a buff supercilium (always more marked than in any European Stonechat), and all ages/sexes have white tail-sides at the base, visible in flight. Furthermore, young have a slightly spotted breast. – On the Canaries, might be confused with Fuerteventura Stonechat, but latter has clean whitish chin and throat in all plumages, and ♂♂ an obvious white supercilium, narrower white half-collar and more restricted, duller red on underparts, while ♀♀ have a more obvious collar and supercilium effect, paler (whitish-buff) breast, and longer and thinner bill. Also, overall shape is slimmer, less rotund, with a flatter crown.

S. r. rubicola, 1stS ♂, Italy, May: unlike ad, black of head and upperparts often less uniform, with pale tips and fringes, even in summer. Not all young ♂♂ are as easy to age as this bird, which has clear moult limits between noticeably worn juv primary-coverts and outer greater coverts and remiges, and fresh and darker inner greater and smaller coverts. (D. Occhiato)

S. r. rubicola / hibernans, 1stS ♀, Netherlands, Apr: moderately worn ♀ in spring, with browner and streakier upperparts, grey-mottled throat but no black on head, no white half-collar nor bright rufous-orange chest of ♂. It seems perfect match for *rubicola* (e.g. limited rufous below), although impossible to be sure. Diffuse streaking on rear flanks more common in European than Eastern Stonechat. Aged by worn and bleached juv primary-coverts, contrasting with fresher greater coverts. (C. van Rijswijk)

AGEING & SEXING Ageing requires close inspection of moult pattern and feather wear in wing. Sexes differ, although some 1stW require close looks and can still be difficult. Seasonal variation primarily due to gradual wear of feather-tips, most pronounced in ♂. – Moults. Complete post-nuptial and partial post-juv moults generally follow close after breeding. Post-juv renewal usually includes head, body, lesser, median and variable number of inner greater coverts and possibly a few tertials. Pre-nuptial moult apparently absent. – **SPRING Ad** ♂ By Apr acquires all-black head (any remaining pale patches insignificant), and largely uniform black upperparts and wing-coverts, mantle/back and scapulars with some brown admixed. **Ad** ♀ Differs from ♂ by browner and streakier upperparts, lacking any of black head, white half-collar and bright rufous-orange chest. Due to loss of pale fringes dusky throat patch becomes much more prominent. Both sexes safely aged by having only ad and relatively less worn wing-feathers. **1stS** As ad, but primaries, primary-coverts and alula (and perhaps some outer greater coverts and tertials) are juv, visibly more worn and browner, creating clear moult limits. – **AUTUMN Ad** ♂ Compared to breeding plumage, duller due to brownish or buff fresh feather fringes, but retains general pattern with reduced white neck-sides and rufous-orange on chest. **Ad** ♀ Slightly warmer colours than in spring, and throat mottled paler. Both sexes aged by evenly very fresh wing- and tail-feathers. **1stW** ♂ Like fresh ad ♂ or intermediate between that and ♀; flight-feathers, primary-coverts, tail, often some tertials and outer greater coverts juv, relatively weaker textured and rather more worn and browner than ad (also, if inner greater coverts still largely juv, these have only 1–2 all-white feathers, versus 3–4 of ad ♂). Also note greater contrast between browner secondaries and blackish tertials in birds that have moulted all/most tertials. Similarly, those which renewed all greater coverts will still show rather strong contrast between blackish-centred greater coverts and paler grey-brown primary-coverts. Until late winter has slightly paler greyish feather tips above, with broader greyish-brown tips to crown-feathers and is less uniform rufous-chestnut below. **1stW** ♀ Best aged by moult limits as in 1stW ♂ but contrast less sharp. Both sexes also aged by more worn and pointed outer tail-feathers (which have more pointed ends to dark centres and frayed pale edges), while at least until Sep inside of upper mandible more yellowish (blackish-grey in ad). **Juv** Resembles dull ♀ but lacks obvious collar, being spotted and streaked buff on head and back, washed rufous on rump, and finely spotted and barred dark brown on breast and flanks. ♂ shows hint of white wing-panel and slightly darker underparts.

S. r. rubicola, 1stW ♂, Italy, Jan: young winter ♂ *rubicola* tends to attain overall more 'complete' plumage from mid-winter, with black feather bases less concealed. Aged by juv alula, primary-coverts and remiges. (D Occhiato)

S. r. rubicola, ♀, Italy, Jun: with feather wear, some ♀♀ (probably older birds) show exposed dark feather bases on throat, and whitish on neck-sides. Due to heavy feather wear and bleaching, and angle of view, this bird is impossible to age. (D. Occhiato)

BIOMETRICS (*rubicola*) **L** 12–13 cm; **W** ♂ 63–69 mm (*n* 26, *m* 66.2), ♀ 63–66.5 mm (*n* 17, *m* 64.8); **T** ♂ 43–49 mm (*n* 26, *m* 46.0), ♀ 44–48.5 mm (*n* 17, *m* 45.7); **T/W** *m* 69.9; **B** 13.1–15.5 mm (*n* 39, *m* 14.4); **BD** 3.3–4.3 mm (*n* 37, *m* 3.7); **BW** 4.0–5.0 mm (*n* 41, *m* 4.5); **Ts** 20.7–22.9 mm (*n* 42, *m* 22.0). **Wing formula: p1** > pc 4.5–9.5 mm, < p2 20–26 mm; **p2** < wt 5–7.5 mm, =6/7 or 7 (56%) or =7/8 or 8 (44%); **pp3–5** about equal and longest (rarely either of p3 and p5, or both, 0.5–1 mm <); **p6** < wt 1–4 mm; **p7** < wt 4.5–7.5 mm; **p8** < wt 7.5–10.5 mm; **p10** < wt 11–14.5 mm; **s1** < wt 11.5–16 mm. Emarg. pp3–6.

GEOGRAPHICAL VARIATION & RANGE Moderate or small variation, with a cline of darker and more saturated plumage towards north-west, near the Atlantic. Two subspecies recognised, but intermediates occur over a diffuse transition area.

○ **S. t. hibernans** (Hartert, 1910) (Ireland, Britain, coastal Portugal, NW Spain, coastal W France; probably this race in W Netherlands; some move to SW Europe or NW Africa, others largely resident). Subtly darker and warmer on average than *rubicola* in fresh plumage, and ♂ has fringes to head and upperparts darker rufous-brown, and underparts slightly darker due to less pale fringes. In worn plumage orange-chestnut of breast and upper flanks deeper and less clearly demarcated from rest of underparts (which are paler/whiter in *rubicola*). White on rump/uppertail-coverts lacking or very limited in fresh plumage (usually evident in many *rubicola*), and less extensive than in worn *rubicola*. Neck patches generally narrower in breeding plumage and obscured by more buffish tips when fresh (broader and whiter, respectively, in *rubicola*). In comparison to *rubicola*, ♀ has deeper black feather-centres and darker rufous-brown fringes to upperparts. Underparts also darker, but much overlap. Grades into *rubicola* in W Europe, and in general birds in S England and along coast of SW Europe are less dark rufous than breeders in Scotland and Ireland. On average a fraction larger. **W** ♂ 65.5–70 mm (*n* 18, *m* 67.5), ♀ 63.5–70 mm (*n* 15, *m* 66.6); **T** ♂ 44.5–50.5 mm (*n* 18, *m* 46.9), ♀ 43–48.5 mm (*n* 15, *m* 46.0); **T/W** *m* 69.3; **B** 13.3–15.8 mm (*n* 25, *m* 14.5); **Ts** 21.6–23.9 mm (*n* 25, *m* 22.7). (Syn. *theresae*.)

○ **S. t. rubicola** (L., 1766) (Europe north to S Norway, Denmark, S Sweden, S Poland, except British Isles and coastal Portugal, NW Spain, W France; breeds east to Ukraine, Kalmykia, Turkey, much of Caucasus to at least 2500 m, at least locally in Transcaucasia, possibly rarely in NW Iran; winters N Africa, Arabia and Middle East, regularly as far east as SW Iran). Treated above. Generally averages slightly paler overall than *hibernans*. In SW & W Europe the variation between *rubicola* and *hibernans* is unstable and inconsistent, and odd *rubicola* with extensive rufous below, closely approaching *hibernans*, can be found far inland and away from the Atlantic coast. Biometrics above. – Note that some ♂♂, especially in Iberia and NW Africa, but rarely elsewhere in S Europe, can approach closely appearance of *maurus* with quite large white rump patch (dark streaks restricted to edges of tail-coverts), large white half-collar and concentrated rufous-orange chest patch. This is presumed to be part of normal variation rather than gene flow from east (but see Illera *et al.* 2008 for apparent proof that gene flow from the east does rarely play a role). (Syn. *amaliae*; *archimedes*; *desfontainesi*; *gabrielae*; *graecorum*; *insularis*.)

TAXONOMIC NOTES The Stonechat complex is comparatively extensively studied (see References) but there are still unanswered questions regarding variation and relationships between the various groups in the Palearctic, mainly due to insufficient studies on the breeding grounds in several key areas and lack of genetic data for some taxa. Geographical variation is complex with as many as *c.* 25 races described, mostly from outside the covered region, and more species than the four recognised here (European Stonechat *S. rubicola*, Fuerteventura Stonechat *S. dacotiae*, Eastern Stonechat *S. maurus*, African Stonechat *S. torquatus*) are possibly involved, with remaining uncertainties primarily in Africa but to some extent also in Asia. The adopted taxonomy is based on examination of fairly long series of specimens of *rubicola*, *hibernans*, *hemprichii*, *variegatus*, *maurus*, *stejnegeri* and *felix*. (See Svensson *et al.* 2012 for a full explanation of the partly altered nomenclature within the complex.) Added to this is the study of 3000+ photographs of birds in the field from across the range. All taxa (except *felix*) were also seen in the field or trapped and examined in the hand. – It eventually became clear that two morphologically diagnosable taxa coexist in NE Turkey, S Caucasus and Transcaucasia (possibly also in extreme W Iran), *rubicola* and *variegatus*, with no intermediates evident, separated also by ecological preferences, and this led us to concur with the taxonomy already proposed by Wittmann *et al.* (1995), Wink *et al.* (in Urquhart 2002a), Illera *et al.* (2008) and Zink *et al.* (2009, 2010) based on genetic analyses. In mtDNA, *maurus* differed from *rubicola/hibernans* by 2.8%. The genetic evidence also showed that *rubicola/hibernans* was sister to the Fuerteventura Stonechat, not to *maurus*, making it difficult to maintain *rubicola* and *maurus* as one species, and Fuerteventura Stonechat a separate species, if paraphyly is to be avoided. Finally, due to an apparently recent eastward range expansion by *rubicola* this now overlaps with *hemprichii*.

S. r. hibernans, ad ♂, England, Apr: darker than ♂ *rubicola*, with more extensive orange-chestnut underparts, smaller white rump patch (not visible here), and neck patches generally narrower and less clean. Aged by entire wing being evenly feathered following post-nuptial moult with broader, blacker and neatly white-fringed primary-coverts. (C. Bovis)

S. r. hibernans, ♀, presumed 1stS, England, Jun: compared to *rubicola*, ♀ in spring and summer on average deeper and darker rufous-buff below, and darker above, though difference often difficult to perceive, and some overlap between the races inevitable. Ageing difficult at this angle of view, but brownish primary-coverts and limited white could indicate 1stS. (D. Mitchell)

S. r. rubicola / hibernans, ♀, Netherlands, Oct: some fresh ♀♀ in autumn have unstreaked and paler rufous rump/uppertail-coverts, and sometimes even more pronounced pale supercilium, recalling Eastern Stonechat. When flanks and/or scapulars conceal coverts, impossible to age by moult, but rather narrow and pointed tail-feathers suggest 1stW. (N. D. van Swelm)

Juv ♂, Spain, Jul: in all races has brown, grey and buff-spotted and streaked, fluffy plumage; some rufous on rump is visible and white wing panel indicates ♂. (A. M. Domínguez)

REFERENCES Barthel, P. (1992) *Limicola*, 6: 217–241. – Corso, A. (2001) *BB*, 94: 315–318. – Flinks, H. (1994) *Limicola*, 8: 28–37. – Grant, K. & Small, B. (2004) *BW*, 17: 154–156. – Hellström, M. (2005) *Vår Fågelv.*, 64 (2): 36–43. – Hellström, M. (2006) *Roadrunner*, 14(3): 42–43. – Hellström, M. & Waern, M. (2011) *BB*, 104: 236–254. – Illera, J. C. *et al.* (2008) *Mol., Phyl. & Evol.*, 48: 1145–1154. – Kirwan, G. M. & Bates, J. M. (2008) *Sandgrouse*, 30: 114–116. – Riddiford, N. (1981) *Ringers' Bull.*, 5 : 120. – Siddle, J. P. (2006) *BB*, 99: 372–374. – Stoddart, A. (1992) *BW*, 5: 348–356. – Svensson, L. *et al.* (2012) *BBOC*, 132: 37–46. – Ullman, M. (1986) *Vår Fågelv.*, 45: 227–229. – Urquhart, E. D. (2002) *Birdwatch*, 120 (Jun): 21–25. – Vinicombe, K. (2005) *Birdwatch*, 160 (Oct): 28–31. – Walker, D. (2001) *BW*, 14: 156–158. – Wink, M. *et al.* (2002) *BB*, 95: 349–355. – Wittmann, U. *et al.* (1995) *J. Zool. Syst. Evol. Res.*, 33: 116–122. – Zink, R. M. *et al.* (2009, 2010) *Mol. Phyl. & Evol.*, 52: 769–773, 57: 481–482.

EASTERN STONECHAT
Saxicola maurus (Pallas, 1773)

Alternative names: Asian Stonechat, Siberian Stonechat, White-rumped Stonechat

Fr. – Tarier oriental; Ger. – Pallasschwarzkehlchen
Sp. – Tarabilla común oriental
Swe. – Vitgumpad buskskvätta

This close relative of the European Stonechat mainly breeds on Asian taiga bogs and wet marshes with lush grass or open fields with tussocks and low bushes. It is a fairly common breeder in Russia, Siberia, Kazakhstan and W Mongolia but also in E Turkey, Caucasus and the Caspian region. Northern and eastern populations winter in N India and Burma, while those from E Turkey and the Caspian region winter mainly in Arabia and the Horn of Africa. The mainly Kazakh/Siberian race *maurus* is a fairly regular straggler to W Europe in autumn, and *hemprichii* has been recorded several times in Scandinavia, the Levant, Middle East and Cyprus.

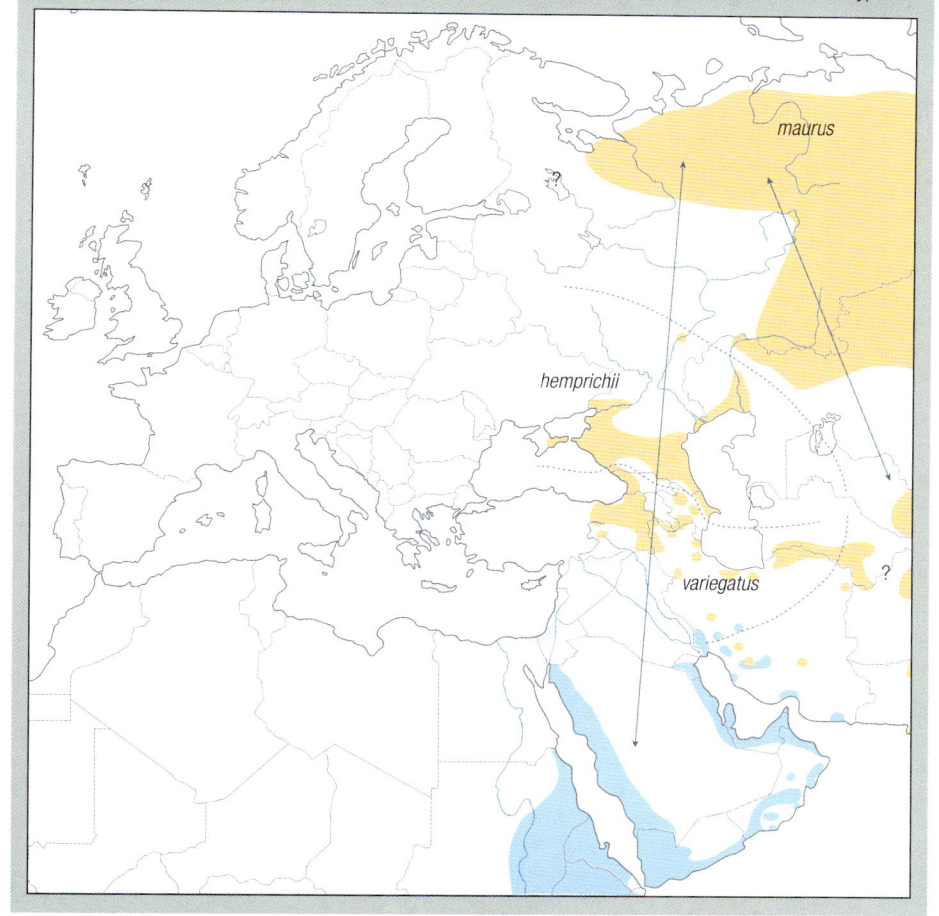

S. m. maurus, ad ♂, Kazakhstan, Jun: race by locality, though relatively pale rufous breast, grading into paler orange upper belly and flanks, with whiter rest of underparts, are typical of *maurus*; also large white neck-side patches. Entire wing-feathers ad; remiges tend to wear more strongly compared to coverts (common in *Saxicola*). (A. Audevard)

IDENTIFICATION Does not differ much from European Stonechat, is by and large similarly sized and shaped, with very similar plumage. Separation requires attention to fine detail. Eastern Stonechat as here understood combines rather varied populations which require separate consideration, although basics are shared, these being: (i) *unstreaked large pale rump patch* (rusty-buff when fresh, white in ♂ or pale buff in ♀ when bleached), and *no prominent dark streaking* even at the edges including *on uppertail-coverts* (though small dark spots or insignificant brown streaks visible only at close range can occur on tail-coverts in some birds); (ii) ♂♂ have more *concentrated rufous-orange breast and whiter belly* (but beware a few tricky birds, which overlap, and that fresh autumn plumage habitually has orange-buff tinge far down upper belly and flanks), (iii) *larger white half-collar patch* on average (a few tricky birds occur), and (iv) in ♂♂ *blackish underwing* due to blackish axillaries and underwing-coverts with only very narrow paler fringes, wearing all black in summer; (v) ♂♂ also have on average blacker upperparts (mantle/back/scapulars) in summer plumage compared to Stonechat, which retain more brown elements even when worn; (vi) ♀♀ have *paler throat* which has no or only very little dark obscured by broad pale tips, hence nearly always stay pale-throated into spring and summer (though more dark-throated ♀♀ occur more frequently in the Caspian region and E Asia), and they are on average (vii) *slightly paler above*, more ochre-brown, less dark earth-brown. Having applied these general characters, it remains to deal with the variation within Eastern Stonechat (ranges given under

Summer-plumaged ♂ Eastern Stonechat, *S. m. maurus* (left, Kazakhstan, Jun) and European Stonechat, *S. r. rubicola* (centre and right, Italy, Jun): separation, especially of summer ♂♂, not always straightforward. ♂ *maurus* with extensive orange below is not dissimilar to many *rubicola*. Examples of *rubicola* with more restricted and orange underparts occur in many S European/NW African populations; the rump can also be almost pure white, with just a few small dark spots on uppertail-coverts. (Left: Y. Belousov; centre: L. Sebastiani; right: D. Occhiato)

S. m. maurus, ♀, presumed ad, Kazakhstan, May: large pale buffish-white and unstreaked rump/uppertail-coverts patch, with open-faced impression due to extensive pale neck-side patches and pronounced supercilium; also orange-buff underparts, concentrated on breast, rather dark-centred feathers above, and substantial whitish inner wing-covert patch. Entire wing-feathers apparently ad. (Y. Belousov)

S. m. maurus, ad ♂, Jordan, Mar: from *hemprichii* by combination of being ad (note primary-coverts), lack of white at base of outer tail-feathers, from *variegatus* only with difficulty on smaller size and less deep brick-red breast patch; from *rubicola* by snowy-white rear underparts, and relatively large/pure white neck-sides and unstreaked rump. (H. Shirihai)

Presumed S. m. maurus, ♀, Israel, Mar: ♀♀ of *variegatus* and *hemprichii*, especially former, often have tail pattern like *maurus*, making identification very tricky. This individual has extreme number of ill-defined streaks on longest uppertail-coverts. Ageing of ♀♀ often difficult without handling, but apparent strong wear to remiges, primary-coverts and tail suggest this bird is young. (Y. Perlman)

Geographical variation & Range). (1) Ssp. *variegatus* (as to nomenclature see Taxonomic notes): the largest race, the size of a Whinchat with almost no overlap in measurements compared to ssp. *maurus* (which is among the smallest of the complex). Also, *variegatus* is the race with darkest colours and strongest contrasts: adult ♂ spring–summer has *very deep reddish-chestnut breast in strong contrast to pure white belly*. In fact, it resembles one of the strongly-coloured African taxa, or the Arabian ssp. *felix* of African Stonechat (which see). In summer, *upperparts become nearly uniformly black*. The clinching feature (though normally very hard to see in the field!) is the limited *pure white on the base of most tail-feathers* (usually none on central pair, and less on outermost). The white on tail-feathers, which only exceptionally reaches outside tips of tail-coverts, usually appears as pointed wedges (pointing towards tail-tip), mainly on inner webs. ♀♀ are much more difficult to identify except in the hand on size. They do not differ much from ♀♀ of *maurus* in plumage, sharing with these the lack of white at base of tail-feathers (exceptionally a hint), but are on average slightly darker on throat and more saturated rufous-buff on breast in fresh plumage. (2) Ssp. *hemprichii*: intermediate in size between *variegatus* and *maurus*. Immediately identified in ♂ plumage by *large white bases to tail-sides*, striking in flight or when tail is spread but can be invisible on perched birds with folded tail. White on tail-feathers ends distally more or less squarely or bluntly (rather than in a sharp point on each

Autumn/winter ♂ *maurus* (top left: ad, Kazakhstan, Sep; bottom left: 1stW, Germany, Oct) versus European Stonechat *rubicola* (top right: ad, Italy, Dec; bottom right: 1stW, Italy, Nov): as result of generally more extensive pale feather-tips on fresh *maurus*, ad ♂ overall closer to 1stW ♂ *rubicola*, while 1stW *maurus* can be more ♀-like. Unlike *rubicola*, fresh autumn ♂ *maurus* is paler and buffier above, with more pronounced supercilium, and large buff unstreaked rump patch. Some *maurus* difficult to identify unless dark axillaries and underwing-coverts seen, or amount of white on inner wing-coverts assessed. 1stW ♂ identified as *maurus* by much concealed black face (*variegatus* develops ♂ plumage quicker), rather wide shaft-streaks and brownish wash to longer uppertail-coverts. 1stW separated from ad by juv feathers in wing. (Top left: R. Pop; bottom left: A. Halley; right two: D. Occhiato)

feather as in *variegatus*), and is lacking only on central tail-feathers. Tail pattern thus recalls that of Wheatear or ♂ Red-backed Shrike, only often shafts and outer webs of feathers are dark, creating a striped pattern on the white panels. Nearly all ♀♀ have a little or a moderate amount of white (or cream-white) at the base of outer tail-feathers (thus clearly less than in ♂) but are for practical purposes inseparable from the other races in the field; confirmation of the presence of white or buff-white on tail base in ♀♀ require handling. (Exceptions found in collections with no white at all on tail were from migration or winter and might in some cases—but not all—have been misidentified *maurus* or *variegatus*.) (3) Ssp. *maurus*: small and slightly paler overall than preceding two races. ♂ has apparently invariably a very little white hidden at the tail base (requires handling to be seen), much less than in *variegatus*, whereas ♀ has no white in tail. – Behaviour similar to Stonechat, thus frequently perches on visible lookout and is easily spotted in open habitats. Flight is rapid with whirring wingbeats. Flicks wings and tail on landing, or when anxious.

VOCALISATIONS Song appears to be very similar to European Stonechat but is said to be often longer (V. Ananian pers. comm.; G. Sangster *in litt.*), though this perhaps requires further confirmation. The song is a rather short twitter of high-pitched, jingling notes on fairly even pitch, delivered rather monotonously. – Calls are also very similar to those of European Stonechat, a sharp, straight whistle *vist* and throaty click or cracking sound which often follows, *vist trak-trak*. Although the full variation within Eastern Stonechat is probably not yet fully known, it has been suggested (Constantine *et al.* 2006; G. Sangster *in litt.*) that a down-bent whistling sound, *viest* or *tseeu*, is characteristic for Eastern Stonechat, whereas a straight or upslurred *tsuist* is typical of European Stonechat.

SIMILAR SPECIES Separating Eastern Stonechat from *European Stonechat* is usually difficult; as many characters as possible should be combined for reliable identification. Several details to note are given both under European Stonechat (Similar species and above under Identification), and are not repeated here. Has on average larger whitish inner wing-coverts patch than in European Stonechat. Also has slightly longer primary projection and slightly longer wings, but shorter bill and tarsus. ♂♂ best told by combination of unstreaked pale rump patch, blackish underwing and large white

Autumn/winter ♀ *maurus* (left images, Kazakhstan: top, Oct; bottom, Sep) versus European Stonechat *rubicola* / *hibernans* (top right: Spain, Sep; bottom right: Netherlands, Oct): top two birds show 'easy' Eastern versus European Stonechat, with former readily identified by characteristic pale coloration, enhanced by broader pale-fringed wing-feathers. Open-faced impression, with broad pale supercilium, and large pale buffish and unstreaked rump/uppertail-coverts. Bottom birds demonstrate that the two forms can be very similar, especially with overall paler and buffier-coloured European (bottom right), which has paler rufous-buff, unstreaked rump/uppertail-coverts. Nevertheless, also has relatively darker, streakier head and underparts patterns, lacking better-developed pale supercilium and throat patch of Eastern. Ageing without handling uncertain, but top right European is apparently ad by some white on inner greater coverts and evenly-feathered wing; others all possibly 1stW, as most wing-feathers seem to be juv. (Top left: Y. Belousov; top right: M. Estébanez Ruiz; bottom left: R. Pop; bottom right: P. van Rij)

Presumed *S. m. maurus*, ♀, Netherlands, Oct: typical autumn vagrant in W Europe, with racial identification especially tricky in ♀ plumages. From European Stonechat by overall paler and more sand-coloured plumage, pale (but here diffuse) supercilium, broad unstreaked orange-buff rump, and long primary projection. Evenly-feathered wing and all-black bill support ageing as ad. (B. van den Boogaard)

S. m. variegatus, ad ♂, E Turkey, Jun: race by locality, also typical impression of large bird, extensive clean white areas and deep brick-red and concentrated breast patch (often darker red-brown than *maurus* or *hemprichii*). Tail pattern shows no visible white bases, but these are normally hidden in this race. Aged by all-black wing-feathers. (G. Bakker)

***S. m. variegatus**, ad ♂, Armenia, Jun*: tail pattern varies, between birds with no or only marginal visible white bases (left, not dissimilar from *rubicola* and *maurus*), to birds with some exposed white beyond uppertail-coverts when tail spread (right). Both have only ad-type wing- and tail-feathers, with no vestiges of brownish immature feathers in rest of plumage. Note dark brick-red concentrated breast patches and very black upperparts, typical of *variegatus*. (V. Ananian)

half-collar, ♀♀ by paler throat, hint of pale supercilium and pale unstreaked rump patch (though beware tricky European Stonechats showing one or more of these characters, too). – Separation from *African Stonechat* can be very tricky, as there is strong similarity between Eastern Stonechat *variegatus* and Arabian ssp. *felix* of African Stonechat (but other subspecies of African Stonechat in NE Africa also resemble *variegatus*), and these meet in winter. That ♂ African Stonechat lacks any white on base of tail-feathers is usually impossible to use in normal encounters. African Stonechat *felix* has about same body size as Eastern Stonechat *variegatus* but has more rounded wing with shorter primary projection. The black bib of *felix* extends a little further down upper breast, and the pale rump patch is on average smaller and less pure white, being more streaked (approaching European Stonechat in this respect). – The Caspian Stonechat *hemprichii* with its extensive white tail-base in ♂ plumage could theoretically be mixed up with S Asian extralimital White-tailed Stonechat *S. leucurus* (not treated), perhaps most likely in winter season, but note that White-tailed is slightly larger and proportionately longer-tailed, more greyish and white, less brown and rufous-buff, further that it has only dark outer webs of outer tail-feathers, while inner webs are largely white (only tips of inner webs faintly brown in some), whereas Eastern *hemprichi* has the dark of outer webs running around tip and continuing somewhat on inner webs (can form a dark hook on each outer tail-feather, at the most a hint of such a hook on r6 in White-tailed).

AGEING & SEXING (*maurus*) Ageing requires close inspection of moult pattern and feather wear in wing. Sexes differ, although some 1stW require close look and can still be difficult. Seasonal variation primarily due to gradual wear of feather-tips, but partly also to moult, most pronounced in ♂. – Moults. Complete post-nuptial and partial post-juv moults generally follow close after breeding. Post-juv renewal usually includes head, body, lesser, median and variable number of inner greater coverts and possibly a few tertials. Pre-nuptial moult partial, apparently confined only to, or at least more common in, 1stW ♂♂ and limited to head, neck and throat.
– **SPRING Ad** Evenly very fresh wing- and tail-feathers, latter with neat rounded tips. Note that greater coverts and primary-coverts stay quite blackish into summer, primaries by then can bleach dark brown suggesting moult contrast to the unwary. **Ad ♂** As in European Stonechat, black head developed in Apr; in May or Jun, upperparts and most wing-coverts become nearly uniform black. Much of orange-buff tinge on upper belly and flanks of autumn plumage worn off, now more whitish, creating better contrast between rufous-orange breast and whitish rest of underparts. Large rusty-buff rump patch bleaches progressively to pure white, unstreaked. **Ad ♀** Differs from ♂ by browner and streakier upperparts, lacking any of black head, white half-collar and bright rufous-orange chest. Both sexes aged by having only ad and relatively less worn wing-feathers. Tends to show hint of pale supercilium. **1stS** As ad, but primaries, primary-coverts and alula (and rarely some outer greater coverts and tertials) are juv, visibly more worn and browner, creating clear moult limits. – **AUTUMN Ad** Tail-feathers have neater edges, more rounded tips and are slightly wider on average. **Ad ♂** Compared to breeding plumage, duller due to brownish or buff fresh feather fringes, but retains general pattern albeit now with reduced white neck-sides visible (partly covered by orange-buff tinge) and rufous-orange on chest invades much of rest of underparts, only paler rufous below breast. The most ♀-like still recognised as ♂♂ by blackish lores, partly black ear-coverts and throat, blackish underwing and much white on secondary-coverts (usually 3–4 white greater coverts, rarely 2 or 5). No contrast between black greater coverts and black base of primaries. **Ad ♀** Slightly warmer than in spring. **1stW** Tail-feathers slightly narrower and more pointed, often with frayed tips near shafts, and with more pointed ends to dark centres. **1stW ♂** Variable, sometimes quite like ♀ or, more often, showing first traces of ♂ plumage with black on lores and ear-coverts, and sometimes on throat, but invariably identified as ♂ by blackish underwing-coverts. Flight-feathers, primary-coverts, tail, often some tertials and outer greater coverts retained juve-

***S. m. variegatus**, ♂, presumed ad, Armenia, May*: ♂♂ of this race in breeding plumage typically have small but dark brick-red patch concentrated on upper breast and quite blackish upperparts. The race is also larger than both European Stonechat and ssp. *maurus* of Eastern. The race *hemprichii* can easily be eliminated by the visibly all-dark tail. (N. Bowman)

***S. m. variegatus**, ad ♀, E Turkey, Jul*: race by locality and date, otherwise there are no useful plumage differences that can separate ♀ *variegatus* from ♀♀ *hemprichii* or *maurus* in the field. Post-nuptial moult still on-going, with outer tail-feathers growing, inferring age. (B. Gocmen)

ROCK THRUSHES AND CHATS

S. m. hemprichii, ad ♂, Russia, May: substantial white basal portion on exposed outer tail-feathers beyond uppertail-coverts diagnostic, permitting identification from similar *variegatus*. Aged by wholly ad blackish wing-feathers, including primary-coverts. (A. Varlamov)

S. m. hemprichii, ♀, Russia, May: race by locality. No plumage feature separates ♀ *hemprichii* from ♀ *variegatus* or *maurus*, but overall pale plumage with whiter underparts and unstreaked rump, and bulkier appearance, separate it from ♀ *rubicola*. Strong feather wear makes it impossible to age this bird. (A. Varlamov)

nile, rather more worn and browner than ad (though note that primary-coverts are often quite blackish; better to look for slight contrast between black moulted greater coverts and subtly browner bases to primaries). Outermost one or two greater coverts with white pattern often have dark area divided, a dark patch on each web with white 'channel' along shafts (this would be most unusual pattern in ad). Note greater contrast between browner secondaries and blackish tertials in birds that have moulted all/most tertials. Until late winter has slightly paler greyish feather-tips above, with broader greyish-brown tips to crown-feathers than ad, and is less uniform rufous-chestnut below. **1stW** ♀ Best aged by moult limits as in 1stW ♂ but contrast less sharp. **Juv** Resembles a dull ♀, being spotted and streaked buff on head and back, washed rufous on rump, and finely spotted and barred dark brown on breast and flanks. ♂ shows hint of white wing-panel and slightly darker buff underparts.

BIOMETRICS (*maurus*) **L** 11–12 cm; **W** ♂ 63–72 mm (n 22, m 68.3), ♀ 64–70 mm (n 13, m 66.7); **T** 44–51 mm (n 33, m 47.3); **T/W** m 69.8; **B** 12.7–14.7 mm (n 33, m 13.9); **BD** 3.2–4.1 mm (n 27, m 3.7); **BW** (prox. edge nostril) 4.0–4.8 mm (n 34, m 4.5); **Ts** 19.8–22.9 mm (n 34, m 21.2). **Wing formula: p1** > pc 5–11 mm, < p2 23–29.5 mm; **p2** < wt 4–7 mm, =6/7 or 7 (96%) or =7/8 (4%); **pp3–5** about equal and longest (rarely either of p3 and p5, or both, 0.5–1 mm <); **p6** < wt 1–5.5 mm; **p7** < wt 4–8 mm; **p8** < wt 8–11.5 mm; **p10** < wt 12–16 mm; **s1** < wt 12–17.5 mm. Emarg. pp3–6. **White** (or buff-white) **at base of rr5–6, inner web** in ♂ 7–15 mm (once 22; exceptionally missing).

GEOGRAPHICAL VARIATION & RANGE Rather marked and complex variation, as can be expected considering partially disjunct range. Three subspecies breed or occur fairly regularly within the treated region, a few more are extralimital, all in Asia, one of which is treated below as it is a very rare vagrant.

S. m. hemprichii Ehrenberg, 1832. 'Caspian Stonechat' (previously known as '*variegatus*'; N Caucasus and Kalmykia plains from lower Don and E Crimea east to lower Volga; winters in Horn of Africa, probably also Arabia). Intermediate in size and plumage between *variegatus* and *maurus* but differs readily in ♂ plumage by large white bases on tail-sides ('Wheatear pattern'). From above, the extensive white area is usually readily visible when the tail is spread. However, this character is rather variable, and some ♂♂ (perhaps mainly in south of range), and especially 1stY, have only very little (at times no) white visible beyond the coverts in field views. Thus, field separation from ♂ *variegatus* is not always straightforward. Note that on perched ♂♂ a certain amount of white is also often visible at base of r6 on the folded undertail, just beyond the tips of undertail-coverts (never so in *variegatus*), or white even shows extensively outside the coverts. Unlike in the other races, ♀♀ usually (though not invariably) have some white at the tail base, but compared to ♂♂, pale area is reduced (especially in 1stY) and is often more cream-white (less pure white), is most pronounced on inner web, and is probably always invisible in the field. To play safe it is best to limit reliable subspecific identification in the field to ♂♂ with sufficiently prominent white in tail. More difficult ♂♂ and probably all ♀♀ require handling to be confidently separated from *variegatus* and *maurus*. – Supporting characters for *hemprichii*: has largest pale rump patch (though much overlap), which appears to

S. m. hemprichii, ad ♂, Kuwait, Mar: partially spread tail revealing substantial white basally on exposed outer rectrices. In Mar, upperparts still partially pale-fringed, and white parts less pure. Aged by entirely ad wing-feathers, including primary-coverts. (R. Al-Hajji)

S. m. hemprichii, 1stS ♂, Israel, Mar: tail has Isabelline Wheatear-like band, with much white on exposed tail-feathers, identifying it as *hemprichii*. Dark areas heavily pale-fringed and white parts quite strongly saturated pinkish-rufous. Aged by many juv wing-feathers, including primary-coverts. (S. R. Waagner)

S. m. hemprichii, variation in winter ♂♂ (left: ad, Kuwait, Dec; right: 1stW, Georgia, Sep): ad ♂ *hemprichii* unmistakable by its narrow Black-eared Wheatear-like tail-band (here 70–80% of exposed outer rectrices white); young ♂ *hemprichii* has less white, and much broader Isabelline Wheatear-like tail-band (c. 60 % white) with dark outer webs reaching far up tail. Ad aged by having entire wing freshly moulted, while young bird shows moult limit in greater coverts, and juv primary-coverts and remiges, and tail pattern typical. (Left: A. Al-Sirhan; right: L. De Temmerman)

S. m. hemprichii, 1stY ♂, Kuwait, Feb: combination of relatively extensive white tail bases, reaching approximately level with tips of undertail-coverts, and aged/sexed as 1stY ♂, indicate *hemprichii*. (A. Al-Sirhan)

Presumed *S. m. hemprichii*, 1stW ♂, Israel, Oct: juv tail-feathers, with relatively extensive white on bases (almost to level of uppertail-coverts), strongly suggests *hemprichii* (*variegatus* of same age lacks or has much-reduced white). Aged by juv remiges and primary-coverts, and narrow and pointed rectrices. Note how black of face is much more visible than in many 1stW *maurus*. (L. Kislev)

Presumed *variegatus / hemprichii*, 1stW ♀, Israel, Oct: photographed during influx of *hemprichii*, this individual lacked any trace of pale tail base. By plumage inseparable from *variegatus* or *maurus*, but overall impression of large, strong bird with stronger bill suggest either *variegatus* or *hemprichii*. Aged by juv remiges, primary-coverts and tail (latter also narrow and pointed). (L. Kislev)

extend further on sides and sometimes to lower back, where admixed with grey (also whitens more rapidly and extensively in worn ♂). In breeding ♂ deep rufous-chestnut breast patch rather restricted, contrasting slightly more with rest of whitish underparts than in *maurus*; white neck-collar broader and deeper at rear, and sometimes almost meets on nape (though much overlap, especially with *variegatus*). Fresh non-breeding ad ♂ has strongly marked blackish head-sides and throat (as in *variegatus* but usually unlike *maurus*), and even 1stW ♂ usually has partially dark cheeks and throat; also, despite buff fringes, neck-side patches more obvious than in *maurus*. Compared to latter, all fresh plumages of *hemprichii* are more extensively fringed pale sandy-buff above, and are whiter on rump and belly. Near-complete overlap in ♀ plumages of *hemprichii* with *variegatus* and *maurus*, with no single feature known to readily separate them in the field (only size and amount of white or pale on tail base helpful with birds in the hand), but while *hemprichii* and *maurus* similarly average medium pale brown above, *variegatus* is often subtly darker, and underparts in *variegatus* average subtly darker and more extensively orange-pink; still much overlap and difference subtle. **W** ♂ 68–76 mm (n 64, m 72.0), ♀ 66–72.5 mm (n 22, m 69.4); **T** 44–52 mm (n 87, m 48.9); **T/W** m 68.6; **B** 13.0–15.5 mm (n 86, m 13.9); **BD** 3.2–4.1 mm (n 86, m 3.7); **BW** (prox. edge nostril) 3.8–5.3 mm (n 83, m 4.6); **Ts** 19.5–22.2 mm (n 87, m 21.0). **Wing formula: p1** > pc 2.5–9 mm, < p2 26–32.5 mm; **p2** < wt 3.5–7 mm, =6/7 or 7 (93%) or =6 (7%); **p8** < wt 7–16 mm; **p10** < wt 13–19 mm; **s1** < wt 14–20 mm. Emarg. pp3–6. **White at base of rr5–6, inner web** in ♂ 21–45 mm, in ♀ 5–25 mm (rarely diffuse or missing). (Syn. *variegatus* in error.)

S. m. variegatus (S. G. Gmelin, 1774) 'ARMENIAN STONE-CHAT' (previously known as '*armenicus*'; NE Turkey, extreme NE Iraq, Transcaucasia, W & S Iran; apparently S slopes of Caucasus; winters Horn of Africa, possibly also Arabia). The largest race, being substantially larger than *maurus* but only somewhat larger than *hemprichii*, and has proportionately subtly shorter legs than both. ♂ averages darker with blacker upperparts and concentrated, darker reddish-chestnut (deep brick-red) chest patch in contrast to pure white belly (though some overlap in colours with *hemprichii* and extreme *maurus*). Pale rump patch somewhat larger than in *maurus* but similar or marginally less extensive than in *hemprichii*. ♂ *variegatus* has only moderate amount of white at base of tail-feathers, always being less than half of entire tail length and often no more than one-third, white tail base as a rule entirely hidden by tail-coverts and not visible in the field (except sometimes in flight or when preening). White bases can be ascertained on handled birds (when tail-coverts can be lifted), usually in the form of narrow white wedges on inner web of many outer tail-feathers (pointing to tail-tip) and mainly on inner webs, much less on outer webs. Those ad ♂♂ with most white approach those 1stY ♂ *hemprichii* with least white in tail, but no overlap when same age compared, so ageing important for any controversial bird. The white on base of inner web of r6 of 1stY ♂ *variegatus* compared to ad is more reduced in size and less sharply defined, and is lacking or nearly so on the outer web. ♀♀ lack white on tail base, or have just a hint of buff-white at the very root of the feathers, just like ♀ *maurus*. Thus, all ages and sexes of *variegatus* can overlap to some degree in tail pattern with *maurus* (but nearly always differ in size), while only known overlap with *hemprichii* is in ♀ plumages. Like *hemprichii*, 1stW ♂ *variegatus* attains more ♂-like plumage in autumn, especially on head and neck (unlike 1stW ♂ *maurus*, which is slower to attain ♂-like plumage). **L** 12.5–14 cm; **W** ♂ 70.5–80 mm (n 94, m 75.0), ♀ 70–76.5 mm (n 43, m 72.8); **T** 47–56 mm (n 137, m 51.2); **T/W** m 69.0; **B** 13.0–16.0 mm (n 120, m 14.5); **BD** 3.3–4.3 mm (n 118,

m 3.8); **BW** (prox. edge nostril) 4.1–5.3 mm (n 104, m 4.7); **Ts** 19.5–23 mm (n 118, m 21.3). **Wing formula: p1** > pc 3–9 mm, < p2 27–35.5 mm; **p6** < wt 1.5–4 mm; **p7** < wt 5.5–9 mm; **p8** < wt 9–12.5 mm; **p10** < wt 13–17.5 mm; **s1** < wt 14–20 mm. **White at base of rr5–6, inner web** in ♂ (once 8) 13–27 mm, in ♀ usually missing but very rarely to 15 mm. (Syn. *armenicus*; *excubitor*.)

S. m. maurus (Pallas, 1773) 'Siberian Stonechat' (Russia, W Siberia east to NW Mongolia, south to Altai, N Kazakhstan and mountains of N Central Asia, south-west to steppe N of Aral Sea, in the south to Afghanistan and NW Pakistan and possibly this race in NE Iran; few pairs also in E Finland; winters mainly S Iran, S Pakistan, India, Burma; scarce but regular migrant in Middle East and Arabia, but apparently only rare straggler to NE Africa; comparatively regular vagrant to W Europe, mainly in autumn). A small and fairly pale race. Tail-feathers dark but pattern differs from Stonechat by presence in ♂ of small white (or buff-white) patch at base, invariably hidden by coverts (see Biometrics). Separation from *hemprichii* in ♂ plumage by lack of extensive white basally on tail-sides; from *variegatus* by distinctly smaller size (only tiny overlap), by comparatively paler colours overall and less white on tail base in ♂ plumage (very few controversial birds as to this); still, separation away from breeding grounds often unreliable without handling. Also, compared to *hemprichii* and especially *variegatus*, the rufous breast patch of breeding ♂ *maurus* averages slightly paler and more orange-tinged (less dark reddish-chestnut), while orange colour is somewhat more extensive, less concentrated on rounded chest patch, and before plumage too worn this patch tends to grade into pale pink-buff lower chest and upper flanks. (Syn. *tschecantschia*.)

Extralimital: ○ **S. m. stejnegeri** (Parrot, 1908) 'Stejneger's Stonechat' (E Siberia from Yenisei and Irkutsk east, C & E Mongolia, NE China, Korea, Japan, Anadyr; winters E & SE Asia; has straggled as vagrant to W Europe). Very similar to *maurus* but also slightly recalls European Stonechat in appearance. Compared to *maurus* has a trifle heavier bill and longer tail, and in ♂ plumage tends to have richer and more extensive rufous underparts and rump, and attains even blacker upperparts in worn summer plumage. When handled the somewhat broader bill base (by c. 15%) can help diagnose the race, but some are intermediate as to this character. Of migrants at Beidahe, E China, 1/4 of examined birds had boldly black-streaked longest uppertail-coverts (Hellström & Norevik 2014), a character not known from ssp. *maurus*. (See also Taxonomic notes.) **W** ♂ 66–71.5 mm (n 22, m 68.8), ♀ 64–69 mm (n 19, m 67.3); **T** 45–52 mm (n 42, m 48.2); **T/W** m 70.9; **B** 13.6–15.5 mm (n 42, m 14.5); **BD** 3.6–4.4 mm (n 42, m 4.0); **BW** (prox. edge nostril) 4.7–5.6 mm (n 42, m 5.2); **Ts** 19.6–22.5 mm (n 42, m 21.2). **Wing formula:** very similar to *maurus*. **White at base of rr5–6, inner web** (only in 65% of ♂♂) 5–15 (once 23) mm.

TAXONOMIC NOTES A certain nomenclatural confusion has long adhered to the stonechats breeding in the Caspian region (Svensson *et al.* 2012). S. G. Gmelin's description of *variegatus* in 1774 was made from birds on the SE slopes of the Caucasus (Samaxi, or Semakha, at 800 m) in an area more likely frequented by the Transcaucasian subspecies until recently known as '*armenicus*', and indeed neither description nor drawing eliminate that taxon. In fact, Gmelin gives a table of measurements which fit only the Transcaucasian taxon. Interestingly, Stegmann (1935) found proof of breeding '*armenicus*' from the very same type locality for Gmelin's *variegatus*. Therefore, the name *variegatus* must be removed from the N Caspian taxon and instead be applied to Transcaucasian breeders. Incidentally, Mlíkovský (2011) came to the same conclusion but based it on unconvincing evidence. The name *hemprichii* has been peculiarly neglected in major handbooks and checklists, being unmentioned in Vaurie (1959), Peters (1964), Urquhart (2002) or Dickinson (2003). Hartert (1910) listed *hemprichii* but, surprisingly, synonymised it with *maurus*. Ehrenberg described a series of stonechats from Nubia and Arabia in winter, which differed from *rubicola* by having white at the base of the tail, and it has been erroneously assumed by some (e.g. Grant & Mackworth-Praed 1947) that these birds are the same as '*armenicus*'. However, as detailed in Svensson *et al.* (2012), the name *hemprichii* is the oldest available valid name for the northern subspecies in the Caspian region. – The general considerations behind the adopted treatment of the Stonechat complex as four different species (European Stonechat *S. rubicola*, Fuerteventura Stonechat *S. dacotiae*, Eastern Stonechat *S. maurus*, African Stonechat *S. torquatus*) have already been briefly mentioned under European Stonechat. Since *variegatus* and *rubicola* come into contact in NE Turkey, Transcaucasia and in parts of the Caucasus, living there sympatrically though generally separated altitudinally, and since they differ consistently and rather clearly morphologically, and no intermediates have been found to date, it seems reasonable and logical to treat them as separate species. In this we follow previous recommendations based on genetic data (Wittmann *et al.* 1995, Wink *et al.* in Urquhart 2002a, Illera *et al.* 2008, Zink *et al.* 2009, 2010). It should be added that *rubicola* is now sympatric with *hemprichii* in E Ukraine and SE Russia following a recent eastward range expansion by the former. – Zink *et al.* (2009, 2010) suggested based on mtDNA that *stejnegeri* was basal to all Palearctic stonechats and not closest relative of *maurus*. However, they included a sample from as far west as Astrakhan in this clade, which seems to be a warning signal, and we prefer to await further studies before evaluating their results. Would their findings stand, *stejnegeri* cannot remain a subspecies within Eastern Stonechat.

REFERENCES Barthel, P. (1992) *Limicola*, 6: 217–241. – Corso, A. (2001) *BB*, 94: 315–318. – Ehrenberg, C. G. (1832) *Symb. Phys.*, pl. 8. – Flinks, H. (1994) *Limicola*, 8: 28–37. – Gmelin, S. G. (1774) *Reise durch Russland*. Part 3: 106–107. St. Petersburg. – Grant, C. H. B. & Mackworth-Praed, C. W. (1947) *BBOC*, 67: 47–48. – Grant, K. & Small, B. (2004) *BW*, 17: 154–156. – Hellström, M. (2005) *Vår Fågelv.*, 64(2): 36–43. – Hellström, M. (2006) *Roadrunner*, 14(3): 42–43. – Hellström, M. & Waern, M. (2011) *BB*, 104: 236–254. – Hellström, M. & Norevik, G. (2014) *BB*, 107: 692–700. – Illera, J. C. *et al.* (2008) *Mol., Phyl. & Evol.*, 48: 1145–1154. – Mlíkovský, J. (2011) *J. Natl. Mus. (Prague)*, 180: 102–103. – Siddle, J. P. (2006) *BB*, 99: 372–374. – Stegmann, B. K. (1935) *Compt. Rend. (Dokl.)*, vol. 3: 45–48. – Stoddart, A. (1992) *BW*, 5: 348–356. – Svensson, L. *et al.* (2012) *BBOC*, 132: 37–46. – Ullman, M. (1986) *Vår Fågelv.*, 45: 227–229. – Urquhart, E. D. (2002a) *Stonechats*. Helm, London. – Urquhart, E. D. (2002b) *Birdwatch*, 120 (Jun): 21–25. – Vinicombe, K. (2005) *Birdwatch*, 160 (Oct): 28–31. – Walker, D. (2001) *BW*, 14: 156–158. – Wink, M. *et al.* (2002) *BB*, 95: 349–355. – Wittmann, U. *et al.* (1995) *J. Zool. Syst. Evol. Res.*, 33: 116–122. – Zink, R. M. *et al.* (2009, 2010) *Mol. Phyl. & Evol.*, 52: 769–773, 57: 481–482.

S. m. stejnegeri, ad ♂, NE China, Sep: virtually inseparable from *maurus*, though larger, bulkier and typically has darker and warmly rufous-tinged plumage. Some have a few characteristic dark streaks on longest uppertail-coverts, but (as here) these can be very limited. Bill generally somewhat stronger (wider), but hard to establish in field. Such birds with solid black in front of eye, and blackish primary-coverts with distinct and narrow white edges around the tips indicate sex and age. (M. Hellström)

S. m. stejnegeri, 1stW, NE China, Sep: sex unknown. This race in fresh autumn plumage is on average darker and more rufous-tinged than *maurus*, often with rump warmest (although some more buff like *maurus*). Uppertail-coverts either unmarked, as in *maurus*, or show dark markings of variable distribution, intensity and shape, and a few can be as strongly marked as European Stonechat. This bird shows moult contrast in greater coverts, but as in *maurus* sexing of 1stW difficult until at least late winter/early spring. (M. Hellström)

AFRICAN STONECHAT
Saxicola torquatus L., 1766

Fr. – Tarier africain; Ger. – Afrikaschwarzkehlchen
Sp. – Tarabilla africana; Swe. – Afrikansk buskskvätta

The widely distributed Stonechat, previously often regarded as one variable species, is now commonly split up into several, and this is followed here. Thus, the African Stonechat gathers all African and one Arabian taxa under one entity. There are more than 15 subspecies described, and they range all the way south to the Cape, but only one is of concern within the region treated here, occurring in SW Arabia where it is sedentary. Habits and appearance much as its European and Asian relatives, keeping to open montane habitats with low vegetation. Often perches in the open, on the lookout for insects.

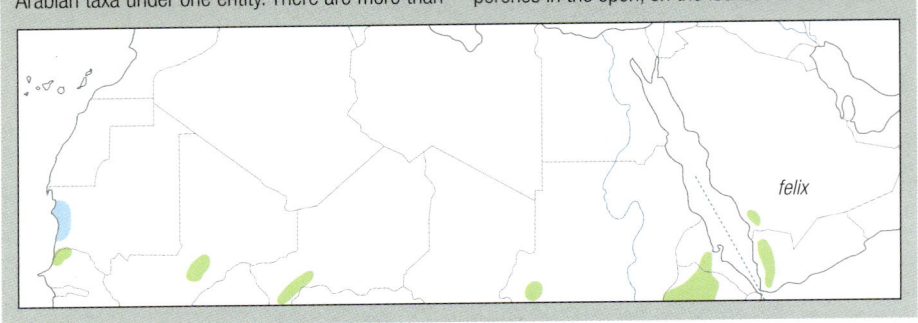

S. t. felix, ad ♂, Yemen, Jan: relatively long tail and shorter wings; tail pattern close to European Stonechat, invariably all dark, and when fresh only has narrow buff fringes. Uppertail-coverts (just visible) are white and chestnut, streaked black. Aged by having ad wing-feathers (same individual as image to the lower left). (H. & J. Eriksen).

S. t. felix, ad ♂, Yemen, Jan: distinctly small, with overall plumage like European Stonechat; paler ginger-chestnut breast patch, with more white at sides. (H. & J. Eriksen)

IDENTIFICATION (*felix*) Although it has *shorter, more rounded wing*, and due to this proportionately *longer tail*, is very similar to Eastern Stonechat of subspecies *variegatus*, i.e. ♂ has strong, dark colours and contrasts, with *concentrated brick-red chest patch* and *white rest of underparts*. *White rump patch* (rusty-buff when fresh) like Eastern Stonechat, but as in several other members of African Stonechat, the patch is rather small and often 'unclean' (thus somewhat approaching European Stonechat ssp. *rubicola*), and there is *often some variable dark and fine rufous streaks* or spots admixed in the white or grey-white ground colour. Upperparts in ♂ *nearly all black*, even when fresh. Differs from both European and Eastern Stonechats in ♂ plumage by subtly *larger black hood*, reaching further down lower throat or even to upper breast. Underwing of ♂ *blackish* like Eastern Stonechat, differing clearly from European Stonechat. ♂ has *very black tail*, often with no white at base, but a few have very little (< 10 mm). *White wing patch usually large* due to many greater coverts being white in adult ♂. ♀ differs from Eastern Stonechat by being rather dark brown above with *dark-streaked ochre-buff rump patch* and usually *darkish-looking throat* (greyish feather-bases partially visible), hence resembles more *rubicola* European Stonechat.

VOCALISATIONS Song appears to be very similar to European Stonechat, a rather short, monotonous high-pitched jingle. – Calls are also similar to those of European Stonechat.

SIMILAR SPECIES African Stonechat of the Arabian subspecies *felix* needs to be separated primarily from Eastern Stonechat, and in all likelihood either of the races *hemprichii* and *variegatus* could occur in the same area on migration or in winter. ♀♀ of these two subspecies have an unstreaked and often larger rump patch, and have mainly pale throat, so separation in the field is probably often possible. More problematic could be distinguishing ♂♂. In spring, *variegatus* and *felix* appear very similar, but latter is a little smaller with subtly more rounded wing (in the hand, note long p1 and short measurement for p1 < p2) with shorter primary projection, and has slightly smaller and less uniform pale rump patch, often being finely streaked dark, at least partly. – ♀ *felix* is very similar to ♀ European Stonechat, both having brownish rump streaked dark. European Stonechat is not known to reach S Arabia in winter, but a straggler could occur and if so would be very difficult to separate.

AGEING & SEXING (*felix*) Ageing not closely studied but seems to be similar to that of Eastern Stonechat. Sexes differ. Seasonal variation primarily due to gradual wear of feather-tips. – Moults. Complete post-nuptial and partial post-juv moults generally follow shortly after breeding (late winter–early spring). Post-juv renewal usually includes head, body, lesser, median and variable number of inner greater coverts and possibly a few tertials. Pre-nuptial moult little known; probably lacking or limited. – For details of ageing and sexing, see Eastern Stonechat, which are thought applicable also for African Stonechat.

BIOMETRICS (*felix*) **L** 12–13 cm; **W** ♂ 64–68.5 mm (n 17, m 65.8), ♀ 62–65 mm (n 11, m 63.5); **T** 45–51.5 mm (n 28, m 48.4); **T/W** m 74.6; **B** 13.2–14.5 mm (n 25, m 13.8); **BD** 3.4–4.1 mm (n 21, m 3.7); **BW** 4.3–5.2 mm (n 23, m 4.8); **Ts** 19.8–22.1 mm (n 24, m 20.9). **Wing formula**: p1 > pc 8–14 mm, < p2 18–23 mm; **p2** < wt 5.5–10 mm, =8/9 or 9 (60%) or =7/8 or 8 (40%); **pp3–6** about equal and longest (sometimes p6 0.5–1.5 mm <); **p7** < wt 2.5–4.5 mm; **p8** < wt 5–7.5 mm; **p10** < wt 7.5–10.5 mm; **s1** < wt 10–12 mm. Emarg. pp3–6. White at base of rr5–6 usually lacking, very rarely in ♂ 5–10 mm on bases of inner webs (never any white in ♀).

GEOGRAPHICAL VARIATION & RANGE Marked and complex variation even when, as here, Afro-Arabian taxa are separated as specifically different from stonechats of Europe and Asia, as African Stonechat. At least 15 subspecies recognised (e.g. in Dickinson 2003), but only one breeds within the covered region, the rest are extralimital and hence not treated. Among the extralimital one or two taxa have been suggested to constitute separate species.

  S. t. felix Bates, 1936 (SW Saudi Arabia, W Yemen; largely resident). Treated above. Within the African Stonechat one of the smaller races.

TAXONOMIC NOTES See European Stonechat and Eastern Stonechat for general considerations underlying our treatment of the Stonechat complex as four species (European Stonechat *S. rubicola*, Fuerteventura Stonechat *S. dacotiae*, Eastern Stonechat *S. maurus*, African Stonechat *S. torquatus*). Wink et al. (*in* Urquhart 2002, *Stonechats*) found that African Stonechat differed genetically (mtDNA) from Stonechat by 4.6–5.7%.

REFERENCES Grant, C. H. B. & Mackworth-Praed, C. W. (1947) *BBOC*, 67: 47–48. – Illera, J. C. et al. (2008) *Mol. Phyl. & Evol.* 48: 1145–1154. – Urquhart, E. D. (2002) *Birdwatch*, 120 (Jun): 21–25. – Wittmann, U. et al. (1995) *J. Zool. Syst. Evol. Res.*, 33: 116–122. – Zink, R. M. et al. (2009, 2010) *Mol. Phyl. & Evol.* 52: 769–773, 57: 481–482.

S. t. felix, ♀, Yemen, Jan: rather pale but overall pattern still like that of European Stonechat, including rump and uppertail-coverts, which are streaked dark on chestnut-brown background. Impossible to age from this angle. (H. & J. Eriksen)

PIED STONECHAT
Saxicola caprata L., 1766

Alternative name: Pied Bushchat

Fr. – Tarier pie; Ger. – Mohrenschwarzkehlchen
Sp. – Tarabilla pía; Swe. – Svart buskskvätta

Unmistakable dark stonechat of arid open land or cultivation in much of Central Asia. Widespread, particularly in lowland areas, from E Iran to Indochina and through S Asia, as well as the Sundas and parts of the Philippines, with northerly populations being migratory, occasionally wandering to the covered region, having straggled to Cyprus, Israel, Saudi Arabia, Oman and the United Arab Emirates.

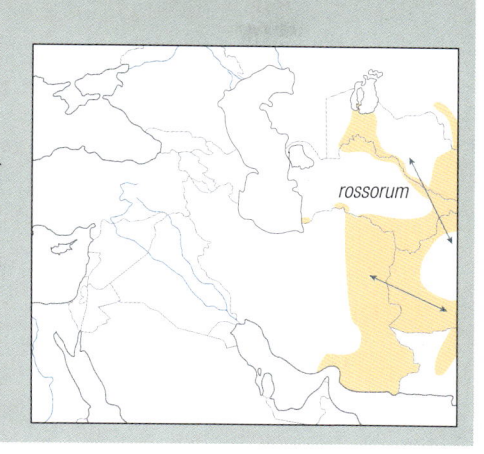

S. c. rossorum, ad ♂, Iran, Apr: shape and size, plus pied pattern make this plumage unmistakable in the covered region. Sex by pied plumage. Rather large white area on belly extends as streaks onto lower breast, typical of western race. Aged by whole wing being evenly feathered, and having extensive and solid white 'shoulder patch' (an extreme example). (C. N. G. Bocos)

S. c. rossorum, ♀, presumed 1stS, Iran, Apr: ♀ readily identified to species and sexed by being in predominantly grey-brown plumage. Rufous-buff rump/uppertail-coverts patch only just visible. Age unclear, but most wing-feathers seem worn and probably juv. (C. N. G. Bocos)

IDENTIFICATION Slightly larger than European Stonechat. Has lively, upright and alert character typical of *Saxicola*, and very dark ♂ has sooty-black plumage, relieved only by longitudinal white innerwing patch, whitish to buff rump and uppertail-coverts patch, as well as buffish-white vent and undertail-coverts. ♀ is, by comparison, a rather dull nondescript chat with predominantly grey-brown plumage (about as dark as ♀ Black Redstart or a little darker still), discreetly streaked darker (can appear plain in the field), with darker brown wings (fringes slightly paler when fresh) and tail. Throat slightly paler, dull buffish, and breast, upper belly and flanks pale fulvous-grey (sometimes tinged rusty and streaked), with cream/buff undertail-coverts, while rump patch is rather small and plain rufous-buff, indistinctly blotched paler. Behaviour and jizz strongly recall European Stonechat, including tail movements and frequent habit of visiting the ground, etc. Song-flight characteristic, descending partially in a glide.

VOCALISATIONS Song a series of rather straight, clearly whistling or almost piping notes, rather staccato and with marked tonal steps both rising and falling in irregular pattern (almost in *Tchagra* fashion), and the song could be rendered 'you would not believe it—no!'. A typical strophe often consists of 4–7 piping or whistling notes, at times with a strained trill added at the end, e.g. *ju, ju see cha see-we zoo chiih zrrreeh*, delivered with a variable pattern and repeated at intervals of a few seconds. – Calls include a repeated, loud and rather plaintive, downslurred *fiiep* (with tail movements), often alternating with subdued short harsh sounds, *che, che, che*. Utters a harsh, scolding *chach* in strong alarm, repeated in long series.

SIMILAR SPECIES ♂ should not be confused with any other species, given obvious chat behaviour, size, structure and black-and-white plumage. However, ♀ requires careful separation from ♀ European Stonechat, but bear in mind that ♀-type Pied Stonechat is nearly unstreaked above or only very indistinctly streaked, lacking the other species' pale inner greater-covert mark, and in comparison has an inconspicuous and unstreaked rufous rump patch, predominantly dark brown tertials with very dull and much narrower fringes, and overall darker brown plumage with a noticeable pale throat, especially in fresh plumage.

AGEING & SEXING Ageing of ♂♂ sometimes possible in the field (♀♀ require closer inspection) by moult pattern and feather wear in wing. Sexes differ markedly. – Moults. Complete post-nuptial and partial post-juv moults in late summer (mostly Jul–Sep). Partial post-juv moult usually includes head, body, lesser, median, a variable number of inner greater coverts, and possibly also a few tertials. Pre-nuptial moult apparently absent. – **SPRING Ad** Both sexes aged by having only ad, relatively less worn, wing- and tail-feathers. **Ad ♂** More strikingly glossy black and white than

S. c. rossorum, 1stS ♀, Israel, Mar: despite being colourless, unstreaked underparts typically tinged buffish-brown, with pale cream-coloured throat; rufous rump/uppertail-coverts just visible. Overall pale and diluted coloration indicative of western race. Aged by juv remiges, alula, primary-coverts and some outer greater coverts. (H. Shirihai)

S. c. rossorum, 1stW ♂, Israel, Oct: considerable variation in black plumage and white shoulder patch in 1stY ♂♂, with this bird moderately advanced in respect of black, and poorly as to white shoulder. Characteristic white rump/uppertail-coverts and lower body. Aged by many juv wing-feathers. (Y. Perlman)

S. c. rossorum, 1stW ♂, Egypt, Oct: like other *Saxicola*, white innerwing patch often concealed below long scapulars, but distinctive by small size, white rump/uppertail-coverts and lower underbody versus otherwise blackish coloration. Aged by juv (extremely worn) remiges, alula and primary-coverts. (K. Gauger)

in autumn due to loss of buff-brown feather fringes. **Ad ♀** Darker and browner above than in autumn, losing most grey-buff fringes through wear, face being flecked buff and brown, and rufous below less obvious. **1stS** As ad, but primaries, primary-coverts and alula (and perhaps some outer greater coverts and tertials) are juv, slightly more worn, bleached browner, and offer moult contrast, but note that fringes by now are very narrow or even completely lost, and relative texture and wear in different feather tracts are difficult to judge. — AUTUMN **Ad ♂** As spring but rump and undertail-coverts have rufous-buff fringes, plumage generally less glossed black with slightly mottled and dull-scaled appearance. **Ad ♀** As spring but has slightly darker brown feather centres contrasting with grey-brown fringes to upperparts, and extensive rufous wash below; narrow greyish-buff fringes to wing-coverts, tertials and inner secondaries. Both sexes aged by evenly very fresh wing- and tail-feathers. **1stW** Much like ad, but both sexes show subtle moult limits (flight-feathers, primary-coverts, tail-feathers, often some tertials and outer greater coverts are juv, relatively weaker textured and rather more worn and browner than respective ad). Black of ♂ has brownish cast and more extensive greyish-brown fringes, with less distinct demarcation between black and white of underparts, and smaller white upperwing patch. 1stW ♀ separated from ad ♀ by moult pattern and wear (see above), otherwise extremely similar. Both sexes have more pointed tail-feathers and, until early autumn, some juv outer greater coverts may have tiny rusty-buff tips. **Juv** Already resembles ad of respective sex, but juv ♂ is blackish-brown, rarely with trace of wing patch, and in both sexes upperparts are spotted paler.

BIOMETRICS (*rossorum*) **L** 12.5–14 cm; **W** ♂ 71–77 mm (*n* 12, *m* 74.1), ♀ 70–76 mm (*n* 10, *m* 73.1); **T** 48–54 mm (*n* 22, *m* 51.5); **B** 12.2–14.9 mm (*n* 22, *m* 13.4); **Ts** 19.1–24.2 mm (*n* 22, *m* 21.9). **Wing formula: p1** > pc 6–12 mm, < p2 24–31 mm; **p2** < wt 5–9 mm, =7 or 7/8 (very rarely =6/7); **pp3–5** about equal and longest; **p6** < wt 1–2 mm; **p7** < wt 4–7 mm; **p8** < wt 7–11 mm; **p10** < wt 10–14 mm; **s1** < wt 11–15 mm. Emarg. pp3–6.

GEOGRAPHICAL VARIATION & RANGE Relatively slight variation, mainly involving size, degree of gloss and distribution of white in ♂, and depth of colour in ♀. Up to 16

S. c. rossorum, 1stW ♂, United Arab Emirates, Sep: least-advanced young ♂ in early autumn, with entire wing (except inner greater coverts) juv. Such plumage rapidly becomes blacker when pale greyish tips wear off, but this 1stW ♂ has paler and more ragged plumage than usual. (T. Pedersen)

subspecies described, with those occurring in or near the covered region as follows.

S. c. rossorum (Hartert, 1910) (S Central Asia to E Iran & NW Indian subcontinent; winters SE Iran, Pakistan; probably this subspecies reaches Middle East as a straggler). Rather large with white on belly of ♂ often extending (as streaks) onto lower breast, while ♀ has upperparts dull grey-brown, with reduced or often no streaking below, and paler rufous rump/uppertail-coverts.

Extralimital: **S. c. bicolor** Sykes, 1832 (N Indian subcontinent; winters mainly India). Somewhat smaller, and ♂ duller black with more restricted white on underparts, and ♀ has much deeper earth-brown upperparts, buff-rufous and darker-streaked underparts, and noticeably deeper rufous rump and uppertail-coverts. Intergrades with previous race where they meet. **W** 66–72 mm (*n* 15, *m* 68.5).

ISABELLINE WHEATEAR
Oenanthe isabellina (Temminck, 1829)

Fr. – Traquet isabelle; Ger. – Isabellsteinschmätzer
Sp. – Collalba isabel; Swe. – Isabellastenskvätta

Arguably *the* wheatear of steppes and arid plains, being common in semideserts in Asia (from S Russia south and east to N China), Isabelline Wheatear reaches west to Bulgaria and Greece, where it dwells on stony or sandy, almost barren or grass-covered open areas, in both lowlands and uplands. Winters in the Sahel of NE Africa and from Arabia east to C India. In much of W Europe it is a very rare but regular visitor in autumn, far less frequent in spring.

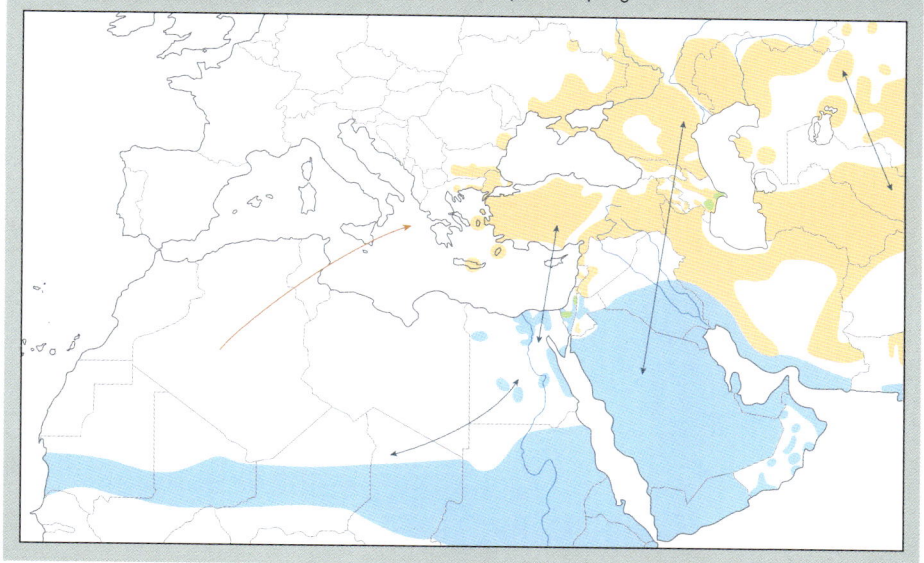

♂, Kuwait, Mar: especially in spring, some can be sexed by clear-cut black lores. Note rather long legs and short-tailed impression (ending well above ground, making any tail-wagging seem freer). Seemingly strong wear to primaries and primary-coverts, and apparent moult limits in greater coverts, might suggest 1stY. (A. Halley)

♂, Bulgaria, May: often, tail pattern is key to identifying the species, best seen when spread—has broad black band encompassing > 1/3 of the exposed rectrices. Black lores in ♂. Ageing always difficult, even in the hand. (C. Bradshaw)

Oman, Mar: in flight, when tail partially folded, black tail-band and white rump can appear almost equal in width; in such circumstances, be aware of possibility of misidentification as Desert Wheatear. (M. Varesvuo)

IDENTIFICATION Basically *fairly uniform sandy* (both sexes) with relatively few contrasting features. Throughout much of region resembles ♀-like plumages of Northern Wheatear and, to lesser extent, several other species. Typically adopts *upright stance*, imparting well-built impression, *longer-legged* but *shorter-tailed* than Northern Wheatear, with terminal tail-band intermediate between that species and Desert Wheatear, and *tail-wagging often conspicuous*. Like rest of genus, mostly blackish bare parts. In flight, but especially on landing, distinctive *broad black tail-band* but *narrower white rump* are exposed, also rather broader-winged and shorter-tailed than Northern Wheatear. Often appears quite jerky and typically forages by dropping to the ground from elevated perch to make rapid dashes after prey, but also performs aerial sallies. After perching on stone, sometimes bobs body forward (accompanied by tail-raising and wing-flicking). Not shy, and often aggressively defends territory. Song, given from low perch or in display-flight (with fluttering wingbeats and splayed tail), highly characteristic in often containing series of loud whistles.

VOCALISATIONS Song noticeably powerful, very rich, highly varied and imitative, with long strophes (often 10–15 sec., pauses brief) usually commencing with some repeated high-pitched *cheeehp* notes, and often including quickly repeated short 'wolf-whistles' in series, *vi-vi-vi-vü-vü-vü-vuy-vuy...* (the so-called 'Harpo Marx whistles'), at times interspersed with downslurred notes, varied sharp, clicking and scrunching sounds, chirps and some tingling notes. Compared to Northern Wheatear, song is longer and more varied, more powerful and contains more high-whistled and nasal notes. In winter, especially prior to migration, a soft or half-volume sub-song is sometimes heard. – Commonest call, also given in territorial disputes or when taking off or landing, a hard tongue-clicking *check* (repeated, especially if excited). Also a more whistling *weet* or *hiiet*. Anxiety sometimes expressed with harsh, strained *cheeh*.

SIMILAR SPECIES In most areas the main confusion species is *Northern Wheatear*. Usually, however, they are separated, at a glance, by jizz, and by Isabelline Wheatear's predominantly pale and more uniform plumage and upperwing, the broader black tail-band and whiter supercilium in front of the eye. However, some will prove less straightforward using these characters alone and should be identified only following careful study of a combination of features. Isabelline is more robust and longer-legged, with a larger, broader head and broader-based and longer bill (although there is overlap in most measurements). Primary projection (equal to or shorter than exposed tertials) is shorter than in Northern Wheatear. Tail, though shorter and fuller, projects further beyond folded wing, which reaches only to tip of undertail-coverts or to much less than halfway down black of tail (in Northern Wheatear, wings being more pointed extend well beyond this point). Isabelline often perches very upright, appearing pot-bellied or even pear-shaped, when its longer legs become even more striking, as its shorter tail (ends well above ground) makes tail-wagging seem 'freer'. Clearly has a broader black tail-band, most easily judged in flight. Whitish supercilium broadest and palest in front of and above eye, short, often buffish and tapering to a point behind. (In Northern Wheatear, supercilium washed buffish and less distinct in front of eye, whitish and more obvious behind; only very few difficult birds as to this character, but always best to check many characters.) Note also that lores vary in both: often dark (darkest in adult ♂), but can be pale. Isabelline has more uniform wings almost concolorous with rest of upperparts (upperwing of Northern Wheatear always has rather striking dark feather-centres, including on coverts), but some variability. Largest (and second-largest) alula-feather very dark brown (almost blackish) and is darkest tract of entire upperparts, usually contrasting strongly with much paler surrounding coverts (alula also dark on Northern Wheatear, but very little contrast due to darker surrounding coverts). However, alula is often hidden by breast-sides, and heavy wing abrasion may obscure this feature and its effect. Tends to have more even sandy/whitish-buff suffusion from breast to vent (on Northern Wheatear warmer, especially in autumn, forming broad breast-band extending to flanks and contrasting with whiter belly), but spring ♀ Northern Wheatear may be equally pale. Underwing-coverts and axillaries very pale, white to pale buffish (greyish-buff to rather blackish on Northern Wheatear), and remiges almost as pale.

Isabelline (top and bottom left: Israel, Mar) versus ♀ **Northern Wheatear** (top right: Finland, Apr; bottom right: Spain, Apr): some are less straightforward to separate, especially in spring, when wings of Isabelline appear contrastingly darker and pale fringes to Northern mostly worn off. Isabelline is overall paler and more uniform, with less contrast between upperparts and wings, thus dark alula (here only partially exposed) is more obvious. Supercilium usually whiter in front of eye, buffier behind (opposite in Northern), less contrasting, smoothly buff-washed underparts, broader tail-band, but smaller whitish rump. Jizz of Isabelline usually comprises heavier impression (compared to *oenanthe* and *libanotica* Northern), upright stance, longer legs, proportionately shorter tail and primary projection than Northern. (Top and bottom left: H. Shirihai; top right: M. Varesvuo; bottom right: C. N. G. Bocos)

This diagnostic difference may be impossible to confirm in the field, but near-translucent underwing of Isabelline often detectable in flight. Race *leucorhoa* of Northern Wheatear is large, but all of the above-mentioned criteria remain valid. – Locally in SW Arabia, Isabelline must also be separated from superficially similar *Red-breasted Wheatear* (which see).

AGEING & SEXING Ageing requires close inspection of moult, feather wear and pattern in wing. Sexes scarcely differ. When using lores for sexing, ageing should precede sexing. Combination of wing- and tail-lengths recommended —see Biometrics. Seasonal variations also limited. – Moults. Complete post-nuptial and partial post-juv moults in summer (mostly Jul–Aug). Post-juv renewal includes all head- and body-feathers, lesser and median coverts, most greater coverts and some tertials, more rarely some tail-feathers, entire alula and innermost secondaries. Partial pre-nuptial moult in winter (Nov–early Mar) includes some head- and body-feathers, a few lesser and median coverts, some inner greater coverts and some or all tertials, more rarely some innermost secondaries. – SPRING **Ad** ♂ Broad pale supercilium and blackish lores well defined. Less warm sandy than in autumn with somewhat greyer cast above, while sandy wash to underparts frequently reaches throat- and neck-sides, and breast and belly washed almost pinkish. Dark parts of tail-feathers and wing-feathers on average subtly darker and are blackish rather than dull brown (beware some overlap or difficult borderline cases). **Ad** ♀ Slightly smaller, duller and paler, and has narrower, less intense, grey-brown to blackish-brown lores, often admixed with paler feathers and never all black, and has on average paler underparts. Dark part of tail-feathers and wing-feathers on average subtly less dark and more brownish than black (but difference subtle, and some overlap). Some birds difficult to sex, even in the hand. **1stS** As ad, and ageing particularly difficult; note that, due to pre-nuptial moult, both ad and 1stS often show moult limits in inner greater coverts and tertials/secondaries. Typically, 1stS has stronger contrast with

Turkey, May: ill-defined darkish lores suggest ♀, but there is overlap between sexes, with some 1stS (and poorly-marked) ♂♂ especially similar. (D. Occhiato)

Bulgaria, Apr: in song-flight, with fluttering wingbeats and splayed tail, showing diagnostic tail pattern. Note quite pale underwing. (B. Nikolov)

browner, more strongly bleached juv primaries and primary-coverts, often some outer greater coverts, and perhaps some tertials and alula. With wear, pattern of juv greater coverts (see 1stW) very difficult or impossible to ascertain. **1stS ♂** tends to have less clear-cut facial markings, approaching ad ♀. – **AUTUMN Ad** Differs from 1stW in being evenly fresh. Tail-feathers slightly broader and blacker than juv ones, with purer white and more prominent pale tips, especially on r6. Plumage generally similar to spring, but has broader pale sandy fringes to wings, and primaries narrowly tipped pale buff. Supercilium more buff in tone. Differences between sexes less clear as lores of ♂ sometimes less solid black. **1stW** Almost identical to ad but face marks even more poorly defined, and most cannot be sexed using loral character. As a rule very difficult to age. Primaries slightly to moderately worn, at most with slight whitish or pale ochre tips. Primary-coverts slightly paler and less fresh than ad ones. Moult limits detectable with practice, e.g. between inner and outermost greater coverts and between tertials (juv ones slightly less intensely coloured with paler shafts and more worn tips than ad-like post-juv feathers). Ill-defined sandy-orange fringes to greater coverts narrower and less smoothly rounded, producing shallow L-shape with buffy step on tip of outer web (detectable with experience). Tail-feathers less broad and browner, while tips of r6 lack or have very little white. **Juv** Soft fluffy body-feathering and much of upperparts faintly mottled. Crown and nape has hint of spotting or scalloping, and throat and chest, too, have faint dark scalloping (not always apparent). Best distinguished from juv Wheatear by tarsus length and tail pattern.

BIOMETRICS L 15.5–17 cm; **W** ♂ 96–105 mm (*n* 12, *m* 101.5), ♀ 96–100.5 mm (*n* 13, *m* 98.2); **T** ♂ 53–63 mm (*n* 11, *m* 59.0), ♀ 53–60 mm (*n* 13, *m* 56.2); **T/W** *m* 57.6; **B** 17.5–21.9 mm (*n* 29, *m* 19.6); **B(f)** 12.6–15.3 mm (*n* 27, *m* 13.9); **BD** 4.2–5.2 mm (*n* 26, *m* 4.7); **Ts** 27.5–33.4 mm (*n* 26, *m* 29.9). **Wing formula: p1** < pc 0–5 mm, < p2 49–59 mm; **p2** < wt 0.5–5 mm, =4/5 or 5 (83%), =3/4 or 4 (15%) or =5/6 (2%); **pp3–4** usually equal and longest (but p4 sometimes 0.5–1 mm <); **p5** < wt 4–5.5 mm; **p6** < wt 10.5–14 mm; **p7** < wt 16–19 mm; **p10** < wt 24–30 mm; **s1** < wt 27–33 mm. Emarg. pp3–5, but on p5 frequently less prominently, and rarely missing.

GEOGRAPHICAL VARIATION & RANGE Monotypic. – Breeds in E Bulgaria, E Romania, Ukraine, Kalmykia, Turkey, E Caucasus, Levant, Middle East, N Arabia, Iran, Pamirs, NW Himalaya, Kirghiz steppes, S Siberia, Altai, Dauria, N Mongolia, Tibet, W China; winters Sahel zone and much of sub-Saharan Africa from Mauritania to Egypt and Horn of Africa, Kenya, Tanzania, further Middle East, S Iran and east to India.

REFERENCES ALSTRÖM, P. (1985) *BB*, 78: 304–305. – BRADSHAW, C. (2000) *BB*, 93: 488–492. – CORSO, A. (1997) *DB*, 19: 153–165. – KÖNIGSTEDT, D. G. W., ROBEL, D. & BARTHEL, P. H. (1992) *Limicola*, 6: 3–22.

Oman, Nov: some birds, especially in autumn, readily separated from Northern Wheatear by generally pale and more uniform plumage and upperwing, broader tail-band and whiter supercilium in front of eye. Relatively strong wear to wing, including primary-coverts, may indicate 1stW. (D. Occhiato)

Germany, Oct: some have more variegated wing, approaching Northern Wheatear, with more contrasting dark feather-centres. Face pattern and buff breast and flanks, contrasting with whiter belly, also atypical and more like Northern. Nevertheless, note contrasting black alula, and shorter primary projection and tail, typical of Isabelline. (A. Halley)

Isabelline (left: Oman, Oct) versus ♀-like Northern Wheatear (right: Italy, Sep): especially in fresh plumage, Isabelline paler and more uniformly patterned (at least until mid/late autumn, Northern tends to have more and often striking dark centres to wing-coverts). They can have rather similar underparts and supercilia, so check for Isabelline's broader tail-band. (D. Occhiato)

Juv, Bulgaria, Jun: faintly mottled above and on throat and chest (not always evident unless seen close), which helps separate it from usually better-marked juvenile Northern, although tail pattern is best feature. (R. Wilson)

RED-BREASTED WHEATEAR
Oenanthe bottae (Bonaparte, 1854)

Alternative names: Buff-breasted Wheatear, Botta's Wheatear

Fr. – Traquet à poitrine rousse
Ger. – Braunbrust-Steinschmätzer
Sp. – Collalba de Botta
Swe. – Svartpannad stenskvätta

Mainly Afrotropical but also inhabits semiarid zones in SW Arabia, where it is a common resident in dry, almost barren or grassy open areas (and local in open riparian woodland on upland plateaux). Red-breasted Wheatear is most closely related to the more familiar Isabelline Wheatear, and within Red-breasted the race *bottae* is paler and somewhat approaches Isabelline (like extralimital NE African *frenata* to some extent also does), but ssp. *heuglini* (of Mauritania to Sudan) appears more different and might be specifically distinct.

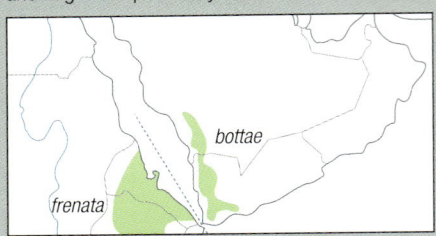

O. b. bottae, Yemen, May: sometimes has slightly broader terminal tail-band than Isabelline, and black on the two central feathers very short. (W. Müller)

O. b. bottae, presumed ♂, Yemen, Jun: mostly resembles Isabelline Wheatear (virtually identical tail pattern), but darker grey crown and blackish mask, emphasized by well-defined short white supercilium, and usually has buffy-orange breast and flanks. Sexes hardly differ, but ♂ tends to have slightly better-marked face. (H. & J. Eriksen)

IDENTIFICATION A large *chunky wheatear* resembling ♀ Northern Wheatear and, especially, Isabelline Wheatear, but is larger than both and much less uniform than latter with *striking dusky greyish-tinged forecrown darkening to near-black forehead, blackish mask and well-defined short white supercilium*. Also to some degree has *brighter, buffy-orange underparts*. Tail pattern closer to Isabelline or has slightly *broader terminal tail-band* (even approaching Desert Wheatear), thus base of inverted 'T' mark very short. Typically adopts upright stance, but especially characteristic is the *stronger bill*. Also long-legged, but tail-wagging less conspicuous. Usually encountered in pairs or alone.

VOCALISATIONS Song, from perch or in flight, consists of long, varied strophes separated by brief pauses, strophes built up by mixture of jumbled, fluty and hard, scratchy notes and much mimicry, especially at outset. – Usual call a hard *chack* or, in alarm, *cheet*.

SIMILAR SPECIES Main risk of confusion is with *Isabelline Wheatear* but Red-breasted is larger, with slightly stronger legs and bill (latter has quite broad base), proportionately shorter wings and primary projection but subtly longer tail, and near-vertical stance. Identification also possible using following characters (but beware that the form in SW Arabia, *bottae*, is somewhat less strongly marked than E African birds): generally larger and heavier, especially bill (giving slightly more *Monticola*-like appearance); greyish (less sandy) crown, mantle and scapulars, variably dusky forehead, more striking white supercilium (most marked in front of eye) and clear-cut, black eye-stripe broadening slightly behind eye to form suggested mask. Lower neck-sides, breast and flanks richer, pale cinnamon-buff (brighter than Isabelline), enhancing notably white-throated impression. – *Northern Wheatear* in ♀-like plumages similar, but is smaller, has different structure, face and wing pattern, and narrower dark terminal band to tail.

AGEING & SEXING Ageing as in Isabelline Wheatear. Sexes scarcely differ, including in size. Rather limited seasonal variation. – Moults. Breeding and moult varies with region and altitude, but in SW Arabia juv noted 2 Jun–19 Aug; ad in early or well-advanced moult 5–13 Sep; generally rather worn by Jan–Feb and extremely worn May–Jun (NHM). – **SPRING Ad** Sexing difficult. ♂ has slightly better-marked face, tends to have blacker eye-stripe and duskier crown, with somewhat richer rufous-buff breast than ♀, but much overlap. **1stS** Best distinguished by having more worn juv primaries and primary-coverts. By Feb virtually no pale fringes on primary-coverts and none on primaries (in ad, primary-coverts still well fringed and less

O. b. bottae, presumed 1stW ♂, Yemen, Oct: especially when fresh, pale fringes to wing, including tertials, better defined than in Isabelline. Darker, greyer crown and white throat contrasting with buffy-orange breast and flanks. Well-defined supercilium habitually appears broken above eye. ♂ tends to have blacker lores. Apparently freshly moulted central tail-feathers suggest age. (U. Ståhle)

O. b. bottae, Yemen, Jan: dullest birds can strongly resemble Isabelline Wheatear, but note still darker crown and whitish supercilium partly broken above eye, but with continuing and widening facial mask, reaching well behind the eye. Not enough of wing visible for reliable ageing, hence the overall dull impression could indicate either 1stY ♂ or ad ♀. (H. & J. Eriksen)

O. b. bottae, ad ♂, Saudi Arabia, Nov: when very fresh with much pale fringing may resemble autumn Northern Wheatear, but told by diagnostic wide black tail-band in flight. Diagnostic partly broken white supercilium, long and at rear widening blackish facial mask, and black on forehead and forecrown already noticeable. Crisply plumaged, evenly fresh ad wing and strong facial pattern infer age and sex. (J. Babbington)

abraded, and primaries darker with very tiny whitish tips). Juv greater coverts more abraded but pattern as described for 1stW (cf. Autumn) more difficult to evaluate at this season. Also check moult limits with fresher inner greater coverts and some renewed tertials. At least some birds tend to be somewhat duller and less strongly saturated orange-cinnamon below, and less pure grey above, with reduced head markings, but much overlap. – **AUTUMN Ad** Following complete post-nuptial moult, evenly fresh with greyish upperparts washed pale sandy-buff, tail narrowly tipped white, and primary tips greyish-white. Fringes of tertials broadly buffish-cinnamon. Mask darker and more extensive in ♂, but differences even less clear than in spring. **1stW** Following partial post-juv moult (including all head- and body-feathers, lesser and median coverts, most greater coverts and tertials) very similar to ad. In the hand, juv remiges, primary-coverts and greater coverts slightly weaker textured and more worn than ad. When fresh, pattern of juv greater coverts differs slightly: cinnamon-buff fringes more ill-defined, and centres have clearer 'step' or 'L' pattern on tips of outer webs (in ad fringes and centres better defined and paler, but dark centre forms less clear and square-cut 'L' shape). Overall duller than ad with less strongly-marked face, especially 1stW ♀, often with slightly paler lores and buffier supercilium, but much overlap. Also less strongly saturated orange-cinnamon below, and drabber and less pure grey base to upperparts. Sex differences difficult to evaluate. **Juv** Soft fluffy body-feathering and, unlike ad, head and mantle spotted rufous-buff with broader rufous-buff fringes to wing-feathers.

BIOMETRICS (*bottae*) **L** 17–18 cm; **W** ♂ 103–106 mm (n 4, m 104.5), ♀ 98–104 mm (n 7, m 101.4); **T** ♂ 61–63 mm (n 4, m 61.8), ♀ 56–63 mm (n 7, m 59.0); **T/W** m 58.5; **B** 19.5–22.6 mm (n 11, m 21.7); **BD** 5.2–5.9 mm (n 11, m 5.5); **Ts** 33.0–35.0 mm (n 11, m 33.9). **Wing formula: p1** > pc 2–6 mm, < p2 46–52 mm; **p2** < wt 3.5–6 mm, =5/7; **pp3–4** (pp3–5) about equal and longest; **p5** < wt 0–2 mm; **p6** < wt 3–7 mm; **p7** < wt 9–13 mm; **p10** < wt 17–26 mm; **s1** < wt 20–28 mm. Emarg. pp3–5 (often less prominently also on p6).

GEOGRAPHICAL VARIATION & RANGE Rather prominent variation, in particular if, as here, so-called Heuglin's Wheatear is included (see below). Within treated range only one subspecies. All populations probably mainly resident.

O. b. bottae (Bonaparte, 1854) (Yemen, SW Arabia).

Slightly larger and duller, less richly coloured (less grey above, not so bright pink-buff below) than extralimital *frenata*. Apparently invariably has white supercilium broken above eye.

Extralimital: ***O. b. frenata*** (Finsch & Hartlaub, 1869) (Eritrea, Ethiopia). Subtly smaller than *bottae* and a little brighter. Supercilium of varying length but unbroken over eye in examined material. **W** ♂ 98–105 mm (n 14, m 102.0), ♀ 95–100.5 mm (n 12, m 97.4); **T** ♂ 56–65 mm (n 12, m 61.0), ♀ 53–64 mm (n 12, m 59.7); **T/W** m 60.5; **B** 18.4–21.0 mm (n 24, m 19.8); **BD** 4.5–5.8 mm (n 19, m 5.2); **Ts** 31.0–36.9 mm (n 26, m 33.9).

TAXONOMIC NOTES Usually considered close to Isabelline Wheatear *O. isabellina*, and these two form a superspecies. The best treatment of so-called Heuglin's Wheatear, *O. b. heuglini* (Sahel region east to Sudan and W Ethiopia), is open to discussion and depends on further research. Considered here part of polytypic Red-breasted Wheatear, together with distinctive *frenata*, but could equally well be seen as a parapatric incipient species without overlap in size or habitat, its darkness and slightly smaller size perhaps warranting specific recognition.

O. b. frenata, presumed ad ♂, Ethiopia, Sep: vagrancy to Arabia of this rather distinctive African taxon should be considered. Generally lighter built than *bottae* with overall brighter plumage, with especially warmer underparts, and diagnostically the supercilium is unbroken above eye. Evenly very fresh ad wing and strong facial pattern infer age and sex. (H. Shirihai)

(NORTHERN) WHEATEAR
Oenanthe oenanthe (L., 1758)

Fr. – Traquet motteux; Ger. – Steinschmätzer
Sp. – Collalba gris; Swe. – Stenskvätta

The most widespread wheatear, breeding in treeless areas with exposed rocks or boulders and short grass, including seashores and uplands, where it is frequently a characteristic bird, perching visibly on top of rocks, stone fences or lower buildings. A summer visitor to Greenland, Iceland, Fenno-Scandia and east across N Russia all the way to the Chukotski Peninsula, in Europe breeding south to the N Mediterranean. Northern Wheatear (in a more international context) also breeds in extreme NE & NW North America. All populations winter in C & S Africa, and thus undertake very long journeys twice a year.

IDENTIFICATION A medium-sized, long-winged bird with characteristic *pure white tail-base and white tail with black inverted 'T' mark*, readily visible in flight. Tail pattern shared with several congeners, so necessary to check other plumage details. ♂ in breeding plumage easily recognised by *ash-grey crown and upperparts* contrasting with long dark wings, *whitish supercilium*, *narrow black mask*, extensive whitish rump and underparts, with *pink-buff breast*. ♀-like plumages (including fresh winter-plumaged ♂) greyish-brown and prone to confusion with Isabelline and Black-eared Wheatears. *Black terminal tail-band moderately and evenly broad* (width c. 1/4 of exposed rectrices). Axillaries and underwing-coverts blackish-grey fringed paler, broader in ♀♀ and young. Strongly territorial and usually encountered alone or in pairs. Mostly forages in series of hops on ground (fewer aerial sallies), perching briefly on raised surface before dashing forward again to secure its prey. Also short runs on flat surfaces, and stance usually half-upright. Often wags tail in rather deep/quick downward and slower upward motion, with quite deep bobbing of head and fore body. Generally sings from elevated perch. Song-flights frequent when newly arrived, rarer later.

VOCALISATIONS Song a fast, scratchy strophe which often commences with or includes the whistling call note *heet* and sometimes imitations, a short warble or chatter with characteristically bright jingling sounds (strophes vary depending on context and also individually). A certain geographical variation noted in that breeders in S Europe have more low-pitched and *Monticola* thrush-like voice, not as high-pitched and scratchy as in N Europe. – Calls include a hard tongue-clicking *chack* or *chack-chack* (or rendered *tuc*) and a straight whistle, *heet*, in contact or when alarmed. The two are often combined and used alternately.

SIMILAR SPECIES Breeding ♂ largely unmistakable, with diagnostic blue-grey tone to crown and mantle, and unmistakable face marks. Main confusion risk of ♀-like plumages is with *Isabelline Wheatear* (which see). – ♀-like birds (♀, fresh autumn ♂ and any first-winter) require careful separation from *Black-eared Wheatear*. Latter usually more buffy or sandy in general coloration, with rather plain, featureless face, lacking obvious dark eye-stripe or distinct whitish or pale buff supercilium (as typical of Northern Wheatear). Further, most Black-eared have a narrow, broken tail-band, with more extensive black on outer web of outermost feather, and least black on second to fourth tail-feathers. In all plumages, Black-eared is longer-tailed, more slender, and frequently flicks wings and tail, while preserving an upright posture. – For differences from similar juvenile Isabelline, Pied and Black-eared Wheatears, see those species.

O. o. oenanthe, ad ♂, **Norway**, Jun: diagnostic ash-grey upperparts and unmistakable face markings; orange-buff below varies in extent and tone, here quite restricted to throat. Aged by evenly blacker ad wing-feathers. (E. Foss Henriksen)

O. o. oenanthe, 1stS ♂, **Netherlands**, Apr: less 'tidy' and less pure grey above than ad ♂, but best aged by more worn, browner wings and tail (juv feathers), especially primary tips and primary-coverts. Note obvious moult limits in median coverts. (N. D. van Swelm)

WHEATEARS

AGEING & SEXING (*oenanthe*) Ageing of ♂♂ usually possible in 1stY but of ♀♀ require close inspection of moult pattern and feather wear in wing. Colour of inside of upper mandible most useful in autumn but hardly practical for other than handled birds. Sexes obviously differ in spring/summer but less clearly so in autumn; biometrics sometimes useful. – Moults. Complete post-nuptial and partial post-juv moults in summer (mostly Jul–Aug). Post-juv renewal includes all head, body, lesser and median coverts and, less frequently, a few innermost greater coverts; rarely 1–2 tertials. Partial pre-nuptial moult in late winter (mostly Jan–Mar), usually involving head/body, much less frequently a few innermost greater coverts and, rarely, some lesser and median coverts or tertials. – **SPRING Ad ♂** Crown to back blue-grey to very pale grey, tinged pale brownish when fresh, and whitish underparts variably tinged sandy-buff or pinkish-buff, sometimes reaching almost to belly. Dark areas of flight-feathers blackish, feathers usually neat and only moderately worn. **Ad ♀** Dark areas of ear-coverts, remiges and tail brown (less contrasting), not black or blackish-brown as in ♂. Overall dull buff-brown above, and breast tinged sandy-yellow to buff-brown. Some are better marked and slightly recall ♂, but less grey above, and face marks more diffuse, less

O. o. oenanthe, 1stS ♀, Netherlands, May: easily separated from ♂ by overall dull buff-brown upperparts, with dark areas of ear-coverts, remiges and tail brown, not black or blackish-brown as in ♂ (some better-marked ♀♀ recall ♂, but always less grey above). Aged by many juv wing-feathers, although note new smallest tertial. (R. Pop)

Ad ♂, probably *O. o. oenanthe*, Spain, Oct: breeding plumage only partially concealed, so despite being to some degree ♀-like at this season, crown and mantle retain vestiges of grey, with diagnostic blacker lores and ear-coverts; note evenly-feathered wing with blacker-centred remiges and primary-coverts. (A. M. Domínguez)

♀-like, Belgium, Sep: probably *O. o. oenanthe*, but in fresh autumn plumage cannot be certainly identified to race. Sexing of all ♀-like plumages (including 1stW ♂) often unsafe without handling, and typically requires accurate ageing. However, rather well-marked dark lores and blackish primaries might suggest 1stW ♂. (K. De Rouck)

like ♂ and many probably not possible to sex in the field (ageing should precede sexing). On average browner-tinged flight-feathers, not as blackish as in ♂ (but some overlap). Warmer brown upperparts without greyish cast. Less dark underwing. **1stW** Very like ad, being ♀-like. Lores and ear-coverts brown to brown-black, supercilium buffish. On average, extent of black on outer tail-feathers greater than in ad. Best aged by in-hand examination of retained juv upperwing-feathers (though moult limits less clear than in spring); in particular note slightly worn primary-tips, less firmly structured primary-coverts, and some show characteristic moult limit in inner greater coverts (juv coverts slightly browner than ad-like post-juv ones). Inside of upper mandible partly yellow or grey (very rarely black), but becomes more ad-like by late autumn. **Juv** Unlike similar ad ♀, body plumage has soft, fluffy feathering, while much of upperparts mottled, indistinctly so on throat/breast and upper flanks. Underparts largely buffy-white. Ear-coverts and lores usually clearly darker.

O. o. oenanthe/libanotica, 1stW, presumed ♀, Turkey, Sep: aged by juv primary-coverts (narrower with rather narrow tips, broadly edged sandy-white with browner centres), as well as apparent moult limit in greater coverts (inner ones renewed). Rather pale lores suggest 1stW ♀. Note species' characteristic black T-shaped tail marking. (V. Karakullukçu)

contrasting. Flight-feathers dark grey-brown, similarly neat as in ad ♂. **1stS ♂** As ad ♂ but has more worn and browner wings and tail, browner cast to grey upperparts, less extensive white forehead, sometimes a less black and contrasting mask, paler underparts, and is overall less evenly and contrastingly patterned. Most reliably aged by clearly bleached, browner juv primaries, primary-coverts and usually some outer greater coverts. Inside of upper mandible (both sexes) largely as ad, but some retain faint yellow tinge. **1stS ♀** At times possible to age by juv primaries, primary-coverts and outer greater coverts, but difference in wear and colour slight and moult limits less obvious than in 1stS ♂, and many birds difficult to age safely. – **AUTUMN Ad** Both sexes evenly fresh (very rarely retains a few unmoulted feathers). **Ad ♂** More ♀-like than in spring, but crown retains vestiges of grey, with diagnostic blacker lores and ear-coverts, and has blacker secondaries and primaries than ♀/1stW. Brown mantle/back has faint greyish cast (whereas ♀ is warmer rufous-cinnamon). Underwing-coverts and axillaries on average darker than in ♀, blackish with only narrow paler fringes. Sometimes a little pure white visible in forehead or foremost supercilium. **Ad ♀** Much

O. o. oenanthe, juv, Hungary, Jun: soft, fluffy body plumage extensively mottled above, on throat, breast and upper flanks. Yellow mouth flanges still present. Base of lower mandible pale. (T. Grüner)

BIOMETRICS (*oenanthe*) **L** 14.5–16 cm; **W** ♂ 93–101.5 mm (*n* 30, *m* 97.8), ♀ 91–98 mm (*n* 29, *m* 94.3); **T** ♂ 52–60 mm (*n* 31, *m* 55.6), ♀ 48–57 mm (*n* 29, *m* 52.7); **T/W** *m* 56.4; **B** 15.0–18.7 mm (*n* 60, *m* 16.7); **B**(f) 10.7–13.5 mm (*n* 21 *m* 11.9); **BD** 3.7–4.7 mm (*n* 21, *m* 4.1); **Ts** 24.7–28.3 mm (*n* 55, *m* 26.7). **Wing formula: p1** < pc 2–7 mm, < p2 51–60.5 mm; **p2** < wt 1–4 mm, =4/5 or 5 (90%) or =3/4 or 4 (10%); **pp3–4** usually equal and longest (but p4 sometimes 0.5–2.5 mm <); **p5** < wt 5–10.5 mm; **p6** < wt 12–17 mm; **p7** < wt 17–22 mm; **p10** < wt 27.5–33.5 mm; **s1** < wt 30–37 mm. Emarg. pp3–4, sometimes also on p5 but if so usually less prominently.

GEOGRAPHICAL VARIATION & RANGE Moderate variation, with northern birds being larger and slightly more rufous-tinged in fresh autumn plumage. Separation requires careful examination of birds of equivalent sex/age and degree of wear. Isolated N African population (*seebohmi*) morphologically quite different and here tentatively treated as a separate species, Seebohm's Wheatear.

○ ***O. o. oenanthe*** (L., 1758) (most of N, C & W Europe south to Pyrenees, France, N Balkans, much of E Europe east through N Ukraine to Siberia except in south; winters sub-Saharan Africa, commonly in Mauritania, N Nigeria, Mali, Egypt, Sudan, Ethiopia and south to Zambia). Described above. (Syn. *integer*; *oenantheoides*; *rostrata*.)

○ ***O. o. libanotica*** (Hemprich & Ehrenberg, 1833)

O. o. libanotica, ad ♂, **Turkey, May**: averages paler, being ash-grey above, with frequently more extensive white forehead than *oenanthe*, while underparts are relatively whiter, but much overlap in these characters. On average, black subterminal tail-band narrower in *libanotica*, best seen here when tail slightly spread. Aged by evenly blacker ad wing-feathers. (R. Armada)

O. o. libanotica, ♀, **Turkey, May**: slightly paler and greyer above, and on average whiter below, with somewhat stronger face pattern like ♂, but many only fractionally paler or inseparable from ♀ *oenanthe*. Note apparently longer bill, typical of this form, especially from Turkey and eastwards. No moult limits in wing visible, while very dark remiges and primary-coverts might suggest ad, but strong feather wear prevents certain ageing. (M. R. Kaleli)

O. o. libanotica, ad ♂, **Israel, Sep**: only head shows some grey ♂-like coloration, with blackish lores and extensive white forehead, which in this case extends to forecrown, characteristic of *libanotica*. However, identification in autumn of *oenanthe* versus *libanotica* is very difficult or impossible. Aged by evenly fresh plumage. (A. Ben Dov)

O. o. leucorhoa, ♂, presumed ad, **Greenland, Jun**: less colourful ♂, although still has more pinkish-orange than most *oenanthe* at same season. Presumed ad by apparently evenly-feathered wings. (R. Martin)

O. o. leucorhoa, ♂, presumed 1stS, **Greenland, Apr**: not all ♂ Greenland Wheatears are as distinctive as this bird, with its rich and extensive rufous-orange underparts. Note apparent moult limits in greater coverts and what appear to be juv primary-coverts, indicating 1stS. (G. Conca)

Presumed *O. o. leucorhoa*, ♂, **Netherlands, Apr**: heavy build, broad tail-band and warmer below. The species' black T-shaped tail marking is shown perfectly here. Ageing difficult without handling. (N. D. van Swelm)

(S Europe south of Pyrenees; Balearics, N Africa, Greece, Bulgaria, E Romania, S Ukraine, Turkey, Levant, Transcaucasia, Iran, Central Asia east to Altai; Tien Shan, Mongolia, Transbaikalia; winters probably in much the same range as *oenanthe*). Averages subtly smaller, longer-tailed and longer-billed than *oenanthe*, and breeding ♂ is paler with upperparts slightly paler ash-grey, underparts whiter with paler breast (cream or cream-buff), while white on forehead is more extensive and purer than in *oenanthe*. Many ♀♀ differ less strongly from ♂♂ than in other races, being to some extent paler above and whiter below, and face pattern often quite well marked with whiter/broader supercilium and even blackish lores. In fresh plumage difficult or impossible to separate from *oenanthe*. Both sexes have on average narrower black tail-band than *oenanthe*. **W** ♂ 93–100.5 mm (*n* 15, *m* 96.9), ♀ 88–95 mm (*n* 12, *m* 92.8); **T** ♂ 54–61.5 mm (*n* 15, *m* 56.7); ♀ 50–56 mm (*n* 12, *m* 52.8); **T/W** *m* 54.9; **B** 17.0–19.0 mm (*n* 27, *m* 18.0); **B**(f) 12.0–14.0 mm (*n* 20 *m* 13.1). **Wing formula: p5** < wt 3–8 mm; **p10** < wt 24–31 mm; **s1** < wt 27–34 mm. (Syn. *argentea*; *grisea*; *nivea*; *virago*.) – Many birds from wide contact zone between this race and *oenanthe* intermediate, and migrants in Africa and S Europe often impossible to assign to subspecies.

O. o. leucorhoa, 1stS ♀, Greenland, Jun: extensive but rather diluted pinkish-cinnamon on cheeks and over entire underparts. Very similar to *oenanthe* and would not be possible to distinguish in the field without the help of the locality. Heavily worn and bleached juv primary-coverts. (R. Martin)

Presumed *O. o. leucorhoa*, ♂ (left) and ♀-like (right), Netherlands, Sep: larger and longer-winged birds, especially those with brighter, pink- or rufous-orange underparts encountered in W Europe, are often assigned to *leucorhoa*, but some *oenanthe* can be confusingly brighter and larger-looking. The ♂ is ad (extensive grey crown and solid black lores) and can be identified as *leucorhoa* by very long primary projection and relatively broad tail-band, while ♀ (probably ad by lack of apparent moult limits in wing) is also probably *leucorhoa* given very rich and extensive cinnamon underparts. (N. D. van Swelm)

O. o. leucorhoa (J. F. Gmelin, 1789) 'GREENLAND WHEATEAR' (Canada, Greenland, Iceland, Jan Mayen, Faeroes; winters N Afrotropics between Senegal and Mali). Rather distinctive but still not always safely separable. Typically larger and longer-winged, with ♂♂ having brown-tinged grey upperparts, and brighter, rufous-orange underparts. ♀/1stW browner and on average rustier on upperparts than in *oenanthe*, and cheeks and underparts deeper and more extensively rusty-cinnamon. Wing shape on average subtly more pointed than in *oenanthe*. **L** 16–17.5 cm; **W** ♂ 98.5–109 mm (*n* 26, *m* 104.4), ♀ 97–107 mm (*n* 19, *m* 101.5); **T** ♂ 54–62 mm (*n* 26, *m* 57.8), ♀ 53–59.5 mm (*n* 19, *m* 56.4); **T/W** *m* 57.3; **B** 16.1–18.5 mm (*n* 24, *m* 17.0); **Ts** 26.5–29.5 mm (*n* 24, *m* 28.0). **Wing formula: p1** < p2 56–64 mm; **p2** =3/4 or 4 (50%) or =4/5 (50%); **p3** longest; **p4** < wt 0.5–2.5 mm; **p10** < wt 28–34 mm; **s1** < wt 32.5–38 mm. (Syn. *schioeleri*.)

TAXONOMIC NOTE Considering morphological distinctness of isolated North African population, ♂♂ invariably having a black throat bib, we have tentatively chosen to treat this as a separate species, Seebohm's Wheatear *O. seebohmi* (which see).

O. o. leucorhoa, ♀-like, Azores, Nov: race deduced by range, extremely long primary projection, extensive pinkish-cinnamon underparts, and overall heavier/chunkier appearance. Sex and age impossible to establish with certainty. (D. Occhiato)

SEEBOHM'S WHEATEAR
Oenanthe seebohmi (Dixon, 1882)

Fr. – Traquet de Seebohm
Ger. – Seebohmsteinschmätzer
Sp. – Collalba del Atlas; Swe. – Atlasstenskvätta

A localised summer visitor to the Atlas Mountains of Morocco and Algeria, and morphologically well differentiated from Northern Wheatear. Although still poorly known, and genetic data are lacking, Seebohm's Wheatear is apparently an incipient species that is treated separately here to highlight its distinctiveness. A relatively short-distance migrant, *seebohmi* winters mainly close to the Atlantic coast, though it has been recorded once each in W Egypt and Gibraltar, and several times in the Libyan Desert. It is a specialist of high steppes and dry, rocky mountain-slopes.

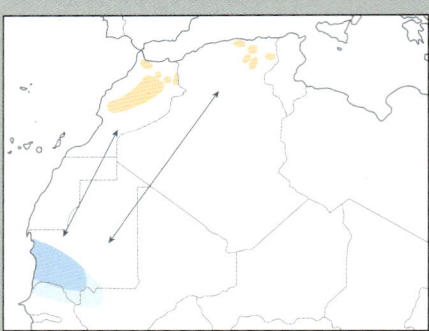

Ad ♂, Morocco, Mar: characteristic are pale feather fringes rather late in spring; note pleasant buff-cinnamon tinge on outer scapulars, a feature which is variable. Black bib and head-sides, and uniformly black-centred (ad) primary-coverts age and sex this bird. (A. B. van den Berg)

Ad ♂, Morocco, Apr: aside from diagnostic black bib, has rather narrow black tail-band, making tail appear proportionately longer, i.e. greater portion exposed beyond wingtips, and shorter primary projection. In Mar and even Apr, can still show some pale tips from fresh plumage, here mainly on greater coverts. The primary-coverts are diagnostically ad. (K. De Rouck)

IDENTIFICATION About same size as Northern Wheatear (especially close in size to southern population of latter) with *slightly shorter, more rounded wing* (exposed primary projection just under equal length of exposed tertials, but is usually somewhat shorter than that, *c.* 3/4 of tertials), sometimes creating a slightly more long-tailed impression. *Narrower black terminal tail-band*, exposing more eye-catching white on tail beyond primary tips, contributes to long-tailed impression. Bill somewhat stronger and legs a little longer than in *oenanthe* Northern Wheatears of N Europe, but about equal to those of *libanotica* in S Europe. Breeding ♂ recognised first of all by *extensive black face/bib that appears to merge narrowly with black of wing-bend*, as well as by *ash-grey upperparts, very broad whitish supercilium and forehead* (white reaching to forecrown), and rest of underparts and rump being whitish, with diagnostic *solid black underwing-coverts*. ♀ almost identical to ♀ Northern Wheatear, especially of southern populations, but rarely also dark-throated morph appears different from any Northern. Tail of both sexes has *typical black 'T'-shaped mark*, similar to Northern Wheatear or slightly narrower, less solid terminal band. Behaviour much like Northern Wheatear.

VOCALISATIONS Song characteristically consists of a few slow, low-pitched phrases often with an inserted 'crumpled paper' sound. Can include imitations. Song usually longer than Northern Wheatear and said to be more measured, melodious and sonorous. A drumming sound, like that of displaying Common Snipe, also described. Perched song may contain notably buzzy elements, but chuckles and rattles also reported, sometimes simultaneously. Similar structure to display-flight song. – Calls probably similar to Northern Wheatear, although *tuc* call apparently softer and less sharp.

SIMILAR SPECIES Unlike *Northern Wheatear*, plumage characters of ♂ Seebohm's (especially adults) less obscured by pale feather fringes in autumn, and therefore retains diagnostic characters year-round. ♀ and some 1stW ♂♂, however, virtually identical to ♀-like Northern Wheatear, and perhaps identified only by the following features, though care is required: (i) averages slighter and slimmer with proportionately shorter wings/longer tail than Northern Wheatear; (ii) the shorter primary projection of Seebohm's is *c.* 3/4 of tertial length (at most just under equal length of tertials), whereas in Northern Wheatear it usually far exceeds tertial length. White rump patch is distinctly larger than in Northern Wheatear (by 10–40%), mainly due to more restricted grey back, reaching only to about midpoint of tertials rather than almost their entire length in Northern Wheatear. Underwing-coverts characteristically more blackish/dusky than in Northern Wheatear. On average, black terminal tail-band narrower and perhaps less solid than in Northern Wheatear. Typically, especially older ♀ Seebohm's tend to have extensive bluish-grey upperparts, warmer cinnamon-buff head-sides, throat and breast, solid dark ear-coverts and lores, and prominent whitish supercilium, but many *libanotica* Northern Wheatears share some or all of these features. In the hand, also compare distance of p5 and p6 to wingtip, which seems to differ between Northern Wheatear and Seebohm's Wheatear, and corresponds to the shorter primary projection and more closely-spaced primary tips in Seebohm's. – In NW Africa, might also be confused with dark-throated ♀ *Maghreb Wheatear* (*O. halophila*). Seebohm's needs also to be distinguished with care from ♀-like plumages and young/winter ♂♂ of *Black-eared Wheatear*. Much less likely to be confused with *Desert Wheatear*.

AGEING & SEXING Ageing requires close inspection of moult and feather wear and pattern in wing. Colour of inside of upper mandible apparently as in Northern Wheatear but unstudied. Sexes differ year-round. – Moults. Few data

♀ (left) and 1stS ♂ (right), Morocco, May: generally rather daintier than Northern Wheatear, and ♂ has diagnostic black throat. Especially in distant views, ♂ also superficially resembles dark-throated western race of Black-eared Wheatear, and also consider possibility of confusion with Maghreb Wheatear, given some overlap in range, at least in non-breeding season. Seebohm's has ash-grey upperparts and well-developed broad whitish supercilium. Separation of ♀ far trickier. Ageing of ♂ also more straightforward, especially due to strong moult contrast in wing; ageing of ♀ impossible. (K. De Rouck)

♀, **Morocco, May**: two images of the same 'pale type' ♀, which is the most frequent in Seebohm's, and most challenging to separate from ♀ Northern. Compared to that species has slighter and slimmer feel, proportionately shorter wings and longer tail, shorter primary projection, typically narrower tail-band, and some ash-grey on upperparts. This bird has rather obscure whitish supercilium. Strong feather wear to primary tips, possible moult limits in wing and juv primary-coverts suggest 1stS. (B. Maire)

♀, **Morocco, Apr**: 'intermediate type' in terms of dusky elements on face. Ash-grey above, and such birds are readily separated from ♀ Northern. (D. Mitchell)

on moult but apparently mostly as Northern Wheatear with timing as *libanotica*, but post-breeding moult slightly earlier, from Jun. – **SPRING Ad** ♂ See above under Identification. **Ad** ♀ Almost identical to ad ♀ (and young ♂) *libanotica* Northern Wheatear, and contrary to previous descriptions nearly always lacks dark bib. Only very rarely do ♀♀ with grey throat occur. Upperparts largely pale grey (with limited sandy-brown suffusion) including on crown. Usually has duskier or almost blackish lores and ear-coverts; whitish supercilium often narrow and short, hardly extending to forehead (less obvious than in ♂, but more so than in young ♀). Also, compared to ad ♂, flight-feathers and tail-band often browner, and upperwing, if pale fringes not worn, less pure grey. Underparts often slightly buffier than in ♂, but can become almost white when worn. **1stS** ♂ As ad ♂ but subtly browner wings and tail, browner cast to grey upperparts, less extensive white forehead, sometimes less solid or only partially dark face, especially bib (which in least-developed birds is almost absent), thus usually appears overall less contrastingly patterned; least-marked 1stS ♂ without black bib usually still separated from any ♀ by greyer upperparts and better-marked mask. Usually readily aged by browner juv primaries and some coverts (including

♀, **Morocco, Apr**: two images of the same 'dark-throated' type ♀, which is unlikely to be confused with Northern or any other species in range. Note pale gap at centre of lower throat. Ageing uncertain without handling, but generally fresh wing-feathers in Apr suggest ad. (D. Mitchell)

all primary-coverts). **1stS ♀** Best aged by moult limits (see 1stS ♂). Almost always has pale sandy-brown upperparts with limited or no bluish-grey cast, while supercilium is much less distinct, dark ear-coverts reduced and often mottled, and lores paler. Quite typically, head-sides and throat to breast tinged buffish-cinnamon. — AUTUMN **Ad ♂** As spring. Unlike Northern Wheatear, pattern only to limited degree obscured by pale fringes, with black of face still extensive and upperparts largely greyish or slightly sandy-grey. **Ad ♀** Much as spring but warmer. **1stW ♂** More ♀-like and rather featureless, but still approaches fresh ad ♂, having greyer upperparts and darker face with exposed black mottling especially on sides and lower throat; warmer below. (Variation poorly known and requires more study.) **1stW ♀** Least-contrasting plumage, with paler face pattern, and browner/duller ear-coverts. Both sexes aged by moult limits, with juv primaries, primary-coverts and outer greater coverts (ad evenly fresh) as Northern Wheatear. **Juv** As juv Northern Wheatear or paler and buffier overall, with weaker dark mottling and scalloping. Juvs of the two are unlikely to come into contact.

BIOMETRICS L 14–16 cm; **W** ♂ 91–101 mm (n 20, m 95.4), ♀ 87–95 mm (n 13, m 92.0); **T** ♂ 54–61 mm (n 20, m 57.9), ♀ 48–57 mm (n 13, m 53.1); **T/W** m 59.5; **B** 16.8–20.3 mm (n 33, m 18.4); **BD** 3.9–4.7 mm (n 32, m 4.3); **Ts** 25.2–29.3 mm (n 33, m 27.4). **Wing formula: p1** < pc 0–4.5 mm, < p2 48–58 mm; **p2** < wt 2–5 mm, =4/5 or 5 (83%) or =5/6 (17%); **pp3–4** equal and longest; **p5** < wt 2–4.5 mm; **p6** < wt 7.5–11.5 mm; **p7** < wt 13.5–17 mm; **p10** < wt 22–25 mm; **s1** < wt 23–30 mm. Emarg. pp3–5 (on p5 sometimes a little less prominently).

GEOGRAPHICAL VARIATION & RANGE Monotypic. — Atlas Mts and adjacent ridges in Morocco, Algeria and presumably W Tunisia; short-range movements in winter to lower altitudes and somewhat further south.

TAXONOMIC NOTE Appears to have diverged quite significantly from allopatric Northern Wheatear, with key differences in vocalisations and plumage, most notably the black bib of ♂. Genetically it appears to be close to identical to Northern Wheatear, a parallel to the situation with Cyprus Wheatear and Pied Wheatear. The split is tentative and based on degree of distinctness and a comparison with other similar cases with diverged allopatric populations. Requires further research.

REFERENCES   Riad, A. M. (1995) *Bull. OSME*, 35: 36–37.

Ad ♂, Morocco, Sep: in fresh plumage following post-nuptial moult, with pale sandy-brown or white tips covering parts of summer plumage, but at least some ♂♂ show some grey on crown, and with much of black face and bib still visible they remain quite distinctive. (R. Short)

Ad ♀/1stW ♂, Morocco, Sep: from this angle, impossible to conclude if this bird is 'dark-throated' ad ♀ (a rare variety) or 1stW ♂. Only black-centred greater coverts visible, which together with solid black lores and partial blackish bib support the first option, but highlights need for further study of this poorly-known wheatear. (B. Maire)

♀, Morocco, Sep: virtually identical to ♀ Northern, including pale supercilium above and behind eye. Note apparent relatively short primary projection (with close-spaced primary tips) and longish tail, and ash-grey tinge on scapulars. Ageing not easy. (B. Maire)

1stW ♀, Morocco, Sep: very fresh, probably at end of post-juv moult, with still rather unsettled feathering of head and mantle. (B. Maire)

PIED WHEATEAR
Oenanthe pleschanka (Lepechin, 1770)

Fr. – Traquet pie; Ger. – Nonnensteinschmätzer
Sp. – Collalba pía; Swe. – Nunnestenskvätta

While rather abundant in S & E Arabia on passage, and as a breeder east of the Caspian Sea, Pied Wheatear is a rare and local breeder or a vagrant over most of the covered region. Throughout its range, the species is often confused with Black-eared Wheatear, chiefly the eastern race *melanoleuca* (with which Pied Wheatear hybridises locally), and the occurrence of a close congener, Cyprus Wheatear, makes this group one of the toughest identification challenges. The breeding range stretches from E Bulgaria and E Romania to Lake Baikal, south to C Iran and NC China.

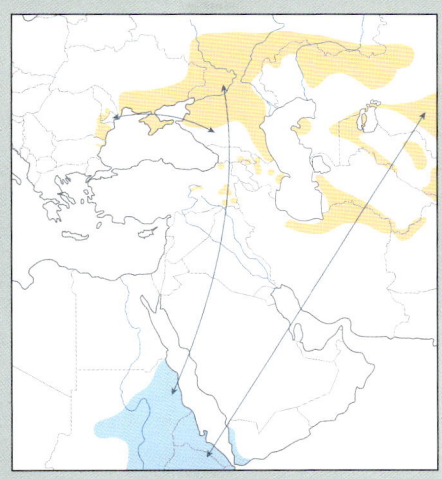

♂, presumed ad, Mongolia, Jun: black underwing-coverts and rather narrow and ragged black tail-band widest at sides. Birds with such extensive black tail-band usually 1stY, but in this case tips are rounded and broad suggesting an ad. (J. Normaja)

Ad ♂, Russian Altai, Jun: with wear during breeding season, buffish and brown fringes lost and pale parts become purer white. Combination of black of face down to upper breast, broadly connected to upperwings on neck-sides, black of wings connected over back, no buffish-pink tinge to undertail-coverts, and no white/grey wing panel typical. Aged by lack of any juv wing-feathers. (M. Vaslin)

IDENTIFICATION A medium-large, fairly slim wheatear with marked sexual dimorphism and an inverted black 'T' against white tail pattern, with black centre and uneven-width terminal band, *outer tail-feathers having more extensive black than those tail-feathers nearer centre* (a pattern shared by Black-eared Wheatear). Some adults, especially ♂♂, have broken black terminal band on each side of centre, white on inner webs reaching tips. ♂ in summer has *black head-sides, throat, chest, wings and back*, whereas crown, nape, and lower breast to vent are mainly white (lower breast below black chest sometimes tinged buff or faintly rusty-ochre when fresh, in autumn–late spring). Black of chest continues uninterrupted into black of wing. (As to rare white-throated morph, var. 'vittata', see below.) *Centre of crown usually sullied or mottled grey-brown*. Note that *undertail-coverts are clean white*, never buff or rusty-tinged. ♀ is dull brown-grey above and on head and chest, and wings are dark brown-grey. Plumage, shape and behaviour very similar to Black-eared Wheatear (particularly ssp. *melanoleuca*), and some ♀♀/young may be close to indistinguishable; see under Similar species and that species. When angle prevents view of *black back* of ♂ Pied, other features must be used, detailed under Similar species. Problematic autumn plumages usually have characteristic *pale feather fringes, imparting scaly impression*, and ♀-like birds are *colder and greyer brown above*. ♂ Pied can also strongly resemble Cyprus Wheatear but differs in longer primary projection (roughly equal to exposed tertial length, whereas projection is only 2/3 in Cyprus), more extensive white on both nape and rump, on average narrower black tail-band, and in having less extensive yellow-ochre and brown-grey tips to white feathers in fresh plumage. Pale 1stW ♀ Cyprus and fresh 1stW ♂ Pied can appear almost identical and require careful assessment of all characters. Separation from most other congeners is less problematic. – Dark-throated birds predominate, while pale-throated morph '*vittata*' is rare (comprising 2–3% or less in museum collections). Among spring/summer ♀♀, dark-throated (normal) morph shows large spectrum of variation as to darkness of throat and breast, and extremes could be labelled 'dark-throated' and 'pale-throated', though division will be arbitrary and diffuse due to variation. Note that pale-throated birds never have as clean white throat as ♀♀ of the '*vittata*' morph.

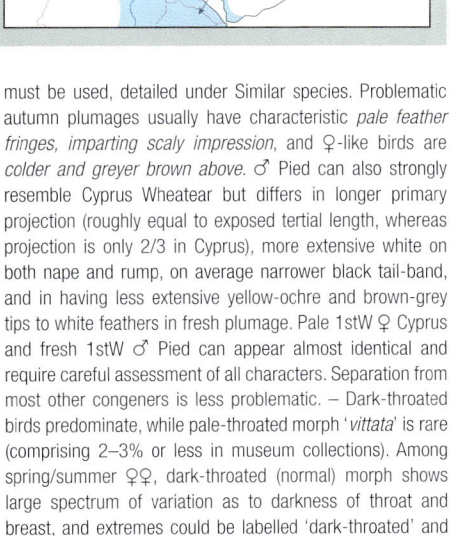

VOCALISATIONS Song often indistinguishable to human ear from Black-eared Wheatear (especially *melanoleuca*) being similarly explosive, twittering and dry, containing many musical, whistled, clicking and trilling notes: e.g. *surtru-shirr-echu*. Strophes sometimes rise in pitch and are often repetitive. There is some variation in intonation, many dialects occur, and at times it resembles Northern Wheatear song. Strophes are perhaps shorter on average and intervening pauses longer than in Black-eared Wheatear, but general performance very similar. Habitually includes brief elements of mimicry of local birds. Song occasionally protracted, often then with more mimicry interwoven and with mechanical, clattering elements (likened to sound of typewriter; Panov 2005). – Main call a frothy *brsche*, at times also throaty clicking *tshak* notes. Alarm is a straight, shrill whistle.

SIMILAR SPECIES Most often confused with Black-eared and Cyprus Wheatears, and advice below focus on these. A possible confusion could also arise with extralimital Gould's Wheatear (see below). – Black-eared Wheatear of eastern race *melanoleuca* has similar structure to Pied: primary projection differs only subtly and on average, thus much overlap; only birds with even longer primary projection than exposed tertials are most likely Pied, and those with clearly shorter are Black-eared Wheatear, whereas most birds fall into the 'overlap zone'. However, western Black-eared race *hispanica* differs more clearly from Pied by having projection *c.* 3/4 of tertial length. Tail pattern nearly identical in Pied and Black-eared, but separates them from most other wheatears.

Ad ♂, Bulgaria, Jun: black of wings connects over back, a feature shared with some of *O. lugens* complex, but unlike that lacks any pale inner webs visible in spread wings. Aged by evenly-feathered wing. (R. Wilson)

HANDBOOK OF WESTERN PALEARCTIC BIRDS

1stS ♂, Kuwait, Apr: many 1stY ♂♂ appear scruffy; in general, retain brownish tips on crown into April or even May. Note distinctive moult limits with especially contrasting juv remiges, primary-coverts and greater coverts. (M. Pope)

Variation in spring/summer ♀♀, Mongolia, Jun: the normal and commonest form, the so-called 'dark-throated morph', includes extensive variation as to the degree of darkness of throat, but this also varies with degree of feather wear. The darker-throated bird on the left shows rather uniformly dark throat and upper breast, whereas the paler-throated bird on the right has paler throat but still streaked and mottled. Such pale-throated birds are clearly differentiated from the white-throated '*vittata*' morph (which see). (H. Shirihai)

Pied adult ♂ can only be confused with Black-eared adult ♂ at angles where colour of back is not visible. Pied has large black bib broadly meeting black or brownish-black mantle and wings, whereas Black-eared has pale mantle and scapulars, and dark throat of black-throated morph of Black-eared is often less extensive and never meets dark of wings (beware apparent connection in some postures, especially in *melanoleuca* Black-eared, which has larger bib than *hispanica*). Additionally, fresh ad ♂ Pied has broad pale fringes to black lesser coverts (almost solidly black on Black-eared Wheatear). First-winter ♂ Pied is generally intermediate between fresh ad ♂ and ♀, and extremely similar to first-winter ♂ (and some more ♂-like ♀♀) *melanoleuca* Black-eared. Some may be identified if dark throat extends broadly onto upper breast (partially mottled), as well as (on some) by definite dark connection between throat patch and scapulars/mantle (in fresh dark-throated Black-eared, pale-fringed throat patch is isolated and restricted, or only fractionally extends onto upper breast, and is more contrastingly separated from the warmer orange-buff breast). Upperparts perhaps diagnostically duskier (or even patchily blacker) and more brownish-grey (lacking warmer tones of some Black-eared Wheatears), typically with more distinct pale fringes, broader on scapulars and affording overall mottled and much less uniform appearance than in Black-eared. In some, crown also mottled unlike in Black-eared. Furthermore, in the hand, has concealed centres of lower mantle feathers black (not

♀, Mongolia, May: in typical worn plumage, identification is less problematic (especially given dark upperparts and blotchy lower breast). Darker patches on sides of lower throat, upper breast and neck highly characteristic, and better developed with wear. Ageing without handling difficult. (T. Langenberg)

♀, dark-throated bird, Bulgaria, May: some ♀♀ paler and have ginger/cinnamon hue to crown and ear-coverts, but rest of upperparts colder and greyer. Note how exposed black feathering of throat extends to upper breast and neck-sides, and rather long primary projection. (B. Nikolov)

Ad ♂, Kuwait, Oct: slim and elongated shape with good primary projection. These characters, together with narrow black tail-band and seemingly much white basally on nape and upper mantle, help to eliminate similar 1stW ♂ or ad ♀ Cyprus Wheatear. Fresh feathers provide safest age criterion. (M. Pope)

Ad ♂, Bulgaria, Sep: either an extreme variation of Pied Wheatear, or, perhaps more likely, a hybrid between this and *melanoleuca* Black-eared Wheatear, considering the very narrow (almost broken) connection between black bib and wing, which is untypical for an ad Pied. Aged by primary-coverts, which are black-centred with whitish fringes that broaden near tips (juv coverts have thinner buffish fringes of more even width). (T. Muukkonen)

white as in Black-eared), even though such central patches may be very small and missing on some feathers. Spring/summer ♀ (worn) Pied typically is dusky greyish-brown above and on head (never has ginger/cinnamon tone of most *hispanica* and some *melanoleuca*), thus usually is clearly darker than Black-eared Pied has extensive, but variable, dark grey-brown or blackish throat reaching neck-sides/scapulars and upper breast, where border usually diffuse or uneven, and typically breast is colder buff, blotched/streaked darker at sides. In spring/summer dark-throated Black-eared Wheatear ♀, throat patch is smaller, not reaching upper breast (or only marginally so in *melanoleuca*), nor connecting with scapulars, and normally has quite conspicuous, well-defined orange-rufous breast-band. Autumn/winter ♀ Pied (fresh) sometimes impossible to separate from Black-eared, but note: typical Pied has neat pale fringes to cold grey-brown crown and mantle/scapulars, while Black-eared usually is rather more uniform and warm darkish brown above (some *hispanica* can have pale rufous tinge, whereas this often is lacking in drabber *melanoleuca*), and usually has only limited and faint paler fringes to scapulars/mantle. Throat and breast patch coloration often essential in identification of ♀♀: both Pied and Black-eared can have pale-tipped dark throat, but note that bib may reach upper breast on Pied but not in Black-eared. Note that bib of dark-throated Pied ♀ has ill-defined paler mottling on lower part, and any whitish lower border to throat/upper breast is indistinct or lacking; in dark-throated Black-eared ♀, throat is rather less grey, and whitish divide more pronounced. Beware, however, that especially on *melanoleuca* Black-eared, the pale area of throat and whitish divide are less contrasting, and bib can appear more extensive. Juvenile Pied Wheatear is often indistinguishable from juv Northern Wheatear (except that tail pattern usually differs) and, especially, from juv Black-eared Wheatear, but often has slightly dusky buff, less scalloped throat and breast. Best identified by accompanying adults. – Very similar to *Cyprus Wheatear*, sharing with it the black back in ♂, but Cyprus differs in several obvious or subtle details: (i) darker, in fresh plumage with more extensive rusty-buff or ochre and dark grey tips to pale areas rendering autumn birds markedly 'swarthy'; (ii) shorter primary projection (about 3/4 of exposed tertials) and subtly shorter tail; (iii) less extensive white rump area (but larger black area on back); (iv) stronger rusty-buff suffusion on underparts even in summer ♂ (when Pied becomes white); (v) on average broader black terminal tail-band (pattern with tail-band broken by white, as often seen in Pied, does not occur); and (vi) subtly less extensive white area on nape, usually not reaching upper mantle (as often in Pied). ♀ Cyprus is much more ♂-like than ♀ Pied

and needs to be separated from either 1stY ♂ Pied or the most dark-throated and ♂-like adult ♀ Pied. Use primary projection, size of white rump patch and rusty-buff saturation of underparts as main separation criteria. – ♂ Pied is superficially similar to *Mourning Wheatear*, but latter has obvious pale areas on flight-feathers (whitish edges to inner webs of primaries and secondaries) readily visible on flying birds, and rusty undertail-coverts. – *Gould's Wheatear* (*O. capistrata*; extralimital) is very similar, and ad ♂♂ of these two are easily confused, but Gould's has near-even-width black tail-band and whiter lower breast, only slightly buff-tinged when fresh, pure white when bleached. On average, crown in Gould's is less grey-mottled on centre than in Pied. For separation of ♀♀, see under Blyth's Wheatear, Taxonomic notes. – Separation from *Northern Wheatear* is by and large as for Black-eared Wheatear (which see).

AGEING & SEXING (normal morph; white-throated morph var. '*vittata*' treated separately below) Ageing after post-juv moult requires close inspection of moult and feather wear in wing. Sexual dimorphism generally obvious in spring and summer, but sometimes less so in fresh plumage. Seasonal

♂, Netherlands, Oct: typical autumn ♂ with pink-buff hue on pale parts and extensive pale tips obscuring summer pattern. Ageing difficult; degree of pale fringing more or less intermediate, or closer to 1stW ♂ than to fresh ad ♂. Moult pattern in wing and shape and pattern of primary-coverts also difficult to precisely match to any particular age. (J. van Holten)

1stW ♂, Sweden, Dec: characteristic pinkish-buff breast, with dark connection between throat and scapulars/wings (via neck-sides), and brownish-grey upperparts, which are often pale-fringed, broadest on scapulars. Aged by moult limits in wing, with juv primary-coverts and outer greater coverts. (S. Hage)

1stW ♂ Pied Wheatear (left: Denmark, Nov) versus dark-throated Eastern Black-eared Wheatear (right: Israel, Sep): most 1stW ♂ Pied identified by dark throat extending to upper breast, or if dark connection between throat and scapulars/wings is confirmed. In contrast, in dark-throated Eastern Black-eared the extensively pale-fringed dark throat at most only reaches upper breast, never onto neck-sides. Pied is more brownish-grey, often distinctly pale-fringed, broader on scapulars with overall mottled or scalloped effect, lacking warmer-toned and plainer upperparts of most Eastern Black-eared. (Left: S. Kristoffersen; right L. Kislev)

1stW ♀, Netherlands, Nov: a 'classic' example with neat pale fringes to cold grey-brown crown and mantle/scapulars, pale drab-grey throat with ill-defined whitish lower border, while breast has slight buffy tinge, forming diffuse breast-band. Clear moult limits among greater coverts establish age. (T. Bakker)

Autumn ♀ Pied (top left and bottom left: Kazakhstan, Sep), Eastern Black-eared (top centre and bottom centre: Israel, Sep) and Western Black-eared Wheatears (top right: UK, Oct; bottom right: Spain, Sep): many ♀♀ cannot be labelled to species with certainty. Separation of Pied from Western Black-eared is less problematic due to warmer or more russet breast and plainer upperparts of latter, and considerably shorter primary projection than Pied. Note that Pied and Eastern Black-eared can look virtually identical. However, Pied is colder and more greyish above with pale-fringed crown and mantle/scapulars (more uniform and warm darkish brown upperparts in Eastern Black-eared). Pied's pale grey throat extends down to upper breast, and whitish lower border is indistinct or absent (throat less grey, and whitish divide usually more pronounced in Eastern Black-eared). Breast-band suffused pinkish-buff, sides sometimes slightly blotched (Eastern Black-eared tends to have better-defined orange-buff band, contrasting with throat and whitish divide). Ageing uncertain without handling. (Top left and bottom left: R. Pop; top centre and bottom centre: A. Ben Dov; top right: M. Goodey; bottom right: G. Reszeter)

variation involves mainly abrasion of pale feather tips, most obvious change noted in ♂♂. – Moults. Complete post-nuptial and partial post-juv moult mainly in Jul–Aug (rarely Jun–early Oct). Post-juv renewal usually involves head, body, lesser and median coverts and often c. 5 innermost greater coverts, more rarely also the innermost tertial. Partial pre-nuptial moult apparently absent or scarce and then very limited (at least no proof of regular partial pre-nuptial body moult). – SPRING **Ad** ♂ Black of face and bib (to upper breast) and upperparts, including wings, broadly connected on neck-sides. Crown usually has brownish-cream feather-tips (enhancing white supercilium), but becomes cleaner (some already pure) white from Apr/May. Only limited orange-buff tinge to white underparts. Some retain a few pale fringes on black throat, rarely also on mantle, scapulars and wings. Tail pattern variable but usually has narrow or broken terminal band and extensive black on outer web of r6. **Ad** ♀ Upperparts uniformly dusky olive-brown (sullied greyish when fresh). Ear-coverts often slightly warmer with gingery-brown tint. Throat/upper breast variably dark, though usually has slightly paler upper throat and chin, a darker lower border, where characteristically bib is faintly streaked and blotched (especially at darker sides), and often a faint, narrow orange-buff wash below this. Some are paler and more uniform on throat, but in other extreme (especially older dark-throated ♀♀) throat can appear quite solidly dark with wear. Tail pattern variable but on average broader and more even-width terminal band than in ♂. **1stS** ♂ Like ad ♂ but often less smart plumage with more pale tips. Wings browner due to retained juv primaries, primary-coverts and perhaps some alula and outer greater coverts (bleached and centred dark brown, not black), forming clear moult limits. **1stS** ♀ Like ad ♀ but primary-coverts duller brown. Juv greater coverts and alula very worn with browner centres contrasting with fresher ad-like post-juv feathers, but moult contrast less obvious than in 1stS ♂. – AUTUMN **Ad** ♂ Compared to 1stW ♂, black upperparts only partially concealed by dusky greyish or white tips (appears paler fringed especially on scapulars and lesser coverts), and black face and neck-sides have only limited pale fringes. Also, unlike 1stW ♂, much of wing-feathers solidly black-centred, including primary-coverts. Also characteristic

♀, Bulgaria, Sep: pale fringes and some greyish tips above and on crown are notable, and uniform whitish throat to upper breast is evident; note hint of dark lateral throat and whitish submoustachial stripes, and rather long primary projection. No moult limits visible in wing, but age still uncertain. (T. Muukkonen)

♀ Pied (left: Denmark, Oct) versus pale-throated Eastern Black-eared Wheatear (right: Israel, Sep): especially with pale-throated birds, overall pattern can be very similar (with white throat patch of Eastern Black-eared extending to upper breast), but overall coloration of Pied is colder/greyer and breast-band blotchier on sides, whereas Eastern Black-eared has broad rufous-buff breast-band. (Left: P. Poulsen; right: L. Kislev)

Juv, Mongolia, Jun: no definitive differences from juv Black-eared Wheatear, but tends to average duller and to be less boldly scalloped. (H. Shirihai)

is greyish-white supercilium bordering cream-brown crown/nape (mottled darker mainly in centre) down to mantle. Pale pinkish-orange breast quite pronounced. **Ad** ♀ Ageing difficult but unlike 1stW ♀ wings evenly fresh, and inside to upper mandible mostly black or blackish-grey. Unlike in spring, typically colder brown-grey above with ill-defined pale mottling (faint scaly fringes) on crown, nape, scapulars and mantle. Rather uniform dusky throat extends to upper breast, and breast-band greyish indistinctly tinged pinkish-buff, generally mottled dark, especially on sides. Chin and upper throat are usually creamy-white. **1stW** ♂ Resembles ad ♂ but usually more extensively fringed and mottled pale on dark areas above and on face, thus, appearance approaches ♀; diagnostic black connection on neck-sides obscured or seldom apparent, being more or less concealed by pale tips. Some already have quite substantial black bases exposed on scapulars/mantle, forming conspicuous dark mottling, and blackness of upperparts may approach ad ♂. Compared to latter, apparently only very rarely has all-black lores. Unlike ad ♂, remiges and all coverts, including primary-coverts, are prominently pale-fringed. Also check for moult limits with retained juv greater coverts, tertials and alula, which are softer, have brown centres and ochre-white fringes, while ad-like post-juv ones are fresher, more firmly structured, have black centres and pale yellowish-white fringes. **1stW** ♀ Extremely similar to fresh ad ♀ and very difficult or impossible to age in the field, but primary-coverts have medium to dark brown centres with thin whitish or ochre-white fringes (in ad dark brown to blackish-brown with very narrow or no pale fringes and tips); juv greater coverts, tertials and alula softer structured with browner centres and ochre-white fringes, while ad-like post-juv feathers are slightly fresher, more firmly structured, with dark brown to blackish-brown centres and pale yellowish-white fringes. At least in early autumn, inside of upper mandible partially greyish-yellow. **Juv** Body plumage recalls ♀, but most of upperparts mottled darker, and indistinctly mottled also on throat/breast, which are dusky buff and contrast with buffy-white underparts; tail-band slightly broader than in ad.

Ad ♂, var. '*vittata*', Mongolia, Jun: the white-throated morph; unlike Black-eared Wheatear, black of ear-coverts connects via neck-sides with sides of mantle, and back is all black. Aged by all-dark wing. (H. Shirihai)

1stS ♂, var. '*vittata*', United Arab Emirates, Mar: in early spring crown still extensively dark-tipped. Aged by pointed and bleached brown juv primary-coverts, and retained juv outer greater coverts and all remiges. (O. Campbell)

BIOMETRICS L 14–15.5 cm; **W** ♂ 90–100 mm (n 73, m 94.3), ♀ 85–97 mm (n 48, m 91.0); **T** ♂ 55–68 mm (n 73, m 60.7), ♀ 56–63 mm (n 48, m 58.8); **T/W** m 64.5; **B** 15.0–17.7 mm (n 112, m 16.4); **BD** 3.5–4.3 mm (n 32, m 3.9); **Ts** 21.5–24.0 mm (n 37, m 22.7). **Wing formula:** **p1** > pc 4 mm, to 3 mm <, < p2 44–52 mm; **p2** < wt 3.5–8 mm, =5/6 (83%), =5 (16%) or =6 (1%); **pp3–4** about equal and longest; **p5** < wt 1–4.5 mm; **p6** < wt 7–12 mm; **p7** < wt 12.5–17 mm; **p10** < wt 21–29 mm; **s1** < wt 22–30 mm. Emarg. pp3–5.

GEOGRAPHICAL AND OTHER VARIATION & RANGE
Monotypic. – E Bulgaria, E Romania, Ukraine, SE Russia, SW Siberia east to Transbaikalia, N Mongolia, N Kazakhstan, Caucasus, Transcaucasia, N Iran, east to Pamirs, Tibet; winters SW Arabia and NE & E Africa. (Syn. *leucomela*; *morio*.) – The white-throated morph var. '*vittata*' is here afforded a separate entry with a brief description and biometrics (although as yet based on scant material) to facilitate future investigations. See Taxonomic notes for alternative taxonomic treatments of var. '*vittata*'.

O. pleschanka var. '*vittata*' (occurs over much of the range of *O. pleschanka* at least between Transcaspia and Tibet but hardly at all in western part; also on migration and in winter in Arabia, Middle East and NE Africa; winter range as normal morph, vagrant to Britain). Differs in both sexes by having white or cream-white throat so that the resultant black eye-stripe forms a prominent mask, continuing on sides of neck and at rear meeting black/dark of shoulders/wings. On average perhaps slightly richer pink-buff on underparts in fresh plumage (but requires confirmation on larger sample).

Var. '*vittata*' or hybrid, presumed 1stS ♂, Kuwait, Feb: variation also includes ♀-like presumed 1stS ♂ (or ♂-like ♀) with blackish ear-coverts, to some extent resembling Northern and pale-throated Eastern Black-eared Wheatears, but has characteristic Pied/Black-eared wing and tail patterns. Since blackish mask is not connected with mantle, a hybrid Pied × Black-eared cannot be excluded. (V. & S. Ashby)

♀, var. '*vittata*', Kazakhstan, Apr: highly distinctive given prominent white throat. Probably ad by less worn wings, without apparent moult limits. (A. Wassink)

♀, presumed ad, var. 'vittata', Kuwait, Mar: distinctive bold white throat; wing pattern and long primary projection essentially ♀ Pied-like. Colour (still blackish in Mar) and shape of remiges and primary-coverts point to ad. (R. Al-Hajji)

Ad ♂, var. 'vittata', United Arab Emirates, Oct: when fresh, the black areas of the ear-coverts and wing-bend are mostly disconnected, or reduced to a fractional, dusky indication of it on the neck-sides. Combination of evenly fresh ad wing and intense black ear-coverts allow inferences on age and sex. (O. Campbell)

W ♂ 91–100 mm (n 18, m 94.8); **T** ♂ 58–68 mm (n 18, m 62.2); **T/W** m 65.7; **B** 15.2–17.3 mm (n 20, m 16.3); **BD** 3.4–4.3 mm (n 18, m 3.9); **Ts** 22.0–24.1 mm (n 19, m 22.8).

TAXONOMIC NOTES Forms superspecies with Cyprus Wheatear and Black-eared Wheatear, and has been considered conspecific with the former and sometimes even with the latter. Breeding ranges largely complementary, with hybrid populations of *melanoleuca* × *pleschanka* in E Azerbaijan (Grabovsky et al. 1992), and intermediate birds of possibly recent but more likely ancient hybrid origin observed on E Caspian Mangyshlak Peninsula (Panov 1986, 2005), possibly also in Bulgaria, Turkey (?) and N Iran. Note that some of these reports may refer to colour morph var. 'vittata'. – It should be noted that not all authorities agree that 'vittata' is a colour morph, this being only one of two traditional treatments. Panov (1986, 1992, 2005) suggested that 'vittata' is the result of ancient hybridisation ('governed by the laws of balanced genetic polymorphism') between Pied and Black-eared Wheatears east of the Caspian Sea, for which he gives some statistical support. See also Grabovsky et al. (1992) regarding hybridisation in Azerbaijan. However, since 'vittata' is extremely rarely recorded in Bulgaria, Romania, Ukraine and Turkey, countries where you would expect mixing of Pied and Black-eared Wheatears to be equally common, and because 'vittata' seems to appear fairly frequently well east of the range of Black-eared, locally even deep into China (Xining, NE Tibet, own observations), hybridisation as sole explanation seems less plausible, or at least not the full reason. Others have even suggested that 'vittata' is a separate subspecies (which it clearly cannot be) or species, but sufficient proof of the latter as yet lacking. – Name 'gaddi' from Iran refers to hybrids in which black eye-stripe reaches shoulders (as in Pied), but with pale mantle and inner scapulars (as in Black-eared).

REFERENCES Ash, J. (1956) *BB*, 49: 317–322. – Dernjatin, P. & Vattulainen, M. (2005) *Alula*, 11: 98–107. – Grabovsky, V. I., Panov, E. N. & Rubtsov, A. G. (1992) *Zool. Zhurn.*, 71: 109–121. – Haffer J. (1977) *Bonn. zool. Monogr.*, 10: 1–64. – Hinchon, G. (2008) *BW*, 21: 121–122. – Panov, E. N. (1986) *Zool. Zhurn.*, 65: 1675–1683. – Stoddart, A. (2008) *BW*, 21: 156–157. – Ullman, M. (1992) *Sandgrouse*, 14: 58–59. – Ullman, M. (1994) *DB*, 16: 186–194. – Wassink, A. (2004) *DB*, 26: 43–45.

1stW ♂, var. 'vittata', Kuwait, Oct: aged by juv remiges, and primary-coverts and outer greater coverts, forming moult contrast with surrounding feathers, while blackish neck-sides best fit fresh young ♂. (R. Al-Hajji)

♀, var. 'vittata', presumed 1stW, United Arab Emirates, Oct: when fresh, white areas typically sullied buff-grey, here notably on throat and supercilium. At first glance more likely overlooked as 1stW ♂ Black-eared *melanoleuca* rather than Pied, but note Pied's pale-tipped upperparts. Tentatively aged by juv remiges (seemingly worn and brown) and moult limit in greater coverts, while obscured blackish neck-sides and ear-coverts identify it as ♀. (O. Campbell)

CYPRUS WHEATEAR
Oenanthe cypriaca (Homeyer, 1884)

Alternative name: Cyprus Pied Wheatear

Fr. – Traquet de Chypre; Ger. – Zypernsteinschmätzer
Sp. – Collalba chipriota; Swe. – Cypernstenskvätta

As its name suggests, this is an endemic breeder to Cyprus, where it is widespread, even occurring in villages and on the outskirts of towns. Often perches on treetops and other high lookouts, even TV aerials in towns. Winters in NE Africa and regularly recorded on passage in Levant March–April and October–November; frequently overshoots to S Turkey in spring. Previously regarded as a race of Pied Wheatear but nowadays treated as a separate species on account of totally different song and sufficiently different morphology.

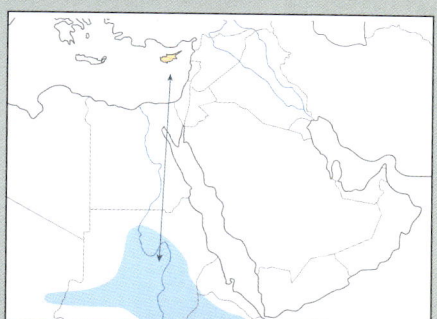

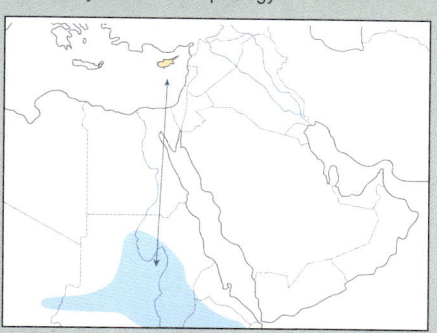

Ad ♂, Cyprus, Apr: more extensive black on upper mantle compared to Pied. This bird could invite confusion with Mourning Wheatear, but note uneven black tail-band pattern with more black on outer feathers. No juv wing-feathers and less worn wings identify it as ad. (A. McArthur)

Ad ♂, Cyprus, May: despite individual variation in both species, ♂ Cyprus, compared to Pied, has on average more extensive black upper breast, and at least until early spring rest of breast and flanks more extensively suffused deeper cinnamon-buff (but whiter ♂♂ also occur). Aged by apparent lack of moult limits in wing. (D. Occhiato)

IDENTIFICATION Rather small and compact wheatear with only *slight sexual variation*. Spring ♂ and fresh autumn birds (both sexes, especially first-winter) are almost identical to Pied Wheatear, while *very dark spring ♀ exhibits bold buffish or white supercilia that may meet on nape*, forming diadem-like marking unique among similar species. Generally darker above than Pied, with *buff-tinged underparts, dark sooty-brown to brown-grey crown and nape*, and *smaller white or buff rump patch* than Pied. Thus, whitish rump patch ends level with tip of shortest or central tertial, but in Pied white area is larger, reaching higher on back, approximately level with tip or base of smallest tertial (young and ♀ of both species tend to have a smaller white area). White on nape extends subtly less far down upper mantle than in Pied, but difference slight and some overlap. *Black tail-band is generally broader* but somewhat variable (overlaps with Pied, and young and ♀♀ have on average wider tail-band than adult ♂ in both). Structural features also useful, especially *smaller size* and *shorter primary projection* with resultant shorter and blunter wingtip (best appreciated in flight) and longer outer primary. In flight, wings often appear short and square. *Shorter bill and tail* also useful, though some experience may be required in their use, particularly with lone individuals. For difficult birds see below under Similar species.

VOCALISATIONS Song diagnostic within genus, a series of hoarse, cicada- or bush-cricket-like buzzing notes totally lacking the clearer warbling and varied character of relatives, a fairly loud and monotonous *bizz-bizz-bizz-...* or *zee-zee-zee-...* Single strophe usually lasts on average eight seconds (varying between 5–15 sec.) but may occasionally last up to one minute. The song may terminate in a few high-pitched, harsh or drawn-out notes. Apparently does not mimic other birds. – Calls include a hard *tchak* or *tsik*, a frizzling *brzü*, rasping *dsed-dsed* and a whistled *djui* or *jüi*.

SIMILAR SPECIES Very similar to *Pied Wheatear* in many plumages, but separable if all characters are carefully assessed in combination. First note size, structure and general plumage differences as detailed under Identification. Furthermore, behaviour and voice also useful in breeding season: songs differ clearly, and Cyprus often uses elevated song posts. The following is an identification summary of the main plumages in relation to Pied. Adult ♂ Cyprus has more extensive black upper breast, sometimes extending

1stS ♂, Israel, Mar: in spring young ♂ is like ad, but has more brown-grey primary-coverts, less solid black face, and brownish tips to mantle and scapulars. (T. Krumenacker)

♂ **Cyprus Wheatear (left) versus Pied Wheatear (right), Israel, Mar**: some overlap in characters, partially because birds are different ages: Cyprus is ad while Pied is 1stS. Normally, Cyprus tends to maintain darker crown and more saturated pink-buff on underparts than Pied, but this difference is less obvious when you compare 1stY. Also, the two can almost overlap in amount of black on breast, while differences in length of primary projection and spacing, size of white rump, and length of p1, can be difficult to judge, thus use a range of features in combination. Cyprus also often appears smaller or more compact, with broader white forehead and supercilium, more solidly dark upperparts, and longer and broader black edges to outer tail-feathers. (H. Shirihai)

♀, **probably ad, Cyprus, May**: generally very similar to ♂, but on average a little darker and duskier, including crown, with bold whitish supercilia that often meet on nape. Upperparts typically sooty-brown to dusky brown-grey (rather than black), and underparts usually tinged buff. Difficult to confirm moult pattern and degree of wear in wing, but best fit is ad ♀. (A. McArthur)

1stS ♀, Israel, Mar: very similar to ad ♀, but duller brown above and black of face to some extent less solid, although age only confirmed by juv primary-coverts and remiges. (H. Shirihai)

to sides, while in (at least early) spring rest of breast and flanks are rich cinnamon-buff (much paler in Pied). In fresh autumn plumage crown/nape blackish-brown diagnostically concolorous with mantle (paler than lower mantle in Pied); in spring (Mar–Apr) centre of crown/nape still tipped dull black (paler grey-brown in Pied). Cyprus has black of back extending to upper mantle (white on Pied). In autumn, mantle/scapular fringes distinctly darker and thinner in Cyprus, and primaries only narrowly tipped pale, while black throat has less extensive, narrower and not as pale fringes as Pied, and upper breast is darker cinnamon-buff. ♀ Cyprus resembles ♂, and in spring has diagnostic dark plumage never found in Pied, while in fresh autumn plumage adult ♀ Cyprus often strongly resembles first-winter ♂. ♀ Pied (all ages) very different: grey-brown above (contrasting with darker wings), with diffuse creamy or buff-white supercilium and variable gingery-brown ear-coverts; chin and throat variably dark, rest of underparts whitish, though some have pale buff breast. — Some first-winter ♂ Cyprus approach dark, fresh adult ♂ autumn Pied (sometimes first-winter, too), but separated by relatively indistinct upperparts fringes, being almost uniform dark brownish from crown to mantle, scapulars only indistinctly tipped paler, tips not as pale as in Pied. Also warmer

Ad ♂, Israel, Oct: characteristic of ad Cyprus (both sexes) is the whitish central throat stripe, and the well-defined whitish-edged wing-feathers. Very similar to ad ♀ Cyprus but mantle and scapulars are blacker (less brownish) and underparts less deeply saturated orange-buff. Unlike the superficially similar fresh ad ♂ Pied, note only faint grey tips on mantle, while crown is uniformly tinged dusky olive-brown (with no pale mottling), making the shorter pale supercilium stand out better. Note also relatively shorter primary projection. (H. Shirihai)

Ad ♀, Israel, Oct: unmistakable due to typical dusky olive-brown crown, and about the same mantle colour. Compared to similarly-fresh ad ♂ (see image left), upperparts and wing-feathers less glossy black, somewhat tinged brownish, and fringes of scapulars, tertials, inner secondaries and greater coverts more buffish-cream, and with both supercilium and underparts warmer tinged. (H. Shirihai)

1stW ♂, Cyprus, Oct: in autumn, age and sex variation in Cyprus Wheatear is complex. Here, more buffish-olive fringing on lower mantle and scapulars (still faint compared to any Pied) and extensive dark facial bib point to ad ♀, but note retained juv primary-coverts and primaries, thus 1stW. Such plumage pattern can only be 1stW ♂ (1stW ♀ Cyprus is much duller—see other images). Also characteristic is extensive whitish tipping on sides of lower throat (sometimes almost making ear-coverts appear as solid black mask). (H. Shirihai)

Autumn Cypress Wheatear (left three, from top to bottom: 1stW ♂, Israel, Oct; 1stW ♀, Cyprus, Oct; ad ♀, Israel, Oct) versus Pied Wheatear (right three: 1stW ♂, Oman, Nov; 1stW ♂, Germany, Oct; 1stW ♂, Kuwait, Nov): separation in autumn quite often tricky. In general, Cyprus is always slightly darker than Pied when same age and sex compared, thus 1stW ♀ Cyprus is most similar to 1stW ♂ Pied, and 1stW ♂ or ad ♀ Cyprus is most similar to fresh ad ♂ Pied, etc. Familiarity with both species' variation and ability to age/sex birds correctly will enable identification. Nevertheless, some features assist separation of most ages/sexes: Cyprus has relatively indistinctly pale-fringed mantle and unmarked crown (lacking Pied's mottled or scalloped effect), appearing more uniform dark brown and dusky, with usually more contrasting pale supercilium, on average larger and more solid dark bib, warmer buff underparts, shorter primary projection, and broader tail-band. (Top left, centre and bottom: H. Shirihai; top right: T. Quelennec; centre right: B. van den Boogaard; bottom right: M. Pope)

buff below, more contrasting supercilium, more extensive blackish bib, while pale fringes to primary-coverts and tips to primaries are narrower. Size, structure, rump patch and tail pattern also useful. First-winter ♀ Cyprus is dullest, with coloration matching first-winter ♂ Pied, and some are so similar that, without in-hand examination, identification is unsafe. Given good views, size, structure, smaller white rump patch, more uniform dusky grey-brown upperparts, including

crown, and more evenly cinnamon-buff pigmented underparts, in combination can help identification, but possibly still inconclusive. – Beware also confusion risk between 1stW ♀ Cyprus and *melanoleuca* Black-eared Wheatear ♀ in autumn, but note Cyprus' decidedly duskier upperparts, better marked supercilium and broader black tail-band.

AGEING & SEXING Ageing requires close inspection of moult, feather wear and pattern in wing. Sexual dimor-

phism less clear-cut than in Pied. Seasonal variation mainly involves abrasion of pale feather-tips, but mainly evident in 1stY. – Moults. Complete post-nuptial and partial post-juv moult in summer (mainly Jul–Aug, rarely to early Oct). Post-juv renewal includes all head, body, lesser and median coverts, usually all greater coverts and most or all tertials, less frequently some tail-feathers and part of or entire alula. Pre-nuptial moult absent (*contra* previous reports suggesting

1stW ♀, Israel, Oct: although sometimes very similar to the dullest 1stW ♂ Pied, a few can be recognised by more solid olive-brown crown (no whitish tips), and shorter but more contrasting pale supercilium. Some variation as to degree of white tips concealing black on throat (in extreme examples as here, extensive whitish throat present) and the extent of rusty-buff on the breast (in some forming hint of breast-band), both variations bearing resemblance to pale-throated morph *melanoleuca* Black-eared ♀. Best means for separation from Black-eared are the duskier upperparts, better marked supercilium and broader black tail-band. (H. Shirihai)

it is partial). Variation in tail pattern by and large as in Pied Wheatear and Black-eared Wheatear ssp. *hispanica*. — **SPRING Ad ♂** With wear, dusky crown and buffish tinge below reduced, whitening as season advances, and usually from May/Jun these become almost pure white. Compared to 1stS ♂ neater with fresher and blacker remiges and primary-coverts, and secondary-coverts and alula lack moult limits. **Ad ♀** Typically ♂-like, but unlike ♂ dusky crown and buffish underparts never white in summer. Crown and nape variably greyish-black to dirty brown, mantle dark brown to blackish-brown, almost solid brown-black when worn. Supercilia may meet on nape, where typically blotchy. Lack of moult limits among secondary-coverts and alula, and remiges and primary-coverts fresher and darker, also help distinguish it from 1stS. **1stS ♂** As ad but note browner juv primaries, primary-coverts, perhaps some alula and outer greater coverts (very worn and weaker structured, and centres dark brown, not blackish), while some are less glossy and browner, approaching ad ♀ in early spring. **1stS ♀** Best aged by moult and wear (as 1stS ♂). — **AUTUMN Ad** Both sexes differ from 1stW in being evenly fresh without discernible moult limits. They also have characteristic narrow central whitish throat stripe, as whitish fringes on throat mostly concentrate there (in 1stW ♂ a wider area of throat is whitish, usually involving much of lower and lateral parts). **Ad ♂** Unlike 1stW ♂ (which generally recalls ad ♀) remiges black, not brown-black, primary-coverts purer black with only very thin greyish-white fringes (in 1stW ♂ slightly broader, buffier and less well defined), greyish-white fringes to wing-feathers narrower and more distinct, and face less extensively tipped white, giving impression of more solid dark bib. Also, better-marked greyish-white supercilium is characteristic (bordering brown crown/nape to upper mantle). Warm buff below. **Ad ♀** Differs from ad ♂ in having dark areas (including ear-coverts and bib) somewhat browner, and warmer buff-brown underparts add to rather darker appearance; supercilium also more sandy-buff. Overall pattern thus rather like 1stW ♂, but wings evenly fresh and primary-coverts blackish-brown with very narrow pale fringes. **1stW ♂** Generally like ad ♂ but more extensive pale fringes and less intense black areas, thus somewhat approaching fresh ad ♀ (ageing should precede sexing). Beware that due to extensive post-juv moult of secondary-coverts and tertials, it is usually difficult to locate moult limits in wing; still, any moult contrast best found between renewed tracts and retained juv primaries, which are slightly to moderately worn, with pale tips partly/mostly worn off; primary-coverts blackish-brown with slightly broader buff-white fringes (fringes at times of uneven width); any retained juv greater coverts, tertials and alula softer structured with dark brown centres, while ad-like post-juv ones fresher, firmer and have black centres, although fringed buffish (not as whitish as in later feather generations). Juv tail-feathers have broader black subterminal bar with less extensive whitish admixed. **1stW ♀** Moult and ageing as 1stW ♂. Typically has duller and more uniform brown upperparts tinged olive-grey, and compared to other plumages browner wing-feathers and more extensive pale fringes to wing- and face-feathers (thus dark bib often extensively concealed), as well as paler buff underparts, with pale buffy supercilia, making it palest of all *cypriaca*, and near-identical to dark 1stW ♂ Pied. **Juv** Differs from similar 1stW ♀ by being overall paler and having most of upperparts distinctly mottled. Throat and breast duskier than rest of buffy underparts with extensive dark scalloping.

BIOMETRICS L 14–15 cm; **W ♂** 83–90.5 mm (*n* 31, *m* 86.5), ♀ 81–87 mm (*n* 23, *m* 83.3); **T ♂** 54–61 mm (*n* 31, *m* 57.5), ♀ 53–59 mm (*n* 23, *m* 55.8); **T/W** *m* 66.7; **black on r1** 31–42 mm; **black on r4** 12–23 mm; **black on r6** 22–44 mm; **B** 15.0–17.2 mm (*n* 53, *m* 16.1); **Ts** 20.9–23.3 mm (*n* 31, *m* 22.1). **Wing formula: p1** > pc 5 mm, to 2 mm <, < **p2** 38–46 mm (cf. 44–52 mm in *pleschanka*); **p2** < wt 5–8.5 mm, =5/6 (rarely =6); **pp3–4** about equal and longest; **p5** < wt 1–2 mm; **p6** < wt 7–10 mm; **p7** < wt 12–15 mm; **p10** < wt 17.5–22 mm; **s1** < wt 20–25 mm. Emarg. pp3–5.

GEOGRAPHICAL VARIATION & RANGE Monotypic. — Restricted to Cyprus. More or less regularly overshoots in spring to S Turkey; winters N & NE Africa, SW Arabia.

TAXONOMIC NOTE Formerly considered a subspecies of Pied Wheatear, but the two differ markedly in song and to a lesser extent in morphology, behaviour, breeding biology and habitat. Randler *et al*. (2012) found that the *O. hispanica–melanoleuca–cypriaca–pleschanka* complex is genetically quite 'uniform', but still maintained species status for *O. cypriaca*, despite lack of any substantial genetic difference from *pleschanka*. It seems that in *Oenanthe*, morphometrics, plumage characters and vocalisations are of even greater importance for determining species status than genetic markers.

REFERENCES CHRISTENSEN, S. (1974) *Orn. Scand.*, 5: 47–52. – FLINT, P. (1995) *BB*, 88: 230–241. – OLIVIER, P. J. (1990) *Sandgrouse*, 12: 25–30. – FÖRSCHLER, M. I. *et al.* (2010) *Bird Study*, 57: 396–400. – RANDLER, C. *et al*. (2012) *J. Orn.*, 153: 303–312. – SLUYS R. & VAN DEN BERG, M. (1982) *Orn. Scand.*, 13: 123–128. – SMALL, B. J. (1994) *DB*, 5: 177–185.

1stW ♀, Israel, Oct: little known 1stW ♀ Cyprus Wheatear—best ID clues are the uniform and dusky olive-brown crown and upperparts, with only indistinct buff fringes on scapulars, bold cream-buff supercilium, short primary-projection and very extensive black tail-band. Since such dull plumage only can occur in 1stW ♀, ageing and sexing is unproblematic. (H. Shirihai)

Juv, Cyprus, May: strongly patterned with distinctly mottled upperparts, and throat to chest duskier with extensive dark scalloping and spotting. (G. Catley)

BLACK-EARED WHEATEAR
Oenanthe hispanica (L., 1758)

Fr. – Traquet oreillard;
Ger. – Mittelmeersteinschmätzer
Sp. – Collalba rubia; Swe. – Medelhavsstenskvätta

Almost anywhere you travel in the Mediterranean area and in Turkey in summer you are likely to see this wheatear on stony slopes and rocky outcrops, with open garrigue, scattered bushes and herbs. The breeding grounds stretch from S France south to Morocco, then east through the Balkans and in N Africa to NW corner of Libya. The range extends in the east to the Levant, Turkey, Transcaspia and W Iran. It is a migrant, wintering south of Sahara.

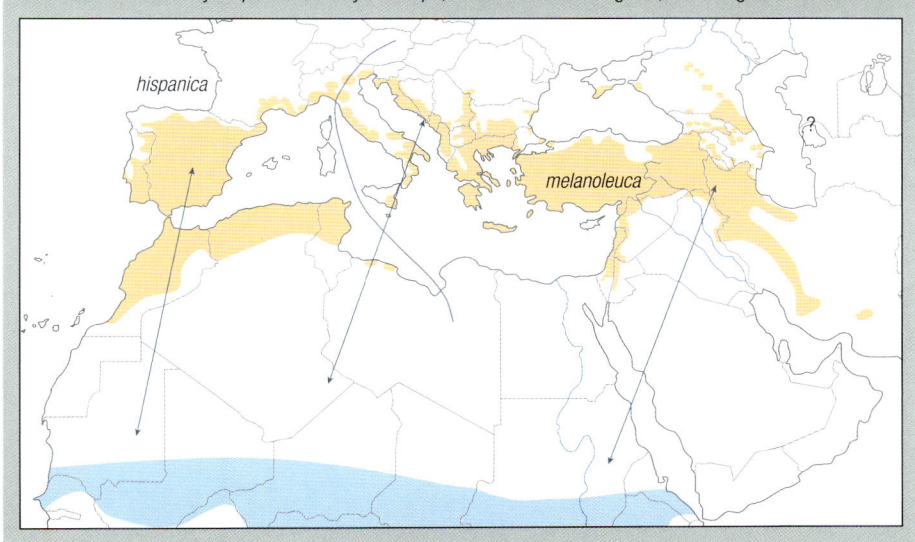

IDENTIFICATION Identified as a wheatear on size, shape, behaviour and *rump/tail pattern, white with inverted black 'T'*, best seen when it flies away or sideways. Two morphs, present in both sexes but most obvious among ♂♂, pale-throated (var. '*aurita*') and dark-throated (var. '*stapazina*'). Slim with *rounded head, smallish bill and longish tail*. Most have blackish or *very dark axillaries and underwing-coverts*. Width of black *terminal tail-band* variable, *most extensive on outermost feathers*, narrowest either side of centre, and in ♂♂ sometimes broken here by white (a pattern shared only by Pied Wheatear). Dark-throated and pale-throated morph ♂♂ (and some older ♀♀) distinctive, whereas most ♀♀ and young require care. Two rather distinctive subspecies complicate matters, Western *hispanica* and Eastern *melanoleuca*. – WESTERN: In dark-throated ♂, *black bib usually rather small* (but varies from very small to almost as large as in Eastern) and *always separated from dark area of wing and scapulars*, quite typically leaving *whitish band on sides of lower throat and upper breast*. Mask in pale-throated morph rather narrow, not extending appreciably onto forehead. *Pale/whitish area across mantle and back* separates it from several congeners with black mantles. ♀ much duller, more brown above, usually with only slightly darker ear-coverts and wings, and (mainly in autumn) often very similar to or inseparable from Pied (greatest problem usually with first-winters), though may have a *russet or ochre tinge to upperparts*, with no, or when fresh very faint and restricted, pale fringes to mantle and scapulars. Darkish (grey) throat usually clearly demarcated from rusty-tinged breast, though ♀ often lacks dark throat. – EASTERN: Very similar to Western but adult ♂ has more brown-grey (rather than buff-brown or sometimes rich cinnamon) *upperparts* when fresh, *becoming purer white* with wear. *Black of head extends slightly onto forehead* above bill base (no black above bill in classic Western), and dark-throated morph has *more extensive black bib, reaching lower throat* and sometimes marginally onto upper breast (in Western bib usually smaller, leaving lower throat and upper breast whitish). Also, mask of pale-throated morph of Eastern somewhat broader, including on lores, and can reach forehead. Note that nearly all first-winter and many first-summer ♂ Eastern lack the black forehead stripe and may appear to have a slightly smaller bib, thus subspecific identification of ♂♂ best attempted after age is established and by combination of several characters. Generally, first-summer ♂ Eastern appears more immature-like or less advanced than same plumages of Western. At all seasons, especially spring/summer, the black scapular area in Eastern is broader than in Western, leaving somewhat narrower pale or whitish area of mantle/back. ♀ Eastern is extremely similar to Western, and probably only a few can be reliably distinguished in the field outside their breeding ranges. Typical Eastern ♀ is duller than typical Western with fewer and duller cinnamon-buff pigments above and on chest (but some overlap). Also, dark-throated ♀♀ are more frequent in Eastern, and in extremes (usually older and worn), bib broader and more pronounced than in equivalent ♀ Western. Further, primary projection (both sexes) in Eastern averages longer than in Western but some overlap, and even other biometrics of limited use. Degree of overlap and individual variation make separation of ♀♀ best restricted to classic individuals, and must often be considered tentative. – Habitually tends to perch on bushes and small trees, less frequently on boulders or buildings (bobs and/or flicks wings, then pumps tail several times). Generally appears quite flighty, foraging by dashing over ground to catch prey and making short aerial sallies. Flight often high between perches and is generally direct and rather agile. Usually encountered singly. Song frequently given in circling song-flight (with jerky undulations on fluttering wings and fanned tail).

VOCALISATIONS Song is a repetitive, near-monotonous series of harsh, clear and chirpy warbling notes, e.g. *chu-chürrche-chuchirr-tri*, slightly falling in pitch (at distance superficially like Short-toed Lark or speeded-up Common Whitethroat). Also a good and varied mimic, capable of including fragments of perfect imitations of species like Barn Swallow, Linnet, Goldfinch, etc. At times perfect mimicry replaces own song, or alternates with it. Chattering subsong

O. h. hispanica, ad ♂, black-throated morph, Spain, Jun: small black bib and rich cinnamon buff-brown upperparts; black of face hardly reaches forehead. Evenly blackish remiges and primary-coverts readily age this bird as ad; by Jun outer primaries can bleach brown. (A. M. Domínguez)

O. h. hispanica, ad ♂, black-throated morph, France, Jun: tail often fanned in display demonstrating relatively little black terminally, except corners and central panel. Aged by less worn blackish remiges. (J. Fouarge)

O. h. hispanica, 1stS ♂, pale-throated morph, Spain, Jun: black mask relatively narrow (cf. *O. h. melanoleuca*), only extending to lowest forehead and hardly reaches above eye. Relatively immaculate plumage in 2CY ♂ (often more so than 2CY ♂ *melanoleuca*). Aged by strong moult contrast in wing. (A. M. Domínguez)

O. h. hispanica, 1stS ♀, pale-throated morph, Italy, Apr: russet tinge to plain upperparts and breast-sides, but otherwise extremely similar to ♀ *O. h. melanoleuca*. Compared to ♀ Northern has poorly-developed dark eye-stripe and pale supercilium, usually darker ear-coverts (though not in this bird), warmer tawny-buff above lacking obvious grey element, narrower and less even tail-band, and slightly shorter primary projection. Moult limits in wing reveal age. (R. Pop)

of harsh, hoarse sounds mixed with chirpy, clear whistles and mimicry. – Calls include a hissing or crackling *brsche*, a Northern Wheatear-like *tshack* and a descending whistle in alarm. Most calls similar to Pied Wheatear.

SIMILAR SPECIES ♀♀ and some first-winter ♂♂ require careful separation from *Northern Wheatear*. Note in Black-eared on average more uneven terminal tail-band with more black at corners and narrower black either side of centre, white sometimes reaching tail-tip (though pattern more similar in those with more extensive black in tail); Northern Wheatear never has very narrow or broken terminal band. Also, in Northern Wheatear crown to back has grey element being duller grey-brown in ♀♀ and first-winters, or rather pale greyish in summer ad ♀, less rufous-ochre, tawny-buff or cinnamon-tinged as in fresh Black-eared. Head pattern rather variable depending on species, age and sex, but any bird with very plain cinnamon-brown head (lacking dark eye-stripe or pale supercilium) is Black-eared Wheatear, not Northern Wheatear. Unlike Northern, Black-eared usually has warmer tawny-buff pigments on breast and sharply demarcated breast-band (bordered by darker sides and whitish belly, and often by dusky throat in dark-throated morph). Dark

O. h. hispanica, ad ♂, black-throated morph, Spain, Sep: evenly fresh, and unlike 1stW ♂, pale fringes disrupt overall pattern less, thus generally closer to spring/summer, with mask and bib almost solid black, while most of wing also black except for pale panel formed by broad rusty-buff fringes to greater coverts. (H. Shirihai)

O. h. hispanica, ad ♂, pale-throated morph, Spain, Sep: when very fresh, black of wing is broken up by quite broad buff-white edges and tips. Also, rufous-buff upperparts and breast more saturated. Aged by being evenly fresh and having more complete ad ♂ plumage with jet-black centres of primaries and primary-coverts. (J. L. Muñoz)

O. h. hispanica, 1stW ♂, pale-throated morph, Sweden, Oct: to some extent ♀-like, but ear-coverts too black for that, while upperparts are washed bright buffish rufous-brown, creating greater contrast with darker wings. Aged by juv extensively pale-fringed primary-coverts and outer greater coverts. (R. Svensson)

underwing-coverts and axillaries are more pronounced in all plumages of Black-eared. Combination of shorter wings (with slightly shorter primary projection) and narrow tail-band in Black-eared gives impression of broader white rump and uppertail-coverts. With experience, Black-eared is subtly smaller and slimmer (at least ssp. *hispanica*) with shorter legs and is more inclined to use bushes or other arboreal look-outs as perches, and often adopts less erect posture. Adult ♂ white-throated morph *hispanica* Black-eared Wheatear and Northern Wheatear have superficially similar blackish masks, clearly separated from blackish wings by pale neck-sides, which in Black-eared are obviously whiter (often tinged pale buffish-brown), always lacking grey-blue suffusion of Northern, and therefore appearing more clean black and white. Note, in strong sunlight, grey cast to mantle of Northern is less obvious so that it can appear quite whitish. Juveniles generally indistinguishable and best identified when seen in company of both parents. Tendency to have more obscure face markings, and different tail pattern compared to Northern Wheatear already evident. – For separation from Pied, Cyprus, Finsch's and Desert Wheatears, see those species.

AGEING & SEXING (*hispanica*) Ageing after post-juv moult requires close inspection of moult and feather wear in wing. Colour of inside of upper mandible useful in autumn for handled birds. Sexes differ, especially in spring/summer and adults. Seasonal variation mainly involves abrasion of pale feather tips, most obvious in ♂♂ and especially in young. – Moults. Complete post-nuptial and partial post-juv moults mainly in Jul–Aug (rarely Jun–early Oct). Post-juv renewal includes all head, body, lesser and median and most greater coverts, while innermost tertials and smallest alula-feather less frequently renewed. Partial pre-nuptial moult in Dec–Feb restricted to some feathers on head and foreparts of body, occasionally also a few innermost greater coverts and tertials, but incidence and extent apparently greater in 1stY than in ad. – **SPRING Ad** ♂ Lores, ear-coverts, upper-wings and axillaries solid black. Throat white or black. Pale upperparts white, variably tinged buff, but especially crown often tinged grey-brown. **Ad** ♀ Lores and ear-coverts brownish to dark brown-grey. Axillaries dark greyish-brown or greyish-black. Upperparts mainly brownish (with some russet tinge), and crown, nape, mantle and upper back lack or have very small white areas. For in-hand identification note frequent presence of small white central spots on most mantle-feathers (safely eliminating any Pied Wheatear). With wear, especially older ♀♀ can develop almost black ear-coverts (and bib in dark-throated morph). **1stS** ♂ Like

O. h. hispanica, autumn ♀♀, (left: England, Oct; right: France, Sep): certain individuals, or at certain angles, especially of western race, can appear superficially like Northern Wheatear, especially if differences in tail pattern not seen, or if only part of the pattern is visible, as here. Smaller size, delicate build, blander face pattern, and more rufous-ochre or cinnamon-tinged plumage provide first clues for fresh Black-eared. Unlike Northern, Black-eared usually lacks obvious dark eye-stripe or pale supercilium. Further, check for shorter primary projection, shorter legs and lack of clear grey tinge to upperparts. The left bird is 1stW due to already not so fresh tips to flight- and wing-feathers; the right bird showing fresh and more ad-like primaries could be ad, but difficult to be sure without handling. (Left: G. Thorburn; right Y. Kolbeinsson)

O. h. hispanica, 1stW ♀, Netherlands, Nov: typically washed buff above and on breast, with on average rather obscure face markings, but otherwise very similar to ♀ *O. h. melanoleuca*, and many impossible to separate. Post-juv median coverts contrast with juv primary-coverts. (M. Prins)

O. h. hispanica, juv, Spain, Jun: buff-brown body plumage is faintly marbled, upperparts being obviously scalloped brown on off-white ground. Very difficult to separate from juv Northern. (R. Aymi)

O. h. melanoleuca, ad ♂, pale-throated morph, Israel, Mar: versus *O. h. hispanica*, buff-brown wash to upperparts, and black of mask extends narrowly across base of forehead, broadening across lores and above eye (but slight overlap with extreme *hispanica*). Aged by blackish remiges and primary-coverts. (D. Occhiato)

O. h. melanoleuca, ad ♂, dark-throated morph, Turkey, Jul: compared to *O. h. hispanica* rather extensive black on forehead. Ad ♂ of both races can have black tail-band very thin or even completely lacking on many feathers. Aged by being evenly feathered and less worn, with blackish remiges and primary-coverts. (S. İmamoğlu)

Ad ♂ black-throated morph *O. h. melanoleuca* (left: Cyprus, Apr) versus *O. h. hispanica* (right: Tunisia, Mar), and young ♂ *O. h. melanoleuca* (centre: Turkey, May): black on *melanoleuca* extends slightly higher on forehead (none or very little in *hispanica*) and black bib reaches marginally down to upper breast, whereas in *hispanica* bib is usually smaller, with lower throat/upper breast whitish. However, many 1stS ♂ *melanoleuca* can be more like *hispanica*, thus important to age birds first. Much variation though, since this 1stS ♂ *melanoleuca* has larger bib than the ad ♂ at left. Ad ♂♂ aged by their evenly blackish ad-type wing-feathers, which are relatively less worn, while 1stS ♂ *melanoleuca* has distinctive moult limits in wing. (Left: J. East; centre: V. Karakullukçu; right: D. Occhiato)

ad ♂ but sometimes less neat or less strikingly patterned (e.g. in some, black throat less solid with some pale tips). Wings bleach browner than in ad ♂, especially juv remiges, primary-coverts and perhaps some alula and outer greater coverts, forming clear moult limits. **1stS** ♀ Like ad ♀ but wings more worn and bleached browner, though moult limits less obvious than in 1stS ♂. – **AUTUMN Ad** ♂ Unlike 1stW ♂, buffish-white wing fringes restricted to tertials, inner secondaries and few inner coverts, while most of wing, including primary-coverts, solid black. Mask (and bib in dark-throated morph) almost solid black, though fine pale fringes evident. Upperparts including crown typically washed buff to grey-brown. **Ad** ♀ Unlike similar 1stW, wing evenly fresh. Plumage often similar to, or inseparable from, 1stW ♂, thus ageing should precede sexing, and in most cases has browner wings, while axillaries and ear-coverts (and perhaps throat in dark-throated morph) are paler/obscured. **1stW** ♂ Rather variable: some rather advanced and recall fresh ad ♂, but many more like ♀, though usually have darker/blacker ear-coverts patch (and bib in dark-throated morph), upperwing and axillaries. Upperparts broadly washed cinnamon-buff or rufous-brown, obscuring whiter summer pattern, while black tail-band averages broader than in ad. For reliable ageing, in particular check for prominently pale-fringed juv remiges and wing-coverts, including primary-coverts, which are also of subtly softer structure, slightly less fresh and dark brown with pale ochre fringes and tips. Juv greater coverts softer textured with browner centres, offering some contrast with fresher ad-like post-juv ones. Juv alula-feathers have dark brown centres, while ad-like post-juv ones are distinctly black-centred. **1stW** ♀ Very similar to ad ♀ and reliably aged only by moult pattern in wings (as described under 1stW ♂), but limits much less contrasting. Generally, the least-marked and dullest plumage, but many difficult to separate from 1stW ♂. **Juv** Unlike ♀-like plumages, body plumage faintly scalloped dark brown on otherwise pale buff-brown head, rest of upperparts and throat/breast, which are never black but like rest of dirty cream-white underparts.

O. h. melanoleuca, 1stS ♂, black-throated morph, Kuwait, Mar: 1stS ♂ *melanoleuca* frequently has rather untidy plumage, especially if many juv wing-feathers are kept, in this case forming subtle moult limits with worn and browner juv remiges, primary-coverts and outer greater coverts. (A. Halley)

BIOMETRICS (*hispanica*) **L** 14.5–15.5 cm; **W** ♂ 85–93 mm (*n* 18, *m* 89.9), ♀ 83–90 mm (*n* 17, *m* 87.3); **T** ♂ 57–64 mm (*n* 16, *m* 61.5), ♀ 54–62 mm (*n* 17, *m* 58.8); **T/W** *m* 67.9; **B** 15.5–18.7 mm (*n* 28, *m* 17.2); **BD** 3.7–4.4 mm (*n* 17, *m* 4.0); **Ts** 21.5–24.3 mm (*n* 30, *m* 23.1). **Wing formula: p1** > pc 5.5 mm, to 3 mm <, p2 39–47 mm; **p2** < wt 2.5–7.5 mm, =5/6 (94%) or 5 (6%) (once =6/7); **pp3–4** usually equal and longest; **p5** < wt

O. h. melanoleuca, 1stS ♂, black-throated morph, Turkey, May: some ♂ *melanoleuca*, especially 1stS, have much duskier brown upperparts, almost like dark ♀♀; they also often have strong orange tinge to breast. Confusion with older dark-throated morph ♀ avoided if bird first aged. Note distinctive moult limits between older remiges and newer tertials. (D. Occhiato)

O. h. melanoleuca, 1stS ♀, pale-throated morph, Israel, Mar: compared to ♀ Northern has relatively weak face pattern, warmer tawny-buff breast contrasting with whitish belly and relatively shorter wings. Separation from ♀ *O. h. hispanica* and Pied Wheatear far more challenging. Evidence of moult limits among greater and median coverts, while primary-coverts, alula and tertials are entirely juv, indicative of 1stY. Nevertheless, the juv remiges are unusually fresh for spring. (D. Occhiato)

WHEATEARS

Spring pale-throated ♀♀, *O. h. hispanica* (left: Spain, Apr), *O. h. melanoleuca* (centre: Israel, Mar) Black-eared and Pied Wheatears (right: United Arab Emirates, Apr): ♀ *melanoleuca* is very similar to *hispanica* and only rarely separable, usually through being clearly duller cinnamon-buff above, but as shown here this does not always hold true due to individual variation or the effect of light etc. ♀ Pied represents another challenge, especially versus *melanoleuca*, but it is distinguishable by longer primary projection, typically greyish-brown upperparts, dark of throat reaching neck-sides and upper breast, breast being colder buff, and blotched or streaked darker at sides (lacking well-defined orange-rufous breast-band). Without closer inspection of moult pattern definitive ageing is impossible. (Left: A. Tate; centre: M. Varesvuo; right: A. Al Ali)

O. h. melanoleuca, 1stS ♀, Israel, Mar: during early spring migration, many ♀♀ are dull grey-brown above, with diffuse paler supercilium, variable dark on lower throat and cinnamon-tinged breast-band. Aged by many juv wing-feathers, with remiges and primary-coverts being worn and abraded. (H. Shirihai)

O. h. melanoleuca, 1stS ♀, Israel, Apr: a warmer-toned individual with rather short primary projection (somewhat like *hispanica*), but is duller/browner above, with paler breast-band, although assigned to race mainly by location. Age by wing being worn and brown with abraded fringes. (T. Krumenacker)

Spring dark-throated ♀ *O. h. melanoleuca* Black-eared (left: Israel, Mar) versus Pied Wheatear (right: Kazakhstan, May) with tricky intermediate (centre: United Arab Emirates, Mar): variation in ♀ *melanoleuca* especially compared to Pied often involves difficult identification challenges due to overlap in characters and some hybridisation. ♀ on left is dark-throated *melanoleuca* Black-eared by browner upperparts and dark of throat not reaching upper breast, instead contrasting with better-defined rufous-tinged breast-sides. In contrast, dark-throated ♀ Pied (right) has colder greyish-brown upperparts, with dark of throat reaching neck-sides and well onto upper breast, where border diffuse, with breast colder grey-brown and some faint blotching on sides. The central bird is more challenging, given its *melanoleuca*-like breast pattern. It is rather dull grey above with blackish-grey of throat reaching upper breast, so despite rufous on breast is probably an atypical Pied, although possibility of a hybrid cannot be eliminated and it is best left unidentified. Evenly-feathered and less worn wing of the *melanoleuca* indicates ad, but age of other two uncertain. (Left: A. Ben Dov; centre: D. Clark; right: C. Bradshaw)

1–3 mm; **p6** < wt 6–11 mm; **p7** < wt 11–17 mm; **p10** < wt 18–26 mm; **s1** < wt 20–26 mm. Emarg. pp3–5.

GEOGRAPHICAL VARIATION & RANGE Two subspecies displaying fairly obvious differences when typical birds are compared. However, in a slightly diffusely-delimited zone stretching from SE France or at least N & C Italy to N Balkans and E Europe, birds with intermediate appearance either dominate or at least occur frequently. The precise border between the races is therefore difficult to define.

O. h. hispanica (L., 1758) 'WESTERN BLACK-EARED WHEATEAR' (SW Europe east to at least S France but probably also extreme NW Italy, in the south in NW Africa from Morocco to NW Libya; winters sub-Saharan W Africa in Senegal, Mauritania, Mali, N Nigeria and SW Niger). Described above, in particular ♂ being on average more ochre-buff (or even pale rufous-tinged) in fresh plumage, and tends to retain ochre-buff colour longer in summer. ♀, too, on average more ochre-tinged, less greyish-brown above, but many very similar to *melanoleuca*. ♂ has rather narrow black mask (pale-throated morph 'aurita') or variably-sized but generally small black bib (dark-throated morph 'stapazina'). Note that some have black bib very small, covering little more than chin, whereas others have rather extensive black bib, approaching or copying size of some *melanoleuca* (bib length 12–25 mm). Black of lores does not extend to forehead above bill except very rarely (and then < 1 mm).

O. h. melanoleuca (Güldenstädt, 1775) 'EASTERN BLACK-EARED WHEATEAR' (S & C mainland Italy, local on Sicily, Croatia, east through Balkans, Crete, Bulgaria, SW Romania, Turkey, Levant south to S Israel, Transcaucasia, W & S Iran; winters Mali and Chad east to Sudan, W Eritrea). Described in rather great detail under Identification. ♂ has more extensive black mask and bib (bib length 20–33 mm), and black frequently (but not invariably) extends onto forehead above bill (by 0.5–2.5 mm), more often in second-year (or older) than in first-year birds. Tends to bleach to more white plumage in summer. Note that first-year ♂♂ are slightly more similar to ssp. *hispanica* than adults, having on average less extensive black mask and bib, and can have slightly more dull colours and less contrast. ♀ on average more greyish-brown, less warm ochre-buff, though some worn summer birds can appear very similar. Differs subtly in biometrics, being slightly larger and longer-winged with longer primary projection on average (but much overlap), and having a little longer p1. **W** ♂ 86–98 mm (n 25, m 91.9), ♀ 83–92 mm (n 22, m 88.1); **T** ♂ 54–64 mm (n 25, m 60.8), ♀ 55–64 mm (n 22, m 58.5); **T/W** m 66.3; **B** 14.6–17.9 mm (n 46, m 16.6); **BD** 3.7–4.2 mm (n 27, m 4.0); **Ts** 21.1–23.8 mm (n 35, m 22.4). **Wing formula: p1** > pc 0–5 mm, < p2 39–50.5 mm; **p2** < wt 3–7 mm, =5/6 (88%) or 5 (12%); **pp3–4** usually equal and longest; **p5** < wt 1–5 mm; **p6** < wt 7–12 mm; **p7** < wt 12–21 mm; **p10** < wt 20–28 mm; **s1** < wt 21–29 mm. (Syn. *xanthomelaena*.)

TAXONOMIC NOTE The relationship between the two forms of Black-eared Wheatear, *hispanica* and *melanoleuca*, and their relationship with Pied Wheatear, is subject to ongoing research. A case could be made for separation of *hispanica* and *melanoleuca* as two species, but intermediates seem to be fairly common in the contact zone, and fairly little study have been performed in this area regarding extent of mixing and genetic composition, although Randler et al. (2012) found very little genetic distance between any of Pied, Cyprus and Black-eared Wheatears (both subspecies). On balance we prefer to continue to treat the two forms as subspecies of the same species.

REFERENCES DERNJATIN, P. & VATTULAINEN, M. (2005) *Alula*, 11: 98–107. – HAFFER J. (1977) *Bonn. zool. Monogr.*, 10: 1–64. – PANOV, E. N. (1986) *Zool. Zhurn.*, 65: 1675–1683. – RANDLER, C. et al. (2012) *J. of Orn.*, 153: 303–312. – ULLMAN, M. (1994) *DB*, 16: 186–194. – ULLMAN, M. (2003) *DB*, 25: 77–97.

O. h. melanoleuca, ad ♂, black-throated morph, Turkey, Sep: evenly fresh and, unlike 1stW ♂, any pale fringes narrower and therefore overall pattern closer to spring/summer, with almost solid black face and much of wing black already in Sep. Note also the lack of broader whitish-buff fringes to primaries and primary-coverts typical of juv. (A. Atıcı)

O. h. melanoleuca, 1stW ♂, dark-throated morph, Egypt, Sep: a few 1stW ♂♂ have well-advanced face patterns, with blacker (though less solid) ear-coverts and bib. Nevertheless, wing-feathers still mainly juv. Note tail pattern with, as far as can be judged, rather broad and more even black band. (E. Winkel)

O. h. melanoleuca, 1stW ♀, possible pale-throated morph, Israel, Sep: drab grey-brown above, with orange-buff breast-band and slightly dusky sides, contrasting with whitish-buff throat and cream-white rest of underparts. Juv primary-coverts, remiges and several greater coverts indicate 1stW. (L. Kislev)

O. h. melanoleuca, ♀, Germany, Oct: grey-brown with sandy-buff tinge on rather plain upperparts and less rufous-tinged breast-band with only moderately darker sides are features of fresh ♀ *melanoleuca*. Slightly exposed dark bases on upper throat may indicate morph involved, while the primary projection seems too short for Pied. Safe ageing difficult. (S. Pfützke)

DESERT WHEATEAR
Oenanthe deserti (L., 1758)

Fr. – Traquet du désert; Ger. – Wüstensteinschmätzer
Sp. – Collalba desértica; Swe. – Ökenstenskvätta

Desert Wheatear breeds in NW Africa and Sinai, as well as locally through the Middle East and east in Central Asia to Tibet and Mongolia, and winters in N Africa and Arabia east to NW India. It inhabits various types of desert and semidesert, as well as dry plains and hillsides, and often associates with Desert Warbler and various specialised larks of the same habitats.

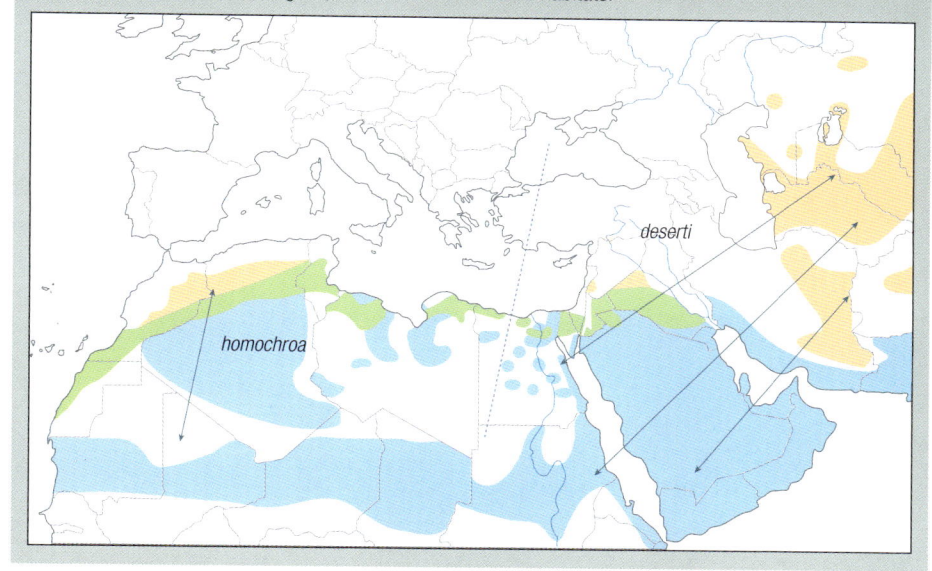

♂, *O. d. homochroa* (above: Tunisia, May) and presumed *O. d. oreophila* (below: Oman, Apr): diagnostic all-black tail, large pale rump, black underwing-coverts, pale coverts and fringes above, and prominent pale supercilium, while lower bird shows extensive white bases to remiges that reach shafts suggestive of *oreophila*. (Above: R. Pop; below: A. Below)

O. d. homochroa, ad ♂, Tunisia, Apr: note black wing forming rather narrow band and in contact with black side of head. Also, tail nearly all dark (diagnostic). By spring, many have lost much of the pale fringes of fresh plumage and appear immaculate. Evenly fresh, blacker remiges and primary-coverts readily age this bird. (D. Occhiato)

IDENTIFICATION Same size as Black-eared Wheatear and of rather similar shape. When alighting after a short flight, the relatively *long tail is frequently pumped*. In all plumages has diagnostically *almost all-black tail, white rump* (often tinged buffish) *and sandy upperparts*. ♂ recalls dark-throated Black-eared Wheatear in general plumage, but is easily separable in good views, especially by *black of face narrowly joining black of wing*, contrasting *whitish 'shoulder'* (lower scapulars) *and wing panel*, large buff-white rump and black tail. ♀ can appear like smaller version of Isabelline Wheatear, being largely pale buffish grey-brown. However, unlike latter has characteristic wing pattern of *paler lesser coverts, pale innerwing panel* and *dark-centred median coverts*, as well as *whiter rear supercilium* and diagnostic tail pattern and action, which should prevent confusion. Typically hops across sandy patches among low scrub, and easily detected in open habitats as frequently perches on bushes (from distance, due to fairly compact build and round head, has rather *Saxicola*-like jizz), often flipping tail up and down to keep balance if there is a wind. Flight low and flitting, usually over short distances, but very agile when chasing prey. Song delivered from low post or in circling song-flight. Often heard in late night or early dawn.

VOCALISATIONS Song has a decidedly melancholic tone (delivered from atop bush, fern or in circling, jerky song-flight, even at first dawn), constantly repeated with very little variation between strophes, a short, plaintive whistle of 2–4 clear descending notes after one initial usually higher-pitched and more stressed, e.g. *trüü-trururu* or *dih-lioo-lioo-lioo-lioo*, sometimes admixed with clicking, squeaky and muffled, rattling notes. Song of ♀ simpler and rarely heard. – Call a squeaky, drawn-out *viieh* and a hard clicking *tsack*. When alarmed near nest often a thin, muffled rattle, *tk-tk-tk-*..., like when a boy puts a cardboard on his bicycle wheel to simulate engine.

SIMILAR SPECIES Tail-and-rump pattern—lacking inverted black 'T' against white so often seen in congeners—distinctive in all plumages, though approached by some Isabelline. Ad ♂ recalls dark-throated Black-eared Wheatear but easily separated by tail pattern, black of face narrowly joining black of wing, pale sandy-white (not black) scapulars and (often) warm buffish-sandy cast to mantle/crown, whitish 'shoulder' (lesser coverts near wing-bend), supercilium and large buffy-white rump. – ♀ most recalls *Isabelline Wheatear*, but usually noticeably smaller and stockier (with shorter legs and bill, more rounded head, slightly longer tail and less upright stance), and tail pattern differs. Plumage differences minor, but easily appreciated: clearly paler lesser coverts contrast with mantle and black-centred median coverts; often obvious pale inner greater coverts and secondary panel; supercilium less marked in front of eye and obviously whiter behind it (reverse in Isabelline which also often has obvious dark lores), and ear-coverts sometimes tinged rusty-brown. – Separated from occasional very pale *Northern Wheatear* by all-dark tail.

AGEING & SEXING Ageing difficult and requires close inspection of moult pattern and feather wear in wing. Sexes differ, especially in spring/summer and ad, but a few are very similar and perhaps inseparable in the field. Seasonal variation mainly involves abrasion of pale feather edges and tips, most obvious in 1stW ♂. – Moults. Complete post-nuptial and partial post-juv moults in summer (mostly late Jun–Aug). Post-juv moult includes all head, body, lesser and median coverts, and three to all greater coverts, perhaps some alula- and tail-feathers and maybe some tertials. Pre-nuptial moult apparently largely absent. – **SPRING Ad ♂** Wings wear to mostly black except pale fringes to inner secondaries, tertials and inner lesser coverts; pale scapular panel also strongly reduced later in season. **Ad ♀** Dull buff-brown above, and head less strongly patterned. Throat usually pale but with wear partially blackish or mottled dark, and older ♀♀ may approach least-advanced 1stS ♂. Unlike in 1stW ♂, bib never solid black, chin and upper throat usually paler, and wings less contrastingly patterned with less obvious pale outer scapulars, rump is often buffier, and if aged first differences easier to appreciate. A simple division into pale-throated or dark-throated ♀♀ obscured by intermediates. **1stS** ♂ As ad, but sometimes less neatly coloured or strikingly patterned, with browner wings (especially bleached juv primaries, primary-coverts and some outer greater coverts), and sometimes has pale tips to black throat-feathers. **1stS ♀** Moult limits less obvious than in young ♂ but diagnostic when seen. Worn primary-coverts centred dull brown with more pointed tips and almost no pale fringes by mid spring, but tips still pale buff (in ad, greyish-black centres and narrow whitish-grey fringes). — AUTUMN **Ad ♂** Unlike 1stW ♂ entire plumage evenly very fresh without discernible moult limits, thus much of wing solid black, and bib almost so. Alula, greater coverts and primary-coverts centred black or blackish, and primary-coverts have very narrow whitish fringes mainly confined to

O. d. homochroa, 1stS ♂, Morocco Feb: aged by strong moult contrast in wing, especially outer greater coverts and primary-coverts are browner. There is some individual variation in upperparts coloration, related to feather wear but also light, making racial discrimination difficult. (M. Varesvuo)

O. d. homochroa, ♀, Tunisia, May: much duller than ♂, dull buff-brown above and head weakly patterned; with wear, dark mottling appears on throat, and wings become more evenly dark. Importance of large buff rump and virtually all-dark tail remains. ♀ less easily aged, especially from single image. (R. Pop)

O. d. deserti, ad ♂, Mongolia, Jun: in Mongolia, many ♂♂ approach race *oreophila* by having black of bib extending backwards towards nape, and some even have faint semi-necklace. Ad by evenly dark wing-feathers, which are not very worn, but by Jun most pale edges have worn off. (H. Shirihai)

O. d. deserti, ♀, presumed 1stY, Kuwait, Jan: some ♀♀ distinctive by virtually unpatterned and diluted sandy-buff plumage. Presumed 1stY by apparently juv remiges and primary-coverts. (D. Monticelli)

O. d. deserti, ad ♂, Oman, Nov: when still fresh, pale tips to black of face and wings obscure pattern, including less sharp division between dark ear-coverts and pale supercilium. Intense black centres to remiges, greater coverts and primary-coverts indicate ad. (M. Varesvuo)

tips. Buffish-white wing fringes narrower, from Dec already restricted to tertials, inner secondaries and a few inner lesser coverts. **Ad ♀** As spring but extensive pale fringes to upperwing (cf. 1stS ♀). **1stW ♂** Resembles ad ♂ but extensive pale fringes may obscure throat patch (sometimes almost entirely), and juv remiges, primary-coverts and some outer greater coverts less intense black with much broader pale buff fringes and bold tips. Juv middle alula has broad ochre fringe and tip (just a small white spot at very tip in ad). Juv feather groups form moult limits with ad-like post-juv lesser, median and inner greater coverts, and sometimes tertials, which are more firmly textured and black-centred (but moult pattern less contrasting than in spring). **1stW ♀** Best told from ad ♀ by moult limits, but these are less marked than in 1stW ♂. Primary-coverts weaker textured and centred dark brownish-grey with rather broad pale buff fringes (in ad, greyish-black centres narrowly fringed greyish-white). **Juv** Similar to 1stW ♀ but has soft, fluffy body-feathering and distinct mottling on head and upperparts. Underparts mainly buffy-cream lacking obvious mottling.

BIOMETRICS (*deserti*) **L** 14.5–15.5 cm; **W** ♂ 89–95 mm (once 86; n 23, m 91.9), ♀ 85–90 mm (n 10, m 88.0); **T** ♂ 57.5–65 mm (n 23, m 60.7), ♀ 56–60.5 mm (n 10, m 88.0); **T/W** m 66.1; **white on base of r1** (disregard black on shaft) 17–25 mm (n 33, m 20.6); **B** 15.2–18.7 mm (n 32, m 17.2); **Ts** 23.3–26.1 mm (n 30, m 25.0). **Wing formula: p1** > pc 4 mm, to 3 mm <, < p2 42–52.5 mm; **p2** < wt 1.5–5 mm; =5 or 5/6 (rarely =4/5); **pp3–4** about equal and longest; **p5** < wt 2–4.5 mm; **p6** < wt 8–11 mm; **p7** < wt 13–16 mm; **p10** < wt 21.5–25 mm; **s1** < wt 21–28 mm. Emarg. pp3–5.

GEOGRAPHICAL VARIATION & RANGE Indistinct and clinal throughout most of range, and size and general plumage prone to individual variation and wear/bleaching, producing many intermediates and much overlap. Westernmost *homochroa* and easternmost *oreophila* are most distinct. We agree with Grant & Mackworth-Praed (1947) and Svensson (1992) that '*atrogularis*' can be included in *deserti*, and recognise just three races, two of which are still quite similar.

○ *O. d. homochroa* (Tristram, 1859) (N Africa east to NW Egypt; some move in winter to Sahel zone, others remain closer to breeding range). By and large same size as *deserti* or very slightly smaller, differing subtly in having pinkish-cinnamon or slightly vinaceous-tinged mantle and scapulars with no or only very slight greyish tinge on crown, nape and upper mantle, appreciable only when series are compared. Chest is pink-buff. White wedge on inner webs of remiges averages narrower than in *deserti*. **L** 14.5–16 cm; **W** ♂ 86–96 mm (n 57, m 91.2), ♀ 82–90 mm (n 31, m 86.1); **T** ♂ 58–68 mm (n 18, m 62.1), ♀ 55–62.5 mm

O. d. deserti, 1stW ♂, Oman, Nov: some young ♂♂ very similar to fresh ad ♂ at same season, but usually show clear moult limits in wing, with contrasting juv remiges, several outer greater coverts and primary-coverts, which feathers are noticeably less fresh, browner and have narrower pale fringes. (D. Occhiato)

O. d. deserti, ♀, presumed 1stW, Oman, Nov: fresh ♀ is trickiest to separate from Isabelline Wheatear, especially if tail pattern not seen well and size cannot be evaluated. Desert usually has better-marked wing pattern with dark-centred feathers and paler lesser coverts, as well as whiter rear. (D. Occhiato)

O. d. deserti, juv, Mongolia, Jun: easily aged by typical 'marbled' or scalloped head and fore-body. All-dark tail best clue to species. (H. Shirihai)

O. d. oreophila, ad ♂ (left) and 1stS ♀, NW India, May: sometimes possible to note that this is larger race, while ♂ often has sides of black of bib reaching nape, whereas ♀ is usually only reliably separated by measurements. ♂ is ad by evenly-feathered and less worn wing, while many juv feathers in wing of ♀ indicative of 1stS. (I. Waschkies)

(n 17, m 58.1); **T/W** m 68.0; **white on base of r1** (disregard black on shaft) 18–28 mm (n 26, m 23.4); **B** 15.4–18.0 mm (n 30, m 17); **Ts** 22.5–26.3 mm (n 32, m 24.3). Wing formula much as in *deserti*.

O. d. deserti (Temminck, 1825) (NE & E Egypt, Sinai, Middle East through N Central Asia to Kazakhstan, Mongolia; winters NE Africa, Arabia, NW India). Differs from similarly-sized or slightly smaller *homochroa* by more greyish-cinnamon or sandy-grey crown, nape, mantle and scapulars, and deeper buff or cinnamon-buff chest. Still, some are near-identical to *homochroa*, and separation away from breeding range and outside breeding season often requires comparison of series rather than of single birds. – Birds in S Caucasus east through Iran, Transcaspia, Altai and Mongolia sometimes separated as '*atrogularis*' on account of being slightly larger and in fresh plumage darker and more pinkish-brown above on crown to mantle, and tendency to have more prominent white edges to inner webs of remiges in ♂♂, but size difference negligible, and plumage differences very slight, clinal and inconsistent, so *atrogularis* included here in *deserti*. White on inner webs of primaries usually separated from shaft by 1–5 mm-wide brownish-black zone, but in a very few white reaches shafts proximally on secondaries and some inner primaries. – Intergrades with *oreophila* in N Afghanistan, NW China and extreme S Mongolia. (Syn. *atrogularis*; *salina*.)

O. d. oreophila (Oberholser, 1900) (Kashmir, Baltistan, Ladakh, Pamirs, through Tibet to Inner Mongolia; winters SE Iran and W Pakistan with some reaching S Arabia, Socotra and Somalia). Somewhat larger than *deserti*, and further differs by tendency in ♂♂ to have hint of black seminecklace, black at side of bib reaching to side of nape. Ad ♂ has white reaching shafts on inner webs of most flight-feathers save the outermost 2–3 long primaries (basal about half of feathers) and division with dark tips rather obtuse and sharp. 1stY ♂ similar but can have narrow dark zone between white on inner webs and shafts, and dark tips less blackish and division less sharp. Often more white at base of tail than in other races. ♀♀ variable as to amount of white or pale on inner webs of flight-feathers, majority being inseparable from *deserti* as to this, whereas others (mainly ad) have more extensive white on secondaries and inner primaries, and should be possible to identify especially in combination with large size. **L** 15.5–16.5 cm; **W** ♂ 96–104 mm (n 40, m 99.6), ♀ 90–98 mm (n 20, m 93.8); **T** ♂ 63–70 mm (n 23, m 60.7), ♀ 59–67 mm (n 16, m 62.7); **T/W** m 66.4; **white on base of r1** (disregard black on shaft) 21–31 mm (once 15; n 36, m 24.8); **B** 16.2–19.9 mm (n 36, m 18.0); **Ts** 24.0–27.3 mm (n 36, m 25.9). **Wing formula: p1** < pc 0–4 mm, < p2 51.5–56.5 mm; **p2** < wt 2–5 mm, =4/5 or 5 (sometimes =5/6); **pp3–4** about equal and longest; **p7** < wt 14–19 mm; **p10** < wt 24–28 mm; **s1** < wt 26–31 mm. (Syn. *montana*.)

REFERENCES GRANT, C. H. B. & MACKWORTH-PRAED, C. W. (1947) *BBOC*, 67: 47–48.

Presumed *O. d. oreophila*, ad ♂, United Arab Emirates, Oct: seminecklace near wing-bend suggests race *oreophila*. Evenly black-centred and white-fringed primary-coverts identify it as ad. (D. Clark)

FINSCH'S WHEATEAR
Oenanthe finschii (Heuglin, 1869)

Fr. – Traquet de Finsch; Ger. – Höhlensteinschmätzer
Sp. – Collalba de Finsch; Swe. – Finschstenskvätta

A Middle Eastern and SW Asian wheatear of open rocky and stony uplands, moderately undulating terrain with boulders and sparse bushes, breeding across Turkey east to N Pakistan. Finsch's Wheatear appears slightly heavier and more compact than many of its similar-looking congeners. It is a short-distance migrant, regularly moving south to the Levant, e.g. Israel, and largely winters in areas used by Black-eared Wheatear *melanoleuca* for breeding, but is generally very local and scarce in its winter range, which reaches south to N Arabia and N Egypt.

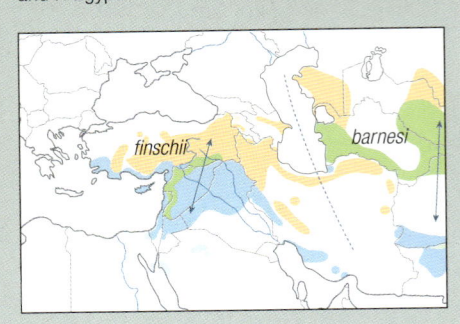

Ad ♂, Turkey, May: superficially resembles several other black-and-white wheatears, but has pale (with wear pure white) mantle meeting white back/rump, black foreparts broadly connected to black wings, and rather broad, even-width tail-band. Lack of moult limits in wing indicate ad. (G. Reszeter)

♂, Turkey, May: in song flight, revealing all of species' diagnostic marks; this bird has slight dusky wash to crown and mantle retained from fresh plumage. Note white centre of back and very dark remiges. (R. Armada)

IDENTIFICATION Stocky, large-headed wheatear with a rather short stubby bill. *On landing, typically flicks and cocks tail and 'hugs ground'*. Generally, ♂ recalls dark-throated Eastern Black-eared, and ♀ is like ♀/first-winter Pied/Eastern Black-eared, but in particular recalls ♀ Gould's (extralimital, but see p.342) and NW African Maghreb Wheatear. ♂ superficially resembles several other black-and-white wheatears, but readily separated by *narrow buff-white mantle narrowly meeting white back/rump*, crown usually tinged greyish (can have pink-buff tinge; becoming whiter with wear), *black of throat extends to upper breast and connects broadly with black wings*, and neat black inverted 'T' on tail forms *moderately broad tail-band of even width*. Note that some ♂♂ have narrow black bridge across whitish back, or at least a tight 'waist-line' on the white where black either side comes close, and such birds are best assessed from rear. Even so, the white back might be largely concealed by scapulars and tertials on a perched bird, but note *pointed shape of white nape* forming a wedge onto mantle (square in Mourning and Pied), and check pattern on flying bird. ♀ is brownish-grey and off-white, and highly variable as to number of exposed blackish throat-feather centres, but unlike several congeners may show *thin whitish wing-bars, darker greater covert panel* and *rusty-tinged ear-coverts*.

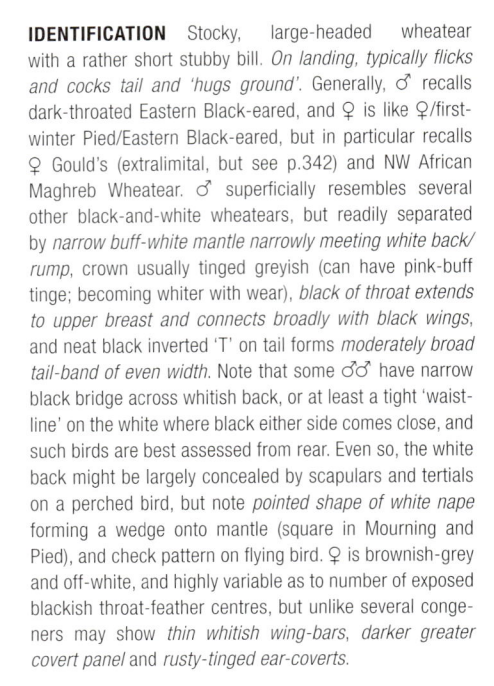

1stS ♂, Jordan, Mar: with wear pale areas become purer white, enhancing rather large black bib, which broadly connects with black of wings. When wings tightly folded, however, diagnostic 'channel' of white from mantle to back/rump is hidden. Aged by juv remiges and primary-coverts. (H. Shirihai)

VOCALISATIONS Varied and melodious song (uttered year-round, but mostly at onset of breeding), and each ♂ has its own variations. Very different from Black-eared Wheatear, consisting of short, clear whistled notes at leisurely pace, some whistles being bent (glissando) interspersed by harsh, scraping, clicking and even 'electric', sparkling notes, e.g. *tk-tk siiih-va chi-chi trrüü-u.....chu-chu kretruü-chiiih-a* or *chik-chik-chik-cheeoowee chrr-ch*, etc.; in song-flight gives much longer, almost continuous song (then sometimes more resembling Black-eared). Less mimicry than several congeners. Another song type is shorter, more hurried and thin, and often preceded by rasping sounds or calls (both sexes). – Calls include *tsit* or *zik*, often repeated, a harsh pebble-clinking *tsheck-tsheck*, and a rasping *bsheh* in alarm.

SIMILAR SPECIES ♂ best separated from *Eastern Black-eared Wheatear* (dark-throated morph) by larger black bib, broadly connected to black of wings, while tail pattern, size, structure and behaviour are also distinctive. Black extends further onto upper breast than in dark-throated Eastern Black-eared, which furthermore has pale neck-sides. Underwing in ♂ Finsch's has striking contrast between black coverts and silvery flight-feathers (less striking in Eastern Black-eared), and black tail-band evenly broad, not narrow either side of centre and becoming broader at corners. – ♂ can be confused with ♂ *Pied Wheatear* if angle is less favourable so that back looks all black (and, as pointed out above, remember that some Finsch's have narrow black bridge where mantle meets back). Still, once seen well from behind, white mantle wedge, or complete white back, of Finsch's will be obvious. Note also conclusive even-width black tail-band lacking broader black on outermost feathers (except insignificantly so). – A major pitfall, usually ignored in the literature, is the similarity of ♀ Finsch's to ♀ Gould's Wheatear (extralimital, but may be recorded in E Iran as vagrant) and extremely pale ♀ Blyth's Wheatear (breeds in much of Iran, winters Arabia); see latter species for both. – ♀ Finsch's separated from other similar congeners, most importantly ♀ *Pied Wheatear*, by more uniform and colder brown-grey upperparts (also lacks Pied's fine pale fringes on mantle and scapulars even when fresh), often has rusty-tinged ear-coverts, contrastingly dark greater coverts bordered by whitish wing-bars, often grey lesser coverts and quite broadly pale-fringed primary-coverts, but these differences only obvious in fresher plumage (i.e. autumn to early spring). Also when fresh, pale feather-tips partially conceal dark throat, forming variable but usually quite characteristic pattern, but note that dark-throated ♀ Pied can

WHEATEARS

1stS ♀, Turkey, May: extensive blackish throat and ear-coverts (recalling ♂) characteristic of dark-throated ♀. Note species' rusty-tinged head, subtle pale wing-bars and darker greater coverts panel. With wear, some ♀♀ also show whitish-grey on mantle. Aged as juv by strongly bleached brown remiges, primary-coverts, alula and outer greater covert. (G. Reszeter)

♀, presumed 1stS, Turkey, May: despite strong wear and bleaching, characteristic greyish upperparts and rusty-tinged ear-coverts and crown still visible, but only limited blackish throat feathering. Wings too worn to age this bird by moult pattern, but degree of wear to primaries point to 1stY. (G. Reszeter)

develop rather extensive dark bib, similar to Finsch's. Tail pattern (with broader and more even-width terminal band) and near lack of orange-buff on breast also useful at most seasons. Behaviour and calls also separate the two. — ♀ *Eastern Black-eared Wheatear* tends to be warmer and browner above. Tail and wing patterns, shape and behaviour are also different, mostly as described for separation from Pied. — ♀ may resemble ♀ *Maghreb Wheatear* and to some extent also *Arabian Wheatear* (Finsch's may occur sympatrically with these two in winter in W Egypt and around Persian Gulf, respectively). However, Finsch's is greyer above than Maghreb, and paler than (and never as deep brown as) Arabian; it also lacks the rusty-orange ventral region, and dark underparts streaking of latter (which see). Note that not all Maghreb have obvious buff-pink on vent/undertail-coverts, and conversely that some Finsch's can show limited warm coloration (normally vent is white). Also beware that Maghreb can show rather similar whitish wing-bars, rusty-tinged ear-coverts and a dark throat, and some may prove inseparable in the field (although note calls and behaviour). Four additional supportive characters could assist with difficult birds: (i) unlike Finsch's, the fresh primary-coverts of Maghreb (stronger in first-winter) usually have broader pale tips and narrower fringes (pale fringes in Finsch's tend to be of more even width and do not form spots at tips); (ii) at least

Ad ♂, Israel, Oct: typically appears rather large-headed and short-tailed. Even at this angle white back suggested by deep white wedge down mantle. Age obvious by very black wing, unlike similar 1stW, fresh primary-coverts lack white fringes and tips. (L. Kislev)

1stW ♂, Cyprus, Oct: when fresh, ♂ often has buffy-grey wash to crown (enhancing whitish supercilium). Unlike ad ♂, primary-coverts have pale fringes concentrated on tips, whereas juv primary tips are less pure white and bold than autumn ad. (A. McArthur)

Ad ♀, Israel, Dec: much individual variation in ♀ plumages, but darkest birds typically ad. Here nearly solid back throat, even when very fresh in Dec, and wing-coverts are blackish. Fresh ad remiges and primary-coverts confirm age. (A. Ben Dov)

♀, presumed 1stW, Israel, Nov: highly distinctive by grey upperparts, almost clean white underparts, rusty tinge to ear-coverts, quite prominent pale supercilium and wing-bars. Age uncertain, although primary-coverts apparently juv. (A. Ben Dov)

♀ **Finsch's** (top left: Israel, Nov), **Maghreb** (top right: Morocco, Jan), **Arabian** (bottom left: Yemen, Jan) and presumed **Gould's Wheatears** (bottom right: India, Jan): the four birds depicted here represent particularly similar individuals—many others are easier to identify. Finsch's: cold olive-grey above but double pale wing-bars, especially strong on greater coverts, and rufous-tinged ear-coverts. Maghreb: dullest ♀ can be virtually identical to Finsch's, but primary projection averages longer and bill usually thinner and more pointed, while primary-coverts usually have much broader pale tips and narrower fringes (more even width in Finsch's), and at least some show paler, somewhat translucent remiges in flight. Arabian: usually darker overall with rather well-developed diffuse streaking below, rusty-orange undertail-coverts and decurved bill tip. Gould's: proportionately long tail but short wings, rather uniform upperparts and wings, and buffish-rufous tinge to face and breast, contrasting with white belly; elimination of fresh Pied rests on lack of pale tips to upperparts and fairly short primary projection. (Although probably Gould's, an extremely pale ♀ Blyth's cannot be excluded.) (Top left: A. Ben Dov; top right: R. Schols; bottom left: H. & J. Eriksen; bottom right: S. Singhal)

some Maghreb show paler, somewhat translucent remiges in flight (most reduced in young ♀), somewhat approaching pattern of Finsch's; (iii) bill of Maghreb is also usually markedly thinner and more finely pointed, whereas in Finsch's bill is broader and appears stubbier; and (iv) the primary projection averages longer in Maghreb.

AGEING & SEXING Ageing requires close inspection of moult, feather wear and pattern in wing and is generally only possible with ♂♂. Sexual dimorphism striking. Limited seasonal variation, mostly in dark-throated ♀♀. — Moults. Complete post-nuptial and partial post-juv moults in late summer (mostly Jul–Aug, some finish early Sep). Post-juv renewal involves all feathers of head and body, lesser and median coverts, most or all greater coverts, perhaps some tail-feathers and usually some tertials, as well as (usually) innermost (more rarely entire) alula-feathers. Pre-nuptial moult absent. — SPRING **Ad** ♂ Aged by being evenly feathered with relatively less worn and blacker primaries and primary-coverts. Creamy-white or pale buff on crown and nape strongly reduced with wear, mantle to back pure white. Creamy-white underparts tinged buffish when less worn, but wear whiter. **Ad** ♀ Unlike 1stS ♀ lacks moult limits in wings, and primaries and primary-coverts less worn and darker. Much of throat and upper breast dark grey or

♀, presumed 1stY, Israel, Feb: raises tail and 'hugs' ground on landing, producing typical posture, but tail can be flicked even further upwards. Faint breast-streaking variable feature of many ♀♀, like Arabian Wheatear. Strong contrast between remiges and blacker tertials suggests moult limits of young bird. (H. Shirihai)

WHEATEARS

Dark-throated ♀ Finsch's (left: Israel, Dec), Maghreb (centre: Tunisia, Nov) and Blyth's Wheatears (right: Oman, Dec): all three can appear dark-throated and extremely similar, if not identical. Finsch's could marginally overlap on migration with Maghreb (in N Africa) and Blyth's (in Arabia), and vagrants outside normal ranges will prove challenging. Some Finsch's (like here) show similar buff undertail-coverts patch to ♀ Maghreb, or even Blyth's, in which case identification will rest on Finsch's on average shorter primary projection, usually visibly stouter bill, and, at least when very fresh, much-reduced pale supercilium. The Maghreb is ad, lacking diagnostic white tips to primary-coverts of 1stY ♀, but also the typically even-width fringes of Finsch's; at least some Maghreb show paler, somewhat translucent remiges in flight. Dark-throated Blyth's also extremely difficult to separate, but usually has darker upperparts, clearer lower border to dark bib (which unlike other two extends to upper breast), and often has more ragged tail-band. (Left: A. Ben Dov; centre: A. Torés Sánchez; right: H. & J. Eriksen)

♀, Israel, Jan: throat patch intermediate. Note faintly rusty-tinged ear-coverts and quite prominent whitish wing-bar. Degree of wear to at least some tracts in wing suggest 1stY, but certain ageing impossible. (L. Kislev)

blackish by summer (especially in older birds), though usually less solid or intense black than ♂, but rest of plumage typically dull ♀-like: upperparts grey/sandy-grey, forehead and ear-coverts often tinged rufous-brown (some have pale supercilium, mainly above and behind eye). See also under Similar species. **1stS** As ad, but note browner juv primaries, primary-coverts and often some outer greater coverts, with moult contrast enhanced (against fresher post-juv wing-coverts). In general, 1stS ♀ has less black on throat than ad ♀, but much variation. – AUTUMN **Ad ♂** Unlike 1stW ♂, wings evenly fresh and black. Pale primary tips well marked, and uniform black primary-coverts may have only a tiny white spot at very tip. Black areas of face and wings indistinctly tipped or fringed paler. Whitish areas of forehead to back variably but conspicuously washed drab buffy-grey, not clean white as spring ♂. **Ad ♀** Unlike 1stW ♀ wings evenly fresh and primary-coverts often slightly darker with narrower paler fringes. Plumage much as in spring, but blackish bases of throat partially or wholly concealed by greyish-white tips (chin, centre of throat and submoustachial stripe typically mottled white with dark or greyer lower-throat crescent, which with dark lateral throat encloses pale area of chin/upper throat). Variable diffuse greyish-buff streaks on chest, otherwise predominantly cold drab greyish above, with diffuse whitish supercilium, some buff or gingery tones to forehead, fore supercilium and ear-coverts. **1stW ♂** After post-juv moult in late summer as ad ♂ but aged by broad pale fringes to primary-coverts. Juv greater coverts, tertials and alula blackish-brown with buffish fringes, while ad-like post-juv feathers are fresher with black centres and, at most, tiny whitish fringes on tertials, and some have broad pale fringes to black bib. **1stW ♀** Extremely similar to ad ♀ and often impossible to separate from it except when handled and with experience. Differs in late summer/early autumn by slightly broader paler fringes and tips to primary-coverts, and often first traces of irregular wear to pale fringes on primaries and tertials. Sometimes possible to establish (subtle) moult difference in wing between juv and post-juv coverts; also tends to have pale throat/chest with fewer dark bases exposed, but much individual variation. **Juv** Unlike ♀, underparts mainly deep buffy-cream and upperwing extensively fringed rufous (compare juv congeners without obviously scaled body plumage, which appear uniform in field). ♂ usually clearly greyer above with darker eye-stripe.

BIOMETRICS (*finschii*) **L** 15–16 cm; **W** ♂ 84–94 mm (n 28, m 89.5), ♀ 82–87 mm (n 13, m 84.5); **T** ♂ 55–64 mm (n 27, m 58.7), ♀ 53–58 mm (n 13, m 55.2);

Juv, Iran, May: upperwing broadly fringed pale rufous and lacks obvious darker scaling to body-feathers; probably ♂, being greyer above, whiter below, and having black centres of flight-feathers. (C. N. G. Bocos)

T/W m 65.5; **black on r1** 33–41 mm (n 32, m 36.6); **black on r4** 10–18 mm (n 31, m 13.8); **B** 15.5–19.3 mm (n 40, m 17.5); **BD** 4.0–4.7 mm (n 24, m 4.3); **Ts** 24.0–27.0 mm (n 27, m 25.5). **Wing formula: p1** > pc 1.5–7 mm (very rarely to 0.5 mm <), < p2 38–46 mm; **p2** < wt 3.5–7 mm, =5/6 to 6/7 (very rarely =5); **pp3–4** about equal and longest; **p5** < wt 0.5–1.5 mm; **p6** < wt 4–7 mm; **p7** < wt 10–13 mm; **p10** < wt 18–23 mm; **s1** < wt 20–25 mm. Emarg. pp3–6, although on p6 sometimes less distinctly, rarely lacking.

GEOGRAPHICAL VARIATION & RANGE Only slight variation. Much overlap between *finschii* and eastern *barnesi*, and effects of age/wear and presence of intermediates mean that only typical individuals are identifiable.

○ **O. f. finschii** (Heuglin, 1869) (Turkey, Syria, Lebanon, Transcaucasia, NW & SW Iran; short-range movements, wintering S Turkey, Levant, N Arabia, Kuwait). Described above.

○ **O. f. barnesi** (Oates, 1890) (NE & E Iran, S Turkmenistan, Afghanistan, north to S Kazakhstan; winters perhaps mainly Arabia, Gulf States, S Iran, but poorly known). Very similar to ssp. *finschii*, but ♂ has vent/undertail-coverts more deeply tinged rufous pink-buff. It is also on average tinged more pinkish cream-buff on crown to mantle and on underparts in fresh plumage (paler sandy-cream in *finschii*), but both wear to white in summer. ♀ *barnesi* averages warmer above with isabelline wash and rustier ear-coverts; also buffier below, with much less exposed black throat-feathering, except in summer when differences are obscured. Although it is true that it averages larger than *finschii* this is still very subtle and of little help. Claims that it has broader terminal tail-band could not be convincingly confirmed. **W** ♂ 85–94 mm (n 19, m 90.8), ♀ 85–91.5 mm (n 14, m 88.0); **T** 55–65 mm (n 33, m 60.5); **T/W** m 67.5; **black on r1** 35–47 mm (n 18, m 40.0); **black on r4** 9–17 mm (n 19, m 13.6); **B** 16.3–19.5 mm (n 31, m 17.7); **BD** 4.0–4.6 mm (n 17, m 4.3); **Ts** 23.9–26.8 mm (n 21, m 25.2).

RED-RUMPED WHEATEAR
Oenanthe moesta (M. H. C. Lichtenstein, 1823)

Fr. – Traquet à tête gris
Ger. – Fahlbürzel-Steinschmätzer
Sp. – Collalba culirroja; Swe. – Berberstenskvätta

This rather large wheatear is patchily distributed in deserts and semideserts of N Africa (from extreme NW Mauritania to NW Egypt), and even more so in the Middle East (S Syria to SW Iraq), where it is either sedentary or nomadic. Red-rumped Wheatear typically frequents sandy and stony hillsides with some bushy cover, and throughout its range is strongly associated with the availability of colonial rodent holes for nesting sites.

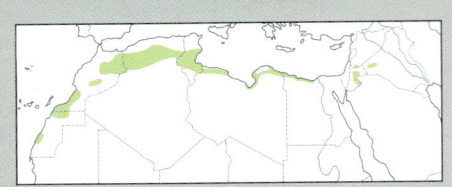

1stW ♂, Tunisia, Dec: lower back and rump cream-buff, contrasting with mostly blackish tail that when spread shows restricted but bright reddish base—diagnostic in combination with whitish wing panel and drab crown/nape. 1stW with pointed and bleached primary-coverts as most obvious vestige of juv plumage. (T. Lindroos)

VOCALISATIONS Unusual song, consisting of long series of strained and often disyllabic, rather low-pitched 'vibrating' notes, series rising and falling with half-tone steps, or both in same strophe, e.g. *churrrr chorror churrurr chirrirr*. There is variation, however, and some use more coarse or rattling notes than others. Whereas some strophes are machine-like and extremely unusual for a bird, others can recall (vaguely) song of Redwing. Both sexes sing, and in courtship pairs duet with one starting as the other finishes. – Varied calls include a repeated, drawn-out *prrriht* when agitated, and various clicking *kwit* and *tsyak* notes. Sometimes a short purring *prrt* can be heard (recalling Little Bustard), or doubled, *prrt-prrt*.

Ad ♂, Morocco, Jan: unlikely to be confused with any other wheatear, especially by black face merging with slate-tinged black dorsal areas, dirty grey-white crown, and dark wings prominently fringed whitish-buff. Evenly fresh blacker remiges and primary-coverts readily age it as ad. (M. Schäf)

IDENTIFICATION Large and also large-headed and robust, with long legs, a fairly strong bill and distinctive tail pattern (often fans tail widely). ♂ unmistakable given *black face connected to dull, slate-tinged or blackish dorsal areas*, which isolate *grey-white crown and hindneck*. Black wings with well-defined whitish-buff fringes, forming narrow *pale double wing-bars and ill-defined innerwing panel* (strongly reduced with wear). Breast and belly whitish. Rear body has diagnostic pattern: *lower back and rump whitish, upper- and undertail-coverts, vent and tail-base bright rufous, contrasting with blackish outertail*. ♀ also distinctive: largely drab sandy-grey upperparts and paler below, with Bluethroat-coloured tail, ill-defined but obvious *pale rufous-orange crown and hindneck*. The diagnostic upperwing pattern and rear-body coloration resembles ♂ but is inconspicuously pigmented, and pattern poorly defined. Pale-winged appearance in flight is also rather distinctive (remiges silvery below, contrasting with very dark coverts).

SIMILAR SPECIES The rufous or cinnamon rump and tail base readily distinguish it from most other wheatears. *Desert Wheatear* may have warm buff rump and uppertail-coverts, and ♀ *Hooded Wheatear* a buff to rufous-buff rump and tail, but both are otherwise easily separated from Red-rumped. – Could be confused with *Kurdish* and *Persian Wheatears* as they share similar rufous rear body and tail-base, but Red-rumped is more robust, with longer bill and legs (not as sleek and Black-eared Wheatear-like as Kurdish and Persian). Kurdish and Persian are also less prone to perch on ground (favouring rocks, large boulders and bushes) and have less black in tail, but note following: ♂ Red-rumped has dark face, wings and upperparts similar only to ♂ Kurdish

1stW ♂, Morocco, Jan: clear moult limits with juv remiges, primary-coverts and some outer greater coverts, which are brownish with very thin off-white fringes and tips; otherwise mostly like ad. (M. Schäf)

1stW ♂, Morocco, Feb: characteristic posture of this species in typical habitat of sandy and stony areas with some bushy cover. Most of exposed wing-feathers are juv. (D. Occhiato)

♀, Morocco, Jan: some ♀♀ rather more russet overall, especially rufous-orange crown and hindneck; note marked upperwing pattern. Primary-coverts concealed, preventing assessment of age. (M. Schäf)

1stW ♀, Morocco, Jan: rufous-orange crown and hindneck often give round head a capped appearance. Sometimes basal tail-feathers are more exposed, when pattern can recall Kurdish and Persian Wheatears. Juv primary-coverts. (M. Schäf)

Wheatear, but lacks latter's diagnostic tail pattern and basal white tail corners. Fringes to wing-coverts in Red-rumped form distinct wing-bars, wholly unlike Kurdish, and in worn plumage note black mantle and scapulars, with usually whiter head and paler underparts. ♀ Red-rumped is similar to both sexes of Persian Wheatear, and to some ♀ white-throated Kurdish, but has rusty-buff crown, nape and head-sides, pale buffish-white wing panel, and usually a broader black tail-band, leaving much narrower rufous base.

AGEING & SEXING Ageing requires close observation of moult, feather wear and pattern in wing. Sexes strongly differentiated but rather limited seasonal variation. – Moults. Complete post-nuptial moult in ad, and partial post-juv moult in young, mostly Jul–Aug. Post-juv moult involves all head, body, lesser and median coverts, most or all greater coverts, usually some tertials and perhaps some tail-feathers, occasionally also the smallest alula and innermost secondaries. Pre-nuptial moult absent. – SPRING **Ad** Both sexes differ from 1stS by being evenly feathered with relatively less worn and darker primaries and primary-coverts. **Ad ♂** Crown and hindneck white, mottled grey or buff-brown, thus supercilium less well defined. Mantle blacker. Largely black tail has restricted rusty bases to outer feathers. **Ad ♀** Crown and hindneck has variable cinnamon or rufous-buff tone, and mantle a greyer tone, both sometimes slightly contrasting. Paler below, some with slight rusty-buff wash to breast, but more invariably on vent and undertail-coverts. Tail like ♂ but at least some have even more rufous on bases, producing clearer 'T'-shaped pattern, though terminal band often diffuse. **1stS** Like ad, but usually has somewhat browner juv primaries and perhaps some outer greater coverts and all primary-coverts. Overall, especially in ♂, less evenly patterned with browner tone to grey and black areas. Note however that due to the rather pale flight-feathers also in ad, these wear rather heavily in winter, and difference between ages not always evident. – AUTUMN **Ad** Both sexes differ from 1stW by evenly very fresh wings without discernible moult limits. **Ad ♂** Largely as in spring, though greyish-brown crown and hindneck, enhancing whitish supercilium and contrasting less with slate-grey mantle. Black of lower face and throat tipped pale greyish. Rump darker (mostly rufous with some greyish). Uniformly fresh pale fringes give wing neat pattern. Primary-coverts have dark grey or blackish centres, and well delimited and broad greyish-white (sometimes slightly ochre) fringes and tips. **Ad ♀** As spring, but rufous parts brighter, and mantle (less grey) and crown more concolorous. Brighter wing fringes form more prominent linings and panel almost as in ad ♂. Primary-coverts on average slightly firmer structured, darker-centred and fringed greyish (less buff) than 1stW ♀. **1stW ♂** Variable, some approaching ad, but especially earlier in season black of lower face often has more extensive pale tips, while crown often tinged rufous, and more brown-grey above. Clear moult limits exist due to juv remiges and primary-coverts and perhaps some greater coverts and tertials; primary tips less fresh, and primary-coverts have pale grey-brown centres and relatively poorly delimited off-white fringes and tips. Juv tertials (if not moulted) and secondaries softer with slightly worn tips, greyish-brown centres and ill-defined pale fringes (contrasting with fresher, more firmly structured ad-like post-juv ones, which have dark grey or blackish centres and better delimited pale fringes). Juv greater coverts (sometimes a few outer retained) have slightly browner centres and more buff or rufous (less greyish-white) fringes than fresher ad-like post-juv coverts. **1stW ♀** Very similar to ad ♀ but note moult limits in wings (as 1stW ♂ but take into account sexual differences in colours). **Juv** Underparts mainly buffy-cream (partial black or dusky throat in juv ♂), lacking rufous on vent/undertail-coverts. Upperwing-feathers extensively fringed rufous. Upperparts mainly dusky buff-brown (paler in ♀) lacking any obvious mottling.

BIOMETRICS Rather peculiarly, there is a larger size variation among ♂♂ than is usual within any one taxon; there seems to be little correlation for this with geographical range, very large ♂♂ being found, e.g., both in Morocco and Tunisia. – **L** 15.5–18 cm; **W** ♂ 85–100 mm (n 54, m 92.7), ♀ 85–93 mm (n 15, m 87.6); **T** ♂ 55–73 mm (n 52, m 63.5), ♀ 58–69 mm (n 14, m 61.4); **T/W** m 68.7; **B** 17.0–22.4 mm (n 68, m 19.8); **BD** 4.0–5.7 mm (n 45, m 4.9); **Ts** 26.1–31.0 mm (n 67, m 28.5). **Wing formula:** $p1 >$ pc 3–10 mm, $< p2$ 35–47 mm; $p2 <$ wt 3–6.5 mm, $=5/7$; $pp3–4(5)$ longest; $p5 <$ wt 0–1.5 mm; $p6 <$ wt 1–5 mm; $p7 <$ wt 6–12 mm; $p8 <$ wt 9–16.5 mm; $p10 <$ wt 15.5–22.5 mm; $s1 <$ wt 16–25.5 mm. Emarg. pp3–5 (sometimes faintly also on p6).

GEOGRAPHICAL VARIATION & RANGE Monotypic. – N Africa, Sinai, Levant north to Syrian Desert, N Arabia; sedentary or nomadic. (Syn. *brooksbanki*; *theresae*.)

TAXONOMIC NOTES Some authors recognise '*brooksbanki*' for Middle Eastern birds, claimed to be on average less black on mantle, more sooty grey-black than N African birds, with paler, usually pink-buff undertail-coverts, and bill is said to be somewhat heavier. These claimed differences are generally indistinct or impossible to confirm on material examined by us, and there seems to be much overlap in all of them. – Breeders on the Tiznit Plateau in S Morocco ('*theresae*') have been separated as being slightly darker and more saturated in colour, but like Vaurie (1955) we fail to see this as anything other than a subtle clinal tendency.

REFERENCES VAURIE, C. (1955) *Amer. Mus. Novit.*, 1731: 1–30.

1stS ♀, Tunisia, Apr: some ♀♀ much duller with less rufous, especially with wear in spring/summer. Juv primary-coverts are very worn and abraded. (D. Occhiato)

Juv ♂, Morocco, May: upperparts rather uniform dusky buff-brown, upperwing extensively fringed rufous, and underparts tinged buffy-cream, usually with partially black or dusky throat in juv ♂. (A. Faustino)

KURDISH WHEATEAR
Oenanthe xanthoprymna (Hemprich & Ehrenberg, 1833)

Alternative names: Red-tailed Wheatear, Kurdistan Wheatear

Fr. – Traquet kurde; Ger. – Rostbürzel-Steinschmätzer
Sp. – Collalba culirroja; Swe. – Kurdstenskvätta

Also known as Red-tailed Wheatear, a rather inaccurate name if you consider that in most plumages the species has a black-and-white tail. Often treated as conspecific with Persian Wheatear *O. chrysopygia*, and the two are certainly closely related, but here they are afforded separate species status considering rather clear plumage differences and limited hybridisation. Despite its attractive plumage, Kurdish Wheatear is an elusive species that often goes undetected, especially in summer, owing to its shy and unobtrusive behaviour and its subdued colours that blend perfectly into its grey, rocky habitat. A summer visitor mainly to SE Turkey and Iranian Kurdistan, wintering in NE Africa and Arabia.

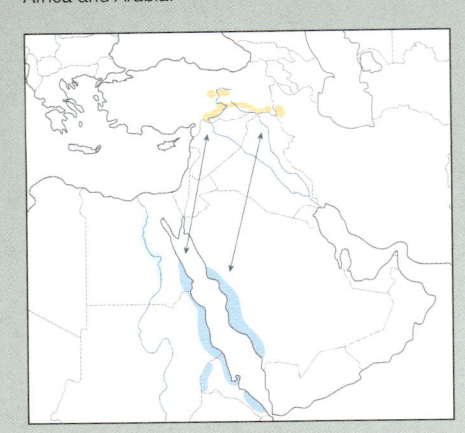

♂, presumed ad, Israel, Mar: rufous wash on rump/uppertail-coverts, with diagnostic white basal corners between rufous area and broad black tail-band (can be difficult to see on folded tail). Two outer greater coverts on right (but not left) wing seem shorter and duller, as if juv, otherwise primary-coverts, overall plumage and extensive white in tail best fit ad ♂. (T. Krumenacker)

♂, presumed ad, Israel, Mar: readily identified by blackish lower face and bib, narrow buffish-white supercilia, dingy grey upperparts with darker wings, and drab white breast and belly, but best confirmed by rufous wash on posterior parts and tail pattern. Presumed ad given evenly fresh blacker remiges and primary-coverts. (T. Krumenacker)

IDENTIFICATION A medium-sized wheatear with rather short and broad-looking tail. ♂ plumage characteristic, *dingy grey above with blackish lower face and bib*, bordered above by *narrow whitish supercilium*. Breast and belly drab white, with *slight pinkish-rufous wash on rear flanks, vent, rump and on upper- and undertail-coverts*. Has black inverted 'T' on tail, leaving *diagnostic white basal corners* (difficult to see on perched bird). ♀ variable: some are rather dark-throated (more so with wear), resembling ♂ but usually a little duller and browner, having very little rufous visible away from vent, while others with pale throats (especially when fresh), *rusty-brown ear-coverts* and *mottled pale brown-grey throat* are very similar to Persian Wheatear, but usually have *white basal tail corners* (Persian: invariably rufous), blacker under-wing-coverts, *rufous on rear body hardly extending to belly or flanks*, and different face pattern. Differences between Kurdish and Persian Wheatears are discussed further under latter. Might also be confused with Red-rumped Wheatear within narrow area of overlap in winter, and—perhaps at distance—dark-throated Eastern Black-eared Wheatear (cf. Black-eared Wheatear).

VOCALISATIONS Song rather softly trilling, strophes generally low-pitched and short (but can be extended when in the 'right mood', spacing between songs also varies depending on excitement). Often consists of only 5–7 syllables usually with last stressed, e.g. *tru-ee tri-tru-tre churr*. Song can recall both Black Wheatear and a distant *Monticola* thrush. – Call (mainly in alarm?) a low-pitched squeaky, whistling note *ihp*, more discreet than the similar call of Northern Wheatear. There is also a buzzing or rasping *brzeeh* (or *chehk*) from anxious birds, or in close encounters with others.

SIMILAR SPECIES ♀ with pale throat easily confused with either sex of *Persian Wheatear*, which see. – ♂ could be confused with dark-throated *Eastern Black-eared Wheatear* at distance, but mantle coloration, reddish rump and under- and uppertail-coverts, and black of throat meeting black of wings offer good distinguishing marks. – ♂ *Pied Wheatear* is readily separated by many of the same characters as mentioned in connection with Black-eared, and unlike Kurdish its mantle and back are blackish in adult/summer (not greyish) and distinctly pale-fringed when fresh. Both Eastern Black-eared and Pied Wheatears also differ considerably in tail pattern (tending to have an uneven band with more black at the corners). – Red-rumped Wheatear in ♂ plumage, which has superficially similar black face, is unlikely to be confused given completely different tail and wing pattern with hint of whitish wing-bars, while duller ♀ Red-rumped is easily eliminated by much broader black tail-band, and by usually being cinnamon/rufous overall rather than having gingery tones on head restricted to ear-coverts; note also general structure (Red-rumped deeper-bellied, broader-winged and has relatively heavier bill).

AGEING & SEXING Ageing often difficult and usually requires close inspection of moult and feather wear, and resulting wing pattern. Sexes usually differ, but a few are very nearly alike. Seasonal variation involves mainly abrasion of feather tips, with most marked changes in 1stY ♂ plumage and face of ♀. – Moults. Complete post-nuptial moult of ad in summer (Jun–Sep), and partial post-juv moult at the same time. Post-juv renewal involves all head, body, lesser and median coverts, most (inner) or all greater coverts, and usually the innermost tertial and innermost alula, more rarely all tertials and innermost secondaries. Although not known to undertake any pre-nuptial moult, an ad collected in Egypt in Mar had certainly replaced one innermost tertial in winter (NHM). – **SPRING Ad** Both sexes differ from 1stS by being evenly feathered, with relatively less worn and darker primaries and primary-coverts. Base of outer tail-feathers apparently as a rule pure white. **Ad ♂** Throat and face black, dark part of uppertail near-black, crown drab grey.

1stS ♂, Turkey, May: aged by worn, browner and bleached outer wing (including visible alula, primary-coverts and primaries); note moult limit in greater coverts. Proof that young ♂ can develop pure white inner tail corners. (D. Occhiato)

WHEATEARS

With wear becomes more strongly patterned, with well-marked black bib, darker wings (being almost black on carpal area) enhancing contrast with greyer mantle/scapulars and small upperwing-coverts. Supercilium usually narrower and whiter than in ♀. Base of outer tail-feathers pure white (or at the most with very faint orange tinge on outer webs when fresh). **Ad** ♀ Usually paler above, slightly more brownish on crown and greyish or dirty whitish-cream below, often with less rufous tinge on lower belly compared to ♂. Dark portion of uppertail brownish-black rather than blackish (but difference subtle, and some overlap). Throat and face in dark-throated morph usually dull blackish or dark sooty-grey with some white tips (primarily on sides of chin and lores), at times bib blacker and approaching ♂ plumage closely (especially when heavily worn). **1stS** As ad but usually possible to discern more worn juv remiges and some coverts (including all primary-coverts). ♂ less evenly patterned compared to ad, with browner tone to dark areas. Moult limits between brown juv primaries, pale-tipped primary-coverts and some outer also pale-tipped greater coverts against much darker post-juv inner greater, median and lesser coverts and perhaps some tertials (most readily seen in ♂). At least majority, possibly all, 1stS ♀♀ have rufous base to outer tail-feathers, not pure white. 1stS ♂ has mainly white tail-base but with variable rufous tinge; the rufous tinge may bleach to white in summer. — **AUTUMN Ad** Both sexes differ from 1stW in having evenly very fresh wings, with no discernible moult limits. Primary tips fresh or, at most, slightly worn, and primary-coverts fresher with darker centres. At least majority, perhaps all, have white base to outer tail-feathers. **Ad** ♂ Largely as spring, though crown and hindneck tinged brown, and whitish supercilium now tinged buff. Rest of upperparts and fresh pale fringes to wings tinged slightly brownish (less grey than in spring). Black of lower face and throat slightly tipped pale greyish. Base to outer tail-feathers pure white. **Ad** ♀ As spring, but paler in general, with pale fringes obscuring dark face/throat, and wings more extensively fringed paler. **1stW** ♂ As fresh ad ♂ though blackish throat usually slightly mottled, and white on tail-feathers edged pale rufous on outer webs, while upperparts are duller/browner (less greyish). Retains some slightly worn juv feathers (including all primary-coverts, remiges and tail-feathers); juv primary-coverts have paler and browner centres and usually more prominent pale tips than ad; juv greater coverts (if any retained) have brown-grey centres and ochre fringes, the latter clearly broader and slightly whitish at the tip (contrasting with fresher ad-like post-juv ones which

♀, presumed ad, Kuwait, Mar: a pale-throated bird, with characteristic rusty-brown tinge to ear-coverts and bright rufous rear body. Pure white at basal tail corners excludes similar Persian Wheatear and indicates ad. (A. Halley)

♀, Turkey, May: rufous on rear body rarely extends to belly and flanks, but in this unusually richly-coloured ♀ it extends well onto breast, flanks and even throat-sides, like Persian Wheatear. Ageing impossible, although white on tail-base might indicate ad. (D. Occhiato)

♀, presumed ad, Turkey, May: a pale-throated greyer bird; with wear rusty-brown on ear-coverts enhanced. Apparently ad with evenly-feathered wing, and relatively less worn and darker primaries and primary-coverts. (D. Occhiato)

Ad ♂, Israel, Nov: very much like spring/summer ♂, but paler, upperparts less pure grey, some pale feather tips in black of face, and wings extensively fringed buff. Uniformly fresh and broadly pale-fringed wing-feathers indicate ad. (H. Shirihai)

1stW ♂, Israel, Dec: 1stW ♂ acquires ad-like plumage early, displaying combination of grey-tinged upperparts, whitish supercilium, black bib and sides of neck and breast, and rufous-tinged ventral area. Quite worn wing indicate age. (A. Rinot)

335

♀, **dark-throated variant, presumed ad, Israel, Jan**: dark-throated ♀♀ rather scarce and poorly known. Throat darkens with wear, especially from winter, albeit never becoming as solid or intense black as ♂. Such ♀♀ are still paler above. Blackish-centred primary-coverts and broad rufous-buff tips to greater coverts suggest ad. (H. Shirihai)

♀, **dark-throated morph, presumed ad, Cyprus, Nov**: dark throat largely obscured by pale tips. Ageing difficult, but quite substantial white tail corners suggest ad, as do fresh primary tips. (A. McArthur)

1stW ♀, probably pale-throated variant, Israel, Nov: some 1stW, especially ♀♀, lack white on tail and can be difficult to separate from ♀ Persian. However, relatively broad black tail-band almost reaches tip of longest uppertail-coverts, while obscure dark eye-stripe but better-developed pale supercilium are important additional features. Newly-moulted inner greater coverts, but rest of greater coverts, remiges and primary-coverts are juv feathers. (Left: O. Horine; right: A. Ben Dov)

have blackish centres and more even-width rufous-brown fringes and tips). Juv tertials and secondaries less fresh, especially at fringes and tips, with paler centres and less rufous fringes than ad-like post-juv ones. If smallest alula replaced it appears clearly darker-centred with a fresher pale fringe than larger juv feathers. **1stW ♀** As ad ♀; only safely separated by moult limits (less obvious than in 1stW ♂), but averages duller, and as a rule (possibly invariably) has rufous inner tail corners, not white. **Juv** Recalls ♀ but has loose body-feathering and reduced rufous-orange on rear body, though upperwings extensively fringed cinnamon. Combination of rufous tail-coverts and vent and lack of any obvious mottling on body distinguish it from all other juv wheatears.

BIOMETRICS L 15–16 cm; **W** ♂ 89–98.5 mm (n 13, m 95.0), ♀ 86–95.5 mm (n 8, m 90.9); **T** ♂ 57–63.5 mm (n 13, m 60.3), ♀ 53.5–63 mm (n 8, m 58.6); **T/W** m 63.8; **B** 17.5–20.8 mm (n 20, m 18.6); **Ts** 22.4–25.6 mm (n 19, m 24.1). **Wing formula: p1** > pc 0.5–6 mm, < p2 39–49 mm (once 36); **p2** < wt 4.5–9 mm, =5/6 (75%) or =6 or 6/7 (25%); **pp3–4**(5) longest; **p5** < wt 0–2.5 mm; **p6** < wt 6–9 mm; **p7** < wt 11–14.5 mm; **p8** < wt 14–19 mm; **p10** < wt 19–24 mm; **s1** < wt 22–28 mm. Emarg. pp3–5.

GEOGRAPHICAL VARIATION & RANGE Monotypic. – SE Turkey, NW Iran, probably NE Iraq; winters Sinai, Egypt, Sudan, Arabia. Hybridises with Persian Wheatear in a rather restricted montane area of NW Iran (Kherman Shan). (Syn. *cummingi*; *hawkeri*.)

TAXONOMIC NOTES Relationships between *xanthoprymna* and *chrysopygia* were studied in NW Iran (Chamani et al. 2010), showing a fairly limited area of hybridisation (see above). Hybrids described from NW Iran have been named '*cummingi*', whereas those from Afghanistan and NW India are called '*kingi*'. Relationships between Kurdish Wheatear and congeners still unclear and indicate a perhaps closer link to Mourning and Finsch's Wheatears than to Persian Wheatear, thus—if confirmed—further refuting the view that Kurdish and Persian are conspecific.

REFERENCES Aliabadian, M. et al. (2007) *Mol. Phyl. & Evol.*, 42: 665–675. – Chamani, A. et al. (2010) *African Journal of Biotechnology*, 9: 7817–7824.

Juv, Turkey, Aug: combination of dusky head and upper breast, some rufous-orange on rear body (mostly on uppertail-coverts and vent) and lack of any obvious mottling on body distinguish it from all other juv wheatears in region. (R. Bonser)

PERSIAN WHEATEAR
Oenanthe chrysopygia (De Filippi, 1863)

Alternative name: Red-tailed Wheatear

Fr. – Traquet de Perse; Ger. – Kaukasussteinschmätzer
Sp. – Collalba afgana; Swe. – Persisk stenskvätta

Also known as Rufous-tailed Wheatear and often regarded as conspecific with Kurdish Wheatear, although *chrysopygia* merits specific status based on its distinctive plumage and on recent genetic evidence that it is apparently more closely related to Mourning and Finsch's Wheatears than to Kurdish. There is a limited hybridisation zone between Persian and Kurdish but no evidence of intergradation. Persian Wheatear is a summer visitor to Iran and eastern Transcaucasia east to Afghanistan and Tajikistan, south to the Iranian Zagros Mountains, and it winters in E Arabia and east to NW India.

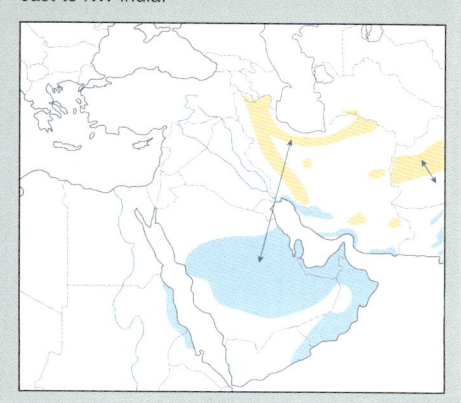

IDENTIFICATION Sexes virtually alike and generally very similar to pale-throated ♀ Kurdish Wheatear, being overall greyish-tinged, *relatively plain brown-grey above* and *grey-white below* (no black throat), with warm *reddish-buff vent and undertail-coverts* extending variably to flanks and belly. *Base of outer tail rusty* (not white), not contrasting with rump, and tail has typical inverted black 'T'. Also, characteristically has *warm buff-brown tone to ear-coverts/lower cheeks*, greyer cast to neck-sides and nape, sometimes extending to crown (mostly in ♂♂), and *blackish eye-stripe, especially pronounced on lores* (and further emphasized by ill-defined greyish-white supercilium). Note also relatively paler, cream-buff upperwing fringes (when fresh almost wholly concealing browner, less black, centres), forming *Isabelline Wheatear-like uniform upperwing. Underwing-coverts mainly whitish*, at most with some black mottling (rather than being solidly blackish as in Kurdish Wheatear). Further differences between Kurdish and Persian Wheatears are discussed under Similar species. Other confusion risks perhaps involve ♀ Red-rumped Wheatear, within narrow area of overlap in winter, and Isabelline Wheatear seen at a distance.

VOCALISATIONS Apparently quite similar to Kurdish Wheatear, but no detailed study available. The song is short but persistently repeated for long periods, rather softly warbling or trilling, a little lark-like. It can also recall a distant Robin or Rufous-tailed Scrub Robin, or even a Blackstart, with variable slow or rapid twittering or trilling phrases and clear notes, and may utter hoarse, rasping sounds in the strophes or between them; usually lacks mimicry. – Calls include a low grating *grat-grat*, a dull, muffled, *zrrp-zrrp*, further a rasping *chairz* in alarm and constant chacking notes, *tchek tchek tchek…*, and this may be repeated in a fast series as a rattling snore *thrrr*.

SIMILAR SPECIES Slightly different from *Kurdish Wheatear*: both sexes largely pale uniform brown-grey above and dirty buffish-white below and on underwing, and both have dark eye-stripe. Lacks dark throat and underwing-coverts of Kurdish, and especially ♂♂ tend to have characteristic pale greyish neck-sides (often continuing onto crown) and extensive cinnamon rear flanks and lower belly. Black tail-band of Persian averages slightly narrower in all plumages, and at all ages has rufous tail-bases. Some young Kurdish Wheatears (both sexes), however, may possess rusty tail-base, too, and others with rusty tail-base may be hybrids ('*cummingi*'). Thus, other features acquire importance, e.g. face pattern (especially dark eye-stripe) and the relatively narrower black tail-band with rufous tip (when fresh). Additional characters, e.g. upper and underparts in general and underwing coloration should also be used. – Superficially recalls much heavier *Isabelline Wheatear* (both have extensive buff upperwings and dark lores), but is distinctly greyer above with conspicuous rufous at rear. – Might also be confused with larger ♀ *Red-rumped Wheatear* but this is easily eliminated by large size, much wider black tail-band, gingery tones on much of head rather than restricted to ear-coverts and different overall structure (Red-rumped deeper-bellied, broader-winged and has relatively heavier bill).

AGEING & SEXING Ageing requires close inspection of moult and feather wear or pattern in wing. Sexes hardly differ. Seasonal variation also limited. – Moults. Extent and timing of moult mostly as in Kurdish Wheatear (which see). – SPRING **Ad** ♀ almost inseparable from ♂ though usually notably duller with less grey around neck/head, weaker dark eye-stripe and more limited rufous below. Differs from 1stS by being evenly feathered with relatively less worn, darker primaries and primary-coverts. **1stS** As ad but usually has clearly more worn, brownish juv remiges and some coverts (including all primary-coverts); also overall duller (especially ♀). – AUTUMN **Ad** Much as in spring, but has fresh sandy-

Ad ♂, Kuwait, Nov: seasonal and sexual variation very limited. Probably ♂ given purer grey plumage, especially on neck and supercilium, with better-marked blackish lores and whiter throat. However, not all ♂♂ as distinctive as this. Aged by very fresh and blackish remiges. (T. Quelennec)

1stS ♂, Kuwait, Mar: very similar to pale-throated ♀ Kurdish Wheatear, but base of outertail rusty, invariably lacking white corners of latter. Dark eye-stripe, especially on lores, emphasized by ill-defined paler supercilium. Strong-billed. (R. Al-Hajji)

1stW ♀, Oman, Oct: paler birds very similar to Kurdish Wheatear and difficult to identify, especially if exact tail pattern or darkness of underwing-coverts not seen. Pattern and condition of primary tips, primary-coverts and tail-tip indicate 1stW; sexed as ♀ by rather bland face (lacking darker lores), less pure grey upperparts and head. (A. Audevard)

buff (less greyish) tips above and below, and broad fringes to wing-coverts and tertials. Unlike 1stW, wings evenly very fresh without discernible moult limits. Primary tips fresh or at most slightly worn, and primary-coverts fresher and firmer with darker centres. **1stW** Only safely separated from fresh ad by juv wing-feathers, including all primary-coverts, perhaps several outer greater coverts (but some moult all), remiges and tail-feathers, and moult limits as in Kurdish. **Juv** Differs from ad in characteristic loose body-feathering, but also in reduced rufous-orange on rear body, though upperwing-feathers extensively fringed cinnamon. Combination of rufous tail-base, like ad, and lack of any obvious mottling on body distinguish it from all other juv wheatears.

BIOMETRICS L 14.5–15 cm; **W** ♂ 89–99 mm (n 24, m 93.7), ♀ 85–93.5 mm (n 14, m 90.0); **T** ♂ 55–65 mm (n 23, m 59.4), ♀ 54–62 mm (n 14, m 57.4); **T/W** m 63.5; **black on r1** 25–37 mm (n 37, m 31.7); **black on r4** 11–18 mm (n 37, m 14.4); **B** 17.0–19.8 mm (once 16.5; n 39, m 18.6); **Ts** 23.6–26.2 mm (n 14, m 24.8). **Wing formula: p1** > pc 0–7 mm (rarely to 0.5 mm <), < p2

Presumed ad, Iran, Jan: especially note facial pattern including dark eye-stripe, pale greyish upperparts, whitish throat and large area of rufous in tail. Seems to have subtly larger head and stronger bill than Kurdish Wheatear. (A. Ouwerkerk)

41–48 mm; **p2** < wt 3.5–7.5 mm, =5/6 or 6 (rarely =6/7); **pp3–4** longest; **p5** < wt 1–3 mm; **p6** < wt 6–9.5 mm; **p7** < wt 11–15.5 mm; **p10** < wt 21–27 mm; **s1** < wt 23–29 mm. Emarg. pp3–5, although rarely also a hint on p6.

GEOGRAPHICAL VARIATION & RANGE Monotypic. – S Armenia, Azerbaijan, much of Iran except extreme west and central plains, Afghanistan; local in S Turkmenistan, W Pakistan; winters in E Arabia and east to NW India. – See also Taxonomic note under Kurdish Wheatear. (Syn. *kingi*.)

REFERENCES ALIABADIAN, M. et al. (2007) *Mol. Phyl. & Evol.*, 42: 665–675. – CHAMANI, A. et al. (2010) *African Journal of Biotechnology*, 9: 7817–7824.

♀, presumed ad, Kuwait, Mar: often strongly recalls ♀ Kurdish, but note slightly narrower black tail-band and invariably rufous tail-base. Also, dark eye-stripe and pale greyish neck-sides obvious, as is strong bill. (A. Halley)

Oman, Mar: very difficult to separate from ♀-like Kurdish Wheatear, but note strong bill. Also, dark eye-stripe and light supercilium offer good clues. (H. & J. Eriksen)

Presumed ad ♂, Iran, Jun: in very fresh plumage, during active post-nuptial moult, but still has characteristic broad rufous tail-tip. Locality in N Iran helps eliminate Kurdish Wheatear. (A. Sadr)

BLYTH'S WHEATEAR
Oenanthe picata (Blyth, 1847)

Alternative names: White-bellied Wheatear, Eastern Pied Wheatear

Fr. – Traquet varié; Ger. – Elstersteinschmätzer
Sp. – Collalba variable
Swe. – Vitbukig orientstenskvätta

This is one of three forms constituting the so-called Variable Wheatear (or Eastern Pied Wheatear). As detailed under Taxonomic notes, Blyth's Wheatear is treated here as a separate species. It occurs over N & E Iran extending into Turkmenistan, Afghanistan and N Pakistan. In most areas north and east of Iran it overlaps and often hybridises with either of Gould's or Strickland's Wheatears, the two other members of the 'Variable Wheatear complex' (both extralimital). Blyth's Wheatear is a regular winter visitor to coastal S Iran, E Arabia, and occasionally wanders further west. It favours open rocky country with bushy cover, often along riverbanks or in foothills with scant vegetation, in winter extending to even drier areas.

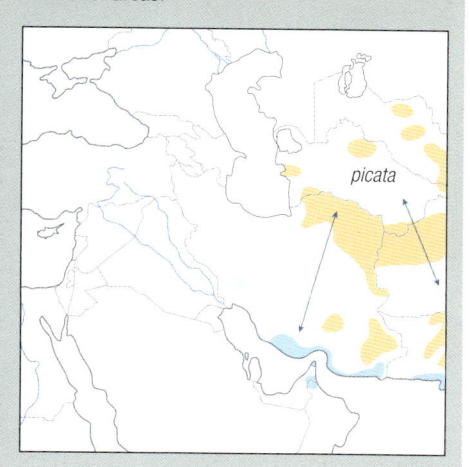

IDENTIFICATION The size roughly of Black-eared Wheatear, with contrastingly black-and-white (♂) or dusky grey-brown plumage (♀, young). ♂ can only be confused with superficially similar Hume's Wheatear, but has nearly always some cinnamon tones to vent and undertail-coverts (Hume's: pure white). ♀ resembles ♀-like plumages of Pied and Finsch's Wheatears (less so Arabian) but is generally darker sooty grey-brown than either. Most ♀ Blyth's are distinctive in approaching ♂ with dusky appearance to head, upperparts and breast, but some are paler overall ('pale-throated') and easily confused with ♀ Gould's Wheatear. – All plumages share white tail base with inverted 'T' mark and *moderately broad to rather narrow tail-band*, somewhat midway between Northern and Pied Wheatears, with a tendency to have *rather extensive dark edges to outer rectrices*. Tail rather long. Vent and undertail-coverts variably coloured, cream-white, pink-buff or rather saturated rufous or cinnamon-tinged. Bill comparatively short, and all bare parts black. Highly territorial, even in winter, and usually encountered alone or in pairs, perching prominently. *Typically makes deep bobbing or curtsying movements*, with slight tail fans, especially when alarmed and nearly always upon landing. Sings from perch or in short display-flight with some hovering and spread tail.

♂ (left) and ♀, Iran, May: the white-bellied member of so-called 'Variable Wheatear complex', and the only one that breeds in the treated region. The ♂ looks like a Pied but without the white cap. (A. Talebui Gol)

VOCALISATIONS Typical wheatear song, a short or moderately long outburst of warbling, whistling and scratchy notes. Compared to most other wheatears has a more varied and complex song than most (all?) Palearctic congeners (Panov 2005). Song said to be quieter and less piercing than Isabelline Wheatear, but can be long and complex, mixing trills and much mimicry, low-pitched calls and Robin-like 'trembling whistles', metallic sounds and chirrups (Roberts 1992) and, sometimes, quite explosive phrases. It has been claimed that in contrast to other members of the 'Variable Wheatear complex', Blyth's Wheatear song often has a very wide frequency range (Panov 2005; though sonograms show similar range to Gould's Wheatear) with more notes of long duration, many with overtones. In areas of overlap, songs of the different forms have been noted to be more similar (Panov 2005). Song of ♀ probably similar but rarely heard and quieter. – Calls typical of genus, e.g. commonly a clicking *chek-chek*, and in alarm a harsh and scratchy *brshe* and brief rattles; also a soft whistle *plew* or disyllabic *plew-wit* much like Isabelline Wheatear (Rasmussen & Anderton 2005).

SIMILAR SPECIES ♂ differs from *Hume's Wheatear* by subtly smaller size (also lacks large-headed look and has weaker bill), and black areas have less metallic gloss. Demarcation between black and white on upper breast on average runs slightly lower, in particular on sides near wing-bend (Hume's: subtly higher up on breast and more straight division, but many appear similar). Hume's invariably has white or cream-white vent/undertail-coverts, whereas Blyth's has a faint or rather obvious pink-buff or cinnamon tinge to these feathers. Hume's is further typically more confiding, at least in E Arabia. Fresh first-year (but not adult) ♂ Blyth's also has diagnostic whitish fringes and tips to primary-coverts (lacking in Hume's, or these are only very narrow and diffuse pale grey). – ♀ Blyth's Wheatear is distinctive in approaching its ♂ plumage, especially by its dusky or even blackish head, upperparts and throat/breast, but some are paler overall, especially young and fresh pale-throated birds (or hybrids), and are then closer to ♀ *Gould's Wheatear* or ♀ *Pied Wheatear*. In fresh plumage, ♀♀ of both Blyth's and Gould's lack Pied's diagnostic fine pale fringes above, the upperparts being uniform dull grey-brown or olive-brown. Wings are also more uniform, less conspicuously pale-fringed. Most also tend to have a cold grey throat and upper breast (with slight pink-brown cast) gradually merging with greyish orange-buff or dull pinkish-buff rest of breast and upper belly, contrasting only with whitish-cream lower belly (in fresh ♀ Pied, buffy/orange tinge quite pronounced on breast, but throat paler/whiter). In worn plumage, both Pied and Blyth's are similarly uniform above, but most ♀ Blyth's are dusky or blackish to

1stY ♂, Iran, Jan: superficially resembles Hume's Wheatear, but smaller with a slighter bill and less glossy black plumage, while demarcation between white and black on breast is lower and better defined. Often shows some buffish-pink on undertail-coverts. Aged by juv primary-coverts with whitish fringes, here barely visible. (H. & J. Eriksen)

♀, **dark-throated, United Arab Emirates, Feb**: dark-throated ♀♀ most frequent in *picata*, here an extreme example being quite ♂-like. Blackish bib becomes more solid with wear, extending well onto upper breast. (H. & J. Eriksen)

♀, **pale-throated, Oman, Jan**: compared to fresh ♀ Pied Wheatear, overall impression is more uniform, with cold grey throat and upper breast (often with slight pink-brown cast), gradually merging into buff-brown below, and dusky-white belly. ♀ Finsch's has more contrasting pale fringes, especially whitish wing-bars, emphasized by darker greater coverts panel. ♀ Arabian Wheatear has pinkish-cinnamon undertail-coverts, dusky-streaked breast and flanks, and broader tail-band. (H. & J. Eriksen)

medium grey-brown (Pied: dull umber-brown). As in fresh plumage, pale Blyth's lacks Pied's rather pronounced buffy/orange tinge on breast, throat and chest, being more uniform pale greyish pink-brown with only indistinctly warmer breast, but much individual variation in both species. Extent of dark on throat and upper breast variable but usually more extensive than in dark-throated Pied, often reaching onto upper breast, but again some individual variation. Blyth's has a broader, more evenly broad black tail-band (broader on outermost feathers in Pied, but some overlap). The two are sometimes best separated by call (Blyth's *check-check*; Pied gives frizzling *brsche*), structure (Pied slimmer with proportionately longer primary projection) and behaviour (Blyth's typically makes deep bobbing). – ♀ *Finsch's Wheatear* is very similar to ♀ Blyth's Wheatear. Finsch's, however, has more patterned wing with thin whitish wing-bars, darker greater coverts panel, and more contrasting pale fringes to many wing-feathers, as well as different calls, but deep bobbing behaviour is similar, yet differs in frequency. Upperparts of Finsch's usually distinctly paler, less grey-brown or dusky than in Blyth's, but palest examples more similar. Beware that extent of dark on throat and breast in Blyth's Wheatear is highly variable and can be as extensive as ♀ Finsch's, and in both species becomes especially extensive with wear and age. – Duller ♀ *Persian Wheatear* (but for diagnostic chestnut tail base) and, especially, *Kurdish Wheatear* (which has rather similar tail pattern) may appear superficially similar to a ♀ Blyth's Wheatear, particularly in poor views, but even in the dullest Kurdish, cinnamon-buff upper- and undertail-coverts and vent/rump are still distinctive (belly of Persian/Kurdish also usually warmer cinnamon than breast). – ♀ *Arabian Wheatear* might prove confusable with extremely pale ♀ Blyth's, but former is eliminated by subtly smaller size and more compact jizz, on average slightly more prominent pinkish-cinnamon undertail-coverts/vent, presence of narrow dusky streaks on breast and flanks, and by broader tail-band.

AGEING & SEXING Ageing possible by use of moult and feather wear or pattern in wing. Rather complex plumage variation given moderate to strong sexual dimorphism and two ♀ plumage types. Fairly limited seasonal variation (wear of feather tips) with most obvious change in dark-throated ♀♀. – Moults. Complete post-nuptial and partial post-juv moult in summer (mainly Jun–Sep). Post-juv renewal includes all head, body, lesser and median coverts, most or all greater coverts, usually some tertials, and some or (rarely) entire alula. No pre-nuptial moult apparent. – **SPRING Ad** ♂ Aged by being evenly feathered with relatively fresher remiges, primary-coverts and dark areas of tail blacker (or blackish-brown). No pale tips or fringes to primary-coverts. **1stS** ♂ Readily aged by very worn and dull retained juv primaries, primary-coverts, often tertials and perhaps some outer greater coverts, alula and darker areas of tail (ad has primary-coverts black or black-brown without pale fringes). Pale primary-covert tips and fringes wear considerably from Mar/Apr, after which these may disappear through abrasion, but condition and ground colour of primary-coverts and primaries always conclusive. **Ad** ♀ Differs from 1stS ♀ in being evenly feathered with relatively fresher primaries and primary-coverts (latter with only indistinct greyish fringes), and dark areas of tail darker brown, less faded. Two ♀ plumage types, pale-throated and dark-throated, with latter more frequent. Blackish bib of dark-throated ♀ more solid with wear. **1stS** ♀ Much as ad ♀ but has more worn and duller brown juv primaries, primary-coverts, often some tertials, and perhaps some outer greater coverts, alula and darker areas of tail, forming moult contrast, but less contrasting than ♂, thus ageing more difficult. Primary-coverts also still narrowly fringed pale buff, but these wear off later in season. Dark-throated 1stS ♀ frequent, but blackish bib often less solid and complete compared to ad, though much variation. – **AUTUMN Ad** ♂ Reliably aged in being

♀ **Blyth's/Gould's Wheatear (left: India, Nov) versus Pied Wheatear (right: Sweden, Oct)**: unlike Pied, in all ♀-like plumages, Blyth's, Gould's and Strickland's Wheatears have notably shorter primary projection, and in fresh plumage lack Pied's pale-fringed mantle and scapulars. The colder/greyer upperparts and whiter underparts of the bird on left best match Gould's, but it could be an extremely pale Blyth's. Ageing of autumn ♀♀, especially in the field, is difficult. (Left: S. Singhal; right: S. Hage)

WHEATEARS

Blyth's/Gould's, ♀, pale-throated, presumed 1stW, Iran, Jan: plumage best matches Gould's, but it could be an extremely pale Blyth's; by range, Blyth's is more likely. Plumage similar to pale-throated ♀ Finsch's, but wing pattern more uniform. Contrast between fresher/new tertials and worn/browner remiges and primary-coverts indicate young bird (M. Hornman)

Gould's Wheatear *O. capistrata*, ad ♂, Iran, May: rather similar to ♂ Pied and Arabian Wheatears, and both sexes of Mourning. From latter two separated by lack of white flashes in remiges (visible when wing spread). From worn Pied Wheatear, Gould's differs mostly in jizz (being slightly bulkier), behaviour and calls, and further by having on average smaller white cap and slightly broader black bib and tail-band, and most have buffish-pink undertail-coverts and ventral area. Aged by homogenously feathered wing. (C. N. G. Bocos)

evenly fresh without pale fringes to primaries and primary-coverts, which are also blacker. Plumage much as in spring, but tinged pale pinkish-buff on flanks, vent and undertail-coverts (sometimes more deeply cinnamon), and washed or blotched grey-brown on crown. **1stW ♂** As fresh ad but aged by buff or whitish fringes to primary-coverts and tips to primaries, and dark areas of remiges, primary-coverts and tail-feathers browner. Often some or all tertials and several outer greater coverts juv and distinctly fringed buff. At least some 1stW ♂♂ have whitish spots above lores, but this also shown (to lesser extent) by fresh ad. Juv/1stY of both sexes, while slightly shorter-winged, also has on average broader tail-band. **Ad ♀** Aged by being evenly fresh with indistinct pale fringes to primary-coverts (greyer and less sharply defined) and almost no pale tips to primaries. Remiges, tail-feathers and primary-coverts fresher. Ad tertials and greater coverts lack broad rufous fringes of juv feathers. Plumage as in spring but pale tips on throat obscure bib in dark-throated morph. **1stW ♀** Unlike similar ad ♀ has slightly more worn and browner juv primaries, primary-coverts (latter also have bold pale buff fringes, unlike in ad), often some tertials and outer greater coverts (usually distinctly fringed pale rufous-buff). Also, alula and darker areas of tail browner, but moult limits less obvious than in ♂; differences between pale-throated and dark-throated birds somewhat less obvious due to broad pale fringes. **Juv** Much as ♀. Soft, fluffy feathering, indistinctly scalloped with faint dark subterminal marks and paler tips. Extensively rufous-fringed upperwing-coverts.

BIOMETRICS L 15 cm; **W** ♂ 90–97 mm (n 15, m 93.3), ♀ 84–92 mm (n 12 m 88.2); **T** ♂ 60–67 mm (n 15, m 64.5), ♀ 58–65 mm (n 12, m 61.3); **T/W** m 69.3; **B** 16.2–18.5 mm (n 26, m 17.3); **BD** 3.7–4.3 mm (n 23, m 4.1); **Ts** 23.4–26.0 mm (n 25, m 24.5). **Wing formula: p1** > pc 3–6.5 mm, < p2 38.5–44.5 mm; **p2** < wt 5.5–10 mm, =6/7 (rarely =6); **pp3–4** (5) longest; **p5** < wt 0–1.5 mm; **p6** < wt 3–6 mm; **p7** < wt 9–14 mm; **p10** < wt 18–23 mm; **s1** < wt 20–25 mm. Emarg. pp3–5 (6) (emarg. on p6 variable: rather short but obvious in > 50%, indistinct or virtually lacking in rest).

GEOGRAPHICAL VARIATION & RANGE Monotypic. – N, E & SE Iran, Central Asia including S Turkmenistan, Tashkent region, Pamirs, N Afghanistan, N & C Pakistan, NW India; in winter either short-range movements or migrates to Arabia, S Iran, Pakistan, NW India. – Fairly extensive but local overlap and ongoing hybridisation (e.g. in S Uzbekistan, N Afghanistan, parts of N Pakistan) with the closely related Gould's and Strickland's Wheatears, producing intermediates closer to one or other of these two, and such birds at times difficult to identify. Blyth's Wheatear is expanding its range northward in Hindu Kush, NE Afghanistan, entering the breeding range of Strickland's Wheatear; and in the Surkhandariinzky district, Uzbekistan, the ♂ population (*fide* Panov 2005) consists of 84% Gould's, 13% Strickland's and 3% Blyth's Wheatears, and the ♀ population > 90% Gould's. At least some apparently dark-throated ♀ Gould's Wheatears may in fact be hybrids with Blyth's Wheatear, but dark-throated Gould's ♀♀ surely occur, as such birds occur in pure populations, and are usually distinguishable by their paler grey, sandier upperparts (these parts usually much darker in Blyth's). For further details see Panov (1992, 2005).

TAXONOMIC NOTES There can hardly exist a more controversial taxonomic case in W Palearctic ornithology than the 'Variable Wheatear complex', i.e. how to best deal with the closely related taxa *picata*, *capistrata* and *opistholeuca*, Blyth's, Gould's and Strickland's Wheatears. Long treated as colour morphs of the same species, but also proposed as separate species (e.g. by Ticehurst 1922 and Zarudny 1923) or subspecies (Panov 1992, 2005). Neither

Gould's Wheatear *O. capistrata*, 1stW ♂, India, Nov: retained juv primary-coverts have whitish fringes and tips, and undertail-coverts can be substantially tinged pinkish-buff, in contrast to rest of whitish underparts washed faintly cream-buff (but never as contrasting as in Mourning or Arabian Wheatears). Unlike Mourning and Arabian, this taxon lacks the white remiges panel in flight. (C. N. G. Bocos)

Presumed Gould's Wheatear, 1stW ♀, pale-throated, India, Feb: best fits Gould's, but could be an extremely pale Blyth's. Following characters are useful: rather uniform upperparts and wings, subtly brown-tinged ear-coverts and breast, and contrasting white belly. Juv primary-coverts with whitish tips often a specific character of this plumage. (K. Malling Olsen)

Gould's Wheatear *O. capistrata*, 1stW ♀, pale-throated, India, Dec: overall pale plumage, especially central throat and breast areas. Aged by juv primary-coverts and outer greater coverts. (R. Suvarna)

of these interpretations is perfect. If they were morphs they would all share large parts of the range, and hybrids of all shades would not occur as they do now. Against the separate species theory is perhaps the high degree of hybridisation in parts of the range (though theoretically could reflect poorly developed—but existing—barriers against hybridisation with other closely related species). Regarding the subspecies proposal, it should be noted that it is quite unusual for geographically connected subspecies to differ so distinctly as these three do, especially in ♂ plumage, and it is difficult to explain how races of the same species could have evolved so differently, even if isolated from each other for a long time before coming into fresh contact. Also, subspecies rarely coexist in certain areas like these three do, and hybridisation does not appear to be completely unhindered where they meet, the three phenotypes in their pure form seemingly being too common in relation to intermediates (according to Panov 2005, only 10–13% birds are of hybridogenous type in areas of overlap). Of the three models, all with their shortcomings, species status for all three might still be the best solution, and is adopted here. – Future field research needs to establish whether offspring of mixed pairs are as fit as young of pure pairs. Future molecular studies should sample pure populations of the three species, e.g. in WC Iran for *picata* and in Badakhshan, NE Afghanistan, for *opistholeuca*; for *capistrata*, it is possible that nowhere does its genotype persist in the pure form, because its original breeding range is now claimed to be entirely occupied by hybridogenous populations of *capistrata* × *opistholeuca* (Panov 1992, 2005). This claim also requires confirmation via further molecular and field research. – Until the 1980s, *opistholeuca* was thought to breed also in Jordan/Syria, but recent work (Shirihai et al. 2011) clearly shows these birds to be a different dark form, known as Basalt Wheatear, sometimes treated as a subspecies of Mourning Wheatear or, as here, as a separate species *O. warriae*.

For completeness a summary of characters for the two extralimital closely related species in the 'Variable Wheatear complex' is given here:

GOULD'S WHEATEAR *O. capistrata* (Gould, 1865) (Turkmenistan & S Kazakhstan to N Afghanistan and N Pakistan; recorded in E Iran; altitudinal winter movements or shorter migrations) has Pied Wheatear-like plumage, thus ♂ has an off-white or pale grey-brown cap; paleness of cap varies greatly, partially due to individual variation but some with dusky, heavily-streaked cap often thought to be hybrids. Whitish lower breast and rest of underparts with faint cream-buff tinge in some (though undertail-coverts often slightly more pale pinkish-buff when fresh), otherwise black or blackish-brown, and black of throat reaches upper breast. ♀ predominantly dull grey-brown above, on average slightly paler than ♀ Blyth's. Ear-coverts warmer brown, sometimes tinged rusty-brown. Most ♀♀ are pale-throated (unlike Blyth's) with chin to chest mostly greyish-buff or tinged pale pinkish-buff, becoming whiter on rest of underparts. Rare dark-throated ♀ has more restricted bib, usually terminating slightly higher on upper breast and is less solid than in Blyth's, but some overlap, although most are readily separated from dark-throated Blyth's by paler upperparts. **W** ♂ 90–95.5 mm (n 14, m 93.1), ♀ 88–94 mm (n 13, m 90.7); **T** ♂ 61–68 mm (n 12, m 65.1), ♀ 60–65.5 mm (n 13, m 63.0); **T/W** m 69.8; **B** 16.4–18.5 mm (n 26, m 17.5); **Ts** 23.5–25.9 mm (n 27, m 24.8).

STRICKLAND'S WHEATEAR *O. opistholeuca* (Strickland, 1849) (W Pamirs to N India, N & E Afghanistan, usually above 1400 m; altitudinal winter movements or shorter migrations; apparently unrecorded within the covered region) is in ♂ plumage almost all black, with only lower abdomen, vent and undertail-coverts apart from tail-base white; in a very few, however, almost solidly black even on vent. ♀ more strongly recalls ♂ than in the other two species but usually somewhat greyer or browner, being dull sooty brown-grey to olive-brown above, with underparts more variable. Unlike preceding taxa rather dark below including most of belly. Darkest birds show blackish-brown underparts, almost matching ♂ but browner and less solid from lower breast; in palest birds, especially 1stW ♀, underparts a mixture of medium to dusky buffish-grey with variable amounts of white on lower belly. Face is usually darker with less rusty-brown ear-coverts, and chin/throat only sometimes slightly paler (more often in fresh plumage). Crown usually slightly paler/browner than head-sides. Apparently a dark-throated ♀ also occurs rarely in this species, with much darker throat/upper breast and warm greyish-brown belly. **W** ♂ 87.5–96 mm (n 24, m 92.4), ♀ 87.5–92 mm (n 13, m 89.8); **T** ♂ 60–69 mm (n 24, m 65.2), ♀ 60–68 mm (n 13, m 63.4); **T/W** m 70.6; **B** 16.0–19.3 mm (n 37, m 17.6); **Ts** 22.4–25.7 mm (n 37, m 24.1).

REFERENCES ALIABADIAN, M. et al. (2007) *Mol. Phyl. & Evol.*, 42: 665–675. – HARTERT, E. (1909) *Falco*, 5: 33–36. – MAYR, E. & STRESEMANN, E. (1950) *Evolution*, 4: 291–300. – PALUDAN, K. (1959) *Vidensk. Medd. Dansk naturh. For.*, 122: 203–213. – PANOV, E. N. (1992) *BBOC*, Suppl., 112A: 237–249. – RICHARDSON, C. (1999) *Sandgrouse*, 21: 124–127. – SHIRIHAI, H., KIRWAN, G. M. & HELBIG, A. J. (2011) *BBOC*, 131: 270–291. – TICEHURST, C. B. (1922) *Ibis*, 64: 151–155. – VAURIE, C. (1949) *Amer. Mus. Novit.*, 1425: 1–47. – ZARUDNY, N. A. (1923) *Izv. Turkestansk. Otd. Russ. Geog. Obshch.*, 16: 65–72. (In Russian.)

Strickland's Wheatear *O. opistholeuca*, 1stW ♂, India, Feb: plumage very nearly identical to Basalt Wheatear. The two mainly separated by structure, with Strickland's having proportionately longer tail, but shorter and rounder wings. Also, only a hint of white tips to primary-coverts further support separation from Basalt Wheatear. Fine whitish fringes and tips to brownish juv primary-coverts indicate age. (K. Malling Olsen)

Strickland's Wheatear *O. opistholeuca*, ♀, India, Jan: often quite distinctive by overall dark grey and slightly blotchy plumage, especially below. Age impossible to assess. (G. Ekström)

MOURNING WHEATEAR
Oenanthe lugens (M. H. C. Lichtenstein, 1823)

Fr. – Traquet deuil
Ger. – Schwarzrückensteinschmätzer
Sp. – Collalba núbica; Swe. – Sorgstenskvätta

A desert specialist of Egypt and the Middle East, where it is a resident or partial migrant. A fairly common inhabitant of dry wadis, stony ravines or boulder-strewn flat deserts with scant vegetation, the apparently disjunct Iranian population *persica* is mainly migratory, wintering widely in Arabia, but more commonly east of the Arabian Peninsula. It is a tame bird and nearly always permits close approach (or even comes to investigate observers). The taxonomy of this wheatear is complicated and still incompletely known. Further study much needed.

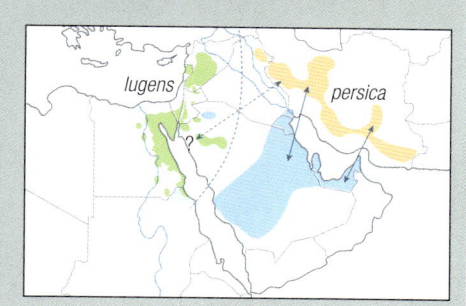

O. l. lugens, ad, S Jordan, Mar: at rest, when extensive whitish wing flashes cannot be seen, best separated from worn ♂ Pied Wheatear by more restricted white cap, rufous-tinged vent, and very different behaviour, shape and calls. Almost uniform black primary-coverts readily age this bird. (H. Shirihai)

O. l. lugens, ad, Israel, Oct: when still fresh, central cap can be variably darker, occasionally with some whitish tips to lesser coverts and throat. However, some ad already have remiges bleached browner at this season, although lack of white tips to primary-coverts age this bird. (H. Shirihai)

IDENTIFICATION A medium-sized wheatear, rather like Black-eared Wheatear and clearly smaller than Hooded Wheatear, with more 'standard' proportions. (Further clues for in-hand separation are given under Ageing & sexing and Geographical variation & range.) Unlike in Maghreb and Arabian Wheatears, *both sexes are similarly patterned black and white* and generally very much resemble a worn ♂ Pied Wheatear (i.e. black of face and wing is connected via the black back). However, unlike Pied has diagnostic combination of characteristic *pinkish-buff to rufous ventral region* and *prominent whitish flash in open wing*, as well as tendency to have broader black tail-band. The prominence of the white wing flashes in ssp. *lugens* is because the white portions covering wide parts of the inner webs of all remiges often reach to shafts. – The Iranian race *persica* (which might breed further west than known) is extremely similar to *lugens*, having *undertail-coverts often strongly rufous-buff* (although some have weaker cream-buff tinge), but has *distinctly less prominent whitish flash in open wing*, as well as a *broader black tail-band*. However, when assessing wing flash, beware of extensive age- (and to lesser extent sex-) related variation in *lugens*, in which retained juvenile remiges have much less white, approaching the amount of white in *persica*. The *black throat tends to extend further onto upper breast* in *persica* (especially at sides), while *in fresh plumage the whitish cap is often tinged more brownish* (sometimes even dusky or blackish-brown). Still, beware extensive individual variation in both these traits. With practice it may be possible to perceive that *persica* is subtly larger than *lugens*. Song given from perch and in flight.

VOCALISATIONS Song seems to differ somewhat from Maghreb Wheatear in being clear and melancholic with short, well-spaced strophes, at times a rather monotonous and repetitive soft sweet babble (recalling Blue Rock Thrush or a distant lark), e.g. descending *tritra-cheerr-tru...*, but also often inserts more subdued raspy *tchek* notes, and song can be more complex, jingling and rolling with some inserted rasping or crackly notes almost like in Siberian Rubythroat or White-throated Robin, and more protracted strophes do occur when agitated. Occasionally includes some good mimicry, e.g. a Linnet-like nasal *tett* as last note in the strophe, but in general shows comparatively little variation. – Variety of calls, including a Common Crossbill-like *chüp-chüp* in take-off. Also a straight whistle, *iht*, usually followed by the Crossbill-like clicking notes (much like in Northern Wheatear), *iht, chüp*. In alarm a low harsh or rasping *zeep* or *drzee*.

SIMILAR SPECIES For separation from *Maghreb Wheatear*, see that species. – For comparison with similar *Arabian* and *Gould's Wheatears*, see those species (extralimital Gould's briefly treated under Blyth's Wheatear, Taxonomic notes). – Could be confused with ♂ *Pied Wheatear* but has shorter black bib hardly reaching to breast (in Pied longer bib reaches to upper breast), and readily identified by extensive whitish wing flashes (visible in flight), lacking in Pied. White cap is restricted to head and nape (reaches upper mantle on Pied). When fresh, normally has uniform mid-grey crown and hindneck (autumn Pied has crown usually darker, resulting in more obvious white supercilium), and also lacks extensive pale fringes to mantle/scapulars and throat of Pied. Also note the larger area of white on rump in Mourning compared to Pied (extending further onto lower back), and the rufous or paler orange-buff tinge on vent/undertail-coverts (wearing paler) (in fresh Pied, lower breast/belly buff-tinged with white vent, though generally whole underparts much whiter in spring), and black terminal tail-band more or less uniform in width (Pied: uneven width with more extensive black corners). In flight, blackish underwing-coverts contrast more strongly with pale remiges.

AGEING & SEXING Most birds can be easily aged in most seasons by moult, feather wear and pattern in wing, and by the pattern of the primary-coverts. Sexes alike as to plumage (though see Biometrics). Seasonal variation limited. – Moults. Complete post-nuptial and partial post-juv moults in summer or early autumn (Jun–Sep). Post-juv moult includes all head, body, lesser and median coverts, most or all greater coverts, and usually the innermost tertials, less frequently all tertials and alula, even some tail-feathers. Pre-nuptial moult absent. – **SPRING Ad** Relatively less worn and lacks traces of immaturity (cf. 1stS). Sexes alike; ♀ has black areas fractionally duller in some birds, and extent of white/grey in remiges is somewhat reduced, but neither can be used as reliable sexing criteria. **1stS** Aged by more worn and browner juv remiges and primary-coverts (the latter may still retain the diagnostic small white tips, at least until Mar/Apr). Also, less intensely coloured, often with some pale tips to black feathering. Juv remiges have reduced white (or pale grey) inner edges, especially in young ♀♀ of the respective taxa. – **AUTUMN Ad** Evenly very fresh wings without discernible moult limits; primary tips fresh or, at most, slightly worn, and primary-coverts only slightly worn with blacker centres and lack bold white tips of 1stW. Plumage mainly as in spring, though often has variable, pale buffish-grey tinge to white underparts and crown. **1stW** Similar to ad but has diagnostic bold white tips to primary-coverts, poorly marked and slightly to moderately worn pale primary tips, and moult limits with slightly browner and looser-structured juv greater coverts, tertials and alula (if retained). Note that both primaries and primary-coverts are quite dark brown-

O. l. lugens, ad, Israel, Jul: extensive whitish wing flashes, here an extreme example and thus ♂, while narrow black tail-band is of even width. (A. Ben Dov)

black, not much paler than in ad. **Juv** Strongly patterned: crown to mantle and scapulars, cheeks and neck-sides sandy-buff with diffuse darker grey-brown mottling, breast sandy-buff, in strong contrast to blackish flight-feathers and greater coverts. Highly distinctive are broad buffy-white upperwing fringes, forming distinct double wing-bars. Sometimes has indistinct paler sandy-cream supercilium. When moult to 1stW starts (Jun–Jul, finished Aug–Sep), first lores, upper cheeks and ear-coverts become black (with faint buff mottling), forming contrasting mask, and then mantle/back (broadly tipped buff-white).

BIOMETRICS (*lugens*) **L** 15–16 cm; **W** ♂ 86–98.5 mm (*n* 40, *m* 94.0), ♀ 85–93 mm (*n* 14, *m* 89.3); **T** ♂ 54–66.5 mm (*n* 40, *m* 61.3), ♀ 54–63 mm (*n* 14, *m* 58.5); **T/W** *m* 65.3; **black on r1** 27–38 mm (*m* 32.4); **black on r4** 7–15 mm (*m* 11.0); **black on r6** 7–15 mm (*m* 10.6); **B** 16.7–20.4 mm (*n* 53, *m* 18.7); **Ts** 24.3–27.3 mm (*n* 54, *m* 25.6); **rump patch** 30–51 mm (*m* 40.6). **Wing formula: p1** > pc 0–7 mm, < p2 39–50 mm; **p2** < wt 4–7.5 mm, =5/6 or 6 (96%) or =5 (4%); **pp3–4** longest; **p5** < wt 0.5–4 mm; **p6** < wt 6–10 mm; **p7** < wt 13–16 mm; **p8** < wt 16–21 mm; **p10** < wt 22–29 mm; **s1** < wt 23–30 mm. Emarg. pp3–5 (rarely very indistinctly on p6).

O. l. lugens, 1stW, Israel, Oct: aged by white tips to not quite black primary-coverts, while rather extensive whitish wing flashes are almost certainly indicative of ♂. Rufous-tinged undertail-coverts shine through. (H. Shirihai)

O. l. lugens, 1stW, Israel, Oct: rufous vent and undertail-coverts typical. At this season, easily aged by broadly white-tipped primary-coverts (juv feathers). (H. Shirihai)

O. l. lugens, juv, Israel, Jul: highly distinctive with prominent broad buff wing-bars, faint darker grey-brown streaks above, and dark ear-coverts. (B. Nikolov)

GEOGRAPHICAL VARIATION & RANGE Moderate variation. The Mourning Wheatear is closely related to Arabian, Maghreb and Basalt Wheatears (which see), forming a superspecies with these, and apparently is also closely related to 'Abyssinian Black Wheatear' (*O. lugubris*) and 'Schalow's Wheatear' (*O. schalowi*) in E Africa (latter two extralimital and not treated).

O. l. lugens (M. H. C. Lichtenstein, 1823) (E Egypt & Levant; resident or makes short-range movements). Described above. Black tail-band about as wide as in Maghreb Wheatear (or on average subtly narrower), markedly narrower than in *persica*. White basally on inner webs of remiges (forming 'wing flashes') in ad that are most extensive within the complex, covering whole (or nearly whole) of inner webs to over half of feather length, and outer edge of white portions runs rather diagonally across the inner web from shaft towards the feather edge, thus creates wider and more solid white flashes on spread wings in flight (see photos). However, beware of considerable age-related variation in pattern of juv primaries (usually kept through 1stY), these having much less white on inner webs than in ad, often divided from shafts by extensive dark zones, at times producing a pattern identical to *persica*. Further, ♀♀ of both age categories have on average smaller white flashes than ♂♂, but variation and overlap render this an unreliable sexing tool. Reliable separation of *lugens* from *persica* outside the breeding season or away from the breeding regions probably only possible with ad ♂♂ using the white wing pattern and supported by a broad tail-band and long wing length. In the hand, and when viewed from below, the white base on p1 is rather extensive and usually well-defined. For Biometrics, see above.

O. l. persica (Seebohm, 1881) (NE Iraq & Iran, possibly also N Syria and adjacent parts of W Iraq; winters Arabia, possibly also in NE Africa). Very similar to *lugens*, lacking sexual dimorphism and being black and white with usually strong cinnamon or rufous-buff tinge to undertail-coverts, differing only as follows: (i) black tail-band broader, usually > 16 mm; (ii) white area basally on inner webs of remiges much reduced, white never reaches the shafts, and edge of white portions is rather straight, running diagonally across the inner webs for a long distance, resulting in more restricted wing flashes (but beware of variation in *lugens* as related above); further, the white base on p1 (seen from

O. l. persica, ad, Kuwait, Sep: virtually identical to *lugens*, with on average a trifle longer and more pointed wing, broader tail-band, darker crown when fresh and deeper rufous-buff undertail-coverts. However, extensive overlap with *lugens* in all characters. All-black primary-coverts diagnostic of ad. (R. Al-Hajji)

below) is rather small and ill-defined or often lacking. In fresh plumage, (iii) cap on average tinged more brownish. **W** ♂ 90.5–99 mm (*n* 23, *m* 96.2), ♀ 88–99 mm (*n* 12, *m* 92.5); **T** ♂ 59–66 mm (*n* 23, *m* 63.4), ♀ 58–66 mm (*n* 12, *m* 61.5); **T/W** *m* 66.1; **black on r1** 37–46 mm (*m* 40.8); **black on r4** 12.5–26 mm (*m* 17.6); **black on r6** 13–22 mm (*m* 17.2); **B** 16.9–20.6 mm (*n* 35, *m* 18.7); **Ts** 22.9–27.3 mm (*n* 35, *m* 25.1); **rump patch** 21–50 mm (*m* 38.5). **Wing formula: p1** > pc 6.5 mm, to 2 mm <; **p7** < wt 11.5–14 mm; **p10** < wt 20–25 mm; **s1** < wt 23–29 mm.

TAXONOMIC NOTES Forms superspecies with closely related *O. halophila* (Maghreb Wheatear; N Africa east at least to Libya, local NW Egypt), *O. lugentoides* (Arabian Wheatear; S Saudi Arabia, Yemen, Oman) and *O. warriae* (Basalt Wheatear; basalt deserts of SE Syria, formerly also adjacent part of Jordan). Morphologically all four differ rather clearly; *halophila* and *lugentoides* show clear sexual dimorphism, the others not. The recurrently advocated split of Mourning Wheatear, Maghreb Wheatear, Arabian Wheatear and the extralimital Abyssinian Black Wheatear *O. lugubris* and Schalow's Wheatear *O. schalowi* has recently been tested by Förschler *et al.* (2010), using both genetic and morphometric analyses. The three last-mentioned proposed splits (the Arabian and African taxa) received strong support in this study, and these have been followed here. However, genetic divergence was low between *lugens* and *halophila* (0.2–1.0%), suggesting that the latter is best kept as a subspecies of *lugens* if genetic distance (mtDNA) based on a few samples (in this case three *halophila* and five *lugens*) is given conclusive weight for taxonomic evaluations. Added to this are reports of presumed intermediates between *halophila* and *lugens* (Guichard 1955), or even a cline within *halophila* with progressively more ♂-like ♀♀ in the east (Baha el Din & Baha el Din 2000; cf. also *HBW*), which therefore seemingly argue against splitting *halophila*. However, note that both these references apparently underestimated natural variation and the occurrence of a darker ♀ morph. We, at least, have been unable to find any confirmed ♀ *halophila* with fully ♂-type plumage, nor have we seen a tendency for ♀♀ to become darker in the east. Furthermore, certain differences in song and habitat selection exist between *halophila* and *lugens*, while genetic differences are almost non-existent between commonly split *Oenanthe* species like Pied and Cyprus Wheatears (Randler *et al.* 2012), making the genetic argument potentially less important. There are therefore rather strong reasons to also split *halophila* and treat it as a separate species, a policy we have adopted. – Also for *persica* a case could be made for a split from *lugens*. Always known to be morphologically very similar to *lugens*, but recently found to be more distinctive than previously thought (especially by wing shape, size and shape of white wing flash,

O. l. persica, 1stW, Kuwait, Jan: as with *lugens*, sexes virtually alike but ageing usually straightforward using primary-coverts pattern—clear whitish tips diagnostic of 1stY. Even birds in hand are difficult or impossible to separate outside the two subspecies' breeding ranges. (D. Monticelli)

1stS, possible *O. l. persica*, NE Jordan, Mar: rather dark crown and seemingly broad tail-band. Clear whitish-tipped primary-coverts diagnostic of 1stY. (H. Shirihai)

and broader black tail-band; Shirihai *et al.* 2011), Förschler *et al.* (2010) showed that genetic divergence of *persica* from *lugens* was fairly substantial (1.2–2.2%) and suggested that specific status is warranted. Major handbooks (e.g. Vaurie 1959, Cramp *et al.* 1988) suggest that Syrian breeders are *lugens*, being widespread there. However, we consider that available proof of this is inadequate. We are aware of just four specimens from Syria that could be breeders (two each in NHM and AMNH), all four young ♂♂ and assigned to *lugens* but showing *persica*-like traits, or traits intermediate between *lugens* and *persica*, and since all are juv moulting to 1stW they could either be genuine *persica* or represent an intermediate population between the two. We also examined photographs of live Mourning Wheatears from Syria, mostly from Palmyra, which seem to show typical *persica* characteristics, but these are mostly from winter (see photos). Since *persica* is migratory, all Syrian records dated between Dec and Mar could have been wintering away from their usual range in E Arabia. Further, we found several specimens with classic *persica* characters collected in winter in S Sinai, Egypt, and S Israel (TAUZM, including three ad ♂♂). The combined evidence seems to suggest that breeders in Syria are either (i) *O. l. persica*, (ii) an undescribed *persica*-like taxon, or (iii) an intermediate population between *lugens* and *persica*, or that such birds at least occur regularly as far west as S Sinai and S Israel in winter. Without further information from currently war-torn Syria we cannot properly assess the correct range and taxonomic rank for *persica*, hence leave it as a subspecies of *lugens*.

REFERENCES ALIABADIAN, M. *et al.* (2007) *Mol. Phyl. & Evol.*, 42: 665–675. – BOON, L. J. R. (2004) *DB*, 26: 223–236. – FÖRSCHLER, M. *et al.* (2010) *Mol. Phyl. Evol.*, 56: 758–767. – GUICHARD, K. M. (1955) *Ibis*, 97: 393–424. – BAHA EL DIN, S. & BAHA EL DIN, M. (2000) *Sandgrouse*, 22: 109–112. – KHOURY, F. *et al.* (2010) *Sandgrouse*, 32: 113–119. – RANDLER, C. *et al.* (2012) *J. of Orn.*, 153: 303–312. – SHIRIHAI, H. (2012) *BBOC*, 132: 226–235. – SHIRIHAI, H., KIRWAN, G. M. & HELBIG, A. J. (2011) *BBOC*, 131: 270–291.

Presumed *O. l. persica*, Syria, Feb: brownish cap, rather broad black tail-band and much-reduced white on flight-feathers highly suggestive of *persica*, or might indicate an intermediate between *persica* and *lugens*. Apparently ad by lack of obvious moult limits. (N. Martinez)

MAGHREB WHEATEAR
Oenanthe halophila (Tristram, 1849)

Fr. – Traquet halophile; Ger. – Berbersteinschmätzer
Sp. – Collalba bereber; Swe. – Maghrebstenskvätta

A close relative of Mourning Wheatear, and together with this and the closely related Arabian and Basalt Wheatears, they form a superspecies. Usually found in barren rocky deserts and semi-deserts, on mountain plateaux and in valleys with bushes and some acacia cover. Not shy, often easy to approach, but can also suddenly take off and be gone for long periods. Its eastern border is found in NW Egypt, allegedly; there are some local populations in NW and NC Egypt, but breeding range does not seem to reach as far east as the Nile.

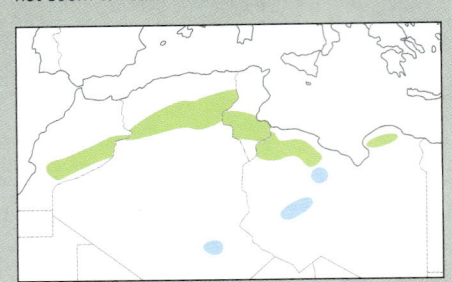

Ad ♂, Morocco, Oct: unlike Mourning Wheatear, clear sexual dimorphism, although ♂ plumage is quite variable. Pinkish-buff undertail-coverts. Aged by evenly-feathered wing including almost uniform black primary-coverts. (T. Peral)

IDENTIFICATION Much like Black-eared Wheatear in size and shape. Clear sexual dimorphism. ♂ is black and white and generally very similar to Mourning Wheatear, while ♀ Maghreb has markedly different plumage, being mainly *brownish-grey and white* with only limited black and is confusable with several ♀ congeners (surprisingly so with ♀ Finsch's Wheatear). Important to check the *pinkish-buff undertail-coverts*, and the *pale flash in open wing* (although both strongly reduced compared to Mourning). The pink-buff on vent and undertail-coverts is highly variable in both sexes, from being very diluted pinkish-white or cream-buff to light orange or rufous-buff (in some birds it is reduced to a few feathers or is partially hidden by whiter outer feathers of the undertail-coverts). The pale wing flash (formed by basally paler grey-white edges to inner webs of remiges) is normally ill-defined and sullied grey, and pale edges usually do not reach the shafts (but occasionally almost so basally on secondaries and inner primaries), thus only forming *smaller, more proximally positioned and less solid white upperwing flashes* (see photograph). ♀♀ are dimorphic as to throat colour, and are also highly variable as to darkness of neck-sides and wings, with the odd extreme dark ♀ (with very dark, almost black throat and sides), which could be mistaken by inexperienced observer as a ♂, still possible to sex correctly by the *greyish-brown back*. Also quite characteristic in Maghreb Wheatear is the pale *cap of fresh ♂ often being extensively sullied dirty grey-brown*, especially on centre of crown and hindneck, although forehead, crown-sides and upper mantle are usually somewhat paler (thus crown generally tends to be less whitish than in Mourning). Further in both sexes, *white areas of plumage are lightly suffused greyish pink-cream*. Both sexes also have inverted black 'T' on tail typical of most wheatears, blackish underwing-coverts and whitish tips to primary-coverts (especially distinctive in young and if not heavily worn).

VOCALISATIONS Song based on own impressions and several recordings (e.g. from Morocco, Algeria, Tunisia and Libya) perhaps slightly different from Mourning Wheatear (*lugens*) being a little slower or even staccato-like, a little drier and more mechanical, yet can include a recurring clear yodelling *yalala* with slightly falling pitch and more warbling trills. At times more extended song, being more varied, jumbled and richer, a continuous sweet, unhurried warbling (alternately ascending and descending), with each strophe sounding like a ringing trill mixed with brief whistles and metallic calls. Appears to show stronger individual variation than Mourning *lugens*, but this requires confirmation. Some appear more imitative. Subsong also consists of quiet warbles. Song of ♀ lower-pitched and slower. – Calls, highly varied according to context include a quiet *tchek*, high-pitched and repeated *hiihp* notes (in excitement) and, in alarm, sometimes a hard, rapid *trac-trac-trac-trac-...*

SIMILAR SPECIES ♂ Maghreb is likely to be confused only with *Mourning Wheatear* (sexes alike), especially in areas where both allegedly mix in winter (e.g. in W Egypt), or where they could occur as vagrants. Thus, ♂ Maghreb hardly differs from Mourning, except in flight by narrower and greyer (more subdued, not so pale and contrasting) basal area to flight-feathers (notably strongly reduced on primaries) and by usually rather indistinct (sometimes almost lacking) cream-pink wash to ventral region (never as deep buff, rufous or cinnamon as in many Mourning). Also tends to have pale areas of lower back, rump and belly less pure white (greyish pink-cream in some), while crown and hindneck when fresh are less pure whitish, being dirty ash-brown. – ♀♀/young: though unlikely to overlap, ♀ Maghreb can be very similar to ♀ *Finsch's Wheatear*, which see. – Further pitfalls: both *Gould's* (and some extremely pale ♀ *Blyth's*) *Wheatear* and, especially, *Arabian Wheatear* are similar to Maghreb Wheatear but both are well separated geographically. – Just as with Mourning, ♂ Maghreb could be confused with worn ♂ *Pied Wheatear*. Further, ♀ Pied, and even ♀-like plumages of *Black-eared Wheatear*, and also ♀-like *Northern Wheatear* and *Seebohm's Wheatear*, could resemble ♀ Maghreb, though, usually, the paler remiges in flight and the cream-pink vent/undertail-coverts of Maghreb should exclude all four. Supporting details to note are offered by the different upperparts and breast colorations, and details in tail pattern, which all support separation of Maghreb from the former two (♀ Maghreb is usually paler and sandier above, when fresh lacks pale fringes or the strong buffy breast-band of these species, and has a more evenly broad terminal tail-band). With experience, subtle difference in size, shape and behaviour will help too.

1stW ♂, Tunisia, Jan: less immaculately patterned, with less solidly black upperparts. Such difference reflects individual variation, but is also age-related. Age by white fringes to juv primary-coverts (with wear mostly restricted to tips). (P. Alberti)

♂, Morocco, Apr: in ♂ whitish panel on spread wing rather restricted (compared to Mourning), especially on primaries, with white bases usually visibly wider and purer on secondaries. (P. Adriaens)

Ad ♂, Tunisia, Nov: especially whiter ♂♂ are very similar to Mourning, but pinkish-buff hue on ventral region usually more diluted, and subterminal tail-band averages wider. Aged by distinctive pale primary tips, while primary-coverts lack obvious whitish tips. (A. Torés Sánchez)

WHEATEARS

AGEING & SEXING For reliable ageing, moult and feather wear, in particular the pattern of the primary-coverts, should be checked closely. Sexes differ, but seasonal variation rather limited, involving mainly abrasion of feather-tips, most evident on dark throat of ♀♀. – Moults. Complete post-nuptial and partial post-juv moults in summer or early autumn (Jun–Sep). Post-juv moult includes all head, body, lesser and median coverts, most or all greater coverts, and usually the innermost tertials, less frequently all tertials and alula, or even some tail-feathers. Pre-nuptial moult absent. – SPRING **Ad ♂** Almost pure black-and-white plumage. Compared to ♀ has more pronounced still weak (and somewhat variable) rusty-buff tinge to vent/undertail-coverts (often rather difficult to see). Crown can be pure white or washed dirty white. Wings and tail relatively less worn and lack any signs of immaturity. Primary-coverts dark brown-black with no or only very tiny paler tips in some (cf. 1stS). **Ad ♀** Predominantly pale buffy grey-brown above, ear-coverts sometimes have warm

Ad ♂ (left) and dark-throated ♀, Morocco, Apr: extremely dark-capped ♂ with strong white supercilium, while ♀ is very dark-throated, the dark continuing into blackish wings, with dark grey upperparts. Both are ad given lack of whitish tips to primary-coverts. (T. Grüner)

Ad ♀, dark-throated, Tunisia, Nov: ♀ dimorphic as to throat pattern, with much individual variation, some of it age-related. When heavily worn, dark areas become blacker and general pattern can mirror ♂, but back always grey-brown, never black as ♂. This bird is at least 2ndW due to only very thin whitish fringes to primary-coverts. (R. Altenburg)

1stS ♀, Tunisia, Apr: a dark-throated type ♀, but bib less solid than previous bird; remiges, tertials, primary-coverts and outer greater coverts are juv. (D. Occhiato)

brown tone. ♀♀ apparently dimorphic, and black on face also strongly dependent on wear and possibly age (though questionable whether ♀♀ with much black are invariably older). Thus entire face, including chin and throat, variably blackish with wear, and may become almost (though never as intensely) black as ♂, while others lack any black on face. Note that black-throated/faced ♀♀ still have normal ♀-like remainder of plumage, never becoming pure black in wing, and mantle/back and scapulars are never black (though in very rare extremes can have sooty smudging). Rest of underparts white or buffy-white with orange-buff or buff vent, though entire underparts become dirty white with wear. Compared to ♂, pale basal area on remiges rather indistinct. Primary-coverts uniform, lacking pale tips, or only show fine traces of such (cf. 1stS). **1stS ♂** As ad ♂ but until early spring and sometimes summer often has diagnostic prominent white fringes to primary-coverts, mostly restricted to tips (note also browner juv remiges, primary-coverts and perhaps some outer greater coverts and tertials), and may also show pale tips to black throat-feathers and have less pure white crown. **1stS ♀** As ad ♀ but moult limits as in 1stS ♂ (but less contrasting, more difficult to judge), with similar white tips to primary-coverts (though these appear to wear off earlier). Pale fringes and tips to some tertials, median and greater coverts often still evident (creating wing-bars). Dark-throated ♀ tends to develop less intense dark/blackish throat than ad ♀♀. – AUTUMN **Ad** Aged most safely (both sexes) by being evenly fresh without moult limits in wings or by lack of extensive pale fringes to primary-coverts (cf. 1stW). Pale primary tips well marked, remiges fresh or at most only slightly worn. **Ad ♂** Variable buffish-grey tinge to underparts and crown. **Ad ♀** As in spring but upperwing more extensively fringed/tipped pale, and underparts washed off-white/buffish, especially undertail-coverts, any dark throat wholly or partially concealed, often appears very pale brown, though from Oct/Nov dark-throated birds already have quite well-developed bib. **1stW ♂** Similar to ad ♂, but primary-coverts diagnostically fringed and tipped duller whitish, and pale primary tips weakly developed, being slightly to moderately worn. Also has distinct moult limit in wing: few retained juv outer greater coverts, tertials and alula weaker structured and centred blackish-brown with prominent pale buffish fringes contrasting with newly-moulted ad-like post-juv ones (often including most inner greater coverts), which are firmer textured, have black centres and when fresh, at most, tiny whitish fringes. **1stW ♀** As fresh ad ♀ and best aged by moult limits as 1stW ♂, but pattern less contrasting. Distinctly pale-fringed primary-coverts (a diagnostic ageing character that can be used in the field). In ad, fringes much narrower and tips reduced but, when still fresh, moderately broad; in both ages pale tips form line of white spots, but these smaller and more spaced in ad. Juv greater coverts moderately worn, with dark brown centres, narrow ochre outer fringes and broad and bolder ochre-white tips, especially at shafts; ad-like post-juv greater coverts fresher, with dark grey centres, narrow greyish-brown outer fringes and small whitish tips. In particular young ♀♀ have much-reduced white basally on flight-feathers, forming rather weakly developed flashes in flight. **Juv** Body plumage shows clear sexual variation, with ♂ appearing as juv Mourning, while ♀ is generally ♀-like and lacks any black on face or throat, being mottled pale grey-buff or buff-brown, especially on ear-coverts. Compared to juv ♂, feather-centres of lower mantle and scapulars paler and less contrasting, but like juv ♂ has obvious double wing-bars. Note extensive sandy-buff fringes to tertials.

BIOMETRICS L 14–15.5 cm; **W** ♂ 87–96 mm (n 32, m 92.0), ♀ 83–92 mm (n 39, m 87.2); **T** ♂ 57–65 mm (n 32, m 61.0), ♀ 54–63 mm (n 39, m 57.8); **T/W** m 66.3; **black on r1** 30–36.5 mm (m 34.0); **black on r4** 8–17 mm (m 11.2); **black on r6** 9–13 mm (m 10.6); **B** 15.8–19.6 mm (n 68, m 17.8); **Ts** 22.0–26.6 mm (n 67, m 24.8); **rump patch** 30–43 mm (m 36.6). Wing formula: p1 > pc 0–6 mm (rarely to 0.5 mm <), < p2 40–48 mm; p2 < wt 3.5–8 mm, = 5/6 or 6; **pp3–4** longest; p5 < wt 0.5–2.5 mm; p6 < wt 5–8.5 mm; p7 < wt 11–15 mm; p10 < wt 20–25.5 mm; s1 < wt 23–28 mm. Emarg. pp3–5 (rarely faintly also on p6).

GEOGRAPHICAL VARIATION & RANGE Monotypic. – N Africa from Morocco to at least Libya, probably also regularly NW & NC Egypt; sedentary or winters near breeding range. No evidence of longer or easterly winter movements.

TAXONOMIC NOTE See extensive notes under Mourning Wheatear.

REFERENCES ALIABADIAN, M. *et al.* (2007) *Mol. Phyl. & Evol.*, 42: 665–675. – BOON, L. J. R. (2004) *DB*, 26: 223–236. – FÖRSCHLER, M. *et al.* (2010) *Mol. Phyl. & Evol.*, 56: 758–767. – GUICHARD, K. M. (1955) *Ibis*, 97: 393–424. – BAHA EL DIN, S. & BAHA EL DIN, M. (2000) *Sandgrouse*, 22: 109–112. – SHIRIHAI, H., KIRWAN, G. M. & HELBIG, A. J. (2011) *BBOC*, 131: 270–291.

1stW ♀, Morocco, Dec: combination of being fresh, young and pale-throated produces such plumage, at palest end of spectrum. Warm brown tones to ear-coverts and whitish supercilium above and behind eye. Primary-coverts broadly tipped white diagnostic of 1stY. (M. Varesvuo)

BASALT WHEATEAR
Oenanthe warriae Shirihai & Kirwan, 2012

Fr. – Traquet de basalt; Ger. – Basaltsteinschmätzer
Sp. – Collalba del basalto; Swe. – Basaltstenskvätta

Formerly treated as a colour morph of Mourning Wheatear (after initially having been identified as Strickland's Wheatear *O. opistholeuca*), more recently described as a separate taxon in its own right, and here split at species level for consistency within the genus *Oenanthe*. As the nearly all-black plumage indicates, it is restricted to the localised dark basalt lava patches in the Syrian deserts, and was formerly also found in Jordan, in similar habitat, but recently appears to have become very rare or extinct from there. It seemingly does not mix much or overlap with neighbouring populations of Mourning Wheatear, its closest relative, but more research is desirable.

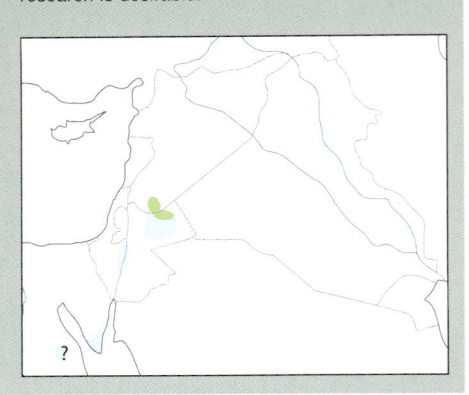

1stS ♂, NE Jordan, Apr: aged and sexed by combination of juv remiges and primary-coverts (with only tiny white spots on tips of latter), and glossy black rest of plumage. (H. Shirihai)

IDENTIFICATION Same size as Mourning Wheatear, its very close relative. The nearly *all-black coloration* with glossy bluish sheen, strongest on upperparts, *leaving only base of tail and vent white*, resembles the more familiar (but usually distinctly larger) Black Wheatear, or young White-crowned Wheatear, the latter being initially black-crowned. Thus, it differs from other forms in the Mourning Wheatear complex by its unique extensive black coloration but also by the *white instead of rufous-buff undertail-coverts* (at most pale sandy-cream or dusky white, probably due to soil staining), and it also has a *distinctly smaller white rump patch*. Within the Mourning Wheatear complex has the *least developed white flashes in the open wing*, and this also varies with age and sex, adult ♂♂ having a limited flash but first-year ♀♀ practically none (just a greyish translucence in certain lights). There is a very slight sexual dimorphism, ♀ being slightly less intense and pure black and having on average less extensive pale inner edges to remiges, although evaluation of this depends on age (see below) and is often difficult to apply in the field (even in the hand). With in-hand examination there is a tendency for outermost primary to be shorter (about equal to primary-coverts); outermost primary is also all dark greyish below (lacking the white tip or white basal areas of Mourning).

Underwing-coverts are as body, contrastingly darker than rest of underwing.

VOCALISATIONS Field observations in NE Jordan and Israel suggest that the song of Basalt Wheatear is quite different from that of Mourning, being continuous and varied, lacking deep and clearer flute-like sounds, being lower-pitched, more 'complicated' and improvised, clicking and scratchy notes admixed with higher-pitched warbling ones. Such a song was recorded by A. Ben Dov in Jan 2016, and it is not believed to be a subsong since so far all songs heard have been of this continuous type rather than the briefer strophes with pauses practised by Mourning. A creaky straight call, *cheep*, often quickly repeated, was recorded on the same occasion. Obviously further research on possible vocal differences within the complex is highly desirable.

SIMILAR SPECIES Very similar to *Black Wheatear* but is subtly smaller and has longer primary projection. See under that species. – Can be confused with first-year *White-crowned Wheatear*, this being completely or almost completely black-crowned (latter category only having one or two white feathers on entire crown, which can be overlooked), but note different tail pattern with practically all-white outer tail-feathers, black subterminal marks on outertail (if any) restricted to narrow isolated patches. – Separation from the near-identical extralimital ♂ *Strickland's Wheatear* is another complicated challenge; for basic information on Strickland's Wheatear, see under Blyth's Wheatear, Taxonomic notes.

AGEING & SEXING Ageing possible using moult and feather wear or pattern in wing, and by pattern of the primary-coverts. Seasonal variation limited. Sexes generally very similar in plumage but some birds at least can be sexed if correctly aged first using combination of general plumage pigmentation, amount of white on the inner webs of remiges (♂♂ have on average purer and more extensive white inner edges to primaries especially) and by some biometrics (see below). – Moults. Not studied, but based on trapped birds and the few available specimens it seems to apply complete post-nuptial and partial post-juv moults in late summer or first half of autumn. Post-juv renewal includes all head, body, lesser and median coverts, most or all greater coverts, and usually some of the tertials and alula; replacement of some tail-feathers seems irregular. Pre-nuptial moult absent or limited (some birds in spring had very fresh secondary-coverts and tertials which apparently had been renewed recently). – **SPRING Ad** Generally only moderately worn, or even quite fresh until Apr, especially the primaries which are very black. Wings also lack any obvious moult limits or any other remnants of immaturity (cf. 1stS). ♀ has black areas generally duller, less strongly glossed metallic bluish compared to ad ♂, and extent of white/grey in remiges reduced, only small amount visible on inner webs of fully stretched open upperwing (in ad ♂ clear-cut and purer white inner-web edges on stretched wing—see in-hand photograph in Shirihai et al. 2011). **1stS** Aged by more worn and browner retained juv remiges and primary-coverts, with most usually still showing the diagnostic small white primary-coverts tips, at least until Apr. Also, both sexes are less intensely coloured and glossed, with especially the black areas of 1stS ♀ tinged brown. Juv remiges have reduced white/grey flashes, and especially young ♀ lacks any visible white edges on the inner webs of fully stretched upperwing. Most 1stS ♂♂ have some white visible on the stretched upperwing, approaching the amount in ad ♀, but some lack any visible white. – **AUTUMN Ad** Very fresh wings without discernible moult limits. Primary tips fresh or, at most, slightly worn, and primary-coverts only slightly worn and still blacker and glossier, and lack bold white tips of 1stW. Plumage and amount of white on the inner webs of remiges generally as described in ad spring. **1stW** Similar to ad but has diagnostic bold white tips to primary-coverts, and birds with retained juv alula and tertials have also contrasting bold white tips to these feathers. (White tips progressively reduced with wear and sometimes moult.) Slightly to moderately

1stS ♂, S Syria, Feb: aside of virtually all-black plumage, lacks rufous on undertail-coverts of Morning, and pale wing panel is restricted to diffuse edges of inner webs. Readily separated from juv (black-capped) White-crowned Wheatear by tail pattern and smaller size. Near-identical to Strickland's Wheatear: check rather broad and even-width black tail-band, and relatively small white rump. Primary-coverts retain some whitish tips as line of spots, characteristic of worn 1stY. (N. Martínez)

worn pale primary tips, and moult limits with less intensely black (or browner) and looser-structured retained juv greater coverts (if any). Rest of plumage and amount of white on inner webs of remiges generally as in 1stS (see above). **Juv** Very distinctive compared to juv of, e.g., Mourning Wheatear, being generally smoky and browner overall with grey-brown upperparts and dusky buff-brown undersides, often with darker/blacker ear-coverts and bib, and warmer buff wing fringes (see in-hand photographs in Khoury et al. 2010).

BIOMETRICS L 15–16 cm; **W** ♂ 93–99 mm (*n* 4, *m* 97.0), ♀ 92–94 mm (*n* 3, *m* 93.3); **T** ♂ 61–65 mm (*n* 4, *m* 63.9), ♀ 62.5–64.5 mm (*n* 3, *m* 63.2); **T/W** *m* 66.6; **black on r1** 41–44 mm (*m* 42.0); **black on r4** 15–19 mm (*m* 17.0); **black on r6** 11.5–20 mm (*m* 16.1); **B** 18.0–19.8 mm (*n* 10, *m* 18.9); **Ts** 24.5–27.0 mm (*n* 10, *m* 25.9); **rump patch** 28–35 mm (*m* 30.8). **Wing formula: p1** > pc 4.5 mm, to 0.5 mm <, < p2 46.5–49 mm; **p2** < wt 4–6 mm, =5/6; **pp3–4** longest; **p5** < wt 1.5–2.5 mm; **p6** < wt 8–9 mm; **p7** < wt 13–15 mm; **p10** < wt 23–25 mm; **s1** < wt 25–28 mm. Emarg. pp3–5 (rarely very indistinctly also on p6).

GEOGRAPHICAL VARIATION & RANGE Monotypic. – Restricted to basalt deserts in NE Jordan (possibly now extinct) and adjacent S Syria; in winter southward short-range movements and occasionally straggles to other parts of Middle East and Egypt. Declining.

TAXONOMIC NOTES Basalt Wheatear was recently described as a subspecies of Mourning Wheatear (ssp. *warriae*; Shirihai et al. 2011) rather than a colour morph as had been customary previously. Khoury et al. (2010) noted that *warriae* has a unique dark juv plumage, while Shirihai et al. (2011) hinted at possible differences in song compared to Mourning Wheatear. To our knowledge there is no firm proof of overlap in breeding ranges, and interbreeding between Basalt and Mourning appears extremely rare (Khoury et al. 2013). Here we treat Basalt Wheatear as a separate species, since this is the most consistent taxonomy for the genus *Oenanthe* as a whole. (See comments under Mourning Wheatear.) – There is a most surprising record of a ♂ Basalt Wheatear paired with a ♀ Finsch's Wheatear in SE Turkey in spring 2011. Interestingly, a molecular study of the genus (Aliabadian et al. 2007) revealed that Finsch's and Mourning are closely related genetically.

REFERENCES ALIABADIAN, M. et al. (2007) *Mol. Phyl. & Evol.*, 42: 665–675. – BOON, L. J. R. (2004) *DB*, 26: 223–236. – FÖRSCHLER, M. et al. (2010) *Mol. Phyl. Evol.*, 56: 758–767. – KHOURY, F. et al. (2010) *Sandgrouse*, 32: 113–119. – KHOURY, F. et al. (2013) *Sandgrouse*, 35: 134–137. – SHIRIHAI, H., KIRWAN, G. M. & HELBIG, A. J. (2011) *BBOC*, 131: 270–291.

1stS ♀, Israel, Mar: diagnostic narrow white inner webs to remiges, which in this 1stY ♀ end well short of shafts and are mostly or only visible on underwing. Note white vent and undertail-coverts. Dark brown-grey plumage looks very black due to light and contrast. (H. Shirihai)

1stS ♀, Israel, Mar: same bird as image to the left, here showing true more dark brown-grey and less glossed black plumage usually feature of ♀♀. White-tipped primary-coverts diagnostic of 1stY, and an important identification mark. Note pure white ventral region, lacking rufous pigmentation of Mourning. (H. Shirihai)

1stS ♂, SE Turkey, May: remarkably, this bird was paired with a ♀ Finsch's Wheatear, in spring 2011. Note that head and body plumage of 1stY ♂ hardly differs from normal ♀ coloration. (E. Yoğurtcuoğlu)

Basalt Wheatear (top left: Israel, Jan), ♂ Strickland's Wheatear (top right: India, Nov), young White-crowned Wheatear (bottom left: Morocco, Feb) and Black Wheatear (bottom right: Spain, Sep): when perched with folded wings, Basalt and ♂ Strickland's differ only by Basalt's proportionately longer wings and shorter tail; thus wingtip of Basalt usually reaches closer to inner edge of tail-band (which band averages more even and broader than Strickland's), while longer primary projection shows more and narrower-spaced primary tips. When still fresh, 1stW differ in pattern of primary-coverts; Basalt has distinctive small white tips, while Strickland's has thin edges that extend along outer webs. Identification of Basalt (and ♂ Strickland's) also requires careful elimination of larger and heavier all-black-headed young White-crowned, and Black Wheatears. Note distinctly longer primary projection in Basalt compared to Black, while White-crowned's black plumage is always glossy, and it is the only species of the four lacking a solid black tail-band; also, both Black and White-crowned have much stronger bills. All birds here 1stY. (Top left: H. Shirihai; top right: C. N. G. Bocos; bottom left: D. Occhiato; bottom right: H. Shirihai)

ARABIAN WHEATEAR
Oenanthe lugentoides (Seebohm, 1881)

Alternative name: South Arabian Wheatear

Fr. – Traquet d'Arabie; Ger. – Arabsteinschmätzer
Sp. – Collalba arábe; Swe. – Arabstenskvätta

One of several rather similar black-and-white wheatears, Arabian Wheatear is endemic to S Arabia, where it is mainly resident and occurs in barren rocky hills and valleys with sparse vegetation and scattered trees, but also among juniper and along coasts. The taxonomic status of *lugentoides* is according to some unresolved and requires further study, as with the rest of the *Oenanthe lugens* complex. However, its breeding range is widely isolated from the other members and it is fairly distinctive, and is therefore treated as a species here.

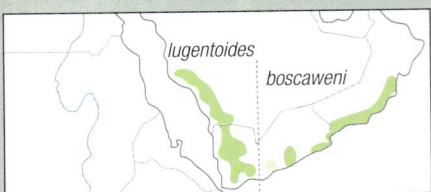

O. l. boscaweni, ♂, Oman, Nov: in flight has much less white on remiges than Mourning, restricted to primary bases, and has distinctly shorter white rump patch than that. (T. Langenberg)

O. l. boscaweni, ad ♂, Oman, Nov: differing from Mourning by being sexually dimorphic, thus only ♂ is very similar to Mourning. Arabian smaller but stockier, with on average more extensive black on throat/upper breast, neck-sides and mantle. White on nape reaches less far back, with broader black subterminal tail-band, and orange-buff vent often deeper coloured. Aged by evenly-feathered wing and black, ad-like primary-coverts. (D. Occhiato)

IDENTIFICATION Small but rather stocky wheatear. ♂ has *black throat, face, mantle, scapulars, wings and inverted black 'T' to tail, pale* (often white) *crown* and white rump and uppertail-coverts, and white underparts with *rufous buff-coloured ventral region* sharply demarcated from black areas. *Centre of crown is often streaked or washed darker*, usually leaving a *cleaner white supercilium*, forehead and rear end to the pale cap. ♀ largely *dusky grey-brown above* with variably *warmer ear-coverts, often a greyer throat* and characteristic grey-and-white *streaky or blotchy lower breast*, with variable *pinkish-orange vent/undertail-coverts*. Variable *white or grey primary flashes*, especially prominent in ♂, but mainly restricted to basal parts of feathers and most visible on primaries. No overlap in characters with *lugens* race of Mourning Wheatear, the principal confusion risk (see Similar species for details). In many respects also closely recalls Maghreb Wheatear, with which there is also no overlap. Note in ♂ dusky centre of crown even in worn plumage, *fairly small white rump patch, broad black terminal band on tail*, on average subtly stronger bill and shorter legs in Arabian.

VOCALISATIONS Song, which most recalls the *lugens* race of Mourning Wheatear, is a short, explosive bubbling, rather low-pitched warble with some rather scratchy elements in the strophes, e.g. *dweet twa weedle dit shree-zwie chu-chu*… In one recording contains varied musical whistles (perhaps mimicry). Subsong is a slightly extended and more varied, low warble with a thin reedy quality. – Has several calls; a rasping *kahk*, and a scolding *shrrrr*; but these two related or just minor variations of the same. A hash clicking *chzak-chzak* (like two stones being knocked together) in alarm. Also a mournful, high-pitched whistle *seek* noted, and low harsh *zeeb* or *dree*.

SIMILAR SPECIES Regarding ♂♂, the only real confusion risk is between ♂ Arabian and *lugens* race of Mourning Wheatear. They are extremely similar (but no overlap in characters if details are seen well), ♂ Arabian mainly differing in being smaller and more thickset with on average more extensive black on throat/upper breast, neck-sides and mantle, somewhat less extensive white on nape and distinctly narrower white rump/uppertail area than in *lugens* (in Arabian white between back and dark area of central rectrices is almost 50% shorter than in *lugens*). Arabian also often shows a dark-streaked greyish-white cap and has broader black terminal tail-band, especially on the outer tail-feathers, which makes the distance between the tail-band and wingtip appear shorter. Orange-buff vent-patch often darker, but much overlap. Furthermore, in flight also has much smaller white area on flight-feathers ('wing flash'), especially on second-

O. l. boscaweni, 1stW ♂, Oman, Nov: often shows dark streaking on greyish-white cap. Aged by white fringes to primary-coverts, which with wear become mostly restricted to tips; also browner juv remiges. (T. Lindroos)

O. l. boscaweni, ♀, Oman, Nov: some ♀♀ distinctive by rather darker upperparts and variably warmer ear-coverts, greyer and streaked throat and subtly streaky breast, with diagnostic pinkish-orange ventral region. Apparent moult contrast in greater coverts, and already somewhat worn wing, could suggest 1stW, but ageing must be considered tentative. (D. Occhiato)

O. l. boscaweni, ♀, Oman, Nov: a slightly more worn and washed-out bird (compared to image to the left), but still has diagnostic pinkish-orange vent/undertail-coverts and thin, diffuse streaking below. Ageing difficult as degree of feather wear and moult pattern does not match usual cycle. (A. Audevard)

aries. – ♂ Arabian could also be confused with worn ♂ Pied Wheatear or ♂ Gould's Wheatear (*O. capistrata*; extralimital, see under Blyth's Wheatear, Taxonomic notes), but both lack conspicuous whitish primary flashes of ♂ Arabian and have different jizz. Pied also has longer primary projection and lacks prominent orange-rufous wash to vent and undertail-coverts of Arabian. – ♀ Arabian differs clearly from its ♂, as in Maghreb Wheatear but unlike in other species of the Mourning Wheatear complex, being predominantly dusky to dull grey-brown with warmer brown ear-coverts, diffusely grey-streaked underparts and sometimes fine dark shaft-streaks on crown and rump/uppertail-coverts. The ♀ also lacks ♂'s white primary patch, at most having a semi-translucent greyish panel in flight. – ♀ Finsch's Wheatear lacks ♀ Arabian's diagnostic pinkish-orange wash to ventral region (at most only a fraction of this visible in Finsch's), and both ♀ Finsch's and ♀ Blyth's lack streaked underparts and usually have no pale-based remiges. – The same characters also readily eliminate ♀-like Pied Wheatear, Eastern Black-eared Wheatear and even Northern Wheatear, all of which can appear superficially similar to ♀ Arabian, especially given poor views and atypical birds. – Both sexes of Maghreb Wheatear share many characteristics with corresponding plumages of Arabian, but they are well separated geographically.

AGEING & SEXING Ageing requires close observation of moult and feather wear or pattern in wing. Sexes well differentiated. – Moults. Complete post-nuptial and partial post-juv moults in late summer (Jun–Sep). Post-juv moult includes all head, body, lesser and median coverts, 5–10 greater coverts, less often innermost tertials and some tail-feathers. Pre-nuptial moult absent. – **SPRING Ad ♂** White crown invariably has sooty brown-grey feather-centres over much of its area except laterally, borders being near-white. Orange-rufous wash to vent/undertail-coverts often still strong. Wings jet-black with no moult limits. **Ad ♀** Warm pigments on ear-coverts vary and may extend to forecrown, and with wear become paler with faint shaft-streaks. Supercilium ill-defined but sometimes whiter above and behind eye. Dark throat and upper breast vary but wear duskier with partially exposed blackish bases. If not heavily worn, faintly streaked lower breast still evident and may extend to flanks or is almost confined to breast-sides (and breast often tinged slightly yellowish-buff). Cinnamon vent and undertail-coverts obvious. Dark wings (but not black), with pale bases to remiges rather indistinct, and when not heavily worn has pale fringes to tertials, remiges and some coverts. No moult limits in wings. **1stS** Aged by relatively more worn and browner retained juv remiges, primary-coverts and often a few outer greater coverts and tertials, otherwise as respective ad. Moult limits more obvious in ♂, which is often less solidly coloured than ad ♂, with some pale tips to black feathers (including sometimes even on primary-coverts tips, if not heavily worn) and streakier/duskier crown. Juv remiges have reduced white/grey on inner webs in both sexes. – **AUTUMN Ad ♂** Largely as in spring but often has crown more washed (rather than streaked) grey, and black areas (including wings) more solid black. **Ad ♀** Head browner (warm ear-coverts coloration often extends more evenly onto crown) and throat creamier with dark bases wholly or partially concealed. Upperwings more extensively fringed and underparts washed off-white or buffish, especially on vent/undertail-coverts. Both sexes differ from 1stW by evenly very fresh wings without discernible moult limits. Noticeable pale primary tips fresh or at most slightly worn, and primary-coverts only slightly worn and lack diagnostic broad white tips and fringes of 1stW. **1stW ♂** Similar to ad ♂ but primary-coverts have well-marked whitish fringes and tips, and pale primary tips slightly to moderately worn. Generally, black areas tend to be sooty/charcoal-grey, and flight-feathers browner. Moult limits less obvious than in spring (juv greater coverts, tertials and alula, if retained, have blackish-brown centres and prominent pale rufous/buffish

O. l. boscaweni, Oman, Nov: some ♀♀ distinctly paler, with much-reduced streaking below and paler throat, inviting confusion with several migrant wheatears, but for the almost diagnostic pinkish-orange vent/undertail-coverts. Safe ageing not possible at this angle. (D. Occhiato)

O. l. boscaweni, ♀, Oman, Jan: distinctive pinkish-orange vent/undertail-coverts and faint streaking on chest. Impossible to age at this angle. (H. & J. Eriksen)

Presumed *O. l. lugentoides*, ad ♂ (left) and ♀ (right), Yemen, Dec: generally darker than *boscaweni*, and especially ♀ is more heavily streaked below, with deeper orange undertail-coverts; black in ♂ is more extensive on breast and white areas average less pure. Both these birds are thought to be ad based on lack of visible juv wing-feathers. (A. Al-Sirhan)

fringes, while ad-like post-juv ones are fresher, firmer, have black centres and shafts, and when fresh, at most, a tiny whitish fringe to tertials and greater coverts, especially outer ones). **1stW ♀** As fresh ad ♀ and best aged by moult limits (see 1stW ♂ though less contrasting and more difficult to detect). Note that especially young ♀♀ have considerably reduced white/grey on inner webs of flight-feathers. **Juv** As juv Maghreb Wheatear, thus sexual difference obvious early, ♂♂ having blackish bib and very dark upperparts.

BIOMETRICS (*lugentoides*) **L** 15–16 cm; **W** ♂ 87–90 mm (n 12, m 88.7), ♀ 80.5–91 mm (n 11, m 84.4); **T** ♂ 56.5–60 mm (n 12, m 58.4), ♀ 53–63 mm (n 12, m 57.4); **T/W** m 66.9; **black on r1** 42–50 mm (m 45.8); **black on r4** 14–22 mm (m 18.9); **black on r6** 15–26 mm (m 20.7); **B** 17.0–19.8 mm (n 23, m 18.4); **BD** 4.0–5.0 mm (n 24, m 4.5); **Ts** 22.1–25.6 mm (n 24, m 23.5); **rump patch** 10–24 mm (m 16.5). **Wing formula:** p1 > pc 4–9.5 mm (once 1.5 mm), < p2 30–40 mm; **p2** < wt 5–9 mm, =6/7 (rarely =6 or 7, exceptionally =7/8); **pp3–4** (5) longest; **p5** < wt 0–2 mm; **p6** < wt 1.5–5 mm; **p7** < wt 6–10 mm; **p10** < wt 16–20 mm; **s1** < wt 18–22 mm. Emarg. pp3–5 (sometimes faintly on p6).

GEOGRAPHICAL VARIATION & RANGE Two races recognised, differing only slightly. All populations resident or make only short-range winter movements.

    O. l. lugentoides (Seebohm, 1881) (Saudi Arabia to W & C Yemen) is described above.

    ○ ***O. l. boscaweni*** Bates, 1937 (E Yemen to Dhofar, W Oman) differs only marginally, mainly by paler plumage of ♂, underparts slightly whiter with paler undertail-coverts, and is whiter on lower back to uppertail-coverts. Differs also in more extensive white in remiges, and by black on chin extending only to lower throat (to chest in ♂ *lugentoides*). Black on r1 somewhat less extensive. ♀ as ♀ *lugentoides* but paler, being more drab-grey above, and rufous on ear-coverts reduced. Underparts whiter with paler drab wash on breast and less streaked below. Both sexes average smaller than *lugentoides* but much overlap. **W** ♂ 81–92 mm (n 8 m 85.9), ♀ 77–83 mm (n 3, m 79.7); **T** 52–59 mm (n 10, m 55.1); **T/W** m 66.1; **black on r1** 34–40 mm (m 37.2); **B** 16.8–19.3 mm (n 11, m 18.4); **Ts** 22.4–25.4 mm (n 11, m 24.3).

TAXONOMIC NOTE The Arabian Wheatear *O. lugentoides* is closer genetically to the so-called 'Abyssinian Black Wheatear' *O. lugubris* (extralimital) together with its close relatives *schalowi* and *vauriei* (both also extralimital) than to the Mourning Wheatear of the Middle East or its close relative Maghreb Wheatear of N Africa (Schweizer & Shirihai 2013). This would render Mourning Wheatear polyphyletic if *lugentoides* is kept as a race of Mourning. Hence, Arabian Wheatear is treated here as a separate species.

REFERENCES Boon, L. J. R. (2004) *DB*, 26: 223–236. – Förschler, M. *et al.* (2010) *Mol. Phyl. & Evol.*, 56: 758–767. – Schweizer, M. & Shirihai, H. (2013) *Mol. Phyl. & Evol.*, 69: 450–461. – Shirihai, H., Kirwan, G. M. & Helbig, A. J. (2011) *BBOC*, 131: 170–191.

Juv, Yemen, May: loose body-feathers and reduced rufous-orange on rear body, though still distinctive by overall dusky plumage; distinctly different from juv Mourning. (W. Müller)

HOODED WHEATEAR
Oenanthe monacha (Temminck, 1825)

Fr. – Traquet à capuchon;
Ger. – Kappensteinschmätzer
Sp. – Collalba monje; Swe. – Munkstenskvätta

A large long-billed, slim and 'flighty' wheatear of Middle Eastern deserts (occurring from NE Sudan and E Egypt to Sinai, S Israel, C Saudi Arabia and Oman, Iran and S Pakistan), where it frequents both mountains and plains with some bushes and *Acacia*. Often occurs around nomad camps and domestic mammals. Essentially resident but performs some short movements.

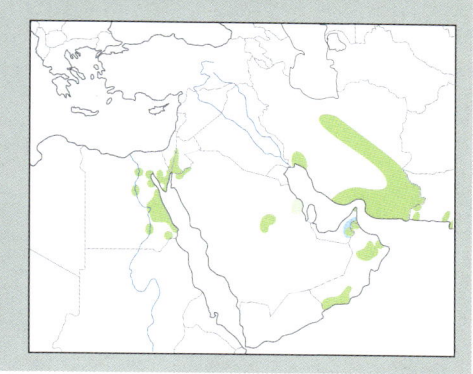

IDENTIFICATION *Medium-large wheatear with longish bill, wings and tail.* Long-bodied with flat head, short legs (often invisible when squatting). ♂ plumage black and white. *Forehead to nape white. Black face to mid/lower breast merges with black mantle and wings. Belly and tail largely white. Black central tail-feathers but diagnostically lacks prominent tail-band* (sometimes has narrow but clearly broken terminal band, or just scattered black spots on tail corners). Black underwing-coverts contrast with pale flight-feathers. ♀ has buffish-grey or drab brown upperparts, darker but *conspicuously pale-fringed wing-coverts, tertials and secondaries*. Underparts, lower back and rump paler, tinged buffish-white; *tail as in ♂, but central area dark brown, pale area tinged rufous-buff*, especially at base, and inverted 'T' often better developed. Habitually perches prominently on bushes, atop *Acacia* trees and on fences, and performs long, sometimes high, flights, chasing larger flying insects.

VOCALISATIONS Song often comprises long strophes with continuous *Acrocephalus*-like warbling or chattering, even stuttering slow series or scratchy or warbling notes, sometimes interspersed with varied hard knocking, clicking and crackling sounds, and occasionally much mimicry. – Variety of other whistles and clicks, a harsh *tsak* or *zak*, and a *wit-wit* in alarm or contact. In territorial disputes a rattle.

SIMILAR SPECIES Structure and plumage usually sufficient for separation from any other black-and-white wheatear (see e.g. Mourning Wheatear). Adult ♂ unmistakable. Long wings can conceal white belly on distant perched bird, which may then recall *White-crowned Wheatear* (similar tail pattern), but Hooded has more white on head and longer bill. ♀'s long bill, pale rust-coloured rump and uppertail-coverts, ill-defined tail-band and extensively pale-fringed tertials, upperwing-coverts and secondaries (forming elongated innerwing panel) are useful features, in addition to size, structure and behaviour.

AGEING & SEXING Ageing requires close observation of moult and feather wear in wing. Sexes well differentiated, but seasonal variation rather limited. – Moults. Complete post-nuptial and partial post-juv moults in summer (mainly Jun–Sep). Post-juv moult includes all head, body, lesser and median coverts, most or all greater coverts, and usually smallest alula, less frequently 1–3 tertials, while some or all tail-feathers and entire alula can also be replaced. Pre-nuptial moult absent. – **SPRING Ad ♂** General description under Identification. Aged by being evenly feathered with relatively less worn, blacker primaries and primary-coverts. **Ad ♀** Unlike 1stS ♀ lacks moult limits in wings, and primaries and primary-coverts less worn and darker. With wear, wings become darker than rest of greyer upperparts, rump to base of outertail paler creamy-buff, and underparts dirty grey-white with buff-washed breast-sides, flanks and undertail-coverts. **1stS ♂** As ad ♂ but usually shows clear moult limits with more heavily worn and distinctly bleached juv remiges, primary-coverts and perhaps some outer greater coverts. Often less evenly coloured, with more whitish tips to black areas, which can be slightly browner, especially bib. **1stS ♀** Best aged by moult limits as in 1stS ♂, but moult pattern much less contrasting; otherwise as ad ♀. – **AUTUMN Ad ♂** Much as in spring, but when very fresh black of throat and especially breast, as well as mantle, narrowly tipped greyish, while white areas sometimes have slight buffish-cream tone. Primary-coverts black, at most with narrow paler fringes (if any) and tips. **Ad ♀** Amount of buff/russet varies, but generally slightly brighter than in spring. Also, fringes to remiges, especially secondaries and tertials, and upperwing-coverts highly contrasting. Primary-coverts firmer and slightly darker than in 1stW ♀. Both sexes also differ from 1stW by evenly very fresh wing-feathers, without discernible moult limits. **1stW ♂** As fresh ad ♂ but more extensive whitish tips to black of face, mantle and especially breast, with browner bases. Distinctive pale-fringed juv remiges, some tertials and coverts, forming moult limits with fresher ad-like post-juv coverts, which have black centres and, at most, narrow whitish fringes. Retained primary-coverts centred dark grey-brown with well-marked whitish fringes and tips. Some conspicuously variegated due

Ad ♂, Israel, Mar: spread long tail showing diagnostic pattern being nearly all white. Age inferred by black wings and very restricted black on tips of outer tail-feathers (amount of black on tips thus partly age-related). (H. Shirihai)

Ad ♂, Israel, Mar: large, slim, black-and-white wheatear, with broad neck, large head, long bill and short legs; tail pattern almost unique—mostly white with black central feathers, and at most a narrow, broken terminal band. No moult limits or juv feathers in wing. (D. Occhiato)

♀, presumed ad, Israel, Mar: drab-brown above with conspicuously pale-fringed darker upperwings; lower back to upper tail-coverts and entire underparts pale buff-white, warmest on under tail-coverts. Tail has similar pattern to that of ♂ but the pale area is cinnamon, the dark parts are dark brown rather than black. Note also characteristic jizz with slim, long bill and bland facial pattern. Whole wing seems evenly feathered ad. (T. Krumenacker)

to more limited or unfinished post-juv moult, almost midway in appearance between ♂ and ♀, due to larger number of retained ♀-like juv feathers (including tail, smaller coverts and tertials). Crown often yellowish-buff, and white areas slightly dirty yellow-brown. **1stW ♀** Separation from ad ♀ sometimes less straightforward, especially birds which moulted most secondary-coverts, tertials and alula. Retained juv primary-coverts slightly softer with paler, browner centres than ad ♀; juv greater coverts, tertials and alula (if retained) have dark brown centres (contrasting with recently moulted post-juv ones which are firmer with darker greyish-brown centres); juv greater coverts also fringed pale ochre (pale greyish-brown in ad-like post-juv feathers). Juv central alula has moderately worn tip and diffuse greyish-brown outer fringe (post-juv feather has fresher and sharply delimited pale greyish-brown fringe that whitens with wear). Juv tail-feathers have slightly more extensive dark brownish marks at tips (except central tail-feathers). **Juv** Like ad ♀ but crown, upperparts and lesser coverts faintly mottled cream-buff. Head-sides and chin to breast-sides narrowly scaled dull black. Upperwing, especially coverts and tertials, extensively fringed pinkish-cinnamon. Tail mostly as ♀.

1stS ♀, Israel, Mar: characteristic shape and colour even in flight. Note large pale rump and rufous tail, in which inverted blackish 'T' can be just visible (more when partly spread). Moult limits in wing (including tertials), with worn and bleached primaries, indicate 1stY. (M. Huhta-Koivisto)

1stS ♂, Israel, Mar: even as late as well into spring retains variable amount of white fringing and tipping, including on the juv remiges and primary-coverts. Note also more solid dark terminal band or tail in 1stY compared to ad ♂. (H. Shirihai)

1stW ♂, Israel, Dec: flight-feathers, primary-coverts and tertials juv, brown-grey and already worn and bleached but still with narrow whitish fringes. The pale areas of the juv outer tail-feathers sometimes tinged rufous-buff, affording hint of ♀-like coloration in young ♂. (H. Shirihai)

BIOMETRICS L 16.5–18 cm; **W** ♂ 100–108 mm (*n* 21, *m* 105.4), ♀ 98–104 mm (*n* 12, *m* 100.6); **T** ♂ 68–75 mm (*n* 21, *m* 72.0), ♀ 67–73 mm (*n* 12, *m* 70.1); **T/W** *m* 68.8; **B** 19.6–24.7 mm (*n* 33, *m* 22.0); **BD** 3.9–4.6 mm (*n* 32, *m* 4.2); **Ts** 21.4–24.0 mm (*n* 32, *m* 22.9). **Wing formula: p1** > pc 4 mm, to 2 mm <, < p2 50–57.5 mm; **p2** < wt 3–6 mm, =4/6; **p3** (pp3–4) longest; **p4** < wt 0–1.5 mm; **p5** < wt 3–5.5 mm; **p6** < wt 10–16 mm; **p7** < wt 16–22 mm; **p10** < wt 26–32 mm; **s1** < wt 28–34 mm. Emarg. pp3–5.

GEOGRAPHICAL VARIATION & RANGE Monotypic. – Egypt along Nile and eastwards, Levant, Arabia, S Iran, Afghanistan; in winter makes only shorter movements, if any.

1stW ♀, Israel, Nov: young ♀ very similar to ad and aged only by close check of moult and of retained juv primary-coverts, and sometimes moult limits in secondary-coverts and tertials. In fresh plumage, underparts and ear-coverts are warmer tinged. Note also the characteristic obvious pale fringes over most of wing typical of species. (A. Ben Dov)

HUME'S WHEATEAR
Oenanthe albonigra (Hume, 1872)

Fr. – Traquet de Hume
Ger. – Schwarzkopf-Steinschmätzer
Sp. – Collalba de Hume; Swe. – Svartvit stenskvätta

Yet another black-and-white wheatear, this one closely copying the pattern of Blyth's Wheatear. Compared to that, it has a somewhat more southerly and westerly range, being found in arid SW Asian deserts and is local in NE Arabia and from E Iraq and S Iran east through S Afghanistan to Pakistan. It prefers barren, rocky hills and montane valleys with scant vegetation. Largely resident, though some evidence of winter movements, e.g. to S Iraq and Kuwait.

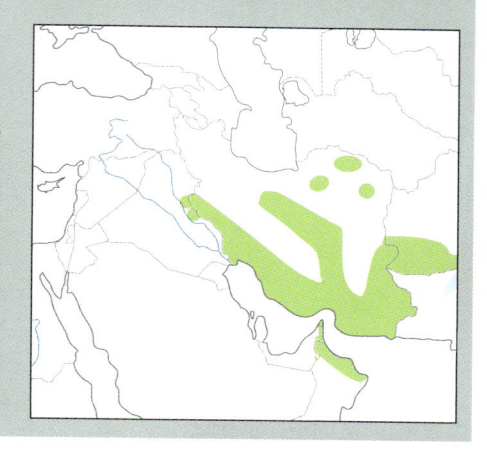

Presumed 1stW, Oman, Feb: inverted black 'T' on tail, all-black head and white belly provide diagnostic combination. Seems to have juv (slightly brownish) remiges and primary-coverts, which would indicate 1stW. (H. & J. Eriksen)

Ad, Oman, Oct: black head, down to upper breast, black upperparts in striking contrast to white below. All wing-feathers uniform glossy black making this bird an ad, possibly ♂ (though sexes nearly identical). (A. Audevard)

IDENTIFICATION *Large, somewhat plump, piebald wheatear with rather large, rounded head. Sexes alike. Black head, throat and upper breast, mantle, scapulars, wings and inverted 'T' on tail, while the rest (diagnostically much of breast and entire belly) is pure white,* sharply demarcated from black areas. Combination of all-black head and *long white rump patch* (invading lower back) diagnostic. Bill and legs strong, legs appearing short. ♀ similar but duller and juvenile even more so, being matt blackish-brown. In E Arabia, migratory white-bellied Blyth's Wheatear (of former Variable Wheatear complex) is main confusion risk, with White-crowned and Hooded Wheatears only superficially similar.

VOCALISATIONS Song a loud cheerful and pleasant *chew-de-dew-twit* or *chiroochiri-chirrichirri* on a rising scale (with a Blackcap- or bulbul-like quality), thus less discordant and harsh than most wheatears. Each strophe is often preceded by soft introductory *dze*, *kooi* or *pepeepoo* notes. Can sing for several minutes with only brief pauses between strophes. ♀ never positively recorded singing. – Contact call a repeated short, sharp whistle, given 3–4 times, and a harsh monosyllabic grating *chit-tit-tit* or *ti-ti-te*. Also a clicking *chak-chak-chak* in alarm.

SIMILAR SPECIES All plumages separable using following characters. Combination of lack of white crown but presence of inverted 'T' on tail distinguish it from most potential confusion species, e.g. adult *White-crowned* and *Hooded Wheatears*. White breast and belly separate it from all species or forms with similar black heads, e.g. juvenile/immature White-crowned Wheatear, Strickland's Wheatear (extralimital), Basalt Wheatear and Black Wheatear, all of which have black bellies. – Superficially resembles only white-bellied *Blyth's Wheatear* but notably is larger and heavier, especially given 'bull-headed' appearance, with longer and stouter bill, longer wings, larger feet and more rounded head. Black parts glossier, and black of throat reduced in size, hardly extending to breast, and black of mantle barely reaching onto back. Upper edge of white rump (actually on back) slightly rounded (more square in Blyth's Wheatear). Black underwing-coverts less contrasting (due to darkish primaries). Less active and more confiding than Blyth's, remaining perched for longer periods. Extreme variation among Blyth's Wheatear hybrids (with either of Gould's or Strickland's) could potentially produce somewhat similar plumage patterns, but even so, size, behaviour and other plumage details should prevent confusion.

AGEING & SEXING Ageing requires close observation of moult and feather wear in wing. Sexes hardly differ (but see Biometrics), and seasonal variation limited. – Moults. Complete post-nuptial and partial post-juv moults in late spring–summer (late Apr–early Sep, probably mostly Jun–Aug). Post-juv moult includes all head, body, lesser and median coverts, 6–10 greater coverts, the smallest alula, some or rarely all tertials, and (less frequently) a few tail-feathers. Pre-nuptial moult absent. – **SPRING Ad** ♂ Often has pronounced gloss to head, mantle and wings. **Ad** ♀ Similar but usually slightly duller, although difference too subtle for field use, and some in collections appear identical. **1stS** Like ad but both sexes somewhat less glossed black with more heavily worn and browner juv primaries, primary-coverts and perhaps some outer greater coverts and tertials. On some, tiny pale tips to primary-coverts still visible (lacking in ad). – **AUTUMN Ad** Much as spring but can have browner or duller black body-feathers. Best aged on being evenly fresh without moult limits or obvious pale tips to wing-feathers (cf. 1stW). Primary tips fresh or, at most, slightly worn. **1stW** Like autumn ad, but black areas sooty/brownish, lacking obvious sheen. Juv primary-coverts, greater coverts and alula have small, diffuse pale tips (subsequently worn off, but in some retained into winter and even beyond). Also, juv primary-coverts and remiges, and greater coverts and alula (if retained), slightly more worn and browner than ad-like post-juv feathers; primary tips moderately worn. **Juv** Body-feathering loose and fluffy. Otherwise as ad but much duller, greyer brown-black, less glossy and bluish. All wing-coverts profusely fringed pale ochre.

BIOMETRICS L 16–16.5 cm; W ♂ 102–110 mm (*n* 16, *m* 105.1), ♀ 96–104 mm (*n* 14, *m* 99.6); T ♂ 64–71 mm (*n* 16, *m* 68.1), ♀ 63–68 mm (*n* 14, *m* 65.5); T/W 65.2; B 18.7–20.9 mm (*n* 23, *m* 19.8); Ts 24.5–27.8 mm (*n* 21, *m* 25.9). Wing formula: p1 > pc 4–9 mm, < p2 42–53 mm; p2 < wt 5–9 mm, =6 or 6/7; pp3–4(5) longest; p5 < wt 0–2 mm; p6 < wt 4–9 mm; p7 < wt 10–18 mm; p10 < wt 22–29 mm; s1 < wt 24–34 mm. Emarg. pp3–5 (sometimes faintly on p6).

GEOGRAPHICAL VARIATION & RANGE Monotypic – NE Arabia to N Oman, S Iran, S Afghanistan, W & N Pakistan north-east to Gilgit and Kashmir; mainly resident, sometimes making shorter movements, but information scant.

REFERENCES RICHARDSON, C. (1999) *Sandgrouse*, 21: 124–127.

Oman, Nov: readily distinguished from most other wheatears by lack of white crown, but having white breast and belly. However, separation from Blyth's requires care that black parts are glossier, black of throat hardly extends to breast, while Hume's is notably larger, with more round-headed appearance and longer and stouter bill. Ventral region white. (P. Alberti)

1stW, Oman, Nov: subtly browner-tinged remiges and primary-coverts, latter with hint of paler tips, confirm this as a young bird. (D. Occhiato)

WHITE-CROWNED WHEATEAR
Oenanthe leucopyga (C. L. Brehm, 1835)

Alternative name: White-crowned Black Wheatear

Fr. – Traquet à tête blanche
Ger. – Saharasteinschmätzer
Sp. – Collalba yebélica
Swe. – Vitkronad stenskvätta

A striking mainly black wheatear of Saharan and Arabian deserts, usually in rocky or cliff-edged wadis, where it attracts the observer's attention with its loud and complex song full of mimicry. White-crowned Wheatear is often seen at the edge of settlements and in Bedouin encampments, where it may warn the occupants of snakes. Mainly resident, from E Morocco to Egypt, south to Mauritania and Ethiopia; also found in Sinai east to N & C Saudi Arabia.

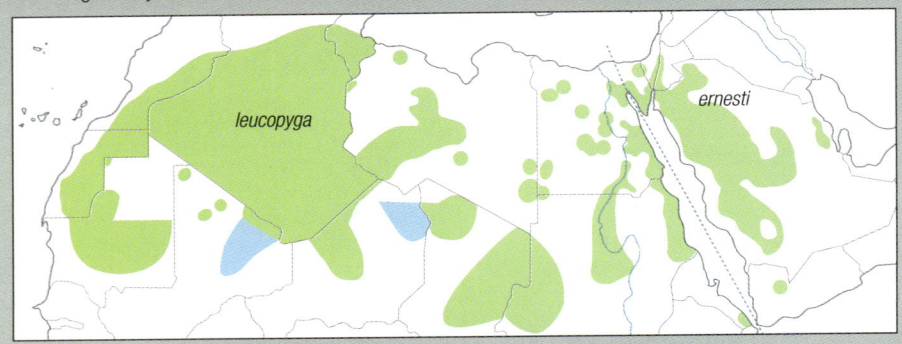

O. l. ernesti, ad, Israel, Mar: white cap can appear small in some angles. Beware that in some postures tail is completely folded and dark central feathers conceal the white tail-sides. (P. Alberti)

O. l. leucopyga, ad, Morocco, Feb: glossy black with white vent and rear, including tail that in principle has only central feathers black. White cap sometimes retains a few black feathers. Aged by being evenly feathered with no moult limits in wing. (D. Occhiato)

IDENTIFICATION A largely *glossy black* medium-sized wheatear. *Adult has conspicuous white crown*, whereas in immature the crown is *black or largely so*. Forehead black. Lower back and belly to tail white, except black central tail-feathers, thus *outer parts of tail either entirely white or have only small black patches near tips* of outer feathers, a pattern shared only with Hooded Wheatear. Structure and plumage only superficially resemble Black Wheatear, Basalt Wheatear and Strickland's Wheatear (extralimital; see Blyth's Wheatear, Taxonomic notes), White-crowned having *rather long bill and wings* but proportionately *slightly shorter legs*. Bold and confident at times but can also be wary, taking cover early.

VOCALISATIONS An expert mimic. Loud and powerful sustained song with expert imitations of other birds ('the Hill Myna of the desert'), given by both sexes year-round, consisting of repeated short phrases, descending mellow whistles (glissando) and other clear notes, rather deep and low-pitched, e.g. *si-sü-so-trui-trü* or *tchu-tchu-tchee, tchu-ichee, tchu-tchu-tcheu*…, sometimes in song-flight with fluttering wings and fanned tail. Song at times recalls Mourning or Black Wheatears, a lark or a *Monticola* thrush. Individual variation emphasized by mimicry, including other birds, human whistles or mammalian sounds. Also has a chattering, softer subsong. – Calls varied, too, including a shrill whistled *hiit*, high cracked *bizz* and hoarse, throaty, grating *tschreh* or *dzzrsh*.

SIMILAR SPECIES The following applies to all ages and sexes. Similar-sized *Black Wheatear* confusable with young White-crowned lacking white in cap but latter's tail almost all white, apart from black central feathers and occasionally has small dark spot near tips of other tail-feathers (Black: invariably a solid evenly broad tail-band). Beware that first-years normally show a narrow, broken terminal tail-band often with white spots intermixed; rarely some have broader black tips, but still rather ragged tail-band compared to pattern of Black. White-crowned is also slimmer than Black with proportionately longer wings and tail, and finer bill. Black also lacks gloss to dark plumage (first-years are even sooty-brown, not black), and black of underparts often extends beyond legs, unlike in White-crowned. In flight, Black has paler flight-feathers, whereas White-crowned has all-dark upper- and underwings. – ♂ *Hooded Wheatear* shares tail pattern, but structure different and has all-white belly to lower breast. – Black-capped young White-crowned resembles *Basalt Wheatear* of Syria, and also *Strickland's Wheatear* (*O. opistholeuca*), but is larger and glossier than both, lacks solid black tail-band, has larger area of white on vent (reaching lower belly), and usually shows more white in tail when perched. It also lacks whitish-grey tone to remiges (the others have variable pale inner webs to flight-feathers, creating slight pale flash in flight).

AGEING & SEXING Ages separable during 1stY, chiefly involving age-related development of white cap. Sexes alike in plumage (though see Biometrics for rather marked size difference in ssp. *leucopyga*), and seasonal variation rather limited. – Moults. Complete post-nuptial and partial post-juv moults in spring or summer, thus quite extended, generally late Mar–early Sep. Post-juv moult includes all head, body, lower and median coverts, four to all greater coverts, more rarely 1–3 tertials and some or entire alula. Different assumptions exist concerning cap development,

O. l. ernesti, ad, Israel, Feb: combination of largely black with white cap and rear parts, as well as tail pattern, make plumage unmistakable. This bird has clean cap with no black feathers or tips. Aged by lack of moult limits in wing. (L. Kislev)

WHEATEARS

but based on field studies in Eilat/Negev deserts, Israel, and specimens, it seems that, independent of sex, during post-juv moult some 1stW acquire one or up to several white crown-feathers, though crown essentially black until first complete post-nuptial moult (to 2ndW plumage), when number of black feathers in cap is c. 50% or less (admixed white feathers spread at variable rate); following second complete post-nuptial moult (to 3rdW plumage), few or no black feathers in otherwise white cap; older ad often acquires several isolated black feathers. Most predominantly black-capped 1stS do not breed, though this requires confirmation. Pre-nuptial moult in late winter absent. — **SPRING Ad** Cap usually all white, but sometimes some black feathers retained, and outer tail may have variable black subterminal marks. Blue gloss to most black tracts. Primary tips moderately worn. ♀ often slightly duller and less glossy than ♂ but differences too subtle to be used in the field. Both sexes lack moult limits. **1stS** Black cap lacks or has only some isolated white feathers. Also, compared to ad has less gloss in black plumage, especially noticeable on juv primaries and primary-coverts (becoming browner in summer), and perhaps some retained outer greater coverts. Some still show very faint pale tips to primary-coverts (lacking in ad). — **AUTUMN Ad** Crown all white or very nearly so. Plumage evenly fresh and glossy. Remiges, primary-coverts, greater coverts and alula black or dark brownish-black, and primary tips very fresh or at most slightly worn. **1stW** Like ad, but crown always all dark or has only a few isolated white feathers, and glossy tones confined to breast, mantle and coverts. Moult limits, especially in fresh plumage, difficult to detect. Juv primaries moderately worn at tips. Some juv wing-feathers when still very fresh have faint pale fringes but these wear off rapidly (primary-coverts slightly more worn, paler and browner than ad, with very faint whitish fringes at tips). Juv greater coverts, tertials and alula (if retained) slightly more worn, paler and browner than ad-like moulted ones. Tail pattern varies: some have very few black subterminal marks, but most have more, though only visible on open tail or if observed from below. Very rarely a near-complete broader tail-band. **Juv** Less deep black than ad. Usually lacks any white on crown. Black parts often have some very indistinct, narrow pale fringes, including to tertials and secondary-coverts. Tail pattern as 1stW.

BIOMETRICS (*leucopyga*) **L** 16.5–17.5 cm; **W** ♂ 103–112.5 mm (*n* 15, *m* 109.3), ♀ 93–105 mm (*n* 13, *m* 98.8); **T** ♂ 65–75 mm (*n* 15, *m* 70.9), ♀ 60–70 mm (*n* 13, *m* 64.7); **T/W** *m* 65.2; **B** 17.7–23.2 mm (*n* 26, *m* 20.5); **BD** 4.0–5.3 mm (*n* 17, *m* 4.6); **Ts** 23.9–27.6 mm (*n* 27, *m* 25.9). **Wing formula:** p1 > pc 2–9 mm, < p2 42–55 mm; **p2** < wt 5–10 mm, =5/6 or 6 (rarely =6/7); **pp3–4**(5) longest; **p5** < wt 0–3 mm; **p6** < wt 6–11 mm; **p7** < wt 11.5–18 mm; **p10** < wt 24–31 mm; **s1** < wt 25–33 mm. Emarg. pp3–5 (sometimes faintly on p6).

GEOGRAPHICAL VARIATION & RANGE Very slight variation with two very similar races. All populations mainly sedentary.

○ ***O. l. leucopyga*** (C. L. Brehm, 1835) (N Africa from Morocco to Suez). Described above. (Syn. *aegra*.)

○ ***O. l. ernesti*** Meinertzhagen, 1930 (Sinai, S Levant, NW Arabia; possibly S Iraq). Very similar to *leucopyga* but subtly larger on average with proportionately slightly longer tail and stronger bill. Oddly, the larger size seems to be a function of ♀♀ being equally large as ♂♂ in this race, since in specimens no small ♀♀ were found and sexes had very nearly identical size. Note large overlap in bill size between the two subspecies. When series are compared, *ernesti* is often glossier bluish-black, but much overlap and indistinguishable on this trait alone in the field. Amount of black in tail-band claimed as useful for subspecific identification, but is a function of age-related plumage variation, younger birds having more black, rather than geographical differences. **W** 102–112 mm (*n* 20, *m* 106.6); **T** 66–75 mm (*n* 20, *m* 69.2); **T/W** *m* 69.2; **B** 19.2–23.8 mm (*n* 19, *m* 21.6); **BD** 4.0–5.7 mm (*n* 19, *m* 5.0); **Ts** 24.1–28.4 mm (*n* 20, *m* 26.1).

O. l. leucopyga, juv, Morocco, Apr: lacks white cap and has less glossy black plumage than ad; black parts often show some very indistinct, narrow pale fringes, here mainly noticeable on rear belly. Like ad tail it is mainly the central tail that is black, but outer tail-feathers can have variable black subterminal marks. (R. Armada)

O. l. ernesti, 1stY, Jordan, Mar: young have crown all back (or have few isolated white feathers—see image below left); gloss generally reduced, and in close views moult limits visible (note browner juv largest alula and primary-coverts). (H. Shirihai)

O. l. leucopyga, 1stY, Canary Is, Jan: some young birds already acquire several white feathers on crown (even partial cap); characteristic strong moult limits in wing created by juv coverts (browner), and variable black subterminal marks on tail. (R. Smith)

O. l. leucopyga, 1stY, Morocco, Mar: young with more extensive black subterminal marks (like here) in tail might initially suggest Black or Basalt Wheatears. (D. Monticelli)

BLACK WHEATEAR
Oenanthe leucura (J. F. Gmelin, 1789)

Fr. – Traquet rieur; Ger. – Trauersteinschmätzer
Sp. – Collalba negra; Swe. – Svart stenskvätta

Large, bulky, mainly black or dark wheatear of the Iberian Peninsula, where it is the only species with such plumage, and of NW Africa, where its range overlaps with the quite similar White-crowned Wheatear. Resident in much of Iberia, except parts of the north and east, and from Western Sahara and S Morocco apparently as far east as extreme NW Libya. Range was perhaps broader in the past, with records east to Italy, Bulgaria and Egypt. It shuns deserts, preferring rocky, well-vegetated slopes in dry country. Behaviour is typical of wheatears, often seen singly or in pairs. Territorial birds use regular lookout posts and are often easy to locate, but still, in places, it can be elusive.

O. l. riggenbachi, presumed 1stS ♂, Tunisia, Apr: similar to *leucura* but on average matter black, and tail-band slightly broader. Some difficult to age and even sex, but here rather brownish-black plumage suggests ♂, and quite heavily worn tail (seemingly juv) indicates 1stS. (D. Occhiato)

IDENTIFICATION Black (♂) or dark sooty-brown to dusky grey-brown (♀) wheatear with contrasting white lower back, rump, vent and most of tail. Note that *undertail-coverts are white* (unless soiled or stained), not rufous-buff or cinnamon. *Tail has* diagnostic *inverted black 'T' pattern, with almost even-width, narrow to moderately broad terminal band.* In flight shows somewhat paler (dull silver-white) bases to underwing. General size and plumage similar to dark-crowned (immature) White-crowned Wheatear, as well as to (slightly smaller) geographically well-separated Basalt Wheatear and extra-limital Strickland's Wheatear (for which see Blyth's Wheatear, *O. picata*). In Europe immediately identified by plumage and range, but in North Africa requires careful assessment. Song given from ground or in undulating song-flight.

VOCALISATIONS Song a coarse but partly clear, loud, rapid twittering, slightly lower-pitched than other species of *Oenanthe* in same region, and has somewhat Rock Thrush-like quality, or can resemble Garden Warbler due to its deep warbling voice and rather even pitch and pace. – Variety of throaty and harsh call notes, e.g. a rolling *cherrr* and thick *chrett-chrett*. Also a whistled *piüp* or *psiiep*.

SIMILAR SPECIES The black ♂ is confusable with any young White-crowned Wheatear lacking white in cap, but is slightly bulkier with larger head and heavier bill. Bold black terminal tail-band is nearly always solid, but rarely narrow white intrusions can occur on third outermost feather in northern birds (White-crowned has extensive white in outer-tail, with white on several feathers even in young birds with darkest tail-band). Also note lack of strong metallic bluish gloss to dark areas of plumage even in ♂♂ (first-years and some ♀♀ are sooty- or grey-brown, not black, and cause less problems), while black of underparts often extends beyond legs, which further separates Black from White-crowned. In flight, Black has variably paler flight-feathers, often visible as a silvery sheen on underside, whereas White-crowned has all-dark upperwing and underwing, and somewhat slower wing action. Beware of rare young White-crowned with confusingly extensive tail-band, but all other characters remain valid and obvious. – Although widely allopatric and rare and local, Basalt Wheatear could theoretically cause severe identification problems due to virtually identical ♂♂ plumages and only slightly smaller size. (♀♀ are less of a problem since in Black they are usually distinctly browner and duskier grey-brown looking.) Check primary projection first, Black having roughly 2/3 of exposed tertial length (Basalt: primary projection about equal to tertial length). Then note width of black tail-band, which averages slightly narrower in Black (though birds in West Sahara and partly S & C Morocco have on average broader tail-band, just as Basalt Wheatear, but these are the least likely to ever occur together with Basalt). These two characters combine to produce a third: the distance between wingtip and upper edge of the tail-band on perched birds is greater in Black than in Basalt; in the former the gap between the two is at least twice the width of the black tail-band (Basalt: wingtip reaches tail-band, or almost so, gap being less than width of tail-band). Their wing patterns are very similar, but on average Black has more even and subdued pale inner webs (dull silver-white, no sharp border), whereas Basalt has more sharply defined whitish edges to inner webs (rather obvious in ad ♂, limited and hardly noticeable in young ♀). Thus, although prominence of pale bases to remiges varies with sex and age, even in least marked Basalt there is still to some degree a characteristic pale 'wing flash' effect in flight, enhanced by dark trailing edge, with typical contrast between pale translucent inner webs and dark shafts—mostly visible from below. Note also that first-year Basalt (before late autumn), with its retained juv primary-coverts, often still shows white tips/spots to these feathers (also often on alula and tertials), lacking in Black Wheatear. In the hand, note that Black Wheatear has longer outermost primary (p1), this being considerably longer than tips of primary-coverts (Basalt: equal to tips of primary-coverts or only insignificantly longer).

AGEING & SEXING Ageing requires close inspection of moult pattern and feather wear in wing, but can be quite difficult in ♂♂ due to black colour and similarity between ad and juv feathers. Sexes differ in plumage (and see Biometrics), whereas seasonal variation is rather limited. – Moults. Complete post-nuptial and partial post-juv moult in summer (mostly Jun–Sep; may start mid Jul and finish early Oct). Post-juv moult includes all head, body, lesser and median

O. l. leucura, 1stS ♂, Spain, Feb: a large, bulky (strong-billed and strong-legged) and mostly black wheatear; note browner (juv) remiges and primary-coverts. (C. N. G. Bocos)

O. l. leucura, ♀, presumed ad, Spain, Feb: note long bill, strong feet and bulky shape. ♀ usually readily distinguished by browner plumage. Remiges, primary-coverts and rest of coverts rather evenly fresh suggesting an ad (ageing very difficult as ad and juv feathers similar). (C. N. G. Bocos)

WHEATEARS

O. l. riggenbachi, ♂, presumed 1stY, Morocco, Feb: slightly less glossy black with browner remiges and coverts suggesting a young ♂. Tail-band a little broader than *leucura*. (D. Occhiato)

O. l. riggenbachi, juv, Morocco, Apr: note soft, fluffy body-feathers with some paler grey-brown fringes, and prominent yellowish mouth flanges remaining on this recently fledged young. (R. Armada)

O. l. riggenbachi, ♀, presumed ad, Tunisia, Apr: sooty-brown overall, rather than near black of ♂, but not all ♀♀ are as obviously different as this. Typical clear-cut black T-shaped tail pattern, while paler grey-fringed remiges are also typical. All wing-feathers seemingly ad. (D. Occhiato)

coverts, three to all greater coverts, less frequently some or all tertials and tail-feathers, as well as smallest or occasionally entire alula. Pre-nuptial moult absent. – **SPRING Ad** Dark and fairly well-kept primaries, dark primary-coverts. Sexes differ in that ♂♂ are completely black, whereas ♀♀ are dark grey (can appear almost blackish) or sooty-brown with some rufous tones and varying amount of diffuse streaking or blotching below. **1stS** Aged by more worn and browner juv primaries, primary-coverts and perhaps some outer greater coverts. Also, ♂ tends to be browner, less pure black. Still, some are difficult to age reliably due to individual variation. – **AUTUMN Ad** In fresh ♂, black plumage on head and back and wings slightly glossed, at least until late winter. Face and underparts in ♀ profusely tipped/fringed rufous-brown when fresh (wear to mostly dull brown). Both sexes generally differ from 1stW in being evenly fresh without discernible moult limits in wings, though beware of some variation in darkness, odd ♂♂ having slightly browner primaries and coverts, which can appear as 1stW. Both tail-feathers and primary-coverts average more broad and round-tipped than in young. **1stW** Similar to ad but ♂♂ less pure black and less glossed, and young ♀♀ have duller brown cast in general. Best aged by moult limits, but difficult in the field (and even in the hand!): primary tips and primary-coverts slightly less fresh, weaker textured and paler/browner than in ad, and primary-coverts are subtly pointed (still rounded!) with broader but faint brownish-buff fringes and tips. Also, compared to ad-like post-juv feathers, juv greater coverts and alula (if retained) slightly more worn, browner and have broader and paler buff fringes and tips (pale fringes and tips subsequently largely wear off). **Juv** Soft, fluffy body-feathering. ♂ has black parts slightly duller than in subsequent plumages; cheeks and chin to breast have faint, paler grey-brown fringes to feather tips. ♀ even browner, with sandy- or rufous-brown tinge to underparts, and lesser and median coverts have rufous-buff fringes.

BIOMETRICS (*leucura*) **L** 17–19 cm; **W** ♂ 96.5–102 mm (n 16, m 98.9), ♀ 93–98 mm (n 12, m 95.5); **T** ♂ 62–70 mm (n 16, m 66.3), ♀ 60–69 mm (n 12, m 65.3); **T/W** m 67.6; **B** 19.0–23.3 mm (n 28, m 21.1); **BD** 4.4–5.5 mm (n 28, m 5.0); **Ts** 26.3–29.5 mm (n 28, m 27.5). **Black tail-band:** on **r6** (measured parallel with shaft; disregard inner 5 mm of inner web) **max.** 7–18 mm (n 23, m 11.7), **min.** 6–13.5 mm (n 23, m 9.8); on **r4 max.** 8–16 mm (n 29, m 12.2), **min.** 0–13 mm (n 23, m 8.2). **Wing formula: p1** > pc 4.5–11 mm, < **p2** 37–41 mm; **p2** < wt 6.5–9 mm, =6/7 or 7; **pp3–5** longest (p5 sometimes < wt 1 mm); **p6** < wt 2–4 mm; **p7** < wt 6–11 mm; **p10** < wt 15–23 mm; **s1** < wt 18–24 mm. Emarg. pp3–6.

GEOGRAPHICAL VARIATION & RANGE Slight variation with two moderately differentiated races. Given much overlap and individual variation, only fresh, preferably adult birds separable and most are probably indistinguishable in the field. See also Taxonomic notes as to valid names.

O. l. leucura (J. F. Gmelin, 1789) (Iberia; mainly resident). Described above. See Biometrics. Distinguished mainly by its on average slightly narrower tail-band in both sexes, where white sometimes narrowly penetrates black on r4 (rarely rr3–5). Ground-colour of ad ♂ deep black (usually with a little gloss when fresh, more so on average than in following race, though never as bluish and metallic as *O. leucopyga*); claimed to be darker than *riggenbachi* but difference very slight and often impossible to discern. 1stY ♂ less pure black than ad, duller, thus general tone somewhat approaches that of ad ♂ *riggenbachi*. ♀ is saturated deep brown or dark chocolate-brown with pronounced rusty tinge, being darker and more rufous than *riggenbachi*, with fringes especially to wing-coverts and underparts much less greyish sandy-tinged.

O. l. riggenbachi Hartert, 1909 (N Africa in Morocco east to NW Libya, south to Western Sahara; resident). Separated from *leucura* in ♂ plumage by combination of subtly duller, less glossy black feathering and on average slightly broader and more solid black tail-band. The width of the tail-band seems clinal, increasing subtly from north-east to south-west, but differences small, and much individual variation obscures this pattern. ♀ distinctive, being paler overall, less dark chocolate-brown with pronounced rusty tinge as in *leucura*, and many have distinctly sandy-tinged fringes especially to wing-coverts and underparts (latter also generally more streaked or blotched). **W** ♂ (once 92.5) 95–103 mm (n 58, m 98.6), ♀ 90–98 mm (n 20, m 94.3); **T** ♂ 64–73.5 mm (n 58, m 67.2), ♀ 59–67 mm (n 20, m 63.8); **T/W** m 68.0; **B** 19.5–23.6 mm (n 90, m 21.2); **BD** 4.7–5.5 mm (n 74, m 5.1); **Ts** 24.1–28.9 mm (n 95, m 26.7). **Black tail-band:** on **r6** (measured parallel with shaft; disregard inner 5 mm of inner web) **max.** 9–23 mm (n 72, m 14.5), **min.** 4.5–16.5 mm (n 59, m 11.7); on **r4 max.** 8–20 mm (n 96, m 14.5), **min.** 4–15.5 mm (n 59, m 10.7). (Formerly *syenitica*; see Taxonomic notes.)

TAXONOMIC NOTES It has been shown (Shirihai et al. 2014) that the holotype of *syenitica* Heuglin, 1869, a name previously attached to the N African population of Black Wheatear, collected in June at El Kab, 'Nubia', N Sudan (or possibly S Egypt), and held at NMW, is probably an overlooked taxon related to Basalt Wheatear *O. warriae* (Shirihai et al. 2011), and thus belongs to the Mourning Wheatear *O. lugens* complex. Vaurie (1959) synonymised *riggenbachi* Hartert, 1909, described from four specimens (AMNH) from Rio de Oro, Western Sahara, and claimed to have broader tail-band, with *syenitica* from rest of N Africa. We have examined these and agree; with such a small sample, only one of which is adult, it is impossible to judge how much individual variation is causing the slight differences noted. Until more material is available *riggenbachi* is best synonymised with birds elsewhere in N Africa. With *syenitica* not applicable to this population, *riggenbachi* becomes the oldest available name for these. – Historically, the range of Black Wheatear was perhaps once more extensive, with old records from Italy, Bulgaria and Egypt, but apparently most such claims lack documentation.

REFERENCES Shirihai, H., Kirwan, G. M. & Helbig, A. J. (2011) *BBOC*, 131: 270–291. – Shirihai, H. et al. (2014) *Zootaxa*, 3785: 1–24.

O. l. riggenbachi, Morocco, Apr: juv (right) with an adult. Note tail patterns and some grey on inner webs to remiges, still wings rather all dark in this species. (R. Armada)

BLACKSTART
Oenanthe melanura (Temminck, 1824)

Fr. – Traquet à queue noire; Ger. – Schwarzschwanz
Sp. – Colinegro común; Swe. – Svartstjärt

In many desert areas Blackstart is among the few birds likely to be encountered, often sharing its habitat only with Sand Partridge *Ammoperdix heyi*, Little Green Bee-eater *Merops orientalis*, Desert Lark, Rock Martin, White-crowned Wheatear and Scrub Warbler. But Blackstart is often the most easily seen of these, as it frequently assumes prominent lookouts. Resident in rocky desert with acacia and other scrub from Mali to the Horn of Africa, and from Sinai to C Arabia and Yemen, as well as SW Oman. The three rather distinct geographical populations may even represent separate species.

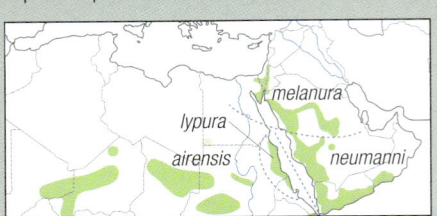

O. m. melanura, Israel, Mar: habitually hops on ground and perches on rocks or bare stumps, continuously flicking or fanning its tail, or occasionally spreading wings. (P. Alberti)

O. m. melanura, Israel, Mar: essentially drab ash-grey, with conspicuous black tail. Unless clear moult limits are visible in wing, ageing usually very difficult or impossible (except when handled). (H. Shirihai)

IDENTIFICATION A slim, inquisitive, desert-loving chat, with head and upperparts *drab ash-grey*, a slightly long-legged appearance and fairly small bill. Most conspicuous is the *black tail, which characteristically is flicked and fanned*. May show a faint cream-coloured supercilium (mainly in front of eye), darker loral stripe and variable whitish eye-ring and brownish ear-coverts, but head otherwise rather uniform. Wings, when fresh, have dark-centred tertials (with ill-defined but obvious pale grey fringes), a *paler secondary panel* and *blackish alula*. Underparts mostly grey-white with variable cream-buff tone to throat and belly, and darker grey flanks. At all ages, eye, bill and legs are black. Flight rather buoyant, but usually only over short distances. Hops frequently on ground, constantly flicking and fanning its black tail (as if signalling), at times accompanied by a flick or half-spread of the drooped wings.

VOCALISATIONS Song has slightly melancholic ring and comprises brief phrases of 2–3 well-spaced (4–5 sec.) deep, clear fluting syllables, the whole song lasting *c.* 30 seconds, *chürre-lu...... trü-troo...... chürlü...... chür-chur...*, with accent on first part of each and often admixed with short, scratchy warbles; song varies slightly, individually and regionally, and is usually uttered from prominent vantage, e.g. atop an acacia. – Commonest call a brief, soft, fluty outburst (a phrase from the song), e.g. *cher-u*, *chairee-churee* or *cheerarie*; also a crackly alarm, *bshreh*, *tschirr* or *tzeetch*, or a high-pitched metallic *vih* when anxious. – Vocalisations probably differ to some extent between main forms, but few data. W African *ultima* (which is closest to *airensis*) reported to have quite pleasant song, comprising liquid notes and some scratchy sounds typical of wheatear or like a Corn Bunting with more piercing, higher-pitched components. In S Arabia *neumanni* has more trilling song than *melanura*, with emphasis on first and last notes, *che-lulu-we*.

SIMILAR SPECIES Although it is normally true that no other W Palearctic passerine is essentially evenly grey with a contrasting black rump and tail, making the Blackstart unmistakable, it may be useful to keep in mind that ♀ White-throated Robin can very rarely lack any visible orange on sides and is then greyish and white with a blackish-looking tail. However structure and habits should soon separate them, and the shorter bill and neck, and more rounded head of Blackstart renders it a 'kinder' look. Also, the throat is more concolorous with rest of head, lacking the contrasting whitish bib of ♀ White-throated Robin.

AGEING & SEXING (*melanura*) Ageing requires close inspection of moult pattern and feather wear in wing. Sexes similar in plumage (see Biometrics). – Moults. In most of the covered range undergoes complete post-nuptial and partial post-juv moults, mostly Jun–Sep, but Jul–Nov in south, especially *neumanni* and *airensis*. Post-juv renewal usually includes head, body, lesser, median and variable number of inner greater coverts and possibly a few tertials. – **SPRING Ad** Aged by having only ad wing- and tail-feathers. ♀ tends to be rather browner above in direct comparison with ♂, but both sexes slightly brown-grey when heavily worn. **1stS** Differs from ad by having some retained juv outer greater coverts and tertials (slightly more worn, tinged browner with sharper buffish-grey fringes), forming slight contrast with greyer secondary-coverts renewed in post-juv moult. All primaries, primary-coverts and tail-feathers more worn and bleached brownish-grey, but moult limits less clear than in some other wheatears. – **AUTUMN Ad** Aged by being evenly feathered, lacking moult contrast. **1stW** Aged by juv primaries and primary-coverts (weaker textured, browner and less fresh than ad); primaries also lack (or have only slight) diffuse silvery-whitish tips of ad, and primary-coverts have narrower and sharper, less pure grey fringes (greyer, broader, more diffuse and fresher in ad). Often shows moult limits with juv tertials, larger alula-feather and outer greater coverts (slightly browner centres and narrower, sharper and buffier fringes) but, again, not always easily detected and some may moult all or most, or conversely none or few greater coverts and tertials, and are therefore more difficult to age. At least in early autumn iris slightly duller, being less blackish-brown than ad (requires perfect light). **Juv** Much like ad but generally browner, with pale throat and paler fringes to wing-coverts and tertials, resulting in less uniform appearance. Gape usually yellowish.

BIOMETRICS (*melanura*) Sexual size difference seems tiny or non-existent, so sexes lumped here. **L** 15–16 cm; **W** 76–86 mm (*n* 26, *m* 81.0); **T** 56–65 mm (*n* 26, *m* 60.9); **T/W** *m* 75.2; **B** 15.3–17.3 mm (*n* 25, *m* 16.6); **BD** 3.5–

O. m. melanura, Israel, Nov: apart from the jet black tail a rather feature-less greyish bird. Evenly fresh wing suggests ad. (A. Ben Dov)

Presumed *O. m. lypura*, Awash, N Ethiopia, Jan: smaller and short-tailed race with reduced grey hue, upperparts being pale greyish-brown, underparts cream-coloured, while ear-coverts nearly always are warmer rufous-tinged. Variation of Blackstart in NE Africa needs further work; still, birds in Eritrea and N Ethiopia are best regarded as race *lypura*. Ageing difficult or impossible, as usual with Blackstarts if no moult limits are visible. (C. N. G. Bocos)

O. m. neumanni, Oman, Nov: again, note slightly darker plumage and proportionately shorter tail in this race. Primary-coverts appear ad-like. (M. Schäf)

O. m. neumanni, 1stW, Oman, Nov: distinctly darker and subtly more bluish-tinged than *melanura* with proportionately shorter tail. 1stW by moult limits between new greater coverts and tertials, and relatively worn juv primaries and primary-coverts. (D. Occhiato)

4.3 mm (*n* 24, *m* 3.8); **Ts** 20.9–24.6 mm (*n* 26, *m* 23.3). **Wing formula: p1** > pc 4–10 mm, < p2 32–39 mm; **p2** < wt 5–8 mm, =6/7 (83%), or =7 (17%); **pp3–5** about equal and longest (p5 sometimes < wt 1 mm); **p6** < wt 2–5 mm; **p7** < wt 6–9.5 mm; **p8** < wt 9–13 mm; **p10** < wt 13–16 mm; **s1** < wt 15–17.5 mm. Emarg. pp3–6.

GEOGRAPHICAL VARIATION & RANGE Rather slight and clinal variation, mostly concerning body coloration, with three races described in the covered region and a fourth found immediately south of the treated range. Further races have been described elsewhere in Africa. All populations thought to be resident.

O. m. melanura (Temminck, 1824) (S Israel, W Jordan and Sinai to Saudi Arabia south to about Riyadh and Medina). Described above. Generally the greyest, with upperparts drab ash-grey (when fresh has very slight creamy or sandy hue), and pale ash-grey (with slightly buffish tinge) on throat and breast, merging with whiter belly; body-sides and vent have rather restricted buffish-grey suffusion.

O. m. neumanni (Ripley, 1913) (S Saudi Arabia to S Oman). Differs from *melanura* in being distinctly darker and plainer lead-grey, and slightly more compact (shorter total length), but wing and tail lengths are similar. Intergrades with *melanura* over much of Arabia, some being intermediates, although Bundy (1986) described the two forms as being well segregated ecologically. **L** 14–15 cm; **W** 75–86 mm (*n* 26, *m* 81.2); **T** 57–65 mm (*n* 26, *m* 60.9); **T/W** *m* 75.1; **B** 14.3–17.3 mm (*n* 26, *m* 16.4); **Ts** 21.6–24.8 mm (*n* 25, *m* 23.3). **Wing formula: p2** < wt 6–11 mm, =6/7 or 7 (70%), or =7/8 (30%); **p10** < wt 14–18 mm; **s1** < wt 14–19 mm. (Syn. *erlangeri*.)

O. m. lypura (Hemprich & Ehrenberg, 1833) (SE Egypt to N Eritrea). Slightly smaller but distinctly less greyish than *neumanni*, with pale greyish-brown upperparts, occasionally with purer grey small upperwing-coverts, and is almost uniformly pale sandy-cream below with very subdued greyish tinge. When fresh, underparts tinged pale pinkish-cinnamon, slightly warmer on sides. **L** 13–14 cm; **W** 73–82 mm (*n* 24, *m* 77.7); **T** 51–62 mm (*n* 24, *m* 56.8); **T/W** *m* 73.1; **B** 15.1–17.7 mm (*n* 23, *m* 16.3); **Ts** 20.5–23.6 mm (*n* 21, *m* 22.0). **Wing formula: p10** < wt 11–18 mm; **s1** < wt 11.5–19 mm.

Extralimital: *O. m. airensis* (Hartert, 1921) (Aïr Mts, N Niger, east to Darfur in Sudan). Relatively small and close to *lypura* but has head and upperparts drab olive-brown, any grey hue being strongly reduced. Pale sandy-brown or buff breast enhances greyish-cream throat and whitish-cream belly centre, and is often slightly warmer on vent and sides. **W** ♂ 77–84 mm (*n* 7, *m* 80.1), ♀ 75–77 mm (*n* 7, *m* 76.1); **T** ♂ 52–55 mm (*n* 7, *m* 53.0), ♀ 45–52 mm (*n* 7, *m* 49.0); **B** 15.9–17.9 mm (*n* 14, *m* 16.8).

TAXONOMIC NOTES Relationships among the various forms have received only limited attention, and potentially more than one species is involved. It is possible that *melanura* and *neumanni* represent two rather distinct races of a separate species in the Arabian Peninsula (the former widespread in arid deserts, the latter in moister well-vegetated habitats of the SW); *airensis* could be another distinct species, possibly including *ultima* (E Mali to Niger), while the races *lypura* and *aussae* (E Ethiopia to N Somalia) are perhaps closest to the *melanura* group, but could also represent a third species—they appear intermediate between the *airensis* and *melanura* groups, and on present knowledge do not appear to naturally belong with either. Further study of DNA and vocalisations of the different groups is clearly required. – Recently two genetically-based phylogenetic studies (Outlaw *et al.* 2010, Sangster *et al.* 2010) showed that the Blackstart, traditionally placed in the genus *Cercomela*, is nested within the genus *Oenanthe*. The simplest way of avoiding a non-monophyletic *Oenanthe* is, as suggested by Sangster *et al.* (2010) and Aliabadian *et al.* (2012), to include *Cercomela* within it (rather than splitting *Oenanthe* into three or more smaller genera), and this recommendation is followed here.

REFERENCES ALIABADIAN, M. *et al.* (2012) *Mol. Phyl. & Evol.*, 65: 35–45. – OUTLAW, R. K., VOELKER, G. & BOWIE, R. C. K. (2010) *Mol. Phyl. & Evol.*, 55: 284–292. – SANGSTER, G. *et al.* (2010) *Mol. Phyl. & Evol.*, 57: 380–392.

O. m. airensis, Niger, Oct: compared to race *melanura* of Middle East, this is a slightly smaller and drabber olive-brown race with a sandy-buff cast, and thus is decidedly less grey. (U. Liedén)

WHITE'S THRUSH
Zoothera aurea (Holandre, 1825)

Alternative names: Mountain Ground Thrush, Northern Scaly Thrush

Fr. – Grive dorée; Ger. – Erddrossel
Sp. – Zorzal dorado de Siberia; Swe. – Guldtrast

A large exotic-looking thrush that breeds in the extreme north-east of the treated area in dense, damp taiga. Very secretive and shy, it 'freezes' at first sign of danger and relies on its well-camouflaged plumage. Spends much time on the ground, but seeks cover in trees when disturbed. There are two ways to find it, neither simple: be extremely lucky to flush a vagrant on a W European island or shore in autumn, or go to the Siberian taiga in early June and listen at very first light for a peculiar drawn-out, thin and straight, monosyllabic whistle, slowly repeated. It migrates to SE Asia for the winter.

Scotland, Oct: one of the most attractive thrushes, which on size and shape in the covered region must be separated from Mistle Thrush, but has bold black vermiculations above and below. Ageing without handling or close observation inadvisable, especially if moult limits cannot be detected in greater coverts. (H. Harrop)

IDENTIFICATION A quite *large and elongated* thrush, with long wings and a *long neck*, a *proportionately small head* but long, *strong bill*. The bird may appear rather nondescript darkish at a distance but proves to have beautifully-patterned plumage if seen close: upperparts *olive-brown and greyish*, underparts off-white, with breast, throat and lower flanks tinged buffish-olive, the *whole plumage* except wings and tail *covered with black scale-shaped marks* or coarse vermiculations. There is usually a diffuse but reasonably conspicuous *pale eye-ring* around the *large, dark eye*. The head-sides are fairly plain but there is a characteristic *dark stripe at side of throat* and *dark patch on rear ear-coverts*. Legs are strong and brownish-pink. The most striking feature is on the underwing, which has *broad white and black bands* over its length (much like in American *Catharus* thrushes). The upperwing has a hint of 'bands' too, but these are much darker and more diffuse. The tips of the otherwise largely buff-olive primary-coverts are dark and form a characteristic *dark patch on the closed wing near its edge*. When flying away after having been flushed, note *white outer corners on the tail*.

VOCALISATIONS The song is unique among species in the covered region, a monotonously repeated clear, straight whistle, slowly repeated in late evening, by night or in the early dawn, *weeee… weeee … weeee...*, with pauses between each note of 4–5 sec. Sometimes a whistling note is replaced by a very thin, high-pitched *ziiiih*, inserted among normal notes without breaking the rhythm. Song carries far, often 1 km or more, yet does not appear very loud even at close range. The singing bird is extremely difficult to spot before it is flushed. – Very silent apart from song, and calls rarely heard. May utter a subdued straight whistle very much like a note from the song, not audible far. In anxiety gives a fine, sharp, drawn-out piping *ziiih* (Jahn 1942), like the alternate note in the song, and like many other thrush species. Said to have a snoring or churring alarm call near nest (*BWP*; Clement & Hathway 2000). Apparently lacks 'chucking' or clicking calls of most thrushes.

SIMILAR SPECIES Size, shape and general coloration give it a general resemblance to *Mistle Thrush*, but latter has extensively white underwing, is spotted black below rather than vermiculated, has near-uniform upperparts (lacking black vermiculations) and frequently gives its dry rolling flight call. Juvenile Mistle is streaked buff above and may be more of a confusion risk, but even a juvenile White's has a bolder pattern above of blackish and buff blotches. Note also that White's has strongly patterned flight-feathers in all plumages (including juv), notably with secondaries being dark-based but pale buffish on their outer part, and rather well-marked darker tips. The same pattern is mirrored on the primaries, only is less clear-cut. – In flight could be confused with ♀ or immature *Siberian Thrush*, both have dark-and-white banded underwing, but latter is smaller with a plain, dark brown back and rump, and coarsely marked vent and undertail-coverts. (White's has centre of belly, vent and central undertail-coverts unmarked white.) – Problems of another magnitude occur in SE Asia, where several closely related and similar taxa must be eliminated (outside the scope of this book). Luckily, none is migratory or likely to straggle to our region. See Clement & Hathway (2000) and Rasmussen & Anderton (2012). For handled vagrants always check number of tail-feathers, which in White's Thrush should be 14. All other taxa within the White's–Scaly Thrush complex have 12 tail-feathers with one exception: Scaly Thrush ssp. *horsfieldi* (Indonesia) has also 14 tail-feathers, but this is a resident form, which will hardly be found outside its range.

AGEING & SEXING (*aurea*) Ages differ very slightly in autumn, rarely also in spring, but differences small and require handling or close observation, and even then some are indeterminate. Sexes inseparable by plumage, and differ only slightly in size. – Moults. Complete post-nuptial moult of

1stW, Denmark, Jan: large bulky body, small head and very long bill. Often a dark patch on ear-coverts. Apparent moult limits and seemingly asymmetrical moult in greater coverts indicate 1stY. (K. Aaen)

— 362 —

England, Oct: in flight, diagnostic broad white and black bands over entire underwing, while upperwing also has characteristic banding, with darker primary-coverts and white outer corners to tail. (C. Galvin)

Russia, June: here seen in typical posture singing its peculiar whistling song, drawn-out straight notes with long pauses, from the top of a spruce in the Siberian taiga (V. & S. Ashby)

1stW, Scotland, Sep: large dark eye with thin pale eye-ring. Dark patch on ear-coverts here fairly obvious, and shared only by some Mistle Thrushes. Moult limit among greater coverts, with renewed ad-like and longer innermost covert. (S. Arlow)

ad in late summer (late Jul–mid Sep). Partial post-juv moult in summer (Jul–Aug) does not involve flight-feathers, tertials or primary-coverts. Often no greater coverts are moulted either, but some birds replace a few inner. Apparently no pre-nuptial moult in winter. – **ALL YEAR Ad** All greater coverts uniformly patterned and of similar length. Many have several central greater coverts buff-tipped in the shape of an even margin or rounded patch (rather than having tips wedge-shaped), but some have wedge-shaped tips to most greater coverts. All greater coverts have black centres and buff tips separated by stripe of olive-brown. Tail-feathers overall rather pointed but still fairly broad, tips rounded (finely pointed tip only at shaft). Outer tail-feathers often rather dark, white tips with clear-cut borders. **1stY** Very similar to ad, sometimes inseparable, although a contrast might be detectable between a few inner greater coverts being moulted, slightly longer and paler olive-brown, with pattern described under ad, and outer juv greater coverts that are shorter, slightly darker and more rufous-tinged (but frequently no juv greater coverts moulted). Juv greater coverts have no olive-brown stripe between black centre and buff tip. Tail-feathers on average somewhat narrower, tips more pointed. Outer tail-feathers not quite as dark, and white tips often diffusely bordered. **Juv** Differs from 1stW in being both streaked and boldly spotted (black below, buff and blackish above) rather than neatly vermiculated.

BIOMETRICS L 27–30 cm; **W** ♂ 157–178 mm (n 36, m 164.3), ♀ 155–173 mm (n 28, m 160.9); **T** 94–111 mm (n 42, m 103.1); **T/W** m 65.4; **B** ♂ 28.5–32.6 mm (n 36, m 30.2), ♀ 27.0–31.5 mm (n 28, m 29.5); **Ts** 31.0–36.6 mm (n 41, m 34.7). **Wing formula: p1** < pc 5–12 mm, < p2 87–95 mm; **p2** < wt 3–9 mm, =5 or 4/5 (=5/6); **pp3–4** about equal and longest; **p5** < wt 6–14 mm; **p6** < wt 19–29 mm; **p10** < wt 47–57 mm; **s1** < wt 53–62 mm. Emarg. pp3–4, often faintly also on p5. Note: 14 tail-feathers instead of the usual 12 of most Palearctic passerines.

GEOGRAPHICAL VARIATION & RANGE Monotypic. – Extreme E Russia on slopes of SW Urals, much of Siberia, Kazakh Altai, N Mongolia; winters SE Asia). The largest form in the complex. Ground colour somewhat paler than in smaller relatives in SE Asia due to more golden-buff than rufous patterning. Legs are more pink-brown than yellow-grey. (Syn. *toratugumi*; '*varia*'.)

TAXONOMIC NOTE Monotypic, with the large taiga-living *aurea* treated here as a separate species from the many similar-looking and closely related taxa in SE Asia (mainly of Scaly Thrush *Z. dauma*). For a definite evaluation of the relationships within the entire White's Thrush complex, and the status of some allopatric or more distinct taxa, a combination of additional genetic and behavioural research is desirable. We agree with Rasmussen & Anderton (2005), who regard taiga-living *aurea* as a separate species, being larger and differing clearly in song and habits, and somewhat in morphology from the rest of the *dauma* complex.

REFERENCES Jahn, H. (1942) *J. f. Orn.*, 90: 163–164. – Khan, A. A. & Takashi, M. (2006) *Forktail*, 22: 117–119. – Klicka, J., Voelker, G. & Spellman, G. M. (2005) *Mol. Phyl. & Evol.*, 34: 486–500. – Voelker, G. & Klicka, J. (2008) *Mol. Phyl. & Evol.*, 49: 377–381.

SIBERIAN THRUSH
Geokichla sibirica (Pallas, 1776)

Fr. – Grive de Sibérie; Ger. – Schieferdrossel
Sp. – Zorzal siberiano; Swe. – Sibirisk trast

As its name implies, a real Siberian taiga species, occurring only rarely in Europe as a vagrant. Shy and secretive in its breeding forests, and not easy to get to grips with, at least not until the song has been learnt. When seen, the ♂ proves to be very neat and attractive. Migratory, spending the winter in SE Asia, from NE India and Vietnam to Indonesia. Spring migration occurs mainly in April and May, but the last birds do not reach their Siberian territories until mid June.

G. s. sibirica, ad ♂, E China, May: blacker-tinged face and crown to otherwise mostly dark slate-grey plumage and very striking white supercilium. Not visible here is white patch on centre of belly and bold white barring on ventral region. Aged by evenly-feathered wing that is edged uniformly dark blue-grey. (M. Parker)

IDENTIFICATION A medium-large, dark thrush, with characteristic underwing pattern similar to White's Thrush and the American *Catharus* species: *broad white and black bands* cover the length of the wing (black replaced by olive-grey in ♀), sometimes visible when the bird takes off. Adult ♂ is a beautifully *dark slate-grey* bird with a *blackish head*, striking *white supercilium* and some *white on central belly and undertail*. In good light the dark grey has a bluish tinge, especially on the rear upperparts. Outer tail-feathers tipped white. First-winter ♂ is similar but has variable amount of off-white or brownish-white on chin, throat, sides of head and breast. Furthermore, supercilium is often not pure white or as distinct, and wings are tinged olive-brown. The ♀ is *dark olive-brown* with *buffish supercilium* (frequently spotted dark, and can be broken-up or irregular), *ochraceous-buff breast and flanks, coarsely marked or 'scaled' olive-grey*, whereas belly and vent are white with some olive-grey blotches. Side of head coarsely patterned with diffuse dark stripes on ochraceous-buff ground, the *lores* being *dark* and other stripes running on the side of the throat and rear ear-coverts. Flight-feathers edged rufous-brown. Legs are pale brown.

VOCALISATIONS The song somewhat resembles that of Eyebrowed Thrush (and simpler variants of extralimital Pale Thrush *T. pallidus* of E Asia; not treated), one or two fluty notes immediately followed by a brief, subdued high-pitched twitter, *chewluh* (*sirr*)... *truih* (*zrizri*)... *churly* (*sirr*)... *cheely* (*sirr*)..., etc. Strophes are uttered at a moderately quick pace, with pauses of 1–2 sec. between each. Compared to Eyebrowed, the voice is more high-pitched and less mournful, fluty notes are fewer or shorter in each strophe (shortness often best character to note), the tempo is often slightly quicker, and the high-pitched twitter is rarely omitted (more frequently so in Eyebrowed). – On migration (often at night) a 'delicate vibrating *seep* or thin *tseee*' (Kolthoff 1932), apparently not unlike Redwing. Also credited with a shorter Song Thrush-like *zit* (Andrew et al. 1955). Various chuckling or squawking calls when flushed or in anxiety. Alarm at nest said to be rattling (Panov 1973).

G. s. sibirica, 1stS ♂, E China, May: as ad ♂, but lead-grey plumage less clean and solid, with pale elements on throat and head-sides; supercilium also less pure white and less well demarcated, while underparts often display pale shaft-streaks and some coarse brown-grey smudges. Brownish-grey juv remiges, primary-coverts and most outer greater coverts, contrast with renewed lead-grey feathers. (R. Schols)

SIMILAR SPECIES The adult ♂ is characteristic with its bluish-grey plumage. A first-winter ♂ seen at a distance in poor light could be confused with a *Dusky Thrush*, both having prominent pale supercilium, but note much greyer upperparts on Siberian and different underwing pattern (Dusky has orange-tinged wing-coverts and no black-and-white bands). – In flight, a ♀ or immature Siberian could perhaps be confused with *White's Thrush*, both having broadly dark-and-white banded underwing, but White's is a larger and more elongated bird with pale olive-brown and black-vermiculated upperparts, including rump, and unmarked white vent and central undertail-coverts (Siberian has plain upperparts, and vent coarsely marked black and white). – In Himalaya (outside range of this book), ♀ must be separated from quite similar ♀ *Pied Thrush* (*G. wardii*; not treated), which has paler, yellowish bill, all-white vent/undertail-coverts, and a longer and more prominent supercilium, reaching to sides of nape.

AGEING & SEXING (*sibirica*) Ages differ during 1stY in ♂, but in ♀ usually not. Sexes separable after post-juv moult. – Moults. Complete post-nuptial moult of ad in late summer (mid Jul–early Sep). Partial post-juv moult in summer (Jul–Aug) does not involve flight-feathers, tertials or primary-coverts. Often no greater coverts are moulted either, but some birds replace a few inner. Apparently no pre-nuptial moult in winter. – **ALL YEAR** **Ad** ♂ Largely dark bluish-grey, head darkest (near black), prominent supercilium pure white, chin, throat and breast bluish-grey. All flight-feathers and coverts uniformly dark grey edged blue-grey, greater coverts sometimes have small white tips. Narrow stripe on centre of belly white, undertail-coverts blue-grey, broadly tipped white. **1stY** ♂ Similar to ad ♂, but differs by varying amount of off-white, ochraceous-buff and olive-brown admixed with bluish-grey of throat and head-sides. Supercilium generally not pure white, and is usually less neatly outlined. Contrast between olive-tinged brownish-grey or duller lead-grey juv flight-feathers, primary-coverts and all or at least outer greater coverts, and moulted brighter bluish-grey scapulars, lesser and median coverts, and often a few inner greater coverts. Breast and flanks sometimes have white shaft-streaks and some coarse brown-grey barring. **Ad** ♀ Upper-parts olive-brown, or less commonly have a slight bluish-grey tinge on rump, back and scapulars, and an olive-grey sheen on mantle. (Rarely even more bluish-tinged, with olive-blue tinge also on flanks and edges of some flight-feathers.) Tail-feathers generally slightly broader, with more rounded tips, in

G. s. sibirica, 1stS ♀, E China, May: overall dark olive-brown above with buffish spotty supercilium, and dusky underparts coarsely barred olive-brown. Often difficult to separate from ad ♀, but has rather strongly contrasting dark tips to primary-coverts and large pale tips to (juv) greater coverts. (M. Parker)

G. s. sibirica, ♀, presumed 1stS, E China, May: unlike imm ♂ has variegated head pattern, overall brownish upperparts and heavily-barred underparts; obvious buff-tinged breast and apparent moult limits in greater coverts suggest 1stY ♀. (M. Parker)

ad than in 1stY, but ambiguous intermediates not uncommon. Outer webs of primary-coverts medium olive-brown, offering moderate contrast with rest of primary-coverts. Pale tips to median and greater coverts usually small, pale buff and somewhat diffusely outlined. – Unless a blue-tinged variety, or a 1stY with conspicuous ochraceous wedge-shaped tips to many secondary-coverts, very difficult to age, even in the hand. **1stY ♀** Very similar to many ad ♀, and often difficult to age. On average more tinged ochraceous on breast and throat-sides. Outer webs of primary-coverts pale rufous-buff, contrasting rather strongly with darkish tips. Pale tips to median and greater coverts on average larger, deeper ochraceous-buff and more distinctly bordered. Tail-feathers on average narrower and more acutely pointed, but some overlap. **Juv** Rather similar to 1stY ♀ but has narrow ochre or buff shaft-streaks above (largest on scapulars), including on lesser and median coverts.

BIOMETRICS (*sibirica*) **L** 20–21.5 cm; **W** ♂ 115–127 mm (*n* 18, *m* 121.6), ♀ 113–126 mm (*n* 21, *m* 118.7); **T** 75–89 mm (*n* 15, *m* 82.8); **T/W** *m* 68.4; **B** 20.0–24.8 mm (*n* 38, *m* 22.8); **Ts** 27.0–29.8 mm (*n* 13, *m* 28.6). **Wing formula: p1** minute, < pc 5–13 mm, < p2 64–75 mm; **p2** < wt 1–5.5 mm; **p3** longest; **p4** < wt 1.5–4 mm; **p5** < wt 10–13 mm; **p6** < wt 16.5–21 mm; **p10** < wt 31–40 mm; **s1** < wt 36–44 mm. Emarg. pp3–4.

GEOGRAPHICAL VARIATION & RANGE Only two subspecies, neither of which breeds in the covered region. Rather well differentiated in ad ♂ plumage, more similar in other plumages. Ssp. *sibirica* has straggled to W Europe. No known record of genuine vagrancy by *davisoni* to Europe.

Extralimital races: ***G. s. sibirica*** (Pallas, 1776) (Siberia W to Angara and W Baikal; winters SE Asia) is described above. Ad ♂ invariably has narrow white patch on centre of belly, and broad white tips to undertail-coverts (broadest tips on longest undertail-coverts usually 5 mm or much more).

G. s. sibirica, ♀, China, May: In flight, note the dark-and-white banded underwing, a pattern it shares with White's Thrush—a distant ♀ Siberian could be confused with that species. Nevertheless, note Siberian's smaller size and more compact shape, its more barred undertail-coverts (near white in White's), and also, if upperparts seen, these are plain (not vermiculated as in White's). (T. Townshend)

Flanks slightly paler blue-grey than breast and often appear patchy due to whitish feather-shafts and sometimes a few slightly darker tips.

G. s. davisoni (Hume, 1877) (Sakhalin, part of Japan, Kurils; winters SE China) differs in being darker in both sexes. Ad ♂ much blacker above, and white on belly, vent and undertail-coverts either lacking or more restricted (broadest tips on longest undertail-coverts usually not more than 2 mm). Flanks rather uniformly dark slate-grey. Only 1stY ♂ can cause some problems. **W** ♂ 115–127 mm (*n* 18, *m* 121.6), ♀ 113–126 mm (*n* 21, *m* 118.7); **T** 75–89 mm (*n* 15, *m* 82.8). (BWP.)

TAXONOMIC NOTE The Siberian Thrush was traditionally

G. s. sibirica, 1stW ♂, Scotland, Oct: some fresh young ♂♂ are very ♀-like in head and underpart-pattern, and also have grey above washed-out and less pure. Nevertheless, lead-grey tinge above usually still obvious (as here) to infer ♂, as does the lack of strong ochraceous-buff hue on face. Supercilium rather pure white supports sexing, too. Much of the wing still contrastingly retained juv, tinged brown-grey. (R. Somers Cocks)

placed together with White's Thrush in the genus *Zoothera*. However, it has been established genetically (Klicka *et al.* 2005, Voelker & Outlaw 2008) that *Zoothera* is polyphyletic and comprises two distinct clades, that containing Siberian Thrush widely separated from that with White's Thrush. The best solution seems to refer the clade containing Siberian Thrush to the available genus *Geokichla*.

REFERENCES Andrew, D. G., Nelder, J. A. & Hawkes, M. (1955) *BB*, 48: 21–25. – Klicka, J., Voelker, G. & Spellman (2005) *Mol. Phyl. & Evol.*, 34: 486–500. – Kolthoff, K. (1932) *Medd. Göteb. mus. zool. avd.*, 59 (5, Ser. B, 3[1]). – Panov, E. N. (1973) *Ptitsy yuzhnogo Primor'ya*. Novosibirsk. – Voelker, G. & Outlaw, R. K. (2008) *J. Evol. Biol.*, 21: 1779–1788.

HERMIT THRUSH
Catharus guttatus (Pallas, 1811)

Fr. – Grive solitaire; Ger. – Einsiedlerdrossel
Sp. – Zorzalito colirrufo; Swe. – Eremitskogstrast

The Hermit Thrush is a native of North America, where it breeds from Alaska and over much of Canada and the USA south to Baja California, Texas and Maryland, and winters in the S USA, Mexico and N Central America. It is a rare vagrant to the British Isles, mainly in autumn, but also once in spring, with records from at least three other European countries.

1stW, Ireland, Oct: the smallest North American thrush often shows some rufous tones, mainly on uppertail—an important feature compared to other *Catharus*. Also note narrow but quite distinct lateral throat-stripe and pale eye-ring. Aged by juv greater coverts and rectrices, the latter having pointed tips. (C. Batty)

IDENTIFICATION Slightly larger than Thrush Nightingale but marginally smaller than other *Catharus*, and *up to 20% smaller than any Palearctic thrush*. Crown to mantle and wings olive-brown, with faint tawny tone in some lights, and *rump to tail varyingly washed rufous-chestnut, contrasting with tertials and back*. Face often paler, with *buff-mottled lores, dull white or buff eye-ring*, brown-mottled cheeks, and *narrow but distinct lateral throat-stripe of black spots*. Breast and flanks washed buff to pale brown-grey, *former quite heavily (but sparsely) marked with bold and large black-brown spots*, while flanks are usually unspotted (at most a few faint olive-brown marks); vent and undertail-coverts off-white. Rufous fringes to primaries form contrasting panel, unlike most congeners. As in other *Catharus*, underwing patterned white-dark-white, distinctive in flight. Bill dark brown or horn-coloured with pale flesh base to lower mandible, large eye generally blackish and legs pale flesh. Frequents understorey of woodland but not shy and can be encountered on open grass near cover.

VOCALISATIONS The commonest call, and that most likely to be heard in Europe, is a low *chuck* or *chup*, softer than Blackbird, which can be given as a double note. Also a whining or miaowing, rising, 'querying' *zhweeeh*, to some ears not unlike a common call of Grey Catbird *Dumetella carolinensis*. In flight a clear, plaintive *peew*.

SIMILAR SPECIES The smallest of the North American thrushes, its general structure and upperparts coloration recall *Thrush Nightingale* or even *Rufous-tailed Robin*, but Hermit Thrush (like other *Catharus* thrushes) is larger and more compact with a shorter tail, heavier spotting on chest, while its habit of cocking and slowly lowering the tail, sometimes accompanied by lateral flicking of the wings, is also distinctive. These movements may be shared by other *Catharus*, but are never as emphatic as in Hermit. – Given prolonged views, readily distinguished from similar *Catharus* by virtue of contrast between chestnut rump to tail and rest of upperparts (unlike *Veery* and *Swainson's Thrush* in which upperparts are concolorous dull rufous or olive, respectively). However, in some races this pattern is less evident, and note that the other species may possess a hint of this character. In addition, unlike most congeners, Hermit has a combination of bolder markings on chest and often prominent eye-ring (still narrower than that of Swainson's). With less distinctive birds concentrate on size/shape and tail movements. For separation from *Grey-cheeked Thrush* see that species. Although not proven to have occurred in the covered region (but a potential vagrant), *Bicknell's Thrush* can be similar to Hermit Thrush in size and in tail colour but has less contrast between tail and back, and greyer underparts, lacks the more prominent pale eye-ring of Hermit, has an orange-yellow basal half to the lower mandible, darker legs and relatively shorter primary projection.

AGEING & SEXING Ageing difficult in the field and requires close inspection of moult, feather pattern or wear in wing. Sexes alike in plumage (but see Biometrics). – Moults. Complete post-nuptial and partial post-juv moult in late summer to early autumn (mostly Jul–Oct). Post-juv moult including all head, body and lesser coverts, some or all median and usually a few (up to 5) inner greater coverts. Pre-nuptial moult absent. – **SPRING Ad** Compared to 1stS has evenly feathered secondary-coverts (no moult limits) and less worn primaries, primary-coverts and tail-feathers, plumage still rather fresh. Remiges bleach duller, while crown and back wear to more dull olive, and black marks on chest become more exposed than in fresh plumage. **1stS** Ageing as in autumn (see below) but moult contrast in outer median coverts or inner greater coverts usually evident, and most still retain buffish tips to juv greater coverts (though sometimes worn off). Plumage usually less 'tidy' than in ad. – **AUTUMN Ad** Compared to spring, upperparts warmer buffish-brown, rufous of lower back to tail and outer remiges brighter. Evenly very fresh wing- and tail-feathers (without moult limits) with remiges, secondary-coverts and primary-coverts firmly textured and glossy greyish-black; also, lacks distinct buff tips to greater coverts of 1stW (though occasionally has faint pale spots when very fresh). **1stW** As fresh ad, but juv primary-coverts, remiges, tertials and tail-feathers slightly less fresh and weaker textured, and tail-feathers more tapered (tips on average narrower and more pointed); p1 rather broader and longer. Moult limits among greater and median coverts often obvious (with juv feathers looser, slightly worn and paler, fringed buffier, contrasting with recently moulted inner ad-like feathers), and has characteristic pale buff tips and shaft-streaks to juv median and greater coverts, albeit often heavily abraded by Oct; on greater coverts bold buff tips create spotty wing-bar (unlike ad, but in some populations less evident on juv feathers or, conversely, some fresh ad have very thin and indistinct pale tips). **Juv** Occurs on breeding grounds and unlikely to reach the treated region.

BIOMETRICS (*faxoni*) **L** 16–18 cm; **W** ♂ 93–100 mm (*n* 20, *m* 96.7), ♀ 89–94 mm (*n* 21, *m* 92.4); **T** ♂ 63–71 mm (*n* 20, *m* 67.0), ♀ 60–66 mm (*n* 21, *m* 63.6); **B** 16.7–18.7 mm (*n* 41, *m* 17.7). **Wing formula: p1** < pc 3 mm (up to 5 mm <; Pyle 1997), to 1 mm >; **p2** < wt 0.5–1 mm, =3/4 or 4 (or as short as =5/6; Pyle 1997); **p3** longest; **p4** < wt 1–2 mm; **p5** < wt 5–6 mm; **p6** < wt 11–13 mm. Emarg. pp3–5. (*BWP*.)

GEOGRAPHICAL VARIATION & RANGE Up to 13 races recognised, often arranged in three groups. One race, *faxoni* (of so-called 'northern group'), has been claimed among those birds which have reached W Palearctic.

C. g. faxoni (Bangs & Penard, 1921) (C Yukon, Mackenzie and NE British Columbia E to S Labrador and Nova Scotia; winters Central & South America). Characterised by warm or rich brown upperparts (with moderate russet tinge), flanks washed quite extensively greyish-brown. Bill base broad.

1stW, England, Oct: some are less distinctive, with the rufous uppertail obscured or less obvious. In such cases, focus on bolder breast markings and prominent eye-ring. Aged by buff tips and shaft-streaks to juv median and greater coverts (two innermost of latter have been moulted). (L. Gwynn)

Mexico, Apr: some show less bold and sparser dark spotting on breast. Rufous tail often visible from below. (H. Shirihai)

Presumed 1stS, USA, Apr: in spring wears duller above, head-sides often slightly greyer, with dark lateral throat-stripe more obvious; drabber upperparts form greater contrast with rufous panel in remiges. Hint of rufous tips to greater coverts and contrasting pattern in primary-coverts suggest 1stY. (C. Manville)

SWAINSON'S THRUSH
Catharus ustulatus (Nuttall, 1840)

Fr. – Grive de Swainson; Ger. – Zwergdrossel
Sp. – Zorzalito quemado; Swe. – Beigekindad skogstrast

Swainson's Thrush breeds in North America from Alaska, across N Canada and south to New Mexico and Maryland, wintering from Mexico south through Central and South America, and the West Indies. Different races appear to occupy segregated wintering areas. It is a vagrant to W Europe, mainly to Britain in autumn, but has occurred east to Finland and even in Ukraine.

Ireland, Oct: can for a moment resemble Song Thrush, but aside from being smaller, spots are restricted to breast and upper flanks (rather than covering virtually entire underparts). Ageing impossible from this image. (S. Cronin)

Presumed 1stW, Scotland, Oct: unlike similar Grey-cheeked Thrush has cold-toned upperparts and flanks. Especially note larger, teardrop-shaped and more regular but browner spots on warmer lower throat and breast. Buffy eye-ring bolder. (H. Harrop)

1stW, France, Oct: drab greyish-brown to olive-tinged upperparts, broad pale eye-ring, conspicuous pale buff loral stripe, buff-tinged, weakly mottled cheeks, and narrow dark lateral throat-stripe. Quite heavily spotted black-brown breast, but sides only faintly dappled, leaving belly to undertail-coverts white. Aged by (faint) moult limit among greater coverts (inner half post-juv, longer and greyer-edged). (V. Legrand)

IDENTIFICATION Arguably the 'typical' *Catharus* thrush, in general pattern closer to larger Song Thrush: upperparts, wings and tail drab greyish-brown or olive, though western birds usually appear more reddish-brown. *Face has conspicuous pale buff loral stripe and broad eye-ring*, brown-mottled and weakly buff-tinged cheeks, and *narrow, black-spotted lateral throat-stripe*. Throat pale buff, becoming *warmer on breast and greyish olive-brown on flanks*, with *quite heavily spotted black-brown breast but sides only faintly dappled*, leaving belly, vent and undertail-coverts white. Has typical *underwing pattern in flight shared by most Catharus thrushes with broad white-dark-white bands*. Bill dark horn-brown with basal lower mandible pale buffish flesh-coloured. Large eye usually blackish. Legs greyish-flesh. Frequents woody habitats, often forages on ground and usually solitary; tail movements rarely obvious.

1stS, Canada, Jun: wears rather duller and greyer above, with dark spotting below more exposed and heavier. However, species' characteristics of buffier and better marked eye-ring and loral stripe still evident. Ageing not easy, but subtle contrast between feather generations discernible between heavily-worn primaries and tertials versus renewed inner greater coverts. (O. Samwald)

VOCALISATIONS Often silent, apparently calling less frequently than many European thrushes. Commonest call an emphatic, sonorous *whit!* or *pwip!* Sometimes heard is a totally different high-pitched, nasal rather odd *qui-brrrrr* (or just *brrrr*). In flight a short, high-pitched squeaky *heep*.

SIMILAR SPECIES Small thrush, which for many European observers has general form and appearance of a nightingale (see Hermit Thrush). Readily differentiated from Song Thrush by being smaller with smaller spots that are restricted in area (reaching down breast and upper flanks, rather than across virtually the entire underparts), and in flight by its marked underwing pattern. – Main confusion risk is *Grey-cheeked Thrush*, but Swainson's has distinctly buffier and better-marked eye-ring and loral stripe (greyer or off-white and less well marked in Grey-cheeked, but in some confusingly better developed), and face and throat are more buff-tinged (Grey-cheeked: off-white, or pale buff at most). Swainson's Thrush may cock its tail, but Grey-cheeked never does. – Tail movements like *Hermit Thrush* (but cocks and lowers tail more slowly and far less often than that species), from which it usually can be distinguished by the lack of contrast between tail to rump and rest of the upperparts (although some geographical variation in both species may obscure these differences), as well as the buffier ground colour to the face and underparts, and usually bolder facial pattern with more distinct pale eye-ring and loral stripe. – Similar-sized *Veery*, especially the western race *salicicolus*, is close in tone to more russet-backed Swainson's, but Veery lacks strongly-spotted breast (and spotting rarely extends below upper breast), lacks an obvious eye-ring, has contrasting clean white belly and flanks, and is relatively longer-tailed.

AGEING & SEXING Ageing as in Hermit Thrush. Sexes indistinguishable on plumage. Seasonal variation limited. – Moults. Moult pattern and timing mostly as in Hermit Thrush, but with on average more secondary-coverts replaced in post-juv moult. – **SPRING Ad** With wear appears rather duller, but dark spotting below more exposed and heavier. Aged by lack of moult limits amongst secondary-coverts, and has less worn primaries, primary-coverts and tail-feathers; by and large plumage less affected by wear. **1stS** Aged as in autumn, but moult limits in wing often more contrasting. Buff wing-coverts spotting usually retained, though sometimes wears off, and otherwise averages more worn, especially tail-feathers. – **AUTUMN Ad** As spring but usually more russet on upperparts, even in more olive-brown populations. Best aged by evenly very fresh wing- and tail-feathers. No buff tips to greater coverts, at most tiny tips visible when very fresh. **1stW** Recalls ad, but retains juv feathers in wings and tail, with pale buff tips or shaft-streaks on greater coverts creating spotty wing-bar. In the hand, juv primary-coverts, remiges, outer greater coverts and tail-feathers are weaker textured and slightly less fresh; p1 rather broader and longer. Look also for moult limits among greater and median coverts (juv coverts are slightly looser, paler, worn and fringed buffier, and differ from darker and fresher, ad-like recently moulted inner ones). Buff tips to secondary-coverts sometimes weak, obscured or can wear off by late autumn. **Juv** Occurs only on breeding grounds and unlikely to reach W Palearctic.

BIOMETRICS (*swainsoni*) L 16.5–19.5 cm; W ♂ 94–107 mm (*n* 17, *m* 101.9), ♀ 93–102 mm (*n* 21, *m* 97.5); T ♂ 62–72 mm (*n* 14, *m* 66.5), ♀ 58–66 mm (*n* 18, *m* 62.4); B 15.2–18.6 mm (*n* 37, *m* 16.8). **Wing formula:** p1 < pc 0–5 mm (3–8 mm; Pyle 1997); p2 < wt 1.5–4 mm, =3/5; p3 (pp3–4) longest; p4 < wt 0–2 mm; p5 < wt 7–12 mm; p6 < wt 13–18 mm. Emarg. pp3–4. (BWP.)

GEOGRAPHICAL VARIATION & RANGE Four or six races generally recognised, of which one—*swainsoni*—is known to have reached W Palearctic. These races are usually considered to form two groups: olive-backed (including *swainsoni*) over much of range, and russet-backed, restricted to the Pacific littoral. In addition to the obvious differences in upperparts coloration, the russet-backed group tends to have narrower and less contrasting 'spectacles', and less obvious breast markings.

C. u. swainsoni (Tschudi, 1845) (C & E Canada, northernmost North America; winters Mexico and south to Argentina). Treated above.

Mexico, Apr: typical view of this essentially arboreal thrush, when characteristic buffish ground colour to face and underparts, bold throat pattern, and pale eye-ring and loral stripe are notable. (H. Shirihai)

GREY-CHEEKED THRUSH
Catharus minimus (Lafresnaye, 1848)

Alternative name: Gray-cheeked Thrush (American spelling)

Fr. – Grive à joues grises; Ger. – Grauwangendrossel Sp. – Zorzalito carigrís; Swe. – Gråkindad skogstrast

The only one of its genus to breed in Eurasia, in extreme NE Siberia, Grey-cheeked Thrush is also widespread from Alaska south to NW British Columbia and NW Ontario. It winters in N South America and on Trinidad, and vagrants have reached W Europe, where, as is typical of this genus of thrushes, most records are from Britain.

England, Nov: dull grey-olive upperparts, greyish face, mottled lores and cheeks paler than crown, and eye-ring generally narrow and partial, plus rather well-defined, black-spotted lateral throat-stripe. Breast quite heavily and densely spotted black-brown, but lower breast and flanks merely dappled dusky-brown. Ageing uncertain, though no moult limit among (buffish orange-tipped juv) greater coverts visible. (D. Stewart)

1stW, Shetland, Nov: slightly plainer greyish face with subdued light mottling, lacking any indication of dark loral line or a conspicuous white eye-ring. Note also rather whitish and clean fore-underparts, limited black streaking on throat-sides (hardly forming stripes) continuing into densely spotted breast, and grey-tinged flanks. Juv greater-coverts tipped buff, forming vague wing-bar. (R. Nason)

1stW, Ireland, Oct: note densely spotted breast (spots rather smaller, too) and paler and greyer cheeks versus Swainson's Thrush. From smaller Hermit Thrush by lack of tawny uppertail, and usually better-developed pale eye-ring (though hardly here). Some young retain most greater coverts and when still fresh the buff tips form wing-bar. (S. Piner)

Presumed 1stS, USA, Jun: in spring and summer generally wears greyer above, but vestiges of buff-tipped juv greater coverts can remain as apparently here. (R. Pop)

IDENTIFICATION The largest *Catharus*, with dull, faintly grey-olive upperparts, *greyish face, finely mottled off-white lores and cheeks* paler than crown, almost concolorous, *narrow and partially broken eye-ring* (can sometimes be slightly stronger and more complete), and a narrow, well-defined, black-spotted lateral throat-stripe. *Breast-sides and flanks buff- to ochre-grey, and breast quite heavily spotted black-brown*, but flanks merely dappled dusky brown, while throat, belly and undertail-coverts are white. Underwing pattern consists of broad white-dark-white bands (like other *Catharus*). Bill dark brown with base of lower mandible greyish-flesh, eye large and black, and legs dusky flesh. Frequents lightly wooded areas. Does not cock tail.

VOCALISATIONS Gives a low *what?* or *chuck* when wary. On migration (often at night) gives low-pitched, shrill piping, slightly drawn-out note with slight or marked drop in pitch at end, *wee-cher* or *chee-wer*, sometimes slightly recalling a Blyth's Pipit call but less buzzing.

SIMILAR SPECIES Principal confusion risk is with Swainson's Thrush (which see). General separation of *Catharus* thrushes from Song Thrush is also discussed under Swainson's. Beware that slight chestnut tinge on tail, especially in ssp. *minimus*, might cause confusion with several congeners. – From Hermit Thrush differs by slightly larger size, lacking that species' distinctive tail-cocking, stronger contrast between the chestnut tail and rest of upperparts, and details of face pattern. – For differences from Veery, see that species. – Should it ever prove to occur in the W Palearctic, Bicknell's Thrush would require extremely careful separation from the present species. Bicknell's is subtly smaller with shorter exposed primaries (part visible beyond tertial tips), and tends to have more chestnut-toned uppertail (contrasting subtly with the olive-brown rest of upperparts), a warmer and plainer face (i.e. less grey, head-sides more streaked/mottled with brown and olive tones), has more buff on the throat but greyer underparts with olive-grey flanks, more reddish (less grey-toned) primaries, an orange-yellow basal half to the lower mandible, and darker legs. Bicknell's has only a vague pale eye-ring. However, most of these characters (especially in young/autumn birds) are subject to individual variation and show some overlap.

AGEING & SEXING Ageing and moult strategy mostly as in Hermit Thrush. Sexes indistinguishable on plumage. Seasonal variation limited. – **SPRING Ad** Wears even greyer above and whiter below, with heavier dark spotting. Told from 1stS by lack of moult limits in wings, with primaries, primary-coverts and tail-feathers less worn; overall, plumage relatively fresher. **1stS** Moult limits in wing often more contrasting due to wear and bleaching. Buff wing-coverts spotting usually retained, but sometimes wears off. – **AUTUMN Ad** Has slightly more olive upperparts. Best aged by being evenly very fresh, without discernible moult limits in wing- and tail-feathers. No or only very small buff tips to greater coverts. **1stW** Resembles ad, but breast spotting more smudged and tends to merge into rows, central breast more likely to have pale yellow ground colour. Base to lower mandible pale flesh. Most reliably differs from ad by juv feathers in wing and tail (with moult limits among greater and median coverts), and by prominent spotty pale buff shaft-streaks forming wing-bar on juv greater coverts (tips sometimes worn off by late autumn); p1 more rounded, longer and broader. **Juv** Occurs only on breeding grounds and unlikely to reach W Palearctic.

BIOMETRICS (*minimus*) **L** 17–19 cm; **W** ♂ 101–111 mm (*n* 18, *m* 106.7), ♀ 98–107 mm (*n* 13, *m* 102.3); **T** ♂ 67–76 mm (*n* 17, *m* 71.7), ♀ 64–70 mm (*n* 13, *m* 67.4); **B** 16.9–18.7 mm (*n* 28, *m* 17.8). **Wing formula: p1** < pc 2–9 mm (3–12 mm; Pyle 1997); **p2** < wt 1.5–4 mm, =4 or 4/5; **p3** longest; **p4** < wt 1–2 mm; **p5** < wt 7–9 mm; **p6** < wt 14–18 mm. Emarg. pp3–4. (*BWP*.)

GEOGRAPHICAL VARIATION & RANGE Monotypic. – NE Russian Far East, Alaska, Canada; winters N South America. – We follow AOU in splitting Bicknell's Thrush *C. bicknelli* (Ridgway, 1882) as a separate species, although acknowledging that some doubt still exists about this. Under such an arrangement, Grey-cheeked Thrush consists of just two weakly differentiated subspecies, *minimus* and *aliciae* (Baird, 1858). Variation between them is entirely clinal, and some recent authorities have considered *aliciae* a synonym of *minimus*, a policy followed here.

REFERENCES Evans, W. R. (1994) *Wilson Bull.*, 106: 55–61.

VEERY
Catharus fuscescens (Stephens, 1817)

Fr. – Grive fauve; Ger. – Dickicht-Musendrossel
Sp. – Zorzalito rojizo; Swe. – Rostskogstrast

The Veery breeds in North America across Canada south to Tennessee and Maryland, and winters mainly in S & E Amazonian Brazil and N Bolivia, although it has recently been observed several times in Cuba at this season. Vagrants have reached W Europe in autumn, most frequently seen in Britain, with the odd record in Sweden. Often seen feeding on ground in woodland.

1stW, England, Oct: combination of uniformly warm tawny-brown upperparts and poorly-developed dark spotting below distinguishes Veery from other *Catharus* species. Juv greater coverts still very fresh, with buff tips forming wing-bar. (J. Atkinson)

1stW, England, Nov: note general contrast between bright tawny-brown upperparts and dusky-white underparts. All visible greater coverts are juv, with buff tips forming 'tatty' wing-bar. (D. Morrison)

1stW, Scotland, Oct: the brightest *Catharus* as to upperparts (even this rather dull individual), but least marked below, somewhat recalling a *Luscinia*. Greyish face with indistinct pale eye-ring. Dark markings on throat weakly developed, while breast is tinged yellow-buff, with dull olive- to tawny-brown spots but rest of underparts largely unmarked. Greater coverts juv. (H. Harrop)

VOCALISATIONS Commonest call a somewhat harsh, clearly downslurred, whistled *whree-uh* or *chee-op*. Variations include an almost even-pitched *cheep*, a more subdued *chuck* and low rattling sound in anxiety.

SIMILAR SPECIES Due to it having the most strongly rufescent upperparts of any *Catharus*, Veery is equally likely to be confused with either of the *Luscinia* nightingales, especially Thrush Nightingale given that both have indistinctly spotted breasts, as with any of its congeners. Given a reasonable view, however, Veery should be distinguished by its bulkier, rather neckless form, faintly indicated wing-bars (in first-winters) and well-marked lateral throat-stripe, as well as different structure and behaviour. – The combination of uniformly warm upperparts and lack of well-marked dark spotting below distinguish Veery from other *Catharus*

IDENTIFICATION The *brightest and least marked below* of the *Catharus* thrushes, with a slightly longer tail, and thus most closely recalls a *Luscinia* nightingale. *Upperparts warm tawny-brown* and tail slightly brighter, inclining to chestnut. Face pale grey-buff, with prominent dark eye, but has *indistinct pale eye-ring* (visible only at very close range). Off-white upper loral stripe (narrow and flecked), a pale buff area behind cheeks, narrow, dark brown-spotted lateral throat-stripe and *indistinct submoustachial stripe* are other head markings. Breast cream-buff, with *restricted* dull olive- to tawny-brown *spots, chiefly on upper breast*. Flanks *white suffused grey*, lightly dappled but appearing uniform at distance, and *rest of underparts pure white*. Underwing has broad white-dark-white bands (like other *Catharus*). Bill dark brown or horn-coloured with pale flesh-coloured base to lower mandible; legs pale pink-brown. Frequents woods and thickets, and habits most like those of Swainson's Thrush.

species, although Pacific races of Swainson's Thrush are rather similar (Veery has whiter underparts with contrasting grey flanks), while Grey-cheeked Thrush is olive-grey above and has a greyish face, strongly marked breast and browner flanks. – Hermit Thrush should also prove distinctive, having (contrasting) chestnut coloration restricted to the rump/tail, and being rather smaller with better-marked breast spots. – Wood Thrush (an extremely rare vagrant to the covered region) also has warm chestnut upperparts, but is a larger bird with a prominent white eye-ring and very bold and large black underparts spotting.

AGEING & SEXING Ageing, moult pattern and timing of moults mostly as in Hermit Thrush. Sexes alike in plumage, and seasonal variation limited. – **SPRING Ad** With wear upperparts become much duller, though still bright tawny in eastern races, and breast spotting blotchier, with grey of flanks and whitish eye-ring reduced. Lack of moult limits in wing-feathers and less abraded primaries, primary-coverts and tail-feathers best ageing clues. **1stS** Most retain pale wing-coverts spotting, but otherwise similar to ad. Moult limits in wing as in autumn but often more contrasting due to wear and bleaching of juv feathers. Plumage in general less fresh. – **AUTUMN Ad** Warmer and more russet tones at this season. Separable from 1stW by being evenly very fresh; no buff tips to greater coverts, or only very limited in some. **1stW** Resembles ad, but retains juv primary-coverts, remiges and outer greater coverts, with bright cinnamon-buff tips or shaft-streaks to latter (forming narrow wing-bar, but sometimes worn off by late autumn). Occasionally has buff shaft-streaks to retained juv outer median coverts. In the hand also check for moult limits among greater and median coverts (as in Swainson's Thrush); p1 on average more rounded, longer and broader than in ad. **Juv** Occurs only on breeding grounds and is unlikely to reach W Palearctic.

BIOMETRICS (*fuscescens*) **L** 17–19.5 cm; **W** ♂ 99–105 mm (*n* 19, *m* 102.4), ♀ 93–101 mm (*n* 13, *m* 96.9); **T** ♂ 66–73 mm (*n* 19, *m* 69.4), ♀ 62–69 mm (*n* 13, *m* 65.2); **B** 16.3–18.6 mm (*n* 30, *m* 17.4). **Wing formula: p1** < pc 2–9 mm; **p2** < wt 2–4 mm, =4 or 4/5; **p3** longest; **p4** < wt 0.5–2 mm; **p5** < wt 5–7 mm; **p6** < wt 10–14 mm. Emarg. pp3–5. (*BWP*; Pyle 1997.)

GEOGRAPHICAL VARIATION & RANGE Five races proposed but only four (*fuscescens*, *fuliginosus*, *salicicolus* and *subpallidus*) generally recognised, and these differ only rather slightly and almost entirely clinally, ranging from those in the east with tawny-rufous upperparts (*fuscescens* and *fuliginosus*) to those in the west (*salicicolus* and *subpallidus*) with duller, darker and more olive-grey upperparts, a slightly broader and darker lateral throat-stripe, and darker underparts spotting. No information on racial identification of most W Palearctic records, but presumably most are *fuscescens*.

C. f. fuscescens (Stephens, 1817) (SE Canada & NE USA; winters SC Mexico). Described above.

Ad, Canada, May: in spring/summer, upperparts wear duller but faint tawny hue remains; as in autumn, underparts spotting weak and restricted to upper breast, with buffy tinge. No moult limits visible, and entire wing appears evenly feathered. Primary-coverts have rounded tips, as have tail-feathers. (H. J. Lehto)

YEMEN THRUSH
Turdus menachensis Ogilvie-Grant, 1913

Fr. – Merle du Yémen; Ger. – Jemendrossel
Sp. – Zorzal yemeni; Swe. – Jementrast

Endemic to N Yemen and the Asir range of SW Saudi Arabia, this unobtrusive-looking thrush is a resident of montane woodland, near cultivation and in higher-elevation wadis, where it requires dense, lush and green thickets for cover. It is usually found above 1500 m, but sometimes slightly lower, and has been recorded at 2900 m. The total population is estimated to number no more than 5000 pairs, and the bird is considered Vulnerable due to ongoing habitat loss.

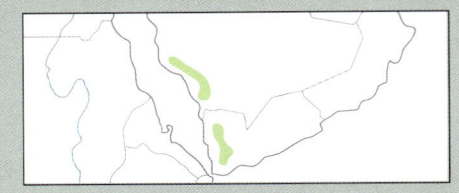

IDENTIFICATION A rather shy and skulking bird, fractionally smaller than a Blackbird, appearing *long-tailed*, the long tail being enhanced by the rounded wings with *short primary projection*. Plain olive-brown or grey-brown above, dusky buff-white or pale brown-grey below with *prominently dark-streaked throat*, the streaking becoming more diffuse on upper breast. *Lores, eye-surround and ear-coverts rather dark* brown. Flanks very slightly and diffusely tinged warmer brown or rufous, or has a few irregular patches of rufous. *Underwing bright orange*. Undertail-coverts diffusely blotched pale and dark grey. The only more remarkable feature of this plain-looking bird is its *strong, brownish-yellow bill*. A narrow and hardly noticeable orbital ring is reddish. Legs are yellowish or dull greyish faintly tinged yellowish-brown. Sexes very similar if not inseparable. First-winter has a darker bill than adult, buff-tipped greater coverts and sometimes some pale shaft-streaks on median coverts and scapulars.

VOCALISATIONS The song is variably described as vaguely recalling Blackbird or Golden Oriole (but much less pleasing, lacking the deep fluty tone), and Redwing, with a series of similar notes falling slightly in pitch, *chi-che-che-chu*. Like many other *Turdus* species, each strophe usually ends with some added squeaky, higher-pitched notes. – Calls include Blackbird-like *chuck*, singly or slowly repeated; a fast chattering series of notes in alarm; a thin, high-pitched alarm *ziih*; and a noisy, metallic or 'electric' *tchr-tchr-tchr-tchr-tchr-...*, which can perhaps recall Fieldfare (Bowden 1987; Magnus Ullman *in litt.*).

SIMILAR SPECIES Paler below than ♀ Blackbird, with stronger bill, which is all yellowish in adults. Also has shorter primary projection and a 'pale-and-dark pattern' on undertail-coverts.

AGEING & SEXING Ages differ during 1stW, and often also in 1stS if seen close. Sexes appears to be alike (although general size, and colour and size of bill might prove helpful; insufficient material examined). – Moults. Complete post-nuptial moult of ad in late summer or autumn. Partial post-juv moult in summer does not involve flight-feathers, tertials, primary-coverts or all or most greater coverts. Apparently no pre-nuptial moult in winter. – **ALL YEAR Ad** ♂ Upperwing uniform olive-brown, no pale tips or wedge-shaped marks on secondary-coverts or scapulars. Bill strong and all yellowish. **Ad** ♀ Like ad ♂ (and many cannot be separated) but on average appears to have a slightly weaker and duller yellowish bill. (Castell *et al.* 2001 suggested that ♂♂ are slightly darker overall and have brighter yellow bill.) **1stW** All or most greater coverts tipped buff-white. Sometimes a few pale shaft-streaks remain on unmoulted juv scapulars, lesser or median coverts. Bill largely brown. **Juv** Undescribed.

Ad, Yemen, Jan: dark greyish upperparts, buffy-brown to grey underparts, prominently dark-streaked throat, more diffuse on breast, and brownish-yellow bill and legs render species unmistakable. Aged by bright yellowish bill and legs, reddish orbital-ring, reddish sepia-brown iris, and evenly fresh wing-feathers. (H. & J. Eriksen)

Ad, Yemen, Jan: in some ad, dark streaking is restricted to throat. Note very short primary projection, making tail appear very long. Face rather plain. Bare-parts coloration and evenly-feathered wing identify this bird as ad. (H. & J. Eriksen)

1stW, Yemen, Oct: aged by juv greater coverts (outer few), which are shorter and tipped buff-white, and by duller and more reddish bill and legs and, if seen close, olive-tinged iris. Otherwise, plumage much like ad. (U. Ståhle)

Yemen, presumed 1stY, Jan: variation poorly known and it is unclear if such boldly-marked throat and breast reflects age or sex variation. However, partial wing-bar in greater coverts should be indicative of 1stY. (P. Bison)

BIOMETRICS (Only scant material available for examination.) **L** 25–27 cm; **W** ♂ 120–128 mm (n 7, m 124.3), ♀ 116–121 mm (n 3, m 118.0); **T** ♂ 105–117 mm (n 7, m 112), ♀ 105–109 mm (n 3, m 107.3); **T/W** m 90.4; **B** ♂ 26.0–28.2 mm (n 7, m 27.1), ♀ 25.5–27.3 mm (n 3, m 26.4); **Ts** 30.0–32.5 mm (n 10, m 31.4). **Wing formula: p1** > pc 1–5 mm, < p2 48–57 mm; **p2** = wt 14–16.5 mm, =9 or =8/9; **p3** < wt 2–4 mm; **pp4–5** equal and longest; **p6** < wt 1.5–3 mm; **p7** < wt 5–10 mm; **p8** < wt 12–16 mm; **p10** < wt 19–23 mm; **s1** < wt 22–27 mm. Emarg. pp3–6, in most less clearly also on p7.

GEOGRAPHICAL VARIATION & RANGE Monotypic. – N Yemen and adjacent SW Saudi Arabia; resident.

REFERENCES Bowden, C. G. R. (1987) *Sandgrouse*, 9: 87–89. – Castell, P. *et al.* (2001) *Sandgrouse*, 23: 49–58.

RING OUZEL
Turdus torquatus L., 1758

Fr. – Merle à plastron; Ger. – Ringdrossel
Sp. – Mirlo capiblanco; Swe. – Ringtrast

A true mountain bird, breeding on rocky alpine slopes or in steep-sided gorges within boreal forests. There its simple, strong but rather monotonous, fluty song can be heard in spring. Monotonous maybe, but also finely tuned to its remote or barren habitats. Although normally a shy bird, it is fairly easy to see in its largely treeless breeding terrain, but is surprisingly 'invisible' on migration to S Europe and NW Africa. Then, its presence is more often detected by its hard clicking calls, given from cover. Returns early in spring, often in mid March and April.

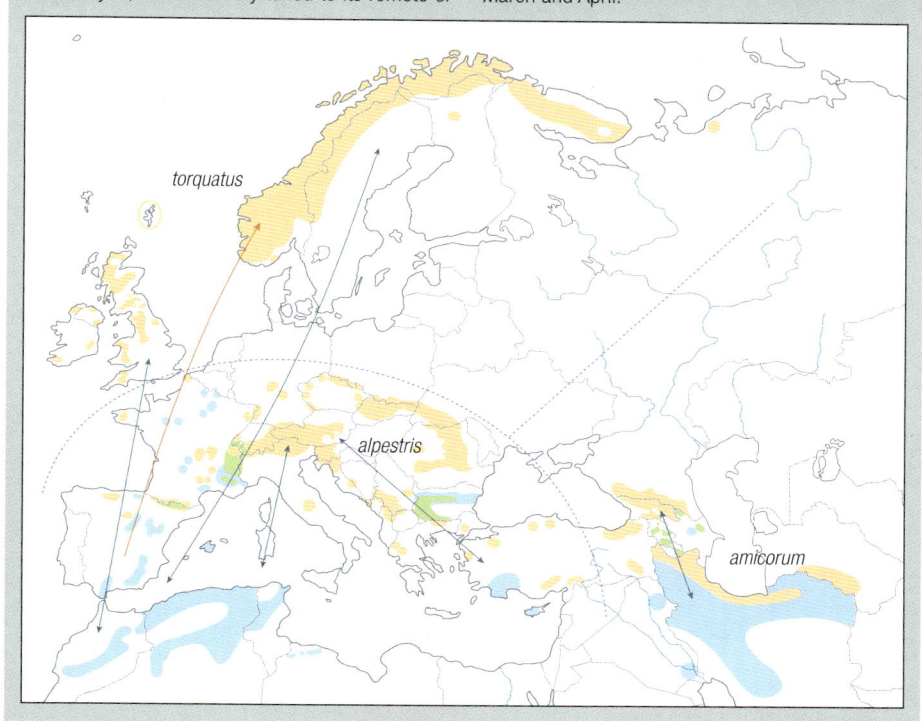

IDENTIFICATION Similar to a Blackbird, being all dark and of similar size, only perhaps slightly slimmer with longer, more pointed wings, and the long tail is on average less rounded, more square. *Dark brown* (appearing blackish) or *brown-grey* in all plumages, adults having a striking feature in their *white transverse patch* or crescent *on the dark breast*. The pale crescent is brown-tinged or lightly scalloped and less contrasting in first-year birds, and can even be predominantly brown or lacking in the young ♀. *Underparts have whitish feather fringes* creating a scaly pattern, but their prominence varies, young and ♀ having more, as do C European birds (*alpestris*) compared to Fenno-Scandian or British (*torquatus*). Compared to Blackbird, adults have *paler wings*, often appearing paler than body in flight (Blackbird uniformly black or dark grey on both body and wings). This paleness is created by off-white outer edges to upperwing-coverts and some flight-feathers. The same difference is sometimes visible in first-year ♂, although more subdued. The young ♀ has even less pale wings, in practice showing very little difference (still some) compared to a flying Blackbird. The paleness of the wing is slightly less obvious in N & W Europe, whereas it is more obvious in S Europe and very striking in Asia. Adults have an attractive dark-tipped *yellow bill*, most prominently on ♂♂, whereas the ♀ has duskier tip. Legs are dark.

VOCALISATIONS The song is both simple in structure and somewhat variable, both between different birds and when the repertoire of one individual is studied, as well as to some extent geographically ('dialects'). However, it is generally recognised by its, for a thrush, simple and repetitive strophes, loud voice and melancholy ring. Strophes are amply spaced, as if the bird is listening for what its competitors achieve. Individuals can alternate between strophes like *trruh trruh trruh*, *che-wee che-wee che-wee che-wee* and *chüll chüll chüll*. There may be added a squeaky twitter after each strophe, but this is more subdued and usually inaudible at > 300 m, or even at less. Rarely utters slightly more complex strophes, like *ter-li ter-li ter-li chu chu*. – Call a very hard and dry clicking *tac*, which may be repeated in short series if the bird is alarmed. It is drier and more 'stony' than the corresponding calls of Blackbird or Fieldfare, and slightly less deep. Other calls include a somewhat Fieldfare-like squeaky *zierk* and a very fine, high-pitched alarm *ziiieh*, in common with many other *Turdus*.

SIMILAR SPECIES Can hardly be confused with any other species than *Blackbird*, at least not within the covered region. For differences see Identification. Note that first-winter ♀ Ring Ouzel can lack any trace of a paler breast crescent, hence important on all-dark Ring Ouzels to see the pale scaly pattern on the underbody, and to note the calls.

AGEING & SEXING (*torquatus*) Ages differ slightly during 1stY, but ageing requires close views or handling. Sexes often separable after post-juv moult, but difference slight and frequently ambiguous. Sexing more reliable from 1stS, but even then can be difficult. – Moults. Complete post-nuptial moult of ad in late summer (Jul–early Sep). Partial post-juv moult in summer (Jul–Sep) does not involve flight-feathers, tail-feathers or primary-coverts. Some inner greater coverts are also moulted, but usually 3–6 outer retained; rarely none or all greater coverts moulted. Rarely one or two innermost tertials replaced. No pre-nuptial moult in winter. – **SPRING Ad** ♂ Head, throat and most of underparts very dark brown (actually never black, although can appear so in the field), initially

T. t. torquatus, ad ♂, Finland, Apr: quite dark, with few pale fringes below, these being already almost worn-off from late spring. White breast crescent relatively broad, and when abraded becomes cleaner white, usually with pointed ends (in ♀ normally blunt-ended). Aged by evenly-feathered greater coverts with uniform greyish-white edges and tips. (J. J. Nurmi)

T. t. torquatus, ad ♀, Scotland, Jun: although some are quite similar to ♂, most ♀♀ have ground colour of plumage slightly paler and browner, and tend to retain obvious scaly pattern (even in summer), as well as brownish-tinged breast patch. Bill on average duller or darker (though not here). Greater coverts apparently ad-like, without obvious moult limit. (S. Garvie)

T. t. torquatus, 1stS ♂, **Finland, Apr**: unlike ad, outer half of greater coverts are juv, these being contrastingly shorter with more distinct (broader and bolder) whitish edges and tips; inner coverts moulted and have narrower, irregular and pale-speckled edges. Similar to ad ♂ due to darkness of plumage, but prominent pale edges remain, especially below, while white breast crescent is rather clean and wide; bill on average less pure/bright yellow, approaching ♀. (M. Varesvuo)

T. t. torquatus, 1stS ♀, **Netherlands, Apr**: size and colour of pale breast-patch best fit ♀, as do browner ground colour and extent of pale edges above and below, which also match pattern in *torquatus* at this season. Seemingly two generations of greater coverts confirm age. (A. Ouwerkerk)

T. t. torquatus, 1stW ♂, **Finland, Dec**: moult limit in greater coverts reveal age. Once aged as 1stY, sexing easier: size and purity of white breast patch (although in mid winter still brown-tinged) best fit 1stW ♂, supported by blacker ground colour and number of pale edges above and below. (P. Komi)

T. t. torquatus, 1stW ♀, **Netherlands, Oct**: rather brownish plumage and highly muted breast patch (densely barred cinnamon-brown and hardly visible!) and pale wing panel best match 1stW ♀. Contrast between inner moulted and mostly juv outer greater coverts. (L. van Loo)

with fine paler fringes; pale fringes progressively wear off so that head, throat, belly and flanks become uniformly dark in Jun, sometimes earlier. Pale breast patch usually nearly pure white, only a few traces of brown tips in some, reaching far onto side of chest, usually ending in pointed wedge (Malmhagen 2012). Belly and flanks dark brown with fine whitish fringes forming discreet scaly pattern. Tail-feathers rather broad, tips not acutely pointed. All greater coverts uniformly edged and tipped greyish-white, border between edges and centres often irregular and faintly speckled dark; greater coverts sometimes largely barred dark. Bill mostly bright yellow with variable amount (usually restricted) of dark on tip and around nostrils. **Ad ♀** Similar to ad ♂ (and a few cannot be reliably separated) but ground colour of head and throat on average slightly paler brown, and generally has slightly more prominent pale fringes. Pale breast patch on average less pure white, often more dirty white with more brown tips remaining, and as a rule not reaching as far up on chest-sides as in typical ♂, ending more bluntly (Malmhagen 2012). Belly and flanks have whitish fringes forming rather obvious scaly pattern. Tail-feathers and greater coverts as in ad ♂. Bill as in ♂, only on average a little duller or darker. (Clearly more difficult to sex ad than often implied, and only the neatest ♂♂ can be picked out on quite white breast crescent and unmarked dark brown rest of plumage, often supported by shape and size of breast crescent. If

T. t. torquatus, 1stW ♂, **France, Oct**: aged as 1stW by pointed tips to tail-feathers; ♂ by dark general plumage, rather white-looking prominent breast crescent with pointed ends. Note silvery-white underwing due to whitish outer webs to secondaries and their coverts. (V. Legrand)

ageing proves impossible, most ad ♀♀ and 1stS ♂♂ will look the same.) **1stS ♂** Very similar to ad ♀, but differs by usually having a varying number (often 3–7) of juv outer greater coverts unmoulted, contrasting with new inner ones by shorter length and more distinct whitish edges and tips to outer web (whitish edges not continuing on inner web), and by lacking speckled or irregular border to pale edges (but often a hint of a dark spot inside tip at shaft, and sometimes pale outer edge enhanced by dark subterminal bar). Innermost unmoulted juv greater coverts sometimes have whitish shaft-streak near tip. Tail-feathers on average narrower and more acutely pointed (if not too abraded to judge). Bill with at least as much yellow basally as ad ♀. (Exact shape of breast crescent less useful age criterion than in ad.) **1stS ♀** Like 1stS ♂, and often inseparable, but typical birds sexed on very dark, brown-tinged breast-patch, only faintly paler than surrounding feathers. – AUTUMN Sexing and ageing much as in spring (which see), but plumage fresh with prominent white fringes to dark feathers of underparts (broad on belly and flanks), and brown tips to white feathers on breast patch. Sexing of ad is, if anything, more difficult than in spring, but see above for shape of breast crescent in ad. **Ad** As in spring, all greater coverts uniformly patterned (see above for details), and tail-feathers broad with rounded tips. **1stW** Contrast exists between moulted and unmoulted greater coverts (see Spring for details), and tail-feathers

T. t. alpestris, ad ♂, Czech Republic, Jun: combination of strikingly large white wing panel, and still relatively extensive pale edges on underparts typical for this subspecies. Aged as ad by evenly ad-patterned greater coverts, and sexed as ♂ by dark head and extensive white crescent with sharply pointed tips. (D. Jirovsky)

T. t. alpestris, ♀, presumed 1stS, Germany, Jul: paler and browner than ♂, and even in summer has obvious pale scaling below and blunt-ended brownish-tinged breast patch. General colours and brown-sullied crescent indicate 1stS, but accurate ageing impossible at this angle. (T. Grüner)

T. t. alpestris, ad ♀, Switzerland, Mar: browner above, and has comparatively small blunt-ended breast patch with distinct brown vermiculations, as well as slightly duller (less pure white) wing panel, equal ♀; evenly-feathered ad-type wing. Compared to top right image, white scaling below still very obvious. (A. Gygax)

T. t. alpestris, presumed 1stS ♂, Switzerland, Apr: moult limits and wear difficult to assess, and therefore not easy to separate ad ♀ and 1stS ♂. Nearly pure white breast patch with pointed end, very broad whitish fringes below, pale wing and all-yellow bill best fit young ♂ *alpestris* at this season. (R. Aeschlimann)

rather narrow with pointed tips. Many 1stW ♀♀ immediately recognised on having very dark brown breast patch, as dark as surrounding breast (only of different tint). **Juv** Rather similar to 1stW ♀ but has narrow buff-white shaft-streaks above (largest on scapulars), including on lesser and median coverts.

BIOMETRICS (*torquatus*) **L** 24.5–27.5 cm; **W** ♂ (once 136) 138–149 mm (*n* 61, *m* 143.1), ♀ 133–143 mm (*n* 48, *m* 138.8); **T** ♂ 102–118 mm (*n* 60, *m* 108.2), ♀ 99–110 mm (*n* 31, *m* 103.0); **T/W** *m* 75.1; **B** 24.0–26.1 mm (*n* 34, *m* 25.1); **Ts** 32.0–34.9 mm (*n* 50, *m* 33.4). **Wing formula: p1** minute, < pc 8–18 mm, < p2 78.5–89 mm; **p2** < wt 5–8 mm; **pp3–4** equal and longest (though often p4 0.5–3 mm <); **p5** < wt 5–8.5 mm; **p6** < wt 18–24 mm; **p7** < wt 26–31.5 mm; **p10** < wt 39–47 mm; **s1** < wt 44–51 mm. Emarg. pp3–5.

GEOGRAPHICAL VARIATION & RANGE Three subspecies, rather well differentiated morphologically and geographically. Main differences are number of pale feather edges on lower parts and on upperwing.

T. t. torquatus L., 1758 (N Fenno-Scandia, Britain, Ireland; winters S Europe, NW Africa). Described above.

T. t. alpestris, 1stW, presumed ♂, Turkey, Oct: juv outer greater coverts, large alula and primary-coverts contrast with inner freshly moulted greater coverts. Sexing not as straight-forward: darkness of ground colour, size and purity of pale breast patch both best fit 1stW ♂, whereas blunt-ended breast patch is more suggestive of ♀; still, this is not an infallible sex criterion. (Z. Kurnuç)

T. t. alpestris, 1stW ♀, Spain, Oct: the dullest plumage, browner above with highly obscure pale breast patch, while pale scaling below narrower and greyer. Moult limits in greater coverts: many inner moulted, longer and greyer with dark-barred effect; the few outer juv feathers are distinctly browner, with thin but well-defined whitish-buff fringes. (A. M. Dominguez)

Rather dark, with insignificant to moderately broad pale fringes below in fresh plumage. Has darkest upperwing of the three races, relatively speaking.

T. t. alpestris (C. L. Brehm, 1831) (mountains in C, S & E Europe, W & C Turkey; winters Mediterranean region, N Africa). Has more prominent pale edges on both upperwing and underbody, with broad pale fringes and central streaks on belly and undertail-coverts. Ad ♂ has about as prominent pale fringes below as young ♀ *torquatus*. About same size as *torquatus*. Sexes nearly equal in size. **W** 135–148 mm (*n* 32, *m* 140.0); **T** 97–111 mm (*n* 32, *m* 103.9); **T/W** *m* 74.4; **B** 22.5–27.7 mm (*n* 31, *m* 24.9); **Ts** 30–35.7 mm (*n* 32, *m* 33.0).

T. t. amicorum Hartert, 1923 (E Turkey, Caucasus, Transcaucasia, N Iran, SW Turkmenistan; winters SW Asia). Even darker below than *torquatus*, with hardly any pale feather edges on belly even in fresh plumage, and breast crescent is both larger and on average purer white. Also, upperwing has very prominent whitish edges, creating a strikingly pale panel when folded. All other plumages are similarly somewhat more neat and contrasting than corresponding ones of *torquatus*. Slightly more long-tailed and long-billed. **L** 25–28 cm; **W** ♂ 138–149 mm (*n* 22, *m* 143.4),

T. t. amicorum, ad ♂, Georgia, Apr: dark underparts with hardly any pale edges on belly, even compared to *torquatus* (of same age/sex and season); breast crescent, however, larger on average and purer white. Upperwing has very prominent whitish edges, forming large solid panel. Aged by evenly-feathered ad-patterned greater coverts and primary-coverts. (D. Tipling)

T. t. amicorum, 1stS ♂, Georgia, May: despite being young, pale fringes below are very narrow. Large pure white breast patch, but bill not pure yellow. All greater coverts are post-juv, but apparent differences in pattern and length of inner three, amplified by their placement, form 'false' moult limits. Tentatively aged by apparently juv primary-coverts and degree of wear to primaries. (S. J. M. Gantlett)

T. t. amicorum, ♀, presumed ad, Georgia, Apr: compared to ♂ browner with on average somewhat smaller, brownish-tinged and ill-defined pale breast patch, while pale scaling below is broader, in particular on chin and throat. More extensive white wing panel in this race (and *alpestris*) makes assessment of moult limits in greater coverts more difficult than in *torquatus*. As far as can be judged at this angle, all wing-feathers uniform of ad type. (D. Tipling)

♀ 133–145 mm (*n* 19, *m* 140.4); **T** ♂ 106–119 mm (*n* 22, *m* 112.4), ♀ 100–111 mm (*n* 13, *m* 106,1); **T/W** *m* 78.0; **B** 24–28.3 mm (*n* 37, *m* 26.1); **Ts** 31.3–35.4 mm (*n* 37, *m* 33.7). (Syn. *orientalis*.)

REFERENCES MALMHAGEN, B. (2012) *Roadrunner*, 20: 34–39.

T. t. amicorum, 1stW ♂, United Arab Emirates, Jan: despite being fresh, pale scaling below very narrow, making underparts appear very dark compared to *torquatus*, and especially well differentiated from *alpestris*. Breast patch relatively large and clean, with very pointed ends suggesting ♂, while darkness of head and mantle also better fit ♂. Moult limits in greater coverts age it as 1stY. (N. Moran)

T. t. amicorum, juv, Georgia, Jun: narrow buff-white shaft-streaks above on scapulars and smaller secondary-coverts; juv greater-coverts with contrastingly and well-defined pale edges to outer webs, which do not continue over inner web. (R. Pop)

(COMMON) BLACKBIRD
Turdus merula L., 1758

Fr. – Merle noir; Ger. – Amsel
Sp. – Mirlo común; Swe. – Koltrast

One of the best-known birds of all, and loved for its truly beautiful song, the Blackbird thrives just as well in gardens as in remote forests. It is also one of the characteristic birds of towns, being content with backyard trees and lawns. This versatility has made it one of the commonest birds in Europe; according to a rough estimate there could be as many as well over 50 million pairs. It is a hardy bird that winters wherever the ground is mainly snow-free. Some northern birds migrate south to warmer areas, but many stay in towns, where they take advantage of slightly higher average winter temperatures and food provided in gardens and parks.

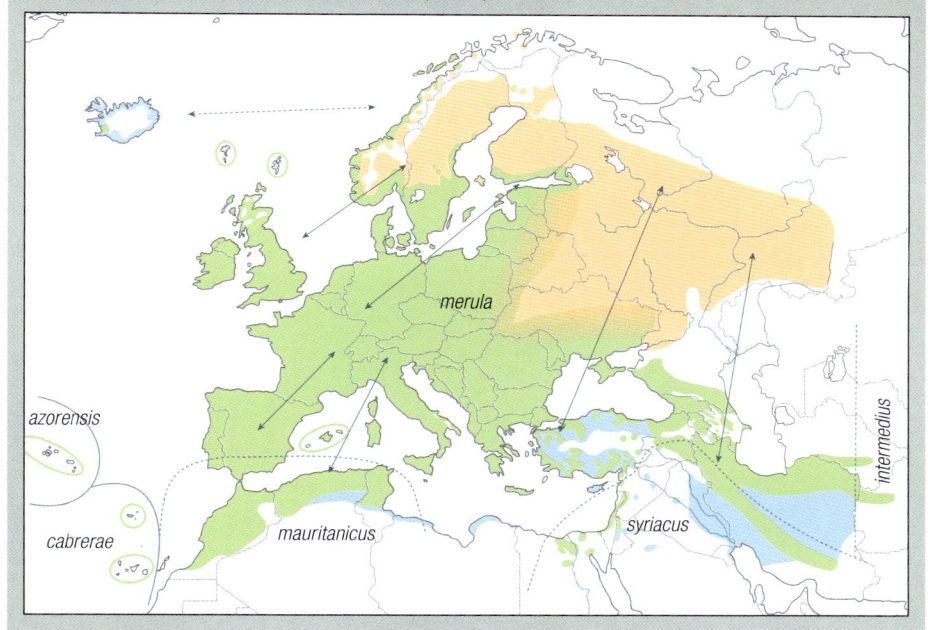

T. m. merula, ad ♀, Finland, Apr: ♀ predominantly dark brown-grey, but markings vary individually, especially greyish-white and blackish streaks on throat, and amount and strength of rufous and dark spots on breast. Some are more greyish, others warmer and browner. Bill coloration also variable. This is a rather strongly patterned and brighter ♀. Aged by evenly-feathered and rather fresh wing. (M. Varesvuo)

IDENTIFICATION A medium-large thrush with *fairly long and well rounded tail*. As the name implies, it is indeed above all a *black* bird, but the ♀ is brownish-grey, and the ♂ has a *bright yellow bill* in spring and summer (can have an orange tint), and a similarly *yellow narrow orbital ring*. Otherwise, the ♂ is completely black. The dark brown-grey ♀ has a *greyish-white throat with bold blackish streaking*, and the breast is usually slightly rufous-tinged and diffusely mottled dark. ♀ plumage is actually rather variable, some being more greyish, others more warm brown, and the plumage can look rather patchy and 'untidy'. Note that there is no pale area on the underparts of the ♀ (except for the ground colour of the throat); belly, vent, undertail-coverts and underwing-coverts are all dark grey. The strong legs are dark in both sexes. The ♀ bill can have some pale yellow-brown basally, but mainly it is rather dark brown. In winter the ♂ bill darkens somewhat, and the orbital ring shrinks and is hardly perceptible. Juvenile similar to the ♀ but readily recognised by its buffish or whitish wedge-shaped shaft-streaks on upperparts, most prominently on scapulars, mantle and secondary wing-coverts, and by its rufous-tinged and coarsely blotched underparts. The Blackbird often raises the tail after landing, e.g. on a lawn or on a television aerial on a rooftop, and then lowers it slowly. Twitches wings and tail when agitated. *Flight is fast but a little jerky* with somewhat uneven bursts of wingbeats and unsteady flight path. When feeding on the ground it either *hops quickly* or *walks* short stretches *in a hunched posture* interrupted by pauses when perches more upright, bends forward to watch the ground intensely, looking for movements by earthworms.

VOCALISATIONS The song is most pleasing to the human ear, improvised fluty, melodic figures in a mellow alto voice, large tonal steps bridged with *glissandi*, seemingly infinitely varied in the details. The general impression being that the song is 'written in the major key', but with 'sincerity'. The rather brief strophes, with some soft twitter appended, are well spaced. Sings from high post, preferably at dawn and dusk but at all times of day. As to song, can normally only be confused with Mistle Thrush (in the rare case that a particular Blackbird has a more monotonous and desolate-sounding song, but Mistle sings a little quicker, has a somewhat harder voice and utilises a narrower tonal range) and, at long range, Golden Oriole (same pitch and mellow, fluty tone, but easily told when closer). Sometimes has a varied subsong, often delivered in earliest spring while feeding on the ground or when perched in a low thicket, full of clever mimicry of other birds. – Has rich repertoire of calls and uses them frequently. In startled alarm a few hard chuckling sounds that turn into a hammering, metallic, high-pitched series, often when the bird takes off for better cover or before going to roost, *chuck chuck-chuck-chuk-kli-kli-kli-kli-kli-kli-*... There is some slight geographical variation, southern populations generally having a higher-pitched metallic, clicking sound. Single *chuck* notes, or slower series of same call, are used in milder anxiety. A very fine, high-pitched whistling *tsiiiih* is usually a warning from a hidden Blackbird when a cat or a Sparrowhawk *Accipiter nisus* is spotted. Once the Blackbird decides to mock a cat in the open it uses nerve-racking, drawn-out series of the metallic clicking sounds, *kli-kli-kli-*... From migrants a soft *srri* with audible 'r' is heard, often from the Nordic late autumn sky by night.

SIMILAR SPECIES Within the treated region, can hardly be confused with any species other than Ring Ouzel. But beginner, watching his or her garden lawn, needs to eliminate Starling, both being blackish birds with yellow bill feeding on the ground. Note that Blackbird has a much longer tail, which it often raises above the back, that it walks rather warily in

T. m. merula, ad ♂, Finland, Apr: ♂'s simple coloration and shape make its identification unproblematic within the treated region. Note bright yellow bill and orbital-ring, contrasting with jet-black head. Aged by evenly-feathered and rather fresh wing. (M. Varesvuo)

T. m. merula, 1stS ♂, Finland, May: unlike ad ♂, clear moult limits between juv alula, primary-coverts and remiges, which are brownish and worn, contrasting with newer and blackish greater coverts. Plumage also scruffier, and black of head browner. (M. Varesvuo)

T. m. merula, 1stW ♂, Finland, Nov: young birds when fresh, in non-breeding season, appear duller, with browner hue, forming mottled effect below. Bill rather dark. Best aged by clear moult limits, with new and blackish inner block of greater coverts forming clear contrast with outer ones, alula, primary-coverts and remiges, which are juv and browner. (M. Varesvuo)

brief bursts, with 'frozen' stops, or that it hops with both feet together (the short-tailed Starling rushes or walks energetically with jerking head). – Most Ring Ouzels have a striking pale breast patch, and even if this is missed or is lacking (as in some first-winter ♀♀) they generally have pale-edged feathers on upperwing, which creates a pale-winged impression when a bird takes off, whereas Blackbirds invariably have all-dark wings. As first-winter Ring Ouzels have slightly darker wings than adults important also to note the pale scaly underbody, lacking in Blackbird, and to note the harder, drier calls.

AGEING & SEXING (*merula*) Ages differ slightly during 1stY, but ageing requires close views or handling. Sexes separable after post-juv moult, but difference in 1stW not always obvious, and ageing requires care. – Moults. Complete post-nuptial moult of ad in summer and early autumn (Aug–early Oct in N Europe, onset sometimes much earlier in W & S Europe). Partial post-juv moult in summer (Jul–Sep, sometimes Oct) does not involve flight-feathers or primary-coverts. Rarely a few tertials and tail-feathers also moulted. A variable number of inner greater coverts are also moulted, but usually a few outer retained; rarely none or all greater coverts moulted. Some birds, probably a minority, appear to undergo a very limited partial pre-nuptial moult in late winter, only involving parts of head, body and (exceptionally) some secondary-coverts. – **SPRING Ad ♂** Entire plumage black, including uniformly black wing, thus all greater coverts evenly dark. Tail-feathers rather broad, tips rounded. Bill orange-tinged yellow. **Ad ♀** Dark brownish-grey above and on belly to undertail, warmer brown-grey (sometimes with rufous tinge) on breast. Throat off-white boldly and densely streaked blackish. Greater coverts uniformly dark brown, tail-feathers shaped as ad ♂. Bill variable, rarely all dark, more often with some yellow or yellow-brown, quite extensively yellow-brown in some. **1stS ♂** Resembles ad ♂, but aged much as in autumn (which see); however, shape of tail-feathers less easy to assess due to wear. Bill initially dark, generally lacking any obvious yellow, but progressively attains colour and becomes ad-like in this respect by (Dec) Jan–Apr (individual variation). **1stS ♀** Like ad ♀, and often inseparable in the field, but typical birds aged by visible contrast among greater coverts between browner unmoulted juv and greyish-olive new. Note also that ad type greater coverts usually are subtly barred darker (growth bars), whereas juv type greater coverts are uniform. If still controversial, check primary-coverts, which are a bit narrower and more pointed at tips compared to broad and well rounded at tips in ad. – **AUTUMN Ad ♂** Plumage like in spring (though now freshly moulted). Bill invariably yellow at first, but may attain darker 'stains' Sep–Jan (Feb). **Ad ♀** Plumage like spring. Bill variable, sometimes all dark but often has a little yellow or yellow-brown. **1stW** Sexes differ like in ad, although both have dark bill (in ♂, all dark through Nov, with small yellowish patches appearing from Dec or Jan, or even later). Aged (apart from bill colour in ♂♂) by having slightly browner primaries, primary-coverts and a variable number of unmoulted juv outer greater coverts, latter also often tipped paler brownish-white, contrasting with new darker (black in ♂, grey in ♀) rest of wing-feathers (easiest told on ♂, more subtly in ♀). Tail-feathers on average narrower and more pointed than in ad (beware of a few ad with slightly narrower and more pointed tail-feathers, though outer 2–3 pairs usually have more square-ended tips than 1y). **Juv** Resembles 1stW ♀ but has narrow buff-white shaft-streaks above (largest on scapulars), including on lesser and median coverts. Underparts rufous-tinged coarsely blotched darker.

BIOMETRICS (*merula*, Sweden) **L** 25–27.5 cm; **W** ♂ 126–140.5 mm (*n* 60, *m* 133.2), ♀ 123–135 mm (*n* 33, *m* 128.2); **T** ♂ 98–112 mm (*n* 60, *m* 105.5), ♀ 96–110 mm

T. m. merula, 1stS ♀, Netherlands, Mar: moult contrast among greater coverts in ♀ often less obvious than ♂, but in close views it is sometimes possible to detect browner juv outer greater coverts and primary-coverts. This a rather dull and uniform ♀ (no grey or rufous pigments). (N. D. van Swelm)

T. m. merula, juv, Germany, May: readily aged by buff-white shaft-streaks above and on smaller secondary-coverts. Unlike other plumages, underparts coarsely blotched and barred dark on more rufous-tinged background, while facial marks are more pronounced than in 1stW ♀. (M. Schäf)

T. m. mauritanicus, ad ♂, Morocco, Oct: proportionately longer tail but blunter/shorter wing characterise this N African race, as well as typically glossier slate-black plumage. Aged by evenly-feathered and fresh wing. (A. B. van den Berg)

T. m. mauritanicus, 1stY ♀, Morocco, Oct: longer and slimmer tail also evident in this ♀; note characteristically grey tinge in plumage and rather more extensively yellow bill, but much overlap with other races. Aged by juv alula, primary-coverts and remiges, which are very subtly contrastingly browner. (A. B. van den Berg)

(n 33, m 103.2); **T/W** m 79.7; **B** 24.0–28.6 mm (n 60, m 26.3); **Ts** 31.1–35.0 mm (n 61, m 33.4). Breeders in Continental Europe average subtly smaller. **Wing formula: p1** small, < pc 1–9 mm, < p2 59–71 mm; **p2** < wt 8.5–17 mm; **pp3–5** about equal and longest (though sometimes p3 0.5–3.5 mm <); **p6** < wt 3–9 mm; **p7** < wt 9–19 mm; **p8** < wt 17–24 mm; **p10** < wt 27–34 mm; **s1** < wt 29–37 mm. Emarg. pp3–6.

GEOGRAPHICAL VARIATION & RANGE Many subspecies described, but variation generally slight and mostly clinal. After comparison of series in several museum collections we have concluded that some described races are synonyms, impossible to uphold following a 75% rule of distinctness. Apart from those sufficiently dissimilar subspecies listed below there are several more, but they are all extralimital and not known to have occurred within the treated area.

T. m. merula L., 1758 (Europe, except possibly in southwestern parts, Ukraine, Crimea, W, C & NE Turkey, Caucasus, Transcaucasia, N Iran; winters W & S Europe to Middle East, northern birds more migratory, southern and western largely resident). Described above. Birds in S Europe have on average fractionally shorter wings and tail than in Fenno-Scandia, but differences slight and clinal. No detectable plumage differences noted, hence best included in *merula*. (Syn. *aterrimus*; *insularum*; *mallorcae*; *pinetorum*; *ticehursti*.)

○ **T. m. syriacus** Hemprich & Ehrenberg, 1833 (SE Turkey, Levant, N Egypt, Iraq, W & S Iran; resident). Very similar to *merula* but somewhat smaller. Said to be more greyish in both sexes (♂ 'slate-coloured rather than black'), but this cannot be confirmed when comparing long series. A very subtle race. **W** ♂ 124–134 mm (n 12, m 129.5), ♀ 120–128 mm (n 12, m 125.3); **T** ♂ 99–108 mm (n 12, m 102.8), ♀ 96–104 mm (n 12, m 100.5); **T/W** m 79.8; **B** 24.9–29.4 mm (n 23, m 26.5); **Ts** 30.5–33.9 mm (n 24, m 32.5). Wing formula very similar to *merula*.

T. m. intermedius (Richmond, 1896) (Central Asia from Afghanistan eastwards, but winters as far west as Iraq and Iran to NW India). The largest race within covered area. Proportionately long-tailed. ♀ darker and greyer than in *syriacus* and *merula*. **L** 27–30 cm; **W** ♂ 135–149 mm (n 16, m 139.5), ♀ 128–140 mm (n 16, m 135.9); **T** ♂ 109–130 mm (n 17, m 118.9), ♀ 111–121 mm (n 14, m 115.8); **T/W** m 85.5; **B** 26.2–30.3 mm (n 29, m 28.2); **Ts** 30.7–35.0 mm (n 31, m 33.4). (Syn. *brodkorbi*.)

T. m. mauritanicus Hartert, 1902 (NW Africa, possibly also coastal S Iberia and E Canaries; resident). Very similar to *merula* but has slightly longer tail, further enhanced by a slightly blunter wing, and stronger tarsi. Ad ♂ is on average slightly deeper and glossier black; ♀ is darker and greyer, with a yellower bill. **W** ♂ 122–132 mm (n 12, m 127.4),

T. m. azorensis, ad ♂, Azores, Oct: a small race with short/blunt wings, and proportionately short tail but longer bill; ♂ also has deep, glossy black plumage. (D. Occhiato)

♀ 116–130 mm (n 12, m 124.6); **T** ♂ 102–118 mm (n 12, m 110.1), ♀ 100–116 mm (n 12, m 107.9); **T/W** m 86.4; **B** 24.8–28.7 mm (n 24, m 26.9); **Ts** 31.2–36.0 mm (n 24, m 34.5). **Wing formula: p1** < pc 0–4 mm or > 3; **p6** < wt 1.5–7 mm; **p7** < wt 7–14 mm; **p10** < wt 20–25 mm; **s1** < wt 24–28 mm. (Syn. *algira*.)

T. m. cabrerae Hartert, 1901 (Madeira, W Canaries; resident). Rather small but proportionately strong-billed. ♂ deep, glossy black, ♀ dark greyish with densely dark-streaked throat. **L** 24–26 cm; **W** ♂ 124.5–133 mm (n 10, m 127.8), ♀ 118–125 mm (n 10, m 122.5); **T** ♂ 94–105 mm (n 10, m 100.2), ♀ 91–100 mm (n 10, m 96.2); **T/W** m 78.5; **B** 24.9–27.7 mm (n 20, m 26.1); **Ts** 31.4–33.7 mm (n 20, m 32.4). **Wing formula: p1** < p2 55–64 mm; **p6** < wt 3–6 mm; **p7** < wt 10–14.5 mm; **p10** < wt 24–30 mm; **s1** < wt 27–33 mm. (Syn. *agnetae*.)

T. m. azorensis Hartert, 1905 (Azores; resident). Coloration much as *cabrerae*, but is even smaller and more blunt-winged, with a proportionately shorter tail; also longer tarsi, especially in ♂. **L** 23.5–24.5 cm; **W** ♂ 121–128 mm (n 11, m 124.5), ♀ 114–126 mm (n 10, m 119.5); **T** ♂ 90–98 mm (n 11, m 93.5), ♀ 86–96 mm (n 10, m 89.9); **T/W** m 75.6; **B** 24.7–27.6 mm (n 21, m 26.0); **Ts** ♂ 33.0–36.0 mm (n 11, m 34.5), ♀ 31.0–35.0 mm (n 10, m 33.1). **Wing formula: p1** < pc 0–6 mm or > 3.5, < p2 50–57 mm; **p7** < wt 7–12 mm; **p10** < wt 19–26 mm; **s1** < wt 24–28 mm.

T. m. azorensis, ad ♀, Azores, Oct: characteristic dark greyish hue above and stronger facial pattern, especially the densely dark-streaked throat. (D. Occhiato)

EYEBROWED THRUSH
Turdus obscurus J. F. Gmelin, 1789

Fr. – Merle obscur; Ger. – Weissbrauendrossel
Sp. – Zorzal rojigrís; Swe. – Gråhalsad trast

Distributed over much of the Siberian taiga, from east of the Ob basin to Kamchatka, the Eyebrowed Thrush is fairly numerous and easy to see on a birdwatching trip to Siberia. It straggles very rarely but annually to W Europe, especially in autumn. Normally though, it migrates south and south-east to winter between NE India and SE China. It is rather shy and wary, and not easy to get a close look at in its vast, remote breeding forests, and normally the best you can hope for is a distant view of a singing bird in the top of Siberian pine or larch.

Ad ♂, Thailand, Feb: tiny white chin and almost uniform grey throat/upper breast. Blackish lores and lead-grey cheeks, as well as white supercilium, white mark below eye and at base of lower mandible, but throat by and large unstreaked unlike in ♀ and young ♂. Underparts extensively sullied deep orange-red. Best aged by uniform tawny-brown greater coverts (lacking moult contrast). (A. Jearwattanakanok)

IDENTIFICATION About the size of a Song Thrush but perhaps slightly slimmer with a proportionately little longer tail. Basically a brown-and-grey bird with orange-brown breast and flanks, and white belly. ♂♂ are generally the handsomest, but individual variation makes sexing quite difficult. Neatest birds (which are usually adult ♂♂) have *lead-grey head and neck* with *blackish lores* and dark grey upper cheeks, *distinct white supercilium*, a *white mark below the eye* and distinctly set-off white chin; a *dark tawny-brown mantle and back*; and orange-brown or even *pale tomato-red breast and flanks*. Adult ♀♀ are similar but have a fraction less neat head pattern and usually a certain amount of white on throat with dark streaking. Dullest birds (which are generally first-year ♀♀) have brownish head and neck with only very slight grey element, indistinct and short off-white supercilium, white-streaked cheeks, both chin and upper throat dusky white with much dark streaking; and dull brown-buff breast and flanks with only slight orange element. First-years recognised by having a number of unmoulted outer greater coverts edged and tipped pale brown-grey in contrast to inner more tawny- or rufous-edged. Tail dark, often with tiny white tips to outer feathers. Underwing pale grey-brown. Bill dark with yellow base to lower mandible. Legs yellowish-brown. Juvenile very different, resembling more a young Song Thrush or Redwing, is pale buff on breast and rusty-ochre on flanks with dense blackish rounded or transverse spotting (much fainter spots on throat and belly), immediately recognised as a juvenile thrush by its buffish wedge-shaped shaft-streaks on upperparts, most prominently on scapulars, mantle and secondary wing-coverts. Behaviour as in most *Turdus*. Flight is fast and mostly direct and purposeful. When perched often holds head high, bill pointing slightly upwards.

VOCALISATIONS Song consists of a simple strophe or outburst of a few fluty notes, repeated with 2–3 sec. pause for long spells (at times slightly more hurried). The tone has a desolate ring and the voice is rather deep. Fluty strophes are usually alternated with brief, subdued, high-pitched twitters or a few squeaky notes. The song resembles several other thrushes of the taiga, perhaps mostly Siberian Thrush, but is separated by the deeper and more mournful voice, its generally slower pace, often slightly longer and more complex strophes, and by tendency to sometimes omit the high-pitched twitter after each strophe. Among common fluty strophes, often repeated or recurring in the song, the following serve as examples: (i) *chu-ee chu-ee ... chill-chevu*; (ii) *chuh-chu-chu*; (iii) *chip-chille-vuh*; and (iv) a shorter *cherluh*. – Call a hard clicking *tak*, often repeated a few times, doubtfully separable from similar calls of Black-throated and Red-throated Thrushes. Also a fine, slightly rolling *ts(r)iih*, a little reminiscent of Blackbird. In alarm utters a clicking series, *chik-chik-chik-...*, energetic and with 'electric' tone, halfway between Redwing and Song Thrush.

SIMILAR SPECIES Adults and first-years are fairly distinctive once combination of orange breast and flanks, and greyish head and neck with a white supercilium, seen. If you clearly see a whitish supercilium on a mid-sized thrush (but much of the rest of the bird is covered), other options are limited to *Dusky Thrush*, *Naumann's Thrush*, *Redwing* and extralimital *Grey-sided Thrush* (*T. feae*; not treated). You need to see more of the bird to clinch the identification, but if it just flies away, remember that the three first-mentioned species have orange or reddish underwing, whereas Eyebrowed has dull greyish-white. – Juvenile resembles a juvenile *Song Thrush* or *Redwing* with its bold spotting below but should be possible to separate by deeper rusty-ochre tinge confined to flanks with white central belly (Redwing and Song Thrush lightly yellowish-buff over breast but lack marked ochre tinge on flanks), less spots on white vent and undertail-coverts, and lack of contrasting head pattern of young Redwing, which develops dark cheeks and white supercilium early. Underwing greyish-white (with modest buff tinge at most), lacking orange or reddish tinge of Song Thrush and Redwing, respectively.

AGEING & SEXING Ages differ slightly during 1stY, but ageing requires close views or handling. Sexes similar and sometimes inseparable due to (sometimes under-rated) individual variation. Sexing should only be attempted with positively aged birds. – Moults. Complete post-nuptial moult of ad in summer and early autumn (Jul–early Sep). Partial post-juv moult in summer (Jul–Sep) does not involve flight-feathers, tertials or primary-coverts. A variable number of inner greater coverts are also moulted, but usually 5–7 outer retained; rarely none or—apparently—all greater coverts moulted. No pre-nuptial moult in winter. – **SPRING Ad ♂** Usually identified by combination of uniform dark tawny-brown greater coverts (lacking contrasting outer greater coverts with pale edges and whitish tips), rather broad tail-feathers with fairly rounded tips (usually requires handling or good photographs), blackish lores, very dark grey upper cheeks and eye surround, and more or less complete dark grey necklace, usually leaving only a small whitish chin-spot. Some have a few white feathers on central throat, forming an irregular patch in the otherwise uniform dark grey, but throat should not be largely streaked and blotched dark and white (a pattern typical of ad ♀/2y ♂). Note also distinct white supercilium and white mark below eye and to base of lower mandible; a rather deep tomato-red tinge to orange-red of breast and flanks, and much yellow on lower mandible. **Ad ♀** Resembles ad ♂ but differs by dark grey rather than blackish lores; cheeks finely streaked white; grey on head and neck faintly brown-tinged, less pure lead-grey; somewhat larger whitish chin with ill-defined border to grey throat, almost invariably with a few scattered dark streaks, and some white invading upper throat, streaked dark; slightly duller orange-brown breast and flanks; and slightly duller yellow bill. A few advanced ♀♀ and retarded ♂♂ can be

♂, presumed 1stS, Mongolia, May: being ♂ (grey throat and blackish lores) and probably young (primary-coverts apparently juv) explain rather dull appearance. However, no moult limits in wing with buff-tipped greater coverts discernible, which could infer extensive post-juv moult, or even 1stY that has not moulted the coverts, but by May has lost the buff tips through wear. (H. Shirihai)

very similar and require careful assessment. (Note that some specimens differ from the above, showing characters of opposite sex, but after careful examination of long series and assessment of plumage characters against wing length, we regard these as wrongly sexed.) **1stS** Resembles ad, and some indistinguishable in the field, but usually recognised by retained outer juv greater coverts being paler, edged greyish and sometimes tipped whitish, in contrast to inner dark tawny-brown. Also, tail-feathers somewhat narrower and tips more pointed (unless too abraded to be judged). Apparently some 1stY moult all greater coverts and become difficult or impossible to age from ad. – Sexing difficult as in ad. Birds positively aged as 1stS that have neat head pattern with dark lores, much grey on neck and only small white patch on chin should always be ♂♂. Those 1stS with dullest plumage (head and neck more brown than grey, entire throat off-white and streaked, breast and flanks dull orange-brown) should be ♀♀. – **AUTUMN Ad** Similar to spring but entire plumage fresh. All greater coverts uniformly greyish-brown. Tail-feathers broad, tips rather rounded. **1stW** As spring. Generally, all or many outer greater coverts retained juv feathers, pale-edged and pale-tipped (either evenly broad tips or wedge-shaped ones). Tail-feathers on average

Ad ♀, China, May: like ♂ but head generally duller grey with dusky-brown loral line and cheeks finely streaked white; also larger whitish chin/throat, nearly invariably bordered at sides with streaky lateral throat-stripe. Duller orange-brown breast and flanks also characterise ♀ at this season. Round-tipped and more greyish-fringed primary-coverts indicate ad. Note apparent differences in length of inner three greater coverts, forming 'false' moult limit often found in thrushes. (M. Parker)

1stW, presumed ♂, Japan, Oct: aside from two innermost coverts, greater coverts are juv with large buff spots at tips. Being 1stW, sexing could be attempted: note greyish-tinged head, bold white supercilium and blackish lores, plus rather vivid orange breast and flanks, and mostly grey throat (only small white patch and no streaking) all strongly suggest a ♂. (T. Shimba)

1stW, presumed ♂, Finland, Nov: all or most outer greater coverts are juv, with pale edges and bold tips; tail-feathers seem rather narrow and acutely pointed, making this bird 1stW. Probably a ♂ given strong face markings, greyer head and upper breast; mostly off-white and streaked throat is typical of young ♂, but much overlap with young ♀. Vagrants usually unmistakable once greyish head and neck, white supercilium, and orange breast and flanks, are seen well. (M. Varesvuo)

1stS ♀, Thailand, Mar: except innermost feather, greater coverts are juv (tipped whitish). Once aged as 1stS, sexing is possible: note brownish (rather than grey) head and neck, with entire throat off-white, and rather dull orange-brown breast and flanks, all suggestive of ♀. (I. Waschkies)

narrower and often acutely pointed. However, several birds with pointed, narrow tail-feathers have all greater coverts of ad type, presumably due to a complete moult of these (rather than being ad with uncharacteristic tail-feathers), and such birds are very difficult to age (or sex!). **Juv** Has narrow buffish shaft-streaks above (largest on scapulars), including on mantle, lesser, median and inner greater coverts. Underparts densely and boldly spotted dark, breast yellowish-buff, flanks rusty-ochre. Centre of belly, vent and undertail-coverts are largely unspotted white.

BIOMETRICS L 20.5–22.5 cm; **W** ♂ 121–137 mm (n 59, m 127.4), ♀ 118–128 mm (n 41, m 123.1); **T** ♂ 79–93 mm (n 59, m 85.6), ♀ 76–89 mm (n 41, m 81.6); **T/W** m 66.8; **B** 20.4–23.5 mm (n 36, m 22.0); **Ts** 28.5–32.7 mm (n 36, m 30.4). **Wing formula:** p1 minute, < pc 3.5–14 mm, < p2 68–81 mm; **p2** < wt 3–8 mm; **pp3–4** equal and longest (though sometimes p4 0.5–1 mm <); **p5** < wt 5.5–11 mm; **p6** < wt 17–23 mm; **p7** < wt 22–29 mm; **p10** < wt 37–43 mm; **s1** < wt 40–46.5 mm. Emarg. pp3–5.

GEOGRAPHICAL VARIATION & RANGE Monotypic. – Siberia from east of Ob to Okhotsk and Kamchatka, south to Altai and N Mongolia; winters E India, SE Asia.

AMERICAN ROBIN
Turdus migratorius L., 1766

Fr. – Merle d'Amérique; Ger. – Wanderdrossel
Sp. – Zorzal robín; Swe. – Vandringstrast

A widespread and common American thrush, named 'Robin' by the first British immigrants due to its red breast. It has straggled many times to W Europe, apparently aided by prevailing strong westerly winds in autumn. But it is also a strong flier which undertakes long annual migrations, for instance from Canada to Mexico and back, a fact indicated by its scientific name.

Ad ♂, USA, May: unmistakable in spring and summer plumage, having jet-black head with striking white 'eyelids'; yellow bill also obvious and mostly lead-grey upperparts contrast with deep orange-rufous breast, belly and flanks; all of the greater coverts uniformly edged grey. (A. B. van den Berg)

IDENTIFICATION Roughly Blackbird size and Fieldfare shape, but has subtly shorter bill, proportionately. Easily identified by *dark reddish breast and belly* (some ♀♀ and young ♂♂ are slightly duller and paler reddish-brown), and by dark grey or *black head* with short *white marks above and below eye* (white 'eyelids'). The *throat is white with dense, bold black streaking* (but a few ♀♀ can have the chin and central throat rather unstreaked). Upperparts plain greyish, sometimes with a brownish cast. Adult ♂ usually has crown, nape and at times upper mantle blackish. Centres of tertials and uppertail are also dark grey or blackish. Adult ♀ is sometimes quite similar, difficult to separate, but more often is less black, more dark grey on nape and sometimes crown. The *outer tail-feathers are white-tipped, underwing reddish* or orange-buff. Bill either *all yellow* (adult summer), or brownish-yellow with dark culmen and tip; in winter bill darkens to brownish-grey. First-winter is slightly duller on average (but beware individual variation), and usually has unmoulted outer greater coverts edged paler and finely tipped whitish (adult has uniformly grey-edged coverts; cf. Ageing & Sexing). Behaviour much as Fieldfare, thus commonly seen feeding on lawns in parks and gardens, walking or hopping briefly, then stopping, looking for earthworms or insects. When feeding on ground adopts typical thrush posture with raised head and drooping wings.

VOCALISATIONS Has several calls, many of which somewhat recall various European species. Commonly gives deep, emphatic *chuck chuck chuck* (rather like Blackbird but softer, almost like Redwing) or in alarm a faster and high-pitched, metallic *chik-chik-chik-...* (vaguely recalling Song Thrush alarm). Alarm notes can also fall in pitch, *chi-chi-chek-chek-chek*. The flight note is a lisped, high-pitched *see-lip* (Farrand et al. 1983) or *ssip* (Dunn et al. 1999), on recordings apparently mainly trisyllabic, squeaky *cheep-cheep-chep*, rather reminiscent of flight call of Fieldfare.

SIMILAR SPECIES A straggler in Europe should not cause any serious identification problems as there are no other similarly-sized birds with such a red breast and dark head with white 'eyelids'. (A few relatively similar species occur in E Asia, Africa and South America, but none is likely to ever occur naturally in Europe.) The only thinkable pitfall might be Eyebrowed Thrush, itself a rare straggler to Europe from Siberia, which can show a superficial similarity to the dullest possible American Robin. Note, however, greyish upperparts (not brown) on American Robin, and darker legs.

AGEING & SEXING (*migratorius*) Ages differ slightly during 1stY, but ageing generally requires close views. Sexes very similar (more than generally maintained), still often separable after post-juv moult; difference in 1stW very slight and requires great care and consideration of age. – Moults. Complete post-nuptial moult of ad in late summer and early autumn (Aug–Sep). Partial post-juv moult in summer (Aug–Sep, sometimes Oct) does not involve flight-feathers, tertials or primary-coverts. A variable number of inner greater coverts (rarely all) and sometimes 1–2 tertials are also moulted. No pre-nuptial moult in winter. – **SPRING Ad ♂** Head largely jet-black or dark blackish (except chin and throat), the black sometimes invading upper mantle. Well-defined short white patches above and below eye (white 'eyelids'). White throat densely streaked black, streaks sometimes partially merge, especially at lower edge. Breast, belly and flanks deep tomato-red (dark orange-rufous). All greater coverts uniformly edged grey (but rarely have some indistinct white tips, though never wedge-shaped). Tail-feathers rather broad, tips rounded. Uppertail blackish. Bill all yellow. **Ad ♀** Resembles ad ♂ (some difficult to separate!) but is on average a little duller, with slight brownish tinge to grey upperparts. Head in many very nearly as black as in ♂, but at least some have dark grey crown and nape, rather than black; pale 'eyelids' are usually less distinct and pure white, and underparts are often slightly paler, more orange-brown with more pale or brownish mottling even in spring (though some acquire quite dark rufous underparts and are hard to separate on this from ♂). Uppertail usually dark grey, on average somewhat less black than in ♂. Bill yellow like ♂ or yellowish with slightly darker tip and culmen. **1stS ♂** Similar to ad ♂, but differs in slightly browner primaries (often heavily abraded tips from c. May), primary-coverts and a varying number of brownish unmoulted juv outer greater coverts, contrasting with new greyer rest of wing-feathers. Tail-feathers on average narrower and more pointed (with worn tips from late spring). **1stS ♀** Like ad ♀, but differs in visible contrast among greater coverts, between browner unmoulted juv (outer) and greyer new (inner). – **AUTUMN** Ageing and sexing much as in spring, but

Presumed ad, England, Oct: the only American *Turdus* to have occurred in the treated region has similar posture to Fieldfare. Ageing and sexing rarely straightforward, but wing appears uniform and pattern of greater coverts more ad-like (diffuse pale tips to some coverts not unusual in fresh plumage). Broad grey edges to primary-coverts and blackish lores suggest fresh ad ♂. (G. Thoburn)

plumage now fresh with some pale fringes to dark feathers. Unmoulted juv outer greater coverts and tertials often have small whitish spots on tips of outer webs. Note that some birds moult 1–2 tertials and have very tiny pale spots on remaining juv greater coverts, so ageing using presence of these spots requires care. **Juv** As in most juv thrushes has narrow buff-white shaft-streaks above (largest on scapulars), including on lesser and median coverts. Underparts rich orange or rufous-tinged and boldly spotted dark. Found only on breeding grounds and unlikely to occur in Europe.

BIOMETRICS (*migratorius*) **L** 22–25 cm; **W** ♂ 126–138 mm (*n* 16, *m* 131.6), ♀ 120–137 mm (*n* 11, *m* 128.2); **T** ♂ 94–106 mm (*n* 15, *m* 98.3), ♀ 85–100 mm (*n* 11, *m* 93.8); **T/W** *m* 74.2; **B** 21.3–24.3 mm (*n* 27, *m* 22.9); **Ts** 31.5–36.2 mm (*n* 27, *m* 33.3). **Wing formula: p1** minute, < pc 3–12 mm, < p2 65–77 mm; **p2** < wt 4–10 mm; **p3** < wt 0.5–4 mm; **p4** longest (sometimes pp3–4 equal); **p5** < wt 1.5–6 mm; **p6** < wt 8–12 mm; **p7** < wt 17–23 mm; **p10** < wt 31–36 mm; **s1** < wt 36–40 mm. Emarg. pp3–5, sometimes faintly also on p6.

GEOGRAPHICAL VARIATION & RANGE Several subspecies described based on variation in colour saturation, amount of white on tail corners, and size. There is also some individual variation within taxa. Only one subspecies known to have reached W Palearctic as a straggler, *migratorius*, hence only this treated in detail above. However, vagrants are generally 1stY, which are more difficult to label subspecifically.

T. m. migratorius L., 1766 (Alaska, much of Canada except extreme east, NE USA). Described above. Usually ad ♂ has blackish head and variably dark grey or black nape, with clear border to grey mantle. A few birds presumably of this race have a little dark grey or black invading upper mantle (like in following race), but apparently never extensively. Amount of white on tips of outer tail-feathers individually variable.

? T. m. nigrideus Aldrich & Nutt, 1939 (easternmost Canada, Newfoundland). Claimed to differ in ad ♂ plumage, being 'more consistently and extensively black' above, and deeper rufous-red below, less orange-tinged (Phillips 1991), and throat-streaking is said to be denser, tending to coalesce. However, not all from this range are blackish on the mantle, and as to claimed differences in the underparts, these are just vague and variable tendencies, and it remains questionable whether this race is sufficiently distinct. Biometrics are nearly identical. Until the race has been re-evaluated using a large sample of breeding-season material it is perhaps better to treat birds with very extensive black mantles as 'var. *nigrideus*'.

REFERENCES PHILLIPS, A. R. (1991) *The known birds of North and Middle America*. Part 2. Denver.

1stW ♂, England, Jan: fresh young ♂ to some extent duller than fresh ad ♂, and very similar to fresh ad ♀, thus ageing should precede sexing. In this case, ageing straightforward due to distinctly patterned juv greater coverts; also unlike ad, primary-coverts edged narrowly and less pure grey. Overall plumage and blackish lores strongly suggest ♂. In all non-breeding plumages, bill brownish-yellow with dark culmen and tip. (I. Fisher)

1stW, presumed ♂, England, Dec: aged by pale-tipped juv greater coverts, but somewhat intermediate coloration makes certain sexing impossible. However, closer to young ♂ than young ♀. Fresh young birds often duller, with some brownish tones to grey upperparts, and black of head replaced by dark streaks, while underparts are paler, more orange-brown with more pale tips. (G. Reszeter)

1stW ♀, England, Nov: the dullest plumage. All greater coverts appear to have been replaced during post-juv moult, but thin buff edges to juv primary-coverts, and moult contrast among tertials indicate 1stW. The weak facial pattern and paler underparts make this bird a young ♀. Confusion perhaps possible with Eyebrowed Thrush, but has predominantly greyish upperparts (not brown) and grey (not yellowish) legs. (R. Lortie)

NAUMANN'S THRUSH
Turdus naumanni Temminck, 1820

Fr. – Grive de Naumann; Ger. – Rotschwanzdrossel
Sp. – Zorzal de Naumann; Swe. – Rödtrast

This is one of the least known of a quartet of closely related Siberian taiga-living thrushes. Visitors from W Europe usually have little problem finding a Black-throated Thrush in Siberia in summer. Running into a Dusky or a Red-throated Thrush requires that you move further east or south-east and have a good guide. But to spot a Naumann's on a regular birding trip means that you are either skilled or blessed with luck. At least so it seems, given how few birdwatchers return from Siberia with this species on their trip lists. It breeds mainly in SC Siberia, reaching north roughly to 65°N. Only a few times has it straggled to Europe in autumn/winter.

Ad ♂, Japan, Mar: rufous-red supercilium, throat, breast and neck-sides. Rufous-red blotches also on flanks and undertail-coverts (lacking in Red-throated Thrush, which has these parts cleaner whitish-grey). Brown-grey upperparts with varying amount of rufous-red fringes (especially when fresh). Rufous also extensive on rump, uppertail-coverts and outertail. Aged by evenly-feathered ad greater coverts, and by bright rufous plumage. (K. Kimura)

IDENTIFICATION In size and shape rather like a Song Thrush, thus a mid-sized thrush with moderate-length tail. Flight direct and agile with short glides, much like Song Thrush. In flight from below, note *orange-brown underwing*, just as in Song Thrush. As other thrushes, feeds mainly on ground. Mostly shy or wary, taking off early and moving some distance before taking a new perch. Despite being very closely related to Dusky Thrush, its appearance differs greatly in having a good deal of rusty-red in plumage, especially on breast, flanks, rump and tail. All plumages similar, but adult ♂ is on average brightest and most extensively and deepest red, whereas immature ♀ is dullest and has least red. Adult ♂ is extensively *rufous-red on underparts*, all *feathers tipped white* when fresh (generally not worn off until spring or summer, first on throat and breast), and *supercilium and neck-sides of neck* also *rufous-red*, encircling grey cheeks. Often has black spots on sides of throat forming stripes, but a very few lack any dark spots. Only centre of belly is largely white, and throat may be pale rufous or off-white (variation not fully understood). Upperparts are brown-grey with varying amount of rufous-red on scapulars and rump/uppertail-coverts, at times also on mantle and back. Adult ♀ is very similar, only on average less deeply and extensively reddish below, invariably has black-spotted lateral throat-stripes and upper throat, and usually lacks any obvious rufous on brown-grey mantle and scapulars (back, rump/uppertail-coverts rufous-tinged in some). Wing-feathers variably edged rufous. Immature ♂ is either similar to adults or clearly less reddish below (lower breast and flanks off-white with some ill-defined grey and rufous blotches or streaks), but invariably differs in having many outer or all greater coverts unmoulted, being duller and paler-edged. Also, note that upperparts often have some rufous admixed like the adult ♂, at least on scapulars and rump/uppertail-coverts. Immature ♀, finally, is very similar to immature ♂ (though on average a little duller), but differs by uniform brown-grey upperparts lacking any rufous (except sometimes on uppertail-coverts). Juvenile is similar to young ♀ but is recognised by its buffish wedge-shaped shaft-streaks on upperparts, most prominently on scapulars, mantle and secondary wing-coverts.

VOCALISATIONS The song is still poorly known, the few existing descriptions being partly contradictory, partly based on a misidentified bird (*BWP*). The only known published recording is by Veprintsev & Leonovich (1986) and is just 9 sec. long. However, two unpublished recordings from Yakutsk, one by P. Alström and one by E. Hassel, have been made available to us, and the description here is based on these. The full song is relatively similar to that of Dusky Thrush, being rather primitive in structure and contains clear and loud notes, several of which are 'bent' (glissando, diphthongs). Notes or phrases are often repeated, and interpolated by brief pauses. Part of a song can be rendered *chaauuh-chi, chaauuh-chi... chip-chuuah... chip chip chaauuh...*, etc. The song appears to lack rolling or harsh notes, all are clear, whistling or squeaky. Delivered from elevated song perch at moderate pace. – Most calls apparently rather Fieldfare-like, including common series of rather harsh, shrill, *chik-chik-chik-chik-...*, and high-pitched, squeaky *giieh*, and variations of this, in flight.

SIMILAR SPECIES Main confusion risk is with young ♀ *Red-throated Thrush*, which can be surprisingly similar to young ♀ Naumann's, but note on Naumann's at least some ill-defined rufous blotches among grey on lower breast, flanks and undertail-coverts (Red-throated: off-white breast and flanks with only diffuse greyish streaks or blotches, no rufous, and usually largely uniform whitish undertail-coverts), and on average more rufous tinge to tail (but many similar in last respect). – Must also be separated from closely related *Dusky Thrush* (treated here as separate species), of similar size and shape, and both have orange underwing. Note on Naumann's extensive rufous portions on tail in adult (Dusky: all-dark tail), and rufous blotching of underparts, but lack of rufous edges, or large rufous patch, on closed wing (Naumann's has plain brownish upperwing). Upperparts never blotched blackish, as in most Dusky ♂♂, is brown-grey, or brown-grey with some reddish admixed (♂). Supercilium often rather ill-defined and tinged rufous (Dusky: prominent and whitish). – Naumann's and Dusky Thrushes reputedly hybridise and produce intermediates in area of contact. It is therefore best to avoid positive identification of birds with mixed characters, e.g. those looking like Naumann's but having partially blackish (or very dark rufous-brown) spotting or blotching on breast and flanks, or extensive vivid rufous edges on upperwing.

AGEING & SEXING Ages differ slightly during 1stY, but

Ad ♀, Japan, Feb: average less deeply and extensively reddish below than ♂, but black lateral throat-stripe and patch usually better developed. Note lack of obvious rufous in brown-grey upperparts, and rufous in rump/uppertail-coverts strongly reduced. Due to similarity with young ♂, ageing should precede sexing: note ad-like greater coverts and primary-coverts. (T. Shimba)

Ad ♀, Korea, May: even if correctly aged, sexing is not always straightforward. Ad mainly by no moult limits in greater coverts. Rufous centres to scapulars could suggest young ♂, but apparent lack of any rufous on rump/uppertail-coverts and throat better fit ♀. (A. Audevard)

1stY ♂, Japan, Feb: some 1stY ♂♂ in winter/spring develop very colourful plumage, like ad ♂. However, moult assessment (all primary-coverts and some outer greater coverts being juv) shows this bird to be 1stY. (T. Shimba)

ageing requires close views or handling. Sexes separable after post-juv moult, but separation in 1stW not always easy or even possible unless at close range. – Moults. Complete post-nuptial moult of ad in summer and early autumn (Jul–Sep), before migration. Partial post-juv moult in summer (Aug–Sep) does not involve flight-feathers, tertials or primary-coverts. Median and a varying number of inner greater coverts (usually 1–4, rarely 0–6) are also moulted. Apparently no pre-nuptial moult in winter. – **SPRING Ad ♂** Identified by combination of uniform ochraceous-brown greater coverts (lacking contrasting outer greater coverts with pale edges and whitish tips), broad tail-feathers with rather rounded tips, extensively bright rufous chin, throat, breast, upper belly, flanks, vent and undertail-coverts, all feathers tipped white when fresh and to variable extent still in spring, *and* rufous feather-centres visible on otherwise greyish upperparts, notably on scapulars and rump/uppertail-coverts, sometimes also (but less prominently) on mantle and back. Individual variation as to colour of throat, some being dark rufous (rare), others pale rufous or off-white; variation also as to presence of black spots on throat-sides, some having none, others just as distinct as ♀/imm ♂. **Ad ♀** Differs from ad ♂ in lacking any rufous feather-centres on brown-grey scapulars and mantle, and by having on average slightly less deep and extensive rufous on underparts (but overlap, and only extremes can be separated on underparts coloration alone). Invariably black-spotted throat-sides and upper throat, the throat being either strongly rufous as in a or paler, more off-white with rufous tinge. **1stS ♂** Resembles ad ♀ and less brightly coloured ad ♂, but shows contrast between moulted inner greater coverts (brown, without pale tips) and retained juv outer greater coverts (edged pale brown, innermost tipped whitish); the contrast is difficult to see on those with only 0–2 inner coverts renewed, but paler edges and tips on greater coverts reveal age. Most have rufous feather-centres on scapulars and rump/uppertail-coverts, less obvious than on ad ♂ but usually visible at close range (very few lack any rufous above). Throat usually mainly rufous, spotted black at sides. Tail-feathers narrow, tips often rather acutely pointed. **1stS ♀** Resembles 1stS ♂ but invariably lacks any rufous on upperparts, is usually somewhat plainer below, with less rufous, and has invariably off-white chin and throat with prominent black streaking on sides. Some individual variation, the most brightly coloured being inseparable from least brightly coloured 1stS ♂. – **AUTUMN Ad ♂** Similar to spring but whole plumage fresh. All greater coverts uniformly greyish-brown with only subtly paler edges. Tail-feathers broad, tips rather rounded. **Ad ♀** As spring. All greater coverts uniformly patterned. Tail-feathers as in ad ♂. **1stS ♂** As spring. All or many outer greater coverts retained juv feathers, pale-edged and pale-

1stS ♂, Mongolia, May: this falls within normal variation of Naumann's Thrush, although perhaps possibility of a hybrid *naumanni × eunomus* cannot be completely excluded given dark breast-band (some *naumanni* can show hint of this, but never prominent). Apparent moult limit among greater coverts, and presence of juv primary-coverts indicative of 1stY; combination of bright and extensive rufous-mottled flanks, rufous-tinged throat and narrow and spotty black lateral throat-stripe are features of young ♂. (S. Leveelahti)

1stW ♂, Finland, Nov: extensive rufous blotching on underparts not always as extensive and pure as this. Naumann's can require careful separation from Dusky Thrush, but has mostly plain brownish upperwing (lacking bright rufous patches) and much rufous on rump and uppertail, which are features of ♂ Naumann's (tail invariably all dark in Dusky). Aged by juv primary-coverts and most greater coverts, while relatively substantial rufous in scapulars and on flanks confirm the bird as young ♂. (M. Varesvuo)

tipped. Tail-feathers on average narrower and often acutely pointed. **1stW ♀** As 1stW ♂, although on average plainer and duller. **Juv** Resembles 1stW ♀ but has narrow buff-white shaft-streaks above (largest on scapulars), including on lesser, median and inner greater coverts.

BIOMETRICS L 21–24 cm; **W** ♂ 125–138 mm (n 27, m 130.4), ♀ 126–136 mm (n 11, m 129.7); **T** ♂ 83–98 mm (n 27, m 88.2), ♀ 84.5–92 mm (n 11, m 87.6); **T/W** m 67.6; **B** 21.0–24.4 mm (n 36, m 22.6); **Ts** 28.5–34.0 mm (n 38, m 31.6). **Wing formula: p1** minute, < pc 7–15 mm, < p2 75–80 mm; **p2** < wt 3.5–6.5 mm; **p3** longest; **p4** < wt 0.5–2 mm; **p5** < wt 5–8.5 mm; **p6** < wt 17–22 mm; **p10** < wt 35–43 mm; **s1** < wt 41–46 mm. Emarg. pp3–5.

GEOGRAPHICAL VARIATION & RANGE Monotypic. – C Siberia from Angara and middle Yenisei east to upper Lena and Okhotsk, south to Baikal and Stanovoy; winters in China and Korea.

TAXONOMIC NOTES Naumann's Thrush was previously treated together with Dusky Thrush *T. eunomus* as one species—and the two are certainly closely related—but we prefer to regard them as separate species based on distinctly different plumage and widely sympatric ranges apparently without common and unhindered interbreeding. Recently, the BOU decided to split them (Knox *et al.* 2008, BOU 2009). Hybridisation in a fairly substantial contact or overlap zone in Siberia has been inferred from the occurrence of birds with intermediate characters. Although not denying that a certain amount of hybridisation might take place, we are not aware of any records of observed interbreeding, and the noted variability in plumage characters may partially reflect historic gene flow and lack of selection against this variability. Birds with intermediate characters also occur far from the contact zone. For references, see e.g. Hartert (1910), Johansen (1954), Glutz von Blotzheim & Bauer (1988), Stepanyan (1990), Clement & Hathway (2000) and *HBW* (2005). In our opinion, Naumann's and Dusky Thrushes differ very clearly in sufficiently many ways to be treated as different species, and they are unlikely to merge. Until more is known of the nature and frequency of the purported mixing we prefer to regard this as a stable but subordinate feature, much in the same way as Pine Bunting and Yellowhammer mix in an apparently stable zone of overlap.

REFERENCES BOU (2009) *Ibis*, 151: 224–230. – Clement, P. (1999) *Limicola*, 13: 217–250. – Johansen, H. (1954) *J. f. Orn.*, 95: 319–342. – Knox, A. *et al.* (2008) *Ibis*, 150: 833–835. – Veprintsev, B. N. & Leonovich, V. (1986) *Birds of the Soviet Union: A Sound Guide*. Passeriformes: Thrushes. Moscow.

1stW ♂, China, Dec: less bright overall plumage than previous bird, but rufous-orange on breast and flanks, greyish head, well-defined dark lores and lateral throat-stripe, weak supercilium, plain upperparts and wing, but rufous uppertail, identify species, age and sex. (D. Thirunavukkarasu)

1stW, presumed ♂, China, Oct: aged by juv primary-coverts and some outer greater coverts. Relatively substantial rufous in supercilium, throat, neck-sides, lower-breast and flanks are better match (but not conclusive) for 1stW ♂, whereas lacking rufous on rump/uppertail-coverts more usual in ♀. Thus, sexing tentative only. (M. Andrews)

1stW ♀, China, Dec: dullest plumage, with restricted rufous-orange on body-sides and head, upperparts and wing brownish, and no rufous on rump area, but has well-developed lateral throat-stripe and patch, as well as rather plain wing; most (visible) wing-feathers juv. (D. Thirunavukkarasu)

Presumed hybrid Naumann's × Dusky Thrushes, ad ♂, Taiwan, ?Dec: freshly moulted, and very likely a hybrid Naumann's × Dusky Thrushes. Most features good for ad ♂ Naumann's, but has Dusky-like wing with extensive rufous, and on the breast dense rufous and partly black spotting, forming a hint of the two cross-bars typical of Dusky. (O. Apple)

DUSKY THRUSH
Turdus eunomus Temminck, 1831

Fr. – Grive à ailes rousses; Ger. – Rotflügeldrossel
Sp. – Zorzal eunomo; Swe. – Bruntrast

The vast Siberian taiga has proven ideal habitat for the evolution of a variety of thrushes. One of these is Dusky Thrush, the English name of which can appear a bit unfair: the older ♂ is a very attractive bird with a distinctive pattern. This species breeds mainly in N and NE Siberia, and its range reaches all the way from the middle Yenisei and Lena rivers (roughly above 60°N) to the edge of the tundra on the Taymyr, to Anadyr, Kolyma and Kamchatka peninsulas. It has straggled many times to Europe in the autumn or winter. Both arboreal and terrestrial habits.

Ad ♂, Japan, Mar: in this plumage, unmistakable, with large and bright copper-red wing, black upperparts (while still fresh with brown fringes), almost unbroken black breast-band, and prominent whitish supercilium, neck-side and throat. Aged by evenly-feathered wing with extensive rufous parts including much of primary-coverts. (S. Rooke)

IDENTIFICATION In size and shape, very similar to Naumann's Thrush (which see for a general description of jizz and behaviour), only differing very subtly in a proportionately larger size but shorter tail. Plumages can be divided into three categories: (i) adult ♂ with its strikingly beautiful plumage, (ii) adult ♀/immature ♂, which resemble the old ♂ but are a little duller, and (iii) the often more discreet immature ♀. Adult ♂ has extensive *bright rufous* ('copper-red') *edges on the wing forming a large unbroken patch*, and *much black on crown, nape, mantle and scapulars* (fresh feathers fringed brown, in summer wearing to near all black; some individual variation though, and a few have less black regardless of abrasion). Further, has a nearly unbroken *black breast-band* (few exceptions), a *prominent whitish supercilium*, very *dark* and unpatterned *cheek patch* (few exceptions), and dark bill with *yellow base to lower mandible*. The immature ♂ is similar but differs mainly in having generally *much less rufous on the wing* (adult ♂ has whole outer web of inner two tertials, and much of median and all of greater coverts rufous, immature ♂ only the outer edges of tertials and greater coverts, and median coverts largely dark). Bill is like in adult ♂ or a little darker. Other characters differ only slightly and not consistently. Adult ♀ is very similar to immature ♂ but has *generally no obvious black admixed on olive-brown upperparts* (usually at the most a hint of smaller dark grey centres in some; exceptions very few and may be due to mistake when sexing), median coverts are mainly dark brown-grey with only small portion rufous, if any, and greater coverts have darkish centres and limited rufous at edges only. Although nearly all ♀♀ differ from ♂ in having less rufous on outer edges of tertials, the odd birds overlap. Generally rather limited yellow at base of lower mandible. Immature ♀, finally, is generally clearly more streaked and duller, lacking any clean rufous, and has throat buffish with much streaking. Breast-band poorly developed, and streaking on flanks rather diffuse. Bill rather dark. Juvenile is similar to young ♀ but is readily recognised by its buffish wedge-shaped shaft-streaks on upperparts, most prominently on scapulars, mantle and secondary wing-coverts. Differs in flight from larger and longer-tailed Fieldfare by *orange*, not white, *underwing* (although slightly paler buff in young ♀).

Ad ♀, Japan, Mar: main difference from ad ♂ is darker and less extensive rufous panel on closed wing, and less black feather-centres on upperparts. Note that rufous on tertials is usually limited to outer edges, on primary-coverts absent or, as here, just hinted, and on median coverts mainly on outer webs. Aged by evenly-feathered wing, all greater coverts moulted to ad type. (H. Iozawa)

VOCALISATIONS The song is rather primitive in structure, much like in Fieldfare or Black-throated Thrush, brief squeaky notes, uttered singly or a few in combination, relieved occasionally by one or more rather characteristic deep, drawn-out, fluty notes, at times becoming glissando and falling in pitch, with subdued twitters interspersed, *chu-vü peeeh, peeuh…* (*trixizri*)*… chu-che peeh, peeuh, peeah…* (*trixizri*)*…tru-zi peeeuuh…* (*sisisiri*)*…*, etc. (Partly based on an unpublished recording by V. Arkhipov, Chukotka, Siberia, Jun 2007.) Some recordings also include high-pitched short whistled notes. Very similar to song of Naumann's Thrush, and safe separation difficult or impossible. Some recordings feature a squeaky *kve-kve* (or *kve-kve-kve*) uttered for long periods, and this may have song function, too. The song is delivered from elevated perch at moderate pace. Little variation, and not so pleasing to the human ear (descriptions in the literature mention a more complex and pleasing song variation, but this either refers to subsong rather than territorial song, or requires better verification). – Most calls resemble the voice of Fieldfare, including common series of hard, rather harsh notes, *chuk-chuk-chuk-chuk-…*, a high-pitched, squeaky *giieh*, and variations, in flight. A thin, ringing *srree* has been noted from migrants. The normal call is said to be similar to Redwing (Seebohm, cited by Hartert 1910), but this might be the same as *srree*.

SIMILAR SPECIES Main confusion risk is closely related *Naumann's Thrush* (treated here as separate species). Separated in most plumages by rufous edges on upperwing forming large bright reddish-brown patch when wing is folded; black blotches or crescent-shaped marks on breast forming complete or partial breast-band, and similar black or dark grey blotches on flanks; blackish or diffusely black-spotted upperparts in ♂; largely blackish or dark grey-brown uppertail (not largely rufous); prominent white or cream-white supercilium. Young ♀ is less characteristic, but note that patches on flanks are dark grey, not reddish, and that a pale supercilium is fairly obvious. Often a hint of a dark breast-band (absent in Naumann's). – Dusky Thrush and Naumann's Thrush apparently hybridise and produce intermediates in area of contact. It is therefore best to avoid positive identification of birds with mixed characters, e.g. those that look like Dusky but have very little rufous on upperwing, or more rufous than blackish spotting on breast and flanks, or with a lot of rufous on rump and uppertail.

AGEING & SEXING Ages differ slightly during 1stY, but ageing requires close views or handling. Sexes separable after post-juv moult, but separation in 1stW not always easy or even possible unless at close range. – Moults. Complete post-nuptial moult of ad in summer and early autumn (Jul–Sep), before migration. Partial post-juv moult in summer (Aug–Sep) does not involve flight-feathers, tertials or primary-coverts. Median and a varying number of inner greater coverts (usually 2–4) are also moulted. Apparently no pre-nuptial moult in winter. – SPRING **Ad ♂** Extensively bright rufous-brown ('copper-red') on upperwing, broad outer edges to tertials (whole outer web of inner two feathers rufous, and sometimes an isolated rufous patch also on inner webs), and all median coverts largely rufous, and greater coverts extensively rufous (often entire outer webs), the rufous colour merging to form unbroken area on folded wing, but somewhat less extensive on a few, involving about half of outer webs. (Some have small pale patch at base of outer primaries, but this variable and appears unrelated to age or sex.) Primary-coverts often bright rufous with only small black tips (diagnostic), sometimes with more black, including partially dark outer webs. Crown to back either near all-black or at least dark grey-brown with blackish feather-centres. Supercilium prominent, pale cream-white. Jet-black blotches or crescent-shaped marks on breast and flanks (usually forming complete breast-band and irregular blackish sides), spots on lower flanks and vent variably tinged dark chestnut (this thought to be within normal variation and not to indicate hybrid origin). Throat often (but not invariably) unstreaked cream-white. Ear-coverts ('cheek patch') typically blackish, but some have patch admixed with a little brown and grey. Tail-feathers rather broad, tips rounded. **Ad ♀** Differs from ad ♂ by lacking extensive and bright rufous on wing, the rather narrow pale edges to greater coverts and tertials being reddish tawny-brown or dark rufous. Dull rufous primary-coverts extensively tipped dark grey, and entire outer webs of primary-coverts also grey (only rarely a hint of rufous on base of outer webs approaching ♂ pattern). Upperparts usually uniform olive-brown (exceptionally some diffuse dark grey or blackish feather-centres on mantle; never extensively blackish above when abraded). Throat generally finely streaked or spotted dark except at centre. **1stS ♂** Resembles ad ♀, but rufous edges on upperwing are brighter rufous (generally narrow as in ad ♀, but a few have slightly more extensive edges), and upperparts usually have much black admixed. Outer greater coverts unmoulted (juv feathers), dark-centred with duller pale brown edges and whitish tips, versus moulted rufous-edged inner. Primary-coverts like ad ♀ but tips blacker and more contrasting. Tail-feathers narrow. **1stS ♀** Somewhat plainer than preceding plumages, resembles ad ♀ but dark spotting on breast and flanks less prominent, dark colours brown-grey, never blackish. As 1stS ♂, outer

Ad ♀, China, Apr: some ♀♀ are duller than any ♂, with less extensive and less contrasting dark blotching on breast and flanks, and with upperparts duller with diffuser and less black centres. Also, less solid and bright rufous on wing (none visible on primary-coverts, practically none on median coverts, and tertials only rather narrowly edged rufous). Aged by having all ad-like greater coverts and tertials. (J. Martinez)

1stS ♂, Hong Kong, Mar: moult limit in greater coverts (at least inner two post-juv), and retained juv primary-coverts, clearly indicative of 1stS. Bold, solid black breast-band and large blackish blotches on flanks, bright rufous edges to renewed greater coverts, broad blackish bases to scapulars and mantle typical of young ♂, although safe separation from boldest patterned ad ♀ could be challenging if bird not aged first. (M. & P. Wong)

1stS ♀, United Arab Emirates, Mar: distinct moult limit among greater coverts (inner two have been replaced) and juv primary-coverts indicate 1stY, while rather plain brown upperparts and poorly-developed blotches on breast and flanks are typical of ♀. Rufous edges to new greater coverts dull and restricted to outer webs, while dark face pattern less pure black. (I. Boustead)

Presumed ad ♂, Japan, Oct: uniformly ad greater coverts and primary-coverts (latter extensively rufous with black tips) indicate fresh ad. Similar to spring ad ♂, but broad blackish feather-centres above and below partially concealed. Feather-centres essentially blackish. Rufous of wing solid and contrasting, with more than half of tertials this colour. Compared to summer, blotchy breast-bands more broken. (T. Shimba)

greater coverts unmoulted, contrasting in colour with inner moulted. Primary-coverts all dark. Tail-feathers narrow. — AUTUMN **Ad** ♂ Similar to spring but whole plumage fresh, all dark feathers tipped or fringed paler. All greater (and median) coverts edged uniformly bright rufous, often tipped buff-white. White fringes on black breast-feathers often make beast-band incomplete. Tail-feathers broad, tips not acutely pointed. **Ad** ♀ As in spring but slightly paler overall due to pale fringes to all dark feathers. Upperparts usually uniformly olive-brown. All greater coverts uniformly patterned. Tail-feathers broad, tips not acutely pointed. **1stW** ♂ Similar to ad ♀, but differs in slightly brighter rufous edges to tertials and greater coverts, and in having some diffuse blackish feather-centres above. Outer greater coverts duller brown with whitish tips, contrasting with rufous-edged inner. Tail-feathers on average narrower and often acutely pointed. **1stW** ♀ Like ad ♀ (lacking any bright rufous on wing) and 1stW ♂ (having contrast among greater coverts and narrow, pointed tail-feathers), but usually recognised by less distinct spotting below. Tail-feathers narrow, tips acutely pointed. **Juv** Resembles 1stW ♀ but has narrow buff-white shaft-streaks above (largest on scapulars), including on lesser, median and inner greater coverts.

Ad ♀, Japan, Nov: as in spring, differs from ad ♂ by lack of extensive rufous panel on closed wing and less prominent blackish feather centres on upperparts. As always, important to start with ageing, considering that some 1stW ♂♂ are so similar. Moulted and uniformly fresh greater coverts, and neat primary-coverts, indicate ad. Typical for ♀ are the dark primary-coverts lacking rufous. (T. Shimba)

BIOMETRICS L 21–24 cm; **W** ♂ 125–139 mm (n 48, m 132.2), ♀ 122.5–135 mm (n 30, m 129.2); **T** ♂ 80–96 mm (n 48, m 87.9), ♀ 80–89 mm (n 30, m 85.5); **T/W** m 66.4; **B** 20.5–24.3 mm (n 78, m 22.5); **Ts** ♂ 30.3–34.3 mm (n 62, m 32.0), ♀ 29.0–32.3 mm (n 15, m 31.3). **Wing formula:** p1 minute, < pc 6.5–17 mm, < p2 71–84 mm; **p2** < wt 2.5–7.5 mm; **p3** longest; **p4** < wt 0.5–2.5 mm; **p5** < wt 4–10 mm; **p6** < wt 14.5–23 mm; **p10** < wt 36.5–45 mm; **s1** < wt 40–48 mm. Emarg. pp3–5.

GEOGRAPHICAL VARIATION & RANGE Monotypic. — N Siberia, in west from Middle Yenisei and through Lena roughly from 60ºN to the edge of the tundra on Taymyr, east to Anadyr, Kolyma and Kamchatka peninsulas; winters in S China and Korea. (Syn. *fuscatus*; *turuchanensis*.)

TAXONOMIC NOTE See under Naumann's Thrush for comment concerning the relationship between that species and Dusky Thrush.

REFERENCES CLEMENT, P. (1999) *Limicola*, 13: 217–250.

1stW ♂, Belgium, Jan: juv primary-coverts and most of greater coverts (except inner two) identify this bird as 1stW. Despite being young, the relatively substantial rufous on replaced greater coverts, some diffuse blackish feather-centres above, and rather strong dark markings on head, breast and flanks best fit ♂. (J. Fouarge)

1stW ♀, Japan, Jan: the dullest plumage variation with predominantly olive-brown upperparts with hardly any exposed darker feather-centres and very limited rufous restricted to edges of inner greater coverts. Note very obvious moult limit in greater coverts. Young ♀ of Black-throated Thrush (which it somewhat resembles) separated on patterned rather than plain greyish wing. (H. Iozawa)

1stY ♂, suspected hybrid Dusky × Naumann's Thrush, S. Korea: note mix of Dusky-like head pattern (bold whitish supercilium and throat, and blackish breast-bands), but mostly Naumann's-like dark rufous-brown blotches on lower breast and flanks. Some Dusky have a few rufous blotches on body-sides, but not as extensive as this bird, while solid rusty rump/uppertail-coverts is more Naumann's-like (though tail appears to be grey-brown like Dusky). (N. Moores)

RED-THROATED THRUSH
Turdus ruficollis Pallas, 1776

Fr. – Grive à gorge rousse; Ger. – Rotkehldrossel
Sp. – Zorzal papirrojo; Swe. – Rödhalsad trast

This elegant thrush breeds in the taiga over a large area in SC Siberia from the Altai in the west to Transbaikalia in the east, south into N Mongolia and north to about 60°N. It is closely related to the Black-throated Thrush (and these two were formerly often considered one species, 'Dark-throated Thrush'), but Red-throated has a more southerly and easterly distribution, and a preference for slightly more elevated areas. The two can apparently interbreed in the zone of overlap, inferred from birds with intermediate plumage, but details of actual hybridisation are scant. Red-throated Thrush winters in a large area of S Central Asia and SE Asia. It has straggled several times to W Europe, but not nearly as frequently as Black-throated Thrush.

Ad ♂, Mongolia, May: immaculate deep rufous-red bib makes this plumage unmistakable. Note also bright rufous outer tail-feathers. Aged by evenly-feathered wing. (T. Langenberg)

IDENTIFICATION A rather *large, slim* thrush with *fairly long tail*. Adult ♂ has *deep rufous-red throat and breast* with a *distinct border to off-white rest of underparts* (although undertail-coverts are sometimes partly rufous-red too). The red further invades much of cheeks and supercilium. The entire *upperparts are plain brownish-grey*, but crown-feathers have darker centres, and the *outer tail is bright orange-brown*, more visible from above when spread. Bill is dark with extensive *yellow base* to lower mandible. Adult ♀ and young ♂ are similar, and compared to the adult ♂ less neat, have some white on chin and throat, black streaking on sides of throat and sometimes across breast, and often a little off-white admixed in red breast; they also have some grey blotches on more dirty white belly and flanks. Young ♀ is even duller with more streaked lower throat/breast, and rather greyish-tinged upper belly and flanks, often with some indistinct darker grey streaking. Juvenile is readily recognised by its buffish or whitish wedge-shaped shaft-streaks on upperparts. It is ochraceous-tinged on sides of breast and somewhat pale buff on flanks. Breast is densely blotched dark, flanks more faintly marked. Behaviour as most other *Turdus* thrushes. In flight, note *reddish or orange-brown underwing* and outertail. – Indisputable ♂ hybrids with Black-throated have throat a mixture of blackish and rufous where black is prominent (not just rufous throat with some black marks on lateral throat, which is normal pattern for imm ♂ Red-throated). Note also that it is common for pure Red-throated to have darkish shade of uppertail, or dark patches on otherwise rufous uppertail.

VOCALISATIONS The song has previously been poorly documented and hence not well known. On existing commercial recordings (Schubert 1979, Mongolian Altai; Veprintsev & Veprintseva 2007, Sayan), the song is probably a misidentified Eyebrowed Thrush (although a brief sequence of correct song can be heard on the last-mentioned source after 1 min. 23 sec.). The real song, studied by LS in NC Mongolia and SE Altai, is rather primitive, most like that of Fieldfare, well-separated brief outbursts of a few squeaky or chattering notes, e.g. *chuk-chuk-chuk ... skweeh ... zvik-zvik-zvik ... chuk-chuk ... djieh ... chek-chek ... quee-qua ... chuk-chuk-chuk ...*, etc. Apparently there are no fluty or 'pleasing' notes in the song. Whether it differs much from song of Black-throated Thrush requires further study. – Calls are mainly hard clicking notes given singly or in series, a little like Blackbird but fuller, perhaps best described as mixture of Blackbird and Redwing calls. They are very similar to those of Black-throated Thrush, perhaps identical. Alarm call rather like Redwing, series of high-pitched, metallic, clicking notes, *chik-chik-chik-...* Other calls include squeaky, high-pitched notes of unknown function.

SIMILAR SPECIES Main risk of confusion is between first-year ♀♀ of this species and *Naumann's Thrush*. Both have reddish outertail and orange-brown underwing, and have basically same plumage pattern and size. Important to note that classic Red-throated has no rufous streaks or spots on upper belly or flanks, only grey (Naumann's has mixture of rufous and grey streaks), and usually unpatterned dusky white undertail-coverts (Naumann's has diffusely rufous-based feathers giving at least slightly blotchy appearance). Very rarely, birds have been claimed to be crosses between Red-throated and Naumann's, but those examined by us invariably seem to be extreme variations of immature Naumann's. Such birds have rather pure grey upperparts and well-developed rufous gorget, but also rufous streaks or

Ad ♂, Mongolia Jun: below rufous-brown bib, rest of underparts are off-white with some diffuse grey blotches; bib colour also reaches cheeks and supercilium. Tail-feathers have rather broad, rounded tips. (J. Normaja)

Presumed ad ♀, China, May: compared to summer ♂, plumage overall less neat, but diagnostic rufous bib is still obvious (though note pale fringes and black streaking on throat-sides); belly and flanks heavily blotched grey. Aged by ad-like greater coverts and primary-coverts, although inner three greater coverts are longer and appear fresher. (M. Parker)

THRUSHES

Ad ♀, Mongolia, May: ♀ plumage somewhat variable, but contrast between rather solid breast patch and paler throat common, the latter only sparsely streaked rufous and always with rather bold black lateral streaking. Age inferred by uniform greater coverts and broad and rounded primary-coverts, but also by rufous supercilium and quite solid rufous breast. Note also the rather dark bill. (T. Lindroos)

1stS ♂, Mongolia, Jun: some 1stY ♂♂ have advanced plumage with uniform rufous face and bib lacking dark streaks on throat and are only separated from ad ♂♂ by moult limits among greater coverts (visible innermost three contrastingly fresher). (H. Shirihai)

Presumed ad ♀, Mongolia, May: typical plumage of ad ♀ in spring with white-tipped but broadly rufous-centred feathers forming reddish gorget across breast beneath off-white throat with black lateral stripes. As far as can be judged no moult limits among greater coverts. (T. Langenberg)

blotches on flanks and undertail-coverts. – In poor light and less than good views can be confused with *Black-throated Thrush*, which has nearly identical size and general plumage, but usually possible to get a glimpse of reddish colour on outertail or on throat and breast, where Black-throated is just grey or black.

AGEING & SEXING Ages differ slightly during 1stY, but ageing requires close views or handling. Sexes separable after post-juv moult, but separation in 1stW not always easy or even possible unless at close range. – Moults. Complete post-nuptial moult of ad in summer and early autumn (mainly Jul–Sep), before migration. Partial post-juv moult in summer (Jul–Aug) does not involve flight-feathers, tertials or primary-coverts. Median and a varying number of inner greater coverts (usually 2–3, sometimes four) are also moulted. Apparently no pre-nuptial moult in winter. – **SPRING Ad ♂** Dark rufous-brown bib without white admixed on chin, throat and breast, and same colour on supercilium and fore cheeks. Sometimes a few black spots on sides of throat (apparently not age-related). Belly to shortest undertail-coverts off-white with a few diffuse grey blotches; longest undertail-coverts

Ad ♂, Russia, winter: very similar to spring with prominent rufous-brown bib, but some rufous feathers finely tipped white before wearing off later in spring. Note uniform fresh wing, with only ad-type greater coverts and tertials. (V. Ivushkin)

1stW ♀, Russia, Feb: young ♀♀ have dullest plumage, with quite subdued pattern on throat and breast, and often no real bib. Ageing should precede sexing, but this not always straightforward, as this example shows. Here, a young bird without any white-tipped coverts or tertials signalling ad, but a closer look reveals that about half of the outer greater-coverts are unmoulted juv, shorter and subtly browner than inner greater-coverts. (V. Ivushkin)

1stW ♂, South Korea, Oct: in autumn, both 1stW ♂ and ad ♀ share this plumage, hence ageing should precede sexing. Moult limit among greater coverts, which are mostly juv, as are primary-coverts (but latter difficult to judge). Being young, only ♂ is a possibility. (A. Audevard)

Baikal to upper Lena; winters in S Central Asia, N India and SW China.

TAXONOMIC NOTE Closely related to Black-throated Thrush *T. atrogularis*, with which it is often lumped as one single species, mainly due to occurrence of presumed hybrids or at least intermediates. However, we feel that Red-throated and Black-throated Thrushes are distinct enough in a number of ways to afford them separate species status. Apart from the obvious difference in throat and tail colour—black/grey versus red—Red-throated has proportionately slightly longer wings and tail, and the two forms prefer slightly different habitats and ecological niches, at least in part. Considering that the overlap area between the two taxa is fairly large it can further be questioned whether not intermediate birds ought to be more common than they seem to be, if the two were just subspecies of the same species. There also seems to exist an average difference in song between Red-throated and Black-throated Thrushes (Arkhipov *et al.* 2003), although the comparison was based on an unreliable example of Red-throated Thrush. Therefore, vocalisations of the two species require further study before taxonomic conclusions can be drawn based solely on their voices.

REFERENCES Arkhipov, V. Y., Wilson, M. G. & Svensson, L. (2003) *BB*, 96: 79–83. – Clement, P. (1999) *Limicola*, 13: 217–250.

same or sometimes partly rufous. Entire upperparts plain brownish-grey (no contrast among greater coverts). Tail-feathers rather broad, tips rounded. **Ad ♀/1stS ♂** Differs from ad ♂ by having throat either paler rufous than in ad ♂ or more whitish with rufous streaking or blotches. Invariably has dark streaks or spots on throat-sides. Retains more white fringes on rufous breast. These two separable when handled, or generally when seen close, on width and pointedness of tail-feathers, and on slight contrast between inner greater coverts (being discreetly cross-barred and often slightly longer) and outer (plainer-looking, often with wedge-shaped indentation at tip of innermost unmoulted, after white spot has been abraded). **1stS ♀** Somewhat plainer than preceding plumages, resembles ad ♀ but breast more streaked or spotted rufous (thus breast patch appears less full). As 1stS ♂, many outer greater coverts unmoulted, contrasting subtly in colour with inner moulted. Tail-feathers narrow and pointed. – **AUTUMN Ad ♂** Similar to spring but whole plumage fresh, all rufous feathers finely tipped white. All greater coverts edged uniformly grey. Tail-feathers broad, tips rather rounded. **Ad ♀/1stW ♂** Differ from ad ♂ by having paler chin and throat (off-white with rufous admixed or pale rufous), and stronger dark streaking on sides of throat. Also, slightly more prominently white-tipped rufous feathers on breast. Upperparts a trifle more brown-tinged on average than in ad ♂. These two separated when aged, which is easier than in spring: 1stW has moult contrast among greater coverts, innermost unmoulted covert frequently with white wedge-shaped spot at tip. Tail-feathers on average narrower and often acutely pointed. **1stW ♀** Like in spring. Unmoulted greater coverts as in 1stW ♂. Tail-feathers narrow, tips acutely pointed. **Juv** Resembles 1stW ♀ but has narrow buff-white shaft-streaks above, most prominently on scapulars, mantle and median wing-coverts. Breast ochraceous-buff densely spotted dark. Flanks lightly spotted yellowish-buff.

BIOMETRICS L 24–25.5 cm; **W** ♂ 132–146 mm (n 30, m 139.2), ♀ 130–143.5 mm (n 12, m 135.3); **T** ♂ 90–103 mm (n 30, m 97.7), ♀ 88–104 mm (n 12, m 94.0); **T/W** m 70.0; **B** 21.6–26.0 mm (n 40, m 23.4); **BD** 6.0–7.0 mm (n 30, m 6.5); **Ts** 30.2–35.0 mm (n 42, m 32.9). **Wing formula: p1** minute, < pc 6–16 mm, < p2 77–90 mm; **p2** < wt 3.5–9 mm; **p3** longest; **p4** < wt 0.5–2.5 mm; **p5** < wt 4–8 mm; **p6** < wt 15–22 mm; **p7** < wt 23.5–30 mm; **p10** < wt 37–45 mm; **s1** < wt 41–48 mm. Emarg. pp3–5.

GEOGRAPHICAL VARIATION & RANGE Monotypic. – NW China, S Altai, W & NC Mongolia, C Siberia north of

1stW ♀, England, Oct: a quite dull ♀; age confirmed by moult, with most of greater coverts being juv, showing whitish and well-defined fringes. (V. Tiwari)

1stW Red-throated Thrush (left: England, Oct) versus Naumann's Thrush (right: Hong Kong, Nov): former typically has no rufous on body-sides, only variable grey streaking, whereas Naumann's has mix of rufous and grey blotching, and undertail-coverts of Red-throated usually dusky white (blotched rufous in Naumann's). Some 1stY Naumann's (mostly 1stW ♀♀) are less rufous below, but even poorly-marked birds have well-developed rufous gorget, and some rufous on flanks and undertail-coverts. Both birds 1stW, but sexing not easy, and given amount of rufous on foreparts could be either well-marked ♀ or duller ♂. (Left: R. Chittenden; right: M. & P. Wong)

BLACK-THROATED THRUSH
Turdus atrogularis Jarocki, 1819

Fr. – Grive à gorge noire; Ger. – Schwarzkehldrossel
Sp. – Zorzal papinegro; Swe. – Svarthalsad trast

Black-throated Thrush is probably the best-known and commonest of a quartet of closely related Siberian taiga-dwelling species. It is particularly close to Red-throated Thrush, but has a more northerly and westerly distribution than that species, with a breeding range reaching just into Europe west of the Urals. It shows a preference for overgrown streams in the taiga, but on the whole has a wide habitat tolerance. Black-throated Thrush has straggled many times to W Europe, most frequently in late autumn or winter, often detected when feeding on windfalls in garden.

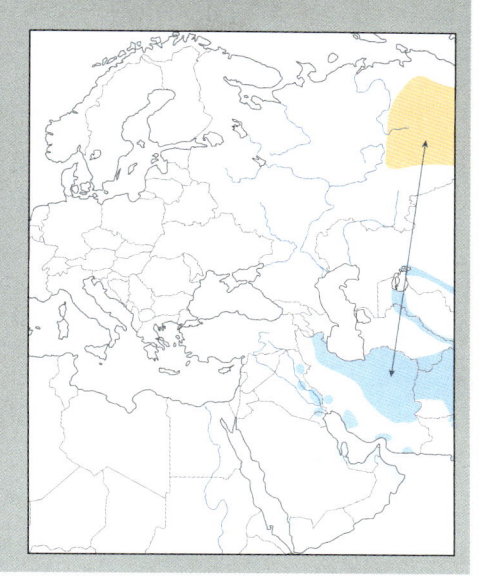

IDENTIFICATION Rather *large* with *fairly long tail*. Very similar to Red-throated Thrush, only has slightly shorter wing and tail, and any rufous-red (except on underwing) is replaced by black. Adult ♂ has *jet-black throat and breast* with a *distinct border to off-white rest of underparts* (although some undertail-coverts sometimes partly or largely tinged rufous-brown). Black also on cheeks and supercilium (rare variety 'relicta' has most of head blackish in worn plumage; see below). *Upperparts plain brownish-grey*, but crown-feathers have darker centres, and *outer tail is blackish*. Bill is dark with extensive *yellow base* to lower mandible. Adult ♀ and young ♂ are rather similar but are less neat, have some white on chin and throat, clearly most in ♀, whereas ♂ usually has chin and throat densely streaked black. Young ♀ is the dullest with more streaked lower throat/breast, and rather greyish-tinged upper belly and flanks, often with some indistinct darker grey streaking. Juvenile readily recognised by its buffish or whitish wedge-shaped shaft-streaks on upperparts. It is ochraceous-tinged on sides of breast and somewhat pale buff on flanks. Breast is densely blotched dark, flanks more faintly marked. Behaviour as Red-throated Thrush and most other *Turdus* thrushes. In flight, note *orange underwing*. – Rarely hybridises with both Red-throated and Naumann's Thrushes, hybrids often identified by rufous tinges to upperwing or undertail, and diffuse rufous streaking or spotting on breast and flanks.

VOCALISATIONS The song is composed of well-spaced series of simple, fairly Fieldfare-like chattering sounds, *chip-chip-chip*, *chet-chet-chet-chet-chet*, *chip-chip-chip*, etc., now and then relieved by a richer, low-pitched and slightly husky warbling (recalling elements of Blackbird song), *trro-uu trre-vee* and similar. In some ♂♂, or for periods, the simple chattering sounds dominate the song, but others use huskier warbling than chattering notes (Arkhipov *et al.* 2003; M. Strömberg unpubl. recording; Rogacheva 1992). High-pitched, feeble twittering after the stronger notes seems lacking. (The claim in *BWP* that the song resembles Song Thrush is thought to be based on recording of a misidentified bird.) – Common calls include a chattering series similar to that in the song, *chet-chet-chet-*…, or rather deeper and harder, *chuck chuck chuck*, recalling Ring Ouzel or Blackbird but slightly drier than latter, and a squeaky *kyeek* or *gvieh*, quite close to calls of Fieldfare. Single *gyack*, uttered in mild anxiety, recalls Redwing.

SIMILAR SPECIES The young ♀ could be confused with first-year ♀ Red-throated Thrush, which sometimes has very little rufous in plumage, but note dark or grey tail (Red-throated: rufous outertail) and a rather cold grey cast on breast and flanks (Red-tailed: at least a faint reddish hue on

Ad ♂, Netherlands, Apr: unmistakable due to contrasting jet-black face and bib (cheeks variably greyer). Belly off-white with any faint blotching or streaking subdued. Aged by evenly-feathered wing. (A. Ouwerkerk)

Ad ♂, Russian Altai, Jun: dark-headed morph (var. 'relicta'), which in worn plumage has black of head extending solidly onto crown, nape and even sometimes the mantle. Evenly rather fresh wing-feathers, and ad-like primary-coverts, provide clues to ageing. (M. Waern)

1stS ♂, England, Mar: some 1stY ♂♂ show advanced plumage (though black bib is still not so solid), thus best aged first, by moult limits in wing (warmer brown outer greater coverts and primary-coverts are juv). (M. Walford)

sides of head and breast). – Another problem is presented by hybrids (or at least birds looking like intermediates) between Black-throated and Red-throated, or between Black-throated and Naumann's Thrushes. Birds should be recorded as intermediates, rather than one of the two species, if they display, e.g., a rufous bib with a blackish tinge, or look like a perfect Black-throated but for partly rufous tail-feathers, or if they have faint red patches on the black-streaked breast and flanks. However, it should be noted that a large proportion (perhaps up to 25% of adults, slightly less of immatures) of seemingly pure Black-throateds have partially dark rufous-tinged or rufous-streaked undertail-coverts, especially the longest ones, and this is apparently not a sign of hybridisation, or at least not recent interbreeding.

AGEING & SEXING Ages differ slightly during 1stY, but ageing generally requires close views or handling, at least in spring. Sexes separable after post-juv moult. – Moults. Complete post-nuptial moult of ad in summer and early autumn (mainly Jul–Sep), before migration. Partial post-juv moult in summer (Jul–Aug) does not involve flight-feathers, tertials or primary-coverts. Median and a varying number of inner greater coverts (usually 2–4) are also moulted, and very rarely odd tail-feathers too. Apparently no pre-nuptial moult in winter. – **SPRING** Ad ♂ Black bib usually from about Apr without white admixed, on chin, throat and breast, and black extends also to supercilium and fore cheeks. Belly to shorter undertail-coverts usually off-white with only few diffuse grey blotches; longest undertail-coverts same or partially or largely dark rufous-brown. Entire upperparts plain brownish-grey. Edges of secondaries, tertials and greater coverts usually smoothly tinged olive-ochre, only slightly paler than centre of feathers. No moult limit among greater coverts. Tail-feathers rather broad, tips rounded. (As to largely black-headed variants, see Geographical variation. Such birds usually also have whiter, entirely unstreaked belly and brighter yellow bill-base.) **Ad ♀** Black breast and lower throat, or at least a black band across breast. Black streaking on sides of throat. Those with most black approach ad ♂ plumage but have at least chin and upper throat whitish narrowly streaked black. Wing-feathers uniformly greyish as in ♂, lacking moult limits. Tail-feathers rather broad, tips rounded. **1stS ♂** Differs from ad ♂ by having black bib admixed with some white, or throat and breast densely streaked black on white ground, rather than being all black. Differs from ad ♀ by having darker, more streaked chin and throat (but can be quite similar as to bib). Sometimes slight contrast between inner greater coverts (being discreetly cross-barred and often slightly longer) and outer (plainer-looking, often with wedge-shaped indentation at tip of innermost unmoulted, after white spot at tip has been abraded). Generally, edges of secondaries, tertials and outer greater coverts are greyish-white with fairly obviously contrasting darker centres (not as warm olive-ochre as usual of ad). Tail-feathers on average narrower with more pointed tips, but latter worn and rather heavily abraded. **1stS ♀** Somewhat plainer than preceding plumages, resembles ad ♀ but breast more streaked or spotted dark (rather than having black breast patch). As 1stS ♂, many outer greater coverts unmoulted, contrasting subtly in colour with inner moulted. Tail-feathers narrow and pointed. – **AUTUMN Ad ♂** Similar to spring but whole plumage fresh, all black feathers finely tipped/edged white, more so on chin/throat than on breast. All greater coverts uniformly grey and smoothly edged rufous-ochre or warm olive-brown. Tail-feathers broad, tips rather rounded. **Ad ♀** As spring but white tips to black feathers make breast pattern more diffuse. All greater coverts uniformly grey edged warm brown. Tail-feathers broad, tips rather rounded. **1stW ♂** As in spring but white tips to black feathers make breast pattern more diffuse. Moult contrast among greater coverts sometimes visible, innermost unmoulted covert frequently with white wedge-shaped spot at tip. In general, juvenile secondaries, tertials and greater coverts are edged more greyish-white, less warm brown than in ad, making edges appear more contrasting. Tail-feathers on average narrower and tips acutely pointed, becoming

Ad ♀, Denmark, Nov: since young ♂ plumage can be quite similar to ad ♀, ageing should precede sexing. Here, evenly-feathered wing infers ad, while much white on throat and submoustachial area only fit ♀. Note reasonably solid dark breast patch and neck-sides in contrast to whitish throat (only sparingly streaked). (J. Salomonsson)

Ad ♂, Sweden, Feb: similar to spring but when fresh in winter black bib-feathers finely tipped white, obscuring pattern. Note evenly fresh ad wing-feathers. (J. Stenlund)

Presumed ad ♀, Belgium, Dec: impression of moult limits in greater coverts is false and caused by disorder to feathers. All greater coverts and tertials (ad type) edged pale brown with moderate contrast to centres; throat/breast pattern not unlike many 1stW ♂, although darkest ad ♀ can also have rather solid breast patch and finely black-streaked chin and throat. Note hint of rufous-tinged vent and undertail-coverts, which is quite frequent and not an indication of hybridisation with Red-throated Thrush. (V. Legrand)

abraded from late autumn. **1stW ♀** Like spring. Unmoulted greater coverts as in 1stW ♂. Tail-feathers narrow, tips acutely pointed and worn. **Juv** Resembles 1stW ♀ but has narrow buff-white shaft-streaks above, most prominently on scapulars, mantle and median wing-coverts. Breast tinged buff densely spotted dark. Flanks pale buff-white lightly spotted. Very similar to juv Red-throated Thrush.

BIOMETRICS L 22.5–25 cm; **W** ♂ 131–142 mm (n 28, m 136.3), ♀ 126–137 mm (n 24, m 131.9); **T** ♂ 90–102 mm (n 28, m 94.6), ♀ 86–97 mm (n 24, m 91.8); **T/W** 69.7; **B** 21.3–26.3 mm (n 49, m 23.2); **Ts** 30.5–34.5 mm (n 26, m 32.5). **Wing formula: p1** minute, < pc 8–13 mm, < p2 74–84 mm; **p2** < wt 4–7 mm; **p3** longest; **p4** < wt 0.5–2 mm; **p5** < wt 3–8 mm; **p6** < wt 15–20 mm; **p7** < wt 22–27 mm; **p10** < wt 35–43.5 mm; **s1** < wt 40–47 mm. Emarg. pp3–5.

GEOGRAPHICAL VARIATION & RANGE Monotypic. — Extreme E European Russia in Ural region, W & C Siberia east to upper Lena, south to Altai; winters from SE Iraq through Iran to Afghanistan, N Pakistan and W Himalayas. — Predominance of dark-headed (in worn plumage black also on crown, nape and sometimes upper mantle) and otherwise very brightly coloured ♂♂ in Altai and Sayan taiga constitutes probably just a local colour morph, var. '*relicta*', or may merit separation as separate race (Zarudny & Koreyev 1903, Portenko 1981). (Syn. *vogulorum*, being largely inseparable from *atrogularis*.)

TAXONOMIC NOTE See note under Red-throated Thrush *T. ruficollis*, with which it is often lumped as single species. Here treated as separate species on account of different plumage, slight differences in proportions, and possibly also consistent differences in song. Judging from specimens with mixed characters, hybridisation between *atrogularis* and *ruficollis* occurs, but such birds seem to be relatively few in relation to large area of contact between the two, indicating a predominance of assortive mating. Hybrids with Naumann's Thrush seem even rarer, unsurprisingly so considering the limited overlap in range between the two.

REFERENCES Arkhipov, V. Y., Wilson, M. G. & Svensson, L. (2003) *BB*, 96: 79–83. — Clement, P. (1999) *Limicola*, 13: 217–250. — Portenko, L. A. (1981) *Trudy Zool. Inst. Akad. Nauk SSSR / Proc. Zool. Inst. Acad. Sci. USSR*, 102: 72–109. (In Russian.) — Zarudny, N. A. & Koreyev, B. P. (1903) *Orn. Monatsber.*, 11: 129–130.

1stY ♂, Kazakhstan, Feb: some 1stY ♂♂ mirror ad ♂ plumage, but in general tend to have more extensive white tips to black feathers, making breast pattern more mottled. Duller young ♂ to some degree approaches 'advanced' ad ♀, making it highly advisable to attempt ageing first: note moult limits among greater coverts, with inner three freshly moulted (lacking thin bold white tips of juv coverts), and primary-coverts are juv. (A. Isabekov)

1stY ♀, Netherlands, Jan: an extremely streaky ♀, with hardly any breast-band. Predominantly dark-streaked birds are often 1st autumn ♀, but can be reliably aged (and sexed) only by moult and feather wear in wing. It appears that all tertials and some inner greater coverts are missing or growing, with most coverts juv having whitish and well-defined fringes; primary-coverts typical of juv with thin buffish-white fringes. (R. Pop)

1stW ♀, England, Jan: given substantial variation in autumn/winter, fresh ♀♀ should be aged by moult pattern in wing. Moult limit among greater coverts (visible in other images of same individual) and juv primary-coverts confirm this bird's age. (G. Jenkins)

1stW ♀ Black-throated Thrush (Norway, Jan) versus 1stW ♀ Red-throated Thrush (Russia, Feb): these two can appear almost identical in 1stW plumage and only differ when seen close; Red-throated has some rufous streaks or spots on breast and rufous-edged outer tail-feathers (latter invisible here; dullest 1stW ♀ Red-throated at times shows visible rufous only when tail is spread). Traces of rufous-buff on undertail-coverts and especially on vent-side and lower submoustachial area apparently not unusual for Black-throated, and do not necessarily indicate hybridisation with Red-throated. Both are 1stW by moult limits in wing and general dull plumage pattern. (Left: B. de Bruin; right: V. Ivushkin))

FIELDFARE
Turdus pilaris L., 1758

Fr. – Grive litorne; Ger. – Wacholderdrossel
Sp. – Zorzal real; Swe. – Björktrast

The Fieldfare used to be confined to N Europe in the breeding season, but in winter it has always been well known elsewhere, visiting much of Europe, and often seen in flocks feeding on berries in parks, churchyards and gardens. In fairly recent times, it has spread south and west, and now breeds locally in Scotland, the Netherlands, Belgium, over the whole of Germany and much of France. As its name implies, it is often seen in open habitats, feeding on fields, pasture and park lawns. At times breeds in loose colonies, all parent birds joining in noisy and fierce defence of their nests and young against Magpies and other predators.

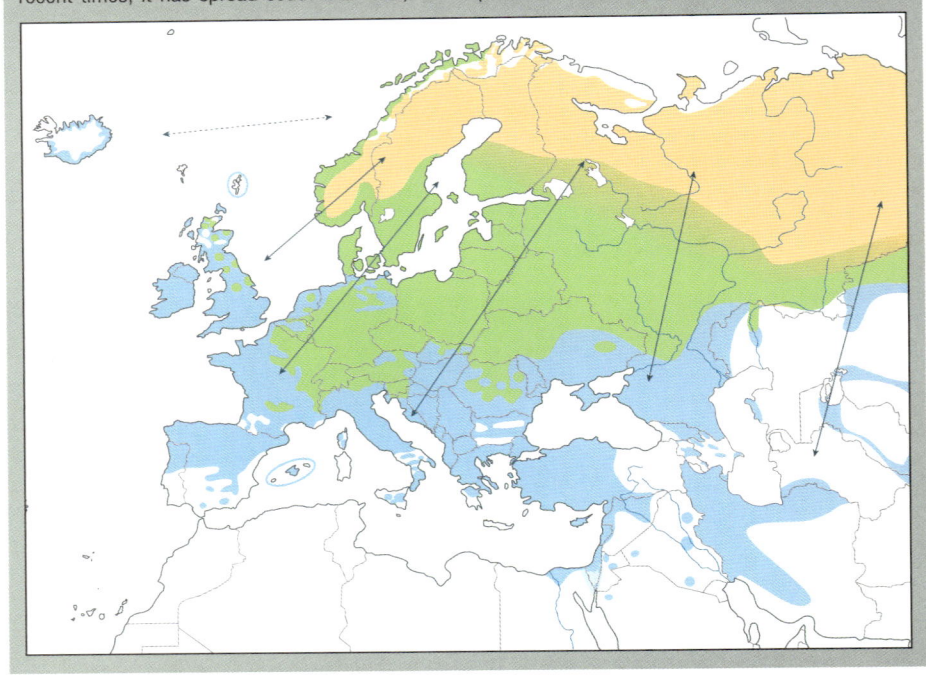

IDENTIFICATION One of the largest thrushes, strongly built and rather long-tailed. An attractive bird with characteristic plumage. Sexes are very similar and can only be separated when seen close—if then (see below). Head, nape, lower back and rump light grey, mantle and upperwing tawny- or chestnut-brown, wing- and tail-feathers dark. Underparts buff-white with an *ochre or even pale rufous tinge on breast*, the *breast and flanks heavily spotted black*, spots crescent- or arrow-shaped. On sides of neck spots often merge into a larger blackish patch, and forecrown is usually spotted or streaked black. The *bill is bright yellow with a black tip*. In flight best characters are size, *long tail* and pure *white underwing*, last feature only shared with Mistle Thrush. When seen on the ground a little of the white underwing often shows near the bend of wing (especially if the wings are not tucked into their wing-pockets properly, as so often when feeding). Juvenile readily recognised by its buffish or whitish wedge-shaped shaft-streaks on upperparts. Flight is direct and less jerky or undulating than some smaller thrushes: series of wingbeats are relieved by short pauses with folded wings, but flight path is still rather straight. When feeding on the ground, behaviour is like Blackbird (which see) or other thrushes, with short walks and stops to intently watch the ground. Vocalisation, once learned, is also important for identification.

VOCALISATIONS Has two song types, neither of which can be said to be particularly pleasing to the human ear. The normal, relaxed territorial song (often heard, but not always perceived as song) is delivered from a high perch and consists of a steady but moderately paced flow of varied chattering call notes and squeaky sounds, given in a staccato, unstructured fashion with brief pauses between each combination of notes or 'sounds'. It is easy to miss that this is a song; it can pass for just agitated calls or 'conversation' between the birds as they often breed colonially and the song is so unobtrusive. The other song variety fits better with our preconception of bird song, as it is longer and continuous, and often louder. The 'long song' is mainly used in agitation and in flight, in a display performance, while chasing a rival male, or a Magpie which came too close to the nest for comfort (thus a parallel to the two song types of White/Pied Wagtail). It is a faster and louder stream of chattering, twittering and squeaky notes, mixed with series of harsh *tscha-tscha-tscha-...* or *chre-chre-chre-...* sounds. – Calls include the hoarse chattering *chre-chre-chre* and variations, and in flight a high-pitched and characteristic squeaky *giieh* or *giih*, often doubled, *gih-gieh*.

SIMILAR SPECIES The combination of white underwing, chestnut-brown mantle and light grey nape and rump prevents confusion with other species. Juveniles readily recognised by rufous mantle, grey rump and heavily dark-spotted ochraceous-buff breast.

AGEING & SEXING Ages differ slightly during 1stY, but ageing requires close views. Sexes often separable after post-juv moult, but difference slight and frequently ambiguous unless birds are handled (very rarely difficult even then). – Moults. Complete post-nuptial moult of ad in late summer (mid Jul–early Sep). Partial post-juv moult in summer (Jul–Sep) does not involve flight-feathers, tertials (except very rarely) or primary-coverts. Some inner greater coverts are also moulted, but usually 4–7 outer retained. (Very rarely no greater coverts moulted, and once just two.) No pre-nuptial moult in winter. – **SPRING Ad ♂** Best sex character is extensive and rather rounded black feather-centres on forecrown, becoming progressively more visible with wear through spring. (A very few atypical birds have more pointed black centres, closely resembling some ♀♀.) Supporting characters are: mantle deeper chestnut, sometimes with diffuse dark (even blackish) centres; darker uppertail, frequently blackish; deeper and more extensive rufous-ochre breast (some overlap with ♀, though); on average more prominent and blacker crescent-shaped markings on sides of breast and flanks (though beware odd

Ad ♂, Finland, Jan: sexes very similar and rarely separable. Aged by uniformly greyish-edged (outer) and tipped greater coverts. Bright plumage supports both ageing and sexing: blackish uppertail, deep rufous-ochre breast with prominent black crescent-shaped markings on flanks; spots on neck-sides merge into large blackish patch, mantle deep chestnut with diffuse blackish centres, and bill bright yellow with black tip. Most important character for sexing is extensive and rather rounded black feather-centres on forecrown, though mostly concealed in this fresh bird. (M. Varesvuo)

♀♀ with similarly heavy such markings!); and on average heavier black markings on undertail-coverts. Age by tail-feathers usually being rather broad, tips rounded (but a few have narrower feathers!), and all greater coverts uniformly edged and tipped chestnut, or, more often, edged chestnut and tipped greyish. (Beware that outermost 1–2 greater coverts of ad habitually are subtly duller brown, not so vividly chestnut, which may appear as a moult contrast.) Bill bright yellow with black tip. **Ad** ♀ Like ad ♂ (and some difficult to tell without handling) but forecrown narrowly streaked black with wider grey margins, black lacking rounded shape of ♂ (some are quite similar, though). Supporting characters are: mantle slightly duller tawny-brown, not deep chestnut, and lacking blackish centres; olive-brown, less dark upper-tail; on average slightly paler ochraceous breast lacking extensive rufous tinge (but many are similar); and on average less prominent markings on flanks and undertail-coverts (but see above about exceptions). Tail-feathers and greater coverts as ad ♂. Bill as in ad ♂ or on average slightly darker. **1stS** ♂ Very similar to ad ♂, but differs in having a variable number of juv outer greater coverts unmoulted, contrasting with new inner by narrow paler brown, less chestnut, edges and usually white tips (unless too worn). If only few inner greater coverts moulted then innermost unmoulted juv greater coverts frequently have whitish tip wedge-shaped or even have a narrow whitish shaft-streak. Tail-feathers on average narrower and more acutely pointed (if not too abraded to judge, as often from Jun onwards). **1stS** ♀ Like ad ♀, but has retained juv outer greater coverts, and narrower and more pointed tail-feathers. – AUTUMN Sexing and ageing much as in spring, but plumage fresh. **Juv** Resembles 1stW but has narrow buff-white shaft-streaks above (largest on scapulars), including on lesser and median coverts. Typical ♂♂ can already be recognised by blackish, ♀♀ by very plain olive-brown, tail-feathers.

BIOMETRICS L 25–27 cm; **W** ♂ 138–153 mm (once 158 mm; n 50, m 147.2), ♀ 137–152 mm (n 49, m 142.9); **T** ♂ 95–113 mm (n 50, m 104.1), ♀ 94–108 mm (n 49, m 101.9); **T/W** m 71.0; **B** 22.3–26.1 mm (n 71, m 23.9); **BD** 6.4–7.9 mm (n 66, m 7.1); **Ts** 31.0–34.9 mm (n 73, m 33.1). Wing formula: p1 minute, < pc 8–18 mm, < p2 80–98 mm; **p2** < wt 3.5–8 mm, =4/5 (57%), =5 (32%) or =5/6 (11%); **pp3–4** about equal and longest (p4 sometimes to 3.5 mm <); **p5** < wt 4–11 mm; **p6** < wt 17.5–24.5 mm; **p10** < wt 41–52 mm; **s1** < wt 45–54 mm. Emarg. pp3–5.

Ad ♂, Finland, Apr: with wear in spring, broad black feather-centres on forecrown become progressively larger and more visible, broadest in ♂♂; deeper chestnut mantle, extensive rufous-ochre breast and blackish uppertail are other ♂ characters. In spring, especially if some primary-coverts and greater coverts are concealed, ageing can be difficult, but overall fresh plumage is ad-like. (M. Varesvuo)

Ad ♀, Finland, Apr: rather fresh wing-feathers of same generation (indicative of ad), and overall rather dull plumage with narrowly-streaked black crown, duller tawny-brown mantle and paler ochraceous breast with smaller dark markings on flanks (strongly suggesting ♀). (M. Varesvuo)

Finland, Oct: pure white underwing-coverts stand out on otherwise dark and long-tailed bird. Ageing and sexing not feasible in flight, though pointed tips to tail-feathers might indicate 1stW and if so rather a ♂. (M. Varesvuo)

Ad, presumed ♂, Netherlands, Oct: ad-patterned greater coverts and primary-coverts (latter with rounded tips and rather broad grey fringes to outer webs). Some ad, apparently mostly ♀, can have juv-like dark bill, with hardly any yellow. Nevertheless, overall plumage best fits fresh ad ♂, especially quite visible blackish centres to mantle and very broad black blotches on flanks. (N. D. van Swelm)

GEOGRAPHICAL VARIATION & RANGE Monotypic. – N Europe south to C & SE France, Austria and N Ukraine, in west from Britain, east through taiga to C Siberia, Transbaikalia; short to medium-range migrant, southern birds being mainly resident, N European breeders moving to at least S Baltic but commonly to W & C Europe, less commonly S Europe.

1stY ♂, Netherlands, Jan: unlike ad, some outer greater coverts are juv (shorter, pale-tipped, less rufous), contrasting with new inner coverts; primary-coverts more pointed, with rather thin pale grey fringes. The rather bright plumage, especially rufous-buff chest, strong blackish markings below and quite deep rufous-brown upperparts, all strongly indicate a ♂. (N. D. van Swelm)

♀, presumed ad, Netherlands, Feb: aged and sexed by combination of no apparent moult limit in greater coverts, and by ad-type primary-coverts (more rounded tips); also overall dull ♀-like plumage, with narrow and diffuse crown-streaking, smaller, less bold underparts markings on mostly white, cream and some buff ground colours, including breast. (R. Pop)

1stY, presumed ♀, England, Jan: moult limit in greater coverts (three innermost clearly longer and warmer brown, having been replaced), and pointed juv primary-coverts. Overall dull ♀-like plumage with narrow and diffuse crown streaking, less bold underparts markings on mostly white, cream-coloured and pale buff ground colours. (P. Smith)

Juv, Sweden, summer: note the characteristic scaly bars on the sides of the body, and narrow pale shaft-streaks above (most obvious on the scapulars), typical of juvenile plumage. (B. Gibbons)

… THRUSHES

SONG THRUSH
Turdus philomelos C. L. Brehm, 1831

Fr. – Grive musicienne; Ger. – Singdrossel
Sp. – Zorzal común; Swe. – Taltrast

Many people know the song of the Song Thrush without being aware of it. It is quite strong, well broken and often repeats phrases. You could say that it 'proclaims' more than it sings—the song is not a fluent melody like that of Blackbird. Some have the bird in their garden, as in Britain and over much of W Europe, but further north and east in Europe it is a shy bird of woodlands. In winter, to evade snow and ice, Song Thrush migrates southwest and departs those areas where it is a 'wild' and shy bird, whereas it is mostly resident in countries which know it as a garden or park dweller.

pleasing. Typical is the repetition a few times of a certain tonal figure, a momentary pause, followed by a completely new tonal figure repeated a few times, etc. The song is quite variable, but generally recognised by its structure, with 'proclaiming', paused presentation and frequent use of metallic and very high-pitched, sharp notes. (In Fenno-Scandia, reports of unusually early arrival of Thrush Nightingale usually refer to Song Thrush.) A typical phrase could be *kokleevih-kokleevih-kokleevih*, *philip philip philip*, *chuvey chuvey*, *tix-ix-ix tix-ix-ix tix-ix-ix*, *pih-e titititititi*, *tru-tru-tru-tru*, etc. Sings from high post, often top of tree, mainly at dusk and dawn, but all day, e.g. often after rain. – Call given in flight on migration, by day or during night, is a short, high-pitched clicking note, *zip*, softer than Robin or any of eastern *Emberiza* buntings, but reasonably similar. Birds in the Caspian area described as giving a disyllabic *zilip* (Schütz 1959). No proof for a drawn-out Redwing-like call on migration (*contra* BWP), but alarmed perched birds may give a thin *ziiih* from cover. Alarm near nest is a fast, high-pitched, scolding series of sharp, 'electric' notes, *tix-ix-ix-ix-ix-…* Young beg with high-pitched *tschip*, often run into small series, *tschip-tschip-tschip*.

SIMILAR SPECIES Similarly-sized *Redwing* is eliminated by lack of buff-white supercilium or submoustachial stripe, and lack of reddish-brown flanks. – *Mistle Thrush* is similarly dark-spotted on pale underparts, but is a considerably larger and more elongated bird. Also, Mistle Thrush has more rounded, less arrowhead-shaped dark spots on underparts, which continue similarly strong and rounded further onto belly and flanks than in Song Thrush. It also has pure white underwing and white corners to longer tail, usually visible as it flies off. Note also the more strongly patterned folded wing, cheeks and ear-coverts on Mistle Thrush (Song Thrush is plainer brown on wings and head-sides). – Juvenile plumages of all thrushes are more similar than adults, but juvenile Song Thrush should be recognised by size and shape, plain brown head pattern and rich buff breast. Compared to *Mistle Thrush*, has brown crown and nape (Mistle: more tinged greyish), plainer brown wing with ochraceous-buff double wing-bars (Mistle: darker and more variegated wing with yellow-brown panel formed by pale edges to tertials and greater coverts, and large buff tips to tertials).

AGEING & SEXING (*philomelos*) Ages differ slightly during 1stY, but ageing requires close views. Sexes do not differ in plumage and only very slightly in size. – Moults. Complete post-nuptial moult of ad mainly in summer (late

IDENTIFICATION A small and compact thrush, rather short-tailed, about same size as Redwing (a fraction smaller, if anything), which it resembles in shape and behaviour. Sexes similar. Song Thrush is recognised by: brown or *olive-brown upperparts*, wing-coverts with ochraceous-buff tips usually forming two discreet wing-bars; *ochraceous-buff breast and flanks*, more whitish throat, belly and vent, and most of *underparts spotted black*, spots *arrowhead-shaped* on throat and breast, more rounded on flanks. Lacks prominent pale supercilium, has only buffish-white eye-ring and hint of a short buff supercilium above lores. *Bill mainly dark brown*, base of lower mandible yellowish-brown (not clear yellow like Redwing). Cheeks irregularly patterned brown, grey and whitish. Streaks on sides of throat a little denser, seem to form a small dark patch in some (but smaller and less prominently than on Redwing). Juvenile readily recognised as a young thrush by its buffish wedge-shaped shaft-streaks on upperparts, and as a Song Thrush on size and shape, and lack of long buff supercilium or ochraceous-red tinge on flanks. Song Thrush is often seen feeding on lawns in gardens and parks, especially under bushes and hedges, where it runs or hops, then stops and 'freezes', hops again, etc. Stance upright when alert. Flight is a little jerky and undulating, similar to Redwing. In flight note pale *orange-brown underwing*.

VOCALISATIONS Song loud and usually well structured, exclamatory and staccato-like, rather than fluent, soft and

T. p. philomelos, presumed ad, Netherlands, Oct: small, compact and short-tailed thrush, with plain olive-brown upperparts, except pale-tipped wing-coverts, and mostly cream-white black-spotted underparts. Relatively limited ochraceous-buff wash on breast and flanks, and buffish-white eye-ring. Greater coverts appear ad, while shape of tail-feathers and age of primary-coverts would require handling to be established. (A. Ouwerkerk)

Jun–early Sep). Partial post-juv moult in summer (Jul–Sep, occasionally later) does not usually involve flight-feathers, tertials or primary-coverts. Some inner greater coverts also moulted, but usually 5–8 outer retained. Very rarely a few tail-feathers also replaced. No pre-nuptial moult in winter. — SPRING **Ad** All greater coverts uniformly edged brown or olive-brown, rather uniformly tipped yellow-buff, pale spots on tips small and rounded, at times even missing entirely (especially on innermost and outermost greater coverts). Tail-feathers rather broad, tips rounded. Primaries still rather fresh. **1stS** Similar to ad, but generally differs by having a variable number of juv outer greater coverts unmoulted, contrasting with new inner by slightly shorter length, and by having either triangular whitish spots or, in summer, notches at the tips (where now-abraded pale spots used to be). Tail-feathers on average narrower and more acutely pointed (if not too abraded to judge). Primaries usually heavily worn at tips. — AUTUMN **Ad** All greater coverts uniform, edged olive- or tawny-brown with small diffusely rounded paler tips on outer web being buff and not touching shafts. Tail-feathers rather rounded at tips (but may have fine extending tips at shaft). Longest primary-coverts with rather rounded tips and comparatively uniform brown, edged olive-brown. **1stW** Outer 5–8 greater coverts unmoulted (juv feathers), innermost of which are usually shorter and have prominent and

T. p. philomelos, presumed ad, Belgium, Feb: in certain lights some appear richer brown above. Spots below predominantly arrowhead-shaped, but more rounded on lower flanks and upper breast. Lateral throat-stripe poorly developed, especially compared to Redwing. Probably ad by apparently uniform greater coverts, which are tipped yellow-buff; primaries relatively fresh. (R. Debruyne)

T. p. philomelos, 1stW, Netherlands, Oct: mainly dark brown bill, with yellowish-brown base to lower mandible, and legs pale flesh-brown. Aside from inner two, greater coverts are juv feathers, which are shorter and have prominent wedge-shaped ochraceous-buff to rufous-yellow tips; moulted feathers have smaller, more diffuse yellow-buff spots. (N. D. van Swelm)

T. p. philomelos, Netherlands, Oct: in flight a small and rather compact thrush in usually hurried, somewhat jerky flight. Note pale orange-brown underwing coverts. (A. Ouwerkerk)

well-marked wedge-shaped ochraceous-buff spots (even slightly tinged rufous) at tip of outer web and frequently extending to shafts, contrasting with inner moulted, which have smaller or more diffuse and more yellow-buff spots. On average, new greater coverts edged a fraction more olive-brown, less rufous, but this is not always obvious or the case. Tail-feathers rather narrow and acutely pointed at tips. Longest primary-coverts have somewhat narrower, less rounded tips, and more contrast than in ad, edged ochre or pale brown with rufous tinge. **Juv** Recognised by narrow buff-white shaft-streaks above (largest on scapulars), including on lesser and median coverts.

BIOMETRICS (*philomelos*) **L** 19–21.5 cm. Sexes similar size. **W** 111–125 mm (n 90, m 117.2); **T** 72–85 mm (n 90, m 78.9); **T/W** m 67.4; **B** 19.1–24.3 mm (n 68, m 21.5); **Ts** 29.0–33.9 mm (n 69, m 31.7). **Wing formula:** p1 minute, < pc 5–13 mm, < p2 64–72 mm; **p2** < wt 2–5 mm; **pp3–4** longest or differing 1–3 mm; **p5** < wt 3–7.5 mm; **p6** < wt 13–19 mm; **p10** < wt 30–37 mm; **s1** < wt 33–40 mm. Emarg. pp3–5.

T. p. clarkei, 1stS, England, Feb: a subtle race, only very faintly darker/browner (less olive-tinged) above than *philomelos*, with underparts often slightly richer ochraceous-buff, especially breast. Greater coverts are juv, with prominent rufous-buff spots. (G. Thoburn)

GEOGRAPHICAL VARIATION & RANGE Three subspecies have occurred within the treated region, two of them fairly distinct in coloration, whereas the third is more subtle. Size differs little. Eastern end of cline in Siberia reputedly slightly larger (ssp. *nataliae* Buturlin, 1929), but extralimital and not studied by us.

T. p. philomelos C. L. Brehm, 1831 (Europe except Britain, W Siberia, N Mongolia, Caucasus, Transcaucasia, N Iran; winters W & S Europe, rarely NW Africa, breeders in W & S Europe mainly sedentary or make only short movements). Described above. – Birds in NW Continental Europe are intermediate between this and *clarkei*, but perhaps closest to *philomelos*. Vaurie (1959) reported larger spring birds from Zagros, Iran (**W** 118–128 mm, *n* 11, *m* 122), but unknown whether these were migrants or local breeders.

○ **T. p. clarkei** Hartert, 1909 (Britain except extreme north-west; winters England, Ireland, SW Europe, but also partly resident). Of same size but on average subtly shorter-winged and longer-tailed, and very faintly darker brown above with slightly less olive cast. Underparts often a little richer ochraceous-buff, especially on breast. Many cannot be separated from *philomelos* when handled alone, without direct comparison, and even then some are doubtful. A subtle race. Sexes differ < 2% in size. **W** 113–122 mm (*n* 25, *m* 116.7); **T** 76–83 mm (*n* 25, *m* 80.8); **T/W** *m* 69.1. – Birds in Scotland slightly darker, tending towards *hebridensis*, sometimes separated as *catherinae* Clancey, 1938, but difference slight and not consistent. (Syn. *catherinae*; *ericetorum*.)

T. p. hebridensis Clarke, 1913 (Outer Hebrides, Skye; apparently winters Britain). Somewhat larger and clearly darker than preceding two races, and streaking and spotting of underparts decidedly heavier, especially on breast, adding to the dark appearance. Brown above with very limited olive cast, similar to *clarkei* but even more pronounced. **W** 113–127 mm (*n* 23, *m* 120.1); **T** 74–89 mm (*n* 23, *m* 82.7); **T/W** *m* 68.9; **B** 21.2–24.5 mm (*n* 23, *m* 22.5); **Ts** 31.9–34.6 mm (*n* 23, *m* 33.2).

T. p. clarkei, presumed ad, Scotland, Apr: Scottish birds are slightly darker and often have heavier spotting below, tending towards *hebridensis*. Here a heavily marked bird, almost like Mistle Thrush, but spotting more arrowhead-shaped. Also has browner (less greyish-tinged) crown and nape, and plainer wing with ochraceous-buff wing-bars. Ageing frequently difficult, but in this case has ad-like greater coverts, with small pale buff tips to outer webs. (S. Round)

T. p. clarkei, juv, England, Jul: note fluffy, soft feathers and rufous-buff wedge-shaped shaft-streaks on upperparts, broadest on scapulars. Greater coverts, too, have prominent wedge-shaped rufous-buff tips to outer webs. (S. Round)

T. p. hebridensis, Outer Hebrides, Scotland, Jul: noticeably darker, with heavier streaking and spotting below, chiefly on breast and flanks, adding to generally almost pipit-like appearance. (R. Robinson)

REFERENCES McGowan, R., Clugston, D. & Forrester, R. (2004) *BW*, 17: 71–75. – Schütz, E. (1959) *Die Vogelwelt des südkaspischen Tieflandes*. Stuttgart.

REDWING
Turdus iliacus L., 1758

Fr. – Grive mauvis; Ger. – Rotdrossel
Sp. – Zorzal alirrojo; Swe. – Rödvingetrast

One of the smallest *Turdus* thrushes, with its stronghold in boreal forests, often in the subalpine birch zone, in N Europe. There, it is one of the rather few abundant songbirds to be heard (others being Brambling and Willow Warbler). Like the Fieldfare, it has expanded its range southwards in recent times. Redwings winter in W, C and S Europe, and once its drawn-out, high-pitched 'sucking' call is learnt you will notice—especially on misty nights—that it is a common migrant in October, even over large towns, where they are probably attracted (or confused?) by the many lights.

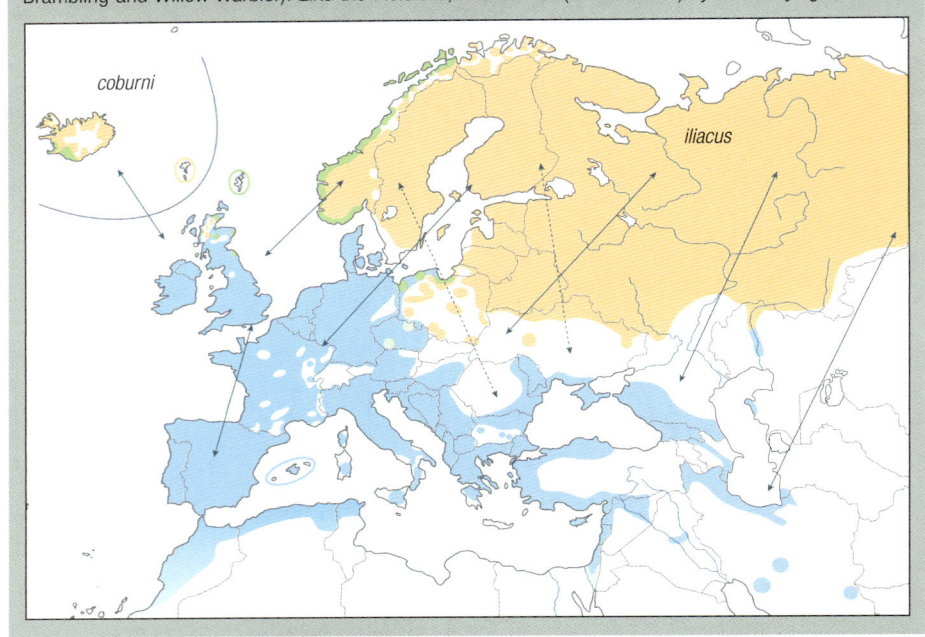

T. i. iliacus, Finland, Apr: in flight note diagnostic rufous-red underwing and well-marked face pattern separating from similar-sized Song Thrush. (M. Varesvuo)

IDENTIFICATION A small, compact, rather short-tailed thrush, about the same size as Song Thrush, which it also resembles in shape and behaviour (except that it has marginally shorter legs and often appears to have a proportionately slightly larger head). Sexes similar. Redwing has two obvious features which readily separate it from Song Thrush: a *long, prominent buff-white supercilium*, which runs from base of bill to nape; and *rufous-red flanks*. Additional characters are a *whitish stripe at the bottom of the cheeks* (submoustachial stripe), *dark lores and crown* (enhancing the prominence of the supercilium), rather uniform brown upperparts and cheeks, a bright *yellow base to lower mandible* (mainly in adults in summer), and *strong dark streaking on whitish throat and breast*, often running together on throat-sides to form a large blackish patch, but nearly unstreaked white belly. In flight note *rufous-red underwing*. Juvenile readily recognised as a young thrush by its buffish wedge-shaped shaft-streaks on upperparts, and as a Redwing by the long buff supercilium, ochraceous tinge on flanks and pale-tipped inner tertials (Song Thrush: uniform olive-brown tertials without pale tips). Flight is a little jerky and undulating, again similar to that of Song Thrush.

VOCALISATIONS The song is quite variable, but nearly always instantly recognisable by its tone and pattern, a fast series of a few loud similar notes on a falling or rising scale, or other variants, immediately followed by a short, high-pitched or squeaky twitter. A very common variant is descending (the twitter omitted in the following renderings), e.g. *chirre-chürre-charre*, or ascending *tra-tro-trü-tree*, short and mournful *truii-traee*, clear ringing *till-ill-ill-ill-ill*, or even a Common Rosefinch-like (!) *vedye-vedyuh*. Each bird has only one song type, and often one type dominates in a region. However, a strict pattern of local 'dialects' does not prevail, and from time to time you can hear two or even more song types competing in the same valley. In spring, roosting flocks of migrants can give quite a noisy chorus of both song and chattering squeaky sounds. – The call given in flight on migration, also at night from high up (referred to above), is a drawn-out 'strained whistle', *seeef*, which can be described as 'inhaling' or 'sucking', with a slightly hoarse quality. Perched birds in cover reveal themselves with a slightly nasal *gack* with almost ventriloquial quality. Alarm near nest a hoarse rattling scold, *tret-tret-tret-tret-...*

SIMILAR SPECIES The combination of small thrush-size and general thrush-shape, buff-white supercilium and submoustachial stripe, rufous-red flanks and in spring yellow bill-base should exclude all other species. Very rarely the red on flanks is covered by the wing or partly missing, or is difficult to see in poorer light, but the other characters should be sufficient.

AGEING & SEXING Ages differ slightly during 1stY, but ageing requires close views. Sexes do not differ in plumage and only very slightly in size. – Moults. Complete post-nuptial moult of ad in summer (mid Jul–Sep, rarely to mid Oct). Partial post-juv moult in summer (Jul–Sep) does not usually involve flight-feathers, tertials or primary-coverts. Some inner greater coverts also moulted, but usually 4–7 outer retained. Rarely a few tertials and tail-feathers also replaced. No pre-nuptial moult in winter. – **SPRING Ad** All greater coverts and tertials uniformly edged brown-grey. Tail-feathers rather broad, tips rounded. Primaries still rather fresh. **1stS** Similar to ad, but generally differs by having a variable number of juv outer greater coverts unmoulted, contrasting with new inner in having triangular whitish spots or notches at the tips (where now-abraded whitish tips used to be). Often also all or inner 1–2 tertials unmoulted, with similar wedge-shaped whitish spots or notches at tip of outer web. Tail-feathers on average narrower and more acutely pointed (if not too abraded to judge). Primaries usually heavily worn at tips. – **AUTUMN Ad** All greater coverts uniform, edged rufous or tawny-brown and tipped ochre-buff (not white). Tertials same, or sometimes have pale buff or whitish tips to outer webs of inner two, but these are narrow, not wedge-shaped. Tail-feathers rather rounded at tips. **1stW** Outer 3–6 greater

T. i. iliacus, ad, Finland, Apr: size and shape Song Thrush-like, but easily distinguished by prominent, long and well-defined white supercilium, and rufous-red flanks. Whole wing very fresh still in Apr, and no traces of pale tips to shorter tertials or greater coverts (nor of indentations where such tips once were) make this an ad. (M. Varesvuo)

THRUSHES

T. i. iliacus, 1stS, Spain, Feb: tertials are juv and have characteristic whitish wedge-shaped pale tips. Greater coverts difficult to judge as to age, appearing as ad type, but may still be atypical juv. (E. Winkel)

T. i. iliacus, ad, England, Feb: apart from rufous-red flanks, note prominent whitish supercilium enhanced by dark eye-stripe and lateral crown-stripe, and dark-streaked whitish throat and breast. All greater coverts are ad and uniformly brown-grey (no rufous-buff tips), and tertials are of ad type, too. (M. Lane)

T. i. iliacus, 1stW, Netherlands, Dec: due to wear, pale spots on outer greater coverts are already reduced, but on juv tertials there are still clear and wedge-shaped pale tips (especially on innermost), making this bird 1stW. (J. van den Bosch)

T. i. iliacus, 1stW, Finland, Oct: juv greater coverts (six outer) are shorter and have prominent wedge-shaped buff-white spots at tips, mostly concentrated on outer webs. (M. Varesvuo)

coverts unmoulted (juv feathers), often edged duller, slightly less rufous-brown, and at least inner ones have prominent wedge-shaped buff-white spots at tip of outer web. Inner two tertials often also have triangular buff-white tips on outer web (but some have narrower white tip only). Tail-feathers pointed at tips. **Juv** Has narrow buff-white shaft-streaks above (largest on scapulars), including on lesser and median coverts; pale-tipped inner two tertials; and on at least some ochre-buff tinge on flanks, signalling where the rufous-red of ad will later develop.

BIOMETRICS (*iliacus*; Sweden) **L** 19.5–21.5 cm; **W** ♂ 112–125 mm (*n* 59, *m* 118.8), ♀ 111–121 mm (*n* 50, *m* 116.6); **T** 73.5–86 mm (*n* 109, *m* 79.1); **T/W** *m* 67.2; **B** 20.4–24.6 mm (*n* 33, *m* 21.6); **BD** 5.4–6.9 mm (*n* 43, *m* 6.2); **Ts** 26.7–29.8 mm (*n* 43, *m* 28.5). **Wing formula: p1** minute, < pc 8–14 mm, < p2 66–83 mm; **p2** < wt 2.5–5.5 mm, =4/5 (72%), =5 (27%) or =5/6 (1%); **pp3–4** about equal and longest (though sometimes either p3 or p4 0.5–2 mm <); **p5** < wt 4.5–8 mm; **p6** < wt 15–20 mm; **p10** < wt 33–40 mm; **s1** < wt 36–43 mm. Emarg. pp3–5.

GEOGRAPHICAL VARIATION & RANGE Only two subspecies described, fairly well separated by differences in colour and size.

T. i. iliacus L., 1758 (Europe, except Iceland and Faeroes; winters W & S Europe, N Africa, SW Asia). Described above. Underparts rather clean whitish with well-defined streaks. (Syn. *musicus*.)

T. i. coburni Sharpe, 1901 (Iceland, Faeroes; winters W Europe). Larger (though bill about same), darker and more olive-tinged above, less brown, and more buff below, less white. Streaks on underparts tend to be warmer brown and blur together, are less well-defined. **L** 21–23 cm; **W** ♂ 120–131 mm (*n* 15, *m* 124.9), ♀ 118–127 mm (*n* 12, *m* 121.5); **T** 76–90 mm (*n* 27, *m* 83.2); **T/W** 67.4; **B** 20.5–24.5 mm (*n* 25, *m* 21.9); **Ts** 29.4–32.1 mm (*n* 26, *m* 30.7).

T. i. coburni, 1stW, Faeroes, Mar: note darker upperparts, and generally more heavily-streaked and a little duskier underparts, with streaks usually poorly defined. Moult limit in greater coverts make this bird 1stW. (S. Olofson)

T. i. coburni, 1stW, Iceland, Apr: some are very similar to *iliacus* and identified only by range, as here, though upperparts still a little duller and darker. Moult limit in greater coverts reveals it to be 1stW. (J. Larsen)

T. i. coburni, juv, Iceland, Jul: well-marked facial pattern, especially long buff-white supercilium, and pale-tipped inner tertials. Fluffy plumage and buff-white shaft-streaks to upperparts are other features of this plumage. Note that juv tertials lack bold wedge-shaped tips, tips being instead more rounded and ad-like. (S. Kunttu)

MISTLE THRUSH
Turdus viscivorus L., 1758

Fr. – Grive draine; Ger. – Misteldrossel
Sp. – Zorzal charlo; Swe. – Dubbeltrast

Although probably well known by name, rather few have a close familiarity with this large European thrush. In N Europe it is a shy forest bird that prefers coniferous woods interspersed with smaller fields, often in mature pine forests on rocky or sandy soils. In S Europe mostly a bird of mountains and alpine meadows. And in Britain and Ireland frequently a garden or park bird year-round—thus quite a variety of preferred habitats. Migrates south and south-west in autumn from northern parts of its range, but returns early in spring, as soon as the snow is gone.

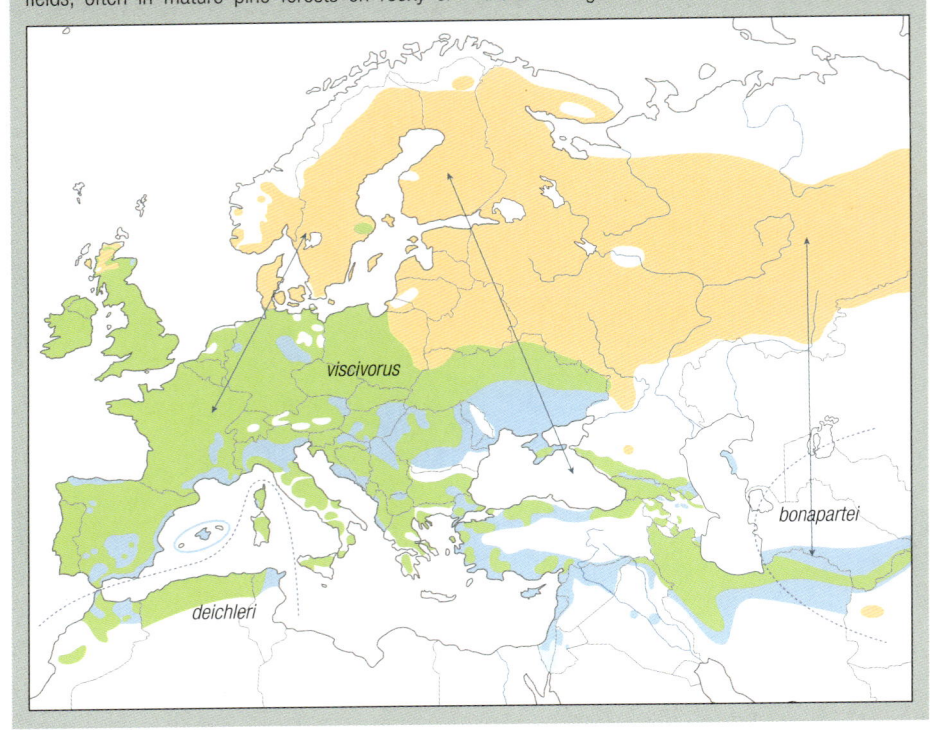

T. v. viscivorus, presumed ad, Finland, May: larger, and usually longer-tailed and stronger-billed than Song Thrush, with more extensive and densely patterned spots below, reaching further back, and generally more variegated wing pattern (especially pale panel on folded secondaries, and pale-edged primary-coverts); also bolder facial pattern. Probably ad by uniform greater coverts (no moult contrast) and by less worn, round-tipped primary-coverts (ad-like). (M. Varesvuo)

IDENTIFICATION A *large, rather slim-shaped* thrush with *long wings* and long tail. Stance on ground upright when alert. Shy and wary, takes off early, quickly gains height and heads for nearest wood. Flight fast and typically *shallowly undulating*. Sexes similar. When perched superficially looks like a large, slim Song Thrush but on closer check recognised by: (i) proportionately *longer tail with white corners* (visible at least in flight); (ii) more variegated wing pattern with *pale-edged tertials*, generally a *pale panel along folded secondaries*, and pale-edged primary-coverts and greater coverts forming other pale patches (wing on average more variegated in first-years, slightly more uniform in adults); (iii) underparts with more *even dark spotting*, reaching far down on belly and flanks, and *spots rounded* (except on throat); (iv) slightly *darker bill* (often appearing rather short); and (v) different head pattern with *two hinted dark vertical stripes on pale cheeks*, one below the eye, the other at rear of ear-coverts (a hint of same pattern is sometimes visible in Song Thrush, but never as prominently). Like Song Thrush has narrow pale eye-ring, most visible behind eye. Underparts variably tinged ochraceous-buff on breast and flanks, but can wear nearly all white. On sides of breast sometimes has olive-brown patch. In flight, note *pure white underwing*. Juvenile recognised as a young thrush by its buffish wedge-shaped shaft-streaks on upperparts, and as a Mistle Thrush on general size and shape, prominently buff-tipped tertials, and same variegated wing pattern as at other ages (see above).

VOCALISATIONS Song resembles that of Blackbird, varied, loud and fluty, and could be confused with it (especially as some have more mellow, Blackbird-like tone), but is generally recognised by more monotonous, desolate ring (due to smaller tonal range and more half-tone steps, as if in the minor key), hard voice, faster pace and shorter pauses between each strophe, and less use of high-pitched concluding twitter (often none): *churichuruh… churutree… churi-churoo… trichüvutru… churuvütritru…*, etc. Sings from top of tree, often at dusk and dawn, but commonly also at midday when other thrushes are silent, even on days with fresh breeze. – Normal call in flight (also on migration), sometimes also by perched bird, is completely different from that of other thrushes, a dry, rolling rattle *zerrrrrr* with wooden or mechanical sound. Alarm rather resembles that of Fieldfare, a noisy scolding rattle, but drier, *zret-zret-zret-zret-…* In contact when breeding, rarely utters a low *kewk*, likened to call by Coot (Hollom in *BWP*).

SIMILAR SPECIES Separated from similarly patterned *Song Thrush* by different song and calls, larger size, longer and white-tipped tail, white underwing, more extensive spotting (spots rounded) on belly and flanks, more variegated wing and head patterns (see under Identification). Juvenile from juvenile Song Thrush by buff-tipped tertials and ochraceous-brown edges to tertials, secondaries and greater coverts, and by white-tipped outer tail-feathers. – Could theoretically be confused with *White's Thrush*, but this has scaly black marks below, not rounded spots, is brightly marked in golden-brown and black above, too, not just on underparts, and has broad white and black stripes on underwing.

AGEING & SEXING (*viscivorus*) Ages differ slightly during 1stW, but ageing requires close views. Ageing in spring frequently difficult (from Mar onwards) due to bleaching of greater coverts, making moulted and unmoulted feathers very similar. Sexes do not differ in plumage and only very slightly in size. – Moults. Complete post-nuptial moult of ad in summer (mainly Jul–early Sep). Partial post-juv moult in summer (Jul–Sep, occasionally later) does not usually involve flight-feathers, tertials or primary-coverts. Some inner greater coverts also moulted, but usually 4–7 outer retained. Apparently no pre-nuptial moult in winter. – SPRING **Ad** All greater coverts and tertials uniformly edged pale brown or off-white. Tail-feathers rather broad, tips rounded, feathers still rather fresh. **1stS** Similar to ad and often impossible to separate with certainty, but a few differ in having faint contrast among greater coverts, outer being edged paler, inner browner.

T. v. viscivorus, Sweden, Mar: variegated and boldly-patterned head with two vertical dark stripes on pale cheeks. Ageing in spring/summer without handling often impossible. (M. Vaslin)

T. v. viscivorus, 1stW, Spain, Oct: pale panel on folded secondaries. Pale-edges of tertials, primary-coverts and greater coverts more pronounced in 1stY. Aged by moult contrast with juv coverts (outer); tips of remiges already quite worn. (A. M. Domínguez)

Tail-feathers narrower and more heavily abraded at tips. — AUTUMN **Ad** All greater coverts and tertials uniformly edged yellowish-brown. Tail-feathers rather broad, tips rounded. **1stW** Similar to ad, but generally differs by having a variable number of juv outer greater coverts unmoulted, edged very pale brown (nearly off-white), contrasting with new inner edged darker, yellowish-brown. (Both old and new greater coverts have similarly-shaped whitish tip to outer web.) Tail-feathers narrower and more acutely pointed, but sometimes difference hard to discern. **Juv** Has narrow buff-white shaft-streaks above (largest on scapulars), including on lesser and median coverts.

BIOMETRICS (*viscivorus*) **L** 26–28.5 cm; **W** ♂ 151–165 mm (*n* 17, *m* 156.5), ♀ 145–160 mm (*n* 14, *m* 152.9); **T** ♂ 101–114 mm (*n* 17, *m* 107.4), ♀ 99–110 mm (*n* 14, *m* 104.3); **T/W** *m* 69.0; **B** 22.6–25.6 mm (*n* 32, *m* 24.1); **Ts** 31.8–35.2 mm (*n* 31, *m* 33.1). **Wing formula: p1** minute, < pc 7–18 mm, < p2 84–95 mm; **p2** < wt 3–9 mm; **pp3–4** about equal and longest (though rarely p4 0.5–1 mm <); **p5** < wt 4–8 mm; **p6** < wt 15–25 mm; **p10** < wt 46–56 mm; **s1** < wt 52–60 mm. Emarg. pp3–5, rarely also a short, less distinct emarg. on p6.

GEOGRAPHICAL VARIATION & RANGE Mostly clinal variation, affecting both size and colour, but differences are rather small and complicated by much individual variation

T. v. viscivorus, Denmark, Dec: compared to Song Thrush, note long tail with diagnostic white corners (usually visible in flight, or as here when feeding). Wings, too, are long. (T. Ravn Kristiansen)

T. v. viscivorus, Denmark, Mar: in typical, rather shallowly undulating flight, mainly white underwing-coverts are apparent, shared only by Fieldfare; white edges to tail just visible. (J. Stenlund)

T. v. viscivorus, juv, Scotland, Jul: buffish-white wedge-shaped shaft-streaks above and prominently buff-tipped tertials. Both wings and face have highly variegated patterns. Being juv, note fluffy feathers and seemingly not fully grown bill. (J. Anderson)

and effects of bleaching. Many subtle or questionable variations have been given subspecies status in the past but are not recognised here.

T. v. viscivorus L., 1758 (much of Europe, W Siberia, Turkey, Caucasus, Transcaucasia, NW Iran; most northern breeders winter SW Europe, possibly also N Africa, southern and western resident or short-range migrants to W & S Europe). Described above. Large and fairly heavily spotted on underparts. (Syn. *bithynicus*; *hispaniae*, *jordansi*, *jubilaeus*; *loudoni*; *precentor*; *reiseri*; *tauricus*; *uralensis*. All fit well with ssp. *viscivorus* when applying a slightly wider subspecies concept involving a 75% rule, not trying to describe and name every subtle variation, mostly of clinal nature. Sometimes birds of EC Europe and NW Balkans are labelled '*reiseri*—but see below—or other names, but they fall within variation of *viscivorus*.)

○ **T. v. deichleri** Erlanger, 1897 (Corsica, Sardinia, NW Africa; resident). Subtly paler and greyer above, and slightly smaller. Birds in Africa tend to have on average very slightly longer bills. A subtle and not entirely consistent race. **L** 25–27.5 cm; **W** 143–159 mm (*n* 24, *m* 152.2); **B** 22.7–26.5 mm (*n* 24, *m* 25.3); **Ts** 30.8–35.4 mm (*n* 24, *m* 32.8). (Syn. *reiseri*.)

T. v. bonapartei Cabanis, 1860 (NE Iran, Central Asia;

T. v. deichleri, presumed 1stS, Morocco, Feb: subtly paler and greyer above and, at least in NW African birds, facial marks appear weaker, whereas bill can look longer; note extremely prominent pale panel in folded secondaries. Probably 1stS by moult contrast with juv outer greater coverts. (D. Occhiato)

T. v. deichleri, Morocco, Apr: here the often rather small head, long neck and plump body is evident. Paler and greyer upperparts on average than in *viscivorus*. Ageing impossible. (G. Conca)

T. v. bonapartei, India, Apr: this race, breeding in Central Asia and wintering in S Asia, is generally larger, paler and greyer above than *viscivorus*, with appearance of smaller and less dense spots on whiter underparts, though some overlap. (N. Devasar)

winters S Central Asia, possibly N Arabia). Generally larger, paler and greyer than *viscivorus*, pale tips to outer tail-feathers tend to be more extensive, and, often, large pale brown wedges on inner web of r6 add to pale impression of outertail. Wing slightly more rounded with emarg. p6 in ad (somewhat variable, but generally more prominent than in *viscivorus*), but only rarely in 1stY. **L** 26–30 cm; **W** ♂ 164–172 mm (*n* 12, *m* 167.3), ♀ 158–170 mm (*n* 9, *m* 163.6); **B** 24.7–27.7 mm (*n* 22, *m* 26.7); **Ts** 32.8–36.8 mm (*n* 22, *m* 34.6). **Wing formula: p1** < p2 85–103 mm; **p2** < wt 5–10 mm; **p5** < wt 2–5 mm; **p6** < wt 9–23 mm.

Presumed *T. v. bonapartei*, 1stS, Pakistan, Mar: ssp. *bonapartei* based on range, but differences from *viscivorus* only slight and not easy to appreciate in the field (subtly larger and averages a little greyer above; in flight has more white in tail). Rather unusually, this bird seems to have retained all juv greater coverts being whitish-edged, thus is 1stS, and relatively worn flight-feathers support this. (I. Shah)

CETTI'S WARBLER
Cettia cetti (Temminck, 1820)

Fr. – Bouscarle de Cetti; Ger. – Seidensänger
Sp. – Cetia ruiseñor; Swe. – Cettisångare

Commonly, the loud and explosive song is the only proof to many that this species exists! It is a secretive bird of dense reeds and bushes growing on or near wet ground, revealing its presence with a short outburst of its song. This rufous-tinged warbler, which acquired its name from an Italian ornithologist living in the 18th century and who wrote about the birdlife of Sardinia, is mainly a resident species in Europe but more migratory further east. It has expanded its range north-west in recent decades, and since the 1970s has established a small but healthy population in mainly S England.

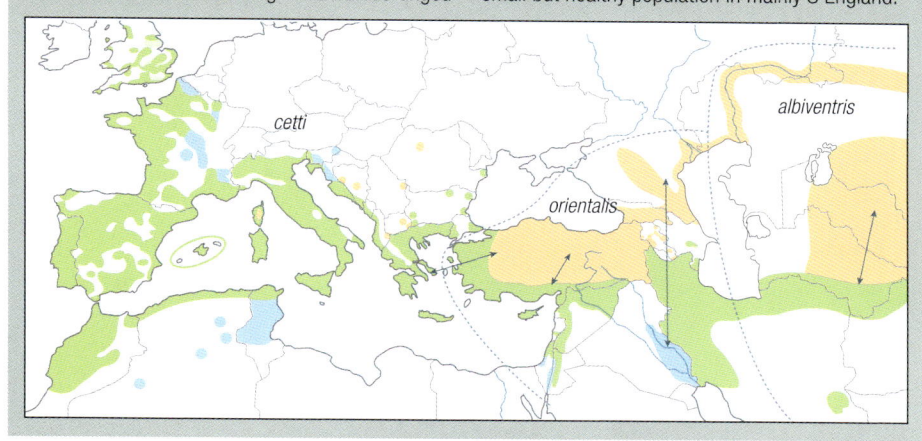

IDENTIFICATION Obviously the song provides the best and easiest means of identification. But once seen well it is fairly easy to identify even without vocal support. To see the bird, however, takes some luck and knowledge, as it is an unobtrusive bird which is quick to take to cover if visible at all. Among warblers, Cetti's is *medium-large*, with *short, rounded wings* and a *fairly long tail* which is often *cocked up and spread* a little. The tail consists (peculiarly for a European passerine) of just ten tail-feathers, but to compensate for this lower than normal number, they are unusually broad and rounded at tips. Cetti's Warbler has *dark rufous-brown upperparts* and *greyish-white underparts*, lacking any striking plumage features. There is *a narrow, short, greyish-white supercilium*, an *off-white eye-ring*, and the throat is diffusely paler than the rest of underparts, but apart from that few characteristics. The *undertail-coverts* are sometimes exposed, and these are *short* in relation to members of *Acrocephalus* or *Locustella*, and are *rufous-tinged with off-white tips*, a rather distinctive feature (although whitish tips are largely missing in juveniles). The rather short pointed bill is mainly dark with some pink on the lower mandible. Legs are rather strong and pinkish. – The eastern race *albiventris* is usually appreciably larger and paler. Underparts and sides of head are more whitish than grey with only a moderate brown-grey tinge on flanks and vent, and this makes supercilium and eye-ring more difficult to discern and gives the bird a paler and 'open-faced' appearance compared to western birds. Also, upperparts are less dark rufous, more greyish-cinnamon or pale tawny.

VOCALISATIONS The striking song, given mostly at dusk and dawn and in the morning hours, but which can be heard at night (preferably in its later part) and less intensely at other times of day, is one of the most characteristic and easily recognised among European warblers, a brief (2.5–4 sec.) explosive outburst of loud notes, often opening with a short, loud and summoning phrase followed by a fleeting pause (to raise our attention!), then a rapid series of repeated disyllabic notes, concluded by a few more notes that seem a little hesitant: *chit chit-chet!... cetti-cetti-cetti-cetti, chet, chet!* or similar. Song delivered from hidden perch. Songs well spaced, at high intensity every 10–30 sec., but in daytime much sparser. There is some geographical variation in the phrasing, but this does not seem to coincide well with named races. No dramatic difference noted between day and night songs, only latter often shorter and more stereotyped (cf. *BWP*). – Among calls, that most often heard is a strong metallic *chik!*, in anxiety repeated and often becoming a metallic trill, *chik-ik-ik-r'r'r'r'r'r*, which has been likened to similar call of Wren. A soft upward-inflected *huit* has been described (Ireland in *BWP*) but not heard by us.

SIMILAR SPECIES Although something of a personality as to shape, habits and movements, it can still be confused with other species if observation is short or made in poor conditions. The general size and uniform warm brown colour above invite confusion with *Nightingale*. Note uniform rufous colour to entire upperparts (Nightingale: contrasting vividly reddish-brown uppertail) and smaller size. Undertail-coverts of Cetti's are reddish-brown, finely tipped whitish (Nightingale: uniform pale buff). – Separation from the group of uniform brown and similarly-sized *Acrocephalus* (*Reed*, *Marsh* and *Blyth's Reed Warblers*) is best made by structure and behaviour. Cetti's has a rounded head and rather short neck and bill, but a generally appreciably longer tail, which it frequently cocks and semi-spreads. Primary projection is also very short in Cetti's, c. 1/3 of tertial length, considerably longer in Reed and Marsh, and somewhat longer in Blyth's Reed. Again, the rufous-tinged and white-tipped undertail-coverts are a useful clue if the view is favourable. – Differs from the likewise uniformly dark brown *Savi's* and *River Warblers* by the rounded head, long, broad and often somewhat spread tail, and above all the short and rounded wings (both *Locustella* species have long pointed wings, more or less lacking emarginations).

AGEING & SEXING (*cetti*) Ages very similar and often alike. Sexes very similar or alike as to plumage, but ♂ larger than ♀ with only small overlap. – Moults. Complete post-nuptial moult of ad in late summer (Jul–Sep; sometimes from late Jun). Partial post-juv moult coinciding with moult of ad, sometimes involving head, body, median coverts, one or two inner tertials (Ginn & Melville 1983; J. Blasco-Zumeta, see refs.) and 4–5 inner greater coverts. Rarely all tertials and a few to many secondaries and primaries renewed, too, but never

C. c. cetti, ♂, Italy, Apr: Rarely seen singing in the open as here. Note short and rounded wing shape, greyish-white throat and chest, and narrow whitish supercilium. Greyish ear-coverts and neck-sides contrast with rusty-brown upperparts. Sex based on singing. Freshness of wing in Apr might indicate ad, but best not to attempt ageing without handling. (D. Occhiato)

C. c. cetti, presumed 1stS, Spain, Mar: typical compact shape and posture, with short bill in relation to head-size. Apparently mostly juv wing-feathers, judging by the abraded tips to remiges, and tail seemingly also mostly juv (possibly only two replaced feathers). Sexes alike with regard to plumage. (C. N. G. Bocos)

C. c. cetti, ad, Italy, Sep: very short-winged and warm-toned warbler, and some (especially in spring/ summer or in SW of range) are rustier above, making them even less mistakable. Supercilium can be rather subdued, but here is relatively pronounced and whitish-grey, and together with reddish-tinged upperparts and extensively pure grey underparts indicate ad. Evenly fresh ad-like wing also assists ageing. Longish broad tail is constantly cocked nervously. (D. Occhiato)

C. c. cetti, presumed ad, Spain, Sep: compared to *Acrocephalus* or *Locustella* warblers, undertail-coverts, if seen well, are characteristically tipped off-white; greyish-tinged cheeks sometimes noticeable. Ageing difficult without closer inspection of wing, though visible tracts are ad-like. (M. Varesvuo)

outer five primaries or primary-coverts. Partial pre-nuptial moult not well known and often absent. In a few moulting birds (Mar), some tail-feathers and innermost tertials were growing (Bub & Dorsch 1988). – **SUMMER–AUTUMN Ad** Upperparts distinctly rufous. Underparts greyish-white with rufous tinge to flanks and vent. Cheeks and neck-sides slightly tinged warm brown, throat with hint of olive-yellow. Tips of primaries and tail-feathers first worn, then these feathers moulted, finally uniformly fresh. Iris rufous-brown. Sexes very similar and often alike. A tendency for ♂♂ to have richer rufous upperparts, greyer neckside, throat and breast, and to have better marked supercilium then ♀♀ (D. Bigas, pers. comm.), but much overlap and requires practice and direct comparison, and only useful for extremes. **1stW** Often inseparable from ad in the field (at times even in the hand) but tends to have darker iris (brown hard to see), and cheeks, neck-sides and throat are colder greyish-tinged or off-white. Rarely a faint moult contrast discernible between new inner tertials and greater coverts and outer unmoulted (J. Blasco-Zumeta, see refs.). **Juv** Upperparts tawny-brown with some rufous tones. White of underparts slightly tinged yellowish-buff. Flight-feathers and entire plumage uniformly fresh. Feathers of nape, vent and undertail-coverts rather loose and greyish-tinged.

BIOMETRICS (cetti) **L** 12.5–14 cm; **W** ♂ 60–66 mm (n 25, m 63.2); ♀ 53–61 mm (n 12, m 56.3); **T** ♂ 56–67 mm (n 25, m 61.6); ♀ 52–59 mm (n 12, m 55.2); **T/W**

C. c. cetti, presumed 1stW, Italy, Aug: note very short primary projection. Contrasting greyish-white throat/chest often important for identification, while whitish supercilium (narrow but prominent in most) is an important facet of the 'personality' of the species. Off-white eye-ring, bill rather short and pointed, legs pinkish. Longish tail often partially spread. Much of wing and tail already quite worn and abraded (juv) indicate 1stW. (D. Occhiato)

m 97.6; **B** 12.7–15.6 mm (n 34, m 14.0); **BD** 2.8–3.6 mm (n 34, m 3.3); **Ts** 20.0–22.7 mm (n 31, m 21.5). **Wing formula: p1** > pc 8.5–12.5 mm, < p2 13–18.5 mm; **p2** < wt 7–11 mm, =ss or <ss (66%), =p10 or 10/ss (39%), or =9 or 9/10 (5%); **p3** < wt 0–3 mm; **pp4–5** about equal and longest (though rarely p5 0.5–1 mm <); **p6** < wt 1–4 mm; **p7** < wt 4–6 mm; **p10** < wt 6.5–8 mm; **s1** < wt 7–9.5 mm. Emarg. pp3–6, sometimes faintly also on p7. – Wing length from a much larger sample of live birds in Camargue, S France (Mester 1975): **W** ♂ 59–66 mm (n 609, m 62.2), ♀ 50–61 mm (n 1179, m 55.8). In a Greek study (Alivizatos et al. 2011) the following values were obtained: **W** ♂ 58–70 mm (n 59, m 62.1), ♀ 53–63 mm (n 65, m 56.2).

GEOGRAPHICAL VARIATION & RANGE Clinal variation, birds from west to east becoming larger and less rufous, more pale and greyish-tinged brown above, and paler, more whitish below. Three races within covered area, which intergrade to some extent where they meet. There is much variation within eastern *albiventris*, especially as to size, birds of Kirghiz Steppe being particularly large and pale, and more detailed research might yield another subspecies there.

C. c. cetti, presumed 1stW, Italy, Sep: a duller individual, with indistinctly paler underparts; whitish supercilium usually more pronounced above and behind eye. Short pointed bill is pinkish-based. Given early date, and already brown-tinged and worn greater alula, possibly 1stW (ad alula glossier and blacker-centred). (D. Occhiato)

C. c. orientalis, presumed ad, Turkey, May: this race averages paler and less rufous overall, and at least some individuals (like this bird) are more greyish-brown, approaching *albiventris*. Not heavily worn plumage in May could indicate ad. (K. Malling Olsen)

C. c. orientalis, Israel, Dec: note dull upperparts of this race; also, when fresh (autumn/winter) underparts and supercilium usually cream-white, being only slightly sullied grey, just a little on the neck-sides. (H. Shirihai)

C. c. orientalis, juv, Israel, May: essentially tawny-brown above and yellowish-buff below; body-feathers have a loose, fluffy quality. Pinkish-yellow mouth flanges are still obvious. (L. Kislev)

C. c. orientalis, Iran, Jan: race *orientalis* averages duller rufous above, with slightly greyer flanks and vent, though still not as dull above and pale-bellied as *albiventris*, thus intermediate between *cetti* and *albiventris*. Difficult to age without handling. (E. Winkel)

C. c. cetti (Temminck, 1820) (W & S Europe east to W Black Sea, NW Africa; western breeders mainly resident, eastern short-range migrants). Treated above, being deep rufous above with only faint olive-grey cast on crown and nape; rump somewhat more vividly rufous-tinged. Uppertail dark brown. Throat whitish, breast washed brownish-grey, flanks same but tinged more rufous. Vent and undertail-coverts ochre or rufous-brown, longer feathers broadly tipped whitish. See also Biometrics. – Breeders in NE Italy and Balkans sometimes separated ('*sericea*') tending towards *orientalis*, but represent subtle cline and better not formally named. (Syn. part of *sericea*; see also *orientalis*.)

○ **C. c. orientalis** Tristram, 1867 (Turkey, Levant, SE Russia, Caucasus area, NW & W Iran; northern breeders make short-range movements southwards or to lower elevations). Small like *cetti*, but somewhat paler and less rufous upperparts, and less vividly rufous-tinged on flanks and vent, more greyish-brown (intermediate between *cetti* and *albiventris*). Tail on average proportionately very slightly shorter than in the other two races. **W** ♂ 61–68 mm (n 18, m 63.9), ♀ 57.5–61 mm (n 11, m 59.1); **T** ♂ 55–65 mm (n 18, m 61.6), ♀ 51–59.5 mm (n 11, m 55.3); **T/W** m 95.6; **B** 12.6–15.2 mm (n 28, m 14.0); **Ts** 20.0–23.5 mm (n 29, m 21.4). **Wing formula: p6** < wt 0–1 mm; **p7** < wt 2–3.5 mm. – Breeders in W Turkey sometimes separated ('*sericea*') being more like *cetti*, but represent subtle cline and better not formally named. (Syn. part of *sericea*; see also *cetti*.)

C. c. albiventris Severtzov, 1873 (N & E Iran, Central Asia, in north-west to NW Kazakhstan; winters Afghanistan, Iran, Pakistan). Somewhat variable size, but generally large. Long-tailed like *cetti*. Underparts pale, whitish, with flanks and vent light grey-brown. Whitish supercilium broad or washed out and not well-marked due to general paleness of head-sides. Upperparts rather pale rufous-brown or tawny with greyish cast. **L** 13.5–15 cm. **W** ♂ 64–74 mm (n 24, m 67.7), ♀ 58–67.5 mm (n 12, m 62.0); **T** ♂ 61–71 mm (n 23, m 65.7), ♀ 57–66 mm (n 12, m 60.8); **T/W** m 97.3; **B** 12.7–16.0 mm (n 36, m 14.5); **Ts** 20.0–24.5 mm (n 35, m 22.0). Wing formula as *orientalis*, thus has slightly more rounded wing than *cetti*. (Syn. *cettioides*.)

REFERENCES ALIVIZATOS, H. et al. (2011) *Ring. & Migr.*, 26: 74–76. – BLASCO-ZUMETA, J. www.javierblasco.arrakis.es. – BUB, H. & DORSCH, H. (1988) *Cistensänger, Seidensänger, Schwirle u. Rohrsänger*. NBB 580: 25. Wittenberg Lutherstadt. – MESTER, H. (1975) *Ardeola*, 21 (Suppl.): 421–445. – VILLARÁN, A. (2000) *Butll. GCA*, 17: 1–9. – ZAVIALOV, E. V. & TABACHISIN, V. G. (2007) *DB*, 29: 303–305.

C. c. albiventris, Kazakhstan, Aug: typical posture, hunched down with tail held high, inquisitive and cautious, a species more often heard than seen, dwelling in dense vegetation. Note long and rather broad tail, rufous-tinged upperparts and off-white underparts. Facial pattern also characteristic with off-white rather prominent but not so long supercilium, enhanced by rufous eye-stripe. (G. Dyakin)

ZITTING CISTICOLA
Cisticola juncidis (Rafinesque, 1810)

Alternative names: Fan-tailed Warbler, Fan-tailed Cisticola

Fr. – Cisticole des joncs; Ger. – Cistensänger
Sp. – Cistícola buitrón; Swe. – Grässångare

A rather secretive warbler, which betrays its presence mainly by its monotonous 'zitting' song, delivered in undulating display flights over grassy meadows or fields. Although originally an inhabitant of natural grass fields such as open savannas and marshes, it has adopted to suitable habitats provided by man like cereals, sugar cane and lucerne fields. A resident, confined to the warmer southern and south-western parts of Europe. The nest is built on or near the ground in dense vegetation, and each male often has at least two females, frequently as many as four (one noted to have had 11).

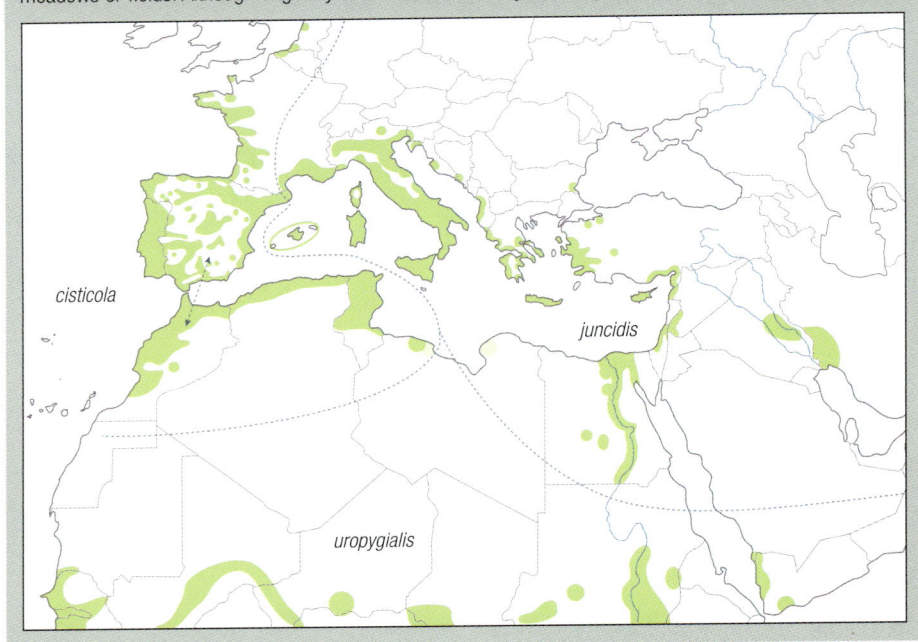

IDENTIFICATION A Zitting Cisticola in song-flight is easily identified: *one of Europe's smallest birds*, no larger than a Goldcrest, in *undulating wandering flight* high up and in broad circles over a field, uttering a persistently repeated shrill, sharp note. It is a 'round' compact little bird with *short, rounded wings* and *short tail*, which it *holds spread during the song-flight*. With the right light and background the *dark tail with white trailing edge* is evident. The difficulties are all about getting a fair view of the bird when perched, and then to safely separate it from Graceful Prinia, scrub warblers, Sedge Warbler and other streaked 'little brown jobs'. Size is a great help. It has actually a smaller body than the two crests, so although the length from bill to tail-tip does not indicate it, it could well be called the smallest bird in Europe. Note *fine, pointed bill, pale and rather plain brownish side of head* without field marks (but darker, streaked crown), coarsely black-streaked back, a *cinnamon- or ochre-tinged streaked rump, cinnamon-tinged flanks* and a short, 'spiky' *tail*, which folded becomes quite narrow (and can look a little longer than it really is) and when spread shows diagnostic *black subterminal patches and white tips on all but the central pair of feathers*. Legs are rather long and pale pinkish. The bill is usually pinkish with dark culmen, but in the breeding season bill of ♂ darkens and acquires variably extensive dark tip to lower mandible. In worn summer plumage, ♂ usually acquires darker crown than ♀. The general plumage colour varies somewhat geographically, birds in SW France, Iberia and NW Africa being slightly darker and greyer brown than those of the rest of the Mediterranean basin, which are paler and more brightly cinnamon-tinged.

VOCALISATIONS The characteristic song, nearly always delivered in daylight in undulating song-flight at a height of 5–30 m, is a persistently repeated shrill or sharp single note, *dzip... dzip... dzip... dzip... dzip...*, etc., one note per c. 0.7 sec. given at each undulation, song bouts lasting for half a minute or more. There is some slight individual and possibly geographical variation in the sound, but not of a magnitude to be confusing. – Calls are basically of two types, one a short ticking *zip* or *chip* which in its variations can recall the song but generally is simpler and not as loud, the other being a series of ticking or clicking sounds run into

C. j. juncidis, 1stS ♂, Greece, Apr: typical perched posture; compact with very short wings and narrow tail, which has diagnostic black bands and whitish tips. Upperparts marked by pale and dark tramlines, while bill rather fine and pointed. Ageing difficult and often impossible, but the renewed tertials among otherwise mostly juv wing-feathers and freshly moulted tail indicate age. In spring, a blacker bill and gape are suggestive of sex, as is contrastingly darker crown. (D. Tipling)

C. j. juncidis, 1stS ♂, Italy, Jul: sometimes hops or runs over open ground. Saturated rufous-ochre upperparts, especially the virtually unstreaked rump, warm ochre-buff underparts and pale and almost unstreaked neck are characteristics of *juncidis*. Alula, primaries, primary-coverts and tail appear to be juv, while some greater coverts have been replaced and the renewed scapulars indicate a 1stS at this season. Bill colour and blackish gape infer sex. (D. Occhiato)

a metallic trill, *pt-pt-pt-pt-*… There are numerous other variations described but they are either not so important to know or reflect 'observer variation'.

SIMILAR SPECIES Separated from *Graceful Prinia* by smaller size and much shorter tail, generally has coarser black streaking above and, even in darker brown SW European/NW African population, an ochre-tinged rump and overall warmer brown colours. – *Saharan Scrub Warbler* has rather uniform upperparts, finely streaked underparts (Zitting Cisticola: unstreaked), a buffish supercilium and a longer tail. – *Grasshopper Warbler* is considerably larger, sleeker and slim-shaped with olive-brown tones, and has dark-centred undertail-coverts. – *Rusty-rumped Warbler* is as large as Grasshopper Warbler, but its whitish tips to outer tail-feathers and streaked reddish-brown rump could invite confusion if only a brief glimpse is obtained. Note size, longer primary projection and more prominent pale supercilium of Rusty-rumped. – *Sedge* and *Aquatic Warblers* are immediately eliminated by their large size, slimmer shape, long primary projection, prominent pale supercilia and uniform brown tail.

AGEING & SEXING (*juncidis*) Ages and sexes similar in plumage, though a few 1stY with incomplete wing moult can be separated. Also, a few extremes may be possible to sex with practice (see below). – Moults. Complete post-nuptial moult of ad rather variable in timing, and is protracted (especially in ♂♂), mainly Jul–Oct (rarely late May–early Nov). Moult of secondaries usually irregular. Post-juv moult, coinciding with moult of ad, is either partial or complete. Although apparently most juv moult completely, quite a few arrest moult and replace only some inner primaries. Partial pre-nuptial moult Feb–Apr seems variable in extent. – **SPRING** ♂ Sexes very similar but sometimes separable, ♂ having broader and more prominent streaks on crown and on average subtly paler head-sides (more contrasting head pattern), and darker bill, often with dark outer half of lower mandible. In worn breeding plumage crown can appear rather solidly dark. ♀ Streaks of crown generally thinner than in ♂, and head-sides finely streaked and a little darker, creating a more diffuse head pattern without the appearance of a darker crown, even in worn breeding plumage. Bill pinkish with dark culmen all year. – **AUTUMN–WINTER** Ad and 1stY usually inseparable once young have concluded a complete moult. Some young leave a variable number of juv primaries unmoulted, and these can then readily be aged through 1stY.

BIOMETRICS (*juncidis*) **L** 9.5–11 cm; **W** ♂ 47.5–54 mm (*n* 20, *m* 50.7); ♀ 44–51 mm (*n* 16, *m* 47.8); **T** ♂ 37–45 mm (*n* 21, *m* 42.3); ♀ 34–46 mm (*n* 16, *m* 39.2); **T/W** *m* 81.7; **B** 10.2–12.5 mm (*n* 33, *m* 11.5); **BD** 2.7–3.5 mm (*n* 16, *m* 3.1); **Ts** 17.0–19.8 mm (*n* 33, *m* 18.3). **Wing formula:**

C. j. juncidis, presumed ad, Turkey, May: unique undulating song-flight, often in wide circles over a field, when bird may be high up and tricky to spot. Short tail often looks contrastingly dark. Prolonged and complex breeding and moult cycles, though start of primary moult already in May, as here, strongly indicates ad. (M. Varesvuo)

C. j. juncidis, presumed ♀, Mallorca, Apr: pale and dark tramlines above, while rufous-ochre in plumage is feature of this race. As post-juv moult can be complete, 1stY is inseparable from ad thereafter unless retaining a few juv feathers. Relatively paler bill, and seemingly narrowly-streaked crown best fit ♀. (M. Schäf)

C. j. juncidis, presumed ♂, Mallorca, Oct: when fresh and if seen well, may appear quite brightly coloured. Note pointed bill, pale head-sides with contrasting darker-streaked crown, coarsely black-streaked mantle/back, cinnamon-tinged rump and flanks, narrow, 'spiky' shortish tail and very short wing, all of which identify species. Legs rather long and pinkish. Broad and contrasting black streaks on crown best indication of ♂. (S. Garvie)

p1 > pc 7–12 mm, < p2 12–21 mm; **p2** < wt 4–7 mm, =7/8 or 8 (55%), =8/9 or 9 (27%), =9/10 or 10 (14%) or =7 (4%); **p3** < wt 1–2.5 mm; **pp4–6** about equal and longest (sometimes p4 and/or p6 0.5–1 mm <); **p7** < wt 0.5–3 mm; **p8** < wt 2–5 mm; **p10** < wt 4.5–7.5 mm; **s1** < wt 5–8 mm. Emarg. pp3–6, often faintly also on p7.

GEOGRAPHICAL VARIATION & RANGE Rather slight and basically clinal variation, Atlantic and W Mediterranean birds being somewhat darker and colder brown. A certain amount of individual or local variation makes pattern less clear, and some described subspecies thought too subtle and inconsistent to be upheld. Two races kept, representing main 'themes'. All populations mainly resident.

C. j. juncidis (Rafinesque, 1810) (SC & SE Europe, from SE France including Balearics, Corsica, Italy including Sardinia & Sicily, east to Turkey, Levant, Iraq, Kuwait, SW Iran, Egypt). Treated above. Compared with ssp. *cisticola*, warmer ochre-buff on underparts and brighter rufous-ochre on often almost unstreaked rump (especially in fresh plumage). Throat usually more contrastingly white than in

C. j. juncidis, presumed 1stW, Egypt, Oct: note complex moult contrast among tertials and remiges, with very worn outer primaries indicating 1stW, but due to regional and age-related variation, it is difficult to relate this safely to age. Tail-bands rather indistinct in this individual. (M. Heiß)

C. j. cisticola, presumed 1stS ♂, Spain, Mar: birds in SW Europe and NW Africa are to some extent darker and duller, and less rufous-ochre-coloured. The almost uniform dark crown and paler face, blacker bill and gape suggest ♂. Moult contrast/limits in wing, especially among greater coverts and tertials may indicate a young bird. (A. Torés Sánchez)

cisticola. Hindneck often paler ochre-rufous with more diffuse streaking, creating a hint of a broad, paler necklace. (Syn. *neuroticus* of Middle East; this must be regarded as a synonym as the type and a paratype in AMNH cannot be separated from a series of *juncidis*. Series of '*neuroticus*' from Syria and Iraq are a trifle paler above, and often not quite as brightly tinged rufous-ochre on flanks as typical *juncidis*, but many appear identical, and reasons for upholding this race are weak.)

C. j. cisticola (Temminck, 1820) (W France, Iberia, NW Africa). More greyish-brown above, boldly streaked, and rump is by comparison less brightly rufous-tinged. Underparts rather cold greyish buff-brown, still with flanks somewhat warmer ochre-tinged, although lacking bright ochre or almost rufous tones of fresh *juncidis*. Nape generally nearconcolorous with crown, not appearing as a pale necklace. **W** ♂ 48–56 mm (*n* 25, *m* 52.0), ♀ 46–53 mm (*n* 15, *m* 49.0); **T** ♂ 36–45 mm (*n* 25, *m* 40.6), ♀ 33–46 mm (*n* 15, *m* 38.7); **T/W** *m* 78.3; **B** 10.6–13.7 mm (*n* 40, *m* 11.9); **Ts** 17.7–19.7 mm (*n* 38, *m* 18.9). **Wing formula: p1** < p2 15–22.5 mm; **p10** < wt 5–9 mm.

○ **C. j. uropygialis** (Fraser, 1843) (SW Arabia, Sahel belt from Mauritania to Ethiopia in north, to Senegal and N Kenya in south). Subtly paler than *juncidis*. Pale subterminal

C. j. cisticola, Spain, Nov: some fresh birds in autumn can be rather buffish, but still not as russet as *juncidis*. Once moult is complete (as here) an evenly fresh bird can be ad or 1stY, and sexing is also impossible. (S. Fletcher)

patches on tail-feathers cinnamon rather than white. No seasonal variation. A weak race. **W** 45–54.5 mm (*n* 12, *m* 48.4); **T** 34–41 mm (*n* 11, *m* 38.4); **T/W** *m* 79.4; **B** 10.8–11.8 mm (*n* 11, *m* 11.4); **Ts** 17.1–19.3 mm (*n* 12, *m* 18.5). (Syn. *perennia*.)

C. j. cisticola, juv, Spain, Sep: rather loose, fluffy body-feathers, in particular visible at flanks and ventral area, and clear yellow gape reveal age (J. Hawkins)

CRICKET WARBLER
Spiloptila clamans (Cretzschmar, 1826)

Alternative names: Cricket Longtail, Scaly Longtail, Scaly-fronted Longtail, Cricket Prinia, Scaly Prinia, Scaly Warbler, Scaly-fronted Warbler

Fr. – Prinia à front écailleux; Ger. – Schuppenkopfprinie
Sp. – Prinia charlatana; Swe. – Sahelsångare

This rather *Prinia*-like warbler is the sole member of the genus *Spiloptila*. It is rather widespread in the N Afrotropics, mainly in the Sahel zone, breeding commonly from S Mauritania and N Senegal east to Sudan and W Eritrea. Largely resident, but makes some movements apparently governed by local rains. Has recently (2009) been found to breed also in Western Sahara. The species inhabits the sub-Saharan dry thorny scrub vegetation and grassy areas, with some northward extension into more arid Saharan habitats or some southward spread into savanna country.

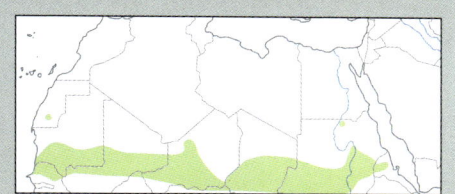

Presumed 1stS ♂, Western Sahara, Mar: ageing uncertain, but primary-coverts (just visible) apparently juv, and iris dull-coloured, suggesting young ♂; black on crown and extensive grey neck-sides also point to ♂. Note black-tipped tail-feathers creating sparsely barred impression. (D. Monticelli)

IDENTIFICATION An unmistakable small warbler, best identified by its *long-tailed*, *Prinia*-like shape with *boldly streaked black-and-whitish crown* (forming delicate scaly pattern when seen close due to fine pale tips to dark feathers, especially obvious when fresh and in ♂♂), and rich *pinkish-brown* or cinnamon *upperparts* often contrasting with grey-tinged neck-sides. Upperwing appears spotted or *chequered black-and-white* (black-centred greater and median coverts fringed white forming double wing-bars). Also, *blackish tertials and secondaries broadly fringed whitish and pale cinnamon*, primaries more narrowly fringed white. Underparts uniform cream-white or pale buff. The long tail is well graduated and brown-grey with *bold buff-white tips, subterminally banded black* (best seen from below when tail raised); *three black bands are formed on undertail* when tail is folded. Wings very short and rounded. Bill has narrow blackish area on culmen and tip, contrasting (especially in breeding ♂♂) with pinkish rest of bill. Legs pinkish-yellow. Sexes are often separable (see below), dark markings in ♀♀ being duller, neck-sides more pinkish (not grey), and all colours slightly less contrasting. Other seasonal plumage variation limited, but cap of juvenile pink-buff with faint darker streaking. Rather social, and often encountered in parties. Often hops or sidles through vegetation, moving restlessly, but can spend extended period on the ground when feeding. Usually not so shy and therefore easy to locate. Flight is straight and low on whirring wings with slim silhouette, enhanced by the long tail.

VOCALISATIONS As its name suggests, often utters a cricket-like trilling sound from exposed perch, which appears to be its song, *zirrr zirrr zirrr zirrr...* Also has long series of a quickly repeated, shrill note, rather like a squeaking, un-oiled machine, *chi-chi-chi-chi-chi-chi-...* – Calls include various high-pitched whistles, repeated slow or fast, sometimes mixed with trills that can recall Scrub Warbler.

SIMILAR SPECIES Within the covered region, confusion possible only with Scrub Warbler (either Levant or Saharan), which is similarly long-tailed as Cricket Warbler. However, Cricket Warbler has different face, wing and tail patterns, and differs especially by the double white wing-bars enhanced by the black-centred greater and median coverts. Cricket Warbler also lacks the blackish eyestripe of Scrub Warbler, the faintly-streaked throat, breast and upper flanks, and it lacks the warmer-coloured flanks. Note also in Scrub Warbler the more uniformly blackish and less graduated tail. – Although Graceful Prinia does not overlap in range with Cricket Warbler, and therefore should not create a problem, it is reassuring to know that it is readily distinguished by plumage, despite some similarity in tail pattern. Note in Graceful Prinia the lack of a bold wing pattern found in Cricket Warbler.

AGEING & SEXING Ageing difficult, often even impossible. Sexes separable (at least with distinctive birds, above all with ad). Seasonal plumage variation small and largely related to wear. – Moults. Complete post-nuptial moult in early summer. Post-juv moult variable in extent, from partial to complete, starting in summer and lasting into autumn. Photographic evidence suggests that a substantial number of 1stY retain remiges, several greater coverts and primary-coverts. Pre-nuptial moult in winter apparently absent, or (in some populations) partial and rather limited. – **ALL YEAR Ad** Both sexes are evenly fresh (or worn), including wing-feathers. However, ad is inseparable from any 1stY which has moulted completely. **Ad ♂** Streaks on crown bold, broad and black, giving darker-crowned impression than in ♀, feathers finely fringed/tipped white forming discreet scaly pattern, and centres of greater and median coverts also black, fringed/tipped pure white, median forming strongly contrasting pattern. Neck-sides grey or bluish-grey. Mantle, scapulars and back (sometimes rump) are deep cinnamon-buff or pinkish-brown. Pale area on bill deep orange-pink,

1stW ♂, Western Sahara, Jan: some ♂♂ are richer vinaceous-pink above. Diagnostic upperwing pattern of ♂ looks variegated black and white, with double wing-bars. Moult limit in greater coverts and tertials indicative of age. (A. Lees)

iris orange red-brown, orbital ring reddish. **Ad ♀** Streaks on crown narrower and browner (hardly forming scaly pattern), pale ground colour buff-grey, not white (or only partially white and cream); often appears paler-crowned than ♂. Centres of greater and median coverts more brown-grey than in ♂ and form less bold pattern. Neck-sides with no or limited grey, while mantle, scapulars and rump are less bright cinnamon-buff or pink-brown. Pale area on bill duller pink, iris and orbital ring as in ♂ or slightly less bright. **1stY** If post-juv moult is incomplete, should be easily aged by some retained juv wing-feathers, but apparently indistinguishable from ad if moult is complete. Indeed, many 1stY seem to retain remiges, several greater coverts and primary-coverts, and these are noticeably browner, more worn and have thinner and less pure white fringes. At least at an early stage, young birds should have, to some degree, duller bare parts, including more olive-tinged iris, but this has not been studied on live birds (only from photographs, and sample small). **Juv** Largely sandy-buff, and note the looser, fluffy quality of juv plumage. The least colourful and weakest-patterned plumage, lacking blackish-streaked crown of post-juv, but wing pattern approaches ♀. Note also olive-tinged iris, dull orbital ring and bill base.

Presumed ad ♂, Western Sahara, Jul: the strongly rounded tail is evident, as each tail-feather is subterminally black-tipped. All greater coverts seem uniformly fresh, and reddish bill and iris indicate ad ♂. (F. López)

♀, presumed ad, Western Sahara, Jan: duller version of ♂ with weak-patterned head bearing less bold black-speckled forecrown, and duller and reduced grey area on neck-sides. The iris being tinged rather deep orange-brown could indicate ad rather than 1stY. Note both orange-red orbital ring and prinia-like shape. (C. N. G. Bocos)

1stS ♂, Western Sahara, Mar: some young ♂♂ are more like ad ♂ (e.g. as here with very bold head pattern), but still have substantial number of juv greater coverts, all or at least most remiges and primary-coverts (browner and narrowly pale-fringed). Iris dull orange-brown, which also confirms age. (V. Legrand)

Presumed 1stW ♀, Western Sahara, Jan: combination of rather dull plumage with speckled forecrown pattern and near complete lack of grey tinge on neck-sides, but also duller iris, and scruffy appearance in general, in part due to moult (also noticeable among forecrown feathers) suggest young ♀. (A. Lees)

BUSH AND REED WARBLERS

BIOMETRICS L 11–12 cm; **W** ♂ 47–50 mm (*n* 15, *m* 48.3); ♀ 44–49 mm (*n* 12, *m* 46.0); **T** ♂ 55–64 mm (*n* 15, *m* 60.1); ♀ 48–58 mm (*n* 12, *m* 52.0); **T/W** *m* 119.4; **B** 11.0–12.5 mm (*n* 27, *m* 17.4); **BD** 3.0–3.4 mm (*n* 25, *m* 3.2); **Ts** 16.0–18.0 mm (*n* 26, *m* 17.4). **Wing formula: p1** > pc 11–14 mm, < p2 11–15 mm; **p2** < wt 3.5–5 mm, =7/ss; **p3** < wt 1–2 mm; **pp4–6** about equal and longest; **p7** < wt 1–3 mm; **p8** < wt 2.5–4.5 mm; **p10** < wt 4–6 mm; **s1** < wt 4–6 mm. Emarg. pp3–7.

GEOGRAPHICAL VARIATION & RANGE Monotypic. – Western Sahara, S Mauritania, N Senegal across Sahel to Sudan and W Eritrea; mainly resident or makes only short seasonal movements. – Some slight variation in birds examined at NHM, but no indication of clear geographical variation over its wide African range.

TAXONOMIC NOTE Sometimes included in *Prinia* (e.g. by Vaurie 1959), but has 12 (rather than ten) tail-feathers and brighter, more contrasting colours, especially in wing, quite different from any *Prinia*.

REFERENCES Amezian, M. *et al.*. (2011) *DB*, 33: 229–233.

♂, presumed 1stW, Western Sahara, Jan: from mid-winter some young birds may already show clear sexual dimorphism; note the very bold back marks on crown, denoting ♂. The combination of evenly fresh plumage with dull orbital ring and iris suggest that it is a young bird that has already moulted completely. (D. Brown)

Juv, presumed ♂, Western Sahara, Mar: juv has somewhat less colourful and more weakly-patterned plumage; note also olive-tinged iris, dull orbital ring and bill base, and rather fluffy quality typical of juv plumage. Subdued blackish-speckled forecrown, greyish neck-sides and bolder black wing markings seem to indicate ♂. (R. Gwóźdź)

Juv, presumed ♀, Western Sahara, Jan: compared to image on left, even duller plumage with weakly-patterned head and browner upperparts and wing markings, most likely indicating a young ♀. (V. Legrand)

GRACEFUL PRINIA
Prinia gracilis (M. H. C. Lichtenstein, 1823)

Fr. – Prinia gracile; Ger. – Streifenprinie
Sp. – Prinia grácil; Swe. – Streckad prinia

Graceful Prinia is generally a common resident of open or semi-open habitats with low scrub from Egypt south to Somalia, also in parts of Arabia and in lowlands from S Turkey to the Indian subcontinent. In the covered region, it is perhaps best known in the Levant, and in Israel is highly adaptive, even breeding in saltmarshes in the Dead Sea depression, while in coastal areas it is a garden bird as well as being typical of shrubby and bushy fallow fields.

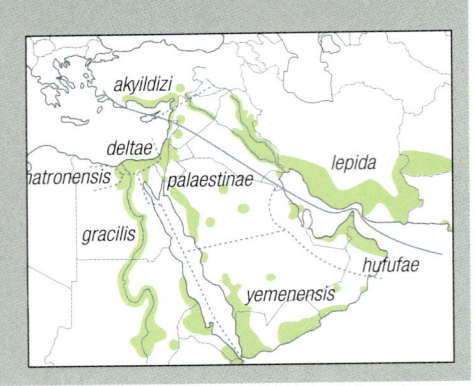

P. g. deltae, ♂, presumed 1stS, SW Israel, Mar: long, noticeably graduated and strikingly barred tail, typical of *gracilis* group, while race assigned by dull sandy-brown and moderately broad-streaked upperparts, underparts tinged olive-grey with slight pale brown or buff tinge, mostly to breast and flanks. Black bill feature of ♂ during courtship period, as are brighter straw-yellow or orange legs. Impossible to distinguish ad from completely moulted 1stY. However, other images of same bird show remiges and primary-coverts to be old and worn (thus reasonably juv), making it 1stS. (H. Shirihai)

IDENTIFICATION A small, *dull-coloured Prinia*, with *short rounded wings* and a *long, narrow tail* (which is frequently cocked, jerked or side-switched). Draws attention by its often-insistent, monotonous calls. Often rather skulking but certainly not shy. *Head rather plain* but has indistinct pale eye-ring and supercilium (sometimes with brown-buff loral area), and slightly paler rear ear-coverts. *Upperparts sandy-brown, streaked dull black-brown on crown, nape and mantle*, but markings decrease in intensity on back and rump, with dark wings fringed buff-grey (including tertials). *Tail structure and pattern diagnostic: noticeably graduated and faintly barred, each feather having a dull black subterminal mark and white tip*, creating striking pattern which is arguably the species' most obvious feature and should quickly identify it over much of the region. Underparts generally whitish-cream with pale buffy flanks, and some have indistinct blotches on breast-sides. Bill rather long, brown with flesh-coloured lower base (but all black in breeding ♂), and legs yellowish-flesh. Often hops or sidles through vegetation. Flight straight and low on whirring wings with slim silhouette, especially the long tail.

VOCALISATIONS Song a frenzied grinding repetition of high-pitched and disyllabic notes, *srrlip-srrlip-srrlip-srrlip-*... or *zerrrit-zerrrit-zerrrit-zerrrit-*..., given from perch or in flight. Occasionally accompanied by rapid, short *brrrp* wing-snap. – Commonest call an explosive *tlipp* or metallic rattling trill, *srrrrrrrt*, often repeated, but various variations occur, like shorter *srrt-t* or more high-pitched and fine *zirrrrr*. Alarm a sharp *tsiit*, harder than Tree or Red-throated Pipits.

SIMILAR SPECIES The sole Western Palearctic representative of a diverse Afrotropical and S Asian genus, thus, in the covered region, confusion possible only with Zitting Cisticola or Scrub Warbler. – In addition to obvious plumage features, *Scrub Warbler* has a much more striking face pattern, especially the blackish eye-stripe, bolder-streaked crown, faintly-streaked throat, breast and upper flanks, warmer flanks, and tail solely blackish and less graduated. Also, Graceful Prinia is readily distinguished by voice, behaviour and habitat, as Scrub Warbler usually inhabits very dry areas and spends much time on the ground. – *Zitting Cisticola* should pose few real problems either, due to the overall much drabber and duller plumage, and note distinctive pipit-like dark and pale 'tramlines' on mantle and scapulars, russet rump, distinctly longer tail, and different behaviour and vocalisations. Some caution is required, however, if faced with a short-tailed juvenile Graceful Prinia, which might suggest Zitting Cisticola, but plumage still distinctive like adults.

AGEING & SEXING Ageing generally impossible after post-juv moult. Sexes virtually alike (except bill colour when breeding), and seasonal plumage variation largely related to wear. Sexes much the same size in most subspecies. – Moults. Both post-nuptial and post-juv moult complete, although some young may leave some feathers unmoulted. Moult usually through summer into autumn (but some southern populations breed almost year-round and moult accordingly). Pre-nuptial moult apparently absent or (in some populations) partial and rather limited. – **SPRING Ad** Sexing often possible: ♂ in courtship period, chiefly Feb–May, has black bill, deeper orange-brown iris, tends to have brighter straw-yellow or orange legs and brighter/cleaner underparts. With wear, upperparts streaking becomes more pronounced and narrower, is overall paler and greyer, underparts wear whiter, and heavily-worn wings and tail have much-reduced pale fringes/tips. **1stS** Apparently indistinguishable from ad following complete post-juv moult. A few trapped in spring with duller bare parts and even some worn wing-feathers (remiges, primary-coverts and tertials), but unclear if these are 1stS and, if so, when they fledged; some spring birds had renewed body-feathers, 1–3 tertials and a few or all tail-feathers, but these too could not safely be aged. – **AUTUMN Ad** As spring but iris duller tawny-brown (the only safe ageing criterion once moult completed). Moulting birds may still

P. g. deltae, ♂, SW Israel, Mar: long, slender tail often cocked and waved. Uneven-length tail-feathers and dark subterminal spots near tips produce 3–4 bars. Bird on right shows typical jizz, while left-hand image shows some fine streaking on face and breast, more frequent in ♂ and especially in *deltae*, which is often warmer below, too. Sexing ad during spring easy by ♂'s black bill. From winter/spring (or even earlier), all ages/sexes have variable brownish-orange iris. (H. Shirihai)

BUSH AND REED WARBLERS

P. g. deltae, ♀, SW Israel, Mar: unlike breeding ♂, ♀ is comparatively pale-billed during breeding season, like non-breeding ♂. Also duller and less clean underparts compared to ♂ during courtship period. At this season, ♀ more secretive and elusive, and less vocal. (H. Shirihai)

P. g. palaestinae, ♂, E Israel, Mar: very similar to *deltae*, but differs subtly by pale greyish (worn) to buffy (fresh) sandy-brown upperparts, moderately broad-streaked black-brown, with paler breast and flanks (some ♂♂ almost snowy-white below). No observable moult limits in wing and orange-brown iris at this season do not establish age, but black bill equals ♂. (D. Occhiato)

show unmoulted, extremely worn post-breeding ad remiges and some coverts, and even tail or body-feathers. **1stW** At least until early autumn iris olive-grey. Those in moult have some juv feathers, but once moult completed only a few still possible to age by iris colour. **Juv** Like ad but is less heavily streaked above and cleaner below, with paler face, rustier fringes to wing-feathers and weaker black tail-tip. Iris dull olive grey-brown.

BIOMETRICS (*palaestinae*) Sexes of the same size. **L** 11–12.5 cm; **W** 41–48.5 mm (*n* 26, *m* 43.8); **T** 50–65 mm (*n* 24, *m* 57.8) (and see below); **TG** (r5 < r1; only 10 rr) 21–34 mm (*n* 24, *m* 28.9); **T/W** 131.9; **B** 10.2–13.1 mm (*n* 24, *m* 12.2); **Ts** 16.7–20.7 mm (*n* 24, *m* 18.4). **Wing formula**: **p1** > pc 9–13 mm, < p2 8.5–10 mm; **p2** < wt 5–7 mm, < ss; **p3** < wt 1–2.5 mm; **pp4–7** about equal and longest (sometimes p4 and p7 < wt 0.5–1 mm); **p8** < wt 1–1.5 mm; **p10** < wt 3–5.5 mm; **s1** < wt 3–6.5 mm. Emarg. pp3–6 (sometimes faintly on p7, too). – Size of live birds in Israel (*n* 185): fresh ♂ 53–68 mm (*m* 61.9), worn 51–64 mm (*m* 56.2); fresh ♀ 51–64 mm (*m* 57.9), worn 51–60 mm (*m* 55.4) (HS).

GEOGRAPHICAL VARIATION & RANGE Two rather distinct subspecies groups involved. In total, eight subspecies

P. g. palaestinae, juv–imm, E Israel, May: broad black subterminal tail-band of *gracilis* group obvious here. As species breeds as early as late winter, young can appear almost ad-like, or mostly ♀-like later in spring. Wings comprise mostly soft and loose juv feathers, while duller orange-brown iris confirms bird is just a few months old. Bill already greyish-black, suggesting juv ♂. The species can even breed at such young age. (L. Kislev)

P. g. deltae, C Israel, Nov: typical appearance in autumn/winter, when sexing impossible (bill colour is same), as is ageing. Warmer olive-brown saturation below separates it from *palaestinae* of Rift Valley. (A. Ben Dov)

P. g. natronensis, ♂, Wadi el Natrun, N Egypt, Apr: breeders in Wadi el Natrun are very similar to *gracilis* of the Nile Valley, only differing by on average slightly longer bill and tarsus. Black bill confirms sex as ♂, but dull olive iris and orbital ring may indicate a young bird. (V. Legrand)

P. g. gracilis, juv, Abu Simbel, Egypt, Jul: juv less heavily streaked above and cleaner below than ad, with paler face, rustier-fringed wing and weaker black tail-band; iris dull olive grey-brown. (R. Armada)

P. g. yemenensis, S Oman, Nov: darker and browner above, with broad, ill-defined dark streaking; washed greyish-buff below. Combination of active wing moult and especially the few unmoulted remiges best fit young bird with delayed moult. However, there are limited data on time of breeding and moult for this population, so best left un-aged. Note very ad-like deep orange-brown iris, but can develop this colour very fast during first few months. (A. Audevard)

recognised within the covered region and one extralimital (not treated or examined). Plumage variation within each group clinal, and the various forms intergrade with much overlap. All populations are sedentary.

LEPIDA GROUP: small with relatively longer tail (and bill and tarsus shorter), with poorly-developed tail-bands (subterminal bars narrower and fainter), generally darker brown upperparts with broader, ill-defined streaks and rather obvious breast-side markings. Mainly breeds in moist habitats in Turkey eastwards.

P. g. akyildizi Watson, 1961 (S Turkey). Small and short-legged but long-tailed. Warm yellowish-brown above with quite bold dark brown streaking. Warm dull buffish-yellow below. Tail-bands narrow and washed out. Scant material available for examination. **W** 40–43 mm (*n* 4, *m* 41.9); **T** 53–69 mm (*n* 4, *m* 62.5); **T/W** 149.0; **B** 10.4–11.8 mm (*n* 4, *m* 11.3); **Ts** 15.7–17.1 mm (*n* 4, *m* 16.5).

P. g. lepida Blyth, 1844 (NE & SW Syria, Iraq, S Iran east to NW India). Very similar to *gracilis* but differs by proportionately longer tail with more washed out and narrow tail-band. There is a cline from west to east as birds become subtly larger and paler with thinner and paler streaks on upperparts, but differences very minor, with much variation and overlap when long series are compared, and considered insufficient grounds for separation of western birds as '*irakensis*'. Differs from *palaestinae* apart from tail pattern by much paler grey-brown upperparts with much finer streaking, from *akyildizi* apart from tail pattern by whiter underparts and colder grey-brown finely streaked upperparts (*akyildizi*: boldly streaked

P. g. hufufae, Bahrain, Nov: rather pale, with narrower, diffuse and duller streaking above, whiter below, while tail-band is better developed and among boldest and blackest. At this season, both sexes have pale bills, making sexing impossible. In active wing moult with several unmoulted remiges and primary-coverts, best matching young bird with delayed moult. Deep red-brown iris does not signify ad, as earlier-hatched offspring could also develop this. (A. Drummond-Hill)

United Arab Emirates, Jan: usually treated as *P. g. hufufae*, but at least some individuals show narrower black tail-band, and these two birds also have narrow dark streaking above, and relatively longer tail approaching *lepida*, which occurs on other side of Persian Gulf. Ageing uncertain, but at least right-hand bird has dull iris, symmetric growth bars on uppertail and what appear to be juv primary-coverts, which suggest 1stY (i.e. later-hatched offspring). (M. Barth)

BUSH AND REED WARBLERS

P. g. akyildizi, Turkey, Jan: being part of the *lepida* group, the tail has poorly-developed tail-band (dark subterminal bars narrower and fainter). Other characteristics of this group include the relatively longer narrow tail, shortish bill, generally darker/browner upperparts with broader, ill-defined streaks, and rather obvious breast-side markings. (V. Pek)

yellowish-brown upperparts). **W** ♂ 41–48.5 mm (*n* 33, *m* 44.2), ♀ 41–46 mm (*n* 18, *m* 43.4); **T** ♂ 57–73 mm (*n* 32, *m* 64.5), ♀ 52–70 mm (*n* 18, *m* 61.8); **T/W** *m* 144.4; **B** 10.5–13 mm (*n* 51, *m* 11.5); **Ts** 16.3–18.9 mm (*n* 48, *m* 17.4). (Syn. *irakensis*.)

GRACILIS GROUP: characterised by larger size, strongly-developed subterminal tail-band, paler upperparts with finer streaks and obscure body-sides streaking, and inhabits drier habitats of S Middle East and NE Africa.
○ *P. g. natronensis* Nicoll, 1917 (Wadi el Natrun, N Egypt). A very subtle race, extremely similar to or almost indistinguishable from *gracilis* and could be synonymised with latter, but for the on average slightly longer bill and tarsus. **W** 43.5–47 mm (*n* 12, *m* 45.0); **T** 48–65 mm (*n* 11, *m* 57.7); **T/W** *m* 127.9; **B** 12.3–14.3 mm (*n* 12, *m* 13.3); **Ts** 18.0–20.6 mm (*n* 12, *m* 19.3).
P. g. deltae Reichenow, 1904 (Nile Delta and east to SW Israel). Resembles *gracilis* but is slightly smaller (subtle!) and darker above, with a little bolder streaking, and underparts darker, more tinged olive-grey with slight hint of pale brown. Bill and tail on average proportionately slightly longer than *gracilis*. **W** ♂ 41–47 mm (*n* 17, *m* 43.9), ♀ 40–44.5 mm (*n* 8, *m* 42.2); **T** 50–64 mm (*n* 24, *m* 58.2); **T/W** *m* 133.9;

B 12.0–13.5 mm (*n* 26, *m* 12.7); **Ts** 17.2–19.7 mm (*n* 25, *m* 18.4). (Syn. *adamsoni*.)
P. g. gracilis (M. H. C. Lichtenstein, 1823) (Nile Valley to N Sudan, W Red Sea coast, south to Ethiopia). Noticeably pale, greyish-sandy above with very narrow brownish-black streaks, buffy abdomen and vent, and cream-white throat and breast. **W** ♂ 43–49.5 mm (*n* 23, *m* 46.3), ♀ 42–47 mm (*n* 10, *m* 44.3); **T** 52–64 mm (*n* 33, *m* 57.5); **T/W** *m* 126.9; **B** 11.5–13.3 mm (*n* 36, *m* 12.3); **Ts** 17.0–20.8 mm (*n* 33, *m* 18.4). – Birds along W Red Sea coasts ('*carlo*') described as being slightly darker with subtly bolder upperparts streaking, but difference, if any, minute, many are inseparable and best included here. (Syn. *carlo*.)
○ *P. g. palaestinae* Zedlitz, 1911 (N Sinai, Israel except in SW, S Syria, Jordan & NW Arabia). Extremely similar to *deltae* but is on average paler, more whitish, below. Biometrics above.
○ *P. g. hufufae* Ticehurst & Cheesman, 1924 (Bahrain & Al Hufuf oasis, NE Saudi Arabia, United Arab Emirates, N Oman). Very like *palaestinae* or marginally paler, with narrower and duller streaks on upperparts, and is whiter below, while subterminal tail-band better developed than in previous races, being relatively bolder, broader and blacker. – Birds of United Arab Emirates and N Oman ('*carpentieri*') have been separated on account of being darker and bolder streaked above, but if anything the opposite is true; the main theme evident from examined material is of marked individual variability, including birds with thinner dark subterminal tail-band, and best to keep these within *hufufae*. **W** 41–47 mm (*n* 15, *m* 44.3); **T** 52–66 mm (*n* 14, *m* 61.4); **T/W** *m* 138.5; **B** 12.0–13.4 mm (*n* 15, *m* 12.8); **Ts** 17.3–20.2 mm (*n* 14, *m* 18.6). (Syn. *anguste*; *carpentieri*.)
P. g. yemenensis Hartert, 1909 (coastal SW Saudi Arabia, Yemen, S Oman). The darkest race of this group, upperparts browner (slightly warmer when fresh) with moderately broad, ill-defined dark streaks; also washed greyish-buff below, and tail-band broadest of all races. Compared to *gracilis*, upperparts darker and colder grey-brown with bolder black streaking; underparts more dusky grey-white, less buffish. **W** ♂ 43–50 mm (*n* 15, *m* 46.1), ♀ 43–48.5 mm (*n* 14, *m* 45.5); **T** 50–63 mm (*n* 29, *m* 56.9); **T/W** *m* 124.5; **B** 11.8–13.5 mm (*n* 29, *m* 12.8); **Ts** 17.5–21.5 mm (*n* 28, *m* 19.2).

REFERENCES GALE, S. W. & MCMINN, S. D. (1989) *BB*, 82: 78–79. – PILCHER, C. W. T. (1997) *Sandgrouse*, 19: 65–67. – SCHEKKERMAN, H. (1991) *DB*, 13: 100–102.

P. g. akyildizi, Turkey, Feb: subterminal dark tail-band narrower, and dark marks often fainter at their edges, often not even reaching edges. (D. Kökenek)

P. g. akyildizi, ♂, Turkey, May: among the *lepida* group, this race has broader and darker streaks above on brownish-olive background with pale pink-buff suffusion, and is also rather rich buff on breast and flanks, with proportionately longer tail. Black bill indicates ♂, but age impossible to ascertain. (D. Kökenek)

P. g. lepida, ♀, Kuwait, Mar: long tail with washed-out and narrow tail-band; overall cold grey-brown above with strongly reduced streaking, and predominantly whitish underparts. Unlike breeding ♂, ♀ is pale-billed at this season. (G. Brown)

LEVANT SCRUB WARBLER
Scotocerca inquieta (Cretzschmar, 1830)

Fr. – Dromoïque du desert; Ger. – Wüstenprinie
Sp. – Prinia desértica; Swe. – Snårsångare

A resident of low scrub, this warbler is typically found on rocky slopes and sandy plains, usually in desert and semi-desert, often keeping well hidden, although its far-carrying calls make the bird easy to follow. It breeds from NE Egypt, W Arabia and the Levant east to S Kazakhstan, Transcaspia and W Pakistan. Closely related to the Saharan Scrub Warbler, the two have commonly been united as one species, but are here treated as separate species (see Taxonomic notes).

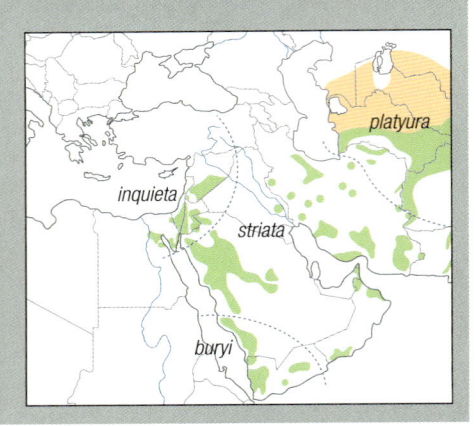

S. i. inquieta, Israel, Mar: head pattern quite variable, but always has blackish streaks on crown and pale supercilium enhanced by dark eye-stripe. Amount of white and silvery-grey in crown/face is especially variable, as is the degree of the streaking on breast and body-sides and how developed the lateral throat-stripe is (this bird is a heavily-streaked example). Note also short, thin but stout bill. Ageing impossible without handling. (H. Shirihai)

IDENTIFICATION A small, pale-plumaged warbler with a proportionately large head and conspicuously *long, contrastingly blackish tail* that is *constantly raised and cocked sideways*. Can appear to have almost mouse-like behaviour, keeping in cover and moving quickly and nervously. Upperparts dull sandy greyish-brown with *dark brown streaks on crown*. A *pale supercilium* is enhanced by distinctive *dark eye-stripe*. At close range short, rounded wings have dusky alula and centres to tertials, while fresh *tail has off-white tips to outer feathers* and slightly paler sooty-brown central rectrices. Mostly creamy-white below, with *breast-sides, flanks and undertail-coverts suffused pink-buff*, and is variably but *diffusely streaked dark on throat, breast* (mostly sides) *and upper flanks*. Short, fine but stout bill is mostly grey-brown with slightly paler base to lower mandible. Longish legs (that appear 'set far back') are yellowish-brown. Somewhat like a miniature *Turdoides* babbler, in Israel typically found in the most desolate areas, inhabited by few other birds. Jumps and hops rapidly on ground, and habitually feeds around tiny shrubs and stones (occasionally skulking). Flight on whirring wings is rather weak but quite direct, usually very low and over short distances. Usually encountered in *very vocal groups* of pairs or family parties.

VOCALISATIONS Rich vocabulary, noisy. Song is a combination of a few whistling notes that fall markedly in pitch, *see-seh-soo-suu*, sounding 'enchanted', and various rolling trills and twitters, often subtly downslurred at the end, e.g. *tcher-terererrrrrr*, *see-suu-soo*, *tsri-srü* or *trsirr-vey-vuy-vuu*. – Calls variable, often melodious and complex: frequently commence with explosive notes (like slow, distant Cetti's Warbler), sometimes including short disyllabic whistles, *te-te te-te te-te*; also *wii-wew* or *tche-tchu*, and a high-pitched slow whistling series with each note lower in pitch than the previous, *sii sü suh soh*, recalling distant call of Common Sandpiper but with Blue Tit-like quality. Other variations include a clear, high-pitched disyllabic *te-he* and stammering *tett-tett-tett-tett-...* Alarm an often-repeated, dry descending, clear trill, *te-terrrrr* or *peurrrrrrrl* with almost Greenfinch-like ring.

SIMILAR SPECIES Perhaps most easily confused with *Graceful Prinia*, although they are unlikely to occur in the same habitat. – *Desert Warblers* (of either species) share the same habitat and frequently forage on open ground like Scrub Warbler, and could be confused given only a brief view. However, unlike Scrub Warbler, both Desert Warbler species have a wholly plain face, with unstreaked crown and throat/breast, and lack the blackish eye-stripe, but do have a conspicuously patterned tail (which is never cocked). – Although not a serious pitfall, due to their widely allopatric ranges, Levant Scrub Warbler differs from *Saharan Scrub Warbler* by being slightly darker and more greyish-tinged, with longer pale supercilium and more prominent black eye-stripe extending well behind eye. Underparts average subtly darker, diffusely streaked on throat, breast-sides and upper flanks. Bill perhaps marginally stouter (but much overlap), base of lower mandible yellowish.

AGEING & SEXING (*inquieta*) Ageing difficult and requires close inspection of moult pattern and feather wear in wing, usually only possible in the hand. Sexes alike and seasonal variation limited. – Moults. Complete post-nuptial moult and mostly partial post-juv moult protracted and variable, depending on breeding dates, often occurring in late spring–summer. Post-juv moult includes all head- and body-feathers, but usually only a few upperwing-coverts. Pre-nuptial moult probably mostly absent, at most partial. – SPRING **Ad** For a general description see Identification. Bleaches paler above, with crown becoming more heavily and sharply streaked, supercilium and underparts whiter with more prominent streaking apparent on throat and breast, but pale fringes to wings and tail heavily abraded, thus wings relatively darker. Iris looks dark in the field, but in the hand

S. i. inquieta, Sinai, Egypt, Dec: often hops rapidly across open ground, and perches on rocks, adopting typical posture, with conspicuously long, contrastingly blackish tail, which is constantly cocked upwards and sideways. This bird has a strikingly patterned head, especially the whitish feathering on crown. (R. van der Weijde)

BUSH AND REED WARBLERS

S. i. inquieta, Israel, Feb: a bird with somewhat weaker head pattern, as well as much weaker lateral throat-stripe and even unstreaked body-sides. Body-sides on the other hand are more suffused pink-buff. Legs appear to be set far back on body. Note typical very dark, narrow tail with white-tipped feathers. (H. Shirihai)

S. i. striata, United Arab Emirates, Mar: this race has darker and greyer upperparts, with densely streaked dark crown, brighter and more extensive cinnamon-buff belly and flanks, and boldly dark-streaked throat and breast. (M. Barth)

dull yellowish-grey, brownish-cream or dark brown; unknown if these variations refer to specific age or sex. **1stS** At least some differ from ad by more heavily-worn juv feathers in wing and perhaps tail. — AUTUMN **Ad** Aged by being evenly feathered, lacking moult contrast. **1stW** Usually retains juv remiges, primary-coverts and tertials, and some or all alula and greater coverts (subtly weaker-textured and soon less fresh than ad; primary-coverts also narrower with narrower pale fringes on outer webs), but moult limits not always easy to detect, especially if all greater coverts retained. Also, outer juv tail-feathers (if kept) have distinctly narrower and more diffuse pale tips than ad. Some, however, reportedly undertake complete moult and are then indistinguishable from ad. Unknown if juv/1stW has darker iris than ad. **Juv** Similar to ad, but has more diffuse and greyer crown streaking, less obvious eye-stripe and supercilium, and very little underparts streaking.

BIOMETRICS (*inquieta*) Very slight sexual size difference, if any. **L** 10–11.5 cm; **W** 43–50 mm (n 27, m 47.6); **T** 45–52 mm (n 26, m 48.9); **T/W** m 102.5; **TG** (r5 < r1) 7–10 mm; **B** 10.7–13.4 mm (n 25, m 11.7); **BD** 2.6–3.3 mm (n 22, m 3.0); **Ts** 17.6–20.0 mm (n 25, m 18.7). **Wing formula: p1** > pc 12–14 mm, < p2 10–13 mm; **p2** < wt 4.5–8 mm, =ss (rarely <ss or =10 or 10/ss, exceptionally =9/10); **p3** < wt 0–1.5 mm; **pp4–6** longest, rarely pp3–7; **p7** < wt 0–3 mm; **p8** < wt 1–4 mm; **p10** < wt 3–6.5 mm; **s1** < wt 3.5–7 mm. Emarg. pp3–5 (often less prominently on p6 and p7).

GEOGRAPHICAL VARIATION & RANGE Four subspecies within the covered region, recognition mainly based on ground colour and amount and prominence of streaking on head and breast, but variation is clinal, and many birds are intermediate where ranges meet, thus difficult to label. All populations mainly sedentary.

S. i. inquieta (Cretzschmar, 1830) (NE Egypt, Sinai, NW Saudi Arabia, Levant). Warm drab-brown above, crown finely but distinctly streaked blackish, a little less obviously on nape, more diffusely and weaker on mantle, whereas back is nearly or entirely unstreaked. Facial pattern quite distinct with cream-white or off-white broad supercilium enhanced by blackish lores continuing as eye-stripe and a narrow black stripe on lateral crown. Underparts buff-white with very fine streaks on throat and upper breast. Biometrics given above.

S. i. striata (Brooks, 1872) (WC Arabia, Oman, SW & SE Iran to NW India). Rather similar to *platyura*, which breeds north of it, but differs in rather obvious rufous-cinnamon tinge to breast-sides and flanks, and by much more obvious

S. i. striata, United Arab Emirates, Feb: a rather weakly-patterned bird, appearing almost unstreaked. The poorly-marked plumage might also be partially due to the bird's probable young age (already very worn in Feb, with apparently juv remiges). (M. Barth)

S. i. striata, Oman, Apr: at least in Oman, some birds assigned to this race to some degree approach *buryi*, though streaking and warm rufous-brown tinge below are not as well developed as in true *buryi*, and iris is dark (not whitish-grey). (H. & J. Eriksen)

and extensive streaking on throat. Upperparts similar, although *striata* is slightly warmer grey-brown, less greyish, and crown is subtly more heavily streaked, streaks being blackish rather than dark brown. **W** ♂ 50–54.5 mm (*n* 12, *m* 52.1), ♀ 47.5–51 mm (*n* 8, *m* 49.5); **T** ♂ 47–54 mm (*n* 12, *m* 50.5), ♀ 48–53 mm (*n* 8, *m* 49.9); **T/W** *m* 98.3; **B** 10.7–13.1 mm (*n* 24, *m* 12.0); **Ts** 18.2–20.5 mm (*n* 24, *m* 19.5). – Birds of W Arabia and in Oman separated as '*grisea*', but insignificant difference when series compared. (Syn. *elaphrus*; *grisea*.)

S. i. buryi Ogilvie-Grant, 1902 (Yemen, S Arabia). The darkest race with heaviest streaking. Upperparts quite dark grey-brown with rufous tinge, black streaking heavy on crown and nape, diffuse on mantle and back. Underparts dusky white on throat with prominent streaking, whereas breast to vent is warmer brown-white with rufous streaking, Flanks prominently tinged cinnamon-rufous. **W** 49–54 mm (*n* 12, *m* 50.9); **T** 48–54 mm (*n* 12, *m* 51.3); **T/W** *m* 100.9; **B** 11.6–13.1 mm (*n* 12, *m* 12.2); **Ts** 18.0–19.8 mm (*n* 12, *m* 18.9).

○ **S. i. platyura** (Severtsov, 1873) (Transcaspia, south to NE Iran & N Afghanistan). Differs from *inquieta* by paler greyish drab-brown and much less streaked upperparts, streaking confined to crown and nape (or just reaching upper mantle very faintly). Throat also less streaked. In *inquieta*, crown is much more prominently black-streaked, and diffuse streaking usually reaches onto back; upperparts also warmer brownish, and throat thinly but clearly streaked (throat-streaking in *platyura* very subdued). Differs from *striata* by much plainer and more whitish underparts (*striata* has rufous-tinged breast-sides and flanks, and fine but well-marked throat-streaking) and by less clearly streaked crown/nape. Sexes of same size. **W** 46–53 mm (*n* 37, *m* 49.7); **T** 42–55 mm (*n* 36, *m* 49.2); **T/W** *m* 99.1; **B** 10.5–12.9 mm (*n* 25, *m* 12.1); **Ts** 18.5–21.0 mm (*n* 24, *m* 19.6). – Birds in southern part tend to be subtly darker and stronger streaked ('*montana*'), but available material insufficient to evaluate validity. (Syn. *montana*; *turanica*.)

S. i. platyura, NE Iran, Apr: birds in Transcaspia and NE Iran differ from *striata* by much plainer and more whitish underparts and by greyer-tinged drab-brown upperparts with reduced streaking, streaking mainly confined to crown and nape. (W. Müller)

S. i. buryi, Yemen, Jan: highly distinctive race, being darkest and more heavily streaked (broadly/densely on crown to upper mantle). Upperparts dusky brownish-olive, underparts distinctly suffused deep rufous-cinnamon and rather heavily streaked. Note hint of pale iris, which appears intermediate between Saharan Scrub Warbler and other races of *inquieta*. (H. & J. Eriksen)

TAXONOMIC NOTE The relationship between the two species of Scrub Warblers, the Levant Scrub Warbler and Saharan Scrub Warbler, may require further investigation before their taxonomy can be conclusively resolved. They have previously usually been combined into one species, but considering the gap in distribution between them, the rather clear morphological differences, some differences in vocalisations and the genetic difference both in nuclear and mitochondrial DNA, with e.g. a cytochrome b mtDNA distance of 11.5% (U. Olsson, unpubl. 2014), they clearly merit separate species status, a policy here adopted.

REFERENCES ALSTRÖM, P. et al. (2011) *Ibis*, 153: 87–97. – BERGIER, P., THÉVENOT, M. & BERG, A. VAN DEN (2013) *DB*, 35: 107–121.– FREGIN, S. et al. (2012) *BMC Evol. Biol.*, 12: 157. – GALE, S. W. & MCMINN, S. D. (1989) *BB*, 82: 78–79.

SAHARAN SCRUB WARBLER
Scotocerca saharae (Loche, 1858)

Fr. – Dromoïque du Sahara; Ger. – Saharaprinie
Sp. – Prinia desértica del Magreb
Swe. – Berbersångare

The westerly relative of the Levant Scrub Warbler, of which it is sometimes treated as a subspecies. However, they differ markedly in DNA and morphology, and are separated by a wide range gap, thus are treated here as two species. Resident from Morocco east to Libya. Habits and habitat preferences are much the same as Levant, both being found in arid semi-deserts or deserts with scattered scrub.

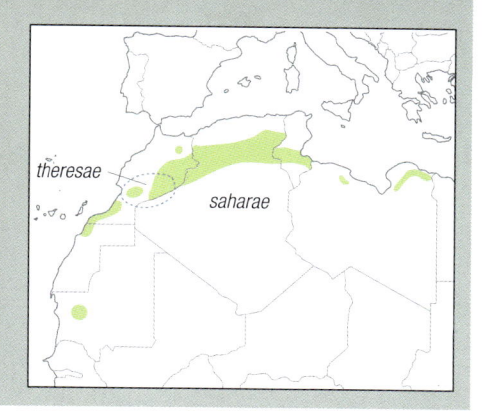

S. s. saharae, Morocco, Apr: in typical posture with long, contrastingly dark tail that is often cocked. Plumage very similar to Levant Scrub Warbler, but note pale iris and the relatively shorter bill. Also, crown has only faint or much narrower dark brown streaks, cream-white supercilium is shorter and only faintly extends behind eye and is slightly buffish in front; dark eye-stripe narrower. (P. Kuhno)

IDENTIFICATION General structure, jizz and behaviour similar to Levant Scrub Warbler, a *small, rather rotund warbler with a long contrastingly dark tail that is often cocked*. It also has *short rounded wings* and skulking behaviour. Saharan Scrub Warbler differs from Levant Scrub Warbler by being subtly smaller (much overlap!), having *paler eyes* (in adult yellowish-grey to sulphur-yellow, rather than 'dull dark'), slightly longer legs, and a relatively *shorter, more pointed* (narrower and more delicate) *bill* (though some overlap in both tarsus and bill sizes) with *more orange-tinged base* to lower mandible. Plumage, although somewhat variable, is usually distinctive, with *crown marked only by faint dark brown streaks*, extending to about nape; *cream-white supercilium shorter and only faintly extends behind eye*; becoming more *buffish above lores* and pale pinkish-buff ear-coverts; dark *eye-stripe also generally thinner and appears shorter* or almost absent behind eye. Those in easternmost Morocco to Libya (*saharae*) noticeably *pale above*, the upperparts and wing edgings being predominantly sandy-brown to isabelline with pinkish-grey overtones (generally more pinkish in fresh plumage, isabelline when worn); *underparts generally paler*, mostly creamy-white suffused pink-buff on breast-sides and flanks, which are only weakly streaked (mostly on throat and breast-sides).

VOCALISATIONS Differences from Levant Scrub Warbler are reportedly quite marked (Bergier et al. 2013), but detailed studies apparently lacking. In S Morocco, song recorded as a short, fast warble, both scratchy and musical, and involving three main phrases with two call types interspersed. Described as 'tremulous and musical' and (in Algeria) as 'pleasant and melodious'. – Some call types, however, are very similar to Levant Scrub Warbler, e.g. the ringing trill is given in contact between group members, and a high-pitched staccato *tic* is also similar to calls of the other group. A characteristic rather low-pitched *prrruuu* call (presumably mainly uttered as alarm) in Saharan Scrub Warbler has no real counterpart in Levant Scrub Warbler, or is at least much higher-pitched (Bergier et al. 2013).

SIMILAR SPECIES Potential confusion with *Desert Warbler*, *Zitting Cisticola* and *Graceful Prinia* are covered under Levant Scrub Warbler. The latter is not known to overlap with Saharan Scrub Warbler, and the differences between them are highlighted under Identification.

AGEING & SEXING Not closely studied but presumably similar to Levant Scrub Warbler, which see. Like that species, iris coloration and its relation to age requires further study.

BIOMETRICS (*saharae*) No detectable sexual size difference. **L** 10–11 cm; **W** 44.5–48.5 mm (*n* 30, *m* 46.7); **T** 44–54 mm (*n* 30, *m* 47.5); **T/W** *m* 101.7; **TG** 6–11 mm; **B** 10.2–12.5 mm (*n* 28, *m* 11.2); **BD** 2.7–3.3 mm (*n* 28, *m* 3.0); **Ts** 18.3–20.9 mm (*n* 29, *m* 19.3). **Wing formula: p1** > pc 12–15 mm, < p2 9–12 mm; **p2** < wt 4–6 mm, =ss (sometimes <ss); **pp3–6** about equal and longest (p3 sometimes < wt 0.5–1 mm); **p7** < wt 0.5–1.5 mm; **p8** < wt 1–2 mm; **p10** < wt 2.5–5.5 mm; **s1** < wt 3.5–5.5 mm. Emarg. pp3–6(7) (on p7 rather weak and can be missing).

GEOGRAPHICAL VARIATION & RANGE Two rather well-separated subspecies differing clearly in saturation and amount of streaking. Exact ranges not known in detail, and there may be intermediary populations where the two are in contact. All populations sedentary.

S. s. saharae (Loche, 1858) (N Western Sahara, S & NE Morocco, Algeria to Libya). Paler overall than Levant Scrub Warbler and much paler and less heavily streaked than *theresae*, pale drab or sand-brown above with faint pinkish hue, unstreaked or with very fine streaking only on upper mantle (apart from crown/nape). (Syn. *harterti*.)

S. s. theresae Meinertzhagen, 1939 (interior SC Morocco, S Atlas Mts). Differs from both *saharae* and Levant Scrub Warbler by being darker above, with heavier

S. s. saharae, Tunisia, Feb: some young birds in bleached plumage confusingly attain whiter and longer supercilium, and dark eye-stripe and crown-streaking may be enhanced. Nevertheless, note unstreaked underparts and plain upperparts. (R. Pop)

S. s. saharae, Western Sahara, Mar: rather plain individual, with prominent streaks only on crown. Note also the relatively large pale tips to outer tail-feathers of this taxon just visible here. (V. Legrand)

streaking on crown and throat/breast, and having much deeper pink-cinnamon on flanks. Note that supercilium is rufous-buff, not whitish. In fresh plumage tinged rufous-buff, especially around face (with rufous-pink supercilium in front of and above eye, but off-white or greyish and rather inconspicuous behind). Scant material available for examination. **W** 45.5–49.5 mm (n 5, m 46.9); **T** 45.5–51 mm (n 5, m 47.7); **T/W** m 101.7; **B** 10.2–11.7 mm (n 5, m 11.2); **Ts** 17.9–19.3 mm (n 5, m 18.5).

TAXONOMIC NOTE See under Levant Scrub Warbler for reasons to recognise Saharan Scrub Warbler as a separate species.

REFERENCES Alström, P. et al. (2011) *Ibis*, 153: 87–97. – Bergier, P., Thévenot, M. & van den Berg, A. (2013) *DB*, 35: 107–121. – Fregin, S. et al. (2012) *BMC Evol. Biol.*, 12: 157.

S. s. theresae, SW Morocco, Feb: rather well-marked head pattern and rather deep pink-brown on flanks and fore supercilium are features of this race. Some faint streaking below makes well-marked individuals quite similar to Levant Scrub Warbler, but unlike latter has pale iris and ochre-buff forepart of supercilium. (D. Monticelli)

S. s. theresae, EC Morocco, Mar: some are distinctly warmer, being overall rufous-buff, with rufous-pink supercilium before and above eye (but off-white or greyish and rather inconspicuous behind); underparts tinged deep pink-brown. (Q. Marcelo)

S. s. theresae, juv, SW Morocco, Apr: rather similar to ad, but has pale grey iris and obvious yellow gape. Underparts in this race are often streaked rather heavily, with dark brown stripes on lower throat and breast. (T. Svensson)

PALLAS'S GRASSHOPPER WARBLER
Locustella certhiola (Pallas, 1811)

Alternative name: Rusty-rumped Warbler

Fr. – Locustelle de Pallas; Ger. – Streifenschwirl
Sp. – Buscarla de Pallas; Swe. – Starrsångare

This is no more than a rare vagrant to Europe from Asia, but it has occurred here in autumn on so many occasions that it has become a well-known and sought-after rarity, no longer regarded as an 'unlikely' species. Being a *Locustella*, it is a skulking bird of dense cover and one that mostly keeps low in bushes and weeds. Within its breeding range in Siberia and N Mongolia, it generally occurs in wet meadows within forest clearings or glades, in reedbeds with scattered bushes or in fields, where seeks shelter at the bottom of low-growing *Salix*. It winters in SE Asia from NE India eastwards, returning in late May or, mainly, in June.

L. c. rubescens, C Mongolia, Jun: diagnostic white tips to tail-feathers, which on some birds and when very fresh can be large. However, this bird does not show the characteristic bold white tips to inner webs of tertials—probably due to angle of view. Ages alike. (H. Shirihai)

L. c. centralasiae, W Mongolia, May: misleadingly superficial resemblance to Sedge Warbler, including rather pronounced pale supercilium. However, has stronger-streaked upperparts and clear black uppertail-coverts spotting (diagnostic). From other angles, check for tail-tip and tertial patterns. Ages alike at this season. (J. Normaja)

IDENTIFICATION Resembles a Sedge Warbler, with similar size, shape and general coloration, including a *rather prominent pale supercilium* visible on most (but still not quite as long or well marked as Sedge), but differs in slightly darker brownish breast-sides and flanks, *streaked uppertail-coverts* (Sedge: nearly uniform), *dark distal parts of tail-feathers with white tips* except on central pair (white tips visible unless feathers heavily abraded) and blackish-centred tertials with distinct pale brown margins, the *margin on inner webs being paler and wider near tips, often forming a white spot* there, especially obvious on inner two tertials. Compared with Sedge Warbler, legs are usually *paler and pinker*, less brown, and the bill appears fractionally shorter and stouter. Like most *Locustella*, the tail is well rounded, and the undertail-coverts are very long, longer than outermost tail-feathers (Sedge: tail squarer, still moderately rounded; longest undertail-coverts shorter than outermost tail-feathers). Rump rufous or tawny, unstreaked or has only a few narrow or diffuse streaks. Undertail-coverts pale buff or (in *rubescens*) more saturated olivaceous-ochre, longest often with greyish central shade (but hardly fully developed streaks) and whiter tips. Some have a rather pale greyish cast on nape.

VOCALISATIONS Song differs from that of congeners in that it resembles more a brief phrase of an *Acrocephalus*, thus is not a monotonous insect-like reeling but a rhythmic strophe comprising a few motifs, the brief, hurried song being repeated with little variation. A common variant can be rendered *zerr-zerr-zerr seo-seo-seo char zer-zer-zer sweeh-sweeh-sweeh-sweeh*. The song is not very strong, but the *sweeh* notes are louder and carry far. Song is mainly given at dusk and dawn, but can also be heard in the middle of light summer nights, and briefly in daylight. Song-flights have been reported, but much less practised than in Sedge Warbler. – Usual call is a sharp, metallic *chick*. Other calls include a purring *tri-trrr* and short clicking *tk*, latter sometimes repeated in series.

SIMILAR SPECIES For differences from *Sedge Warbler*, see Identification. – In autumn can be confused with a vagrant first calendar-year *Lanceolated Warbler*, but latter is smaller. – Same size and shape as *Grasshopper Warbler*, but this is slightly duller and plainer, lacking the rather distinct pale supercilium of Pallas's, the contrastingly rufous-ochre rump and the dark and white tips to outer tail-feathers.

AGEING & SEXING (*rubescens*) Ages generally readily separable in autumn. Sexes alike in plumage and size. – Moults. Post-nuptial moult after breeding (Jul–Aug) either partial or complete; to what extent it is complete is not known. Only a few actively moulting birds in summer seen, but based on these, and judging from the fresh primaries of many autumn adults, a complete post-nuptial moult might be quite common. Post-juv moult in late summer very limited, and some might migrate in juv plumage. Complete moult of ad usually in late winter (Feb–Apr), but geographical or individual variation occurs, and some moult earlier, immediately after reaching winter quarters (Sep–Nov). Latter category can hardly have moulted completely, or even extensively, on breeding grounds too, but details poorly known. Pre-nuptial moult of 1stY in late winter apparently complete. – **SPRING** Ages alike. – **AUTUMN Ad** Upperparts usually rather cold brown (even olive-brown), underparts dusky white tinged rufous. Freshly moulted birds often have fine, minute dark spots across breast or on lower throat (difficult to see in the field, though). **1stCY** Easily aged as more rufous-tinged above and yellowish below, some being saturated ochre-yellow on all underparts. Many have prominent dark streaking on lower throat and upper breast; those with fewest streaks have these forming a faint gorget or necklace.

BIOMETRICS (*rubescens*) L 13–14 cm; W 62–73 mm (n 72, m 66.7); T 49–58.5 mm (n 19, m 53.6); T/W m 79.4; B 14.5–16.3 mm (n 18, m 15.7); Ts 21.3–23.6 mm (n 19, m 22.4). **Wing formula:** p1 < pc 0–2 mm or > 0–4 mm, < p2 31–36 mm; **p2** < wt 2.5–5 mm, =4/5 (rarely =4 or 3/4); **p3** longest; **p4** < wt 2–3 mm; **p5** < wt 2.5–4 mm; **p6** < wt 5–8 mm; **p7** < wt 7–11 mm; **p10** < wt 14–19 mm; **s1** < wt 16.5–22 mm. Emarg. p3.

GEOGRAPHICAL VARIATION & RANGE Slight variation.

L. c. rubescens, Russia, Jul: based on date probably an ad breeding. Note species characters of slightly darker brownish breast-sides and flanks, and blackish-centred tertials with bold white tips to inner webs (diagnostic, if present). (P. Parkhaev)

Three races within treated area, rather similar and clinally intergrading. The following is based on the large collection in ZISP, supplemented by material in NHM, AMNH, MNHN, ZMUC and NRM. Individual variation, and variation by age, complicate identification of single birds. Streaking of rump and uppertail-coverts varies individually in all three races. One extralimital race in E Asia (*minor*), not treated.

L. c. rubescens Blyth, 1845 (W & partly C Siberia, C & N Mongolia, W Transbaikalia; winters SE Asia from NE India eastward). Treated above. The darkest race, being rather deep rufous-brown above with an olive-grey tinge, dark streaks less contrasting due to dark ground colour. Often strong rufous-buff uppertail-coverts and lower flanks, but some are less saturated, more like other races in this respect. Underparts usually tinged olive-grey, especially at sides. – Breeders in S Siberia, N Mongolia and W Transbaikalia ('*sparsimstriata*') sometimes separated and said to be intermediate between *rubescens* and *certhiola*; sometimes synonymised with ssp. *certhiola*, but morphologically very close to *rubescens* and best included therein. A very few have dark streaking on mantle subdued, a feature not found in classic *rubescens*. **W** 62–73 mm, n 38, m 66.7. (Syn. *sparsimstriata*.)

L. c. certhiola (Pallas, 1811) (SE Transbaikalia, E Mongolia, NE China; winters SE Asia). Somewhat paler above, not so dark 'dusky' rufous-brown, more ochre-rufous, and is more heavily streaked, streaks being slightly broader and sharply defined. In worn plumage, crown becomes quite dark. Undertail-coverts ochraceous-buff, tipped paler, usually without darker central shade. **W** 59–71 mm (n 53, m 64.2); **T** 46–54 mm (n 15, m 49.9); **T/W** m 78.6; **B** 13.9–16.7 mm (n 15, m 15.0); **Ts** 20–22.0 mm (n 13, m 21.1). Wing formula similar to *rubescens*, but **p2** =5 or =5/6 in *c*. 30%, and **p1** < p2 28.5–34.5 mm.

L. c. centralasiae Sushkin, 1925 (Central Asia, W Mongolia; winters SE Asia, though details poorly known). Resembles *certhiola* but a fraction smaller and subtly paler and more tinged ochraceous-tawny on average, except on nape and hindneck, which are pale greyish (streaked dark). Dark streaks of upperparts more brown-black, less pure black, and rather narrow and less well defined. Crown apparently never wears very dark like in *certhiola*. On average broader pale edges to tertials in fresh plumage. **W** 59–70 mm (n 35, m 64.0); **T** 48–53 mm (n 12, m 50.7); **T/W** m 77.2; **B** 13.3–15.4 mm (n 35, m 14.8); **Ts** 20.2–22.0 mm (n 13, m 21.3).

REFERENCES   Madge, S. (1999) *Birdwatch*, 80: 32–35. – Millington, R. (1998) *BW*, 11: 387–389. – Riddiford, N. & Harvey, P. (1991) *BW*, 4: 324–326. – Roselaar, C. S. *et al.* (2006) *DB*, 28: 273–283.

L. c. rubescens, juv/1stW, Russia, Sep: readily separated from ad by fresh plumage, often with rather typical warm tone. Characteristic curved wing of *Locustella* with white outer edge, and long and 'bulky' undertail-coverts. Note the few exposed blackish spots on scapulars and uppertail-coverts. Legs strong and bright pinkish (M. Vaslin)

Juv/1stW, Scotland, Oct: easily aged by being fresh, with rufous-tinged upperparts and faint ochre-yellow tinge below. Note gorget-like pattern formed by prominent dark streaking on lower throat/upper breast. Distinctive blackish spots on scapulars and uppertail-coverts. However, face pattern is rather poorly developed, with rather indistinct pale supercilium. (M. Breaks)

1stW Pallas's Grasshopper Warbler (left: Scotland, Oct) versus Sedge Warbler (right: Faeroe Is, Oct): despite belonging to different genera, these two species can at times resemble one another. Former has more graduated tail, curved wing shape and very long undertail-coverts, but these structural differences are not visible in these images. In most, the tertials of Pallas's Grasshopper have distinct white spotty tips on inner webs (invariably lacking in Sedge), bold black streaks on uppertail-coverts (in Sedge more or less uniform rufous-tawny, or at most narrow dark shaft-streaks; example above shows maximum streaking). Crown is more evenly dark-streaked (lacking Sedge's tendency to have darker lateral crown-stripes). Pallas's Grasshopper has diagnostic small bold white tips to rectrices, but they can be reduced or hardly visible, and in some very fresh Sedge there are thin pale (even whitish) tips. Legs of Pallas's Grasshopper paler and pinker, never brown or grey like most Sedge, but some unusually pale-legged Sedge can be more yellowish pink-brown. Both birds are juv/1stW, judging by fresh evenly-feathered body- and wing-feathers. (Left: A. Banwell; right: S. Olofson)

Juv/1stW, Faeroe Is, Oct: diagnostic small white tips to tail-feathers. However, due to individual variation and angle of view, these are often hardly visible, even in fresh birds. (J. Ravnborg)

LANCEOLATED WARBLER
Locustella lanceolata (Temminck, 1840)

Fr. – Locustelle lancéolée; Ger. – Strichelschwirl
Sp. – Buscarla lanceolada; Swe. – Träsksångare

This, the smallest *Locustella*, breeds across the taiga belt in NE Russia and east over much of Siberia and the Far East. Recently it has started to expand further west and is now an annual summer visitor in Finland. It is found in overgrown marshes or in wet parts of more open taiga with rich undergrowth. As with other members of the genus, it is usually shy and skulking, keeping mostly to cover near the ground. The majority winters in SE Asia, returning in late May and June. However, judging from the regularity of occurrence in Europe, a small part of the population might have developed an alternative strategy of wintering in S Europe or Africa.

Finland, Jun: streaks on breast and upper flanks can be thin and sharp, but always become broader and bolder posteriorly. In this individual the black central streaks on undertail-coverts almost extend to longer coverts. (J. Peltomäki)

IDENTIFICATION As with other members of the genus, in summer usually first noticed by its song. Behaviour much as Grasshopper Warbler. A small warbler (clearly smaller than a Sedge Warbler) with proportionately rather short, compact tail and short, stout bill. Primary projection moderately long, just over half of tertial length. It has brown and off-white or buffish colours and is noticeably dark-streaked both above and below. Undertail-coverts long and 'bulky', reaching close to tip of tail. *Upperparts* brown, *boldly streaked black*. Tertials and greater coverts *black-centred, tertials typically with thin, sharply defined and nearly evenly broad pale brown margins*. There is a very short, indistinct paler supercilium in some, but usually much less obvious than in Grasshopper Warbler. *Underparts* dusky white or off-white tinged yellowish-buff, flanks and undertail-coverts tinged greyish-rufous or olive-brown, *with variable number of well-defined, short black streaks*, apparently invariably *on lower throat/upper breast* and on shorter undertail-coverts, but usually (especially in adults) also on lower breast and flanks. Thick-based, short bill is dark but has pinkish cutting edges and lower mandible. Legs pinkish. Important to note *black-centred tertials* (*solidly black up to paler margins*) with comparatively *narrow, distinctly demarcated pale brown margins*. Also useful to note *short, well-marked black central streaks to shorter undertail-coverts* (whereas longer undertail-coverts either are unmarked or appear to be so in normal side-on views).

VOCALISATIONS Song resembles those of both Grasshopper Warbler and River Warbler, a drawn-out insect-like, mechanical reeling, being quite high-pitched and 'electric' in tone, *zezezezezezezezeze…*, with a hint of chuffing as in River Warbler, only finer, faster and higher-pitched. Like Grasshopper, the song is mainly given at dusk into early night, and at dawn (but can also be heard throughout light summer nights, and briefly in daylight). As an interesting detail of singing behaviour, the lower mandible vibrates during song, whereas in Grasshopper Warbler it is still (Tjernberg 1991). – Call is a compound, 'throaty clicking' *chk*, in alarm often repeated in long series at moderate tempo, then somewhat recalling Garden Warbler alarm.

SIMILAR SPECIES Main confusion risk is *Grasshopper Warbler*, both on its Siberian breeding quarters and with a vagrant Lanceolated in autumn in Europe. On breeding grounds of Lanceolated you are likely to see adults, which are normally much more profusely and distinctly streaked black over most of underparts than any Grasshopper Warbler (but beware of rare examples of Grasshopper with quite densely dark-spotted throat!). Lanceolated occurring in W Europe in autumn are usually first-years and have reduced streaking on breast and flanks, thereby resembling more Grasshopper. However, note on Lanceolated smaller size (although overlap), shorter-looking tail and shorter primary projection (just over half of tertial length), proportionately shorter and stronger-based bill, more distinct black streaking on upperparts (including crown and uppertail-coverts) and on lower throat/upper breast than >95% of autumn Grasshopper Warblers. Further, in field encounters, undertail-coverts differ, long ones of Lanceolated appearing unstreaked and rufous-tinged with paler tips, shorter ones rufous-brown with short distinct blackish streaks (Grasshopper: usually all undertail-coverts have long but rather diffuse central streaks). Best character of Lanceolated is the blackish-centred tertials with narrow, neatly demarcated pale brown margins, dark centres being solidly dark right up to the pale margin (Grasshopper has diffusely demarcated pale outer edges grading into progressively darker centres, paler edges which are broader basally and generally more pointed distally).

AGEING & SEXING Small seasonal and age-related plumage differences, partly obscured by individual variation. Sexes alike in plumage and size. – Moults. Complete or extensive moult of both ad and 1stY generally in late

China, May: the smallest *Locustella* is noticeably dark-streaked both above and below. Black streaking may cover most of lower throat/upper breast on dusky-white background. Poorly-developed face pattern with indistinct pale supercilium. Bill rather short and stout; legs bright pinkish. (G. Ekström)

China, May: short tail and primary projection afford compact impression. Underparts can have highly variable number and shape of dark streaks, but always sharply demarcated on breast and upper flanks, and broader and even bolder posteriorly. Especially note pattern to black centres of tertials and central streaks on undertail-coverts. Fresh-feathered wing, but both ages moult completely in winter. (G. Ekström)

Young (juv/1stW) Lanceolated (left: Scotland, Sep/Oct) versus Grasshopper Warbler (top right: France, Sep; bottom right: Scotland, Sep): young autumn Lanceolated often has reduced streaking on breast and flanks, whereas some Grasshoppers can be strongly marked. Correct identification of Lanceolated (smaller size not always easy to detect) depends on following: Lanceolated has shorter primary projection (c. 1/2 or less than tertial length), sharper and better-spaced black streaking on upperparts, which can be almost as strong on crown and uppertail-coverts, and lower throat/upper breast. Black-centred tertials almost evenly fringed with thin but sharply-defined pale margins that become whiter near tips, whereas Grasshopper has pale outer edges and dark centres more diffusely demarcated, the fringes more evenly rufous. Undertail-coverts of Lanceolated have short, well-marked black central streaks, virtually always restricted to shorter coverts; in Grasshopper usually all undertail-coverts have long rather diffuse central streaks. All birds juv/1stY by evenly-feathered body- and wing-feathers. (Top left: H. Harrop; bottom left: R. Nason; top right: A. Audevard; bottom right: M. Breaks)

winter (Feb–Apr), but timing and extent of replacement of flight-feathers variable, and no clear pattern known. Partial post-nuptial moult in late summer involves head, body and some wing-coverts plus sometimes (perhaps regularly) a variable number of primaries (often a few central, rarely only some inner, or both inner and central), then arrests moult for migration. Moult is resumed in winter quarters with replacement of all or remaining wing-feathers. Post-juv moult in late summer variable and limited. Pre-nuptial moult of 1stY either complete or partial, leaving some juv flight-feathers unmoulted. – **SPRING** Ageing impossible, as both ad and 1stY apparently can occur with two generations of flight-feathers. – **AUTUMN Ad** Plumage worn, especially tips of outer primaries, and tertials. Note that a few start to moult head and body, and rarely odd tertials and tail-feathers, while still in breeding area, and then look fresh. Plumage on average less yellow-tinged, colder brown above and dusky white below, and lower throat, breast and flanks usually heavily streaked black. **1stCY** Uniformly fresh. On average more warm rufous-tinged above and yellowish below, and breast and flanks much less streaked, some having only lower throat streaked plus a few indistinct streaks on lower flanks.

BIOMETRICS L 11–12 cm; **W** 53.5–61 mm (n 20, m 57.5); **T** 41–47 mm (n 20, m 44.6); **T/W** m 77.6; **B** 11.4–13.2 mm (n 20, m 12.4); **Ts** 17.0–19.0 mm (n 19, m 17.9). **Wing formula: p1** < pc 0–2 mm or > 0–1 mm; **p2** < wt 0–2.5 mm, =3, 3/4 or 4 (sometimes =4/5); **p3** longest; **p4** < wt 1–2.5 mm; **p5** < wt 2.5–5.5 mm; **p6** < wt 4 5–7 mm; **p7** < wt 6–8.5 mm; **p10** < wt 9–14 mm; **s1** < wt 11.5–16 mm. Emarg. p3.

GEOGRAPHICAL VARIATION & RANGE Monotypic. – SE Finland, Russia, Siberia east to Kolyma, Kamchatka, Ussuriland, Manchuria, in taiga south to N Mongolia, Altai; winters SE Asia. – Breeders on Sakhalin, S Kurils, Hokkaido and N Honshu sometimes separated ('*hendersonii*') on account of subtly larger size and less streaked plumage. Judging from scant material available to us this is a variable and insufficiently distinct race, and we follow *BWP* in treating the species as monotypic. (Syn. *gigantea, hendersonii*.)

REFERENCES ALSTRÖM, P. (1989) *Vår Fågelv.*, 48: 335–346. – DAVIS, P. (1961) *BB*, 54: 142–145. – MADGE, S. (1999) *Birdwatch*, 80: 32–35. – MILLINGTON, R. (1998) *BW*, 11: 387–389. – RIDDIFORD, N. & HARVEY, P. V. (1992) *BB*, 85: 62–78. – TJERNBERG, M. (1991) *DB*, 13: 11. – VERBELEN, D. & DE SMET, G. (2003) *DB*, 25: 221–234.

Juv/1stW, England, Oct: this well-marked fresh young bird displays all of the species' diagnostics: short primary projection, boldly-streaked black upperparts, black-centred tertials with thin, sharp and almost evenly broad pale margins, and buff-tinged underparts with extensive well-defined black streaks, especially on throat/upper breast. (P. Baxter)

(COMMON) GRASSHOPPER WARBLER
Locustella naevia (Boddaert, 1783)

Fr. – Locustelle tachetée; Ger. – Feldschwirl
Sp. – Buscarla pintoja; Swe. – Gräshoppsångare

Along with Savi's Warbler, this is the most widespread *Locustella* species in Europe—which does not mean that it is often seen and well known. All *Locustella* warblers are discreet and skulking birds which skilfully keep hidden in grass, thickets and reeds. But for their songs most would be missed. This is true also for Common Grasshopper Warbler, which in N Europe is a bird of lakeside meadows with well-developed tussocks, overgrown pastures or young forest plantations, but in S Europe is often found at higher altitudes on alpine meadows with tall weeds and low scrub. It winters in tropical Africa, returning in April–May.

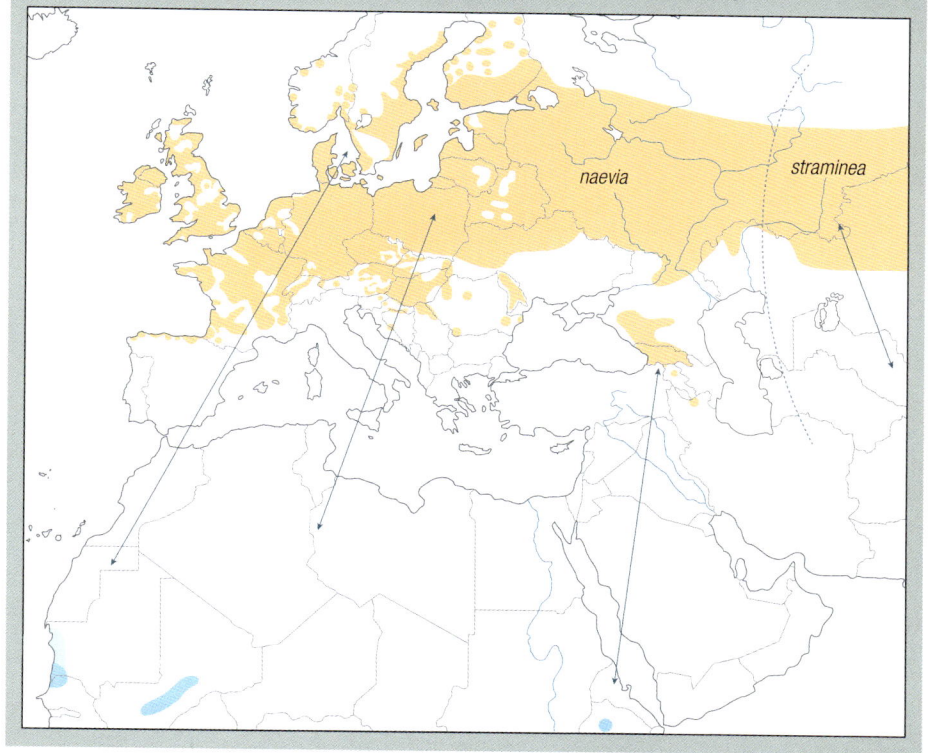

L. n. naevia, Estonia, May: showing typical posture, with plumage melting into the surrounding vegetation. Note rather well-marked dark streaking above, while supercilium is subdued. Dark-centred tertials have broad outer borders, narrowing towards tip, and borders at times diffusely outlined creating less of a contrast to moderately dark centres. Could be either ad or 1stY, as both ages moult completely in winter. (M. Varesvuo)

IDENTIFICATION Usually noticed by its characteristic song. Medium-large warbler (roughly as Sedge Warbler) with olive-brown and off-white or buffish-grey colours and noticeably dark-streaked upperparts. Secretive and *skulking habits*, will run quickly on the ground in deep grass like a mouse and only take low flight to find better cover if absolutely necessary. The edge of the folded wing is rather rounded (typical of *Locustella*), and there is a fair primary projection, just less than tertial length. *Tail long, broad and well rounded at tip*, but when kept folded often appears rather pointed. *Undertail-coverts very long, reaching close to tip of tail.* When (seldom!) seen sitting still, often assuming crouching, horizontal posture, note dark tawny upperparts with faint olive cast visible on most, *boldly streaked blackish on mantle and back*, and dark-centred tertials, greater coverts and scapulars. Head, nape and rump have finer and somewhat less obvious streaking or spotting, and uppertail-coverts are almost uniform brown. The supercilium is *poorly marked* and *greyish- or buffish-white*. There is a fair degree of colour variation, with a tendency to form two—or even three (Harvey & Small 2007)—morphs. In particular the underparts are variable, either dusky white or warmer 'dirty' yellow-white, either unmarked or with *some diffuse brown streaks on flanks* and dark spots on throat/upper breast (rarely densely and quite boldly spotted on lower throat). Bill dark but has diffuse pinkish-yellow base to lower mandible. Legs pinkish. Crucial to see *rather broad and diffuse rufous or olive-brown outer edges to tertials* generally *lacking sharp borders to their dark centres*. Also useful to note *long and ill-defined darkish central streaks to all undertail-coverts*.

VOCALISATIONS Song is a peculiar, prolonged insect-like, mechanical, dry reeling, *svir'r'r'r'r'r'r'r'r'r'r...*, high-pitched and maintaining the same note for minutes, varying only slightly in volume when the bird turns its head relative to the listener. The song is mainly given at dusk into early night, and at dawn (but can also be heard throughout light summer nights, sometimes in daylight). It sounds almost like an old-fashion wrist-watch alarm or some kind of a cricket, and it carries far, on calm nights over several hundred metres. – Usual call is a sharp, metallic *psitt* or more 'electric' *zeck*. Can give brief fragments of song in anxiety or excitement.

SIMILAR SPECIES In normal encounters in Europe, needs to be separated from *Sedge Warbler*, but this has much stronger whitish supercilium, more obvious dark streaking on

L. n. naevia, England, Apr: note less contrasting dark streaks on faintly tinged olive tawny-brown upperparts. Crown, nape and rump have finer and less obvious streaking or spotting, and uppertail-coverts almost uniform. Poorly-marked buffish-white supercilium. Any streaking on 'dirty' yellow-buff underparts usually insignificant. Fresh-feathered plumage and wing, but still impossible to age. (S. Round)

(sides of) crown, unstreaked and more vividly rufous-tinged yellow-brown rump, and unstreaked undertail-coverts. – In autumn can be confused with a vagrant first calendar-year *Lanceolated Warbler*, but latter smaller, has shorter primary projection (just over half of tertial length), proportionately shorter and stronger-based bill, more distinct black streaking on upperparts and lower throat/upper breast than >95% of autumn Grasshopper Warblers, sometimes flanks also distinctly streaked, and undertail-coverts differ, long ones of Lanceolated appearing unstreaked and rufous-tinged with paler tips, shorter ones rufous-brown with short distinct blackish streaks or diamond-shaped spots. Best character of Lanceolated is the blackish-centred tertials with narrower, neater pale olive-brown margins, slightly darker-centred than on Grasshopper Warbler, and in particular uniformly dark right up to the distinct, pale edge, which is similarly broad around entire tertials (Grasshopper has diffusely demarcated pale edges broader basally). – Pallas's Grasshopper Warbler differs in its more prominent pale supercilium and, unsurprisingly, rusty rump that is almost unstreaked (whereas upper-tail-coverts are well streaked). When flushed often possible to see diagnostic dark distal parts and fine whitish tips to most tail-feathers (except central pair). When a perched bird is seen close look for the white spots on inner web tips of all or some inner tertials.

L. n. naevia, Finland, Jun: long and ill-defined darkish central streaks to most of undertail-coverts clearly visible in this typical view with slightly cocked tail. The smoothly rounded wing edge, long and 'bulky' undertail-coverts, boldly streaked back, diffusely streaked head with very subdued paler supercilium combine to identify the species. (P. Parkkinen)

L. n. naevia, juv, Wales, Jul: this image shows nicely how juv wing- and tail-feathers very rapidly get abraded, and demonstrates that some birds are much plainer with obscure streaking above and none below. Still the better-marked mantle, and lack of mottled underparts prevent confusion with superficially similar River Warbler. (D. Tromans)

L. n. naevia, 1stS, Russia, Jun: many worn and bleached brown juv primaries, primary-coverts and longest tertial, contrasting with fresher rest of wing-feathers. Some birds are more heavily streaked above (enhanced due to wear), but also on flanks, which could invite confusion with Lanceolated Warbler. However, note tertial pattern with broad olive-brown fringes narrowing at tip, also strong bill, longer primary projection and lack of obvious breast-side streaks of Lanceolated. (I. Ukolov)

AGEING & SEXING (*naevia*) Small seasonal and age-related differences in size and plumage, partially obscured by individual variation. Sexes alike in plumage and size. – Moults. Complete moult of ad generally in late winter (Feb–Mar, sometimes Apr). Partial post-nuptial moult in late summer, generally only of head, body and some wing-coverts, not involving flight-feathers and primary-coverts, but in a few cases some tertials, tail-feathers and inner secondaries moulted as well, rarely also a few inner primaries (but indisputably *complete* full moult before migration as yet undocumented). Records of complete moult in Sep just after arrival to winter quarters might represent alternative strategy or resumed moult, which was arrested during migration. Post-juv moult in late summer controversial (its occurrence has been questioned), and in any case very limited. Pre-nuptial moult of 1stY in late winter either complete or (*fide* J. King, pers. comm.) 'in *c*. 90% arrested before complete, spring birds retaining varying number of juv wing-feathers'. However, among specimens collected in spring, rather few have two generations of wing-feathers indicating that a substantial proportion of 1stY moult completely. – **SPRING F.gr.** Plumage fresh, all wing-feathers of same generation. **1stS** One or a few brown and worn retained juv wing-feathers, contrasting with fresher rest of wing-feathers. – **AUTUMN Ad** Plumage worn, especially tips of outer primaries, and tertials. Note that a few start to moult head and body, and rarely odd tertials and tail-feathers, while still in the breeding area, which then look fresh. Irrespective of the tendency to fall into colour morphs, plumage on average less yellow-tinged, more olive-brown above and dusky-white tinged rufous below (but apparently some overlap with 1stCY). **1stCY** Uniformly fresh. On average more rufous-tinged above and yellowish below (but some overlap, or occurrence of intermediates).

BIOMETRICS (*naevia*) L 13–14 cm; **W** 59–68 mm (*n* 106, *m* 64.4); **T** 48–61 mm (*n* 45, *m* 53.9); **T/W** *m* 83.6; **TG** 12–21 mm; **B** 13.2–15.2 mm (*n* 21, *m* 13.9); **Ts** 18.8–21.0 mm (*n* 19, *m* 19.8). **Wing formula: p1** > pc 0–3 mm or < 0–2 mm, < p2 31–37 mm; **p2** < wt 1–4.5 mm, =3/4, 4 or 4/5 (equal frequency for all three; rarely =5); **p3** longest; **p4** < wt 1–2.5 mm; **p5** < wt 3.5–5.5 mm; **p6** < wt 5–8.5 mm; **p7** < wt 7–10 mm; **p10** < wt 12.5–16.5 mm; **s1** < wt 15–18 mm. Emarg. p3; sometimes a faint emarg. on p4, too.

GEOGRAPHICAL VARIATION & RANGE Clinal variation, European populations being slightly larger, shorter-tailed and on average more diffusely streaked above, Asian populations slightly smaller, longer-tailed, more distinctly streaked, and

Presumed *L. n. naevia*, ad, Kuwait, Aug: an unusually dark bird with extensive and contrasting black streaking all over, making it strongly resemble Lanceolated Warbler, but note tertial pattern typical of Grasshopper Warbler. Heavily worn primary tips, and with moult contrast between newly-replaced head and body, some tertials and greater coverts. Also, unlike fresh young birds, plumage on average less yellow-tinged, more olive-brown above and dusky-white below. Legs are strangely dark pink on this bird, perhaps just an aberration. (P. Fågel)

sometimes paler. Intermediate populations in E Russia and W Siberia. The following is based on examination of the large collection at ZISP, supplemented by material in NHM, AMNH, MNHN, ZMUC and NRM.

L. n. naevia (Boddaert, 1783) (W & C Europe, Fenno-Scandia, W & C Russia, Kalmykia, W Caucasus, Transcaucasia, NW Kazakhstan; exceptionally NE Turkey; winters sub-Saharan W Africa, but eastern breeders might winter E Africa). Treated above. Rather large, dark rufous-brown with olive tinge above and, relatively speaking, less contrasting dark streaks. Wing rather pointed. – Birds of E Turkey and Caucasus ('*obscurior*') represent only a subtle clinal variation towards *straminea*, in our view insufficiently distinct to be treated separately. (Syn. *obscurior*.)

L. n. straminea Seebohm, 1881 (easternmost Russia, W & C Siberia, N & E Kazakhstan, Altai, Tien Shan; winters Indian subcontinent). Slightly smaller than *naevia* with a slightly blunter wing (**p2** =4, 4/5, 5 or 5/6, all about equally common; **p1** < p2 27–35.5; **p10** < wt 8–15 mm; **s1** < wt 11–16 mm), and proportionately longer tail (**T/W** *m* 86.1), with **TG** 15–25 mm. Has on average subtly paler feather-edges above in reasonably fresh spring plumage, often creating a trifle more contrast with more distinct streaking, but many are close to *naevia*. (On the whole it is hardly 'more olive-grey above' in corresponding plumages than *naevia*, something claimed by several authors, but is on average just as olive-brown, although admittedly odd birds show a tendency towards this difference.) Underparts, especially flanks, paler buff-grey or light yellowish, less saturated than most *naevia* (though some overlap). Often, individual plumage variation is such as to make only measurements and structure useful. **L** 12–13 cm; **W** 53–65 mm, ♂ generally > 57 mm, ♀ < 60 mm (*n* 74, *m* 58.8); **T** 45–59 mm (*n* 71, *m* 50.6); **B** 12.7–14.3 mm (*n* 17, *m* 13.3); **Ts** 17.5–19.7 mm (*n* 17, *m* 18.6). – Birds of Zaisan in NE Kazakhstan, W Mongolia, Sayan Mts and adjacent areas of SC Siberia ('*mongolica*') stated to have narrower dark streaks above than *straminea*, but those examined rather had broader streaks, or were similar. Also said to be greyer, less olive above, but impossible to detect any consistent such difference. (Syn. *mongolica*.)

REFERENCES Harvey, P. V. & Small, B. J. (2007) *BB*, 100: 658–664. – Millington, R. (1998) *BW*, 11: 387–389. – Ruttledge, R. F. (1955) *BB*, 48: 235–236.

L. n. naevia, juv/1stW, Scotland, Oct: body and wing fresh, tinged yellowish-rufous above and subtly yellowish-cream below. Again, note relatively broader, less clear-cut rufous-olive outer edges of tertials which are subtly broader at the base and very narrow at the tip. Black streaking above is indistinct, and below practically lacking. Also useful are overall size, fair primary projection and larger bill. This bird has evenly fresh juv wing. (H. Harrop)

L. n. naevia, 1stW, Spain, Oct: unlike ad, very fresh and fringed rufous on wings, and slightly tinged yellowish below. Typical tertial pattern with basally broader and slightly diffuse pale rufous-brown edges. (C. N. G. Bocos)

RIVER WARBLER
Locustella fluviatilis (Wolf, 1810)

Fr. – Locustelle fluviatile; Ger. – Schlagschwirl
Sp. – Buscarla fluvial; Swe. – Flodsångare

Among the *Locustella* warblers treated in this book, four species possess a peculiar reeling, insect-like song—Grasshopper, Lanceolated and Savi's Warblers, and this species. River Warbler breeds in damp thickets like river edges and marshes with bushes and scattered trees, but occurs in a variety of habitats provided there is sufficient vegetation *c.* 2–4 m high, and dense enough to provide cover and shade. Its main range is in E Europe on lowland plains with rather high summer temperatures. River Warbler winters in tropical E Africa, returning in May–early June.

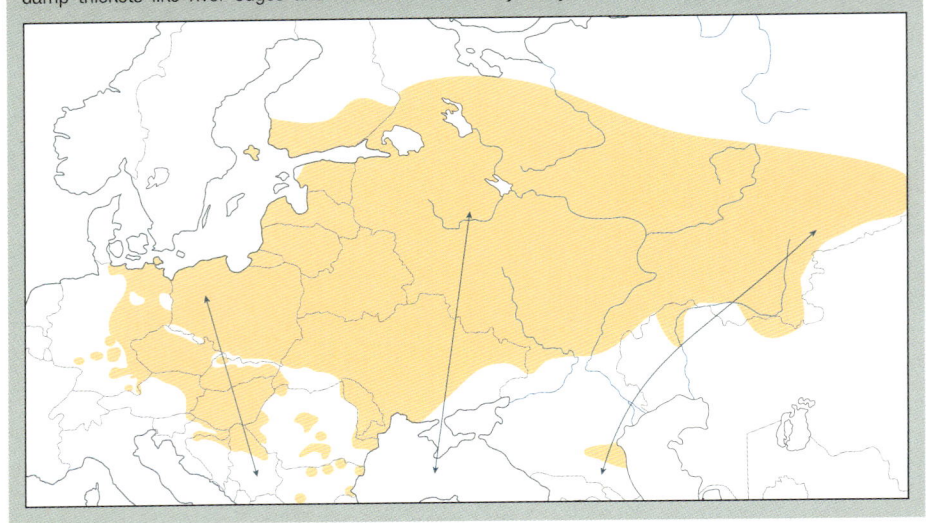

IDENTIFICATION Usually first noticed by its song (see below), which provides the easiest means of identification. Shy and wary, keeping mostly in cover near the ground and often skulking, except singing birds, which take a more elevated perch. Rather large and long-winged with proportionately short bill. Primary projection exceeds tertial length (roughly by 125%). Tail fairly long, broad and well rounded. Upperparts including wings *plain olive-brown*, but outer edge of folded wing (outer web of outermost primary) is paler brown. Underparts are dusky white with olive-brown flanks. *Undertail-coverts also olive-brown, broadly tipped whitish* forming coarsely barred pattern and in good views a striking field mark. Note that there is a fair amount of individual variation in plumage colours, some being darker grey, others paler and more tinged olive-brown. Most obvious field mark regardless of plumage variation is *lower throat/upper breast being either diffusely or prominently streaked olive-brown*. Sides of head mottled olive-brown and dusky white. Often a hint of a short, ill-defined paler supercilium. Legs pale pinkish. Bill rather dark-looking. Song posture with wide open bill pointing obliquely upwards and tail folded down, lower mandible vibrating while singing, distinctive.

VOCALISATIONS Song, mainly given from dusk, during night, in early dawn and variably into morning or briefly at midday, resembles that of a bush-cricket or the sound of a sewing machine, a pulsating or 'chuffing' *dzre-dzre-dzre-dzre-dzre-dzre-dzre-...*, not spinning or buzzing on one note like Grasshopper Warbler. When heard at close range, you will usually notice a peculiar high-pitched metallic jingling background noise, created by fine, shrill notes interwoven in the strophe. Strophes long when undisturbed, even extending over an hour without a break! After a brief pause, new strophe begins with single, less complex notes, but singer soon assumes full song. – Calls include a sharp, metallic *zick-zick* and single *zeck* or *tshick*. These resemble other *Locustella* and are hardly species-specific. A more purring *zrr zrr* has been noted in contact.

SIMILAR SPECIES Can be confused with *Savi's Warbler* of its eastern subspecies *fusca*, although the latter generally favours tall-grown reedbeds (where River Warbler is very unlikely to occur). Eastern Savi's can have an undertail-coverts pattern approaching that of River, being pale rufous-brown with diffusely paler tips, at times even contrastingly whitish (River: rather dark olive-brown coverts with broad and distinctly demarcated white tips). The upper breast, too, of some *fusca* can be diffusely mottled grey, further contributing to the similarity with River Warbler, but latter still much more prominently streaked dark over lower throat and upper breast, and sides of head are more variegated and distinctly mottled dark (Savi's: more uniformly rufous-brown sides of head). Note also olive-brown breast-sides and flanks in River, whereas Savi's is basically rufous-tinged. Primary projection longer in River than Savi's, 125% against broadly equal with tertial length. – In poor views could be confused with a variant *Grasshopper Warbler* with boldly spotted lower throat, but once seen well, uniform upperparts of River, and different undertail pattern, should be evident. – From extralimital *Gray's Warbler* (*L. fasciolata*; vagrant) from E Asia by smaller size (Gray's is of Great Reed Warbler size) and more prominently streaked lower throat/upper breast (Gray's: upper breast uniform or sometimes diffusely mottled grey, but never streaked). Undertail-coverts of Gray's are peculiarly short for a *Locustella* and are rufous-yellow or orange-brown, either uniform or rarely have diffusely paler tips. – From extralimital *Middendorff's Grasshopper Warbler* (*L. ochotensis*; not treated), from E Asia, of which some first-years can have streaked lower throat/upper breast, and which further resemble River in having plain upperparts and olive-brown flanks, by (i) yellowish tinge to underparts, (ii) tail-feathers having narrow white tips and (iii) unstreaked chin/upper throat forming a yellowish-white bib. Rump slightly rufous-tinged (River: olive-brown as rest of upperparts). Note also wing formula of River with no emarginations and second primary forms wingtip; in Middendorff's, third primary clearly emarginated, and second falls between third and fourth, or equals fourth. – A third extralimital E Asian warbler should be mentioned, *Oriental Reed Warbler* (vagrant), which to a surprising extent can recall River Warbler in plumage, with some birds almost as streaked on lower throat and upper breast, and plain grey-brown upperparts. However, the uniform pale brownish-white undertail-coverts of Oriental Reed should immediately separate it from River once seen well.

AGEING & SEXING Ages similar but some 1stY possible to separate if seen close. Sexes alike in plumage and size. Claimed slight sexual differences in plumage have not been possible to confirm on long series of sexed specimens. – Moults. Complex moult pattern, not yet fully studied. Some

Russia, May: on breeding grounds, only likely to be observed when singing. Here a moderately-streaked individual. Upperparts, head and body-sides brown but tinged olive-grey, while wings are fringed more brownish (even rufous-brown). Legs usually dull pinkish red-brown. Age classes inseparable after winter moult. (O. Khromushin)

BUSH AND REED WARBLERS

Finland, Jun: typically, only throat and chest are streaked, some birds rather heavily so, and greyer/browner above (less olive), as here. Undertail-coverts brown with broad pale tips, but this is often not as striking or obvious as on this bird. (A. Juvonen)

birds, presumed to be only ad, moult some outer primaries (apparently frequently from outside inwards) in autumn on stopover in NE Africa, then moult completely, including recently renewed outer primaries, the primaries now from inside outwards, in winter (Jan–Mar) in SE Africa. 1stW is believed to moult just once, completely in SE Africa along with ad. In spring, all those reaching Europe are equally fresh, without moult contrast. The arrested primary moult in autumn might not involve all ad, but perhaps represent individual variation. Post-juv moult in late summer very limited if occurs at all (Müller 1981). – **SPRING** Ageing impossible due to both ad and 1stY having complete moult in winter. – **AUTUMN Ad** Plumage slightly worn, especially tips of outer primaries, and tertials. Plumage on average more tinged olive-brown above and dusky-white below. In late autumn–early winter at least a few moult some outer primaries. **1stCY** Uniformly fresh, including tips of outer primaries. On average more warm rufous-tinged brown above and sometimes faintly tinged yellowish below, but some variation and overlap in coloration. Apparently only one complete moult of flight-feathers in winter, hence autumn birds never show moult contrast in primaries.

Czech Republic, May: typical shape, with long, pointed wings lacking primary emarginations but having rounded outer edge, and long, bulky undertail-coverts. Note olive-tinged grey-brown tones above and rather weakly patterned plumage below. Diagnostic dark-streaked throat and mottled breast, and typically white-tipped olive-grey undertail-coverts, are clearly visible. (L. Mráz)

Finland, May: typical jizz and predominantly olive-brown upperparts are characteristics of the species. This individual has an unusually poorly-streaked throat and breast. Note primary projection equalling or exceeding tertial length. Visible here are features of the genus *Locustella*: pale outer edge to the outermost long primary, and rounded wing edge. (A. Below)

1stW, Iceland, Oct: prior to winter moult. Even some young birds in autumn can be confusingly poorly streaked on throat, breast and head-sides. However, the mostly olive-/grey-tinged head and body are characteristic of the species; primary projection exceeds tertial length. Overall fresh wing typical of 1stW in autumn. (Y. Kolbeinsson)

United Arab Emirates (left) and Israel, May: in Middle East at migration stopover sites, may adopt different postures and show variable behaviour depending on habitat, but when terrestrial often typically *Locustella*-like (left), but could also at first glance look almost pipit-like when resting in acacia (right). Nevertheless, diagnostic *Locustella*-structure and dark-streaked throat and breast, and pale-tipped olive-grey undertail-coverts (giving barred effect) confirm identify. (Left: M. Barth; right: R. Mizrachi)

1stW River Warbler (left: Netherlands, Sep) versus Savi's Warbler (right: Ukraine, Aug): separation is especially problematic between poorly-marked River and Savi's with relatively stronger-marked foreparts (often young, or of the eastern race *fusca*); Savi's can also have undertail-coverts pattern approaching River. However, most River separated reliably by being more prominently streaked or mottled dark, at least on head-sides and lower face (in Savi's more uniform rufous-brown). River also usually has darker olive-brown undertail-coverts, with bolder white tips (in Savi's bases are rufous-brown, and whitish-cream tips are more diffuse). Predominantly olive-brown or olive coloration in River, but more rufous-tinged in Savi's. Primary projection longer in River. Both these birds are juv/1stW by evenly fresh body- and wing-feathers. (Left: T. Luiten; right: L. Mráz)

BIOMETRICS **L** 13.5–15 cm; **W** 72–80 mm (n 24, m 76.1); **T** 51–61 mm (n 23, m 56.7); **T/W** m 74.4; **B** 13.3–16.4 mm (n 22, m 15.0); **Ts** 19.4–22.3 mm (n 21, m 21.2). **Wing formula**: **p1** < pc 0–7 mm; **p2** longest (or =p3); **p3** =p2 or 0.5–3 mm <; **p4** < wt 3–6.5 mm; **p5** < wt 5.5–10 mm; **p6** < wt 8.5–13 mm; **p7** < wt 10–16 mm; **p10** < wt 17–23 mm; **s1** < wt 20–26 mm. No emarg. of primaries, or only a very faint suggestion on p3 in some.

GEOGRAPHICAL VARIATION & RANGE Monotypic. – C Europe from S Finland, E Germany, Austria and Bulgaria east to middle Ob in W Siberia; winters SE Africa.

REFERENCES Müller, H. (1981) *Falke*, 28: 258–265. – Normaja, J. (1994) *BW*, 7: 192–195. – Pearson, D. J. & Backhurst, G. C. (1983) *Ring. & Migr.*, 4: 227–230.

1stW, Kazakhstan, Aug: some young in early autumn are very dusky-looking and heavily streaked below (here just visible), thereby appearing overall darker. Note long primary projection and lack of any emarginations of long primaries. (F. Karpov)

SAVI'S WARBLER
Locustella luscinioides (Savi, 1824)

Fr. – Locustelle luscinioïde; Ger. – Rohrschwirl
Sp. – Buscarla unicolor; Swe. – Vassångare

The species epithet of the scientific name, *luscinioides*, means 'nightingale-like', a very apt description. However, the resemblance only refers to the bird's general appearance. Unlike the Nightingale, it lives in dense, tall reedbeds at lakesides, and has a reeling insect-like song, just like many other *Locustella*, rather different from the pleasing song of the Nightingale. Savi's Warbler has a scattered but wide distribution in C and S Europe, occurring wherever its favourite habitat is found. It is a summer visitor, wintering in Africa south of the Sahara and returning in April or May.

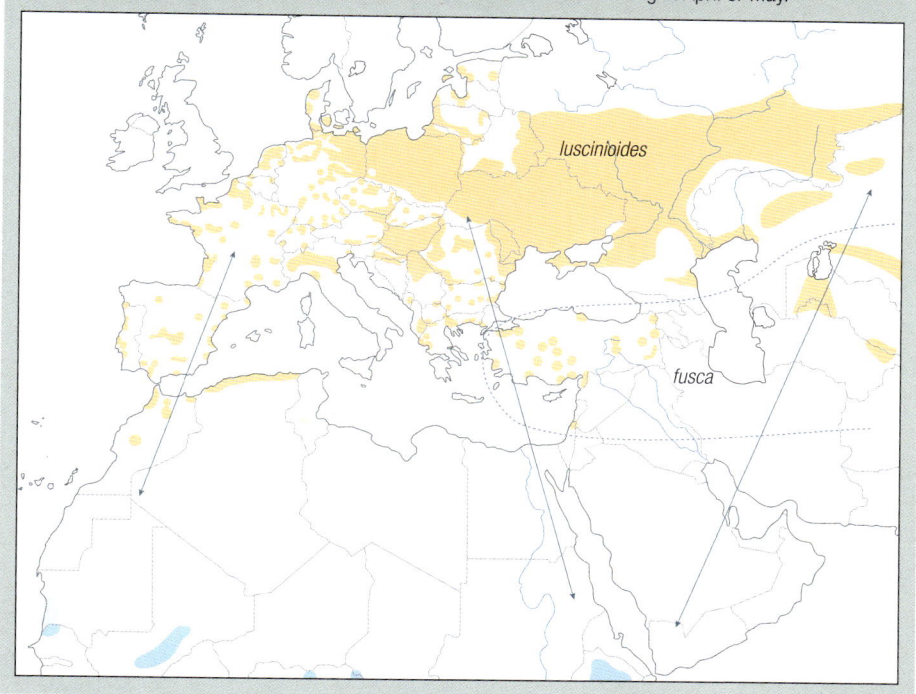

with rest of upperparts in Savi's but usually contrastingly brighter rufous-ochre rump in reed warblers. It is useful also to note the length and relative 'bulkiness' of the tail in Savi's, whereas the tail of the reed warblers is narrower and better-proportioned. – For separation from *River Warbler*, see that species. – *Cetti's Warbler* is similarly plain rufous above, and is frequently seen in reeds (although it prefers a mix of dense cover, with stands of reeds just one component). Cetti's has a shorter neck and smaller bill than Savi's. Note also the narrow, better-marked supercilium of Cetti's. – *Nightingale* is rather similar to Savi's, as mentioned above, but has a much brighter rufous tail, darker legs and heavier bill.

AGEING & SEXING (*luscinioides*) Small seasonal and age-related differences. Sexes alike in plumage and size. – Moults. Timing and extent of remiges moult in both ad and 1stY variable, apparently partially depending on population and subspecies but also on timing of breeding, in some northern and eastern populations generally in autumn (Sep–Nov) immediately after migration (Stresemann & Stresemann 1970), but in southern and western populations may start and at least locally often be completed before migration, mainly moulting remiges in Jul–Sep (Neto & Gosler 2006); odd birds in NC Europe may also moult more extensively before migration (Stein 2011); moult of remiges can also often be suspended to be concluded on stopover (Lynes 1925, Neto et al. 2006) or after migration. Judging from some very fresh spring birds, moult could also take place later, in midwinter (but ageing of these difficult). Moult sequence of primaries is somewhat unusual in that often any of pp5–7(8) is shed first, whereupon moult proceeds centrifugally, or is more conventionally descendent, from p10 outwards. Post-juv moult in late summer either non-existent (late growth of body-feathers notwithstanding) or in particular in southern populations variably extensive, and may even involve some or many outer primaries, all tertials and greater coverts. Pre-nuptial moult of ad and 1stY in winter variable but is either partial, leaving some or all primaries and primary-coverts unmoulted, complete, or suspended to be finished on stopover on spring migration (Neto et al. 2006). Birds with three feather generations among remiges sometimes occur in spring. The result of this plastic and irregular moult strategy is that few

IDENTIFICATION A plain *rufous-coloured* warbler seen in reedbeds can be one of the reed warblers (*Acrocephalus*) or the Savi's Warbler. Savi's has *uniform upperparts* (lacking a paler, brighter rufous-ochre rump), rather darker *dusky rufous underparts*, especially flanks and undertail-coverts, proportionately *longer and broader tail often looking bulky and more rounded*, and a more *rounded wing with pale edge* (outer web of long outermost primary paler brown-white). There is a *narrow and indistinct pale supercilium*. *Undertail-coverts usually uniform rufous-buff*, but sometimes with diffusely paler—even whitish—tips. Legs and bill are rather dark, legs being brownish-pink, bill dark grey, sometimes with a paler, yellowish or pinkish base to lower mandible. In certain lights and angles, the uppertail can be seen to be densely and narrowly cross-barred darker (effect created by so-called growth bars). Eastern birds are more greyish-brown above, less rufous. In summer, most easily noticed and identified by typical song.

VOCALISATIONS Song is a drawn-out, monotonous, insect-like, mechanical reeling, *churrr'r'r'r'r'r'r'r'r'r'r...*, lower-pitched and harder and drier in tone than song of Grasshopper Warbler, which it otherwise most resembles. The song opens with a few hard, metallic call notes that soon merge into the reeling, full song. It is mainly given at dusk and nocturnally, but can also be heard after dawn and briefly throughout the morning. – Usual call is a sharp, metallic *pvitt*. Another call is a chattering series, *chet-et-et-et-...*, almost like a Graceful Prinia in tone.

SIMILAR SPECIES The similarity with some *Acrocephalus* species has already been dealt with under Identification. The best means of separation from these is rump colour: uniform

L. l. luscinioides, Switzerland, Jun: typical posture, singing from dense reeds. Undertail-coverts usually broadly but diffusely pale-tipped; otherwise rather unmarked underparts. Also typical facial jizz, with narrow and tapering pale supercilium, emphasised by thin dusky eye-stripe. (H. Shirihai)

L. l. luscinioides, Netherlands, Apr: uniform dull rufous-brown upperparts (feature of *luscinioides*), bland facial appearance with only hint of short pale supercilium, affording unique look. Chunky body and broad rounded tail; faint growth bars on uppertail. This bird has apparently skipped winter moult, but complicated moult strategy makes it difficult to age. (R. Messemaker)

Israel, Mar: plumage appears neither dull enough for typical *fusca*, nor as warm as classic *luscinioides*—most birds are difficult or impossible to assign to race during migration. This bird has mostly older primaries, but without handling difficult to determine if these are juv or ad feathers. (J. Normaja)

Netherlands, May: ssp. *luscinioides* by range, but plumage more like *fusca* (best explained by degree of individual variation and overlap). This bird also demonstrates that some Savi's have supercilium that is stronger and whiter above and behind eye. In spring, evenly fresh birds could be either age. (H. Gebuis)

birds can be confidently aged without a full appreciation of amount of wear and much experience. – SPRING **Presumed ad** Plumage uniformly fresh or worn, all wing-feathers of same generation. (Note that in this category could be some young birds that kept all juv primaries and therefore show no obvious moult contrast.) **1stS** Wing showing contrast between several brown and worn retained juv wing-feathers, usually 2–5 inner primaries, contrasting with fresher rest of wing-feathers. – AUTUMN **Ad** Plumage either heavily worn, in active wing moult with heavily worn unmoulted remiges, or (especially in S Europe) has completely fresh wing just prior to migration. Some birds migrate after suspending moult with variable number of moulted central primaries, and some inner (usually) and outer (rarely) retained primaries heavily worn and brown. Upperparts rufous grey-brown, slightly greyer and duller on average than in young birds. Iris deep rufous-brown or at least warm brown (some slight overlap with advanced 1stCY). **1stCY** Either uniformly fresh, although tips of tail-feathers can show first signs of wear, being a little 'frayed', or some outer primaries moulted, but these only moderately fresher than retained inner juv primaries. On average more vividly rufous-tinged above. Iris dark grey-brown or brownish-grey, less rufous-tinged than in ad (but slight overlap depending on early or late hatching). Any streaking or mottling of breast would indicate 1stCY.

BIOMETRICS (*luscinioides*) **L** 13–15 cm; **W** 66.5–76 mm (n 40, m 70.9); **T** 51–61 mm (n 42, m 56.4); **T/W** 79.8; **B** 14.0–16.5 mm (n 40, m 15.4); **Ts** 19.5–22.1 mm (n 40, m 20.9). **Wing formula:** p1 narrow and pointed, < pc 0.5–4 mm, < p2 40–45 mm; **p2** longest (rarely =3), tip pointed; **p3** < wt 0.5–3 mm; **p4** < wt 3.5–6 mm; **p5** < wt 3.5–6 mm; **p6** < wt 6–9 mm; **p7** < wt 9–13 mm; **p10** < wt 17–21 mm; **s1** < wt 20–24 mm. Emarg. p3, but slight and hardly noticeable in many.

GEOGRAPHICAL VARIATION & RANGE Only slight and clinal variation. Two races.

L. l. luscinioides (Savi, 1824) (Europe including Russia; W European breeders winter in W & C sub-Saharan Africa from Senegal to perhaps W Chad and south to N Ghana, E European breeders in C & E sub-Saharan Africa, mainly Sudan, Eritrea and Ethiopia). Treated above. Dark rufous-brown above and rufous-buff below with greyish tinge on breast-sides. Throat and centre of breast usually paler, pale brownish-white. Lower flanks, lower belly and undertail-coverts usually vividly rufous-tinged, latter sometimes with diffuse whitish tips. – Breeders of SE Ukraine and SE Russia east to N Caspian Sea sometimes separated ('*sarmatica*') on account of duller colours. This seems to be a subtle clinal variation between *luscinioides* and *fusca*, being closer to the former, and on available material, admittedly scant, in our view not distinct enough to be upheld as a separate taxon. (Syn. *sarmatica*.)

○ **L. l. fusca** (Severtzov, 1873) (Turkey, Central Asia; winters sub-Saharan Africa in Sudan, along Red Sea coast, Ethiopia). On average very slightly smaller but proportionately longer-tailed and longer-billed than *luscinioides*, and subtly paler more greyish-brown above (appreciable in direct comparison). In fresh plumage the upperparts are at most warm tawny (but not rufous), soon bleaching to a more greyish-brown. Underparts on average paler, more pinkish-buff than rufous-tinged. Note that 1stY is a little warmer brown above, more like *luscinioides*. Birds in E Europe (C & SE Russia) can be slightly duller than typical *luscinioides*, tending towards *fusca* (see above). Rarely, birds in Middle

Presumed L. l. luscinioides, Upper Volga, Russia, May: locality within the range of *luscinioides*, but coloration of this bird close to *fusca*, showing the degree of (clinal) variation and overlap between races. This bird appears to show moult limits in wing, with older feathers (presumably juv) very worn. Being of eastern origin, it is not unusual for 2ndY or older birds to arrest moult, but without handling difficult to be sure. (O. Khromushin)

L. l. fusca, during spring/summer (left: Mongolia, Jun; right: United Arab Emirates, Mar): sometimes eastern locality in combination with generally more greyish-brown upperparts, lacking any obvious warm tawny or rufous tinges, can help identify subspecies *fusca*. Paler underparts would further support identification. (Left: H. Shirihai; right: M. Barth)

East in late Sep have heavily worn inner 3–4 primaries, and bleached ss1–6, but rest of wing new; these birds with suspended (or arrested?) moult could be 2ndY, though firm proof that they are exclusively this age still lacking. **W** 67–73 mm (*n* 16, *m* 70.7); **T** 57–64 mm (*n* 16, *m* 60.1); **T/W** *m* 85.0; **B** 15.1–17.0 mm (*n* 16, *m* 16.0); **Ts** 20.0–22.0 mm (*n* 16, *m* 21.1).

REFERENCES Lynes, H. (1925) *Ibis*, 68: 344–416. – Mead, C. J. & Watmough, B. R. (1976) *Bird Study*, 23: 187–196. – Müller, H. E. J. (1981) *Falke*, 28: 258–265. – Müller, H. E. J. (1982) *Falke*, 29: 242. – Neto, J. M. & Gosler, A. G. (2006) *Ibis*, 148: 39–49. – Neto, J. M., Gosler, A. G. & Perrins, C. M. (2006) *J. Avian Biol.*, 37: 117–124. – Stein, H. (2011) *Berichte Vogelw. Hiddensee*, 21: 81–85. – Stresemann, E. & Stresemann, V. (1970) *J. Orn.*, 111: 237–239. – Thomas, D. K. (1977) *Ring. & Migr.*, 1: 125–130.

L. l. luscinioides, juv/1stW, Italy, Aug: oddly enough, can briefly resemble Cetti's Warbler. However, latter readily eliminated by smaller size and different jizz, with shorter neck and weaker bill; Cetti's also has stronger face pattern and much shorter primary projection. Warm flanks coloration but unstreaked breast separate Savi's from River Warbler. Aged by being vividly rufous above, while iris is dull olive or grey-brown, and note abraded tail-feather tips (ad much fresher with whitish edge). (D. Occhiato)

Presumed ssp. *fusca*, juv/1stW, Israel, Oct: walks, rather than hops, when on ground (unlike *Acrocephalus*). Rather fresh plumage, but worn tips to primaries and tail-feathers, and olive iris, provide ageing clues. Being a young in autumn but still rather dull with no obvious rusty tinge suggests *fusca*. (A. Ben Dov)

L. l. fusca, juv/1stW, Kuwait, Aug: some eastern birds are noticeably more greyish-brown and olive-tinged above, with hardly any rufous, even when fresh; also noticeably whiter below, sometimes with dark streaks on upper breast, potentially inviting confusion with River Warbler. (M. Pope)

MOUSTACHED WARBLER
Acrocephalus melanopogon (Temminck, 1823)

Fr. – Lusciniole à moustaches; Ger. – Mariskensänger
Sp. – Carricerín real; Swe. – Kaveldunssångare

A bird of patchy distribution in S Europe, frequenting marshes and ponds with lush vegetation and scattered bushes. In vast reedbeds often found near small patches with open water, broken reeds and stands of reedmace (*Typha*). Not always easy to separate from Sedge Warbler, but characteristic elements in the song help. Unlike Sedge Warbler resident or only short-range migrant, wintering in marshes along Mediterranean coasts and rivers, or in N Africa and the Middle East. Eastern populations are more migratory than those in west.

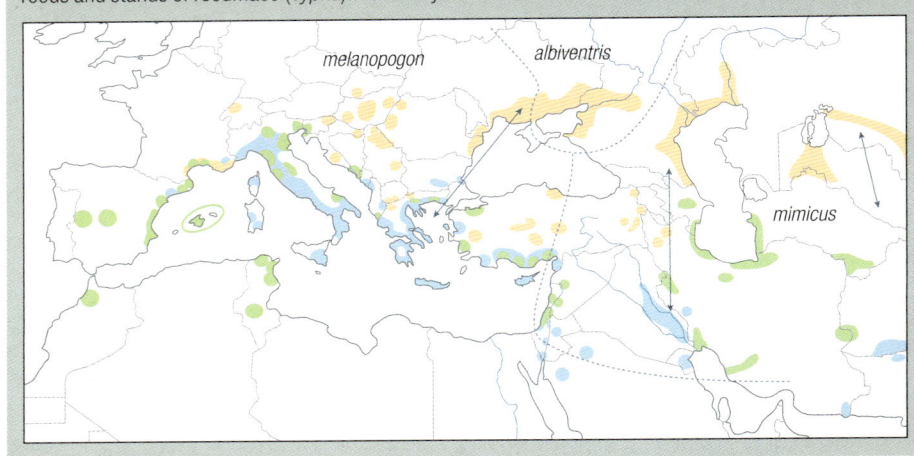

IDENTIFICATION Very similar to Sedge Warbler, and in particular the eastern race *mimicus* of Moustached resembles Sedge and requires close views and care. Same size and general plumage pattern as Sedge, with *streaked back* and prominent *pale supercilium*, but differs structurally by more *rounded wings* with *shorter primary projection* (about 1/3 to 1/2 of tertial length) and more *rounded tail*. European populations (ssp. *melanopogon*) are rather distinct due to *quite dark* (often blackish-looking) *crown, very prominent white supercilium ending abruptly and squarely at rear* (yellowish-buff and tapering rather narrowly in Sedge) and *deep rufous tinge on upperparts and flanks*. Note also darker or *duskier ear-coverts* in Moustached with *a more clearly defined dark lower edge* (the 'moustache'!), whereas in Sedge the ear-coverts are paler and browner, and usually merge indistinctly into the pale throat-sides (although there is some variation in this, making the character less than absolutely reliable taken alone). The more Sedge Warbler-like eastern race *mimicus* is less uniformly dark on crown (browner with dark streaks on centre of crown), less vividly rufous above and on flanks, but differs in same way as European birds from Sedge Warbler in structure and darker and better-defined ear-coverts. Also, throat more contrastingly white, and flanks more tinged pinkish-brown, not so yellowish-buff. *Legs and bill are darker* on average in all populations of Moustached than in Sedge, often being dark grey and can look black. Hence a bird which looks 'promising' for Moustached on plumage but which has pinkish or yellowish legs is probably not one. Supplementary characters to note are habit of *often cocking its partially spread tail* or twitching it upwards and sideways, and frequently feeding low down, often using broken reeds at the very edge of the water. See also call below.

VOCALISATIONS The song can be described as a mixture of Sedge and Reed Warbler songs, a 'talkative' but hurried series of harsh and whistling, mainly monosyllabic sounds. It is by comparison livelier and a little faster than Reed, but not as fast as that of Sedge, lacking the accelerations and crescendos of repeated harsh notes of the latter. It is also rather softer and quieter than both, never audible as far, and is best recognised by recurring *series of softly whistled notes on slightly rising pitch*, not unlike similar series of Nightingale, *lu lu lo lo le le lee lee*. – Call a muffled but hard clicking, *chrek*, rather like European Stonechat but less loud. Related to this are a shorter *trr* or *trk*, and a clicking or rolling series, *trk-trk-trk-trk-*…

SIMILAR SPECIES Separation from *Sedge Warbler* is treated under Identification. Although the majority do not present severe problems, one should be aware of the existence of more difficult birds that require all characters be checked and used in combination. – *Pallas's Grasshopper Warbler* can appear reasonably similar, with streaked back and a rather prominent pale supercilium, but differs on several points, notably the distally dark and white-tipped outer tail-feathers and usually prominently streaked lower rump and uppertail-coverts (Moustached: unstreaked).

AGEING & SEXING (*melanopogon*) Insignificant seasonal and age-related differences. Sexes alike in plumage. – Moults. Complete post-nuptial moult of ad in Jul–Sep (Oct). Rather uniquely among related warblers, juv, too, moults completely at the same time. Both age groups have a partial spring moult. – **SPRING** Ageing impossible. – **AUTUMN Ad** Recognised in late summer by remaining worn

A. m. melanopogon, Spain May: even in late spring plumage, most birds of this race are still rather obviously warmer and darker than any Sedge Warbler, making prominent whitish supercilium and throat quite contrasting. Note also diagnostic proportionately shorter wings/primary projection. Due to moult strategy, ageing impossible. (Q. Marcelo)

A. m. melanopogon, Spain, May: distinctly bold dark and white facial markings, with almost blackish crown, white supercilium and dark eye-stripe, and note well-developed dark moustachial stripe; also warmer-coloured breast-sides and flanks. (Q. Marcelo)

A. m. melanopogon, Italy, Nov: rather Sedge Warbler-like, but note blackish-looking crown, contrasting with rather broad and square-ended white supercilium. Also differs by proportionately shorter wings and longer-looking tail, which often is raised and waved. By late autumn, ad and young have already completed moult and are alike. (D. Occhiato)

A. m. melanopogon, Italy, Sep: some are noticeably rufous above and appear to have almost solid dark crown and ear-coverts, accentuating almost snow-white supercilium and throat. Extensive and warm-toned flanks. (D. Occhiato)

flight-feathers, but once last primary has been shed usually impossible to age. **1stCY** A bird in active wing-moult with reasonably similar fresh both old and new primaries is young.

BIOMETRICS (*melanopogon*) **L** 12–13.5 cm; **W** ♂ 55–65 mm (*n* 14, *m* 58.8), ♀ 53–62 mm (*n* 13, *m* 57.2); **T** 46–56 mm (*n* 27, *m* 50.8); **T/W** 87.6; **B** ♂ 13.8–15.6 mm (*n* 14, *m* 14.8), ♀ 12.7–15.4 mm (*n* 13, *m* 14.2); **Ts** 19.5–21.7 mm (*n* 15, *m* 20.7). **Wing formula:** p1 > pc 4–10 mm, < p2 19–25 mm; **p2** < wt 4–7.5 mm; **pp3–4** longest; **p5** < wt 1–2.5 mm; **p6** < wt 2.5–4.5 mm; **p7** < wt 4–6.5 mm; **p10** < wt 7.5–10 mm; **s1** < wt 8.5–12 mm. Emarg. pp3–5, sometimes faintly also on p6.

GEOGRAPHICAL VARIATION & RANGE Slight clinal variation noted, birds from west to east becoming less dark and rufous, more pale and olive-tinged brown above. On the whole, eastern populations are very slightly larger than western ones. At least two distinct races (a third claimed, apparently rather subtle).

A. m. melanopogon (Temminck, 1823) (Europe, NW

A. m. melanopogon / mimicus, Cyprus, Nov: breeders of Cyprus usually appear midway between the two races, but this bird is closest to *melanopogon*. Typical bold head pattern, with dark cap and ear-coverts, forming clear contrast with clean white supercilium and throat. Compared to Sedge, legs and bill on average darker grey and can look black. Ageing impossible following completion of post-nuptial/juv moult. (A. McArthur)

Africa, W & C Turkey; winters coastal areas around Mediterranean Sea). Quite dark rufous above, crown in worn plumage often appearing near-black (at least laterally), and breast and flanks rather strongly rufous-tinged. See also Biometrics. – Birds in SE Europe and W & C Turkey tend towards ssp. *mimicus*.

? **A. m. albiventris** Kazakov, 1974 (Sea of Azov, NE Black Sea and lower Don; thought to winter Middle East). Claimed to be darker and greyer above than *mimicus* (see below), and whiter on underparts, but differences apparently slight, and not accepted by Stepanyan (1990). No material examined.

A. m. mimicus (Madarász, 1903) (E Turkey, Caucasus, Transcaucasia, Levant, NW Caspian coast, lower Volga east to Kazakhstan, Uzbekistan; winters E Mediterranean Sea, S Middle East, Persian Gulf, Pakistan). Slightly paler and less vividly rufous than *melanopogon*, more tinged olive-grey, crown being slightly paler and more clearly streaked. Also, overall size slightly larger on average. Birds of Cyprus intermediate between this race and *melanopogon* but closest to *mimicus*. **W** ♂ 60–65 mm (*n* 14, *m* 62.9), ♀ 59.5–63 mm (*n* 15, *m* 61.3); **T** 50–58 mm (*n* 28, *m* 54.1); **T/W** *m* 85.9; **TG** 6–11 mm; **B** 14.0–16.0 mm (*n* 25, *m* 14.8); **Ts** 19.5–

A. m. mimicus, Israel, Dec: superficial Sedge Warbler-like impression, with duller coloration, especially the paler and streakier crown, and paler upperparts and flanks, with less rufous, but main diagnostics still hold, e.g. shorter primary projection, unstreaked rump and hint of dark moustachial stripe; often cocks its tail up and sideways. (L. Kislev)

Moustached Warbler (top left: Israel, Nov; bottom left: Italy, Nov) versus Sedge Warbler (top right: Spain, Aug; bottom right: Italy, Sep): in fresh plumage, when comparing European population of Moustached (*melanopogon*) to Sedge, the two species are easily separated (bottom two photographs). Moustached has different jizz, with shorter/rounder wings, and its tail is often cocked upwards and sideways. The darker and more solidly patterned cap, cheeks and body accentuate the prominent white supercilium and clear white throat. Moustached is also warmer and more rufous above. Nevertheless, the two can be very similar, in particular when comparing a duller example of race *mimicus* to duller Sedge (upper two photographs). In such cases, check for shorter primary projection (c. 1/3 to 1/2 of tertial length in Moustached, versus 2/3 to 3/4 in Sedge), less streaked crown and well-developed blackish lateral crown-stripes, while the supercilium and throat are whiter and better defined. In addition, upperparts and flanks are slightly warmer in Moustached, which lacks the often contrasting rufous rump/uppertail-coverts of Sedge, and bill and legs are darker/greyer. (Top left: T. Muukkonen; bottom left: D. Occhiato; top right: Q. Marcelo; bottom right: D. Occhiato)

A. m. mimicus / melanopogon, Israel, Nov: probably *mimicus* given paler head and flanks, but rufous above somewhat approaches *melanopogon*. Paler brown median crown-stripe more obvious in eastern populations, and therefore resembles Sedge Warbler. Nevertheless, unlike Sedge, prominent whitish supercilium ends abruptly and squarely posteriorly. (L. Kislev)

22.0 mm (n 27, m 20.7). **Wing formula: p6** < wt 1–3.5 mm; **p7** < wt 3–5.5 mm; **p10** < wt 8.5–11.5 mm; **s1** < wt 10–13 mm. Complete post-nuptial moult apparently usually later than in *melanopogon*, perhaps most commonly in winter quarters, finishing in Nov, but may also start on breeding grounds, then be suspended and finished in winter. (Syn. *caspicus*.)

REFERENCES Bradshaw, C. (2000) *BB*, 93: 29–38. – Madge, S. (1992) *BW*, 5: 299–303. – Melling, T. (2006) *BB*, 99: 465–478. – Stepanyan, L. S. (1990) *Conspectus Orn. Fauna USSR*. Moscow. – Vinicombe, K. (2002) *Birdwatch*, 118 (Apr): 22–25.

AQUATIC WARBLER
Acrocephalus paludicola (Vieillot, 1817)

Fr. – Phragmite aquatique; Ger. – Seggenrohrsänger
Sp. – Carricerín cejudo; Swe. – Vattensångare

A rare bird with a restricted range, unusual migration route, and requires damp short-grass meadows and flooded riversides. From its known strongholds in E Poland, Belarus and SW Russia, in late summer Aquatic Warbler migrates due west and then south through W Europe to reach its main wintering grounds, which have recently been located in the swamps of Senegal. Spring migration is poorly known but may take a slightly more easterly course through Europe. Perhaps the best place to become acquainted with this rare warbler is the vast Bierbza marshes of E Poland, which supports a population of *c.* 2000 singing males, arriving from late April to early May.

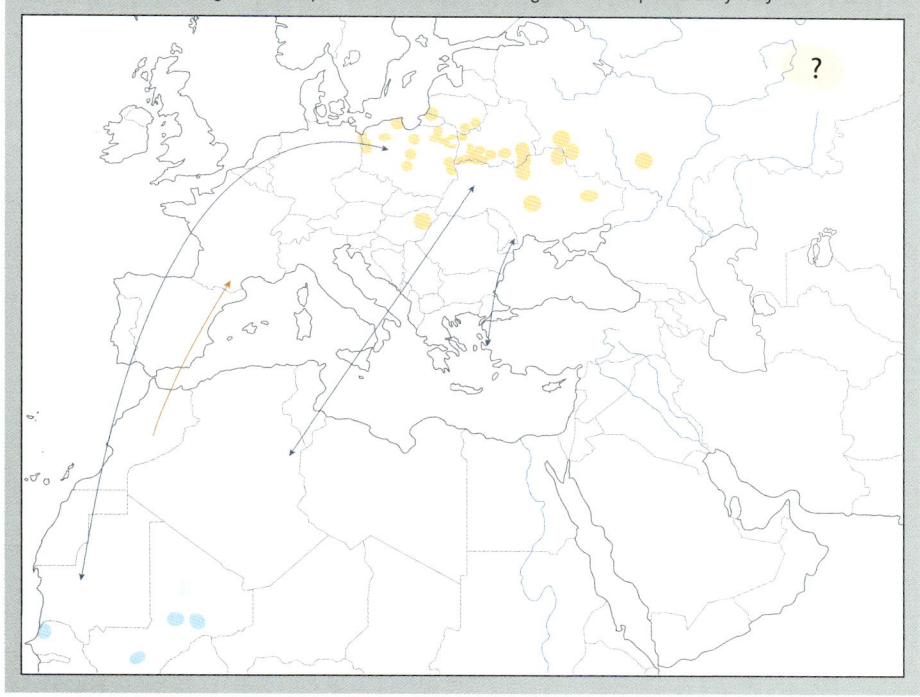

Poland, Jun: typically stretches neck when singing. If diagnostic pale median crown-stripe is not visible, other features useful for separation from Sedge Warbler include the prominent buff-and-black mantle stripes, pale lores (invariably dark in Sedge) and finely streaked flanks and brownish-buff neck. Also, primary projection shorter than in Sedge. (M. Matysiak)

IDENTIFICATION Very similar in shape and size to Sedge Warbler, differing primarily in *shorter bill* and *shorter primary projection* (c. 1/2–2/3 of exposed tertial length), making the bird appear slightly more compact but tail at least sometimes proportionately a little longer. Other characters are the *broader and better-marked buffish supercilium* and presence of *a distinct pale buff median crown-stripe*. It also has *prominent buffish lateral mantle-stripes*, lacking in Sedge. Supplementary characters to note are the moderately *streaked rump* (Sedge is unstreaked) and overall *less dark rufous, more pale yellowish-ochre upperparts*. Between the pale supercilia and pale crown-stripe run *blackish lateral crown-stripes*, which end before forehead. *Legs* are *pale pinkish*. Adults are finely streaked on lower throat, upper breast and usually on flanks (very few exceptions unstreaked), whereas juveniles are unstreaked or have just a few indistinct blotches on lower throat/upper breast.

VOCALISATIONS The song, delivered mostly in evenings until dusk, but also in the morning and sporadically at other times, is like a 'sleepy', uninspired Sedge Warbler, the voice being similar but the strophes brief, uncomplicated and lacking crescendos or long series of harsh notes. Most songs start with a creaky or churring *trrrrr*, immediately followed by a rapid series of clear whistling or nasal and squeaky notes, *trrrrrrrr-didididi…*, *trrrrr-lülülülü…*, etc. This simple song can be maintained for long periods but sometimes, if agitated, strophes become slightly more complex, containing a few more elements, increasing the similarity with Sedge Warbler, *trrrrr-jojojojo-terr… churrrrr-ju-chi-chi-chi-errr…*, etc. Song usually given from perch on low weed, but occasionally also in short song-flight. Posture includes peculiarly stretched neck and tail pointing down. – Call is a hard clicking *chek*. Alarm is the same, or a similar churring trill as in the song, *trrrrr*. Some variations of both calls heard.

SIMILAR SPECIES Mainly confused with *Sedge Warbler*, but differs in slightly shorter bill, overall paler and more sandy-tinged plumage (although can be difficult to evaluate in certain lights), much paler and better-defined and prominent buffish-yellow median crown-stripe (some young Sedge have rather pale central crown but this never forms a distinct and unstreaked yellow-buff stripe), pale lores (Sedge: invariably dark), prominent yellow-buff lateral mantle-stripes, more yellow-brown (less rufous) and distinctly streaked rump and uppertail-coverts, and, in adults, finely streaked breast and flanks. Tail subtly more rounded, and tail-feathers slightly narrower and more pointed.

AGEING & SEXING Small but often useful age-related differences in summer–autumn. Sexes alike in plumage. –

Poland, Jun: one of the streaked and striped *Acrocephalus* species, and unmistakable if the diagnostic head pattern is seen well, sporting a well-marked straw-coloured median crown-stripe. Following complete winter moult, ageing impossible. (M. Matysiak)

Poland, Jun: with feather wear in summer streaking becomes even more bold and prominent. Note in profile often round-headed and short-billed impression. Although pale supercilium broad and contrasting with dark lateral crown, eye-stripe can be less distinct, as here. (M. Matysiak)

1stW, Belgium, Sep: striped crown pattern diagnostic. Note that black lateral crown-stripes do not reach forehead, and that pale crown-stripe is rufous-buff in its forepart. Underparts of this young autumn bird typically unstreaked yellowish rufous-buff. (K. De Rouck)

1stW, Belgium, Sep: predominately sandy-buff and rufous-tinged young birds like this are unmistakable, this colour also invading upperparts. Again, note the unusually short bill for an *Acrocephalus*. Overall fresh plumage ages this bird as 1stW. (K. De Rouck)

Moults. Complete post-nuptial moult of ad, and complete moult of 1stY, variably started just before or during autumn migration, when limited to head and body, some wing-coverts and (in ad) sometimes r1 and odd tertials, finished early winter (Oct–Dec) with flight-feathers, possibly simultaneously for both age groups. Apparently no post-juv moult. Both age groups have a partial pre-nuptial spring moult of variable extent, apparently usually rather restricted. – **SPRING** Ageing impossible. – **AUTUMN Ad** Worn, especially tips of outer primaries, tertials and tail-feathers. Head and body generally also worn, but some replace these partially while still in Europe. Underparts rather whitish if still unmoulted. Breast-sides and flanks almost invariably finely and distinctly streaked, but very rarely nearly unstreaked. A darkish loral 'smudge' in some, but others have quite pale lores as in 1stCY. Wings tinged greyish-olive if largely unmoulted. **1stCY** Uniformly fresh plumage, including primaries, tertials and tail, but some have slightly frayed tips to tail-feathers from Sep. Tertials broadly edged buff. Underparts tinged warm yellowish-buff. Breast and flanks usually unstreaked, but a few have fine streaks on lower flanks and some spots on lower throat/upper breast, and exceptionally are prominently streaked on breast and flanks just as ad. Lores invariably pale. Wings tinged ochre.

Young/1stW Aquatic (left: Sweden, Aug) versus Sedge Warbler (right: Italy, Sep): even at long range the paler, more sandy-tinged upperparts of Aquatic with distinctive blackish mantle stripes are striking. The well-defined but rather narrow pale median crown-stripe is usually detectable in closer views and at certain angles. However, some young Sedge have rather pale central crown, but this is usually finely streaked. Lores of Aquatic also pale and unmarked, at most shaded darker (i.e. lacking dark loral line of Sedge). Tail-feathers narrower and more pointed compared to Sedge. Primary projection of Aquatic generally shorter than Sedge, but some birds are closer in length (as here). On average broader and better-marked buffish supercilium in Aquatic. Legs paler in Aquatic and mostly pinkish. Both are juv/1stW by fresh body- and wing-feathers. (Left: L. Lundmark; right: D. Occhiato)

BIOMETRICS L 11.5–13 cm; **W** ♂ 59–68 mm (*n* 30, *m* 63.5), ♀ 59–65.5 mm (*n* 16, *m* 62.7); **T** ♂ 44–52.5 mm (*n* 29, *m* 47.3), ♀ 44–50 mm (*n* 16, *m* 46.0); **T/W** *m* 74.5; **TG** 6.5–12 mm (*n* 29, *m* 8.9); **B** 11.8–13.8 mm (*n* 46, *m* 13.0); **BD** 2.9–3.5 mm (*n* 22, *m* 3.2); **Ts** 19.0–21.6 mm (*n* 36, *m* 20.3). **Wing formula: p1** < pc 0–4 mm, < p2 34–40 mm; **p2** < wt 0–2 mm, =3/4 (53%), =3 (40%) or =4 (7%); **p3** longest (or p2 = p3); **p4** < wt 1–2.5 mm; **p5** < wt 2.5–6 mm; **p6** < wt 5.5–9 mm; **p7** < wt 7–11 mm; **p10** < wt 12–18.5 mm; **s1** < wt 15–21 mm. Emarg. p3, in some ad also faintly on p4.

GEOGRAPHICAL VARIATION & RANGE Monotypic. – C & E Europe, mainly Poland, Belarus and SW Russia, more local elsewhere; winters Senegal, Mali and adjacent areas. (Syn. *A. aquaticus*.)

REFERENCES Gorman, G. (2002) *Alula*, 8: 62–65. – Schulze-Hagen, K. (1989) *Limicola*, 3: 229–246. – Vinicombe, K. (2003) *Birdwatch*, 134 (Aug): 22–24. – Walbridge, G. (1991) *BW*, 4: 237–241.

1stW, Spain, Aug: unlike ad, underparts usually unstreaked, but sometimes a few indistinct marks on breast-sides and lower flanks. Note distinctive mantle, scapulars and head patterns. The sharply pointed tail-feathers are typical. (J. Sagardía)

1stW, England, Sep: a strongly marked young bird. Rather fresh, evenly-feathered wing easily separates it from ad. Again, note very sharply pointed tail-feathers. (P. Leigh)

SEDGE WARBLER
Acrocephalus schoenobaenus (L., 1758)

Fr. – Phragmite des joncs; Ger. – Schilfrohrsänger
Sp. – Carricerín común; Swe. – Sävsångare

Widely distributed in N, C and SE Europe, and further east, usually breeding in wet meadows with sedges (*Carex*) and willows (*Salix*) bordering lakesides, ponds, water-filled pits and rivers, but is also found in drier habitats such as young plantations, along hedges or overgrown ditches, etc. Unlike Reed Warbler, it avoids tall, dense, extensive reedbeds, but is occasionally found at their edges or where scattered bushes and openings with sedges break the conformity. Often noticed by its loud, energetic song, at times delivered in a short song-flight. A long-distance migrant which winters south of the Sahara, returning in late April and May.

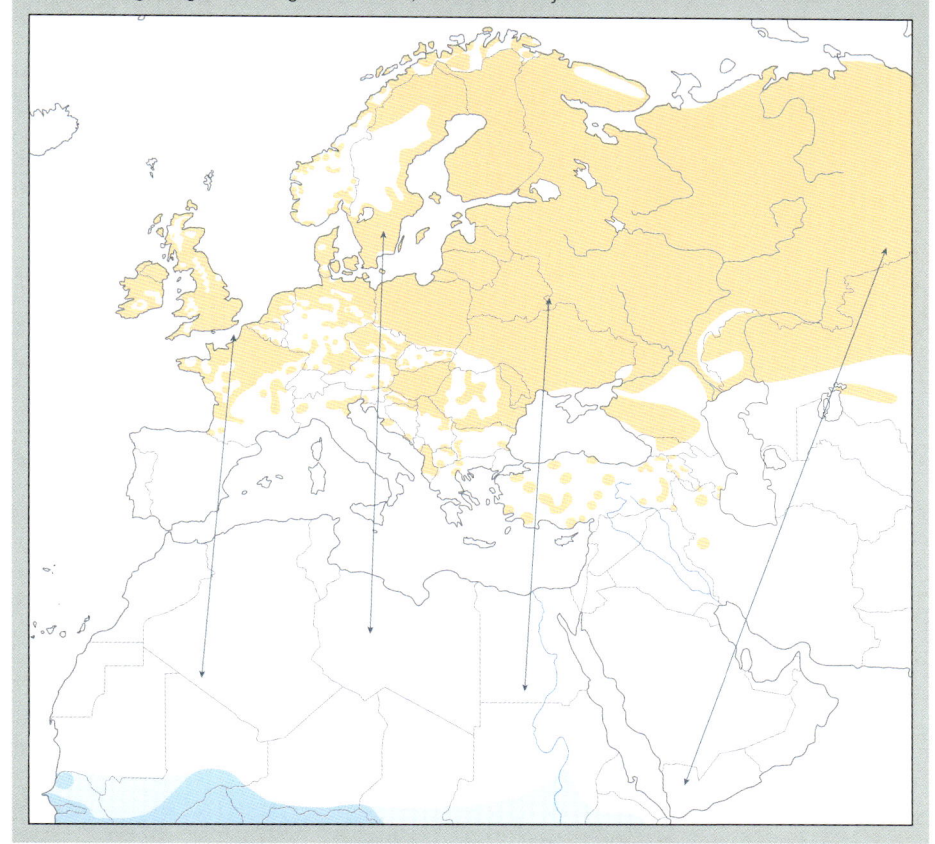

Latvia, May: with wear, prominent cream-white supercilia bleach whiter, as do the (unstreaked) underparts, with especially throat and belly becoming cleaner. Supercilium further enhanced by dark eye-stripe and lateral crown-stripes, leaving paler brown median crown-stripe. With wear, mantle and scapulars become duller. (M. Varesvuo)

IDENTIFICATION A small, slim warbler of brown and buff colours. Best field marks are *streaked back* and *prominent pale supercilium*. For safe separation from Moustached and Aquatic Warblers important to note combination of several details: (i) *supercilium is cream-coloured* or yellowish-buff (can bleach more whitish), enhanced by *laterally rather dark crown-sides* and a *dark eye-stripe*, and *tapers rather narrowly at rear*; (ii) *crown* is a shade *paler brown along centre*, streaked dark, but this is never a prominent yellowish median crown-stripe (beware of some juveniles with a paler centre than most); (iii) *cheeks* are generally fairly *pale brown*; (iv) *primary projection is rather long* (*c.* 3/4 to equal tertial length); (v) *rump is an unstreaked bright tawny-rufous*, whereas *back is duller and darker brown with some subdued dark streaking*. Legs a rather *pale pinkish-brown*. Underparts of adults are unstreaked and cream-coloured except whitish throat, whereas juveniles usually have a little dark spotting on lower throat and breast.

VOCALISATIONS The song is quite loud, prolonged and energetic, often delivered from cover or from a barely visible, low song-post, but now and then—especially during morning hours—in a short song-flight. Although the tempo varies, compared to Reed Warbler the song can be much faster and more agitated, containing typical accelerations and crescendos of repeated harsh notes. It can continue for longer spells with rather uninspired sequences of repeated, grating phrases, short trills or long series of harsh notes, e.g. *zetre-tret, zetre-tret, zetre-trit, zezeze-tirrrrrr zret-zret-zret-zret-zrezrezrezre...* etc., then suddenly pick up speed and 'inspiration', and burst into an impressive flourish, *zezezezezrizrizrizri-psit-trutrutru-pürrrrrrerrrrr vi-vi-vi lülülü zetre zetre*. Sings at almost any hour but usually rests in afternoons and early evening, with peaks at dusk to midnight and again at dawn and sunrise. – Call a hard clicking *tsek!*, usually audibly harder and sharper than any Reed Warbler call (instead, more like Blackcap). Alarm is the same or a harsh, somewhat muffled, rolling *cherrrr*. More subdued variations of both calls heard.

SIMILAR SPECIES Can be confused with *Moustached Warbler* (especially the eastern race *mimicus*) but differs in longer primary projection, more tapering, pointed end to supercilium, less rufous tinge to upperparts, breast-sides and flanks, and paler legs. That the cheeks are slightly paler and less greyish, with usually a much less pronounced dark moustachial stripe, is a supplementary character (but subject to variation, and should not be used alone). Details in behaviour and calls differ as well (see above and under Moustached). – Similar to *Aquatic Warbler* but a little darker brown

England, May: often uses exposed branches of bushes and low trees as song posts. Note head pattern and rufous-tinged rump/uppertail-coverts. Birds returning to Europe having completed winter moult are impossible to age. (M. Stanley)

Netherlands, May: even in late spring still relatively fresh. Primary projection not much shorter than tertial length in this bird. Supercilium and throat, although pale, not as whitish and well-marked as in Moustached Warbler. (P. van Rij)

Ad, Denmark, Jul: in summer ad easily separated from juv by worn and slightly duller plumage, emphasising whitish supercilium and dark lateral crown-stripes. Unstreaked and bright tawny-rufous rump and uppertail-coverts. Primary projection here rather short (2/3 of tertial length). (K. Dichmann)

1stW, Netherlands, Jul: dark lateral crown-stripes, though streaked rather than solidly dark, and paler centre to crown can appear obvious on some birds and at certain angles. Aged by being evenly fresh. (N. D. van Swelm)

and duller above, with less obvious pale central crown-stripe, and no hint of ochre-yellow lateral mantle-stripes. Further, the rump is unstreaked tawny-rufous (Aquatic: tawny-ochre and distinctly streaked dark). – *Pallas's Grasshopper Warbler* can appear reasonably similar, with streaked back and a rather prominent pale supercilium, but differs in having distally dark, white-tipped outer tail-feathers, a shorter supercilium and usually prominently streaked lower rump and uppertail-coverts (Sedge: unstreaked).

AGEING & SEXING Small seasonal and age-related differences. Sexes alike in plumage. – Moults. Complete post-nuptial moult of ad, and complete moult of 1stY, in winter, possibly simultaneously and often early (Oct–Dec), but timing variable and quite a few moult later (finishing Mar–Apr) or in two stages, suspending in Dec–Jan for a movement further south. Apparently no post-juv moult, or only a very restricted one. Both age groups have a partial pre-nuptial spring moult, but only affecting (majority of) birds which moulted wings and tail early in winter; late moulters of wings and tail do not need, or have time for, a pre-nuptial moult. – **SPRING** Ageing impossible. – **AUTUMN Ad** Worn, especially tips of outer primaries, tertials and tail-feathers. Head and body generally also worn, but some replace these partially while still in Europe. Tertials edged off-white or cream-white (if not too abraded). Breast almost invariably unstreaked, but very rarely has a few small distinct spots or streaks. **1stCY** Uniformly fresh plumage, including primaries, tertials and tail, but some have slightly frayed tips to latter. Tertials broadly edged buff. Breast nearly always speckled with dark brown rather rounded spots (sometimes ill-defined and difficult to discern in the field).

BIOMETRICS L 11.5–13 cm; **W** ♂ 64–70 mm (n 24, m 67.8), ♀ 61–68 mm (n 18, m 65.4); **T** ♂ 46–53 mm (n 24, m 49.5), ♀ 43–52 mm (n 18, m 47.6); **T/W** m 72.9; **B** ♂ 13.0–14.8 mm (n 23, m 14.3), ♀ 12.4–14.8 mm (n 17, m 13.9); **Ts** 19.3–22.1 mm (n 42, m 20.7). **Wing formula: p1** < pc 1–5 mm; **p2** < wt 0–3 (4) mm; **p3** longest (p2 sometimes equal); **p4** < wt 1–4 mm; **p5** < wt 3.5–6 mm; **p6** < wt 5.5–10 mm; **p7** < wt 7.5–12 mm; **p10** < wt 14–18 mm; **s1** < wt 16–21 mm. Emarg. p3, sometimes faintly also on p4.

GEOGRAPHICAL VARIATION & RANGE Monotypic. – N & C Europe from Britain, France and Fenno-Scandia east to Yenisei in Siberia and to Balkhash area in E Kazakhstan, south to Turkey; winters sub-Saharan Africa.

Juv, England, Jul: recently fledged juv usually has buff and dark areas more boldly patterned, including stronger lateral crown-stripes and paler central crown than ad, thus potentially inviting confusion with Aquatic Warbler. However, juv Sedge lacks Aquatic's black and ochre-yellow tramlines on mantle and dark-streaked rump and uppertail-coverts, as well as the solidly patterned head. (G. Thoburn)

1stW, Israel, Oct: breast typically speckled dark. Separated from most congeners by rather long primary projection; whitish-buff supercilium (tapering narrowly at rear) bordered by dark eye-stripe and lateral crown-stripe, with paler central crown that is streaked dark, upperparts faintly dark-streaked, but brighter tawny-rufous rump is unstreaked (or only weakly streaked). Leg colour varies greatly, from pinkish-brown (as here) to bluish-grey (see Aquatic Warbler). (L. Kislev)

PADDYFIELD WARBLER
Acrocephalus agricola (Jerdon, 1845)

Fr. – Rousserolle isabelle; Ger. – Feldrohrsänger
Sp. – Carricero agricola; Swe. – Fältsångare

This smaller version of Common Reed Warbler, with similar plumage and habits, shares a preference for vast reedbeds at lakesides, although it seems to like nearby open patches with broken, flattened reeds best. (It has very little association with 'paddyfields'.) Mainly an Asian species, breeding only in SE Europe, and is only rather local and rare there, Paddyfield Warbler winters in S Asia, mainly in India east to Assam. However, young birds sometimes appear in W Europe in late summer and early autumn, it being practically an annual visitor in Britain.

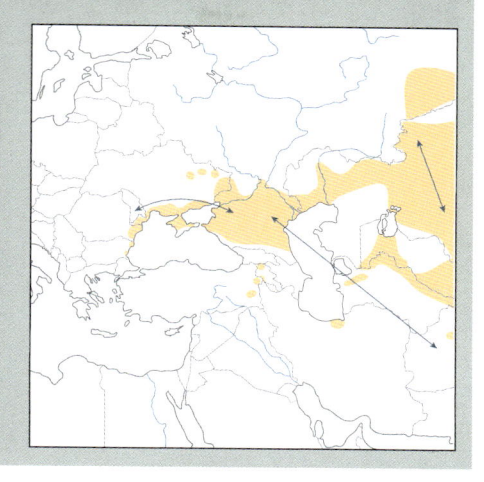

Turkey, Jun: smaller than Common Reed Warbler, with shorter/rounder wings, but proportionately longer tail and shorter bill. Long, prominent whitish supercilium, amplified by fairly dark eye-stripe and dark crown, bicoloured (black-tipped and pink-based) bill, and paler iris. After completing winter moult, impossible to age. (V. Legrand)

Bulgaria, May: typical jizz and pale supercilium still prominent despite being slightly broken behind eye. Also whitish undertail, often a characteristic of species. Some ad can have quite dark bill. (B. Nikolov)

IDENTIFICATION The Paddyfield Warbler is easily mistaken at a quick glance for a Reed Warbler, as the slightly *smaller size* of Paddyfield is not always apparent. They have similar warm tawny colours with a *rufous-tinged rump*, *unstreaked back*, and they share the same habitat: Paddyfield is also a true reed-dwelling species. However, once a better look is secured, it does appear smaller with quicker and more restless movements. Structurally, it has *shorter and more rounded wings*, making the *slightly longer tail* appear more obvious. The *bill is a little shorter* too than in Reed Warbler, making the head look more rounded. A common encounter with a singing bird is of a small and lively bird clinging to a reed, swaying from side to side with jerky movements, singing energetically with raised crown-feathers, the *crown then appearing darker than rest of plumage*. After a while it will make a short sally, only to 'crash-land' on a new reed lower down. In certain lights it can look rather drab grey-brown differing from Reed Warbler mainly by its *more contrasting head pattern*. Once seen well you should notice (i) the rather *long and prominent whitish supercilium* reaching from forehead to behind the eye (very few exceptions among young birds, which may have a more indistinct one), enhanced by *dark eye-stripe* and *dark crown*, especially laterally; (ii) the pink-based but *dark-tipped lower mandible* (a few exceptions among young have poorly-developed dark tip), making the bill look more *bicoloured* and *black-tipped*; and (iii) the *paler eye* (again, young birds can have a somewhat darker brown iris, not that different from Reed Warbler). To these add (iv) a *paler brown uppertail* with pale (!) shafts distally; and (v) a generally *slightly paler colour* overall. There is a fair amount of colour change between freshly moulted birds that are more rufous and cinnamon-tinged, and worn ones, which are more grey-brown and lack all warm tones, and the extremes can appear quite different. Legs and toes usually look rather *pale pink*, paler than on Reed Warbler, but toes are slightly more greyish, and claws pencil grey.

VOCALISATIONS The song is a fast, prolonged chattering warble of moderate strength. It most resembles a subdued song of Marsh Warbler, sharing with this species the high speed and varied performance with many harsh and squeaky notes, and some mimicry of other birds. If you are familiar with Marsh Warbler song you will usually be able to pick out Paddyfield's 'lower volume', higher-pitched voice, and more even, fast pace with fewer imitations. Its song generally lacks both the tempo changes and sudden crescendos of Marsh, and the *Prinia*-like trills and characteristic repeated, harsh *zi-zaaih* notes of that species are also absent. Still, standing at the edge of a lakeside reedbed, the chorus of singing birds can be quite confusing, and apart from Marsh, at least Sedge and Moustached Warblers, sometimes even a spirited Reed Warbler, will need eliminating before you can note a certain Paddyfield. – Calls include a 'fat', clicking note, *chlek*, a drier and sharper clicking *chik*, and a drawn-out, slurred or squeaky, nasal *cheehr*.

SIMILAR SPECIES The most common identification problems have been treated under Identification and Vocalisations. An additional pitfall is *Booted Warbler* in autumn, which can resemble some dull Paddyfield. A few young Paddyfield Warblers lack the characteristic rufous rump/uppertail-coverts, are confusingly dull overall, and may have a less prominent pale supercilium than most; these require care and a check of all characters to be safely separated from Booted. Note the shorter primary projection (roughly half tertial length or a little less) of Paddyfield, and its paler brown flight-feathers (the outertail especially is paler). Also, tertials of Paddyfield have more contrasting pale outer edges (before becoming too worn), and eyes are usually visibly paler. Additional clues are slightly longer legs, longer and more rounded tail in Paddyfield, narrower tail-feathers (especially the central pairs, and distally), generally slightly more prominent supercilium (broad also over eye, where Booted often has a hint of a 'waistline' indention), and distally rather pale

Turkey, Jun: long whitish supercilium enhanced by hint of dark lateral crown. Buffy body-sides, but whiter throat. Note also comparatively short primary projection, short bill with dark tip and pale brown iris, all important differences from the average Reed Warbler sharing the same habitat. (E. Yoğurtcuoğlu)

1stW, Scotland, Aug: note short primary projection, whitish supercilium emphasised by dark lateral crown and eye-stripe, and generally pale plumage with cinnamon tinge (a character of moderately fresh individuals). Legs pale pink, outer part of hind toe and claws greyer. Aged by fresh wing; note less contrasting bill colours, another characteristic of young. (S. McElwee)

1stW, Scotland, Oct: a rather rustier bird. Warmer body contrasts with whiter throat/chest, extending onto sides of throat and as greyish shade behind ear-coverts. Correct identification relies on accurate assessment of head pattern, short primary projection, short bill, and bill and leg colours. (S. McElwee)

1stW, Scotland, Sep: a rather dull/pale individual, with long supercilium. With wear, dark crown sometimes tinged greyish, and especially young birds have less distinctly two-coloured lower mandible. This bird already has medium grey iris. Aged by still moderately and evenly fresh plumage. (H. Harrop)

shafts to outer tail-feathers and sometimes tertials (Booted invariably dark-shafted). For in-hand identification, or in good photographs, note also that Paddyfield has a little shorter and narrower outermost short primary (p1), and 3–4 strong and well-curved rictal bristles each side of gape (Booted: longer and broader p1, and 4–6 thinner and straighter bristles).

AGEING & SEXING Moderate age-related differences, but these often clouded by more marked seasonal and individual variation. Sexes alike in plumage. – Moults. Complete post-nuptial moult of ad in winter quarters, usually soon after autumn migration, mainly Sep–early Nov, but a few start already in late Jul–Aug, and others later (Nov–Dec; presumably mainly young), hence spring birds display a varying degree of abrasion to wing-feathers. Juv apparently invariably moults body and wing-coverts on breeding grounds (though extent variable), and then undergoes complete moult in winter quarters. As to complete moult, it is reasonable to assume that late-moulting birds are mainly young. Both age groups have a partial spring moult, but judging from the wear of some birds this might be very restricted or inhibited in some. – **SPRING** Sexes and ages probably inseparable, but iris colour may sometimes prove useful, birds with darker brown eyes presumably 1stS, although this requires study. – **AUTUMN Ad** Flight-feathers with abraded tips (no wing-moult, or moult just started), body and head mainly worn and greyish-brown, old tertials lack pale outer edges. Iris rather light brown of varying shade. Tip of bill rather blackish. **1stCY** Flight-feathers fresh. (Moult of primaries presumably invariably starts in late autumn or early winter.) Tertials fresh with contrasting paler outer edges. Iris medium or rather dark brown, or olive grey-brown. Bill colours less contrasting, tip not so dark.

BIOMETRICS L 12–13 cm; **W** ♂ 55–62.5 mm (n 49, m 58.9), ♀ 54–60 mm (n 34, m 56.9); **T** ♂ 48–60 mm (n 48, m 53.9), ♀ 48–57 mm (n 32, m 51.1); **T/W** m 90.9; **TG** 5–11 mm; **B** 13.0–16.0 mm (n 103, m 14.4); **BD** 2.3–3.5 mm (n 66, m 2.8); **Ts** 20.1–22.6 mm (n 70, m 21.5). **Wing formula: p1** > pc 0–5.5 mm, < p2 26–32 mm; **p2** < wt 3–6.5 mm, =6/7 or 7 (55%), =5/6 or 6 (30%), or =7/8 or 8 (15%); **pp3–4** longest, **p5** sometimes too, or < wt 0.5–3 mm; **p6** < wt 1–5 mm; **p7** < wt 3–7 mm **p10** < wt 7.5–12.5 mm; **s1** < wt 9–14 mm; **notch p2** 11–15 mm, < ss 2–9.5 mm; **notch p3** 9–13 mm, = p9/ss, but often faint only. Emarg. pp3–5, but on p5 sometimes only faintly, and in 1stW rarely lacking entirely.

GEOGRAPHICAL VARIATION & RANGE Monotypic. – E Europe west and north of Black Sea, E Turkey, S Russia, SW Siberia, Kazakhstan, W Mongolia, Central Asia; winters extreme SE Iran, S Pakistan, Indian Subcontinent. – There is a fair amount of variation in plumage colours related to age and season, and we have been unable to confirm the distinctness of described subspecies. Even if there is a slight tendency for breeding birds in the north and west of the range to be slightly darker and less rufous-tinged, this is only a rather insignificant difference which does not hold for anything near 75%. Numerous breeders in Central Asia appear identical to E European ones, and vice versa. (Syn. *capistratus*; *septimus*. Note that the former is sometimes referred to as *brevipennis*, but within *Acrocephalus* this name is preoccupied by the Cape Verde Warbler.)

REFERENCES Bradshaw, C. (1997) *BB*, 90: 142–147. – McAdams, D. & Jännes, H. (2000) *Birdwatch*, 99 (Sep): 26–30. – Schulze-Hagen, K. & Barthel, P. H. (1993) *Limicola*, 7: 1–34. – Votier, S. (1997) *Birdwatch*, 64 (Oct): 27–32.

Paddyfield Warbler (top left: Scotland, Jun; bottom left: England, Oct) versus Booted Warbler (top right: Estonia, May; bottom right: Germany, Oct): Paddyfield has proportionately longer tail and very short primary projection (slightly shorter than 1/2 of exposed tertial length, as opposed to c. 2/3 in Booted). Dark eye-stripe better-marked than Booted, while more prominent and generally evenly broad supercilium extends well behind eye, whereas in Booted it tends to narrow over fore edge of eye, thereby appearing to have a 'waistline'. Iris often paler than Booted, which always has all-dark eye. Legs of Paddyfield slightly longer, and tail-tip rounder, with central feathers often noticeably narrower. Pale (not dark) shafts to outer tail-feathers and tertials. Also, Paddyfield's undertail-coverts are longer than in Booted, the bill is stouter (typically shorter/finer in Booted) and head flatter (Booted has domed head, more *Phylloscopus*-like). (Top left: H. Harrop; bottom left: R. Tidman; top right: U. Paal; and bottom right: A. Halley)

BLYTH'S REED WARBLER
Acrocephalus dumetorum Blyth, 1849

Fr. – Rousserolle des buissons; Ger. – Buschrohrsänger
Sp. – Carricero de Blyth; Swe. – Busksångare

Whereas Blyth's Reed Warbler is a rarity in W and C Europe, it is a very common bird over much of S Russia and S Siberia, and is frequently seen on passage through Central Asia. It is not invariably found in close proximity to water like many of its relatives, nor is the species fond of reedbeds. Typically, it occurs at forest edges or glades with undergrowth and lush meadows with bushes and scattered broadleaf trees. Since this is a common habitat, there is no shortage of suitable territories. It arrives late to its NW Russian and Finnish breeding sites, in late May or early June, which is due to its long migration; the species winters mainly in India and Sri Lanka.

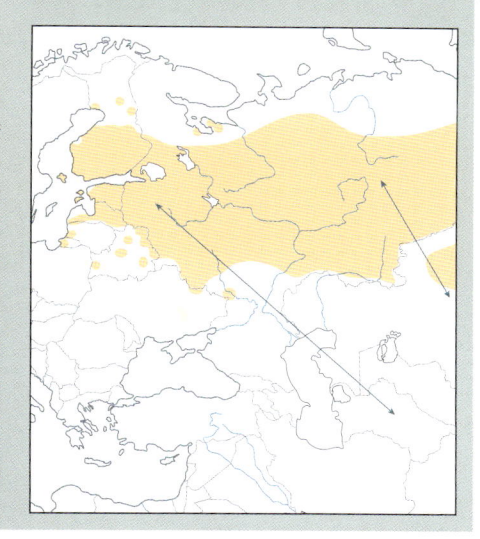

Denmark, Jun: rather bland-looking, but several features permit field separation from two close congeners, Common Reed and Marsh Warblers. Note shorter primary projection (usually up to six primary tips visible in tightly folded wing), especially differing versus long-winged Marsh. Pale supercilium slightly more pronounced. Upperparts greyish-brown with rather plain wing. Birds in spring impossible to age. (E. Ødegaard)

IDENTIFICATION Roughly same size and shape as Reed Warbler, and similarly *unstreaked brown* without a very prominent pale supercilium. Spring birds usually sing, making identification reasonably easy, but silent birds need to be carefully checked for full set of characters to eliminate both Reed and Marsh Warblers. Bill fine and pointed, similar to Reed (not short and stout like Marsh, although difference subtle), giving head a slim, pointed look. *Wing slightly more rounded and a little shorter*, proportionately, than both other species. Often crouches with slightly raised bill and tail (the 'banana posture'), but this is also practised by Reed and Marsh, only perhaps less frequently. Persistently twitches and semi-spreads tail while moving through canopy. Important to note colours for reliable identification: adults are *greyish-brown above* with only subtle olive tinge in some (lacking the rather bright greenish hue of most spring Marsh) and *only moderately warmer brown rump* (lacking bright rufous tones of Reed); *wing is rather uniformly brown* in most adults, lacking prominent dark centres and contrasting bright edges to tertials, primary-coverts and alula (but slightly more contrast shown by some juveniles). *Lower mandible nearly always has a darkish smudge near tip*, lacking in the other two species. *Legs* of adults are dull pinkish-grey or brown-grey and can *look rather dark* in the field, usually marginally darker than in Reed and clearly darker than Marsh. Young have greyish legs with pale yellow soles. Supercilium similar to Reed or slightly longer and more pronounced, reaching just behind the eye in most. Young in autumn are surprisingly similarly warm brown with rufous-tinged rump in all three species, making identification even more challenging. The somewhat shorter, more rounded wing and shorter primary projection, in combination with dark-shaded lower mandible tip, rather dark legs, rather prominent supercilium and comparatively uniform wing might suffice, but note that first-years have on average slightly more contrasting tertials and alula than adults.

VOCALISATIONS Full song is loud and quite characteristic in its slow, 'reciting' pace and in that it repeats phrases exactly several times, sometimes as many as 10–12, more often 4–8, even complex ones, nearly always interspersed by two or three (at times more) dry, clicking call notes, *chek-chek*. Mimicry of other birds masterly, but less striking than in Marsh Warbler. More characteristic is recurring 'scale exercises' with clear or whistled notes and marked scale steps, such phrases repeated several times identically, *loh-lü-lii-ah* (three rising notes but last falling) or similar. Sometimes moderately increases speed of delivery, and sometimes omits the clicking notes, or omits most repetitions of phrases; such songs often heard from migrants ('loud sub-song'), but in south of range may represent an habitual song variant (Marova *et al.* 2010). Still, very different from full Marsh Warbler song, and separation should be unproblematic given some experience. (More difficult to separate such slightly faster and more varied songs from Large-billed Reed Warbler, see below, and this requires careful documentation and comparison.) – Calls; either the clicking note from the song, *chek*, or a drawn-out, slurred trill, *chrrrt*.

SIMILAR SPECIES The most common identification problems, separation from *Reed Warbler* and *Marsh Warbler*, have been treated under Identification and Vocalisations. An additional pitfall is the eastern race *fuscus* of Reed Warbler, which is sometimes more greyish-brown and less brightly rufous-tinged above than race *scirpaceus* of Reed, thus approaching Blyth's Reed in appearance. Here structural differences help, *fuscus* having a more pointed and longer wing than Blyth's Reed, and in fresh plumage exposed primaries of *fuscus* are usually darker grey with fine whitish tips (Blyth's Reed has paler grey-brown primaries without prominent paler tips). – Young *Paddyfield Warbler* separated from Blyth's Reed by smaller size, usually more contrasting bill pattern with a more pronounced dark tip to lower mandible (although some are

Finland, Jun: moderately worn spring migrant with characteristic cold greyish-tinged brown upperparts. This bird has a paler bill than most, but there is still a dark smudge on lower mandible. Supercilium pronounced in front and reaching a little behind eye. Wing is rather uniform, and note the relatively short primary projection with three primaries emarginated, the outer one falling well short of the secondary tips. (D. Occhiato)

Lithuania, Jun: differs from Marsh and Reed Warblers by shorter primary projection, plainer olive-brown wing, and on average longer and better marked supercilium reaching well behind eye. This bird is noteworthy by lacking a clear dark tip on lower mandible. Note also the tick under the lower mandible near gape. (V. Jusys)

Sweden, Jun: in Jun, primary tips can be substantially worn. Nevertheless, rounded wing shape is visible, with three emarginated primaries (versus one in Marsh and nearly all Common Reed Warblers), emarginations falling well inside tips of secondaries (opposite or outside in Marsh and Common Reed). Note also shortness of outermost long primary. (M. Hellström)

Juv, Finland, Aug: recently fledged juv usually has brighter and cinnamon-buff tones, with more obvious dark parts to wing, especially darker-centred tertials, forming bolder pale fringes. Supercilium rather poorly developed, and dark smudge on lower mandible obvious but restricted to outermost tip at this stage. (J. Normaja)

quite similar), longer and more prominent pale supercilium (very few exceptions), darker eye-stripe, and more contrast between dark centres and bright brown outer edges of tertials. — For separation of extralimital but extremely similar Large-billed Reed Warbler, see Taxonomic note.

AGEING & SEXING Moderate age-related differences in autumn. Sexes alike in plumage. — Moults. Complete post-nuptial moult of ad generally in autumn stopover areas (e.g. in NW India) prior to reaching its winter quarters proper, usually mid Sep–Nov (Dec), but a few start already in Aug (Gaston 1976). No regular post-juv moult on breeding grounds, but autumn moult might commence just prior to or during migration. Both age groups have a partial spring moult of variable extent. — SPRING Ageing apparently impossible. It has been suggested (BWP) that a few birds with odd retained older flight-feathers could be 1stS, but this has not been studied closely, and no such birds encountered by us, so perhaps affects only a tiny percentage. — AUTUMN Ad Flight-feathers with abraded tips. Upperparts mainly worn and greyish-brown, rump only subtly paler and warmer brown. **1stCY** Flight-feathers fresh. Upperparts tawny-brown, rump tinged rufous.

BIOMETRICS **L** 12.5–13.5 cm; **W** 59–67 mm (n 215, m 62.7); **T** 45–56 mm (n 197, m 51.5); **T/W** m 82.0; **B** 15.0–18.2 mm (n 201, m 16.7); **B**(n) 7.2–10.4 mm (n 184, m 8.9); **BD** 2.7–3.7 mm (n 118, m 3.3); **BW** 2.8–4.1 mm (n 195, m 3.5); **Ts** 20.2–23.5 mm (n 196, m 22.1); **HC** 5.2–7.2 mm (n 194, m 6.1); **MC** 4.0–6.4 mm (n 181, m 5.3). (Bill-width measured at outer edge of nostrils.) **Wing formula:** **p1** < pc 3 mm, to 2.5 mm >, < p2 30–37 mm; **p2** < wt 3–6.5 mm, =6/7 or 7 (56%), =5/6 or 6 (42%), or =7/8 or 8 (2%); **pp3–4**(5) longest, **p5** sometimes too, or < wt 0.5–3.5 mm; **p6** < wt 1–6 mm; **p7** < wt 4–9 mm; **p10** < wt 8.5–14 mm; **s1** < wt 10.5–16 mm; **notch p2** 10.5–15.5 mm, < ss 0.5–8.5 mm; **notch p3** 8–12.5 mm, = pp6/10, but often only faint, and can be missing entirely in 1stW. Emarg. pp3–5 in ad, but sometimes weaker on p5, in 1stW pp3–4, sometimes a hint also on p5, and very rarely only distinct on p3 and weaker on p4.

GEOGRAPHICAL VARIATION & RANGE Monotypic. — S Finland, N Baltic States, NE Belarus, N Ukraine, Russia, S Siberia east to upper Lena, south to N & SE Kazakhstan, N Tien Shan; winters Indian subcontinent, Sri Lanka, Burma. Southern breeding range in Central Asia needs to be better established. — (Syn. *A. montanus*.)

TAXONOMIC NOTE Considering the great similarity between Blyth's Reed Warbler and extralimital Large-billed Reed Warbler *A. orinus* (apparently rare, still little-known species, breeding locally in mountains of S Central Asia and apparently wintering in Burma and Thailand, possibly also India; Svensson et al. 2008, Timmins et al. 2009, Ayé et al. 2010, Arkhipov et al. 2011), a description of the latter is given here as a service to researchers in Asia and in museum collections. Obviously, Large-billed Reed Warbler has been overlooked in the past due to this similarity:

LARGE-BILLED REED WARBLER *A. orinus* (Oberholser, 1905) (NE Afghanistan, SE Tajikistan, possibly adjacent areas; presumed to winter Burma, Thailand and apparently N India; reported also from C & S India but neither in midwinter nor confirmed records). Monotypic. Size and plumage as Blyth's Reed Warbler (possible subtle average difference that Large-billed Reed is more saturated buff below gives little guidance, and paler lores and possibly slightly more subdued supercilium than in Blyth's Reed are difficult to use in field), differing subtly mainly in structure and bare-parts coloration. Bill shape in dorsal view usually broader, more wedge-shaped and has 'fuller' tip, lacking more attenuated tip of Blyth's Reed (but a few *dumetorum* approach this), and lower mandible ventrally more rounded, less 'keeled'. Bill longer, usually obviously, but there is slight overlap in extremes of the two species. Tarsi and claws longer, especially hind claw, but slight overlap noted in both measurements. Shape of hind claw thinner and more pointed, *dumetorum* typically having rather thick and blunt hind claw in lateral view (but some overlap in shape, too). Lower mandible usually all pinkish-yellow, lacking the dark smudge near tip present in >95% of *dumetorum* (but a few have faint hint of darker patch). Iris colour seems to average slightly lighter and more olive-grey (less dark and warm brown). Proportionately longer tail but shorter primary projection. Wing marginally blunter on average; p2 never =5/6 (as in 20% of *dumetorum*). Song intermediate between Blyth's Reed and Marsh Warblers, repeating phrases as Blyth's Reed but less exactly and fewer times, is a little faster and more varied and includes a few 'freer' passages in Marsh Warbler style. Contains many trilling sounds (with 'rolling r'). **W** 59–65.5 mm (n 27, m 62.2); **T** 50.5–60.5 mm (n 28, m 55.3); **TG** 6–8 mm (m 6.9); **T/W** m 88.9; **B** 17.3–20.6 mm (n 29, m 18.7); **B**(n) 8.3–10.7 mm (n 27, m 9.9); **BD** 3.0–4.1 mm (n 26, m 3.5); **BW** 3.4–4.2 mm (n 27, m 3.8); **Ts** 22.6–24.1 mm (n 28, m 23.4); **HC** 6.7–8.5 mm (n 28, m 7.3); **MC** 5.6–7.1 mm (n 27, m 6.3). **Wing formula:** **p1** > pc 3 mm, to 1.5 mm <, < p2 27.5–33 mm; **p2** < wt 3.5–6.5 mm, =6 or 6/7 (50%), =7 or 7/8 (46%), or =8 or 8/9 (4%); **pp3–4** longest, **p5** sometimes longest too, or < wt 0.5–2 mm; **p6** < wt 2–4 mm; **p7** < wt 3.5–6 mm; **p10** < wt 9–12 mm; **s1** < wt 11–14 mm; **notch p2** 13–15.5 mm, < ss 3.5–9 mm. Emarg. pp3–5 in ad, only exceptionally slightly weaker on p5; sometimes a hint of emarg. also on p6. (Bill-width measured at distal edge of nostrils.) — Since each measurement taken alone is hardly statistically significant for

1stW, Scotland, Nov: especially later in season, young can wear to more greyish-olive and identification is more challenging, e.g. versus duller *fuscus* Common Reed. Shorter primary projection, dark-shaded lower mandible tip, rather prominent supercilium and uniform wing offer useful clues. Note Blyth's Reed's colder underparts, less sullied buff than Common Reed. (H. Harrop)

species determination, a combination of two or more gives more reliable result. By plotting lengths of HC (x axis) and B (y axis) on a graph, two well-separated clusters are recovered (Svensson et al. 2010). The combination of four measurements into a 'multiple character value' (MCV) gives 97.5% separation of the two species: add lengths of Ts, HC and B, and subtract s1 < wt. **MCV** in orinus is 34.3–40.2 (n 24, m 36.9), in dumetorum 27.3–34.9 (n 192, m 31.8) (Svensson in prep.).

REFERENCES Arkhipov, V. et al. (2011) PLoS ONE, 6(4): e17716. – Avé, R., Hertwig, S. T. & Schweizer, M. (2010) J. Avian Biol., 41: 452–459. – Bensch, S. & Pearson, D. (2002) Ibis, 144: 259–267. – Bradshaw, C. (2001) BB, 94: 236–245. – Ellis, P., Jackson, P. & Suddaby, D. (1994) BW, 7: 227–230. – Gaston, A. J. (1976) Ibis, 118: 247–251. – Golley, M. & Millington, R. (1996) BW, 9: 351–353. – Harrap, S. (1989) BW, 2: 318–324. – Helbig, A. & Seibold, I. (1999) Mol. Phyl. & Evol., 11 (2): 246–260. – Marova, I. M., Ivanitskii, V. V. & Veprintseva, O. D. (2010) Biology Bulletin, 37: 1–15. – Pearson, D. J., Kennerley, P. R. & Bensch, S. (2008) BBOC, 128: 136–138. – Svensson, L. (in prep.) BBOC. – Svensson, L. et al. (2008) J. Avian Biol., 39: 605–610. – Svensson, L. et al. (2010) Ibis, 152: 323–334. — Timmins, R. J. et al. (2009) BirdingASIA, 12: 42–45. – Timmins, R. J. et al. (2010) Forktail, 26: 9–23. – Votier, S. & Riddington, R. (1996) BW, 9: 221–224.

1stW, Norway, Oct: young often have confusingly warmer brown upperparts and stronger-patterned tertials than ad, and can resemble Common Reed and Marsh Warblers. This bird is still separated by bill colour and relatively short primary projection, while the plain brown primaries lack prominent paler tips. Note deep emargination on p3 and p4, but only hint on p5 (quite normal for 1stY). (J. Bell)

1stW, Sri Lanka, Dec: Blyth's Reed Warbler can sometimes resemble Sykes's or Olivaceous Warblers, especially the former. However, Blyth's Reed is slightly darker grey-brown above, has different bill shape, more rounded tail, more curved outline to folded primaries, and relatively longer undertail-coverts. This bird has rather unusual broad pale loral area. (H. Shirihai)

Large-billed Reed Warbler *A. orinus*, West Bengal, India, Apr: a potential vagrant to Iran, this species is morphologically very similar to Blyth's Reed Warbler, but differs markedly in genetics and song, but also in its longer bill, tarsus and claws. (S. K. Sen)

Large-billed Reed Warbler *A. orinus*, Thailand, Mar: compared to Blyth's Reed, the longer, stronger bill is usually appreciable, especially the 'swollen' outer part; also lower mandible is usually entirely pale flesh, having only hint of dark smudge if any. Note the rounder wings with even shorter primary projection, making the tail appear proportionately long. Legs generally brownish-grey. (P. Round)

Large-billed Reed Warbler *A. orinus*, Afghanistan, Jun: bill shape in dorsal view (though not shown here) has a broader, more wedge-shaped appearance with convex outline. Note that iris is on average lighter and more olive-grey than Blyth's Reed. Birds in spring seemingly wear olive-grey above and white below, further obscuring differences from the near-identical Blyth's Reed Warbler. (S. N. Mostafawi/WCS Afghanistan)

MARSH WARBLER
Acrocephalus palustris (Bechstein, 1798)

Fr. – Rousserolle verderolle; Ger. – Sumpfrohrsänger
Sp. – Carricero políglota; Swe. – Kärrsångare

The Marsh Warbler inhabits dense stands of herbs, tall weeds, nettles and low bushes. It follows that—despite its name—it is not dependent on water. It is a master of mimicking other birds' sounds, combining them to form its own attractive song, which is mainly heard by night and at dusk and dawn. It migrates on a broad front to tropical E Africa in late summer and early autumn, making something of a loop by returning in spring—in late April and through May—along a more easterly route, being very abundant on the Arabian Peninsula, before entering Europe from the south-east.

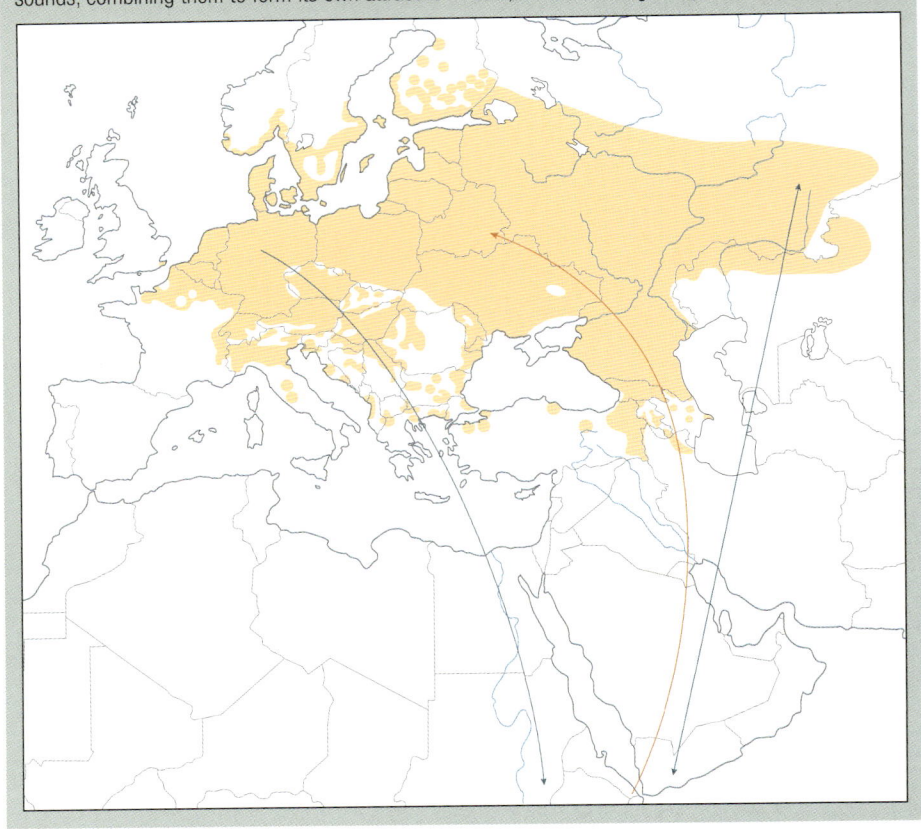

IDENTIFICATION Very similar to Reed Warbler in size and general shape, and is similarly *unstreaked above* and lacks a prominent pale supercilium, having only a short and indistinct hint of one. Further, both Marsh and Reed have pinkish-yellow lower mandible without any dark tip. However, Marsh differs in spring in a decidedly *greenish sheen to medium grey-brown upperparts*, especially on mantle (though beware more grey-brown impression to abraded birds in summer), *rump only slightly warmer ochre-brown* (not brightly rufous or cinnamon like Reed). Underparts largely yellowish-cream with *whitish throat*, which can be rather striking on a singing bird. Bill on average *a little shorter and stouter* than in Reed (some overlap, but often a useful clue), which renders the bird a slightly more 'kind' appearance with *rounder head*. Legs of adults on average a shade *paler pinkish-brown* than Reed (but some overlap). Note also long, pointed wing with *longest primary projection* of the two (though some overlap), and on average darker primaries with more obvious white tips in fresh plumage (again, some overlap). Young in autumn are frustratingly similar, Marsh as well as Reed being warm brown above with rufous-tinged rump, and identification in the field must often rest on a combination of structural differences, and realistically is sometimes impossible (rarely even in the hand!). There is a tendency for young Marsh to be slightly paler brown above, with a slight hint of olive in the mantle, with less strong rufous tones on the rump, which may at times be visible during close observations in good light. Legs of young Marsh are even paler and more pinkish than in adults (only rarely with a faint grey tinge), and this is a supplementary character (young Reed: grey-brown, sometimes with lead-grey sheen, thus usually darker and greyer than Marsh), as are the claws, which are rather uniformly straw-coloured (young Reed: more bicoloured, grey-brown dorsally with paler, yellowish underside). In spring and early summer, song is best means of identification.

VOCALISATIONS Famous for its accomplished song, consisting (entirely!) of mimicry of other birds. Almost 200 different species identified among imitations, but each ♂ uses on average 75 species, and some of these only rarely. The song is a fast-flowing mixture of many easily identified European bird sounds like Blue Tit alarm, Bee-eater *Merops apiaster* flight call, various Chaffinch calls, Blackbird alarm and call, Magpie alarm, Barn Swallow alarm, Linnet flight call, Quail *Coturnix coturnix* song, Jackdaw calls, Common Gull *Larus canus* call and alarm, Greenfinch calls, etc., mixed with unfamiliar and often more harsh or trilling calls which

Netherlands, Jun: bill on average slightly shorter and stouter and head rounder than Common Reed, while legs are on average paler pinkish-brown, with subtly shorter yellowish-brown claws. However, some variation and overlap. Lower mandible invariably pinkish-yellow lacking any dark. (N. D. van Swelm)

Switzerland, Jun: shares with its closest congener, Common Reed, poorly-developed short pale supercilium, but Marsh has longer primary projection and in spring/summer typically has uniform greenish-olive hue above, with rump scarcely brighter (lacking rufous or cinnamon of Reed). Underparts tinged yellowish-cream with whiter throat. As it moults later in winter, still in Jun wing is fresher than most Common Reed, with usually more conspicuous pale primary tips that are more evenly spaced (up to eight visible). In this image, it is possible to appreciate that the emargination on p3 falls well outside secondary tips. (H. Shirihai)

BUSH AND REED WARBLERS

Sweden, May: in different habitats, circumstances and light conditions, like on migration, impression of species can differ from singing bird on breeding grounds. However, most diagnostic features are useful: pointed wing with long primary projection and fine pale primary tips, and buffish cream-tinged underparts. Depending on light, olive tinge of upperparts not always obvious. Compared to Common Reed head rounder, and bill averages shorter and stouter. (S. Persson)

are also imitations, of African birds, which the young ♂ Marsh Warbler has heard during its first autumn migration and winter. Although the song's pace varies, and can include 'thoughtful' or 'absent-minded' repetitions of a few calls at slower speed (vaguely recalling Blyth's Reed Warbler), it soon picks up its usual high speed and bravura, exploding in a variety of mimicked sounds, now and then intermixed with some almost *Prinia*-like dry trilling sounds, *prri-prri-prrü-prri-*…, and very characteristic hoarse, drawn-out *ti-zaaih, ti-zaaih*. Song differs clearly from Reed Warbler by its speed, variation and abundant imitations. Some Reed Warblers mimic a little more than most, and can mix spells of more energetic and varied song from time to time, but once you have listened a minute or more, Reed is recognised for its 'lecturing' performance and overall more monotonous song. – Calls resemble those of Reed, either a 'muffled clicking' *chrek* or *tret*, sometimes repeated in long series, or a dry, slurred trilling, *chrrreh*.

SIMILAR SPECIES Separation from *Reed Warbler* in the field is treated under Identification and Vocalisations, but see also Biometrics for help with tricky handled birds or specimens. – In certain lights, a pale *Melodious Warbler* is also a possible pitfall. Not all Melodious are strongly lemon-yellow below; this is either due to variation or to light effects related to reflections from surrounding vegetation. Still, prolonged views will generally reveal the more yellow underparts of Melodious, and note the pale loral area, brownish-grey legs and more square-ended tail. – Due to its greyish-green upperparts, adults may resemble *Blyth's Reed Warbler*, but the longer and more pointed wing of Marsh (primary projection about as long as exposed tertials, or even slightly longer) in combination with slightly shorter bill without a greyish 'smudge' at tip of lower mandible should prevent a mistake. Further, leg colour in Marsh is on average slightly paler and more pinkish. – Theoretically, Marsh could be confused with *Garden Warbler*, especially on migration, when habitat choice is less indicative of species. Note pinkish-tinged pale brown legs of Marsh (Garden: grey), entirely pinkish-yellow lower mandible (Garden: entire bill mainly grey, base of lower mandible light grey), bright yellow gape (Garden: pink), and paler and more yellowish-white underparts (Garden: most of underparts sullied olive-grey).

AGEING & SEXING Slight age-related differences in autumn. Sexes alike in plumage. – Moults. Partial post-nuptial moult of ad starts soon after breeding, mainly Jul–Aug, with partial replacement of head and body, generally

Unstreaked *Acrocephalus* warblers in spring, Marsh (top left: Russia, Jun), 'Caspian Reed' *A. scirpaceus fuscus* (top right: United Arab Emirates, Mar), Common Reed *A. s. scirpaceus* (bottom left: Netherlands, May), and Blyth's Reed Warbler (bottom right: Finland, May): all of these are very similar and require careful study. In spring, Marsh is greenish-tinged above, but this is not always obvious. Still, Common Reed usually readily separated by bright rufous upperparts, in particular rump. Primary projection in Marsh longest, with primary spacing more even and wider; also, pale tips to primaries as a rule more obvious than in Common Reed. However, in all of these characters, *fuscus* Reed is intermediate. Bill of Marsh generally stouter, less narrow-based than both Reeds, but much overlap. Finally, Blyth's Reed, unlike rest, has near-diagnostic dark-shaded lower mandible tip (often less easily seen than here), on average better-developed supercilium, notably shorter primary projection (usually no more than six visible primary tips) and hardly any contrast between wing and rest of upperparts. Leg colour often supportive: legs of Marsh typically paler, and those of Blyth's often greyer, though beware of changing impression due to light as demonstrated above. (Top left: O. Khromushin; top right: P. Arras; bottom left: L. van Loo; bottom right: T. Muukkonen)

Marsh Warbler (left) **and 'Caspian Reed Warbler'** *A. s. fuscus*, **United Arab Emirates, May**: separation of Marsh from *fuscus* Reed can be extremely difficult, even impossible, without close and prolonged views, as well previous experience. On migration, they often give brief views, while primary projection, spacing and pattern of pale tips do not always clearly separate the two, and the degree of contrast in the tertial fringes and leg colour (pinkish-brown with more yellowish pigments in Marsh) only weakly differentiate them. Nevertheless, as shown here, Marsh when fresh is more saturated buffish-cream below and tinged more uniformly olive above (versus the rufous-ochre rump and uppertail-coverts of *fuscus*). (Left: O. Campbell; right M. Barth)

of limited extent, then head/body-moult suspended and continued after migration on stopover or in northerly winter quarters, usually Oct–Nov. Whether there is any partial post-juv moult on breeding grounds, or at onset of autumn migration, has been questioned and is apparently unresolved. (Young with growing feathers in Aug claimed by some to only finish growth of first set of juv-feathers.) Both age groups have a complete winter moult in Afrotropics, mainly Feb–Mar (Dec–Apr). – **SPRING** Ageing impossible. – **AUTUMN Ad** Flight-feathers with abraded tips. Upperparts mainly worn and greyish-brown, rump only subtly paler and warmer brown. **1stCY** Flight-feathers fresh. Upperparts tawny-brown, rump tinged rufous.

BIOMETRICS L 12.5–14 cm; **W** ♂ 66–75 mm (n 66, m 69.4); ♀ 65–70.5 mm (n 20, m 68.0); **T** 48–56.5 mm (n 48, m 52.9); **T/W** m 77.3; **B** 14.3–17.5 mm (n 78, m 15.8); **BD** 2.8–3.8 mm (n 41, m 3.3); **BW**(n) 3.5–4.9 mm (n 38, m 4.2); **Ts** 21.5–23.5 mm (n 45, m 22.3); **HC** 5.0–6.8 mm (n 39, m 5.9). **Wing formula: p1** < pc 0–6 or 0.5–1 mm >, < p2 36–46.5 mm; **p2** < wt 0–4 mm, =3/4 (55%), =4 (31%), =4/5 (7%), or =3 or 5 (7%); **p3** longest (sometimes p3 = p2); **p4** < wt 1–3 mm; **p5** < wt 3–6 mm; **p6** < wt 5.5–10 mm; **p7** < wt 8–12 mm; **p10** < wt 14–18 mm; **s1** < wt 17.5–22 mm; **notch p2** 8–11.5 mm, =7 or 7/8 (64%), =8 (22%), =6 or 6/7 (14%) or, only in ad, =8/9 or =9

Ad, Denmark, Jul: from mid/late summer ad becomes worn and duller and greyer brown (less olive) above, and whiter below. Typical rounded head shape and rather short bill, with yellowish tinge to predominately pinkish-brown legs. Largely yellow-brown claws short. (S. Kristoffersen)

(5%). Emarg. p3. – For difficult birds, calculate ratio of **notch p2/W**: 12.2–16.9 mm (once 17.3) (n 86, m 14.4), pay attention to depth of notch on p2, and its position in relation to rest of primaries, to colour of legs and claws, hind claw length, and upperparts colour. Another and more complex formula is given in Svensson (1992). Since SE European migrants have been reported to require a partially different identification technique, see also Wilson *et al.* (2001).

GEOGRAPHICAL VARIATION & RANGE Monotypic. – Europe except west, south-west and north, from Britain, France, S Fenno-Scandia east to W Kazakhstan, Caucasus, Transcaucasia, south to E Turkey and NW Iran; winters SE Africa. (Syn. *laricus*; *turcomana*.)

REFERENCES Harrap, S. (1989) *BW*, 2: 318–324. – Helbig, A. & Seibold, I. (1999) *Mol. Phyl. & Evol.*, 11: 246–260. – Leisler, B. (1972) *J. f. Orn.*, 113: 366–373. – Walinder, G., Karlsson, L. & Persson, K. (1988) *Ring. & Migr.*, 9: 55–62. – Wilson, J. D. *et al.* (2001) *Ring. & Migr.*, 20: 224–232.

Juv./1stW, Kuwait, Aug: aged by fresh wing. Long, pointed wing with very long primary projection, evenly and widely spaced primaries, and on average darker remiges. Warm tawny-brown tone above, approaching same-age Common Reed, but paler and not quite as rufous-tinged as Common Reed. Also characteristic is relatively well-developed pale fringing on wing, especially tertials. However, majority are so similar that one must check bill length, wing formula and claws. (M. Pope)

BUSH AND REED WARBLERS

Unstreaked *Acrocephalus* in autumn, 1stW, Marsh (top left: Netherlands, Aug), 'Caspian Reed' *A. scirpaceus fuscus* (top right: Kuwait, Sep), Common Reed *A. s. scirpaceus* (bottom left: Italy, Sep) and Blyth's Reed Warbler (bottom right: Kazakhstan, Sep): field identification (without handling) of young in autumn very difficult. Marsh has relatively shorter and blunter bill but longer primary projection, while both Reeds have the opposite, but experience and use of as many features in combination will facilitate identification (though many must still be left unidentified). Again, main problem is with 'Caspian Reed', which in some respects appears intermediate between Common Reed and Marsh. Some populations of *fuscus* are particularly long-winged, with greyish hue above, often quite like Marsh. Nevertheless, most young Marsh tend to show a touch olive above, with less strong rufous tinge to rump than both Reeds. Legs of young Marsh often visibly paler, more yellowish-pink, with paler claws, which are often subtly shorter (in young Reed claws darker and mostly grey-brown). Experienced observers may find that separation of Blyth's Reed from the others, especially Marsh, is less problematic by its shorter primary projection, more pointed/longer bill with greyish 'smudge' at tip of lower mandible and more uniform wing. (Top left: D. F. Asselbergs; top right: M. Pope; bottom left: D. Occhiato; bottom right: A. Wassink)

Juv/1stW, Netherlands, Aug: some young appear duller and more olive-tinged above (especially when not in direct sunlight). Typical head and bill shapes, the latter being subtly shorter than in Common Reed Warbler. Note also very long primary projection. Finely pale-tipped primaries and contrasting tertial edges also often aid identification as Marsh. Aged by evenly fresh plumage. (D. F. Asselbergs)

Young Marsh/Reed Warbler, Israel, Sep: rather pointed wing and long primary projection and short hind claws better fit young Marsh. Also, legs and claws appear fairly pale, matching Marsh, as does bill structure. However, being 1stW, one expects young Marsh to show paler primary tips than this bird. Upperparts coloration fits young of both species. Borderline cases such as this one are best left undetermined, and the present bird could be an odd *fuscus* Reed. Aged by being evenly fresh. (L. Kislev)

(COMMON) REED WARBLER
Acrocephalus scirpaceus (Hermann, 1804)

Alternative names: Eurasian Reed Warbler, European Reed Warbler

Fr. – Rousserolle effarvatte; Ger. – Teichrohrsänger Sp. – Carricero común; Swe. – Rörsångare

From mid-April onwards at the nearest reed-fringed pond, you can generally hear a gentle, chatty song from the reeds, delivered at a steady pace, *tret tret trit tret trut trit…*: the Reed Warbler. It is a brown and buff-white bird, which attaches its basket-shaped nest to a few old reed stems and spends most of its time in the 'reed jungle'. European birds are summer visitors. In late summer, when the young are fledged, they migrate, like most insectivorous passerines by night, to sub-Saharan Africa and winter mainly in a belt from Guinea and Mali to Ethiopia and south to Botswana, returning to breed in Europe in April and first half of May. There are also a few, apparently entirely resident, populations in Arabia and N Africa, and a numerous and widespread, mostly resident, population in C & S Africa.

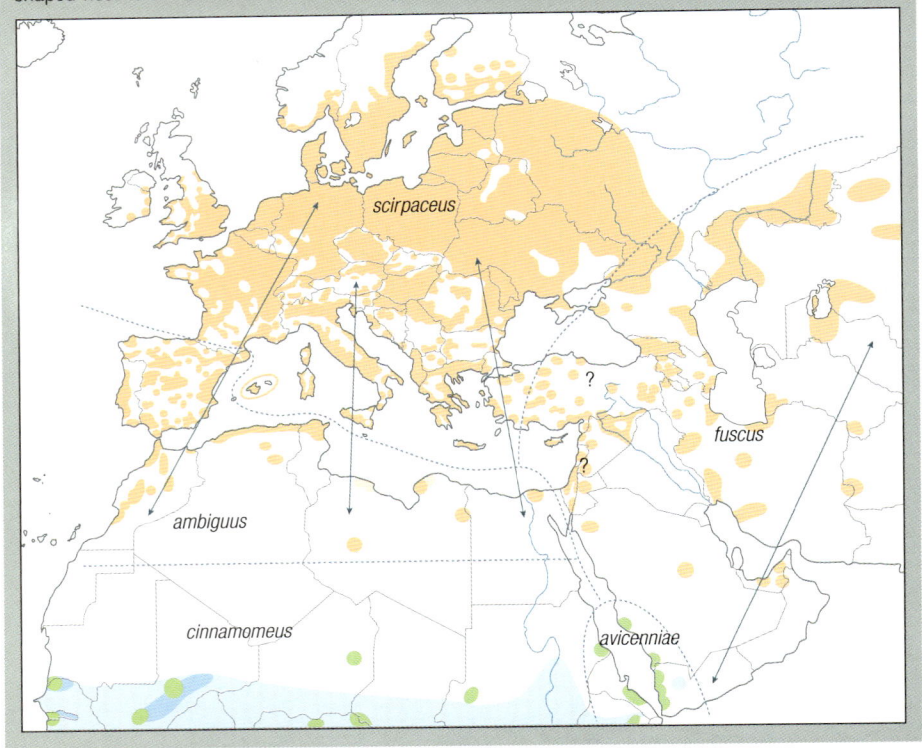

A. s. scirpaceus, England, May: sometimes in overcast conditions can appear duller, more olive grey-brown, with barely indicated cinnamon-brown rump, as here. Fortunately, at this season, Marsh Warbler is still readily separated by song and habitat choice. Here, unlike Marsh, note the relatively shorter primary projection with more tightly bunched tips, and emargination on p3 closer to secondary tips (usually falls nearer wingtip in Marsh). (I. H. Leach)

IDENTIFICATION A medium-large, *slim-shaped* warbler with *warm brown upperparts* and *yellowish-buff and whitish underparts*, lacking any striking plumage features. Often seen climbing reed-stems with hopping movements, or perched in crouched posture with head lower than tail. The *fine pointed bill* and *flat forehead* renders the head an elongated look. Constantly on the move, restless and active, briefly surveying the area from atop a reed but soon disappearing into the depths and safety of the reedbed. On migration sometimes found in gardens, thickets and hedges, so is not necessarily only seen in reeds. Can be confused with several close relatives, and even with other warblers. Easily identified by its song, but also by careful examination and confirmation of (for European breeders) *rather long primary projection* (about equal to exposed tertials, or slightly less), *warm brown crown and mantle* with only a faint olivaceous wash in some lights (unlike Marsh, lacking any obvious purer green hue), slightly *rufous- or cinnamon-tinged rump* (Marsh is less rufous on rump, more subdued ochre), *long bill* (longer on average than Marsh), and *all-pale pinkish-yellow lower mandible* (like Marsh; Paddyfield has dark-tipped lower mandible, Blyth's Reed usually at least a dark smudge at tip). Note also *no more than a hint of an indistinct and short paler supercilium*, usually invisible at some distance. Legs are *pinkish-brown or pale brown-grey* but impression varies, and some age-related differences are involved, too; can look fairly pale, but at other times quite dark. Also, slight geographical variation seems to exist in that N African resident breeders have rather dark grey legs and contrasting yellow soles (Jiguet et al. 2010). – Young Reed and Marsh are very similar, both having practically identical cinnamon or tawny-brown upperparts, with slightly paler and brighter rufous-tinged rump, and some are indistinguishable unless seen close and structural differences, as well as leg and claw colours, are well documented. – Eastern race *fuscus* variable but frequently slightly paler, less cinnamon above, more greyish-tinged, especially on crown and nape, and less cream-coloured below, more whitish. Many, in particular adults, have on average more prominent whitish tips and inner fringes to tail-feathers than *scirpaceus*. – African and Arabian resident populations have similar song and plumage to *scirpaceus* but considerably more rounded wing with shorter primary projection. They also seem on average slightly warmer buff including on the supercilium.

A. s. scirpaceus, Switzerland, May: ssp. *scirpaceus* when still fresh, and especially in sunlight, has upperparts tinged rich brown with in spring diagnostic rufous-cinnamon rump and uppertail-coverts (versus Marsh Warbler). Also separating from Marsh Warbler in spring is song and generally habitat. Legs usually rather dark, but in bright sunlight can appear misleadingly paler, as here. Following winter moult, ageing impossible in spring. (H. Shirihai)

BUSH AND REED WARBLERS

A. s. scirpaceus, ad, France, Aug: at this season, easily separated from juv by worn plumage (especially primary tips, tertials and tail), and reddish-tinged (rather than olive-grey) iris. Plumage bleaches greyer rufous-brown in summer. (A. Audevard)

A. s. scirpaceus, 1stW, France, Sep: young typically fresh and warmer tawny or rufous above (rump and uppertail-coverts even brighter) and often warmer cream-buff below than ad. Wing fringed light rufous-cinnamon, especially coverts, tertials and alula. Subdued but still obvious buffish-white eye-ring. (A. Audevard)

A. s. scirpaceus, 1stW, Italy, Aug: some young are slightly duller and more uniform above, approaching *fuscus*. Both races vary somewhat, and migrants rarely possible to identify. Young Marsh eliminated by long pointed bill, quite dark legs, not very long primary projection, and rather warm buff-brown underparts. (D. Occhiato)

this and the on average slightly longer and more pointed bill should be helpful. When obvious white-tipped tail-feathers are present in *fuscus* (far from on all!), this is also a helpful clue. – Resembles *Blyth's Reed Warbler*, again especially compared to eastern race *fuscus* of Reed due to the on average more grey-brown upperparts. Notice on Blyth's Reed more rounded wing with shorter primary projection (*c.* 2/3 of exposed tertials or slightly less), and in adult Blyth's Reed plainer brown wings, by and large lacking obvious dark feather-centres. Legs of Blyth's Reed are on average slightly darker pinkish-brown or grey-brown than Reed (but overlap). One should also mention the great resemblance of some N African populations, e.g. in Tibesti, Chad, but surely elsewhere as well, to Blyth's Reed Warbler (Fry *et al.* 1974, Dowsett-Lemaire & Dowsett 1987). Latter is colder brown, lacking rufous rump and warm ochre-red upperparts of fresh Reed, and underparts are more yellowish buff-white, flanks not tinged rufous-buff as in Reed. It also has a darker tip to

VOCALISATIONS The song, given mostly at dusk and dawn and in morning hours, but less intensely at other times during day (Reed is not a nocturnal singer like Sedge and Marsh Warblers), is rather chatty and almost 'lecturing' in its steady, slow pace of mainly mono- or disyllabic notes, repeated 2–3 times, now and then with an inserted clear whistling phrase or imitation of other birds, e.g. *tret tret tret tiri tiri trü trü tie tre tre* [whistling:] *vi-vü-vu tre tre trü trü tiri tiri...*, etc. Song delivered from perch, low down or atop reed; short song-flights exceedingly rare. Eastern race *fuscus* sings the same or very similarly, sometimes a fraction slower and more varied and 'interesting', possibly with more frequent use of mimicry. – Usual calls are all 'gruff' or 'muffled', an unobtrusive *che*, or with 'r', *chrek* or *tret*, sometimes repeated in long series, *chre chre chre...* In alarm a hoarse, dry, slurred trilling *chrrreh*, similar to Marsh Warbler. Apparently no corresponding call to normal hard, tongue-clicking call of Marsh and Sedge Warblers.

SIMILAR SPECIES Separation from *Marsh Warbler* is treated under Identification and Vocalisations, and under that species. See also Biometrics. Note that eastern race *fuscus* of Reed is often larger, slightly paler and duller grey-brown above, and therefore more similar to Marsh Warbler. However, they usually have greyer, less green tinge on mantle, and

A. s. ambiguus, Spain, May: Iberian and N African breeders differ very slightly from breeders elsewhere in Europe by having shorter, more rounded wings, and slightly more brown-buff underparts and often more subdued supercilium. Tail and bill are subtly shorter than in *scirpaceus*, but this often difficult to assess in the field. (A. Manglano)

A. s. ambiguus, Spain, Jun: during summer, ad wears duller and greyer above, and whiter below. Versus *scirpaceus* of C & N Europe, short-winged jizz perhaps most obvious, but also note rather large-headed impression. More rounded wing shape than *scirpaceus* is reflected in its shorter migrations. (R. Herrero Viturtia)

— 453 —

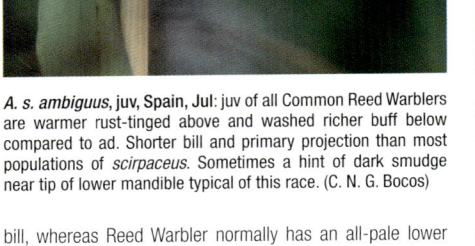

A. s. ambiguus, juv, Spain, Jul: juv of all Common Reed Warblers are warmer rust-tinged above and washed richer buff below compared to ad. Shorter bill and primary projection than most populations of *scirpaceus*. Sometimes a hint of dark smudge near tip of lower mandible typical of this race. (C. N. G. Bocos)

A. s. fuscus, Kazakhstan, May: race based on locality, while long pointed bill, rather long primary projection and fine contrasting whitish primary tips support identification as Reed Warbler. The race *fuscus* often moults later in winter and is fresher in spring. Regularly more greyish-tinged upperparts and has relatively reduced rufous on rump, although this is slightly variable and depends on light and feather wear; the bird here looks quite warm-coloured. (A. Belyaev)

A. s. fuscus, 1stW, Israel, Sep: typical young *fuscus* with rather cold colours and long primary projection. Legs have lead-grey cast and claws are even darker grey, eliminating young Marsh Warbler. Note dull (olive-tawny, less rufous) rump and uppertail-coverts. (L. Kislev)

'Levantine *A. s. fuscus* type', ad, E Israel, Sep: by locality and perhaps plumage colours best fits *fuscus*, with short primary projection indicative of a short-range migrant and probably a local breeder. Such birds can even invite confusion with Blyth's Reed Warbler. (L. Kislev)

bill, whereas Reed Warbler normally has an all-pale lower mandible. – From *Paddyfield Warbler*, which shares favourite habitat of Reed, by slightly larger size, longer primary projection (Paddyfield: only about half tertial length), usually slightly darker brown upperparts with less contrasting tertial edges, darker flight-feathers, longer bill lacking dark tip to lower mandible, and usually darker iris. – A few particularly greyish *fuscus* with whitish underparts can be surprisingly similar to *Olivaceous Warbler* (ssp. *elaeica*), with quite similar size, shape, facial expression and even a hint of contrasting paler edges to tertials, flight-feathers and greater coverts, but separated in usually having slightly more cream-buff tinge on flanks, more rounded tail-tip, more pointed wing with longer primary projection (Olivaceous has blunter wing with three primaries emarginated), and longer undertail-coverts. Legs are basically pale grey with a faint pink tinge in Olivaceous, whereas in Reed they are pinkish-brown (often with a greyish cast in young birds). – A local but real problem of separation occurs in the Red Sea region and in N East Africa, where shorter-winged ssp. *avicenniae* can co-exist in mangroves with Olivaceous Warbler (ssp. *pallida* or claimed *alulensis*). These two can appear very similar, but note in Reed generally slightly darker brown upperparts, slightly longer and more pointed bill (tip tends to be subtly downcurved), tarsi and toes by and large uniformly pinkish-brown (toes not darker and greyer), different song and lack of tail-dipping habit. – Theoretically, Reed could be confused with *Savi's Warbler*, another species living exclusively in dense reed-beds, but there are clear differences. Note Savi's uniform dark brown upperparts lacking a paler and brighter rufous rump, rather dark underparts sullied rufous-grey (especially on flanks and undertail-coverts), a well-curved edge of folded wing, its outer edge slightly paler than rest, and somewhat broader, more rounded tail. – *Cetti's Warbler* is plain rufous-brown above and whitish below but separated by shorter bill, rounder head, short rounded wings with very short primary projection (only about 1/3 of tertial length), long, broad tail often held cocked and spread, and longer more prominent whitish supercilium.

AGEING & SEXING (*scirpaceus*) Slight age-related differences during 1stY, or a little longer. Sexes alike in plumage. – Moults. Partial post-nuptial moult of ad starts either soon after breeding, mainly mid Jul–Aug, or a little later, on migration, with partial replacement of head and body, generally of limited extent, but a few moult more extensively before migration, involving also tertials and central tail-feathers; instances of complete moult (including wings and tail) noted in SW Europe. Partial post-nuptial moult grades into complete moult without clear division, or with short suspension between, flight-feathers moulted in winter quarters rather soon after autumn migration, birds in active wing-moult Sep–Nov (Dec). Partial post-juv moult on breeding grounds (some of head, body and wing-coverts) followed by complete moult in winter quarters, apparently with similar timing to ad. Partial pre-nuptial moult, mainly Jan–Mar (Apr), affecting head, body and some wing-coverts. – **SPRING** Ageing on plumage criteria impossible, but with practice many can be aged using iris and leg colour (Walinder et al. 1988). **Ad** Iris hazelnut-brown or cinnamon-brown. Legs pinkish-brown without bluish tinge. **1stS** Iris olive-brown or olive-grey with a narrow reddish or cinnamon pupil. Legs brownish with variable bluish-grey tinge. – **AUTUMN Ad** Flight-feathers have abraded tips. Upperparts, unless already moulted, bleached and worn, tinged greyish-brown. Iris colour as in spring. Legs lack blue-grey tinge, pinkish-brown, sometimes with yellow cast. **1stW** Flight-feathers and entire plumage uniformly fresh. Upperparts tawny-brown, rump tinged rufous. Iris rather dark grey or grey-brown faintly tinged olive. Legs rather dark greyish-brown with bluish cast, pink or yellow tones subordinate.

BIOMETRICS (*scirpaceus*) **L** 12.5–14 cm; **W** ♂ 61–70 mm (*n* 52, *m* 66.4); ♀ 61–68 mm (*n* 32, *m* 64.3); **T** ♂ 48–59 mm (*n* 28, *m* 52.3); ♀ 49–54 mm (*n* 22, *m* 51.1); **T/W** *m* 78.6; **B** 15.0–18.5 mm (*n* 100, *m* 16.8); **BD** 2.7–3.8 mm (*n* 51, *m* 3.2); **Ts** 21.0–24.0 mm (*n* 64, *m* 22.6); **HC** 6.0–8.0 mm (*n* 34, *m* 6.9). **Wing formula: p1** < pc 0–3 or 0.5–2 mm >, < p2 34–44 mm; **p2** < wt (0) 0.5–3.5 mm, =4 or 4/5 (73%), =3/4 or 3 (26%) or =5 (1%); **p3** longest (rarely pp3–4 or); **p4** < wt (0) 0.5–3.5 mm; **p5** < wt 2–6.5 mm; **p6** < wt 4–9 mm; **p7** < wt 7–11 mm; **p10** < wt 12–17 mm; **s1** < wt 14–19.5 mm; **notch p2** in ad 10.5–13.5 mm, =9/ss or < ss 0–2 mm (70%), =9 (13%), =8/9 (10%) or =8 (2%); in 1stCY 9.5–12 mm, =8/9 or 9 (69%), =8 (15%), =9/10 (9%), or =7/8 (7%). Emarg. p3, very rarely faintly (at times even prominently) also p4 (especially ad ♀). – For birds difficult to separate from Marsh Warbler, calculate ratio of **notch p2/W**: 15.4–21.7 (once 14.8) (*n* 136, *m* 18.2), pay attention to depth of notch on p2, colour of legs and claws, and upperparts. Another and more complex formula is given in Svensson (1992). As SE European migrants have been reported to require a partially different identification technique, see also Wilson *et al.* (2001).

'Levantine *A. s. fuscus* type', juv/1stW, Israel, Jul: young of breeders in Israel are overall pale, less dark brownish or russet above (cinnamon on rump rather subdued) and less warm below. Compared to *fuscus* from Kazakhstan, wing and tail is *c.* 6% shorter, still the bill is proportionately long; wing also notably rounded (with p2 mostly =4/5, sometimes =5, rarely =5/6, clear emargination also on p4), with shorter primary projection. Some even show slight dusky smudge on lower mandible. (H. Shirihai)

Presumed *A. s. avicenniae*, Somalia, May: the so-called Red Sea or Mangrove Reed Warbler is a generally pale and short- and round-winged *Acrocephalus*, resembling Blyth's Reed Warbler. Facial expression (especially the weak supercilium) and bill shape, as well as quite rufous-tinged underparts identify the bird as a Reed Warbler and not an *Iduna* species. (C. Cohen)

GEOGRAPHICAL VARIATION & RANGE Slight clinal variation, birds from west to east becoming larger and less rufous, paler and greyish-tinged brown above, and paler, more whitish below, while southern birds are more round-winged due to being short-range migrants or residents. There is much variation within, and overlap between, all subspecies. Recently breeders in Iberia and Morocco have been separated on account of different leg colour, shorter bill and rounder wing. Five extralimital subspecies in Africa, not treated.

A. s. scirpaceus (Hermann, 1804) (most of Europe except much of Iberia; W Turkey; northern populations winter Afrotropics, southern too, or make somewhat shorter movements). Treated above, being warm tawny- or cinnamon-brown with faint olive cast on crown to back and rather saturated cream-buff on breast to vent in fresh plumage. See also Biometrics. – Some birds in SE Europe tend towards ssp. *fuscus*, being a little greyer-tinged above. (Syn. *strepera*.)

○ *A. s. ambiguus* (A. E. Brehm, 1857) (Iberia, apparently French Pyrenees, N Africa in Morocco but possibly (though still unconfirmed) east through Algeria to Egypt, more hypothetically also N Arabian coast of Red Sea; largely resident or makes short-range movements in winter). Differs genetically from *scirpaceus* and has on average slightly shorter and more rounded wing, and rarely a faint or even prominent emarg. of p4 (apart from emarg. on p3). Also has darker grey legs (with contrasting yellowish soles) and slightly shorter tail and bill. In fresh plumage perhaps on average warmer buff, including supercilium. Sexes very nearly same size. Moults completely on or near breeding grounds in late summer–early autumn. **W** 60.5–67.5 mm (*n* 33, *m* 63.5); **T** 44–55 mm (*n* 31, *m* 50.2); **T/W** *m* 78.9; **B** 14.9–17.3 mm (*n* 33, *m* 16.0); **Ts** 21.0–22.9 mm (*n* 31, *m* 22.0). **Wing formula: p1** < p2 32–41 mm; **p2** < wt 0.5–3.5 mm; **p5** < wt 2–5.5 mm; **p6** < wt 4–8 mm; **p7** < wt 5.5–10.5 mm; **p10** < wt 10.5–16 mm; **s1** < wt 13–18 mm. (See also Taxonomic notes.) (AMNH, NHM, MNHN, NRM.)

A. s. fuscus (Hemprich & Ehrenberg, 1833) 'CASPIAN REED WARBLER' (SE Russia, C & E Turkey, Cyprus, Levant, Caucasus, Transcaucasia, N Iran, W Kazakhstan; winters E Africa). A variable race (which might even contain one or two undescribed taxa), either warm brown above and very close to *scirpaceus* (doubtfully separable in the field, as indeed is the holotype, ZMB 260, 'Arabia'), or has a slightly paler and more greyish-tinged crown and nape, less yellow-buff, more whitish underparts, a slightly longer wing, and frequently whitish-tipped and whitish-fringed outer tail-feathers. (A claimed proportionately shorter p2 could not be confirmed by

Presumed *A. s. avicenniae*, Somalia, May: short primary projection, rather dull brown or olive greyish-brown above like Marsh Warbler, and rather restricted cinnamon on rump. Underparts have faint yellowish-buff hue; whitish supercilium narrow and short. Note extremely worn tail on right-hand image suggesting retained juvenile feathers. (C. Cohen)

us when assessing birds from whole range, but may be true when only birds from Levant, S Turkey and Cyprus are examined.) Differences often require series of this and *scirpaceus* for comparison, or much practice, before appreciable. Some birds noted on migration, presumably from Central Asia (Pearson *et al.* 2002), even greyer above and rather different from *scirpaceus* (even recalling Olivaceous Warbler, ssp. *elaeica*). Breeders in Levant small like *scirpaceus*, smaller than *fuscus* of SE Russia or Central Asia, and often have a long bill but slightly blunter wing with a hint of emarg. of p4 (cf. Morgan 1998). Song perhaps on average slightly slower and more varied, using more mimicry than *scirpaceus*. **W** ♂ 64–73.5 mm (*n* 20, *m* 69.8), ♀ 63–69 mm (*n* 14, *m* 66.6); **T** ♂ 52–59.5 mm (*n* 21, *m* 56.1), ♀ 49–55.5 mm (*n* 14, *m* 52.7); **T/W** *m* 79.6; **B** 16.0–18.3 mm (*n* 37, *m* 17.2); **Ts** 20.5–23.6 mm (*n* 35, *m* 22.5); **HC** 5.8–8.0 mm (*n* 15, *m* 7.1). **Wing formula**: p1 < p2 37.5–45 mm; **p2** < wt 1–3.5 mm, =4 or 4/5 (89%), =3 (7%), or =5 (4%). Complete post-nuptial moult in winter quarters, variable as to timing, often in Ethiopia and Sudan at same time as *scirpaceus* (Sep–Dec), but some populations wintering further south in Africa moult later than in *scirpaceus*, commonly Jan–Mar (Apr), hence many *fuscus* in spring return with fresher primary tips, these being tipped whitish, contrasting against dark rest of feathers.

A. s. avicenniae Ash *et al.*, 1989. 'RED SEA REED WARBLER' (N Somalia, possibly also S Red Sea coasts; resident or short-range migrant to E Africa). Small with much shorter and rounder wing, plumage rather like Marsh Warbler, dull brown or olive greyish-brown above, with rather restricted cinnamon on rump; whitish underparts with slight yellowish hue, but no buff. Only two specimens examined, one of which in heavy moult (Oct, Somalia) (**W** 60 mm, **T** 51 mm, **T/W** *m* 85.0, **B** 15.0–16.9 mm, **Ts** 20.3–21.2 mm; **HC** 5.9–6.3 mm; **wing formula**: p2 < wt 3.5 mm, =6/7; **p7** < wt 4 mm; **p10** < wt 8 mm; **s1** < wt 10 mm; emarg. p3 and faintly also on p4). Theoretically very similar to a small Blyth's Reed Warbler, but combination of short bill and primary projection, much less prominent emargination of primaries, and paler yellowish-white underparts (*dumetorum*: usually tinged olive-grey, sometimes with faint yellowish-buff hue, on flanks and breast-sides), should separate *avicenniae*.

TAXONOMIC NOTES Extralimital African *baeticatus* (Vieillot, 1817), in which the Red Sea form *avicenniae* is sometimes included (though not here), is often treated as a separate species, 'AFRICAN REED WARBLER'. However, we include all African taxa in Reed Warbler, following Dowsett-Lemaire & Dowsett (1987), who showed that *baeticatus* responds fully to playback of song of *scirpaceus*, and vice versa, songs which sound very similar or identical, and that morphology generally makes them inseparable except for the more rounded wing of *baeticatus*. This difference is an expected and natural adaptation to their resident life in the Afrotropics and is not thought important for taxonomy. That they have fairly recently become separated is indicated by the only moderate difference in mtDNA of 1.9–2.4%, much less than between several other traditionally recognised species in the genus *Acrocephalus* (Helbig & Seibold 1999). A similar-sized difference was found between *scirpaceus* and *fuscus*. It should be noted that intra-taxon differences within *fuscus* were almost of the same magnitude. Leisler *et al.* (1997) showed that, if anything, *avicenniae* and *scirpaceus* are sister taxa, more recently evolved than *baeticatus*, and all are sister to *fuscus*. This would make a division in 'Eurasian' and 'African Reed Warblers' polyphyletic, but support for the results were partially rather weak, and more research is needed; also, sampling of additional African populations is desirable. – Recently, breeders in Libya and W Egypt have been claimed to include both 'African Reed Warbler' (subspecies unknown) and *avicenniae* (or a taxon close to *avicenniae*), based on DNA (Hering *et al.* 2009).

REFERENCES AMEZIAN, M. *et al.* (2010) *Ardea*, 98: 225–234. – DOWSETT-LEMAIRE, F. & DOWSETT, R. J. (1987) *BBOC*, 107: 74–85. – FRY, C. H., WILLIAMSON, K. & FERGUSON-LEES, I. J. (1974) *Ibis*, 116: 240–346. – HARRAP, S. (1989) *BW*, 2: 318–324. – HELBIG, A. J. & SEIBOLD, I. (1999) *Mol. Phyl. & Evol.*, 11: 246–260. – HERING, J. *et al.* (2009) *Limicola*, 23: 202–232. – JIGUET, F., RGUIBI-IDRISSI, H. & PROVOST, P. (2010) *DB*, 32: 29–36. – JIGUET, F. *et al.* (2015) *Ibis*, in press. – LEISLER, B. *et al.* (1997) *J. of Orn.*, 138: 469–496. – MORGAN, J. (1998) *Ring. & Migr.*, 19: 57–58. – PARKIN, D. T. *et al.* (2004) *BB*, 97: 276–299. – PEARSON, D. J., SMALL, B. J. & KENNERLEY, P. R. (2002) *BB*, 95: 42–61. – WALINDER, G., KARLSSON, L. & PERSSON, K. (1988) *Ring. & Migr.*, 9: 55–62. – WALTON, C. (2012) *Ring. & Migr.*, 27: 94–98. – WILSON, J. D. *et al.* (2001) *Ring. & Migr.*, 20: 224–232.

Unidentified warbler, possibly ssp. *avicenniae* or related little-known taxon, Red Sea coast, Saudi Arabia, Jul: *Iduna*-like bill shape and flat forehead, plain pale ochre-brown above, pale pinkish buff-white below. Shows affinity with *Acrocephalus* warblers by the rather clear graduation of tail and no visible white on tail-sides. Head pattern 'open-faced', having pale lores and very poorly marked dark eye-stripe. Seems to have emargination of both p3 and p4, and at least p3 clearly falls well inside tips of secondaries. Compared with presumed *avicenniae* from Somalia (see images on p. 455) the long bill and flat head shape of this Arabian bird is especially striking. This variation is difficult to explain without further research. (P. Roberts)

Subspecies unknown, Western Desert, Egypt, May: these birds are closest to *avicenniae* genetically. They are small-sized (including bill, at least true for the right hand bird) and blunt-winged. Both were trapped at the same location, but the degree of variation in coloration (partially affected by light) and bill shape is striking. On current knowledge and away from the breeding grounds, it is unclear if such birds could be separated reliably from either *ambiguus* or *avicenniae*, even if handled! (J. Hering)

CAPE VERDE WARBLER
Acrocephalus brevipennis (Keulemans, 1866)

Alternative names: Cape Verde Cane Warbler, Cape Verde Swamp Warbler

Fr. – Rousserolle du Cap-Vert; Ger. – Kapverdenrohrsänger Sp. – Carricero de Cabo Verde; Swe. – Kapverdesångare

Endemic to the Cape Verde Islands, the Cape Verde Warbler appears like a cross between Reed and Clamorous Reed Warblers, perhaps with a touch of Nightingale. Found in natural reeds and in similar man-modified habitats such as thickets on wetter ground, sugar cane, mango and coffee plantations interspersed with fruit trees and in dense gardens. It occurs mainly on the island of Santiago, with small recently re-discovered or re-established populations on São Nicolau and Fogo. The total population has been estimated at just *c.* 500 pairs, and it is considered Endangered. Human developments and disturbance, and the effects of drought on its habitat, are the main threats.

IDENTIFICATION A fairly large, plump *Acrocephalus* warbler with *long bill, large head* and rather long neck but *short wings and tail*. However, the short, rounded wings mean the tail still appears fairly long. The *feet are long and strong* (as so often on island passerines) and rather dark brown-grey. The bird lacks any striking plumage features, the *upperparts being dark olive-brown* and *underparts greyish-white*. Although underparts are mainly greyish with somewhat whiter upper throat and central belly, many in fresh plumage show a *faint cinnamon tinge to flanks and vent. No pale supercilium.* Lower mandible yellowish. *Sides of head and neck*, and often sides of breast, *tinged greyish*, sometimes appearing subtly striped or mottled. Unlike most of its relatives in the genus *Acrocephalus*, the rump is not brighter cinnamon or rufous, although juveniles are overall warmer tawny-brown above. Due to its size and build, movements are often rather slow and deliberate. It is not always a skulker in dense reeds but is sometimes observed feeding in the canopy of trees, then moving rather quickly and agilely.

VOCALISATIONS The song is often likened to that of Common Bulbul or Nightingale, but could also be compared with Eastern Orphean Warbler, being loud and slightly explosive, delivered in staccato fashion, with some notes or phrases repeated a few times. Song could be rendered *chow, put dud-loo put dud-loo dee-de-lo dee-de-lo dee-de-lo.* – Usual calls are a throaty, 'thick', disyllabic *ker-chow*, a short, sharp *chik*, often repeated in short series, and a deeper, slightly nasal or guttural *gok*.

SIMILAR SPECIES Although confusion with other warblers is rather unlikely most of the year, a stray *Isabelline Warbler* in early spring or late summer is theoretically possible on the Cape Verdes, but much plumper body and somewhat larger size of Cape Verde Warbler should prevent misidentification. Add to that the shorter more rounded wings, the more greyish cast on sides of head and neck, the heavier bill, and stronger legs.

AGEING & SEXING Very slight age-related differences in winter and spring. Sexes alike. – Moults. Complete post-nuptial moult of ad in winter, apparently late Jan–Mar, but possibly slightly earlier, at least some finishing full moult exist in Feb. Partial post-juv moult extended and variable due to long breeding period (breeding generally Aug–Nov), usually partial but may be complete, mainly Jan–Apr. Apparently no partial pre-nuptial moult in either age group. – **WINTER–SPRING Ad** Plumage rather dark greyish olive-brown above and greyish-white below. **1stS** Plumage warm tawny-brown

Presumed ad, Santiago, Cape Verde Is, Mar: rather plump and large-headed, with fairly long bill, but lacks any striking plumage features, though white-throated appearance often noticeable. Note also very short primary projection making tail look long. Appearance of evenly feathered wing and red iris suggest ad. (D. Occhiato)

Presumed ad, Santiago, Cape Verde Is, Apr: often very tame and easy to approach with notably slow movements in vegetation. Typical jizz with large-headed impression and long bill. Often sullied greyish or brown on underparts making white-throated appearance distinctive. Other images of both birds reveal evenly-feathered wing, thus probably ad. (H. Shirihai)

Presumed ad, Santiago, Cape Verde Is, Apr: distinctly long tail, which can appear tinged whitish below (mostly on outer feathers). Birds often visit canopy and are typically observed from below. Evenly-feathered wing, as far as can be judged, suggests ad. (H. Shirihai)

above, underparts tinged cinnamon-buff on breast, flanks and vent. – **SUMMER–AUTUMN** Plumage differences between ad and young much less obvious due to wear and bleaching.

BIOMETRICS **L** 15–16.5 cm; **W** 62–69 mm (n 25, m 66.2); **T** 55–65 mm (n 25, m 60.4); **T/W** m 91.3; **TG** 11–19 mm; **B** 18.5–21.3 mm (n 25, m 19.8); **BD** 3.4–4.0 mm (n 44, m 3.2); **Ts** 25.5–27.5 mm (once 24.5; n 25, m 26.6). **Wing formula:** **p1** > pc 11–17 mm, < p2 16–21 mm; **p2** < wt 7–10.5 mm, =10 or 10/ss (46%), ≤ss (42%), or =9 or 9/10 (12%); **p3** < wt 1–3.5 mm; **pp4–5** (pp4–6) longest; **p6** < wt 0–1 mm; **p7** < wt 1–4.5 mm; **p8** < wt 2.5–6 mm; **p10** < wt 6–10 mm; **s1** < wt 8–12 mm. Emarg. pp3–6, sometimes faintly also on p7.

GEOGRAPHICAL VARIATION & RANGE Monotypic. – Cape Verde Is; resident.

REFERENCES Donald, P. & Taylor, R. (2004) *Malimbus*, 26: 34–37. – Hazevoet, C. J., Monteiro, L. R. & Ratcliffe, N. (1999) *BBOC*, 119: 68–71. – Hering, J. & Fuchs, E. (2009) *BB*, 102: 17–24.

Ad, Santiago, Cape Verde Is, Mar: distinctly whitish throat, very short pale supercilium above lores, bicoloured bill and dusky grey-brown underparts; very short wings hardly reach beyond tail-base. Note reddish iris, but probably exaggerated by strong light. (S. Piner)

Presumed 1stY, Santiago, Cape Verde Is, Mar: very short primary projection makes tail look long. Note rounded head shape and strong and bluntly-tipped bill, almost recalling a robin or nightingale. Red-brown iris indicates young age. (S. Gatto)

Juv, Santiago, Cape Verde Is, Apr: untidy buffish-cinnamon feathers below (uniform greyish in ad). Legs long, strong and tinged dark grey. (H. Shirihai)

CLAMOROUS REED WARBLER
Acrocephalus stentoreus (Hemprich & Ehrenberg, 1833)

Fr. – Rousserolle stentor; Ger. – Stentorrohrsänger
Sp. – Carricero estentóreo; Swe. – Papyrussångare

Clamorous Reed Warbler replaces and partly overlaps with Great Reed Warbler in S Middle East, where it inhabits undisturbed reedbeds and where, on migration, both Great Reed and Basra Reed Warblers might also occur. Nonetheless, the characteristic far-carrying, almost frog-like song of Clamorous Reed Warbler is very typical of marshy habitats such as fishponds in Israel's Hula Valley. It is a widespread resident or partial migrant in papyrus or reed swamps from Egypt through NE Africa and east across Asia to New Guinea and extreme NE Australia.

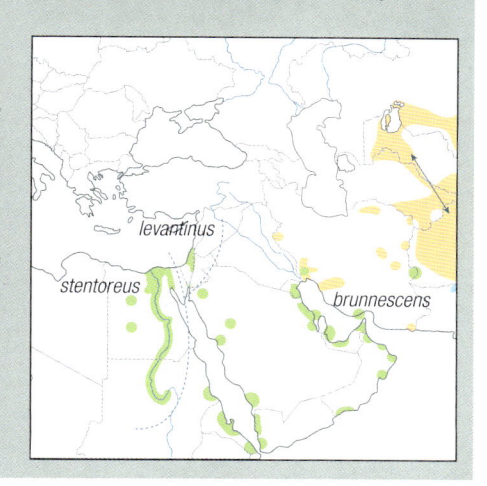

A. s. stentoreus, Egypt, Mar: large unstreaked *Acrocephalus*, which mostly requires separation from Great Reed Warbler. Note longer pointed bill, which sometimes appears slightly decurved; wing and primary projection short, leaving proportionately long tail. Following completion of post-nuptial and post-juv moults, ad and 1stY indistinguishable. (T. Small)

A. s. stentoreus, Egypt, Apr: typically has long tail that is well rounded and fairly broad. Face pattern rather weak. Also, note white-throated appearance. (D. Occhiato)

IDENTIFICATION A medium-large, unstreaked, bulky *Acrocephalus* very similar to Great Reed Warbler. Generally brownish-olive above with rufous cast to rump, being predominantly cream-coloured to greyish-buff or tawny-buff below, with warmer sides (but much variation in race *levantinus* of Israel/Jordan, which has dark and pale morphs). Subspecies *stentoreus* and *levantinus* of Clamorous Reed Warbler differ from Great Reed as follows: (i) *bill longer and pointed*, with a finer tip, less deep base and sometimes appears to have a slightly decurved distal part; (ii) wing and *primary projection shorter*, 1/2 to 2/3 of visible tertial length with only c. 6 primary tips visible beyond tertials (Great Reed has longer wing and longer primary projection, latter about equal to tertial length with 8–9 exposed tips); (iii) *tail proportionately longer and deeply rounded* (shorter and less graduated in Great Reed); (iv) jizz is thus somewhat more elongated (owing to narrower dagger-like bill and longer tail often held drooping), with more smoothly tapered head profile, rather longish and less rounded crown compared to Great Reed. Other finer differences include *rather weaker face pattern* with *shorter, more indistinct supercilium* that is often fainter or lacking behind eye (more pronounced and longer in Great Reed, normally extending behind eye, and emphasized by darker eye- and vague lateral crown-stripes). Clamorous Reed is usually warmer and darker overall, with restricted (often near-absent) whitish area on mid belly and throat to chin (Great Reed paler and suffused more rufous-brown above in ssp. *arundinaceus* or greyish-olive in *zarudnyi*, and more extensively pale below). Legs greyish or dull bluish-grey, whereas they are predominantly pale brown or mouse-grey in Great Reed (some overlap though). Usually sings from cover, does not habitually show up at top of reed like Great Reed.

VOCALISATIONS Song loud and powerful and similarly gruff like Great Reed Warbler, and the two can at times be difficult to separate; strophes more broken up and song structure tentative or varied and 'loose'. A common song is slightly accelerating and increases in loudness, often ending with characteristic loud phrase containing raucous and inserted shrill notes, e.g. *track, track, track karra-kru-kih karra-kru-kih chivi-trü chivi-chih*. Does not repeat same note 3–4 times and whole phrases again and again in a more rigid pattern like Great Reed Warbler. – Contact calls include a loud, deep *track* or *chak*, and a hard rolling *karrrk*.

SIMILAR SPECIES In Levant, separation of Clamorous Reed from *Great Reed Warbler* (both ssp. *arundinaceus* and *zarudnyi*) relatively easy due to Clamorous Reed's combination of different structure and plainer face pattern (see above), especially if dealing with darker birds. Further east, however, race *brunnescens* of Clamorous Reed is proportionately slightly longer-winged and shorter-billed with a less graduated tail and has colder brown upperparts (dull olive-brown or greyish-olive) and paler, whiter underparts with less greyish-olive on body-sides, and is therefore less easily separated from Great Reed (especially ssp. *zarudnyi*). However, most Clamorous Reeds are still separable by the relatively longer and slimmer bill, and shorter primary projection, as well as slightly longer, more graduated tail, with colder, greyer appearance and shorter (but often strikingly whitish) supercilium. Furthermore, complete post-nuptial moult of *brunnescens* is in September–November, thus many young and adults are then in heavy moult, while Great Reed migrating through, e.g., Arabia usually show no active moult at this period. In spring (e.g. in Arabia, where *brunnescens* breeds and Great Reed occurs on passage), *brunnescens* is more worn, chiefly on primary tips, than Great Reed (which mainly undertakes complete moult in winter quarters). – Race *brunnescens* of Clamorous is very similar to *Oriental Reed Warbler*, but see latter for differences. – For separation from *Basra Reed Warbler*, which is not always easy, see that species. – Thus, especially in the Middle East, observers should remember that separation of the four taxa mentioned above can be extremely difficult and sometimes impossible, demanding prolonged and careful observations. Like

A. s. stentoreus, ♂, Egypt, Apr: bright orange gape obvious in this ♂ when singing. This race is relatively paler brown above than pale morph *A. s. levantinus*. Face pattern among blandest of genus, with rather indistinct short supercilium, often fainter or lacking behind eye. Legs greyish, often dull bluish-grey. (V. Legrand)

A. s. levantinus, Israel, Nov: typical crouched posture perched on a reed. An intermediate-coloured bird of this dimorphic race; compared to *stentoreus* from Egypt, relatively darker and more brownish-tinged (less tawny-olive). (A. Ben Dov)

A. s. levantinus, Israel, Apr: an extremely pale bird of the pale morph; here several months after moult with some bleaching and wear, contributing to its paleness. Such pale birds are basically identical to *stentoreus* from Egypt. (L. Kislev)

A. s. levantinus, Israel, Nov: very distinctive dark morph, with dark earth-brown and slight russet casts above, and entire underparts deep cinnamon-brown to buff-brown (throat and breast often slightly paler). Face marks considerably reduced, supercilium on average very obscure or lacking. Such birds are unmistakable. (L. Kislev)

A. s. brunnescens, Kyrgyzstan, Jun: slightly longer-winged with longer primary projection, and stouter/shorter-billed, while overall plumage less rufous, generally colder and duller than western populations, and less easily separated from Great Reed (especially *zarudnyi*). Face pattern also to some extent better marked, approaching Great Reed. Tail less graduated and less deeply rounded. (M. Westerbjerg Andersen)

Great Reed, unlikely to be confused with any of the smaller *Acrocephalus* due to overall size, structure and voice.

AGEING & SEXING Moult and feather wear can be used for ageing prior to completion. Iris colour especially useful in autumn. Sexes alike, and seasonal plumage variation limited to some wear and bleaching. – Moults. Complete post-nuptial moult of ad rather prolonged and variable, *stentoreus* and *levantinus* mostly Sep–Nov (even to Dec or later); *brunnescens* earlier, mostly Aug–Oct (Nov). Post-juv moult also complete, mostly soon after fledging, but timing probably even more extended than in ad. Races *stentoreus* and *levantinus* essentially resident and moult in or near breeding areas; *brunnescens* recorded moulting on breeding grounds and on migration. – **SPRING Ad** Less worn in early spring but as breeding starts soon bleaches paler, upperparts becoming more greyish-olive with slightly warmer rump and uppertail-coverts, underparts and supercilium becoming whiter, breast and flanks less heavily suffused rufous or buff, and pale fringes and tips to wing-feathers reduced. **1stS** Indistinguishable from ad by plumage. Iris in many still differs (see 1stW). – **AUTUMN Ad** Generally worn or has started moulting, at least in early autumn; iris dark sepia-brown. **1stW** In early autumn easily separated from ad by mostly fresh juv plumage, including flight- and tail-feathers; in late autumn (or earlier), as fresh ad but iris olive-grey, becoming browner in late autumn. **Juv** Brighter, more rufous-cinnamon upperparts and fringes to wing- and tail-feathers, with rump even warmer. Underparts and supercilium deeper tawny-buff even than fresh ad, and entire plumage has soft, fluffy appearance.

BIOMETRICS (*stentoreus*) **L** 17–19 cm; **W** ♂ 78–82 mm (n 13, m 79.8), ♀ 73–81.5 mm (n 13, m 77.0); **T** ♂ 70–77 mm (n 13, m 74.2), ♀ 67–76 mm (n 13, m 70.8); **T/W** m 92.5; **TG** 11–19.5 mm (m 14.8); **B** 23.5–27.4 mm (n 26, m 25.6); **BD** 4.9–5.8 mm (n 25, m 5.2); **Ts** 26.0–29.8 mm (n 26, m 27.9). **Wing formula: p1** < pc 1.5–8 mm, < p2 35–47 mm; **p2** < wt 3–6 mm, =5/6 or 6 (83%), or =6/7 or 7 (17%); **pp3–4** longest (sometimes pp3–5); **p5** < wt 0–3 mm; **p6** < wt 3–7 mm; **p7** < wt 5–8 mm; **p8** < wt 7–9 mm; **p10** < wt 11–18 mm; **s1** < wt 14–18 mm; **tert. tip** < wt 12–19 mm; **notch p2** 15–20 mm, < ss. Emarg. pp3–4.

GEOGRAPHICAL VARIATION & RANGE Up to 11 races recognised, most of them extralimital and hence not treated. Of the three subspecies occurring within the covered range, *brunnescens* is rather different in structure, seemingly forming a link to closely related Great Reed Warbler.

A. s. stentoreus (Hemprich & Ehrenberg, 1833) (Egypt east to Suez Canal, along Nile south to N Sudan; mainly resident, or makes shorter movements in winter). Described above.

A. s. levantinus Roselaar, 1994 (N Israel, possibly

A. s. brunnescens, United Arab Emirates, Mar: shorter bill and stronger supercilium versus western populations; also colder (dull olive-brown or greyish-olive) above, and paler, whiter below. In spring, in Arabia, where *brunnescens* breeds and Great Reed occurs on passage, former is diagnostically more worn, chiefly on primary tips. (C. Tiller)

Unstreaked large *Acrocephalus*, Clamorous Reed (top left: pale morph, *levantinus*, Israel, Dec), Clamorous Reed (top right: *brunnescens*, United Arab Emirates, Feb), Great Reed (centre left: probably *arundinaceus*, Greece, Apr), Great Reed (centre right: *zarudnyi*, Kuwait, Apr), Oriental Reed (bottom left: China, May) and Basra Reed Warblers (bottom right: Kuwait, Apr): in attempting to separate these species, especially outside their normal ranges, it is essential to check as many features as possible, including moult and feather wear, and trapping birds is sometimes required. Pale morph of western populations of Clamorous Reed differs from Great Reed Warbler in having longer, slimmer and more pointed bill, which often appears slightly decurved; its primary projection is distinctly shorter, but tail proportionately longer and more graduated. It also has considerably weaker face pattern, and overall coloration warmer, especially underparts. Both *brunnescens* Clamorous Reed and Oriental Reed Warblers have intermediate structure, with bill length and shape approaching Great Reed, but have shorter wings and primary projection. Their identification can be highly problematic, as well as telling them from longer-winged *zarudnyi* Great Reed and Basra Reed. Note latter's combination of very long primary projection and longish, very slim bill, somewhat approaching *brunnescens* Clamorous Reed, as well as colder and overall slimmer appearance. (Top left: L. Kislev; top right: T. Varto Nielsen; centre left: M. Schäf; centre right: K. Haataja; bottom left: J. S. Hansen; bottom right: M. Pope)

A. s. brunnescens, Kuwait, Aug: despite wide range occupied by *brunnescens*, variation is poorly known. These Kuwaiti birds appear to have very long bills, approaching western populations of Clamorous Reed. Very few have been trapped and they require further investigation. These images also perfectly capture differences between worn bird (left) and recently fledged and evenly fresh juv (right). Ageing the worn bird is not straightforward: the level of wear to primary tips and tail best match ad, but impossible to eliminate possibility that it is an early-hatched juv that is already heavily abraded, and the dark olive-grey iris supports latter assumption. (A. Al-Sirhan)

A. s. brunnescens, Oman, Jan: this individual has rather long bill, much like western populations. Most *brunnescens* when fresh are less rufous, generally olive-brown or greyish-olive above, and paler below, with greyish-olive body-sides. Some show faintly warmer brown rump. (H. & J. Eriksen)

Variation of *A. s. brunnescens* in Kuwait in non-breeding season (left: Feb; right: Nov): left bird colder-toned, mostly olive grey-brown above, and quite whitish below, while right bird is warmer, rustier in colour. Birds wintering in Kuwait could come from large area, so variation is perhaps not just individual. Furthermore, note how some *brunnescens* (especially duller/greyer individuals), in certain lights and angles (when bill appears pointed and straight), can look almost identical to Basra Reed. Safest separation by much shorter primary projection but longer and rounder tail of *brunnescens*. (A. Al-Sirhan)

also adjacent Jordan; mainly resident). Resembles *stentoreus* but has proportionately slightly longer tail and bill, and is distinctly darker and warmer-coloured. Two morphs, dark and pale, the latter approaches ssp. *stentoreus* but is on average darker and duller brown with buff-olive tinge reduced, albeit with some overlap, while dark morph is very distinctive in being dark earth-brown with slight russet cast to upperparts, and entire underparts (including throat and belly) deep cinnamon-brown to buff-brown, with facial marks much reduced, supercilium on average very obscure and warmer, or even virtually lacking. **W** ♂ 79–86.5 mm (*n* 14, *m* 83.8), ♀ 78–84 mm (*n* 12, *m* 81.1); **T** ♂ 73–85 mm (*n* 15, *m* 80.1), ♀ 70–82 mm (*n* 12, *m* 76.3); **T/W** *m* 94.9; **TG** (11) 13.5–21.5 mm (*n* 24, *m* 15.5); **B** 24.5–29.4 mm (*n* 27, *m* 26.8); **BD** 4.0–5.3 mm (*n* 27, *m* 4.7); **Ts** 26.1–31.5 mm (*n* 28, *m* 28.5).

A. s. brunnescens (Jerdon, 1839) (coastal Red Sea, S & E Arabia, Iran and in Central Asia north to Fergana basin, E Kazakhstan and east to Mongolia; perhaps E Iraq; northern populations winter Gulf States, S Iran east to India, Nepal). Has longer and more pointed wings with longer primary projection than *stentoreus*, and although the less graduated tail is moderately longer, too, it is still shorter proportionately. Also, tarsus longer and bill somewhat shorter and stouter, while overall plumage is less rufous, generally a little colder and duller (see Similar species). **W** ♂ 87–93 mm (*n* 12, *m* 89.1), ♀ 82–91 mm (*n* 13, *m* 86.4); **T** 74–82 mm (*n* 12, *m* 78.3), ♀ 70–82 mm (*n* 13, *m* 75.3); **T/W** *m* 87.6; **TG** 10.5–16 mm (*m* 11.6); **B** 21.2–25.6 mm (*n* 26, *m* 23.8); **BD** 4.7–6.1 mm (*n* 26, *m* 5.5); **Ts** 27.6–31.6 mm (*n* 26, *m* 29.6). **Wing formula: p1** < p2 45–53 mm; **p10** < wt 15–19 mm; **s1** < wt 18–21 mm; **tert. tip** < wt 15.5–25.5 mm.

REFERENCES VAN DEN BERG, A. & SYMENS, P. (1992) *DB*, 14: 41–48. – LAIRD, W. (1992) *BB*, 85: 83–85. – WALLACE, D. I. M. (1973) *BB*, 66: 382–385.

A. s. brunnescens, Oman, Jan: in fresh plumage, a rather rufous- and buffish-toned bird. Whiter underparts with limited buffish-olive body-sides, and rather strong whitish supercilium. From similarly cold and dull *zarudnyi* Great Reed, note shorter primary projection, and narrower, less stout bill. (H. & J. Eriksen)

GREAT REED WARBLER
Acrocephalus arundinaceus (L., 1758)

Fr. – Rousserolle turdoïde; Ger. – Drosselrohrsänger
Sp. – Carricero tordal; Swe. – Trastsångare

The epitome of the large unstreaked *Acrocephalus*. Although rather similar to several congeners, its far wider distribution affords Great Reed Warbler a much greater familiarity among birdwatchers. It comes into contact with Clamorous Reed Warbler in a few places, e.g., the Levant and parts of Central Asia, and these two and two other species form a challenging group for birders seeking to identify them in the field. Great Reed Warbler is a strictly migrant breeder in reedbeds and other tall, dense waterside vegetation, from continental Europe and S Scandinavia east to Kazakhstan, and from NW Africa to N Iran. It winters in sub-Saharan Africa, south to the Cape.

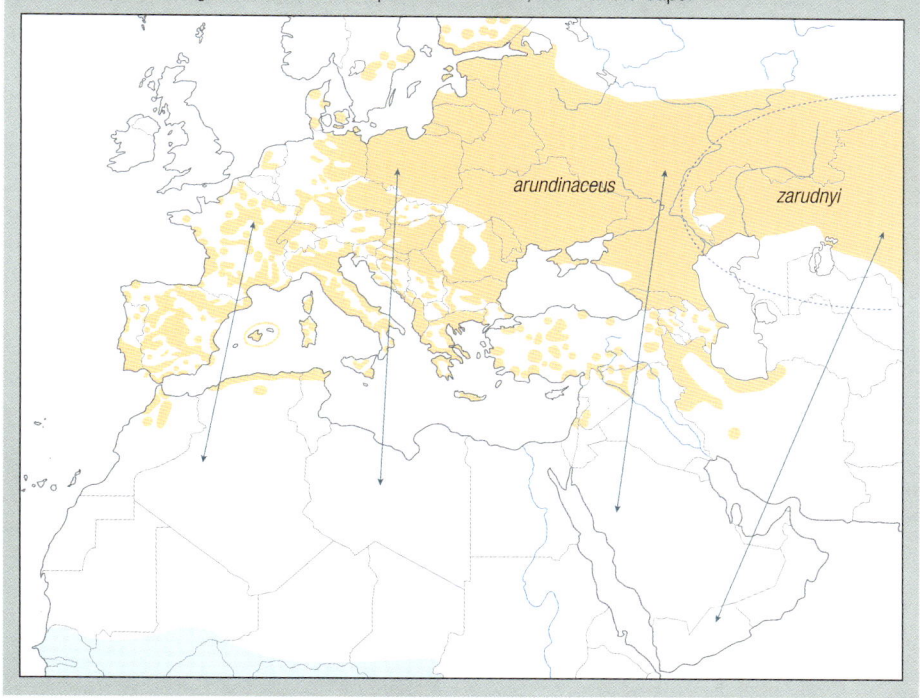

A. a. arundinaceus, Turkey, May: Typical erect posture when perched on twig in dense lake-side reed vegetation. Some are more short-billed than majority. Degree of whitish tips to tail (mostly outer feathers) also variable, but although diffuse rather extensive here. (R. Armada)

A. a. arundinaceus, Greece, Apr: typically tinged tawny and rufous on olive-brown upperparts, with rump and uppertail-coverts even warmer. Note proportionately long primary projection, which separates it from Clamorous Reed. Following completion of pre-nuptial moult, ad and 1stY indistinguishable. (R. Pop)

IDENTIFICATION A large, thickset unstreaked *Acrocephalus* warbler, with a *moderately long, blunt-tipped and deep-based bill* and comparatively long legs. Also *proportionately long-winged with long primary projection* (about equal to exposed tertials, with 8–9 exposed primary tips) and *short, only moderately graduated tail*. Face pattern comparatively well-marked among unstreaked *Acrocephalus*, with more *prominent* (though somewhat variable) *supercilium* usually extending well behind eye, emphasized by darker eye-stripe and lateral crown-stripe, giving the bird a rather stern look. Mainly warm olive-brown above, with rump and uppertail-coverts still warmer and ochre- or rufous-tinged. Wing-coverts and inner flight-feathers have pale rufous fringes. Rather *whiter below*, only *slightly deeper buffish on body-sides*. When seen close, on adults note fine shaft-streaks on sides of throat and across lower throat/upper breast. Bill mainly dusky horn above and on tip, with bright pinkish or flesh-coloured base to lower mandible. *Legs predominantly pale brown or greyish*. On breeding grounds the loud song is usually the earliest and best indication of its presence. Typically 'crashes' through vegetation sluggishly and deliberately, and gait is distinctly less nimble than smaller *Acrocephalus*; sometimes hops powerfully on ground, almost thrush-like. Flight rather laboured and jerky over short distances; tail often held below level of body, and it seems to crash into cover on landing. Often takes erect song-post clinging to reed, and this confident bird frequently perches in open.

VOCALISATIONS Song is loud and guttural, usually a rhythmic series of *karra-karra-karra krie-krie-krie trra-trra kih-e kih-e*, or *karre-karre-keet-keet dree-dree-dree trr trr trih*, occasionally interspersed with piping or creaking notes, and even mimicry (less harsh and tentative in structure than Clamorous Reed Warbler, habitually repeating most notes or phrases 2–4 times). – Commonest calls a deep, throaty *chack*, a harsher *trrak* and deep, croaking *churrr* in alarm.

SIMILAR SPECIES Readily distinguished from smaller unstreaked *Acrocephalus*, e.g. 50–60% larger than *Reed Warbler*, with proportionately larger and deeper-based bill, longer wings, more striking face pattern, and longer-looking tail (also song and calls strikingly different from latter and other similar species). – Thus, confusion much more likely with *Clamorous Reed Warbler* (especially ssp. *brunnescens*), *Basra Reed Warbler* and *Oriental Reed Warbler* (which see).

AGEING & SEXING Can be aged by moult and feather wear prior to completion of moult, iris colour being especially reliable in autumn. Sexes alike, and seasonal plumage varia-

A. a. arundinaceus, Denmark, May: unstreaked chunky *Acrocephalus*, with strong blunt-tipped and deep-based bill. Often raises crown-feathers when agitated. Note long primary projection (equal to tertials) and relatively shorter and less rounded tail. Plain brown upperparts, but rump and uppertail-coverts slightly brighter/warmer. (H. Brandt)

A. a. arundinaceus, juv/1stW, Italy, Aug: many fresh young in autumn are a little paler above, tinged tawny-brown to cinnamon-brown. In early autumn, this age easily separated by mostly fresh/juv body- and wing-feathers, and by dark olive-brown iris. (D. Occhiato)

A. a. arundinaceus, juv/1stW, Italy, Aug: some young are duller and can be washed extensively warm buff below, any whitish being restricted to throat and central belly, while pale supercilium varies in length and degree of buffish tinge. Evenly fresh plumage at this age, with dark olive-brown iris. (D. Occhiato)

A. a. zarudnyi, ad, Kuwait, Sep: unlike young birds (at least until early autumn or arrival in Africa, prior to winter moult), entire plumage, especially tail and wingtips, heavily worn. Iris dark sepia-brown. Plumage also often bleached greyer above and white below, especially noticeable in eastern birds. (M. Pope)

tion limited to effects of wear and bleaching. – Moults. Partial post-nuptial moult of ad on breeding grounds (ad renewing odd flight- and tail-feathers recorded, but exceptional), at same time partial post-juv moult, Jul–Aug, then both suspend and resume, and start flight-feather moult, during migration stopover in N sub-Saharan Africa (Sep–Dec) or complete south of equator (Dec–Jan), in both cases performing complete moult. For convenience all moults in the tropics could be termed 'complete pre-nuptial', although some birds moult in autumn. – **SPRING** Generally fresh while still on migration, but those that largely moulted flight-feathers in previous autumn (rather than in midwinter) already slightly to moderately worn. Subsequently bleach paler, with upperparts duller, less saturated rufous-yellow, and underparts and supercilium whiter, with breast and flanks less heavily suffused, and fringes and tips to wing-feathers reduced. 1stS usually not distinguishable on plumage wear as this is more dependent on moult timing than age, but for some iris colour still differs (see 1stW). – **AUTUMN Ad** Worn, often heavily, at least until early autumn or until reaching Africa. Fine shaft-streaks present on sides of throat and across lower throat/upper breast (though very faint in some ♀♀). Iris dark sepia-brown, bright olive-brown or reddish-brown. Legs on average paler, more warm brown or pinkish (Kennerley & Pearson, 2010). **1stW** In early autumn easily separated by mostly fresh juv plumage, including flight- and tail-feathers. Throat and upper breast unstreaked. Iris dull dark olive-grey. Legs on average greyer and slightly darker than ad (Kennerley & Pearson 2010). **Juv** As 1stW but brighter, more cinnamon upperparts, with warm-coloured underparts (except cream-white central belly), and has soft, fluffy appearance.

BIOMETRICS (*arundinaceus*) **L** 18–20 cm; **W** ♂ 91–102 mm (*n* 21, *m* 97.9), ♀ 88–98 mm (*n* 13, *m* 92.6); **T** ♂ 70–83 mm (*n* 21, *m* 76.4), ♀ 70–77 mm (*n* 13, *m* 73.4); **T/W** *m* 78.6; **TG** 6–12.5 mm (*m c.* 10); **B** 19.9–24.4 mm (*n* 36, *m* 22.5); **B**(f) 15.0–18.3 mm (*n* 31, *m* 16.9); **BD** 5.2–6.2 mm (*n* 32, *m* 5.7); **Ts** 25.5–30.9 mm (*n* 35, *m* 28.6). Wing formula: **p1** < pc 3–10 mm, < p2 54–64 mm; **p2** < wt 0–3 mm, =3 or 3/4 (92%), or =4 (8%); **p3** (or pp2–3) longest; **p4** < wt 1–4.5 mm; **p5** < wt 6–9 mm; **p6** < wt 9–13 mm; **p7** < wt 12–16 mm; **p10** < wt 23–27.5 mm; **s1** < wt 26.5–32 mm; **tert. tip** < wt 20–31 mm; **notch p2** 10–16 mm, =6–8 (8/9) [9].

GEOGRAPHICAL VARIATION & RANGE Two races often recognised, but note individual variation and extensive overlap between them (e.g. worn ad in autumn largely inseparable).

○ ***A. a. arundinaceus*** (L., 1758) (NW Africa, Europe, Turkey, Levant, Transcaucasia, east to Caspian Sea, NW Iran; winters Afrotropics to S Africa). Treated above. Connected with following race by a wide zone of either intermediate birds or both types occurring together in the same area.

○ ***A. a. zarudnyi*** Hartert, 1907 (SE Russia east of Volga, SW Siberia, N Central Asia, W Mongolia; winters E & S Africa). A fraction larger than *arundinaceus* on average and subtly paler and colder greyish, less rufous, being essentially greyish-olive above (indistinctly paler and browner on rump and uppertail-coverts), and pale parts are much whiter, including supercilium, throat, upper breast and belly, while flanks and undertail-coverts are still cream-buff. This said, difference sometimes tiny, and wear of *arundinaceus* makes it more grey-and-white and often more difficult to separate from *zarudnyi*. Also, the two races are connected by wide zone of intergradation. **W** ♂ 94–102 mm (*n* 13, *m* 98.2), ♀ 93–100 mm (*n* 8, *m* 96.8); **T** ♂ 72.5–81 mm (*n* 13, *m* 76.8), ♀ 71–79 mm (*n* 8, *m* 75.3); **T/W** *m* 78.0; **B** 20.8–24.3 mm (*n* 22, *m* 22.76); **B**(f) 15.5–18.5 mm (*n* 21, *m* 17.1); **BD** 5.2–6.3 mm (*n* 21, *m* 5.7); **Ts** 28.5–30.9 mm (*n* 21, *m* 29.6).

TAXONOMIC NOTE Oriental Reed Warbler and Basra Reed Warbler were formerly treated as subspecies of Great Reed Warbler, but most authorities now consider them as separate species, a policy also followed here. Basra Reed Warbler is a W Palearctic breeder and therefore treated in full, whereas Oriental Reed Warbler is an extremely rare vagrant to the covered region.

REFERENCES COPETE, J. L. *et al.* (1998) *J. of Orn.*, 139: 421–424. – LEISLER, B. (1977) *Ibis*, 119: 204–206. – PEARSON, D. J. (1975) *Ibis*, 117: 506–509.

A. a. zarudnyi, Kuwait, May: can be notably paler than *arundinaceus*, being essentially colder and tinged greyish-olive above (less rufous), usually with whiter (less buffish) supercilium, throat, upper breast and belly, and cream-buff body-sides. However, all features of *zarudnyi* represent cline towards *arundinaceus*, making racial designation best based on combination of locality and plumage colours. (G. Brown)

A. a. zarudnyi, juv/1stW, Kuwait, Sep: locality in combination with overall paler and greyer plumage, yet fresh, point to the subtly differentiated *zarudnyi* (here an extremely pale young bird, emphasised by strong light and some feather wear). Aged by mostly fresh juv body- and wing-feathers; also dark iris. (J. Tenovuo)

BASRA REED WARBLER
Acrocephalus griseldis (Hartlaub, 1891)

Fr. – Rousserolle de Basra; Ger. – Basrarohrsänger
Sp. – Carricero de Basora; Swe. – Basrasångare

As befits its name, Basra Reed Warbler is a summer visitor to reedbeds and thickets near water mainly in S and C Iraq, north to Baghdad. Due to severe habitat destruction in the Basra region in the 1980s and 90s the species apparently responded by expanding into new areas, and has recently (since 1995) been proven to breed in Kuwait (albeit perhaps not annually), in SW Iran (since 2006), in N Israel (2006–08) and probably also in Saudi Arabia, where it is occasionally recorded in large numbers on passage at both seasons. Arrives on the breeding grounds from April, from wintering areas in E & SE Africa, and vagrants have been recorded west to Cyprus and Syria, with a single claim from Egypt.

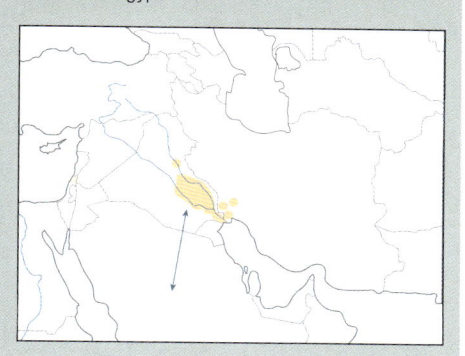

Kuwait, May: more delicate than Great Reed, with equally long primary projection, but more slender bill with purer pink at base and lower mandible. Tail rather short and squarer. At this season, usually tinged rather cold olivaceous-brown above. Uppertail often darker than in Great Reed and Clamorous Reed Warblers. Head well marked with whiter supercilium that is variable in extent behind eye (non-existent above); often enhanced by dark eye-stripe, and white eye-ring also often very obvious. (P. Fågel)

IDENTIFICATION Decidedly smaller and slimmer than other unstreaked large *Acrocephalus*, being *medium-sized* (between Reed and Great Reed Warblers) with moderate to longish and *slender, pointed bill, long wings, long primary projection* (about equal to exposed tertial length, often longer), and *rather short tail*. Upperparts uniform cold *olivaceous brown-grey*, slightly greener and darker than ssp. *zarudnyi* of Great Reed Warbler. Tail rather dark, lacking warm rufous tone of ssp. *arundinaceus* Great Reed (but suffused pale rufous-brown when very fresh). *Whiter below* than Great Reed with *restricted cream-yellow tinge to flanks*, pure white throat and no faint streaks on lower throat/upper breast. Head well marked with a narrow but prominent, *long, whitish supercilium* (cream or buffish on most other larger reed warblers), reaching at least to rear edge of eye, sometimes well behind, enhanced by *quite pronounced dark eye-stripe*; a white eye-ring is also noticeable. Furthermore, note *only moderate tail graduation*. Heavily built legs and toes are dull grey. Gait and general behaviour as Great Read, but *less bulky and clumsy*.

VOCALISATIONS Song is best described as a cross between Reed Warbler and Great Reed Warbler, a sequence of chirping and squeaky notes with slight variation at moderate ('calmly talking') pace, more subdued than Great Reed, less grating and guttural, still with some more raucous and squeaky notes than given by Reed, thereby hinting at a slightly larger bird, e.g. *chuk chivvy cheep churrac cheep chivy chuk cher cheep chivip-chivip chivy chu-chic...* – Main call (of migrants) a rather harsh, nasal *chaarr* (louder than similar note of Reed Warbler). Reportedly also gives a *trak* or *chack*, less deep than Great Reed Warbler.

SIMILAR SPECIES Field identification rather difficult and requires care and, preferably, experience with all relevant forms. Considerably smaller than *Great Reed Warbler* and best separated from it by proportionately more slender, finely pointed bill which appears longer. Also differs in stronger facial pattern, slightly darker flight- and tail-feathers (tail especially is dark, when fresh), and underparts on average whiter. Legs also look greyish (not greyish-brown as Great Reed). However, ssp. *zarudnyi* of Great Reed (especially those from extreme east of range) is to some extent closer to Basra Reed and can appear equally cold above and paler below with emphasized face markings. Some intermediate-looking birds could prove very puzzling, and require in-hand examination for positive identification! – Separation from *Clamorous Reed Warbler* (*stentoreus* and *levantinus*) is rather straightforward, especially given usually clear structural differences. Basra Reed has proportionately far longer wings and primary projection, with a relatively shorter, less graduated tail, its bill is marginally shorter and straighter, and overall plumage is paler and less saturated, especially below, with considerably stronger facial marks and dark tail contrast. However, race *brunnescens* of Clamorous Reed (Iran, Arabia) appears to be an overlooked pitfall, especially as it is relatively long-winged (versus *stentoreus*), short-billed and has a less graduated tail and colder coloration (upperparts sometimes greyish-olive like Basra Reed) with paler underparts, and is thus less easily separated. Given prolonged close views, however, Basra Reed should still differ structurally, especially the longer primary projection, with 7–8 (rather than in *brunnescens* usually six, rarely seven) exposed primary tips, which are more widely and evenly spaced, and its less graduated tail. Also, Basra Reed has better-marked face pattern than any Clamorous Reed, and in particular has a longer and whiter supercilium. Furthermore, Basra Reed moults later than *brunnescens*, thus adults are unlikely to be seen moulting in autumn (unlike all ages of Clamorous Reed), and in spring all ages of Basra Reed are fresher—but much experience and preferably direct comparison is required to evaluate this, and individual variation exists in both species. – Another confusion risk is the quite similar *Oriental Great Reed*, which see. – Furthermore, bear in mind that even experienced observers

Kuwait, May: here the paler and greyer upperparts are more obvious. Note long rather thin bill with much pink basally, flat-crowned outline and strong legs, which are dull grey with flesh tinge. Prominent whitish eye-ring broken by dark eye-stripe. Following pre-nuptial moult, ad and 1stY indistinguishable. (A. Al-Sirhan)

Iran, Apr: well-marked individual with longer whitish supercilium, clearly reaching behind eye and enhanced by dark eye-stripe, which breaks whitish eye-ring only thinly (in all other relevant taxa, the dark loral line breaks the pale eye-ring more broadly and diffusely). Can show faint streaks on breast, but more subdued than Great Reed. (W. Müller)

Ad, Kuwait, Sep: ad plumage in early autumn often bleached greyer above. Unlike young birds (at least until early autumn or before arriving in Africa, prior to complete pre-nuptial moult), most of ad plumage, especially wingtips, heavily worn. Ad replacing occasional remiges and rectrices outside Africa known but are rare. (J. Tenovuo)

may occasionally mistake, at least initially, a large, dark eastern (ssp. *fuscus*) Reed Warbler as Basra Reed Warbler.

AGEING & SEXING Much as Great Reed Warbler. – Moults. Both ad and 1stY moult flight-feathers in winter quarters, often on sub-Saharan stopover in Sep–Nov, but some suspend and finish later and further south in E Africa, maybe (Dec)Jan–Mar. Pre-nuptial moult in Mar limited to head and body. – SPRING Generally very fresh until early spring (when upperparts perhaps somewhat brighter, especially mantle, with even darker tail), but in Apr–May bleaches paler, with upperparts duller and colder, and underparts and supercilium whiter, breast and flanks much less saturated, and fringes and tips to wing-feathers narrower. Ageing of 1stS based on more olive-grey tone to iris requires much practice and generally in-hand examination (iris of ad apparently dark rufous-brown in all or most). – AUTUMN Ad Worn at least until early autumn or until arrival in Africa. Iris dark sepia-brown. **1stW** Fresh, with juv remiges and tail, at least until early autumn. Iris olive-grey. **Juv** Like ad, but has soft, fluffy body-feathering, and brighter upperparts with slightly pale cinnamon underparts. Face markings less distinct.

BIOMETRICS L 15–16 cm; **W** ♂ 81–84.5 mm (n 10, m 83.1), ♀ 78–85 mm (n 11, m 81.8); **T** ♂ 60–67 mm (n 10, m 63.3), ♀ 56–64 mm (n 11, m 60.6); **T/W** m 75.1; **TG** 4–8.5 mm (m 6.2); **B** 20.1–23.8 mm (n 21, m 22.0); **B**(f) 14.6–18.2 mm (n 19, m 16.6); **BD** 3.9–4.7 mm (n 20, m 4.3); **Ts** 23.5–26.8 mm (n 21, m 25.1). **Wing formula: p1** < pc 2.5–7.5 mm, < p2 45–54 mm; **p2** < wt 0–3 mm, =3/4 or 4 (89%), or =3 (11%); **p3** longest (sometimes pp2–3); **p4** < wt 1.5–3 mm; **p5** < wt 4–7.5 mm; **p6** < wt 7–10 mm; **p7** < 9–14 mm; **p10** < 17–21 mm; **s1** < 20–24 mm; **tert. tip** < wt 17–24 mm; **notch p2** 11–15 mm, =pp7–10.

GEOGRAPHICAL VARIATION & RANGE Monotypic. – W, C & S Iraq, Kuwait, SW Iran; has bred N Israel; probably breeds Arabia; winters E & SE Africa. (Syn. *babylonicus*.)

TAXONOMIC NOTE Long treated as a subspecies of Great Reed Warbler, but nowadays commonly regarded as a separate species based on distinctive structure, plumage and song.

REFERENCES AYÉ, R. (2006) *DB*, 28: 304–306. – VAN DEN BERG, A. & SYMENS, P. (1992) *DB*, 14: 41–48. – PERLMAN, Y. & SHANNI, I. (2008) *Alula*, 14: 46–47. – TENOUVO, J. (2006) *BW*, 19: 66–68.

Juv/1stW, Kuwait, Jul: fresh young often tinged brownish (less olive) above. However, unlike Great Reed, note slenderer, finely pointed bill, usually better marked whitish eye-ring, slightly darker uppertail, and whiter underparts. Legs greyish, not paler and mostly brownish-grey as Great Reed. Separation from Clamorous Reed easy if structural features correctly evaluated, but *brunnescens* Clamorous Reed differs less, including having more similar coloration and face pattern. (M. Pope)

1stW, Kuwait, Oct: slightly warmer tawny-tinged upperparts often characterise fresh young. Unlike Great Reed and Clamorous Reed, note longer whitish supercilium, reaching clearly behind eye, and proportionately very long primary projection. (A. Al-Sirhan)

THICK-BILLED WARBLER
Iduna aedon (Pallas, 1776)

Fr. – Rousserolle à gros bec; Ger. – Dickschnabelsänger
Sp. – Carricero picogordo; Swe. – Tjocknäbbad sångare

A summer visitor to S Siberia east to NE China, wintering in SC and SE Asia, Thick-billed Warbler inhabits taller thickets and reedbeds with scattered trees and bushes, but is not dependent on the presence of water. It is a very rare vagrant to N Europe, with most records from Shetland (Britain), but once recorded as far south as Egypt. Most vagrants have appeared in autumn.

Presumed 1stW, Scotland, Oct: quite unique jizz, particularly short, stubby bill, rounded head with hint of rufous cap, and short primary projection but proportionately long tail. Dark olive iris suggests 1stW, but unknown if remiges are juv (seemingly not contrasting enough with fresh-looking tertials). (H. Harrop)

IDENTIFICATION A bulky unstreaked warbler more reminiscent of an *Acrocephalus* than an *Iduna*, distinctive by its *relatively short and somewhat stubby bill* (diagnostic), and combination of *very plain, almost featureless and more rounded head*—lacking pale supercilium and dark loral stripe—and almost *uniformly pale lores* giving an open-faced appearance. Typically has *short wings and short primary projection*, and *tail is proportionately very long and markedly graduated*. Otherwise rather plain, with largely uniform fulvous olive-brown upperparts (though often has warmer brown crown, contrasting with greyish head-sides, and rump/uppertail-coverts and fringes to remiges often slightly more rufous, too). Pale ochre-white underparts rather strongly suffused buffish on breast-sides and flanks, whereas quite long undertail-coverts are deeper ochre. Diffuse eye-ring cream-white. Bare parts: upper mandible dark horn-brown and lower mandible entirely yellow-horn or flesh; legs grey or bluish-pink. General impression is actually shrike-like, especially given long tail and stubby bill, which is reinforced by the species' preference for more bushy areas and glades.

VOCALISATIONS The loud warbling song is very different from Great Reed Warbler, being fast, highly melodious and varied (lacking repetitive structure or deep guttural elements), and as it also contains much expert mimicry it can recall Marsh Warbler but is louder and lacks the typical harsh *zi-cheh zi-cheh…* of latter. Song often introduced by repeated harsh clicking calls, *chek, chek, chek…* – Calls include the harsh clicking *chek* and doubled *tack-tack* (usually much less deep than in equally large *Acrocephalus*, more like smaller *Iduna* and *Acrocephalus* species), but sometimes a louder, harsh, fast-repeated *chok-chok* and a hard *tschurrr*.

SIMILAR SPECIES The overall impression, especially from structure and behaviour, is quite unlike any other large *Iduna*, *Hippolais* or *Acrocephalus*. In the covered region might be confused with *Great Reed* or *Clamorous Reed Warblers*, both of which are bulkier and differ diagnostically in having a dark tip to the longer, less stubby bill. Great Reed also has better-marked head pattern (with obvious pale supercilium, dark lores and eye-stripe), longer wings and a proportionately shorter, less graduated tail. Clamorous Reed superficially shares some of the structural and plumage features, but its bill dimensions are strikingly different. – Separation from *Oriental Reed Warbler* is dealt with under that species. With experience, and reasonable views, Thick-billed Warbler has a very different jizz from all the above species, which is useful once learnt.

AGEING & SEXING For ageing, moult and feather wear can be used prior to completion of moult. Iris colour apparently most useful in autumn. Sexes alike including being very similar in size, and plumage variation rather limited (due to wear and bleaching). – Moults. Complete post-nuptial moult of ad and partial post-juv moult in late summer–winter, protracted, often starting with head, body, tail partly or wholly, and some wing-coverts on breeding grounds, then suspended during migration until resuming moult immediately upon arriving in wintering areas, where most feather renewal including most of flight-feathers occurs: ad finishes earlier (Aug–Oct) than 1stW (Oct–Jan). Post-juv moult includes head, body and apparently sometimes a few tail-feathers on breeding grounds, followed by moult in winter quarters of remiges and rest of plumage not yet moulted. Occasional complete post-juv moult in autumn has been

Presumed 1stW, China, Oct: stubby bill enhanced by lack of dark tip to lower mandible. Also typical are pale (greyish) lores and bland face. Underparts rather strongly suffused buffish, and long undertail-coverts deeper ochraceous. Same individual as next to the right. (J. Martinez)

Presumed 1stW, China, Oct: rather short rounded wings, and tail proportionately very long and seemingly well graduated. Moult and ageing can be quite complex: the iris here appears reddish-brown (suggesting ad), but narrow and pointed tail-feathers already abraded is probably stronger indication of age. (J. Martinez)

China, May: at times somewhat shrike-like due to very long tail, large head and stubby bill. Ad and most young birds have a complete moult between late summer and winter, thus in spring most birds should be evenly feathered and impossible to age. (B. van den Boogaard)

Mongolia, May: on migration often feeds on the ground. Again, note hint of rufous darker cap, subtly contrasting with paler, more greyish-tinged lores, head- and neck-sides. At this season, most are impossible to age. (H. Shirihai)

Ad, Siberia, Jul: just after breeding, but before post-nuptial moult has started or advanced much, ad are frequently quite worn and tatty. This is not to be confused with the slightly fluffy plumage of recently fledged young birds, which are in fresh, neat plumage. Note characteristic strong, thrush-like bill, rather unique among Palearctic warblers. (M. Henderson)

claimed (*BWP*), but proof not entirely convincing and further study required. Partial pre-nuptial moult of head, body, some secondary-coverts and tertials, occasionally central tail-feathers, in winter quarters (Mar–Apr/May). – **SPRING** Entire plumage fresh Jan–Feb but worn Mar–Apr onwards. **Ad** With wear bleaches paler, upperparts duller and flight-feathers, tail-feathers and wing-coverts browner (pale fringes and tips reduced), and underparts paler with less buff wash to breast and flanks. **1stS** Indistinguishable from ad if winter moult complete, but some retain juv inner primaries and/or outer secondaries, and a few primary- and secondary-coverts, which are then contrastingly heavily worn. Perhaps iris colour is useful as with other similarly-sized warblers (see e.g. Great Reed Warbler). – **AUTUMN** In early autumn, ad is worn with strong moult contrast, or in active moult, while 1stW is fresh with less contrasting moult limits. **Ad** Fresh plumage (from Oct/Nov onwards). A browner bird, less olive-tinged than in spring. **1stW** Much like fresh ad, but in early autumn most flight-feathers, primary-coverts, some or most secondary-coverts and some tail-feathers are still juv, being subtly narrower, fresh or only moderately worn, becoming gradually slightly abraded from Sep. Apparently differs in iris colour as in Great Reed Warbler (which see), but not closely studied. Upperparts more rufous, with buffish-cinnamon wash to underparts. **Juv** Very fresh plumage, slightly softer and 'looser' than in ad, and tail-feathers slightly narrower on average. Unlikely to reach the treated region.

BIOMETRICS L 17–19 cm; W ♂ 77–86 mm (*n* 30, *m* 80.9), ♀ 74–84 mm (*n* 19, *m* 78.7); T ♂ 78–91 mm (once 74; *n* 30, *m* 83.8), ♀ 77–88 mm (*n* 18, *m* 82.1); **T/W** *m* 103.6; **TG** 15–31 mm (*n* 34, *m* 22.3); **B** 17.4–20.3 mm (*n* 51, *m* 19.0); **BD** 5.1–6.3 mm (*n* 52, *m* 5.5); **BW** 4.8–5.9 mm (*n* 26, *m* 5.2); **Ts** 25.5–29.6 mm (*n* 37, *m* 27.6). **Wing formula: p1** > pc (1.5) 3–11 mm, < wt 25.5–31.5 mm; **p2** < wt 7–10 mm, =7/8 or 8 (50%), =6/7 or 7 (42%), or =8/9 or 9, or shorter (8%); **pp3–4** (5) about equal and longest; **p5** < wt 0–2 mm; **p6** < wt 2.5–5 mm; **p7** < wt 5.5–8 mm; **p10** < wt 13–16 mm; **s1** < wt 15.5–20 mm; **notch p2** < ss. Emarg. pp3–5 (on p5 rarely less distinct).

GEOGRAPHICAL VARIATION & RANGE Monotypic. – Siberia east to Amur, N Mongolia, NE China; winters India, SE Asia. – Sometimes an easterly race ('*rufescens*'; altered to '*stegmanni*' when placed in *Acrocephalus*) separated on account of smaller size and subtly darker and more rufous-tinged upperparts, but plumage differences not possible to confirm when reasonably long series from same season were compared, and size differed only very subtly (1–2% average difference; very slight in subspecies taxonomy) with large overlap, hence not followed. (Syn. *rufescens*; *stegmanni*.)

TAXONOMIC NOTES Formerly kept in separate genus, *Phragamaticola* (but see below), or lumped with *Acrocephalus*, but genetic evidence has recently shown it to be closely related to *Iduna* and *Hippolais*, being sister species and basal to the group containing among others Booted, Sykes's, Olivaceous and Isabelline Warblers (Fregin *et al.* 2009). Since inclusion in *Iduna* avoids a monotypic genus it is preferred here. – See Pittie & Dickinson (2013) for reason why *Phragamaticola* is a junior synonym of *Arundinax*.

REFERENCES Dickinson, E. C. & Gregory, S. M. S. (2006) *Zool. Med. Leiden*, 80: 169–178. – Fregin, S. *et al.* (2009) *Mol. Phyl. & Evol.*, 52: 866–878. – Pittie, A. & Dickinson, E. C. (2013) *Zool. Bibliography*, 2: 151–166.

ISABELLINE WARBLER
Iduna opaca (Cabanis, 1851)

Alternative name: Western Olivaceous Warbler

Fr. – Hypolaïs obscure; Ger. – Isabellspötter
Sp. – Zarcero bereber; Swe. – Macchiasångare

A close relative of Olivaceous Warbler, but restricted to S Iberia and coastal parts of NW Africa east to at least Tunisia. A summer visitor, arriving early to Morocco (late March and early April) but in Spain usually not until late April–early May. Winters in W Africa between Senegal and N Cameroon. Breeds in riverine forests, tall maquis (mainly Iberia), orchards, palm groves and lush gardens.

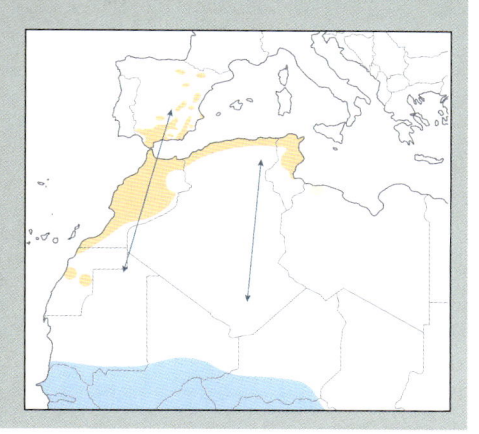

Morocco, Mar: rather long-tailed, but unlike eastern congeners does not 'dip' the tail, except a shallow 'trembling' while singing. Whitish lores and short supercilium. Safe elimination of *reiseri* Olivaceous can be difficult, relying on plain wing, stronger bill and overall longer tail and large size, still is sometimes impossible without hearing the song or handling. (T. Bedford)

Morocco, Mar: the most featureless *Iduna*, isabelline- or pale drab-brown above, with almost uniform wing (no paler secondary panel) and cream-white below (faint grey-brown hue on body-sides). However, has quite long strong bill with obvious pinkish-yellow lower mandible. Following pre-nuptial moult, ad and 1stY indistinguishable. (V. Legrand)

IDENTIFICATION Usually noticed, and to the experienced ear identified, by its song. Agile and very active, keeps well hidden in canopy, flight is fast and direct. Medium-sized, slim-shaped warbler with nondescript features, being isabelline-brown ('milky-tea') above and off-white below with a faint grey-brown hue on breast-sides and flanks, lacking any particular features except a quite *long and strong bill* with pinkish-yellow lower mandible. The bill is 'swollen' and has almost invariably *very slightly convex sides* to its outer half if seen from below, and this can on rare occasions be ascertained in the field (but bill shape is more often impossible to assess). Proportionately it is a *rather long-tailed* bird, but surprisingly it *does not dip or otherwise move its tail* like its relatives (apart from a fine 'shivering' while singing). The longish tail can appear rather full in its outer parts, with slightly rounded sides. That the outer tail-feathers are faintly edged and tipped off-white is not always easy to see. Legs strong. Pale supercilium short and poorly marked, lores pale recalling facial expression of Melodious Warbler. Although the upperparts often appear pale brown in sunlight, they have a faint olive sheen to the brown, at least before too worn, which is easier to see in overcast weather. In normal light no pale panel is visible on the folded wing.

VOCALISATIONS Song differs from Olivaceous Warbler (which see) by being less scratchy and monotonous, lacking the cyclic structure, being more varied and including nasal notes (recalling Icterine Warbler). It is delivered at a slightly slower pace, sounding 'better pronounced' as if 'talking' (thereby recalling Reed Warbler, but lacking the habitual repetition 2–3 times of most notes of that species). Strophe-length varies with mood, often 15–30 sec. Each strophe commonly opened with a few clicking notes, *chek chek chek...* – Call is a tongue-clicking, 'fat' *chek*, very similar to Olivaceous, possibly a trifle louder but doubtfully separable. Short series of calls, and churring *cher'r'r'r'r* can also be heard when on territory.

SIMILAR SPECIES Needs to be separated chiefly from Olivaceous Warbler in NW Africa (*reiseri*), which locally may breed sympatrically but usually occurs in drier habitats away from the coastal zone (desert oases, palm groves, etc.), and from Melodious Warbler in Iberia. Very similar to Olivaceous *reiseri* but has different song, does not move the tail (Olivaceous frequently 'dips' its folded tail), is a trifle larger and has stronger, more 'thick-tipped', 'swollen-looking' bill (*reiseri*: more finely pointed outer bill). Also, the closed wing

Morocco (left: May; right: Mar): both these birds have uniform wings concolorous with upperparts, relatively long tail, long 'swollen' bill and strong legs. However, note typical isabelline coloration of right-hand bird in shade, compared to much paler and greyer-looking bird exposed to direct sunlight. (Left: A. Faustino; right: V. Legrand)

Summer Isabelline (left: Spain, Jun) and Olivaceous Warblers (right: *reiseri*, Morocco, May): song is safest means of identification, with calls useful too, while behaviour (especially tail movements) also provides clues. However, use of structural and plumage features may be necessary if faced with a silent bird. Note Isabelline's proportionately longer tail and less pointed bill. (Left: Q. Marcelo; right: O. Jonsson)

Ad, Spain, Jul: even in worn plumage, ad is separated from *reiseri* Olivaceous by overall larger size, chunkier jizz, stronger bill, overall duller colour, and lack of paler secondary panel. (K. Drissner)

appears uniform (often paler buff-brown secondary panel on *reiseri*), and overall colour is slightly darker (*reiseri* is a fraction paler brown above). On average more extensive and well-marked white tips to outer tail-feathers in *reiseri* than in Isabelline Warbler. – From Melodious, which has similar call and in certain lights can appear 'colourless' below, by much slower song, lack of pale secondary panel, slightly paler and browner upperparts, and not as pale lores. – There is also a remote risk of confusion with Reed Warbler, but this has rounded tail and stronger rufous tinge above, especially on rump. Also, Reed has thinner bill, on average darker legs, somewhat longer primary projection with darker primaries and often contrasting fine whitish tips to these (Isabelline has shorter projection and paler, more uniform primaries).

AGEING & SEXING Insignificant seasonal and age-related differences. Sexes alike in plumage and size. – Moults. Complete moult of both ad and 1stY generally in mid or late winter (Dec–Mar, sometimes Nov), but some spring birds are quite worn indicating that some moult earlier in autumn. – **SPRING** Ageing not possible. – **AUTUMN Ad** Worn, especially tips of outer primaries, and tertials. Note that some moult head and body while still in breeding area. **1stCY** Uniformly fresh.

BIOMETRICS L 13–14.5 cm; **W** ♂ 64.5–74 mm (*n* 58, *m* 70.2), ♀ 63.5–71.5 mm (*n* 22, *m* 68.1); **T** ♂ 52.5–61.5 mm (*n* 58, *m* 57.6), ♀ 51–58 mm (*n* 23, *m* 54.9); **T/W** *m* 81.6; **B** 16.4–19.5 mm (*n* 96, *m* 17.7); **BD** 2.8–3.8 mm (*n* 92, *m* 3.3); **BW** 4.6–5.6 mm (*n* 94, *m* 5.0); **Ts** 21.0–24.2 mm (*n* 94, *m* 22.6). **Wing formula: p1** > pc 4–9.5 mm, < p2 25–31.5 mm; **p2** < wt 4–8 mm, =7 or 7/8 (61%), =6/7 (33%) or =8 (6%); **pp3–5** about equal and longest (though in particular p5 often 0.5–1.5 mm <); **p6** < wt 1.5–4.5 mm; **p7** < wt 5–7.5 mm; **p8** < wt 7.5–10.5 mm; **p10** < wt 11–16 mm; **s1** < wt 13–18 mm. Emarg. pp3–5; p6, too, has slight emarg. in *c.* 50%.

GEOGRAPHICAL VARIATION & RANGE Monotypic. – Spain, Morocco, N Algeria, N & E Tunisia, possibly also NW Libya; winters W Africa from Senegal to N Cameroon. Has straggled to Sweden (once), Egypt (once) and Greece (once, although last record, based on museum specimen, might involve a labelling mistake).

TAXONOMIC NOTES Recently separated as independent species from the Olivaceous Warbler complex based on differences in morphology, vocalisations and behaviour (cf. above), also on substantially different DNA, and apparently no interbreeding with parapatric *reiseri* of Olivaceous Warbler (Roselaar in *BWP*, Helbig & Seibold 1999, Svensson 2001, Ottosson *et al.* 2005). – Helbig & Seibold (1999) showed that *opaca* and three other *Hippolais* taxa were closer to *Acrocephalus* and suggested subgenus *Iduna*. Dickinson (2003) adopted this but used *Iduna* as full genus. Latter policy found strong support in Fregin *et al.* (2009) using both mitochondrial and nuclear data for near-complete Acrocephalidae phylogeny, and is followed here. Still, its consequence that two extremely similar species, Olivaceous and Upcher's Warblers, appear in different genera seems odd.

REFERENCES FREGIN, S. *et al.* (2009) *Mol. Phyl. & Evol.*, 52: 866–878. – HELBIG, A. & SEIBOLD, I. (1999) *Mol. Phyl. & Evol.*, 11: 246–260. – OTTOSSON, U. *et al.* (2005) *J. of Orn.*, 146: 127–136. – SVENSSON, L. (2001) *BW*, 14: 192–219.

Juv/1stW, Spain, Aug: aged by seemingly fresh plumage. Typical head and bill shapes in profile. Faint short supercilium and hint of pale eye-ring are the only features to note. (L. Barrón)

BUSH AND REED WARBLERS

OLIVACEOUS WARBLER
Iduna pallida (Hemprich & Ehrenberg, 1833)

Alternative name: Eastern Olivaceous Warbler
Fr. – Hypolaïs pâle; Ger. – Blasspötter
Sp. – Zarcero pálido; Swe. – Eksångare

From thickets in a Greek grove or an Algerian oasis, from the tall *Robinia* trees in the Topkapi Palace garden in Istanbul, or from riverine forests in Uzbekistan you hear the same scratchy and squeaky, repetitive song of this dull grey-brown and off-white songster. A summer visitor, arriving in late March to May from wintering grounds in C and E Africa, populations at the southern edge of the range might be resident.

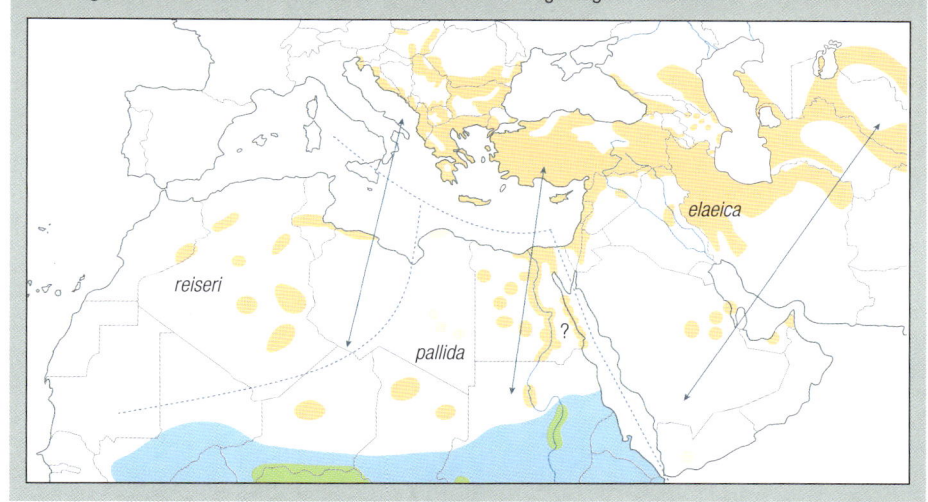

I. p. elaeica, Israel, Mar: some *elaeica* can initially suggest Sykes's Warbler, obviously due to being similarly brown above and whitish below. Note that *elaeica* has slightly longer primary projection and lacks Sykes's usually dark-smudged lower mandible tip. Also, Olivaceous' pale secondary panel and whitish secondary tips are absent in Sykes's. (D. Occhiato)

I. p. elaeica, Israel, Mar: typical coloration, with olive greyish-brown upperparts and pronounced pale secondary panel; underparts off-white, sullied slightly creamy-buff at sides. Proportionately rather short, square-ended tail but longish primary projection, although these features are not always obvious and require field experience with entire group. Following pre-nuptial moult, ad and 1stY indistinguishable. (A. Ouwerkerk)

IDENTIFICATION Usually noted and identified by its characteristic song. A lively bird that mostly keeps in the canopy of trees, hopping about, stretching for insects and suddenly flying off in direct, fast flight to next tree. Medium-sized, about as large as a Willow Warbler, slim, grey-brown above and off-white below with a creamy tinge on breast-sides and flanks, and (in *elaeica*, and less obviously in *reiseri*) with a hint of paler edges to long tertials and secondaries forming a slight *pale panel on the folded wing* in fresh plumage. Identification clinched when *repeated downward 'dips' of the tail* are seen, the *folded* tail being quickly 'pumped' down—'dipped'— from horizontal and back (each tail-dip often accompanied by a call). In many birds with reasonably fresh plumage, fine pale tips to secondaries form a whitish narrow mark on the folded wing (*elaeica*). Reed Warbler-like facial pattern with pale eye-ring and short pale supercilium, and Reed Warbler-sized bill, and as in that species *lower mandible invariably all pinkish-yellow*. Outer tail-feathers tipped and edged whitish, especially on inner webs, although the pale edges may be worn off or difficult to see in the field due to being slightly sullied brown-grey, especially in young birds. Legs greyish with variable pink tinge, though toes generally greyer.

VOCALISATIONS Song is a rather monotonous 'babbling' strophe, with hoarse, scratchy and squeaky nasal notes, and only few whistles, these often recurring in high-pitched parts of song (and can, perhaps surprisingly, recall song of Budgerigar *Melopsittacus undulatus*). Recognised by being cyclically repeated, with notes going up and down the scale in a repetitive pattern, lower notes scratchy, higher squeaky or whistling. Notes tend to blur together, articulation being poor. Very little variation geographically from E Morocco to Kazakhstan. – Call a tongue-clicking *chek*, slightly 'thick' with a nasal undertone, often repeated when feeding, usually given when moving in canopy and in synchrony with tail movements. Sometimes calls run together in subdued, fast, short series, *cher'r'r'r*, or somewhat slower, *chet-et-et-et-*….

SIMILAR SPECIES For separation from very similar sibling species in Europe and Africa, Isabelline Warbler, see that species. – In Middle East and Central Asia, the main pitfall is the strong similarity between Olivaceous and *Sykes's Warbler*. Sykes's does not dip its tail downwards, at most twitching it up or sideways (only marginally downwards). Further, Sykes's lacks a paler secondary panel on the closed wing (other than a very faint one in certain lights), does not have a whitish mark at the tips of the folded secondaries, is more uniformly sandy grey-brown above, with duller brown exposed primaries (slightly darker and greyer in *elaeica* Olivaceous) and has slightly shorter primary projection than in *elaeica* (2/3 of visible tertial length rather than 3/4). Olivaceous invariably has an all-pinkish lower mandible, whereas Sykes's often has a faint dark smudge near the tip. Still, separation of *pallida* Olivaceous in NE Africa and S Levant from a vagrant Sykes's would be a real challenge as *pallida* compared to *elaeica* has paler brown primaries, shorter primary projection and more uniform wing. Song and tail movements differ, but silent birds must be observed very carefully, and some are probably inseparable in the field. – Ssp. *elaeica* must be separated in Turkey and SE Central

I. p. elaeica, Turkey, May: note whitish tips and edges to outer tail-feathers. Also obvious are pale secondary panel and general coloration. (G. Thoburn)

— 471 —

Asia from *Upcher's Warbler*, but latter is slightly larger, has darker uppertail and darker exposed primaries, and the white tips on outer tail-feathers are purer white and usually more prominent (rarely though, the tail pattern can overlap). Look also for average differences in prominence and shape of pale secondary panel (if not abraded), this being narrower in Olivaceous and running along nearly whole length of secondaries/longest tertial, not forming a rather prominent whitish oval on the centre of the secondaries (though exceptions do occur, with Upcher's having pale edges to entire length of secondaries). Apart from having a pale secondary panel, Upcher's usually also has pale-edged greater coverts (in Olivaceous only seen in some fresh first-winter birds). Also, Upcher's has a slightly fuller tail, waving tail-movements with slightly spread tail and is not known to frequently practise repeated 'tail-dipping' with closed tail, if at all. Even after paying attention to all these characteristics a few birds can remain difficult to separate, but note more well-defined pale supercilium in front of eye in Upcher's, slightly more diffuse in Olivaceous. – The 'tail-dipping' habit is shared with *Olive-tree Warbler*, and this can lead to a mistake when the smaller size and finer bill of Olivaceous are not evident. Note that Olivaceous invariably has a pink tinge to its greyish legs, not all-grey legs, and has paler grey cheeks and very faint grey cast on breast-

I. p. elaeica, Israel, May: beware that differences in jizz and tail movement do not always easily eliminate Upcher's. Also, due to wear the characteristic pale fringes to secondaries in *elaeica* can be lacking, and similarly the white tips to secondaries can be absent. Further, if wings are not tightly folded the shorter primary projection is difficult to evaluate. Still, note shorter and slenderer tail being not so dark above, and also pinkish (less greyish) legs. (H. Shirihai)

I. p. elaeica on spring migration (Israel; left: Mar; right: May): can have very different jizz to birds observed on breeding territories, but the usual clues still work, like wing structure and general colour, tail colour and tail movements, bare-parts coloration, as well as calls. (Left: K. Malling Olsen; right: A. Ben Dov)

sides and flanks alone. Greater coverts are plainer grey with only slightly paler edges in Olivaceous. The shorter primary projection of Olivaceous should clinch the identification. – There is a real risk of confusion in the Red Sea region and N East Africa, where either ssp. *pallida* or claimed ssp. *alulensis* can co-exist with *Common Reed Warbler* of ssp. *avicenniae*. Note slightly shorter, straighter and more blunt-tipped bill in Olivaceous, often pinkish tarsi with greyer toes, and different song. Habit of dipping folded tail should also help. – A remote risk of confusion when only the song is heard is with *Masked Shrike*, whose song is similarly scratchy and monotonous, but deeper and even more monotonous in pitch, and easily separated once learnt. – Confusion has been reported also with aberrantly singing *Garden Warbler*, which occasionally can have a hoarse voice, but this species has shorter, thicker and darker bill, and longer primary projection, aside from many other differences.

AGEING & SEXING (*elaeica*) Insignificant seasonal and age-related differences. Sexes alike. – Moults. Complete moult of both ad and 1stY generally in midwinter (Dec–Feb, sometimes Nov), but many autumn migrants (30%) arrive, e.g. in Kenya, with interrupted moult (Pearson *in BWP*), thus moult may start earlier in autumn at stop-over sites. Furthermore, spring birds are quite heavily worn already in May soon after arrival, indicating that some might moult earlier in autumn. – **SPRING** Ageing impossible. – **AUTUMN Ad** Worn, especially tips of outer primaries, and tertials. Note that some moult head and body while still in breeding area, which then look fresh. **1stCY**

I. p. elaeica, ad, Oman, Sep: worn tips to exposed primaries, tertials and tail; head and body appear fresh, probably due to partial moult in breeding area, which is not unusual. Due to feather wear, wing panel less clear. Obvious pale eye-ring and short (almost absent) pale supercilium. Greyish pink-brown legs also support ageing. (H. & J. Eriksen)

I. p. elaeica, juv/1stW, Greece, Jul: aged by uniformly fresh plumage, and soon after fledging bill is often still slightly shorter than ad. Compared to Booted and Sykes's, note slightly darker flight-feathers, with more contrasting pale fringes. Also, upperparts are more grey-tinged than brown. (S. Garvie)

Uniformly fresh. **Juv** As 1stCY but soon after fledging bill often slightly shorter, and plumage has somewhat warmer buff tinge.

BIOMETRICS (*elaeica*) **L** 13–14 cm; **W** 62–72 mm (n 209, m 66.8); **T** 46–56.5 mm (n 209, m 51.6); **T/W** m 77.2; **B** 14.1–17.5 mm (n 208, m 15.9); **BD** 2.6–3.6 mm (n 183, m 3.2); **BW** 3.6–4.9 mm (n 190, m 4.2); **Ts** 19.0–22.4 mm (n 182, m 20.9). **Wing formula: p1** > pc 1.5–7.5 mm, < p2 26.5–33 mm; **p2** < wt 2.5–7 mm, =6 or 6/7 (66%), =5/6 (20%), =7 (8%) or =5 or 7/8 (each 3%); **pp3–4** about equal and longest; **p5** < wt 0.5–2 mm; **p6** < wt 2–6.5 mm; **p7** < wt 4–9.5 mm; **p8** < wt 6.5–11.5 mm; **p10** < wt 10–15.5 mm; **s1** < wt 12–18 mm. Emarg. pp3–5; rarely (in only c. 5%) p6, too, has very slight emarg.

I. p. elaeica, juv/1stW, Israel, Aug: beware that some young *elaeica* in autumn can closely resemble Upcher's Warbler, to the extent that calls and behaviour are required to confirm identification; differences in size and shape are slight and not always obvious. Nevertheless, note rather short tail and near lack of any secondary panel. Aged by evenly and only moderately worn plumage. (A. Ben Dov)

I. p. elaeica, juv/1stW, Kuwait, Sep: some young are tinged sandier and can have diagnostic wing markings (whitish panel and secondary tips) subdued. Rather long supercilium could invite confusion with Sykes's, but *elaeica* lacks the latter's dark-smudged lower mandible tip, and primary projection also too long for Sykes's. (A. Al-Sirhan)

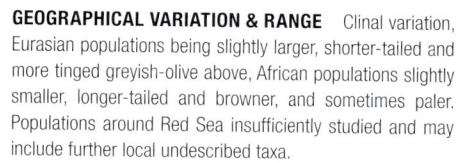

GEOGRAPHICAL VARIATION & RANGE Clinal variation, Eurasian populations being slightly larger, shorter-tailed and more tinged greyish-olive above, African populations slightly smaller, longer-tailed and browner, and sometimes paler. Populations around Red Sea insufficiently studied and may include further local undescribed taxa.

I. p. elaeica (Lindermayer, 1843) (SE Europe, Levant and N Middle East, Caucasus, Central Asia; winters sub-Saharan Africa, mainly E Africa). Largest (cf. Biometrics), with proportionately shortest tail, has most obvious olive-grey cast above in fresh plumage, and as a rule most pronounced pale secondary panel on closed wing. In fresh plumage sometimes has pale tips to secondaries, forming thin whitish mark on folded wing. Flight-feathers rather dark grey when not too heavily worn or bleached. Underparts pale, white with only faint creamy tinge on sides. (Syn. *tamariceti*, which is too subtle and variable to uphold; Svensson 2001.)

I. p. pallida (Hemprich & Ehrenberg, 1833) (Egypt, mainly in Nile Valley; presumably N Sudan, possibly also Red Sea coast of SW Arabia, Yemen and Egypt; short-range migrant to Sudan and adjacent E Africa). Slightly smaller than *elaeica* with a slightly blunter wing (**p2** =6/8; **p1** < p2 24–30.5 mm; **s1** < wt 11–15 mm), and proportionately longer tail, is browner above with rump slightly tinged ochre. In fresh plumage paler and warmer tawny-brown, when worn more greyish and slightly darker. Folded wing has only faintly paler panel, often invisible in the field. **W** 59–66.5 mm (n 43, m 63.1); **T** 44–55 mm (n 44, m 50.6); **T/W** m 80.2; **B** 14.3–16.6 mm (once 17.0; n 43, m 15.6); **Ts** 19.9–22.7 mm (n 43, m 21.1). Wing formula: **p1** > pc 2–8.5 mm, < p2 24.5–30.5 mm; **p2** < wt 3.5–6.5 mm, =6/8; **p6** < wt 1–4 mm; **p7** < wt 3.5–6.5 mm; **p8** < wt 5–9 mm; **p10** < wt 9–13 mm; **s1** < wt 11–15 mm. – Populations around Red Sea in mangroves are possibly this race, but could belong to *alulensis* (see below) or constitute a separate undescribed race.

I. p. pallida, Egypt, Apr: unlike *elaeica*, has proportionately longer tail, browner upperparts, and folded wing has only indistinct pale panel, or lacks it alltogether. Best identified on typical song. (Mattias Ullman)

Presumed *I. p. pallida*, Egypt, Apr: though ssp. *elaeica* is likely to migrate through Egypt at this season, appearance of short/small bill, seemingly moderate primary projection and brownish upperparts with indistinct pale secondary panel strongly suggest *pallida*. (D. Monticelli)

I. p. pallida, Egypt, Dec: a worn/scruffy bird that is apparently in active body moult. Note proportionately long tail. It is far more likely that a Dec bird in Egypt is local *pallida*; ssp. *elaeica* winters in E Africa north to C Sudan. (E. Winkel)

I. p. reiseri, Morocco, May: note square tail with white-tipped outer feathers of *Iduna/Hippolais*. Also note pointed tip to bill, which is not as broad as in Isabelline Warbler. Also unlike Isabelline has obvious pale sandy-buff secondary panel. Ageing in spring not possible. (B. Maire)

I. p. reiseri (Hilgert, 1908) (EC & SE Morocco, Algeria, W Tunisia south of coastal mountains, Libya, N Niger, Lake Chad region; northern birds winter in Sahel zone of W Africa or south to Nigeria, Niger, Mali, southern birds mainly resident). Similar to *pallida* in size and has similar wing formula, but slightly paler (olive-) brown above with a usually more obvious pale sandy-buff secondary panel. Furthermore, it has proportionately somewhat longer tail, and usually much pure white on the tips and inner webs of outer tail-feathers, at least in ad. Bill rather broad-based but finely pointed, slightly shorter on average than in *pallida*. Primary moult as in *elaeica*, thus either entirely in winter quarters (probably some ad and all 1stY) or started in stop-over south of Sahara, then suspended for migration to winter quarters and concluded in latter (Salewski et al. 2005); however, moults start on average earlier than in *elaeica*, usually from Oct. **W** 60.5–70 mm (*n* 46, *m* 65.5); **T** 48–60 mm (*n* 45, *m* 53.6); **T/W** *m* 81.8; **B** 14.3–16.2 mm (*n* 43, *m* 15.3); **Ts** 19.2–23.0 mm (*n* 45, *m* 21.1). **Wing formula: p1** > pc 3–9.5 mm, < p2 24.5–32 mm; **p2** < wt 3–6.5 mm, =6/8; **p6** < wt 0.5–2.5 mm; **p7** < wt 4–7.5 mm; **p8** < wt 6–9.5 mm; **p10** < wt 9–13 mm; **s1** < wt 11–15 mm. – Birds in the south (Saharan Algeria, N Niger, Chad) tend to be subtly smaller and still paler sand-coloured, described as '*lenaeni*', but form part of a faint cline with many overlapping characters and at least half inseparable, therefore included in *reiseri* here. Birds from southern region: **W** 59.5–65 mm (*n* 19, *m* 62.6); **T** 46–55 mm (*n* 19, *m* 51.2); **T/W** *m* 81.7; **B** 14–15.8 mm (*n* 18, *m* 14.9); **Ts** 18.5–21.4 mm (*n* 19, *m* 19.8). **Wing formula: p1** > pc 3–9 mm, < p2 23–29.5 mm; **p6** < wt 0.5–3 mm; **p7** < wt 2–5.5 mm; **p8** < wt 5–8 mm; **p10** < wt 9.5–12 mm; **s1** < wt 10.5–13.5 mm. (Syn. *lenaeni*.)

? *I. p. alulensis* (Ash, Pearson & Bensch, 2005) (coastal N Somalia; apparently resident; apparently also in mangroves in coastal E Egypt, Saudi Arabia and Yemen; Baha El Din in prep.). Described as being 'small and short-winged like *laeneni*' (*laeneni* treated here as southern cline of *reiseri*) but a little darker and greyer, in plumage closely approaching *elaeica*, which it is also genetically closest to. However, this new subspecies was apparently described from only four available specimens. Two of these (in USNM) were examined by LS. Both were in heavily worn plumage, thus comparatively dark and greyish, and very close to similarly worn *pallida*. One biometrically inseparable from average *pallida*, the other had 1 mm shorter primary projection than lower limit of known *pallida* variation, but no other measurements separated it. **W** 59, 63 mm; **T** 49, 50 mm; **B** 14.5, 15.8 mm; **Ts** 21.4, 21.7 mm. **Wing formula: p1** > pc 4.5, 5 mm, < p2 26, 27 mm; **p6** < wt 2, 3 mm; **p7** < wt 4, 5 mm; **p8** < wt 5, 7 mm; **p10** < wt 9, 10 mm; **s1** < wt 10, 11.5 mm. The mtDNA placed *alulensis* closer to *elaeica* than to *pallida*, which remains the main reason for its recognition. We recommend that this proposed new subspecies is better documented before it is unhesitantly accepted, and to facilitate safer identification of single birds.

TAXONOMIC NOTES An analysis of morphology, song, behaviour and DNA of all taxa of the *I. pallida* complex *s.l.* (Ottosson et al. 2005) showed that, whereas *opaca* differed clearly in all respects, the other four taxa shared song and behaviour, and had very similar morphology and DNA. Ssp. *elaeica* differed slightly more from the three African taxa, but genetic differences were still within what is normally found within species. (Cf. also Isabelline Warbler; Roselaar in *BWP*, Helbig & Seibold 1999, Svensson 2001.) – As to generic taxonomy, see note under Isabelline Warbler.

REFERENCES Ash, J. S. & Pearson, D. J. (2002) *BBOC*, 122: 222–228. – Ash, J. S., Pearson, D. J. & Bensch, S. (2005) *Ibis*, 147: 841–843. – van den Berg, A. B. (2005) *DB*, 27: 302–307. – Fregin, S. et al. (2009) *Mol. Phyl. & Evol.*, 52: 866–878. – Helbig, A. & Seibold, I. (1999) *Mol. Phyl. & Evol.*, 11 (2): 246–260. – Jiguet, F. (2005) *BW*: 18: 262–263. – Ottosson, U. et al. (2005) *J. of Orn.*, 146: 127–136. – Salewski, V., Herremans, M. & Stalling, T. (2005) *Ring. & Migr.*, 22: 185–189. – Shirihai, H., Christie, D. A. & Harris, A. (1996) *BB*, 89: 114–138. – Svensson, L. (2001) *BW*, 14: 192–219.

I. p. reiseri, Tunisia, May: similar to *pallida*, but has proportionately somewhat longer tail, is slightly paler overall, and has purer white tips and inner edges on outer tail-feathers, but this usually requires handling to be appreciated. (R. Pop)

I. p. reiseri, Morocco, Apr: rather delicate bill with pointed tip, pale sandy-buff secondary panel and pale mark at tip of bunched secondaries all combine to exclude Isabelline Warbler. (A. B. van den Berg)

Poorly known population described as *I. p. alulensis*, Somalia, May: among the poorest-known taxa of the complex, doubtfully separable from ssp. *pallida* and apparently uniquely confined to mangroves, where it overlaps with *avicenniae* Common Reed Warbler. Separated from latter by song, bill shape and bill length, better-defined white supercilium, but most importantly the new outer rectrix has the white tip extending along inner edge. (C. Cohen)

Poorly known population presumed to be so-called *I. p. alulensis*, Egypt, May: identification of these birds based on song, locality and habitat choice (mangroves), but also on overall smaller size and delicate jizz, seemingly different proportions (especially the relatively long tail and short wings), greyer upperparts. Note also fairly well-defined pale supercilium. Outer tail-feather has obvious white outer edge and tip. However, in all of these features there is overlap with especially *pallida*. (V. Legrand)

BOOTED WARBLER
Iduna caligata (M. H. C. Lichtenstein, 1823)

Fr. – Hypolaïs bottée; Ger. – Buschspötter
Sp. – Zarcero escita; Swe. – Stäppsångare

The 'typical' warbler of the vast Kirghiz Steppe, being content with the lowest possible scrub, often spirea and pea species. But also found in low brushwood on meadows where the steppe meets the southern fringes of the taiga, and locally in suitable habitat in more arid areas of the south, S of Aral Sea and Lake Balkhash. Has expanded northwest in recent years, utilising overgrown pasture land as far as SE Finland. Migrant, wintering in India.

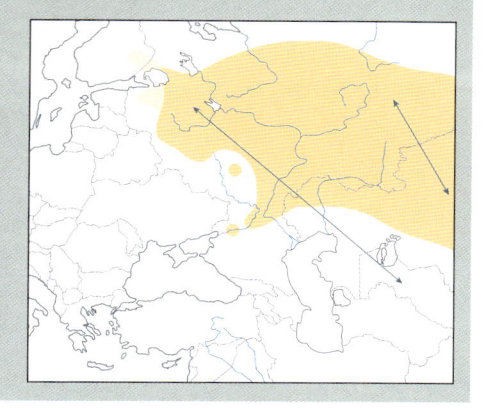

Kazakhstan, Jun: superficially *Phylloscopus*-like jizz, augmented by pale supercilium and hint of dark eye-stripe. Bill invariably dusky on outer part of lower mandible. Also note brownish upperparts, generally lacking an obvious paler secondary panel; at most, tertials appear dark-centred. Legs mainly brownish-pink. (J. Normaja)

IDENTIFICATION Small to medium-sized, rather nondescript brown and off-white warbler, in May–Jun often noticed by its song. Impression of structure rather variable, generally appearing compact with fairly short tail and bill, not unlike a *Phylloscopus*, but at times looking more slim, long-billed and long-tailed. In fresh plumage rather dark and warm brown above, and sides of *breast and flanks tinged ochre-buff*. In worn plumage paler and plainer greyish-brown above and off-white below, more like Sykes's Warbler. Pale *supercilium* rather distinct and often *reaches 2–3 mm behind eye*, enhanced by fairly dark brown sides to forecrown, dark loral spot and hint of dark eye-stripe. Generally, no obvious paler secondary panel. Tertials, greater coverts and uppertail usually appear 'contrasty' with dark centres and paler edges (and when seen well, thin shafts to these feathers can appear near-black). Bill of variable length, usually *fairly short* but longer-billed birds occur. Lower mandible pinkish-yellow with a variably prominent *dark smudge near the tip*, and some have rather extensive dark outer lower mandible. Often twitches or flicks the tail 'nervously', upwards or sideways, at times also slightly downwards (though nothing like the repeated 'tail-dipping' of Olivaceous). When flicking tail *often twitches wings simultaneously*, rather like a Willow Warbler, a habit not seen in Sykes's Warbler. Outer tail-feathers tipped and edged paler, though these may wear off or be difficult to see in the field. Legs pinkish with a pale grey cast, toes often slightly greyer/darker.

VOCALISATIONS The rather subdued song is a very fast warbling twitter that opens tentatively, gathers speed, pitch

Russia, Jun: a compact, small-sized warbler, with short bill (dark-smudged lower mandible tip), rather brownish upperparts. Whitish below with light ochre-buff sides. Pale supercilium rather distinct, often reaching clearly behind eye. Nervously flicks tail (like Willow Warbler) with downwards and sideways movement. Legs predominantly pinkish, but toes slightly darker and greyer. (P. Parkhaev)

India, Apr: pale supercilium prominent but here short, ending just over eye. Lateral crown often appears darker due to angle of view and light. Unlike Sykes's Warbler, overall jizz recalls *Phylloscopus*, augmented by on average shorter bill and tail. Note characteristic pink-brown tarsi with darker and greyish toes (the 'boots'). (N. Devasar)

Russia, May: fairly compact, with rounded head and short white-sided tail. In worn plumage, paler and plainer greyish-brown above, and off-white below, with diluted buff sides. As here, pale supercilium sometimes a little shorter. Pale fringes to outer tail-feathers distinct, but often difficult to see and can wear off during summer. (A. Levashkin)

Presumed 1stW, Netherlands, Oct: despite quite worn plumage, probably not ad, but young with already quite worn and untidy plumage. Some Booted, as this one, can appear less compact, have quite plain and pale grey-brown upperparts, and a slightly longer bill with rather pale lower mandible, thus inviting confusion with Sykes's. In rare borderline cases handling may be required to establish species. (C. van Rijswijk)

and strength, and usually ends abruptly and emphatically. The notes burst out in a seemingly disordered pattern (like 'simmering water'), and the rather guttural and nasal voice is 'trembling with energy'. Since notes are not repeated in sequence, at least not obviously, no real resemblance occurs with Sedge Warbler, helping to eliminate Sykes's Warbler. Common strophe length 2–4 sec. No mimicry. – Call a harsh, tongue-clicking *chek* or *chrek*, surprisingly similar to the call of European Stonechat or Moustached Warbler, and more 'compound' and throaty than the *zett* call of Sykes's.

SIMILAR SPECIES Very similar to *Sykes's Warbler* (which see), and silent birds not always safely separable. Song and call are best means of separation once learned, together with on average shorter bill and tail. Note that although typical Booted is slightly warmer brown and also darker than Sykes's, with rather shorter bill and tail, recalling a *Phylloscopus* in shape, some are both plainer and paler brown above with a longer bill and quite pale lower mandible. In general, prominence and length of pale supercilium is a good clue, but some Sykes's match this. Legs on average slightly darker pinkish-brown than in Sykes's. – From *Olivaceous Warbler* by absence of repeated downward movements ('dips') of the tail, slightly smaller size, usually shorter wings and bill, and vocalisations. – A more remote risk of confusion is *Paddyfield Warbler*, of which some young birds in autumn can appear reasonably similar in coloration, and even appear to have paler outer tail-feathers. Paddyfield has a longer tail, a longer and more prominent supercilium, and, above all, better-marked dark eye-stripe than Booted. Best character is very short primary projection in Paddyfield (often slightly less than half length of exposed tertials against almost two-thirds in Booted), and its somewhat paler brown flight-feathers (especially outertail). Tertials have more contrasting pale edges in Paddyfield, unless too worn. Also, the eye of Paddyfield can be seen to be a little paler and warmer brown than in Booted, which has an all-dark eye.

AGEING & SEXING Insignificant seasonal and age-related differences. Sexes alike in plumage. – Moults. Complete moult of ad in winter quarters (Oct–Dec), but may start replacing body-feathers earlier. Unclear if all 1stY undergo complete moult. – **SPRING** Ageing apparently impossible, although birds with heavily abraded primary tips probably 2ndY. Still, such birds surprisingly few if all 1stY leave primaries unmoulted, and feather wear often difficult to categorise. – **AUTUMN Ad** Worn, especially tips of outer primaries, and tertials. **1stW** Uniformly fresh. **Juv** Warm rufous-brown above, and tertials and secondary edges rufous-buff, recalling a young Reed Warbler.

BIOMETRICS L 11–12.5 cm; **W** 56–65 mm (n 259, m 60.9); **T** 40–51 mm (twice 52) (n 258, m 47.1); **T/W** m 77.3; **B** 12.0–15.6 mm (only rarely > 14.5; n 248, m 13.5); **BD** 2.4–3.3 mm (n 221, m 2.8); **BW** 3.2–4.3 mm (n 232, m 3.7); **Ts** 17.5–21.3 mm (n 214, m 19.8). **Wing formula: p1** > pc 2–9.5 mm, < p2 21–29 mm (once 20); **p2** < wt 2–7 mm, =6/8 (92%), =6 (4%) or =5/6 or ≤8 (4%); **pp3–5** about equal and longest (though in particular p5 often 0.5–2.5 mm <); **p6** < wt 1–5 mm; **p7** < wt 2.5–8 mm; **p8** < wt 4.5–10 mm; **p10** < wt 7.5–13.5 mm; **s1** < wt 9–15 mm. Emarg. pp3–5; p6, too, clearly emarg. in *c.* 25%, less obviously in *c.* 55%, and totally lacking in *c.* 20%.

GEOGRAPHICAL VARIATION & RANGE Monotypic. – SE Finland, Estonia, Russia, SW Siberia, Kazakhstan; winters E India. – Ssp. '*annectens*', described from Russian Altai, is a synonym of *I. caligata* (large part of type series examined; Svensson 2001). Another synonym is '*scita*', a junior name for *caligata*. (Syn. *annectens, scita*.)

TAXONOMIC NOTE See comments under Sykes's Warbler.

REFERENCES Aalto, T. & Dernjatin, P. (2006) *Alula*, 12: 2–8. – Shirihai, H., Christie, D. A. & Harris, A. (1996) *BB*, 89: 114–138. – Svensson, L. (2001) *BW*, 14: 192–219. – Svensson, L. (2003) *BW*, 16: 470–474.

1stW, France, Oct: already in Oct, some young can commence moult when on migration (unmoulted feathers are juv and less worn than ad). Obvious pale eye-ring but short supercilium. Outer tail-feathers edged whitish. Tertials have dark shafts, and despite being worn still show better contrast than Sykes's. (R. Armada)

1stW, Scotland, Aug: apart from considering separation from Sykes's Warbler, also bear in mind possibility of confusion with Paddyfield Warbler, versus which Booted has shorter tail but relatively longer primary projection, less prominent dark eye-stripe, usually shorter supercilium typically with 'waistline' over eye, further contrasting pale edges to tertials, and all-dark eye. (H. Harrop)

Juv/1stW, Turkey, Sep: compared to very similar Sykes's Warbler, note prominent, almost *Phylloscopus*-like supercilium, warmer/browner upperparts, somewhat longer primary projection (though not easy to judge on these images), shorter bill and tail, and slightly darker pinkish-brown legs. From *elaeica* Olivaceous by lack of whitish wing panel, shorter primary projection, smaller size, whitish supercilium generally reaching well behind eye, and dark-smudged lower mandible tip. (E. Yoğurtcuoğlu)

Juv/1stW, Scotland, Aug: again, most Booted Warblers have slight indention on supercilium ('waistline'), just over fore part of eye, thus supercilium narrows then broadens again. Such pattern also occurs in Paddyfield, albeit much less commonly, and it therefore constitutes a useful subsidiary feature in their separation. (H. Harrop)

SYKES'S WARBLER
Iduna rama (Sykes, 1832)

Fr. – Hypolaïs rama; Ger. – Steppenspötter
Sp. – Zarcero de Sykes; Swe. – Saxaulsångare

Widespread and locally abundant in sand or clay deserts with saxaul trees or tamarisks, mainly in the West Turkestan range, from NE Middle East to E Central Asia, in the north reaching the N Caspian Sea (Volga Delta), Lake Balkhash and Dzungarian basin, in the south to United Arab Emirates, N Oman, Iran and N Pakistan. Migrant, wintering in India, Pakistan and E Arabia. A rare vagrant to Middle East and even to W Europe, mainly in autumn

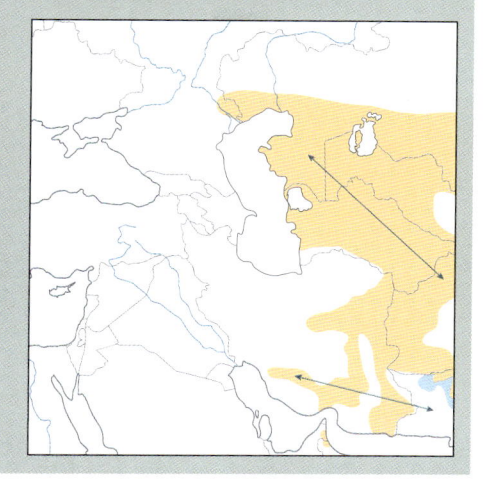

SIMILAR SPECIES Very similar to *Booted Warbler* and separable only with prolonged views and familiarity with variation and problems involved. Song is safest means once learned. Call can help, too. Habitat often gives clue, Sykes's prefers vegetation of 2 m or higher, whereas Booted can make do with knee-high scrub. Sykes's has on average longer bill and tail (some overlap), and shorter primary projection (much overlap), but despite overlap structure is a good start. Sykes's is on average slightly paler above, being plainer greyish-brown without much contrast on tertials or wing-coverts (though wear makes the two species more similar), and underparts are whiter (Booted often has rather obvious ochre-buff tinge on sides). Lower mandible has only a tiny grey smudge near tip (Booted more extensive dark tip in many). Sykes's supercilium is as a rule short and less well marked, generally fading at rear edge of eye, or fractionally further back (Booted has more distinct supercilium frequently reaching 2–3 mm behind eye), but a few Sykes's have just as long supercilium, so no single plumage character is conclusive, all must be used in combination. Does not seem to move its wings when it twitches its tail, unlike Booted often does, which behaves more like a Willow Warbler in this respect. – Very similar also to *Olivaceous Warbler* but note absence of repeated downward 'dips' of the tail, which is

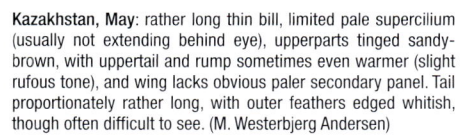

Kazakhstan, May: rather long thin bill, limited pale supercilium (usually not extending behind eye), upperparts tinged sandy-brown, with uppertail and rump sometimes even warmer (slight rufous tone), and wing lacks obvious paler secondary panel. Tail proportionately rather long, with outer feathers edged whitish, though often difficult to see. (M. Westerbjerg Andersen)

United Arab Emirates, Apr: these two images show birds that to some extent approach Booted, which is not unusual. Across bulk of Sykes's range, they can often appear very similar without the benefit of vocalisations or biometrics. It is very easy to be confused by bill length, the supercilium and contrasts in plumage—all of which depend on individual variation, light conditions, angle of view, and other circumstances. Note extensive whitish tips and edges to outer tail-feathers. (H. & J. Eriksen)

IDENTIFICATION Medium-sized, slim, rather nondescript warbler, usually noticed by its song. Keeps mostly hidden in the canopy of low trees and bushes, but is rather easily seen due to the open, arid habitat it frequents, e.g. when it sings from an exposed perch or in fast, direct, low flight from one isolated tree to another. Sandy-brown ('milky-tea') upperparts and off-white underparts. The uppertail and rump can appear slightly warmer rufous-tinged in flight, recalling Reed Warbler, but this impression disappears as soon as the bird alights. Pale *supercilium* indistinct and short, generally *not extending* appreciably *behind eye*, but a few are longer just like Booted Warbler, the most similar species. The entire *upperparts are very plain and unpatterned*, and there is no obvious paler secondary panel. In fresh plumage and in strong light, the upperparts may appear to have a pale olive tinge to the brown. Fairly long and thin bill, lower mandible pinkish-yellow, usually with a faint dark smudge near the tip, although this can be difficult to see in the field or be nearly absent. Primary projection short. *Tail* generally appears *rather long*, and the bird often twitches or flicks it 'nervously', upwards or sideways, at times also a little downwards (though nothing like the repeated tail-dipping of Olivaceous). Unlike Booted it *does not flick its wings* simultaneously when the tail is twitched. Outer tail-feathers generally tipped and edged paler, though these may be absent, worn off or difficult to see in the field. Legs pinkish with a pale grey cast, toes often slightly greyer/darker. Young birds, in fresh, slightly paler and warmer brown plumage, reported to have slightly more greyish legs.

VOCALISATIONS Song resembles that of Booted Warbler but is louder and includes more scratchy and hard notes, some of which are repeated in twos or threes, creating a somewhat structured pattern and a vague resemblance to Sedge Warbler song, enhanced by some interwoven whistles and brief trills. Alternative song in early dawn consists of staccato-like, 'squeezed', chirping notes delivered at slow pace, *zeeh... zree... zeeh... tsretsre... zrih...*, etc. – Calls: (1) a fairly characteristic 'full' *tslek*, perhaps most recalling Bluethroat, and (2) a drier tongue-clicking *zek* (or *zet*), not unlike the call of Lesser Whitethroat or other small *Sylvia* warblers, less throaty or muffled than the *chek* of Booted. When agitated or alarmed emits a churring trill, *zerrr*.

Kazakhstan, Jul: pale supercilium only sometimes extends behind eye. In certain lights upperparts can look darker, more grey-brown (less sandy). Typically, bill fairly long and thin, with lower mandible rather contrastingly pinkish-yellow; in the field, dark-smudged lower mandible tip is usually not detectable (cf. Booted Warbler). On average, primary projection proportionately shorter, but tail generally appears rather longer than Booted. (O. Amstrup)

BUSH AND REED WARBLERS

Spring/summer Sykes's Warbler (top left: Turkey, May) versus Booted (top right: Finland, Jun) and Olivaceous Warblers (bottom left: *elaeica*, Israel, Mar; bottom right: *pallida*, Egypt, Apr): as these images clearly show, separation of Sykes's from Booted, and both these from Olivaceous, ranks among the most difficult tasks in W Palearctic field ornithology. Sykes's usually has shorter supercilium (but can rarely, as here, extend behind the eye), on average longer bill and tail, slightly shorter primary projection, paler upperparts, plainer wing, and whiter underparts. Dusky smudge near tip of lower mandible is feature of both, but more likely to go undetected in Sykes's. Elimination of Olivaceous: Booted/Sykes's have proportionately shorter primary projection and, at least compared to *elaeica*, plainer and browner upperparts and wing (no pale secondary panel). Dark-smudged lower mandible tip lacking in Olivaceous, while Booted/Sykes's do not practise repeated downward dips of tail so typical of Olivaceous. Reliable identification often achieved only after prolonged observation and an awareness of the difficulties involved. (Top left: E. Yoğurtcuoğlu; top and bottom left: D. Occhiato; bottom right: V. Legrand)

proportionately slightly longer than in Olivaceous, as well as shorter primary projection, and song. Supporting character in Central Asia (where ssp. *elaeica* occurs) is plain brown upperparts and wing (*elaeica* is more tinged olive-grey, has hint of pale secondary panel and in fresh plumage darker grey primaries with at times fine whitish tips).

AGEING & SEXING Moderate seasonal and age-related differences due to wear and bleaching. Sexes alike in plumage. – Moults. Complete moult of ad apparently immediately after arriving in winter quarters (Aug–Sep). With this timing, 1stY is unlikely to moult completely. However, examined specimens in active moult few, and some may moult later (Oct–Dec). – **SPRING** Ageing apparently impossible. – **AUTUMN Ad** Worn, especially tips of outer primaries, and tertials. Plumage on average somewhat darker and greyer than young birds. **1stW** Uniformly fresh at first, but progressively becomes moderately abraded on edges of flight-feathers and tertials in autumn. Plumage on average slightly paler and warmer brown above. **Juv** Warm rufous-brown above, slightly paler overall than juv Booted, especially tips to primaries and tail-feathers, but otherwise quite similar. Vent/undertail-coverts, nape and sometimes rump can show more loose feathering where barbs not completely interlocking.

BIOMETRICS L 11.5–13 cm; **W** 57–66 mm (n 241, m 61.5); **T** 49–57 mm (once 47) (n 241, m 52.7) (for only live ad **T** measured 52–61, m 57; Wilson 2011); **T/W** m 85.6; **Ts** 18.5–21.9 mm (n 226, m 20.4); **B** 13.6–16.5 mm (only

Kazakhstan, May: separated (with great care) from Booted by shorter supercilium, on average longer bill and tail, but shorter primary projection. Also by generally paler upperparts, plainer wing, and whiter underparts. Dusky smudge near tip of lower mandible limited to thin line at very lower edge, thus unlike Booted, which usually has more extensive and visible dark smudge. Sykes's can be very similar, or almost inseparable from Olivaceous, but identified by proportionately shorter primary projection and, at least compared to *elaeica*, plainer sand-brown upperparts and wing. (M. Westerbjerg Andersen)

Ad/1stW, Scotland, Sep: thin rather long bill and fairly short supercilium typical of the species. Level of wear to primary-tips and tail best fit ad, but impossible to eliminate the chance that it is an early-hatched juv, already heavily abraded. (G. Thomas)

BUSH AND REED WARBLERS

Juv/1stW, Scotland, Aug: note how different light can change impression of coloration in same bird, although characteristic pale and plain plumage remains evident. In the field, the dusky smudge near tip of lower mandible is often invisible or almost so, unlike in Booted, which usually has more extensive and visible dark tip to lower mandible. In strong light, a Sykes's can appear to lack any smudge (right image), whereas in overcast light (left) it can be more obvious, even approaching the bill pattern typical of Booted. Typical short supercilium seen in most Sykes's. Aged by fresh plumage. (Left: D. Preston; right: A. Wilson)

7% < 14.5; n 236, m 15.1); **BD** 2.5–3.3 mm (n 220, m 2.8); **BW** 3.2–5.0 mm (n 228, m 4.0). **Wing formula: p1** > pc 3.5–10 mm, < p2 20–27 mm; **p2** < wt 3.5–8 mm, =7/9 (rarely =7 or 9, or shorter; juv may have =6/7); **pp3–5** about equal and longest (though either of p3 and p5 often 0.5–1.5 mm <); **p6** < wt 0.5–4 mm; **p7** < wt 2–6.5 mm; **p8** < wt 4–9 mm; **p10** < wt 6.5–12 mm; **s1** < wt 9–13.5 mm. Emarg. pp3–5; p6 clearly emarg. too, in c. 45%, less obviously in c. 50%, and totally lacking in c. 5%.

GEOGRAPHICAL VARIATION & RANGE Monotypic. – NE Middle East to E Central Asia, in the north reaching to N Caspian Sea, Lake Balkhash and Dzungarian basin, in south to United Arab Emirates, N Oman, Iran and N Pakistan; winters Indian subcontinent, Pakistan, W Himalaya. – Ssp. 'annectens', described from SE Kazakhstan, is merely a synonym for Booted Warbler *I. caligata* (large part of type series examined, inseparable from *caligata*; Svensson 2001), and therefore is not 'closer to *rama*' as suggested by Castell & Kirwan (2005).

TAXONOMIC NOTES Previously often regarded as a subspecies of Booted Warbler, but recently separated on account of apparently sympatric ranges without known interbreeding, and different song, morphology, habitat and DNA (Helbig & Seibold 1999, Svensson 2001). – For population on southern side of the Persian Gulf, differences in egg pattern and nest structure have been pointed out (Castell & Kirwan 2005), but we regard song, plumage and biometrics of this population more important, all being typical of Sykes's Warbler. – The population in the extreme NW (W Ural River, Ryn Desert; not formally described and named) is apparently (based on a few available specimens) somewhat darker above and has duskier bill, in these respects approaching Booted Warbler; merits closer study. – For generic taxonomy, see note under Isabelline Warbler.

REFERENCES CASTELL, P. & KIRWAN, G. M. (2005) *Sandgrouse*, 27: 30–36. – HELBIG, A. & SEIBOLD, I. (1999) *Mol. Phyl. & Evol.*, 11: 246–260. – SHIRIHAI, H., CHRISTIE, D. A. & HARRIS, A. (1996) *BB*, 89: 114–138. – SVENSSON, L. (2001) *BW*, 14: 192–219. – WILSON, K. (2011) *BW*, 24: 386–391.

Juv/1stW Sykes's Warbler (top left: Scotland, Sep) versus Booted (top right: Turkey, Sep), Olivaceous (bottom left: Israel, Aug) and Blyth's Reed Warblers (bottom right: Kazakhstan, Sep): in addition to slightly more compact and short-tailed Booted Warbler, other potential confusion risks when faced by a Sykes's include *elaeica* Olivaceous; remember Sykes's shorter primary projection, plainer and browner upperparts, in addition to behaviour (no tail-dips when moving in canopy). Compared to superficially similar Blyth's Reed, note Sykes's more square-ended tail, which if spread should be diagnostically tipped and edged paler if still fresh; also has flatter/broader bill which lacks dark smudge near tip of lower mandible. (Top left: D. Curtis; top right: E. Yoğurtcuoğlu; bottom left: A. Ben Dov; bottom right: P. Palmen)

UPCHER'S WARBLER
Hippolais languida (Hemprich & Ehrenberg, 1833)

Fr. – Hypolaïs d'Upcher; Ger. – Dornspötter
Sp. – Zarcero lánguido; Swe. – Orientsångare

Breeds in Turkey, Caucasus, Middle East and Central Asia on dry, stony hillsides or in river valleys with thorny thickets, maquis, groves, orchards and open woods of holm oaks, pistachio, etc., also in deserts with tall saxaul or tamarisks. Summer visitor, arriving in April–early May from wintering grounds in E Africa. To get to know this species you need to go at least to SC Turkey, where many birdwatchers have seen it for the first time around Gaziantep or in the Taurus Mountains west of Adana.

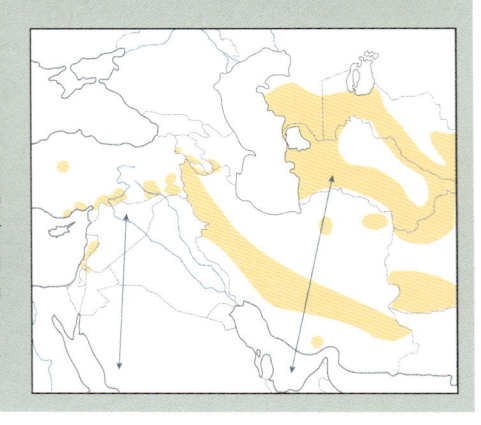

IDENTIFICATION   In many ways resembles the eastern race of Olivaceous Warbler (*Iduna pallida elaeica*), and the two are easily confused, but is slightly larger, in size between that species and Olive-tree Warbler, although closer to Olivaceous. Further differs in having slightly *stronger bill* than Olivaceous (though some are fairly similar), a *slightly larger and often more rounded head*, and somewhat thicker (at times rather short-looking) neck, giving slightly different, front-heavy jizz. It also has longer primary projection and subtly *darker primary tips and uppertail* than Olivaceous (light-dependent). Folded wing has, before too worn, a distinct *whitish secondary panel*, typically concentrated on central part of secondaries and longest tertial, and *often seems almost oval in shape* (Olivaceous has more evenly whitish edges along nearly entire secondaries, forming more of a narrow panel, but no rule without an exception: odd Upcher's also have pale edges along whole length of their secondaries). Typically lacks contrasting pale edges to outer greater coverts (like in Olive-tree Warbler). Both species can have white tips to secondaries forming a white mark on the folded wing. Also, in Upcher's *outer tail-feathers are edged and prominently tipped pure white*, the white being well marked and broad at tips, not diffuse and sullied grey-brown as in Olivaceous. Supercilium generally rather short albeit *well marked, especially in front of eye* (Olivaceous: usually rather diffuse outline to supercilium in front of eye) ending at rear edge of eye (or slightly behind eye). Supercilium more prominent in shade than in sunlight. Often appears large-eyed. Legs greyish-pink, if anything a shade less grey than Olivaceous. The *tail is often semi-spread and waved sideways* or vertically, a habit peculiar to Upcher's (and Olive-tree), or 'dipped' downwards, rather like Olivaceous, but movements are then less frequent, usually slower and the tail is often kept slightly spread. Another reasonably typical habit of Upcher's (and Olive-tree) is the *gliding on outstretched wings* and *slightly raised and semi-spread tail* over a few metres before landing on a new perch.

VOCALISATIONS   Song is a very useful means of separation from Olivaceous Warbler once learnt, a rather guttural and monotonous babble, slightly less strong than Olivaceous (despite being a larger bird), mechanically repeating brief segments (5–10 times, or even more) and includes nasal notes much like Icterine Warbler. Thus, although it broadly resembles the song of Olivaceous, the notes are neither as scratchy, nor as slurred as in that species. Inspired songsters can produce a more pleasing and varied song than Olivaceous, at least partly. – Call a tongue-clicking *chak* (at times slightly harder and louder than Olivaceous, but frequently more subdued). Alarm is a series of slightly throaty clicking notes, *chek chek chek…* or, less often heard, a churring *zerrrr*.

SIMILAR SPECIES   For differences from *Olivaceous Warbler* see Identification, and note that these two can have virtually the same general plumage (Upcher's, despite some claims, is not always greyer above), though darker uppertail and wing-feathers, and more contrasting oval-shaped white secondary panel concentrated on the centre of the folded secondaries are generally obvious. Still, essential to be careful and if possible study tail movements, structure and song as well. – *Olive-tree Warbler* can be eliminated by its larger size, much longer and stronger bill with orange-yellow lower part (more pinkish-yellow in Upcher's), longer primary projection, darker primaries, usually more prominent (but similarly rounded and 'centred') white secondary panel, pale-fringed outer secondaries, greyer (less pink-tinged) legs, and different vocalisations. – In brief views could perhaps recall *Barred Warbler*, which often occurs in the same habitat, and is of similar size and also has white tail corners, but note on Barred shorter, stouter bill, larger head, pale wing-bars and several other details once seen well.

AGEING & SEXING   Moderate seasonal and age-related differences. Sexes alike in plumage. – Moults. Complete moult of ad protracted, often Aug–Jan, sometimes interrupted in late autumn. Whether 1stY moults on average later, and sometimes incompletely, suspected but not known. – **SPRING** Ageing apparently impossible, but birds with

Turkey, May: slightly larger than *elaeica* Olivaceous, with slightly more rounded head and stronger bill. Tail is typically fuller and contrastingly darker than rest of upperparts; also edged and prominently tipped white on outer feathers. Dull greyish-brown tone above contrasts with darker remiges (especially tips). Whitish secondary fringes forming diagnostic white panel which usually is strongest on the centre of the secondaries (at times forming oval patch). (D. Occhiato)

Bahrain, May: example of bird with less well-marked supercilium. Legs greyish-pink, and note broad dark tail that is typically often semi-spread and waved sideways or vertically. Also note diagnostic whitish fringes to secondaries, concentrated in centre, forming diffuse but rather prominent oval panel. This bird has untypically pale-fringed outer greater coverts, recalling Olive-tree Warbler, but it may be an artefact due to light. In spring/summer all are fresh following complete winter renewal and cannot be aged. (A. Drummond-Hill)

United Arab Emirates, May: here gleaning for insects with unusual long-necked jizz and untypically strong-looking bill, almost recalling Olive-tree Warbler. Note full tail, with prominently white-tipped outer tail-feathers. (D. Clark)

Upcher's Warbler (top left: Turkey, May; bottom left: United Arab Emirates, May) versus *elaeica* Olivaceous Warbler (top right: Greece, May; bottom right: Israel, May) in fresh spring/summer plumage: behaviour and impressions on breeding grounds (top two images) versus migration (bottom two) can be different in both species. On breeding grounds, Upcher's movements and jizz are highly characteristic, making it more readily identifiable; however, on migration, Upcher's makes less frequent and less vigorous tail movements. Nevertheless, 'classic' Upcher's should be visibly larger and heavier than *elaeica* Olivaceous, with slightly more front-heavy jizz and stronger bill; head also larger, more rounded, sometimes with fuller-necked appearance. Upcher's has on average longer primary projection, its tail is generally fuller/broader and clearly darker above than dull grey-brown upperparts. Overall note key discriminator: if edges to secondaries have whitish edges concentrated at centre, forming diffuse but rather prominent oval panel, bird is Upcher's, whereas narrow pale edges along most of secondaries, reaching tip, very strongly indicative of Olivaceous. Both can show white tips to folded secondaries, but more often in Olivaceous. (Top left: D. Occhiato; bottom left: G. Schuler; top right: H. Roberts; bottom right: A. Ben Dov)

much pure white on outer tail-feathers, and the white tips sharply demarcated, are believed to be ad. — AUTUMN **Ad** Worn, especially tips of outer primaries, and tertials. White secondary panel lacking due to abrasion, or only shows as faint remnant. **1stCY** Uniformly fresh. Secondary panel rather prominent, tinged olive-grey. Entire plumage slightly tinged buff or cream.

BIOMETRICS L 14–15 cm; **W** ♂ 75–80 mm (*n* 28, *m* 77.5); ♀ 73–78 mm (*n* 19, *m* 75.3); **T** 57–65 mm (*n* 48, *m* 60.4); **T/W** *m* 79.0; **B** 16.3–19.5 mm (*n* 74, *m* 17.6); **BD** 3.2–3.8 mm (*n* 21, *m* 3.6); **BW** 3.7–5.0 mm (*n* 21, *m* 4.5); **Ts** 21.2–23.3 mm (*n* 35, *m* 22.0). **Wing formula: p1** ~pc (< pc 0–3 mm or 0.5–1.5 mm >), < p2 38–45 mm; **p2** < wt 2.5–5 mm, =5/6 (67%), =6 (18%), =6/7 (10%) or =5 (5%); **pp3–4** about equal and longest; **p5** < wt 0.5–3 mm; **p6** < wt 3.5–7 mm; **p7** < wt 7–10 mm; **p10** < wt 14–18 mm; **s1** < 15–20.5 mm. Emarg. pp3–4; p5, too, has slight emarg. in many.

GEOGRAPHICAL VARIATION & RANGE Monotypic. — Turkey except west and north; Levant, Middle East east over much of Iran and S Central Asia north to SC Kazakhstan and SW Pakistan; winters E Africa. (Syn. *magnirostris*.)

REFERENCES FRY, C. H. (1990) *BB*, 83: 217–221. – SHIRIHAI, H. (1987) *BB*, 80: 473–482. – SHIRIHAI, H., CHRISTIE, D. A. & HARRIS, A. (1996) *BB*, 89: 114–138. – SMITH, K. D. (1959) *BB*, 52: 31. – ULLMAN, M. (1989) *BW*, 2: 167–170.

Kuwait, Apr: some Upcher's are less striking and can resemble *elaeica* Olivaceous when size of bill is not apparent. Here, adding to problems, due to harsh light and background reflection, plumage appears superficially sandier than normal. However, the following features identify it as Upcher's: strong whitish secondary panel but only moderately paler fringes to greater coverts, strong bill with nearly all-yellowish lower mandible, and (just discernible) uppertail darker than back. Moreover, primary projection is relatively long. (K. Haataja)

Ad, Ethiopia, Sep: overall very worn and bleached. Due to wear, diagnostic secondary panel is absent. Thus, separation from *elaeica* Olivaceous must concentrate on larger size and heavier appearance, with long legs and strong bill. (H. Shirihai)

Juv/1stW Upcher's Warbler (top left: Turkey, Jul; top right: United Arab Emirates, Sep) versus *elaeica* Olivaceous Warbler (bottom left and right: United Arab Emirates, Sep): species-specific overall jizz and coloration of both species discernible. Upcher's appears larger and heavier with slightly more front-heavy jizz and stronger bill; head rounder, with fuller and clearly darker uppertail. Light conditions and angle of view strongly influence appearance, while, on migration, Upcher's tail movements are often subdued, and some Olivaceous can partially spread their tails. (Top left: A. de Jong; top right, bottom left and right: H. Roberts)

Juv/1stW, United Arab Emirates, Jul: pay attention to characteristic jizz, broad and partially fanned tail that is obviously darker than dull grey-brown upperparts, and whitish edges concentrated on centre of secondaries, forming prominent oval panel. (T. Pedersen)

Juv, Israel, Jul: when wings very fresh, in particular in juv, secondary fringes are tinged buffish. Note broad, partially fanned and contrastingly dark tail, affording distinctive jizz even to juv. (A. Ben Dov)

OLIVE-TREE WARBLER
Hippolais olivetorum (Strickland, 1837)

Fr. – Hypolaïs des oliviers; Ger. – Olivenspötter
Sp. – Zarcero grande; Swe. – Olivsångare

A large *Hippolais* with a deeper and throatier song than its congeners and other close relatives. Confined to SE Europe and Turkey, it is a summer visitor that arrives mainly in May from winter grounds in SE Africa. Often occurs in olive, almond or pistachio groves, orchards or open maquis and scrub with scattered oak trees on lower coastal mountains or grassy hills.

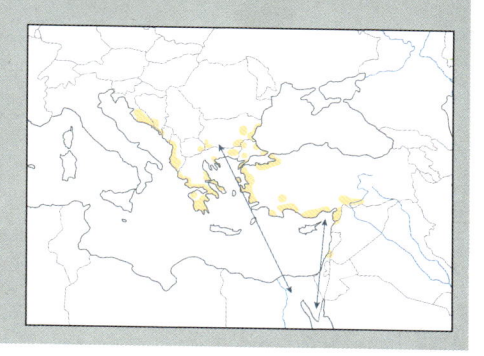

Turkey, May: although being a rather long-tailed species, long primary projection and angle of view can, as here, make tail look shorter and shape more compact. Still, note all characteristics for Olive-tree: strong bill with orange tinge to yellow lower mandible, dark wing- and tail-feathers, former with prominent white secondary panel and hint of paler edges to outer greater coverts. (D. Occhiato)

Turkey, Jun: Olive-tree Warbler is a large rather chunky warbler, heavy-fronted with a strong bill with orange-tinged yellow lower mandible and strong grey legs. Typically has whitish secondary panel and also whitish-edged outer greater coverts, and long, pointed wings, primary projection being about equal to tertial length. (H. Shirihai)

IDENTIFICATION Like with so many other warblers living a rather secretive life in the canopy of trees and larger bushes, usually noted and identified by its characteristic song. Wary, keeps in cover much of the time, hopping about rather sluggishly. Flaps tail sideways or vertically in leisurely, wavering fashion, much like Upcher's Warbler, and can practise downward-directed 'dipping' tail movements, at least briefly but sometimes for longer periods (cf. Olivaceous Warbler). In flight, note the large size, long pointed wings, fairly long tail and grey upperparts, apart from straight, purposeful flight. Perched birds told by grey and white colours with quite *dark flight-feathers, very long, strong and pointed bill* with *orange-tinged yellow lower mandible* and cutting edges, a rather distinct but short, narrow grey-white stripe above the dark lores, whitish eye-ring (but no supercilium at all behind eye), and long primary projection equalling visible tertial length (or very slightly shorter). Folded wing has, before too worn, an *obvious whitish secondary panel* (like in Upcher's often rather rounded, white usually mostly confined to central parts of edges to tertials and secondaries, though some variation), but also as a rule quite *pale-edged dark outer greater coverts* making the wing a little more contrasting than in other related species. *Legs rather strong and dark grey*, nearly always lacking visible pink tones. *Ear-coverts usually rather dark greyish*, slightly darker on average than in its relatives. Whitish underparts have a variable amount of *dusky grey on breast-sides, flanks and undertail-coverts*, latter two often slightly blotched. Outer tail-feathers edged and narrowly tipped white.

VOCALISATIONS Song quite deep-voiced and throaty, the raucous phrases recalling both Fieldfare and Great Reed Warbler, or can be described as a strongly amplified Olivaceous Warbler played at too slow a speed. The strophes are rather monotonous, cyclically repeated (notes going up and down in a recurring pattern), but are also irregular and jerky, and delivered at a moderate, staccato-like pace. – Call is a variably loud, deep tongue-clicking *chuk*, when emphatically uttered is remarkably loud. Alarm a throaty, nasal chatter *kerrekekekek*, like 'giant Blue Tit'.

SIMILAR SPECIES From *Upcher's Warbler* (in approximate order of usefulness) by: (i) longer primary projection, very nearly equalling visible tertial length (c. 2/3 to 3/4 of

Israel, May: greyish plumage but quite dark flight-feathers and tail, latter often leisurely dipped, fanned/waved or (as here) lifted upwards. Chunky appearance, large head with flat forehead, long pointed wings, obvious whitish secondary panel and diagnostic clear white edges to tertials and greater coverts. (A. Ben Dov)

Israel, May: that the lower mandible is all pinkish-yellow without any dark is less evident in some views, but length of bill should invariably be evident. Short grey-white loral stripe joining whitish eye-ring also often obvious, but no supercilium behind eye. Very long primary projection, and outer tail-feathers edged and tipped white. (A. Ben Dov)

tertial length in Upcher's); (ii) stronger all-grey legs; (iii) on average slightly darker grey cheeks, breast-sides and flanks; (iv) longer and stouter bill invariably yellow or orange on lower part (sometimes more pinkish-grey in Upcher's, but some are similarly yellowish); (v) darker greater coverts with more obvious pale edges on outer coverts; (vi) larger size; (vii) darker primaries; (viii) sometimes more obvious white secondary panel; and (ix) different song. – Can resemble *Olivaceous Warbler* due to shared habit of repeated 'tail-dipping' (practised at least sometimes by Olive-tree) if clearly larger size is not evident, but separated by longer and more pointed wings (primary projection in Olivaceous shorter, shorter even than Upcher's), longer and stronger bill, more obviously pale-edged greater coverts, and generally more obvious and differently shaped pale secondary panel. Although both have scratchy-voiced cyclically repeated song, Olive-tree has so much deeper and throaty voice and slower staccato-pace that confusion is impossible with some practice. – If observation is only fleeting can recall a *Barred Warbler*, which often shares same habitat, and is similarly large and long-tailed, but note short bill and much more white on tail corners in Barred. – If only song is heard, can be confused with *Masked Shrike*, but this species has even more monotonous and scratchy song on more even pitch, is more staccato-like in performance, and is not quite so loud and deep-voiced.

Ad, Turkey, Jul: here just prior to start of autumn migration, an ad showing heavily worn primaries and tertials. Upperparts bleached paler and browner, and white secondary panel almost totally lacking due to feather wear. However, large size, bill size and colour and general jizz and behaviour should identify. (Ö. Necipoglu)

Olive-tree Warbler (left: Israel, May) versus Upcher's Warbler (Saudi Arabia, Apr) in spring plumage: Although Olive-tree Warbler is larger and tinged purer grey (less drab) above than often very similar Upcher's, these size and shade differences not always easy to perceive, and other characters become important. Note that—with few exceptions—all greater coverts in Upcher's are uniformly edged pale brown, whereas in Olive-tree the outer coverts are more or less whitish-fringed and contrasting. Olive-tree has stronger bill but there is overlap, hence best to also note that in Olive-tree yellowish lower mandible is slightly orange-tinged and lacks a dark tip or smudge (Upcher's: yellowish, sometimes with faint pinkish tinge, frequently with a dark smudge near tip). Wing- and tail-feathers average darker grey in Olive-tree, in particular wing-feathers as can be clearly seen here. Primary projection in Olive-tree is 3/4 of or equals tertial length, is more like 2/3 or slightly less in Upcher's. Head patterns very similar, though supercilium averages a little longer in Upcher's, extending a little behind eye in some, rarely so in Olive-tree (but slightly here). These two birds can also demonstrate that leg colour can be very similar. (Left: D. Occhiato; right: J. Babbington)

Juv/1stW, Scotland, Aug: unlike Upcher's Warbler (which is intermediate in size between Olivaceous and Olive-tree Warblers) has longer primary projection, longer, stronger and all-grey legs, and longer, stouter bill with yellowish-pink lower mandible. Tertials and greater coverts darker with more obvious pale edges, and white secondary panel can be more obvious at this age. Larger than Olivaceous Warbler, with longer more pointed wings, longer bicoloured bill, more obviously pale-edged tertials and greater coverts, and often (but not here) better-developed pale secondary panel. (H. Harrop)

AGEING & SEXING Moderate seasonal and age-related differences. Sexes alike in plumage. – Moults. Complete moult of both ad and 1stY generally in mid or late winter (Dec–Mar, sometimes Nov). – **SPRING** Ageing impossible. – **AUTUMN** **Ad** Worn, especially tips of outer primaries, and tertials. Upperparts grey, sometimes with slight olive hue. White secondary panel lacking due to abrasion, or only shows as faint remnant. **1stCY** Uniformly fresh. Secondary panel rather prominent, but edges not white as in spring, more buffish olive-grey. Entire plumage slightly tinged buff or cream, less pure (olive-) grey. Underparts have less obvious dark centres on flanks and undertail-coverts.

BIOMETRICS **L** 15.5–17 cm; **W** 84–90 mm (n 32, m 87.4); **T** 62–71 mm (n 32, m 66.0); **T/W** m 75.6; **B** 18.0–21.4 mm (n 30, m 19.6); **BD** 3.7–4.9 mm (n 25, m 4.2); **BW** 4.8–6.0 mm (n 26, m 5.4); **Ts** 21.3–24.5 mm (n 22, m 23.0). **Wing formula:** p1 minute, many < pc; **p2** =4 or 4/5 (rarely =5); **p3** longest (sometimes pp3–4); **p4** < wt (0) 0.5–3 mm; **p5** < wt 4–7.5 mm; **p6** < wt 7–12 mm; **p7** < wt 10.5–15 mm; **p10** < wt 19–23 mm; **s1** < wt 22.5–27 mm. Emarg. pp3–4; p5 has very slight emarg. in a few.

Juv/1stW, Israel, Aug: plumage rather similar to fresh ad in spring, and readily identified by same suite of features. Especially note larger size, chunky jizz and wing pattern, with obvious whitish secondary panel and clear edges to tertials and greater coverts, though less pure white, more buffish/greyish and narrower than in spring. (H. Shirihai)

GEOGRAPHICAL VARIATION & RANGE Monotypic. – Europe from Bulgaria and Greece to Thrace, S Turkey, N Levant; winters mainly S Africa, some E Africa.

REFERENCES HARROP, H. R., MAVOR, R. & ELLIS, P. M. (2008) *BB*, 101: 82–88. – SHIRIHAI, H. (1987) *BB*, 80: 473–482. – SHIRIHAI, H., CHRISTIE, D. A. & HARRIS, A. (1996) *BB*, 89: 114–138.

ICTERINE WARBLER
Hippolais icterina (Vieillot, 1817)

Fr. – Hypolaïs ictérine; Ger. – Gelbspötter
Sp. – Zarcero icterino; Swe. – Härmsångare

'Icterine' is derived from a Latin word meaning 'yellowish', and is thus an appropriate name. An excellent diurnal singer of tall, dense woods with undergrowth, it is the north-eastern counterpart to the Melodious Warbler of SW Europe. The loud song containing nasal notes and the ability to mimic other birds generally draw the attention, as the bird keeps mainly to the treetops, thereby often denying closer scrutiny. A summer visitor, arriving from its S African winter quarters in late April–May (in Scandinavia and Finland not until mid or late May).

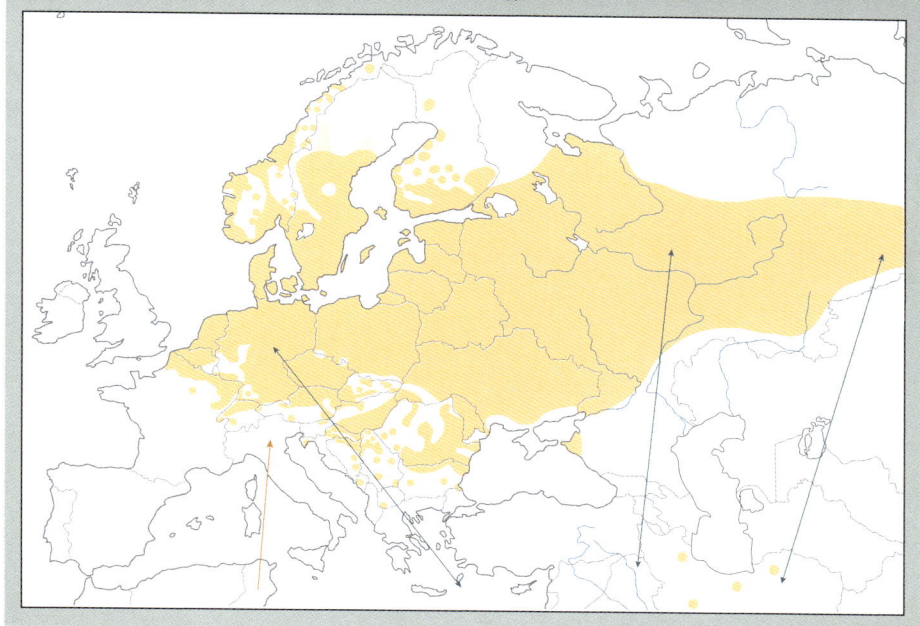

Scotland, May: long pointed wings make tail appear relatively shorter, while long primary projection with usually seven primaries visible clearly separates Icterine from Melodious Warbler. Ageing not possible in spring. (H. Harrop)

IDENTIFICATION Medium-sized warbler with square-ended, fairly short-looking tail and rather large head. Seen in profile, can have a flattish forecrown, or the crown-feathers are raised to give the bird a steep forehead. Upperparts greenish-grey with a generally obvious *whitish secondary panel* formed by edges to long tertials and secondaries. (On more abraded summer birds, the panel is less prominent.) Underparts typically rather *saturated lemon-yellow*, though odd birds, mostly young but a few bleached adults, may be paler below with yellow element less obvious or nearly lacking. Primary projection long, usually c. 3/4 of, or equal to, *tertial length*. Accordingly, in flight the wings look pointed and long. Pointed, strong bill with pinkish-yellow lower mandible. Face pattern 'open' with *quite pale lores* and lacks any dark in front of eye. Tail square-tipped with very slightly paler outer edges (often difficult to see). Legs strong, slate-grey, or have slight bluish cast.

VOCALISATIONS Song strong and pleasing, variably comprising skilful mimicry of other birds (some excel in this, others copy less, or are less accomplished), fast warbling and sharp notes mixed and delivered at varying pace, alternating between leisurely with brief pauses to flowing and energetic, many motifs repeated a few times (could, due to this, even recall Blyth's Reed Warbler). Best recognised by recurring interwoven, shrill, nasal notes, *gih-e gih-e…*, carrying far in dense woodland, and also the trisyllabic call (cf. below) included in the song. – Most typical call in spring–summer is a cheerful and musical trisyllabic, quick *tey-te-**dwee***. Also heard a single harsh, clicking *chek*, or when alarmed short series, rather like in Melodious, *te-te-te-te-tek*. In anxiety a softer whistling ***dwuee***, related to trisyllabic call (last syllable), a bit like Willow Warbler. In alarm also a hoarse, fast series *veh-veh-veh-veh-…*, somewhat like Common Whitethroat or even Red-backed Shrike in tone. Young beg with hoarse *teehv, teehv*. Less vocal on passage.

SIMILAR SPECIES Greatest resemblance is to *Melodious Warbler*, which see. On Icterine, note longer primary projection and more pointed wingtip (making tail appear shorter) often flatter forecrown, and more olive-grey upperparts with more visible pale (actually whitish, not brownish-yellow) secondary panel. Yellow underparts mean there is a risk of confusion not only with Melodious, but also with *Willow Warbler* and *Common Chiffchaff*, both of which have finer, darker bills and a dark eye-stripe, enhancing the length and prominence of their yellowish-white supercilia. – Odd Icterines are less yellow below, being more cream-coloured,

Netherlands, May: rather large-headed and short-tailed, but wings pointed with very long primary projection. When alarmed, forecrown can be raised, forming steep forehead. Also characteristic are pointed, strong bill with pinkish-yellow lower mandible, and noticeable whitish secondary panel and tertial fringes. Underparts variably tinged lemon-yellow, most evident on throat, upper breast and lores. Legs grey. (R. van Rossum)

Finland, May: lemon-yellow wash below is most intense on throat and lores. In this image, crown appears flattish, neck thick and head square-looking. Note bland face lacking dark eye-stripe, which often leads to species being described as 'open-faced'. (M. Varesvuo)

Finland, May: some Icterine are notably less yellow below, more cream-coloured, and can be confused with several other long-winged warblers (e.g. Marsh and Garden), but note all-pale lores, yellowish throat, whitish secondary panel, and whitish-edged tertials and wing-coverts, as well as grey legs. (M. Varesvuo)

Ad, Turkey Jul: here just prior to departure for winter in Africa, an ad with by now worn primaries and primary-coverts. Upperparts bleach duller, and due to feather wear pale secondary panel is variably reduced. Can also bleach strongly below, to duller yellow with whiter areas, but this bird is still quite yellow. (O. Bulut)

perhaps inviting confusion with *Marsh* and *Garden Warblers*, but Marsh has a dark loral smudge (not all-pale lores as Icterine), uniform wings (Icterine usually has a whitish wing panel) and pinkish legs (Icterine: grey), and Garden has a shorter and thicker greyish bill, and is more 'dingy' olive-grey below.

AGEING & SEXING Insignificant seasonal and age-related differences. Sexes alike in plumage. – Moults. Complete moult of both ad and 1stY in Africa in late winter (mid Dec–mid Apr). Post-juv moult in Europe either very limited (parts of head and body, odd wing-coverts) or lacking. – SPRING Ageing impossible. All birds return in quite fresh plumage. – AUTUMN **Ad** Worn, especially tips of outer primaries, and tertials. The whitish secondary panel is of variable prominence due to wear, usually nearly absent. A few birds are very bleached in worn autumn plumage and show hardly any yellow on underparts. **1stCY** Uniformly fresh. Some are paler below, yellowish-cream rather than lemon, although most have lemon tinge to lores, throat and upper breast, much as ad. Edges to long secondaries and tertials paler buffish-yellow, forming paler panel.

BIOMETRICS L 12.5–14 cm; **W** 73–84 mm (*n* 151, *m* 78.1); **T** 49–55 mm (*n* 16, *m* 53.2); **T/W** *m* 66.3; **B** 15.5–17.8 mm (*n* 15, *m* 16.4); **Ts** 19.9–21.3 mm (*n* 14, *m* 20.6). Undertail-coverts < tail-tip 11–17 mm. **Wing formula: p1** 0–3 mm > or < pc, < p2 42–48 mm); **p2** =4/5 (=4); **p3** longest; **p4** 0.5–2 mm < wt (rarely = wt); **p5** < wt 3–6 mm; **p6** < wt 10–15 mm; **p10** < wt 20–24.5 mm; **s1** < wt 22–27 mm. Emarg. pp3–5, though p5 less distinct in many.

GEOGRAPHICAL VARIATION & RANGE Monotypic. – Scandinavia south to N France, N Balkans, east through S Finland, Russia, Ukraine, Crimea, W Siberia to Ob; winters S Africa north to Uganda, Kenya. (Syn. *alaris.*).

REFERENCES SMOUT, T. C. (1960) *BB*, 53: 225.

Juv/1stW, United Arab Emirates, Sep: ageing in autumn unproblematic, with uniformly fresh birds invariably young. Young usually pale creamy below, with yellowish most noticeable on lores, and at least part of throat and upper breast. Whitish secondary panel and tertial fringes useful, but usually fringes are less pure white, reducing the contrast. Diagnostic combination of long primary projection, bluish-grey legs and clearly bicoloured bill. Forecrown feathers often erected, forming characteristic steep forehead. (I. Boustead)

Juv/1stW, Scotland, Sep: despite somewhat diluted appearance, yellowish tinge on face usually evident. Note slightly less prominent secondary panel. An important feature against young Melodious Warbler is the longer primary projection; also bluish-grey legs. Fresh plumage ages this bird. (A. Ash)

Juv/1stW, Madeira, Sep: some young Icterine can appear very similar to Melodious Warbler, due to light or wear making the whitish secondary panel appear very subdued, increasing risk of confusion. However, very long primary projection and rather bold pale tertial fringes confirm identification as Icterine. (O. Krogh)

Juv/1stW, England, Sep: typical posture, exhibiting long primary projection and strongly patterned tertial fringes, as well as large head and clearly bicoloured bill. Legs grey, and face pale and rather featureless. (B. Porter)

MELODIOUS WARBLER
Hippolais polyglotta (Vieillot, 1817)

Fr. – Hypolaïs polyglotte; Ger. – Orpheusspötter
Sp. – Zarcero políglota; Swe. – Polyglottsångare

The song of Melodious Warbler, the south-western counterpart of Icterine Warbler of N and E Europe, is perhaps rather more energetic than 'melodious'. Melodious Warbler accepts more open habitats and lower vegetation on average than its northern relative, provided it is dense, although there is some overlap in this respect. A summer visitor which arrives in SW Europe north to N France from its African winter quarters in late April–May.

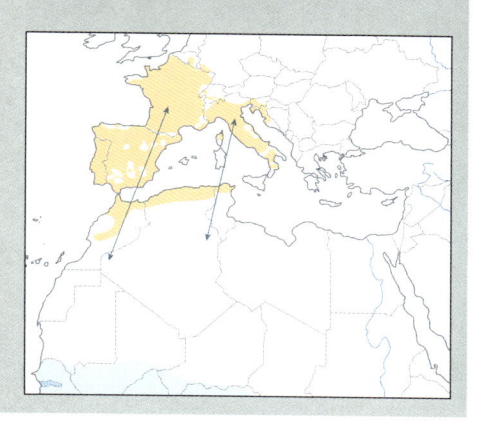

Italy May: bland face and bicoloured bill very similar to Icterine Warbler. Tail appears longer than latter, mainly due to shorter wings and primary projection. Also unlike Icterine, legs are normally a little paler and warmer in colour, usually brown-grey. Upperparts duller, tinged brownish-olive (less greyish), while pale secondary panel and tertial fringes are buff-brown and unremarkable. (D. Occhiato)

Italy, May: compared to Icterine, secondary panel and tertial edges slightly more subdued due to yellow-buff rather than white fringes. Furthermore, rather greyish legs and greenish cast above could invite confusion with migrant Icterine (Melodious is typically brownish above, but here scapulars and nape seem greenish-grey). An important clue to correct identity is the shorter primary projection, about 2/3 of tertial length (about equal length in Icterine). (D. Occhiato)

VOCALISATIONS Song a babbling warble of varying speed and length, often including drawn-out phrases at lightning-fast pace, recognised by combination of the speed and frequently interwoven chattering *trrt-trrt* calls or longer alarm-like *tr'r'r'r'r'r'rt*. Can recall a Mediterranean *Sylvia* song (e.g. in the Subalpine complex). When mimicry is included by a songster perched in lower vegetation it can be mistaken for a Marsh Warbler at first. At times the bird slows and uses more nasal or shrill notes, which might resemble both Icterine and Olivaceous Warblers (which see). Calls on territory are variety of tongue-clicking *chet*, *tre-te-te-tü*, drawn-out series of *te-te-te-te-te-...* or the quite characteristic sparrow-like, chattering *tr'r'r'r'r'r'rt*. Usually silent on passage.

SIMILAR SPECIES Most similar to Icterine Warbler, but fractionally smaller with usually more rounded head (Icterine often has slightly flatter forecrown), and Icterine has less brown, more greyish-tinged upperparts, more obvious whitish secondary panel, and longer primary projection (3/4 of, or equal to, visible tertials). As a function of this, primary emarginations on folded wing on Icterine fall opposite secondary tips, and visible primary tips are more spaced near the wingtip than at the base (on Melodious, emarginations fall inside secondary tips, and visible primary tips are near-evenly spaced). Melodious also has slightly shorter undertail-coverts than Icterine. Note also difference in pre-nuptial moult, Icterine never moulting any tertials or inner secondaries, whereas this occurs rather regularly in Melodious (D. Bigas, pers. comm.). – Yellow underparts could theoretically lead to confusion with Icterine, as well as with *Willow Warbler* and *Common Chiffchaff*, both of which have finer, darker bills and a dark eye-stripe, enhancing the length and prominence of their yellowish-white supercilia. Beware, however, paler-coloured Melodious (rare!), or those on which the yellow underparts are difficult to see (mainly due to light conditions), appearing more cream-coloured below, inviting confusion with *Marsh* and *Isabelline Warblers*. These can be eliminated by their vocalisations, Marsh also on pinkish legs and long primary projection, and Isabelline by slightly larger size and paler brown upperparts.

IDENTIFICATION Like many warblers, usually noticed and identified by its fast, energetic song. Spends most of its time in thickets and foliage, keeping well hidden, perching in the open only briefly. Medium-sized, slim warbler with *greenish-tinged brown upperparts*, generally uniform wing, lacking obvious paler secondary panel (less abraded birds might show a hint, created by paler olive-buff fringes), and characteristic *yellowish* (buffish-yellow in young) *underparts* (rarely paler below, but then has at least pale lemon throat and upper breast). Important to establish that *primary projection is not decidedly long*, usually *about half* or *two-thirds of visible tertial length*. On flying bird this manifests as rounded, not too long wings. The bill is rather long with pinkish-yellow lower mandible. Face pattern 'open' with *quite pale lores* and lacks any dark in front of eye. Mid-length tail is square-tipped and has very slightly paler outer edges (often difficult to see). Legs strong, brown-grey.

AGEING & SEXING Insignificant seasonal and age-related differences. Sexes alike in plumage. – Moults. Complete

France, May: greenish-tinged brown upperparts, with only weak indication of pale secondary panel. Underparts and face variably tinged lemon-yellow, on average stronger than in Icterine. Here, compared to Icterine, note wing structure (shorter primary projection), while primary emarginations fall well inside secondary tips, and primary tips are near-evenly spaced. (A. Audevard)

moult of ad in Africa in autumn or early winter (Sep–Nov, sometimes Aug–Dec). Whether 1stY invariably moults completely is possibly questionable; some spring birds are heavily worn. Pre-nuptial moult of ad in late winter includes body, generally innermost two tertials and often one or a few inner greater coverts (Gargallo 1995). It can also include central and longest tertials and some innermost secondaries (D. Bigas, pers. comm.). – **SPRING** Ageing not possible. – **AUTUMN Ad** Worn, especially tips of outer primaries, and tertials. No paler secondary panel due to wear. Rarely, pale birds with reduced yellow occur, being bleached and worn. **1stCY** Uniformly fresh. Some paler below, yellowish-cream rather than lemon. Edges to long secondaries and tertials paler brownish-yellow, forming indistinct paler panel.

BIOMETRICS L 12–13.5 cm; **W** 62–71 mm (n 47, m 66.2); **T** 48–54 mm (n 34, m 50.9); **T/W** m 76.2; **B** 14.5–17.3 mm (n 30, m 15,7); **Ts** 19.5–21.5 mm (n 30, m 20.4). Undertail-coverts < tail-tip 15–23 mm. **Wing formula: p1** > pc 2.5–8 mm, < p2 27–32 mm; **p2** =6/7 or 7 (rarely =5/8); **pp3–4**(5) about equal and longest; **p5** < wt 0–2 mm; **p6** < wt 2–5 mm; **p10** < wt 11–15.5 mm; **s1** < wt 12.5–17 mm. Emarg. pp3–5; p6 has slight or no emarg. in majority, but equally strong in a very few.

Netherlands, Jun: characteristic yellowish underparts can be strongly subdued, as here, being predominantly cream-coloured. Such paler Melodious can be confused with both Marsh and Isabelline Warblers. However, Marsh has longer primary projection and different face pattern (darker lores and suggestion of eye-stripe), and Isabelline has slightly larger size and paler, longer bill and plainer brown upperparts. (R. Schols)

Melodious (top left: Italy, May; bottom left: England, Sep) versus Icterine Warbler (top right: Finland, May; bottom right: England, Aug), in fresh spring/summer (top two images) and fresh juv/1stW plumages (bottom two): at both seasons, first establish Melodious' shorter primary projection and weaker pale secondary panel, which is well-developed and whiter in fresh Icterine. All other differences are subject to individual variation and some overlap, thus no more than supportive: Melodious tends to have overall longer-tailed appearance, slightly shorter undertail-coverts, more brownish-olive upperparts, yellower-tinged underparts, and paler brown-tinged grey legs. Song and relative position of primary emarginations may be employed in borderline cases, but usually the two species are readily separable. Following pre-nuptial moult, both species cannot be aged (spring/summer), but in autumn fresh birds are juv/1stW. (Top left: D. Occhiato; top right: M. Varesvuo; bottom left: G. Thoburn; bottom right: A. Banwell)

BUSH AND REED WARBLERS

Ad, Scotland, Aug: overall very worn and bleached, with greyer upperparts and whiter underparts, thus more similar to Icterine. Consequently, separation must concentrate on shorter primary projection, and when wing is slightly drooped, but still folded, it can be possible to establish that primary emarginations fall well inside secondary tips, and that primary tips are near-evenly spaced. Also, p1 is clearly longer than primary-coverts (Icterine: about equal). (H. Harrop)

Juv, Spain, Jul: when freshly fledged often saturated bright greenish above and yellowish below (slightly buffish-yellow in young). Shorter primary projection and near-evenly spaced primary tips compared to Icterine. Yellow underparts could suggest Willow Warbler or Chiffchaff, but both readily eliminated by their finer, darker bills and dark eye-stripe, enhancing much longer yellowish supercilium. (A. M. Domínguez)

GEOGRAPHICAL VARIATION & RANGE Monotypic. – SW & C Europe from N France through Iberia to NW Africa, east to SW Germany, Switzerland, Italy, W Balkans; winters tropical W Africa from Senegal to Cameroon.

REFERENCES BERMEJO, A., DE LA PUENTE, J. & PINILLA, J. (2002) *Ardeola*, 49: 75–86. – GARGALLO, G. (1995) *Vogelwarte*, 38: 96–99.

Juv/1stW, England, Sep: due to angle and strong light, appears to have almost Icterine Warbler-like prominent whitish secondary panel. Consequently, Icterine must be eliminated by shorter primary projection and relatively longer tail. Although a little wear of wings and tail is already evident, this is still rather limited and presents no difficulty for ageing. (G. Thoburn)

Juv/1stW, France, Aug: young have in general the dullest plumage, often characterised by whitish underparts with creamy-yellow only on face and throat. Separated from young Icterine by shorter primary projection, lack of obvious secondary panel (here stronger than usual), more rounded head, and shorter undertail-coverts. (T. Quelennec)

Juv/1stW, Scotland, Aug: duller Melodious can be confused with several small *Acrocephalus*, including Marsh, Blyth's Reed and *fuscus* Reed Warblers. However, if seen well, bland face of Melodious lacking dark eye-stripe in combination with rather square tail and short undertail-coverts help to eliminate these. Same-age Marsh is also warmer tawny in autumn (only worn ad similarly greyish). Melodious should have at least some yellowish on lores, lower cheeks, throat and upper breast (never on any *Acrocephalus*). Possible confusion with Isabelline Warbler must be taken into account, but this is usually paler brown above. Aged by being evenly fresh. (H. Harrop)

MARMORA'S WARBLER
Sylvia sarda Temminck, 1820

Fr. – Fauvette sarde; Ger. – Sardengrasmücke
Sp. – Curruca sarda; Swe. – Sardinsk sångare

Named after one of Napoleon's generals, an Italian who was also a naturalist, this small Mediterranean warbler favours low garrigue and similar vegetation, breeding mainly on Corsica and Sardinia, as well as on several smaller islands close to Italy. Largely resident, though small numbers winter in NW Africa, in Algeria to NW Libya, and on Sicily. Claimed to have bred in mainland Spain, France, Italy and on Crete, but none of these records is well substantiated. Documented records exist of spring overshoot records as far north as Scotland, and a vagrant has also been recorded in Egypt.

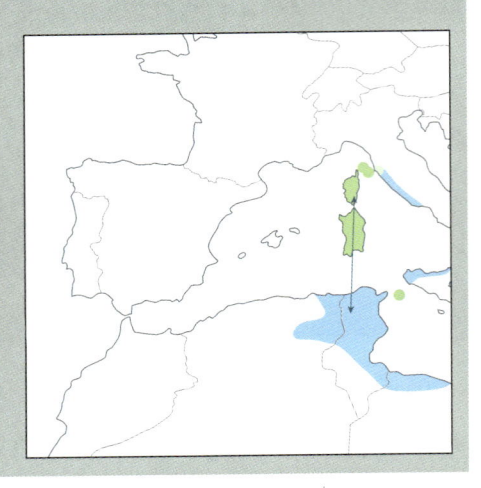

Ad ♂, Corsica, Apr: a small, long-tailed *Sylvia*, though less slim and delicate than Balearic Warbler. Overall tinged slate- or bluish-grey; throat almost as uniform grey as rest of underparts (in fresh plumage often finely flecked white). Buff on belly limited. Less bright base to subtly shorter bill and duller and colder reddish orbital ring. Diffusely darker on lores. Aged on basis of being less worn, with bluish-grey fringes to ad primary-coverts and remiges. (L. Sebastiani)

IDENTIFICATION A small, *long-tailed Sylvia*, recalling Balearic Warbler but larger and proportionately slightly shorter-tailed than latter, being somewhat closer to Sardinian Warbler in these features. Due to size and shape, less likely to be confused with Dartford Warbler than Balearic Warbler. ♂ *has slaty bluish-grey upperparts and wing fringes*, usually with sootier head, mainly due to *slightly blacker lores, forehead and ear-coverts*; throat also grey (only subtly paler than rest of head), finely *freckled white* on upper throat in fresh plumage; rest of underparts also grey, uncommonly with a faint pale buff wash on flanks and paler mid belly. ♀ *browner olive-grey above* (contrasting with greyer crown, rump and uppertail-coverts) *and buffier below*, virtually lacking any darker mask or bluish cast to upperparts and wings. First-winter ♂ recalls ♀ or is intermediate between the two. For all, graduated *tail* is slim and *lacks obvious whitish edges and tip*. Orbital ring red, fairly thin and moderately prominent; iris reddish-orange in adult, but browner/olive in first-autumn birds. Spiky bill has extensive dull pinkish-yellow lower base and darker culmen and tip. Legs pale orange-brown. Typically skulks in cover, moving with cocked tail, which is waved up and down and from side to side, and usually forages near or on ground, though may also make aerial sallies for insects. Tolerates close approach and is not shy, despite nervous behaviour.

VOCALISATIONS Song is a soft, high-pitched warble that sometimes is given in Common Whitethroat-like song-flight. Commences with a short high-pitched section or a few 'explosive' notes followed by a 'bouncing', very fast warble, more varied and pleasant than Balearic Warbler song, often gradually slightly falling in pitch and trailing off in strength, and therefore at times can recall song of Short-toed Lark, *sri-we-chio-trip-trip-chet-chet-chet-sivi-tsrevi-sivui* and variants. Can also somewhat recall Stonechat song. – Commonest call is a short and hard but guttural *chrk* or *tchrek*, often repeated from cover and has a slightly explosive quality. In alarm also gives a dry *tk-tk-tk-tk-tk*, sometimes in long series.

SIMILAR SPECIES Marmora's Warbler is occasionally recorded in the Balearics on migration, and both Marmora's and Balearic Warbler wander to mainland Spain, hence *Balearic Warbler* is the greatest confusion risk. Both song and calls differ (song of Marmora's clearly more liquid and varied with some clearer notes, whereas that of Balearic is distinctly similar to Dartford Warbler and can even recall alarm of Sardinian Warbler; main call of Marmora's is a gruffy, hard *chrk*, in Balearic a more rolling *tr'ret*), but to a less experienced ear it may be easier to concentrate on relative size and structure. Marmora's more closely resembles Sardinian Warbler, while Balearic is nearer Dartford. In particular, Marmora's appears shorter-tailed than Balearic. Bill colour useful: Balearic has a brighter, more intensely orange-toned base and at times a more diffuse and restricted dark tip to lower mandible (rarely almost absent), but much overlap as to dark tip. In breeding season, Marmora's tends to have thinner and less prominent duller red orbital ring (Balearic invariably has very swollen and bright red orbital ring). Furthermore, throat is fairly dark grey freckled white (or homogenously grey with wear, unlike Balearic, which is mainly uniform pale greyish-white, producing pale-throated appearance). In juvenile and young autumn birds, leg colour is helpful too, as Marmora's has browner legs (Balearic: more orange-coloured). – Juvenile and first-winter Marmora's and Balearic share many differences for separation from same plumages of *Dartford Warbler*, but Marmora's is particularly distinctive due to its more Sardinian-like structure and bare-parts coloration (diagnostically darker, lacking any obvious reddish or yellowish colorations characteristic of Dartford). Also, plumage-wise Marmora's virtually lacks distinctive pink-red wash of younger Dartford (and, to some extent, Balearic), especially on underparts. In Marmora's, the lores are dark and concolorous with rest of head, while in Dartford they are often distinctly paler. – Given size and structure of Marmora's, it may be necessary to eliminate ♀ *Sardinian Warbler*, which has a grey hood, brownish mantle and purer whitish throat, a longer bill (with base of lower mandible pale grey, not greyish-pink as in Marmora's), and well-patterned tail with white corners. It also has more prominent reddish orbital ring, and reddish or pinkish-orange eye-ring, and very different voice. – Marmora's has wandered once to Egypt and, therefore, confusion is possible with ♀-like *Cyprus Warbler*, which is bulkier with a shorter, fuller tail (with Sardinian-like pattern), longer primary projection, deeper-based bill, darkish chevrons on the undertail-coverts, browner upperparts and narrower, whiter fringes to the tertials (slightly broader, greyer and buffier in most Marmora's), with better-defined pale fringes to wing-coverts (almost uniform in Marmora's). The movements and voices of the two species are also very different.

AGEING & SEXING Ageing often possible with close inspection of moult and feather wear in wing, tail pattern and iris coloration, but note that young can moult many secondary-

Corsica, May: amount of buffish below varies, and some show more yellowish-orange (less pinkish) bill-base—in these aspects then confusingly similar to Balearic Warbler. Bright orange-brown eye suggests ad, but ageing unsure without assessment of moult and feather wear, as some 1stY develop ad-like bare-parts coloration in spring. Sexing must also remain uncertain in such cases, but visible features favour either 1stS ♂ or ♀ of any age. (S. Pfützke)

SYLVIA WARBLERS

Presumed ad ♀, Sardinia, May: combination of duller, less pure grey plumage, indistinct blackish mask, brown-tinged underparts, and that bird is very likely ad (by apparently less worn and greyer fringes to primary-coverts and remiges) strongly suggest a ♀. As ageing is not always possible without handling, and ad ♀♀ can be very similar to some young ♂♂, it is best to be cautious with ageing/sexing. (A. Comas)

♂, presumed 1stS, Sardinia, May: very similar to ad ♂ but note strong moult contrast with worn brownish remiges (indicating 1stS), although silvery-grey outer webs still present, and iris almost as bright as ad, but primary-coverts not visible, thus ageing requires caution. Note short, slightly chunky bill, with a pale pinkish-straw base, and cold red orbital ring. Legs have atypically strong colour. (L. Sebastiani)

coverts and tail-feathers, and can rapidly acquire ad-like iris colour. Sexes rather similar (but can be differentiated with experience: correct ageing often essential, especially with birds of intermediate appearance and in autumn). Limited seasonal variation. – **Moults.** Complete post-nuptial and partial post-juv moults after breeding in late summer (mostly Jul–Sep). Post-juv moult involves all head- and body-feathers, lesser and median coverts, most or all greater coverts, alula, tertials and some tail-feathers. Partial pre-nuptial moult in late winter to early spring; usually includes some head and body-feathers, tertials, and perhaps a few secondary-coverts and tail-feathers. – **SPRING Ad** Both sexes best aged by having only ad (=fresher) wing-feathers. **Ad ♂** With wear entire upperparts become darker, tending to be pure bluish-grey, and black mask enhanced. Underparts are more solidly patterned, with grey sides and throat more uniform medium grey, almost concolorous with breast, sometimes with fine white speckles visible on upper throat. Flanks and lower belly often concolorous grey, rarely with very faint buff cast. **Ad ♀** Wears slightly greyer than in fresh autumn plumage, but mantle, wing-coverts, tertials and remiges fringed browner, and underparts paler, with grey and buff admixed, while head also paler (due to restricted dark on lores). **1stS** Not always possible to separate, though majority have heavily worn and browner remiges and primary-coverts, with faded buffish-brown (rather than mainly grey) fringes; also usually tinged more buff and brown, and duller throughout, with reduced frontal mask. Juv tail-feathers retained in some, when extremely faded and worn, with distinctly less pure white edges and tips. Iris nearly ad-like in most, very rarely with olive cast, but unreliable for ageing. – **AUTUMN Ad ♂** Compared to spring plumage slight pale earth-brown tone to mantle and scapulars and may show vestigial whitish submoustachial stripe, while underparts variably suffused—usually extensively—grey with some buff on flanks and, most notably, breast-sides and vent; wing-coverts, tertials and remiges mainly fringed grey. Also head generally less dark with reduced mask. **Ad ♀** More extensively tinged brown on mantle and scapulars than ♂, and underparts paler with reduced dark on body-sides and some pale pinkish-buff admixed; wing-feathers fringed less pure grey. Note that ad ♀ strongly resembles 1stW ♂, though some, particularly older ♀♀, can strongly approach ad ♂, too. Both sexes aged by evenly very fresh wing- and tail-feathers (remiges and primary-coverts centred glossy greyish-black), while r6 usually has broader, whiter fringes to outer webs and tips than in other plumages; also, orbital ring and iris as in

1stS ♂, England, Jun: some young ♂♂ have strong bluish-grey plumage and deep reddish eye, approaching ad ♂, and are separated from ad only by moult limits in wing, with retained juv primary-coverts (shown better on other images of same bird). Compared to Balearic Warbler, note stronger build, to some extent approaching Sardinian Warbler. (D. Stewart)

♂ Marmora's (left: Sardinia, Jun) and Balearic Warblers (right: Mallorca, Mar): separation can be very difficult, especially of silent vagrants. With experience, it is often possible to appreciate Marmora's slightly stronger build, with jizz somewhat approaching Sardinian Warbler, while Balearic is more reminiscent of Dartford. Note Marmora's slightly darker grey throat (white tips here already mostly worn off), whereas Balearic usually gives more pale-throated impression. Base of bill differs in that Marmora's is paler pinkish, while Balearic is brighter and more orange-toned. Also, orbital ring of Marmora's is less bright red, more dull pink. These images highlight the characteristics that separate the species, but there is much individual variation (mostly age-related) and overlap between the sexes, e.g. some Marmora's (especially ♀ or young ♂) can be quite substantially sullied pinkish-buff below. (Left: J. Coveney; right: R. Martin)

— 491 —

spring or perhaps slightly less bright. **1stW** Like fresh ad but sexual dimorphism less obvious; both sexes less pure grey, more extensively buff and rusty-brown above, and underparts paler with more extensive whitish-buff wash. Mask reduced and paler, mainly blackish-grey in ♂ and often indistinct or almost absent in ♀ (in both barely reaching forehead). Best aged by juv remiges and primary-coverts, which are softer textured and browner, fringed buffish-brown (primaries virtually lack pale tips of ad); juv tail-feathers browner, very narrowly fringed and indistinctly tipped sandy-buff (white in ad) but these can be replaced by ad-like feathers at any time. Iris darkish olive-brown in nestlings (clearly darker and duller than ad), gradually acquiring reddish-orange tone in 2ndCY; orbital ring dull orange-red. **Juv** Soft, fluffy body feathering is dark earth-brown above, concolorous with head, but paler and buffier below.

BIOMETRICS L 12.5–13.5 cm; **W** ♂ 54–61 mm (n 18, m 56.9), ♀ 52–56 mm (n 13, m 54.2); **T** ♂ 57–65 mm (n 18, m 61.3), ♀ 57–61 mm (n 13, m 59.2); **T/W** 108.6; **TG** 6–12 mm (n 27, m 9.4); **B** 11.2–13.2 mm (n 30, m 12.2); **BD** 2.6–3.2 mm (n 27, m 3.0); **Ts** 18.7–21.2 mm (n 29, m 19.8). **Wing formula: p1** > pc 2.5–7 mm, < p2 21.5–25 mm; **p2** < wt 3.5–6 mm, =7/9 (rarely =9); **p3** < wt 0.5–1.5 mm; **pp4–6** longest (or p6 0.5–3 mm <); **p7** < wt 1.5–5.5 mm; **p10** < wt 6–8.5 mm; **s1** < wt 7–9.5 mm. Emarg. pp3–6 (usually fainter on p6).

GEOGRAPHICAL VARIATION & RANGE Monotypic. – Corsica, Sardinia, and on other, smaller islands close to Italy and, at least formerly, on Zemba off Tunisia; largely resident, though small numbers winter in NW Africa, Algeria to NW Libya, and on Sicily.

TAXONOMIC NOTES Marmora's Warbler was previously considered polytypic, including ssp. *balearica*, but *balearica* (Balearic Warbler) was fairly recently elevated to allospecies status based on its differing vocalisations, coloration, size, structure and mtDNA (Shirihai *et al.* 2001). Here, it is regarded as a valid biological species based on the same evidence and on a recent study (Voelker & Light 2011), which showed *balearica* to be sister not to *sarda* but rather to *undata*. Elevating it to species is consistent with our treatment of other similar cases. Taken together, the differences exceed what is normal for subspecies of the same species.

REFERENCES Shirihai, H. *et al.* (2001) *BB*, 94: 160–190. – Voelker, G. & Light, J. E. (2011) *BMC Evolutionary Biology*, 11: 163.

♂, presumed ad, Corsica, Oct: when fresh, generally less pure grey above, underparts variably faintly suffused brown, dark mask reduced, and sometimes shows some white tips on throat (especially on centre). Degree of dark grey plumage best fit ♂ at this season; wing seems evenly fresh and mainly grey-fringed, while red orbital ring suggests an ad, but iris is less deep orange-brown than usual. (G. Jenkins)

♂, presumed 1stW, Tunisia, Dec: young ♂ in autumn often shows some brownish in bluish-grey upperparts, contrasting with streaky paler chest and warmer (pinkish-brown) belly. Best aged first, by juv remiges and primary-coverts with brownish fringes. Iris olivaceous-ochre and orbital ring duller red. Unlike Balearic Warbler, note well-defined black tip to otherwise pale pinkish lower mandible (Balearic usually has more diffuse and smaller dark tip to brighter orange-toned bill). (P. Casali)

1stW, Corsica, Sep: in early autumn (often at end of post-juv moult) many 1stW are typically rather 'scruffy' with less pure grey plumage, and generally rather ♀-like. Variably tinged brownish on mantle and wing fringes, underparts sullied brownish-grey, with usually dull red-brown orbital ring and olive iris, while juvenile tail-feathers are either fully retained or only partially moulted. Due to individual variation, many young birds are difficult to sex: in this case, overall purer greyish upperparts and head suggest a ♂. (D. Cottridge)

Juv, Corsica, Aug: the soft, fluffy body-feathers are characteristic. Upperparts almost dusky brown, but underparts paler and rusty-tinged grey. Red orbital ring still very undeveloped and dull. The locality helps when eliminating a juv Balearic. (F. Jiguet)

BALEARIC WARBLER
Sylvia balearica Jordans, 1913

Fr. – Fauvette des Baléares; Ger. – Balearengrasmücke
Sp. – Curruca balear; Swe. – Balearisk sångare

Restricted as a breeder to the Balearic Islands (except Menorca), where it can be abundant in low garrigue and so-called matorral vegetation, often favouring suitable habitats near sea level. Breeders at higher altitude generally probably descend in winter, but it is otherwise resident. In the past, frequently reported to breed in SE or E Spain, but the only definite evidence of presence on the mainland involves birds on autumn dispersal and it is not even certain that the species winters there.

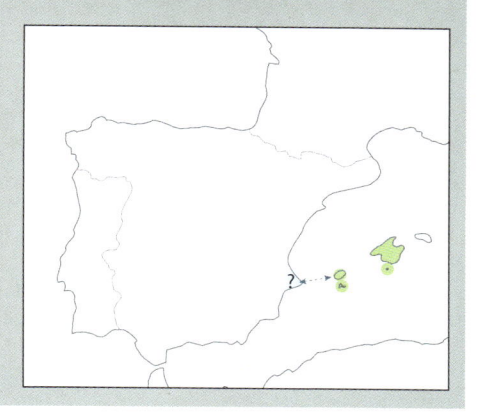

IDENTIFICATION Strongly recalls partly sympatric Dartford Warbler in size and structure, while *plumage closely resembles allopatric Marmora's Warbler*. Markedly *small-bodied and long-tailed* (tail only slightly shorter than in Dartford), with short wings (and correspondingly tiny primary projection). ♂ has *lead-grey upperparts* (subtly paler grey than Marmora's), a *hint of a blackish mask* (less contrasting than Marmora's, though much overlap), and *underparts greyish tinged dusky buffish-grey to warm buffish-pink*, mainly on flanks and belly (but buff hues can also invade lower breast). All year, throat paler than Marmora's, more whitish-grey, creating *distinctive white-throated impression*, markedly contrasting with darker lores and breast (unlike darker grey and finely white-speckled throat of Marmora's). ♀ rather similar (slightly less marked sexual dimorphism than in Marmora's), but *invariably saturated more pale pinkish-buff on grey-brown breast and body-sides*; also averages browner above and *virtually lacks any suggestion of a dark mask* or bluish cast to upperparts. First-winter ♂ recalls ♀. As in Marmora's, graduated *tail lacks obvious whitish edges and tip.* Orbital ring on average more swollen and *prominent*, particularly in ♂, *being bright red*. Iris as in Marmora's, reddish-orange or warm ochre in adult, darker and duller brown in first-autumn birds. *Pointed, thin bill is on average longer* than in Marmora's, with *base of lower mandible more brightly yellow-orange* (Marmora's: duller pinkish-straw), with blackish tip of bill on average more restricted and less sharply demarcated (but much overlap in this respect between them). Legs pale brownish-orange. Like Marmora's, less skulking than Dartford Warbler, and behaviour identical to the former.

VOCALISATIONS Song of Balearic differs from Marmora's, having a more grating Dartford-like tone, and is clearly less liquid and varied. It is a fast, short, repetitive song, the first half often 'winding' up and down, and somewhat recalling alarm of Sardinian Warbler, while second half is more of a fast trilling twitter, simpler and drier than song of Marmora's, e.g. *tri-trui-trui-trui-tritritritritritre*. – Normal call *trt* or *tr'ret* or *terrip*, dry and with audible 'r', almost rolling. Alternatively, a very short high-pitched *vit*, sometimes doubled in excitement, *vit-vit*. Alarm-call similar to Marmora's (mostly consisting of rapidly-repeated, 'stuttering' contact calls).

SIMILAR SPECIES Greatest confusion risk, especially if dealing with vagrants, is with largely allopatric *Marmora's Warbler*, and this pitfall is treated under that species in some detail. – *Dartford Warbler* is also a significant problem, particularly juveniles and first-winters. Balearic Warbler is almost as long-tailed as Dartford and of similar size. Juve-

♂, presumed ad, Mallorca, May: a small, long-tailed *Sylvia* almost identical in plumage to Marmora's. Differs from Marmora's by whiter throat and warmer-tinged underparts, and by deep yellowish-orange lower mandible with on average smaller blackish tip than Marmora's. Also has brighter red orbital ring. Another average difference is slightly longer bill of Balearic. Rather fresh and evenly-feathered wing (including grey-fringed primary-coverts) suggest ad. (F. López)

Variation in spring/summer ad ♂♂ (left: Apr; right: Jun), Mallorca: left bird more similar to Marmora's in being uniform lead-grey (but note characteristic more whitish throat, bill length and pattern, and bright red orbital ring, which identify it to species). The right-hand bird is, however, more typical in its brown hue to flanks and belly, very long tail and bill and orbital ring colours. The relatively fresh wings, lacking any visible moult limits (with broad grey fringes to primary-coverts and remiges) age both as ad. (Left: I. Merrill; right: B. Ramis)

nile Dartford is best separated from juvenile Balearic by their different calls, otherwise they are near-identical except in bill coloration. Juvenile Balearic has the blackish tip to lower mandible paler, less sharply demarcated and more restricted (in some almost absent); juvenile Dartford, however, has a very obvious blackish tip. Juvenile Dartford is also buffier above and below, especially on chin and throat; Balearic has a slightly whiter throat and darker face (especially lores). In particular, less colourful first-winter ♀ Dartford can have the otherwise diagnostic pinkish reddish-brown underparts rather subdued; conversely, some ♀ Balearic can have rather pinkish-grey and buffish-washed underparts. However, Balearic still has a diagnostic whitish-grey throat and darker breast, ear-coverts and lores; Dartford on the other hand completely lacks a pale-throated appearance and has a whitish-spotted pinkish-brown throat and chin. Furthermore, the eye-ring comprises virtually only grey and white feathers in Balearic, but mostly pinkish-buff ones in Dartford. Calls are also very different.

AGEING & SEXING Much as Marmora's Warbler (which see), except some minor differences in timing and extent of moult (for instance, Balearic undertakes a slightly but significantly more extensive post-juv moult). Also note that Balearic is less reliably aged by iris colour, as some 1stW have nearly ad-like iris in early autumn. However, others can be separated by their duller and less deep orange iris; a few, with mostly olive-brown iris, are easily aged as 1stW.

♀, Mallorca, Mar: ♀♀ are typically browner above and paler pinkish-brown below, but sexual dimorphism is not always well-marked. Note the species-specific white-throated impression. Ageing in this case unsure: the less worn plumage, lack of discernible moult limits in wings and deeper reddish eye all suggest ad, but the very worn outer tail-feather suggests otherwise. (R. Martin)

Mallorca, Apr: some birds can develop warm pinkish buff-brown belly, which at first glance could suggest Dartford Warbler. Such pigmentation indicates ♀, and the less worn wing and tail, dull grey (less bluish) upperparts and less obvious dark loral area support this assumption. However, ageing (which is impossible from this photo) should precede sexing. (R. Steel)

1stS ♂, Mallorca, Jun: young ♂♂ are very similar to ad, usually differentiated only by the worn and slightly brown-tinged juv primaries and primary-coverts, subtly contrasting with the renewed bluish grey-fringed secondary-coverts and tertials; very fresh innermost greater covert probably also renewed. Iris is duller, while colour of bill and orbital ring very typical. (B. Ramis)

♀, presumed ad, Mallorca, Dec: the browner upperparts and paler pinkish-buff underparts, with darker, greyer head and tail are features of fresh ♀♀. Compared to Marmora's, note Dartford Warbler-like jizz (with shorter wings and long, slender tail). The apparently evenly-fresh plumage and wing, round-tipped tail-feathers and bright eye suggest an ad. (F. Jiguet)

BIOMETRICS L 12–13 cm; **W** ♂ 49–54 mm (n 14, m 51.2), ♀ 49–52 mm (n 12, m 50.5); **T** ♂ 53–62 mm (n 13, m 57.9), ♀ 53.5–60 mm (n 12, m 56.8); **T/W** m 112.7; **TG** 7.5–16 mm (n 22, m 10.8); **B** 12.5–13.9 mm (n 27, m 13.0); **BD** 2.5–3.2 mm (n 21, m 2.9); **Ts** 17.5–19.5 mm (n 27, m 18.5). **Wing formula: p1** > pc 3–6 mm, < p2 19–23 mm; **p2** < wt 4–6 mm, =7/10 (rarely =10); **p3** < wt 0.5–1.5 mm; **pp4–6** longest (or p6 to 2 mm <); **p7** < wt 1.5–5 mm; **p10** < wt 6–7.5 mm; **s1** < wt 6–9.5 mm. Emarg. pp3–6 (usually fainter on p6).

GEOGRAPHICAL VARIATION & RANGE Monotypic. – Balearic Is (except Menorca); resident; vagrants may reach the Spanish coast.

TAXONOMIC NOTE See under Marmora's Warbler for an explanation of the treatment of Balearic and Marmora's as two different biological species.

REFERENCES Voelker, G. & Light, J. E. (2011) *BMC Evolutionary Biology*, 11: 163.

Early 1stW plumage, Mallorca, Aug–Sep: in early autumn many young birds are typically 'scruffy', with dull greyish plumage (left bird: presumed ♂) or still mostly brownish plumage (right bird: presumed ♀) and only slightly paler underparts; also only very dull red-brown orbital ring, olive iris, pinkish lower mandible and tarsi, and worn pointed tail-feathers. Note predominantly dark juv tail-feathers. (Left: R. McIntyre; right: L. Olsson)

Juv, Mallorca, Aug: note fluffy and buffish brown-tinged plumage. Bill long but has still not attained bright yellow-orange colour basally. (D. Rickards)

DARTFORD WARBLER
Sylvia undata (Boddaert, 1783)

Fr. – Fauvette pitchou; Ger. – Provencegrasmücke
Sp. – Curruca rabilarga; Swe. – Provencesångare

The Dartford Warbler is, despite the periodic and quite marked fluctuations that beset those populations at the northernmost extremities of its range, among the hardiest of *Sylvia*, as it is resident in Britain and N France. Its range stretches south to N Africa, with populations on most of the W Mediterranean islands. There is relatively little winter emigration, although those at higher altitudes move lower and some south to the northern edge of the Sahara. It largely favours low thickets of gorse, dense maquis, well-vegetated heaths and very open pine woodland with a dense understorey.

S. u. dartfordiensis, 1stS ♂, England, Apr: those in S England to NW France tend to be darker brown-grey above and deeper red-brown below, but such differences are slight and not always apparent. Worn, brown juv primaries and primary-coverts age this bird as 1stY, but eye colour is already ad-like. (R. Steel)

IDENTIFICATION A small and rather secretive *Sylvia* with a *long, slim tail that is proportionately the longest of the genus* and is frequently held cocked (or waved sideways). The very attractive ♂ is lead-grey to earth-brown above and *deep vinaceous reddish-brown below* (Preussian Red, Dark Vinaceous-Purple; Ridgway 1912), with *whitish flecking mainly on chin and upper throat*, and has restricted whitish belly. ♀ is slightly duller, generally more brownish above and less dark, more *pinkish-red or orange-brown on underparts* with *more extensive whitish-buff wash* and broader pale belly patch. *Tail graduated with indistinct pale edges and tips* to outermost rectrices. Wings very short and rounded, and narrowly fringed bluish-grey (mainly in adult ♂♂) or grey admixed with or predominantly pale rusty or buffish-brown (mainly ♀♀), mostly on tertials. Young are like duller versions of adults. Bill thin and pointed with contrasting dark culmen and tip, base of lower mandible being pinkish. Orbital ring mostly reddish-orange in adult, but duller brownish-olive or orange-brown in juvenile/first-winter. Iris and legs orange-brown. Typically skulks in dense low cover, but when seen well, perched atop a bush, can be seen to be a rather long-legged and large-headed warbler. Restless, its jerky movements are typical of several small *Sylvia* species, but perhaps undertakes rather longer flights between cover than close congeners; flight generally appears skipping and whirring.

VOCALISATIONS Song is given year around, although mainly in breeding season. A short vigorous and rather grating and monotonous warble with a few inserted sharp whistling or piping notes. Compared to quite similar song of Balearic Warbler has more of the clear whistling notes, which are rather evenly spread through the strophe. Compared to song of Sardinian Warbler is more monotonous and grating, less varied. – Frequently gives its most typical call, a long but softly harsh grating *chaihr*, repeated in long series, often with an added prolongation at the end, *chaih-ehr*; sometimes also uttered in series when alarmed. It also has a rather Western Subalpine Warbler-like clicking *tuck*, sometimes uttered in hurried series, *tek-ek-ek-ek-ek*. Another alarm call resembles that of Sardinian Warbler, but unlike latter initial note always differs from following notes, e.g. *chrrü-ter'r'r'r'r'r*.

SIMILAR SPECIES Greatest risk of confusion is with *Balearic Warbler* and, to a lesser extent, with *Marmora's Warbler* (see those species). – Much less likely to be confused with *Western Subalpine Warbler*, which has orange-red rather than dark rufous-red underparts (♂), longer wings and a shorter, less graduated tail (with distinct white edges and tips) that is rarely waved or cocked. Western Subalpine has also a white submoustachial stripe, rufous-brown fringes to the wing-feathers and an unmarked throat. – Confusion with similarly-sized and similarly-shaped *Tristram's Warbler* is also possible in North Africa (where Tristram's breeds). The tail is similarly long, although proportionately not quite as long and slim-looking as in Dartford. By plumage, they should be readily separated: Dartford lacks Tristram's rufous wing panel, distinct white eye-ring and richly-patterned tail (similar to Spectacled Warbler). Moreover, Tristram's Warbler is somewhat paler vinaceous red-brown below, usually with a clear whitish submoustachial stripe (lacking in Dartford), but throat can have some white tips when fresh (reminiscent of pattern in Dartford). Further, they have different calls.

AGEING & SEXING Ageing possible with close inspection of moult and feather wear in wing, tail pattern and iris coloration, but note that young can moult many secondary-coverts and tail-feathers, and can rapidly acquire ad-like iris colour. Sexes separable, with rather marked age- and sex-related variation, but seasonal variation limited to moderate. – Moults. Complete post-nuptial and partial post-juv moults in late summer (mostly Jul–Sep). Post-juv moult involves all head- and body-feathers, lesser and median coverts, most or all greater coverts, alula, tertials and some tail-feathers, with innermost secondaries often replaced too; limited (but often protracted) winter to early spring replacement involves some head- and body-feathers, tertials, and perhaps a few

S. u. dartfordiensis, ad ♂, England, Apr: in W Europe often found in heath dominated by gorse, which in full spring bloom forms a vivid background to this rather dark, small and elusive warbler, which can otherwise be difficult to spot if not singing from the top of the bush. Despite being a colourful ♂ there is an earth-brown wash above, typical of *dartfordiensis*. Overall fresh, including ad primary-coverts, inferring age. (R. Chittenden)

S. u. dartfordiensis, ♀, presumed 1stS, England, Jun: at least some ♀♀ are decidedly duller than ♂♂, with more diluted pinkish-red or orange-brown underparts, and more extensive pale belly. Very short wings and long slender tail should always help identify this species. Primary-coverts apparently worn and brown juv forming moult contrast with renewed greater coverts, and eye colour duller, all suggesting a 1stS. (B. Baston)

HANDBOOK OF WESTERN PALEARCTIC BIRDS

secondary-coverts and tail-feathers. Partial pre-nuptial moult occurs, but period somewhat variable. – **SPRING Ad ♂** With wear entire upperparts become more uniform dark grey (brown tones almost non-existent in ssp. *undata*); below more solid red-brown, except on throat, which is finely but distinctly speckled white. Wing-feathers modestly abraded and lacking moult contrast, and pale primary tips virtually lost. Iris a bright reddish-orange. **Ad ♀** Also wears slightly greyer above, but at least mantle and fringes of wing-coverts, tertials and remiges are browner. Pink-red underparts paler and admixed with buffish-white. Iris deep orange-brown. Best aged by having only ad (and relatively less worn) wing-feathers. **1stS** Majority have heavily worn and browner remiges and primary-coverts, with faded buffish-brown fringes; also, usually less pure grey (♂) or duller with more buff and brown tones (♀) above and on wing-coverts. Juv tail-feathers, if retained, extremely faded and worn. Obvious moult limits between post-juv moulted tertials and worn primaries. Iris near ad-like in most, very rarely retains olive cast, but colour generally unreliable for ageing. – **AUTUMN Ad** Both sexes aged by evenly very fresh wing- and tail-feathers

S. u. dartfordiensis, ♀♀, England, May: ♀ is generally duller than ♂ (see images on p. 495), but some young ♂♂ can be less colourful, especially in NW Europe where both sexes are typically browner above; still, these two should safely be considered ♀♀. Without close inspection of wing, ageing is unsure. (M. Bennett)

S. u. dartfordiensis, ♂, England, Nov: in fresh autumn plumage overall browner above, including head, and less deeply saturated rufous below partly due to broad white tipping on throat and centre of chest. However, in this case still sufficiently richly coloured below to permit reliable sexing as ♂. Age is unsure without handling. (M. Fidler)

S. u. dartfordiensis, ♀, presumed ad, England, Oct: when very fresh underparts extensively flecked white, especially on chin and throat, sometimes even coalescing to produce white-throated impression. Probably ad, given apparently evenly fresh plumage, but also has bright eye colour and ad tail, the latter blackish with rounded tips that are broadly white. (S. Ashton)

S. u. dartfordiensis, ♀, presumed 1stW, France, Sep: overall dull plumage with relatively pale and dull orange-brown flanks and chest (white central throat even continues onto centre of breast and belly) help to sex the bird as ♀, and it is tentatively aged by having seemingly undergone extensive post-juv moult in wing and tail (older feather seem to be juv). (A. Audevard)

(with remiges and primary-coverts glossy greyish-black); also, orbital ring and iris perhaps slightly less bright than in spring, but still distinctive. Both also show minute, shiny whitish-grey primary tips, and tail has clear-cut, narrow whitish fringe to r6. **Ad ♂** Upperparts and head largely slate ash-grey, with pale olive-brown tone mainly on mantle and scapulars (in *dartfordiensis* entire upperparts tend to be dark earth-brown), and dark vinaceous red-brown chin and throat tipped extensively white (chest and flanks similar, but coloration less intense and only indistinctly tipped whitish). **Ad ♀** Usually distinguished from ♂ by more extensive brown tones above, with reduced bluish-grey on head and mantle, and has paler underparts (profusely washed whitish), whereas wings have broader buff or rusty-brown fringes. Note: ad ♀ often resembles 1stW ♂ (ageing should precede sexing). **1stW** Both sexes duller and more

S. u. undata, ♂, presumed ad, Spain, Mar: a quite distinctive *Sylvia* when it raises crown-feathers, twitches its long tail and shows its attractive ♂ summer plumage. With wear, most ♂♂ in S France and much of Iberia become immaculate uniform bluish-grey above and solid red-brown below except throat, which is finely speckled white. Still rather fresh, with bluish-grey fringes to primary-coverts and remiges, suggesting an ad. (S. Fletcher)

— 496 —

SYLVIA WARBLERS

S. u. undata, ad ♂ (left) and ♀, Spain, Sep: two birds at the end of the post-nuptial moult, at least one outer tail-feather still growing. Note sexual differences, with the ♀ usually visibly duller, with less pure grey upperparts, and being pale below with paler and almost unmarked throat, lacking the white dotty stripes or fine flecks on dark reddish ground of ♂. (H. Shirihai)

S. u. undata, 1stW ♂, Spain, Oct: especially without handling, ageing and sexing can be tricky. Assuming this bird is 1stW (seemingly juv remiges and primary-coverts, dull eye colour), the rather neat plumage with purer greyish-tinged head, neck and exposed greater covert centres, and deeper red-brown chin, throat-sides and flanks indicate a ♂, thus almost certainly 1stW ♂. (A. Tóres)

extensively buff and rusty-brown above, while underparts are paler with more extensive whitish-buff elements (in some young ♀♀ largely concealing pinkish-orange coloration). Best aged by juv primary-coverts and remiges, which are softer textured and slightly browner, fringed buffish-brown (primaries virtually lack pale tips of ad); moult limits between moulted dark greyish alula and juv primary-coverts often evident. Juv outermost tail-feathers browner, very narrowly fringed and indistinctly tipped sandy-buff (white in ad), but these can be replaced by ad-like feathers at any time. Iris darkish olive-brown to yellowish-brown in nestlings (clearly darker and duller than ad), subsequently orange cast is more evident and therefore approaches ad (thus best to rely on moult pattern for ageing); orbital ring duller orange-red. **Juv** Lacks orange-brown coloration below, being mostly whitish, extensively washed buff, and is warmer greyish-brown above. Entire plumage very fresh with typical fluffy appearance.

BIOMETRICS (*undata*) **L** 12.5–14 cm; **W** ♂ 53–58 mm (n 16, m 55.1), ♀ 51–54.5 mm (n 12, m 52.7); **T** ♂ 58–69 mm (n 16, m 64.7), ♀ 60–66.5 mm (n 12, m 62.6); **T/W** m 118.0; **TG** 10–16 mm (n 28, m 12.8); **B** 11.0–12.7 mm (n 28, m 11.9); **BD** 2.7–3.4 mm (n 20, m 3.1); **Ts** 17.8–20.2 mm (n 28, m 19.3). **Wing formula: p1** > pc 4–7 mm, < p2 18–22 mm; **p2** < wt 4–8 mm, =8, 8/10 or =10; **p3** < wt 0.5–2 mm; **pp4–6** longest; **p7** < wt 1.5–3 mm; **p8** < wt 2.5–6 mm; **p10** < wt 6–9 mm; **s1** < wt 6–9.5 mm. Emarg. pp3–6 (rarely fainter on p6).

GEOGRAPHICAL VARIATION & RANGE Three races treated here, but racial differentiation slight and sometimes inconsistent, and races intergrade with intermediate (clinal) populations, hence many birds cannot be safely identified outside breeding season. Throughout its range mainly resident, generally making only local or shorter altitudinal movements.

S. u. dartfordiensis Latham, 1787 (S England, NW France). Smaller and on average a little darker brick-red below (less pink-red) than *undata*, and upperparts a trifle darker and more brown-tinged dull grey. White spots on throat less obvious than in *undata*. **W** ♂ 51–56 mm (n 21, m 53.1), ♀ 50–54 mm (n 12, m 51.8); **T** ♂ 56–68 mm (n 21, m 62.4), ♀ 58–64 mm (n 11, m 60.7); **T/W** m 117.4; **TG** 8.5–15 mm (n 28, m 11.6); **B** 11.0–13.7 mm (n 31, m 12.6); **BD** 2.7–3.4 mm (n 21, m 3.0); **Ts** 18.0–19.9 mm (n 26, m 18.9). – Birds in Brittany ('*aremorica*') appear very similar or the same when series compared, and are best included here. (Syn. *aremorica*.)

S. u. undata (Boddaert, 1783) (S France, N & E Spain, Menorca, Italy). Treated above. Has rather dark greyish upperparts with moderate brown hue, being subtly greyer above than otherwise similar (but marginally smaller) *dartfordiensis*. Grades into smaller and darker *toni* in W Iberia.

S. u. undata, 1stW, Italy, Nov: fresh ♀♀, especially young in autumn, often show extensive rufous-tinged upperparts and renewed greater coverts, and have dull orange-brown flanks and chest. However, this bird could be a very dull young ♂, given the greyish head and rather broad rufous bases to throat-feathers. Aged by distinctly worn and brown juv primaries and primary-coverts, but eye coloration is again ad-like! (D. Occhiato)

– Impossible to see any difference between breeders in S France or E Iberia with those of Corsica, hence proposed '*corsa*' not acknowledged. (Syn. *corsa*; *naevalbens*.)

S. u. toni Hartert, 1909 (Portugal except in north, SW Spain, coastal NW Africa). Subtly smaller than *undata* with deeper and warmer brick-red underparts in ♂♂, and white spots on throat on average smaller. Upperparts very subtly

S. u. undata, juv, Italy, Sep: warmer greyish-brown above with extensively pale pinkish-washed underparts, otherwise entire plumage typically fluffy in appearance. Yellow gape-flanges still present. Iris dark olive grey-brown typical of juv. (D. Occhiato)

S. u. toni, ♂, presumed ad, Portugal, May: a subtle race, differing only by on average darker and duller grey upperparts and darker red underparts, but differences from and border with *undata* obscured. Wing seems of evenly feathered ad type to suggest age. (D. Monticelli)

darker than in other races. **L** 12–13 cm; **W** ♂ 51–55 mm (n 15, m 53.2), ♀ 50–55 mm (n 14, m 52.1); **T** ♂ 56–64 mm (n 15, m 61.0), ♀ 56–64.5 mm (n 12, m 60.2); **T/W** m 114.7; **TG** 9–15 mm (n 22, m 11.4); **B** 11.0–13.4 mm (n 25, m 12.1); **BD** 2.7–3.1 mm (n 14, m 2.9); **Ts** 17.8–19.8 mm (n 22, m 18.9). (Syn. *maroccana*; *tingitana*.)

TRISTRAM'S WARBLER
Sylvia deserticola Tristram, 1859

Fr. – Fauvette de l'Atlas; Ger. – Atlasgrasmücke
Sp. – Curruca del Atlas; Swe. – Atlassångare

To see this endemic to NW Africa, you need to visit well-vegetated mountains from N Morocco to Tunisia, preferably in April to June, and search out bushy areas within park-like woodland between 1400 and 2500 m altitude. Tristram's Warbler is a short-range migrant that in winter either chiefly moves to lower elevations or migrates short distances, although some move further south, reaching W Libya, S Morocco and even Mauritania.

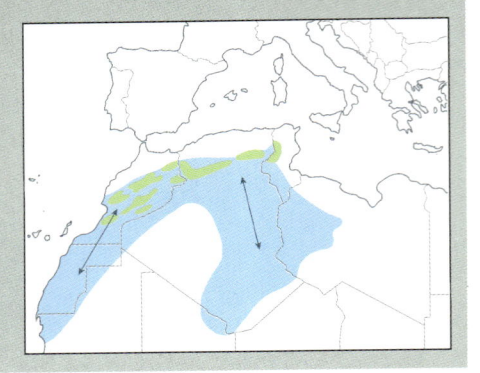

Ad ♂, Morocco, Feb: short-winged and rather long-tailed, with mainly grey upperparts, purer on head, contrasting with bright rufous on wing and underparts, as well as tiny whitish submoustachial and broken but prominent eye-ring. Still rather fresh, evenly-feathered wing and deep red-brown iris confirm that this bird is ad. (R. Burri)

IDENTIFICATION In many ways, appears intermediate between Spectacled, Western Subalpine and Dartford Warblers, being a *relatively small, quite long-tailed Sylvia*. Adult ♂ *has pure grey head*, nape and mantle, contrasting with a *bright rufous wing-panel, extensive deep terracotta-red underparts* (feathers of throat and upper breast finely and diffusely tipped whitish when fresh) with whitish belly and whitish fringes to undertail-coverts, and very *short whitish and poorly-defined submoustachial stripe*. ♀ similar but is less pure grey above and more orange-brown below, while first-winter ♀ is clearly paler below, mostly whitish with just a hint of orange-buff. *White eye-ring broad and Spectacled Warbler-like*, and iris and orbital ring are mostly reddish-orange in adults, but duller and brownish-olive in juvenile/first-winter. Legs yellowish-orange. *Long, slim-looking tail often recalls Dartford Warbler*, but in particular has *clear affinities with Spectacled in jizz and behaviour*. Generally rather skulking, and movements restless and nervous, although is not particularly shy. Occasionally forages on ground, but generally keeps to cover. Frequently raises, waves or flicks tail.

VOCALISATIONS Song resembles several related small *Sylvia*, notably Balearic, Dartford and Western Subalpine Warblers (but even a distant Sardinian Warbler comes into the picture), a hurried, metallic, chattering warble that is somewhat prolonged (longer in flight than when perched) and interspersed with fluting whistles and clicking call-like notes. With practice it may be possible to recognise a lack of loudness, an abundance of interspersed high-pitched whistling notes and the 'winding' structure of grating, quickly repeated notes familiar from Dartford Warbler song. Compared to Western Subalpine Warbler (which theoretically could breed in vicinity), the song is faster, weaker and higher-pitched, consisting mainly of 'whining' whistle-sounds and the disyllabic clattering call *terret*. – In contact gives either a sharp *tsack* like Western Subalpine, or a near-disyllabic *tsrek* or clearly disyllabic *terret*. When alarmed repeats the *tsrek* call rather hurriedly in a series, when at times also gives a high-pitched descending *weeee*.

SIMILAR SPECIES Given the geographical range and several distinctive plumage features (see above), in most instances Tristram's Warbler should be possible to identify reliably. However, sometimes with poorly marked young ♂♂, but chiefly with ♀♀, Tristram's versus *Spectacled Warbler* can represent a true challenge for field observers. When trying to separate such dull and confusing plumages, it is important to bear in mind that many ♂ Spectacled Warblers lack their species' diagnostic dark facial mask, while many Tristram's (including young ♂♂) have strongly reduced rufous on throat and belly. Thus, at times the two could show no more than vague tendencies in their features to assist separation in the field: Tristram's usually appears to have a slightly longer and slimmer tail, partly due to the latter being more constantly and sharply cocked. Tristram's tends to have a greyish head that is better demarcated from sandier mantle, and more evenly buff pinkish-orange throat and breast, at least down to breast-sides, as opposed to Spectacled's purer white-throated impression (in those ♂♂ that lack the dark central throat patch). Furthermore, the two species share the characteristic rufous wing panel with sharp dark tertial centres, but wing panel of Tristram's is often less extensive and duller (with slightly more exposed dark centres and narrower rufous fringes). Also, unlike many Spectacled, Tristram's tends to have the lesser coverts at least partly greyish (rather than pale sandy-brown), contrasting with the sandier median and greater coverts (almost concolorous in Spectacled). ♀ Tristram's seems to have predominantly greyish lores, rather than whitish as in more pale- and uniform-headed ♀ Spectacled; and orbital ring is dull orange-brown to brownish-orange (greyish-brown or dull yellowish-brown in Spectacled, thus on average duller and less prominent). As additional supportive structural differences, note the following: Tristram's tends to have slightly longer primary projection (5–6 exposed primary tips; in Spectacled usually 4–5), and also on average more evenly-spaced tertials (Spectacled tends to have shorter distance between longest and central feathers). Nevertheless, some problematic birds are likely to be intermediates in all of the above-mentioned characters, and their calls will be the only decisive clue: Tristram's commonest call is a short, muted *tsrek* or a disyllabic *terret* (often softer with sparrow-like quality), while Spectacled has a fast, soft, drawn-out rattlesnake-like *zrrrrrrrr*. – Compared to *Western Subalpine Warbler*, Tristram's lacks an obvious white submoustachial stripe, has a rufous wing panel (lacking in Western Subalpine), shorter primary projection, and blackish

♀, presumed ad, Morocco, Feb: superficially similar to Spectacled Warbler, including the broken, broad white eye-ring and rufous wing panel with dark tertial centres, but unlike Spectacled, many ♀ (chiefly ad) Tristram's show substantial red-brown on throat, chest and flanks. Not heavily worn and evenly-feathered wing suggests ad, though iris colour seems less advanced. (D. Occhiato)

and longer and slimmer tail, which is also much more readily cocked. – From *Dartford Warbler*, mainly separated by quite different wing pattern (particularly the bright rufous panel of Tristram's) and whitish eye-ring. With practice, all are easily differentiated by behaviour and calls.

AGEING & SEXING Ageing possible after close check of moult and feather wear in wing, tail pattern and iris coloration, but 1stYs can moult many secondary-coverts and tail-feathers, and can acquire ad-like iris colour. Sexes differ, with rather marked age- and sex-related (or individual) variation, but also seasonal variation, chiefly due to wear. – Moults. Complete post-nuptial and partial post-juv moults after breeding, in late summer (mostly Jul–Sep). Post-juv moult involves all head- and body-feathers, lesser and median coverts, all or at least most greater coverts, alula and tertials, along with some tail-feathers (more rarely some innermost secondaries or entire tail renewed). Partial pre-nuptial moult (Jan–Apr) includes some greater coverts, tertials, rectrices and head and body-feathers, more rarely some innermost secondaries. – **SPRING Ad** Compared to 1stS, both sexes usually have clearly less worn primaries and primary-coverts, and in general plumage is more 'tidy'. Pre-nuptial moult limits usually do not complicate main ageing features, but some birds can have two generations of tertials. Iris deep brownish-orange. **Ad ♂** With wear, crown to mantle become

1stS ♂, Tunisia, Apr: many young ♂♂ are somewhat intermediate between ad ♂ and ♀ plumages, especially having more pinkish-buff tinge below, with duller and more restricted rufous wing panel. Contrastingly worn and brown juv primaries and primary-coverts, moult limit among greater coverts, and duller eye and orbital ring, confirm age. The still rather colourful plumage indicates a ♂. (D. Occhiato)

Ad ♂, Morocco, Dec: pure grey head, deep ochre-rufous iris, bright red orbital ring, rufous underparts including throat irregularly white-tipped and mottled, and (evenly-feathered) rufous wing fringes, plus broad but slightly broken white eye-ring readily identify the species, and provide the age/sex. (R. Messemaker)

almost uniform pale bluish slate-grey and underparts more evenly pinkish rufous-brown, with little or no whitish flecking. Iris bright reddish-orange. **Ad ♀** With wear buff-brown (less grey) above, but crown clearly greyer, and underparts dark vinaceous or buffish rufous-brown (paler and more restricted rufous elements than ♂). **1stS** Juv primaries and primary-coverts more worn, with faded buffish-brown fringes, and moult limits easy to detect in ♂. Iris colour perhaps less advanced, slightly duller than in ad on average. Generally speaking, the plumage state of both sexes, respectively, is less 'tidy', duller and less pure grey above and more buffish below, with paler rufous wing panel. – **AUTUMN Ad** Evenly very fresh wing- and tail-feathers (with remiges and primary-coverts firmly textured and glossy greyish-black, and exposed primary tips thinly silvery white). Iris bright reddish-orange to brownish-orange. **Ad ♂** Grey of head and mantle tinged buffish-brown. Vinaceous rufous-brown underparts profusely washed whitish-cream through whitish tips of fresh feathers, especially on central areas. Whitish chin and submoustachial stripe quite pronounced but diffuse. **Ad ♀** Head and upperparts browner, and broad pale feather-tips conceal rufous-buff ground colour, producing impression

Tristram's Warbler (left: Morocco, Jan) and Spectacled Warbler (right: Israel, Jan) in non-ad ♂ plumages: many ♂ Spectacled Warblers lack the diagnostic dark mask (but prominently present on this bird), while many young ♂ Tristram's have much reduced rufous below. Thus, note Tristram's tendency to have greyish head that is better demarcated from browner mantle, further some pinkish-orange present on throat down to breast-sides, whereas Spectacled gives white-throated impression (still, note that many ♂♂ develop pinkish-grey lower throat patch with wear). Both species have rufous wing panel and dark tertial centres, but panel of Tristram's is often smaller, duller and less solid. Tristram's appears slightly longer-tailed, partly because the tail is more constantly and sharply cocked, and can also show slightly longer primary projection, with more evenly-spaced tertial tips. (Left: M. Schäf; right: A. Ben Dov)

of paler throat and extensive pale belly patch. **1stW** More extensively buff-brown above (♂ has much less pure grey head; ♀ is drab grey-brown throughout), and underparts are predominantly whitish-buff. ♂ approaches ad ♀, while ♀ recalls corresponding plumage of Spectacled Warbler. Juv primary-coverts and remiges slightly less fresh, browner and fringed buffish-brown (primaries lack obvious pale tips of ad). Iris dark greyish-olive in nestlings, in 2ndCY acquiring ad coloration, but even more advanced individuals in late autumn can have warmer olive-brown hue. Also, juv tail-feathers (if retained) slightly abraded, having less clear and less extensive white fringes (more extensively sullied pale brown). Moult contrast more obvious in ♂: moulted ad-like tertials clearly darker and blackish-centred compared to juv ones. Unmoulted alula, at least usually largest one, clearly paler brown and, when present, provides obvious contrast with ad-like blackish smaller ones. **Juv** Largely resembles ♀ but is buffish off-white below, profusely washed greyish-buff on flanks and breast; also warmer greyish-brown above.

BIOMETRICS Sexes very nearly the same size. **L** 11.5–12.5 cm; **W** 52–58 mm (n 33, m 54.7); **T** 51–59 mm (n 33, m 55.2); **T/W** m 101.0; **TG** 4–11 mm (n 27, m 7.0); **B** 11.0–12.9 mm (n 32, m 11.8); **BD** 2.6–3.5 mm (n 31, m 3.0); **Ts** 17.4–19.7 mm (n 33, m 18.4). **Wing formula: p1** > pc 1–5.5 mm, < p2 22.5–27.5 mm; **p2** < wt 2–5 mm, =6, 6/8 or =8; **pp3–5** about equal and longest; **p6** < wt 0.5–2.5 mm; **p7** < wt 2–4.5 mm; **p8** < wt 4–6 mm; **p10** < wt 6.5–9.5 mm; **s1** < wt 7.5–10.5 mm. Emarg. pp3–6 (often fainter on p6).

GEOGRAPHICAL VARIATION & RANGE Monotypic. – NW Africa, in mountains between Morocco and Tunisia; short-range migrant, moving to lower elevations or only shorter distances, although some move further south, reaching W Libya, S Morocco, Western Sahara and Mauritania. – Two subspecies often recognised, but we have found '*maroccana*' Hartert, 1917 (Morocco, NW Algeria) only subtly and inconsistently differentiated from *deserticola* (rest of Algeria & Tunisia); the small series of the former seems on average indistinctly darker and more richly coloured, with less extensive rufous in wings and less white in outer tail-feathers, but with much overlap. Appears not to pass a 75% rule. – A third subspecies was proposed by Meinertzhagen ('*ticehursti*') based on just one specimen. As detailed in Shirihai *et al.* (2001), this represents an aberrant bird rather than a race with an unknown breeding range. (Syn. *maroccana*; *ticehursti*.)

1stW ♂, Morocco, Dec: Tristram's tends to have a greyish head better demarcated from browner mantle, and often shows a more evenly pinkish-orange throat and underparts. Note slightly longer primary projection than Spectacled (here, six exposed primary tips). Ageing by moult pattern difficult, but degree of feather wear to primaries and primary-coverts, dull iris, and overall plumage colours suggest 1stY ♂. (G. Di Liddo)

1stW ♂, Morocco, Dec: a least-advanced young ♂, in almost ♀-like plumage, but has exposed dark grey crown-feathers and rufous-brown bases to some throat-feathers. Apparent moult limit in greater coverts (though difficult to be sure of), apparent retained juv primary-coverts, and dull eye further confirm age as 1stW. (T. Kolaas)

Tristram's (left: Tunisia, Dec) versus Spectacled Warbler (right: Canary Is, Sep) in fresh ♀-like plumages: a very challenging identification problem is posed by these two. Tristram's tends to have greyish head better demarcated from warm pale brown upperparts, brown reaching down to uppertail-coverts. Note pale pinkish-orange chin and throat in Tristram's, versus purer white-throated appearance of Spectacled. Tristram's also tends to appear slightly longer-tailed. A supportive but slight difference lies in the on average slightly redder orbital ring of Tristram's rather than more greyish-brown in Spectacled. Moult difficult to analyse in these birds, but degree of feather wear in primaries and primary-coverts, iris colour and overall plumage suggest both are 1stY ♀♀. (Left: P. Alberti; right: A. Guerra)

SPECTACLED WARBLER
Sylvia conspicillata Temminck, 1820

Fr. – Fauvette à lunettes; Ger. – Brillengrasmücke
Sp. – Curruca tomillera; Swe. – Glasögonsångare

Diminutive *Sylvia* mainly restricted to the W Mediterranean and NE Atlantic islands, but also occurs on Cyprus and in the Levant, and has recently been discovered breeding in SE Turkey. An inhabitant of garrigue and other low scrub, usually on mountain sides and in rather arid regions, but on islands also at lower altitudes. Spectacled Warbler is resident in many parts, but S European populations move to N Africa in winter, and vagrants have reached NW Europe.

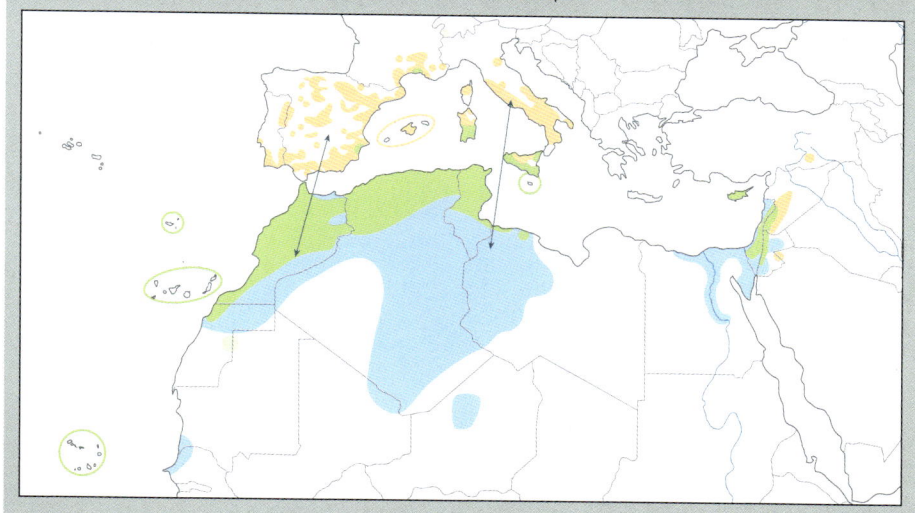

IDENTIFICATION The plumage largely recalls Common Whitethroat, but Spectacled Warbler is *a distinctly smaller and slimmer Sylvia*, rather like Western Subalpine Warbler in size and structure (though latter has marginally shorter tail). *Short, rounded wings* and *comparatively long tail with striking white edges*. Head rather rounded, with high crown and spiky bill. ♂ distinctive when seen well: *dark greyish head and even darker* (blackish) *loral mask*, prominent eye-ring forming *white 'spectacles'*, warm brownish mantle/scapulars, *extensive and solid rufous wing panel* (with dark centres to greater coverts largely invisible), partly visible very neat, *clean black tertial centres*, largely dark tail, *white throat* with rather *dark grey diffuse lower centre*, and saturated pinkish-buff or *pink-rufous underparts*. ♀ more uniform, being grey-brown above, with pale lores, duller underparts, only slight mask, but still has characteristic *rufous wing panel* and *dark tertial centres*. Iris dark sepia-brown (adult) to dark olive-grey (first-winter). Narrow orbital ring mostly black (adults and first-winter ♂) or dark yellowish-brown (first-winter ♀), and legs pinkish- to reddish-yellow, or pale ochre. Like other small *Sylvia*, typically skulks, and movements sprightly and restless, or hops on ground with raised tail, which may be waved, and forages low in scrub, only rarely sallying aerially. Brief undulating song-flight, but usually sings from cover or briefly atop bush.

VOCALISATIONS A short musical, chattering warble nearly always opened by one or a few drawn-out desolate-sounding clear whistling notes on different pitch (at times recalling alternative song of Crested Lark!), which usually makes the song quite characteristic among several otherwise very similar-sounding *Sylvia* species. The high-pitched initial note is then followed by 3–8 rapid, short notes on similar lower pitch, forming a rather discreet warble. – Call a high, buzzing, drawn-out rattling *zrrrrrrr* with fizzling quality, sometimes slowing slightly towards end, *zrrrrr-r-r-r*. Short related *zre* also heard, often run together in anxious series, *zre zre zre zre...*

SIMILAR SPECIES Possible confusion with either of the Desert Warblers or Tristram's Warbler is discussed under those species. – Non-♂ plumages of Spectacled Warbler and Western Subalpine Warbler might be confused, especially dullest Spectacled with Western Subalpine with richest rufous wings (and many of the same characters listed below are also valid for separation of Moltoni's Warbler and Eastern Subalpine Warbler). Spectacled has (i) solid rufous wings, broadest on secondaries and tertials, but also extending onto all greater and median coverts, and often scapulars (Western Subalpine can have markedly sandy-buff or even rusty-brown edges to most wing-feathers, but fringes are narrower and only form small wing panel, which unlike Spectacled is confined to inner greater coverts, secondaries and longest tertial); (ii) darker primary tips and alula, and well-marked and acutely pointed black centre to central tertial (Western Subalpine: centre less clear-cut and more rounded at tip); (iii) shorter, more rounded wings with only 4–5 closely-spaced primary tips visible (Western Subalpine: 6–7 widely-spaced primary tips); (iv) proportionately longer tail and more rounded crown, producing a more slender overall impression than Western Subalpine, which although small, appears more compact and 'solidly' built, with a relatively short tail and flatter crown. Also useful are: (v) the very dark uppertail of Spectacled, contrasting with the paler upperparts (Western Subalpine: uppertail browner grey and nearly concolorous with back); and (vi) prominent white eye-ring being subtly broader above eye (Western Subalpine: narrower ring of uniform width). Furthermore, Spectacled prefers low, sparse bushes and frequently hops on ground, raising and cocking its tail (Western Subalpine prefers taller, denser vegetation and trees, and moves tail less freely), although habitat preferences vary, especially on migration. They differ also in bare-parts coloration, but this is often difficult to evaluate confidently in the field. Finally, Spectacled typically gives a long drawn-out rattling *zrrrrrr* and Western Subalpine a hard *tak* (rather like Lesser Whitethroat, thus a striking difference). However, Moltoni's Warbler has a hard, rattling *trrrr*, which

♂, presumed ad, Cyprus, May: a small, slim and short-winged *Sylvia*, with rather distinctive ♂ plumage: grey head with darker loral area; especially with wear, intense pinkish-cinnamon below; contrasting white throat with dark grey lower throat patch in most; large and solid rufous wing panel. Wing still not heavily worn, and together with overall rich plumage colours and deep red-brown iris, suggest the bird is an ad. (D. Occhiato)

♂, presumed ad, Cyprus, Apr: rather typical pose, and note grey tinge between white throat and pink-cinnamon breast. Lores dark grey. With wear, the white 'spectacles' can become narrower and not so prominent (as here). However, the extensive rufous wing panel with the comparatively small contrasting black tertial centres are still prominent. Rather fresh evenly-feathered wing and deep red-brown iris suggest this ♂ is ad. (D. Jirovsky)

1stW ♂, Spain, Nov: young ♂ is like ad ♂ but upperparts often suffused warmer brown. Throat usually pure white (grey on lower throat still largely concealed), contrasting with buffish-pink rest of underparts. The apparently rather worn and already bleached brownish inner primaries contrasting with darker and fresher renewed outer ones, plus moult limits in tail, show the bird to be 1stW. Note that the iris is already bright orange-brown like in ad. (B. Rodríguez)

1stW ♂, Santiago, Cape Verde Is, Sep: post-breeding and post-juv moults generally occur earlier in the Cape Verdes, but both period and extent of moult vary with rains. This mainly fresh ♂ has apparently retained inner primaries and some secondaries; such clear moult limits between new outer primaries and very heavily-worn inner (juv) primaries in Sep, supported also by dull olive iris, conclusively age the bird as 1stW. (G. Conca)

can recall Spectacled, but Moltoni's call is usually shorter and more Wren-like. Eastern Subalpine has either a slightly 'compound' *t'ret* or a doubled *te-tet*. – Despite the strikingly larger size of Common Whitethroat, ♀-like plumages are superficially similar to Spectacled Warbler, although note that Spectacled is usually more delicate with a finer, narrow-based bill, being not only smaller but importantly has quicker movements and moves the tail more freely (Common Whitethroat usually heavy and slow, and moves tail very little). Spectacled has drawn-out rattling calls lacking in Common Whitethroat, which utters a nasal, buzzing *wheed-wheed-wheed* and prolonged displeased, hoarse *chaihr*. Spectacled further differs in its shorter primary projection, this being c. 1/3 length of the exposed tertials, with only 4–5 tips visible (Common Whitethroat: just over 1/2 length of tertials with 6–7 visible tips). If not too heavily worn and bleached, Spectacled also has more extensive and solid rufous wing panel, with the greater coverts, especially, lacking visible exposed dark centres (Common Whitethroat: dark feather centres of nearly all secondary-coverts and tertials visible, reducing visual impact of rufous panel); in Spectacled the solid rufous area usually extends further over much of the secondaries and is broader on the tertials (leaving on average narrower and more sharply defined dark centres on the latter). Generally, darker tail contrasts more obviously with paler upperparts and, seen from below, has better-demarcated pale pattern (Common Whitethroat has rather weakly-patterned tail with sandy edges to outer feathers and from above appears virtually concolorous with rest of upperparts). There are some subtle bare-part differences, e.g. legs in Spectacled are brighter (slightly darker or duller in Common Whitethroat), and bill on average has better defined and darker tip and culmen (though dark tip sometimes restricted).

AGEING & SEXING Ageing requires close check of moult and feather wear in wing, tail pattern and iris coloration, but 1stY can moult extensively and can acquire ad-like iris colour (also, some ad are less advanced in this respect). Sexes differ, with rather marked age- and sex-related or individual variation (also some seasonal variation, chiefly due to wear). – Moults. Complete post-nuptial and partial post-juv moult in late summer (mostly Jul–Sep in Mediterranean region, but in Canaries & Madeira May–Jul, and much earlier in Cape Verdes, Jan–Jul). Post-juv moult involves all head- and body-feathers, lesser and median coverts, usually all greater coverts, and some tertials and tail-feathers, less

1stS, presumed ♀, Cyprus, May: some young ♂♂ in spring are somewhat intermediate between ad ♂ and ♀ plumages, while some (usually ad) ♀♀ can approach ♂ plumage, but head is never as pure grey and they usually lack any obvious dark loral area (as here). Rather abraded juv primaries and brown and pointed primary-coverts infer age. (D. Occhiato)

1stY, Fuerteventura, Canary Is, Feb: primaries and tail appear to be juv and already rather worn, and iris is rather dull olive-brown, suggesting age. Given 1stY, the amount of grey visible on head, and the apparent blackish feathers emerging in the loral area, suggest less advanced ♂, but the overall dull plumage also matches some ♀♀. (H. H. Larsen)

commonly also some innermost secondaries and even some outer primaries. Partial pre-nuptial winter moult (mostly Nov–Apr) includes some greater coverts, tertials, tail-feathers and head and body (more extensive moult in 1stS). – SPRING **Ad** In both sexes rufous wing fringes slightly abraded and less bright, with slightly more exposed dark centres than in fresh plumage. Remiges wear browner (though relatively less worn than 1stS), and pale primary tips wear off. On the whole, ad plumage more 'tidy' than 1stS. Iris can be brighter than in autumn, but distinction from 1stS unreliable. **Ad ♂** With wear, grey of crown becomes purer, and dark loral mask and white eye-ring more pronounced. Upperparts colder sandy grey-brown, often with enhanced rusty-buff scapulars. Underparts more intensely pink, with variable dark pinkish-grey or grey central throat patch (sometimes creating ill-defined broad whitish submoustachial stripe), grey at times 'bleeding' onto upper breast. **Ad ♀** Head slightly tinged greyish, with duller and colder rest of upperparts, and underparts usually whiter. **1stS** Usually impossible to age in the field (since more heavily worn and bleached juv

1stS ♀, Cyprus, Apr: plumage generally dull with predominantly grey-brown upperparts (head hardly greyer), paler less saturated underparts, smaller and duller rufous wing panel, heavily worn browner remiges and slightly olive-tinged iris reliably age and sex this bird. (G. Reszeter)

♀-like plumages of Spectacled Warbler (left: Israel, Mar) and Common Whitethroat (right: Finland, May): both have superficially similar rufous wing panel. Spectacled's smaller size and more delicate shape with finer bill, as well as more terrestrial habits and nervous movements, are not always obvious. When separation is uncertain, important to check the following closely: Spectacled has shorter primary projection which often creates longer-tailed impression, further has larger and more solid rufous wing panel (especially on the greater coverts, which lack visible dark centres on folded wing). (Left: S. Arlow; right: S. Leveelahti)

primary tips and primary-coverts hard to judge). Rufous wing panel often less solid and duller. Underparts less deeply and extensively pinkish-buff, and dark grey lower throat patch of ♂ on average smaller; in ♂ grey of head sometimes less pure, and much reduced in ♀. Juv tail-feathers (if retained) distinctly more worn. Iris colour similar to ad. – AUTUMN **Ad** Both sexes aged by evenly very fresh wing- and tail-feathers (remiges and primary-coverts firmly textured and glossy greyish-black); exposed primary tips thinly silvery white. Iris dark to medium olivaceous-brown. Outer tail-feathers usually have larger, purer white fringes and tips compared to juv feathers. **Ad ♂** As spring but grey of head less pure and upperparts suffused warmer rusty-buff. Wings have broader rufous fringes (mostly concealing blackish centres). Underparts pale buffish-pink, especially sides. Iris dark sepia-brown, less-advanced birds tinged more olivaceous. **Ad ♀** Head and upperparts browner (head with hardly any grey tones). **1stW** Resembles ad except that juv primary-coverts and remiges are slightly less fresh (softer textured and mainly dark brownish-grey, less glossy). Pale primary tips more abraded, but rarely possible to assess in the field. Juv tail-feathers slightly abraded, with less pure and less extensive white pattern (more extensively sullied buff and brownish), but some replace several or all tail-feathers in post-juv moult. Compared to ad, in ♂ grey of head tinged slightly buff-brown, and loral mask less intense. Underparts less saturated pinkish-buff and more restricted, almost confined to breast. ♀ has upperparts generally paler and more

♀, 1stW, Spain, Sep: young ♀ has dullest autumn plumage, but note strong contrast between paler and more uniform upperparts and still rather obvious rufous wing panel. White eye-ring less contrasting due to whitish lores, and whiter throat is less obvious due to paler underparts. The rufous wing panel mirrors the pattern of Common Whitethroat, but the panel in Spectacled is more solid, with more sharply-demarcated dark centres especially to innermost tertials, and primary projection is considerably shorter. (A. Guerra)

uniform, and underparts whiter. Both sexes have less solid and duller rufous wing panel. Moult contrast more obvious in ♂: moulted ad-like greater coverts and remiges (usually tertials) appear blackish-centred, compared to browner juv feathers. Some, with more extensive post-juv moult, exhibit moult limits among primaries and secondaries; unmoulted alula-feathers more sandy-brown compared to ad-like blacker ones (at least the smaller feathers often moulted). Iris mostly dark olive-brown, but only typical ad separable using iris colour. **Juv** Largely resembles ♀, but buffier below and slightly warmer above, and entire plumage very fresh. Tail and iris as 1stW.

BIOMETRICS L 11.5–12.5 cm; **W** ♂ 52–60 mm (n 30, m 57.1), ♀ 52.5–57.5 mm (n 23, m 55.3); **T** ♂ 45–55 mm (n 28, m 51.4), ♀ 47–53 mm (n 23, m 50.0); **T/W** m 90.2; **B** 11.0–13.2 mm (n 38, m 12.2); **BD** 2.6–3.3 mm (n 20, m 3.0); **Ts** 17.0–19.5 mm (n 20, m 17.9). **Wing formula: p1** > pc 1.5–4 mm, < p2 25–30.5 mm; **p2** < wt 2–4 mm, =5/7 or =7 (very rarely =5); **pp3–5** about equal and longest (p5 rarely to 2 mm <); **p6** < wt 1–3 mm; **p7** < wt 3.5–5 mm; **p8** < wt 5–7 mm; **p10** < wt 7–10.5 mm; **s1** < wt 7.5–12 mm. Emarg. pp3–6 (on p6 sometimes faint).

1stW ♀, Cyprus, Dec: with experience, even the dullest 1stW ♀ should be reliably identified by its dainty jizz, together with the solid and large rufous-orange wing panel, well-defined dark tertial centres, and well-developed white throat and eye-ring. Primary-coverts and primaries appear to be juv and already worn, and the eyes are olive. (N. Fenech)

Spectacled (left: Israel, Oct) versus Western Subalpine Warbler (right: Spain, Sep) in ♀-like autumn plumages: these two species, even in their least-marked plumages, should be readily separated on jizz and calls. However, silent birds and those with less straightforward plumage, including migrants, can require careful checking. Most important is Spectacled's solid rufous wing panel (Subalpine can have markedly rusty-brown wing edges, but fringes narrower and never form a clear, solid and large panel). Note Spectacled's shorter primary projection (only 4–5 closely-spaced primary tips; Subalpine has 6–7 widely-spaced primary tips). Also useful is the darker uppertail of Spectacled, and broader white eye-ring above eye (uniform width in Subalpine). Ageing using moult pattern is difficult for these birds, but the degree of feather wear to primaries and primary-coverts, iris colour and overall plumage pattern suggest both are 1stY ♀♀. (Left: L. Kislev; right: J. L. Muñoz)

GEOGRAPHICAL VARIATION & RANGE Monotypic. – S Europe from Iberia to Italy, Canary Is, Madeira, Cape Verdes, NW Africa from Morocco to Tunisia, Levant, locally SE Turkey; nearly all European populations winter in N Africa from Western Sahara to W Libya, with some reaching Sahel in a belt from Mauritania to Chad; populations in N Africa, S Sicily, Cyprus and much of Levant are resident. – Claimed racial differences bridged by individual variation. Most-saturated ♂♂ with prominent blackish loral area, large grey centre to lower throat and extensive and deep pink-rufous breast and flanks equally found in Canary Is, Madeira and Sardinia (sic!), with very similar birds in Morocco. On average, ♂♂ in Algeria, Tunisia, Spain, S France and mainland Italy are subtly paler pink-rufous below, but difference from darker birds mentioned above still very slight and there is much overlap. Amount of white on chin in ♂ averages slightly less extensive in breeders on Madeira and Canary Is, and most extensive in Cape Verdes and NW Africa, again with much variation and overlap. Size is nearly the same in all populations, and with noted variation and similarity in plumage colours the species is best regarded as monotypic. (Syn. *bella*; *extratipica*; *orbitalis*.)

REFERENCES SHIRIHAI, H., HARRIS, A. & COTTRIDGE, D. (1991) *BB*, 84: 423–430. – SMALL, B. (1996) *BB*, 89: 275–280.

Juv, Cyprus, Jun: like a plain-coloured ♀ but with soft, fluffy body-feathers and with blander head pattern. Iris rather dark still. (A. Al-Sirhan)

WESTERN SUBALPINE WARBLER
Sylvia inornata Tschusi, 1906

Fr. – Fauvette passerinette; Ger. – Maghrebgrasmücke
Sp. – Curruca carrasqueña occidental
Swe. – Rostsångare

A migratory *Sylvia* of the W Mediterranean region, when breeding the attractive Western Subalpine Warbler occupies a range of habitats from laurel garrigue and maquis to open oak woodland, and favours *Acacia* savannas in winter. Regularly overshoots its breeding range to N Europe. What was until recently regarded as one species, the Subalpine Warbler, breeding from Iberia and Morocco east to C Turkey, has been the subject of intense study which has now led to the separation of three different species within this superspecies, the other two being Eastern Subalpine Warbler and Moltoni's Warbler.

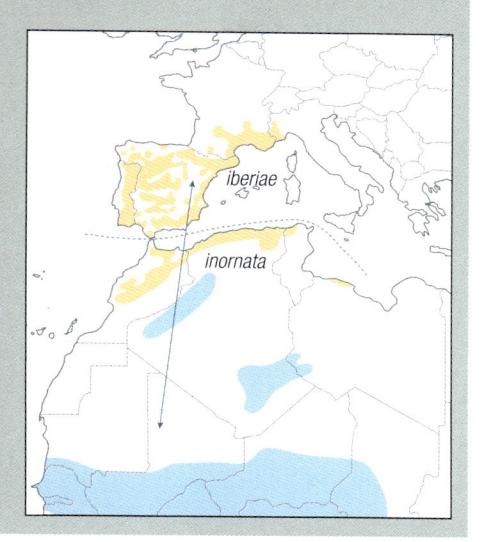

S. i. iberiae, ad ♂, Spain, Apr: rather evenly deep peach red-brown below (like ♂ Dartford Warbler), lacking marked contrast between throat/chest and rest of underparts. Bold but narrow/short white submoustachial stripe also characteristic. Wing-coverts and especially tertials notably pale-fringed. With wear becomes purer bluish slate-grey above. Not heavily worn wing, including broadly greyish-edged primary-coverts, and overall rich coloration, confirm this bird is ad. Deep reddish orange-brown iris is an age criterion, but also a specific tendency of Western Subalpine (iris of ad ♂ Eastern usually duller and less reddish). (J. L. Muñoz)

IDENTIFICATION A rather dainty *Sylvia* with a *fairly short tail but long wings*. ♂ has bluish-grey head and upperparts, a *narrow white submoustachial stripe* and *orange-red throat to lower breast*, becoming whitish on centre of belly, vent and undertail-coverts. *Flanks* frequently *similarly dark orange-red*, or subtly paler orange. Wings slightly darker than mantle with largely greyish-fringed coverts; *tertials stand out due to their more solid dark centres and sandy-buff fringes*. Tail dark grey with white outer edges and white tips to some outer feathers. Some (older) ♀♀ show some pinkish or *pale red on underparts* and even a *fairly well-marked white submoustachial stripe*, but most are rather featureless, buffish brown-grey above *with a slightly greyer head*, only slightly darker brown-grey tail, browner wings (*duller version of ♂ tertial pattern*) and *extensively whitish underparts*. Young autumn birds strongly approach ♀, whereas first-summer ♂ usually acquires adult ♂ pattern. Autumn plumage of adult poorly known, but orange-red underparts of ♂ appear much subdued due to whitish-tipped feathers, creating more pinkish-red impression and confusing similarity with spring ♂ Moltoni's Warbler. The rather thin, relatively short bill has blackish culmen and tip, and paler greyish or dull yellowish lower base. *Orbital ring red*, brighter in ♂♂, and outside this usually a *narrow whitish or cream-coloured eye-ring* in ♀♀. In adults, eye variably tinged ochre or orange-red (often in ♂), and legs fleshy yellow-brown. Lacks obvious raising or wagging of tail of many similar-sized congeners, but tail often flicked. Carriage rather horizontal. Often moves higher in trees than congeners, but also uses low scrub and sometimes takes prey on ground. Song either given from dense cover, infrequently from an exposed perch and at times in parabolical song-flight.

VOCALISATIONS Song rather like Sardinian but on average a little longer, and tempo more hurried, less even and distinctly higher-pitched, a little weaker, lacks any grating notes and is less rattling. Can be likened to a fast Linnet with similar rippling and bouncing quality, and containing many dry, call-like notes *tett-tett-tett*. Differs from Moltoni's Warbler by more varied content and slower pace, lacking the latter's scratchy and 'electric' quality. Differs from Eastern Subalpine Warbler (with difficulty) mainly in somewhat slower pace and being a little more varied. – Contact call a hard dry *tett* (like Lesser Whitethroat), used in fast series in alarm, *tett-ett-ett-ett...*, sometimes accelerating after the start, *tett tett tet-et-et-et-et*. Strongly alarmed birds can utter a repeated harsh *tsheh* in long series. Calls with mixed-in hoarse notes, *tscheh-tett-tscheh-tett-tett-...*, also uttered in anxiety. A nasal *ehd* or *cheh*, loosely repeated, is sometimes heard near nest.

SIMILAR SPECIES The greatest confusion risk is with extremely similar and closely related *Eastern Subalpine Warbler*, and ♀♀ and first-winters of the two can be inseparable in the field, while even adult ♂♂ can be so similar in colour that one must leave them unidentified in areas where both occur (mainly Algeria, Tunisia, Italy). Western is on average subtly smaller, and ♂♂ are slightly warmer orange-red below, this being more extensive and continuing onto lower breast and flanks, and is almost as saturated on flanks as on throat/upper breast (Eastern: slightly more bluish-tinged brick-red, less warm orange, and this colour usually ends rather abruptly on upper breast, creating contrast with nearly white or very pale red lower breast/belly and flanks). Note, however, that a few Eastern have slightly less clear-cut contrast, being tinged pale brick-red on flanks, paler than on breast but not as pale as in majority. While the white submoustachial stripe in Western is on average narrower than in Eastern, which usually has a broad white stripe, there is some overlap and the character should only be used as supporting evidence. On average, adult ♂ Western has a slightly redder iris than Eastern, which usually has a more ochre or olive-ochre iris (but probably some slight overlap). In the hand, and rarely in the field when a bird is

S. i. iberiae, ♀, presumed 1stS, Spain, May: taxon identified by location. ♀ often has some pinkish orange-red on underparts, but this cannot be used to separate taxa within the Subalpine Warbler complex, unless call is heard or tail pattern checked. Such ♀♀ often show a rather clear whitish submoustachial stripe. Posture typically rather horizontal and seldom engages in obvious tail-cocking or waving of similar-sized congeners, e.g. Sardinian Warbler. Note the relatively short bill with blackish culmen and tip, and mostly pinkish grey or yellow hue on lower base. Rather abraded, brownish and pointed (probably juv) primary-coverts, while remiges are also quite worn and fringed brownish, and iris is dull brownish-orange with an olive hue suggesting 1stY. (C. N. G. Bocos)

preening, the tail pattern can be seen, differing in adults of both sexes between the two species in that Western has only a small and square white tip on the penultimate tail-feather (and rather limited amount of white on outermost), whereas adult Eastern has a long narrow white wedge along shaft on inner web of penultimate (and often extensive white on outermost). Some Eastern even have a narrow white wedge on the third outermost tail-feather. The tail pattern is less developed in juveniles (and is usually kept partly or wholly during first-winter), though still often possible to see a hint of a pale wedge in Eastern; however, some juveniles might be inseparable, even in the hand. (Note that Western Subalpine and Moltoni's Warblers have very similar tail patterns.) The call of Eastern often differs in being not a simple dry clicking *tett* as in Western but a more 'fat' *chep* (Italy) or a compound *t'ret* or disyllabic *te-tet* (Balkans, Turkey). – *Moltoni's Warbler* is another challenging pitfall to be aware of, the third species of the Subalpine Warbler superspecies. Adult ♂ Moltoni's should be possible to separate by the paler and buffy salmon-pink colour below (rather than vividly orange-red; only very rarely do dark Moltoni's approach Western Subalpine in underparts coloration, and invariably differ by pink-red rather than orange hues), but this is true only in spring and summer; autumn adult ♂ Western Subalpine becomes more pinkish-red due to white-tipped underparts feathering, thus approaching the appearance of Moltoni's. Iris colour of adult ♂ Moltoni's is usually more ochre-yellow, less reddish-orange, than typical Western Subalpine (but some overlap). ♀♀ and young are by and large inseparable on plumage. Both sexes of Moltoni's differ by their Wren-like trilling call, *trrrr*. The song of Moltoni's is faster and more scratchy and 'electric' in tone, not so Linnet-like and less of a pleasant warble. – Confusion could arise with a vagrant *Ménétries's Warbler*. While ♂ Ménétries's should be separable by its dark lores and forecrown (thus at least a hint of dark cap), darker uppertail and all-white throat, ♀♀ and young are extremely similar to corresponding plumages of Western Subalpine, but when seen well usually have a slightly or clearly darker uppertail with more white on outer edges and corners, and throat is invariably white (pale reddish, warm buff or pink-buff centre to throat in Western Subalpine). Note also more contrast in tertials between dark centres and pale buff-brown edges in Western Subalpine, more uniform grey-brown tertials with more diffusely paler edges in Ménétries's. Ménétries's often utters a rattling call, *trrrr'r'r*, which with familiarity is distinctive from Western Subalpine's series of dry clicks, *tet-et-et-et-et*. – A theoretical confusion risk is between ♀ or first-winter Western Subalpine and ♀ *Sardinian Warbler*, especially of race *momus* (although Western Subalpine and *momus* are hardly ever likely to come into contact). Western Subalpine has longer primary projection (almost exceeding tertial length and usually comprising 6–7 widely- and equally-spaced tips, whereas in Sardinian usually no more than five closely-spaced tips are visible). Also, Western Subalpine has a shorter, less graduated tail which is rarely waved. ♀ Sardinian has brownish upperparts, a darker grey head, much darker tail, obvious warm buff to brown body-sides and purer white throat, while Western Subalpine usually displays a white submoustachial stripe, whiter eye-ring and pinkish grey-brown base to the lower mandible. Also, they usually clearly differ in vocalisations. – Differences from *Rüppell's* and *Spectacled Warblers* are covered under those species.

AGEING & SEXING Ageing requires experience of moult and feather wear, and patterns in wing and tail. Iris coloration is most useful in autumn but always very tricky to assess. Sexes differ in ad and breeding plumages. Rather marked age- and sex-related variation. A certain seasonal variation in ♂. – Moults. Complete post-nuptial and partial post-juv moults in late summer (mostly Jul–mid Aug). Post-juv moult involves all head- and body-feathers, lesser and median coverts, usually all or most greater coverts, and perhaps some tertials, less commonly innermost and, more rarely,

S. i. inornata, ad ♂, Morocco, Apr: overall paler, with warmer and purer orange-tinged underparts characteristic of the NW African race. Combination of less heavily worn wing, including the rounded and broadly-edged primary-coverts, overall rich plumage coloration, and deep red-brown iris are indicative of an ad. (G. Conca)

S. i. inornata, ♀, presumed 1stS, Morocco, Apr: ♀♀ in spring often very similar to migrant *iberiae*, but at least some can perhaps be identified by the more yellowish-orange hue to the underparts. Beware of confusion with least-advanced young ♂♂, though very subdued red orbital ring inside white eye-ring and pale underparts best fit ♀. Apparent moult limit in tail (central pair renewed) and possibly retained brownish and worn juv primary-coverts suggest 1stS. (T. Bedford)

S. i. iberiae, ad ♂, Spain, Sep: fresh autumn ♂♂ are highly variable (see image above), some having peachy red-brown underparts strongly concealed by broad whitish tips, forming pinkish-orange patches on breast-sides, while flanks are mainly light orange-buff. Predominately grey head and rest of upperparts (helping to eliminate ♀). Aged by evenly very fresh wing with silvery-white primary tips, broad and greyish-fringed primary-coverts, and deep orange-cinnamon iris confirm age as ad. (J. L. Muñoz)

SYLVIA WARBLERS

S. i. iberiae, presumed ad ♀, Spain, Sep: in autumn, some rich-coloured ad ♀♀ are very similar to 'advanced' 1stW ♂♂, making ageing a priority. Note apparently evenly fresh wing-feathers, mainly pink-brown (rather than white) eye-ring and extensive white on outer tail-feathers indicating ad. Brownish-olive iris does not exclude ad ♀. Due to fresh pale tips to underparts, these appear pinker, not dissimilar to Moltoni's Warbler (nevertheless, has rusty tone to throat and upper breast). (H. Shirihai)

S. i. iberiae, 1stW ♂, Spain, Aug: some young autumn ♂♂ approach dullest ad ♂♂, especially in the greyer upperparts and already substantial rufous below (essentially brick-red on throat, more orange on flanks). Correct sexing first requires correct ageing: note juv primary-coverts and olive-tinged iris; though (unusually) whole alula was replaced in post-juv moult. Unlike most 1stW ♀♀, young ♂♂ have reddish orbital ring. (S. Fletcher)

central feathers, but tail-feathers usually not renewed. Partial pre-nuptial winter moult (mostly Dec–Feb), in ad includes some tertials and tail-feathers, rarely a few inner greater coverts and secondaries, but usually no primaries, whereas in 1stY moult is more extensive, including some body-feathers and wing-coverts, alula, all tertials and tail-feathers, often some innermost secondaries and up to six primaries. – **SPRING** Ageing of both sexes requires correct assessment of age of post-nuptial primaries and primary-coverts, where ad-type feathers are firmer, less worn and have darker centres and greyish to brownish-grey fringes. Iris more variable in ♂, often as dark olive-grey as 1stS or nearly so, thus only ♀ with deep orange iris reliably aged using iris coloration. **Ad ♂** With wear becomes purer bluish slate-grey above, with darker, more solidly reddish-orange throat, chest and flanks, and more clearly-defined white submoustachial stripe. Iris dull orange-brown or ochre (but somewhat variable). **Ad ♀** Head wears slightly greyer and throat is often tinged pale pinkish-orange, with pale submoustachial stripe quite pronounced. Some older ♀♀ develop deeper reddish throat (even upper breast) but colour is still duller and blotchier than in ♂. Iris deep orange. **1stS** Less advanced and less neatly coloured, with upperparts tinged brownish, and reddish colour below less intense and more restricted. Best aged by juv primaries and primary-coverts (which are distinctly worn and browner and fringed buffish to greyish-brown). Unmoulted juv outer tail-feathers (if any) noticeable for their duller and less pure white fringes and tips, and are generally a little paler and very worn. Iris dark grey-brown to dull orange-brown; iris of ♀ grey-brown, only rarely orange-brown, but ♂ can have duller orange/orange-brown iris, thus approaching ad; iris never deep orange in 1stS. At all ages, orbital ring and, usually, eye-ring pink-orange in ♂ (in ♀, orbital ring usually duller than in ♂, and eye-ring mostly white or pale orange). – **AUTUMN Ad** Both sexes aged with practice by ad-type outer tail-feathers with pure white tip and edge on r6 and pure white square tips on r5 or rr4–5. Supporting evidence of ad from evenly very fresh wing- and tail-feathers (if moult interrupted, some very worn inner secondaries diagnostic). Remiges and primary-coverts firmly textured and glossier greyish-black, with greyer wing edgings in ♂; tips of exposed primary-coverts silvery-white. Ad usually has greyish-brown (or orange-brown) iris and dull orange orbital ring (some 2ndCY ♀♀ very similar to 1stW). **Ad ♂** As spring, but grey of head and upperparts

less pure (being slightly olive-buff on crown, ear-coverts and mantle/scapulars). Underparts variable, most have throat and breast dull reddish or pinkish-orange (less obviously orange than in spring), finely tipped whitish, often appearing somewhat patchy on breast-sides, while flanks are mainly light ochre-buff. Beware that some can appear rather pinkish-red due to fresh feathers being finely white-tipped, inviting confusion with Moltoni's Warbler. (Variation of this plumage poorly known.) **Ad ♀** Upperparts brown tinged olive-grey, but typically purer grey on head and nape. Throat cream-buff (sometimes slightly pink-orange), breast (mostly sides) and flanks variably buffish-cream, with rest of underparts cream-white. **1stW** Sexes almost alike, largely as ad ♀ but usually more uniform dull brown above, with paler underparts. 1stW ♂ more greyish (especially neck-sides and supercilium), with more orange-buff body-sides and throat (thereby enhancing white submoustachial stripe), thus general plumage more like ad ♀ than 1stW ♀. Aged by juv primary-coverts and remiges, and some outer greater coverts (softer and slightly less fresh-looking or less glossed, being mainly centred dark brownish-grey, with rufous-brown fringes). Pale primary tips more abraded. Moulted ad-like greater coverts have darker greyish centres with brownish-grey fringes, forming moult limits, clearer in ♂♂; similar moult limits discernible between tertials if some replaced. Juv tail-feathers slightly browner and paler with ill-defined pale brownish-cream or greyish-white fringes and tips to outermost feathers. Iris dark greyish or brownish-grey, and orbital ring brownish-yellow. **Juv** Largely as ♀ but upperparts uniform greyish-buff, below less pure white, tinged buffish-cream especially on breast and throat.

BIOMETRICS (*iberiae*) Practically no size difference between sexes. **L** 11.5–12.5 cm; **W** 55–63 mm (n 56, m 58.9); **T** 46.5–58 mm (n 56, m 52.5); **T/W** m 89.2; **B** 10.9–13.0 mm (n 52, m 11.9); **BD** 2.6–3.5 mm (n 50, m 3.0); **Ts** 17.0–19.9 mm (n 51, m 18.5). **Wing formula: p1** > pc 0–4 mm (rarely to 2 mm <), < p2 26–36 mm; **p2** < wt 1–3.5 mm, =5/6 or 6 (83%), or =4/5 or 5 (17%); **pp3–5** about equal and longest (p5 sometimes < wt 0.5–2

S. i. iberiae, 1stW, presumed ♂, Spain, Oct: many young in autumn are intermediate in their pigmentation and cannot be sexed with certainty, but in this case the greyer-tinged head-sides and reddish orbital ring indicate 1stW ♂, though not conclusively. Juv primary-coverts and remiges, and moult limits in tertials, age this bird, as does the sepia-olive iris. (E. Padilla)

S. i. iberiae, 1stW ♀, Spain, Sep: following post-juv moult, young ♀ has dullest plumage, with duller, brownish-buff (and more uniform) upperparts and head; also, moulted greater coverts are generally duller. Orbital ring usually at most dull yellowish-brown (not red) and lores whiter. In early autumn the plumage is still fresh; juv primary-coverts weaker textured and buffish-fringed, while primary tips already slightly worn and are not whitish. (J. L. Muñoz)

S. i. iberiae, juv, presumed ♀, Spain, Jul: in general plumage like ♀ but has soft, fluffy body-feathers. Some juv can be sexed, and here the very dull plumage and faint and yellowish-brown orbital ring suggest ♀, but impossible to be certain due to overlap in characters. Whitish marks on tips of many flight-feathers either aberration or due to staining. (K. Mauer)

mm); **p6** < wt 2–5 mm; **p7** < wt 3.5–8 mm; **p8** < wt 5.5–8.5 mm; **p10** < wt 8.5–13.5 mm; **s1** < wt 10–15.5 mm. Emarg. pp3–5 (in ad sometimes faint also on p6). **White tip r5** (inner web) 0–6 mm (n 40, m 2.4; once 12 mm on one side in aberrant ad ♂ Algarve, Portugal), square-shaped or obtusely wedge-shaped. In juv/1stW absent or reduced to very thin greyish-white tip.

GEOGRAPHICAL VARIATION & RANGE Two subspecies recognised, differing only subtly.

○ **S. i. iberiae** Svensson, 2013 (Iberia, France, W Liguria and W Piedmont in extreme NW Italy; wintering grounds poorly known, but mainly reach to S Morocco, C & S Algeria and presumably from Senegal and Mauritania east to Niger). Described above. Throat and breast, and usually also flanks, rather deep orange-red or bright rust-coloured. White submoustachial stripe of variable prominence, often narrow and pointed at both ends, but very rarely has a broader and rather prominent one (although often still with pointed ends). Breeders in S Iberia said to approach *inornata* of NW Africa, but only a faint clinal tendency, and on present evidence best included in *iberiae*.

S. i. inornata Tschusi, 1906 (N Africa, from Morocco to NC Libya; presumed to be resident or makes only local movements). Very similar to *iberiae* but differs subtly in being slightly more yellow-tinged below in both sexes, noticeable when series are compared. Thus, ♂ is more warmly orange below, and ♀ has more saturated ochre-buff tinge on breast and flanks. Still, many are difficult to separate without the support of season and locality, or with series of known subspecies for comparison. Sexes of very nearly same size. **W** 56–63 mm (n 57, m 59.6); **T** 48–56 mm (n 57, m 52.9); **T/W** m 88.8; **B** 11.1–13.0 mm (n 53, m 12.0); **BD** 2.7–3.5 mm (n 48, m 3.1); **Ts** 17.2–19.6 mm (n 54, m 18.6). **Wing formula:** similar to *cantillans* but **p7** < wt 3–7 mm; **p8** < wt 5–10 mm. **White tip r5** (inner web) 0–7 mm (n 45, m 2.1); when > 4 mm usually still squarely rounded, not forming pointed wedge, but two had hint of narrow white wedge near shaft.

TAXONOMIC NOTES It has long been known that some breeders within the widely distributed superspecies Subalpine Warbler differ in calls, notably those breeding on the Balearics, Sardinia and Corsica having a Wren-like rattling call, unlike the short clicking note of most other populations (e.g. Savi 1828). Orlando (1937, 1939) described plumage differences of Sardinian breeders and named them *moltonii* (although *subalpina* has priority; see Taxonomic notes under Moltoni's Warbler). Gargallo (1994) and Shirihai et al. (2001) further noted differences in moult, plumage and vocalisations, and partly based on and inspired by this, Brambilla and co-workers published a series of studies on biogeography and genetic differences (e.g. Brambilla et al. 2008a,b,c, 2010), which has clarified their relationships. They also made an extensive survey of the true breeding range of *subalpina* (Brambilla et al. 2006), demonstrating that this taxon is widely distributed in mainland northern Italy, not just confined to some Mediterranean islands, and in fact breeds sympatrically with another Subalpine Warbler taxon there. This other Subalpine Warbler, differing in morphology, vocalisations and genetics, is numerous and widespread, and is best viewed as the *cantillans* Pallas described in 1764 with the loose type locality 'Italy' (Svensson 2013a). Whereas most taxonomists now agree that *subalpina* deserves species rank based on the above-mentioned results, becoming Moltoni's Warbler, views differ on how to best treat S Italian *cantillans* in relation to remaining populations. As Brambilla et al. indicated (but lacked morphological or other supporting evidence), the S Italian breeders are genetically closer to E European *albistriata* than to Iberian–French breeders (recently named *iberiae*; Svensson 2013a). S Italian breeders are often inseparable from *albistriata* using plumage and biometrics, sharing with this the specific pattern of white wedges on the penultimate tail-feathers, and available studies of vocalisations reveal only small differences in calls and apparently none in song. S Italian *cantillans* thus groups clearly with *albistriata*, and is more distant from *iberiae* of France and Iberia. The natural arrangement then, supported by all evidence, is to divide the Subalpine Warbler complex into three species, Western and Eastern Subalpine Warblers and Moltoni's Warbler, an arrangement also adopted here.

REFERENCES Baccetti, N., Massa, B. & Violani, C. (2007) *BBOC*, 127: 107–110. – Brambilla, M. et al. (2006) *Ibis*, 148: 568–571. – Brambilla, M. et al. (2008a) *Mol. Phyl. & Evol.*, 48: 461–472. – Brambilla, M. et al. (2008b) *J. Avian Biol.*, 21: 651–657. – Brambilla, M. et al. (2008c) *Acta Orn.*, 43: 217–220. – Brambilla, M. et al. (2010) *J. of Orn.*, 151: 309–315. – Gargallo, G. (1994) *BBOC*, 114: 31–36. – Orlando, C. (1937) *Riv. Ital. Orn.*, 7: 213. – Orlando, C. (1939) *Riv. Ital. Orn.*, 9: 147–177. – Savi, P. (1828) *Ornitologia Toscana*. Vol. 2. Nistri, Pisa. – Shirihai, H., Gargallo, G. & Helbig, A. (2000) *BW*, 13: 234–250. – Stoddart, A. (2014) *BB*, 107: 420–424. – Svensson, L. (2013a) *BBOC*, 133: 240–248. – Svensson, L. (2013b) *BB*, 106: 651–668. – Trischitta, A. (1922) *Atti Soc. It. Sci. Nat. Milano*, 61: 121–131.

S. i. iberiae, juv ♂, Spain, Jun: some juv ♂♂, even while post-juv moult is in progress, can be sexed by stronger/warmer orange-red underparts; also note whitish submoustachial stripe and reddish orbital ring. Yellow gape-flanges still present. (J. Sagardía)

EASTERN SUBALPINE WARBLER
Sylvia cantillans (Pallas, 1764)

Fr. – Fauvette de Balkans; Ger. – Weißbartgrasmücke
Sp. – Curruca carrasqueña oriental
Swe. – Rödstrupig sångare

The easternmost species in a trio comprising the Subalpine Warbler superspecies. It breeds in W Turkey, Greece and much of the Balkans, but also commonly in mainland Italy, mainly south of a line from Marche to Lazio, including on Sicily. It also breeds thinly in N Italy, thus sympatrically with Moltoni's Warbler. It is an attractive warbler of garrigue and open woodland with rich undergrowth. In late summer it migrates mainly south-west, like all Subalpine Warblers, through Tunisia and Algeria to the trans-Saharan Sahel region of W Africa, but some apparently winter further east in Africa. In spring many birds take a more easterly route and pass through the Levant.

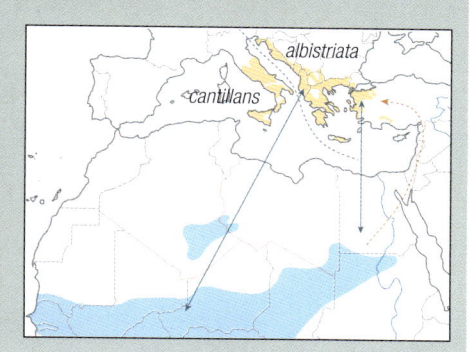

adults) have some *pinkish-red on underparts* and even *a hint of a white submoustachial stripe*. ♀♀ and young are normally inseparable from Western Subalpine unless tail pattern is examined (in the hand or on the rare occasion when a bird preens, spreading its tail; see above). Eastern can also easily be confused with Moltoni's Warbler, the two having sympatric ranges in N Italy, and with which it commonly mixes on migration and in winter. Eastern Subalpine has either a monosyllabic clicking, full *chep* call (S Italy) or a slightly more compound, almost or clearly disyllabic *t'ret* or *te-tet* (Balkans, Turkey), obviously different from the rattling Wren-like *trrrr* of Moltoni's and possibly subtly different from the drier clicking *tett* of Western Subalpine. In autumn, ♂ Eastern Subalpine in post-juvenile plumage usually also differs from Moltoni's in having emerging dark brick-red on throat and breast (whitish tips partly obscuring this), and contrasting whitish flanks and belly, while Moltoni's has either underparts all cream-white with no pink, or underparts paler and more uniformly salmon-pink. (However, autumn plumage variation in adult still poorly known.) The fine bill has blackish culmen and tip, and greyish-horn or dull yellowish lower base (possibly on average duller yellowish-horn than in Western Subalpine, which can be a little brighter yellow). Colours of bare parts do not differ much from the other Subalpine Warblers, *orbital ring being red*, and outside this usually a *narrow whitish or pale buff eye-ring* in ♀♀. In adults, eye variably tinged ochre or orange-cinnamon (perhaps on average subtly less orange-red than in Western Subalpine), and legs fleshy yellow-brown. Behaviour much as the other two Subalpine species.

VOCALISATIONS Song very similar in structure to that of Western Subalpine, being a hurried, varied warble with many scratchy and clicking notes mixed with clear whistles and softer sounds. In comparison, the song is more hurried than Western Subalpine, as fast and energetic as Moltoni's Warbler, but differs from latter by its less scratchy and 'electric' quality. Separation of all three requires much practice and a good ear, and not all cannot be separated even then. – Contact call either (apparently usually in S Italy) a monosyllabic clicking, slightly 'fat' *chep* (not as dry as Lesser Whitethroat, more like call of Red-necked Phalarope or even Radde's Warbler), probably not separable from call of Western Subalpine without comparison or much practice, sometimes used in series as alarm, *chep-ep-ep-ep...* Alternative call (appears to dominate in Balkans and Turkey) is a slightly compound, scratchy *t'ret* or more clearly disyllabic *te-tet* or *terret*. A different call of unknown function is a plain-

S. c. cantillans, ad ♂, Italy, May: in central and southern Italy, breeding populations are genetically distinctive from other forms in the Subalpine Warbler complex, but they are almost identical in morphology to *albistriata* of Balkans and Turkey. This Italian breeder in May is safely an Eastern Subalpine of race *cantillans*. Compared to *albistriata* it tends towards Western Subalpine in having subtly paler grey upperparts and paler, less contrasting brick-red throat. (D. Occhiato)

IDENTIFICATION Similar to Western Subalpine Warbler, and not always separable in the field, but is on average very slightly larger, and adult ♂ has *a little darker and colder reddish-brown or brick-red* (less warm orange-red) *throat and breast*, with on average *more obvious contrast between the red-brown and the very pale rufous or almost whitish flanks and belly*. Upperparts of ♂ are *ash-grey* like in the two closely related species, on average a little darker lead-grey in the eastern race than in the Italian population. ♂ Eastern Subalpine has on average *a broader and more prominent white submoustachial stripe* than Western Subalpine and Moltoni's, but there is some overlap; it is a supporting but not infallible character, and only for Eastern Subalpines with a very broad stripe (broad also at both ends) should one place weight on it. Note that a few Eastern Subalpine ♂♂ in spring and summer have a slightly stronger rufous tinge on flanks than is typical, then approaching Western Subalpine (rarely even strong similarity!). They still differ by tail pattern (useful for adults of both sexes, sometimes also for juv), Eastern Subalpine having *a narrow white wedge inside shaft of penultimate tail-feather* (r5; sometimes also on r4), the other two species a square white tip to this feather (or feathers). There is also on average more white on r6 in Eastern Subalpine (but some overlap). ♀♀ are rather variable as to underparts colours, some being whitish with only a faint cream-coloured tinge on breast and flanks, others (only

S. c. albistriata, ad ♂, Israel, Mar: normally, best clue to Eastern Subalpine Warbler ♂ is strong contrast between dark brick-red (less orange) throat/upper breast and rest of underparts, but beware that some ad ♂♂ (particularly in the race's eastern range) can have wider red area below, rather than restricted to the throat/breast. Taxon still revealed by deep brick-red colour and broad white submoustachial stripe. Age inferred by grey-fringed and rounder-tipped primary-coverts; at least tertials were replaced more recently during partial pre-nuptial moult. (H. Shirihai)

tive *cheeh*.

SIMILAR SPECIES The two most obvious risks of confusion are with *Western Subalpine Warbler* and *Moltoni's Warbler* (which see and above under Identification). – ♀♀ and young are very similar to *Ménétries's Warbler*, and some must be left unidentified unless call is heard or different tail pattern seen well. Ménétries's is a subtly smaller bird with slightly paler sand-brown upperparts that offer stronger contrast to darker grey uppertail, but differences are small and require care and experience. Note that Ménétries's also has on average slightly more uniform grey-brown tertials with more diffusely pale edges than in Eastern Subalpine, which displays more obvious contrast. – A common relative sharing same or similar habitats is *Sardinian Warbler*, and a fleeting view could be confusing. Once seen well the, in general, darker plumage and sturdier build of Sardinian is usually evident. Note especially the darker body-sides (grey in ♂, brown in ♀), the darker uppertail and invariably pure white throat of Sardinian. The strikingly dark cap of ♂♂ or darkish ear-coverts of ♀♀ are often evident too. – Pale legs and brick-red orbital ring separate from invariably dark-legged *Lesser Whitethroat*. – Smaller size and daintier shape, shorter tail, lack of rufous wing panel and presence of brick-red orbital ring readily separates from *Common Whitethroat*. – For differences from *Rüppell's Warbler*, see that species.

AGEING & SEXING As in Western Subalpine Warbler, ageing requires experience of moult and feather wear, and patterns in wing and tail. Iris coloration most useful in autumn, but frequently tricky to assess. Sexes differ at least in ad and breeding plumages. Rather marked age- and sex-related variation. A certain seasonal variation in ♂. – Moults. Complete post-nuptial and partial post-juv moults in late summer (mostly Jul–mid Aug). Post-juv moult involves all head- and body-feathers, lesser and median coverts, usually all or most greater coverts, and perhaps some tertials, less commonly innermost and, more rarely, central feathers, but tail-feathers usually not renewed. Partial pre-nuptial winter moult (mostly Dec–Feb), in ad includes some tertials and tail-feathers, rarely a few inner greater coverts and secondaries, but usually no primaries, whereas in 1stY moult is more extensive, including some body-feathers and wing-coverts, alula, all tertials and tail-feathers, often some innermost secondaries and up to six inner primaries. – **SPRING** Ageing of both sexes requires correct assessment of age of post-nuptial primaries and primary-coverts, with ad-type feathers firmer, less worn and having darker centres and greyish to brownish-grey fringes. Iris of ♀ more variable than in ♂, often as dark olive-grey as 1stS or nearly so, thus only ♀ with

S. c. albistriata, 1stS ♂, Israel, Mar: light bluish grey upperparts with clear-cut pale tertial fringes are often enough to give the first identification clue, before the typical dark brick-red throat and upper breast become visible. Heavily abraded primaries and primary-coverts infer age. (H. Shirihai)

deep orange iris reliably aged using this feature. **Ad ♂** With wear becomes purer bluish slate-grey above, with darker, more solidly rufous or brick-red throat and chest, and white submoustachial stripe more clearly defined. Iris dull orange-brown or ochre. **Ad ♀** Head wears slightly greyer and throat is often tinged pale reddish, with pale submoustachial stripe quite pronounced. Some older ♀♀ develop deeper reddish throat (even upper breast) but always duller and blotchier than in ♂. **1stS** Less advanced and less neatly coloured, with upperparts tinged brownish, and reddish colour below less intense and more restricted. Best aged by juv tail-feathers (often a few, sometimes whole tail juv), primaries and primary-coverts (which are distinctly worn and browner and fringed buffish to greyish-brown). Unmoulted juv outer tail-feathers noticeable for their duller and less pure white tips, r5 having only diffuse paler wedge or oval patch, or no pale pattern at all, and are generally very worn. Some have moulted several inner primaries, many or all tertials and some secondaries, but frequency of such extensive pre-nuptial moult not known. Iris dark grey-brown to dull orange-brown; ♀ grey-brown, only rarely orange-brown, but ♂ can have duller orange/orange-brown iris, thus approaching ad; iris never deep orange in 1stS. At all ages, orbital ring and,

usually, eye-ring pink-orange in ♂ (in ♀, orbital ring usually duller than in ♂, and eye-ring mostly white or very pale orange). – **AUTUMN** Both sexes aged with practice on whole tail moulted and ad-type outer tail-feathers with extensive pure white portion on r6 and pure white long narrow wedge on r5 (or rr4–5). Supporting evidence of ad from evenly very fresh wing- and tail-feathers. Remiges and primary-coverts firmly textured and glossier greyish-black, with greyer wing edgings in ♂; tips of exposed primary-coverts silvery-white. Ad usually has greyish-brown (or orange-brown) iris and dull orange orbital ring (some 2ndCY ♀♀ very similar to 1stW). **Ad ♂** As spring but grey of head/upperparts less pure (being slightly olive-buff on crown, ear-coverts and mantle/scapulars). Underparts variable, most have throat and breast dark rufous-red tipped whitish, often appearing patchy on breast-sides, while flanks are mainly whitish or light rufous-buff. **Ad ♀** Upperparts suffused sandy-buff, but typically greyer mainly around supercilium, neck and nape. Throat cream-buff (sometimes slightly pink-red), breast (mostly sides) and flanks variably buffish-cream, with rest of underparts cream-white. **1stW** Sexes can be similar, largely as ad ♀, but usually more uniform sandy-brown above with paler underparts. More often 1stW ♂ differs slightly, being

Breeding ♂ Western Subalpine (ssp. *iberiae*, left: Spain, Jun), Eastern Subalpine (ssp. *albistriata*, centre: Turkey, Mar) and Moltoni's Warblers (right: Italy, May): Western Subalpine Warbler is characterised by reddish-orange underparts (typically almost equally strong all over), leaving only central belly white. Eastern Subalpine is darker and colder red-brown below, the colour being often solid only on throat and upper breast, contrasting with rest of paler underparts, and tends to have somewhat broader white submoustachial stripe. Moltoni's Warbler has salmon-pink underparts, without any orange. Further, tail pattern is often essential (in the hand, and rarely in the field): most strikingly between Eastern (white wedge on r5 diagnostic!) and the other two as soon as juv outer tail-feathers have been replaced. (Left: J. Sagardía; centre: E. Yoğurtcuoğlu; right: R. Parmiggiani)

S. c. albistriata, presumed ad ♀, Israel, Mar: some ♀ Eastern Subalpine Warblers develop ♂-like (but more subdued) vinaceous reddish-brown throat/breast, making a hint of a white submoustachial stripe appear, and accordingly separation between such ad ♀♀ and 1stS ♂ can at times be difficult. Although the quite grey head and upperparts might indicate a ♂, the largely white eye-ring, the diffuse submoustachial stripe and the overall reasonably fresh wing all fit better with ad ♀. (H. Shirihai)

more greyish (especially neck-sides and supercilium), with better-developed orange-buff breast and throat (enhancing white submoustachial stripe), thus general plumage more like ad ♀ than 1stW ♀. Aged by juv primary-coverts and remiges, and some outer greater coverts (softer and slightly less fresh-looking or less glossed, being mainly centred dark brownish-grey, with rufous-brown fringes). Pale primary tips more abraded; moulted ad-like greater coverts have darker greyish centres with brownish-grey fringes, forming moult limit, clearer in ♂; similar moult limits discernible between tertials if some replaced. Juv tail-feathers slightly browner with ill-defined pale brownish-cream or off-white fringes and tips to outermost feathers (and r5 either has hint of pale narrow wedge inside shaft, or no pale pattern). Iris dark greyish or brownish-grey, and orbital ring brownish-yellow. **Juv** Largely as ♀ but upperparts uniform greyish-buff, below less pure white, tinged buffish-cream especially on breast and throat.

BIOMETRICS (*albistriata*) **L** 12–13 cm; **W** ♂ 59–68 mm (n 92, m 63.2), ♀ 58.5–66.5 mm (n 34, m 62.5); **T** ♂ (48) 50–59 mm (n 92, m 54.0), ♀ 50–58 mm (n 34, m 53.8); **T/W** m 85.7; **B** 11.0–13.9 mm (n 113, m 12.7); **BD** 2.5–3.5 mm (n 98, m 3.2); **Ts** 18.1–20.4 mm (n 116, m 19.1). **Wing formula: p1** < pc 3.5 mm, to 2 mm > (m < 0.7), < p2 32.5–40 mm; **p2** < wt 0–3.5 mm, =5 or 5/6 (65%), =4 or 4/5 (31%) or =3 or 3/4 (4%); **pp3–4** about equal and longest (rarely pp2–4); **p5** < wt 0.5–2.5 mm; **p6** < wt 1.5–5.5 mm; **p7** < wt 4.5–8.5 mm; **p8** < wt 7–11 mm; **p10** < wt 10–16 mm; **s1** < wt 11.5–17.5 mm. Emarg. pp3–5 (in ad sometimes faint also on p6). **White tip r5** (inner web) of post-juv 10–32 mm (n 96, m 16.5; in c. 2% < 10 mm, but then either aberrant or possibly not correct taxon), narrowly wedge-shaped with almost parallel sides. In juv/1stW unmoulted juv tail-feathers, usually restricted to diffuse greyish-white wedge or oval-shaped patch.

GEOGRAPHICAL VARIATION & RANGE Two very similar subspecies, their separation warranted due to subtle but consistent differences in plumage, size, vocalisation and mtDNA. Whether wintering grounds of the two races differ is not known. The combined wintering area is W Sahel, in Niger, N Nigeria, Chad, but since it is recorded as far west as C Morocco (Mansour, W High Atlas; NRM) the winter range could extend much further west; records in early Mar from S Algeria may be migrants.

S. c. cantillans (Pallas, 1764) (mainland Italy, commonly from Umbria, Marche and Lazio and south, including Sicily, thinly north to Tuscany, Emilia-Romagna; winters presumably in sub-Saharan Sahel in W Africa). Inseparable from *albistriata* except for by fractionally smaller size and on average shorter

S. c. albistriata, 1stS ♀, Greece, Apr: combination of clearly pointed and abraded juv primary-coverts, worn primary tips, dull olive-brown iris and overall dull plumage identify this individual as 1stS ♀. Away from breeding grounds, dull ♀♀ cannot be subspecifically identified, but note rather long primary projection, pointed wings, stronger build and larger, longer-billed impression, plus relatively broader white submoustachial stripe. (G. Schuler)

S. c. albistriata, ad ♀, Israel, Mar: on migration some birds can look confusing due to unusually rufous-tinged wing pattern, but note vinaceous reddish-brown hue on throat to flanks, and relatively broad white submoustachial stripe, which readily separate this ad ♀ from e.g. Spectacled Warbler or Common Whitethroat. Reddish orbital ring and longish primary projection further separate it from Spectacled Warbler. Evenly-feathered, rather fresh wing with broad, round-tipped primary-coverts and quite greyish head make this ♀ ad. (P. Dernjatin)

S. c. albistriata, ad ♂, **Turkey, Aug**: in autumn, some ad ♂♂ have only limited reddish-brown below (strongly concealed by broad whitish tips), creating a pinkish wash which can invite confusion with Moltoni's Warbler. However, the few throat-feathers with more reddish-brown help to identify such birds. Note also the rather strong build, longish bill, long primary projection and broad white submoustachial stripe. The complete post-nuptial moult of this bird further helps to eliminate Moltoni's. (O. Bulut)

S. c. albistriata, 1stW ♂, **Turkey, Aug**: young ♂♂ in autumn are similarly variable as ad ♂♂, but just as latter some show dark reddish bases to many feathers of throat and breast, providing an identification clue. Locality is of course also helpful in this case when taxon is determined. Dark olive iris ages this bird as 1stW. (Ö. Necipoglu)

primary projection, subtly paler grey upperparts and less dark brick-red throat/breast in ♂♂, this being slightly more warm orange with on average less strong contrast between throat and upper breast and flanks (though many inseparable). It also has a more monosyllabic clicking call, a 'fat' or compound *chep* rather than *t'ret* or *te-tet*. **W** 58–65.5 mm (*n* 21, *m* 61.1); **T** 50–58 mm (*n* 21, *m* 53.6); **T/W** *m* 87.8; **B** 11.3–13.5 mm (*n* 21, *m* 12.6); **BD** 2.8–3.3 mm (*n* 21, *m* 3.1); **Ts** 18.0–20.0 mm (*n* 20, *m* 19.2). **Wing formula: p1** < pc 2.5 mm, to 1.5 mm >, < **p2** 31–37.5 mm; **p6** < wt 2.5–5 mm; **p7** < wt 5–8 mm; **p8** < wt 7–9.5 mm; **p10** < wt 10–13.5 mm; **s1** < wt 11–15 mm.

○ ***S. c. albistriata*** (C. L. Brehm, 1855) (Trieste, Balkans, Greece, W & SC Turkey; winters in sub-Saharan Sahel in W Africa, at least from C Morocco to Chad, but claimed also from S Egypt and possibly N Sudan). Described above. Very slightly larger on average than *cantillans* (though much overlap), and ♂ has on average subtly darker lead-grey upperparts and darker brick-red throat and chest. Call is a slightly gruffy or 'compound' *t'ret* or disyllabic *te-tet*.

TAXONOMIC NOTE See under Western Subalpine for general comments on the Subalpine Warbler superspecies and its taxonomy, plus full references.

REFERENCES Brambilla, M. *et al.* (2006) *Ibis*, 148: 568–571. – Brambilla, M. *et al.* (2008a) *Mol. Phyl. & Evol.*, 48: 461–472. – Brambilla, M. *et al.* (2008b) *J. Avian Biol.*, 21: 651–657. – Brambilla, M. *et al.* (2008c) *Acta Orn.*, 43: 217–220. – Brambilla, M. *et al.* (2010) *J. of Orn.*, 151: 309–315. – Cade, M. & Walker, D. (2004) *BW*, 17: 202–203. – Pennington, M. (2009) *BW*, 22: 241–245. – Shirihai, H., Gargallo, G. & Helbig, A. (2000) *BW*, 13: 234–250. – Svensson, L. (2013a) *BBOC*, 133: 240–248. – Svensson, L. (2013b) *BB*, 106: 651–668. – Trischitta, A. (1922) *Atti Soc. It. Sci. Nat. Milano*, 61: 121–131.

S. c. albistriata, 1stW, ♂, **Turkey, Jul**: sex based on grey head and mantle (young ♀♀ are invariably sandy greyish-brown above). Note also slight vinaceous-red hue to breast-sides further indicating sex. In early autumn a few young like this bird have greyer tarsus. Orbital ring usually dull olive-brown (not red), whereas white eye-ring is prominent. Dark olive-brown iris confirms that this bird is 1stW. (K. Dabak)

S. c. albistriata, 1stW ♀, **Israel, Sep**: example of the dullest autumn plumage, including grey-brown upperparts and pale underparts totally lacking white submoustachial stripe. Orbital ring unremarkable being dull olive-brown, and iris dark. Such dull plumage is almost identical to 1stW ♀ Ménétries's Warbler, but this species usually waves its tail, which is also contrastingly darker above than rest of upperparts (in *albistriata* both have same colour). Note also sharply defined pale brown tertial fringes, not diffuse as in Ménétries's. (H. Shirihai)

MOLTONI'S WARBLER
Sylvia subalpina Temminck, 1820

Fr. – Fauvette de Moltoni; Ger. – Moltonigrasmücke
Sp. – Curruca subalpina; Swe. – Moltonisångare

Of the three closely related Mediterranean warblers constituting the Subalpine Warbler superspecies, this one breeds in the centre, flanked by the others. Moltoni's Warbler occupies several of the large W Mediterranean islands, except Sicily, and recently its range has been found to also include a large part of the N Italian mainland. Although separating the three species in the field can be a daunting task, a Moltoni's Warbler can often be distinguished by its slightly paler and more uniform buffish salmon-pink underparts, and both sexes have a characteristic call.

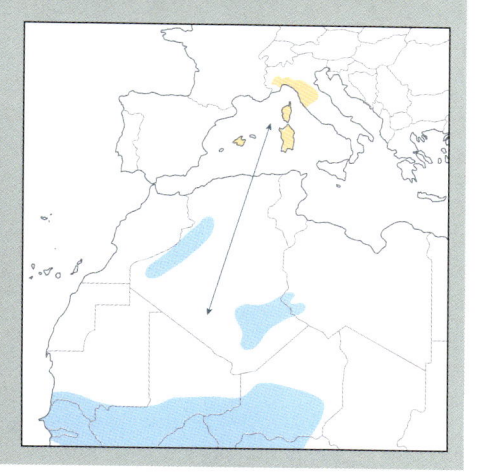

♂, presumed ad, Italy, May: salmon-pink underparts and relatively thin and short white submoustachial stripe are characteristic of ♂. Unlike Western and Eastern Subalpine Warblers, Moltoni's often appears much fresher in spring as pre-nuptial moult is frequently complete, although there is much individual variation in timing (some returning with limited pre-nuptial moult). Deep orange-brown iris indicates ad rather than 1stS ♂, but inconclusive as a few of the latter can develop similar bright iris. (D. Occhiato)

IDENTIFICATION Very similar in size and structure to Western Subalpine Warbler, with only on average subtly longer wings and thereby proportionately *shorter tail*. ♂ Moltoni's has a *rather pale and uniform buffish salmon-pink colour on throat, breast and flanks* ('old rose' or pink-buff); only rarely becomes a little deeper pink-red, which in certain lights may approach the colour of some Eastern Subalpine Warblers, but normally a well-seen ♂ Moltoni's in spring and summer is readily identifiable. The *white submoustachial stripe* is *usually narrow* and pointed at both ends, but a few have a slightly broader and more prominent stripe. ♀♀ and young are extremely similar to corresponding plumages of both Western and Eastern Subalpine, and are often inseparable by plumage from Western. However, their *calls differ and are quite characteristic*, in Moltoni's *a brief Wren-like fast rattle*, rather than the monosyllabic dry clicking of Western Subalpine or similar yet subtly different call of Eastern Subalpine. Many Moltoni's moult some or all flight-feathers in winter, thus such birds *return in spring with much neater wings* than do Western Subalpine, which only moults completely in late summer and returns with rather worn wings, mainly visible through the abraded primary tips and tertial edges. In the hand, the slightly longer and more pointed wing of Moltoni's may be a useful clue, sometimes evident by very short p1 and long distance p1 < p2 (see Biometrics).

Note that, in autumn, adult ♂♂ of both Western and Eastern Subalpine Warblers may appear more pinkish-red below due to white-tipped fresh plumage, inviting confusion with Moltoni's Warbler. Still, both often have some patchy darker, rust-red feathers visible on sides of throat and upper breast, but individual variation of adult ♂ plumages still imperfectly known. As to tail pattern, note that by and large Moltoni's and Western Subalpine Warblers both have the same small and squarish white tip to r5, while both differ from Eastern Subalpine, which has narrow white wedge inside shaft on r5. For further differences from closely related Eastern Subalpine Warbler, see under that species and Similar species. Song usually given from dense cover but infrequently also from an exposed perch or in parabolical song-flight.

VOCALISATIONS Song compared to Western Subalpine Warbler is faster, more high-pitched and scratchy (can almost recall Serin), not as varied and pleasing a warble, slightly more monotonous and 'electric' in tone. Sometimes the Wren-like trilling call is inserted in the song. The song is similarly fast as Eastern Subalpine, but scratchier and more high-pitched. – Call characteristic, a brief, dry rattling or trilling *zrrrr* (rather like a soft Wren or alarm of Ménétries's Warbler, and in tone fairly similar to a very short Spectacled Warbler call).

SIMILAR SPECIES The confusion risk with *Western Subalpine Warbler* is treated under that species (Similar species). – Moltoni's Warbler could easily be confused with *Eastern Subalpine Warbler* in other plumages than adult ♂, and safe separation seems to require handling. In the hand, Eastern Subalpine in all post-juvenile plumages (and sometimes in juveniles, too) differs in having a long, narrow white wedge inside shaft on tip of r5 (sometimes a narrow but shorter wedge on r4), whereas Moltoni's has a rather square, short white tip on r5 (sometimes rr4–5) if any. Adult ♂♂ of course differ clearly in that Moltoni's has rather uniform buffish salmon-pink rather uniform underparts (as a rule including flanks and upper belly) compared to Eastern Subalpine's cold, dark brick-red throat and upper breast, but generally contrastingly whitish or at least considerably paler rest of underparts. The white submoustachial stripe is on average more prominent in Eastern Subalpine than in Moltoni's, and often has rather broad ends, but it may be prudent to caution against using this too blindly; there is some overlap in the prominence of the stripe between all three species of the Subalpine complex. Eastern Subalpine is a subtly larger bird than Moltoni's, but the difference is slight and usually difficult or impossible to evaluate in the field. –

Ad ♂, Italy, May: second outermost tail-feather (r5) shows the small square white tip that eliminates Eastern Subalpine Warbler. Tertials and innermost five primaries were replaced in winter, the rest in preceding autumn, thus only moderate moult contrast. This supports both age as ad (young birds that moult partially tend to prioritise replacing the outermost primaries) and the species identification. (D. Occhiato)

1stS ♂, Italy, May: young Moltoni's (aged as young despite complete winter moult) often less advanced than ad ♂, having the pink underparts most intense on throat/upper breast, but much variation means this can only be used to support ageing. 1stY Moltoni's often moults completely in winter; fresher smallest tertial was replaced in late winter. Note also dull grey-brown iris. (D. Occhiato)

Ad ♀, Italy, Jun: although this ♀ appears to show hint of salmon-pink underparts of Moltoni's, this criterion cannot be used to safely separate ♀♀ of the Subalpine Warbler complex. This bird seems to have performed almost complete winter moult, albeit some odd feathers were not replaced (including the longest tertial); ad ♀ usually shows orange-brown iris. (D. Occhiato)

Although generally widely allopatric, a straggling *Ménétries's Warbler* is similar enough to create problems, not least with first-winters or autumn adult ♀♀. Note in Ménétries's the contrastingly darker uppertail with generally (but not invariably) more extensive white tips to outer tail-feathers, and more concolorous tertials with only diffusely and subtly paler edges (rather than the more contrasting dark tertials with narrow pale edges of Moltoni's). – Other pitfalls, also relevant for the identification of Moltoni's Warbler, are dealt with under Western Subalpine, Similar species.

AGEING & SEXING Ageing requires experience of moult and feather wear, and patterns in wing and tail. Iris coloration is most useful in autumn but frequently difficult to assess, at least in the field. Sexes differ in ad and breeding plumages. Rather marked age- and sex-related variation. A certain seasonal variation in ♂. – **Moults.** Extent of post-nuptial moult in ad in late summer (mostly Jul–mid Aug) variable, the majority moult completely, others perform a partial moult and suspend, while others do not replace any flight-feathers before migration. Post-juv moult involves all head- and body-feathers, lesser and median coverts, usually all or most greater coverts, and perhaps some tertials, less commonly the innermost and, more rarely, central feathers, but tail-feathers usually not renewed. Pre-nuptial winter moult (mostly Dec–Feb), in ad complete in about 1/3 or less, but some have a partial moult (showing mix of new and old primaries) and the remainder return in spring with all or nearly all flight-feathers somewhat abraded, these apparently not renewed since previous late summer. 1stYs nearly always moult completely in winter, or rarely leave several, usually inner, secondaries or odd primaries unmoulted. – **SPRING Ad** Differs from 1stS only if winter moult was partial (rather frequent) and only when unmoulted remiges or tail-feathers can be judged to be post-nuptial (depends on feather wear, colour and pattern). Plumage in general rather fresh. **Ad ♂** With wear becomes purer bluish slate-grey above, with darker, more solid salmon-pink throat and chest. Iris dull orange-brown or ochre. **Ad ♀** Head wears slightly greyer, and throat often becomes tinged pale pinkish, with pale submoustachial stripe quite pronounced; some older ♀♀ develop deeper or more extensive pink throat (even upper breast) but still duller and more blotchy than ♂, and rest of plumage as normal ♀. Iris more variable than ♂, often as dark olive-grey as 1stS or nearly so, thus only ♀ with deep orange iris reliably aged as ad using this feature. **1stS** Less advanced or neatly coloured, with upperparts tinged brownish, and any pink below less intense and more restricted. If any juv flight-feathers or primary-coverts are retained, then reliably aged, otherwise as a rule indistinguishable from ad. Iris dark grey-brown to dull orange-brown (never deep orange as some ad ♂♂); iris of ♀ often grey-brown, only rarely orange-brown, but iris of ♂ often more advanced, dull orange or orange-brown, thus approaching ad. – **AUTUMN Ad** Both sexes aged as ad, with practice, either by lack of summer moult of flight-feathers, these being more worn than comparatively fresh ones of juv/1stW, or by partial and suspended moult of flight-feathers before migration. Quite a few moult completely and can then usually be aged by ad-type outer tail-feathers, with pure white edge on r6 and pure white square tips on r5 or rr4–5. Those that moult partly usually retain some outer primaries, primary-coverts, secondaries and tail-feathers, making ageing easy due to worn state of these feathers. Ad usually has warm greyish-brown (or orange-brown) iris and dull orange orbital ring (some 2ndCY ♀♀ very similar to 1stW). **Ad ♂** As spring but grey of head/upperparts less pure and underparts variable, most have pale pink with faint buff tinge on chin/throat and much of breast and flanks, but extensively and finely tipped whitish. **Ad ♀** Plumage much as in Western Subalpine Warbler. **1stW** Sexes similar, much as ad ♀ but usually more uniform sandy-brown above with paler underparts. For a general description see Western Subalpine Warbler. Aged by juv primary-coverts and remiges, and some outer greater coverts (softer and slightly less fresh-looking or less glossed, being mainly centred dark brownish-grey, with warm brown fringes). Pale primary tips more abraded. Moulted ad-like greater coverts have darker greyish centres with brownish-grey fringes, forming moult limit, clearer in ♂; similar moult limits discernible between tertials if some replaced. Juv tail-feathers slightly paler and browner with ill-defined pale brownish-cream or greyish-white fringes and tips to outermost feathers. Iris dark greyish or brownish-grey, and orbital ring brownish-yellow. **Juv** Largely as ♀ but upperparts uniform greyish-buff and underparts less pure white, tinged buffish-cream, especially on breast and throat.

BIOMETRICS No sexual size difference. **L** 11.5–13 cm; **W** 58–64 mm (*n* 80.0 *m* 60.9); **T** 49–57 mm (*n* 80, *m* 52.2);

1stS ♀, Italy, May: unlike ad, note combination of brownish upperparts, darker and more olive-brown iris, not so pure white eye-ring, and seemingly having undergone complete pre-nuptial moult (in late autumn/winter, which is more frequent in young). (D. Occhiato)

Ad, presumed Moltoni's Warbler, probably ♀, Italy, Sep: a certain number of ad Moltoni's moult completely in autumn, as this bird has. Note pale buffish-pink breast-sides and flanks, but as all ♀♀ in the complex can appear like this when fresh, the specific identity is questionable. Rather clean white throat and breast, rather subdued red orbital ring and thin and broken white eye-ring, best fit ♀, as does the dull iris. (D. Occhiato)

1stW ♂, presumed Moltoni's Warbler, Italy, Sep: young ♂♂ in autumn are generally very similar to ♀♀, but if first aged correctly (by dark olive iris, outer greater coverts juv) they can often reliably be sexed, using the more greyish-tinged upperparts, whiter lores, reddish orbital ring and more pinkish-buff body-sides. Besides some pinkish elements below, in general young of the Subalpine Warbler complex cannot be identified to species, unless trapped, when biometrics and tail pattern come into play, or if diagnostic contact/alarm call is heard. (D. Occhiato)

T/W m 85.8; **B** 11.0–13.5 mm (n 79, m 12.1); **BD** 2.5–3.4 mm (n 68, m 3.0); **Ts** 17.5–19.9 mm (n 76, m 18.6). **Wing formula: p1** < pc 0–2.5 mm, or to 2 mm >, < p2 30–39 mm; **p2** < wt 0.5–4 mm, =5/6 or 6 (75%), =4/5 or 5 (23%) or =4 (2%); **pp3–4** about equal and longest (very rarely pp3–5); **p5** < wt (0) 0.5–3 mm; **p6** < wt 2–6 mm; **p7** < wt 4–9 mm; **p8** < wt 7–9.5 mm; **p10** < wt 9–14 mm; **s1** < wt 10–16 mm. Emarg. pp3–5 (in ad sometimes faint also on p6). **White tip r5** (inner web) 0–9 mm (n 70, m 3.1), square-shaped or obtusely wedge-shaped, exceptionally a short but more pointed wedge. In juv/1stW white tip absent, or restricted to very thin greyish-white tip.

GEOGRAPHICAL VARIATION & RANGE Monotypic. – Mallorca and Cabrera in Balearics; Corsica & Sardinia, E Liguria, part of Piedmont, Emilia-Romagna, Tuscany, Lombardy; winter range is in W Sahel, documented from NW Senegal, Mauritania, E Burkina Faso, N Benin and N Nigeria (B. Piot, in litt., Dowsett, in litt.) extending at least to N Cameroon (specimen in MNHN) and north to Ahaggar Mts, S Algeria (specimen in ZFKB); a winter record from Tunis (specimen in ZMA/NL) might indicate that some winter north of the Sahara. (Syn. *moltonii*.)

TAXONOMIC NOTES See Western Subalpine Warbler for general comments on the taxonomy of the Subalpine Warbler superspecies. See also Svensson (2013a,b). – Temminck's original description of *subalpina* from 1820 based on one ♀ 'in spring plumage' obtained 'near Turin' (the only specimen known at the time, hence the holotype) as a bird with 'a beautiful vinaceous colour below'. We (like Baccetti *et al.* 2007) regard the name *subalpina* valid and place *moltonii* Orlando, 1937, as a junior synonym. Temminck's *subalpina* type is furthermore depicted on a colour plate showing the typical small square white tips to the outer tail-feathers. Although not perfect in all details, this original description is not inferior to numerous others of its time.

REFERENCES BACCETTI, N., MASSA, B. & VIOLANI, C. (2007) *BBOC*, 127: 107–110. – BRAMBILLA, M. *et al.* (2006) *Ibis*, 148: 568–571. – BRAMBILLA, M. *et al.* (2008a) *Mol. Phyl. & Evol.*, 48: 461–472. – BRAMBILLA, M. *et al.* (2008b) *J. Evol. Biol.*, 21: 651–657. – BRAMBILLA, M. *et al.* (2008c) *Acta Orn.*, 43: 217–220. – BRAMBILLA, M. *et al.* (2010) *J. of Orn.*, 151: 309–315. – DE SMET, G. & GOOSSENS, T. (2002) *DB*, 24: 80–88. – FESTARI, I., JANNI, O. & RUBOLINI, D. (2002) *DB*, 24: 88–90. – GARGALLO, G. (1994) *BBOC*, 114: 31–36. – GOLLEY, M. (2007) *BW*, 20: 459–463. – ORLANDO, C. (1937) *Riv. Ital. Orn.*, 7: 213. – ORLANDO, C. (1939) *Riv. Ital. Orn.*, 9: 147–177. – SAVI, P. (1828) *Ornitologia Toscana*. Vol. 2. Nistri, Pisa. – SVENSSON, L. (2013a) *BBOC*, 133: 240–248. – SVENSSON, L. (2013b) *BB*, 106: 651–668. – TRISCHITTA, A. (1922) *Atti Soc. It. Sci. Nat. Milano*, 61: 121–131.

1stW ♀, presumed Moltoni's Warbler, Italy, Sep: unlike 1stW ♂, upperparts are duller and browner (practically no grey), while lores are less pure white, orbital ring dull brown, and underparts lack any obvious pinkish (only breast-sides and flanks are orange-buff). Unless the bird calls, in such plumage specific identification is impossible. (D. Occhiato)

1stW ♀, presumed Moltoni's Warbler, Corsica, Aug: at this season all three species of the Subalpine Warbler complex can be encountered in the C Mediterranean region, and especially young birds look extremely similar, differing subtly in size (eastern birds slightly larger than western ones). Only calls (a Wren-like churr indicates Moltoni's) and tail pattern can help—the very few juv that replace the tail to ad-like feathers with narrow white wedges on the penultimate rectrices can be identified as Eastern Subalpine Warbler. Usually, silent birds must be left unidentified. (F. Jiguet)

MÉNÉTRIES'S WARBLER
Sylvia mystacea Ménétries, 1832

Fr. – Fauvette de Ménétries
Ger. – Tamariskengrasmücke
Sp. – Curruca de Ménétries
Swe. – Östlig sammetshätta

Birders familiar with old field guide images of a pink-flushed, black-headed and rather striking *Sylvia* must frequently have been disappointed by their first experience of this species, as the form most likely to be encountered by W European birders, *rubescens*, is not so dark-headed and has very little pink on the underparts, even in spring plumage.

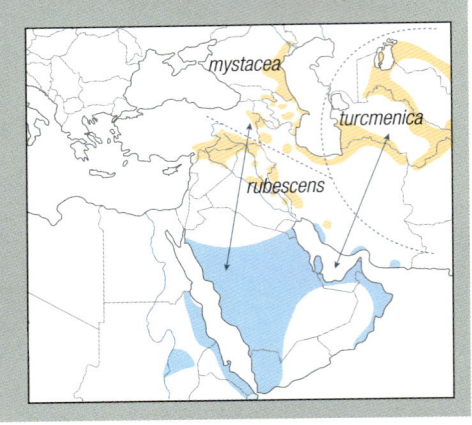

S. m. rubescens, ad ♂, Turkey, May: race based on locality, but note pale pinkish cast to whitish underparts (here at extreme end of spectrum and further enhanced by puffed-out throat). This bird has rather restricted pinkish base to lower mandible, but ill-defined greyish outer webs to tertials are typical. Aged by being not heavily worn, with bluish-grey fringes to ad primary-coverts and remiges, and deeper/purer red-brown iris. (D. Occhiato)

IDENTIFICATION A medium-sized *Sylvia* with broadly similar plumage to Sardinian Warbler. ♂ has *blackish forecrown, lores and ear-coverts*, becoming progressively *paler* (more dark grey than blackish) *on rear crown*, merging with *medium grey rest of upperparts*. Almost *blackish uppertail*, which when fanned shows *extensive white outer corners and edges*. Darkish *wings with broad, ill-defined tertial edgings*. Underparts have very *pale pinkish cast*, mainly on lower throat and upper breast (though amount of pink varies, being more saturated in northern and eastern races, in which whitish submoustachial stripe is better developed). Eye-ring usually pinkish-buff or whitish, and orange-brown iris and reddish orbital ring quite pronounced. ♀ *nondescript, sandy-grey above*, including head (which can be dark grey in older ♀♀), with whitish lores. Eye-ring whitish or has some pale pinkish-buff feathers (orbital ring compared to ♂ paler, and iris often duller). *Tertials diffusely patterned*, and underparts *white, partly tinged pale buff, mainly on body-sides*, while *very dark uppertail is most striking character*. Rather broad bill with pale pinkish-horn base to lower mandible. Legs dull orange-brown. Behaviour and habits much like Sardinian Warbler, but *raises, waves and often fans tail*. Has *chattering or rattling calls*, like Rüppell's Warbler or miniature Barred Warbler (though compared to latter has nasal tone and is feebler, higher-pitched and subtly faster).

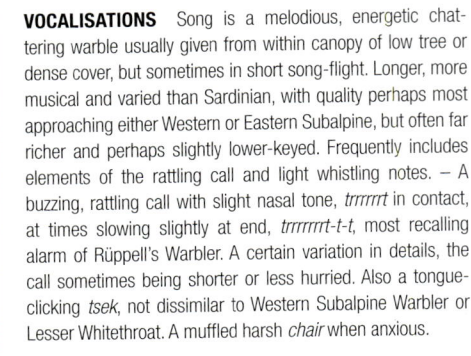

VOCALISATIONS Song is a melodious, energetic chattering warble usually given from within canopy of low tree or dense cover, but sometimes in short song-flight. Longer, more musical and varied than Sardinian, with quality perhaps most approaching either Western or Eastern Subalpine, but often far richer and perhaps slightly lower-keyed. Frequently includes elements of the rattling call and light whistling notes. – A buzzing, rattling call with slight nasal tone, *trrrrrrt* in contact, at times slowing slightly at end, *trrrrrrrt-t-t*, most recalling alarm of Rüppell's Warbler. A certain variation in details, the call sometimes being shorter or less hurried. Also a tongue-clicking *tsek*, not dissimilar to Western Subalpine Warbler or Lesser Whitethroat. A muffled harsh *chair* when anxious.

SIMILAR SPECIES On breeding grounds unlikely to be confused, as greatest potential confusion risks are any within Subalpine Warbler complex or Sardinian Warbler, but little overlap with these, except on migration with Eastern Subalpine and Sardinian race *momus*. Tail of Ménétries's is constantly raised, vigorously waved sideways and up and down (in Eastern Subalpine movements largely absent and in Sardinian much less free or regular), and all three differ in calls. – From ♀-like *Eastern Subalpine*: Ménétries's has the tail diagnostically much darker (in Eastern Subalpine browner, almost concolorous with upperparts), with more contrasting white edges and extensive square tips (particularly noticeable when spread or seen from below; Eastern Subalpine has narrow white wedge on p5, if any), and Ménétries's also lacks sharply defined and narrow pale fringes to the dark-centred tertials (of Eastern Subalpine), although differences may be obscured by wear and moult. Ménétries's has a rather broad-based bill with the pale area on the lower mandible horn-pink (mainly straw or horn with limited pinkish tinge in Eastern Subalpine). – The palest race of Ménétries's, *rubescens*, and the Levant race, *momus*, of Sardinian Warbler are separable, as Ménétries's usually has an ill-defined dark grey rear crown, whereas nearly all ♂ Sardinian have the dark crown more intensely black and more sharply demarcated from nape. Many Ménétries's have pinkish-red on the throat and breast, but others are whitish-grey and similar to *momus*; however, in close views, Ménétries's almost invariably has broad, ill-defined greyish tertial fringes (whiter and narrower, contrasting with blackish centres, in Sardinian). Most Ménétries's have whitish-sandy or very pale buff flanks (in Sardinian, body-sides more intensely grey, enhancing white throat). Ménétries's usually has a broader bill-base, which is typically pinkish-brown, whereas in Sardinian the bill is slenderer and the pale area

♂ Ménétries's Warbler, *S. m. rubescens* (left: United Arab Emirates, Feb) versus Sardinian Warbler, *S. m. momus* (right: Israel, Dec): these two races are the palest taxa of their species, very similar, and their breeding ranges are not far apart. Remember that although ♂ Ménétries's often shows some (diagnostic) pinkish-red on throat/breast, others virtually lack any, being mostly whitish-grey and similar to Sardinian; also, although ♂ Ménétries's tends to show a more ill-defined blackish crown (as here), some can have an almost Sardinian-like intense black crown, well demarcated from grey upperparts. Therefore, confirm Ménétries's broad, ill-defined greyish tertial fringes (whiter and narrower, contrasting with blackish centres, in Sardinian). Age of Ménétries's unsure; Sardinian is ad with evenly-fresh feathers and deep red-brown iris. (Left: D. Clark; right: L. Kislev)

SYLVIA WARBLERS

♂, presumed 1stS, ssp. unknown, United Arab Emirates, Feb: safe separation from Sardinian Warbler relies on pinkish cast to underparts, broad, ill-defined tertial edges and pale area on lower mandible being pinkish horn-coloured. Following pre-nuptial moult both ad and 1stY show moult limits in wing (note renewed inner greater coverts and tertials, contrasting with worn primaries and primary-coverts) and ageing often difficult, but strong wear to retained feather tracts in Feb and completely renewed tail better fit 1stY. Some ♂♂ are intermediate in amount of pinkish below and impossible in winter to ascribe to race. (M. Barth)

♂, presumed 1stS, ssp. unknown, United Arab Emirates, Apr: typical tail-cocking posture. Also typical is for young ♂♂ to have very obscure head pattern, and only limited buff-cream or pinkish wash below. However, variation in all races prevents racial labelling. High degree of moult contrast and feather wear in wing, and olive-brown iris best fit a young bird. (H. Roberts)

is almost invariably grey. ♀-like plumages of Ménétries's are separable from ♀ Sardinian based on their (i) paler upperparts, contrasting strongly with the dark uppertail (upperparts browner in Sardinian, contrasting only slightly with tail); (ii) largely white underparts with limited buffish-yellow on sides (extensive warm buff-brown body-sides and contrasting white throat in Sardinian); (iii) much less obvious greyish head in Ménétries's; (iv) grey-brown tertial centres and ill-defined sandy-grey fringes (centres much darker and fringes well defined, narrower and buff-brown to whitish-cream in most Sardinian); (v) more contrasting dark alula (in Sardinian, the similarly dark alula almost matches surrounding tracts). – ♀-like *Rüppell's Warbler* is larger, more robust and has a proportionately longer, heavier bill, longer primary projection and better-patterned tertials and greater coverts (with well-defined whitish fringes to tertials). – ♀-like *Cyprus Warbler* is further characterised by the warmer olive-brown body-sides, vent and undertail-coverts (with diagnostic whitish-cream fringes/tips). Ménétries's is also readily separated from both these last-mentioned species by its nervous tail cocking and, in relation to Cyprus, very different call.

AGEING & SEXING Ageing requires a close check of moult and feather wear in wing, and sometimes iris coloration

1stS ♂, ssp. unknown, Kuwait, Mar: typical terrestrial behaviour and raised tail. Ragged rear crown delimitation of blackish cap is often a feature of young ♂♂ at this season. The apparently whitish throat, pale cream-buff flanks and pale grey upperparts best match *S. m. rubescens*, but given substantial individual variation it is impossible to eliminate other taxa. Striking moult limits in wing, with very worn and bleached primary-coverts and remiges, and recently moulted tertials. (M. Pope)

S. m. turcmenica, ♂, possibly ad, Uzbekistan, May: rather similar to *S. m. mystacea* in being extensively saturated pink below, but taxon based on locality. Beware potential confusion with Subalpine Warblers, especially Moltoni's Warbler: note Ménétries's darker grey forecrown, more contrastingly dark uppertail (not visible here) and square white tip to r6 (rather than wedge-shaped), and in the field note wavy tail-cocking and voice. Probably ad given brightness of underparts. (G. Baker)

S. m. mystacea, variation in ♂ plumage (NE Turkey, May): the subspecific/sexual characters of this race include duller grey upperparts, throat/upper breast almost entirely pink, with well-developed whitish submoustachial stripe, and paler pink breast. Nevertheless, although both birds were photographed at same site, note differences in amount of pink below due to individual variation and age differences: the left-hand bird is ad (evenly and less worn with purer orange-brown iris, but note misleading winter renewal of inner greater coverts and tertials), whereas right-hand bird is 1stS (rather strong moult limits in wing, with contrasting juv remiges and primary-coverts, and iris colour). The right-hand bird is paler both above and below, and could not be safely differentiated from *S. m. rubescens* if seen away from breeding grounds. (D. Occhiato)

S. m. mystacea/turcmenica, ♂, presumed 1stS, Kuwait, Mar: extensive and rather deep pinkish underparts suggest one of the eastern races. However, their separation during migration is extremely difficult, although the underparts here approach *turcmenica*. Ageing is also unsure without handling: following winter moult, both age classes can show strong moult contrast in wing, although the dull iris colour is better match for 1stY at this season. (A. Al-Sirhan)

S. m. mystacea/turcmenica, 1stS ♂, Kuwait, Mar: faced with a ♂ outside the breeding range, correct ageing can aid subspecific designation. Here, combination of young age (strong moult contrast in wing, with juv primaries and primary-coverts, and duller iris) with deep pink throat/upper breast suggest one of the eastern races. Furthermore, note similarity to *albistriata* Eastern Subalpine Warbler; except darker lores and forecrown, all other traits match both, though the whitish chin/upper throat would probably show some brick-red bases in Subalpine. Stronger contrast with black uppertail, tail pattern and tail-cocking, plus vocal differences will be the chief features to use in such cases (C. Tiller)

is useful too. Tail pattern, however, hardly differs between juv and ad. Sexes differ in ad and in spring (sexual dimorphism attained following first pre-nuptial moult), with seasonal variation evident in ad ♂. – **Moults.** Complete post-nuptial and partial post-juv moult in late summer (mostly Jul–Sep). Post-juv moult involves all head- and body-feathers, lesser and median coverts, usually all or most greater coverts, and perhaps some tertials (those moulting more extensively can replace some primaries and primary-coverts, some secondaries and entire tail). Pre-nuptial winter moult (ad and 1stY) partial (mostly Dec–Mar). Winter moult includes some head- and body-feathers, lesser and median coverts, also variable number of greater coverts, tertials, tail-feathers and inner secondaries, rarely some primaries and primary-coverts (extent greater in 1stY). – **SPRING Ad** Generally plumage less 'scruffy' than 1stS; primaries and primary-coverts much less heavily worn, and moult limits following pre-nuptial moult less obvious than in 1stS (due to relatively minor differences between post-nuptial and pre-nuptial feathers). Iris brighter (especially in ♂) than autumn. **Ad ♂** Blackish forecrown more solid, upperparts purer grey and salmon-pink throat more intense (except in *rubescens*). **Ad ♀** Variable number of blackish-grey crown-feather centres partly exposed. In both sexes, remiges browner and whitish primary tips wear off. **1stS** As ad of each sex, but juv primaries and primary-coverts more heavily worn and browner, with buffish-brown

1stY ♂, Oman, Feb: a least-advanced young ♂, with almost ♀-like plumage, but has characteristic darker forepart of head (exposed bases to forecrown and loral-feathers), and greyish upperparts. Note typical diffuse ash-grey fringe of freshly moulted tertial. Aged by juv remiges and primary-coverts, and dull eye. (B. Herren)

Ad, presumed ♂, Oman, Oct: aged by being evenly fresh, with very fresh primaries and primary-coverts, the former still with pale tips, and rather bright orange-brown iris. Greyer head, with darker lores and some dark centres on head-sides, even in fresh plumage, suggest ♂. Resemblance to ♀ Sardinian Warbler afforded by grey head and white-throated appearance, but focus on diffuse pale tertial fringes (due to the angle, the edges look confusingly better-defined) and pinkish-horn base to the lower mandible. (H. & J. Eriksen)

1stW ♀, United Arab Emirates, Sep: ♀♀ in fresh plumage are often very pale and greyish, with the most striking feature being the contrastingly blackish uppertail. Note frequent habit of cocking tail. Juv primary-coverts, primaries (tips already slightly worn) and outer greater coverts, plus olive iris, age this bird. (H. Roberts)

♀, presumed 1stS, United Arab Emirates, Mar: typically rather uniform pale greyish-brown above, with rather diffuse pale tertial fringes (important), and characteristically contrasting blackish tail; also note slightly pinkish base to lower mandible. Rather heavily worn primary tips and primary-coverts, and rather dull iris colour best match a 1stY, but are not conclusive for ageing. (H. Roberts)

Autumn/winter ♀♀ Ménétries's Warbler (left: United Arab Emirates, Nov) versus Eastern Subalpine (ssp. *albistriata*, centre: Greece, Sep) and Sardinian Warblers (ssp. *momus*, right: Israel, Dec): unlike Subalpine, Ménétries's has a contrastingly blacker uppertail (Subalpine browner, almost concolorous with upperparts), with more diffuse pale tertial fringes (Subalpine usually sharply-defined, narrow pale fringes). Ménétries's appears to have broader-based bill, which often has slight pinkish base to lower mandible (mainly yellowish-grey with little pink in Subalpine). Ménétries's also cocks its tail vigorously and calls differently. ♀ Ménétries's is closer to Sardinian (identical tail patterns), but has decidedly paler upperparts, forming stronger contrast with dark uppertail (upperparts in Sardinian browner and darker), and underparts largely cream-white with some buffish-yellow on sides (in Sardinian extensively sullied buff-brown with clearer white throat). Compared to Sardinian, head of Ménétries's much less obviously greyish and tertial fringes more ill-defined (most Sardinian have narrower, sharply-defined fringes). Sardinian has usually pale bluish-grey tinge to base of lower mandible (no pink). Ménétries's and Subalpine Warblers are 1stW by dull plumage, juv primary-coverts and olive-coloured iris; age of Sardinian unsure as some young acquire almost ad-like plumage and iris. All are 1stW by dull plumage, juv primary-coverts and olive-coloured iris. (Left: S. Jahan; centre: P. Petrou; right: D. Kalay)

♀, Turkey, Jun: already in May/Jun this highly terrestrial *Sylvia* can be very worn, with the only striking (almost diagnostic) feature being the often-cocked, contrastingly blackish tail. As both age classes can show strong moult limits in wing following winter moult, and both can be strongly worn and bleached at this season, age is unsure. (V. Legrand)

fringes. Iris colour perhaps less advanced. Overall plumage state less 'tidy', and some young ♂♂ have patchy hood. — AUTUMN Ad ♂ Compared to spring, grey of upperparts tinged pale sandy-brown, and blackish crown duller and less clearly demarcated; pale salmon-pink throat and breast generally less intense, being tipped whitish (reduced or lacking in *rubescens*). Ad ♀ As spring but in general suffused warmer/buffier (less greyish above) with few or no exposed dark feather-bases to crown. Both sexes aged by evenly very fresh wing- and tail-feathers; exposed primary tips silvery-white. Iris has at least some reddish-orange or yellowish tones. 1stW Sexes almost alike, largely as ad ♀. Upperparts uniform with poorly patterned head. Some 1stW ♂♂, however, have greyer head with reddish orbital ring (greyish in 1stW ♀). Best aged by juv primary-coverts, remiges and perhaps some outer greater coverts (which are softer and slightly less fresh or less glossed, with dark brownish-grey centres and rufous-brown fringes), and pale primary tips virtually lacking; inner ad-like greater coverts have darker centres with brownish-grey fringes, creating moult limit, obvious in ♂, and also discernible in tertials once some replaced. Juv tail-feathers more worn. Iris uniform dark olive. **Juv** More uniform greyish or yellowish-buff above than ♀, including head (except pale lores), and whitish-cream below.

BIOMETRICS (*rubescens*) Sexes nearly of same size. **L** 12.5–13 cm; **W** ♂ 57.5–63.5 mm (*n* 15, *m* 60.2), ♀ 56–62 mm (*n* 13, *m* 59.5); **T** ♂ 51–58.5 mm (*n* 15, *m* 54.4), ♀ 50–59 mm (*n* 13, *m* 54.2); **T/W** *m* 90.7; **B** 11.5–13.9 mm (*n* 28, *m* 12.6); **BD** 2.9–3.3 mm (*n* 22, *m* 3.2); **Ts** 17.5–19.7 mm (*n* 26, *m* 18.7). **Wing formula: p1** > pc 0–5 mm, < p2 25.5–31.5 mm; **p2** < wt 2–5.5 mm, =6 or 6/7 (57%), =7 (24%) or =5 or 5/6 (19%); **pp3–4** (5) about equal and longest; **p5** < wt 0–4 mm; **p6** < wt 1–6 mm; **p7** < wt 3.5–7 mm; **p8** < wt 6–8.5 mm; **p10** < wt 8–12 mm; **s1** < wt 9–14 mm. Emarg. pp3–5 (6).

GEOGRAPHICAL VARIATION & RANGE Three subspecies generally recognised, two of which differ only subtly. Exact ranges difficult to define, and some intergradation where they meet. Scant material pertaining to definite breeders (in particular ♀♀) available in museums, limiting quality of evaluation of geographical variation. All three races move south and south-west in winter, but details concerning where each race winters are poorly known. The combined winter range covers the Red Sea and E Egypt, E Sudan, coastal Eritrea, NW Somalia, Yemen, much of Saudi Arabia, Oman, Persian Gulf and S Iran.

S. m. rubescens Blanford, 1874 (SE Turkey, N Syria, Iraq, W Iran). Differs from the following two races in being paler and more uniform. ♂ is pale grey above with light sandy suffusion when fresh, and white below, with pale pink often restricted to chest and flanks, leaving clear white throat, while crown averages blacker; a few ♂♂ lack almost any pink below in spring (presumably 2ndCY). ♀ drab grey-brown with pale sandy tinge above and mostly dirty white or cream underparts (body-sides paler).

S. m. mystacea Ménétries, 1832 (NE Turkey, Caucasus, Transcaucasia, Azerbaijan, NW Iran). Differs from *rubescens* in ♂ plumage by darker grey upperparts (slate-grey, darkest of all subspecies), suffused pale brown when fresh. Lower throat deep salmon-pink, sometimes contrasting with paler pink upper breast and pinkish-grey flanks, sometimes more uniform with breast. Crown dull black extensively tipped dark grey at rear. ♀ scarcely differentiated from *rubescens*, but usually slightly darker above and warmer sandy-buff below. On average slightly shorter bill and proportionately marginally longer tail. Wingtip slightly more rounded. Frequently difficult to separate from *turcmenica*, the latter being only doubtfully separable. **W** ♂ 56.5–62 mm (*n* 14, *m* 59.8), ♀ 57.5–58.5 mm (*n* 4, *m* 57.9); **T** ♂ 52–58 mm (*n* 13, *m* 54.8), ♀ 52.5–56 mm (*n* 4, *m* 54.1); **T/W** *m* 91.9; **B** 11.5–12.7 mm (*n* 17, *m* 12.2); **BD** 2.9–3.4 mm (*n* 16, *m* 3.1); **Ts** 17.5–19.6 mm (*n* 18, *m* 18.6). **Wing formula: pp3–5** about equal and longest; **p6** < wt 1–3 mm; **p7** < wt 4–5.5 mm; **p8** < wt 5.5–7 mm; **p10** < wt 7.5–10.5 mm; **s1** < wt 8.5–12.5 mm.

○ **S. m. turcmenica** Zarudny & Bilkevich, 1918 (Turkmenistan, N & E Iran east from C Elburz, Transcaspia, W Kazakhstan, Aral Sea region). Closely resembles *mystacea*, with similarly rounded wing and subtly longer tail, also in that ♂ has pink throat and often upper breast (though on average slightly paler pink), but is less dark grey above, more like *rubescens* in this respect. Upperparts tinged sandy-brown when fresh (as in all subspecies). Note that chin and uppermost throat in ♂ often are whitish. ♀ indistinguishable from or intermediate between other races. Slightly larger than the other two races, but much overlap. A subtle and somewhat variable race, and wear and individual variation often make identification difficult away from breeding grounds. **W** ♂ 57–64 mm (*n* 21, *m* 61.2), ♀ 58–63 mm (*n* 3, *m* 60.7); **T** ♂ 54–59 mm (*n* 21, *m* 56.8), ♀ 53–59 mm (*n* 3, *m* 56.3); **T/W** *m* 92.9; **B** 11.5–13.2 mm (*n* 24, *m* 12.4); **BD** 3.0–3.8 mm (*n* 24, *m* 3.2); **Ts** 17.5–19.4 mm (*n* 24, *m* 18.6). **Wing formula: pp3–5** about equal and longest; **p6** < wt 0.5–3 mm; **p10** < wt 7.5–10.5 mm; **s1** < wt 10–13.5 mm.

S. m. turcmenica, juv, Iran, Jun: soft, fluffy body-feathers, but otherwise resembles ♀, albeit tinged greyish-buff above, head marginally greyer (suggesting ♂) and cream-white below. Like ♀, blackish uppertail is most striking feature, along with buffish-brown wing-fringes. Subspecies based on locality and date. (E. Winkel)

SARDINIAN WARBLER
Sylvia melanocephala (J. F. Gmelin, 1789)

Fr. – Fauvette mélanocéphale
Ger. – Samtkopf-Grasmücke
Sp. – Curruca cabecinegra; Swe. – Sammetshätta

Widespread and generally the most common *Sylvia* species in dense scrub, maquis and related wooded habitats of the Mediterranean region. Most Sardinian Warblers are resident, but some populations move south or to lower altitudes in winter, some even reaching sub-Saharan Africa, where they are more catholic in their habitat usage. Often noticed as it quickly flutters off low in jerky flight and disappears into dense cover, but soon reveals its identity by its characteristic loud, rattling call like from a wooden rattle.

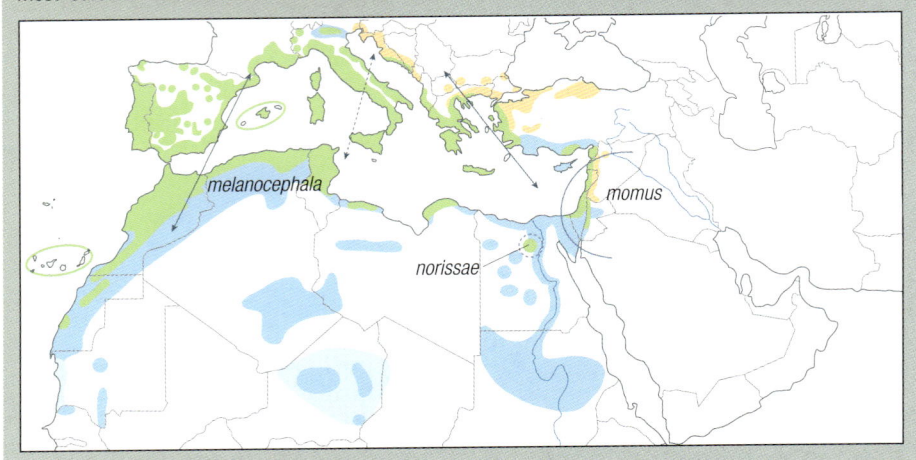

IDENTIFICATION A rather compact and large-headed *Sylvia*. Size varies from quite small (e.g. *momus*) to medium-sized (*melanocephala*), but there is only limited variation in plumage. Short-winged and has a markedly long and full, rounded tail with white tail-sides and broad white tips to most feathers except central pair. ♂ has black hood extending to ear-coverts, *sharply demarcated from striking white throat* and dark to pale *grey upperparts, blackish-centred tertials with well-defined pale fringes*, whitish belly and *greyish wash to flanks and breast-sides*; prominent *orbital ring red or reddish-orange*, eye-ring mainly buff to pinkish-orange. ♀ has dark grey head with slightly darker ear-coverts, contrasting with almost pure white throat, warm olive-brown to dark earth-brown upperparts, whitish-cream belly with buffish-grey to olive-brown flanks, and duller orbital ring and eye-ring. Iris reddish-orange in adults, duller and a little darker in young. Bill quite strong and moderately long, with predominantly *bluish-grey lower mandible* contrasting with blackish culmen and tip. Legs reddish flesh-brown. First-years generally as adults, though white on outer tail-feathers generally more dusky and ill-defined, especially in ♀♀. Typically skulks, with usual views being in whirring flight as bird speeds between patches of cover; rather restless, often cocking tail. Presence usually betrayed by loud, obvious calls.

VOCALISATIONS The song is arguably the epitome of the *Sylvia* chatter, not least because the species is common and widespread; a rather explosive and fast, brief warble, consisting of harsh rattling strophes delivered at high speed, with some whistled notes and traces of the characteristic rattling calls often interwoven. – Most commonly used contact call a hard monosyllabic clicking *tseck*, but more characteristic and when more alarmed is a loud rattle often compared to the sound of traditional wooden rattle used for hare-hunts, *trr-trr-trr-trr-trr*, or, less often, with disyllabic repeated elements, *te-tra, te-tra, te-tra*.

SIMILAR SPECIES In many areas learning the plumages (and vocalisations) of this species will prove extremely helpful in recognising other, apparently similar congeners. It is usually ♀-type plumages that cause the most significant confusion risks, but even ♂♂ require reasonable views to be safely recognised. The potential pitfalls offered by Ménétries's Warbler (race *rubescens*), but to a lesser extent also by Cyprus Warbler, Western and Eastern Subalpine Warblers, and Rüppell's Warbler are discussed separately under each of these species.

AGEING & SEXING (*melanocephala*) Ageing requires a close check of moult and feather wear in wing, and iris coloration is also often useful. Tail pattern differs between juv and ad, but some ♂♂ require handling and careful assessment, and some (many?) young replace outer parts or whole tail early and become largely inseparable from ad thereafter. Sexes differ, but seasonal variation insignificant. – Moults. Complete post-nuptial moult and generally partial post-juv moult (rarely complete, more likely in ♂♂ than ♀♀), mostly Jul–Sep. Post-juv moult includes all of head and body, lesser, median and greater coverts, tertials and many (at times all) tail-feathers. Pre-nuptial winter moult partial, involving some secondary-coverts, tertials, inner secondaries and tail-feathers, apart from head and body. – **SPRING Ad** Both sexes have orbital ring brighter than in autumn, brick-red or deep reddish-orange, and iris is slightly brighter too, mostly deep reddish-orange or brownish-orange. Beware of winter-moulted tertials, as these create moult patterns similar to post-juv. Best aged by fresher ad-type wing-feathers (especially look for less worn primaries and primary-coverts). **Ad ♂** With wear and bleaching, crown becomes blacker and rest of upperparts purer grey. Grey of body-sides often more sharply demarcated. **Ad ♀** Due to more abraded and exposed blackish feather-bases, head often much darker than duller brown upperparts; also more sharply-demarcated olive-brown body-sides. Black-capped ♀ (rare) separated from ♂ by brownish upperparts and brown tinge to flanks. **1stS** Juv remiges and primary-coverts (if any) more worn, with buffish-brown fringes, and moult limits easily detected in ♂. Iris colour often less advanced. Note that pre-nuptial moult can complicate ageing. – **AUTUMN Ad** Both sexes aged by evenly very fresh wing- and tail-feathers (remiges and primary-coverts glossy greyish-black). Iris

S. m. melanocephala, ad ♂, Italy, May: generally the largest and darkest race, with upperparts and body dark grey; black of head often not sharply delimited at nape. Note sharply-demarcated white throat and striking red orbital ring and orange-brown eye. With wear, whitish tertial fringes are reduced. The broadly-edged grey primary-coverts and remiges, and deep red-brown iris, age this bird as ad. (D. Occhiato)

S. m. melanocephala, ad ♂, Spain, Dec: still in fresh plumage with still prominent whitish tertial-fringes. Due to individual and geographical variation, some ♂♂ of this race are paler grey overall (especially in drier areas), but always have extensive grey body-sides. Combination of evenly-fresh wing-feathers, including grey-edged primary-coverts and remiges age this bird as ad. (M. Varesvuo)

much as spring (cf. 1stW). **Ad ♂** Compared to spring, mantle and scapulars slightly brownish, somewhat less pure grey, and underparts variably suffused grey, with some buff on sides and vent. Remiges mainly fringed grey. White tips to rr5–6 extensive, well defined and 'square' (edge c. 90° to shafts). **Ad ♀** As spring, but most have medium to dark slate-grey heads, slightly tinged olive-brown. Upperparts pale to dark earth-brown tinged olive-buff, and body-sides extensively buffish olive-brown. White tips to rr5–6 much as in ad ♂, only on average slightly less extensive. **1stW** As fresh ad but juv primary-coverts and remiges softer textured and browner, fringed buffish-brown (and primaries lack obvious narrow pale tips of ad). Moult limits (more obvious in ♂) usually present between renewed tertials and unmoulted secondaries and, especially in birds that have moulted more extensively, contrast among primaries or primary-coverts also useful. Some 1stW undergo complete moult and are inseparable from ad by plumage. Iris dark grey or dull olive in nestlings, acquiring ad coloration in 2ndCY, but by late autumn differences from ad often small. Orbital ring dull orange-red. Check for any juv outer tail-feathers and note difference in pattern described under juv. **Juv** Largely as ♀, but ♂ has

S. m. melanocephala, 1stW ♂, Spain, Sep: sex inferred by rather dark cap, greyish upperparts and flanks. Heavily-abraded brown juv primary-coverts and remiges contrast strikingly with renewed other wing-coverts, inner secondaries, alula and tertials. Already in late autumn, many young ♂♂ can acquire almost ad-like iris colour, but still usually slightly duller brown. (H. Shirihai)

S. m. melanocephala, ad ♀, Morocco, Nov: ♀ typically has dark grey head, contrasting with white throat, olive-tinged rich brown upperparts and body-sides. Red orbital ring often less prominent, while eye colour is sometimes as striking as in ad ♂. Evenly-fresh wing and deep red iris suggest an ad. (P. Komi)

S. m. melanocephala, ♀, presumed ad, Spain, Apr: an almost black-headed ♀ (uncommon), still readily sexed by brownish-tinged upperparts and less pure and dark grey body-sides. Brick-red orbital ring often brighter than in autumn. The brownish-orange iris, generally fresh plumage and evenly ad-like wing feathers suggestive of age. (S. Fletcher)

greyer head, while in ♀ this is more concolorous with upperparts (and darker ear-coverts more pronounced). Underparts creamy, with extensive brown wash to chest and flanks. Eye-ring mainly cream-buff and whitish. Orbital ring pale orange-yellow (redder in ♂). Outer juv tail-feathers usually clearly different from ad, having whitish fringes and tips more dusky, diffuse and reduced in extent (some even sullied pale brown), but difference most obvious in ♀♀ as most ♂♂ are closer to ad pattern with purer and more extensive white portions, though generally separable on white tips being more wedge-shaped, not 'square'. (Note: many juv replace some or all rectrices during early post-juv moult, becoming inseparable from ad on this character alone.)

BIOMETRICS (*melanocephala*) **L** 13–14 cm; **W** 56.5–63.5 mm (n 51, m 60.1); **T** ♂ 57–67 mm (n 30, m 60.9), ♀ 54–63 mm (n 20, m 59.1); **T/W** m 99.9; **B** 12.3–15.1 mm (n 42, m 13.7); **BD** 3.2–4.0 mm (n 39, m 3.5); **Ts** 19.0–22.0 mm (n 42, m 20.4). **Wing formula: p1** > pc 1–9 mm, < p2 20–28 mm); **p2** < wt 3–8 mm, =6/7 or 7 (50%), =7/8 or 8 (45%) or =8/9 or 9 (5%); **pp3–5** about equal and longest (p5 sometimes < wt 0.5–1 mm); **p6** < wt 1–3.5 mm; **p7** < wt 3–5.5 mm; **p8** < wt 4.5–7.5 mm; **p10**

S. m. melanocephala, ad ♀, Italy, Dec: some ♀♀ show such striking head pattern even when fresh, despite this being partly obscured by pale tips. Note combination of brownish upperparts, extensive grey-brown body-sides, pure white throat and clear pale fringes to tertials. Evenly fresh ad-like wing-feathers and bright/deep eye colour and developed orbital ring at this season safely age this ♀. (G. Falcone)

< wt 7–10.5 mm; **s1** < wt 8.5–12.5 mm. Emarg. pp3–6 (p6 sometimes a little less prominently).

GEOGRAPHICAL VARIATION & RANGE Moderate variation with a general tendency for birds to become slightly smaller and paler towards the south and east, although much local variation. Three subspecies recognised here but a few more claimed by others. Within *melanocephala* there is a slight clinal variation, birds becoming smaller and darker in the south (darker despite general tendency mentioned above), and Canary Islands population ('*leucogastra*') tends to be small and whitish below with rather grey upperparts, but local deviations from this pattern occur. Much individual variation and overlap in characters, and on the whole and when long series compared, *leucogastra* best included in *melanocephala*. *S. m. momus* in Middle East is small, paler and proportionately shorter-tailed, while fairly distinct *norrisae* formerly bred in Egypt, but is apparently now extinct.

S. m. melanocephala (J. F. Gmelin, 1789) (Canary Is, S Europe including Balearics and Sicily, NW Africa east to Libya, Turkey; largely resident, but birds of Bulgaria, Thrace and N Turkey move south; found in winter in N Egypt and along the Nile, and odd records known from, e.g., Mauritania, Senegal, Niger and Mali). Treated above. Generally larger and darker than *momus*. ♂ has black of head glossy, though often not sharply delimited from mantle. Body-sides extensively washed dark grey (or suffused pinkish olive-buff). ♀ dark earth-brown above, while underparts are extensively washed dark brown-buff on sides. Wing > 56.5 mm, tail usually also (*m* for both *c*. 60 mm), and T/W ratio *c*. 100; p2 mostly =6/8. (Syn. *carmichaellowi*; *leucogastra*; *pasiphae*; *valverdei*.)

S. m. momus (Hemprich & Ehrenberg, 1833) (NW Sinai to Lebanon; possibly Syria; resident or short-distance movements to Sinai and Egypt). On average smaller and paler than *melanocephala*. ♂ has less glossed hood, more sharply delimited from paler lead-grey upperparts. Tertials have paler, more sharply demarcated, even whiter fringes, and underparts are whiter with rather limited pale grey and/or buffish-pink wash on sides. ♀ paler overall than *melanocephala*, upperparts being olive-brown and underparts also paler with pale buff or olive-brown sides. Wing and tail usually < 60 mm (mean for wing *c*. 57, tail *c*. 54), and T/W ratio *c*. 96; p2 mostly =6/7 or =7. **L** 12–13 cm; **W** ♂ 55–60 mm (*n* 15, *m* 57.4), ♀ 54–59 mm (*n* 13, *m* 56.1); **T** ♂ 52–59 mm (*n* 15, *m* 55.2), ♀ 51–56 mm (*n* 13, *m* 53.4); **T/W** *m* 95.8; **B** 11.5–13.7 mm (*n* 27, *m* 12.8); **BD** 3.1–

S. m. melanocephala, 1stW ♀, Spain, Dec: young ♀♀ are similar to fresh ad ♀♀, but (as shown here) there is clear moult limit between heavily-abraded brown juv primary-coverts and remiges and the other wing-coverts, alula and tertials, which were renewed in autumn. Eye colour, however, is already ad-like, and orbital ring well developed. (M. Varesvuo)

S. m. melanocephala, juv, Cyprus, April: soft, fluffy body-feathers, especially at vent, otherwise largely as ♀; the slightly greyer head suggests a juv ♂; pale reddish-brown orbital ring also fits a recently-fledged juv ♂. Underparts typically washed creamy-buff and grey. (A. McArthur)

S. m. melanocephala, ♂ (left, Feb) and ♀ (right: Dec), Canary Is (Fuerteventura): some populations in the Canaries, now referred to ssp. *melanocephala*, require further study, given complex variation, many being subtly smaller and purer grey above with well-demarcated black head, somewhat approaching ♂ *momus*. Left bird shown here appears to be evenly fresh with ad feathers and ad-like iris colour. Right bird has moult limit in alula and rather worn primary tips, and is probably 1stS. (Left: T. Holmgren; right: B. Winkel)

3.5 mm (*n* 26, *m* 3.3); **Ts** 18.1–20.3 mm (*n* 24, *m* 19.2).
Wing formula: p1 > pc 1.5–6 mm, < p2 23–27 mm; **p2** < wt 2–6 mm, =6/7 or 7 (50%), =7/8 or 8 (45%) or =8/9 or 9 (5%); **pp3–5** about equal and longest; **p6** < wt 0.5–3 mm; **p7** < wt 1.5–5 mm.

†? ***S. m. norissae*** Nicoll, 1917 (Lake Birket Qarun, Faiyum, N Egypt; considered resident; apparently now extinct). Slightly paler and browner, less dark and grey compared to *momus*, and often has pale pinkish-buff flush on underparts (less on throat and centre of belly) when fresh. Smaller and paler than *melanocephala* with a much better-defined cap in ♂. **W** ♂ 56–59 mm (*n* 11, *m* 57.9); **T** ♂ 52–60 mm (*n* 11, *m* 56.6); **T/W** *m* 97.9; **B** 12.1–14.0 mm (*n* 12, *m* 13.0); **BD** 3.0–3.7 mm (*n* 12, *m* 3.3); **Ts** 19.0–20.7 mm (*n* 11, *m* 19.9).

TAXONOMIC NOTES Races *momus* and *melanocephala* are allopatric and sufficiently divergent morphologically to perhaps justify species status, but genetic divergence is not as large as in other distinctive cases within *Sylvia*, and vocalisations are similar (Shirihai et al. 2001); therefore, on present evidence, they are best treated as subspecies of the same biological species. – Situation in the Canaries is especially complex, where '*leucogastra*' was synonymised with *melanocephala* by Vaurie (1959) and others, but recognised by Williamson (1968) and Cabot & Urdiales (2005). The latter authors also erected the very weakly defined '*valverdei*' for birds in Western Sahara based on just five specimens; the claimed differences in size seem in this case particularly insignificant based on such a short series (e.g. wing shorter than in Moroccan *melanocephala* by 0.88 mm). Although admitting variation both within *melanocephala* and '*leucogastra*', Cabot & Urdiales (2005) did not propose further splits, and we do the same, only at a different level. All Atlantic & NW African populations might require further study, and pending this we prefer to lump them into *melanocephala*. Most breeders on the Canaries somewhat approach *momus* in that ♂ has paler grey upperparts and well-demarcated black hood, but wing is rather short or intermediate; also, note proportionately longer tail and longer and more pointed bill than in *momus*; birds on Gran Canaria are especially distinctive, while those on Palma are darker and less contrasting above, approaching *melanocephala*. Practical subspecies taxonomy cannot always describe every subtle variation within a mosaic-like pattern!

REFERENCES Cabot, J. & Urdiales, C. 2005. *BBOC*, 125: 230–240.

S. m. momus, ♂, presumed ad, Israel, Jan: this distinctively smaller and paler race is characterised in ♂ plumage by its paler and purer grey upperparts and on average whiter and more sharply-defined tertial fringes, while black cap is more sharply demarcated, too. Usually visibly lighter underparts, with rather limited pale grey on flanks, and white throat is less pronounced than in *melanocephala*. Evenly fresh wing and ad-like bright iris colour infer age (but beware that 1stY ♂ of southern populations can attain this colour already after few months). (H. Shirihai)

S. m. momus, ♂, presumed 1stY, Israel, Feb: sharply-delimited tertial fringes and rather pale grey upperparts. Some ♂♂, especially young, can show less solid black hood even in early spring. Slightly worn remiges edged greyish and ad-like, while tertials apparently renewed in pre-nuptial moult. Iris rather dull brown, and orbital ring also dull, which combination suggests a young ♂ that moulted completely in autumn. (H. Shirihai)

S. m. momus, 1stW ♂, Israel, Nov: differs from ad ♂ by olive-brown iris and dull reddish orbital ring, and by retained juv wing-feathers including primary-coverts and largest alula. Beware that some *momus* ♂♂ have paler nape or even poorly defined dark cap at rear, rendering similarity with Ménétries's Warbler (of race *rubescens*), still told by well-defined whitish tertial fringes, greyer flanks and greyish lower mandible. (D. Kalay)

S. m. momus, 1stW ♀, Israel, Nov: ♀♀ of this race are usually paler than *melanocephala*, being olive-brown above with buff-brown flanks; note rather short primary projection. Otherwise, shares with *melanocephala* combination of well-defined pale tertial fringes, bluish-grey lower mandible base, grey head and brownish mantle, blackish uppertail and warmer flanks with white throat. Dull olive-brown iris and dull reddish orbital ring, as well as retained juv primaries, primary-coverts and alula infer young age (1stW ♀ unlikely to moult completely in autumn). (D. Kalay)

CYPRUS WARBLER
Sylvia melanothorax Tristram, 1872

Fr. – Fauvette de Chypre; Ger. – Schuppengrasmücke
Sp. – Curruca chipriota; Swe. – Cypernsångare

The Cyprus Warbler has the most restricted breeding range among the *Sylvia* species; as its name suggests the species is confined to Cyprus in summer, but winters in the Levant and extreme NE Africa along the Red Sea littoral. Its unusual migration route takes it due south across the Mediterranean in autumn, returning via the Levant in spring. Unlike the other Cypriot breeding endemic, the wheatear, Cyprus Warbler is only a very rare spring overshoot to S Turkey (the wheatear is regular and even common in late March and early April there).

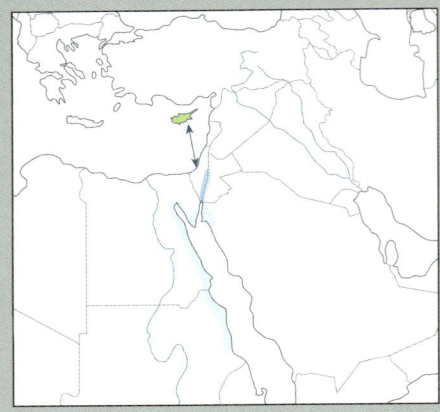

Ad ♂, Cyprus, Apr: surely one of the passerines of the region with the most striking appearance. Superficially recalls Sardinian Warbler, but note diagnostic blackish blotching below, and white submoustachial stripe. Immaculate plumage and less heavily-worn wing, including the primary-coverts and remiges, plus deep red iris age this bird as ad, and probably an older bird. (D. Occhiato)

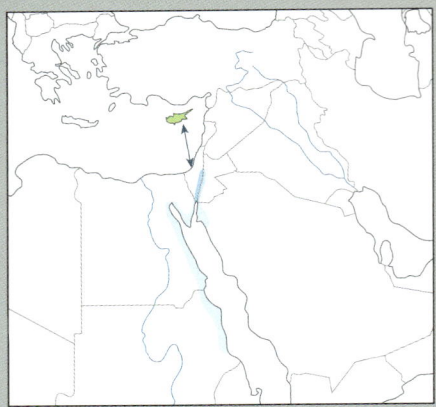

1stS ♂, Israel, Mar: some 1stS ♂♂ are highly distinctive given their heavily-abraded brown juv primary-coverts and remiges, versus the renewed wing-coverts, alula and tertials, forming striking moult limits. Iris like ad in many, or slightly duller, but orbital ring diluted. (M. Høegh Post)

IDENTIFICATION In structure rather like Sardinian Warbler but has slightly shorter tail and longer wings. The striking ♂ has *black hood*, off-white underparts with *striking blackish blotching*, *white submoustachial stripe*, *lead-grey upperparts* and *whitish fringes to tertials* and wing-coverts. ♀-type plumages less distinctive, although some adult ♀♀ may approach ♂ plumage, but generally are *greyish-olive above* with *reduced blotching below*, *dark on head* often *mostly confined to ear-coverts*, and has paler wings with greyer fringes; whitish *submoustachial stripe ill-defined*, and flanks and vent duskier olive-grey. Plumages vary greatly, and young ♀ Cyprus with much-concealed or reduced mottling poses the greatest challenge to identify. Pale fringes to median and greater coverts often appear as a whitish wing-bar. Tail black with white outer edges. Eye-ring consists mainly of blackish-grey and white feathers. Orbital ring reddish, but much duller in young. Iris colour varies from reddish-brown (adults) to olive-brown (juvenile), and legs from orange-flesh to duller brown. Rather secretive, although its frequent calls and, when glimpsed, energetic tail-waving behaviour may betray its presence.

VOCALISATIONS Song a rather chirruping chatter of rattling notes, in part recalling Sardinian Warbler but seems more guttural, tentative, lower-pitched and generally lacks all or most interwoven high-pitched whistling notes. Often commences tentatively with one or two single *chirp* notes before the hurried strophe follows. – In relaxed contact often a rather subdued, repeated clicking *tsek*. A louder, churring, harsh *tchret* repeated in slow staccato (cicada-like) series, *tchret-tchret-tchret-tchret-…*, is also used as contact call (voice differs slightly between sexes) and is quite different to other *Sylvia* calls once learnt. In alarm, a prolonged dry, chattering rattle in two speed variations, more guttural and scratchy in tone than corresponding alarm of Sardinian; often a bouncing *ze-ze-ze-ze*, slower than most other *Sylvia* species. At times emits more complaining, squeaky notes singly or in succession, *pieh-pieh-pieh-…*

SIMILAR SPECIES Distinguishing ♂ Cyprus from ♂ Sardinian Warbler is not always straightforward, especially as both are rather skulking. Unlike Sardinian, ♂ Cyprus has whitish fringes to wing-coverts, tertials and undertail-coverts, and darker grey upperparts. With good views, most (including ♀♀) will not prove problematic, but young ♀ Cyprus lacking markings below can usually be reliably identified only by the scaly undertail-coverts fringes, diffuse pale wing-bars and whiter tertial fringes. Cyprus also lacks Sardinian's warmer, browner upperparts and body-sides, and has a greyer crown and less contrasting white throat. Immature ♀ Cyprus usually lacks reddish or orange eye- and orbital rings of Sardinian, and has pinkish-flesh lower mandible. They also differ in some calls. – ♀-like *Rüppell's Warbler* (chiefly first-year ♀) may recall equivalent age and sex of Cyprus which, however, is smaller, has a shorter bill and shorter wings (six closely-

♀, presumed ad, Cyprus, Mar: greyer olive-brown above with some (highly variable, but diagnostic) neat dark blotches on whitish underparts, some dark on head, and pale fringes to wing-coverts, and especially to tertials. Whitish submoustachial stripe ill-defined. Moderately worn and evenly-feathered wing (except renewed central tertial), and deep red-brown iris suggest an ad. (J. East)

♀, presumed 1stS, Israel, Mar: most young ♀♀ in spring show some (diagnostic) dark mottling below, but otherwise can be tricky to identify. The dull greyish olive-brown upperparts, contrasting with blackish uppertail, and diffuse pale fringes to wing-coverts are also characteristic of this plumage. Impossible to assess moult pattern at this angle, but the reduced spotting below and dull olive-brown iris suggest a 1stY. (K. Malling Olsen)

♀, presumed 1stS, Cyprus, Apr: some ♀♀ (mostly older birds, but also some young) can develop larger spots below in spring, sometimes even more extensive and dense than here. The apparently worn wing best fits a young ♀, as does the duller iris and rather subdued orbital ring. (D. Jirovsky)

♂, ad-like, Cyprus, Dec: fresh plumage, with black spots of underparts largely concealed by broad whitish tips (though note pied undertail-coverts). Obvious whitish fringes to tertials and greater coverts; also diagnostic is (partial) white eye-ring. Fresh wing-feathers, including whitish-edged primary-coverts and remiges, indicate an ad. However, almost complete lack of exposed black blotching below could suggest a 1stY that moulted completely (post-juv moult variable, and generally more extensive in ♂♂, sometimes complete or almost so). (A. McArthur)

♂, ad-like, Israel, Feb: in late winter (when species is already migrating north), the underparts of most ad ♂♂ are still less prominently spotted, spots being concealed by broad whitish tips. Diagnostic distinct whitish fringes to wing. Such an evenly fresh bird could be an ad or 1stY that moulted completely. (H. Shirihai)

spaced primary tips as opposed to seven widely-spaced usually found in Rüppell's), and proportionately slightly longer tail. Cyprus also has warmer underparts with diagnostically scaled undertail-coverts (the latter never sufficiently prominent in ♀ Rüppell's to create confusion). Also, darker upperparts and usually less striking wing-bar. – ♀-like *Eastern Subalpine* and *Ménétries's Warblers* usually do not present significant confusion risks. Both lack the undertail-coverts pattern of Cyprus, and are paler below and on average subtly greyer (or buffish sandy-grey) above. Eastern Subalpine has a browner tail barely contrasting with upperparts (contrastingly blackish in Cyprus, but less so than in Ménétries's). Eastern Subalpine shares with Cyprus paler and more contrasting edges to tertials (plainer with diffuse fringes in Ménétries's), but differs from it in being longer-winged and shorter-tailed (Cyprus is closer to Ménétries's in these respects). Cyprus constantly moves its tail in a distinct vertical swinging motion, like Sardinian but more energetic. Compared to Ménétries's, tail-cocking is less free, and sideways movement much less

Ad ♀, Cyprus, Nov: grey-olive upperparts, duskier flanks and ventral region (characteristic), and slightly indicated dark blotching below (diagnostic). New tertials characteristically have clear-cut dark centres. In this plumage, dark of head mostly confined to ear-coverts. In ♀, combination of evenly fresh ad-like wing-feathers and warm sepia-brown iris at this season age this bird as ad. (A. McArthur)

1stW ♂, Cyprus, Nov: highly variable plumage, with many being intermediate between the sexes, or attaining only partial ♂ plumage in first autumn; especially, the crown is often blotchy, usually with darker ear-coverts. Underparts spotting also sparse, and upperparts often extensively tinged brown. Aged by dull iris and orbital ring; this bird has apparently moulted almost entire wing (though outermost greater coverts might be juv). (H. Shirihai)

1stW ♂, Cyprus, Nov: note underparts spotting, contrasting tertials, dark uppertail. Least-advanced young ♂ has head mixture of grey and emerging black, with limited lingering brown patches from juv plumage, and underparts spotting also usually much sparser and more irregular, but leaving throat mostly white. (A. McArthur)

prominent. Thus, in gait and action Cyprus differs markedly from Eastern Subalpine and Rüppell's. – Possible confusion with *Marmora's Warbler* (they could meet in N Africa) is discussed under that species.

AGEING & SEXING Ageing sometimes possible after close check of moult and feather wear in wing, and iris coloration can be useful too. Tail pattern often differs between juv and ad, but with some overlap. Sexes differ clearly in ad, whereas 1stW exhibits less obvious sexual dimorphism. – Moults. Complete post-nuptial moult of ad, and partial or (rarely) complete post-juv moult, in late summer (mostly Jun–Sep). Extent of post-juv moult variable, in general more extensive in ♂♂ (in a few complete, or almost so), including all head, body, most or all wing-coverts, some outermost primaries and innermost secondaries, and always all tertials, usually most or all tail-feathers, and alula. Pre-nuptial moult partial, usually involving some tertials, typically including central or innermost feathers, and, more rarely, some greater coverts, and also some tail-feathers and head/body. – **SPRING Ad** Both sexes usually have reddish orange-brown iris (although duller and less prominent in some ♀♀). Due to winter moult, limits between old and new feathers can resemble those of 1stS, but ad has only fresher ad-type wing-feathers. **Ad ♂** Upperparts bleach sootier grey, and abrasion of feather-tips below creates bolder blackish-mottled pattern (often coalescing

1stW ♀, Cyprus, Nov: many young ♀♀ are quite featureless, being largely off-white below with dusky brown body-sides, lacking any exposed dark spots on throat or breast. However, note the cold greyish olive-brown upperparts, contrasting strongly with blackish uppertail (visible at least on right bird), and whitish fringes to tertials. Finally, look for the diagnostic pale fringes (diffuse chevron-like pattern) to the warm olive-grey undertail-coverts. Age further confirmed by juv primary-coverts and outer greater coverts, and remiges (browner and already worn primary tips). Iris largely greyish-olive and orbital ring dull brownish-red. (A. McArthur)

1stW ♀, Cyprus, Nov: some young ♀♀ in autumn are more sandy-tinged above (less olive or dusky), and due to distinctive bold white fringes to tertials they can resemble Eastern Subalpine Warbler. Warmer body-sides and white-tipped undertail-coverts are the best eliminating clues, but also the shorter primary projection and relatively longer and fuller tail. Dull iris and orbital ring indicate age. (H. Shirihai)

Juv, N Cyprus, Aug: soft, fluffy body-feathers and small-headed appearance due to not fully developed plumage, characteristic dusky head and upperparts, and underparts typically washed brown-grey, deepest on vent and undertail-coverts. Swollen mouth flanges still evident. (A. Öztürk)

SYLVIA WARBLERS

Spring ♀♀ of Cyprus (top left: Cyprus, Apr), Sardinian ssp. *momus* (top right: Israel, Mar), Eastern Subalpine ssp. *albistriata* (bottom left: Cyprus, Apr), Rüppell's (centre bottom: Cyprus, Apr) and Ménétries's Warblers (bottom right: United Arab Emirates, Mar): some young ♀ Cyprus show no dark mottling below and can be tricky to identify, though very few young ♀♀ remain unmarked in their first spring. The dull greyish olive-brown upperparts, with blackish uppertail, and diffuse pale chevrons on buffy undertail-coverts are the best characteristics. Especially liable to confusion with ♀ ssp. *momus* Sardinian, but latter usually has clearer greyer head, warmer upperparts and flanks, and pinkish-orange (not white) eye-ring. In optimal views, ♀ Eastern Subalpine *albistriata* and Rüppell's are distinctive in having better-demarcated and whiter submoustachial stripe, usually longer primary projection, and clearer-cut pale tertials and secondary-coverts fringes than ♀ Cyprus. Rüppell's is also usually visibly larger/heavier and stronger-billed, while Eastern Subalpine is often proportionately shorter-tailed with browner uppertail barely contrasting with upperparts like Cyprus. Eastern Subalpine and Rüppell's also lack the duskier underparts and diagnostically scaled undertail-coverts of Cyprus. The young ♀ Rüppell's, which lacks any exposed dark bases to throat-feathers, has some pinkish-orange on throat, possibly from fruit-eating, which could invite confusion with Eastern Subalpine. ♀ Ménétries's usually does not present a pitfall for Cyprus, being essentially greyer or buffish sandy-grey above, paler below and lacks the undertail-coverts pattern of Cyprus. The first four ♀♀ are 1stS, but the Ménétries's is probably ad. (Top left: J. Buckens; top right: R. Bisp Christensen; bottom left and centre: G. Reszeter; bottom right: D. Clark)

into larger sooty patches). Note that some advanced 1stS ♂♂ may be indistinguishable from ad ♂ due to complete post-juv moult. **Ad ♀** Acquires blacker head contrasting with greyer upperparts, while dark underparts markings become denser and more complete than in fresh autumn plumage. Some older ♀♀ stronger marked and approach ♂, but have normal (greyish-olive, less lead-grey) upperparts coloration. **1stS** Retained juv remiges and primary-coverts more worn and browner, with faded buffish-brown fringes, forming strong moult limits with newer ad-like tertials, some inner secondaries and outer primaries. Many have iris much like ad, but others, usually ♀♀, have darker olive-brown iris. Some 1stS distinctive due to only partly-developed ad plumage. Note that rarely some 1stS (probably primarily ♂♂) may look as ad due to complete post-juv moult. – **AUTUMN Ad** Both sexes aged by evenly very fresh wing- and tail-feathers (remiges and primary-coverts glossy greyish-black). Tail-feathers glossy black with large white fringes and tips. Pale grey primary tips usually quite obvious. Iris sepia-brown, but some ♀♀ have slightly olive tinge, although usually still have dominant reddish or orange-brown cast. **Ad ♂** Often has paler rear crown when very fresh, but this still contrasts with dark lead-grey upperparts. Underparts more sparsely/narrowly spotted (feathers have broader whitish tips). **Ad ♀** Upperparts warmer olive-brown and underparts spotting more concealed. **1stW** Best aged by some juv primary-coverts and remiges (textured softer and browner), and juv primaries, if retained, lack obvious pale tips of ad. Moult limits usually present between moulted tertials and unmoulted secondaries (but contrast much less obvious in ♂). Some birds moult more extensively (then contrast among primaries or primary-coverts acquires importance). In both sexes, rare juv tail-feathers have less pure or sharply-defined pale tips and fringes, and such feathers are clearly more worn. Iris largely greyish-olive to earth-brown, though some have slight warm sepia or even reddish cast (especially ♂♂) and approach typical ad. **1stW ♂** Intermediate between ad ♂ and ♀, but most recall former. Crown varies from cold greyish-black to dark grey with patchy black spots. Ear-coverts usually darkest part of cap, almost solidly black. Upperparts dusky grey, extensively tinged buff-brown. Wings show distinct moult limits between ad-like blackish-centred tertials, greater coverts and usually outermost primaries, and juv rest of remiges and primary-coverts (browner centres and buff-brown fringes). Underparts spotting usually much sparser than in ad: throat white, with blackish centres largely concealed; body-sides dusky grey-olive, and central underparts have fewer/smaller exposed blackish centres. **1stW ♀** Upperparts cold greyish olive-brown, contrasting strongly with blackish tail. Tertials fringed cream-white, presence of whitish wing-bars and at least subtly dark-marked undertail-coverts important for identification. Below largely whitish-cream with dusky buff-brown body-sides, sometimes with a few dark brown throat and breast spots. **Juv** Sexes alike. Generally resembles 1stW ♀, but has more fluffy feathers throughout. Compared to juv ♀ Sardinian Warbler averages subtly paler and plainer (Walton 2015).

BIOMETRICS L 12–13 cm; **W** ♂ 57–64.5 mm (n 35, m 60.0), ♀ 55.5–62 mm (n 25, m 58.7); **T** ♂ 49–61 mm (n 35, m 55.1), ♀ 50–56 mm (n 24, m 53.0); **T/W** m 91.1; **TG** 3.5–8 mm; **B** 11.9–14.0 mm (n 59, m 13.1); **BD** 2.9–3.8 mm (n 25, m 3.4); **Ts** 18.1–20.0 mm (n 31, m 19.2). Wing formula: $p1 >$ pc 0.5–4 mm, $< p2$ 28–32 mm; $p2 <$ wt 0.5–3.5 mm, =5/6 or 6 (65%), =4/5 or 5 (25%) or =6/7 (10%); $pp3$–5 about equal and longest (p5 sometimes 0.5–2.5 mm <); $p6 <$ wt 2–5 mm; $p7 <$ wt 4–7 mm; $p8 <$ wt 6–8 mm; $p10 <$ wt 8–11.5 mm; $s1 <$ wt 10–13 mm. Emarg. pp3–6 (p6 sometimes a little less prominently).

GEOGRAPHICAL VARIATION & RANGE Monotypic. – Restricted to Cyprus; a few are resident, but majority winters in S Israel, SW Jordan, extreme NW Saudi Arabia, S Sinai and near west coast of Red Sea from Egypt to N Eritrea.

REFERENCES WALTON, C. (2015) *Sandgrouse*, 37: 16–21.

RÜPPELL'S WARBLER
Sylvia ruppeli Temminck, 1823

Fr. – Fauvette de Rüppell; Ger. – Maskengrasmücke
Sp. – Curruca de Rüppell; Swe. – Svarthakad sångare

This lovely *Sylvia* (the ♂ surely is among the most attractive members of this genus) breeds only in dry hilly maquis country with open woodland in Greece, S & W Turkey and the N Levant, wintering in NE sub-Saharan Africa. Vagrants have reached as far north and west as Britain and even the Faeroes. Its normal migratory pattern describes a loop: in autumn, most move south across the E Mediterranean, being commonly seen in Egypt, but rather few pass through the Levant. Being a much more common passage migrant in spring in the Levant, obviously the majority take a more easterly route than in autumn.

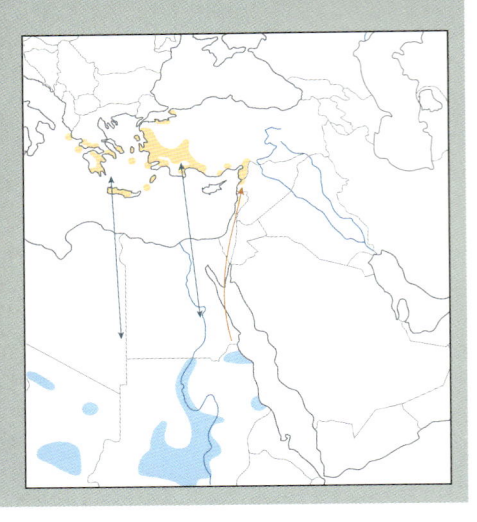

Ad ♀, Israel, Mar: with wear, especially older ♀♀ show variable hint of blackish bib and whitish submoustachial stripe. Note characteristic whitish-fringed greater coverts and tertials, and also rather long and pointed bill with down-curved tip. The apparently not heavily worn and evenly-feathered wing, and rather warm brown iris, confirm age. (G. Ekström)

IDENTIFICATION Size and proportions much like Common Whitethroat, but plumage nearer Sardinian, Cyprus and Ménétries's Warblers, albeit notably more *robust with longer wings and a shorter tail* than any of these three. Long bill (slightly decurved) differs from any similarly plumaged *Sylvia*. Primary projection often nearly the same as tertial length. ♂ unmistakable in good views: *black hood and bib*, prominently enhancing *white submoustachial stripe*, contrasting with mainly *pure grey upperparts*. Greater coverts and tertials dark-centred fringed whitish, dark tail with white sides, and whitish underparts with greyish sides. Indistinct eye-ring, usually a mixture of white and grey. Orbital ring red or dark orange-red, iris ochre to reddish-brown. Legs reddish-flesh. ♀ and young less distinctive, although *older or worn ♀♀ have a variable blackish bib* and, even in poorly marked individuals, *a slight whitish submoustachial stripe and the characteristic whitish-fringed greater coverts and tertials*. Greyish-brown above (darker head in spring in worn or older ♀♀). Almost pure white below, with limited greyish flanks, and duller bare-parts colours. Distinctive behaviour can be judged with experience, Rüppell's Warbler being far less restless than most other *Sylvia* species, and *never conspicuously waves or cocks tail* (but sometimes flicks both wings and tail). It is also less skulking than its principal confusion species.

VOCALISATIONS Song a rather musical, if jerky, warble, being rather similar to Sardinian Warbler but more continuous and subtly slower ('better articulated'), with rather more whistled notes interwoven and is often perceived as more musical than Sardinian. Frequently recognised by call-like chattering, almost 'bouncing' notes which comprise a good part of each strophe. – A hard *tak* or *chet* in contact, which is often quickly repeated in chattering, sparrow-like series, *chet-et-et-et-et-et*; also even faster rattling series of *trrrr* notes in alarm.

SIMILAR SPECIES Adult ♂ should offer few identification problems, but ♀-like birds (including 1st-winter ♂♂) could be confused with an unusually dull *Sardinian Warbler*, although latter is marginally smaller, shorter-winged and longer-tailed, with less obvious paler tertial fringes and, especially, plainer wing-coverts, and has browner upperparts and body-sides. Most Rüppell's have a longer, slightly decurved bill, and a white submoustachial stripe, though this may be hard to see on some birds with least dark smudges on throat. Rüppell's never has Sardinian's buffish-cream or pale pinkish-orange eye-ring; its eye-ring instead mainly consists of white and grey feathers. Note that Rüppell's tail pattern approaches Western Subalpine or Moltoni's Warblers. – ♀-like plumages of *Eastern Subalpine Warbler* may also present a trap for the unwary (they have similar general structure and well-marked tertial fringes). However, Rüppell's is larger (being more robust, clumsy and Common Whitethroat-like) with a longer bill, and has diagnostic white fringes to greater coverts, darker tertial centres (fringes usually more pronounced and whiter), greyer upperparts and body-sides, a darker head and more contrasting dark tail that lacks the deep narrow white wedges inside shaft of the penultimate feather of adult Eastern Subalpine. Furthermore, the pale base to the lower mandible is greyer. –

Ad ♂, Turkey, May: a remarkably handsome *Sylvia*, especially in spring when plumage is immaculate, with strongly-patterned black head and bib, and white submoustachial stripe. By Apr/May, whitish primary tips have worn off and often pre-nuptial renewal of tertials forming strong moult limit. Such immaculate plumage and not heavily-worn wing in May, including primary-coverts and remiges with greyish fringes, and bright ochre iris, age bird as ad. (D. Occhiato)

Ad ♀, Israel, Mar: when still fresh in early spring the mottled blackish bib is largely concealed (but in few it is always lacking, even in worn plumage later in the summer), though there is usually still a hint of whitish submoustachial, and in this bird the blacker cap already started to emerge through wear. Note the characteristic jizz of elongated overall shape, rather strong bill and longish tail, and the diagnostic whitish-fringed greater coverts and tertials. Evenly feathered wing and orange-brown iris confirm the age as ad. (H. Shirihai)

SYLVIA WARBLERS

In the Middle East, *Common Whitethroat* of the race *icterops* (especially dull ♀♀) could be confused (they have similar size and structure), but *icterops* has rufous wing-feather fringes, different tail pattern and pinkish lower mandible. The two are also easily separated by call. – Unpatterned and dark-eyed ♀-like Rüppell's (see above) could be confused with *Lesser Whitethroat* or *Western* and *Eastern Orphean Warblers*. Diagnostically, all those species have dark grey legs, rather than reddish flesh-brown as in Rüppell's. They also lack Rüppell's well-defined pale tertial and greater coverts fringes, being more uniform with broader and diffusely paler fringes. – Separation from *Cyprus* and *Ménétries's Warblers* is covered under those species.

AGEING & SEXING Ageing requires close scrutiny of moult and feather wear in wing and tail, and iris coloration is useful supporting criterion. Sexual dimorphism well developed in spring, and obvious in ad at all seasons, but almost non-existent in 1stW. – Moults. Complete post-nuptial and partial post-juv moult after breeding (mostly Jul–Sep). Post-juv moult includes all head, body, lesser and median coverts, most or all greater coverts and some tertials; some even replace inner secondaries and very rarely a few outer primaries. Partial pre-nuptial moult (late autumn–early winter) involves head, body, tertials, and some

1stS ♂, Israel, Mar: young ♂♂ are often striking in having heavily-abraded brown juv primary-coverts and most remiges, the latter forming moult contrast with most other wing-coverts, secondaries, alula and tertials having been replaced in post-juv or pre-nuptial moult. However, overall plumage very ad ♂-like but for usually mottled black cap and bib and less pure grey upperparts. Iris of this bird already tinged ochre but still with olive-brown hue. Note that on migration this *Sylvia* habitually forages close to the ground, and often will hop for extended period across open ground. (H. Shirihai)

1stS ♀, Israel, Mar: note characteristic longish, slightly decurved bill, and hint of whitish submoustachial stripe. Could be confused with dull ♀ Sardinian Warbler, though latter is usually visibly smaller and proportionately shorter-winged and longer-tailed, as well as markedly browner above and on flanks, with plainer wing-coverts. Eye-ring of Rüppell's mostly white and grey (never buffish-cream or pale pinkish-orange like Sardinian). Could be mistaken for ♀ Eastern Subalpine, but Rüppell's is noticeably larger and more robust, with longer bill and diagnostic white fringes to greater coverts; also, has greyer upperparts and body-sides. There is also an even greater confusion risk with poorly-marked ♀ Cyprus Warbler. Again, Rüppell's is usually visibly larger/heavier and stronger-billed, with better-demarcated and whiter submoustachial stripe, while has longer primary projection and clearer-cut pale tertial and secondary-coverts fringes than Cyprus; Rüppell's also lacks Cyprus's warmer underparts and scaled undertail-coverts. Juv primary-coverts and primaries have faded paler fringes, contrasting with post-juv and winter-moulted secondary-coverts and tertials; iris still olive-brown like 1stW in autumn. (H. Shirihai)

Ad ♂, Turkey, Aug: just as in spring, fresh ♂ in autumn still shows diagnostic black crown and bib (with white submoustachial stripe), but these are less solidly patterned, with some paler grey tips. Note broad ash-grey greater coverts fringes and primary tips. Eye-ring and iris are a little duller in autumn. (F. Yorgancıoğlu)

innermost greater coverts, more rarely secondaries or tail-feathers (less extensive in ad). – **SPRING Ad** Iris and orbital ring brighter than in autumn, and iris largely reddish-brown or bright ochre, though some, mainly younger ad ♀♀, have duller yellow-brown iris. Whitish primary tips abraded, and remiges browner than in autumn (beware of moult limits due to winter moult, as these can confuse ageing, due to their recalling moult patterns of 1stS); best aged by having only ad and relatively less worn wing-feathers. **Ad ♂** Black of head and bib, and grey of upperparts, on average more solid and purer (but advanced 1stS equally neat). All primaries and all or at least inner (more sheltered) primary-coverts edged lead-grey. **Ad ♀** Acquires variable (usually mottled) darker crown and throat. Older ♀♀ can develop almost ♂-like blackish head and throat, but rest of plumage as normal ♀. **1stS** Retained juv remiges and primary-coverts have faded buffish-brown fringes (pale primary tips worn off) contrasting with post-juv and winter-moulted secondary-coverts, tertials and sometimes inner secondaries and (rarely) few outer primaries, which are ad-like, i.e. blackish with lead-grey or greyish-white fringes. Iris much like ad in

— 529 —

many, but others, usually ♀♀, have duller olive yellow-brown iris. Some 1stS distinctive due to less completely developed ad plumage. (1stS ♂♂ often have less complete black head/bib and less pure grey upperparts than ad, but some overlap as to this. Some least-advanced ♀♀ are like 1stW with hardly any dark on crown and throat.) Juv tail-feathers, retained in some, are diagnostic if present. – **AUTUMN Ad** Both sexes aged by evenly very fresh wing- and tail-feathers (remiges and primary-coverts glossy greyish-black). Whitish primary tips broad and well defined. Tail-feathers glossy black with large and well-defined white fringes and tips. Iris largely yellowish-orange to deep brownish-orange, though in some, mainly 2ndCY ♀♀, olive yellow-brown. **Ad ♂** Crown sometimes less intense black than in spring. When still very fresh, bluish ash-grey upperparts tinged buffish-cream on mantle, otherwise as spring. **Ad ♀** As spring, but upperparts warmer buffish grey-brown with dusky greyish-brown crown (dark feather-bases concealed) and, usually, even duskier ear-coverts. Underparts mostly white, but throat (except whitish submoustachial stripe) and flanks tinged pale buffish-grey. **1stW** Both sexes differ from similar ad ♀ by juv remiges and some wing-coverts (weaker and browner with narrower, poorly-demarcated buffish fringes). Diagnostically, primaries lack bold whitish tips of ad (being diffusely tipped buffish at most). Usually only tips of median and greater coverts whiter, forming wing-bars. Juv remiges and primary-coverts usually produce clear moult limits with newly moulted, ad-like (blackish-grey, fringed greyish) greater coverts and tertials. Retained juv tail-feathers browner, with smaller, poorly demarcated pale wedges or tips. Iris invariably dark greyish olive-brown. Usually, ♂ has greyer head and blackish lores (not greyish-white). ♀ is uniform buffish drab grey-brown (less greyish) above. **Juv** Very distinctive. Mostly olive dusky-brown above, including head. Throat and body suffused dusky buffish-cream, with broad greyish-brown breast-band; flanks warmer. Orbital ring mostly medium yellow-brown. Iris greyish brown-olive.

BIOMETRICS L 13.5–14 cm; **W** ♂ 66.5–75 mm (n 28, m 70.3), ♀ 66–74 mm (n 23, m 69.0); **T** ♂ 57–64.5 mm (n 28, m 60.5), ♀ 55–61 mm (n 23, m 58.2); **T/W** 85.3; **B** 13.5–15.9 mm (n 51, m 14.5); **BD** 3.2–4.1 mm (n 28, m 3.6); **Ts** 20.0–21.8 mm (n 25, m 20.9). **Wing formula: p1** < pc 0–5 mm (rarely 0.5 mm >), < p2 37.5–45 mm; **p2** < wt 0.5–2 mm (rarely = wt), =4/5 or 5 (57%), =3/4 or 4 (19%), =5/6 or 6 (13%) or =3 (11%); **pp3–4** (rarely pp2–4) about equal and longest; **p5** < wt 0.5–2.5 mm; **p6** < wt 3–6.5 mm; **p7** < wt 5.5–9.5 mm; **p10** < wt 11–17 mm; **s1** < wt 13–19 mm. Emarg. pp3–5 (p5 sometimes a little less prominently in 1stY).

GEOGRAPHICAL VARIATION & RANGE Monotypic. – Breeds in S & E Greece, W & C Turkey, extreme NW Syria; winters in sub-Saharan Sahel in Chad and Sudan, perhaps also Niger and Mali.

TAXONOMIC NOTE Although named to honour the German ornithologist Eduard Rüppell, Temminck wrote the type description in French and oddly altered Rüppell's name to a perceived French spelling, hence the spelling *ruppeli* of the species epithet in the scientific name. The Code regulating scientific names does not permit what one could see as a natural correction out of courtesy.

REFERENCES Lewington, I. (1992) *BW*, 5: 338–340.

Ad ♀, Israel, Sep: fresh ♀♀ in autumn (ad and young) are rather similar to dullest young ♀♀ in spring, but have warmer upperparts, dusky greyish-brown crown and whitish throat (dark bases largely concealed). Long, strong and slightly decurved bill and long primary projection are often useful clues but not always distinctive. Aged by evenly very fresh wing, and by broad and well-defined whitish primary tips. (H. Shirihai)

Juv, Turkey, Jun: mostly olive dusky-brown above, including head. Throat and body suffused dusky buffish-cream, with slightly dusky-tinged breast and warmer flanks. Buff tertial fringes usually pronounced. Told from rather similarly-plumaged Eastern Orphean Warbler by more obvious and distinct whitish tertial edges. (H. Shirihai)

1stW ♂ (left) and 1stW ♀, Turkey, Sep: following post-juv moult, both sexes attain ♀-like plumage (note juv remiges without white primary tips), and identification is as for 1stS ♀ Rüppell's without dark throat. At least some young can be sexed, e.g. note ♂'s darker and purer grey head (with blacker bases exposed), better-demarcated and whiter submoustachial stripe and darker lore. Despite being the least-marked plumages, pay attention to unique combination of overall jizz, distinctive white throat, long primary projection, blackish uppertail and clearer-cut pale tertial and secondary-covert fringes. (Left: G. Coral; right: F. Yorgancıoğlu)

ASIAN DESERT WARBLER
Sylvia nana (Hemprich & Ehrenberg, 1833)

Fr. – Fauvette naine; Ger. – Wüstengrasmücke
Sp. – Curruca enana; Swe. – Ökensångare

This small *Sylvia* inhabits sandy deserts with scattered scrub and bushes, as well as *Acacia* and *Saxaul* trees, from the Caspian Sea, including N Iran, through Central Asia east to Mongolia. It winters in the NW Indian subcontinent through S Iran to Arabia, the S Levant and the NE African littoral. Vagrants have been recorded as far west as Britain, and in the north-west to Sweden and Finland. Most birdwatchers have probably seen it for the first time in deserts of S Israel, where it is a scarce but regular annual winter guest at least in late November–February.

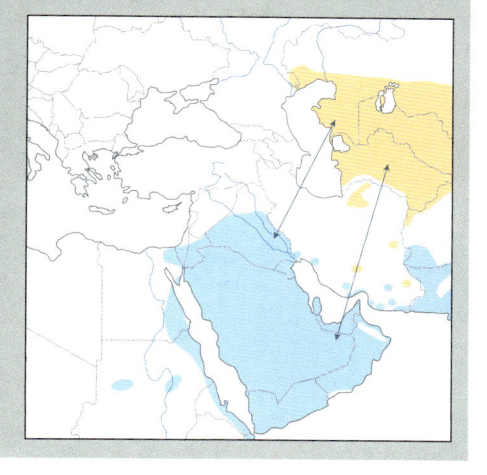

(of which the small and warm-coloured races *minula* and *halimodendri* could be a pitfall), but the latter has no rufous coloration on rump/uppertail or yellow in the bare parts, and their behaviour and vocalisations are also quite different.

AGEING & SEXING Ageing usually only possible in the hand. Sexes alike, and virtually no seasonal variation. – Moults. Complete post-nuptial and partial post-juv moults in late summer (mostly Jul–Aug). Post-juv moult includes most or all of head, body, secondary-coverts and tertials, while some even replace a few inner secondaries and tail-feathers. Partial pre-nuptial moult over extended period (Oct–Mar), usually involving some feathers of head, body, tertials and some innermost greater coverts, more rarely some tail-feathers and/or inner secondaries (generally fewer such replacements in ad). – **SPRING Ad** Upperparts and wings wear duller (more greyish) and underparts whiter; pale primary tips much reduced or absent. Presence of pre-nuptial moult can be difficult to detect, or can create misleading pattern recalling post-juv moult (mostly due to 1–2 fresh inner tertials); ageing should rely on condition and structure of primaries and primary-coverts. **1stS** Virtually indistinguishable from ad, but look for slightly more bleached and worn juv primaries and primary-coverts. In early

Mongolia, Jun: this tiny, round-headed *Sylvia* is characterised by its greyish-sandy upperparts, whitish and cream-buff underparts, with bright orange-rufous rump and uppertail, uniquely bland face pattern (with paler lores), mainly pale yellow lower mandible and bright yellow iris, which all make the species unmistakable. Ageing, even in the hand, is often difficult or impossible. (H. Shirihai)

IDENTIFICATION Tiny, diminutive *Sylvia*, with a short bill, rounded head, short rounded wings (with moderate primary projection), and medium-length tail. Even more compact and dainty than Spectacled Warbler, otherwise these two are rather similarly structured. Plumage curiously unpatterned, characterised by *greyish-sandy upperparts*, with *bright orange-rufous uppertail* (the latter even more distinctive when spread) *and wing* (in Common Whitethroat-fashion, black centres visible on tertials), and largely buffish-cream underparts (slightly deeper on flanks, sometimes also greyish-brown breast-sides). Unlike any other *Sylvia*, combination of large pale loral area and *mainly pale yellow lower mandible, iris and tarsus* very distinctive. In all plumages has *whitish eye-ring* and narrow blackish orbital ring. Restless, *especially energetic and nervous on ground*, where it often forages in almost mouse-like fashion, and typically moves in short sharp hops on sand, often picking insects from below, and moves with fast low flights between bushes. Often very tame, closely approaching observers. In some wintering grounds there appears to be a strong association between Asian Desert Warbler and Desert Wheatear when feeding, with the warbler often following the wheatear.

VOCALISATIONS Song a tinkling, rather short and monotonous warble that commences with a low churring, followed by a loud, clear whistling note turning into a jingling trill that descends slightly at the end, e.g. *eh-cherrrr-si-siiih-sir'r'r'r'r'r*. Slowly repeated with little variation. – Calls or gives milder alarm with the same low churring *cherrrr* that commences the song (can then sound almost like a European Blue Tit), at times with a short initial note added, *eh-cherrrr* (vaguely similar to Grey Partridge song!), squeakier and slower than African Desert Warbler, otherwise similar. Also gives a grating *chee-chee-chee-chee* or *chee-chee-krrrr*.

SIMILAR SPECIES The two desert warblers have never been found to overlap, but theoretically vagrants of either species could be found in the winter range of the other, and accidentals in NW Europe (all thus far have been identified as the present species) require careful separation. Concentrate on the tail and tertial patterns, as African Desert Warbler lacks Asian Desert's distinct blackish central area. Also note greater contrast in Asian Desert between the colder upperparts (sandy drab greyish-brown rather than bright orange-tinged buff-brown as African Desert) and the bright rufous rump/uppertail, the more patterned underparts (largely uniform in African Desert) and the slightly duller bare parts. Voice would probably be of limited use for vagrants (although some Asian Desert Warblers in spring in Europe have sung). – Separation from *Spectacled Warbler* is covered under African Desert Warbler. – Asian Desert Warbler overlaps in SW Asia and Arabia in winter with Lesser Whitethroat

Kuwait, winter: often very tame, and very active on ground, where it often forages in almost mouse-like fashion. Note the attractive lemon yellow iris. Pale-spectacled impression sometimes more evident in some birds or depending on light, as here. Without handling ageing is impossible. (A. Al-Sirhan)

Oman, Nov: when fresh and in certain light, the species is far from appearing dull-coloured, especially if the bright orange-rufous rump and uppertail are evident. Uppertail distinctive when briefly spread, showing rufous central feathers, then dark/blackish-brown ones inside white sides. The pale sandy-orange tertials with darker shafts are just visible here. (D. Occhiato)

Presumed 1stW, Kuwait, Dec: in close views note the darker shafts to tertials, whitish eye-ring and yellow eye. Often has similar jizz and behaviour as Spectacled Warbler. Clear moult limit in tertials, whereas the juv primary-coverts are rather narrow-fringed and pointed, and the primaries perhaps somewhat worn and bleached brownish, indicating a 1stY. (V. Legrand)

spring, whitish primary tips largely worn; sometimes three generations of feathers discernible: some fresher (winter-moulted) ad-like tertials, and perhaps inner secondaries, contrasting with juv remiges and post-juv greater coverts. Juv tail-feathers retained in some, being worn and faded, and usually contrasting with newly moulted ones; compared with ad-like feathers have distinctly less pure, more sandy-whitish wedges. – **AUTUMN Ad** Plumage evenly very fresh without apparent moult limits. Primaries dark blackish-grey with broad bright rufous-buff fringes and whitish-grey tips. Primary-coverts solidly textured. Whitish wedges to outer tail-feathers purer and larger than on juv feathers. **1stW** Extremely similar to ad. Most, however, appear 'untidy', with more worn, paler and browner juv flight-feathers (with narrower, duller fringes), and slightly abraded primaries (whitish tips faded and less bold). Evidence of moult limits in inner wing best observed between newly moulted inner tertial(s) and juv secondaries or outer tertial(s). Blackish central areas on renewed remiges darker and more solidly textured than juv feathers. Primary-coverts also subtly more pointed with on average a little softer texture. Juv outer tail-feathers lack pure white wedges, being mainly off-white or sullied pale buffish-sandy (though some may have ad-like renewed feathers). See Juv for iris colour. **Juv** Occurs only on breeding grounds (May–Aug); resembles ad, but has soft, fluffy body-feathers and wing-coverts, and colder brown upperparts (less sandy), unpatterned head, rufous-buff fringes to wing-coverts and tertials, and more extensively cream-buff underparts. Slightly less pure yellow bare parts; in some, iris less pure yellow than in ad, duller and slightly greyish, mainly in form of narrow ring (still apparent in some 1stW even in late autumn, but usually hard to detect). However, most young rapidly develop pure yellow iris like ad.

BIOMETRICS L 11–12 cm; **W** ♂ 57–61.5 mm (*n* 13, *m* 58.9), ♀ 53.5–57.5 mm (*n* 12, *m* 55.4); **T** ♂ 49–51 mm (*n* 13, *m* 50.3), ♀ 47–50 mm (*n* 12, *m* 48.3); **T/W** *m* 86.3; **B** 11.2–12.8 mm (*n* 25, *m* 11.9); **BD** 2.7–3.5 mm (*n* 25, *m* 3.0); **Ts** 18.4–21.8 mm (*n* 25, *m* 19.7). **Wing formula: p1** > pc 0–4 mm; < p2 24.5–32 mm; **p2** < wt 1.5–3.5 mm, =5/7 (very rarely =7 or 7/8); **pp3–5** about equal and longest (though p5 often 0.5–1.5 <); **p6** < wt 1–3.5 mm; **p7** < wt 3–6 mm; **p10** < wt 7–11 mm; **s1** < wt 7–12 mm. Emarg. pp3–5 (p6 sometimes also with faint emarg.).

GEOGRAPHICAL VARIATION & RANGE Monotypic. – Breeds NE Iran, Transcaspia (including lower Volga) through Central Asia east to S & W Mongolia, NW China; winters NW India, S Iran, Arabia, Red Sea and S Levant. (Syn. *theresae*.)

TAXONOMIC NOTE Asian and African Desert Warblers were formerly treated as conspecific, but were recently separated on the basis of their different songs, plumages and wholly allopatric ranges.

Oman, Dec: typical posture before diving back into cover. Note diffuse dark centres and sharp black shafts to tertials, and contrasting rufous uppertail-coverts and central tail-feathers, latter also black-shafted. (M. Römhild)

Presumed 1stS, Israel, Feb: rather untidy plumage, with worn and scruffy tail-feathers and quite worn tertials, of which the central ones are missing. Primary-coverts also appear to be juv, being narrowly fringed and rather narrow, suggesting a 1stY. (H. Shirihai)

AFRICAN DESERT WARBLER
Sylvia deserti (Loche, 1858)

Fr. – Fauvette naine; Ger. – Saharagrasmücke
Sp. – Curruca sahariana; Swe. – Saharasångare

In contrast to Asian Desert Warbler, the African Desert Warbler is largely resident or only partly migratory, occurring in desert regions of S Morocco to W Libya and S Mali, Mauritania and, perhaps, Niger. Possibly nomadic; there have been several records of short-distance vagrancy, the species having reached the Cape Verdes, Malta, Portugal and off Sicily.

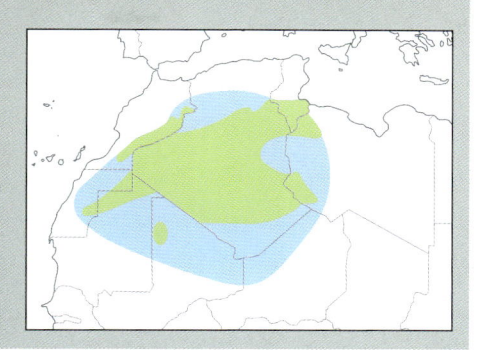

IDENTIFICATION Very small, diminutive *Sylvia* with a short bill, rounded head, short rounded wings, moderately long tail, and creeping, mouse-like behaviour just like Asian Desert Warbler. Also very similar to that species in general plumage, but has *orange-sandy buff-brown upperparts* (lacking any grey tones), thus *appearing almost concolorous above*, with *rufous-orange rump/uppertail creating only moderate contrast with rest of upperparts*, including warm-coloured wing-feather fringes. *Tertials plain*, shafts at most slightly but never conspicuously darker than pale sandy webs. *Paler whitish below* (though often extensively tinged pale pinkish-buff), with more limited buffish-rufous wash on sides. Head pattern often more obvious, with usually *whiter* (rather than greyish) *lores and pale supercilium* and rear ear-coverts. Also differs from Asian Desert by tail pattern: the central feathers lack prominent dark centres, being at most dark orange-brown, and the dark portions never extend to adjacent inner parts of both webs; black centres on rr2–4 usually confined to inner web, and buffish-rufous colour extends (more deeply) over entire outer webs, while white on rr5–6 is more extensive, creating a broader and deeper white wedge on r5 than in Asian Desert. Bare parts as Asian Desert, but bill base and legs more pinkish-yellow; blackish bill tip is less extensive, more diffuse or entirely absent. Nervous and energetic, moving constantly (frequently on the ground), flicking wings and tail, secretive nevertheless (sometimes refusing to fly from deep cover) and can be difficult to see, although not shy and may approach observers.

VOCALISATIONS Song differs clearly from Asian Desert in lacking the clear trill falling in pitch at the end, instead can be somewhat reminiscent of Common Whitethroat, with similar jerky rhythm and scratchy warble type of song. The strophe often opens with similar call-like subdued trill as Asian Desert, but the song then becomes more complex, less clearly trilling and more variable, sometimes with extended and distinctly more complex warbling than Asian Desert Warbler. – Contact calls include a rattling *krrr*, and a harsh, sparrow-like *ch-ch-ch-ch-ch-ch-ch* in alarm, both similar to corresponding calls of Asian Desert Warbler.

SIMILAR SPECIES Overlaps with *Tristram's* and *Spectacled Warblers* in NW Africa, which have rather similar size and structure, share some habits, and are therefore confusion risks. Especially first-winter of both Tristram's and Spectacled can be very dull and superficially similar in plumage to African Desert Warbler, but they have conspicuous rufous

Morocco, Jan: compared to Asian Desert Warbler more gingery above, with by and large plain tertials. Also purer white below, with often whiter lores, but blackish bill-tip averages smaller, or the dark is more diffuse. Confusion also possible with Tristram's and Spectacled Warblers, especially duller 1stW of latter, but they should be eliminated by their rufous wing panels, with usually striking dark centres to tertials, and lack of Desert Warbler's contrasting rufous-orange rump and uppertail. (M. Schäf)

Malta, Apr: both Asian and African Desert Warblers should be considered when faced with a vagrant in S Europe. With wear, especially from spring, often less bright and less gingery above, and therefore more similar to Asian Desert, though still more orange-tinged brown above, forming less contrast with rufous-orange rump and uppertail. Note also plainer wing, especially tertials. (N. Galea)

Lanzarote, Canary Is, Oct: tail pattern also shows fewer black bases or centres, including the almost complete lack of prominent dark shafts to central feathers. Note plain rufous-brown tertials. (D. Pérez)

Morocco, Feb: often very tame and active on ground, here around a desert bush, in mouse-like fashion. Some birds in strong light (and with wear) can appear paler and sandier, with hardly any orange tones. Note lack of prominent dark shafts to central tail-feathers. Also, unlike Asian Desert Warbler, the tertials are plain, too. (R. Schols)

Juv, Morocco, May: even juv has gingery-tinged upperparts. At this young age almost lacks any yellow on bare parts, with iris more cream-coloured, and bill and legs flesh-toned. Unlike ad, plumage is fresh in May. (B. Maire)

wing panels, with extensive rufous fringes to tertials, which have striking dark centres. They also lack the contrasting rufous-orange rump and uppertail of African Desert, and close views will also reveal their very different bare-parts colorations. Vocalisations also differ.

AGEING & SEXING Much as Asian Desert Warbler. Note that in worn plumage, upperparts become duller and more yellowish-sandy, below almost pure white (thus, differences in general pigmentation from Asian Desert become less distinct). – Moults. Complete post-nuptial and partial post-juv moults generally earlier than in Asian Desert, and post-juv moult seems more extensive, even involving (some) primaries (Shirihai *et al.* 2001), but still little studied.

BIOMETRICS L 11–12 cm; **W** ♂ 57–61.5 mm (n 13, m 58.9), ♀ 53.5–57.5 mm (n 12, m 55.4); **T** ♂ 49–51 mm (n 13, m 50.3), ♀ 47–50 mm (n 12, m 48.3); **T/W** m 85.4; **B** 10.3–12.8 mm (n 24, m 11.8); **BD** 2.6–3.4 mm (n 23, m 3.0); **Ts** 17.5–21.4 mm (n 27, m 19.0). Wing formula: **p1** > pc 0–6 mm, < p2 24.5–31 mm; **p2** < wt 0.5–4 mm, =5, 5/7 or 7; **pp3–5** about equal and longest (though p5 rarely 0.5–1.5 <); **p6** < wt 2–3 mm; **p7** < wt 3.5–6 mm; **p10** < wt 6–10 mm; **s1** < wt 8–11 mm. Emarg. pp3–6 (on p6 sometimes not quite as prominent).

GEOGRAPHICAL VARIATION & RANGE Monotypic. – Western Sahara, SE Morocco, Algeria mainly south of Atlas Mts, S Tunisia, W Libya; resident or nomadic, making only more local movements.

TAXONOMIC NOTE See Asian Desert Warbler.

ARABIAN WARBLER
Sylvia leucomelaena (Hemprich & Ehrenberg, 1833)

Alternative name: Red Sea Warbler

Fr. – Fauvette d'Arabie; Ger. – Akaziengrasmücke
Sp. – Curruca árabe; Swe. – Arabsångare

This interesting warbler breeds in S Israel and on both sides of the Red Sea and the Gulf of Aden, where it inhabits denser *Acacia* stands within savanna-like habitats. Behaviour may be equally useful in identifying this species as its plumage, with deliberate movements in the canopy of acacias and constant downward dipping of its long-looking tail.

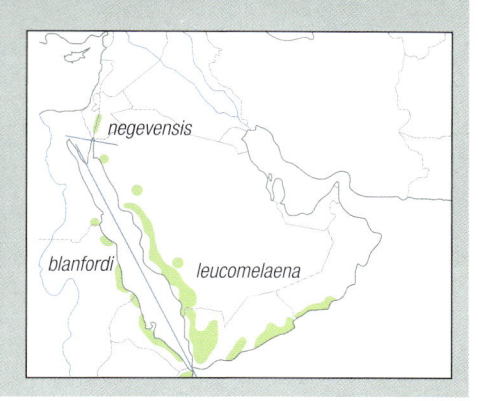

IDENTIFICATION A medium to large, rather bulky *Sylvia* with a *long, full tail* and *short-looking wings*. As to plumage, note the *dark hood, some whitish feathers in eye-ring* (at times many, creating full whitish eye-ring) and *dark graduated tail with characteristic white spotting at tips* if seen from below. Upperparts *brownish-grey* and throat and rest of underparts whitish. *Tertials have well-defined whitish-grey fringes*. *Rather rounded head*, notably shorter bill (than Eastern Orphean Warbler), and *sluggish, rather deliberate movements*, particularly the *repeated downward dips of the tail*. Lacks restless, sprightly, tail-cocking habits of smaller *Sylvia*. Bill has pale bluish-grey tone to base and greyish-black culmen and tip, while legs are dark slate-grey, orbital ring largely blackish and iris, which usually looks blackish in the field, is dark brown or greyish-brown. Very secretive but sometimes perches upright in the open (posture somewhat recalling a bulbul), flying relatively high between treetops. Usually in pairs.

VOCALISATIONS A bubbling, loud far-carrying warble (with both rather melodious and harsh elements), which may recall Blackbird or Blackcap due to prominent inclusion of mellow, fluty notes, rather deep voice and slow rhythm; given year-round. Differs from Eastern Orphean Warbler by lack of repetition of notes and mellower, less scratchy, hard voice. – Commonest call, in contact, is a harsh, peculiarly bulbul-like, hard clicking *chack, chack*, often repeated several times, sometimes in bulbul-like long series when alarmed. Thus, rattling alarm call differs from e.g. Barred Warbler in being less dry and clicking, more nasal and 'compound' in tone, vaguely recalling Garden Warbler alarm (but stronger).

SIMILAR SPECIES Most serious risk of confusion is with *Eastern Orphean Warbler*, given similar size and rather obvious blackish hood, especially in first-summer ♂♂; other plumages are rather well differentiated in overall pattern. Rather easily separated by Arabian Warbler's shorter, less deep-based bill, shorter wings and primary projection (1/3 of tertial length or less; 1/2 or more in Eastern Orphean), more lumbering movements, frequent dips of tail below body-line, darker, more graduated tail with reduced white (in adults forming diagnostic spotting from below; indistinct in juveniles). Usually some white in eye-ring (spring Eastern Orphean may have largely white iris, but never a white eye-ring), and much broader and more obvious pale tertial fringes (plain tertials with narrow greyish fringes in Eastern Orphean).

AGEING & SEXING Ageing can be based on moult and feather wear in wing, but requires handling or close observation. For some birds, tail pattern and iris coloration useful.

S. l. negevensis, ♂, presumed ad, Israel, Feb: note dark rounded head, proportionately long tail and short, rounded wings. Some (usually fresh ad ♂♂) have complete whitish eye-ring, resembling White-spectacled Bulbul. Makes frequent downward dips of the tail. Greyish upperparts and quite dark hood, with well-marked complete white eye-ring, and extensive white in outer tail-feathers best fit ad ♂. However, evenly rather fresh, ad-like wing also matches a 1stY that has moulted completely (rare). (H. Shirihai)

♂ Arabian Warbler *S. l. leucomelaena* (left: Oman, Dec) and Eastern Orphean Warbler *S. c. crassirostris* (right: Turkey, May): at first glance these two species are superficially similar, but given reasonable views the shorter primary projection, frequent dips of longer tail, broader and more obvious pale tertial fringes, and often more white in eye-ring should identify Arabian Warbler. 1stW ♂ Arabian often striking due to heavily-abraded brown juv primaries, these contrasting with fresher most other wing-coverts, secondaries, alula, tertials and even primary-coverts. Purer greyish upperparts best fit ♂. Eastern Orphean is also 1stS ♂ by combination of blackish head and purer grey upperparts, iris colour (mostly dark with only few white dots) and complete pre-nuptial moult. (Left: J. Niemi; right: D. Occhiato)

S. l. leucomelaena, ♂, presumed ad, Oman, Feb: especially in ♂♂, blackish hood becomes better-marked with wear. On the other hand whitish feathers in eye-ring are often fewer due to wear or when in moult, but some, even ad ♂♂, show very few white feathers at any season. Note characteristic dark graduated tail, with ad ♂-like white tips. Greyer-tinged upperparts also characteristic of ♂. Pale bluish-grey tone to bill-base and greyish-black culmen and tip. Impossible to age, but all visible wing-feathers seem ad-like. (D. Forsman)

S. l. leucomelaena, presumed ad ♀, Oman, Dec: subtly browner cap with a little darker ear-coverts, and hood contrasts less sharply with browner-tinged upperparts, as well as less pure white underparts, the latter tinged slightly buffish-grey, all suggest a ♀, especially as the bird appears to be ad. However, tertials appear newly moulted, hence ageing and sexing are a bit unsure. (J. Niemi)

Sexes separable to some degree. Insignificant seasonal variation due to feather wear. – Moults. Complete post-nuptial moult, and post-juv moult (usually partial, possibly occasionally complete), in late summer (mostly Jul–Sep). Extent of post-juv moult extremely variable, from just body-feathers, median, lesser and some inner greater coverts, to entire plumage except a few secondaries or innermost primary-coverts, but commonest are birds moulting all tertials, greater coverts and tail-feathers and some, but not all, primaries, secondaries and primary-coverts; some probably replace entire plumage (firm proof still lacking). Partial pre-nuptial moult (also variable) involves a few inner secondaries and greater coverts, and some tertials and tail-feathers, as well as body-feathers, median and lesser coverts. – **SPRING Ad** In both sexes, notice moult limits following pre-nuptial renewal, but this should not affect main ageing clues, as post-nuptial feathers never possess characteristic juv features of 1stS. Iris greyish-brown, with profuse, irregular and diffuse whitish-grey dots (often with pale greenish hue). **Ad ♂** Blackish head becomes more intense and sharply demarcated, upperparts duller and remiges browner with reduced and duller grey-brown fringes (pale primary tips virtually worn off). White in outer tail slightly reduced through wear, and white eye-ring often strongly so, especially in summer. **Ad ♀** Not always readily separated from ♂, but tends to have browner head with darker ear-coverts, and hood contrasts less sharply with brownish upperparts. Underparts slightly more buffish-grey than in ♂. Same effect of wear as in ♂ but plumage often even duller with more strongly reduced white eye-ring. **1stS** Difficult to separate in the field: juv remiges and primary-coverts on average more worn, browner with faded buffish-brown fringes, and juv tail-feathers (if any) have much-reduced pale fringes and tips. In general, body plumage more heavily bleached and unticy, crown often duller and eye-ring variable, but number of white feathers forming eye-ring reduced or wholly absent in some 1stS ♀♀. Minority apparently moult completely, and if so indistinguishable from ad. Some (especially ♂♂) have diffuse whitish dots in iris, approaching ad. – **AUTUMN Ad ♂** Compared to spring, slightly blacker ear-coverts and ill-defined rear crown merging into paler grey-brown upperparts. Remiges and primary-coverts centred blackish and fringed greyer, and pale grey primary tips obvious. White-flecked eye-ring feathers and pure white fringes and tips to outer tail-feathers. **Ad ♀** Often less clearly separated from fresh ad ♂ but in most hood is browner, darker ear-coverts more contrasting and crown merges more gradually into upperparts. Fringes to remiges and wing-coverts less grey. Underparts usually less pure white and eye-ring less complete or narrower. White in outer tail-feathers as in ♂ or slightly less extensive. **1stW** As fresh ad, but sex differences less obvious; both have browner crown and upperparts, and incomplete white eye-ring (notably 1stW ♀). Extremely variable: those that apparently moult entire plumage indistinguishable from ad, unless they retain uniform dark brown iris (lacking flecked whitish iris of ad, but from late autumn some young develop some flecking). Most, however, show distinct moult limits with some unmoulted secondaries and inner primaries being clearly browner and fringed less pure grey than freshly moulted outer primaries and innermost secondaries, including tertials. Innermost juv primary-coverts or central secondaries are usually contrastingly browner and fringed less pure grey than newly moulted ones. Those with limited moult still distinctive by rather clear moult limit in greater coverts: innermost moulted ones fresher, fringed greyish with more blackish centres than juv outermost ones (which are loosely textured and browner, with narrow pale brown fringes). Tail usually replaced completely but, if not, outermost juv tail-feathers diagnostic by much-reduced whitish-buff tips and very narrow fringes to outer web. **Juv** Soft and fluffy body-feathers. Upperparts almost uniform earth-brown (some greyish olive-buff). Head mainly sooty-brown, ear-coverts and lores blacker. Lacks white in eye-ring but has diagnostic small, bold white spot just above upper rear corner of lores. Wing-coverts and tertials fringed pale rusty buff-brown. Throat and belly clean white, while body-sides and usually vent are pale buffish. Typically has greyish-yellow lower mandible and yellow gape. Tail and bare parts as 1stW.

BIOMETRICS (*leucomelaena*) **L** 14.5–16 cm; **W** ♂ 66–76.5 mm (*n* 19, *m* 70.9), ♀ 64–71.5 mm (*n* 15, *m* 68.9); **T** ♂ 62–74 mm (*n* 19, *m* 68.7), ♀ 62–69 mm (*n* 15, *m* 65.9); **T/W** *m* 96.2; **B** 13.7–15.8 mm (*n* 34, *m* 15.0); **BD** 3.6–4.4 mm (*n* 33, *m* 4.0); **Ts** 19.5–23.5 mm (*n* 34, *m* 21.2). **Wing formula: p1** > pc 8–15 mm, < p2 22–26 mm; **p2** < wt 6–10 mm, =9/ss (rarely =9 or 8/9); **pp3–5** about equal and longest (rarely p5 0.5–1 mm <); **p6** < wt 0.5–1.5 mm; **p7** < wt 1–2.5 mm; **p8** < wt 3.5–5 mm; **p10** < wt 7.5–9.5 mm; **s1** < wt 9–12 mm. Emarg. pp3–6 (p7 sometimes also with faint emarg.). **White tip on r6** 3–12 mm (*m* 5.5).

GEOGRAPHICAL VARIATION & RANGE Four subspecies recognised, three of which are found in the covered region, hence treated here. All populations are mainly resident.

○ ***S. l. negevensis*** Shirihai, 1988 (Arava Valley in Israel and Jordan). Subtly larger than *leucomelaena* and slightly greyer above with blacker hood, and shows more obvious sexual dimorphism. Bill somewhat stronger. **W** ♂ 71–75 mm (*n* 12, *m* 72.5), ♀ 67–72 mm (*n* 9, *m* 69.6); **T** ♂ 67–72 mm (*n* 12, *m* 70.0), ♀ 62–72 mm (*n* 9, *m* 66.1); **T/W** *m* 96.1; **B** 15.2–18.0 mm (*n* 18, *m* 16.1); **BD** 3.5–4.4 mm (*n* 32, *m* 4.1); **Ts** 20.2–24.8 mm (*n* 18, *m* 21.7). (Shirihai 1988; ZFMK).

S. l. leucomelaena (Hemprich & Ehrenberg, 1833) (W & S Arabia). Treated in some detail above in the main text.

S. l. blanfordi Seebohm, 1879 (Eritrea, Sudan north to S Egypt). Distinctly smaller, with a paler greyish mantle and better-defined hood at rear. Bill proportionately a trifle stouter, tail a little shorter. Said to have more white in tail, but difference negligible. **L** 13–14.5 cm; **W** ♂ 65–71 mm (*n* 15, *m* 68.1), ♀ 65–68 mm (*n* 8, *m* 66.1); **T** ♂ 57–67 mm (*n* 15, *m* 64.0), ♀ 61–65 mm (*n* 8, *m* 62.9); **T/W** *m* 94.4; **B** 13.3–15.5 mm (*n* 23, *m* 14.4); **BD** 3.6–4.4 mm (*n* 23, *m* 4.1); **Ts** 19.5–23.2 mm (*n* 23, *m* 21.2). **White tip on r6** 2–10.5 mm (*m* 6.1).

REFERENCES Shirihai, H. (1988) *BBOC*, 108: 64–68. – Shirihai, H. (1989) *BB*, 82: 97–113.

S. l. negevensis, juv, Israel, June: soft, fluffy body-feathers with typically earth-brown upperparts, the latter forming only limited contrast with sooty-brown head, but lores and ear-coverts often somewhat blacker. The small white spot above the eye is diagnostic for juv. Wing-coverts fringed pale rusty buff-brown. (H. Shirihai)

YEMEN WARBLER
Sylvia buryi (Ogilvie-Grant, 1913)

Alternative name: Yemen Parisoma

Fr. – Fauvette du Yémen; Ger. – Jemengrasmücke
Sp. – Curruca yemení; Swe. – Jemensångare

Restricted to the highlands of SW Saudi Arabia and Yemen, the poorly-studied Yemen Warbler is an *Acacia* specialist that may be declining due to habitat degradation. It is a mountain-living bird usually found above 1500 m, in vegetated wadis and where there are rockfaces with numerous acacias, sometimes also in trees and dense shrubbery close to cultivation and groves, as well as in juniper stands. The species was formerly placed in the genus *Parisoma* within the babbler family, but recent genetic research has shown it to be a true *Sylvia*.

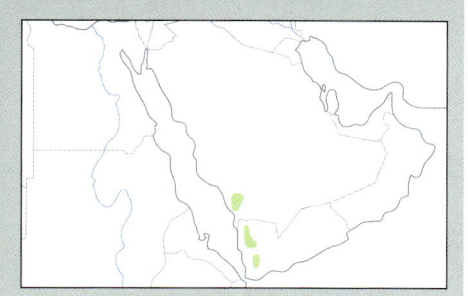

♂, presumed ad, Yemen, Dec: proportionately long tail (usually held half-cocked, and frequently pumped downwards like Arabian Warbler), very short wings and whitish iris (like Eastern Orphean), while warm rufous-brown vent and undertail-coverts are diagnostic. Sexing difficult, but rather grey upperparts, dark hood, clear white-throated appearance, pale iris, and rufous-buff on vent not invading belly is typical of a ♂. Wing seems evenly feathered and ad. (A. Al-Sirhan)

IDENTIFICATION A *rather large and heavy-looking Sylvia* with a relatively long tail and long, slightly curved dark bill. ♂ somewhat *recalls Eastern Orphean Warbler* (which may winter in same area) due to similar size and general colours, but unlike latter has *very short wings* (including short primary projection), structure thus more akin to a babbler, and differs obviously by having *lower belly and vent tinged rufous-ochre* (apparently best developed in ♀). Like ♂ Eastern Orphean has *whitish iris*, and the indistinctly *darker forehead, ear-coverts and lores* might give impression of a darker cap in poor views. Mantle dark grey with rufous (or even purplish) tinge, neck and nape usually slightly paler grey, and wings marginally darker grey. Underparts (apart from rufous-tinged ventral area) mostly greyish-white. Chin and sides of throat can appear diffusely spotted or streaked (mainly an effect of tufted feathers, but some unevenness also in colour). Bill has black tip and pale grey base. ♀ plumage rather more nondescript and browner above, again appearing *darkest on ear-coverts and lores*, and underparts largely cream-grey (faintly mottled whitish), often with paler throat. In all plumages, *uppertail contrastingly darker* (can appear blackish) with fine pale fringes and tips (best seen from below). Tarsus dark grey or brownish-grey. Orbital ring blackish and eye-ring brownish. Rather unobtrusive, frequently seen alone or in pairs. Often hops, and foraging movements *more creeping* than in other *Sylvia*. Gleans prey from low branches and mostly around main trunks of *Acacia* (usually 2–5 m above ground); clings to branches and stems at various angles, and occasionally forages tit-like, even hanging upside-down. *Tail usually held half-cocked, and frequently pumped downwards* (like Arabian Warbler).

VOCALISATIONS Song a prolonged throaty warble, rather slow (but pace somewhat variable) and deliberate with an abrupt start, given from concealed perch, slightly reminiscent of a parrot or Thick-billed Warbler (though slower than latter and hardly contains mimicry), *bi woo woo woo woo eee too-chit too-chit did di chee eeyou-eeyou-eeyou...* (partly based on a recording by P. Davidson/NSA). – Calls unlike other *Sylvia*: in alarm a loud *tsyii*, sometimes repeated, harsh or mewing fluty whistles, *piiuu-ptyii-plii* (Porter et al. 1996), *eeyou* and *wip eeyou*; other calls include *tchah tsuh-tiir-huit*, a rolling *tschee-tschee* followed by high- and low-pitched whistles, while in contact pair members give high-pitched *kiip-kiip-kiip-kiip*, like Lesser Spotted Woodpecker song; also a chattering *chrr-chrr-chrr*.

SIMILAR SPECIES The ♂ might be confused with ♀ *Eastern Orphean Warbler*, but has much shorter primary projection, while tail is proportionately longer with narrower white edges, and is often half-fanned and frequently pumped down. ♀-like plumages wholly distinctive from Eastern Orphean in being decidedly warmer and browner. – Despite a resemblance to *Arabian Warbler*, having similar tail/wing dimensions, similar tail pattern and movements, and also inhabits *Acacia* trees, unlikely to be confused mainly due to long, slightly curved bill and whitish iris of Yemen Warbler. Their jizz is also completely different. Voice offers further clues.

AGEING & SEXING Ageing requires close inspection of moult and feather wear in wing. Tail pattern sometimes useful too. Plumage variation poorly known, and sexes apparently separable mainly in ad, but variation in iris coloration incompletely known. Insignificant seasonal variation. – Moults. Complete post-nuptial and partial post-juv moult in late summer involve body-feathers, median and lesser coverts, some or all greater coverts, and tertials. Timing and extent of partial pre-nuptial moult is poorly known. – **BREEDING Ad** See Identification. **1stS** As respective ad but retains variable number of worn juv remiges and primary-coverts and some or all tail-feathers. Iris colour undocumented. – **NON-BREEDING Ad** As breeding. All remiges and primary-coverts

Presumed ♀, Yemen, Oct: overall rather nondescript, greyish belly rufous buff-tinged and dorsally slightly brown-tinged, with dark on head mainly restricted to lores, and iris not quite as pale as many others suggesting a ♀ (but young ♂ cannot be excluded). Like Arabian Warbler is associated with *Acacia* trees. (U. Ståhle)

Presumed ♀, SW Saudi Arabia, May: judging from the rather featureless plumage but pale iris, this ought to be a ♀. Quite a 'scruffy' and uneven plumage. Still, dark grey lores and eye-surround in combination with strong dark bill and whitish iris render the bird a stern look and 'personality'. (M. Al Fahan)

of same generation, evenly fresh, fringed greyish-brown, with large pale fringes and tips to tail-feathers. Iris white. **1stW** As respective ad but some young ♂♂ approach ♀♀. Differs from ad in juv remiges, primary-coverts and some greater coverts, which are paler and browner with buffier fringes (contrasting with new ad-like feathers, e.g. inner greater coverts and tertials, which are clearly darker). Juv tail-feathers have reduced, greyish-buff fringes and tips (though some moult to ad-like pattern). Underparts more buff-tinged. Iris apparently less white than in ad, some perhaps not becoming white at all. **Juv** Soft, fluffy body-feathers, crown and upperparts have obvious rufous suffusion, and paler underparts with light buff wash mainly to vent. Tail and bare parts as 1stW.

BIOMETRICS (Limited material available. The following is based on the five specimens in NHM, including holotype.) **L** 16–17 cm; **W** 68.5–71 mm (n 5, m 69.5); **T** 68–72.5 mm (n 5, m 70.1); **T/W** m 100.9; **B** 16.0–17.8 mm (n 5, m 17.0); **BD** 4.1–4.6 mm (n 5, m 4.4); **Ts** 22.2–25.5 mm (n 5, m 23.9). **Wing formula: p1** > pc 13–16 mm, < p2 15–19 mm; **p2** < wt 8–10 mm, = ss or < ss; **pp3–6** about equal and longest (rarely p3 and/or p6 0.5–1 mm <); **p7** < wt 0.5–1 mm; **p8** < wt 1.5–3 mm; **p10** < wt 5.5–7 mm; **s1** < wt 7–8 mm. Emarg. pp3–7.

GEOGRAPHICAL VARIATION & RANGE Monotypic. – SW Saudi Arabia, W Yemen; resident.

REFERENCES BROOKS, D. J. (1987) *Sandgrouse*, 9: 90–93.

Presumed ♀, SW Saudi Arabia, Jul: rather similar to the bird on the previous page (bottom), thus reasonably a ♀. Again, note that some birds can appear quite 'scruffy' below. Note heavily abraded central tail-feathers in July. (J. Babbington)

Yemen, Jan: young birds generally tend to be browner. The dark forepart of head and more pronounced pale throat, and rather extensive rufous-brown vent and undertail-coverts, might suggest young ♂. However, our lack of experience with variation in this species prevents safe ageing and sexing. Dark bill-tip is apparently due to shadow. (H. & J. Eriksen)

WESTERN ORPHEAN WARBLER
Sylvia hortensis (J. F. Gmelin, 1789)

Fr. – Fauvette orphée; Ger. – Orpheusgrasmücke
Sp. – Curruca mirlona occidental; Swe. – Herdesångare

Secretive inhabitant of rocky, wooded areas, maquis and orchards, sometimes also oak or pine forests, in the W Mediterranean region east to Switzerland and Italy, and in N Africa east to Libya. Secretive perhaps, but its loud and slightly monotonous song often gives its presence away. It departs as early as July for its winter quarters in sub-Saharan W Africa. Previously regarded as part of a more widely distributed species including the Eastern Orphean Warbler, but nowadays the western constituents are commonly treated as a separate species.

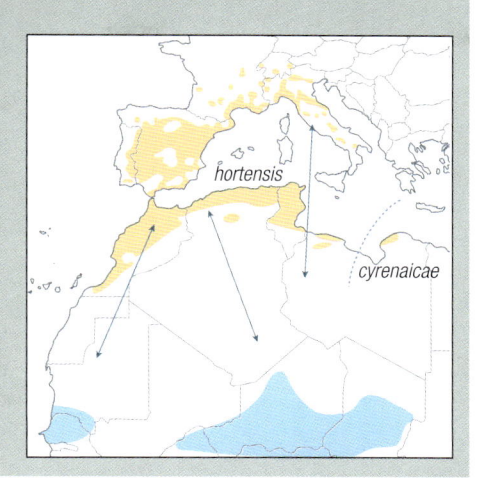

IDENTIFICATION Robust *Sylvia* with *broad-based and long, pointed bill*, largish, rather angular head and *long wings* (6–7 widely-spaced exposed primary tips), often appearing somewhat pot-bellied. *Adult and first-summer ♂♂ have a blackish (or dark grey) hood*, darkest on ear-coverts and contrasting with paler brownish-grey upperparts. Slightly browner ♀-like plumages, including *first-winter ♂, have a greyer, less contrasting hood* but *more obviously contrasting ear-coverts*, and often pale lores. All plumages have *pale underparts with white throat*, and are *tinged pinkish-buff on rear flanks, belly and vent* (somewhat variable). Tertials have indistinct pale grey fringes, and *tail-feathers have extensive white tips/edges on outermost* (amount varies, mostly with age), with characteristic *white narrow wedges on outermost two* (rr5–6). Eye-ring mostly blackish in adult ♂, greyish in first-winter ♂ and grey admixed whitish in ♀. *Iris colour varies from whitish in adults to dark grey-brown in first-winters*, with blackish or brownish-grey eye-ring. Bill has dark culmen and tip, and variably bluish-grey base to lower mandible (shortest-billed birds appear mainly entirely 'dark-billed', longest-billed show more of bluish-grey base). Legs are dark grey. As to difference in tail pattern between Western and Eastern Orphean Warblers, see below under Similar species and under Eastern Orphean. Movements typically clumsy but deliberate. Often rather shy, even on migration, and sometimes keeps high in trees. As to tail pattern, see below under Similar species.

VOCALISATIONS Song a loud and rather monotonous strophe, usually comprising 5–10 syllables, with pairs of rising and falling notes creating 'sawing' pattern (like Ring Ouzel), *tee-ro tee-ro tee-ro...* or *che-wee che-wee che-wee...* Sometimes gives a more complex strophe, but typical repeated sawing notes often included or soon heard. Song is much simpler and more repetitive in structure than Eastern Orphean Warbler (which see), thus readily separable with a little practice. – Call a sharp clicking *check* or *tchak* in contact, very similar to Blackcap. A drawn-out sparrow-like *cher'r'r'r'r'r* in alarm, rather like Barred Warbler. At times when anxious stuttering series of slightly nasal calls also heard, *chet-chet-chet-chet-...*

SIMILAR SPECIES For separation from *Eastern Orphean Warbler* in the field, see that species. Note that in-hand separation from Eastern Orphean on tail pattern is possible for all adults and nearly all young (Svensson 2012), sometimes also in the field. In adult Western Orphean, r6 has long narrow white wedge reaching at least to 2/3 (often to 3/4 or more) along feather length and does not widen much

S. h. hortensis, ♂, Spain, Apr: a robust *Sylvia* with quite strong bill, blackish head and distinctive whitish iris. With wear, blackish hood becomes more well-marked. Note characteristic pinkish-buff suffusion on sides and undertail-coverts. Both age classes return in spring with moult limits in wing, making it difficult without handling to be sure of the age of primaries and primary-coverts. (C. N. G. Bocos)

near tip (adult Eastern Orphean: short, wide or obtuse white wedge, broadening further at tip and proximally reaching no more than 1/3 or 1/2 along feather; very few slightly ambiguous birds found). There is an average difference also in the shape of the white tip to r5 in that this usually has a short but pointed and narrow white wedge near tip (adult Eastern Orphean: square or fan-shaped white tip to r5, but note that a few Western also have rather square white tip to r5); in juvenile Western Orphean, pattern is similar to adult, only sullied brown-grey and more diffuse, and pale tip on r5 is much more diffuse and often oval or rounded (juvenile Eastern Orphean: similar wide wedge as in ad but sullied brown-grey or, rarely, a more narrow wedge, closer to pattern of Western Orphean). – Despite significant size differences, *Lesser Whitethroat* (which is c. 15–20% smaller, much less bulky, more sprightly and nervous, with a shorter, thinner bill) is in overall plumage very similar, differing only in having at least a whitish lower eyelid and often a near-complete narrow whitish eye-ring (♀ or imm Western Orphean has at most a greyish-white upper eyelid), less saturated rear underparts, and a less obviously dark tail with in Europe much narrower white edges. Also, their calls are quite different with experience. – Other dark-headed *Sylvia* species, e.g. Sardinian

S. h. hortensis, ♂, presumed ad, Spain, Jun: diagnostic, long, narrow white wedge on r6. Typical is also blackish or sooty-brown hood diffusely demarcated at rear, and unpatterned pink-buff undertail-coverts. Base of lower mandible almost invariably light bluish-grey. Whitish iris is suggestive of ad, as is fairly moderate wear of wings and tail in June, but impossible to exclude an advanced 1stS. (C. N. G. Bocos)

S. h. hortensis, ♀, presumed 1stS, Western Sahara, Apr: ♀♀ are browner above, with a less dark crown, but usually contrasting darker ear-coverts, and often hint of pale fore supercilium. Some have the underparts as here rather strongly tinged pinkish-buff, making separation from Eastern Orphean easier. Possibly has retained odd outer juv tail-feathers, whereas large alula has been renewed in pre-nuptial moult suggesting 1stS. (T. Svensson)

S. h. hortensis, ♂, presumed 1stS, Portugal, May: cream-yellow iris (like ad) uncommon in 1stS ♂, but can occur. Note abraded and brownish juv primary-coverts, as well as less developed blackish hood (blotchy with slightly darker ear-coverts) and less pure grey upperparts. In 1stS, the brownish-fringed, apparently unmoulted outer primaries are important for separation from Eastern Orphean; rufous-buff rear underparts also important in this respect. (D. Monticelli)

Warbler, are significantly smaller than Western Orphean, with smaller bills, more agile, restless movements, have proportionately longer tails (often cocked or waved) and often red orbital rings, orange-red irides and fleshy pink-brown legs, with broader white areas on the tail. – A relatively unmarked *Barred Warbler* might prove a pitfall for the unwary, being of similar size, but both Orphean Warbler species have a longer bill, darker upperparts, greyer crown, dark ear-coverts, and lack any suggestion of a greyish-white eye-ring or noticeable pale fringes or tips to the wing-coverts, tertials and uppertail-coverts, nor of obvious dark-centred undertail-coverts (although first-winter and first-summer Barred may have these features to some extent obscured).

AGEING & SEXING Ageing possible by checking moult and feather wear in wing, but requires close observation if not handling. Tail pattern and iris colour are also useful to check. Sexual dimorphism most obvious in ad and breeding birds. Little seasonal feather wear. – Moults. Complete postnuptial and partial post-juv moult in late summer (mostly Jun–Aug). Post-juv moult involves most feathers of head and body, lesser and median coverts, usually some inner greater coverts and, less commonly, innermost tertials; no or only few tail-feathers replaced before autumn migration. Postnuptial moult variable in timing and extent: ad starts a partial or near-complete moult, or performs a complete moult, on breeding grounds; those which do not moult completely suspend and resume directly after autumn migration (late Sep–Nov); after suspension in autumn, continued winter moult not entirely complementary, and far less extensive than in Eastern Orphean. Partial pre-nuptial renewal (Feb–Apr) usually includes parts of head and body, small secondary-coverts, some secondaries, tertials, greater coverts and tail-feathers (rarely more extensive). 1stS usually replace all or most tail-feathers in late winter. – **SPRING Ad** For reliable ageing attempt to confirm that only ad wing-feathers are present, with only moderately worn primaries and primary-coverts, and some fresher secondaries and tertials. (Beware, however, that differences in abrasion and colour of primaries do not differ that much between ad spring and 1stS. Also, note that a moult contrast caused by renewed feathers in pre-nuptial moult can confuse the ageing process; also those ad that, very rarely, retain unmoulted odd secondaries can be misidentified as 1stS with extensive pre-nuptial moult.) **Ad ♂** Head wears more solidly black, ear-coverts become concolorous with crown or slightly darker, upperparts greyish-brown and underparts whiter with reduced pinkish-

S. h. hortensis, ad ♂, Spain, Sep: a well-marked fresh ad ♂, but compared to spring crown is less intense black, with hint of pale grey spots on lores. Pale buff-brown wash below, especially on rear flanks, vent and undertail-coverts (the latter with ill-defined paler fringes, cf. Eastern Orphean). Whitish iris in autumn is diagnostic of ad, but also note very fresh pale primary tips. (A. Tamayo Guerrero)

S. h. hortensis, ad ♀, Spain, Sep: sometimes differs from ♂ by not quite so dark hood and paler lores. From 1stW by being overall fresher, including remiges with fine whitish primary tips, and by having primary-coverts that are greyer and neatly pale-fringed, and rather pale iris. Warm buff below, relatively short bill and primary projection separate this species from Eastern Orphean. (J. L. Muñoz)

— 540 —

buff suffusion. In both sexes, primaries slightly browner, with reduced pale grey fringes and whitish tips. Iris always at least partly, sometimes extensively, whitish (in some birds pure white or almost so, especially older birds and mainly ♂♂). Eye-ring mainly black. **Ad ♀** Blackish-grey crown contrasts with blacker ear-coverts; rest of upperparts duller, and below whiter. Iris much as ad ♂. Eye-ring usually a mixture of grey and off-white. **1stS** Juv primaries and primary-coverts more worn and abraded browner with faded buffish-brown fringes. Iris in some mostly olive-brown, lacking whitish spots, but others, apparently mostly ♂♂, almost like ad. Like ad, primaries unmoulted in winter (but unlike Eastern Orphean, which frequently renews some outer). Both sexes show vestiges of immaturity, e.g. ♂ has less completely blackish head (blotchy with clearly blacker ear-coverts) and less pure grey upperparts. Some diagnostically retain juv tail-feathers with easily recognised diffuse and less extensive pale tips, wedges and edges. – AUTUMN **Ad** Both sexes tinged pale buff-brown on lower central throat, chest and extensively on flanks and almost entire belly. (There is a hint of white submoustachial stripe and chin, which still affords white-throated impression.) Pale fringes to diffusely brownish-centred undertail-coverts ill-defined and rather obscure and limited (cf. Eastern Orphean). Separated from 1stW by being overall fresher with glossy blackish grey-brown remiges and primary-coverts, and purer grey fringes and bold whitish primary tips. Tail-feathers darker, outer pairs with more and pure white (mainly on outermost). Iris at least partly whitish; rarely, especially in ♀, yellowish or buffish-cream. **Ad ♂** Compared to spring, crown dark plumbeous-grey (less blackish), with ear-coverts blacker; greyer rear crown merges gradually into mid grey rest of upperparts (tinged pale olive-brown). Eye-ring mainly blackish. **Ad ♀** Usually separated from ♂ by paler crown with dark ear-coverts and paler lores (but some are difficult to sex). Rest of upperparts browner, extensively tinged olive buff-brown. As in spring, eye-ring largely consists of white and grey feathers. **1stW** Both sexes resemble ad ♀; unlike fresh ad, juv primary-coverts and remiges slightly softer in texture and browner, fringed buffish-brown (also, juv primaries lack obvious pale tips of ad). Moult limit usually observable among greater coverts, with innermost fresher ad-like (more solidly textured and fringed greyer) while retained juv outer ones have characteristic looser texture and narrow, faint buffish-brown fringes. Those that replace all greater coverts still exhibit slight difference between these and unmoulted duller brown primary-coverts and remiges, and often also show moult limit in tertials; conversely, some do not replace any greater coverts and are then very difficult to age. Pale tips, wedges and fringes to outermost juv tail-feathers less extensive and sullied brown. Iris diagnostically dark brown, never partly whitish in summer or early autumn, whereas some ♂♂ might start to acquire paler iris from late autumn. Note deeper and more extensive pinkish-buff underparts, contrasting white throat and poorly-developed chevrons on undertail-coverts, or nearly uniform buffish-brown undertail-coverts typical of Western Orphean (some tiny overlap with Eastern Orphean pattern though). **Juv** Like 1stW but has typical soft, buff body-feathers, and ear-coverts only slightly darker than crown. Lores pale with small but obvious whitish upper loral patch, and upper- and underparts profusely washed buffish sandy-grey.

BIOMETRICS (*hortensis*) **L** 15–16 cm; **W** ♂ 73–87 mm (*n* 46, *m* 80.8), ♀ 74–84 mm (*n* 36, *m* 79.2); **T** ♂ 60–72 mm (*n* 47, *m* 66.2), ♀ 59–69 mm (*n* 36, *m* 65.2); **T/W** *m* 82.2; **B** 15.4–19.0 mm (*n* 84, *m* 17.1); **BD** 4.2–5.3 mm (*n* 79, *m* 4.6); **Ts** 21.0–24.0 mm (*n* 78, *m* 22.6). **Wing formula:** **p1** > pc 2–7 mm, < p2 31–41 mm; **p2** < wt 2–7 mm, =5 or 5/6 (63%) or 6 or 6/7 (37%); **pp3–4** (5) about equal and longest; **p5** < wt 0.5–3.5 mm; **p6** < wt 3–6 mm; **p7** < wt 6.5–11 mm; **p8** < wt 9–14 mm; **p10** < wt 14–18.5 mm; **s1** < wt 17–21 mm. Emarg. pp3–5.

GEOGRAPHICAL VARIATION & RANGE Very little variation with a subtle cline of increasing bill length from north to south, and only two subspecies recognised.

S. h. hortensis, 1stW, presumed ♂, Spain, Sep: diagnostic warm pinkish-buff underparts (with only weak chevrons on undertail-coverts) separate from Eastern Orphean, as does relatively shorter primary projection. (Bill unusually long on this bird.) Aged by juv remiges and primary-coverts, fringed brown. Also note moult limit in greater coverts (inner two renewed), while iris is diagnostically dark brown. Some can be sexed; here intense blackish mask and rich buff underparts suggest ♂. (J. Sagardía)

● *S. h. hortensis* (J. F. Gmelin, 1789) (Iberia, S France, local in Switzerland, NW Africa from Morocco to Tunisia, possibly also Tripolitana in Libya; winters in sub-Saharan Sahel from Senegal and Mauritania east to Chad). Described above.

○ *S. h. cyrenaicae* Svensson, 2012 (N Cyrenaica, NE Libya; apparently winters Mali and Niger). Very similar to *hortensis* but differs in having a clearly longer (by 11%) and more obviously two-coloured bill. It also appears to have subtly paler mantle (though this requires confirmation on a longer series than presently available). The following biometrics are based on seven certain breeders and four likely winter birds (10 ♂♂, 1 ♀; AMNH, NHM; Svensson 2012). **W** 79–87 mm (*n* 11, *m* 82.2); **T** 62–68 mm (*n* 11, *m* 65.8); **T/W** *m* 80.1; **B** 18.5–20.3 mm (*n* 11, *m* 19.1); **BD** 4.3–5.3 mm (*n* 10, *m* 4.9); **Ts** 21.3–23.9 mm (*n* 11, *m* 22.9). **Wing formula:** **p1** > pc 1–5 mm, < p2 35–42 mm; **p2** < wt 2–5.5 mm, =5 or 5/7 (once =4/5); **pp3–4** about equal and longest; **p5** < wt 0.5–2.5 mm; **p6** < wt 4.5–6.5 mm; **p7** < w. 8.5–11.5 mm; **p8** < wt 11–19 mm; **p10** < wt 15.5–21 mm; **s1** < wt 18–23 mm.

TAXONOMIC NOTE Western Orphean Warbler considered specifically distinct from allopatric Eastern Orphean Warbler based on differences in plumage (mainly tail pattern, undertail-coverts pattern, underparts coloration), mtDNA and very different song. To this can be added a somewhat different pre-nuptial moult.

REFERENCES SVENSSON, L. (2012) *BBOC*, 132: 75–83.

S. h. hortensis, 1stW, presumed ♀, Spain, Aug: paler overall than previous bird, with slightly more brown-tinged upperparts and more prominent whitish fore supercilium that may indicate a ♀. It has apparently moulted all of the greater coverts and might therefore be tentatively aged as ad ♀, but the mostly dark iris confirms it is 1stW. (Q. Marcelo)

EASTERN ORPHEAN WARBLER
Sylvia crassirostris Cretzschmar, 1830

Fr. – Fauvette orphéan; Ger. – Nachtigallgrasmücke
Sp. – Curruca mirlona oriental; Swe. – Mästersångare

Recently considered a separate species, different from the western form of Orphean Warbler. Eastern Orphean Warbler ranges from Slovenia and Bosnia to Central Asia and Pakistan, and Eastern and Western Orphean are apparently completely allopatric (although occasional contact might be expected in NE Italy or extreme NW Slovenia). Very similar in ecology, plumage and behaviour to Western Orphean Warbler, therefore to correctly diagnose a vagrant in NW Europe, especially in autumn, is a real challenge.

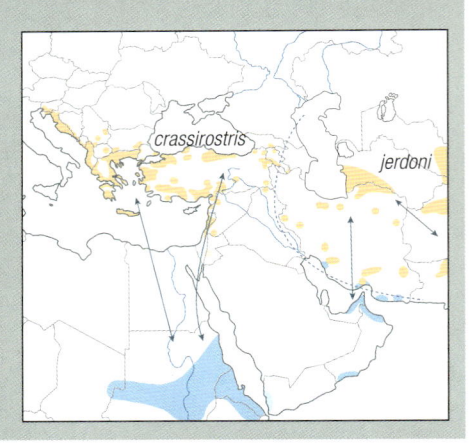

S. c. crassirostris, ad ♂, Israel, Mar: there is individual variation in darkness of hood and paleness of iris. Here a ♂ with quite pale iris but less dark hood. Still, well within variation of ad ♂, and age further supported by moult pattern: new secondaries and innermost primary, and rest of primaries replaced the previous summer. Note whitish and grey underparts lacking any pinkish-buff hues. (H. Shirihai)

IDENTIFICATION Very similar to Western Orphean Warbler: *robust with a notably long pointed bill*, rather large and quite angular head and long wings. Often appears pot-bellied, and movements clumsy and almost 'exaggerated'. ♂ *has a black (or blackish-grey) cap, often darkest on ear-coverts and contrasting with slightly paler grey crown*, the cap being on average better-marked in Eastern than Western Orphean Warbler. ♀ and young are *brownish-grey with a less dark cap* and *more obviously contrasting darker ear-coverts*. *All-whitish underparts, tinged pale buffish-grey (or brown-grey) on flanks* with *dark-centred undertail-coverts*, forming a hint of chevrons or arrowheads. *Iris pale yellowish-cream or whitish in adults*, but darker or blotchy in young. Eye-ring mostly blackish in adult ♂, greyish or admixed whitish in ♀/young. Orbital ring blackish or brownish-grey. Bill has blackish culmen and tip, and pale blue-grey lower base. Legs dark grey. For a useful difference in tail pattern (not always possible to see in the field), see Similar species. Habits as for Western Orphean Warbler.

VOCALISATIONS The song is similarly loud as Western Orphean Warbler, but more varied, rich and pleasing, containing soft trills and series of whistled notes, scratchy sounds, lower-pitched elements and is given in clearly longer and more complex strophes, lacking the simple structure and 'sawing' monotony of Western Orphean. In fact, due to its loudness and explosive start, its song is sometimes mistaken for that of Nightingale, but a prolonged listening will usually soon reveal Eastern Orphean's song to be a little slower, slightly harder, with many dry trilling notes and without the slow series of whistles in crescendo typical of Nightingale (though odd birds seem to mimic these notes from Nightingale song!). – Calls and alarm seem identical to Western Orphean.

SIMILAR SPECIES While Western and Eastern Orphean Warblers are apparently completely allopatric when breeding, and perhaps even on migration and in winter, separation of vagrants in W Europe (especially young or ♀♀) is extremely problematic, especially as *crassirostris*, the race of Eastern Orphean probably most likely to wander west, is most similar to *hortensis*. In normal field views differs from Western Orphean in that adult ♂ is purer grey above, with an often slightly darker cap that is subtly more sharply demarcated at the rear (and much darker and better-marked in eastern race *jerdoni*), and in much cleaner white underparts (any cream-grey or pale brown-grey being limited to the sides). ♀/young are slightly paler and drabber above than Western Orphean, with whiter underparts (flanks, lower belly and vent suffused pale cream-buff or greyish, whereas in Western Orphean these areas are often warmer, more pinkish-brown or pink-ochre) and a slightly darker crown. Common to all is an on average longer and slightly more deep-based bill, with usually more obvious bluish-grey base than in Western Orphean (but much overlap in both length and colour!), and note diagnostic moult limits in first-summer birds. Eastern Orphean further has ill-defined dark centres (with pronounced pale fringes) to most undertail-coverts (these feathers being much plainer in most Western Orphean, with only a few, usually ♂♂, having visible darker centres to the longest coverts alone, the majority having plain whitish or pinkish cream-tinged coverts without darker centres). However, this is difficult to detect in the field, and the ventral area of a few Eastern Orphean, largely ♀♀/young, appears almost plain. Tail pattern of adults and most first-winters also differs, being sometimes possible to see on good photographs or in very close views (and of course easily in the hand; Svensson 2012): in Eastern Orphean the white of r6 usually forms a wide (obtuse) wedge confined to inner web of outer 1/3 or at most 1/2 of feather, and the wedge often widens further near tip (Western Orphean: narrow, long white wedge on inner web of r6 reaching at least 2/3 along feather length, often more, not widening at tip). There is an average difference also in the pattern of white tip on r5 (sometimes also on r4)

S. c. crassirostris, ♀, Turkey, May: poorly developed dark hood and rather dusky iris in spring make this bird a ♀, while only limited pre-nuptial moult (almost entire wing of one generation, just tertials renewed) probably could fit either age. The greyer flanks, dusky grey-brown vent-sides and clearly pale-fringed undertail-coverts help separate the species from Western Orphean. (C. Dogut)

Eastern Orphean Warbler ssp. *crassirostris* (left) versus Lesser Whitethroat, presumably ssp. *halimodendri* (Israel, Mar): despite significant size difference, these two are superficially quite similar. Lesser Whitethroat is distinctly smaller (by c. 15–20%), without Orphean's bulk, and is often more sprightly and nervous. If separation using size is proving problematic, then check for Lesser Whitethroat's (1) shorter, thinner bill; (2) diagnostic broken white eye-ring (usually a single white crescent below the eye, or occasionally above and below, as here); (3) lack of dark centres to undertail-coverts (versus Eastern Orphean) or buffish saturation (Western Orphean), with a less obviously dark uppertail; and (4) calls. The first two points also eliminate Western Orphean. Eastern Orphean is 1stS ♀ by diagnostic moult pattern (outer primaries newer, and contrast between old, worn and brownish juv primary-coverts and winter-moulted alula), rather dark iris, greyer head and darker ear-coverts; ageing the Lesser Whitethroat is trickier without handling, but the less worn primary tips and primary-coverts suggest ad. (D. Occhiato)

S. c. crassirostris, ♂, presumed 1stS, Greece, Apr: Eastern Orphean Warbler tends to be purer grey above and off-white below, less tinged brown and pink-buff than Western. Note also the characteristic dark-centred undertail-coverts. Sex inferred by quite dark hood, while darkish iris and an apparent moult limit in primary-coverts, identify this bird as most likely 1stS. (T. Gaitanakis)

S. c. crassirostris, ♂, presumed 1stS, Israel, Apr: while sex is inferred by quite dark hood, iris is mottled dark indicating 1stS. The problem with a definite ageing of this bird is the moult pattern: primaries except innermost seem very abraded and brown indicating 1stS, but more normal for this age is to replace a varying number of outer primaries, not inner. A good example of variation and remaining difficulties. (H. Shirihai)

in that Eastern usually has a square or fan-shaped white tip, not forming a pointed wedge (Western Orphean: white tip on r5 generally narrow, pointed wedge, only much shorter than on r6, but a few Western have similar blunt wedge-shaped or square white tips as in Eastern); juvenile tail-feathers are similar to corresponding adult, only pale pattern more diffuse and sullied brown-grey, and odd Eastern Orphean are more similar to Western Orphean, requiring even more care or are best left unidentified on this character alone. In summary, when dealing with a vagrant, first-winters should as a rule be left undetermined (unless trapped or clearly showing diagnostic characters), but adults and spring birds showing a full suite of characters may be assigned to species; observers should take into account individual variation and correctly assess age/sex first. – Separation from *Lesser Whitethroat* is covered under Western Orphean Warbler, and *Arabian Warbler* under that species. – Other dark-headed *Sylvia*, including *Sardinian*, *Cyprus*, *Ménétries's* and *Rüppell's Warblers* are considerably smaller, much more agile and restless, and differ structurally, as well as by plumage and bare-part characters. However, a poorly-marked first-year ♀ Rüppell's is more likely to be confused with Eastern Orphean; this pitfall is discussed under Rüppell's Warbler.

S. c. crassirostris, 1stS ♀, Israel, Mar: young ♀♀ are more nondescript due to their darker iris and poorly defined hood but can usually be identified to species by their diagnostic dark-centred and white-fringed undertail-coverts and lack of pinkish-buff hue on sides and vent, often also supported by moult pattern (note many outer primaries renewed during pre-nuptial moult, and typically worn juv primary-coverts just visible). (H. Shirihai)

Eastern Orphean Warbler *S. c. crassirostris* (top left: ad ♂, Turkey, Aug; top right: 1stW, Israel, Jul) and Western Orphean Warbler (bottom left ad ♂, and bottom right 1stW, Spain, Sep) in early autumn: in both ad ♂♂ and 1stW the species differ in similar ways. Eastern Orphean tends to have clearly pale-fringed, dark-centred undertail-coverts (in Western these are plainer, tinged pinkish-brown, with dark centres reduced or lacking), and rest of underparts are cleaner and whiter or at the most washed brown-grey, mostly on sides (in Western these areas are warmer, brownish-buff). Due to the paler underparts, Eastern tends to give less of white-throated impression than the warmer-bellied Western. Eastern also has on average a longer bill (but much overlap), and in ♂♂ the black hood tends to be better demarcated at the rear. Ad of both species aged in autumn by having mostly fresh ad wing-feathers and pale iris, while 1stW birds generally have retained several brownish juv outermost greater coverts, alula and primary-coverts, and dark iris. (Top left: M. Goren; top right: E. Bartov; bottom left & right: J. Matute)

AGEING & SEXING Ageing requires close check of moult and feather wear in wing. Tail pattern and iris colour also useful to note. Sexual dimorphism most obvious in ad and breeding birds. Little seasonal feather wear. – Moults. Much as in Western Orphean, but different moult strategy for 1stY frequently enables spring birds to be aged. Complete post-nuptial moult of ad commonly starts in late summer, then is suspended and resumed after autumn migration. Partial post-juv moult in late summer (mostly Jun–Aug). Pre-nuptial moult (Feb–Apr) frequently more extensive than in Western Orphean, and often includes several tail-feathers, secondaries and outer primaries (or rarely all primaries) plus corresponding primary-coverts, which feathers are thus apparently moulted twice a year. Unknown whether those birds renewing flight-feathers in late winter are ad resuming suspended moult, or are 1stS. If 1stS replaces some flight-feathers in winter it is reasonable to believe that ad, too, moults extensively in winter, either completely or partly. Still, confusing spring birds, presumed ad, have a few inner primaries and inner primary-coverts renewed, these being slightly darker and have neater ash-grey fringes, while outer are older and worn and more grey-brown. The full variation of moult is yet to be established. – **SPRING Ad** Either all primaries and primary-coverts reasonably fresh and lacking moult contrasts, or, less often, primaries slightly worn, rarely with inner ones renewed in pre-nuptial moult in late winter, being darker with ash-grey edges. **1stS** Commonly a moult contrast created by outer 4–7 primaries renewed and fresh, contrasting to inner more abraded and grey-brown primaries. However, some 1stS show no moult contrast, this due to all primaries being retained juv feathers (or rarely all being moulted?). The frequency of limited moult in 1stY is yet to be established. – **AUTUMN** Largely as Western Orphean, but ad post-nuptial moult (before autumn migration) mostly suspended, otherwise similarly complete (only secondaries, sometimes also tail-feathers retained). Partial post-juv moult much as in Western Orphean, but sometimes wing-coverts, and even body-feathers, retained (those with old greater coverts thus usually Eastern Orphean), but differences often insignificant.

BIOMETRICS (*crassirostris*) Sexes do not differ in size. **L** 16–17 cm; **W** 77–84.5 mm (n 66, m 80.5); **T** 61–74 mm (n 67, m 66.6); **T/W** m 82.8; **B** 16.4–19.6 mm (n 66, m 18.1); **BD** 4.2–5.4 mm (n 65, m 4.8); **Ts** 21.5–24.2 mm (n 66, m 22.9). **Wing formula: p1** > pc 6 mm, to 1 mm <, < p2 34–42.5 mm; **p2** < wt 1.5–6 mm, =5/6 or 6 (67%), =4/5 or 5 (30%) or =6/7 (3%); **pp3–4** about equal and longest; **p5** < wt 0.5–4.5 mm; **p6** < wt 5–8.5 mm; **p7** < wt 6.5–12 mm; **p8** < wt 9–14 mm; **p10** < wt 14–19 mm; **s1** < wt 17–24 mm. Emarg. pp3–5 (but sometimes faint or nearly absent on p5).

GEOGRAPHICAL VARIATION & RANGE Moderate variation, with birds tending to be paler and to have better contrasting black cap in ♂ plumage in the east. Two races recognised (a third is intermediate but much closer to *jerdoni*, from which many cannot be separated, and best included therein).

Presumed *S. c. crassirostris*, ad ♂, United Arab Emirates, Sep: around Persian Gulf, two different races can be expected on migration, occasionally also in winter, but this bird appears to fit best with *crassirostris*: hood less blackish-looking, upperparts brown-tinged, bill not as long as in many *jerdoni*. This ♂ has ad-like iris, and moult pattern with partially suspended replacement of secondaries and tail-feathers infers age as ad. (A. Al-Sirhan)

S. c. crassirostris, juv, Turkey, May: typically soft and fluffy feathers and overall buff-brown tinge. Note pale lores, drab sandy-grey upperparts but slightly more greyish crown, while underparts are cream-grey and buff. Yellow gape-flanges still obvious, but bill appears still slightly short of final size. Race by locality. (R. Debruyne)

S. c. jerdoni, ♂, presumed ad, United Arab Emirates, Feb: the eastern race differs in being smarter and more contrasting with paler grey upperparts in strong contrast to very black and distinct hood. Further, underparts tend to be cleaner white. Whole tail seems to be post-juv with very distinct white tips to several penultimate feathers, thus most likely an ad, but definite ageing impossible without handling. (K. Al Dhaheri)

S. c. jerdoni, ad ♂, India, Dec: note deep black hood that reaches nape and ends in a well-defined rear border. Upperparts usually slightly paler grey, and bill and tail average longer. When fresh, pale underparts can be sullied greyish cream-buff, enhancing white throat. Chevron-like pattern to undertail-coverts well developed. This ♂ (by black hood) has whitish iris and evenly very fresh plumage having recently completed its post-nuptial moult. (N. Devasar)

S. c. jerdoni, ♀, presumed ad, India, Apr: combination of greyish head in spring and pale grey iris suggests an ad ♀. Both ad and 1stY undertake partial moults, but with different patterns (as a rule, ad replaces mostly inner remiges, 1stY outermost primaries and tertials). Impossible to be sure if this bird is ad (most likely), or a young ♀ that has moulted completely and developed a pale iris. Note proportionately long bill and tail in this race. (A. Deomurari)

S. c. crassirostris Cretzschmar, 1830 (SE Europe from Slovenia to Turkey, possibly westernmost Transcaucasia, south to Levant; winters around Red Sea south to Sudan, S Saudi Arabia, Oman). Described above. There is a slight tendency for breeders to have shorter bill in Levant (*m* 17.6), somewhat longer in Greece and Balkans (*m* 18.1) and longer still in Turkey (*m* 18.4), but samples are small. (Syn. *helena*.)

S. c. jerdoni (Blyth, 1847) (C & E Transcaucasia, possibly whole Transcaucasia, Transcaspia, Iran, Central Asia north to Tajikistan & Kyrghyzstan, Pakistan, extreme NW India; winters Indian subcontinent, perhaps rarely also west to Persian Gulf). Paler and greyer above, with darker and better-marked cap in ♂♂, in particular at rear towards nape. Underparts whiter with very limited pale cream-buff on flanks; dark chevron-like pattern on undertail-coverts well developed, on average more visible than on *crassirostris*. Also, has longest bill and proportionately longer tail. ♀ is very similar to *crassirostris*, though is slightly paler above and whiter below; older ♀♀ often have solid blackish-grey hood. **W** ♂ 77–87 mm (*n* 20, *m* 81.3), ♀ 76.5–83 mm (*n* 13, *m* 79.2); **T** ♂ 62–73 mm (*n* 20, *m* 67.8), ♀ 62–69 mm (*n* 13 *m* 66.2); **T/W** *m* 83.6; **B** 17.8–21.8 mm (*n* 39, *m* 19.7); **BD** 4.1–5.5 mm (*n* 40, *m* 4.8); **Ts** 21.5–24.0 mm (*n* 34, *m* 23.0). – Birds in Transcaspia and much of Iran sometimes separated ('*balchanica*') on account of their intermediate appearance between *crassirostris* and *jerdoni*. However, they are much closer to *jerdoni* and many (> 50%) cannot be separated from latter. Therefore it seems best to include them in *jerdoni*. (Syn. *balchanica*.)

TAXONOMIC NOTE See Western Orphean Warbler.

REFERENCES Svensson, L. (2012) *BBOC*, 132: 75–83.

S. c. jerdoni, presumed 1stW ♂, United Arab Emirates, Dec: typical long bill of this race. Also note slightly darker grey on head giving hint of hood of ad ♂ being extensive, reaching nape and ending in well-defined border. Combination of rather dark head, greyish-white iris and moult pattern (all or most of primary-coverts, innermost two primaries and outermost secondaries retained juv) age and sex this bird. (H. Roberts)

BARRED WARBLER
Sylvia nisoria (Bechstein, 1792)

Fr. – Fauvette épervière; Ger. – Sperbergrasmücke
Sp. – Curruca gavilana; Swe. – Höksångare

A large *Sylvia* warbler of dry, sunny and open habitats with thorny bushes and dense copses, including open deciduous low-growing woods with glades and undergrowth, and limestone heaths with thorny shrubbery. Perhaps first noticed by its scolding call or low song-flight and scratchy, warbling song. Summer visitor to mainly the eastern parts of the treated region, wintering in E Africa.

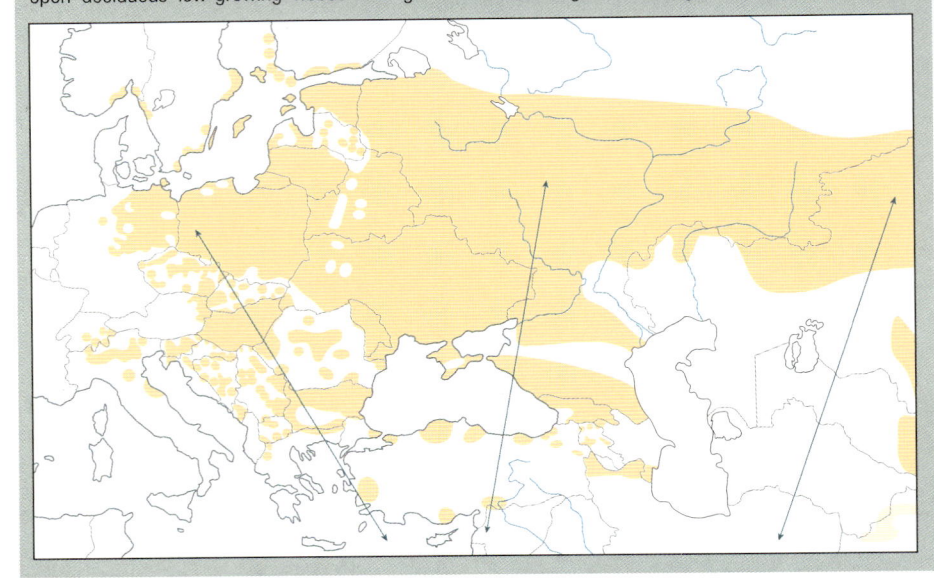

Ad ♀, Finland, May: less worn and greyish-fringed primaries and primary-coverts, yellowish-pink legs and deep yellow iris are features of an ad, while barring is narrower and restricted to small dots on throat, the lores are mostly whitish, strongly suggesting an ad ♀ (plumage is overall like 1stS ♂, but being ad enables the sex to be confirmed). (M. Varesvuo)

IDENTIFICATION Large and stocky as a Red-backed Shrike, and nearly as *long-tailed*, told by its mainly *greyish upperparts* and, in most plumages, *barred underparts*. Narrow, *pale double wing-bars*, *pale-edged tertials* and *whitish tail-corners* are useful features to note, as is the *pale eye* of adults, chrome-yellow in ♂ and duller in ♀. While the adult ♂ is densely barred below like a Sparrowhawk, the barring of ♀♀ and first-summer ♂♂ is less obvious and complete. Bill dark with pale yellow or bone-white base to lower mandible. Keeps mostly hidden in bushes and dense trees, but its loud calls give it away, and not infrequently it will briefly perch in the open or perform a flattish song-flight from one patch of cover to another, often showing its white tail-corners. First-year birds are duller and browner-tinged, and have dark eyes. They can be told by their size and shape, by a hint of the *pale wing-bars* and rather *pale fringes to the tertials*, by some *pale-tipped uppertail-coverts*, dark vermiculations on undertail-coverts and traces of dark barring on the lower flanks in many. Bill is pale greyish with dark tip.

VOCALISATIONS Song is a prolonged warble on fairly even pitch, rather deep-voiced and containing both clear and pleasing, and scratchy or hard notes. Basically, it resembles the song of Garden Warbler, and even with experience it is possible to mix them up if heard only briefly. Barred Warbler includes more hard notes and is less mellow and less low-pitched (thus can also recall flight-song of Common Whitethroat), and is conclusively recognised, in case you hesitate, if excitement call (see below) is uttered together with song. – Perhaps most characteristic call, mostly used in breeding season and in excitement, is a loud House Sparrow-like scolding rattle, slowing at the end, *trrrrr'r'r't't'et*. At other times shorter rattles are used as contact call, singly or doubled, *trrr-trrr*. Single rather subdued tongue-clicking *chak* rarely heard, also.

SIMILAR SPECIES With only a brief glimpse of a bird taking off, can be mistaken for a ♀/immature *Red-backed Shrike* due to similar size and shape, and shared habitat. The shrike has rufous-brown back and lacks white tail-corners. – *Common Whitethroat*, too, occurs in largely similar habitats, but is a paler bird with a more jerky or twisting flight. When seen perched and well, adults are unmistakeable. – First-years can resemble *Garden Warbler* or even *Marsh Warbler*, but note hint of paler, buffy off-white wing-bars and tertial edges, as well as a hint of pale spotting on uppertail-coverts, and dark spots or barring on undertail-coverts, and often on flanks. Bill grey with base of lower mandible pinkish or grey-white, tip often contrastingly darker.

AGEING & SEXING Ages separable in 1stY, at least when seen close or handled. Sexes generally separable in spring within each age-group, but ad ♀ and 2ndCY ♂ can be quite similar. Autumn birds usually not possible to sex. – Moults. Complex post-nuptial moult of ad, among flight-feathers

Ad ♂, Russia, May: a large, stocky *Sylvia*, with long tail and primary projection, while greyish upperparts, barred underparts, whitish double wing-bars and yellow eye make the species unmistakable. Note quite dark primary-coverts, somewhat worn (post-nuptial) primaries and yellowish-pink legs which age this bird as ad, while chrome-yellow iris and dense barring below, with prominently barred flanks reveal it to be a ♂. (P. Parkhaev)

Ad, Finland, May: some birds are difficult to age and sex, even in the hand. Seemingly ad-like primary-coverts (dark-centred and less worn), relatively less worn (post-nuptial) primaries and yellowish-pink legs indicate ad, while bright chrome-yellow iris and sooty-grey head-sides suggest a ♂. However, chin to upper breast is irregularly barred and spotted, leaving throat largely unmarked, with flanks variably and less boldly barred, suggesting either a poorly marked ♂ or well-marked (older?) ♀. (M. Varesvuo)

moulting only primaries and some central tail-feathers in Jul–Aug, whereas secondaries and outer tail-feathers are renewed in Africa after autumn migration (Hasselquist *et al.* 1988). Juv does not replace any flight-feathers or primary-coverts before autumn migration, and in winter generally moults most secondaries, tertials and sometimes a few outer primaries. (A similar moult strategy occurs only in White-throated Robin.) – SPRING **Ad** ♂ Chin, throat, breast, forehead and sides of forecrown densely barred dark. Ear-coverts, lores and forecrown rather dark grey. Flanks prominently and densely barred dark. Axillaries invariably barred, Iris chrome-yellow. Legs grey with some yellow. Nearly invariably fairly fresh primaries and primary-coverts of same generation. (Very rarely ad replaces some outer primaries in winter.) **Ad ♀/1stS ♂** Chin, throat and breast irregularly barred and spotted dark, leaving portions unmarked. Ear-coverts, lores and forecrown brown-grey, uniform with rest of upperparts. Flanks and axillaries variably barred. Iris bright or pale yellow. Legs usually bluish-grey. Ad ♀ has fresh primaries and primary-coverts like ad ♂, whereas 1stS ♂ usually has somewhat or much more abraded (retained juv) primaries and primary-coverts, often with some outer feathers renewed and contrastingly fresher and darker than inner. **1stS ♀** Barring

1stS ♂, Estonia, Jun: worn and bleached brownish juv primary-coverts and, especially, primaries, and purer grey legs age this bird as 1stS, while deep yellow iris and well-developed barring confirm the sex to be ♂. Not as boldly and extensively barred as typical ad ♂, and head-sides hardly darker, all confirming the age and sex. Diagnostic white tail-corners—most rectrices have been replaced in first pre-nuptial moult. (M. Varesvuo)

1stS ♂, Israel, Apr: classic wing pattern with worn and brownish juv primaries and especially primary-coverts, forming clear contrast with fresh greater coverts, tertials and tail. However, this young ♂ is unusually heavily barred below, purer grey above, with a darker head and deep yellow iris. As a rule, always age a bird before attempting to sex it. (A. Ben Dov)

1stS ♀, Finland, Jun: first-spring ♀♀ are often easily aged and sexed, being duller and browner, with only weak barring below, but are still readily identified to species by their size and shape, pale wing-bars, rather pale-fringed tertials, some pale-tipped uppertail-coverts, and at least some dark vermiculations on flanks. (H. J. Lehto)

Ad, presumed ♂, United Arab Emirates, Oct: ageing in autumn often straightforward as ad invariably shows freshly moulted primaries, tertials and greater coverts, with bold white tips and fringes, quite substantial barring below and usually deep yellow iris. Following post-nuptial moult, the bars below are partially concealed and most obvious on flanks. The purer grey upperparts, with stronger white wing markings and regular barring on flanks suggest a ♂. (D. Clark)

of underparts much restricted, throat and axillaries largely unmarked, and whole plumage tinged buffish-brown, crown more brown than grey. Iris pale brown or dull yellowish-grey. Legs bluish-grey without yellowish tinge. Primaries and primary-coverts as described for 1stS ♂. (In monitored breeding populations with close study of many birds of known age, often possible to discern 2ndS, at least among ♂♂, being intermediate between ad and 1stS; however some overlap in characters and requires much practice.) — **AUTUMN Ad** Though somewhat variable, both sexes resemble ad ♀ spring, thus ♂ loses much of its dense barring. Underparts off-white sparsely barred grey. Iris variable but not dark, those with deep chrome-yellow eye likely to be ♂. Pale tips to outer tail-feathers pure white and often prominent. **1stCY** Underparts tinged warm buff, especially on sides of breast and lower flanks. Barring much restricted, mostly confined to lower flanks and undertail-coverts. Rump and uppertail-coverts with some pale spots at tips. Iris brown or dark grey. Pale tips to outer tail-feathers restricted and sullied grey-buff.

BIOMETRICS L 16–18.5 cm; **W** ♂ 86–92 mm (n 27, m 88.7), ♀ 82–91 mm (n 20, m 87.0); **T** 63–74 mm (once 76; n 46, m 69.6); **T/W** m 79.1; **B** 15.9–18.3 mm (n 46, m 17.2). **Wing formula: p1** minute, < pc 5–10 mm; **p2** < wt 0–2 mm, =3 or 3/4; **p3** usually longest (sometimes pp2–3); **p4** < wt 0.5–3 mm; **p5** < wt 3.5–7 mm; **p6** < wt 6–11 mm; **p10** < wt 18–22 (once 26) mm; **s1** < wt 22–27.5 (exceptionally 20–30) mm. Emarg. pp3–4, at times faintly also on p5.

Ad, presumed ♀, Kuwait, Sep: bold whitish primary tips, pure yellow iris and brownish legs age this bird. One secondary is growing but the rest are not visible and are presumably retained and to be replaced in Africa. Brown tinge to mantle, scapulars and crown, weaker wing-bars, and extensive white forehead (compare image on p. 547, bottom right) indicate ad ♀. Being ad, tips to outer tail-feathers are pure white, bolder and quite prominent. (A. Al-Sirhan)

GEOGRAPHICAL VARIATION & RANGE Monotypic. — C & E Europe from Baltic area, Germany and N Italy east through Russia and Turkey to Central Asia; winters southern E Africa, though make long stopover further north, in Sudan to Kenya. – There is a tendency for birds in E Central Asia ('*merzbacheri*') to be very slightly paler above than European birds, but the difference is very faint and inconsistent over large areas. We regard this as insufficient grounds for formal recognition of two subspecies. (Syn. *merzbacheri*.)

REFERENCES HASSELQUIST, D. *et al.* (1988) *Ornis Scandinavica*, 19: 280–286. – LINDSTRÖM, Å. *et al.* (1993) *Ibis*, 135: 403–409.

1stW, England, Sep: 1stY impossible to sex, but usually readily separated from fresh ad by being duller overall with grey eyes (instead of yellow) and are unbarred or only vaguely marked below. In brief views could be confused with Garden or even Olive-tree Warblers, but note hint of at least some dark vermiculations on rear flanks and dark centred undertail-coverts, as well as pale wing-bars and pale-fringed tertials. (A. Allport)

1stW, England, Oct: some 1stW in autumn are browner above and sandier below, and overall featureless, with only poorly marked pale wing-bars, reduced pale-fringed tertials and hardly any visible vermiculations on rear flanks. Such birds could be confused with Garden Warbler, but the just-visible, diagnostic dark centres to undertail-coverts and pale-fringed uppertail-coverts always separate them. (G. Thoburn)

1stW, Greece, Sep: some 1stW are very attractive in being extremely tawny-buff with fine vermiculations below. Note virtually all-juv wing and rather unusual pale olive-grey iris. (L. Stavrakas)

Juv, Estonia, Jul: a recently fledged juv with typical loose, fluffy plumage on head and body, and still not fully grown wings and tail. Prior to autumn migration, juv moults all or most of body-feathers and perhaps some lesser coverts to 1stW, but difference in overall appearance is small. Bleaching or pale brown tips to median coverts, tertials and uppertail-coverts make these more whitish and contrasting. (A. Uppstu)

LESSER WHITETHROAT
Sylvia curruca (L., 1758)

Fr. – Fauvette babillarde; Ger. – Klappergrasmücke
Sp. – Curruca zarcerilla; Swe. – Ärtsångare

A widely distributed medium-sized *Sylvia*, catholic in habitat choice and hence equally well established in forest edges with bushes and open areas in boreal zone in N & E Europe, as in scrub in steppes and deserts on plains, and on alpine slopes in mountains (to 3000 m). Often encountered in hedges, junipers, riparian forests and lush gardens, avoiding only dense woodland. A summer visitor, wintering between E Africa and India. Although it keeps much in cover and has little that sticks out in its appearance, it is frequently noticed in Europe by its characteristic rattling song.

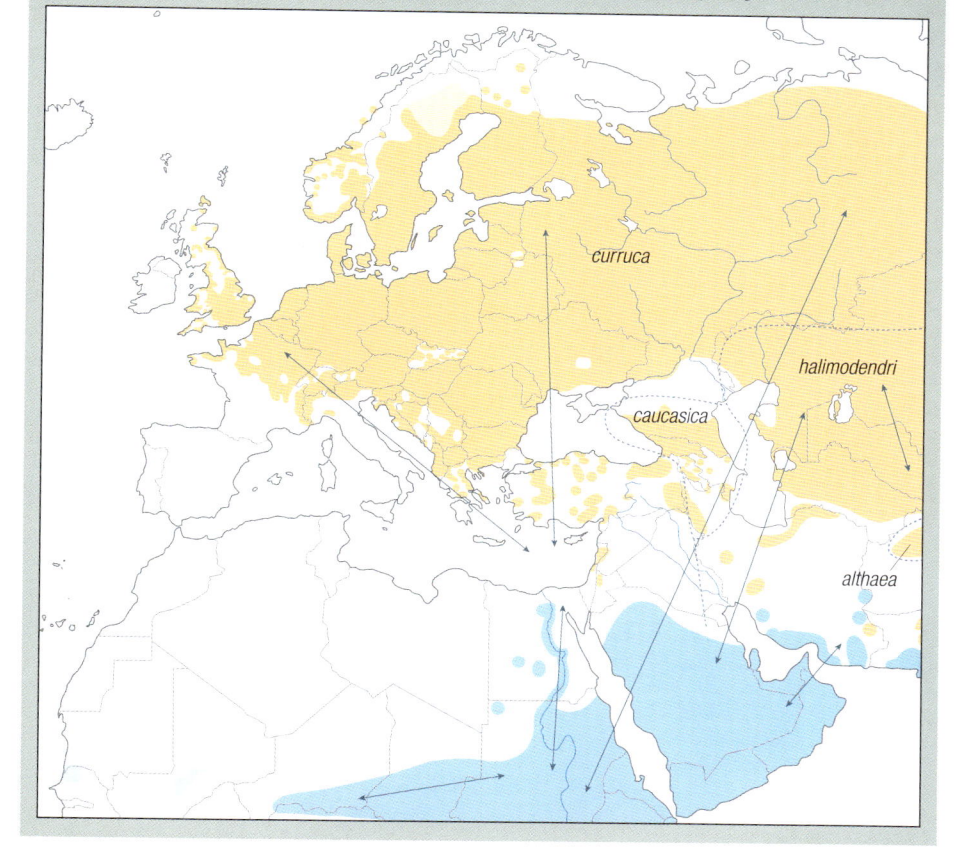

S. c. curruca, 1stS, Netherlands, Apr: this bird has a rather uniform grey head, with ear-coverts hardly any darker, but pinkish-buff flush on breast and flanks enhance white-throated impression. Some young are easier to age, like this one, by rather worn and bleached browner remiges (chiefly primary tips) and primary-coverts, contrasting with the fresh tertials and greater coverts. (R. Pop)

IDENTIFICATION A rather compact, small warbler, appearing almost neck-less and fairly short-tailed, and lacking striking features. Keeps mostly hidden in bushes and low trees, moving at deliberate pace when feeding. For a *Sylvia*, the combination of *dark grey legs* and bill, *brown upperparts* (though crown often grey), *white throat* and *contrasting dark grey ear-coverts* (usually creating a slightly 'masked' effect) is often enough to clinch the identification. The dark tail has a varying amount of white on outer edges. Breast and belly dingy off-white, on many with breast-sides and flanks tinged vinaceous or buff. Often a narrow white partial eye-ring below eye, rarely more complete. In fresh plumage, some (both young and adults) have a hint of a whitish, short supercilium above and in front of eye. Fresh birds, especially paler birds in Central Asia, can have a fairly prominent pale buff secondary panel. Populations in mountains in the Caucasus and in Central Asia are generally subtly larger and darker, and greyer above, with often slightly darker forehead. Eye colour varies, in Europe rather dark nut-brown or chestnut (with a variably prominent pale upper crescent), in Central Asia often somewhat paler, either with a reddish tinge, or more greyish (even bluish-grey).

S. c. curruca, presumed ad, Israel, Mar: small and rather compact, with rather featureless plumage, dark legs and bill, and essentially dull earth-brown upperparts, forming some contrast with greyish crown, and ear-coverts variably darker, contrasting again with white throat. Whitish underparts often have faint pink-buff flush to sides. Narrow, partial white eye-ring, always present below eye (sometimes also above it), and hint of whitish, short supercilium. (D. Occhiato)

S. c. curruca, ad, England, Aug: typical for ssp. *curruca* is dull brown upperparts with hint of olive-grey cast (though this bird looks more earth-brown). Evenly-fresh ad wing-feathers, including greyish-white primary tips, black greater alula and dark-centred primaries, with rather pale grey iris confirming age. (P. Blanchard)

S. c. curruca, 1stW, England, Oct: in fresh plumage, the dull grey-brown upperparts, greyer crown and white throat, with duskier white rest of underparts are evident. Dark uppertail sometimes has varying amount of white on outer edges. Aged by juv carpal-covert, with brown juv secondaries and primaries, contrasting with replaced scapulars and mantle-feathers. Note how neat and ad-like the alula and primaries appear. (G. Thoburn)

S. c. curruca, juv, W Turkey, Jul: as ad, but typical soft and buff-brown feathers, with slightly scruffy head pattern, drab sandy-grey upperparts (buff fringes to coverts and tertials more obvious), and cream-grey and white below. One central tail-feather already replaced (and possibly one outer as well). (H. Yılmaz)

VOCALISATIONS Song of basically two types: (i) In Europe east to W Siberia, Caucasus and in Turkey, the phrase opens with a rather subdued, brief warble, which abruptly runs into a loud rattle or brisk stanza of repeated figures, *tell-ell-ell-ell-ell* or *chiwa-chiwa-chiwa-chiwa*. At some distance, you only hear the rattling or repetitive stanza, and it is convenient in the simplified division here adopted to refer to this type as 'rattling' song. (ii) In eastern populations the initial warble of western birds comprises the entire phrase, being louder and longer, whereas the rattling finale is dropped (or only kept as a brief fragment). The rhythm is jerky, the tone rather scratchy and there is a fair amount of individual variation (still, it can be convenient to lump these variations under the label 'scratchy warble' song). Desert birds (*halimodendri*, *minula*), and those of Central Asian mountains (*althaea*), have very similar song (*contra* Martens & Steil 1997). Intermediate song types between (i) and (ii) are found in a broad overlap zone between these two main groups. Note that (rare) song-flight invariably is of scratchy warble type, even in Europe. — Commonest call is a dry tongue-clicking *chek* (also rendered *zet* or *tett*), often rather subdued and repeated from inside a bush when feeding. To the trained ear this is a little softer than the similar but louder and more emphatic call of Blackcap. It can be more difficult to tell from Western Subalpine Warbler (but then these two hardly occur in the same area). Birds in deserts or arid plains of Central Asia and W China (*halimodendri* and *minula*, respectively) have an alternative fast, nasal, scolding series, *che-che-che-che-che*, vaguely recalling a Blue Tit or distant Great Tit, or variants of this. This latter scolding call, together with the *chek* call, is frequently heard from migrants in Arabia and Middle East, notably in S Israel (in all likelihood involving both *curruca* and *halimodendri*).

SIMILAR SPECIES *Common Whitethroat* occurs in largely similar habitats as Lesser Whitethroat, but is slightly larger, slimmer and longer-tailed, has pink-brown legs and paler iris (pale olive-brown to reddish-ochre), paler-based lower mandible, a more prominent and unbroken whitish eye-ring, and generally shows more white on outertail in flight. In most populations, Common Whitethroat has a bright rufous-ochre wing panel (although this difference is less pronounced in eastern populations of Common and Lesser Whitethroats). — *Western* and *Eastern Orphean Warblers* both have dark legs and bills, and their general plumage coloration is also similar to Lesser Whitethroat, but they are much larger with longer tail, primary projection and bill, the latter often with an obviously paler base to the lower mandible. The head in both often appears quite large with slightly domed rear crown, while Lesser Whitethroat looks comparatively neck-less and small-headed.

AGEING & SEXING (*curruca*) Insignificant seasonal and age-related differences. Sexes alike as to plumage. — Moults. Complete post-nuptial moult of ad in Jul–Sep. Partial post-juv moult of body, all median and some inner greater coverts (rarely all greater) at the same time. Both age-groups have a partial pre-nuptial spring moult, usually involving all tertials and often r1. — **SPRING** Ageing frequently difficult or impossible unless any juv outer tail-feather retained or plumage wear is helpful. **Ad** Tail pattern with a pure white wedge with well-defined dark apical spot on r6. Primaries greyish and reasonably well kept. **1stS** If juv r6 still not moulted, these are sullied brown with diffusely paler (but not pure white) hint of washed-out wedge and no distinct apical spot. Primaries and primary-coverts often tinged brown and quite worn at tips. — **AUTUMN Ad** Plumage fresh with no visible moult limits; r6 rounded at tip and shows much white, r5 white-tipped. Iris brownish-tinged with variable whitish crescent on upper half. **1stW** Plumage rather fresh but may show first signs of slight abrasion to tips, flight-feathers subtly less dark and glossy by comparison; if juv feather retained, r6 more pointed, sullied brown, apical spot diffuse, r5 without white tip. Iris uniformly olive-grey. **Juv** As 1stW but somewhat warmer brown above, and plumage a little softer, especially on nape and vent.

BIOMETRICS (*curruca*) **L** 12.5–13.5 cm; **W** 60–72 mm (*n* 148, *m* 66.5); **T** 51–62 mm (*n* 144, *m* 55.0); **T/W** *m* 82.8; **B** 11.0–13.8 mm (*n* 144, *m* 12.7); **BD** 2.7–3.7 mm (*n* 129, *m* 3.3); **Ts** 17.8–21.2 mm (*n* 126, *m* 19.5). **Wing formula:** p1 > pc 0.5–6 mm, < p2 26.5–36 mm; **p2** < wt 1.5–5.5 mm, =5/6 (72%), =6 (11%) or =6/7 (14%) or =5 or 4/5 (3%); **pp3–4** about equal and longest; **p5** < wt 0.5–3 mm; **p6** < wt 2–6.5 mm; **p7** < wt 5–9 mm; **p10** < wt 11–15 mm; **s1** < wt 12.5–18 mm. Emarg. pp 3–5.

GEOGRAPHICAL VARIATION & RANGE Apparently largely clinal variation, mainly slight, hence often difficult

Presumed *S. c. caucasica*, Turkey, Apr: photographed in SC Turkey, outside the breeding range of *caucasica*, but appears like classic example of this race, presumably on migration. Nevertheless, it could be *curruca* as transition to *caucasica* is gradual. Note, in particular, darker and purer grey upperparts, forming less contrast with the crown (than in *curruca*) and dusky-white underparts, enhancing white throat. Some *caucasica* can be very close to (or inseparable from) *S. c. althaea* in the field. At least the primary tips appear reasonably fresh, suggesting an ad (N. Tez)

Presumed *S. c. caucasica*, 1stW, Turkey, Oct: rather dusky, despite fresh plumage, perhaps suggesting E Turkish origin, but racial labelling without handling can never be conclusive. Aged by retained juv outer greater coverts (fringed rustier) and remiges (with indistinct pale primary tips), and uniform rather dark olive-grey iris. (Ü. Özgür)

S. c. althaea, SE Iran, Apr: in the treated region *althaea* breeds only in E Iran. Most characteristic birds breed in Kashmir, but even at western end of range some birds distinctive, with almost lead-grey head, affording stronger white-throated impression; also slightly larger and bulkier, with stronger bill (although not so visible on this bird) and extensive white on outertail-feathers. (C. N. G. Bocos)

S. c. althaea, SE Iran, Apr: the rather dark upperparts and dusky-white underparts (generally lacking pink-buff suffusion), and heavier jizz, indicate *althaea*, but in general those in western part of this subspecies' range, including Iran, are less distinctive as to plumage, approaching *curruca* and *caucasica* in many respects, rendering it very difficult to identify away from the breeding grounds. (C. N. G. Bocos)

S. c. althaea, 1stS, NE Iran, May: the rather dark upperparts, almost lead-grey crown with darker forecrown, heavier jizz with stronger bill and proportionately rather short tail, and dusky-white underparts (without any pink-buff) are all indicative of *althaea*, and locality and date support this. Already very-worn primary tips, contrasting with fresh tertials, inner secondaries, some tail-feathers, some wing-coverts and alula diagnostic of 1stS. (C. N. G. Bocos)

S. c. blythi, 1stS, NE Mongolia, Jun: generally paler and browner than ssp. *curruca*. Racial identification confirmed by locality, NE of Ulaanbaatar, but outside the breeding range it would be impossible to separate from a brighter *curruca* or duller *halimodendri*. Despite rather subtle differences, *blythi* can often be identified in the hand using wing formula and amount of white in outertail. Heavily worn and bleached browner (juv) primaries, primary-coverts and outer greater coverts, but one new central tail-feather, very typical for 1stS. (H. Shirihai)

S. c. blythi, 1stW, England, Jan: 'Siberian Lesser Whitethroat' is an often-claimed vagrant to W Europe in autumn, but is only very marginally paler and browner above than *curruca*, and wing formula differs only slightly with some overlap. As with this bird, vagrant *blythi* can usually be confirmed only if trapped to check DNA to certainly exclude a duller *halimodendri*. Aged by juv remiges, primary-coverts and tail. (B. Spencer)

to assign single birds to a certain subspecies. Differences between described taxa are small, much more so than often appreciated. Variation especially insignificant in Europe and Siberia, but more complex in southern forms. Mountain-living form 'Hume's Lesser Whitethroat' (*althaea*) constitutes an 'ecological species', but see below under Taxonomic notes regarding its relationship with the other taxa.

S. c. curruca (L., 1758) (Europe, Turkey except extreme east, Levant; winters Egypt, Sudan, sub-Saharan E Africa, S Arabia). Dull earth-brown above, often with faint olive-grey cast. Crown greyish, ear-coverts darker than crown. Sides of breast and flanks often have faint pinkish flush. Outer tail-feathers have limited white, and dark apical spot on r6; juv has r6 sullied brown-grey. See Biometrics above.

○ **S. c. caucasica** Ognev & Bankovski, 1910. 'CAUCASIAN LESSER WHITETHROAT' (Taurus and E Turkey, Caucasus, Transcaucasia, W & NW Iran; wintering grounds poorly known, presumably E Africa and S Arabia). Slightly greyer above, and often has a marginally darker crown and forehead than *curruca*. Also, tarsus and primary projection tend to be a little shorter. Birds with greyest upperparts easily confused with ssp. *althaea* in the field, but note that song in Caucasus region is of rattling type, same or very nearly same as in

Presumed *S. c. halimodendri*, presumed ad, Israel, Mar: this race, a common migrant in Middle East, is best characterised by its brighter plumage than *curruca*, being paler with sandier-brown upperparts. Also note rather short primary projection. However, based on just one photograph like this, it is difficult to eliminate *blythi* (pale *blythi* and dark *halimodendri* are often nearly inseparable even in the hand). The rather evenly-feathered and not heavily-worn wing (including the round and broadly-fringed primary-coverts) indicate an ad. (D. Occhiato)

Presumed *S. c. halimodendri*, ad, Israel, Mar: on spring migration both *halimodendri* and *curruca* pass through the Levant and can at times be difficult to separate. Still, *halimodendri* averages slightly paler and warmer brown above, has blunter wing and proportionately longer tail; also, in ad more extensive white in outertail (though not visible here), and iris is often as on this bird tinged reddish, not as pure pale grey as in most *curruca*. The not-too-worn and rather dark remiges and primary-coverts infer age. (H. Shirihai)

S. c. halimodendri, 1stS, Kazakhstan, May: on the breeding grounds; note overall pale plumage with pale grey head and sandier or pale tawny-brown upperparts, lacking olive-grey cast; also note pinkish-buff flanks. Already very worn (juv) primary tips and pointed primary-coverts, forming contrast with newer post-juv or pre-nuptial greater coverts and tertials. (C. Bradshaw)

curruca. (Reports of scratchy warble type song from Turkey require confirmation or further study.) Pattern of outer tail-feathers same as in *curruca*, thus with restricted white. The eastern limit of *caucasica* in NW Iran, and the western limit of *althaea* in E Iran (E Khorasan?) is not known, or whether they are allopatric, parapatric or sympatric. **W** 62–70 mm (*n* 48, *m* 65.5); **T** 52–60 mm (*n* 47, *m* 55.1); **T/W** *m* 84.5; **B** 11.7–14.1 mm (*n* 49, *m* 12.7); **Ts** 17.5–20.2 mm (*n* 44, *m* 18.9). **Wing formula: p1** < p2 28–35 mm; **p10** < wt 9–14 mm; **s1** < wt 11–16 mm.

S. c. althaea Hume, 1878. 'HUME'S LESSER WHITETHROAT' (mountains of Central Asia, usually at 1600–3000 m, in the west from E Iran east through Afghanistan, NW Pakistan, Kashmir, Tien Shan and W China, north to Tarbagatay; winters S Iran, Pakistan, India, N Sri Lanka). Resembles *caucasica*, but is slightly larger with stronger bill, often a trifle shorter tail but with more extensive white on outertail, generally darker upperparts with often in worn ♂ plumage darker grey forecrown, and often lead-grey, even bluish-tinged (instead of reddish-brown) iris (though iris colour difference requires confirmation from additional material before being used for identification). Underparts are whiter, less warm buffy than in other races, but flanks and sides of breast can be tinged pinkish-brown. Most typical birds, large, strong-billed and dark-crowned, are found in Kashmir and adjacent areas, whereas breeders in e.g. E Iran ('*zagrossiensis*'), Afghanistan and areas north to E Kazakhstan ('*monticola*') are less extreme, more similar to dark *curruca*. Note that imm in autumn has warmer colours with a more brown-tinged back, and is easily overlooked as one of the other races. Song of 'scratchy warble' type, very similar to, or inseparable from, that of both *halimodendri* and *minula* (though often a little louder than both). Reports of a much louder than normal song could be attributed to effect of echo and acoustics in alpine gullies rather than to a real difference. Song-flight appears to be practiced more commonly than among lowland taxa. Ageing possible in autumn, often also in spring, on shape of tail-feathers and amount of pure white in outertail, ad having more rounded tail-feathers and large white wedge on r6 and white-tipped rr5–4(3), 1stY more pointed tail-feathers and narrower off-white wedge on r6 and whitish tip only on r5. **L** 13–14.5 cm; **W** 65–74 mm (*n* 81, *m* 69.3); **T** 51.5–63 mm (*n* 81, *m* 56.1); **T/W** *m* 81.0; **B** 12.3–14.8 mm (*n* 82, *m* 13.6); **BD** 3.0–4.3 mm (*n* 79, *m* 3.6); **Ts** 18.6–21.2 mm (*n* 77, *m* 19.9). **Wing formula: p1** > pc 1.5–8 mm, < p2 26–35 mm; **p2** < wt 3–8 mm, =6/7 or 7 (66%), =7/8 (23%), rarely any of =6, =8, =8/9, or =5/6; **p6** < wt 1–4 mm; **p10** < wt 9–13 mm; **s1** < wt 11.5–15.5 mm. Emarg. pp3–5, sometimes slightly on p6. As to claims of separate species status see Taxonomic notes. (Syn. *monticola*; *zagrossiensis*.)

Presumed ssp. *halimodendri*, S Kazakhstan, Aug: claims of *halimodendri* during autumn migration (that confidently exclude *blythi*) should be viewed very cautiously. However, in this case the sandy or pale tawny-brown upperparts are closer to *halimodendri*. Age unsure: despite rather fresh and dark-centred wing, with notably whitish-grey primary-tips, the greater coverts appear to be juv. Note warm tinge to iris. (Y. Belousov)

○ **S. c. blythi** Ticehurst & Whistler, 1933. 'SIBERIAN LESSER WHITETHROAT' (Siberia east of Irtysh and Ob, north of S Altai, apparently including NC & NE Mongolia, NE China; winters Pakistan, India, western SE Asia). Very similar to *curruca* but on average a trifle paler and browner above (though many are as dark as *curruca*) and has on average slightly blunter wing (**p2** < wt 2–7 mm, =6/7 in 54%, =5/6 in 20%, =6 in 19%, =7 or 7/8 in 13%) and proportionately slightly longer tail (**T/W** *m* 85.3) with more pure white on r6 in many. **W** 61–70.5 mm (*n* 155, *m* 64.9); **T** 51–60 mm (*n* 155, *m* 55.3); **B** 11.2–13.8 mm (*n* 149, *m* 12.7); **Ts** 18.2–21.5 mm (*n* 141, *m* 19.7). **Wing formula: p10** < wt 8–14 mm; **s1** < wt 11–16.5 mm. – Presumably intergrades with *curruca* in west, and birds west of Ob and in Orenburg region difficult to ascribe to subspecies. Apparently also intergrades with *halimodendri* in south-west. Song variable, but at least in eastern part of range of 'scratchy warble' type, or mixture between this and European rattle type, sometimes including rhythmic *rutti-rutti-rutti* part, neither rattle nor warble. A subtle race, which on morphological grounds could be included in *curruca*, but on genetic and vocal grounds best kept separate (see Taxonomic notes below).

S. c. halimodendri Sushkin, 1904. 'TURKESTAN LESSER WHITETHROAT' (deserts or steppe in Central Asia south to E Iran; common migrant in Middle East; winters E & S Arabia, Gulf States, S Iran, possibly also E Africa). Slightly smaller, paler and sandier grey-brown above than *curruca*, lacking olive cast when fresh, has usually much more white on sides of

Presumed ssp. *halimodendri*, presumed ad, Kuwait, Sep: sandy dull brown hue to upperparts suggest this race, but impossible to exclude an extremely pale *blythi*. Ageing also unsure at this angle, it being difficult to evaluate feather age in wing, but the rounded and dark tail-feathers, with dark primaries and alula, suggest an ad. (J. Tenuovo)

S. c. minula, Tarim Basin, Xinjiang, W China, May: the locality, small/delicate appearance including small bill, and warm sandy (slightly rusty-tinged) upperparts confirm the subspecies, rather than *halimodendri*; also note the tiny primary projection and proportionately longish tail that is pale and greyish-tinged. However, as the two to some extent overlap in these characters, separation in the field away from the core breeding ranges is often very difficult. (A. Hellquist)

S. c. minula, Tarim Basin, Xinjiang, W China, May: small and delicate appearance plus very short, fine bill and sandy (slightly gingery) upperparts, and especially the pure white outertail-feathers visible from below, are all typical for this race. Sides of underparts flushed pale cream-buff. Interestingly, this bird has quite brown-tinged head. (A. Hellquist)

proportionately longer tail, has often a little purer grey uppertail-coverts, and a paler, more reddish eye. Blunter wing than preceding races reflected in wing formula. Song invariably of 'scratchy warble' type, apparently indistinguishable from both *minula* and *althaea*. **L** 11.5–12.5 cm; **W** 58–68 mm (n 179, m 62.8); **T** 50–60 mm (n 177, m 54.9); **T/W** m 87.4; **B** 10.6–13.7 mm (n 177, m 12.0); **BD** 2.6–3.9 mm (n 159, m 3.1); **Ts** 17.5–20.7 mm (n 159, m 19.2). **Wing formula: p1** < p2 23–31 mm; **p2** < wt 3–8 mm; =7 or 7/8 (52%), =6 or 6/7 (33%), =8 or 8/9 (14%) or =9 (1%); **pp3–5** about equal and longest (though p5 often to 2 mm <); **p6** < wt 0.5–5 mm; **p10** < wt 7–13.5 mm; **s1** < wt 9–14.5 mm; p6 with slight emargination in some. – The main problem is to separate *halimodendri* from *minula*. Breeders of Tarim Basin and Qaidam Pendi, W China (*minula* s.s.), and 'southern *halimodendri*' (breeders in SW Kazakhstan south to E Iran) have developed very similar morphology, yet are rather well separated genetically (Olsson *et al.* 2013). If true relationships when known define a subspecies, then *minula* is restricted to China, while near inseparable populations known as *halimodendri* breed in Central Asia. Without DNA only birds with W < 65, T < 58, Ts < 20.1, B < 12.6, BD < 3.4 and s1 < 13.5 should be tentatively labelled as 'showing the characters of *minula*'. (For a slightly different interpretation and allowing for subtler subspecies, see Loskot (2005). (Syn. *jaxartica*; *snigerewskii*; *turkmenica*.)

Extralimital races: ○ **S. c. telengitica** Sushkin, 1925. 'CHUIA LESSER WHITETHROAT' (Chuia High Plateau in SE Altai, NW Mongolia; winters presumably India). Resembles *margelanica* with hint of same broad dark mask (see below) but differs in being greyer brown above, and often has pure grey crown/nape and rump/uppertail. Very close to and often difficult to separate from *halimodendri* (although genetically closer to *margelanica*; cf. Olsson et al. 2013). Tend to have slightly longer and subtly more pointed wing and longer tail and tarsus than *halimodendri*: **W** 61–70 mm (n 37, m 65.9); **T** 53–61 mm (n 37, m 56.9); **Ts** 18.2–22.1 mm (n 35, m 20.1). **Wing formula: p2** =6/7 in 59%, =7 or 7/8 in 38%, and =8 in 3%.

S. c. minula Hume, 1873. 'DESERT LESSER WHITETHROAT' (Tarim Basin and apparently also Qaidam Pendi east to Lake Qinghai, W China; winters S Pakistan, NW India, probably also S Iran; claims from Arabia should be further confirmed as they more likely refer to *halimodendri*). A fraction smaller than *halimodendri* (but much overlap in size) with a small bill and reddish- or ochraceous-tinged iris and much pure white on outertail. Upperparts a trifle paler than *halimodendri* on average, though many virtually the same, and sides of underparts tinged creamy-buff (not pink). Wing formula similar to *halimodendri*, although wing on average even blunter. Song as in *halimodendri*, or at least quite similar. **L** 11–12.5 cm; **W** 59.5–64 mm (n 28, m 61.8); **T** 50–57 mm (n 27, m 53.8); **T/W** m 86.9; **B** 10.9–12.5 mm (n 27, m 11.6); **BD** 2.6–3.3 mm (n 25, m 2.9); **Ts** 17.8–20.0 mm (n 25, m 19.0). **Wing formula: p2** =7 or 7/8 in 57%, =8 in 32% and = 6/7 in 11%; **p6** < wt 1–4 mm; p6 with good or slight emargination in many). – Often very difficult to separate from *halimodendri* due to overlapping characters, and only very typical birds (very small, very pale ochre-brown above, proportionately long-tailed with much pure white in outertail, with short p2, etc.) should be identified without the help of DNA. (Syn. *minuscula*.)

S. c. margelanica Stolzmann, 1897. 'GANSU LESSER WHITETHROAT' (E Qinghai, Gansu in NC China, S Mongolia, also apparently N & E Xinjiang, China and NW Mongolia, possibly also SE Mongolia; migrates initially westwards in autumn through Central Asia—originally described from migrants in Fergana—then presumably south to unknown winter quarters, presumed to be S Central Asia and W India; single winter bird known (LS unpubl.) from Jodhpur, BM 1886.7.8.63. Large as *althaea* but is long-tailed and similar to *minula* in coloration, although slightly darker ochraceous-brown above, has a short but strong bill and a more extensive and prominent dark mask usually expanding towards rear, almost in ♂ Penduline Tit fashion. Song loud, irregular and hard, slightly reminiscent of Eastern Orphean Warbler, though still close to song of, e.g., *althaea* and *halimodendri*. **L** 13–14.5 cm; **W** 64–73 mm (n 41, m 68.6); **T** 54–65 mm (n 41, m 60.2); **T/W** m 87.7; **B** 11.1–13.5 mm (n 37, m 12.1); **BD** 3.0–3.6 mm (n 35, m 3.3); **Ts** 19.0–22.0 mm (n 39, m 20.6). (Syn. *chuancheica*.)

TAXONOMIC NOTES The Lesser Whitethroat is widely distributed, yet morphological variation is rather limited. This is probably due to it being an old species that has successfully adapted to changing climate and environments over a long time, most likely involving incidents of secondary contact and mixing of populations after periods of separation during glaciations. A genetic sampling of the entire complex (Olsson *et al.* 2013), with inclusion of several type specimens or typotypical material, recovered six major clades (one of which could be merely a subdivision of *halimodendri*), showing that *minula* and *curruca* are sisters and oldest, while the other four grouped together, with *althaea* being sister to the remaining three. This means that two different pairs frequently grouped close together based on morphology (*curruca* and *blythi*, *minula* and *margelanica*) sit on different branches of the tree, and are not sister taxa. A division into three or four species has been proposed by many authors based mainly on morphology and song, but here we take a different stance. Without full diagnosibility of taxa, with much overlap in plumage and measurements, it seems pointless to split *S. curruca* into fragments based on mtDNA. The recovered phylogenetic tree referred to above does not encourage such a split either. The arguably most distinct Lesser Whitethroat taxon, 'Hume's Lesser Whitethroat' (*althaea*), might be seen as a likely candidate for separate species status, with near-overlap with *halimodendri* in Central Asia, rather distinct morphology and largely separated ecologically (mainly by altitude). However, if cladistic theory is followed and non-monophyletic species are to be avoided, *althaea* cannot be singled out since it is nested in the middle of the *curruca* tree. We therefore recommend that all *curruca* taxa, at least for the time being, are kept in one species. It has obvious practical advantages even though evolutionary history, as far as it has been established, is not indicated by the adopted nomenclature.

REFERENCES LOSKOT, V. M. (2001) *Zoosyst. Rossica*, 10: 219–229. – LOSKOT, V. M. (2005) *Zool. Med. Leiden*, 79: 157–165. – MARTENS, J. & STEIL, B. (1997) *J. f. Orn.*, 138: 1–23. – OLSSON, U. et al. (2013) *Mol. Phyl. & Evol.*, 67: 72–85.

Presumed *S. c. margelanica*, S Kazakhstan, Sep: this far eastern race migrates west in autumn through Central Asia. It is as large as *althaea* but paler and sandier brown, approaching *minula*. If size cannot be judged correctly, impossible to separate from either of these races, or *blythi*, though *margelanica* is slightly darker ochraceous-brown above (i.e. warm brown, but usually not rufous-looking), with faint pink-buff flanks, but most important are the strong bill and more extensive and prominent dark 'mask', at times affording a Penduline Tit face pattern, while the very short primary projection is also evident. Much white on tail-sides, too. Ageing unsure: the rather evenly-feathered and not heavily-worn wing (including broadly-tipped whitish primaries) suggest an ad, but the buffish outer greater coverts and grey iris appear like juv. (Y. Belousov)

COMMON WHITETHROAT
Sylvia communis Latham, 1787

Fr. – Fauvette grisette; Ger. – Dorngrasmücke
Sp. – Curruca zarcera; Swe. – Törnsångare

A widespread and common *Sylvia* warbler of open habitats with dense bushes and scattered trees, often in junipers and brambles on heaths and shores, or in hedges and roses in agricultural land. In the southern part of its range it is often found on mountain-sides. Often uses song post in the top of a bush or on a telephone wire and delivers its brief and scratchy strophe, or now and then bursts out in a longer phrase during a fluttering, short song-flight. Summer visitor, wintering south of Sahara in the Sahel region, but also even further south in Africa.

IDENTIFICATION Medium-sized to large warbler, proportionately rather *long-tailed*, told by its *ochre or rufous* colours in the folded *wing*, its *white throat* which contrasts against the buffy or pinkish breast, its *pale pinkish-brown legs* and, in most plumages, narrow *pale eye-ring*. The adult ♂ is grey (or brown with greyish tinge) on crown and nape, has pinkish-buff breast, and the *eye is orange-brown*, whereas the ♀ is duller with brownish crown and more buffy, less pinkish breast, and the eye is duller yellowish-brown. However, the sexes are fairly similar, and characters vary somewhat, making reliable sexing of all birds impossible or even of most. *Tail-sides white* in all plumages (though duller in imm). Habitat and characteristic anxiety and alarm calls often help to separate this common warbler from its relatives and other species of similar size sharing its territory.

VOCALISATIONS Song a brief, fast strophe with loud but scratchy, gruffy voice, the phrase delivered in a jerky rhythm in a much-repeated pattern, e.g. *cheerioo-che-cheri-che-cheree*, but much variation in finer details. Indefatigable singer when newly arrived. Brief song-flight often practised, and then the song is longer, more varied and pleasant. Some birds sing with 'song-flight voice' for long spells also when perched, which can invite confusion with song of Barred Warbler. – Commonest call, mostly used in mild alarm, is a fast series of slightly hoarse, nasal notes, *weed-weed-weed-weed*. Anxiety is commonly expressed with a hoarse, muffled *chaihr*, sparsely repeated a few times, somewhat recalling Dartford Warbler alarm. Very rarely, when young are just fledged, a discreet clicking *chet* also heard (cf. Lesser Whitethroat), often combined to quick series, *chet-et-et-et*.

SIMILAR SPECIES In N & NW Europe, needs primarily to be separated from *Lesser Whitethroat*, which however has dark grey legs, lacks or usually has only incomplete whitish eye-ring, generally a darker eye and less buff or pink tinge on breast, a more compact build with proportionately shorter tail, and fewer rusty-brown tones on the closed wing (though beware that Common Whitethroat becomes progressively less rusty, and Lesser Whitethroat the opposite, towards east). – In S Europe several more options occur. *Spectacled Warbler* is similar but a smaller and slimmer bird with neater colours, and has a more obvious whitish eye-ring, while the rusty colour on the wing merges into a large uniform patch (dark feather-centres more visible among rusty edges on Common Whitethroat). – ♀ and immature *Western* and *Eastern Subalpine Warblers* or *Moltoni's Warbler* are also similar, but again are daintier birds with a slightly duller wing and different call.

S. c. communis, ad-like ♂, Italy, Jun: note rufous wing panel and white throat. Combination of grey crown and pink-tinged breast usually signals ♂. Bare-parts colorations often important for ageing, the bright orange-brown iris here best fits ad, but without handling it is difficult to decide if very worn, bleached and pointed primary-coverts reflect extreme wear or are juv, and if this bird is 1stY with an already ad-coloured iris. (D. Occhiato)

S. c. communis, 1stS ♂, Italy, Apr: rufous wing panel (greater coverts and especially tertial fringes) together with greyish head, white throat and pinkish suffusion on breast make the specific identification and sexing (as ♂) straightforward. When still not too worn, ageing is also possible, and the combination of very worn, bleached and pointed juv primary-coverts and primary tips, and olive-yellow iris age this bird as 1stS. (D. Occhiato)

SYLVIA WARBLERS

AGEING & SEXING (*communis*) Ageing in the field difficult or impossible, requiring very close examination of moult and wear in wing; ageing in autumn in the hand usually possible, sometimes also in spring if juv outer tail-feathers retained or if three generations of secondary-coverts present. Sexes often separable in spring with help of song, and when pair seen together, but otherwise ad ♀ and 1stS ♂ (even ad ♂!) can be quite similar. Autumn birds frequently not possible to sex. – Moults. In Europe, post-nuptial moult (Jul–Sep) of ad is complete. Juv moults only body and wing-coverts at the same time, and among greater coverts usually just the inner ones (only very rarely all). Both age-groups have a partial moult before spring migration, sometimes including a few central tail-feathers and all tertials. In SE Russia and Central Asia, complete moult takes place after autumn migration, or is started before but suspended to be concluded in winter quarters. – SPRING Ageing very difficult and generally impossible. Any retained juv outer tail-feathers heavily abraded and with sullied buffish-white rather than pure white portions, and with on average more prominent dark shafts, will be signs of 1stS. Similarly, a few birds have three generations of secondary-coverts, the oldest and most worn being juv feathers. **Ad ♂** Typical birds have breast pinkish,

S. c. communis, 1stS ♀, Germany, Jun: duller than ♂, with brownish crown, browner above, less saturated below, and white throat often less contrasting. Strong moult contrast between very worn and bleached brown juv remiges and primary-coverts, and the newer greater coverts and especially tertials, as well as olive iris, age this bird. (S. Pfützke)

S. c. communis, ad ♂, Turkey, Aug: following complete post-nuptial moult in late summer, ♂ has browner head and underparts apart from white throat washed dusky buff-brown. Iris less brightly coloured than in spring/summer. Several secondaries (at least four) have been left to be moulted in winter quarters, which strategy is practised by some Common Whitethroats in parts of Europe and Turkey. (F. Yorgancıoğlu)

S. c. communis, 1stW, Spain, Aug: despite being ♀-like and relatively duller, the rufous wing panel is still quite striking and the whitish throat is also evident. Aged by retained juv remiges and primary-coverts, which are softer in texture and browner, fringed buffish-brown (juv primary tips are cream-buff, and already slightly worn, rather than neat and white as in ad), and iris is diagnostically dark olive. (C. N. G. Bocos)

S. c. communis, 1stW, Turkey, Oct: the most likely confusion species is the much smaller Spectacled Warbler, which can be eliminated by its shorter primary projection, and smaller and more distinct dark centres to tertials and greater coverts. Both remiges and primary-coverts seem to be juv, and primary tips seem slightly worn and tail-feathers narrow and pointed indicating a 1stW. (T. Yılmaz)

crown grey, throat white. Iris orange- or reddish-brown. Less typical birds, difficult to identify, commonly occur, too. **Ad ♀/1stS ♂** Breast creamy-buff or with very faint pink tinge only, crown brown (with slight grey tinge at most), throat off-white. Iris ochre- or yellowish-brown. **1stS ♀** As ad ♀ but on average even browner and duller. – AUTUMN **Ad** Both sexes resemble ad ♀ spring, thus ♂ attains browner head, and pink on breast is mixed with buff. Iris orange-, ochre- or warm nut-brown. Pale tips and edges to outer tail-feathers pure white. **1stCY** Differs from ad in having iris dull brown or olive-brown, and outer tail-feathers rather uniform pale brown (unless already replaced by ad type with pure white portions).

BIOMETRICS (*communis*) L 13.5–14.5 cm; **W** ♂ 71.5–79 mm (n 30, m 75.0); ♀ (67) 69–76.5 mm (n 17, m 72.7); **T** ♂ 58–67 mm (n 30, m 62.5); ♀ 58–65 mm (n 17, m 60.7); **T/W** m 83.4; **TG** 2–9.5 mm (n 17, m 4.4); **B** 12.3–14.7 mm (n 46, m 13.5); **BD** 3.6–4.3 mm (n 43, m 4.0); **Ts** 19.8–22.7 mm (n 45, m 21.3). **Wing formula: p1** < pc 0–6 mm (rarely > 0,5), < p2 40–49 mm; **p2** < wt 0–2.5 mm, =3/5 (=3 or 5); **pp** (2)**3–4** about equal and longest; **p5** < wt 1–3 mm; **p6** < wt 3–7 mm; **p7** < wt 5–10 mm; **p10** < wt 13–18 mm; **s1** < wt 15–21.5 mm. Emarg. pp3–5, on p5 sometimes more faintly.

GEOGRAPHICAL VARIATION & RANGE Clinal variation. From west to east, amount of rufous on wing is reduced, while at the same time size increases a little. Differences are slight and gradual over large areas, and here a simplistic treatment is preferred, with only three formally named subspecies based on colour saturation, which differs more clearly than size.

S. c. communis Latham, 1787 (Europe, W Siberia, NW Kazakhstan, NW Turkey, NW Africa; winters mainly Sahel region between Senegal and Chad, but eastern breeders assumed to winter E Africa). Treated above. Edges of tertials and greater coverts bright rufous or reddish-ochre, and rest of upperbody rather warm grey-brown, not so pure greyish. Eastern birds (SE Russia, W Siberia, NW Kazakhstan) tend to be slightly larger and paler ('*volgensis*', but seems a rather subtle cline). Annual complete moult in late summer, before autumn migration. (Syn. *cinerea*; *hoyeri*; *jordansi*; *volgensis*.)

S. c. icterops Ménétries, 1832 (SC & E Turkey, Caucasus, Levant, Transcaucasia, N Iran; winters E Africa). Has duller and darker wing, tertials and greater coverts narrowly edged yellowish-brown. Ad ♂ has on average purer grey on head and back. Slightly smaller on average than *communis*. Most perform annual complete moult in winter, but some variation as to moult strategy and requires more study. **W** ♂ 69–77 mm (*n* 23, *m* 73.6); ♀ 66–74 mm (*n* 13, *m* 71.7); **T** ♂ 57–65 mm (*n* 23, *m* 61.6); ♀ 57–64 mm (*n* 13, *m* 60.2); **T/W** *m* 83.1; **Ts** 19.5–21.8 mm (*n* 29, *m* 20.9). Wing formula similar to *communis*. (Syn. *traudeli*.)

S. c. rubicola Stresemann, 1928 (E Turkmenistan, E Uzbekistan, SE & E Kazakhstan, NW China, SC Siberia, N Mongolia; winters E & S Africa). Very similar to *icterops* but is clearly larger and on average very slightly greyer. Still, many migrants and wintering birds cannot be separated. Song either like that in Europe or slightly different, with both harder and clearer notes, at times recalling short phrases of Siberian Rubythroat song. **L** 13.5–15.5 cm; **W** ♂ 75–83 mm (*n* 14, *m* 77.5); ♀ 70–78 mm (*n* 12, *m* 74.5); **T** ♂ 60–70 mm (*n* 14, *m* 65.2); ♀ 59–65 mm (*n* 12, *m* 61.8); **Ts** 20.3–22.7 mm (*n* 23, *m* 21.5). **Wing formula: p1** < pc 3–7 mm, < p2 41.5–52 mm; **p5** < wt 1.5–4 mm; **p10** < wt 14–18 mm; **s1** < wt 17.5–22 mm. (Syn. *fuseipilea*.)

REFERENCES Diesselhorst, G. (1971) *J. f. Orn.*, 112: 279–301.

S. c. communis, juv, Sweden, Aug: like ♀ but body-feathers typically soft and fluffy, and has more olive-brown than red-brown iris (but difficult to see here). Chunky shape, pale legs, rufous wing panel and quite contrasting whitish throat help identify the species. (P.-G. Bentz)

S. c. icterops, ♂, presumed 1stS, Israel, Feb: at least in Israel, among the first summer visitors to arrive, often by Feb. Not all are as distinctive as this ♂, with dull greyish-brown upperparts and greyer head. Typically has more yellowish-brown wing panel, less bright rufous than *communis*. Iris colour can support ageing in this race: here still yellowish olive-brown. (L. Kislev)

Presumed *S. c. icterops*, ♀, presumed 1stS, Israel, Mar: ♀♀ of *icterops* also tend to be duller greyish-brown above, but are otherwise inseparable from *communis* unless moult pattern can be analysed. Here, overall plumage looks very fresh, indicating that the bird went through complete pre-nuptial moult recently. Still mostly olive iris suggests, though not conclusively, a 1stY. (L. Kislev)

Presumed *S. c. icterops*, 1stW, Israel, Aug: 1stW of this race is generally indistinguishable from *communis*, but some tend to be slightly duller above, with less rufous wing panel, and by their less extensive post-juv moult. Dark iris confirms age. Sexing of 1stW normally impossible. (D. Kalay)

***S. c. icterops*, ad ♂, Kuwait, Sep**: greyish head and yellowish nut-brown iris, together with partial post-nuptial moult in combination automatically exclude *communis*, and age/sex the bird as ad ♂. However, cannot be separated from race *rubicola* using moult and plumage, but the relatively short primary projection and less pure greyish upperparts should exclude that race. (A. Al-Sirhan)

Presumed *S. c. rubicola*, ♂, possibly ad, Egypt, May: locality and being ♂ on late migration, together with rather cold grey being extensive on upperparts, and above all having extremely long primary projection mean this bird is probably *rubicola*. This gets support from apparent complete pre-nuptial moult in winter (though innermost secondary and one inner greater covert have been left unmoulted). Iris seems yellowish nut-brown, suggesting ad. (D. Occhiato)

Presumed *S. c. rubicola*, 1stW, Oman, Sep: combination of location, rather dull and greyish-tinged upperparts and long primary projection indicate this race, though without handling, *icterops* cannot be excluded with certainty. Wing and tail are juv. (H. & J. Eriksen)

GARDEN WARBLER
Sylvia borin (Boddaert, 1783)

Fr. – Fauvette des jardins; Ger. – Gartengrasmücke
Sp. – Curruca mosquitera; Swe. – Trädgårdssångare

A rather large but featureless and seldom-seen *Sylvia* warbler, most easily detected by its song and calls. It breeds in open woodland, broadleaved or mixed, with tall trees, some undergrowth and preferably access to fresh water. Despite its name, gardens are only used when these are quite large, have tall trees and preferably bushes and tall herbs, and parts that have been left neglected. Although its territory often includes glades or forest edges, it is remarkably secretive and rarely seen in the open. Summer visitor over much of Europe, wintering in sub-Saharan Africa.

IDENTIFICATION *Large* and *plump Sylvia*, basically a *greyish* bird without any striking features. Upperparts grey with faint brownish-olive tinge when fresh, underparts dusky grey-white. The grey on sides of breast and flanks can have a slight yellowish-olive tinge. There is *no pale supercilium* but the eyelids are off-white. Sides of neck are usually paler and purer grey than surrounding parts, creating a hint of a grey necklace, a feature often mentioned in field guides, but unfortunately not always visible and hence not an infallible tool for identification. No white on uniform tail. *Bill and legs grey*, the bill rather *short and stout*, which serves as an important character when eliminating any greyish *Hippolais* or *Iduna* species. Large size and fairly clumsy, heavy movements in the canopy of trees generally obvious and separate it from *Phylloscopus* warblers. Often keeps high up in the canopy of tall trees. Flight direct and fast, though rarely seen.

VOCALISATIONS Song is a prolonged warble on fairly even pitch, rather deep-voiced and containing both clear and fluty and harder notes. It is sometimes likened to the rippling sounds of a distant creak, and its deep, fluty voice can recall the song of a thrush or Black Wheatear. It resembles most the song of Barred Warbler, and even with experience it is possible to confuse them if heard only briefly. However, the song of Garden Warbler is on average slightly lower-pitched, includes fewer hard or scratchy notes, and is more mellow and pleasing. – The alarm call, often uttered near nest, is a quite characteristic prolonged series of throaty, slightly nasal *chek* calls, *c.* 3 per sec, *chek chek chek chek*... The rhythm varies a little depending on level of excitement and on the individual. Single such calls are sometimes uttered in contact.

SIMILAR SPECIES See Identification for some general advice. Other greyish species in wooded habitats include *Spotted Flycatcher*, which is easily eliminated by upright posture, dashing sallies for aerial insects, very short, black legs, and a streaked breast and forecrown. – *Blackcap* occurs in same habitat, has same shape and movements, but has neatly outlined black or brown cap. – First-year *Barred Warbler* is eliminated by more obviously pale-based lower mandible, longer tail, pale-edged tertials, pale-tipped greater coverts and uppertail-coverts, and diffusely dark-spotted undertail-coverts. – *Olivaceous Warbler* and *Isabelline Warbler* both differ by their longer and more pointed bills, with largely pinkish lower mandible, and some pinkish tinge to the greyish legs. – *Marsh Warbler* could theoretically be a confusion risk but has straw-coloured legs, longer, more

Finland, May: no sexual dimorphism in plumage, and ageing impossible once winter moult completed. It may be a rather drab, greyish-brown and largely unpatterned bird, but it compensates for that with its strong and attractive song. Note long and pointed wings, shortish and square-ended tail and rather short, stout bill, as well as paler and purer grey neck-sides. (M. Varesvuo)

Finland, May: some have better-marked pale supercilium and more obvious grey neck-sides, with even bolder whitish tertial fringes, which could invite confusion with poorly-marked juv Barred Warbler. The grey legs and all-dark eye are typical. (M. Varesvuo)

pointed bill with some yellow, and has a short but obvious pale supercilium. Marsh Warbler also has orange-yellow gape, often visible on a singing bird (Garden: pink).

AGEING & SEXING Ages separable in autumn by state of plumage, but this requires close observation or handling. Seasonal changes small and often hard to appreciate without the bird in the hand. Sexes similar. – Moults. Post-nuptial and post-juv moult Jul–Sep both partial, involving some body-feathers and wing-coverts (including inner 1–4 greater coverts). A few ad start moult of flight-feathers in Europe but generally arrest after replacing some tertials, tail-feathers and inner primaries; very rarely wing moult more advanced (possibly complete) before autumn migration. Complete or resumed arrested moult of ad and complete moult of 1stW in winter (Nov–Mar). – SPRING **Ad/1stS** These two usually inseparable being uniformly fresh (or moderately worn) without any moult contrast. **1stS?** A very few presumed 1stS told by a few contrastingly worn unmoulted juv feathers within otherwise fresher plumage. (However, such birds could include ad with aberrantly kept old feathers; more research needed.) – AUTUMN **Ad** Plumage slightly worn, especially tips of tertials, tail-feathers and long primaries. Upperparts brownish-grey rather than tinged olive. Underparts often wear and bleach to whitish with greyish sides. Beware that a few might start a near-complete moult on breeding grounds; look for remaining moult contrast. **1stCY** Plumage evenly fresh. Upperparts greyish-olive. Underparts not so whitish, more tinged warm buffish-olive, especially on sides of breast and lower flanks.

BIOMETRICS L 13–14.5 cm. Sexes of similar size, ♂ only a fraction larger. **W** 74–83 mm (*n* 130, *m* 79.0); **T** 51–59 mm (*n* 43, *m* 54.2); **T/W** *m* 68.2; **B** 12.4–14.3 mm (*n* 19, *m* 13.2). Wing formula: **p1** minute, < pc 1–7 mm, < p2 46–52 mm; **p2** < wt 0–2 mm, =3 or =3/4, or longest (rarely =4); **p3** longest (rarely pp2–3 longest); **p4** < wt 2–4 mm; **p5** < wt 5–8.5 mm; **p6** < wt 9–13 mm; **p10** < wt 18–26 mm; **s1** < wt 20–27.5 mm. Emarg. p3, often also p4.

GEOGRAPHICAL VARIATION & RANGE Monotypic. – Europe east through Russia and N Turkey to W Siberia; winters sub-Saharan Africa. – There is a tendency for birds in Russia, from W Siberia south to Central Asia ('*woodwardi*') to be very slightly paler, greyer and larger than W European birds, but the difference is very slight and not entirely consistent, and the two areas are connected clinally over a large zone of intergradation, including Poland and E Europe to Black Sea. These slight clinal differences are regarded as insufficient grounds for a formal recognition of two subspecies following a minimum 75% rule of distinctness. (Syn. *woodwardi*.)

Finland, May: despite being featureless, can in certain lights appear like some other warblers, here almost recalling Marsh Warbler's jizz, especially due to the yellowish buff-olive reflection from the canopy. The long wings with white-tipped primaries add to this impression. (M. Varesvuo)

Ad, Italy, Sep: all visible wing-feather tracts are ad-like, which during autumn migration are worn, especially noticeable at tips of primaries, contrasting with the recently renewed tertials; upperparts are tinged brownish-grey (instead of olive-brown as in fresher young birds at this season), and underparts often wear and bleach paler (though not so apparent here). (D. Occhiato)

1stW, Italy, Sep: in several respects mirrors fresh spring plumage, and unlike ad in autumn is overall fresh and, at least until Oct, primaries have tiny whitish tips; also note two inner greater coverts are renewed, while most wing-feathers are juv. Also compared to ad, upperparts are warmer toned and underparts slightly richer buffish-olive, especially on sides. (D. Occhiato)

1stW, Spain, Aug: typically tinged more buff-brown than worn ad in summer/autumn when still in Europe. Otherwise wing is mostly juv and usually fresher than ad. Note that the grey neck-side has not developed in a juv yet (other than a vague hint). (S. Fletcher)

BLACKCAP
Sylvia atricapilla (L., 1758)

Fr. – Fauvette à tête noire; Ger. – Mönchsgrasmücke
Sp. – Curruca capirotada; Swe. – Svarthätta

Compared to Garden Warbler, its closest relative, the Blackcap inhabits a wider variety of wooded habitats, including taller and denser macchia, dense spruce forests with a few broadleaved trees mixed in, larger parks and gardens, quite dense broadleaf woods, etc. It is an abundant species with some 25 million pairs in Europe alone, and its strong and clear song is one of the commonest sounds of spring and early summer in forests. While northern and eastern populations are long-distance migrants to SE Africa, W European populations stay in SW Europe or NW Africa, and some move west-north-west from the Continent to winter in Britain and even SW Scandinavia.

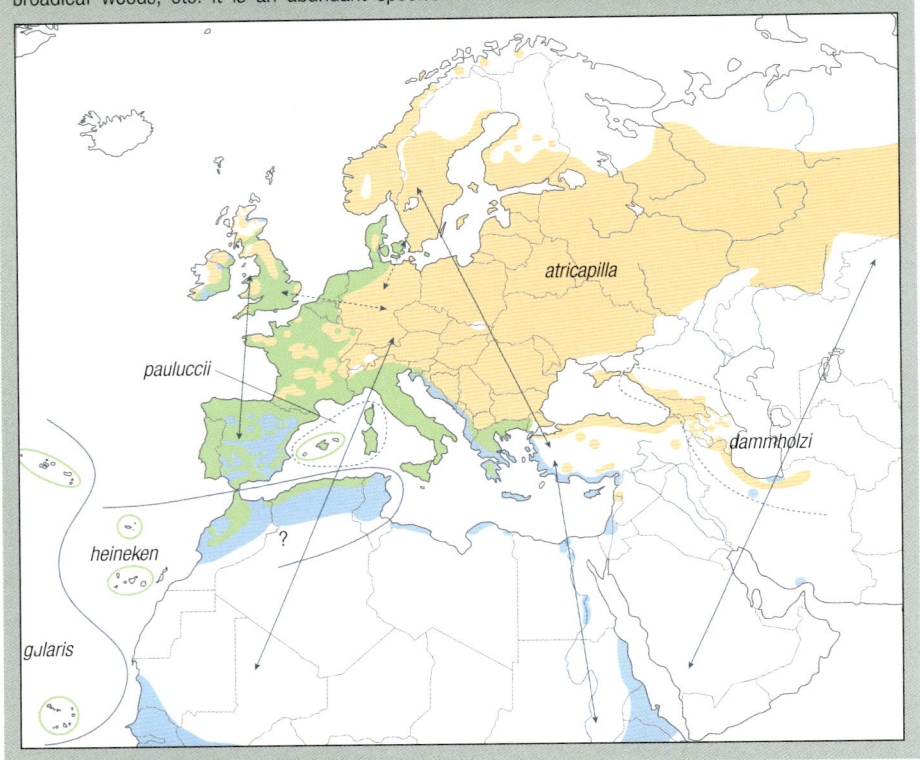

S. a. atricapilla, ♂, presumed ad, Netherlands, Apr: diagnostic black cap is almost always instantly obvious. Ageing, however, is far from easy, even in the hand: here the fresh and pure greyish fringes to remiges and round-tipped primary-coverts indicate an ad. Conversely, the tertials and innermost greater covert are clearly greyer/newer, which is frequent pattern in 1stY. Still, 'presumed ad' seems a fair conclusion. (N. D. van Swelm)

IDENTIFICATION Like the Garden Warbler, basically a *greyish* rather *large Sylvia* characterised by its *black* (♂) or *rufous-brown* (♀ and juv) *cap*. The rest of the plumage is similar to that of the Garden Warbler, with featureless olive-grey upperparts and dusky greyish-white underparts, the only difference from Garden Warbler being that the sides of head and neck are slightly paler and purer grey, especially in the ♂. As in Garden Warbler, no white on uniformly-coloured grey tail. *Bill and legs grey*, the bill rather *short and stout* just like in Garden Warbler. Movements and jizz quite similar to Garden Warbler, but not as shy and secretive, and hence much easier to see. If mimicked in breeding season will often approach to investigate even if you manage only a rather crude rendering of its song. A dark morph can occur in both sexes on the Canaries and Madeira.

VOCALISATIONS Song often starts with a slightly subdued warble of both hard notes and more pleasant ones, not unlike the song of several other *Sylvia* warblers, but soon increases in strength and shifts to its characteristic end, a few clear flute-like notes with a melancholy ring ('minor key'). In midday heat in S Europe, the warbling intro is often dropped, and the fluty end simplified. A variant of this has a high and low note given alternately in drawn-out, rather mechanical series, and this can recall both Orphean Warbler and (vaguely!) Great Tit. A prolonged warbling subsong with a mixture of clever mimicry, hard and whistling notes can sometimes be heard, quite similar to the subsong of Common Whitethroat. – The alarm call, often uttered near nest or young, is a characteristic prolonged series of tongue-clicking *zeck* notes, indicative of great anxiety, with interfoliated drawn-out, hoarse *schree* notes, *zeck zeck zeck schree zeck zeck...* Single *zeck* calls are used in contact, but are obviously more difficult to separate from a number of other warblers.

SIMILAR SPECIES To the beginner the Blackcap needs to be separated from *Marsh* and *Willow Tits*, all three being greyish with black cap. This should not present any problems though, since the cap is smaller on Blackcap (touches upper rim of eye; on the tits it covers the entire eye), the head is proportionately smaller too, and behaviour and calls clearly differ. – Told from the two *Orphean Warblers* by the duskier underparts, all-dark tail (Orpheans: white sides), and well-marked cap at nape (Orpheans: dark grey forehead and ear-coverts, but usually greyish nape). – *Sardinian Warbler*

S. a. atricapilla, Israel, Apr: a simply patterned *Sylvia*, mostly grey or greyish-brown and diagnostically capped black in ♂ (left) or rufous-brown in ♀ (right). In close views and when still not heavily worn, plumage appears less dull, with olive tone above and dusky-buff wash below. Apparent moult limits and strong feather wear in wing suggest both are 1stS, and indeed ♂ still has some rufous in the cap. (E. Bartov)

S. a. atricapilla, 1stS ♂, Israel, May: some young ♂♂ are easily aged if black cap retains a few brown (juv) feathers, while this bird seems to have an intermediate sooty-brown cap rather than black. Also note rather worn wing with heavily-abraded brown juv remiges and longest tertial. Juv primary-coverts have worn and pointed tips. (A. Ben Dov)

S. a. atricapilla, 1stS ♀, Italy, May: generally identical to ♂ but has rufous cap. Garden Warbler is similarly shaped, also with rather short stout bill, but aside from the cap, Blackcap also has head-sides and neck paler and purer grey. Note abraded and brownish juv remiges, primary-coverts and outer greater coverts, and possibly the largest alula and tertials. (D. Occhiato)

has a better-defined black cap, but is a much smaller and slimmer bird with white sides to longer tail, darker flanks, brick-red eye-ring, etc.

AGEING & SEXING Ages separable in autumn, if seen close, by state of plumage; often in spring as well, although some are then dubious, especially ♀♀. Sexes differ after post-juv moult. — Moults. Post-nuptial moult in late Jun–Sep complete. Post-juv moult (can last into Oct) partial, involving some body-feathers and wing-coverts (including all or nearly all greater coverts in about half, and commonly inner 7–8). There is a limited or variable partial pre-nuptial moult of both ad and 1stW in Dec–Mar; most birds of sedentary populations replace very few or no feathers at all, whereas migratory ones replace a varying number of body-feathers, wing-coverts and tertials. — **SPRING Ad ♂** Cap black. Throat, breast and upper flanks grey, moderate brown tinge on lower flanks. All greater coverts and tertials uniform. Tips of flight-feathers fairly fresh. **1stS ♂** As ad ♂, and cap often black as in ad, but a few retain a few brown juv feathers on crown. Generally a slight moult contrast detectable in greater coverts and sometimes tertials. Tips of primaries and tail-feathers rather worn. **Ad ♀** Cap all brown. Breast and flanks warm olive-brown, less grey than

S. a. atricapilla, ♂, presumed ad, Finland, Oct: long pointed wings and mid-length tail give it an almost finch-like shape, especially in flight. Otherwise ♂ plumage is essentially greyish, with tendency to appear purer grey in older birds. Short, stout bill and legs, both being grey. Apparently evenly fresh wing and lack of brown feathers in cap suggest an ad. (T. Muukkonen)

S. a. atricapilla, 1stW ♂, Netherlands, Oct: after post-juv moult many ♂♂ are readily aged and sexed by mixed black-and-brown cap. However, ♂♂ with all-black cap and ♀♀ can be aged only by moult assessment, which is never easy in the field or in photos, and sometimes difficult in the hand. Note the characteristic long pointed wings, but relatively short tail and bill. (H. Gebuis)

S. a. atricapilla, 1stW ♂, Italy, Sep: this ♂ has almost purely black cap (only some traces of brown above lores and eye). Note moult limits due to browner juv primaries, primary-coverts, outer greater coverts and largest alula, contrasting with renewed inner greater coverts and tertials. Also, juv tail-feathers are narrower with pointed tips. Characteristic pure grey head-sides and neck. (D. Occhiato)

♂. All greater coverts and tertials uniform. Tips of flight-feathers fairly fresh. **1stS** ♀ As ad ♀, but often possible to detect a slight moult contrast in greater coverts and sometimes tertials. Tips of primaries and tail-feathers rather worn. — **AUTUMN Ad** Either in final stages of complete moult, with outer primaries and inner secondaries growing, or entirely fresh. Sexing straight-forward by black (♂) or brown cap (♀), but rarely some ♂♂ have brown-tipped black feathers on hindcrown. **1stCY** After post-juv moult a slight moult contrast occurs in greater coverts and sometimes tertials, the outer retained juv greater coverts being more brownish-grey and abraded, edges less tinged olive. Sexing generally possible on colour of cap as in ad, but a very few ♂♂ in Nov (Dec) still have a largely brown cap with just the odd black feather emerging, or a dark brown cap tinged black all over.

BIOMETRICS (*atricapilla*) **L** 14–15.5 cm; **W** 70–80 mm (*n* 77, *m* 75.6); **T** 56–65 mm (*n* 29, *m* 59.4); **T/W** *m* 80.2; **B** 13.3–15.3 mm (*n* 39, *m* 14.2). **Wing formula: p1** > pc 0–6.5 mm; **p2** < wt 5–9 mm, =5/6 (rarely =5 or =6); **pp3–4** about equal and longest (rarely p4 to 1 mm <); **p5** < wt 2–4.5 mm; **p6** < wt 5–10 mm; **p10** < wt 13–20 mm; **s1** < wt 15.5–23 mm. Emarg. pp3–5, on p5 near tip and sometimes slightly less prominent.

S. a. atricapilla, 1stW ♀, France, Sep: rather clear moult limits in wing, with browner and already worn juv remiges, and tail, the latter being narrower with sharply-pointed tips. (A. Audevard)

GEOGRAPHICAL VARIATION & RANGE Clinal and mostly very slight variation, with some more complex variation in resident island populations. Individual birds frequently impossible to name subspecifically without series at hand for comparison, save some very long-billed birds on the Azores.

S. a. atricapilla (L., 1758) (much of Europe, W Siberia, Turkey; winters W & S Europe, rarely S Scandinavia, also SE Europe, Levant and N & E Africa). Treated above. Fairly long-winged, brightly coloured and pale grey below with greyish-white throat, although birds in Britain, the Netherlands, France and Iberia tend to be slightly shorter-winged and duller in colour. (Syn. *riphaea*.)

○ **S. a. dammholzi** Stresemann, 1928 (Caucasus, N Iran; winters E Africa, possibly also Arabia). Very similar to *atricapilla* although on average slightly larger, and is paler and cleaner olive-grey above, less tinged brownish, is whiter on belly, and ♀ is a little greyer, less olive-brown, below, and usually has a trifle paler rufous cap. Still, rather subtle differences. **W** 72–82 mm (*n* 22, *m* 77.7); **T** 57–66 mm (*n* 22, *m* 61.3); **T/W** *m* 79.0.

○ **S. a. pauluccii** Arrigoni, 1902 (Sardinia, Balearics, possibly Corsica; resident or short-range migrant). Quite close to *atricapilla* but, like *dammholzi*, is a little cleaner

S. a. atricapilla, juv, Germany, Jul: juv always mirrors ♀ plumage due to rufous cap, but the brown cap is usually noticeably duller and a little paler rufous, and soft, fluffy body-feathers are typically obvious. (J. Schmitz)

S. a. heineken, 1stS ♂, melanistic form, Madeira, Apr: in this rare morph entire head, neck and breast are blackish and rest of plumage darker/warmer than normal. Young ♂ usually easily aged by having black hood less solid, often with brownish ear-coverts and mottled breast. Note rather clear moult contrast in greater coverts, worn juv primary tips and primary-coverts, and heavily abraded and very pointed juv tail-feathers. Bill is long as so often in island taxa. (H. Shirihai)

S. a. heineken, 1stW ♀, melanistic form, captive-born of Madeira population, Nov: ♀♀ of this dark morph are often distinctive, being overall darker and warmer in colour, with warmer greyish-rufous underparts than normal *heineken*. Rufous cap subdued due to both young age and melanistic morph. (U. Querner)

S. a. heineken, ♂, Madeira, Apr: this race of Madeira, the Canaries and possibly NW Africa averages overall subtly darker and shorter-winged. Due to shorter primary projection, tail will appear slightly longer. Ageing unsure without handling. (H. Shirihai)

olive-grey above, and cap of ♀ usually slightly paler rufous. Bill proportionately long: **B** (Italy, Sardinia, Corsica, Mallorca; *BWP* 1992, own measurements) 14.3–16.2 mm (*n* 24, *m* 15.1). Nevertheless, close to *atricapilla*. (Syn. *koenigi*.)

○ ***S. a. heineken*** (Jardine, 1830) (Madeira, Canaries, apparently NW Africa; resident). About same size as *atricapilla* but has shorter wing and a fairly distinct song dialect. (That it should be smaller and 'markedly darker' than *atricapilla*, as sometimes stated, cannot be confirmed when comparing a fair series from Madeira with European birds; NHM.) A melanistic morph occurs rarely in both sexes, ♂ having whole head, neck and breast black, and rest of plumage much darker than normal morph, ♀ being darker olive-brown. **W** 66–73 mm (*n* 26, *m* 70.0); **T** 55–62 mm (*n* 17, *m* 58.4); **T/W** *m* 82.7; **B** 13.0–15.0 mm (*n* 17, *m* 14.0). **Wing formula: p2** < wt 5–8 mm; **p10** < wt 12.5–17 mm; **s1** < wt 15–19.5 mm. (Syn. *obscura*.)

○ ***S. a. gularis*** Alexander, 1898 (Cape Verdes, Azores; resident). Similar coloration and size to *atricapilla*, but bill proportionately long. Some rather complex variations in size noted within the Azores, but on the whole not regarded as grounds for separation. The following measurements are from San Miguel. **W** 71–79 mm (*n* 16, *m* 74.6); **T** 56–67 mm (*n* 15, *m* 60.7); **B** 14.3–16.3 mm (*n* 16, *m* 15.5). (Syn. *atlantis*.)

S. a. gularis, São Miguel, Azores, ♂ (left, Mar) and ♀ (Nov): similar to *atricapilla* in plumage, but bill proportionately longer. Without handling definite ageing is difficult in both birds. (B. Carlsson)

BROWN WOODLAND WARBLER
Phylloscopus umbrovirens (Rüppell, 1840)

Fr. – Pouillot ombré; Ger. – Umbralaubsänger
Sp. – Mosquitero oscuro; Swe. – Grönvingad sångare

Within the treated region, this resident brownish and green leaf warbler only occurs in W Yemen and SW Saudi Arabia, but it also has a widespread extralimital distribution in E Africa, from Ethiopia to Tanzania. In Arabia it breeds in montane forests, in juniper scrub or in 'green' wadis at altitudes of 1600–2600 m. As is the habit in warm climates, Brown Woodland Warbler breeds in all or most months of the year, which means that moult strategies are not as clear-cut as in a European species, and ageing requires care.

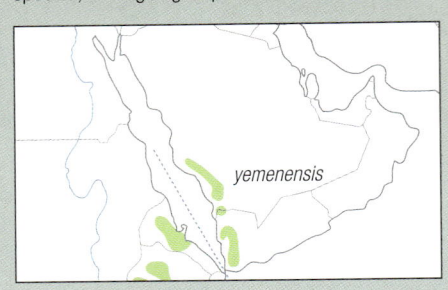

P. u. yemenensis, Yemen, Jan: strong bill with much pink-yellow at base, strikingly bright yellow-green wing in contrast to grey-brown upperparts, and dusky-looking underparts. Supercilium is rather well marked, but particularly so in front of eye. (H. & J. Eriksen)

P. u. yemenensis, Yemen, May: a rather small, compact *Phylloscopus* with relatively distinctive plumage. Especially note rufous-tinged head and bright yellowish-green edges to wing-feathers and tail-sides, while the narrow supercilium and underparts are off-white tinged pink-buff. Ageing impossible without handling. (W. Müller)

IDENTIFICATION Same size or slightly smaller than Common Chiffchaff, which it otherwise resembles in shape and general appearance. However, it is perhaps slightly more compact, short-tailed and 'neck-less', has a shorter, rounder wing, and a proportionately *longer-looking and stronger bill* than Common Chiffchaff. Upperparts brown with faint olive cast and crown often darkest and more rufous. All *edges to wing-feathers* (including greater coverts) *bright yellowish-green*, contrasting in colour and usually appearing as a lighter area against the duller back and scapulars (brightness depending on light and angle). *Tail-sides too have bright yellowish-green edges. Sides of head and most of underparts warm tawny-buff*, strongest on supercilium and sides of head, but also fairly prominent on breast and flanks. Centre of throat and belly paler, more whitish, and undertail-coverts can have a pale yellow tinge. A *dark eye-stripe* helps to enhance the otherwise rather subdued buff supercilium. Supercilium narrow in front of and above eye but widens considerably behind it. Patch near wing-bend and axillaries pale lemon (neither of them normally visible on a perched bird). Bill dark with pale pink-brown or *straw-coloured base to lower mandible* and cutting edges, *tip of lower mandible being dark*, thus many at least have *more bicoloured bill* than is found in its European relatives. Legs dark grey or grey-brown, but can appear blackish in the field.

VOCALISATIONS Song is a fast series of clear, whistling notes that mainly descend the scale, or have last note ascending. Phrases vary somewhat and may be prolonged. Has been likened to Willow Warbler song, presumably due to mainly descending pitch, but is much faster and more varied given inclusion of trilling passages, making it actually more reminiscent of song of Pallas's Leaf Warbler or Canary (M. Ullman *in litt.*). – Call is a short *tee* or *tee-wee* (*HBW*), or a slightly drawn-out, buzzing *dscher*. Also described as a 'descending metallic long *dzziieep*' (Porter *et al.* 1996).

SIMILAR SPECIES Superficially similar to Common Chiffchaff of Siberian race *tristis*, but differs in strong contrast between dark brownish-olive upperparts and pale yellowish-green wings (much stronger contrast between brown and green tones than in any *tristis*). Further, sides of head, breast and flanks darker rufous-buff, and supercilium generally darker buff and therefore less prominent. Largely pale lower mandible on much stronger bill are other differences. – Willow Warbler eliminated by using same characters, and on much shorter primary projection. – Both Bonelli's Warblers separated by their whitish underparts, paler yellowish-ochre or yellowish-green rump, and lack of strong rufous-buff tinge to supercilium and sides of head and neck.

AGEING & SEXING (*yemenensis*) Relatively small differences in plumage between ad and juv. Sexes alike in plumage. – Moults. Complete post-nuptial moult of ad mainly in summer but appears to be variable (due to extended breeding season), and exact timing not studied. Partial post-juv moult (usually in summer, but variable) little studied but does not seem to involve flight-feathers or primary-coverts. Partial pre-nuptial moult mainly in winter. – **SPRING–SUMMER Ad** On average slightly less dark olive-brown above than juv, lacking any yellow below except on underwing and undertail-coverts. Plumage of breeders usually quite worn with frayed tips to flight-feathers. **Juv** Typical birds are slightly darker brown above, have a faint dull yellow cast over much of underparts (not only undertail-coverts), and have slightly 'woolly' and loose feather texture, especially on nape and vent. Tips of flight-feathers fresh at least during first two month after fledging.

BIOMETRICS (*yemenensis*) **L** 10–11 cm. **W** 53–58 mm (n 13, m 56.0; sexes of about same size); **T** 39–46 mm (n 13, m 42.9); **T/W** m 76.6; **B** 11.8–13.0 mm (n 13, m 12.5); **BD** 2.6–3.5 mm (n 13, m 2.9); **Ts** 19.0–21.0 mm (n 12, m 20.0). Wing formula: **p1** > pc 8–13 mm, < p2 13.5–18 mm; **p2** < wt 6.5–13 mm, ≤ ss (70%), =10 or 10/ss (20%), or =9 or 9/10 (5%); **p3** < wt 1.5–4 mm; **pp4–6** about equal and longest; **p7** < wt 0.5–4 mm; **p8** < wt 2.5–5.5 mm; **p10** < wt 6–8 mm; **s1** < wt 6–9 mm. Emarg. pp3–6, sometimes faintly also on p7.

GEOGRAPHICAL VARIATION & RANGE Nine races described, of which only *yemenensis* breeds within the covered region. All the others are extralimital and not treated here, breeding in Africa and are not likely to occur far from their respective ranges. The differences between these races seem mostly slight, and the distinctness of some of them has been questioned (*HBW*).

P. u. yemenensis (Ogilvie-Grant, 1913) (SW Saudi Arabia, W Yemen). Treated above.

P. u. yemenensis, Yemen, Jan: the bill often appears reasonably strong and is typically bicoloured, with rather contrasting pale pink-brown or straw-coloured base to lower mandible and cutting edges. Also note that both pale supercilium and dark eye-stripe tend to be more solid in front of eye. Attractive wing pattern comprising bright yellowish-green edges to most flight-feathers, primary-coverts and some greater coverts, contrasting with greyish tertials. Legs blackish-grey. (H. & J. Eriksen)

ARCTIC WARBLER
Phylloscopus borealis Blasius, 1858

Fr. – Pouillot boréal; Ger. – Wanderlaubsänger
Sp. – Mosquitero boreal; Swe. – Nordsångare

As the German name 'Wanderlaubsänger' implies, this is a real long-distance migrant, together with Willow Warbler one of the longest-migrating species in relation to its size. Breeding in the Arctic taiga, in the west to N Norway, it migrates all the way to SE Asia to winter in Borneo, Sumatra, the Philippines and adjacent areas. To get there it passes north and east of the Central Asian deserts and the Himalayas, a journey of up to c. 13,000 km each way. No small feat considering that it weighs just 10 g and its fuel is merely tiny insects. Due to its long migration, it is perhaps not surprising that the species is the latest of all summer visitors to arrive, often not until midsummer.

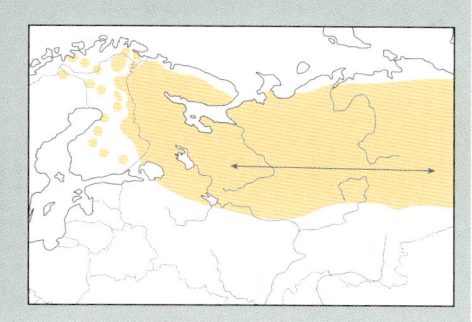

Finland, Jun: among the most robust leaf warblers. Well into summer the greater coverts wing-bar is still obvious, and there can even be a hint of a shorter and duller second bar on median coverts. Note long evenly narrow supercilium. Ageing impossible once winter moult completed. (M. Varesvuo)

IDENTIFICATION Although a small and slender bird, among the so-called leaf warblers (genus *Phylloscopus*) it is one of the largest, roughly the same size as Willow Warbler and Wood Warbler. Being a long-distance migrant it is decidedly long-winged, with a *primary projection at least c. 3/4 of tertial length*. The tail, however, is proportionately short, a little shorter than in Willow Warbler. *Legs are usually pale grey-brown ('horn-coloured') or pinkish-brown*, but rarely are somewhat darker. Best characters apart from size, shape and leg colour are: (i) a rather prominent *white wing-bar* formed by white-tipped greater coverts, and (ii) *long and distinct whitish supercilium* (faintly yellowish in front, whiter at rear), reaching far back towards nape. The supercilium is enhanced by (iii) a *dark eye-stripe* and uniform dark crown, both being olive-grey. Note also (iv) generally *variegated cheeks and ear-coverts*, boldly patterned yellowish-white and dark greenish-grey. Further, (v) the *long bill* which has, compared to most congeners, *paler lower mandible*, much of it pinkish-yellow (only tip variably dark). Two more features add to the character of Arctic Warbler: (vi) the often *grey-blotched lower throat and breast*, and lightly grey-tinged flanks, and (vii) the uniform greyish-green upperparts, *tertials being near-uniform*. At some angles there is an almost 'bronzy' sheen to the otherwise greenish-edged primaries. Underparts are, apart from the grey, pale yellowish-white. When seen well, note that (viii) *supercilia end above lores but do not meet on forehead over bill* (as in most Greenish Warblers of ssp. *viridanus*), whereas (ix) the *dark eye-stripe reaches all the way to base of bill*. The greater coverts wing-bar is usually broader and more prominent throughout its length than in *viridanus*, and has a tendency to show tiny breaks between each feather ('string of pearls', whitish tips mainly confined to outer webs), whereas the wing-bar is more continuous (but shorter as a whole and thinner at each end) in *viridanus*. Rarely the greater coverts wing-bar is absent in summer due to feather wear. When faced with the sometimes tricky separation from Two-barred Warbler, note that (x) dark eye-stripe is uniformly broad through eye (usually pinched-in in front of eye in Two-barred due to more extensive white lower eye-lid), and that in fresh plumage (xi) presence of any median coverts wing-bar is diffuse and yellow-tinged (Two-barred: usually rather distinct and more whitish). Supporting differences are (xii) subtly larger size and more elongated shape, (xiii) short outermost primary ('p1') if visible, and (xiv) never a hint of emargination on p6. (xv) Underparts average darker and duskier, and (xvi) supercilium is usually evenly broad in much of its length (sometimes broadens behind eye in Two-barred). Call from a migrant is particularly useful to note.

VOCALISATIONS Song is a rather monotonous, drawn-out, low-pitched and dryish trill, usually staying on one pitch with a repetition of the same type of note or syllable, *zre-zre-zre-zre-zre-zre-zre-*…, but sometimes changing to a slightly lower pitch halfway, *sre-sre-sre-sre-sre-sre-sre-sru-sru-sru-sru*, if anything increasing slightly in loudness. Can recall song of Cirl Bunting, but is a little 'fuller', more shuttling, less trilling. There is some individual variation, but the song is still usually readily recognised by general structure, pace and 'voice'. Each song lasts c. 2.5–4 sec, often being preceded by a few call notes. – Call characteristic, a short, penetrating, sharp and scratchy note like the call of White-throated Dipper, *dzre* or *zrik*, quite unlike any call of W Palearctic congeners. Anxious birds use a dry trill or churring note, *tr-tr-tr*, in alarm.

SIMILAR SPECIES Due to general coloration, long and distinct supercilium and single wing-bar, commonest problem to European birders is to eliminate *Greenish Warbler* (particularly the western race *viridanus*). Greenish shares the above-mentioned characteristics, but is a smaller and more compact bird with, size-related, quicker and more energetic movements when feeding. It has on average a less distinct and slightly shorter wing-bar than Arctic, but safest separation is to note that supercilia in Greenish (almost) join on forehead (stop short on Arctic), and that the dark eye-stripe is less prominent, as a rule ending before reaching base of bill (reaches it in Arctic). Greenish has on average less boldly patterned cheeks and ear-coverts than Arctic (still, odd birds of both come close), and the supercilium can be rather broad on some Greenish, whereas it seems to be invariably narrow in Arctic. Legs are usually somewhat paler in Arctic, darker in Greenish. The sharp, White-throated Dipper-like call of Arctic will immediately tell it from Greenish, the latter sounding more like a Pied Wagtail. – In several respects, Arctic is far closer in appearance to *Two-barred Warbler*, both having a very prominent wing-bar and supercilium, the latter ending short of the forehead, but Arctic often has grey-mottled or diffusely grey-striped lower throat and breast-sides, a slightly stronger bill and larger head. A subtle average difference is a stronger marked dark eye-stripe through eye, breaking the white eye-ring more prominently in Arctic, whereas Two-barred often has more pinched-in eye-stripe at the front of the eye, white eye-lids coming closer to each other. (See also under Identification for further criteria.) And, again, their calls differ, Two-barred resembling Greenish or Green, not White-throated Dipper. – A far-fetched but realistic risk of confusion is with *Eastern Crowned Warbler* (*P. coronatus*; extralimital), which has straggled a few times from E Asia to W Europe. This resembles an Arctic, but differs in having a hint of a narrow paler central crown-stripe, an even stronger bill and often broader and more obviously yellowish supercilium in front of eye. Both eye-stripe and sides of crown tend to be darker on Eastern Crowned, while underparts are slightly paler, less olive-grey on sides and breast. There is also a tendency for the nape and neck-sides to be slightly greyish-tinged. – After the recent split in three species (cf. Taxonomic notes), severe identification problems exist in E Asia, problems which still require more research to be over-

Finland, Jun: note the proportionately short-tailed impression due to long primary projection. Also note the following combination of characters: prominent white greater coverts wing-bar, long and distinct whitish supercilium, which is less clean in its forepart than at rear and ends above lores (rather than meeting on forehead like in most Greenish Warblers), highlighted by dark eye-stripe and uniform dark crown (concolorous with nape and mantle), and typically mottled head-sides. (J. Tenovuo)

Finland, Jun: legs are pinkish-brown, while the rather long but pointed bill is variably coloured, but in many birds (and compared to most congeners) is typically mainly pale over much of lower mandible, being pinkish-yellow with only tip variably dark. Note characteristic diffusely grey-blotched lower throat and breast, slightly grey-tinged flanks, and rather pleasant greyish-green upperparts. (M. Varesvuo)

Ad, Scotland, Sep: a difficult bird that easily could be overlooked since, due to feather wear, the wing-bar is almost absent. Note the rather unremarkable head pattern, with whiter supercilium that is buffier in front of eye. With wear in autumn, ad becomes greyer above and underparts predominantly clean white or slightly pale yellowish-white at sides. All visible wing-feather tracts are ad-like, which during autumn migration are slightly or moderately worn, mainly noticeable on tips of tertials, tail and longest primaries. (H. Harrop)

come. The three species differ in DNA, song, call and subtly in morphology (see Alström *et al.* 2011 for a summary).

AGEING & SEXING (*borealis*) Small age-related plumage differences only in 1stCY. Sexes alike or very similar in plumage. – Moults. Partial post-nuptial moult in late summer (Jul–Aug) involving body-feathers and some wing-coverts, sometimes also odd tertials and tail-feathers. Complete pre-nuptial moult in late winter (mainly Feb–early Apr). Partial post-juv moult in summer does not involve flight-feathers or primary-coverts. – For both age-classes note peculiar finely pointed tips to flight-feathers ('spikes') unless too worn, also seen in Greenish Warbler and some other Asian species, but usually not in Willow Warbler or Common Chiffchaff. – **SPRING** Sexes very similar, but ♂ appears to be on average very slightly paler on breast and flanks; those with prominently dark olive-grey sides to breast/upper flanks are nearly always ♀♀. Still, much overlap, so this character should be used with caution. – **AUTUMN Ad** Much-abraded tips to primaries and tail-feathers. Ground colour of underparts whitish or pale yellowish-white. Supercilium usually white. **1stCY** Plumage fresh, tips to primaries and tail-feathers pointed but fresh. Underparts tinged yellow, in most birds more so than in ad. Supercilium yellowish.

BIOMETRICS (*borealis*) **L** 11–12.5 cm. **W** ♂ 62–71 mm (*n* 17, *m* 67.3), ♀ 60–65 mm (*n* 14, *m* 62.9); **T** ♂ 40–51 mm (*n* 17, *m* 46.2), ♀ 41–45.5 mm (*n* 14, *m* 43.7); **T/W** *m* 69.1; **B** ♂ 12.7–15.3 mm (*n* 14, *m* 14.0), ♀ 12.3–14.4 mm (*n* 14, *m* 13.2); **Ts** 17.5–19.6 mm (*n* 31, *m* 18.7). **Wing formula: p1** > pc 0–2 mm or < 0–1.5 mm, < p2 29–39 mm; **p2** < wt 4–7.5 mm, =5/6 (75% in ♂♂, 43% in ♀♀), =6 (18% in ♂♂, 43% in ♀♀) or =6/7 (7% in ♂♂, 14% in ♀♀); **pp3–4** about equal and longest; **p5** < wt 0.5–2.5 mm; **p6** < wt 4–7 mm; **p7** < wt 7.5–11 mm; **p10** < wt 13–18 mm; **s1** < wt 14–19 mm. Emarg. pp3–5.

GEOGRAPHICAL VARIATION & RANGE Monotypic, following recommendations by Alström *et al.* (2011). There is a very subtle tendency for birds to become slightly larger and their colours more saturated in the east, but these differences are both very minor and purely clinal over a vast range, hence not sufficient grounds for recognition of more than one taxon. See below regarding separation of two E Asian taxa as separate species. – N Fenno-Scandia, N Russia to N Russian Far East, Alaska; winters SE Asia. (Syn. *flavescens*; *hylebata*; *kennicotti*; *talovka*; *transbaicalicus*.)

TAXONOMIC NOTE Two extralimital Far Eastern taxa formerly treated as races of Arctic Warbler have recently been proposed to constitute separate cryptic species ('Japanese Leaf Warbler' *P. xanthodryas* in Japan except Hokkaido, and 'Kamchatka Leaf Warbler' *P. examinandus* in S Kamchatka, Sakhalin, Hokkaido and Kurils) based on different song, calls, morphology and mtDNA, a proposal here followed. (Alström *et al.* 2011. See also Reeves *et al.* 2008, Saitoh *et al.* 2008, 2010, and some further references therein.)

REFERENCES Alström, P. & Olsson, U. (1987) *Proc. 4th Int. Ident. Meet. Eilat 1st–8th Nov. 1986*: 54–59. – Alström, P. & Olsson, U. (1989) *Limicola*, 3: 269–279. – Alström, P. *et al.* (2011) *Ibis*, 153: 395–410. – Reeves, A. B., Drovetski, S. V. & Fadeev, I. V. (2008) *J. Avian Biol.*, 39: 567–575. – Saitoh, T., Shigeta, Y. & Ueda, K. (2009) *Orn. Science*, 7: 135–142, 8: 1–11. – Saitoh, T. *et al.* (2010) *BMC Evol. Biol.*, 10(35): 1–13.

1stW, Scotland, Oct: unlike ad at same season, plumage is fresh and much greener, supercilium yellowish, and very similar to fresh ad in spring. The underparts, however, are yellower, being less greyish-white than ad. (G. Jenkins)

1stW, Scotland, Sep: some young in autumn are duller and to some degree approach ad coloration, but are still much fresher, with greener edges to remiges. Also note two-toned supercilium with buffier area in front of eye, mottled cheeks, and mainly pale lower mandible. (H. Harrop)

GREEN WARBLER
Phylloscopus nitidus Blyth, 1843

Alternative names: Bright-green Warbler, Green Leaf Warbler

Fr. – Pouillot du Caucase; Ger. – Wacholderlaubsänger
Sp. – Mosquitero del Cáucaso
Swe. – Kaukasisk lundsångare

Closely related to the Greenish Warbler, the Green Warbler has an isolated and rather restricted breeding range in N Turkey, the Caucasus, Transcaucasia, N Iran and adjacent mountain ranges. In Turkey, where most people probably have sought it out on their first encounter, and in Georgia and Armenia, it breeds in lush broadleaved forests on mountain slopes above c. 900 m, often reaching to c. 2500 m. It has been stated to breed in juniper scrub in NE Iran, hence its German name (meaning 'Juniper Leaf Warbler'). In late summer migrates east-south-east to winter in S India and Sri Lanka.

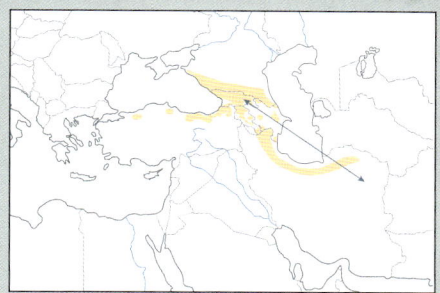

Iran, Apr: the long, broad supercilium usually ends just short of forehead (in most *viridanus* Greenish supercilia meet on forehead), although this can be hard to judge when seen head on. Compared to *viridanus*, note the longer and evenly broad greater coverts wing-bar. (Magnus Ullman)

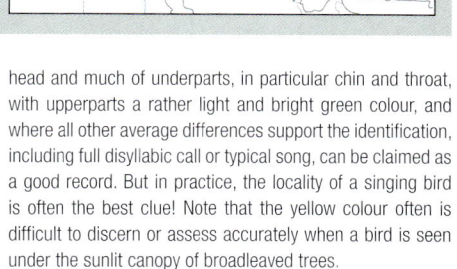

Finland, May: a surprising first record for Finland of this W Asian species, showing well all the characteristic features, like bright yellow and broad supercilium and cheeks, with throat and upper breast also tinged yellow, strong green tone to upperparts, solid and broad whitish wing-bar, and strong bill. Breeding birds at this season cannot be aged due to complete pre-nuptial moult for all birds. (M. Bruun)

IDENTIFICATION By and large of same size and appearance as Greenish Warbler, the Green Warbler differs subtly but consistently in the following ways from ssp. *viridanus* of Greenish (in order of importance): (i) *supercilium, face and whole of underparts pale but bright yellow*, clearly more so than in the most yellow-tinged *viridanus*; although a few *viridanus* Greenish are a little more yellow-tinged than others, they tend to be subtly more yellow on belly and flanks, whereas Green, if anything, is most saturated yellow on supercilium, chin, throat and breast; (ii) *upperparts slightly paler and more brightly green-tinged*, less greyish; (iii) long and broad *supercilium ends short of base of bill* (like in most Two-barred Warblers) and does not continue to forehead (as in most *viridanus* Greenish); (iv) on average *longer and broader lower wing-bar* on greater coverts, usually ending broadly at both ends (wing-bar narrower and short, usually fading off at both ends in *viridanus* Greenish); (v) sometimes *a hint of a short, thin second wing-bar* on median coverts (very rare in *viridanus* Greenish); (vi) on average subtly longer primary projection (but much overlap); (vii) size on average a fraction larger, and wing longer (but much overlap) and appears to have a *larger head* and a *stronger bill*. To these characteristics come subtle but apparently quite useful differences in vocalisation. A bird seen well, close and in good light with bright lemon-yellow supercilium, sides of head and much of underparts, in particular chin and throat, with upperparts a rather light and bright green colour, and where all other average differences support the identification, including full disyllabic call or typical song, can be claimed as a good record. But in practice, the locality of a singing bird is often the best clue! Note that the yellow colour often is difficult to discern or assess accurately when a bird is seen under the sunlit canopy of broadleaved trees.

VOCALISATIONS Song very much resembles that of Greenish Warbler, a high-pitched, string of sharp, 'scratchy' notes in a somewhat jerky, 'hurried' rhythm. Strophes often last 1.5–2.5 sec. Sometimes, the song cannot reliably be separated from Greenish (at least not without use of sonograms, or even then), but many birds include a 'buzzing' or dry trilling element (different from the Wren-like trills sometimes heard from Greenish Warbler), which seems to be species-specific. This dry trilling note, *tserr*, is often initiated by a short, sharp note, thus *tsi-tserr*. A song can include up to three such trilling notes, but more commonly one or two. Another element in typical song (but not used by all singers) is the inclusion of a sweet whistling, short *hooeet*. – Call very like call of Two-barred Warbler (and presumably inseparable from it), a high-pitched, clearly disyllabic *tiss-swee* or *tss-eurr*. Some variants can also be transcribed as the French say '18', *dishuit*.

SIMILAR SPECIES The main criteria to note for separation from *Greenish Warbler* are set out under Identification. As to the very few Greenish with a little stronger yellowish hue on head and underparts than the majority, it is important to note full set of characters, including slightly darker and greyer upperparts and at least a whitish element in supercilium, cheeks, chin and throat, rather than these being entirely lemon-yellow as in *nitidus*. (We are unable to confirm the findings of Albrecht 1984, who reported extensive overlap in plumage characters even in specimens at NHM.) Although frequently difficult to establish precise bill size without handling, size of both head and bill of Green Warbler often appear subtly but clearly larger than in Greenish Warbler (*viridanus*). – Differs from *Arctic Warbler* by yellow of underparts, paler green upperparts, slightly smaller size and lack of greyish tinge or mottling on sides of lower throat and breast.

AGEING & SEXING As for Greenish Warbler.

BIOMETRICS L 10.5–12 cm. **W** ♂ 59–69 mm (*n* 32, *m* 64.3), ♀ 58–64 mm (*n* 15, *m* 61.4); **T** ♂ 43–51 mm (*n* 32, *m* 47.1), ♀ 41–48 mm (*n* 15, *m* 45.2); **T/W** *m* 73.4; **B** 12.0–14.0 mm (*n* 36, *m* 12.8); **BD** 2.5–3.0 mm (*n* 19, *m* 2.7); **BW** 3.3–3.8 mm (*n* 14, *m* 3.6); **Ts** 17.5–19.9 mm (*n* 35, *m* 18.6). **Wing formula: p1** > pc 3–8 mm, < p2 25–31 mm (once 23); **p2** < wt 6–9.5 mm, =7 or 7/8 (52%)

Turkey, May: compared to Greenish Warbler, the greater coverts wing-bar is longer and bolder, and often there is also a hint (as here) of a short, narrow second wing-bar on median coverts. Note compact build with rather large head and strong bill with pinkish-yellow base. (D. Occhiato)

=6/7 (37%) or =8 (11%) (Ticehurst 1938 found a higher proportion having =6/7, 84%, and none with =8); **pp3–5** about equal and longest (though either or both of p3 and p5 sometimes < wt 0.5–1 mm; **p6** < wt 1.5–4 mm; **p7** < wt 5–8 mm; **p10** < wt 10.5–14 mm; **s1** < wt 13–17 mm. Emarg. pp3–6.

GEOGRAPHICAL VARIATION & RANGE Monotypic. – N Turkey, Caucasus, Transcaucasia, N Iran; claimed to possibly breed NW & C Afghanistan (Rasmussen & Anderton 2012) but requires confirmation; winters S India, Sri Lanka. – The claimed greyer plumage of W Turkish birds and those in Caucasus (Albrecht 1984, who largely cited anecdotal notes by others) could not be confirmed by our examination of the same material (NHM), nor is there any indication of this in studied specimens in other collections or live birds, thus the claimed variation remains poorly documented.

TAXONOMIC NOTE See Greenish Warbler for a general overview of the complex.

REFERENCES Albrecht, J. S. M. (1984) *Sandgrouse*, 6: 69–75. – Collinson, M. *et al.* (2003) *BB*, 96: 327–331. – Irwin, D. E. (2000) *Evolution*, 54: 998–1010. – Irwin, D. E., Bensch, S. & Price, T. D. (2001) *Nature*, 409: 333–337. – van der Vliet, R. E., Kennerley, P. R. & Small, B. J. (2001) *DB*, 23: 175–191.

Georgia, May: some birds are duller and slightly greyish-toned above and pale yellowish-white below and thus approach Greenish Warbler in coloration. Such birds outside the breeding range will probably prove indistinguishable from Greenish if not trapped or singing, but note large-headed jizz with strong bill, and prominence of greater coverts wing-bar, plus presence of second wing-bar. (V. Legrand)

1stW, Sri Lanka, Dec: three images of a 1stW bird (prior to the complete pre-nuptial moult; plumage still only moderately worn). Despite the whitish wing-bar being somewhat shorter than usual, it is still broad and solid enough to clinch the identification. Also, yellow tones in plumage in early winter are often somewhat reduced through wear, but even so there are always some remains of pale lemon-yellow on face and vent. In particular, note the rather large head and strong-billed impression compared to *viridanus* Greenish, and also the bold yellow supercilium. (H. Shirihai)

Ad, India, Oct: ad by having quite worn and brownish wing-feathers in Oct, especially the primary-coverts and primaries. Central tail-feathers have been renewed in a partial post-nuptial moult. Prior to complete pre-nuptial moult in late winter, the whitish wing-bar may become somewhat reduced, and especially the inner greater coverts can lose nearly all white. Much yellow on face and underparts. (Y. Krishnappa)

1stW, Turkey, Sep: young in fresh plumage and good light conditions often also show quite vivid yellow supercilium, cheeks and most of underparts, especially throat and breast, with upperparts being bright green, making separation from Greenish less problematic. Note large-headed impression and strong bill. Greater coverts wing-bar prominent, and note hint of second wing-bar on median coverts. (V. Karakullukçu)

GREENISH WARBLER
Phylloscopus trochiloides (Sundevall, 1837)

Fr. – Pouillot verdâtre; Ger. – Grünlaubsänger
Sp. – Mosquitero verdoso; Swe. – Lundsångare

A wing-barred, medium-sized leaf warbler that breeds over a vast area from E Europe deep into Asia, in the west reaching to the Baltic Sea. It is found in both deciduous and mixed forests, often preferring open woodland, edges and mature groves. Primarily a lowland species in Europe, it is also found in mountains in Asia. It spends much time in the foliage of tall trees, so but for its high-pitched, hurried and 'nervous' song, many would go unnoticed. In late summer it starts its long migration to SE Asia. Greenish Warbler returns late in spring, in the far north-western part of its range often not until the end of May or early June.

IDENTIFICATION Slightly smaller than Willow Warbler, more like a small Common Chiffchaff, with *rather short tail and wings*, looking rather compact. Often extremely *active and restless*, changing perch all the time, moving very quickly in canopy, briefly hovering at times. Frequently flicks both tail and wings as it moves. Has a habit of raising crown-feathers when agitated, including while singing. Fairly pale and brightly coloured, never appearing as dusky as many Common Chiffchaffs do. *Upperparts rather pale greyish-green*, including on edges of wing-feathers. *Underparts whitish with faint yellow hue*, this being usually uniform on entire underparts, although chin, throat and centre of breast are often paler; many appear more or less whitish below in the field, the faint yellow hue being hard to discern. In western part of range (*viridanus*), two characters are more important than others to note: (i) *long and prominent yellowish-white supercilium*, reaching from forehead to sides of nape, and (ii) *single, short, narrow and whitish wing-bar*. Note that the supercilium does not end above the lores (as in Arctic Warbler) but *diffusely continues to nostrils* or even goes around forehead; note also that the dark grey-green eye-stripe below supercilium is less prominent (than in Arctic), especially in front of eye (and some *viridanus* have a quite poorly marked eye-stripe). The single *pale wing-bar* is of variable prominence, never very long and distinct, more often quite *thin* and *limited to a few central greater coverts*, thus never reaches the edge of the wing or the tertials. However, other races in Asia have a more prominent greater coverts wing-bar, and rarely display a hint of a second bar on median coverts (which *viridanus* only exceptionally does). *Tertials are largely uniformly grey-green*, lacking the contrasting pale outer edges of Yellow-browed and Hume's Leaf Warblers. Bill is typically proportionately rather strong (though less so than in Green or Arctic Warblers) and has extensive *yellowish base to lower mandible*. *Legs* medium to *dark* grey-brown. Call useful when learnt. Vibrates semi-opened wings while singing from canopy of tree.

VOCALISATIONS Song a brief, high-pitched, string of sharp, 'scratchy' notes in a somewhat jerky, 'hurried' rhythm, often including a half-stop halfway through, tis*lee*-zizi-tis*lee*-zet-zitt*slee*-zette-zit... tis*lee*-zezezi-*tslee*-te*slitt*. Typically, lasts 2–3 sec. There is a fair bit of variation, some birds specialising in a more Wren-like performance, then with a trill inserted, or ending the phrase. Others, in particular those breeding at higher altitude in Central Asia, have a peculiar Coal Tit-like ring to their song, consisting of quick repetition of slightly clearer, less scratchy notes. – Call, frequently uttered both when feeding and between songs, a frothy, faintly disyllabic *t'slee* (other renderings *che'weest* or *wizip*), a little like Pied Wagtail. Possible to confuse with call of Hume's Leaf Warbler, often found in same habitat and range, but Greenish is 'frothier', sharper and more emphatic (Hume's Leaf is a little softer and more variable, and does not resemble Pied Wagtail quite as much).

SIMILAR SPECIES In Europe needs above all to be separated from *Arctic Warbler*, and some differences from this already treated under Identification. Note also in Greenish smaller size and lack of greyish tinge or mottling/diffuse streaking on sides of throat and breast, paler eye-stripe and more uniform cheeks and ear-coverts. – Both *Yellow-browed Warbler* and *Hume's Leaf Warbler* can theoretically be confused with a Greenish if views are less than perfect, but they have much broader wing-bars on greater coverts, prominence enhanced by dark-based wing-feathers in Goldcrest fashion, and they have a short but prominent and white (Yellow-browed) or subdued and pale grey-green (Hume's Leaf) second wing-bar on median coverts. Also, unless plumage is much abraded, tertials in Yellow-browed and Hume's Leaf are contrastingly dark-centred with light outer edges, not more or less uniform grey-green as in Greenish. –

Finland, Jun: single, short and narrow wing-bar, and prominent yellowish-white supercilium, which continues narrowly above bill. Also note slightly smaller and slighter build, characteristic pale greyish-green upperparts, always brightest on edges of wing-feathers. Underparts whitish, but can have faint yellow hue. In spring/summer ageing impossible due to complete winter moult of all age classes. (M. Varesvuo)

Finland, May: legs are sometimes rather dark grey-brown, and as shown here the bill can be quite thin and delicate. Whitish wing-bar often only on four or five greater coverts tips, frequently a little fainter in both ends. (J. Normaja)

For differences from *Green Warbler* and *Two-barred Warbler*, see those species.

AGEING & SEXING (*viridanus*) Small age-related plumage differences usually in 1stCY alone. Sexes alike in plumage. – Moults. Partial post-nuptial moult in late summer (Jul–Aug) involving body-feathers and some wing-coverts, rarely also odd inner tertials and central tail-feathers. Partial post-juv moult (Jul–Aug) affects mainly body and wing-coverts. Complete moult (including wings and tail) of both ad and 1stY in winter (mainly Jan–Mar). – **AUTUMN Ad** Much-abraded tips to primaries and tail-feathers. Odd inner tertials occasionally moulted and fresh. Wing-bar worn and ill-defined, off-white, rarely absent on one or both wings. **1stW** Plumage fresh, including tips to primaries and tail-feathers. Wing-bar well marked, yellowish. **Juv** Like 1stW but plumage more brown-tinged, less green. Grey-green tinge acquired with post-juv moult.

BIOMETRICS (*viridanus*) **L** 10–11.5 cm. **W** ♂ 61–66.5 mm (n 18, m 63.7), ♀ 57–62 mm (n 15, m 59.0); **T** ♂ 42–51 mm (n 18, m 47.9), ♀ 40–47 mm (n 15, m 43.8); **T/W** m 74.8; **B** 11.5–13.3 mm (n 31, m 12.2); **BD** 2.4–3.0 mm (n 18, m 2.7); **BW** 3.3–4.1 mm (n 14, m 3.7); **Ts** 17.2–19.6 mm (n 31, m 18.3). **Wing formula: p1** > pc 5.5–10 mm, < p2 21.5–27 mm; **p2** < wt 6–9.5 mm, =7/8 or 7 (81%), =6/7 (12%) or =8 or 8/9 (7%); **pp3–5** about equal and longest; **p6** < wt 1.5–4 mm; **p7** < wt 5–8 mm; **p10** < wt 10–14 mm; **s1** < wt 12–15.5 mm. Emarg. pp3–6.

GEOGRAPHICAL VARIATION & RANGE Rather well-marked variation with clinal trends. Birds in the west tend to be slightly paler, more greyish-tinged above and have poorly marked wing-bar. Races in Asia in the Greenish Warbler complex often described as forming a 'ring species', but here the ring is broken up because Two-barred Warbler is considered a separate species (see below and that species).

P. t. viridanus Blyth, 1843 (NE Europe, from S Finland, Baltic States, N Germany and NE Poland east through N Belarus and Russia to W Siberia and mountains in Central Asia; winters Pakistan, N India). Treated above. Greyish-green above with a slight brownish tinge in worn birds. Fairly pale plumage overall, underparts being whitish with a variable amount of pale yellow suffusion and only limited greyish tinge on sides in some. Supercilia meet on forehead, or at least reach nostrils. Eye-stripe variable but often comparatively poorly developed, affording facial expression a slightly washed-out quality compared to following races. Single wing-bar (hint of second bar very rare) often narrow and indistinct, diffusely narrowing at both ends, and mostly confined to 3–5 central greater coverts; can be entirely

Finland, Jun: a slightly tricky bird, which shows some features that suggest Arctic Warbler (strong bill, supercilium seemingly ending well short of forehead, rather blotchy cheeks), still diagnostic other ones confirm it is Greenish: short wing-bar fading in both ends, long p1 (much longer than tip of primary-coverts), hint of emargination of p6, as well as rather dark legs. (T. Muukkonen)

Ad, Scotland, Sep: due to wear and bleaching, the wing-bar has here become extremely vague; the upperparts wear greyer and underparts more whitish. Worn ad Greenish in autumn that has completely lost its wing-bar is separated from superficially similar Willow Warbler by smaller body and shorter tail, proportionately shorter primary projection, slightly larger head and bill, and on average longer supercilium at rear. (H. Harrop)

1stW, Scotland, Sep: unlike ad at same season, plumage of young in autumn is fresh and slightly greener, and supercilium is tinged yellowish, thereby to some degree approaching fresh ad in spring. The whitish underparts often appear rather finely yellowish-streaked, while cheeks and sides of breast/neck can have a faint brown tinge. (H. Harrop)

1stW, Scotland, Aug: some young in autumn have just two or three pale-tipped outer greater coverts. Note that tertials are rather uniformly patterned, lacking the contrasting bold whitish edges of fresh Yellow-browed and Hume's Leaf Warblers. Also unlike those species, there is only a greater coverts wing-bar, which is much shorter and less bold (also lacking the dark bar basally on wing-feathers outside tips of greater coverts). (H. Harrop)

Comparison between Arctic (top left: Finland, June), Greenish (top right: Finland, May), Green (bottom left: Apr, Iran), and Two-barred Warblers (bottom right: Mongolia, May): all four share indistinctly-patterned tertials, and, with exception of Two-barred Greenish, have single wing-bars! – Arctic Warbler is relatively large-headed and long-winged; typically, cheeks, lower throat and breast are mottled olive-grey, supercilium is long and distinct (which unlike most Greenish ends above lores); rather long bill has mostly pale lower mandible, with only tip variably dark. – Greenish Warbler (*viridanus*) is slighter and shorter-winged than Arctic; note supercilia almost meeting above bill; typically pale greyish-green above, with brighter wings, and underparts mostly clean whitish; bill typically has extensive yellowish base. – Green Warbler requires careful separation from Greenish, but typical birds in favourable light conditions have bright lemon-yellow face and underparts and longer primary projection; upperparts bright green, and greater coverts wing-bar is on average broader and bolder (albeit some variation). – Finally, vagrant Two-barred Greenish is often first detected by the obvious second (median coverts) wing-bar, but one of the most useful features for *plumbeitarsus* (separating it from Arctic and *viridanus* Greenish) is the very broad, long and even-width greater coverts wing-bar, which includes more feather-tips than in the others; additional characters versus Greenish are the shorter supercilium usually ending well short of forehead and the better-developed dark eye-stripe, generally darker and greener upperparts, and proportionately slightly heavier bill. Beware that the double wing-bars of Two-barred Greenish can invite confusion (at first glance) with the smaller Yellow-browed and Hume's Leaf Warblers, but Two-barred Greenish lacks the contrasting whitish tertial edges and Goldcrest-like dark-based wing-feathers around the bars. (Top left and right: M. Varesvuo; bottom left: Magnus Ullman; bottom right: M. Putze)

missing on worn ad. Lower mandible yellowish-brown with dark tip, latter often obvious but may be nearly missing. Legs often rather dark brown-grey.

Extralimital races, presumed not to occur within treated range:

○ **P. t. ludlowi** Whistler, 1931 (NW Himalaya; presumed to winter India). Poorly differentiated but is slightly larger and more greyish above, less green, than *viridanus*.

P. t. trochiloides (Sundevall, 1837) (C & E Himalayas, SW China; altitudinal or short-range movements in winter). Darker olive-brown above, in particular on crown, markedly tinged olive-grey below, has one well-marked and long wing-bar on greater coverts, and sometimes a hint of a second on median.

○ **P. t. obscuratus** Stresemann, 1929 (mountains of SC China, E Tibet; short-range winter movements). Poorly differentiated, intermediate between preceding and Two-barred Warbler, is a little less dark than *trochiloides*, and more often has two developed wing-bars, the lower one prominent.

TAXONOMIC NOTES The Greenish Warbler complex, comprising apart from the above taxa also the Green Warbler *P. nitidus* and the Two-barred Warbler *P. plumbeitarsus*, is not easy to give a clear and consistent taxonomic treatment. The mainly clinally connected forms are obviously closely related, both morphologically and vocally, most of them grading into the next without sharp borders (like classical geographical races should if there are no barriers), the complication being that they seem to form a ring, but where the two ends (*viridanus* and *plumbeitarsus*) meet they behave as different species. The ring species theory was suggested already by Ticehurst (1938) and was later supported by Irwin (2000) and Irwin et al. (2001) through a genetic analysis and new field research focusing on vocalisations. Since the Green Warbler has an allopatric range, separated from the others by a very wide gap despite no lack of suitable habitat, and since it is in our experience invariably diagnosable on morphology, and shows some vocal differences, it is treated here as a separate species. As to the Two-barred Warbler, this also has long occupied an allopatric range, in its southern part due to habitat destruction, and it overlaps at its north-western end with Greenish Warbler *viridanus* apparently without interbreeding (Irwin 2000), thus is treated here as another separate species. However, recently I. Marova and co-workers reported both frequent (c. 10%) apparent mixed-song types and birds with mixed appearance from the area of contact (Marova et al. 2010), suggesting fairly extensive hybridisation. Requires more study. See also Kovylov et al. (2012).

REFERENCES Alström, P. & Olsson, U. (1987) *Proc. 4th Int. Ident. Meet. Eilat 1st–8th Nov. 1986*: 54–59. – Alström, P. & Olsson, U. (1989) *Limicola*, 3: 269–279. – Bradshaw, C. (2001) *BB*, 94: 284–288. – Collinson, M. (2001) *BB*, 94: 278–283. – Dean, A. R. (1985) *BB*, 78: 437–451. – Irwin, D. E. (2000) *Evolution*, 54: 998–1010. – Irwin, D. E., Bensch, S. & Price, T. D. (2001) *Nature*, 409: 333–337. – Kovylov, N. S., Marova, I. M. & Ivanitsky, V. V. (2012) *Zool. Zhurn.*, 91: 702–713. – Marova, I. et al. (2010) *Proc. 13th Orn. Congr. N Eurasia*, Orenburg. – van der Vliet, R. E., Kennerley, P. R. & Small, B. J. (2001) *DB*, 23: 175–191.

TWO-BARRED WARBLER
Phylloscopus plumbeitarsus Swinhoe, 1861

Alternative name: Two-barred Greenish Warbler

Fr. – Pouillot à pattes sombres
Ger. – Middendorfflaubsänger
Sp. – Mosquitero patigrís
Swe. – Sibirisk lundsångare

Two-barred Warbler, one of several forms in the complex ring species collectively called Greenish Warbler in the broad sense, is treated here as a separate species. It breeds in taiga and wooded areas in SC and E Siberia, in the west to the southern slopes of W Sayan Mountains, where it overlaps with *viridanus* Greenish Warbler, with apparently limited interbreeding. Also found in N Mongolia and N China; rare vagrant to Europe. Behaviour and habitat much as for Greenish Warbler. A lively and active small warbler spending much time high in the canopy of trees, and often first noticed by its song and calls.

1stW, England, Oct: unlike similar Greenish Warbler, longer and broader white greater coverts wing-bar and has a second shorter bar on median coverts, while bill is proportionately slightly stronger. Note tertials are largely uniform, lacking the contrasting whitish edges of Yellow-browed and Hume's Leaf Warblers. Greyish-tinged legs typical. Unlike ad at same season, plumage is fresh. (T. Tams)

IDENTIFICATION Very similar to Greenish Warbler (*viridanus*), differing only subtly in plumage. Main character to look for is a *longer and broader white wing-bar along tips of greater coverts* (rule-of-thumb: clearly broader wing-bar than the tarsus is thick), and *a shorter and narrower second wing-bar on tips of median coverts*. The same size as Greenish (or very subtly larger) with a proportionately *slightly stronger bill*, a marginally *blunter wing*, and has somewhat *darker and slightly brighter green upperparts* (usually with faint brown tinge on crown and mantle), and on average *less yellow-tinged, more dusky-white underparts*. Due to its broadness, the lower wing-bar does not wear off even in worn summer and autumn plumage. Another difference is that the *supercilium usually ends above lores* (thus in most birds short of forehead, shorter than in Greenish; only few exceptions seen with supercilium to forehead like in Greenish), but on the other hand *can reach further back towards sides of nape*. Eye-stripe *slightly darker grey-green* than Greenish, enhancing prominence of supercilium, and *sides of head and ear-coverts on average more variegated and 'patchy'*. Lower mandible is often all yellowish-brown, lacking a dark tip, but a few have a dark smudge on the tip (thus some overlap with Greenish in this respect). Legs same as Greenish, or on average slightly paler brown. Once Greenish has been eliminated, Arctic Warbler needs to be considered, at times equally or more difficult. Greater coverts wing-bar in Two-barred is more solid (will rarely appear as 'string of pearls') and a little broader, and often involves the tips of more coverts. The second wing-bar on median coverts is often short but is more distinctly set off than in those fresh Arctics showing a hint of a second bar, and tips are more whitish (Arctic: diffusely yellowish-tipped coverts). A good indication of Two-barred is the slightly less dark eye-stripe which does not quite reach the base of bill, and which is 'pinched in' in front of eye due to slightly more extensive white lower eye-lid (Arctic: evenly broad and darker eye-stripe reaching base of bill). Also, supercilium often widens behind eye (Arctic: evenly narrow). Underparts average whiter, not so dusky as in Arctic, and ear-coverts are less blotchy (though some overlap). On average legs more grey-tinged than in Arctic, but overlap. See also call, clearly different from the short raspy call of Arctic.

VOCALISATIONS Song to human ear very similar to that of Greenish Warbler, but to experienced observers with a good ear, or by use of sonograms, on average slightly lower-pitched, strophes slightly longer and containing more elements, and often recognised by a more marked 'shuttling' up and down the scale with repetitive phrases, with slightly clearer, less sharp notes. Songs often 3–4 sec long. There seems to be less variation than in Greenish, but this may be due to as yet incomplete documentation of the full range of variation within Two-barred. – Call similar to Greenish but often slightly more complex, more clearly disyllabic or even trisyllabic (like in Green Warbler), *tsi-z'li*, whereas that of Greenish may sound 'near-monosyllabic'.

SIMILAR SPECIES In Europe must first of all be separated from Arctic and Greenish Warblers. Separation from *Arctic Warbler* is generally helped by presence of obvious second wing-bar in Two-barred Warbler, a feature that is very rare in Arctic and at the most present as a hint. Two-barred is also usually subtly smaller and slimmer, and is whiter below lacking strong greyish tinge or mottling/diffuse streaking on

1stW, China, Oct: upperparts generally darker and greener-looking compared to Greenish of race *viridanus*. Supercilium frequently (like here) does not reach forehead, being closer to Arctic Warbler in this respect, but unlike Arctic sometimes (like here) widens behind eye. Note very broad greater covert tips forming the characteristic lower wing-bar. Fresh wing, tail and rest of plumage separate 1stW from worn ad at this season. (I. Fisher)

Ad, China, Aug: after partial post-nuptial moult, with still worn wings and tail. Typically, some greater coverts still show very broad and distinct pale tips. A rather atypical bird showing traits recalling Arctic Warbler (prominent dark eye-stripe, dark-mottled head-sides) but note short primary projection, long p1 and yellower bill than average Arctic. (G. Norevik)

Russia, Jun: on spring migration, all ages are similarly fresh following complete winter moult. Note wing-bars, here rather subdued on median coverts, especially on right-hand bird. Separation from Arctic Warbler is particularly tricky, not least since both species have prominent pale supercilia that end short of forehead. Note in Two-barred moderately dark eye-stripe, cheeks only vaguely mottled and breast only faintly greyish-tinged (Arctic has darker and rather evenly broad eye-stripe, more mottled cheeks with hint of olive-grey, and breast has typical diffuse greyish-olive streaking). Also Two-barred has longer outermost primary reaching well past tips of primary-coverts, although this is rarely visible in field views. (M. Hellström)

sides of lower throat and breast, which is often a characteristic of Arctic. Also, Two-barred has a paler eye-stripe and on average slightly more uniform cheeks and ear-coverts (though some overlap). See under Identification and under Arctic Warbler for full set of differences. – *Greenish Warbler* of ssp. *viridanus* is very similar and perhaps not always separable in the field if views are brief. It has on average slightly paler greenish-tinged grey-brown upperparts, slightly more yellow-streaked or yellow-tinged throat and breast, and a much narrower and shorter single wing-bar. The pale supercilium in Greenish reaches all the way to forehead, thus is longer than in Two-barred. – Differences from *Yellow-browed Warbler* and *Hume's Leaf Warbler* are the identical to those separating these two from Greenish Warbler (which see).

AGEING & SEXING Small age-related plumage differences usually in 1stCY alone. Sexes alike in plumage. – Moults. Partial post-nuptial moult in late summer (Jul–Aug) involving body-feathers and some wing-coverts, rarely also odd inner tertials and central tail-feathers. Partial post-juv moult (Jul–Aug) affects mainly body and wing-coverts. Complete moult (including wings and tail) of both ad and 1stY in winter (mainly Jan–Mar). – **AUTUMN Ad** Much-abraded tips to primaries and tail-feathers. Odd inner tertials occasionally moulted and fresh. Wing-bar worn and ill-defined, off-white, rarely absent on one or both wings. **1stW** Plumage fresh, including tips to primaries and tail-feathers. Wing-bar well marked, yellowish. **Juv** Like 1stW but plumage more brown-tinged, less green. Grey-green tinge acquired with post-juv moult.

BIOMETRICS L 10.5–12 cm. **W** ♂ 57–66 mm (n 13, m 61.2), ♀ 54–62.5 mm (n 17, m 57.6); **T** ♂ 42–54 mm (n 13, m 46.9), ♀ 40–47 mm (n 17, m 43.1); **T/W** m 75.6; **B** 12.0–13.5 mm (n 28, m 12.5); **BD** 2.5–3.3 mm (n 25, m 2.9); **BW** 3.2–4.3 mm (n 23, m 3.7); **Ts** 16.5–19.2 mm (n 27, m 17.9). **Wing formula: p1** > pc 5–10 mm, < p2 19–25 mm; **p2** < wt 6–10 mm, =7/8 or =7 (78%), =6/7 (9%), =8 (7%) or =8/9 (6%); **pp3–5** about equal and longest; **p6** < wt 1–4 mm; **p7** < wt 4.5–7.5 mm; **p10** < wt 9–12 mm; **s1** < wt 10.5–13 mm. Emarg. pp3–6.

GEOGRAPHICAL VARIATION & RANGE Monotypic (treated as a separate species, rather than as a race of Greenish Warbler). – Southern slopes of W Sayan Mts, N Mongolia, SE Siberia east to Okhotsk, Ussuriland, Manchuria, N & E China; winters SE Asia south to Thailand.

TAXONOMIC NOTES See general comments concerning the Greenish Warbler complex under that species. Two-barred Warbler has often been regarded as a subspecies of Greenish Warbler and closely related to its European race (*viridanus*), but is here treated as a full species considering its allopatric range in China (range gap to Greenish ssp. *obscuratus*) and overlap in the Sayan Mts taiga with Greenish *viridanus*, apparently without full-scale hybridisation (although Marova et al. 2010 reported intermediates to comprise c. 10% of birds in the overlap zone, and the frequency of intermediates in the overlap area based on song and morphology comparison was again reported by Kovylov et al. 2012. The reproductive isolation of these two, although far from complete, is presumably helped by the slight but average differences in vocalisation. For example, Kovylov et al. (2012) noted no response from *plumbeitarsus* when song of *viridanus* was played.

REFERENCES Bradshaw, C. (2001) *BB*, 94: 284–288. – Collinson, M. (2001) *BB*, 94: 278–283. – Dean, A. R. (1985) *BB*, 78: 437–451. – Irwin, D. E. (2000) *Evolution*, 54: 998–1010. – Irwin, D. E., Bensch, S. & Price, T. D. (2001) *Nature*, 409: 333–337. – Kovylov, N. S., Marova, I. M. & Ivanitsky, V. V. (2012) *Zool. Zhurn.*, 91: 702–713. – Marova, I. et al. (2010) *Proc. 13th Orn. Congr. N Eurasia*, Orenburg. – van der Vliet, R. E., Kennerley, P. R. & Small, B. J. (2001) *DB*, 23: 175–191.

A tricky pair to separate, Two-barred Warbler (left) and Arctic Warbler (right), Mongolia, May: sometimes, but especially if calls not heard, in the field or in photographs, the two can appear very similar. Beware that this Two-barred shows atypical, rather narrow whitish tips to greater coverts, while it still has an obvious whitish second wing-bar on the median coverts (rather rare in Arctic, and present at the most as a yellowish hint – see bird on the right). Two-barred also has slimmer shape, appearing proportionately larger-headed and thinner-billed than Arctic, and supercilium is slightly wider above eye. Further, Two-barred has a less dark eye-stripe which is slightly 'pinched-in' in front of eye (here also behind eye), and has rather more uniform cheeks. The Arctic Warbler was not heard to call in the field, but in photographs identified on the following combination of characteristics: chunkier jizz, stronger bill with darker tip, and a shorter and 'broken-up' greater coverts wing-bar. Note also narrow and long supercilium, prominent dark loral stripe reaching bill-base, and greyish-tinged and blotchy underparts. (H. Shirihai)

PALLAS'S LEAF WARBLER
Phylloscopus proregulus (Pallas, 1811)

Alternative name: Pallas's Warbler

Fr. – Pouillot de Pallas; Ger. – Goldhähnchen-Laubsänger
Sp. – Mosquitero de Pallas; Swe. – Kungsfågelsångare

This is the smallest of the so-called leaf warblers. It breeds in taiga in southern parts of C and E Siberia, normally winters in S Asia and yet visits W Europe every autumn, often in some numbers. Vagrants often associate with equally small Goldcrests, but frequently stick out in mixed flocks by their much more active and quick movements when feeding. The frequency with which these 'Sibes' (as these visitors from Siberia are often nicknamed) occur in Europe, mainly in late September and October, perhaps indicates that they have now established an alternative wintering strategy in our part of the world.

Mongolia, Jun: often agile or appears nervous, twitching wings, revealing diagnostic yellowish-white rump patch and median crown-stripe, whereas other features are always visible, including prominent supercilium (strongest yellow in forepart) and double wing-bars. Ageing difficult (even with handling). (H. Shirihai)

IDENTIFICATION Very small (Goldcrest-sized), often appearing *large-headed* but *short-billed*. Active and restless when feeding, makes quick moves constantly, can hang upside-down like a small Blue Tit, and hover at tip of branches. Often twitches wings and tail nervously. Very *long and prominent supercilium*, running from forehead to side of nape, *chrome-yellow in its forepart, whitish at rear*. Supercilium strengthened by *quite dark eye-stripe* and *dark greenish-grey crown-sides* (forecrown tinged yellow). A narrow, *pale central crown-stripe* is visible in head-on views. Has *two whitish wing-bars, the lower very broad and striking*, enhanced by dark greater coverts and *dark bases to secondaries*. Tertials dark, tipped white when fresh. Diagnostic, *sharply demarcated, small pale yellow* (rarely white) *rump patch* often concealed in side views, but easily seen on hovering bird, or when flying away. Mantle and back greyish-green, underparts dusky white (with faint yellowish tinge on throat in some). Flight-feathers edged yellowish-green. Bill dark with only insignificantly paler brown base. Legs thin and brownish.

VOCALISATIONS As with Wren in Europe, the song is remarkably loud and striking for such a small bird. Strophes also build up rather like in the song of Wren, with series of notes repeated at high speed, some whistling and clear, others trilling, pitch varying somewhat between series of notes, *tsee yu-yu-yu-tsree, wreecha-wreecha-wreecha seewoo-seewoo-seewoo tsitt, choo-choo-choo-choo-tsee* and the like. The song can also be likened to the song of Olive-backed Pipit (sharing same habitat), but is stronger and more varied. Some birds insert short, sharp notes, *zek, zek...*, between songs or even in the songs between phrases. The function of these is unknown, and the same bird can insert them or drop them. Song delivered from high, exposed song post, often atop a conifer. Rarely heard within treated range. – Call seldom heard; many vagrant Pallas's Leaf Warblers remain very quiet, others will call but do so rarely and rather quietly. The common call is an upwards inflected *twooeet*, recalling Common Chiffchaff (but less loud) and is somewhat squeaky and nasal in tone. Shorter variants can sound monosyllabic, *tweet*.

SIMILAR SPECIES In Europe needs first of all to be separated from *Yellow-browed Warbler*. As the yellow rump patch is not always easy to see, look also for the distinct central crown-stripe (Yellow-browed may have a hint of a paler central stripe on rear crown, but it is never distinct and reaching over whole crown). Most Pallas's Leaf have a vivid yellow fore supercilium (pale yellow at most in Yellow-browed, more often whitish) and broad dark bases to secondaries (narrower and more diffuse in Yellow-browed). Also, bill is shorter and darker than in Yellow-browed. – In Asia must be separated from a few sibling species, recently separated from Pallas's Leaf, mainly based on clear differences in vocalisation (rather slight in plumage), but this falls outside the scope of this book.

AGEING & SEXING Small age-related differences in 1stW. Ageing in spring based on different amount of wear probably applicable but not studied. Seasonal variation slight. Sexes alike in plumage. – Moults. Complete post-nuptial moult in summer (mainly late Jul–early Sep). Partial post-juv moult in summer does not involve flight-feathers or primary-coverts. Partial pre-nuptial moult in winter (both age classes) limited and variable in extent, often involving change of body-feathers, some wing-coverts and odd tertials, sometimes also central tail-feathers. – SPRING **Ad** Rather evenly abraded plumage with no or only limited apparent moult differences, tips of primaries and tail-feathers somewhat worn, feathers dark grey and often slightly glossy. **1stS** Tips of primaries and tail-feathers often much abraded, feathers rather brownish and with reduced gloss. – AUTUMN Ageing frequently difficult, differences slight. **Ad** Freshly moulted, tips to tail-feathers neat and rounded, centres dark and glossy, finely fringed or tipped whitish. **1stW** Some very similar to ad. Typical birds show first sign of abrasion to tips of primaries and tail-feathers, latter often subtly frayed at tips, with on average less gloss and contrasting edges than ad. **Juv** Compared to post-juv plumages lacks most of yellow colours, notably on forecrown, supercilium, wing-bars and rump patch, all these being off-white, with very little yellow tinge only on some. Crown-stripe ill-defined. Upperparts brown-tinged.

BIOMETRICS L 9–9.5 cm. **W** ♂ 51.5–57 mm (n 14, m 53.9), ♀ 47.5–53 mm (n 13, m 49.5); **T** ♂ 38–42 mm (n 14, m 40.6), ♀ 35–41.5 mm (n 13, m 36.4); **T/W** m 75.4; **B** 9.6–11.2 mm (n 28, m 10.3); **BD** 2.0–2.8 mm (n 21, m 2.4); **BW** 2.2–3.3 mm (n 16, m 2.8); **Ts** 15.0–17.5 mm (n 28, m 16.3). **Wing formula: p1** > pc 5–8 mm, < p2 16.5–21 mm; **p2** < wt 6–8.5 mm, =8 or 8/9 (78%), =7 or 7/8 (13%), or =9 or 9/10 (10%); **pp3–5** about equal and

Mongolia, May: tiny (Goldcrest-sized), often giving impression of being large-headed. Note very long and prominent supercilium, with brighter yellow forepart, heightened by dark eye-stripe and appearance of darkish crown-sides. However, also note how median crown-stripe can be obscured, whether due to abrasion or angle of view. The double whitish wing-bars are still striking even on an already worn bird. The yellowish-white rump patch can be invisible at many angles. (T. Lindroos)

Oct, Norway: unlike Yellow-browed Warbler, most Pallas's Leaf show distinct pale median crown-stripe (not always visible at first), brighter yellow fore supercilium and, due to dark crown and eye-stripe, overall head pattern often appears very striking. The yellow rump is often hidden by the wings. Tiny bill. Ageing unsure, especially as post-juv moult usually includes all greater coverts, and most fresh birds show light green tips to inner coverts, which create false moult limit. (C. Tiller)

Scotland, Oct: fresh plumage showing the diagnostic distinct median crown-stripe, brighter yellow forepart to very broad supercilium (further pronounced by dark crown-sides and blackish eye-stripe), well-developed pale wing-bars and tertial edges, and the square yellowish-white rump patch. (H. Harrop)

longest; **p6** < wt 0.5–1.5 mm; **p7** < wt 3–5 mm; **p8** < wt 5–7.5 mm; **p10** < wt 8–10.5 mm; **s1** < wt 9–12 mm. Emarg. pp3–6.

GEOGRAPHICAL VARIATION & RANGE Monotypic. – C & E Siberia from Ob and Altai east to N Okhotsk, Sakhalin, south to N Mongolia; winters S China.

TAXONOMIC NOTE Pallas's Leaf Warbler was previously regarded as a polytypic species, but has recently been split into several (extralimital) species based on differences in vocalisations, morphology and DNA, the extra species being: Gansu Leaf Warbler *P. kansuensis* (C China), Chinese Leaf Warbler *P. yunnanensis* (C & NE China), Lemon-rumped Warbler *P. chloronotus* (Himalayas) and Sichuan Leaf Warbler *P. forresti* (SW & SC China). None of these is thought to be a potential straggler to the area covered by this book and consequently they are not treated herein.

REFERENCES ALSTRÖM, P. & OLSSON, U. (1990) *BBOC*, 110: 38–43. – ALSTRÖM, P., OLSSON, U. & COLSTON, P. R. (1997) *BBOC*, 117: 177–193. – CATLEY, G. P. (1992) *BB*, 85: 491–494. – MARTENS, J. et al. (2004) *J. of Orn.*, 145: 206–222. – SCHUBERT, M. (1982) *Mitteilungen Zool. Mus. Berlin*, 58: 109–128.

Germany, presumed 1stW, Oct: overall duller and less boldly patterned. Aside from the median crown-stripe and brighter yellow fore supercilium, note that the two whitish wing-bars are enhanced by dark bases to greater coverts. This dull bird is probably a 1stW, with apparently renewed inner greater coverts (longer with broader pale tips), but this could also be due to the plumage being disordered. (C. Tiller)

China, Oct: seemingly evenly fresh body plumage, wings and tail, with rounded tips to the rectrices indicating an ad, but again, without handling, definitive ageing is difficult. Here the black bases to secondaries show well, enhancing prominence of greater coverts wing-bar. (S. Fisher)

YELLOW-BROWED WARBLER
Phylloscopus inornatus (Blyth, 1842)

Fr. – Pouillot à grands sourcils
Ger. – Gelbbrauen-Laubsänger
Sp. – Mosquitero bilistado; Swe. – Tajgasångare

Along with Pallas's Leaf Warbler and the Siberian race of Common Chiffchaff, this Siberian autumn vagrant now appears annually in quite surprising numbers in W Europe—surprising when you consider that its nearest breeding area lies some 2500 km from Britain and Western Europe, and its normal winter range is SE Asia. This phenomenon is generally thought to be due to so-called reverse migration, a 180° directional mistake made by a small proportion of mainly inexperienced young birds setting out on their first migration (although other explanations are also possible). Like Siberian Common Chiffchaff, it breeds in taiga from just west of Ural Mountains in the west, and east right across the immense Siberian forests to the Far East.

Presumed 1stS, China, May: effects of feather wear, especially in 1stY, can be rather noticeable in spring, when birds are much duller and 'scruffier'. However, unlike very similar Hume's Leaf, note relatively greener-toned upperparts and brighter fringes to remiges, while the wing-bars and supercilium are still obvious and broad (in Hume's, median coverts wing-bar is usually worn off). The legs are noticeably pale and the bill has a rather substantial yellow-brown base compared to Hume's. Age is suggested by the rather worn primary tips and primary-coverts (both apparently juv). (R. Schols)

China, May: a small, wing-barred leaf warbler mainly characterised by its long and prominent supercilium, which in spring becomes whiter (less yellowish). While in late spring pale tertial edges are worn down to thin edges, the double white wing-bars are still broad. The rather bright greenish upperparts and overall not heavily-worn wings suggest an ad, but ageing without handling is impossible. (R. Schols)

IDENTIFICATION *Quite small*, almost like Pallas's Leaf Warbler, a little smaller than a Greenish Warbler but otherwise of similar proportions to this. Nervous and quick movements when feeding, now and then twitching tail and wings, tail upwards or to sides. Typical wing-barred leaf warbler with one *long white and prominent bar on greater coverts*, usually enhanced by dark surrounding wing (very few exceptions), and a second much *shorter and less obvious* (sometimes broken or in worn plumage even nearly invisible) *white bar on median coverts*. To this is added a *long and prominent white or pale yellowish supercilium*, and rather distinctly *white-tipped and white-edged tertials* (at least in fresh autumn plumage). *Upperparts rather uniformly greenish*, but *crown darker grey-green*. Underparts whitish with variable amount of greyish suffusion on sides of throat, breast and flanks. There is sometimes a hint of a narrow pale central crown-stripe visible on rear crown, but nothing like the distinct crown-stripe of Pallas's Leaf Warbler. Another and rare or local plumage variation has rather obviously less green above, more greyish with just a hint of green; the reason for this is unknown. *Bill rather dark* with variable amount of yellow-brown on base of lower mandible and cutting edges (usually much less pale than Greenish Warbler). *Legs fairly dark*, again somewhat variable, from medium pinkish-brown to quite dark brown-grey.

VOCALISATIONS The song, rarely heard outside breeding range, is a surprisingly high-pitched and low-keyed performance, a few thin, whistling notes repeated in similar fashion, the delivery being rather hesitant and unrhythmic; the best clue is that the familiar call can often be recognised as part of the strophe, or at least notes similar to the call, *tsewees, se-se-wee… sewees*. There is a surprising superficial resemblance to song of Hazel Grouse (the two could theoretically be encountered in the same taiga forest), but that species has a more piercing, less soft tone. Among commoner European birds, the Yellow-browed song recalls somewhat a slowed-down but higher-pitched Short-toed Treecreeper. There is also an alternative song variation, at least in N Mongolia, consisting of a fair number of calls slowly repeated, now and then interrupted by a couple of muffled buzzing notes (local dialect or cryptic different species?). – The call is loud and uttered frequently both by breeders and most migrants, a disyllabic high-pitched and penetrating *tso-weest*, upwards-inflected with much stress on last syllable. The voice can resemble that of Coal Tit, but whereas calls of latter vary somewhat and are slightly less high-pitched, Yellow-browed calls are constant, more drawn-out and 'fizzy' in tone.

Mongolia, May: some Yellow-browed Warblers are overall duller and less green, especially in spring, as this individual. Such birds could be confused with Hume's Leaf, but note the broad and well-marked supercilium and wing-bars, while the cheeks are mottled, thus overall a well-marked bird, unlike the duller Hume's. Also, bill has a rather substantial yellow-brown base to lower mandible (in Hume's bill averages slightly darker). (H. Shirihai)

SIMILAR SPECIES Very similar to closely related *Hume's Leaf Warbler*, and can be inseparable in the field without the help of calls (especially in late spring and summer, when plumage of both more abraded and similarly grey-tinged), but usually differs (at least in autumn) by lack of any buff tinge to supercilium and sides of head, and by whiter and better-defined wing-bars (particularly that on median coverts). In addition, usually the head pattern is a little more contrasting with slightly darker greenish-grey crown-sides and better-defined supercilium due to darker loral mark, whiter supercilium, etc. The more greyish-tinged variety of Yellow-browed mentioned above will invite confusion with Hume's Leaf even more, but note lack of buff or brown hues, and invariably broad and well-marked supercilium and double wing-bars, thus has a pure white, short but distinct, bar also on median coverts. For calls of Hume's Leaf, see that species. – Of similar size to *Pallas's Leaf Warbler* (only fractionally larger on average) but lacks a yellowish rump patch. For more details see that species. – Smaller than *Greenish Warbler* (ssp. *viridanus*), and has much broader and longer lower wing-bar, nearly always a short but clearly visible second wing-bar, white-tipped tertials and dark-based secondaries forming a dark bar next to the lower wing-bar (missing on just a few birds). – Could be confused with *Two-barred Warbler* as this has double wing-bars nearly as prominent as in Yellow-browed, but again note white-tipped tertials, smaller size and different calls of Yellow-browed.

AGEING & SEXING Small age-related differences in 1stW. Ageing in spring based on different wear probably applicable but not studied. Sexes alike in plumage. – Moults. Complete post-nuptial moult in summer (mainly late Jul–Aug). Partial post-juv moult in summer does not involve flight-feathers or primary-coverts. Partial pre-nuptial moult in winter (both age classes) limited and variable in extent, often involving body-feathers, some wing-coverts and odd tertials, sometimes also central tail-feathers. – **SPRING Ad** Rather evenly abraded plumage with no or only limited apparent moult differences, tips of primaries and tail-feathers somewhat worn, feathers dark grey and often slightly glossy. **1stS** Tips of primaries and tail-feathers often much abraded, feathers rather brownish and with reduced gloss. – **AUTUMN** Ageing frequently difficult, differences slight. **Ad** Freshly moulted, tips to tail-feathers neat and rounded, centres dark and glossy, finely fringed or tipped whitish. **1stW** Some very similar to ad. Typical birds show first sign of abrasion to tips of primaries and tail-feathers, latter often subtly frayed at tips, with on average less gloss or contrasting edges than ad. **Juv** Often slight yellow-buff tinge to whitish supercilium and wing-bars. Green of upperparts faintly brown-tinged.

BIOMETRICS L 9–10.5 cm. **W** ♂ 54–61 mm (*n* 24, *m* 57.6); ♀ 51–58 mm (*n* 17, *m* 54.5); **T** ♂ 37–44 mm (*n* 24, *m* 40.7); ♀ 36–41 mm (*n* 16, *m* 38.3); **T/W** *m* 70.5; **B** 9.8–12.0 mm (*n* 41, *m* 10.7); **BD** 2.0–2.7 mm (*n* 27, *m* 2.4); **BW** 2.7–3.3 mm (*n* 16, *m* 3.1); **Ts** 15.8–18.5 mm (*n* 35, *m* 17.3). **Wing formula: p1** > pc 3–9.5 mm, < p2 20–27 mm; **p2** < wt 4–7.5 mm, =7 (48%), =6/7 (31%), =7/8 (19%) or =8 (2%); **pp3–5** about equal and longest; **p6** < wt 1–3 mm; **p7** < wt 4–6.5 mm; **p8** < wt 6.5–9 mm; **p10** < wt 9.5–12 mm; **s1** < wt 11–13.5 mm. Emarg. pp3–6.

GEOGRAPHICAL VARIATION & RANGE Monotypic. – Extreme NE Russia just west of Ural Mts east to E Siberia, Anadyr, south to Altai, N Mongolia and Ussuriland; winters S & SE Asia. In a small area of NE Mongolia, only greyish birds were encountered in 2013 (HS). Whether these represent an undescribed subspecies or were by chance only bleached or 1stS birds is unknown.

TAXONOMIC NOTE Yellow-browed Warbler was formerly regarded as a polytypic species, with *inornatus* living in taiga in the north, and *humei* and *mandellii* in mountain forests or scrub further south. Based on differences in vocalisations, morphology and DNA, the latter two are now generally combined and treated as a separate species, Hume's Leaf Warbler *P. humei*.

REFERENCES ALSTRÖM, P. & OLSSON, U. (1986) *International Bird Identification*, pp. 54–57. – HARRIS, P. (1995) *Ring. & Migr.*, 16: 127. – LUIJENDIJK, T. J. C. (2001) *DB*, 23: 275–284. – SCHUBERT, M. (1982) *Mitteilungen Zool. Mus. Berlin*, 58: 109–128.

Netherlands, Oct: combination of immaculate plumage, evenly very fresh wing-feathers and still broad whitish fringes to primaries suggests ad, but ageing without careful in-hand check of wing, tail, skull and iris is impossible. Note the typical uniformly greenish upperparts and whitish underparts with some greyish suffusion. (A. Ouwerkerk)

Germany, Oct: note hint of narrow and ill-defined pale median crown-stripe, but never as well-developed as Pallas's Leaf. Also, wing-bars are generally less strongly patterned than in Pallas's Leaf (median coverts bar shorter, less obvious and sometimes broken), although both are similarly enhanced by dark bases to secondaries and greater coverts. Uniformly greenish upperparts and clear tertials pattern. Ageing without handling unsure. (M. Schäf)

Netherlands, Oct: note characteristic tertials pattern, as well as rather extensive yellow-brown base to lower mandible and cutting edges. Although there is some overlap in tail-feather shape between ad and 1stW, such narrow and pointed feathers better fit 1stW. Nevertheless, without handling, it is difficult to age with certainty. (A. Ouwerkerk)

Israel, Nov: some birds are overall duller by autumn, but the substantial green element above and slight yellowish tinge to prominent supercilium and greater coverts wing-bar identify to species. Without handling, ageing is again difficult (P. Hytönen)

HUME'S LEAF WARBLER
Phylloscopus humei (Brooks, 1878)

Alternative name: Hume's Warbler

Fr. – Pouillot de Hume; Ger. – Tienshan-Laubsänger
Sp. – Mosquitero de Hume; Swe. – Bergtajgasångare

Thirty years ago, few knew or had even heard of this small warbler that breeds in the mountain forests of Central Asia and S Siberia. This was partly because it was regarded as 'just' a slightly different race of Yellow-browed Warbler, something which only specialists bothered about. But once W European birders started to travel east on a larger scale they noted that it had both entirely different song and calls, and differed in plumage and ecology. This led to its subsequent recognition as a separate species, and it is now known to be a regular autumn vagrant in W Europe, where it was most likely overlooked in the past.

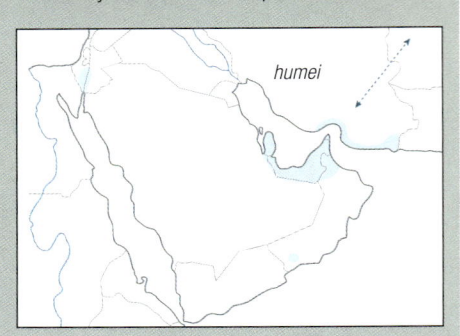

IDENTIFICATION Very similar, and closely related, to Yellow-browed Warbler, with similar size, proportions and habits, being just a fraction smaller. Differs in fresh plumage by slight *buffish tinge to supercilium and sides of head*, lower wing-bar and tertial edges, and by somewhat *less distinctly marked upper wing-bar*, this being a little diffuse, subdued or nearly missing, off-white and often buff-tinged. Underparts often somewhat more yellow-tinged than in Yellow-browed. Green of *upperparts is a fraction more tinged brown-grey*, less bright greenish than normal and fresh Yellow-browed, and *head pattern is less distinct* with slightly paler cheeks and *paler eye-stripe and crown*. Ground colour of tertials and greater coverts is not quite as dark as in Yellow-browed, creating in Hume's Leaf slightly less well-marked contrast with whitish tips and edges. Both bill and legs are on average slightly darker than in Yellow-browed (but some overlap; odd Hume's Leaf have been noted to have slightly paler cutting edges and base to lower mandible). Note that late spring and summer birds are more similar to Yellow-browed in that buff tinges of supercilium, ear-coverts and wing-bars become bleached and whitish, and mantle in both is more greyish. Remember that a bleached summer Yellow-browed still has *prominent* (albeit short) *pure white upper wing-bar*, whereas this is grey-tinged, diffuse, narrow and often broken in corresponding plumage of Hume's Leaf. Unless observation of such birds is sufficiently close and long, reliable identification requires the support of vocalisations.

VOCALISATIONS The song has two components, which may be mixed or given singly for long periods: (i) most commonly heard is a double note, *tiss-yip*, energetically repeated in twos or threes for long periods; (ii) less often heard, but more characteristic, is a drawn-out and peculiarly strained, buzzing note, *bzzeeeeeoo*, faintly downslurred at the end, a little reminiscent of the flight call of migrant Redwings. – Calls are rather variable and sometimes similar to that of Yellow-browed Warbler, but still nearly always easily separable as different, once learnt. Most common call is a disyllabic 'merry' *tsu-viis*, which is not as high-pitched and drawn-out as the call of Yellow-browed, softer and without the frothy 's'. However, some calls are more similar to Yellow-browed, like *weest* and variations. Even though often upslurred it can end with a hint of falling pitch, *tsu-viis(o)*. Less often an almost trisyllabic *se-su-veet* is heard. Note that disyllabic call variants can resemble the call of Greenish Warbler, but latter is sharper with both syllables on same pitch, more like a Pied Wagtail call.

Oman, Feb: upperparts predominantly brown-grey, lacking the bright greenish tones of most (but not all) Yellow-browed in equivalent plumage. Compared to Yellow-browed, note less strongly patterned head, with paler cheeks and less clearly demarcated eye-stripe. Also, typically, lacks any strong yellow in plumage, except greenish-yellow edges to flight-feathers, while supercilium is characteristically cream-tinged. Finally, median coverts wing-bar is typically very short, often involving just one or two, rarely three, coverts (in Yellow-browed usually longer and more complete). (K. Schjølberg)

Israel, Mar: some birds are even duller, being almost totally grey-brown, grey and cream, with the characteristic poorly-developed wing pattern and face markings (diffuse and narrow buff-cream supercilium, and poorly-developed dark eye-stripe). Note some buff wash on flanks, quite typical of some. Bill and legs are on average slightly darker than Yellow-browed. (F. Heintzenberg)

Kazakhstan, May: worn and bleached body plumage and wing (hardly any evidence of greater coverts wing-bar) suggests a 1stS. On some birds, pale base to lower mandible can be large and distinctive, though still smaller than on most Yellow-browed; also rather dark legs. Leg colour and vocalisations often play important parts in separation from Yellow-browed at same season. Beware that in worn plumage the pale outer edges to tertials are lost, thereby even inviting confusion with Greenish Warbler. (M. Vaslin)

Worn Hume's Leaf Warbler (left, Jun) versus Yellow-browed Warbler, Mongolia, May: due to feather wear both become paler and drabber, and closer in overall coloration, especially ad Hume's compared to strongly bleached 1stY Yellow-browed, or with 'grey-type Yellow-browed' (see example under this species). Note how the median coverts bar of Yellow-browed becomes 'dotted' with wear. Another problem is evaluating leg colour under shade, when that of Yellow-browed looks misleadingly darker than Hume's. Best clues at this season are Hume's (1) wing-bars being off-white tinged buffish, like the supercilium, which is bold and narrower; (2) much darker bill with limited pale base; (3) bases to greater coverts and remiges paler; (4) underparts off-white with buffish-cream suffusion, not silky white; and (5) Yellow-browed has substantial greenish on scapulars and fringes to remiges. Without handling neither bird can be aged. (H. Shirihai)

SIMILAR SPECIES For separation from *Yellow-browed Warbler*, see Identification. — Other potential confusion species are the same as given under Yellow-browed Warbler.

AGEING & SEXING (*humei*) Small age-related differences in 1stW. Ageing in spring based on different wear probably applicable but not studied. Sexes alike in plumage. — Moults. Complete post-nuptial moult in summer (mainly late Jul–Aug). Partial post-juv moult in summer does not involve flight-feathers or primary-coverts. Partial pre-nuptial moult in winter (both age classes) limited and variable in extent, often involving body-feathers, some wing-coverts and odd tertials, sometimes also central tail-feathers. — SPRING **Ad** Rather evenly abraded plumage with no or only limited apparent moult differences, tips of primaries and tail-feathers somewhat worn, feathers dark grey and often slightly glossy. **1stS** Tips of primaries and tail-feathers often much abraded, feathers rather brownish and with reduced gloss. — AUTUMN Ageing frequently difficult, differences slight. **Ad** Freshly moulted, tips to tail-feathers neat and rounded, centres dark and glossy, finely fringed or tipped whitish. **1stW** Some very similar to ad. Typical birds show first sign of abrasion to tips of primaries and tail-feathers, latter often subtly frayed at tips, and less gloss or contrasting edges on average than ad. **Juv** Often has slightly buffish tinge to both wing-bars. Upperparts more obviously brown-tinged than in ad.

BIOMETRICS (*humei*) **L** 9–10.5 cm; **W** ♂ 53–60 mm (n 19, m 57.8), ♀ 52–57 mm (n 17, m 54.4); **T** ♂ 38–45 mm (n 19, m 41.4), ♀ 36–40 mm (n 17, m 38.7); **T/W** m 71.4; **B** 10.0–12.0 mm (n 33, m 10.7); **BD** 2.2–2.7 mm (n 18, m 2.5); **BW** 2.5–3.3 mm (n 18, m 2.9); **Ts** 16.7–19.0 mm (n 32, m 17.7). Wing formula: **p1** > pc 4–9 mm, < **p2** 19–25.5 mm; **p2** < wt 5–8 mm, =7 or 7/8 (60%), =8 or 8/9 (37%) or <8/9 (3%); **pp3–5** about equal and longest; **p6** < wt 0.5–2 mm; **p7** < wt 3.5–6 mm; **p8** < wt 5–8 mm; **p10** < wt 8–11.5 mm; **s1** < wt 9–12.5 mm. Emarg. pp3–6.

GEOGRAPHICAL VARIATION & RANGE Two races currently recognised, but only *humei* known to have reached Europe as a vagrant.

P. h. humei (Brooks, 1878) (mountain forests of N India, Pamir, Tien Shan north and east to Altai, Sayan Mts and W Mongolia; winters Pakistan, India, rarely further west to Oman). Treated above.

Extralimital: *P. h. mandellii* (Brooks, 1879) (mountains of C China; winters NE India, SE Asia). So-called 'MANDELLI'S LEAF WARBLER' is in some respects morphologically more similar to Yellow-browed Warbler *P. inornatus* due to its darker lateral crown-stripes and eye-stripe in contrast to bold pale (mainly buffish cream-white) supercilium and fairly well-marked pale

Scotland, Oct: some extremely pale and drab Hume's should be less problematic to separate from Yellow-browed: here, note the thinner, less boldly marked buffish-tinged supercilium and median coverts wing-bar. (H. Harrop)

Kazakhstan, Oct: in fresh plumage and in certain lights, some Hume's can appear more greenish above with the brown element being less obvious, inviting confusion with Yellow-browed Warbler. However, note the more buffish-tinged supercilium, slightly more subdued eye-stripe, and almost totally absent upper wing-bar. Also, the bill and legs are somewhat darker than in Yellow-browed. (Y. Belousov)

Fresh Hume's Leaf Warbler (left: Finland, Nov) versus Yellow-browed Warbler (right: France, Oct): Hume's is generally brownish-grey above (with only limited green), lacking the distinct greenish-yellow tones of many Yellow-browed, and dirtier white below (without strong yellowish hue). However, this Hume's has some buffish-green above and slight yellowish-buff flanks. Hume's also has a narrower and pale buffish supercilium (yellowish in Yellow-browed), the ear-coverts are plain buffish-white and only slightly mottled (darker and strongly mottled in Yellow-browed), while the dark eye-stripe is less prominent. The wing of Hume's is less attractively patterned, with mostly off-white or buffy-white wing-bars, while the median coverts bar is usually indistinct; furthermore, the bright yellow-green wing panel and dark area around wing-bars are less obvious, while the outer edge of longest tertial in (most) Hume's is also whitish (usually yellowish-green in Yellow-browed). Finally, Hume's has on average darker bare parts, with a smaller pale base to lower mandible and usually blackish-brown legs (more fleshy-brown in Yellow-browed). However, both bill and legs vary individually and with light conditions. (Left: M. Bruun; right: M. Vaslin)

Sweden, Dec: Hume's Leaf Warbler occasionally shows rather strongly greenish-suffused upperparts, but head is here still typically greyish, a characteristic of the species. Note the short upper wing-bar, which in Dec can bleach whiter like in Yellow-browed. This individual also has very typical dark bill and legs. (D. Erterius)

crown-stripe (although differs in being more yellow-tinged on underparts, especially lower flanks and vent, and in having less white and therefore less contrasting upper wing-bar). Often some fine greyish streaking on chest. Legs yellow-brown, tip of lower mandible extensively dark. Song is same or very similar to that of *humei*, whereas call usually is more sharp a rapid disyllabic *zeweet*, first syllable very short or almost 'swallowed', the call sounding a little like that of Greenish Warbler. (Syn. *superciliaris*.)

TAXONOMIC NOTES See Yellow-browed Warbler. Race *mandellii* might merit separate species status considering allopatric range, rather different plumage and apparently somewhat different call, but this requires further study.

REFERENCES Luijendijk, T. J. C. (2001) *DB*, 23: 275–284. – Madge, S. (1997) *BB*, 90: 571–575. – Millington, R. & Mullarney, K. (2000) *BW*, 13: 447. – Shirihai, H. & Madge, S. (1993) *BW*, 6: 439–443. – Schubert, M. (1982) *Mitt. Zool. Mus. Berlin*, 58: 109–128.

China, May: as evident here, the Chinese race *mandellii* of Hume's Leaf Warbler (extralimital and unrecorded in our region) is very similar to Yellow-browed Warbler, sharing with it a more greenish and yellowish plumage. It differs by its even in fresh plumage subdued and more yellowish-grey upper wing-bar, its on average less well-marked dark eye-stripe, and—above all—by its different call (see text). (P. Alström)

RADDE'S WARBLER
Phylloscopus schwarzi (Radde, 1863)

Fr. – Pouillot de Schwarz; Ger. – Bartlaubsänger
Sp. – Mosquitero de Schwarz; Swe. – Videsångare

Another frequent Siberian autumn vagrant to Europe, Radde's Warbler is named after the 19th century German ornithologist and explorer of E Siberia, Gustav Radde. This species is something of a personality among leaf warblers with its strong bill and legs, its sturdy head shape and rather different call. It breeds in C and E Siberia in forest edges and glades in taiga, and winters in SE Asia. Still, a few are found each year in W Europe, usually in late September and October, in copses of trees or low vegetation on islands or on coasts. Most individuals of this rather discreet species are detected only because they are trapped at bird observatories.

China, May: this large and robustly built leaf warbler is characteristic due to its steep forehead and strong, stout bill, proportionately rather short wings and long tail that is often raised or twitched. The tail movements, plus the obvious greenish-tinged upperparts and yellow-buff underparts separate the species from Dusky Warbler. Further, the prominent supercilium is tinged rufous-buff in front of the eye but whiter behind (the reverse of Dusky). (B. van den Boogaard)

IDENTIFICATION One of the *largest* leaf warblers, of same size as Willow Warbler (or even a fraction larger), but with a proportionally *larger head* and *stronger feet*, and *strong but short bill*. The *crown usually appears rounded* and broad in its forepart, creating a rather *steep forehead* and, together with the thick and short bill, render the species a characteristic jizz. Rather *long-tailed*, accentuated by the quite *short primary projection*. Often perches with lowered wings and slightly raised tail, and now and then twitches tail upwards and flicks wings. Upperparts rather darkish olive-brown and uniform, in some birds or in certain angles a little more greenish than in others. Underparts somewhat variable but nearly always warmly coloured, *dirty-buff on sides of throat, breast and flanks*, and more *ochraceous-yellow shades on belly and undertail-coverts*. The plumage lacks any other striking pattern other than the *long and prominent supercilium*, which starts above lores (short of nostrils) but reaches far onto sides of nape. It is usually tinged *ochre-yellow or buff in front of eye*, but is *more whitish behind*, and its prominence is enhanced by both a *dark lateral crown-stripe* and a *dark eye-stripe*. Note that it is *often rather diffusely marked in its forepart* bordering the forehead (but a few are better marked). Lower eyelid is white, contrasting rather distinctly. *Legs are pinkish or pale brown*, never dark-looking. Bill has dark culmen and tip, but pinkish-yellow base of lower mandible. Feathering on forehead and above nostrils often erect and 'brushy'. Juvenile much more saturated yellow below and has a yellowish-buff supercilium.

VOCALISATIONS Song a very loud outburst of whistling and trilling notes, each one repeated a few times. Strophes are varied, elements often being combined differently. When less inspired, somewhat resembles songs of Dusky Warbler and Sulphur-bellied Warbler (*P. griseolus*; extralimital, see Vagrants), but is faster and more explosive, and contains more trilling elements. Commonly, each strophe opens with a quick repetition of a few *te* notes, *te-te-te tell-ell-ell-ell-ell... te-te swee-swee-swee-swee-...*, etc. When more agitated, strophes become longer, with shorter pauses between them, and may then recall short strophes by Common Nightingale, e.g. *te-te-serrrrrrr-tyu-tyu-tyu* or *te-te sweerr-pyu-pyu-pyu-tel-el-el-el-el*. – The call is rather subdued, a short, nasal, 'thick' clicking *chrep*, often repeated as the bird feeds and moves around in the vegetation. Rather oddly, it recalls the call of a totally unrelated species, Red-necked Phalarope, which also has a call like 'plucking a violin string'.

SIMILAR SPECIES Distinctive enough if seen well. In less unequivocal views can be confused with *Dusky Warbler*, but this is less obviously greenish-tinged above and much less yellow-tinged below, appearing in the field more grey-brown above and rufous-buff below. Also, Dusky has thinner bill and legs, bill with on average slightly more obvious dark tip to lower mandible, and legs generally darker pinkish-brown, often striking you as 'dark' rather than 'pale' in typical encounters. A rare pitfall is offered by a Radde's with an unusually thin bill, but such birds will still be identified by a combination of all other characters. The call of a migrant Dusky, often uttered and then usually conclusive, is a sharp, tongue-clicking *zeck* like Lesser Whitethroat's, or even harder like Wren's, not at all like the more slurred, nasal clicking of Radde's. When handled or photographed, notice acutely pointed tips to tail-feathers in 1stCY (Dusky: more rounded, like in rest of congeners). – *Booted Warbler* is slightly smaller, has shorter supercilium, lacks yellow tinges on vent/undertail-coverts, has off-white edges to tail, thinner legs, darker toes, etc.

AGEING & SEXING Small age-related differences in 1stW. Sexes alike in plumage. – Moults. Complete post-nuptial moult in late summer (mainly late Jul–Sep). Partial post-juv moult in summer does not involve flight-feathers or primary-coverts. Partial pre-nuptial moult in late winter (mainly Mar) variable in extent, often involving many or all tail-feathers

Mongolia, Jun: proportionately large head, strong but short bill and strong feet (feet always notably pinkish or pale brown). Also, note ochraceous-yellow vent and undertail-coverts. The long, prominent and bicoloured supercilium (often more diffusely outlined in its forepart) is usually tinged ochre-yellow or buff in front, while the lower eyelid is distinctly white. Rather worn primary tips and primary-coverts suggest 1stS, but certain ageing requires handling. (H. Shirihai)

(Williamson 1969); birds moulting tail-feathers in winter might be both 1stW and ad (*contra BWP*). – **SPRING** Ageing according to wear of primaries and tail-feathers requires further study before it can be used. Very fresh birds are presumed to be ad, whereas heavily worn primaries would indicate 1stS. Still, most birds fall in an intermediary category. – **AUTUMN** Ageing frequently difficult, differences slight, but at least many young are readily told by saturated yellowish-brown underparts. However, there is some variation as to amount of yellow, and others are very similar to ad. **Ad** Freshly moulted, tips to tail-feathers neat and rounded (although can have a fine point at the shaft), centres dark and glossy, finely fringed or tipped whitish. On average less tinged olive above, less tinged yellow below. **1stCY** Some are strikingly yellowish-brown below and olive-tinged above, others less so and are then very similar to ad. Irrespective of underparts colour, usually told by acutely pointed tips to tail-feathers, and on first sign of abrasion to tips of primaries and tail-feathers, latter often subtly frayed at tips, and show on average less gloss or contrasting edges than ad.

BIOMETRICS **L** 11–13 cm. **W** ♂ 61–67 mm (n 25, m 64.1), ♀ 56–65 mm (n 14, m 59.8); **T** ♂ 51–58 mm (n 24, m 54.6), ♀ 45–55 mm (n 14, m 50.4); **T/W** m 84.8; **B** 11.5–14.2 mm (n 38, m 12.7); **BD** 3.0–3.9 mm (n 36,

Mongolia, May: some in spring wear greyer above, contrasting with warm yellowish-buff below and greener secondaries panel. Some Radde's lack a bicoloured supercilium (here the supercilium is tinged rufous-buff over much of its length). (M. Putze)

m 3.4); **BW** 3.4–4.3 mm (n 34, m 3.9); **Ts** 20.0–22.9 mm (n 39, m 21.7). **Wing formula: p1** > pc 7–14 mm, < p2 15–22 mm; **p2** < wt 6.5–10 mm, =8/9 or 9 (55%), =9/10 or 10 (25%) or =7/8 or 8 (20%); **pp3–5**(6) about equal and longest; **p6** < wt 0–3.5 mm; **p7** < wt 2.5–6 mm; **p8** < wt 4.5–8.5 mm; **p10** < wt 7.5–12.5 mm; **s1** < wt 9–14 mm. Emarg. pp3–6.

GEOGRAPHICAL VARIATION & RANGE Monotypic. – SC & SE Siberia, E Altai, Baikal area, Transbaikalia, Amur, Sakhalin, Ussuriland, N Korea; winters C & S China, mainland SE Asia.

REFERENCES Bradshaw, C. (1994) *BB*, 87: 436–441. – Johns, R. J. & Wallace, D. I. M. (1972) *BB*, 65: 497–501. – Madge, S. (1990) *BW*, 3: 281–285. – Vinicombe, K. (2004) *Birdwatch*, 148 (Oct): 37–39.

England, Oct: both olive-green and greyish above and dull greyish-yellow below. As in spring, plumage lacks any striking pattern but for the long and prominent supercilium, which reaches back to nape-sides and is usually tinged warm yellow-buff in front of eye, more whitish behind. Supercilium further enhanced by dark lateral crown-stripe and eye-stripe. The rather unusually thin bill of this Radde's could invite confusion with Dusky, but combination of all other characters readily secure the species. (I. H. Leach)

Scotland, Oct: not all young birds in autumn are obviously yellowish-brown below and olive-tinged above, some being not unlike ad, thus it is always best to base ageing on wear and moult in wing and tail rather than on plumage colours. This bird could be either a rather dull young bird, or warmer fresh ad. In terms of species identification, note the striking supercilium, tinged rufous-buff in front of eye, whiter behind, further warmer undertail-coverts and stubbier bill. (H. Harrop)

1stW, England, Oct: Radde's Warbler habitually feeds on the ground. This bird is in typical young plumage in autumn, with characteristic two-coloured supercilium, olive-tinged upperparts and greyish-yellow underparts, and primary-coverts are juv. (G. Thoburn)

DUSKY WARBLER
Phylloscopus fuscatus (Blyth, 1842)

Fr. – Pouillot brun; Ger. – Dunkellaubsänger
Sp. – Mosquitero sombrío; Swe. – Brunsångare

A rare but fairly regular Siberian autumn vagrant to W Europe. On migration it is often found in weeds and low scrub in open dry terrain, but in its Siberian home this species breeds in very wet and damp thickets on open, flat ground, and is a genuine 'marsh warbler' (whereas the real Marsh Warbler is a 'nettle warbler' and not at all fond of marshes). Apart from Siberia, it also breeds in Altai, N Mongolia, Manchuria, N China and east to Sakhalin. Dusky Warbler normally migrates south and south-east to reach its wintering grounds in SE Asia, from Nepal eastwards, but as with several other Asian passerines a number of presumably inexperienced young birds apparently make a 180° mistake on their first autumn migration and end up in Europe instead.

Netherlands, Oct: in autumn, typically appears overall dusky with long, prominent, pale supercilium (better defined and whiter in its forepart than on Radde's). Dusky buff-brown underparts and thin bill further separate from Radde's. The apparently rather freshly moulted and dark-centred wing, with neat pale tips to primaries, and rather rounded tail-feathers, suggest ad, but without handling best left un-aged. (P. Palmen)

Mongolia, May: note delicate proportions, fine bill and thin long legs. Also quite round-winged impression (primary projection short), grey-brown with hardly any green above and mostly dusky grey-brown below. Sides and cheeks often warmer rufous-tinged. It shares with Radde's the long and prominent pale supercilium, which is further enhanced by dark eye-stripe. Unlike in Radde's, the supercilium is distinctly outlined in its forepart. (J. Normaja)

IDENTIFICATION Of same size as Willow Warbler, with similar phylloscopine delicate shape, but proportionately slightly *longer legs* and much *shorter primary projection* (in fact has quite rounded wings). Habits and movements similar to Willow or Radde's Warbler. *Upperparts dark olive-brown and uniform*, although in the field usually appear *grey-brown without any green tinge*. Underparts dusky grey-brown, darkest on sides of breast and flanks, the throat and vent being a little cleaner whitish; there is often a faint *rufous tinge on ear-coverts, sides of throat, breast and flanks*, but no clear yellow on lower flanks, vent or undertail-coverts (as in Radde's). Still, rather variable as to darkness and amount of rufous or buff tones; Ticehurst (1938) very sensibly stressed the variability of this species, both individually and due to wear. Like Radde's, it has a *long and prominent pale supercilium*, usually narrow (but can be broad in its forepart), enhanced by a *dark eye-stripe* and in some lights and angles a hint of a darker lateral crown-stripe. The *supercilium is white in front of the eye but tinged rufous-buff behind* (the reverse pattern compared to Radde's), and unlike in Radde's it is *invariably distinct and well-marked in its forepart*. A very few have the supercilium all white, or slightly whiter at rear. *Thin bill* is rather dark, but has yellow-brown base to lower mandible and paler cutting edges. Legs pinkish yellow-brown but often look surprisingly dark (though never blackish) in the field.

VOCALISATIONS Song a simple series of a loudly repeated note (sometimes disyllabic), delivered in rather calm tempo (thus not as explosive, fast and varied as Radde's). Rarely, two different motifs are used in the same strophe, but then generally only one is repeated, *chill-chill-chill-chill... seewee seewee seewee seewee... suuh-swe-swe-swe-swe...* etc. It is not that unlike Radde's Warbler's song, but the simpler structure and slower tempo in combination with marshland habitat generally identify the species. Also, it does not usually start each phrase with a few short notes (*te-te*) as Radde's so often does, and does not contain fast rattling trills. Much more similar to song of Sulphur-bellied Warbler (*P. griseolus*; extralimital, see Vagrants), but this occurs at higher elevations, in open mountain forests on slopes with scree and boulders. – Call is a sharp tongue-clicking *zeck*, rather like Lesser Whitethroat, but when agitated can be even louder and harder, like Wren.

SIMILAR SPECIES Told from *Radde's Warbler* by finer bill, thinner and usually slightly darker legs, different head jizz, lack of appreciable ochre-yellow tinge on underparts, more rufous and fewer olive tones in rest of plumage, and by tongue-clicking call. Note on handled or photographed birds that tips of tail-feathers are rather rounded, like in Common Chiffchaff or Willow Warbler (young Radde's: acutely pointed). – Superficially similar to *Common Chiffchaff* of Siberian race *tristis*, but this has blackish legs, and unless heavily worn has greenish-tinged edges to wings and tail, and more olive hues on back and rump; further, its supercilium is shorter and less well-marked due to the more subdued, paler eye-stripe. Small patch near wing-bend and axillaries, at times possible to evaluate in the field, are pale yellow in *tristis* but buff-white in Dusky. The call of *tristis* is a straight (or very slightly downslurred) piping *iihp*, very different from short, tongue-clicking call of Dusky Warbler. – *Booted Warbler* is slightly paler grey-brown overall, has shorter supercilium, less dark and prominent eye-stripe, and off-white edges to tail. – *Paddyfield Warbler* is eliminated by having paler brown iris (Dusky: all-dark eye), paler and usually more rufous-buff or ochre tinges on rump, flanks, vent and undertail-coverts, rounded tail-tip, and usually noticeably stronger bill.

AGEING & SEXING Small age-related differences in 1stW. Sexes alike in plumage. – Moults. Complete postnuptial moult of ad either in late summer (mid Aug–Sep) or just after autumn migration (Sep–Oct); frequency of these

Dusky Warbler, variation in spring, Mongolia, May: note that irrespective of feather wear, some Dusky Warblers are warmer yellow-tinged (right) versus more brown-toned overall. The right-hand bird with more buffish supercilium approaches those few Radde's that also have even-toned supercilium. Nevertheless, unlike Radde's, note the less robust jizz with finer bill, thinner legs and better-marked supercilium in front of eye. The monosyllabic tongue-clicking call of Dusky is often best means to separate tricky or briefly-observed birds. (H. Shirihai)

Dusky Warbler, variation in autumn, Hong Kong (left) and Denmark, Oct: note following: (1) chiefly dusky grey-brown, (2) underparts cream-white and extensively suffused fulvous-brown, mainly on breast-sides to undertail-coverts, (3) long supercilium, cream-white in front of eye, becoming buffier above and behind eye, (4) crown colour by and large concolorous with upperparts, (5) dark eye-stripe broad, usually darker or blacker in front of eye, (6) ear-coverts heavily mottled cream and brown. Also characterised by (7) dull flesh-brown slender legs, and (8) bill rather short, fine and pointed with extensive dark culmen and tip. Although Dusky shares with Radde's several structural characteristics, such as (9) relatively longish tail but rather short primary projection, it is (10) usually visibly less large-headed in direct comparison. Both birds apparently 1stW by already abraded primary tips and primary-coverts (retained juv). Juv tail-feathers with pointed and frayed tips, most noticeable on right-hand bird. (Left: M. Hale; right L. Jensen)

alternative strategies not known and requires further study. Partial post-juv moult in summer does not involve flight-feathers or primary-coverts. Pre-nuptial moult in late winter (Feb–Apr), variable in extent, apart from body often involving some or all tertials and odd tail-feathers; claims that it is sometimes complete (Williamson 1969) cannot be confirmed by us. – **SPRING** Ageing according to wear of primaries and tail-feathers requires further study. Note that in worn breeding plumage, it can appear as if outer 3–5 primaries are newer, inner being heavily abraded (implying that these birds are 2ndCY), but this contrast in wear is due to the fact that the outer primaries are less exposed when the wing is closed (due to the wing being so rounded). – **AUTUMN** Ageing frequently difficult, differences slight. **Ad** Freshly moulted, tips to tail-feathers neat and rounded, centres dark and glossy, finely fringed or tipped whitish. On average less tinged olive above, and lacks any yellow below. **1stCY** Some very similar to ad. Typical birds aged by first sign of abrasion to tips of primaries and tail-feathers, latter often subtly frayed at tips and show on average less gloss or contrasting edges than ad. On average more olive cast above and faintly yellow-tinged on centre of belly (but some variation as to coloration).

BIOMETRICS L 10.5–12 cm. **W** ♂ 59–69 mm (n 22, m 62.9), ♀ 53–62 mm (n 20, m 57.7); **T** ♂ 47–58 mm (n 22, m 52.0), ♀ 43–52 mm (n 20, m 47.3); **T/W** 82.3; **B** 11.0–12.7 mm (n 39, m 12.2); **BD** 2.4–3.0 mm (n 36, m 2.7); **BW** 2.7–3.6 mm (n 33, m 3.3); **Ts** ♂ 20.5–23.9 mm (n 20, m 22.1), ♀ 19.0–22.3 mm (n 15, m 20.4).

Wing formula: p1 > pc ♂ 9.5–14 mm, ♀ 8–12 mm, < p2 ♂ 17–21 mm, ♀ 15.5–18 mm; **p2** < wt 6.5–11 mm, =8/9 or 9 (41%), =9/10 or 10 (41%), =8 (10%) or <10 (8%); **pp3–5** about equal and longest; **p6** < wt 0.5–2.5 mm; **p7** < wt 2–5 (6) mm; **p8** < wt ♂ 5–9 mm, ♀ 4.5–7 mm; **p10** < wt 7.5–12 mm; **s1** < wt 9–13 mm. Emarg. pp3–6.

GEOGRAPHICAL VARIATION & RANGE Monotypic (but see below). – Altai, SC Siberia from Ob east to Anadyr, south to C China in N Sichuan; winters foothills of Himalaya, S & SE Asia. (Syn. *altaicus*; *homeyeri*; *mariae*; *robustus*. See Ticehurst 1938 for rationale.)

TAXONOMIC NOTES The darker taxon *weigoldi* Stresemann, 1923 (mountains of E Tibet and SW China), formerly regarded as a race of Dusky Warbler, has recently been referred to extralimital Smoky Leaf Warbler *P. fuligiventer* (Hodgson, 1845) on account of morphology, vocalisations and genetic evidence (Martens et al. 2008). – In 1923 Stresemann separated *robustus* (Sichuan, WC China) as being slightly larger than *fuscatus* and having a dirty yellow-tinged centre of belly (rather than white). However, Ticehurst (1938) in his careful monograph of *Phylloscopus* synonymised *robustus* with *fuscatus*. Later, Vaurie (1959) and Dementiev & Gladkov (1954) held the same view. Martens et al. (2008), however, accepted *robustus*, but based it largely on genetic properties, examined limited material and did not in our opinion provide entirely convincing arguments. – Recently Redkin & Malykh (2011) described a new subspecies from Sakhalin. We have not examined relevant material, but the described differences appear to be rather minor.

REFERENCES Bradshaw, C. (1994) *BB*, 87: 436–441. – Johns, R. J. & Wallace, D. I. M. (1972) *BB*, 65: 497–501. – Madge, S. (1990) *BW*, 3: 281–285. – Martens, J., Sun, Y.-H. & Päckert, M. (2008) *Vert. Zool.*, 58: 233–265. – Redkin, Y. A. & Malykh, I. M. (2011) *Ross. Orn. Zhurn.*, 20: 59–80. – Vinicombe, K. (2004) *Birdwatch*, 148 (Oct): 37–39.

Germany, Nov: like in Radde's, the long, prominent pale supercilium is enhanced by a dark eye-stripe and sometimes by hint of darker lateral crown-stripe, but unlike Radde's the supercilium is white in front of eye and tinged rufous-buff behind. In this individual, also note the dusky rufous-tinged off-white underparts, darkest on sides and warmer behind legs, leaving cleaner whitish belly. (K. F. Jachmann)

Dusky Warbler (left: Germany, Nov) versus Radde's Warbler (right: England, Oct) in autumn: even though Dusky might be instantly separable by its slighter jizz, thinner bill and legs, and often its less brightly saturated plumage, it is advisable to take full notes on any vagrant. Underparts of Dusky almost invariably never show tawny or warm yellowish tones of Radde's, instead are dusky grey or olive-brown with variable buffish suffusion. Dusky also lacks the large-headed appearance and high, rounded forecrown of Radde's. Both have prominent pale supercilium, but in Dusky it is more well-marked and whiter in front of the eye, more rufous-buff behind—the reverse pattern in Radde's, but as shown here this character is not always easy to evaluate. The legs in Dusky average darker. With less straightforward individuals of the two species, calls can be the best means of separation. (Left: K. F. Jachmann; right: G. Reszeter)

WESTERN BONELLI'S WARBLER
Phylloscopus bonelli (Vieillot, 1819)

Fr. – Pouillot de Bonelli; Ger. – Berglaubsänger
Sp. – Mosquitero papialbo; Swe. – Bergsångare

In SW Europe in dry mountain woods and hilly open forests, you are likely to hear this small summer visitor energetically sing its simple but pleasant song. And it is probably by its song that the species' presence is first noted, because it keeps hidden in the foliage most of the time. The song is just a quick, short series of repeated high-pitched notes, but it has an enchanting, joyful ring. This little warbler spends its winters in sub-Saharan Africa, and returns to Europe in mid April and early May. It is nowadays regarded as a separate species from its close relative and counterpart in SE Europe, Eastern Bonelli's Warbler.

can sometimes (though far from invariably) be separated using finer plumage characters, supported by structure and coloration of bare parts: Western Bonelli's is slightly more greenish-tinged on crown and mantle, less brown-grey (but difference is slighter than many believe!). Best criterion is usually slightly different wing pattern, with Western generally having fresher plumage in spring, with better retained *yellowish-green edges to greater coverts, flight-feathers and tertials*. (Eastern has more worn wings with often greyish-white, uneven edges to all tertials and many central greater coverts, these areas often appearing as paler, contrasting patches.) Note that there is some individual variation, with some Western having a hint of the same pattern as Eastern in spring, and that Western breeders gradually develop a more similar wing pattern to Eastern as their wings become bleached and worn in summer. Also, Western often has whitish tips to tertials when these are fresh; it is the *edges to outer webs* that are green, especially basally. – There are some very slight and average structural and other differences too: Western has *slightly shorter and more rounded wing*, making it sometimes *look longer-tailed*, which may serve as a supplementary character (but is too subtle to use alone). Also, the bill is very slightly longer, and has on average a little paler (pink-brown) colour at the base and along cutting edges (Eastern: more often a subtly shorter and darker bill), but differences are slight, and numerous birds are inseparable in this respect. Legs average subtly longer in Western, and their colour is on average a shade paler than in Eastern, but again there is much overlap. (Facial pattern does not appear to differ consistently and is therefore not stressed here, but Western is usually not as rufous-buff-tinged as some Eastern can be.) – Young autumn Western and Eastern Bonelli's are depressingly difficult to separate, since they attain very nearly the same plumage, being olive-brown above with a moderately contrasting slightly paler rump (less contrast on average than in adults), and having rather bright green edges to all wing-feathers. Autumn adults, too, of the two species are very similar, now being in worn and quite similar more greyish plumage. Only the call is still helpful. – For identification in the hand of difficult birds see above regarding plumage colours and moult differences, and below under Biometrics and note especially wing-length, distance p1 < p2, MCV ratio, position of p2 in relation to other primaries, and tendency to have proportionately longer tail, longer legs, longer bill and shorter undertail-coverts in relation to tail-tip.

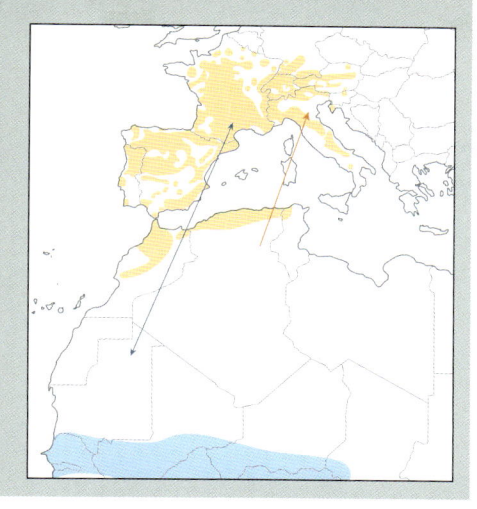

Italy, May: unlike Eastern Bonelli's has slightly longer and stronger bill, while upperparts are suffused slightly brighter brownish-green. Due to fresher edges to greater coverts and tertials at this season, the wing appears more uniformly yellow-green (Eastern has more abraded and greyish-white tertials and central greater coverts). Also note pale cheeks. (D. Occhiato)

IDENTIFICATION The two Bonelli's Warblers, the Western and Eastern, have much in common, and it is often useful to first treat them together in relation to all of the other leaf warblers (mainly genus *Phylloscopus*). They share the following characteristics: (i) a *pale rump*, at least in adults usually *bright yellowish-green* or even ochre-yellow, often noticeably contrasting against grey-brown back; (ii) quite *whitish underparts* with just a faint brownish-grey hue on sides of throat and breast, and some yellowish-green on lower flanks in some; (iii) an *unbroken off-white, narrow eye-ring*, the eye often appearing rather large (dark 'peppercorn'); (iv) rather *pale cheeks and ear-coverts*, tinged yellowish-brown; (v) *bright yellow-green edges to tail-feathers and most wing-feathers*; and (vi) contrasting *tertials, dark-centred with pale outer edges*. Bill is rather strong and has a pale brown base. Legs are pinkish-grey but often look *rather dark*, in between Common Chiffchaff and Willow Warbler in colour. Both Bonelli's species are agile and active, having similar movements to a Willow Warbler, at times twitching the tail in any direction, often paired with quick wing-flicks. Frequently hovers at tip of branches when feeding, thereby exposing the pale rump. – It remains to separate the two twin species, a much harder task. *Safest separation is by call*, or song once this has been learnt. When seen close and well, adults in spring and summer

Switzerland, Apr: note quite whitish underparts, contrasting pale edges to tertials and greater coverts, and off-white, narrow unbroken eye-ring, enhancing large-eyed appearance. This bird is a bit atypical in appearing quite greenish-tinged on crown and back, but this may be due to light and reflections. Ageing impossible due to complete winter moult of both age classes. (R. Aeschlimann)

VOCALISATIONS Song a quick repetition of the same high-pitched note, often 8–12 notes in each strophe, with some moderate variation in pace, pitch and pronunciation, *swee-swee-swee-swee-swee-swee-swee-swee... sresresresresresresresresresre...*, etc. The song recalls most of all song of Wood Warbler, but lack any acceleration, and is simpler in structure and shorter. Compared to song of Eastern Bonelli's, the voice of Western is more twittering or laughing (joyful), and the tone is more 'silvery', or light and clear, rather like Blue Tit. Song length varies depending on excitement of singer, but is often slightly longer than in Eastern Bonelli's. Hardly discernible to the ear, but obvious when compared in sonograms (Bergmann & Helb 1982), is the slower pace under which generally individual notes are uttered. Graphically, the notes form 'V' or 'U' shapes (in Eastern Bonelli's inverted 'V' or 'backslash'). – Common call by adults a distinctly disyllabic *too-eef*, with both syllables equally stressed, superficially at distance recalling Willow Warbler but is quite distinctive when heard close, being much more emphatic, clearly disyllabic and upwards-inflected so that you may think of Greenfinch rather than a leaf warbler. The *too-eef* call is used also as alarm by the parents near

France, Jun: when worn and bleached, separation from Eastern Bonelli's becomes very difficult, often impossible using plumage, making vocalisations the safest means of identification. Compared to images on p. 585, note paler pink-brown legs. (J. Fouarge)

Ad, Spain, Aug: few silent late summer birds can be safely identified, as wear obscures the already small plumage differences between Western and Eastern Bonelli's Warblers. But concentrate at least on eliminating any other leaf warbler; note bland face, unbroken whitish eye-ring and seemingly large dark 'peppercorn' eye. When rump is seen, note its brighter yellow-green tinge. (Left: S. Fletcher; right: C. Zea)

nest. Young birds, and apparently also some adults, have a softer, less up-curled and less disyllabic call, *hweef* (like a loud Common Chiffchaff), and young beg with short metallic notes, *chix*, a little like Robin or a juvenile Song Thrush. In territorial disputes, the adults utter a subdued purring *tr tr tr tr*. Occasionally emits a brief *psit* repeated a few times (Géroudet 1957). Migrants have been noted to utter monosyllabic calls (A. B. van den Berg, *in litt.*), but these seem close to the above-mentioned *psit*, and are rather insignificant (rarely used, common call soon heard).

SIMILAR SPECIES Main risk of confusion is with *Eastern Bonelli's Warbler*, and separation of these two is treated under Identification. – Theoretically could be confused with a northern, paler *Common Chiffchaff* of race *abietinus*, or northerly *Willow Warbler* of race *acredula*, both of which can be rather whitish below, have a rather bright green rump in contrast to duller rest of upperparts, and have greenish-edged wing-feathers, but note on Common Chiffchaff smaller, darker bill and darker legs, and on both species broken eye-ring in two halves, darker cheeks and ear-coverts and more uniform tertials. – *Wood Warbler* is sometimes quite white on underparts, lacking much of the usual yellow throat and upper breast, and such birds can be superficially similar to a Western Bonelli's due to their green-edged wings and contrasting tertials. But Wood Warbler invariably has a

1stW, Spain, Aug: readily aged by the fresh plumage at this season (in contrast, ad is worn). Young autumn Western, however, is very difficult to separate from Eastern Bonelli's as both species are equally fresh and very similarly olive-brown above, with a moderately contrasting paler yellowish-green rump and rather bright green edges to all wing-feathers (C. N. G. Bocos)

prominent dark eye-stripe breaking the pale eye-ring, and much longer primary projection, making it appear short-tailed. Also, legs are paler, a light pinkish-brown.

AGEING & SEXING Small age-related plumage differences usually in 1stCY only. Sexes alike or very similar in plumage. – Moults. Partial post-nuptial moult in late summer (Jul–Aug) involving body-feathers and some wing-coverts, sometimes also odd tertials and tail-feathers. Partial post-juv moult in summer does not involve flight-feathers or primary-coverts. Complete moult including wings and tail takes place early in winter quarters, in late autumn or early winter (mainly late Sep–Nov, rarely Dec; earliest with finished moult 21 Oct). Some moult much later (Feb–Mar), presumably 1stY, but no proof that all are. Judging from freshness of tertials on nearly all birds returning to Europe Apr–May, these feathers are either often moulted twice in winter, or—more likely—their renewal is kept to a late stage of the winter moult. A few return with mixture of new and very worn tertials, sometimes also greater coverts. A very few appear to moult inner primaries in late summer or early winter, arrest, then complete with moult of outer primaries before spring migration (Mead & Watmough 1976; pers. obs.). Exceptionally, spring birds, presumably 1stY, return with their now heavily worn juv flight-feathers unmoulted. – AUTUMN **Ad** Much-abraded tips to primaries and tail-feathers. Rump yellowish, showing good contrast with rest of olive grey-brown upperparts. **1stCY** Plumage fresh, including tips to primaries and tail-feathers. Rump patch variable in prominence, some having more subdued and dull yellowish olive-brown rump. Upperparts, especially back, sometimes tinged warm tawny-brown (even rufous).

BIOMETRICS L 10.5–11.5 cm. **W** ♂ 59–68.5 mm (n 107, m 64.4), ♀ 56.5–66 mm (n 54, m 61.6); **T** ♂ 43–55 mm (n 102, m 48.2), ♀ 42–52 mm (n 53, m 46.1); **T/W** m 74.8; **B** 11.4–13.7 mm (n 185, m 12.7); **B/W** 17.4–22.4 (n 181, m 20.0); **BD** 2.1–3.0 mm (n 159, m 2.6); **Ts** 17.0–19.6 mm (n 101, m 18.5); **Ts/W** 26.5–33.3 (n 94, m 29.5); **UTC** (see below) 11–23 mm (n 158, m 15.9). **Wing formula: p1** > pc 2–8.5 mm, < p2 23.5–33.5 mm (m 28.0, n 181, only rarely > 30 mm); **p2** < wt 3.5–7.5 mm, =6/7 or shorter (82%), =6 (10%), or =5/6 (8%); **pp3–4**(5) about equal and longest; **p5** < wt 0–2 mm; **p6** < wt 2–6 mm; **p7** < wt 5–10 mm; **p10** < wt 9.5–16 mm; **s1** < wt 11–17 mm. Emarg. pp3–5; also p6 often fully emarg. (23%), slightly less prominently (25%), or very faintly (29%); thus, in remaining 23% there is no emarg. of p6. – For difficult birds, calculate 'multiple characters value', **MCV**, as follows (UTC = tail-tip > tip of longest undertail-coverts): add W and p1<p2, and subtract B, UTC, p1>pc, p2<wt. MCV is 39.5–60.3 (n 139, m 52.5); 97.6% had MCV ≤ 58.0. Birds in very abraded or wet plumage, or which have lost longest undertail-coverts, cannot be reliably calculated.

GEOGRAPHICAL VARIATION & RANGE Monotypic. – NW Africa, W Europe from Iberia and France east to C Europe, Slovenia, Italy; winters sub-Saharan W Africa, commonly in Sahel belt from N Senegal to Niger and adjacent areas. (Syn. *montana*.)

TAXONOMIC NOTE Helbig *et al.* (1995) showed that the genetic difference between Western Bonelli's Warbler *bonelli* and Eastern Bonelli's *orientalis*, formerly two subspecies of the same species, was as large as 8.5% (cyt b, mtDNA). Taking into account also that the two forms are allopatric and have very different calls and slightly different songs, with some subtle but consistent morphological differences, and no hybridisation known, they are better treated as two separate species.

REFERENCES GROENENDIJK, D. & LUIJENDIJK, J. C. (2011) *DB*, 33: 1–9. – HELBIG, A. J. *et al.* (1995) *J. Avian Biol.*, 26: 139–153. – MEAD, C. J. & WATMOUGH, B. R. (1976) *Bird Study*, 23: 187–199. – OCCHIATO, D. (2007) *BW*, 20: 303–308. – PAGE, D. (1999) *BB*, 92: 524–531. – SVENSSON, L. (2002) *Birdwatch*, 119 (May): 26–30. – SVENSSON, L. (2002) *Vår Fågelv.*, 61(4): 7–14.

1stW, Shetland, Scotland, Oct: indistinguishable from Eastern Bonelli's using plumage (this bird was confirmed vocally), but reliably separated from other *Phylloscopus*. Especially note the almost pure white underparts and dull grey-brown upperparts, with bright yellowish-green edges to wing- and tail-feathers, unbroken whitish eye-ring and pale supercilium. Aged by fresh plumage. (H. Harrop)

1stW, Mallorca, Spain, Aug: a typical view of a fresh young bird in late summer. Readily noted as one of the Bonelli's by its bright yellowish-green rump and uppertail-coverts, green-edged remiges and tail-feathers in contrast to much duller brown-grey rest of upperparts. Species identification based on locality. (J. J. Bazán Hiraldo)

1stW, Spain, Aug: easily identified as one of two Bonelli's Warbler species based on unbroken pale eye-ring, pale cheeks, and whitish throat and breast in contrast to vivid yellowish-green wing and tail edgings. However, this young Western Bonelli's in mint-fresh plumage is indistinguishable from Eastern Bonelli's on plumage alone. (R. Tidman)

EASTERN BONELLI'S WARBLER
Phylloscopus orientalis (C. L. Brehm, 1855)

Fr. – Pouillot oriental; Ger. – Balkanlaubsänger
Sp. – Mosquitero oriental; Swe. – Balkansångare

This is the SE European counterpart of Western Bonelli's Warbler, until recently regarded as a subspecies, but nowadays commonly treated as a separate species. It breeds in the Balkans, including in Greece, and also in Bulgaria, W & S Turkey and at scattered localities in the Middle East, in dense woods of oak, hornbeam or other deciduous trees, including in tall forests. In autumn it migrates mainly east of the Mediterranean to reach its winter quarters in Sudan and Ethiopia. Returns to the breeding grounds in April. Listen then for its insect-like song.

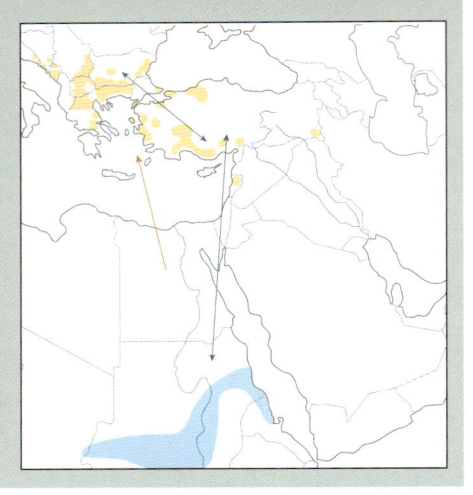

IDENTIFICATION Quite similar to its close relative the Western Bonelli's Warbler, the Eastern Bonelli's Warbler share with it: (i) a *paler rump*, in adults *bright yellowish-green* or *ochre-yellow*, usually contrasting against olive-brown back; (ii) quite *whitish underparts*, off-white with only a faint grey-brown wash on sides of throat and breast; (iii) an *unbroken off-white eye-ring* around dark eye; (iv) quite *pale greyish-brown ear-coverts*; (v) *bright yellow-green edges to tail-feathers and most wing-feathers*; and (vi) contrasting *tertials, dark-centred with pale outer edges*. That the undertail-coverts, vent and lower flanks can have a very faint lemon wash is only visible on handled birds or at very close range. Bill is medium strong and usually *all dark*. Legs are usually *dark grey-brown*. Habits and movements much as in Western. – Safest separation from Western Bonelli's *is by call*, or *song* (although latter requires practice and a good ear). When seen close, adults in spring and early summer can sometimes be separated using finer plumage characters, supported by structure and colour of bare parts: Eastern Bonelli's is slightly less tinged greenish on crown and back, instead more dull brown-grey (but difference is still slight!). Best criterion is usually slightly different wing pattern, with Eastern being on average more worn in spring, in particular on tertials, these having whitish and uneven edges (never smooth and basally green-tinged), and all or many central greater coverts are often abraded, being edged dull greyish-brown or off-white (Western Bonelli's return with much more green-tinged edges to entire wing in spring, although a few are more intermediate). Note that Western breeders gradually develop a more similar wing pattern to Eastern, as they become bleached and worn in summer. As to structure, Eastern has on average a little longer and more pointed wing, rendering the bird a slightly more short-tailed appearance, and the bill is very slightly shorter, and often a little darker, than that of Western (but differences are slight and hardly useful taken alone). Legs are on average a fraction darker and shorter in Eastern (again, much overlap). Still, all these fine structural differences can sometimes in combination translate into a slightly different jizz. – Young autumn Western and Eastern Bonelli's generally cannot be separated as they attain very nearly the same plumage, being olive-brown above with a moderately contrasting slightly paler rump, and having rather bright green edges to all wing-feathers. Autumn adults, too, are very similar, being in worn and quite similar more greyish plumage. Only the call is still very helpful. – For identification in the hand of difficult birds see above regarding plumage colours and moult differences, and below under Biometrics and note especially wing length, distance p1 < p2, MCV ratio, position of p2 in relation to other primaries, and tendency to have proportionately shorter tail, legs and bill, and also longer undertail-coverts in relation to tail-tip.

VOCALISATIONS Song very much resembles that of Western Bonelli's, a quick repetition of the same high-pitched note, but differs to the trained ear by its more mechanical or insect-like and dry tone, *zrezrezrezrezrezrezre*. It is more monotonous than the song of Western, lacking the latter's twittering or 'laughing' quality. If compared with another species it would be Cirl Bunting, which, at least when heard at some distance, produces a similar sound. Often similar song length to Western Bonelli's (or perhaps slightly shorter on average), but both species can prolong the song when exited. When sonograms are compared it is obvious that individual notes are delivered at greater speed, thus more notes on average in songs of similar length (Bergmann & Helb 1982); songs often consist of 14–22 notes. Also, graphically each note forms an inverted 'V' or a sloping mark ('backslash') rather than the 'V' or 'U' shapes of Western Bonelli's. – Call, frequently uttered when feeding and between song phrases, a rather plain, unmelodic *chiff*, not that dissimilar from a begging young House Sparrow *Passer domesticus*. A more subdued but clearly related *isst* is sometimes heard from migrants, and is doubtfully separable from similar call of Western Bonelli's (which see).

SIMILAR SPECIES Main pitfall is obviously Western Bonelli's Warbler, which is treated under Identification, and under that species. Note also that Eastern Bonelli's has on average slightly longer and more pointed wings (and hence shorter-looking tail), but difference is very slight indeed. Also, it has generally paler yellow axillaries und underwing-coverts (although a few have been noted to be vividly yellow as Western), and slightly shorter bill and legs. In spring, a diffusely paler panel or wing patch is formed by worn and bleached tertials and many greater coverts, not seen as prominently in Western. – Other similar species and possible confusion risks are the same as for Western Bonelli's Warbler (which see).

AGEING & SEXING Small age-related plumage differences usually in 1stCY only. Sexes alike or very similar in plumage. – Moults. In general, moults are similar to Western Bonelli's, with the difference that a late winter moult of tertials and greater coverts, as is assumed for *bonelli*, appears to be lacking, judging from the bleached and rather worn state of these feathers when the birds return to Europe/Turkey. No spring bird found with mix of newer and older primaries, as sometimes seen in *bonelli*. – **AUTUMN Ad** Much-abraded tips

Israel, Mar: identified from other migrant leaf warblers by combination of bright yellowish-green rump and edges to wing- and tail-feathers, contrasting with dull grey-brown rest of upperparts. Usually also obvious are the pale (grey-white) tertials and greater coverts edges, and unbroken whitish eye-ring. Rather long and pointed wing, and therefore short-tailed appearance. Due to complete winter moult, ageing is impossible in spring. (H. Shirihai)

Israel, Mar: unlike Western, Eastern Bonelli's is generally drabber and purer brown-grey above. Due to different timing of pre-nuptial moult, Eastern returns in spring on average a more worn state, especially the tertials and some greater coverts being edged duller greyish-brown or off-white. The contrasting almost pure white underparts and face pattern (including unbroken whitish eye-ring) are also obvious in the field. (D. Occhiato)

Western Bonelli's Warbler (top left: Italy, Apr; bottom left: 1stW, Scotland, Oct) versus Eastern Bonelli's Warbler (top right: Israel, Mar; bottom right: 1stW, Kuwait, Aug) in spring (top) and autumn (bottom): this comparison demonstrates how similar these two are and, as described in the text, they are best separated at any time by call, or with practice by song. Only in spring/summer can some birds be identified using plumage. In particular, note how fresh spring Western (top left) is suffused more yellowish-green on brown-grey upperparts than Eastern (top right). Apparently due to late (protracted/extended) pre-nuptial moult, Western returns in spring with fresher, broader and brighter greenish edges to wing-feathers. There are no consistent differences to separate 1stW in autumn, unless they are heard to call. (Top left: D. Occhiato; top right: D. Jirovsky; bottom left: H. Harrop; bottom right: A. Al-Sirhan)

to primaries and tail-feathers. Tertials occasionally partly or completely moulted and fresh. Rump yellowish, showing good contrast with rest of olive grey-brown upperparts. **1stCY** Plumage fresh, including tips to primaries and tail-feathers. Rump patch variable in prominence, some having more subdued and dull yellowish olive-brown rump. Upperparts, especially back, sometimes tinged warm tawny-brown.

BIOMETRICS   L 11–12 cm. **W** ♂ 64.5–73 mm (n 46, m 68.3), ♀ 60.5–67 mm (n 41, m 64.2); **T** ♂ 45–55 mm (n 46, m 48.9), ♀ 43–50.5 mm (n 40, m 45.6); **T/W** m 71.6; **B** 10.6–13.5 mm (n 93, m 12.2); **B/W** 16.5–20.6 mm (n 93, m 18.4); **BD** 2.2–3.2 mm (n 95, m 2.6); **Ts** 16.5–19.0 mm (n 28, m 17.6); **Ts/W** 25.4–29.5 (n 28, m 26.8); **UTC** (see below) 9.5–18 mm (n 83, m 13.8). **Wing formula: p1** > pc 1–7 mm, < p2 28–35.5 mm (m 32.1, n 97, only 3% < 30 mm); **p2** < wt 2.5–6 mm, =5/6 (65%), =6 (20%), or =6/7 (15%); **pp3–4**(5) about equal and longest; **p5** < wt 0–2.5 mm; **p6** < wt 3–7 mm; **p7** < wt 5.5–11 mm; **p10** < wt 11–17 mm; **s1** < wt 12–18 mm. Emarg. pp3–5; p6 never prominently emarginated, but can be less prominently (4%) or only very faintly (19%); thus, in remaining 77% there is virtually no emarg. of p6. – For difficult birds, calculate 'multiple characters value', **MCV**, as follows (UTC = tail-tip > tip of longest undertail-coverts): add W and p1<p2, and subtract B, UTC, p1>pc, p2<wt. MCV is 57.9–72.4 (n 73, m 64.7). 98.6% had MCV >58.0. Birds in very abraded plumage, or which have lost longest undertail-coverts, cannot be reliably calculated.

GEOGRAPHICAL VARIATION & RANGE   Monotypic. – SE Europe from Croatia and Balkans east to W & S Turkey; winters Chad, Sudan and adjacent areas.

TAXONOMIC NOTE   See Western Bonelli's Warbler.

REFERENCES   GROENENDIJK, D. & LUIJENDIJK, J. C. (2011) *DB*, 33: 1–9. – HELBIG, A. J. *et al.* (1995) *J. Avian Biol.*, 26: 139–153. – OCCHIATO, D. (2007) *BW*, 20: 303–308. – PAGE, D. (1999) *BB*, 92: 524–531. – SVENSSON, L. (2002) *Birdwatch*, 119 (May): 26–30. – SVENSSON, L. (2002) *Vår Fågelv.*, 61(4): 7–14.

Kuwait, Aug: young (aged by fresh plumage) Eastern are identical in plumage to Western Bonelli's, and can only be identified within their core ranges and if their calls are heard. Note unbroken pale eye-ring and large dark eye on washed-out side of head. (M. Pope)

WOOD WARBLER
Phylloscopus sibilatrix (Bechstein, 1793)

Fr. – Pouillot siffleur; Ger. – Waldlaubsänger
Sp. – Mosquitero silbador; Swe. – Grönsångare

Among the leaf warblers, to the birdwatcher a group of mostly frustratingly similar species, at least the Wood Warbler stands out as a characteristic bird: large, brightly coloured with an easily identified voice. The Wood Warbler is a common breeder in much of C and N Europe, accepting a variety of wooded habitats—both deciduous and mixed—but perhaps preferring rather tall stands without tall undergrowth, a habitat affording space for its song flight among the tree trunks. It winters in tropical Africa and returns in April and early May.

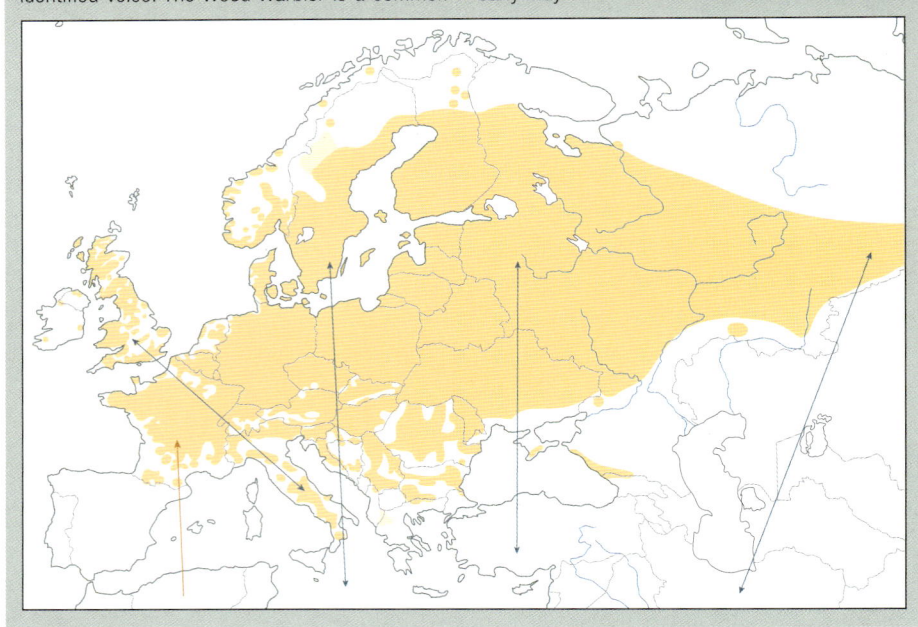

Finland, May: chin, throat and upper breast clear yellow contrasting against white rest of underparts, but this can be reduced in some individuals, which are rather strongly suffused yellowish below. Variation also includes birds with predominantly pale-looking bill. Ageing impossible in spring. (J. Tenovuo)

VOCALISATIONS Song loud, energetically repeated, especially during first days after arrival at breeding site, a 3–4 sec-long series of accelerating sharp, metallic notes, ending in a trill, *zip... zip... zip, zip, zip, zip, zip, zip zip zip-zip-zip-zip-zipzipzipzipzvirrrirrrr*, and likened to the sound of a coin spinning on a marble slab. Often performed in short horizontal song-flight, the concluding trill delivered upon landing, with shivering wings. Interposed between trilling strophes is a peculiar alternative song type, slightly recalling Willow Tit song, a series of soft, melancholy notes, initially intensified and accelerating slightly, *tüh tüh tüh-tüh-tüh-tüh*. – Call a sharp *zip*, like a note from the song. Also a low-pitched fluty *tüh*, like a note from the alternative song, especially in anxiety but also as contact call on migration.

IDENTIFICATION As large as a Willow Warbler, or even slightly larger, with proportionately longer, more pointed wings and therefore seemingly shorter tail. The wing shape results in the longest *primary projection* of all *Phylloscopus*, at least *equalling the tertial length*. Supercilium very broad, long and bright yellow, enhanced by *prominent dark grey-green eye-stripe*. Upperparts bright green, lacking any brown or grey tinges. Tertials and greater coverts very dark, edged yellowish-green or yellowish-white, *showing strong contrast*. Rest of flight-feathers, and tail-feathers, edged yellowish-green. Chin, *throat and upper breast clear yellow, rest of underparts white*. Rarely, birds with reduced green and yellow occur, yet structure and general pattern should still enable identification. Bill strong, dark with pale yellow-brown cutting edges. Feet dark grey-brown with more yellowish soles. Flicks wings and tail when feeding, much like Willow Warbler.

SIMILAR SPECIES Distinctive once long primary projection is seen in combination with broad, long, yellow supercilium and yellow throat contrasting against white belly. One possible confusion species is *Willow Warbler*, mainly due to size and sometimes quite prominent yellowish supercilium, but this has less sharp contrast between yellow throat/upper breast and white lower breast/belly, and has shorter primary projection and less contrast on tertials. – Could also be confused with *Western Bonelli's Warbler* because of similarly white belly and bright green wing pattern, but yellow throat should prevent this. Call is different as well. Very rarely, Wood Warblers with subdued yellow and green colours occur, but

Belgium, Apr: rather large, chunky leaf warbler. Wings characteristically long and pointed, with longest primary projection of all *Phylloscopus*, and thereby appears rather short-tailed. In spring both ages return in evenly fresh and very attractive plumage, with bright greenish upperparts, yellowish face and often pure white rest of underparts. (J. Fouarge)

England, May: some are slightly less saturated and more greyish-tinged, and the extremes (duller even than this) could invite confusion with several congeners. Compared to similar Willow Warbler and both Bonelli's warblers, note the long and pointed wings combined with the broad, long, pale supercilium, hint of slightly yellower throat and whiter belly, and (especially) better-marked pale edgings to tertials and greater coverts. (P. Hobson)

PHYLLOSCOPUS WARBLERS

these still have broad and prominent supercilium and long primary projection. Also note the broken eye-ring in Wood, unbroken in Western Bonelli's.

AGEING & SEXING Small age-related plumage differences usually in 1stCY only. Very slight seasonal variation. Sexes alike in plumage. — Moults. Partial post-nuptial moult in late summer (Jul–Aug) involving body-feathers and some wing-coverts, rarely also odd inner tertials and central tail-feathers. Partial post-juv moult (Jul–Aug) affects mainly body and wing-coverts. Complete moult (including wings and tail) of both ad and 1stY in winter (mainly Feb–Mar). — AUTUMN **Ad** Tips to primaries and tail-feathers abraded, with only narrow traces of yellow-green edges. Odd inner tertials and central tail-feathers occasionally moulted and fresh. **1stW** Plumage fresh, including tips to primaries and tail-feathers, feathers neatly edged yellowish-green. **Juv** Like 1stW, but upperparts slightly brown-tinged, less green.

BIOMETRICS **L** 11.5–13 cm. **W** (69) 70–82 mm (n 144, m 73.9; ♂ averages only very slightly larger than ♀); **T** 45–53 mm (n 31, m 48.8); **T/W** m 64.0; **B** 11.8–14.2 mm (n 30, m 12.8); **BD** 2.5–3.4 mm (n 28, m 3.0); **Ts** 16.8–18.8 mm (n 30, m 17.8). **Wing formula: p1** < pc 0–6 mm (rarely > 1.5), < p2 40–52 mm; **p2** < wt 1–4.5 mm, =4/5 (81%), =4 (18%) or =5 or 5/6 (1%); **p3** longest; **p4** < wt 0.5–2 mm; **p5** < wt 3–7 mm; **p6** < wt 7.5–13 mm; **p7** < wt 11–16.5 mm; **p10** < wt 16–24 mm; **s1** < wt 19–26 mm. Emarg. pp3–4, often also near tip on p5, but usually less well marked.

GEOGRAPHICAL VARIATION & RANGE Monotypic. — Europe except extreme north and south-west, east through S Russia and N Ukraine to W Altai and SC Siberia; found in SE Europe but does not enter Asia Minor; winters tropical W & C Africa, mainly between Liberia and Congo Basin.

Israel, Apr: some are slightly less saturated and more greyish-tinged, and the extremes (duller even than this) could invite confusion with several congeners. Compared to similar Willow Warbler and both Bonelli's warblers, note the long and pointed wings combined with the broad, long, pale supercilium, hint of slightly yellower throat and whiter belly, and especially the better-marked pale edgings to tertials and greater coverts. (H. Shirihai)

Ad, Austria, Aug: Wood Warblers (especially ad) are poorly known in autumn due to their early departure post-breeding. Worn ad is similar to fresh spring plumage, but primaries, tail-feathers and most greater coverts are now rather obviously worn with uneven brown-grey edges. Some undergo limited renewal of coverts or tertials, as here. (G. Loidolt)

1stW, Spain, Aug: fresh 1stW rather like spring ad and is separated from congeners by same key features, i.e. very long primary projection, yellow-throated appearance, well-developed bright yellow supercilium, amplified by dark eye-stripe, prominently edged yellowish-green or yellowish-white tertials and greater coverts, and rather strong bill with extensive pale yellow-brown areas. Mostly juv wing, including primaries, primary-coverts and tail, confirm age. (H. Harrop)

1stW, Shetland, Scotland, Aug: fresh wing with juv primary-coverts confirms age. Face pattern with yellow-washed throat and strong supercilium, long wings and wing pattern, especially the prominently edged greater coverts and tertials, and rather strong and extensively pale bill identify to species. (G. Petrie)

PLAIN LEAF WARBLER
Phylloscopus neglectus Hume, 1870

Fr. – Pouillot modeste; Ger. – Eichenlaubsänger
Sp. – Mosquitero sencillo; Swe. – Dvärgsångare

Although among the most unlikely stragglers to appear in Europe, one bird was caught at a bird observatory on the Baltic Sea near Stockholm, Sweden, on 10 October 1991. It breeds in Iran, S Turkmenistan, Afghanistan and adjacent areas in mountainous forests or scrub of oak, juniper, pistachio, etc. Commonly found above 2000 m, but may breed as low as 1000 m. It winters at low levels in Pakistan, around the Persian Gulf and along the Arabian Sea, so an autumn bird in Sweden is puzzling. The easiest way to see this species is to visit better-wooded hillsides in NW Oman and United Arab Emirates in November–early March.

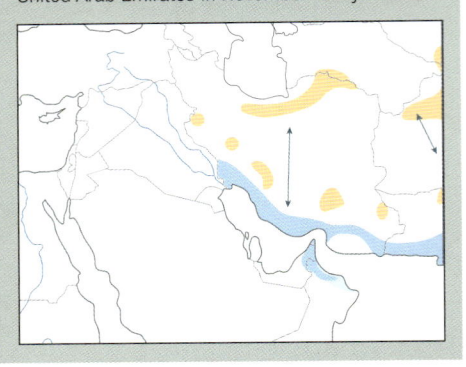

IDENTIFICATION A warm grey-brown and dusky whitish leaf warbler, which needs primarily to be separated from Siberian race *tristis* of Common Chiffchaff, and from Mountain Chiffchaff. *Smallest* member of the genus, approaching Goldcrest in size, but can all the same be difficult to tell from a *tristis* by size alone under field conditions. Although undoubtedly rather compact and at times appearing rather *neck-less*, with *short, rounded wings* and *fairly short tail*, it still often appears *slender and 'pointed'* in shape. Note that the *tail is usually square-cut* with rather sharp corners, thus often lacking the fine central indentation commonly seen on the folded tail of other *Phylloscopus* (although Mountain Chiffchaff has also a rather square tail). By and large has similar tail-movements to the chiffchaffs, with frequent downward dips of the folded tail, but also twitches the semiopened tail sideways. Flight a little quicker and more 'jumpy' than the more purposeful *tristis*, but differences are slight and can only serve as first step in the identification process. When seen close, or handled, Plain Leaf Warbler is not as similar to *tristis* as it looks at some distance: it is more *isabelline-tinged greyish on head, nape, upper mantle* and sides of neck, less warm buff-brown or rufous-tinged. Underparts pale off-white with sandy-buff tinge, strongest on breast and flanks, while undertail-coverts usually appear white. Wings have rather pale yellowish-brown (or even slightly ochre-olive) edges, *on secondaries often* forming *a hint of a pale panel*, which in strong light looks whitish in the field. Wing-bend (and axillaries) buffish, cream-coloured or pale yellowish (never bright yellow). The usually narrow but variably prominent and *long supercilia often meet on forehead*, are *dull off-white* or pale pinkish- or yellowish-buff, whitest at front, often a little blurred and not particularly striking but can be more distinct, at least in front of eye. Compared to Siberian Common Chiffchaff the dark *eye-stripe is usually narrower* and often reaches a little further behind eye. *Fine bill and thin legs are entirely black* (generally including soles), although when seen close the bill can be seen to have narrow paler cutting edges. Rump is rather brighter than rest of upperparts, tinged ochre-brown (not tinged greenish as in reasonably well-kept *tristis*). Very faint traces of yellow can sometimes be seen on centre of breast and undertail-coverts. Some have a slight yellowish-green tinge basally on secondary edges in fresh plumage, but in worn plumage the wing is all brown.

VOCALISATIONS Song differs completely from all other chiffchaffs and relatives in the genus *Phylloscopus*, a brief, quick outburst of whistling and sharp notes, sometimes repeated every c. 3 sec. The song has an opening staccato rhythm vaguely recalling Short-toed Treecreeper, and the high-pitched, squeaky voice recalls—most of all—the ecstatic song variant of Goldcrest, *zit zyt zitteri-seeweeoo*. – Call a rather low-pitched *chep* or *chrep*, perhaps slightly reminiscent of Radde's Warbler (and very different from any chiffchaff). A variation of the call can sound 'bouncing' or almost disyllabic, *chep-pe*. Alarm or irritation expressed with churring *cherrr*.

SIMILAR SPECIES Despite small size, can be tricky to tell from migrant Siberian race *tristis* of *Common Chiffchaff*, which is similarly grey-brown above with blackish legs. Plain Leaf is a smaller, more compact and often more greyish-looking bird with a more subdued supercilium (though variable). Main separating criteria from Siberian Common Chiffchaff are detailed under Identification. – *Mountain Chiffchaff* is another brown-and-white *Phylloscopus* that must be considered. In general, Mountain Chiffchaff is larger and longer-tailed, is darker above, has a stronger bill and legs with longer claws (as first noted by K. Mild, pers. comm.), and has darker tawny-brown crown, appearing particularly dark laterally, making broad whitish forepart of supercilium stand out better (Plain Leaf: greyish-brown crown and dull off-white or yellowish-buff, often slightly more diffuse

Iran, Apr: note characteristic small size, rather long dark legs, short square tail and dark eye-stripe enhancing warm-tinged prominent supercilium. Especially in worn spring plumage on breeding grounds, and in strong light, can look overall quite pale greyish. This bird is collecting nest material. (Magnus Ullman)

Kuwait, Apr, 1stS: frequently as here looks characteristically neck-less, giving a front-heavy and large-headed appearance. Birds in April with heavily worn primaries (as here) and with moult limit in tertials, should be 1stS. (A. Audevard)

United Arab Emirates, Oct: short wings and short, broad and square-cut tail. The fine bill and thin legs can appear all black. Unlike several chiffchaffs, Plain Leaf is typically greyish-isabelline above, with warmer brown cheeks, neck-sides and flanks, almost with a pinkish tinge, and rest of underparts are creamy. Wings have rather pale buff-brown edges (unlike many chiffchaffs, which tend to have green-tinged wing-feathers) often forming hint of a pale panel on secondaries. Note also that supercilia often are broader in forepart. (H. Roberts)

United Arab Emirates, Oct: fluffy plumage with isabelline-tinged greyish-brown head and upperparts, and cream-buff face and underparts. Supercilium rather variable in strength and shape: here strong enough but does not extend to forehead, unlike the bird in the images below. (T. Pedersen)

United Arab Emirates, Oct: note virtual absence of lemon yellow in armpits unlike other taxa within the Chiffchaff complex. General shape of this small warbler is compact with seemingly large head, short, compact neck and rounded wings. (T. Pedersen)

United Arab Emirates, Nov: two images of the same bird, highly distinctive due to pinkish-buff tinge to pale facial parts, breast and flanks. Note also diagnostic thin blackish legs, short wing (with pp3–6 forming wingtip) and short tail (square-ended). (Magnus Ullman)

supercilia). – Extralimital Central Asian *Kashmir Chiffchaff* (*P. sindianus*; see Mountain Chiffchaff) is more similar to Plain Leaf Warbler in its paler sandy grey-brown plumage, but again differs in being slightly larger and longer-tailed, having proportionately longer claws, more prominent rictal bristles, and by subtle plumage differences: very slightly more greyish sand-brown above (Plain Leaf: warmer brown, almost with a pinkish tinge), and more buff-brown on breast, less so on lower flanks (Plain Leaf: usually the opposite pattern, but some individual variation).

AGEING & SEXING Ages and sexes alike in plumage and very similar in size. No appreciable seasonal variation. – Moults. Complete post-nuptial moult of ad in late summer–early autumn (late Jul–early Sep). Partial post-juv moult in summer does not involve flight-feathers or primary-coverts, but may include a few central tail-feathers. Partial pre-nuptial moult (Mar?) of body-feathers, some wing-coverts, tertials and odd central tail-feathers.

BIOMETRICS **L** 9–10.5 cm. **W** 47.5–55 mm (*n* 39, *m* 51.4); **T** 33–42 mm (*n* 26, *m* 38.3); **T/W** *m* 74.3; **B** 9.7–11.0 mm (*n* 25, *m* 10.3); **Ts** 16.2–18.3 mm (*n* 38, *m* 17.2); **HC** 3.5–4.6 mm (*n* 20, *m* 4.2); **MC** 2.5–3.5 mm (*n* 11, *m* 3.1). **Wing formula: p1** > pc 6.5–11 mm, < p2 15.5–19.5 mm; **p2** < wt 4–8 mm, =8/9 or 9 (44%), =9/10 or 10 (38%), ≤10/ss (9%) or =7/8 or 8 (9%); **pp3–6** about equal and longest, though p3 and p6 often to 1 mm (rarely 2 mm) <; **p7** < wt 0.5–4 mm; **p8** < wt 2–6 mm; **p10** < wt 4.5–8.5 mm; **s1** < wt 6–9.5 mm. Emarg. pp3–6 (exceptionally slightly also on p7).

GEOGRAPHICAL VARIATION & RANGE Monotypic. – N, SW & S Iran, S Turkmenistan, Afghanistan, NW Pakistan, local eastern Iraq; winters S Iran, around Persian Gulf including in United Arab Emirates and NW Oman. (Apart from the Swedish record mentioned above there are records claimed from Turkey and Lebanon still to be assessed by national records committees.)

REFERENCES 'BISTER' (2004) *Roadrunner*, 12(3): 28–29. – SVENSSON, L. (2001) *Vår Fågelv.*, 60(4): 19.

Oman, Dec: note distinctly broad supercilium in front of eye that almost appears to run across forehead, but which is somewhat thinner behind eye. Bill dark and more or less thin. When fresh, some birds can appear more olive-tinged and have a hint of warm buff to pale facial parts, flanks and wing edges. (M. Tallroth)

COMMON CHIFFCHAFF
Phylloscopus collybita (Vieillot, 1817)

Fr. – Pouillot véloce; Ger. – Zilpzalp
Sp. – Mosquitero común; Swe. – Gransångare

Widespread and common small warbler over most of Europe, found in a variety of wooded habitats. Although hardly well known outside the ranks of birdwatchers, this is one species that the layman has a chance of learning to pick up on its characteristic song, a sound on which its English name is based, *chiff chaff chiff chaff*..., etc., well-spaced resounding notes in a long series. Birds in SW and W Europe are more or less resident or short-range migrants, but northern and eastern birds are more migratory, wintering in S Europe, N Africa and Middle East. Common Chiffchaff returns early, mainly in February and March (into April in N Europe). Siberian birds sing differently and winter mainly in India, but in small numbers also west to Middle East, and straggle on a regular basis to W Europe.

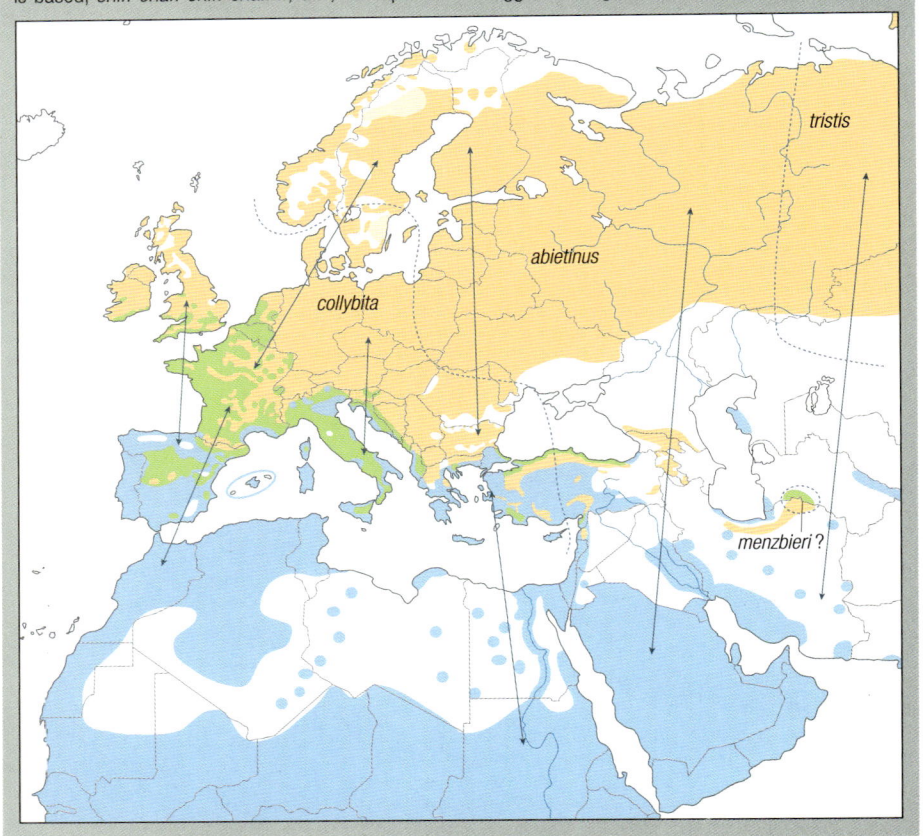

P. c. collybita, **Italy, Apr**: by the time this widespread *Phylloscopus* is settled on its breeding territories in spring, racial identity can be automatically inferred by range, although the plumage may already by then be worn and bleached to prevent plumage confirmation. Rather worn primary-coverts and primaries in May could indicate a 1stS, but without closer inspection ageing is not recommended. (D. Occhiato)

IDENTIFICATION Medium-sized European leaf warbler lacking an obvious wing-bar, thus needs to be separated primarily from Willow Warbler, from which it differs by being slightly more *compact with more rounded head* (and shorter neck), having *shorter, more rounded wings*, and consequently can appear proportionately *longer-tailed*. Primary projection is fairly short, often about 1/2 to 2/3 of tertial length. Important to note *dark legs*, blackish or dark brown-grey, which distinguish 75% of all Common Chiffchaffs from Willow Warbler, but some, especially older birds and breeders in S and C Europe, frequently have 'dark medium-brown' legs, very similar to many Willow Warblers. The *fine bill*, too, is on average quite dark, but some have paler cutting edges and a hint of a paler base to the lower mandible. Compared to Willow Warbler, *pale supercilium and dark eye-stripe are less distinct* giving face a less clear expression, and overall *plumage is on average slightly duskier*, less clean greenish, yellow and white, more *tinged brown-grey and buff*, especially on head, neck and breast. However, bleaching and wear in summer will make colours less brownish, more pure greenish and white. Note also characteristic habit of Common Chiffchaff to *'dip' its folded tail downwards repeatedly* when feeding in the canopy of a tree or bush. (Wings can sometimes be flicked too, but less often and less vigorously, unlike in Willow Warbler where such movements are more obvious.) Call is useful for identification once learnt. – Just as in Willow Warbler there is considerable geographical variation, northern and eastern birds being successively more brown-grey, less greenish above, and buff-white below without yellow apart from on wing-bend and axillaries. Brightest part of upperparts on such more brown-grey birds is often rump, which can strike you as yellowish-green (cf. the two Bonelli's Warblers). Siberian race *tristis* in its most typical form differs in being grey-brown above on crown and mantle (lacking green on these parts) and whitish below with a rusty-buff tinge on supercilium, sides of head and slightly on breast (lacking any yellow except under wing). Edges to flight-feathers and tail-feathers dull greenish in fresh plumage (a bird with very bright yellowish-green edges to wing-feathers in Western Bonelli's Warbler fashion is not a *tristis* irrespective of whether it has a 'straight' call!). In autumn there is often a hint of a paler greater coverts wing-bar due to paler brown tips to these feathers. Some easterly Common Chiffchaffs, possibly of this race or intergrades with race *abietinus* (see below), appear to be more grey and white, lacking the rusty-buff or warm brown elements.

VOCALISATIONS Song in most of range a simple series of single (rarely disyllabic) loud and resounding notes at two (or three) pitches, delivered alternatively (or irregularly) at a moderate (trotting) pace, *chiff chaff chaff chiff chaff chiff chiff chaff*... Songs often last 4–8 sec. Birds newly arrived at selected territory frequently insert subdued stammering notes between strophes, *perre-perre*. Song in extreme north-east (*tristis*; NE Russia, Siberia) differs clearly; see below under Geographical variation & range. – Common call in Europe is a soft whistle, more or less monosyllabic but slightly upwards-inflected ('diphthong') and with stress on final part, *hweet*. When alarmed near nest will use this call repeatedly, sometimes with a slightly harsher voice. There is also a different call that is often transcribed as *sweeoo*, a call which subtly rises, then falls. The function of this call is not well understood; it is heard both in spring and autumn with varying frequency both locally and between years. There is a certain amount of variability also in its details, and some straighter variants may recall common call of subspecies *tristis* (see below). Young (only recently fledged?) often beg with a straight and more piping call, *füt* or *fiet* (again inviting confusion with the call of *tristis*). Call in north-east (*tristis* in Siberia and S Ural area), in Caucasus and Transcaucasia (apparently *abietinus*) and Transcaspia ('*menzbieri*' in N Iran/S Turkmenistan, and local *collybita* in W Turkey/N Levant) differs in being straighter and more plaintive, shrill and piping, slightly recalling Dunnock and has been described as recalling a 'chicken in distress', *iihp* or *peet*. (There may be

W European Chiffchaffs (presumed *P. c. collybita*) in spring, with relatively dull (left: Netherlands, Apr) and rather bright individuals (right: Italy, Mar): just before breeding when plumage is worn, extensive individual variation can render subspecific identification in the field difficult, in particular since the more northern race *abietinus* could still be on passage through areas occupied by breeding *collybita*. These images are meant to reflect the spectrum of variation within *collybita*, which varies from duller olive-grey to brighter greenish and brown above. Duller bird on left is almost identical (even indistinguishable) from *abietinus*. Thus, racial separation is unsure without biometrics and comparison with long series. Same applies to ageing: despite the rather less worn wings suggesting both are ad, ageing without closer inspection of moult and feather wear is inadvisable. (Left: H. Gebuis; right: L. Sebastiani)

subtle but average differences between normal call of *tristis* and latter-mentioned populations though.) A very slight hint of fall in pitch at the end of the *tristis* call is often noted, which accounts for the plaintive tone. – Some birds in late summer and autumn in N Europe, at least in some years, many of which look most like *abietinus*, have been noted to give a *tristis*-like straight call, or a similar call rather like the *füt* call of young described above. It is not known whether these birds are late-hatched young, represent a migrant population of unknown origin, or if an alternative call is about to develop within parts of the ranges of *abietinus* or *collybita*. A cautious approach when using straight calls in autumn for racial identification of Common Chiffchaff is advisable, not only in Turkey and Middle East but also in Europe. Always try to secure a plumage description, just the call is not enough!

SIMILAR SPECIES Some general remarks regarding separation from *Willow Warbler* are given under Identification and under Willow Warbler. Note, apart from the darker legs, different call, unique 'tail-dipping' behaviour, shorter primary projection and the often slightly smaller and more compact look of Common Chiffchaff, its generally better-defined pale eye-ring (partly due to more uniformly dark cheeks and ear-coverts), and on average slightly duskier and less clean-looking plumage (more brown-grey tinges). Common Chiffchaff also usually has thinner and darker bill (but some overlap). Less often adult Common Chiffchaffs of both *collybita* and *abietinus* (but not *tristis*) have paler legs, brown rather than blackish, in practice very like Willow Warbler and inviting confusion, but applying a combination of the other above-mentioned criteria will help, as will any calls. – Many fresh autumn birds (perhaps more frequently among northerly and easterly populations) show a hint of a paler wing-bar, which can resemble *Greenish Warbler*, but any wing-bar on a Common Chiffchaff is more subdued and diffuse, being pale (olive-)brown, never whitish. – In SW France, Iberia and NW Africa needs to be separated from *Iberian Chiffchaff*, but this is more similar to Willow Warbler in plumage (similarly greenish, yellowish and white with no, or practically no, brown tinges). Still, it must be remembered that bleached breeding Common Chiffchaffs in summer can become extremely similar to Iberian, losing any brown or buff tinges in plumage through bleaching (but keeping the yellow and green); and conversely, fresh autumn Iberian can have a little brown on sides of head and breast. Leg colour can be the same in these two, and the slightly stronger and more pale-based bill of Iberian Chiffchaff as compared to Common Chiffchaff is only an average difference. Luckily, Iberian has a downwards-inflected call, *feeu*, quite distinct from the upwards-inflected *hweet* of Common Chiffchaff. – Odd Common Chiffchaffs have a slightly contrasting, somewhat brighter yellowish-green rump, and quite bright yellowish-

Presumed *P. c. collybita*, possibly 1stW, Germany, Oct: the supercilium and the broken white eye-ring are the most obvious features in the otherwise rather subdued face pattern. Also typical are dark legs, and fine, mostly blackish bill. Habitually 'dips' folded tail when feeding (wings sometimes flicked too). Edges to flight- and tail-feathers dull greenish when fresh, whereas upperparts are rather uniform brownish-olive. Retained juv primary-coverts and tail-feathers suggest 1stW. (M. Schäf)

Presumed *P. c. collybita*, 1stW, France, Sep: some are rather yellower in fresh plumage. This bird could even be confused with a young Willow Warbler, but differs immediately by its blackish legs, more diffuse facial pattern, in being overall slightly duskier with more brown-grey, olive and buff pigments (not greenish, yellow and white); also has more compact jizz, with more rounded head and shorter neck. Note shorter primary projection (only c. 60% of tertial length) and consequently longer-tailed appearance. Aged by retained juv primary-coverts and outer greater coverts. (A. Audevard)

green edges to wing-feathers, which might invite confusion with either of the two *Bonelli's Warblers*. Note that the two Bonelli's have unbroken pale eye-ring, contrasting dark and pale-edged tertials, rump almost ochre-tinged bright yellowish-green, bright yellowish-green edges to tail- and wing-feathers, a blander face pattern and whiter underparts. — For differences from *Mountain Chiffchaff*, see that species.

AGEING & SEXING (*collybita*) Small age-related differences in 1stY, but ageing in spring often difficult. Sexes alike in plumage. — Moults. Complete post-nuptial moult in summer (mainly late Jun–Aug). Partial post-juv moult in summer does not involve flight-feathers or primary-coverts, but a few or all tertials are frequently renewed. At least some inner greater coverts moulted, sometimes all. Partial pre-nuptial moult in winter (both age classes) limited and variable in extent, some apparently having no moult, others change body-feathers, some wing-coverts and odd tertials and central tail-feathers (Gargallo & Clarabuch 1995; possibly some *abietinus* in their material). — **SPRING** Ageing frequently difficult and sometimes impossible. The text here refers to the most typical birds. **Ad** Rather evenly abraded plumage with no or only limited apparent moult differences, tips of primaries and tail-feathers somewhat worn, feathers dark grey and often a little glossy. Tips of primary-coverts rather broad

P. c. collybita, juv, Italy, Jun: soft fluffy juv plumage, but overall fresher and often more vividly-coloured than accompanying worn and bleached parents. This is at least true for most juv *collybita*, although the depicted bird here is rather less colourful and could invite confusion with *abietinus* were it not for location. (D. Occhiato)

P. c. abietinus, Finland, Apr: this northern taxon is very similar or indistinguishable from *collybita*, but averages larger, less compact with proportionately slightly shorter bill. It is generally duller, more greyish-green above (less dark olive-brown) and on average slightly less extensively yellow on face and below, with whiter belly. Ageing without handling difficult, but rather less worn primary-coverts and tips of primaries in Apr could suggest an ad. (S. Leveelahti)

P. c. abietinus, Estonia, May: with wear and bleaching on breeding grounds, birds become duller and greyer, approaching *tristis*. To this comes a natural colour variation, some easterly breeders habitually being duller and more grey-brown than others. Ssp. *abietinus* often strikes you as having a more elongated (less compact) jizz, with longer primary projection; also typical is slightly more clear greenish on uppertail-coverts/rump and wing edges. (M. Varesvuo)

and neatly edged greenish. **1stS** Tips of primaries and tail-feathers often much abraded, feathers rather brownish and with reduced gloss; often one or a few central tail-feathers replaced in winter, contrastingly more dark and fresh. Tips of primary-coverts on average slightly more pointed with less neat greenish edges than in ad. Sometimes a slight contrast between outer unmoulted greater coverts, being slightly duller and more greyish, and inner, edged more greenish. — **AUTUMN** Ageing frequently difficult, differences slight. No or only very limited difference in amount of yellow on underparts (unlike in *P. trochilus*). **Ad** Freshly moulted, tips to tail-feathers neat and rounded, centres dark and glossy, finely fringed or tipped whitish. All greater coverts edged greenish. Primary-coverts like in spring. **1stCY** Some very similar to ad. Typical birds told on first sign of abrasion to tips of primaries and tail-feathers, latter often subtly frayed at tips, and on average show less gloss or contrasting edges than ad. Sometimes a faint contrast between inner 1–3 moulted greater coverts edged more greenish, and outer unmoulted being a little duller. Primary-coverts like in spring. A few have moulted one or two central tail-feathers, these being darker grey and glossier than rest, with more rounded tips.

P. c. abietinus, Finland, Oct: variation in autumn can be confusing, when some very fresh, *abietinus* tends to be less greenish-grey but duller olive grey-brown above, while yellow can be restricted to traces on supercilium, eyelids and sides. Ageing impossible. (M. Varesvuo)

P. c. collybita / abietinus, Denmark, Oct: due to substantial overlap between *collybita* and *abietinus*, race of this autumn migrant unsure. A bird with such well-marked head pattern can be confused with Willow Warbler, but in addition to dark legs, short primary projection, thin and dark bill, note on average slightly duskier and less clean-looking plumage. Ageing impossible. (K. Malling Olsen)

P. c. abietinus / tristis, 1stW, Germany, Oct: this bird closely approaches *P. c. tristis* (though note only hint of characteristic rusty-brown hue of *tristis* on supercilium, cheeks and sides of neck, and fore supercilium has a touch of faint yellow) or is midway between *tristis* and *abietinus*, resembling so-called '*fulvescens*'. Aged by moult limits, with newly moulted (longer) inner greater coverts and retained juv primary-coverts. (T. Krüger)

BIOMETRICS (*collybita*) **L** 10.5–11.5 cm. **W** ♂ 56–63 mm (*n* 114, *m* 60.0), ♀ 52.5–60 mm (*n* 43, *m* 56.0); **T** ♂ 42.5–52.5 mm (*n* 114, *m* 47.9), ♀ 41–51 mm (*n* 43, *m* 44.4); **T/W** *m* 79.6; **B** 10.4–13.1 mm (*n* 156, *m* 11.7); **B**(f) 7.5–9.8 mm (*n* 26, *m* 8.3); **BD** 2.0–2.7 mm (*n* 133, *m* 2.4); **Ts** ♂ 17.9–21.2 mm (*n* 105, *m* 19.6), ♀ 17.4–20.0 mm (*n* 42, *m* 18.6); **HC** 4.3–6.2 mm (*n* 28, *m* 5.3). **Wing formula: p1** > pc 3.5–10 mm, < p2 20–26.5 mm in ♂♂, 19–25 in ♀♀; **p2** < wt 4.5–8.5 mm, =7 or 7/8 (58%), =8 or 8/9 (31%), or =6/7 (11%); **pp3–5** about equal and longest; **p6** < wt 0.5–5.5 mm; **p7** < wt 3–8 mm; **p8** < wt 4.5–9.5 mm; **p10** < wt 7.5–12 mm; **s1** < wt 8.5–13 mm. Emarg. pp3–6 (invariably a prominent and rather deep emargination on p6). – See below under ssp. *abietinus* regarding a biometric formula ('MCV') which may support the identification.

GEOGRAPHICAL VARIATION & RANGE Mostly clinal and slight variation, although differences in vocalisations seem to divide the taxa in two groups. Taxonomy for entire range still not fully resolved. Birds tend to be more grey-brown, less green and yellow, in a cline from west to east, and this may involve an incipient species in the east (*tristis*), with a rather distinct song. Vocal variation as to calls does not coincide well with morphological variation, adding to the taxonomic complexity.

P. c. collybita (Vieillot, 1817) (W, C & S Europe, from S Sweden through SW Poland, W Romania and Bulgaria, Thrace; apparently also Asian W Turkey; winters S Europe, N Africa, Middle East). Treated above. On average a little more saturated olive-brown or darkly greenish-tinged, less greyish-green, above, and has slightly more yellow below and on supercilium than *abietinus* (but very subtle with wide overlap!). Also, there is on average a bit more brown or ochre-buff tinge in the plumage when fresh, especially on head, throat, breast and flanks. Worn birds have on average slightly darker crown than usually seen on comparable *abietinus*. Leg colour on average slightly paler medium brown than in *abietinus* or *tristis*, both of which generally have quite dark legs, especially in 1stY. For separation of handled birds from *abietinus*, see latter. – A distributional gap still exists in SC Sweden between *collybita* and *abietinus*, whereas the two intergrade SE of the Baltic. – Breeders in W Turkey described as *brevirostris* Strickland, 1837, but holotype (Cambridge; examined), a Nov bird from Izmir, is inseparable from either '*fulvescens*' (see below under *tristis*) or easterly brownish *abietinus*, whereas true breeders in W Turkey (many handled) do not differ much or at all from *collybita* (contra Watson 1962; Copete & Svensson in prep.). Admittedly, W Turkish birds are somewhat variable, some being more brownish than typical *collybita*, p2 averages *c.* 1 mm shorter (rarely equalling length of p9 or p10), the mtDNA differs by 1.5% (Helbig *et al.* 1996), and the call seems to be invariably straight, not upwards-inflected, but morphological variation also occurs within undisputed range of *collybita*, and inclusion of Turkish '*brevirostris*' is therefore reasonable. Since breeders in E Turkey seem to be *abietinus*, a smooth transient between these two is to be expected in C Turkey. (Syn. *brevirostris*.)

○ *P. c. abietinus* (Nilsson, 1819) (Fenno-Scandia, N Sweden south approximately to River Dalälven; NE Poland, Belarus, Baltic States, Russia except extreme east; Ukraine, E Turkey, Syria, Caucasus, Transcaucasia, where it is generally found below *c.* 1400 m, rarely up to 1800 m, exceptionally even higher; winters S Europe, N Africa, Middle East, N Arabia, Persian Gulf, SW Iran; on migration, appears to be less common in Britain, Netherlands, Belgium, W France, Iberia and Morocco). Differs only very slightly and on average from *collybita* by being a trifle larger, less compact, by having a proportionately slightly shorter bill, and on usually being slightly paler greyish-green above, less dark olive-brown (though some—or even many—are inseparable on colours). Underparts, too, are very similar in the two races, but *abietinus* is on average slightly less extensively yellow, centre of belly being nearly white, while undertail-coverts in both are

Presumed *P. c. abietinus*, presumed 1stW, Germany, Oct: especially in autumn some Chiffchaffs have paler tarsi being greyish yellow-brown, approaching dark-legged Willow Warblers. However, in all other respects this bird is a typical Chiffchaff: short primary projection and tip of p2 level with that of p8, clear emargination of p6, dusky face, indistinct supercilium, clear whitish eye-ring and rather fine, short bill. Overall dull greyish-olive coloration best fits *abietinus*, but lack of brighter/greener wing edges, as well as paler legs are better match for average *collybita*. Seemingly quite pointed tips to tail-feathers suggest a 1stW. (B. van den Boogaard)

Presumed *P. c. menzbieri*, Iran, May: racial designation by the N Iranian locality and month (the least known Chiffchaff population), but seemingly very short primary projection is supportive. Reduced green or olive above, and only limited trace of yellow below characteristic of this form. Ageing is unsure without handling for moult inspection. (C. N. G. Bocos)

invariably tinged yellow. Legs on average slightly darker than in *collybita* (still, some ad similarly medium brown, especially in E Europe). Within *abietinus* there is a trend for birds to become greyer, browner and paler, less green and yellow, towards north and east. Intergradation (or hybridisation, depending on taxonomic view) with '*fulvescens*' (westerly *tristis*) in E Russia and Ural area has been implied and seems to be at hand but is still inadequately studied. **L** 11–12.5 cm; **W** ♂ 59–68 mm (n 88, m 63.7), ♀ 56–64 mm (n 37, m 59.2); **T** ♂ 44.5–55 mm (n 88, m 50.4), ♀ 44–52 mm (n 36, m 46.7); **T/W** m 79.3; **B** 10.3–12.6 mm (n 123, m 11.4); **B**(f) 7.0–9.3 mm (n 28, m 8.0); **Ts** 17.5–20.6 mm (n 125, m 19.2). **Wing formula: p1** > pc 3.5–9 mm, < p2 22.5–29.5 mm; **p2** <wt 5–9 mm, =7 or 7/8 (61%), =6/7 (17%), or =8 or 8/9 (12%); **p6** < wt 1–4 mm; **p7** < wt 4–8 mm; **p8** < wt 6–10.5 mm; **p10** < wt 9–14 mm; **s1** < wt 11–15.5 mm. – For handled birds in N & W Europe sometimes useful to calculate a 'multiple character value' (MCV): (p10 < wt), + (s1 < wt), + (p1 < p2), – Ts, – (B/W ×100). If MCV is **9.0** or more, the bird is most likely *abietinus* (85% likelihood), if less (including negative values) it is probably *collybita* (92%). Overlap: 7.3–10.7. Only 5% of all *abietinus* fell below 8.0, all but one being ♀♀, whereas a mere 3% of

P. c. tristis, Kazakhstan, May: typically a dull brownish Chiffchaff with some rusty-buff tones on supercilium and cheeks, and invading cream-coloured throat, upper breast and flanks. In addition to song and calls, differs from ssp. *abietinus* by much-reduced green and yellow tones, with classic birds lacking any yellow other than on wing-bend. Ageing without handling impossible. (C. Bradshaw)

P. c. tristis, France, Apr: for reliable field identification of *tristis* away from breeding range, it is often helpful to note the voice (this bird's song was recorded). Typically, note overall grey-brown upperparts (lacking any green hue on crown and mantle), whitish-cream underparts with rusty-buff tinge to supercilium, cheeks and sides of neck and breast and clear whitish lower eyelid. The lack of visible green tinge to edges of remiges is presumably an artefact due to the light. As both age groups undertake partial winter moult, the moult limits in greater coverts do not age this bird. (A. Audevard)

collybita exceeded 10.0 (all were ♂♂). Birds with W 64 mm or more, s1 < wt 13.5 mm or more, and p1<p2 27 mm or more are probably invariably *abietinus* irrespective of MCV value. – Breeders in SE Turkey and Syria (latter fairly recently discovered) are closest to this race and best included in it. – Birds in Caucasus and Transcaucasia have been described as a separate race, *caucasicus* (Loskot 1991), on account of claimed darker upperparts and paler, more whitish underparts. They also have a straight rather than upwards-inflected call. Regrettably, the type material, mainly from Caucasus, in St. Petersburg has been held unavailable for an independent evaluation for over 15 years. Breeders in Transcaucasia have been studied by us, though, and these do not differ significantly from *abietinus*, contra Loskot (Copete & Svensson in prep.). The only subtle appreciable difference seems to be a fraction smaller size and shorter wing on average, hardly enough for separation. (Syn. *caucasicus*.)

P. c. tristis Blyth, 1843 'SIBERIAN CHIFFCHAFF' (easternmost Russia, Siberia, N Kazakhstan, N Mongolia; winters S & SE Asia, predominantly India, also S Central Asia, Middle East, Turkey and local S Europe). Siberian Chiffchaff differs from *abietinus*, with which it comes into contact in E Russia, by rather distinctly different song and call, and on much-reduced green and yellow tinges in a considerably more

P. c. tristis (left: Denmark, Oct) versus *tristis*-like presumed *P. c. abietinus* (right: England, Oct): the bird on left represents 'classic' *tristis* (especially note brown and rusty-buff hues), while the greyer and whiter bird with brighter green wing edges is a very likely *abietinus* but which could invite confusion with Eastern Bonelli's Warbler. Recent information suggests that the right-hand bird might emanate from eastern range of *abietinus* (often considered at least partly as intergrades between *tristis* and *abietinus*). There is individual variation in both categories and separation can be very challenging and requires detailed observation, and will benefit from hearing calls. (Left: P. Schans Christensen; right: C. Turner)

P. c. tristis, Italy, Nov: a fairly pale and brightly coloured *tristis*, still typical with brown and rusty-buff hues to head-sides, and any hint of yellow on fore supercilium or upper eyelid insignificant. The rather bright greenish edges to flight-feathers are normal for fresh *tristis*, although can be said to here reach its extreme. Ageing without handling uncertain. (D. Occhiato)

grey-brown and rusty-buff plumage. Song compared to other races is quicker and more complex, with more jilting rhythm and disyllabic or trisyllabic figures joined together in a fluent strophe that fluctuates up and down in pitch. A typical song could be rendered *che-chivy-chu-chovee-chuvo-che-chovee-chivy-chooee-chiv*. As with the European subspecies, agitated ♂♂ often insert a more subdued disyllabic note between strophes, frequently 'sweeter' than European birds and with different pitch to the two syllables, *chew-it*. Call straight and plaintive, a shrill and piping *iihp* (or *peet*). 'Classic' *tristis* breeds from the Yenisei eastwards and lacks any yellow on supercilium, eye-ring, sides of head, throat, breast or undertail-coverts, which instead are pale buff-brown, or in fresh plumage brown with rusty tinge; yellow is confined to underwing (usually pale but bright lemon) and rarely a little on 'thighs' (latter never bright lemon). Frequently has diffuse and faint dark streaking on lower throat and upper breast. As to upperparts, faint olive-green hues are restricted to lower mantle, back, rump and scapulars, and to edges of wing-feathers and tail-feathers. (Note that worn summer ad may show very little or no olive in plumage, being brown-grey above and off-white below.) Western *tristis* (often referred to as '*fulvescens*', from Yenisei west to Ural and Pechora) can look the same, but many differ by slightly brighter green edges to wings and tail, and in that several birds (but not all) display traces of yellow on one or more of fore supercilium, eye-ring, throat and breast, and may have faint olive tinges on crown and mantle. Among birds closest to *abietinus* (sometimes named '*riphaeus*'), some may be hybrids between *tristis* and *abietinus*. Rarely, yellow is missing from underwing (var. '*axillaris*'). For reliable identification of *tristis* away from breeding range important to note call (basically straight, plaintive ring) and a complete record of plumage details. In the hand, note presence of lemon-yellow underwing (variably strong), shorter hind claw than in Mountain Chiffchaff, but longer than in Plain Leaf Warbler. Legs and bill very dark, on average darker than in *abietinus*. On average slightly blunter wing than in *abietinus*, and any with p2 =6/7 is probably not a genuine *tristis*. **W** ♂ 55–68 mm (*n* 44, *m* 62.6), ♀ 54–64 mm (*n* 34, *m* 57.2); **T** ♂ 45–55 mm (*n* 44, *m* 49.8), ♀ 42–51 mm (*n* 34, *m* 45.4); **T/W** *m* 79.7; **B** 10.2–12.1 mm (*n* 87, *m* 11.1); **Ts** ♂ 18.6–21.5 mm (*n* 41, *m* 19.5), ♀ 17.6–20.3 mm (*n* 32, *m* 18.6); **HC** 4.9–6.4 mm (*n* 48, *m* 5.6). **Wing formula: p1** > pc 3–8 mm, < p2 20–30.5 mm (♀♀ ≤26); **p2** < wt 5–8 mm (rarely 4.5–9 mm), =8 or 8/9 (56%), =7 or 7/8 (34%), ≤ 9 (10%); **p6** < wt 0.5–3.5 mm; **p7** < wt 2.5–6.5 mm; **p8** < wt 5–9 mm; **p10** < wt 7.5–12 mm; **s1** < wt 9–14 mm (once 8 mm). (Syn. *axillaris*; *fulvescens*; '*riphaeus*'.)

? *P. c. menzbieri* Shestoperov, 1937 (Kopet Dag in N Iran, S Turkmenistan; possibly also Elburz Mts; presumably mainly short-range movements in winter, but some might reach Persian Gulf). Resembles '*fulvescens*' in lacking all or most yellow tinges below away from underwing, having rather whitish belly and white undertail-coverts, and reduced green above, being more tinged brown. In fresh plumage some buff-brown tinge to sides of head and throat. Rarely a little yellow on supercilium and breast. Song same or similar to that of *collybita* or *abietinus* but call straight like in *tristis*. Differs only subtly from '*fulvescens*' on slightly smaller size and more rounded wing. Scant material available in collections, and accepted here provisionally until better studied. Lectotype and paralectotype (ZMMU) examined and found to be inseparable from *tristis s.l.* on plumage. The border between this race, if at all valid, and *abietinus* in the west needs to be better defined (E Azerbaijan?), but separation from pale and brown-tinged examples of *abietinus* on morphology seems close to impossible. It is difficult to agree with Marova & Leonovitch (1997) that it is 'a well-defined subspecies'.

TAXONOMIC NOTES Genetic distance (cyt b, mtDNA) of Common Chiffchaff *collybita* from Mountain Chiffchaff *P. lorenzii* is 3.6% (Helbig et al. 1996), and species status of latter further supported by morphological differences and ecological separation at same localities. Song of *lorenzii* is very similar to song of *collybita* and *abietinus* (but call more similar to *tristis*). – As outlined in Dean & Svensson (2005), present evidence is still insufficient to indicate whether *tristis* should be regarded as a separate species, or a subspecies of Common Chiffchaff; it is therefore tentatively kept as a race of Common Chiffchaff here. Hybridisation between the two where they meet in E Russia seems—based on song and morphology—to be fairly common (Lindholm 2008, Marova et al. 2009), indicating that they are still at an early stage of separation. Helbig et al. (1996) showed that the genetic difference between *tristis* and *collybita* is 1.7%, which is somewhat less than is normally found between warbler species, but a fair amount between subspecies. – To simply define any Common Chiffchaff with a straight call as *tristis* (van den Berg 2009) almost irrespective of plumage variation seems risky, and we can only agree with the author's statement that 'more study on vocalisations is necessary ... and the use of calls for identification might be limited in the Middle East'. This cautious but sound approach [!] would apply similarly to Turkey and Europe in autumn, where Common Chiffchaffs from distant breeding grounds turn up habitually and need to be carefully scrutinised as to plumage before safe identification can be attempted. – On present evidence, '*fulvescens*' ('westerly *tristis*', W of Yenisei) seems too subtle and variable, and interaction with *abietinus* insufficiently known, to be accepted as a sufficiently distinct race. However, museum material from breeding areas is scant, and collection localities are geographically uneven, and therefore not entirely conclusive; hence, more research could provide a different assessment. – As to the claimed race '*caucasicus*' (Loskot 1991) it should be noted that Helbig et al. (1996) found that '*caucasicus*' differed genetically only 0.2% from W Turkish breeders, here included in *collybita*. Clearly, more research in the entire area under discussion is needed.

REFERENCES VAN DEN BERG, A. B. (2009) *DB*, 31: 79–85. – CLEMENT, P., HELBIG, A. J. & SMALL, B. (1998) *BB*, 91: 361–376. – COLLINSON, J. M. et al. (2013) *BB*, 106: 109–113. – COPETE, J. L. & SVENSSON, L. (in prep.) A review of the taxonomy of Chiffchaff *Phylloscopus collybita* in Asia Minor and Transcaucasia. – DEAN, A. R. (2007) *BB*, 100: 497–501. – DEAN, A. R. (2008) *BB*, 101: 144–150. – DEAN, A. R. et al. (2010) *BB*, 103: 320–338. – DEAN, A. R. & SVENSSON, L. (2005) *BB*, 98: 396–410. – EBELS, E. B. *DB*, 31: 86–100. – GARGALLO, G. & CLARABUCH, O. (1995) *Ring. & Migr.*, 16: 178–189. – HELBIG, A. J. et al. (1996) *Ibis*, 138: 650–666. – DE KNIJFF, P., VAN DER SPEK, V. & FISCHER, J. (2012) *DB*, 34: 386–392. – LINDHOLM, A. (2008) *Alula*, 14: 108–115. – LOSKOT, V. M. (1991) *Vestnik Zoologii*, 3: 76–77. – LOSKOT, V. M. (2002) *Zoosyst. Ross.*, 10 (2): 413–418. – MAROVA-KLEINBUB, I. M. & LEONOVICH, V. V. (1993) *Proc. Zool. Mus. Moscow Univ.*, 30: 147–164. – MAROVA-KLEINBUB, I. M. & LEONOVICH, V. V. (1997) *Contrib. to 1st EU meeting, Bologna.* – MAROVA, I. M. et al. (2009) *Doklady Biolo. Sci.*, 427: 1–3. – MARTENS, J. & MEINCKE, C. (1989) *J. f. Orn.*, 130: 455–473. – WAERN, M. & HELLSTROM, M. (2011) *Vår Fågelv.*, 70: 14–21. – WATSON, G. E. (1962) *Ibis*, 104: 347–352. – VEPRINTSEV, B. N. & LEONOVICH, V. (1986) *Birds of the Soviet Union: A Sound Guide. Sylviidae.* Moscow.

Presumed western *P. c. tristis*, Italy, Dec: so-called 'western *tristis*' (or '*fulvescens*') are often characterised by slightly paler underparts and brighter green edges to wings and tail, and a hint of yellow on fore supercilium, eye-ring, throat and breast (very little here, though), as well as showing faint traces of olive on mantle and crown. However, grey-brown still predominates above with rusty-buff elements on cheeks and flanks. (D. Occhiato)

MOUNTAIN CHIFFCHAFF
Phylloscopus lorenzii (Lorenz, 1887)

Alternative names: Caucasian Chiffchaff, Caucasian Mountain Chiffchaff

Fr. – Pouillot de Lorenz; Ger. – Bergzilpzalp
Sp. – Mosquitero montano del Cáucaso
Swe. – Berggransångare

With its breeding range restricted to mountains in NE Turkey, Armenia, Georgia, Azerbaijan and Caucasus, this close relative of the Common Chiffchaff requires some dedicated and targeted efforts to be seen and studied. It usually breeds at altitudes above 1400 m, sometimes up to 2500 m or even higher, often in willows and scrub at or near the tree-line but commonly also in tall, closed deciduous or mixed forests. Apparently commonly moves south in winter to Iraq, Kuwait and presumably W Iran and to other Gulf States; whether any actually stay all winter at lower altitude in the Caucasus area or in Turkey, as has previously been assumed, is unknown.

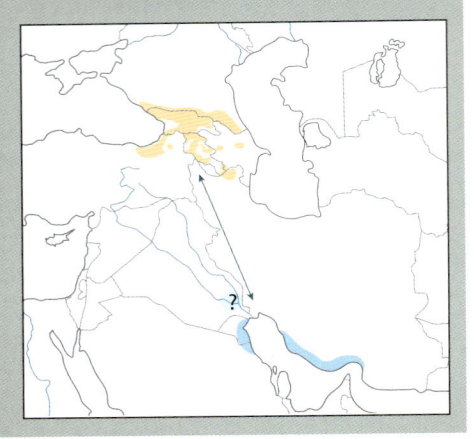

Turkey, May: unlike the very similar Common Chiffchaff, this species has a more compact jizz, and is characteristically tinged brownish above and creamy below. It often appears larger-headed and short-necked, while the short wings create a proportionately long-tailed appearance. Furthermore, it also has relatively broader tail-feathers. Legs and most of bill typically blackish. Ageing impossible in this case. (H. Shirihai)

IDENTIFICATION Normally located by song (Common Chiffchaff-like) and preliminary identified using range, habitat (high altitude, in willows, near tree-line, but also somewhat lower down in closed broadleaved mountain forests), but must be carefully studied to eliminate any local or migrant Common Chiffchaff, in particular of eastern origin. Proportions and size of Mountain Chiffchaff are much as Common Chiffchaff, but Mountain often seems to be short-necked and have a slightly *larger head* with *more rounded crown*, and appears proportionately *long-tailed*. Since the tail-feathers are comparatively broad, especially in their outer part, the whole tail often can appear rather *broad-ended* too, and is quite *square-cut*. Does not seem to practise much tail-dipping like other chiffchaffs when feeding in canopy, but often twitches tail sideways or makes smaller jerky tail movements, rather like Willow Warbler. *Feet* (including tarsi) *appear blackish* (although soles can be yellowish) and *are strong* with *long, blackish claws*. Bill medium strong and blackish with narrow orange cutting edges and only a little yellow-brown at base of lower mandible (can look all black). Upperparts rather dark brownish with *prominent pale supercilia* that *usually meet above base of bill*, more distinct and often whiter in front of eye, on average less distinct and more buffish behind. The supercilium is enhanced by rather *dark brown crown*, uniformly dark brown down to the edge of the supercilium, and by a *darkish eye-stripe*. Ear-coverts and sides of neck rusty-buff, even tinged pinkish-brown, but small area below eye often diffusely whitish. *Breast, flanks and sides of vent usually rusty-buff*, whereas throat is white, at times forming a hint of a white bib. When seen close and well, or when handled, virtually all have a very faint olive tinge to the brown upperparts, most obvious on scapulars, back and on edges of greater coverts and bases of wing-feathers (but none on crown and nape, which are just brown), and there can be a few fine yellowish streaks on buffish breast, but normally underparts are rusty-buff and white, lacking any obvious yellow. (There can also be a very subtle pale yellowish tinge in the buffish-white supercilium and on upper eyelid, but this is not visible in normal field conditions.) Wing-bend and axillaries are off-white *with a pale yellow tinge*, in particular along the fore edge, but some birds have a little stronger yellow here, not that different from Common Chiffchaff (*contra* some statements in the literature).

VOCALISATIONS Song basically very similar to that of Common Chiffchaff, and some birds sound inseparable, a simple series of single notes at two or three alternating pitches. In typical cases, the song differs by being slightly more hurried, higher-pitched and using three rather than two alternating pitches (according to some, creating a vague resemblance to song of 'Siberian Chiffchaff' *tristis*, although this is faster still and has much more complex multisyllabic notes run together in a more fluent strophe). Amplitude between high and low notes more restricted compared to Common Chiffchaff, the individual notes softer and at times slightly 'bent', but voice often less resounding. As with Common Chiffchaff, newly arrived birds often insert subdued stammering notes between strophes, *te-ri te-ri*. – Call is a monosyllabic piping or plaintive whistle, with a very slight hint of a fall in pitch at the end, *pee(eh)*. The call resembles most of all that of *tristis* and some south-eastern Common Chiffchaff populations, and can easily be confused with those, but usually differs subtly to the experienced ear in being less piping, more mechanically whistling ('squeaky gate'), lacking perhaps the full melancholy ring of the *tristis* call.

SIMILAR SPECIES Must first of all be separated from migrant Siberian race *tristis* of *Common Chiffchaff*, which is similarly brown above with blackish legs. Mountain Chiffchaff has a longer, fuller and squarer tail and stronger feet, and often appears more large-headed and bull-necked. It is slightly darker above and has a darker tawny-brown crown, which appears particularly dark laterally, making broad whitish forepart of supercilium more contrasting. Supercilium variable in both, but on average narrower and more evenly buffish (whitish when bleached) in *tristis*, broader and whiter in front even when fresh—and generally more prominent overall—in Mountain, narrowly continuing over bill on forehead (usually not so in *tristis*). Edges of wing-feathers on Mountain Chiffchaff basically brown (insignificant olive hue in some only visible on handled birds or at extremely close range), whereas *tristis* has rather obvious dull olive-green edges to wing- and tail-feathers (unless very heavily worn, and then usually on breeding grounds in Siberia). Also, *tristis* usually has greenish-tinged rump and uppertail-coverts, whereas these parts are much more brownish in Mountain Chiffchaff. On *tristis* you often see a small clear yellow patch

Presumed ad, Georgia, May: note typically prominent buff-white supercilium, which usually reaches to forehead and which can be whiter in front of eye, more buff-tinged behind. Whitish-cream underparts typically washed pinkish-buff on sides, while edges of wing-feathers show tiny amounts of yellowish-olive. Rather clear white eye-ring (broken) and hint of paler cheek patch below eye. Throat typically contrasting whitish and crown warm dark brown. (S. J. M. Gantlett)

on wing-bend near alula (and axillaries are partly rather bright lemon-yellow in 98%), but this would be very rare to see on a Mountain Chiffchaff, which usually has off-white axillaries with some (pale) yellow, and shows more limited (but some) yellow on wing-bend. – Other pitfalls are local or migrant European Common Chiffchaffs, but these are rather obviously more greenish above, more yellowish on supercilium and white on underparts. Frequently they have less blackish, browner legs with smaller feet and shorter claws, and appear less long- and broad-tailed, more narrow-tailed with a notched tip. – *Plain Leaf Warbler* is similarly brown and off-white but is smaller, shorter-tailed and has shorter, more rounded wings, a finer but proportionately longer all-black bill, and has a different, dry clicking call. – Could theoretically be confused with *Dusky Warbler*, although this winters in S Asia (W to Pakistan), vagrants to Europe being rare and the risk therefore remote. Dusky has paler brown legs, often pinkish-brown, and pale-based lower mandible, whereas Mountain Chiffchaff has blackish legs and a darker bill. Dusky generally has a longer supercilium, reaching further back towards nape, a dark and prominent eye-stripe, and has more variegated or 'dotted' pattern on ear-coverts than Mountain Chiffchaff. – Finally, closely related *'Kashmir Chiffchaff'* (*P. sindianus*; extralimital, cf. below) is similar but usually has less prominent, more diffuse supercilium (though can be rather bold above lores), not quite as dark brown crown, lacks nearly any greenish tinge on wings and tail (flight-feathers usually edged sand-brown), lacks the contrasting white throat bib as generally seen in Mountain due to breast being paler brown in Kashmir Chiffchaff, has weaker feet with subtly shorter claws, and has different, finer, more high-pitched and as a rule slightly upwards-inflected call (see Taxonomic notes).

AGEING & SEXING Ages and sexes very similar or alike in plumage. Possible age-related differences in 1stCY based on difference in wear difficult to use before moult pattern has been better established. – Moults. Complete post-nuptial moult (according to the literature, and based on fairly few available specimens) in late summer–early autumn (Aug–Oct), but some birds apparently undergo a complete or at least extensive wing moult in spring, this based on a few observations of trapped birds in late Mar and Apr in Kuwait (Cleere *et al.* 2004), or perhaps replace flight-feathers at both seasons, possibly depending on migrational habits and wintering strategy. Partial post-juv moult in summer does not involve flight-feathers or primary-coverts.

BIOMETRICS L 11–11.5 cm. **W** ♂ 59–67 mm (*n* 23, *m* 62.8), ♀ 55.5–60 mm (*n* 10, *m* 57.5); **T** ♂ 49–57 mm (*n* 23, *m* 53.0), ♀ 45–51 mm (*n* 10, *m* 48.2); **T/W** *m* 84.3; **B** 10.8–12.4 mm (*n* 34, *m* 11.6); **Ts** ♂ 18.2–21.3 mm (*n* 21, *m* 19.9), ♀ 18.2–20.1 mm (*n* 10, *m* 19.1); **HC** 6.3–7.6 mm (*n* 22, *m* 6.9); **MC** 4.4–6.2 mm (*n* 12, *m* 5.1). **Wing formula: p1** > pc 5–11 mm, < p2 18–28 mm (< 23 mm in ♀♀); **p2** < wt 5.5–10 mm, =8 or 8/9 (48%) =9 or 9/10 (28%), =10 (17%) or =7/8 (7%); **pp3–5** about equal and longest; **p6** < wt 0.5–2 mm; **p7** < wt 2.5–5 mm; **p8** < wt 4.5–10 mm; **p10** < wt 7–12 mm; **s1** < wt 8–14 mm. Emarg. pp3–6.

GEOGRAPHICAL VARIATION & RANGE Monotypic. (See Taxonomic notes below.) – NE Turkey, Caucasus, Transcaucasia; winters Iraq, Kuwait, probably also further south around the Gulf and into SW Iran.

TAXONOMIC NOTES Closely related to Common Chiffchaff as evident from both similar morphology and vocalisations. Although apparently a mountain isolate, broadly segregated from Common Chiffchaff by habitat choice and altitude, it comes into contact with the latter in lower alpine forests at *c.* 1400–1800 m (allegedly, rarely even higher, to 1900 m). However, the two seem to be taxonomically well separated, and no hybrids are known. They can invariably be separated on colours and structure, even in the field, and are hence best treated as separate species. – Since a Caucasian subspecies of Common Chiffchaff has been described ('*caucasicus*'; not accepted here), the name Mountain Chiffchaff (rather than 'Caucasian Chiffchaff') for *P. lorenzii* is preferred to avoid confusion. – A close relationship between extralimital *P. sindianus* and *P. lorenzii* is indicated

Presumed ad, Turkey, May: here the characteristic dark brown crown is shown well. Some birds are suffused more overall rusty-brown. Aside of being mostly brownish, note slight greenish tinge to edges of greater coverts, alula and secondaries. Rather less worn wing, chiefly primary-coverts and tips of primaries, in May could suggest an ad. (D. Occhiato)

Georgia, Jun: note very black legs, rounded crown with hint of warm tawny-brown cap-like effect, almost whitish throat but pink-buff breast/throat-sides. Supercilia usually appear to meet on forehead, but some birds (like this) have a much-reduced, narrower and duller supercilium in front of eye. Ageing impossible in this case. (R. Mizrachi)

Georgia, May: some have a very faint greyish-olive tinge to the brown upperparts, most obvious on scapulars, back, edges of greater coverts and bases to wing-feathers. This bird shows well the rather dark rusty-brown crown, and the pink-buff sides to head, neck and breast. Compared with a *tristis* Common Chiffchaff, note the better-marked supercilium reaching to forehead. (K. Haataja)

Juv/1stW, N Caucasus, Aug: young birds slightly less characteristic due to more subdued and washed out plumage characteristics, still uniform brown cap already obvious. Also note the longer, broader and fuller tail, and very dark legs and strong claws. Age inferred by the fluffy and rather loose body-feathering. (I. Ukolov)

Presumed 1stW, Armenia, Oct: rusty-brown above, with brighter rusty-buff cheeks and rear supercilium, but whiter throat and fore supercilium, a diffuse small whitish patch below eye, together with strongly pink-buff body-sides, proportionately short wings and neck, large rounded head and longer and broader tail make the species distinctive even in autumn. Soft-textured juv remiges and already worn tail-tips suggest age. (V. Ananian)

by moderate genetic distance (1.7%; Helbig et al. 1996), but they are morphologically slightly different and widely separated allopatric taxa with partly different vocalisations, and as with Green Warbler *P. nitidus* in relation to Western Greenish *P. trochiloides* are better treated as separate species. Since there is one claimed record of *P. sindianus* in Europe (Gibraltar; Perez 2002) and a few records outside the treated range of this book, but away from the normal breeding and wintering quarters of *sindianus*, a brief description is offered here as a guide.

KASHMIR CHIFFCHAFF *P. sindianus* Brooks, 1880 (Sinkiang Mts, Kashmir and NW Himalaya, west to Pamir and mountains south of Fergana basin, N Pakistan; winters S Pakistan, W India). Differs from Mountain Chiffchaff in being generally subtly paler greyish sandy-brown above, less dark rufous, and by having less prominent and more diffusely marked supercilium. Tail-feathers not as broad and dark as in Mountain resulting in narrower folded tail. Dark bill subtly shorter than in *lorenzii*, but requires handling to be assessed (and much overlap). Further, tarsus shorter and at times slightly paler, and claws also subtly shorter. Somewhat blunter wing, but much overlap. Vocalisations slightly different, too: song is a little slower and perhaps also weaker, but has same 'bent' notes with *tristis* ring, and call is a finer and more high-pitched, soft whistle slightly or clearly upwards-inflected, *weest* or *ho-eest*, usually more similar to call of *collybita / abietinus* than to *tristis* or *lorenzii*. Differs from Siberian Common Chiffchaff ssp. *tristis* on call; generally all-brown wings and tail (any green tinge much restricted and subdued), lacking obvious yellow-greenish edges; paler yellow (or even buffish-cream) wing-bend and axillaries (*tristis*: nearly always vividly lemon-yellow); more uniform sandy-buff lower throat and breast (*tristis* often tends to be diffusely mottled or streaked darker); ear-coverts (and perhaps loral-stripe?) being slightly paler, not as dark as in *tristis*; and somewhat blunter wing (but overlap). Note wing and claw length, and stronger and longer rictal bristles, as compared to Plain Leaf Warbler *P. neglectus* (as first pointed out by K. Mild, pers. comm.). Complete post-nuptial moult in Aug–Oct (but perhaps imperfectly known, as with Mountain Chiffchaff). **W** ♂ 58–62 mm (*n* 16, *m* 59.5), ♀ 53–59.5 mm (*n* 11, *m* 56.0); **T** ♂ 45–51.5 mm (*n* 15, *m* 49.0), ♀ 40–50 mm (*n* 11, *m* 46.3); **T/W** *m* 82.2; **B** 9.8–11.8 mm (*n* 26, *m* 11.1); **HC** 5.0–7.0 mm (*n* 26, *m* 6.0); **MC** 3.9–4.8 mm (*n* 12, *m* 4.5); **Ts** 17.7–19.8 mm (*n* 28, *m* 19 1). **Wing formula: p1** > pc 5–10.5 mm, < p2 15.5–23 mm; **p2** < wt 5–9 mm, =8 or 8/9 (39%), =9 or 9/10 (39%), or =10 or 10/ss (22%); **p3** < wt 0.5–1.5 mm; **pp4–5**(6) about equal and longest; **p6** < wt 0–2.5 mm; **p7** < wt 1.5–5.5 mm; **p8** < wt 3.5–7 mm; **p10** < wt 6–9 mm; **s1** < wt 7–10 mm. (Syn. *subsindianus*.)

REFERENCES CLEERE, N., KELLY, D. & PILCHER, C. W. T. (2004) *Sandgrouse*, 26: 143–146. – CLEMENT, P., HELBIG, A. J. & SMALL, B. (1998) *BB*, 91: 361–376. – HELBIG, A. J. et al. (1996) *Ibis*, 138: 650–666. – MARTENS, J. (1982) *Zeit. zool. Syst. Evol.-forsch.*, 20: 82–100. – PEREZ, C. E. (2002) *Gibraltar Bird Rep. 2001*. – SCOTT, M., SIDDLE, J. & SHIRIHAI, H. (1999) *BW*, 12: 163–167. – SHIRIHAI, H. (1987) *Proc. 4th Int. Ident. Meet. Eilat 1st–8th Nov. 1986*: 60–61.

Kashmir Chiffchaff *P. sindianus*, **India, Aug:** not quite as dark brown as Mountain Chiffchaff, being generally somewhat paler greyish sandy-brown dorsally, with a similarly bold but often more diffuse supercilium often becoming subdued behind eye. It also has slightly weaker feet with shorter claws. It generally resembles Siberian Chiffchaff (*tristis*) and many are unidentifiable unless heard to call or being handled. Marginal difference: wings and tail lack greenish edges of *tristis*. (S. Ghosh)

Kashmir Chiffchaff *P. sindianus*, **India, Apr:** plumage halfway between Mountain Chiffchaff and *tristis* Common Chiffchaff, with a hint of the brown cap of the former (though less dark), but with more diffuse supercilium and head/throat pattern of the latter. Differs also from *tristis* by more sandy-brown edges to remiges, less green-tinged. (A. Hellquist)

IBERIAN CHIFFCHAFF
Phylloscopus ibericus Ticehurst, 1937

Fr. – Pouillot ibérique; Ger. – Iberischer Zilpzalp
Sp. – Mosquitero ibérico; Swe. – Iberisk gransångare

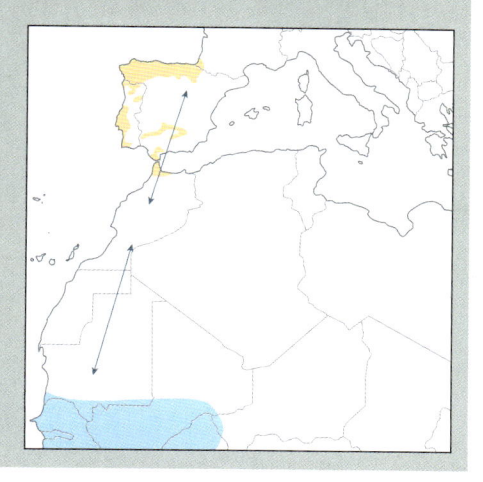

A little-known bird until recently, when it was elevated to full species status from its more obscure existence as a subspecies of Chiffchaff. Since then it has received much interest, and records of several stragglers north of its Iberian breeding range have been noted, even as far north as in Britain, Germany, Denmark and Sweden. Interestingly, although it is one of the two most southerly forms of the Chiffchaff complex and might be expected to be resident or a short-distance migrant, it migrates the furthest south of all chiffchaffs, commonly wintering in tropical West Africa.

streaking on lower throat and breast, giving a quite *yellow-breasted* appearance, but far from all are this bright. Facial pattern intermediate between Common Chiffchaff and Willow Warbler, supercilium and dark loral stripe not as subdued as in some Common Chiffchaffs, but not as prominent as in some Willow Warblers either. Often tinged yellowish-green on sides of neck. *Supercilium characteristically vividly lemon yellow in front of and above eye* (yet more whitish at rear), but note that this can be matched in worn summer plumage by at least some Common Chiffchaffs. In fresh autumn plumage, a very faint grey-brown tinge can often be detected on ear-coverts, sides of throat and breast, and on flanks. *Legs* never as dark as in some northerly Common Chiffchaffs, invariably having *a medium-brown element*, closer to some darker-coloured Willow Warblers. The *bill*, too, is more like in Willow Warbler, on average *slightly stronger* than in Common Chiffchaff, with some *paler brown visible on cutting edges and at the base* of lower mandible.

VOCALISATIONS Main part of song resembles that of Common Chiffchaff in tone, but whole song differs due to the inclusion of two species-specific items: a staccato-like 'stutter', and a softer, drawn-out note repeated a few times. A typical phrase could be *chiff chiff chiff chiff chiff te-te-te-te swee swee swee*, but the order between the last two parts can be altered, also at times by the same bird, or either the softer or the stuttering notes, or both, can be left out, and a common variant is *chiff chiff chiff chiff swee te-te-te-te-te*. A closer analysis of the initial Common Chiffchaff-like part of the song reveals that this differs on average by being slightly faster and drier, not so 'resounding', and in consisting of, by and large, only one type of notes (rather than altering between *chiff* and *chaff*). As with Common Chiffchaff, Iberian often inserts subdued *terre terre* between songs when agitated. Sings from high post, often top or upper part of tree. – Call is a fine, high-pitched whistling downslurred *seeu*, readily separable from upwards-inflected *hweet* of Common Chiffchaff. Theoretically could be confused with a call by Siberian Common Chiffchaff *tristis*, which can fall in pitch very slightly at the end, but Iberian is higher-pitched and normally distinctly more downwards inflected (can even recall call of Reed Bunting). – Beware of hybrids in area of overlap in extreme SW France and W Spanish Pyrenees, or among overshooting birds in NW Europe; most can be recognised by their ability to switch between songs of both parental species, alternatively on intermediate song, or on mixture of Iberian Chiffchaff-like song and Common Chiffchaff call. Others frustratingly will appear and sound just like Iberian, but be revealed as hybrids only by their DNA.

Spain, May: appearance between Common Chiffchaff and Willow Warbler, with intermediate jizz, face pattern, leg colour and length of primary projection. Upperparts moss-green, rather like Willow Warbler, but can have slightly brown-grey wash when worn; underparts whitish with lemon wash to sides of neck, breast, flanks and undertail-coverts. Both pale supercilium (vivid lemon-yellow in forepart, more whitish at rear) and dark loral stripe not as subdued as many Chiffchaffs, but less prominent than most Willow Warblers. (J. Sagardía)

IDENTIFICATION A typical *Phylloscopus* warbler with green, yellow and white colours, and delicate bill and feet. Much a mixture of Common Chiffchaff and Willow Warbler (maybe more than an impression?), having intermediate leg colour and by and large the shape and wing formula of Common Chiffchaff, but the more clean green upperparts and lack of obvious brown and buff colours on head and yellow-streaked breast of Willow Warbler. Just like most of the other members of the Common Chiffchaff group it *habitually dips its folded tail downwards* when moving in the canopy of a tree or a bush (but does not normally semi-open the tail or flick its wings much), a habit which, when seen, readily separates it from Willow Warbler. Primary projection on average a little longer than in Common Chiffchaff but clearly shorter than in Willow Warbler. *Upperparts usually quite bright green* in fresh plumage (as green as a C European Willow Warbler, but not as green as Wood Warbler), especially on mantle to uppertail-coverts, but field impressions—unless the bird is seen very close—are still often of rather brownish-olive upperparts. *Underparts* whitish with lemon streaks on breast, and lemon tinge on flanks and undertail-coverts, *lacking obvious buff and brown hues* (at least in spring and summer). Centre of belly typically white, surrounded by yellow hues. The most vividly-coloured birds have strong lemon tinge or

Spain, May: a less straightforward bird (partly due to wear and bleaching), being less deep green above than normal, more *collybita*-like. However, still separable by rather paler (medium brown) bill and legs, relatively long primary projection, fairly green mantle, scapulars and wing-bend. Entire wing apparently ad, and less worn. (F. López)

SIMILAR SPECIES Separation from Common Chiffchaff and Willow Warbler covered under Identification. Note that bleaching in summer of ssp. *collybita* of *Common Chiffchaff* seems to reduce brown and buff tinges before any lemon-yellow and green become bleached, and thus a few worn summer *collybita* can be extremely similar to Iberian. Duller-than-typical Iberian also occur. Even with close observations, some Iberian cannot safely be separated from Common Chiffchaff without help of vocalisation, biometrics or DNA. – *Willow Warbler*, in particular a small female, can be surprisingly similar, but is usually separated on longer primary projection and lack of habit of repeatedly dipping tail down when moving in canopy like Iberian and other members of the Chiffchaff complex. – *Wood Warbler* has similarly green upperparts (in fact even greener) and yellow throat and breast lacking any buff-brown, but has solidly yellow throat and upper breast, broad all-yellow supercilium, darker olive-grey eye-stripe and all-white lower breast/belly and under-tail-coverts, with much longer wings and primary projection, primaries darker and tertials dark with obvious pale edges (Iberian has near-uniform tertials).

AGEING & SEXING Small seasonal and age-related differences; ages frequently separable in early spring, before becoming too worn, sometimes by help of moult contrast in wing. Sexes alike in plumage. – Moults. Complete post-nuptial moult of ad in summer (mainly Jul–early Sep). Partial post-juv moult in summer does not involve flight-feathers or primary-coverts. Partial pre-nuptial moult of ad in winter appears to be rather limited. In one study (Monteagudo *et al.* 2003), 100% of a small sample (n 12) of 1stY moulted 4–6 outer primaries, innermost secondary (s6, often also s5, rarely more secondaries) and tertials in winter. – **SPRING** Generally no brown or buff tinges in the green on head and sides, except rarely a very faint trace on ear-coverts. Ageing apparently possible for a majority. **Ad** All flight-feathers of same generation, tips of primaries generally only moderately worn until Jun. **1stS** Often a contrast between 4–6 outer primaries, which are newer, less worn and darker grey, and inner browner ones which have heavily abraded tips, at least until early May, after which abrasion makes it more difficult to distinguish between two generations of primaries. Those which have retained all juv primaries are generally heavily abraded. – **AUTUMN Ad** Freshly moulted, appearance much as spring, but frequently has very slight brownish tinge on crown and sides of head, throat and rarely on flanks. Tail-feathers rather rounded at tips. (Material studied small.) **1stW** Basically rather similar to ad, but generally appears to be a little more yellow on breast and flanks, and brown or buff tinges perhaps on average more obvious on head and sides. Tail-feathers often

1stW, Spain, Aug: this bird shows extremely deep yellowish-green tones, very rich not only above but also below. In autumn, 1stW can show bright tones below, always more than ad and 2CY, which are invariably paler on underparts. Signs of feather wear in wing further confirm the bird is 1stW. Legs are too dark for average Willow Warbler. Undertail-coverts appear to be pure yellow. Common Chiffchaff never shows such rich green tones above. Note lack of brown in plumage (except very faintly on cheeks), also unlike Common Chiffchaff. Primary projection is rather longer than Common Chiffchaff but too short for Willow. (A. Fernández)

Presumed ad, Portugal, Aug: upperparts not as green as is typical, still has a greenish sheen. Rather uniform ear-coverts, and predominantly whitish below with lemon streaks on breast-sides, flanks and undertail-coverts. Rich yellow on supercilium, especially fore part, and dark legs. However, in fresh autumn plumage, a faint grey-brown tinge can often be detected on ear-coverts, throat, breast-sides and flanks. The generally pale tones (greyish on head, whiter below) and apparently evenly fresh wing-coverts suggest an ad. (A. Caldas)

rather narrow and slightly pointed at tips, and tips often slightly worn from Sep.

BIOMETRICS L 11–12 cm. **W** ♂ 58–65 mm (n 36, m 61.3; once 56), ♀ 54–59 mm (n 16, m 56.4); **T** ♂ 42.5–52 mm (n 36, m 47.5), ♀ 41–48 mm (n 16, m 44.2); **T/W** m 77.8; **B** 10.4–13.3 mm (n 49, m 12.0); **Ts** ♂ 17.9–20.7 mm (n 34, m 19.7), ♀ 17.6–19.0 mm (n 15, m 18.5). – Onrubia *et al.* (2003) obtained the following data from live breeders: **W** ♂ 60–68 mm (n 44, m 64.3), ♀ 54–62 mm (n 41, m 58.1); **T** ♂ 45–56 mm (n 41, m 50.5), ♀ 41–52 mm (n 34, m 45.3). – **Wing formula:** p1 > pc 2–8 mm, < p2 22–29 mm in ♂♂, 20–25.5 in ♀♀; **p2** < wt 5–8 mm =7/8 or 8 (51% of ♂♂, 80% of ♀♀), =6/7 or 7 (43% of ♂♂, 20% of ♀♀) or =6 (6% of ♂♂); **pp3–5** about equal and longest (p5 sometimes 0.5–1 mm <); **p6** < wt 1.5–6.5 mm; **p7** < wt 3.5–8 mm; **p10** < wt 9–13 mm in ♂♂, 7.5–11 in ♀♀; **s1** < wt 10–15 mm in ♂♂, 9.5–12.5 in ♀♀. Emarg. pp3–6 (on p6 sometimes not deep, still prominent). – For difficult birds, museum workers, who can decide sex from labels, can calculate 'multiple characters value', **MCV**, as follows (mainly helpful for ♂♂): add W, B, p1 < p2, p6 < wt, p7 < wt, p10 < wt, and s1<wt, then subtract T and p1 > pc. MCV is useful for ad ♂♂, and supportive for ♀♀. If MCV is greater than 73.2, the bird is likely to be *ibericus* (only 11% of *ibericus* fell below), if lower it is probably Common Chiffchaff *collybita* (all *collybita* fell below); overlap area for ♂♂ is 71.9–73.2. For ♀♀, 57% of *ibericus* have ≥ 71.0, whereas all *collybita* have < 71.0; overlap area for ♀♀ is 61.1–70.9. Note that heavily worn birds (some in May, many in Jun) can be unsuitable for calculating MCV. On a larger sample one can expect slightly larger overlap (cf. Onrubia *et al.* 2003; see also Gordo *et al.* 2016 for a proposed method to sex many live birds, although this was not based on any material of known sex).

GEOGRAPHICAL VARIATION & RANGE Monotypic. – Extreme SW France, W Spain, E Portugal, N Morocco, N Algeria; apparently winters mainly in sub-Saharan W Africa, though some may stay in NW Africa. (Syn. *biscayensis*.)

TAXONOMIC NOTES Here treated as a separate species based on the results from thorough research over many years mainly by M. Salomon. A. J. Helbig *et al.* demonstrated that the genetic difference between *ibericus* and *collybita* was 4.6% (mtDNA). There is still some hybridisation in area of overlap with Common Chiffchaff, but hybrids are less common than would be expected if no barriers existed, and mainly (only?) hybrid ♂♂ seem to be fertile. Supporting evidence of species status come from the vocalisations and migration habits as related in Svensson (2001) and Perez & Cortes (2003). – Salomon *et al.* (2003) described a new subspecies ('*biscayensis*') from the northern part of the range mainly as being a fraction longer-winged—in ♂ by only 1.28 mm on average. This is not distinct enough to warrant separation. Such minor variations are found in subtle degrees within most widely-distributed continental taxa. The claimed shorter bill and tarsi of the northern population could not be confirmed in our material, in which these measurements co-varied neatly with wing-length, thus were very slightly longer, not shorter. Described differences in habitat selection seem more an effect of availability than choice.

REFERENCES CLEMENT, P. & HELBIG, A. J. (1998) *BB*, 91: 361–376. – COPETE, J. L. (2008) *BB*, 101: 378–379. – GORDO, O. *et al.* (2016) *Ring. & Migr.*, 31: 83–97. – HELBIG, A. J. *et al.* (1996) *Ibis*, 138: 650–666. – HELBIG, A. J. *et al.* (2001) *J. Evol. Biol.*, 14: 277–287. – MONTEAGUDO, A. *et al.* (2003) *Revista de Anillamiento*, 12: 14–17. – MÜLSTEGEN, J.-H., NIEHAUS, G. & SELLIN, D. (1994) *Limicola*, 8: 8–14. – NIETHAMMER, G. (1963) *J. f. Orn.*, 104: 403–412. – ONRUBIA, A. *et al.* (2003) *Revista de Anillamiento*, 12: 18–29. – PEREZ, C. & CORTES, J. (2003) *Gibraltar Bird Report*, 1: 29–31. – SALOMON, M., VOISIN, J.-F. & BRIED, J. (2003) *Ibis*, 145: 87–97. – SLATERNS, R. (2007) *DB*, 29: 83–91. – SVENSSON, L. (2001) *BBOC*, 121: 281–296. – THIELCKE, G. & LINSENMAIR, K. E. (1963) *J. f. Orn.*, 104: 372–402.

Ad, Spain, Oct: primary projection slightly longer than the average Common Chiffchaff, giving the bird a somewhat Willow Warbler-like jizz, but still too short for that species. Note rich green above, yellow supercilium especially above eye and lores, and yellow undertail-coverts—all characteristic of Iberian. The less strong yellow below and evenly fresh plumage (just finishing post-nuptial moult, with p2 still growing) suggest an ad. (E. Padilla)

Common Chiffchaff (top left: Italy, Apr; top right: Italy, Oct), **Iberian Chiffchaff** (centre left: Belgium, Apr; centre right: Spain, Aug) and **Willow Warbler** (bottom left: Netherlands, Apr; bottom right: Greece, Oct): birds on left are in spring–summer (worn), on right in autumn (fresh). Willow Warbler versus Common Chiffchaff: unlike similar but more compact Chiffchaff, Willow Warbler is an 'elegant' bird, slightly larger, longer-winged, and usually stronger-billed; leg colour is paler brown, not blackish-looking. These differences usually suffice, making the longer primary projection of Willow superfluous (c. 3/4 or even equals tertial length, instead of 1/2 to 2/3 as Chiffchaff). The on average bolder face pattern with more contrasting dark eye-stripe of Willow further secures the identification. – Separation of Iberian Chiffchaff from the other two, if not heard, can be very difficult, although it often has slightly richer yellow fore supercilium and yellowish-green hue to ear-coverts, further contributing to yellower-faced look. Important, too, are usually present richer moss-green tones to mantle, scapulars and neck-sides, similar to (or even greener than) Willow Warbler. Also predominantly whitish underparts with lemon streaks on breast-sides and flanks, rich yellow undertail-coverts (not normally present in Common Chiffchaff), and note habit of dipping tail when feeding like Common Chiffchaff, but unlike Willow Warbler (which twitches wings and tail). Last, note Iberian's not very dark legs (invariably with medium-brown element) and primary projection (c. 2/3 to 3/4 of tertial length) midway between the other two species. All chiffchaffs best left un-aged though degree of wear suggests they are 1stY, whereas the autumn Willow is ad. (Top left and right: D. Occhiato; centre left: V. Legrand; centre right: A. Fernández; bottom left: L. van Loo; bottom right: P. Petrou)

CANARY ISLANDS CHIFFCHAFF
Phylloscopus canariensis (Hartwig, 1886)

Fr. – Pouillot des Canaries; Ger. – Kanarischer Zilpzalp
Sp. – Mosquitero canario; Swe. – Kanariegransångare

Visitors to the Canary Islands have long noted some peculiarities about the local chiffchaffs, and these have resulted in the birds being elevated to species status, following genetic studies and a new appreciation of the 'personality' and separate evolutionary history of distinct and isolated populations. As with so many passerines on remote islands at southern latitudes, this is a resident bird. Observers visiting the islands in winter should be aware that there is a small influx of Common Chiffchaffs from mainland Europe, mixing with the local birds.

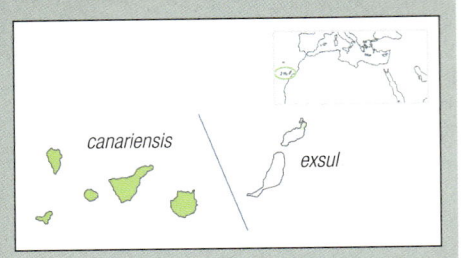

voice and the uneven rhythm and slight acceleration in the normal song. Strophes are somewhat shorter on average than in Common Chiffchaff. Other variations have more even staccato rhythm, still clearly different from Common Chiffchaff on more liquid, metallic voice. There is also a peculiar alternative song, often used in early morning (but function not well understood), a long series of 'complaining', piping notes, now and then interfoliated by a normal song. – Calls are rather variable, more so than in Common Chiffchaff. Perhaps the most common call is quite similar to that of Common Chiffchaff of ssp. *collybita* or *abietinus*, a soft whistling upwards-inflected *hooeet*. There is also a straighter call, *veest* or *heep*, at a distance surprisingly *tristis*-like. Further, a disyllabic call variant is sometimes heard, *veest-ist*, giving almost an echo effect. For local variations between the islands cf. Martens (2013).

SIMILAR SPECIES See Identification. Main risk of confusion is with wintering or migrant *Common Chiffchaff*, but this has slightly longer, more pointed wings, not as long bill and

P. c. canariensis, Canary Is, Apr: a quite variable chiffchaff, with some birds strikingly brownish above and on body-sides, others more greenish and yellow. Note long grey-brown legs and long and well-developed (though typically narrow) supercilium. Unlike Willow Warbler has very short primary projection affording long-tailed appearance. (T. Krüger)

P. c. canariensis, Canary Is, Mar: some birds are streaked more yellow below, but still predominantly washed buff or ochraceous, while upperparts remain rather dark and dusky brown. Typical long-tailed appearance, and long-looking bill with hint of downward-inclined tip. (C. Johansson)

IDENTIFICATION Differs from Common Chiffchaff on the following combination of characters: *song and some calls distinct*; thin but strong, often *long-looking bill with pale cutting edges* and in many (but not all) hint of *downward-inclination at tip*; rather *long legs* that are medium *grey-brown* (about same colour as Willow Warbler) or a little darker; *very short primary projection* (= very rounded wing); upperparts rather *dark greenish with variable brown elements*; underparts usually extensively tinged and streaked yellow in fresh plumage (c. Aug–Mar) with obvious buff or *ochraceous tinge across breast and on flanks*; and *yellowish supercilium narrow* but *rather long* and prominent, vivid lemon-yellow or yellow-buff in front of eye, at times more whitish at rear. As birds become bleached and more worn from April through early summer, they become successively purer yellow and more whitish below, and more greenish above (ochre elements bleach away first). Still, plumage rather variable as to amount of yellow or buff-brown elements. Undertail-coverts pale buffish-yellow. Dips folded tail downwards when moving in canopy like other chiffchaffs, but perhaps not quite as frequently as Common Chiffchaff.

VOCALISATIONS Song somewhat variable but typically loud and 'explosive', has been compared to Cetti's Warbler, although still basically structured like Common Chiffchaff, *chet chit chat chet chit-chet-chit-cheet-chet*. The Cetti's Warbler similarity arises both from the explosive, metallic

P. c. canariensis, presumed 1stW, Canary Is, Sep: some birds are more Common Chiffchaff-like, being more greenish above and yellow-tinged below, and this bird also has a slightly longer primary projection than normal. Still, the plumage colours could just indicate young age and fresh plumage, and the general duskiness is rather typical for Canary Island Chiffchaff. Tips of tail-feathers already abraded. (H. Shirihai)

legs, and the bill appears straighter. Also, Canary Islands Chiffchaff in winter, while still in fresh plumage, is usually darker above and much more ochraceous-tinged on breast and flanks. Might give long-tailed appearance, mainly due to short and rounded wings. (Tail-length in relation to body-size about the same in both species.)

AGEING & SEXING Age-related differences small. Seasonal changes occur due to bleaching. Sexes alike in plumage. – Moults. Complete post-nuptial moult of ad in early summer (May–Jul). Partial post-juv moult in summer poorly studied but does not involve flight-feathers or primary-coverts. Apparently no pre-nuptial moult. – **SPRING** Rather limited brown or ochraceous-buff tinges on underparts due to bleaching and wear, at least from Apr. Ageing not possible. – **AUTUMN Ad** Freshly moulted, as in spring but now with obvious brownish or ochraceous-buff tinge in greenish upperparts, and buff tinge in the yellow on lower throat and breast, and on flanks. **1stW** Not well studied. Basically rather similar to ad, but appears to be more vividly yellow on breast and flanks, but has less ochraceous-buff tinge on flanks.

BIOMETRICS (*canariensis*) **L** 10–11.5 cm. **W** ♂ 50.5–58 mm (*n* 27, *m* 55.1), ♀ 48–56 mm (*n* 13, *m* 52.1); **T** ♂ 46–53.5 mm (*n* 27, *m* 50.2), ♀ 42–51 mm (*n* 13, *m* 46.2); **T/W** 90.4; **B** 11.0–13.8 mm (*n* 45, *m* 12.7); **Ts** 19.0–21.8 mm (*n* 47, *m* 20.4). **Wing formula: p1** > pc 7–12.5 mm, < p2 14–19 mm; **p2** < wt 6.5–10 mm, = p10/s1 or = s1 (40%), < s1 (38%), or =9/10 or 10 (22%); **p3** < wt 0–1 mm; **pp4–5**(6) about equal and longest; **p6** < wt 0–1.5 mm; **p7** < wt 0.5–3.5 mm; **p10** < wt 5.5–9 mm; **s1** < wt 6.5–9.5 mm. Emarg. pp3–6.

GEOGRAPHICAL VARIATION & RANGE Two races, one of which appears to have become extinct in modern times.

P. c. canariensis (Hartwig, 1886) (five W Canary Islands; resident). Treated above.

†? *P. c. exsul* Hartert, 1907 (Lanzarote, apparently previously also Fuerteventura; now presumed to be extinct on both). Differs from *canariensis* by being slightly paler olive-brown above, or brown with only olive on edges to wing-feathers, and on average less yellow below, buff-brown colours dominating even more. Differs slightly in structure by being smaller and having somewhat more pointed wing (e.g. p2 < wt 5.5–8.5 mm versus 6.5–10 mm in *canariensis*) and proportionately slightly longer tail (T/W ratio *m* 93.3 versus 90.2). Plumage colours not much different from some richly-coloured Common Chiffchaffs ssp. *collybita* (which certainly can occur on Canary Is), but differs in much rounder wing with less primary projection.

TAXONOMIC NOTE Previously treated as a race of Common Chiffchaff but in recent years commonly treated as a separate species due to distinct morphology, vocalisation and genetics. Thielcke *et al.* (1978) analysed its song, and Helbig *et al.* (1996) showed that its DNA differed 5% from Iberian Chiffchaff and 3.7% from Common Chiffchaff *collybita*.

REFERENCES Clement, P. & Helbig, A. J. (1998) *BB*, 91: 361–376. – Helbig, A. J. *et al.* (1996) *Ibis*, 138: 650–666. – Thielcke, G., Wüstenberg, K. & Becker, P. H. (1978) *J. f. Orn.*, 119: 213–226.

P. c. canariensis, presumed 1stW, Canary Is, Nov: during late autumn/winter there is a risk of confusion with wintering Common Chiffchaffs, but at this season most (but not all) of the latter do not attain such brownish plumage above and ochraceous below. Also note long-looking bill and tail, but proportionately very short, rounded wings. Apparent moult limit in greater coverts, and retained juv remiges, tail (latter more pointed) and primary-coverts suggest 1stW. (H. Nussbaumer)

Presumed *P. c. canariensis*, Canary Is, Dec: this bird could prove less straightforward to separate from Common Chiffchaff, especially given its short and straight bill, the amount of yellow on supercilium and underparts, and the greenish edges to remiges. Still, Common Chiffchaff can most likely be eliminated by its seemingly short primary projection and long-looking legs. (M. Høegh Post)

P. c. canariensis, Canary Is, Dec: a rather striking bird, easily separated from Common Chiffchaff by heavier, long and marginally decurved bill, also by rather long legs but very short wingtip. Dark brownish upperparts, rather dull underparts and prominent supercilium add to its characteristics. (M. Høegh Post)

WILLOW WARBLER
Phylloscopus trochilus (L., 1758)

Fr. – Pouillot fitis; Ger. – Fitis
Sp. – Mosquitero musical; Swe. – Lövsångare

While House Sparrow, Starling and Feral Pigeon certainly are very common birds in most N European countries, it often comes as a surprise to many outside the ranks of ornithologists that a bird that they have hardly heard of is even more numerous: the Willow Warbler. With European Russia included, there could be as many as a staggering 150 million pairs around in summer of this attractive little bird. Its song is particularly pleasing and can be heard wherever there are trees or tall bushes, from parks and gardens to remote forests on mountain slopes. In autumn the masses move south to winter in S Africa, returning to Europe in April and early May.

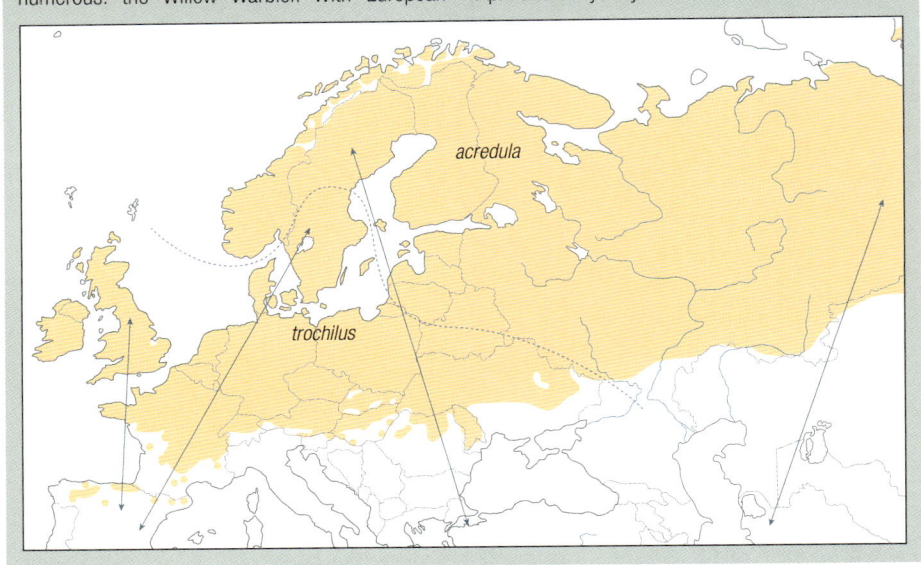

IDENTIFICATION Typical *Phylloscopus* or 'leaf warbler', a small and slender bird with thin bill and legs, being grey, green or brownish above, and whitish and yellow below. There is some geographical variation, northern and eastern birds being on average more brown-grey, less greenish above. In all, brightest part of upperparts is generally rump. Single most obvious field mark is a *yellowish-white supercilium*, enhanced by a *grey-green eye-stripe* below. Legs are *pinkish-brown or yellowish-brown*, only rarely slightly darker brown. *Bill has paler cutting edges* and often a pale base to lower mandible. Note on all but the most grey-brown variants absence of ochre-brown or buff tinges to head, sides of neck or sides of breast (as seen on many Common Chiffchaffs); Willow Warbler is more clean greyish-green and lemon-yellow on these parts. When seen well, it is worth noting that *primary projection* (=part of primaries extending outside tip of longest tertial) *is fairly long*, often *c.* 3/4 of or even equalling tertial length (=visible parts of these three feathers). Juveniles are generally appreciably more yellow-tinged below than adults. At all times, crucial to establish the absence of a pale wing-bar, fairly uniformly greenish tertials, reasonably pale pink-brown legs and pay attention to any calls or song. Active when feeding, displaying irregular wing and tail movements (see Similar species).

VOCALISATIONS Song one of the most pleasing, a softly whistling verse of slightly descending notes, but with some individual variation and often a hint of shift in rhythm and loudness halfway through. Note that the voice is invariably sweet, and most of the notes softly inflected. An average song lasts *c.* 3 sec and can be rendered *sisisi-vuy-vuy-vuy-se-se svi-svi-svi-sieh sesese-seseevuy*. It is a most energetic and tireless singer when newly arrived, delivering a new song about every 12 sec, filling the northern woods with its pleasant melody. – Call a soft whistle, upward-inflected and with first syllable stressed, *hoo-eet*, a bit similar to the call of Common Redstart, only on average a little weaker and more 'anxious-sounding'. It differs from the call of Common Chiffchaff in much of Europe by being more obviously disyllabic and having first, not last, syllable stressed. Agitated or strongly alarmed Willow Warblers give a piping, straighter sound repeatedly, *cheed… cheed… cheed…*.

SIMILAR SPECIES Most members of the genus *Phylloscopus* are very similar and require close and careful scrutiny for safe separation. Vocalisations are generally very helpful, even for migrants. – Most common problem for European birders is to separate Willow Warbler from *Common Chiffchaff* (ssp. *collybita* and *abietinus*). Common Chiffchaff is usually slightly smaller and more compact-looking, less elongated, with smaller or more rounded head and less visible neck than Willow Warbler, although some *abietinus* can be similarly large and slim. The primary projection of Common Chiffchaff is shorter, between just over 1/2 to *c.* 2/3 of tertial length. Note that while Common Chiffchaff habitually dips its folded tail downwards when moving in the canopy of a tree, Willow does this less 'mechanically', and more often twitches both tail and wings quickly, the tail often being slightly spread (not as tightly folded as in Chiffchaff), and although downwards movement is commonest there is often also some slight movements sideways. Also, Common Chiffchaff is on average a little duskier and less clean-looking, with more brown-grey tones, but some come very close. On average, Common Chiffchaff has darker legs, blackish or dark brown-grey against yellowish- or pinkish-brown in Willow Warbler, but there is a little overlap when you take all variation into account. (Incidentally, it is more

P. t. trochilus, Germany, Apr: a bright and strikingly yellow spring bird with lemon-yellow head sides, including supercilium and sides of neck, ear-coverts, breast and flanks, making separation from Chiffchaff easy. Unlike Chiffchaff has pale brown legs, longer primary projection and only three emarginated primaries. Following complete pre-nuptial moult, ageing is impossible. (T. Grüner)

P. t. trochilus, Germany, Apr: a more 'typical' bird with only moderate yellowish wash to face and breast. Note lack of ochraceous-brown or buff tones around head and below which often characterise European Chiffchaffs. Legs can be strikingly pinkish-brown and bill typically pale due to paler base to lower mandible and cutting edges. Also note very long primary projection. (M. Schäf)

common with pale-legged Common Chiffchaff variants than dark-legged Willows.) Common Chiffchaff also usually has thinner and darker bill, and darker cheeks making the whitish lower eyelid stand out better, whereas in Willow the cheeks are paler so that a pale eye-ring is not particularly striking. Conversely, Willow Warbler usually has a better-pronounced dark eye-stripe. – *Iberian Chiffchaff* is quite similar to Willow in plumage, similarly cleanly greenish and yellow, being only a slightly smaller and more compact bird. Iberian often dips its folded tail downwards repeatedly just like Common Chiffchaff, and unlike Willow Warbler, so this is useful to note, as are any calls. – *Wood Warbler* has similar size to Willow but has longer, more pointed wings (longer primary projection), making it look short-tailed, and has more contrasting tertials with paler outer edges (Willow: near-uniform), a more contrasting supercilium and dark eye-stripe, apart from other plumage differences. – Odd Willow Warblers have a slightly contrasting, somewhat brighter yellowish-green rump and this might lead you to think of either of the two *Bonelli's Warblers*. Note that these two, like Wood Warbler, have contrasting pale-edged tertials, and have almost ochraceous-tinged bright yellowish-green on both rump and edges of tail- and wing-feathers, a blander face pattern and, unlike Willow, an unbroken pale eye-ring.

AGEING & SEXING (*trochilus*) Small age-related differences in 1stCY alone. Sexes alike in plumage. – Moults. Uniquely among European passerines, ad has two complete moults per year, one post-nuptial in late summer (Jul–Aug) and one pre-nuptial in late winter. Hence ad appears nearly invariably fresh-plumaged. A very few spring migrants with abraded primaries might be aberrant or odd 2ndCY which have not performed a complete moult, but this might also be due to wintering habits and requires further study. Odd birds return with one or few fresher, more recently replaced inner secondaries and/or odd tertials indicating that moult has very recently been completed. Partial post-juv moult in summer does not involve flight-feathers or primary-coverts, and usually no greater coverts (except one or a few inner in some). – AUTUMN **Ad** Freshly moulted, tips to tail-feathers neat and rounded, centres dark and glossy, finely fringed or tipped whitish. Ground colour of underparts whitish or pale yellowish-white with lemon-yellow streaking on lower throat and breast, rarely a little more yellowish all over. **1stCY** Reasonably fresh plumage, but many show first signs of abrasion to tips of primaries and tail-feathers, latter often subtly frayed at tips, and show less contrasting edges on average than ad. More vividly and uniformly yellow on lower

Presumed *P. t. trochilus*, England, May: compared to Chiffchaff, Willow Warbler is an 'elegant' bird, slightly larger, with more attenuated body, longer wings and subtly stronger bill, and separation will be confirmed by the clearly paler legs. A pale bird which theoretically could be a migrant *acredula*, but such pale birds rarely occur also among *trochilus*. By May many birds have already bleached paler and hardly differ from duller eastern and northern races. (S. Elsom)

P. t. trochilus, juv, England, Jul: like 1stW, but has soft fluffy juv feathers and clear yellowish mouth flanges at gape. Typically washed yellow on lower throat to chest, and has weak supercilium. (G. Thoburn)

P. t. acredula, Finland, Jun: in summer, Willow Warblers in N Europe become rather drabber grey-brown and off-white, with just a pleasant yellowish tinge to lower throat, breast and supercilium. The clear-cut pale supercilium is the most obvious field mark and is enhanced by dark eye-stripe. Typical slender jizz, with obviously pale legs. Seen well, note the fairly long primary projection that exceeds *c.* 3/4 of tertial length. (M. Varesvuo)

P. t. trochilus / acredula, ad, Scotland, Sep: less yellow on face and underparts, with yellow on throat and breast taking form of distinct streaks, are features of ad in autumn. Primaries very fresh after complete moult. Difference in length of greater coverts partly due to feather disorder ('false' moult limits). Typical combination of long supercilium, long primary projection and pale legs. (H. Harrop)

P. t. trochilus / acredula, 1stW, France, Sep: young Willow Warblers can be distinctive, both from fresh ad and Chiffchaffs by being overall more greenish above and yellower on facial areas and underparts. Moreover, the long primary projection, clear-cut supercilium and pale yellow-brown legs and bill-base make identification easy. The primary-coverts and most or all greater coverts are juv (at least the tertials seem to have been replaced during post-juv moult). (A. Audevard)

P. t. trochilus / acredula, 1stW, Spain, Sep: a fresh young autumn bird on migration showing typical jizz with slim shape and long primary projection, and characteristic plumage pattern having well-marked face with prominent supercilium and eye-stripe. Upperparts usually have an olive tinge, and underparts are smoothly lemon-tinged. (H. Shirihai)

throat, breast and flanks than ad, and belly, too, is somewhat more yellowish. Yellow rather even, less restricted to streaking as in ad.

BIOMETRICS (*trochilus*) **L** 11–12.5 cm. **W** ♂ 64–71 mm (*n* 25, *m* 68.5), ♀ 61–68 mm (*n* 16, *m* 63.9); **T** ♂ 45–54.5 mm (*n* 25, *m* 50.7), ♀ 44–51 mm (*n* 16, *m* 47.5); **T/W** *m* 74.1; **B** 10.6–13.0 mm (*n* 36, *m* 11.8); **Ts** 18.3–20.4 mm (*n* 34, *m* 19.2). **Wing formula: p1** > pc 3–6 mm, < p2 31–35 mm in ♂♂, 27–30.5 in ♀♀; **p2** < wt 3.5–7 mm, =5/6 (88%), =6 (8%), =6/7 (3%; only in ♀♀) or =5 (1%); **pp3–4** about equal and longest; **p5** < wt 1–3 mm; **p6** < wt 5–8 mm; **p7** < wt 8–12 mm; **p10** < wt 14–18.5 mm; **s1** < wt 15–19.5 mm. Emarg. pp3–5 (only extremely rarely a hint distally also on p6 in some ♀♀).

GEOGRAPHICAL VARIATION & RANGE Three races, connected by wide zones of intergradation, making many birds difficult to assign to any one particular. Birds tend to be greyer, less green and yellow, and very subtly larger, in a cline from south-west to east.

P. t. trochilus (L., 1758) (Europe except C & N Fenno-Scandia; winters in W & WC Africa in sub-Saharan savannas, thought mainly to occur from S Senegal to Cameroon). Treated above. Morphologically grades into *acredula* in S Sweden, N Poland and through N Ukraine, and numerous birds are impossible to assign to either subspecies (these difficulties being even greater for migrating and wintering birds). On average a little brighter olive-brown or greenish-tinged, less greyish, above, and has slightly more yellow below and on supercilium than *acredula*, but many appear very similar or the same. Note that duller, more brown-grey birds, reminiscent of *acredula*, can rarely occur within the range of *trochilus*. (Syn. *fitis*.)

○ *P. t. acredula* (L., 1758) (Norway, N Sweden, Baltic States, Russia, Siberia; winters C, E & S Africa from Cameroon to S Kenya and south to South Africa). Differs only very slightly and on average from *trochilus* by being less olive-brown above, more tinged greyish, and on being on average less yellow below. A variable race, some of which are quite close to *trochilus*, others much more brown-grey above and pale below, but these variations are only to a certain extent linked geographically. Breast, undertail-coverts, supercilium and sides of neck and ear-coverts invariably tinged yellow (although only slightly in some). Most birds in E Siberia (Yenisei Basin), but also odd birds as far west as Lapland, are quite greyish and dull, lacking nearly all or virtually all green and yellow tinges, and these Siberian birds represent intergrades with the following race. **W** ♂ 67–74 mm (*n* 61,

m 70.0), ♀ 63–68 mm (*n* 11, *m* 66.0); **T** ♂ 48–56 mm (*n* 74, *m* 55.1), ♀ 46–53 mm (*n* 11, *m* 48.4); **T/W** *m* 74.3; **B** 11.1–12.7 mm (*n* 82, *m* 12.0); **Ts** 17.5–20.9 mm (*n* 60, *m* 19.4). **Wing formula: p1** > pc 1–7 mm, < p2 30.5–37 mm in ♂♂, 27.5–32 in ♀♀; **p2** =5/6 (93%), =6 (5%) or =6/7 (1%); **p7** < wt 7–13.5 mm; **p10** < wt 15–19.5 mm; **s1** < wt 16–20.5 mm. (Syn. *eversmanni*.)

P. t. yakutensis Ticehurst, 1935 (Russian Far East: Lena Basin, Yakutia, Anadyr; scant migrant through Middle East; winters E & S Africa). A mere fraction larger and on average even duller than the greyest *acredula*, but differences in both respects are slight, and the two are connected by a zone of intergradation. Differs from typical *acredula* on: (i) no yellow on breast, which instead is diffusely streaked or mottled brown-grey (a few have faint lemon 'stains' in fresh plumage, but any worn breeder should have no yellow; *acredula* has yellow streaking, sometimes faintly, but invariably some, even when breeding); (ii) undertail-coverts white (a few have a faint yellow tinge, but this is not typical; *acredula* invariably tinged pale lemon); (iii) upperparts dull brown-grey with faint olive cast on fore-crown, lower back and scapulars, and frequently a rather obvious green cast on rump (which appears paler than rest) and on edges of most wing- and tail-feathers (rear crown to mantle has generally very little or no green); and (iv) sides of neck and ear-coverts grey-brown without any yellowish-green. Supercilium variable, either entirely off-white, or yellowish on forepart (in front of eye). Since *yakutensis* migrants are in fairly fresh plumage, with more traces of yellow and green compared to when they become worn, they remain difficult to separate from dull and grey variants of *acredula*, and identification should only be attempted with the most typical individuals. Birds which are quite greyish, and which have all-white undertail-coverts, but which lack greyish mottling or diffuse streaking on breast should not be labelled *yakutensis*, since such birds occur rarely throughout the range of *acredula*. **W** ♂ 69–74 mm (*n* 12, *m* 71.7), ♀ 64–70 mm (*n* 9, *m* 66.4); **T** ♂ 49–56.5 mm (*n* 12, *m* 52.9), ♀ 47–53 mm (*n* 9, *m* 49.7); **B** 10.8–12.6 mm (*n* 21, *m* 11.7). **Wing formula: p1** > pc 1.5–5 mm, < p2 22.5–36 mm; **p2** =5/6 (common), =6 or =6/7 (rare); occasionally =5 or 4/5; **p6** < wt 2–9 mm; **p7** < wt 6.5–12.5 mm; **p10** < wt 13.5–18 mm; **s1** < wt 16–20 mm.

REFERENCES ALEXANDER, H. G. (1957) *BB*, 50: 307–308. – HEDENSTRÖM, A., LINDSTRÖM, Å. & PETTERSSON, J. (1995) *Ornis Svecica*, 5: 69–74. – VINICOMBE, K. (2006) *Birdwatch*, 166 (Apr): 22–23.

P. t. trochilus / acredula, presumed 1stW, France, Sep: some young Willow Warblers have slightly less prominent yellow wash on face and below. Note absence of ochrous-brown or buff tinges to these areas (except a hint on ear-coverts) often seen on many Chiffchaffs. Also note slender jizz with thin bill and legs (both mostly pale brown) and well-marked supercilium (enhanced by dark eye-stripe). The primary projection is also fairly long (> 3/4 of tertial length). Much of wing seems to be retained juv feathers. (A. Audevard)

P. t. yakutensis, Yakutia, NE Siberia, Jul: note dull or cold brown-grey upperparts and off-white underparts lacking any green or yellow tinge, including on the undertail-coverts. Separated from *tristis* Common Chiffchaff by longer and more pointed wing, with one emargination less on primaries (three instead of four). Also, legs are brown rather than blackish. (M. Waern)

P. t. yakutensis, Yakutia, NE Siberia, Jul: in head-on view, note tendency to have grey-brown streaking or suffusion across lower throat and breast, most clearly developed on sides. Facial pattern more distinct than on average *tristis* Common Chiffchaff with hint of dark lower edge to brown-tinged cheeks and ear-coverts. (M. Waern)

GOLDCREST
Regulus regulus (L., 1758)

Fr. – Roitelet huppé; Ger. – Wintergoldhähnchen
Sp. – Reyezuelo sencillo; Swe. – Kungsfågel

Wherever there are reasonable stands of conifers, especially of spruce and fir, one can expect to hear the high, thin calls of this, one of the smallest W Palearctic birds. (Firecrest and Wren are similarly small.) It is very widespread, occurring from Britain and N Fenno-Scandia east to N Japan and the Indian subcontinent, and west and south to the Azores and the Mediterranean region. Northerly populations move south and west in winter, when Goldcrests can be found in a wider range of habitats, even occurring in (occasionally huge) numbers in tiny patches of coastal scrub on migration.

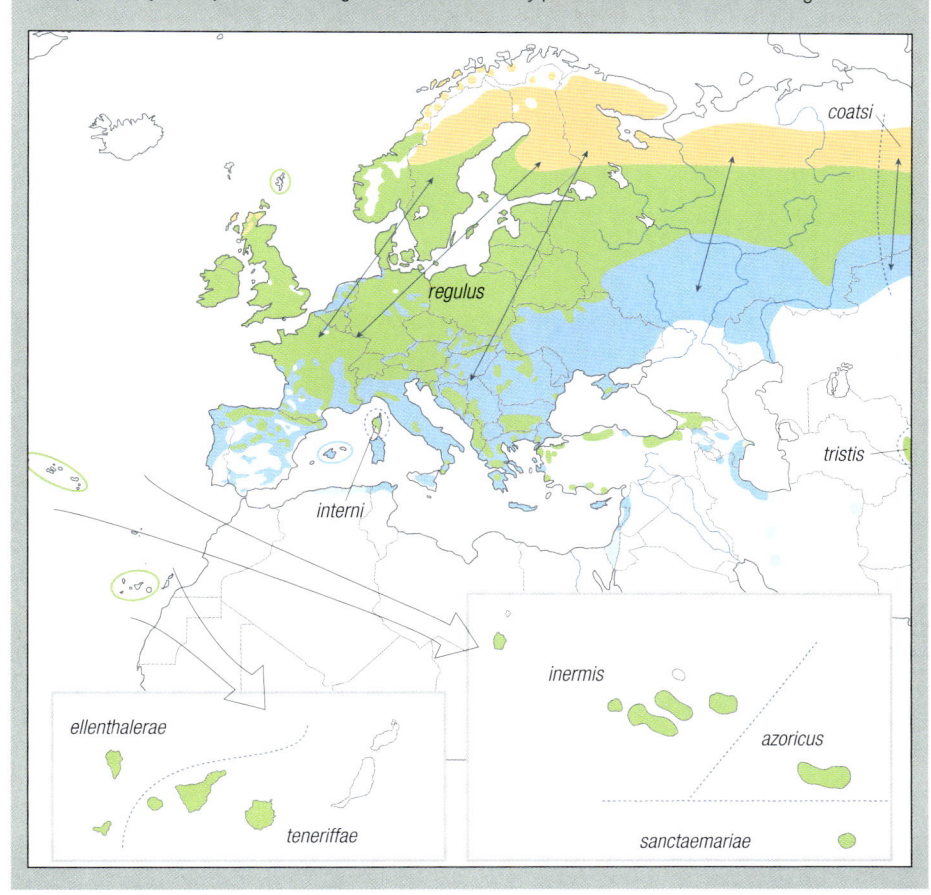

rarely any orange in ♀♀. Bill needle-like and dark (appearing as continuation of thin black moustachial stripes in head-on views, giving sad or surprised facial expression), and legs brownish. Tame but often difficult to detect unless heard, and typically moves restlessly, normally high up in the canopy, though does not dash around among the branches like some of the *Phylloscopus* species; in fact, often identified by constant *small movements with fluttering wings*. Flight thus lighter and slower than even smallest *Phylloscopus* warbler, characteristically forages nervously with constant wing- and tail-flicking, hovers frequently and may creep briefly up and down trunks. Often encountered on migration in small, loose groups, frequently with mixed parties of tits, treecreepers and warblers.

VOCALISATIONS Song a *cyclic* repetition (4–6 times or even a few more) of a thin, quite high-pitched (and therefore often inaudible to elderly people), rhythmic *pit***ee***tilü-pit***ee***tilü-pit***ee***tilü* or *eedle-eedle-eedle*…, rising a little in strength and ending in a Treecreeper-like flourish, *zezesu***zree***o* (or variants; such a complex concluding flourish is invariably absent in Firecrest). Alternative song, rarely heard and apparently only when agitated, lower-pitched, a variable warble of fine trills, soft whistles and mimicry. – Common call, often given in autumn, a thin, high-pitched, faintly sibilant *sree-sree-sree*, of 3–4 syllables, rarely just two notes or uttered singly; much softer and less emphatic than that of Firecrest. Also a shorter, thin *zit*, not unlike some tits, and more drawn-out *sriih*. Fine subdued tongue-clicking *pt, pt*, … also heard.

SIMILAR SPECIES Firecrest is the main confusion species, but confusion also possible with one of the small wing-barred *Phylloscopus*, e.g. *Yellow-browed* or *Pallas's Leaf Warblers*, which often join migrant flocks of crests, and have similar wing pattern. However, separating all of these species is rather easy, given reasonable views, and remember that Goldcrest is commoner over most of the covered region. Both species of vagrant *Phylloscopus* also possess bold, striking yellow-and-green head patterns, but lack the centrally black-demarcated crown patch of the crests (having instead long supercilia, dark crown-sides and, in Pallas's Leaf, an obvious central crown stripe), and thus have much less open-looking faces than Goldcrest. Both warblers, especially Pallas's Leaf Warbler, may hover, like Goldcrest, when Pallas's Leaf reveals its much brighter, yellower rump patch. They also lack Goldcrest's basal white primary patch, among other differences in wing pattern.

AGEING & SEXING Ageing possible during 1stY by moult pattern. Seasonal variation limited to normal and moderate

IDENTIFICATION Significantly *smaller than Common Chiffchaff*, with proportionately shorter neck, wings and tail, thus a *compact, plump-bodied little bird* with *tit-like behaviour*. If descending close enough, the *pale open face with its black peppercorn eye*, and the bold wing pattern may be briefly apparent. The Goldcrest is otherwise mostly *yellowish-green above* and *buff-white below*. Both sexes have an attractive *crown pattern with yellow (♀) or yellow and orange (♂) patches bordered laterally by bold black stripes*. The face is *pale dusky-olive*, and seen head-on the *diffusely pale, almost whitish eye-surround* is characteristic. Striking wing pattern includes *bold whitish greater coverts bar*, emphasized by *conspicuous black patch at base of secondaries* connected to a small white primary patch. Upper (median coverts) wing-bar is narrower and ill-defined, but reasonably distinct when fresh. *Tertials boldly fringed white*, especially prominent at tips, otherwise remiges and coverts are fringed yellowish-green. Upperparts largely pale olive, with yellowish suffusion on *slightly brighter rump*. Ocular region has grey tone, and sometimes faint greyish cast also on side of neck. Underparts off-white with slight buffish-cream tinge, any warm tones most evident on flanks in close views. Sexes almost alike, except for strong orange-red admixed in yellow crown patch of all post-juvenile ♂♂, but only very

R. r. regulus, ad ♂, Finland, Apr: among the smallest bird species within the covered region, being considerably smaller and daintier than Chiffchaff, appearing much like a 'rounded feather tuft'. ♂ has a diagnostic orange centre of crown (most visible when feathers are fluffed up in excitement or display), bordered by black lateral crown-stripes. Evenly quite fresh ad wing-feathers (including dark primary-coverts) and tail infer age. (M. Varesvuo)

R. r. regulus, Turkey, Mar: here in characteristic head-on view with mouse-like curious-looking facial expression. Solid black lateral crown-stripes and possible (but mostly concealed) faintly orange feathers in rear of central crown might suggest a ♂, but could also indicate an advanced/older ♀. Ageing impossible without handling, but rather boldly marked wing suggests ad. (K. Özkan)

R. r. regulus, presumed ad ♀, England, Feb: evenly fresh wing, rather dark primary-coverts and fairly rounded tips to tail-feathers suggest ad, and combination of rather dull plumage for an ad (including matt-yellow central crown surrounded by comparatively narrow dark lateral crown-stripes) suggests ad ♀. (G. Thoburn)

R. r. regulus, 1stS ♀, Denmark, Apr: rather dull and narrow yellow crown and less distinct dark lateral crown-stripes, as well as clearly abraded brown and pointed juv primaries, primary-coverts and tail-feathers identify this bird as 1stS ♀. The species' characteristic wing pattern includes bold pale greater coverts bar emphasised by conspicuous black patch at base of secondaries, and tertials boldly fringed whitish, especially tips. (S. E. Jensen)

R. r. regulus, Finland, Nov/Dec: flight weaker and more fluttering than even the smallest *Phylloscopus* warbler, and the bird may hang down below a branch, or hover while foraging. (M. Varesvuo)

feather wear. Sexes separable by close inspection of crown feathers. – Moults. Complete post-nuptial moult and partial post-juv moult roughly at the same time, generally within timespan of late Jun to mid Oct. Post-juv moult involves head, body, lesser and median coverts, some or all tertials, but no greater coverts. No pre-nuptial moult. – SPRING **Ad** With wear (both sexes), face and hindneck become purer grey and paler, wings and tail browner with narrower and duller greenish edges, and especially whitish tertial tips abraded and narrower or almost completely worn off. Wing-bars reduced, and especially median coverts bar may be almost invisible. Both sexes evenly feathered (cf. 1stS). **Ad ♂** Classic individuals sexed by yellow crown patch containing also some bright and glossy orange-red, especially at rear, though usually invisible on relaxed bird (but may be visible when feathers fluffed up, e.g. in display). In the hand, also check for generally narrower crown-feathers with very narrow whitish/greyish basal area and much orange instead of yellow over exterior part (Svensson 1992). Wing-length is useful supporting character. **Ad ♀** Yellow crown patch lacks visible orange even when crest raised (at least in 99%; exceptionally some advanced ♀♀ can show limited hint of orange with less gloss). In the hand check for somewhat shorter/broader-shaped crown-feathers with rather extensive whitish/greyish basal area and much yellow (on tips reaching as far as distal half) and limited or no orange over rest of exterior. Upperparts on average very slightly paler than ♂ and underparts marginally greyer, but much

R. r. regulus, ♂, presumed ad, Italy, Dec: when fresh and on cold winter days, often puffs out feathers like a ball. When fresh the bright olive-green upperparts and warm tones below are especially attractive. Combination of evenly fresh wing and rounder tips to tail-feathers suggest ad, while intense orange central crown clearly indicate a ♂. (D. Occhiato)

overlap and usually no useful difference. Again, check wing-length as supporting character for birds with hint of orange in crown patch. **1stS** Like respective ad but still has juv remiges, primary-coverts and tail-feathers, latter more worn and abraded browner with faded fringes (moult pattern as 1stW); usually duller overall with more heavily bleached and narrower greater coverts wing-bar (retained juv-feathers). –
AUTUMN Ad Compared to spring, both sexes more intense olive above, with some pinkish-brown below, and wings and tail are blacker with brighter yellowish-green fringes; wing marks more pronounced. Reliably aged by evenly very fresh wing- and tail-feathers. Tail-feathers on average broader with more rounded tips, and at least extremes can be aged according to shape after practice. **1stW** Like respective fresh ad, but for some less advanced birds the sexes are less clearly differentiated using crown features. To separate from ad, check for moult limits produced by retained juv greater coverts, primary-coverts, remiges and more narrow and pointed tail-feathers, which are subtly weaker textured and duller centred (less glossy and blackish) with less pure white, more buffish-cream fringes. **Juv** Soft fluffy appearance, paler bill, plainer head pattern with no yellow or black crown-stripes. Also, has duller upperparts, underparts are less suffused brown, and wing pattern is slightly duller.

BIOMETRICS (*regulus*) L 9–10 cm; **W** ♂ 54–57.5 mm (n 25, m 55.0), ♀ 49–54 mm (n 25, m 52.7); **T** ♂ 38–43 mm (n 25, m 40.1), ♀ 35–40 (n 25, m 38.0); **T/W** m 72.5; **B** 9.6–11.6 mm (n 45, m 10.5); **Ts** 16.1–18.2 mm (n 41, m 17.0). **Wing formula: p1** > pc 6–8 mm, < p2 18–24 mm; **p2** < wt 5–7 mm, =7–8; **p3** < wt (0) 0.5–1 mm; **pp4–5** about equal and longest (sometimes pp3–5); **p6** < wt 1–2.5 mm; **p7** < wt 4–5.5 mm; **p8** < wt 5–8.5 mm; **p10** < wt 8–11 mm; **s1** < wt 9–12 mm. Emarg. pp3–6.

GEOGRAPHICAL VARIATION & RANGE Over a dozen subspecies described, some being extralimital and not treated here. Most continental variation within covered range is clinal and quite subtle, whereas variation in Macaronesia is more marked and complex. – Breeding sometimes claimed in Elburz Mts, N Iran, by a local subspecies ('*hyrcanus*'), but no such population currently known to exist (A. Khaleghizadeh *in litt.*) and either now extinct or was mistakenly based on late-staying migrants from further north, presumably *regulus* of E European Russia or *coatsi* of W Siberia, so '*hyrcanus*' is not accepted here.

R. r. regulus (L., 1758) (Europe, W Siberia, much of Turkey, Caucasus; partly resident in northern parts, some migrating to C Europe in winter). Treated above. A very subtle trend noted to become paler and duller from north-west to south-east but hardly ground for subspecific division. – Birds in Britain & Ireland ('*anglorum*') separated on account of

R. r. regulus, 1stW ♂, France, Oct: in general, 1stW do not differ much from respective fresh ad (here a 1stW ♂ with unusually visible orange feathers on rear crown), but note that wing is not as contrastingly patterned as ad due to many retained juv feathers including all of the greater coverts. Very pointed and already abraded tips to tail-feathers (though difficult to judge here). (A. Audevard)

R. r. regulus, 1stW, presumed ♀, Switzerland, Sep: aged by retained juv remiges and primary-coverts (already not quite fresh) and sharply pointed tail-feathers. Tentatively sexed by the rather dull plumage with indistinct and partly brown lateral crown-stripes, and restricted and matt yellow centre of crown (though dullest young ♂ could be very similar, and safe sexing would require the crown-feathers to be spread). (H. Nussbaumer)

R. r. regulus, juv, Belgium, Jul: soft fluffy body-feathers, largely pale bill, and head lacking yellow and black crown-stripes, but dusky whitish surround to dark eye is still obvious. Pale pinkish-yellow gape-flanges present, typical of recently fledged juveniles. (J. Fouarge)

R. r. azoricus, ♂, São Miguel, Azores, Jul: heavily worn by midsummer, when orange on crown can be highly visible even without feathers being fluffed up. Racial plumage differences are difficult to appreciate on such worn and abraded birds, but note traces of olive-yellow below, and relatively strong bill of this population. (J. Normaja)

R. r. azoricus, ♂, presumed ad, São Miguel, Azores, Oct: here in very fresh plumage with the overall dusky appearance and typical yellowish-olive pigmentations, especially on mantle, neck-side and chest, while bill is also appreciably longer. Tiny visible orange feathering at rear of the yellow crown-stripe makes it ♂, while wing seeming evenly ad to suggest this age. (H. Shirihai)

R. r. inermis, ♂, Flores, Azores, Nov: averaging the darkest-tinged race (though not evident here), being dull olive-green above and often less clean below, with off-white underparts patchily washed olive on belly and flanks, and dull brown on breast; face also rather dusky. Bill rather long. Some visible orange feathers in yellow crown-stripe identify as ♂. Ageing is unsafe. (D. Occhiato)

R. r. inermis, 1stW, presumed ♂, Flores, Azores, Oct: dull olive-green above with light brownish-olive wash below characteristic of this taxon. The whitish surround to the dark eye is often less complete in this race, especially noticeable here. Some young birds are less advanced in their first autumn, being overall duller and less strongly patterned. Crown-stripe rather vividly yellow perhaps indicating a ♂; most wing-feathers are estimated to be juv. (D. Occhiato)

slightly darker and greener upperparts and shorter wing, but differences extremely slight and do not fulfil the 75% rule. – Breeders in NE Turkey, Crimea and Caucasus ('*buturlini*') tend to be very slightly paler and larger, but much overlap in size, and plumage colours close to or overlap with *regulus*, thus we follow Stepanyan (1978) by including these populations in *regulus*. (Syn. *anglorum*; *buturlini*; *sarepta*.)

○ **R. r. interni** Hartert, 1906 (Corsica; resident). Resembles *regulus* closely but is subtly darker, more cold green above, less yellowish-green, and rump is not so brightly yellowish-green. A tendency to have greyish tinge on nape and sides of neck (but diffuse in some). **W** ♂ 53–57 mm (*n* 17, *m* 54.6), ♀ 51–54 mm (*n* 10, *m* 53.0); **T** ♂ 37.5–42 mm (*n* 17, *m* 39.5), ♀ 36–40.5 mm (*n* 10, *m* 38.0); **T/W** *m* 72.1; **B** 10.3–11.6 mm (*n* 27, *m* 11.0); **Ts** (once 15.3) 16.2–18.2 mm (*n* 25, *m* 17.3).

R. r. inermis Murphy & Chapin, 1929 (Flores, Faial, Pico, São Jorge, Terceira in Azores; resident). The darkest race, upperparts a dark dusky olive-green with faint grey-brown tinge, underparts dusky off-white with patchy dirty olive on belly and flanks, and dull brown tinge on breast. Face pattern rather dusky. Very rarely underparts are more yellowish-tinged, then closely approaching *azoricus*. As so often in island populations, bill is proportionately somewhat longer. **W** ♂ 51.5–55 mm (*n* 12, *m* 53.3), ♀ 50–52.5 mm (*n* 12, *m* 51.0); **T** ♂ 37–41 mm (*n* 12, *m* 38.5), ♀ 35–39.5 mm (*n* 12, *m* 37.0); **T/W** *m* 72.4; **B** 11.6–13.2 mm (*n* 24, *m* 12.4); **Ts** 17.0–20.0 mm (*n* 24, *m* 18.4).

R. r. azoricus Seebohm, 1883 (São Miguel, E Azores; resident). Differs from *regulus*, and vast majority of *inermis*, in having in fresh plumage (autumn) underparts strongly tinged yellowish-olive (with dusky-brown tinges on breast), whereas spring birds are paler and whitish, still with traces of olive-yellow. Upperparts as *regulus* or slightly more saturated green. Size and structure as *inermis*, only tail proportionately slightly shorter. **W** ♂ 51.5–55.5 mm (*n* 12, *m* 53.3), ♀ 50–53 mm (*n* 12, *m* 51.1); **T** ♂ 35–40 mm (*n* 12, *m* 37.9), ♀ 34–38 mm (*n* 12, *m* 36.4); **T/W** *m* 71.2; **B** 11.7–13.1 mm (*n* 24, *m* 12.4); **Ts** 17.0–19.2 mm (*n* 24, *m* 18.0).

R. r. sanctaemariae Vaurie, 1954 (Santa Maria, SE Azores; resident). Resembles *inermis* but is paler throughout, upperparts similar to *regulus*, rather dull greyish-green, underparts dusky white with pale yellowish hue and dusky olive diffuse blotching and subtle buffish tinge on breast in some. A tendency to have more whitish eye-surround than *inermis*. Scant material examined. **W** 49–54 mm (*n* 10, *m* 52.8); **T** 38–40.5 mm (*n* 10, *m* 38.9); **T/W** *m* 73.7; **B** 11.3–12.7 mm (*n* 10, *m* 12.2); **Ts** 17.5–19.1 mm (*n* 10, *m* 18.3).

R. r. teneriffae Seebohm, 1883. 'TENERIFE GOLDCREST' (Tenerife, La Gomera, WC Canary Is; resident). Differs from *regulus* in smaller size, duller colours, slightly darker upperparts with less bright moss-green rump, and most obviously in broad black forehead patch merging the two black lateral crown-stripes at front, recalling pattern of Firecrest *R. ignicapilla*. Note also, compared to *regulus*, subtly longer bill and tarsus, reduced white edges/tips to tertials, and slightly deeper pink-buff on rear underparts and flanks. Song a mixture of Goldcrest *regulus* and Firecrest, a rather brief series of sharp, fine notes ending in a simple terminal flourish; does not seem to have the cyclic structure of *regulus*, and the song is slightly simpler, more approaching Firecrest and sometimes even ending with a single note, *see-see-see-see-charr*. Call more finely piping, not as 'lisping' as *regulus*. – Subtly larger with stronger bill than *ellenthalerae*; there is also a slight genetic difference (Päckert et al. 2006). The alleged difference in amount of orange on crown patch in ♂♂ between *teneriffae* and *ellenthalerae* is very subtle and only discriminates c. 60%. **W** ♂ 50–54.5 mm (*n* 19, *m* 51.9), ♀ 47–51 mm (*n* 11, *m* 51.0); **T** ♂ 35–39.5 mm (*n* 19, *m* 37.4), ♀ 33–38 mm (*n* 11, *m* 34.8); **T/W** *m* 71.6; **B** 10.5–12.3 mm (*n* 30, *m* 11.4); **BD** 2.3–2.6 mm (*n* 13, *m* 2.4); **Ts** 17.0–18.5 mm (*n* 23, *m* 17.8).

? **R. r. ellenthalerae** Päckert et al., 2006 (La Palma, El Hierro, W Canary Is; resident). Very similar to *teneriffae*, differing only by being subtly smaller, in DNA (Päckert et al.

R. r. teneriffae, ♂, Tenerife, Canary Is, Sep: the so-called Tenerife Goldcrest is marginally smaller than Goldcrest, differing also in the stronger crown pattern with broader lateral crown-stripes that join in black patch on forehead. Also has on average broader and more contrasting whitish eye patch, moderately developed grey-white base to forehead and underparts being deeper pink-buff. In other images shows orange bases on mid-crown, hence a ♂. (H. Shirihai)

2006) and in that ♂ has the golden-yellow crown patch on average less deeply saturated reddish-orange but more commonly moderately orange-yellow; still, many have more vivid orange (c. 40%) being inseparable from *teneriffae* on this character alone; ♀ crown patch is matt yellow. On average proportionately subtly smaller bill than *teneriffae*. A very subtle race, mainly described on genetic grounds. **W** ♂ 50–53.5 mm (n 12, m 51.1), ♀ 48.5–50 mm (n 11, m 49.1); **T** ♂ 35–38 mm (n 12, m 36.4), ♀ 33–36 mm (n 11, m 34.7); **T/W** m 71.0; **B** 10.1–12.0 mm (n 23, m 11.0); **BD** 2.0–2.5 mm (n 13, m 2.3); **Ts** 16.5–18.1 mm (n 19, m 17.4).

Extralimital subspecies:

? **R. r. coatsi** Sushkin, 1904 (W & SC Siberia south-east to Altai, W Baikal area; short-range migrant, perhaps rarely reaches easternmost parts of covered range in autumn/winter). Similar to *regulus* but allegedly paler above with more yellowish-green wash, and generally broader wing-bars and tertial edges. Said to average larger. No specimens examined by us.

R. r. tristis Pleske, 1892. 'CASPIAN GOLDCREST' (Transcaspia, W Central Asia; resident or short-range movements, possibly reaching Iran). Differs clearly from all other races in lacking or having strongly reduced crown pattern, in larger size and proportionately longer tail. More obvious sexual dimorphism than in other races: ♂ lacks nearly all traces of black marks on crown, these reduced to dark grey shades. ♂ has golden-yellow crown patch (without, or with just hint of, orange), whereas ♀ has no dark grey on crown, and crown patch is small and matt yellow, diffusely merging with olive-grey surround. Duller than *regulus*, limited green above, and dusky buffish-white below. **W** ♂ 55–59 mm (n 18, m 56.3), ♀ 54–57 mm (n 12, m 55.4); **T** ♂ 41–46 mm (n 17, m 43), ♀ 39–44 mm (n 12, m 41.5); **T/W** m 75.8; **B** 10.7–12.4 mm (n 28, m 11.5); **Ts** 16.8–18.9 mm (n 23, m 17.7).

TAXONOMIC NOTES The taxon *tristis* clearly merits more research regarding its best taxonomic status, as the differences in morphology seem rather marked in comparison to all other subspecies. – Sometimes the populations on Canary Is are regarded as a separate species, 'Tenerife Goldcrest', but we feel that this is still based on inadequate evidence. Both plumage and vocalisations much resemble Goldcrest, and genetic differences are apparently moderate and not conclusive.

REFERENCES COLLINSON, M. (2006) *BB*, 99: 306–323. – LÖHRL, H., THALER, E. & CHRISTIE, D. A. (1996) *BB*, 86: 379–386. – PÄCKERT, M. *et al.* (2003) *Evolution*, 57: 616–629. – PÄCKERT, M. *et al.* (2006) *J. Avian Biol.*, 37: 364–380.

R. r. teneriffae, ♂ (left) and ♂/♀ (right), Tenerife, Canary Islands, Sep: this taxon spends considerable time creeping vertically around tree trunks in nuthatch fashion while foraging. Note the contrasting whitish eye-and-forehead mask and very thick black connection of forecrown (best seen on right bird). In the field, the left bird showed orange bases on mid-crown, revealing the sex as ♂, while the right bird never puffed the crown-feathers to allow sexing but the less territorial behaviour suggested that it could be ♀ or perhaps young ♂. (H. Shirihai)

R. r. teneriffae, La Gomera, Canary Is, May: Tenerife Goldcrest is mostly like Goldcrest in crown pattern, but at least some birds tend towards Firecrest pattern, having broader lateral crown-stripes that join on forehead. However, it lacks the black eye-stripe and golden-yellow lower neck-side patches of both forms of Firecrest. (R Debruyne)

R. r. ellenthalerae, La Palma, Canary Is, Nov: generally like ssp. *teneriffae*, differing only in being subtly smaller, in DNA and in that ♂ has duller and on average more orange-yellow crown patch. Ageing is unsure. (H. Nussbaumer)

R. r. teneriffae, juv, La Gomera, Canary Is, May: even in juv plumage the diagnostic broad lateral crown-stripes that join on the forehead are obvious in Tenerife Goldcrest. (R Debruyne)

(COMMON) FIRECREST
Regulus ignicapilla (Temminck, 1820)

Fr. – Roitelet triple-bandeau; Ger. – Sommergoldhähnchen
Sp. – Reyezuelo listado; Swe. – Brandkronad kungsfågel

Equally small as its much commoner relative, the Goldcrest, this attractive species is decidedly rarer and less easily found, especially in the breeding season, when it is easily overlooked, unless one is familiar with its voice. Firecrest breeds in a range of mixed and broadleaf woodlands across much of Europe except in the north, east to Turkey and the Caucasus, and just reaching into NW Africa in the south. Northern and eastern populations are largely migratory, but western and southern populations are more sedentary, although even those usually make local movements in winter.

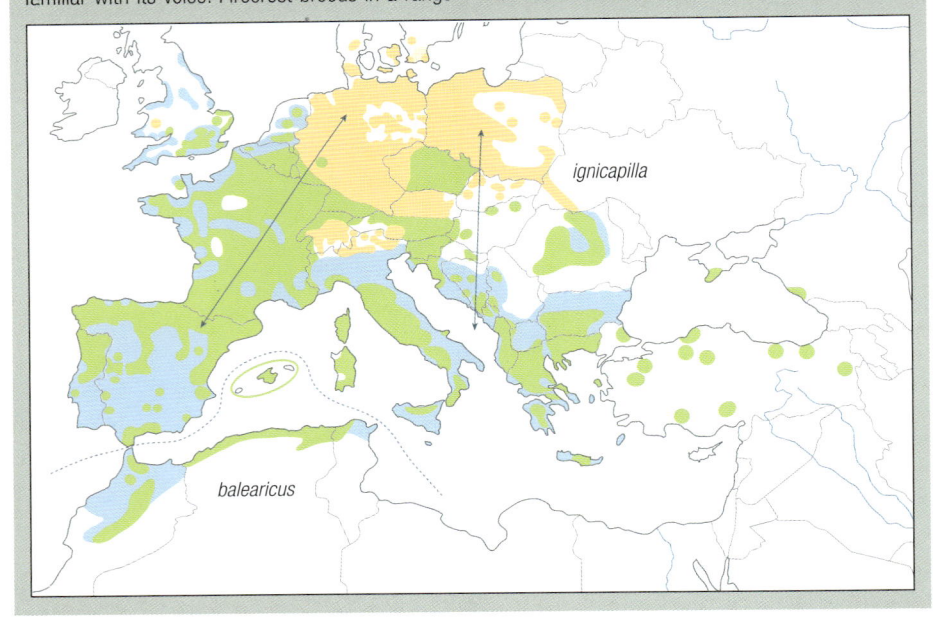

ignicapilla

balearicus

relaxed. Also check for diagnostic *black eye-stripe* and eye-ring, and *off-white upper cheek-patch* just below the eye; dull grey ear-coverts separated from white throat by black moustachial stripe. Has characteristic *ill-defined bronze-coloured lower-neck-side patches* merging with olive-green rest of upperparts (entire upperparts brighter than in Goldcrest). Wing pattern as Goldcrest with double whitish wing-bars, broader and bolder on greater coverts, connecting with white basal primary patch, but overall pattern less bold owing to slightly less contrasting blacker basal remiges and greater coverts. White tertial edges and tips smaller and less well defined than in Goldcrest. Usually appears cleaner off-white below with greyish-ochre wash largely restricted to breast-sides. Needle-like, blackish bill slightly longer than Goldcrest's. Gait, flight and behaviour much like Goldcrest, although probably slightly less restless. Tired migrants can have plumage even more puffed-out than Goldcrest, emphasizing their exceptional beauty. Throughout the year, much less dependent on conifers than its commoner counterpart, but usually found in the canopy of needle trees when singing during the breeding season, and more rarely found in low cover on migration or in winter.

VOCALISATIONS Song may sound like Goldcrest, especially to inexperienced observers, but is a noticeably more uniform or monotonous repetition of insect-like sharp, high-pitched notes (frequently difficult to hear for elderly people) and ending with a single and more stressed note, *zü zu-ze-zi-zi-zi-ziss*; there is usually a slight acceleration in the strophe, lacking Goldcrest's cyclic rhythm and more complex terminal flourish. Firecrest's song in general carries further. – Frequently-heard calls are also fairly similar, but are subtly lower-pitched, sharper and more emphatic: commonly uttered in a series of 3–4, e.g. *ze-ze-zriss*, or single *ze* or *zit*, invariably sharp and 'piercing' in tone.

SIMILAR SPECIES Greatest risk of confusion is with *Goldcrest*, although many features permit easy separation. Most important to confirm is Firecrest's bold white supercilium and almost equally striking black eye-stripe and broader black border to crown patch (these features are all much reduced in juveniles, but still sufficiently well developed to ensure separation from Goldcrest). Colour of central crown patch rather reversed in ♂♂ compared to Goldcrest, i.e. orange-red is at front of crown, not the rear (also visible on relaxed bird, but concealed in ♂ Goldcrest unless feathers are fluffed-up). In addition, Firecrest has an often quite obvious golden-brown ('bronze') shoulder patch (never shown by Goldcrest), a deeper, brighter green mantle, especially in fresh plumage, emphasized by Firecrest being cleaner, brighter white below. On breeding grounds, song is also a useful feature. – Although the *Madeiran Firecrest* is resident on Madeira to where Firecrest normally does not move, the theoretical vagrant Firecrest can be eliminated by having a longer and broader white supercilium, and a black eye-stripe that reaches a little further back. – Given poor views, might be confused with one of the small *Phylloscopus* warblers that wander to the treated region, mainly in late autumn (especially *Pallas's Leaf Warbler*), but the following traits are unique to Firecrest: stark contrast between white supercilium and blacker eye- and lateral crown-stripes, black forecrown patch (much less obvious in juvenile, though), and striking orange forepart to crown patch, bronzy-yellow wash on lower neck-sides, and clearly more striking wing pattern with diagnostic white basal primary patch. Firecrest also lacks the bold yellowish-white rump of Pallas's Leaf.

R. i. ignicapilla, ♂, presumed ad, Belgium, Mar: ♂ is readily sexed by brilliant golden-orange colour to crown patch (yellow only at rear and sometimes narrowly on sides). Unlike its closest relative, Goldcrest, Firecrest has diagnostic bold white supercilium and almost equally striking blackish eye-stripe. Less worn and bright plumage, plus rounded tail-feather tips suggest an ad, but impossible to age with certainty. (J. Fouarge)

IDENTIFICATION Perhaps slightly larger-headed than similarly minute Goldcrest, to which it superficially bears a close resemblance, but is unquestionably more striking and attractive, owing to its strongly-striped head consisting of a *pure white supercilium* (altogether lacking in Goldcrest), *broader black lateral crown-stripes that meet above buff forehead*, and striking *reddish golden-orange* (or yellow in ♀) *central crown patch* that becomes yellower at rear. True colour of crown patch requires good view but unlike in Goldcrest, orange colour in ♂ is also visible when feathers

R. i. ignicapilla, England, Apr: age unsure (wing seems rather evenly fresh like ad, but tips of tail-feathers pointed and abraded like 1stY) and also unclear if it is ad ♀ or 1stS ♂ with limited orange in rear crown. Apart from diagnostic head pattern, also note the characteristic bronze-yellow shoulder patch, merging into the olive-green upperparts (brighter than Goldcrest). (S. Fisher)

R. i. ignicapilla, ad ♀, Germany, Apr: combination of evenly fresh wing (for this time of year, including primary-coverts), and pale yellow central crown patch (with no visible orange) identify the bird as ad ♀. Wing pattern is similar to Goldcrest, with double whitish wing-bars, broad and bold on greater coverts, connected to white basal primary patch, whereas white tertial edges and tips average smaller and less well defined than in Goldcrest.

R. i. ignicapilla, 1stS ♀, Sweden, Apr: combination of lack of visible orange tinge in the crown and less boldly patterned wing (mostly juv, with worn brownish primary-coverts) identify as 1stS ♀. Even in young dull ♀♀, diagnostic strongly striped head pattern is evident, with pure white supercilium and quite dark eye-stripe. Rather clean off-white below. Needle-like blackish bill is slightly longer than that of Goldcrest. (L. Olsson)

AGEING & SEXING Ageing requires close scrutiny of moult and feather wear in wing and tail, and is still not always possible. Seasonal variation limited to feather wear. Sexes separable in close views. – Moults. Complete post-nuptial and partial post-juv in late summer (mainly Jul–Sep). Post-juv moult affects mostly head, body, lesser and median coverts, and some or all tertials. Pre-nuptial moult absent. – SPRING **Ad** With wear (both sexes), head marks become slightly duller, and wing and tail browner with narrower and duller fringes, especially whitish tertial edges become abraded or almost completely wear off, while wing-bars are reduced (median-coverts bar may become almost invisible). Upperparts greyer. Both sexes evenly feathered (cf. 1stS). **Ad ♂** Usually readily sexed by brilliant orange or reddish crown patch, with yellow limited to rear and sometimes narrowly on sides. Wing-length is a useful supporting character. 'Shoulder' (side of nape/neck) on average brighter bronze-coloured than in ♀. **Ad ♀** Crown patch predominately yellow, with little and only very faint orange in a minority. Upperparts and bronze 'shoulder' often slightly duller. Difference in crown pattern often somewhat obscured by wear, making sexing less practicable. **1stS** Like ad but has juv remiges, primary-coverts and tail-feathers, latter being more worn and browner with faded fringes (moult pattern as 1stW); some generally duller compared to ad, with narrower/whiter greater-coverts wing-bar (retained juv-feathers). – AUTUMN **Ad** Both sexes aged by evenly very fresh wing- and tail-feathers (remiges and primary-coverts centred glossy greyish-black). Compared to spring, sexes more intensely and brighter plumaged, and dark and pale areas of wings bolder. **1stW** Much like ad, but sexes differ less clearly in crown features. From fresh ad, check for (subtle!) moult limits produced by juv greater coverts, primary-coverts, remiges and tail-feathers (latter also more pointed), which are weaker textured, duller centred (less glossed and less blackish) and fringed more buffish-cream. **Juv** Clearly differs from ad by soft fluffy appearance and lack of yellow/orange central crown patch, but has blackish eye-stripe, cream-buff to whitish supercilium and partially indicated cheek-mark, features absent in juv Goldcrest.

BIOMETRICS L 9–10 cm; **W** ♂ 51.5–57 mm (n 51, m 54.2), ♀ 48–52.5 mm (n 25, m 50.9); **T** ♂ 36–43 mm (n 51, m 40.1), ♀ 35–39 mm (n 25, m 37.2); **T/W** m 73.7; **B** 10.3–12.3 mm (n 73, m 11.3); **Ts** 15.4–18.6 mm (n 59, m 17.4). Wing formula: **p1** > pc 5–10.5 mm, < **p2** 16.5–22 mm; **p2** < wt 5.5–9 mm, =7/8 or 8 (55%) or =8/9 or 9 (45%); **p3** < wt 0.5–2 mm; **pp4–5** longest; **p6** < wt 0.5–1 mm; **p7** < wt 3.5–5.5 mm; **p8** < wt 5–8 mm; **p10** < wt 8.5–10.5 mm; **s1** < wt 9–11.5 mm. Emarg. pp3–6.

R. i. ignicapilla, ♂, presumed ad, Italy, Nov: apparently very fresh wing and immaculate plumage suggest an ad. Sexing at least is unproblematic—a ♂ based on orange median crown-stripe. Unique head pattern, comprising pure white supercilium (altogether lacking in Goldcrest), broader black lateral crown-stripes that meet above rufous-buff forehead, and diagnostic black eye-stripe. Golden-brown shoulder patch is also obvious. (D. Occhiato)

R. i. ignicapilla, ♂, Italy, Oct: slightly paler and weaker head pattern with paler orange crown-stripe, ill-defined dark eye-stripe and less pure grey cheeks compared to previous bird. The duller appearance could be explained by it being a young bird: pointed and already slightly worn tail-tips indicate 1stW, but remiges appear too dark and fresh for young, and without better view of coverts (or handling) is best left un-aged. (D. Occhiato)

GEOGRAPHICAL VARIATION & RANGE Very modest variation. Only two subspecies treated here, differing only subtly, but two more may be valid. See Taxonomic notes.

R. i. ignicapilla (Temminck, 1820) (C & S Europe south of Baltic except on Balearics; local Turkey; resident in west and south, short-range migrant in north and east). Described above. – Birds of Corsica ('*minor*') were described as being subtly smaller, having a paler and more coldly green back, and slightly more greyish rear ear-coverts. However, none of these criteria hold when longer series are compared, or at least are only faintly suggested. (Syn. *minor*.)

○ *R. i. balearicus* Jordans, 1923 (Balearics, NW Africa). Differs subtly but apparently consistently from *ignicapilla* in having whiter underparts lacking obvious yellowish-green tinge on throat and chest (but can be slightly dusky on upper chest), and by slightly paler and 'cooler' or more greyish-green upperparts (*ignicapilla*: more yellowish-green and slightly darker upperparts). There is a tendency for ♀♀ to have a slight orange tinge on the crown patch, more so than in *ignicapilla*, sexual differentiation thus being a little more difficult (but not impossible) to perform. **W** ♂ 52–56.5 mm (*n* 17, *m* 53.7), ♀ 50–54 mm (*n* 9, *m* 51.9); **T** ♂ 37–42 mm (*n* 17, *m* 39.6), ♀ 36–40 mm (*n* 9, *m* 38.1); **T/W** *m* 73.6; **B** 10.5–12.4 mm (*n* 24, *m* 11.3); **Ts** 16.4–18.0 mm (*n* 20, *m* 17.1).

R. i. ignicapilla, 1stW, presumed ♂, Italy, Oct: some birds are more easily aged—here mostly juv wing-feathers (primaries and primary-coverts notably abraded and browner) give the age—while sex can be more difficult; the very broad and black lateral crown-stripes and bright yellow median crown-stripe might infer a young ♂, still difficult to be sure. (D. Occhiato)

TAXONOMIC NOTES Madeira Firecrest *R. madeirensis*, until recently regarded as a race of the present species, is treated separately here on the basis of different morphology, vocalisations and genetics. – Two more claimed races have been impossible to evaluate due to lack of available specimens: breeders of W Caucasus and W Transcaucasia ('*caucasicus* Stepanyan, 1998') separated by slightly paler and brighter coloration, and those of S Crimea ('*tauricus* Redkin, 2001') on somewhat darker colours. Size for both apparently the same as for *ignicapilla*.

REFERENCES Hartert, E. (1906) *BBOC*, 16: 45–46. – Jackson, C. H. W. (1992) *Ring. & Migr.*, 13: 127. – von Jordans, A. (1924) *J. f. Orn.*, 72: 165–166. – Päckert, M. *et al.* (2003) *Evolution*, 57: 616–629. – Parrot, C. (1910) *Orn. Jahrb.*, 21: 156–157. – Redkin, Y. (2001) *Ornithologia*, 29: 98–104. – Rogers, M. J. (1970) *BB*, 63: 179. – Stepanyan, L. S. (1998) *Zool. Zhurn.*, 77: 1077–1079.

R. i. ignicapilla, ad ♀, Germany, Oct: note that unlike Goldcrest the dark centres to the ad greater coverts are not as intensely black, but nevertheless the dark and rounded primary-coverts and very dark remiges confirm the age, and corroborate that the paler yellow median crown-stripe equals an ad ♀, rather than poorly-marked young ♂. Thus, ageing is sometimes essential to achieve reliable sexing. (T. Krüger)

R. i. ignicapilla, 1stW ♀, Italy, Oct: the least-marked plumage, including wing pattern (most of the visible wing-feathers are juv, especially the dull and more pointed primary-coverts). Again, pure white supercilium and cheek patch, and black eye-stripe and eye-ring, are best means of separation from Goldcrest. (D. Occhiato)

R. i. ignicapilla, juv, Belgium, Jul: blackish eye-stripe, and cream-buff to whitish supercilium and cheeks are still present in juv, and secure this bird's separation from juv Goldcrest. (J. Fouarge)

MADEIRA FIRECREST
Regulus madeirensis Harcourt, 1851

Fr. – Roitelet de Madère; Ger. – Madeiragoldhähnchen
Sp. – Reyezuelo de Madeira; Swe. – Madeirakungsfågel

Restricted to Madeira where it is the only regularly occurring *Regulus* species. This highly localised form has only recently attracted renewed attention from European field ornithologists and taxonomists, and it is now generally regarded as a separate species, rather than as a subspecies of Firecrest, which it resembles most.

Ad, Madeira, Sep: even in post-nuptial moult the species shows very limited seasonal plumage variation, with ad in autumn very similar to spring. (T. Kuppel)

Ad ♂, Madeira, Apr: resembles Firecrest but uniquely has a narrow white bar across forehead, appearing as an extension of the white supercilium. Supercilium moreover is shorter, often creating impression of 'white spectacles'. Median crown-stripe broader and squarer than in other crests and differs in being bright glossy golden-orange or tawny in ♂. Also note patch on neck-sides is brighter ochraceous-yellow and more contrasting. Aged by being evenly rather fresh. (H. Shirihai)

Ad ♀, Madeira, May: ♀ is generally duller than ♂, but separation is still not always easy. Note especially the narrower black lateral crown-stripes and less bold eye-stripe, more dusky whitish supercilium and cheeks, and paler orange-yellow median crown-stripe. Relatively little worn wings and even feathering at this season identify this bird as ad. (M. Danzenbaker)

IDENTIFICATION Closely recalls Common Firecrest in plumage, habits and behaviour, but orange crown patch is duller in ♂♂, *white supercilium much shorter* ending just behind eye (this and white mark below eye give more the impression of expanded eye-crescents), while the *black eye-stripe is correspondingly shorter*. Very useful additional field characters are the slightly but noticeably *longer bill* and very slightly *longer legs*, giving the bird a different 'feel'. Furthermore, upperparts are even brighter green, especially in fresh plumage, contrasting with the blacker basal wing-coverts and *bronzy-golden 'shoulder' (lower-neck-side) patch*, which is also more striking than in Firecrest. The *bolder and more extensive white tertial tips* are closer to Goldcrest than to Firecrest.

VOCALISATIONS Song a hurried series of sharp, fine notes, ending on a more stressed one than it starts. In general, the song is closer to Firecrest than to Goldcrest, although playback studies have revealed that the species does not respond to song of Firecrest from mainland Europe (Päckert *et al.* 2003). – Several short and monosyllabic calls resemble those of Firecrest, but perhaps distinct are slightly more drawn-out and shrill *wheez* and a desolate *peep*, latter at a distance slightly recalling *tristis* Common Chiffchaff. There is also a thin upward-inflected *suueet*, recalling Yellow-browed Warbler at a distance, but at close range this call has a shrill or strident tone, almost like Tree Pipit.

SIMILAR SPECIES The only *Regulus* on Madeira, and therefore it should pose no identification problems given a reasonable view.

AGEING & SEXING As in Firecrest.

BIOMETRICS L 9–10 cm; W ♂ 52–58 mm (*n* 16, *m* 56.2), ♀ 51–55 mm (*n* 15, *m* 52.4); T ♂ 37.5–42 mm (*n* 16, *m* 40.6), ♀ 36–41 mm (*n* 15, *m* 38.1); T/W *m* 72.5; B 11.2–12.9 mm (*n* 27, *m* 12.1); BD 2.4–3.3 mm (*n* 20, *m* 2.8); Ts 18.5–20.6 mm (*n* 28, *m* 19.6). **Wing formula:** p1 > pc 9–12 mm, < p2 15–19 mm; **p2** < wt 7–10 mm, =8/ss; **p3** < wt 1–3 mm; **pp4–5** about equal and longest; **p6** < wt 0.5–2 mm; **p7** < wt 2–5 mm; **p8** < wt 3.5–6.5 mm; **p10** < wt 7–10 mm; **s1** < wt 8.5–10.5 mm. Emarg. pp3–6.

GEOGRAPHICAL VARIATION & RANGE Monotypic. – Madeira; resident.

TAXONOMIC NOTE Previously usually treated as a subspecies of Firecrest, but increasingly considered a separate species in recent years. Morphological and acoustical differences, and genetic divergence, seem sufficient to warrant such recognition.

REFERENCES PÄCKERT, M. *et al.* (2003) *Evolution*, 57: 616–629.

Presumed 1stW, Madeira, Oct: combination of rather dull plumage, and seemingly already worn tips to primaries and tail-feathers (latter also pointed), and appearance of retained juv primary-coverts suggests age; sexing is difficult or impossible outside breeding season and if not seen in pairs. (J. J. Nurmi)

Juv, Madeira, Sep: juv is very similar to same-age Firecrest, but face pattern is generally more obscure, thereby somewhat more resembling juv Goldcrest. (M. Römhild)

VAGRANTS TO THE REGION

This section covers those very rare passerine vagrants that have seldom been recorded in the treated region (see Introduction for geographical coverage of the handbook). In general, until the end of 2016, fewer than ten individuals of each species included here are believed to have reached the covered region unassisted and in a wild state. Records of rare species are made continuously, and for some the limit of ten birds has already been exceeded. Although lists of rare birds often focus on the number of occasions that a certain species has been found, i.e. 'records', we feel that it is more interesting to know the number of birds involved. One record can involve more than one bird.

Descriptions in this section are kept short, and the number of photographs low, to save space and because most of these species are neither part of the normal avifauna of the region, nor even of the entire Palearctic. They have often arrived from distant regions due to extreme weather conditions or imperfect navigation abilities in young and inexperienced birds. Some may even have been partly or wholly ship-assisted, but details of this are understandably largely unknown, and all records listed are thought to involve natural movements of wild birds (i.e. excluding known or strongly suspected introductions and escapes, often referred to as 'category D species'). It is hoped that these descriptions will suffice as a first introduction to identification, as well as summarising known records in the covered region.

The texts are not always of the same length, some East Palearctic species being afforded somewhat more detailed accounts reflecting the anticipated greater interest in their identification and taxonomy. Also, species more challenging to identify due to their similarity to related species have often received more attention than, say, a fairly straightforward Indian Pitta or Banded Martin.

We gratefully acknowledge the valuable help in compiling the list of records for each species and country, largely done by José Luis Copete and Marcel Haas. They have in turn corresponded with the various national Rarity & Record Committees or similar bodies to as far as possible ensure that all records are officially accepted. Others who in various ways helped assemble and check the data, in particular regarding the Middle East, namely Guy Kirwan, Nigel Redman and Magnus Ullman, are also gratefully thanked.

Vagrancy, and its underlying reasons, has always attracted interest both among keen amateur ornithologists (notably by 'twitchers' focusing on rare birds) and by professional ornithologists or scientists trying to understand the mechanisms behind these movements. By listing all of these rare species and their records and numbers within the covered range, an up-to-date tool is provided for further study and analysis. For those wanting to delve deeper into the subject we recommend reading introductory sections in, e.g., Alström *et al.* (1991), Haas (2012) and Howell *et al.* (2014), and a useful paper by Thorup (2004).

EASTERN KINGBIRD *Tyrannus tyrannus* (L., 1758)

Ireland two (Oct 2012, Sep 2013), Britain one (Sep 2016). Breeds throughout much of North America; winters in South America south to Argentina. Often encountered in woodland clearings, around farms, in orchards; often seen near water. **IDENTIFICATION** A large tyrant flycatcher (**L** 18–22 cm) with *relatively small head* but *strong black bill*. Typically has *dull black head*, *slate-grey back* and contrasting *silky-white underparts*, although *breast is tinged grey*. Tail black with a *broad white terminal band*. Wing-coverts and secondaries edged pale grey to white when fresh. Crown-feathers often raised when alert to form *peaked crown*, which then might reveal otherwise concealed *vermilion central crown-patch* (better developed in ♂). Bill and legs black. Often perches in the open atop a lookout. In flight, wings typically look small and narrow. Call a piercing, slightly harsh, high-pitched *dzeet*, also given in series, sometimes mixed with dry chattering notes. **AGEING & SEXING** 1stW paler and duller above than ad, more brownish, lacking orange crown-patch at least until Dec. Also often has scruffy feathers in autumn due to incomplete post-juv moult prior to migration. Otherwise, moults largely take place on the winter grounds. Limited seasonal plumage variation, mainly due to wear and bleaching. Sexes similar. **GEOGRAPHICAL VARIATION** Monotypic.

Peru, Oct (left) and Maryland, USA, Jun (right): note upright posture, small head, with peaked crown, strong, straight bill and square-ended rather long tail. Black head and white-tipped tail are typical, plus inconspicuous white fringing on dark brownish wing and almost clean white underparts. (Left: H. Shirihai; right D. Monticelli)

EASTERN PHOEBE *Sayornis phoebe* (Latham, 1790)

Britain one (Apr 1987). Breeds predominantly in E & C North America; winters S USA and Mexico. Breeds in woodland, farmland and often suburbs if some trees are present. **IDENTIFICATION** Rather large (**L** 16–18 cm), relatively *long-tailed flycatcher* with olive brownish-grey upperparts, and *head is clearly darker*. Pale wing-bars only moderately developed, becoming even more indistinct when worn, but *light tertial edges* more visible. Underparts mostly whitish-cream with a *rather dark olive wash on sides of breast*. In autumn characteristically *sullied yellow below*. Further characterised by *all-black bill*, including lower mandible. Being a phoebe it habitually *pumps tail downwards and spreads it* and, unlike *Empidonax* flycatchers, *lacks distinct whitish eye-ring*. Typically perches in a rather upright position, from which it makes regular sorties to catch airborne insects. Call a sharp, short, loud, slightly metallic *tsyp* or *chip*, which when repeated in series can recall song of Zitting Cisticola! **AGEING & SEXING** 1stW as ad, but a few juv greater coverts (outer) and tertials are usually retained, being distinctly buff-tipped and creating moult limits (none in evenly fresh autumn ad). Rectrices more pointed than in ad. After partial pre-nuptial moult ageing more difficult and usually impossible. Sexes alike, and weak seasonal plumage variation other than that caused by effects of wear and bleaching. **GEOGRAPHICAL VARIATION** Monotypic.

Canada, May: dusky head, and upperparts slightly paler brown-grey, while underparts are whitish with a faint yellow wash, chiefly on belly and undertail-coverts. Note also some faint grey-brown at sides of upper breast and off-white edges to wing-feathers. Whitish double wing-bars subdued. (B. Lasenby)

S USA, Nov: the yellow wash below is deeper and more extensive in autumn. Note also narrow and rather square tail, often pumped down, while bare parts are more or less black. The lack of a white eye-ring separates it from *Empidonax* flycatchers. (B. Taylor Barr)

VAGRANTS TO THE REGION

ACADIAN FLYCATCHER *Empidonax virescens* (Vieillot, 1818)

Iceland one (Nov 1967), Britain one (Sep 2015). This North American tyrant flycatcher is a migrant breeding in NE America and SW Ontario; migration passes through E Mexico and the Caribbean, winters from S Central America to NW South America. Found in open forests, woodland and farmland and often in suburbs with sufficient tall vegetation. **IDENTIFICATION** Medium-sized (**L** 13.5–15 cm), long-winged flycatcher, with flattish forehead but *peaked rear crown*. Ad generally has olive- or greenish-tinged grey upperparts and quite striking wing pattern with *bold double yellowish-white bars*, while underparts are whitish (throat cleanest white), though *breast is olive-grey* and *belly and vent pale lemon*. Also has quite pronounced narrow but complete pale eye-ring, and a *broad-based bill* with convex sides and with dark upper mandible and yellowish lower. Legs grey. Needs to be separated with care from several congeners. Main call is a loud *peez* or *whit!* Also utters a squeaky, almost 'lisping', downslurred *pwee-est*. **AGEING & SEXING Ad** In late summer and early autumn fresh, wing-bars average whiter yellowish (less buff-tinged) and slightly narrower than in 1stW. Underparts average whiter (yellow tinge less prominent). Tips of tail-feathers average broader and more rounded. Still, many are similar and ageing can be difficult. – **1stW** In Sep–Oct starts to show some wear of primary tips. Wing-bars average broader and more buffy-yellow than in ad. Underparts average more yellowish, and tail-feathers slightly narrower and more pointed at tips. Birds likely to reach W Europe are 1stW. After partial pre-nuptial moult in late winter ageing becomes more difficult or impossible. Sexes alike. No obvious seasonal plumage variation. **GEOGRAPHICAL VARIATION** Monotypic.

Presumed 1stW, England, Sep: small greenish- and yellow-tinged flycatcher with thin white eye-ring and broad whitish wing-bars. Note broad bill with yellowish base to lower mandible. Already some wear at tips to remiges, quite broad wing-bars and deeper yellow underparts all indicate 1stW. (D. Monticelli)

ALDER FLYCATCHER *Empidonax alnorum* Brewster, 1895

Iceland one (Oct 2003), Britain two (Oct 2008, Sep 2010), Norway one (Sep 2016). Breeds widely in brushy habitats near bogs and in birch and alder thickets in N North America; winters mainly in W South America. On migration in wide variety of vegetated habitats. **IDENTIFICATION** Medium-sized flycatcher (**L** 14–16 cm) with characteristic combination of *strongly developed pale wing markings* and *yellow-orange lower mandible*. Very similar to Willow Flycatcher (*E. traillii*; not recorded within the treated region), but *bill is slightly shorter*, eye-ring usually *more prominent on upper half*, and has slightly *more olive* (less brown) *upperparts*. Differs from some races of Willow Flycatcher by marginally darker head and better defined light tertial edges and bolder wing-bars, while rather longer primary projection can be supportive. Best identified by voice, with song a typical three-syllable phrase, *free-**bree**-oh*. Main call a loud and explosive, short *chrip* or *prip*, but more characteristic is a buzzing, 'bent' whistle *zwee-oo*. Identification by wing formula requires trapping. **AGEING & SEXING** Ageing often possible by moult pattern in wings and tail; rather complex moults mostly occur just prior to migration and on the winter grounds (strategies appear to differ between populations). At least until mid or late autumn, **1stW** can be safely aged by rather fresh plumage, while still unmoulted **ad** is worn or shows strong moult contrasts. In spring, all ages similar. Birds likely to reach W Europe are 1stW, which tend to have some yellowish-buff wash below, making the throat appear whiter, and often pale wing-bars tinged buff-brown. Sexes alike, and no obvious seasonal plumage variation. **GEOGRAPHICAL VARIATION** Monotypic.

1stW, Norway, Oct: note combination of characteristic prominent light wing markings and largely yellow-orange lower mandible. Difficult to distinguish from Willow Flycatcher (not recorded in the covered region), but at least shorter bill is indicative (and identity of this bird later confirmed by DNA). Brownish-tinged whitish wing-bars and some warm hue in greyish upperparts infer age. (C. Tiller)

1stW, England, Sep: Alder Flycatcher differs from at least some races of Willow Flycatcher by its slightly darker head, shorter bill and bolder whitish wing-bars; sharply defined light tertial edges, more greenish-tinged than brownish upperparts and rather longer primary projection are sometimes supportive, too. Safely 1stW based on rather fresh plumage, but interesting to note variation compared to the Norwegian record. (G. Reszeter)

LEAST FLYCATCHER *Empidonax minimus* (W. M. & S. F. Baird, 1843)

Iceland one (Oct 2003). A North American tyrant flycatcher breeding in deciduous woodland, orchards and parks from W and SE Canada to NW and E USA; winters in Mexico and Central America. **IDENTIFICATION** A *small and compact* migrant tyrant flycatcher (**L** 11–13 cm) with quite *greyish* plumage. Characteristically *large-headed*, with *bold white eye-ring* and short triangular-shaped bill with lower mandible largely pale straw. Rather *short primary projection*. Throat whitish, forming some contrast with grey-washed breast that gradually becomes more yellowish on belly and deeper still on undertail-coverts. Upperparts greyish-olive, although crown and nape grey with little or no greenish tinge often appearing a bit darker and duskier than the purer greyish-olive rest of upperparts. Note also *bold whitish double wing-bars*, lemon-tinged pale remex-edges and narrow, almost square tail. Main call on migration a sharp, piercing *whit* or *pwit*, which is sometimes also given in a long series, c. two calls per second. **AGEING & SEXING** Partial post-juv moult occurs on the breeding grounds, and ad post-nuptial moult commences on the breeding grounds but is completed after autumn migration. Thus, fresher birds in autumn with buff-tinged wing-bars are 1stW, which also tend to have more yellowish wash below. Sexes alike, and no obvious seasonal plumage variation. **GEOGRAPHICAL VARIATION** Monotypic.

Far left: 1stW, Iceland, Oct: quite small and generally drab in colour with olive-grey upperparts, whitish throat, grey-washed breast, and yellowish-tinged belly and undertail-coverts. Note bold white eye-ring and two white wing-bars. Legs and feet are black. Fresh in autumn with subtly buff-tinged wing-bars infer 1stW. (J. O. Hilmarsson)

Left: Mexico, Apr: note short, flattened triangular bill with pale lower mandible, and longish tail. Rather strongly washed yellowish below. (H. Shirihai)

EASTERN WOOD PEEWEE *Contopus virens* (L., 1766)

Azores two (Oct 2015). Breeds in SC & SE Canada and E USA; winters N South America. Breeds in broad-leaved or mixed woodland, often at edges or near glades. **IDENTIFICATION** Rather large (**L** 15–17 cm), *long-tailed and long-winged flycatcher* with greyish upperparts with olive tinge, but *head a little darker. Double pale wing-bars obvious* and contrasting, white in ad, buff-tinged in young, becoming more subdued when worn; also, *light tertial edges* visible in fresh plumage. Underparts mostly dusky, but throat and vent/undertail-coverts whiter, latter with dusky bases creating slightly variegated pattern. Note *dark bill with pinkish-yellow base to lower mandible*. Thin and faint whitish eye-ring only visible at close range. Unlike phoebes or *Empidonax* flycatchers has *no obvious tail movements*. Perches like other American flycatchers high up and well visible in a rather upright posture, making darting sallies after insects. Call either a liquid, short, discreet *tslip* or a high-pitched, squeaky, drawn-out and upslurred *peweee* (as its name suggests). **AGEING & SEXING** 1stW as ad, but upperparts feathering finely buff-tipped, and buff wing-bars average wider. Ad has plainer upperparts (any paler tips very narrow and insignificant) and on average narrower pale olive-grey wing-bars. Rectrices more pointed than in ad. Ad moults completely late summer to early autumn, young birds apparently partly (though some birds seem to suspend and complete in winter). Sexes alike, and weak seasonal plumage variation other than that caused by effects of wear and bleaching. **GEOGRAPHICAL VARIATION** Monotypic.

Azores, Oct: though identifying tyrant flycatchers can be challenging, Eastern Wood Pewee is characteristically greyer overall (rather than having mostly olive-brown upperparts), with longer wings and prominent whitish wing-bars and tertial edgings. Unlike *Empidonax* species it lacks pale eye-ring. (V. Legrand)

Mexico, Apr: note contrasting paler belly and typical bicoloured bill. A hint of a crest can be raised to form peaked crown. Quite a long-tailed species. (H. Shirihai)

INDIAN PITTA *Pitta brachyura* (L., 1766)

Iran one (Nov 1968). The only member of the highly diverse and colourful family Pittidae to have occurred within the treated region, being predominantly found in the Oriental region. The Indian Pitta breeds from foothills of the Himalayas to C & SW India; migratory, with characteristic south-westerly movement at the end of the rainy season, returning to breeding grounds in advance of SW monsoon; winters in C and S India and Sri Lanka. **IDENTIFICATION** Rather large (**L** 17–19 cm) and robust ground-dweller in undergrowth of dense subtropical jungle, a characteristically *rather long-legged and short-tailed, compact and rotund* bird. Told by its striking plumage pattern, with *a bold black stripe through eye* contrasting with *white throat and buff-white supercilium*, and *olive-buff lateral crown-stripes separated by a black stripe along centre of crown*. Upperparts green, *rump and wing-bend iridescent azure-blue*. Underparts mustard-yellow with *pale scarlet central belly and undertail-coverts*. In flight, shows *small white patch on wing*. Bill and eye mostly black, legs pale pinkish-brown. If seen well unlikely to be confused with any other Pitta. Main call a sharp and loud, disyllabic whistle, the second note descending, *weee-aaaahr*. **AGEING & SEXING Ad** Sexes rather similar. Limited seasonal variation in colour. **Juv** Distinctive, has orange-red bill with black band near tip and smoky underparts with scarlet on belly faint and hardly visible. **1stY** Similar to ad, but young still in moult recognised by retained juv remiges, duller green upperparts and diluted or partial red patch on lower belly. **GEOGRAPHICAL VARIATION** Monotypic.

Sri Lanka, Dec: unmistakable, still beware of some similarity with unrecorded African Pitta *Pitta angolensis* (found as far north as in Addis Ababa, Ethiopia), but this lacks Indian's white supercilium and throat, while having larger red ventral patch extending to belly, and different wing pattern. (H. Shirihai)

CHESTNUT-HEADED SPARROW LARK *Eremopterix signatus* (Oustalet, 1886)

Israel one (May 1983). Breeds from NE Kenya to S Sudan, Somalia and Ethiopia, where occurs in dry short-grass in *Acacia* savannas, with movements in response to rains. **IDENTIFICATION** *Small*, compact and dainty (**L** 11.5–12.5 cm), but at the same time *rather long-legged*. The striking ♂ has *a rounded white patch on cheeks* (in Great Tit fashion) narrowly connected with *white on neck-sides and lower nape*. There is also a smaller *white patch on centre of crown*, and all of these white areas contrast with a complex pattern of *dark chestnut-brown and black forepart of 'face', sides of crown and lower neck-sides*. There is also a *dark central stripe on the underparts*, becoming wider and blacker on the belly. Mainly *light grey-brown above* with off-white fringes to most wing-coverts. ♀ to some degree mirrors ♂ plumage, but pattern and colours are much more subdued: especially note rufous-brown supercilium, buffish-white cheeks, contrasting with *darker lower neck-side patches* and a ragged dark central breast-line and blacker belly. Both sexes have a *large pale bill* and pale (pinkish) legs. Beware that 1stY ♂ Black-crowned Sparrow Lark (p. 33) has variable and often abraded browner (almost chestnut-tinged) black areas—which can also be much less extensive than in ad ♂—and such an incompletely patterned ♂ Black-crowned is superficially similar to ♂

♂, *Ethiopia, Aug*: note pure white crown patch and cheeks in ♂ (extent of both varies individually, partly age-related; see main text) surrounded by chestnut pattern connected to the blacker belly by narrow breast-line. Rest of upperparts are pale brown with bold whitish edgings. Bone-coloured bill characteristically heavy with curved culmen. (H. Shirihai)

♀, *Ethiopia, Aug*: pattern vaguely mirroring ♂'s plumage, but much duller and more washed-out with brown-streaked crown and upperparts, and dusky and diffusely streaked head and breast pattern. There is a hint of a rufous supercilium and central breast-line, and blackish blotches on breast and belly usually coalesce to a solid patch in centre. (H. Shirihai)

Chestnut-headed. ♀ Black-crowned Sparrow Lark is much more uniform above than ♀/imm of Chestnut-headed, lacking any obvious dark on the underparts, having a whitish throat and neck-sides, and no rufous in the supercilium. Flight call a sharp *chip* or *chip-up*. **AGEING & SEXING** **Ad** Sexes differ (see above). **1stY** Much like respective ad following complete post-juv moult, but young ♂♂ may to a variable extent have attained the dark plumage of ad. **GEOGRAPHICAL VARIATION** Ssp. *harrisoni* (SE Sudan, SW Ethiopia and NW Kenya) recorded as vagrant in Israel.

BANDED MARTIN *Neophedina cincta* (Boddaert, 1783)

Yemen one (Mar 1982), Egypt one (Nov 1988), Saudi Arabia one (Oct 1996). Resident and intra-African migrant. Scattered populations in open habitats, often near water and in montane grassland and alpine moorland, e.g. Ethiopian highlands (ssp. *erlangeri*). **IDENTIFICATION** Like a large, compact Sand Martin (**L** 15–17 cm). Compared to Sand Martin has *broader and more solid breast-band* and diagnostic *white supra-loral mark*, while usually notably larger overall size serves as additional clue. Note also *whitish underwing-coverts* in flight (Sand Martin brownish). Tail square-ended (somewhat forked in Sand Martin). Utters a squeaky song, which sometimes culminates in a short trill, and also gives loud, short chattering calls. **AGEING & SEXING** **Ad** Sexes alike. Medium brown upperparts relieved only by white eyebrow in front of eye, with predominantly white underparts, from throat to undertail-coverts, except brown broad breast-band, with slightly ragged lower edge, and whitish underwing-coverts contrasting with dark brown flight feathers and under-tail. **1stY** Much like ad following post-juv moult. **Juv** Like ad but has a much less distinct breast-band and cream-coloured to rufous fringes to the upperparts feathers. **GEOGRAPHICAL VARIATION** The likely race to reach the region is Ethiopian *erlangeri*. Previously often placed in genus *Riparia*.

Ethiopia, Sep: only liable to be confused with Sand Martin but has a peculiar white supraloral marking, and on average broader brown breast-band. Its distinctively larger size is usually the first thing noticed in the field. Differs from Sand Martin also by the white underwing-coverts usually visible in flight. (H. Shirihai)

TREE SWALLOW *Tachycineta bicolor* (Vieillot, 1808)

Britain two (Jun 1990, May 2002), Azores eight (four Oct–Nov 2005, two Oct 2007, one Oct 2012, one Oct 2013), Iceland one (May 2012). This transatlantic vagrant is a widespread North American migrant summer visitor; winters mainly in Mexico, Central America and the Caribbean. **IDENTIFICATION** A small, compact swallow (**L** 13–14 cm) characterised by *iridescent blue-green upperparts, white underparts* and a *slightly forked tail*. Has a darker loral mask, and its *paler blue-green rump* may be visible in close views. Often, the *dark crown and head-sides* and *contrasting white throat* are eye-catching. Underwing-coverts grey-brown. In the treated region could only be confused with Common House Martin (which see), if the latter's white rump is not seen. A variety of high-pitched chirps or chatters can be heard from birds in flight, at times a little reminiscent of some calls from Short-toed Lark. **AGEING & SEXING** Ad and 1stY undergo complete post-nuptial and post-juv moults that commence on the breeding grounds and are completed in winter. In autumn, juv and 1stW are duller grey-brown above than ad and may show a hint of a grey breast-band. **GEOGRAPHICAL VARIATION** Monotypic.

Ad, Iceland, May: unlike superficially similar House Martin lacks white rump and has greener-tinged upperparts. (S. Ásgeirsson)

Ad, Cuba, Mar: glossy blue-green above and white below with a thin black eye mask. In flight, note small size but jizz of a chunky swallow with a shallowly forked tail and broad-based triangular wings. Glides more than many other swallows. (H. Shirihai)

1stW, Azores, Oct: young birds in autumn duller with more grey-brown in upperparts, and may show a hint of a blurry grey-brown breast-band merging into whitish (often sullied buffy-cream) rest of underparts. When fresh, tertials and secondaries edged pale. (Left: V. Legrand; right: D. Monticelli)

PURPLE MARTIN *Progne subis* (L., 1758)

Azores two (Sep 2004, Oct 2011), Britain one (Sep 2004). Although declining over much of North America (which could explain its scarcity as a transatlantic vagrant), it is still locally common where suitable nest sites are available. A long-distance migrant, wintering in South America. **IDENTIFICATION** Large and powerful martin (**L** 19–21 cm) with shallowly forked tail. ♂ is *dark, glossy blue* (actually rather little purple hue, and looks blackish from a distance), with paler fringes only on sides of rump and uppertail-coverts. *Underwing-coverts blackish*. ♀ *sullied and mottled greyish below*, with grey-brown face, some limited blue gloss on upperparts, too, and *forehead and hindcollar grey*. Duskier throat can appear a bit streaky and mottled if seen close, and whitish breast is finely dark-streaked. 1stS ♂♂ often develop partial though extensive blue coloration, including below. Typical swallow flight, but often appears heavy-fronted with rather slow wingbeats, and short glides alternate with flapping wingbeats (has even been compared with Starling!).

Ad ♂, USA, Apr: the largest North American swallow. Ad ♂ is an attractive glossy dark purple-blue with brown-black wings and tail. (J. Normaja)

Ad ♀, Chicago, USA, Apr: largely mottled grey, but some steel-blue sheen discernible on crown, mantle and shoulders, and has a darker facial mask (partly created by shadow). Note the slightly forked tail and dusky-white but grey-flecked undertail-coverts. (T. Zurowski)

1stW, which is most likely to reach the treated region, is close to ♀, but duller still, with both *dark streaking and breast-band better developed on whiter ground*; darkish grey-brown above, with more *contrasting whitish-grey forehead and collar, creating darker mask*, but any blue-black sheen above is limited. Calls, mostly given in flight, involve varied rather deep-voiced and twangy *chrep*, often rolled and burry, and buzzes and crackles from birds giving subsong. **AGEING & SEXING** At least in autumn due to suspended moult in both age categories, many ad and 1stY still readily separated, young being fresher and more evenly feathered than any ad ♀ with incomplete moult of flight-feathers. Ad moults remiges from late summer to early winter, whereas young usually only begin their moult once they have reached the winter grounds. In spring, all birds return after complete moult and cannot be aged. **GEOGRAPHICAL VARIATION** In general only moderate geographical variation, although several races recognised. Race to have reached Europe as vagrant has not been established.

1stW, Azores, Oct: note clearly marked pale neck collar, darker mask and contrasting whitish-grey forehead. Also, whitish lightly buff-sullied underparts show some streaking and a breast-band. Darkish grey-brown above, while variable blue-black sheen above is limited. Note also the slightly forked tail. (V. Legrand)

ETHIOPIAN SWALLOW *Hirundo aethiopica* Blanford, 1869

Israel one (Mar 1991). Resident in SE Mali, W Nigeria to Eritrea and south to W Uganda and NE Tanzania; also scattered occurrence in Senegal to Togo. Mostly open areas in montane grassland, broadleaf woodland, *Acacia* and arid savannas, thornbush below 2750 m, and often around habitation. **IDENTIFICATION** Separated from European races of Barn Swallow (p. 118) by *pale orange-buff* (ad) or *cream-white* (young) *throat that extends to breast*, bordered by a *broken black band* (rather narrow) on the upper breast. Tail-streamers shorter than in Barn Swallow of respective age and sex, with more white on central rectrices. Beware that both American Barn Swallows (*erythrogaster*) and East Asian birds (*gutturalis*) habitually have incomplete blue-black breast-band and that young of these have a paler orange throat. Confusion could also arise with Wire-tailed Swallow (which see), especially if this species' diagnostic entirely chestnut crown is not seen (in Ethiopian Swallow any reddish on crown is restricted to forehead). Beware that young Ethiopian has whitish-cream or buff throat (see below), as is usually true (in all ages) of Wire-tailed Swallow. Beware also that, although Wire-tailed has no or only poorly indicated breast-band, there is some overlap between them as to this character. Ethiopian Swallow also lacks the incredibly long and narrow tail-streamers of ad Wire-tailed, but these can be broken or missing due to moult, or not yet fully developed in young. Song a melodious twittering. Also gives various *cheep* calls, or variants of this. **AGEING & SEXING Ad** Sexes alike. Upperparts deep blue, with a restricted reddish forehead, buff-orange throat and clean white underparts, with very narrow broken black breast-band. Tail well forked with extensive white on central rectrices, but rather short or only moderately long 'streamers'. **1stY** Much like ad once post-juv moult concluded. **Juv** Overall duller and browner than ad, with reddish on forehead paler and diffuse, and whitish-cream or buff throat. **GEOGRAPHICAL VARIATION & RANGE** Ssp. *aethiopica* (W & C Ethiopia, Kenya south to Tanzania and west to Senegal) and *amadoni* (E Ethiopia, Somalia and NE Kenya). Both could be potential vagrants to the treated region, but the Israeli record was not identified to subspecies.

Ad, Ghana, Mar: note diagnostic combination of very narrow broken dark breast-band bordering pale orange throat and bib, and a very restricted reddish forecrown. Tail well forked but with rather short 'streamers'. (H. Shirihai)

LESSER STRIPED SWALLOW *Cecropis abyssinica* (Guérin-Méneville, 1843)

Oman one (Dec 1986). Widespread in Africa south of the Sahara except in the south-western part of the continent. Resident or short-range migrant in north and south of breeding range, with seasonal movements in non-breeding season towards centre of range. Frequents grassland near forest edges and open woodland, e.g. in savanna and montane grassland, especially in the vicinity of freshwater rivers and lakes, up to 3000 m altitude. **IDENTIFICATION** Fairly large swallow (**L** 15–19 cm), which shares some features (e.g. red rump) with Red-rumped Swallow, but is easily separated from this species by *very prominent black-streaked white underparts* and an almost *entirely bright rufous head* (only chin and throat whitish and dark-streaked). Squeaky initial notes to song, which quickly becomes more nasal and descending. Also gives a slightly downslurred, wheezy *cheeew*. **AGEING & SEXING Ad** Sexes similar, but ♀ has shorter tail-streamers. Deep blue upperparts with browner flight-feathers, long tail-streamers, a rufous head, white underparts from throat to undertail-coverts, heavily streaked black from throat to belly, small rufous thigh-patches, buffish undertail-coverts and a white band across undertail. **1stY** After post-juv moult much like ad. **Juv** Compared to ad has shorter tail, overall duller plumage, darker crown, paler rump and pale tawny tips to the wing-coverts and some secondaries. **GEOGRAPHICAL VARIATION & RANGE** Ssp. *abyssinica* (E Sudan, Eritrea and Ethiopia) is likely to be the source of vagrancy to the treated region.

Ad, Ethiopia (left: Sep) and Cameroon (right: May): easily separated from superficially similar Red-rumped Swallow by being the only swallow in the treated region with black-streaked white underparts and an entirely rufous head. (H. Shirihai)

ASIAN HOUSE MARTIN *Delichon dasypus* (Bonaparte, 1850)

United Arab Emirates three (Oct 1999, Nov 2001, Feb–Mar 2008). A close relative of Common House Martin *D. urbicum*, and their ranges come into contact locally in Asia. Has a wide distribution primarily in East Asia from Transbaikalia and N Mongolia east to Japan, Korea, south to Tibet, Kashmir, Himalaya and much of China; winters in South-East Asia. Found in similar open habitats as Common House Martin, including near habitation for nesting. **IDENTIFICATION** Resembles Common House Martin but is *smaller and more compact*, with subtly shorter wings and clearly shorter tail, the *tail also being more shallowly forked*. Underparts not pure white, invariably *washed ashy-brown*, though bleached birds in strong light might look white. It has a more *fluttering and 'light' flight*, in which the on average *darker underwing-coverts* may visibly contrast with the somewhat paler rest of underwing (lacking the white coverts of Common House, although these rarely show well in flight). It should be stated that, in all these traits, Asian House Martin is more similar to the Asian subspecies of Common House Martin (*lagopodum*) rendering safe separation difficult. The *white rump patch extends on average a little further onto lower back* (but the longer uppertail-coverts are black unlike in ssp. *lagopodum*). A more subordinate and variable character is a little less gloss on upperparts. In close views, Asian House Martin can be seen to have *more extensive blue-black on sides of head, reaching a little bit further onto sides of chin*, running narrowly across chin, and fractionally darker upperparts overall. Flight calls very similar to Common House Martin but perhaps on average more trilling and 'electric' in tone. **AGEING & SEXING Ad** Sexes similar. Tertials with no or only thin pale fringes. Fair amount of blue gloss above. **1stY** Much like ad after post-juv moult. **Juv** Compared to ad has broader whitish fringes to tertials, and overall duller plumage with less blue gloss above. **GEOGRAPHICAL VARIATION** Polytypic, but the three races described are quite similar and difficult to separate in the field. Ssp. *dasypus* (E Siberia, Kuriles, Sakhalin, Korea, Japan), perhaps the most likely to reach the covered region, is described above; *cashmiriense* (N Pakistan, Himalayas, C China) differs in being subtly paler blue-black above and purer white below, while *nigrimentale* (S & E China; unlikely to reach the treated region) is like *dasypus*, only a little smaller.

Japan, Apr: very similar to Common House Martin but unlike that has relatively short tail, and less clean white underparts, these being faintly brown-tinged. (S. Price)

Presumed 1stW, Hong Kong, Nov: dark brown underwing-coverts a clue but not diagnostic, as many Common House Martins can appear similar, but note combination of compact shape due to shorter and less forked tail and rounder wing, and more extensive dark cheek area that extends over the throat-sides. Underparts never pure white, but impression will depend on light and reflections. Possibly 1stW in advanced complete post-juv moult. (J. & J. Holmes)

GOLDEN PIPIT *Tmetothylacus tenellus* (Cabanis, 1878)

Oman one (Jun 1983). Resident and intra-African migrant, in bushy and wooded grassland in dry country, light *Acacia* savanna and other scrubby areas. Breeds from SE Sudan, NW Somalia and E Ethiopia to NE & C Tanzania; from sea-level to 1800 m. Local movements occur, often in response to rains. **IDENTIFICATION** (**L** 13.5–15 cm.) ♂ especially distinctive and should not be confused with any other species within the covered region, having *deep yellow underparts and supercilium* with a *broad black breast-band*. Upperparts greenish-grey with dark brown streaking from crown to back. Most of the wing-feather tracts are yellow tipped black (except all-yellow lesser and some greater coverts). In flight, especially note the distinctive broad black wingtips. Except for the central rectrices, tail largely yellow (feathers only dark subterminally). ♀ much more like other pipits, thus much less easily identified if seen alone, being *mainly plain buff below* (only some pale yellow on belly and undertail-coverts) with a pale supercilium and dark-streaked grey-brown upperparts. Note *yellow tail-sides and very narrow pale yellowish fringes to flight-feathers*. In flight, the *yellow underwing-coverts* should provide a good clue (♂♂ have virtually the entire underwing yellow, this colour being restricted to coverts in ♀♀). Both sexes have mostly dark bill and pale legs. Perches freely in trees, moving tail up and down. Whistled song, given in flight or from perch, is described as having a weaver-like quality, but is generally silent, except after rains. Sometimes gives a hurried warbling series when descending to ground from a treetop. **AGEING & SEXING Ad** Sexes differ as described above. **1stY** Much like ♀, but may retain some breast spotting, and young ♂ shows yellow in secondaries. Young best aged by moult pattern with at least some retained juv wing-feathers. If aged correctly, young can sometimes be sexed in autumn, too. **Juv** Like ♀ but browner with streaked breast. **GEOGRAPHICAL VARIATION** Monotypic.

♂, Ethiopia, Sep: distinctive and should not be confused with other pipits due to bright yellow underparts and black breast-band, but also because of striking yellow wing panel. (H. Shirihai)

♀, Ethiopia, Sep: very different from the ♂, being mainly plain buff below (just some pale yellow on belly and undertail-coverts) with hint of broad, pale supercilium, yellow tail-sides and pale yellowish and rather narrow fringes to the flight-feathers. In flight, look for the yellow underwing-coverts. (H. Shirihai)

PADDYFIELD PIPIT *Anthus rufulus* Vieillot, 1818

Iran two (Dec 2010). A resident and partially migratory breeder in open scrub, grassland and cultivation, in Central Asia west to C & SE Afghanistan and in S Asia in NW Pakistan and east to the Philippines. **IDENTIFICATION** Belongs to the group of 'large pipits' (**L** 15–16.5 cm) and hence needs to be compared with Richard's, Tawny, Blyth's and Long-billed Pipits, with all of which it theoretically could overlap in range through vagrancy to the south-eastern part of the treated region. Within its size group at the smaller end, with *moderately streaked and not so warm-coloured upperparts*, but the *breast is usually prominently streaked*, while the *loral stripe is quite poorly marked*. Paddyfield has a distinctive and *diagnostic call* that if heard eliminates all the other four large species—see below. In plumage very similar to Richard's Pipit but is *usually visibly much smaller*, with slightly more horizontal carriage. Further, it has a proportionately *shorter tail, smaller bill* and *shorter legs*. Hind claw also *much shorter*. Flight weak and rather fluttering, and it *does not habitually hover before landing* in tall grass like Richard's. Compared to most Richard's, Paddyfield is *less streaked and less warm-toned above*, and can show a narrow dark loral

A. r. rufulus, Sri Lanka, Nov: two birds showing variation from rather streaky plumage with stronger facial markings (left) to more plain-backed and pale-lored (right). Paddyfield Pipit has only recently been recorded in Iran, but it could well be overlooked due to extreme similarity with Richard's Pipit, although fortunately flight calls are clearly different, and Paddyfield lacks Richard's hovering before landing in grassy fields. Most Paddyfields also should average smaller-sized, shorter-legged and shorter-tailed with more pointed bill, and they are usually less heavily streaked above but on average have narrower dark loral line. The other tricky pipit to separate from Paddyfield is poorly known sedentary race *A. c. eximius* of African Pipit of SW Arabia (which see). Compare also Blyth's Pipit. Based on present knowledge, reliable field separation of all these large pipits without calls seems unlikely. (H. Shirihai)

stripe. Shape and carriage more closely resemble Tawny Pipit, although it is *shorter-tailed*, often walking slightly more upright. Also, compared to Tawny, it has *more strongly streaked upperparts, and the lateral throat-stripe and breast-streaking are better developed* (beware that degree of streaking can vary, with overlap between the two due to age-related and individual variation). In Paddyfield, *lores are pale and almost unmarked* or have a rather faint mark, unlike Tawny's distinct dark stripe. Often almost identical in plumage and closer in size to Blyth's, but *dumpier in shape than Blyth's*, with a particularly *short tail*. Frequently the best distinction is the *weaker-streaked and duller upperparts*; unlike Blyth's, Paddyfield Pipit frequently shows a narrow and faint dark loral stripe. Told from Long-billed Pipit by considerably *smaller size* and *shorter bill and tail*, but has *stronger facial markings* and *heavier-marked and richer-coloured upperparts*. Call an emphatic, 'fat' and metallic clicking *chlitt*, often repeated, usually slow but at times quicker, *chlet-et-et-et*. **AGEING & SEXING Ad** Sexes similar. **Juv** Easily distinguished from ad using same criteria as in Richard's and Tawny Pipits, which see. **1stY** Separable if juv primary-coverts can be recognised by their pattern, and the same applies if some juv median and greater coverts are retained; however, many young moult these completely to ad pattern, and ageing in the field is generally not advisable. **GEOGRAPHICAL VARIATION & RANGE** Polytypic, with usually five subspecies recognised. On geographical grounds, vagrancy to Iran is most likely attributable to ssp. *rufulus* (SE Afghanistan, much of Indian Subcontinent, Sri Lanka, adjacent parts of China).

CEDAR WAXWING *Bombycilla cedrorum* Vieillot, 1808

Britain six (Jun 1985, Feb–Mar 1996, Sep 2013, three Jun 2015), Iceland two (Apr–Jul 1989, Oct 2003), Ireland three (Oct 2009, Nov 2012, Jun 2015), Azores two (Oct 2010, Oct 2013). Breeds in North America, in S Alaska, N & E Canada to S USA; winters from S Canada to C Panama, exceptionally in N South America; just like its Eurasian counterpart, the Bohemian Waxwing, it performs winter movements that brings it further south (e.g. non-breeders regularly reach Costa Rica). Found in open habitats where berries are available, and in North America can be highly gregarious on migration and in winter. **IDENTIFICATION** *Smaller* (**L** 15–16.5 cm) and *browner than Bohemian Waxwing*. Unlike Bohemian, *belly is tinged pale yellowish, grading whiter onto undertail-coverts* (in Bohemian, belly is grey and undertail-coverts chestnut). It also lacks Bohemian's yellow (or white) tips to flight-feathers and primary-coverts (missing on secondaries, too). However, it shares with Bohemian the dark facial mask and bib (though bib is small and diffuse), as well as yellow-tipped tail and the waxy red appendages on secondaries. Otherwise a typical *Bombycilla* in having sleek, velvety plumage, being long-winged, rather short-tailed, with a jagged, swept-back crest (variable in size), a short bill and short legs. Main call a piercing, very high-pitched Goldcrest-like *seeeee* or *ssssir*, like a trilled whistle, not loud but penetrating, although easily missed by older ears. **AGEING & SEXING Ad** Sexes rather similar, with a larger number of waxy red tips in ♂ (and increasing with age: 3–9 in ad, 0–7 in 1stY) than in ♀ (1–7 in ad, 0–3 in 1stY), and chin in ♂ glossy black (duller and more brownish-black in ♀). In addition, ♂ has on average broader yellow tail-tip (slight overlap, though). **1stY** As breeding usually occurs later in summer, the partial juv plumage is kept well into autumn, including the dusky streaking on breast, often until Oct–Dec. Primary-coverts also narrow, tapered to a point, relatively abraded and tinged grey-brown. Tail-feathers also narrower with duller and smaller yellow tips. **GEOGRAPHICAL VARIATION** Monotypic.

♂♂, ad (left: USA, June) and 1stS (Mexico, Mar): note diagnostic yellowish-tinged lower body and vent and lack of Bohemian Waxwing's yellow/white tips to outer webs of primaries (but sometimes on fresh primaries, variable number of very tiny white tips visible, as on left bird). Both are ♂♂ by extensive black chin. Reduced number of secondaries with red waxy tips infer young age of right bird. (Left: D. Monticelli; right: H. Shirihai)

1stW, ♂ (left) & ♀, Azores, Oct: lack of yellow/white tips to outer webs of primaries, and also of chestnut undertail-coverts, excludes Bohemian Waxwing (as does yellowish-tinged lower body). Secondaries lack any red waxy tips, inferring young age. Lack of black chin and narrowness of yellow tail-band, together with diffusely streaked underparts, separates young ♀ (right) from young ♂ (left). (Left: D. Monticelli; right: V. Legrand)

NORTHERN MOCKINGBIRD *Mimus polyglottos* (L., 1758)

Britain two (Aug 1982, May 1988), Netherlands one (Oct 1988). Breeds in SE Canada, USA, N Mexico, the Bahamas, the Cayman Islands and the Greater Antilles; mainly a resident, but northern birds may move south during harsh weather. Found in a variety of habitats, including green areas of towns. **IDENTIFICATION** Rather large (**L** 23–26 cm) and chat-like, but plumage pattern like no other bird in the treated region. *Crown, nape and upperparts grey*, contrasting with blackish wings that show *a large white wing patch, narrow double white wing-bars, and whitish tertial and secondary edgings*. Face, throat, and underparts whitish to dusky pale grey, with dusky lores. There is also a *broad white patch across bases of primaries* which is striking in flight, while *blackish tail has white outer rectrices*, which is often the first thing seen on a bird in low escape flight. *Eyes pale yellowish*, bill and legs blackish. Song a slowly recited mix of original and imitative phrases, a varied series of loud, rich or grating phrases, many repeated several times, all with clear pauses between. Calls include a fairly full and loud, sharp *tcheck* or *chek*, and a Starling-like drawn-out grating *keeersch*. **AGEING & SEXING Ad** Sexes similar in plumage. Performs complete post-nuptial moult (Jul–Oct). **1stY** Usually possible to age due to partial post-juv moult of wing (with moult contrast created by some retained juv coverts and most remiges), and in early autumn by still dusky eye. **GEOGRAPHICAL VARIATION & RANGE** Ssp. *polyglottos* (extreme SE Canada, E USA south to S Mexico) is by range the most likely race to have reached Europe.

Texas, USA, Feb: a thrush-like, slender-bodied bird, long-tailed and mainly grey with large white primary patches. Unlikely to be confused with any native bird of the treated region (although a brief distant view of a bird in flight might resemble Great Grey Shrike). Note small head, pale iris and hint of blackish mask. Long-legged. (A. Murphy)

Texas, USA, Feb: note that the characteristic white primary patch differs from that of a Great Grey Shrike by including also primary-coverts. The white patch is often well visible on perched birds, but even more so in flight, when also extensive white on outer tail-feathers is striking. (A. Murphy)

VAGRANTS TO THE REGION

GREY CATBIRD *Dumetella carolinensis* (L., 1766)

Germany two (Oct 1840, May 1908), Channel Islands one (Oct 1975), Ireland one (Nov 1986), Canary Islands one (Nov 1999), Britain one (Oct 2001), Belgium one (Dec 2006), Azores three (two Oct 2010, one Oct 2011). Breeds mainly in E & C USA and S Canada; winters in SE USA to Central America. Prefers dense second growth, hedgerows and forest edges. Rather social, and loose flocks sometimes seen on migration. **IDENTIFICATION** Fairly large (**L** 20–21.5 cm), in structure and size like a slender, long-tailed thrush, and is typically all grey with black cap and deep chestnut undertail-coverts (the latter not always easily seen). Hops on ground with tail cocked, wings flicking, but usually skulking and may venture out of cover only at dawn and dusk. Nevertheless, if seen well unlikely to be mistaken for any other species within the covered region. Song rather varied, reminiscent of the song of some European robins and thrushes, or subsong of a *Lanius* shrike, consisting of an entertaining mixture of nasal, rather harsh, cat-like miaowing sounds, squeaky, grating and clear notes. Most frequent call a miaowing, slightly harsh, somewhat Jay-like *weee* or *wee-eh*. **AGEING & SEXING Ad** Sexes similar, and aged by usually uniformly black crown and evenly fresh wing (after complete post-nuptial moult). Tail-feathers typically blackish, broad and rounded at tips. Red iris. **1stY** Differs subtly from ad on slightly more pointed tail-feathers with looser texture and by Oct already abraded tips. Flight-feathers and primary-coverts very slightly more brownish grey, not blackish with slate-grey edges as in ad (though subtle and often difficult to use). Differs also by greyish-brown to dull reddish-brown iris. **GEOGRAPHICAL VARIATION** Monotypic.

Presumed 1stW, Belgium, Dec: medium-sized, thrush- or babbler-like songbird with long, rounded tail and short wings. Bill is dark, rather narrow and down-curved. Plumage entirely slate-grey but for the diagnostic small black cap, blackish tail and rich rufous-brown vent (hardly visible here). Seemingly abraded wing infers 1stW. (V. Legrand)

Presumed ad, Mexico, Mar: surprisingly, one record in the treated range was made in spring, hence this bird from March is included. Note long legs, and that the long broad, rounded tail often is fanned. Typically secretive but makes vigorous movements, hopping briskly through the undergrowth. Age suggested by broad tail-feathers with rounded tips, and red iris. (H. Shirihai)

BROWN THRASHER *Toxostoma rufum* (L., 1758)

Germany one (autumn 1836), Britain one (Nov 1966–Feb 1967). Breeds mainly in E USA and S & C Canada. Northern populations are migratory, wintering in a variety of scrub, but also in hedgerows and woodland edges, often close to human habitation. **IDENTIFICATION** Rather *large* (**L** 24–29 cm), *long-tailed and long-billed*, and unlikely to be confused with any European bird, also by *reddish-brown head and upperparts*, greyer face, *pale double wing-bars*, and *coarsely dark-streaked pale underparts*. The *yellow eyes* add to its characteristics. Throat often appears pale buff with dark lateral throat-stripes, and undertail-coverts buff. Typically skulking, keeping to thickets and often observed low down or on ground. Song a long series of varied melodious phrases, not unlike subsong of Song Thrush. Call a rather sharp, clicking *tchek*. Also utters a low buzzing churr, *chaaihr*, a bit like a Magpie sound. **AGEING & SEXING Ad** Sexes similar in plumage. Undertakes a complete post-nuptial moult on breeding grounds (Jul–Aug). **1stY** Eyes often grey or brown in early autumn, and iris may attain ad coloration only early in 2nd calendar year. Unlike ad undergoes partial post-juv moult, and hence usually shows moult contrast among greater coverts (retained juv wing-feathers are duller brown and relatively abraded with pale buff tips to coverts, but note that the patterns of the two age categories are still quite similar). When fresh, wing-bars tinged buff (less pure white). **GEOGRAPHICAL VARIATION & RANGE** Two races generally recognised. By range the most likely one to be encountered as vagrant in W Europe is ssp. *rufum* (SE Canada and NC & E USA south to Florida).

New York, USA, Apr: fairly large, long-tailed and slender, enhanced by long, strong and slightly down-curved bill. Unmistakable due to rufous-brown upperparts and wings (latter with double black-and-white wing-bars), yellow eye and bold dark streaking on whitish underparts. The tail is often cocked. Often skulks in shrubby tangles, and heard before seen. (R. Brewka)

RUFOUS-TAILED ROBIN *Larvivora sibilans* Swinhoe, 1863

Britain three (Oct 2004, Oct 2010, Oct 2011), Poland one (Dec 2005), Denmark one (Oct 2012). A summer visitor to its Siberian and E Asian breeding grounds, frequenting damp and dense undergrowth in mixed and coniferous taiga forest, from sea-level to lower valleys of mountains; winters in SE Asia, where mostly found in thickets on floodplains. **IDENTIFICATION** A smallish robin (**L** 12.5–13.5 cm) with rather *unpatterned* plumage, having *uniform olive-brown upperparts* except for *rufous-fringed wing-feathers* (forming hint of chestnut panel) and *rufous-tinged uppertail*. Underparts off-white with *irregular grey-brown scalloping, most prominent on breast*, where sometimes gives impression of darker breast-band (often diffusely set-off from a whitish semi-collar across lower throat). Flanks washed olive-brown or grey, leaving lower belly and undertail-coverts the whitest parts. Facial pattern subdued, includes a *pale eye-ring, short pale supercilium* and dusky loral stripe. Bill usually greyish-brown with some pinkish elements. Pale legs pinkish with grey or brown hue. Black eye appears proportionately large. Shy and retiring, but notable in spring for remarkable sibilant song, a loud metallic and clattering trill or rattle falling in pitch, consistently repeated. Utters thin whistles and hard knocking notes, *tuck, tuck*, when alarmed. **AGEING & SEXING Ad** Sexes similar in plumage. All wing-coverts uniform brown, no visible moult contrasts. **1stY** All or a varying number of outer greater coverts retained juv feathers with paler rufous tips and edges, creating contrast to any inner coverts moulted to uniform ad type. **GEOGRAPHICAL VARIATION** Appears to be best treated as monotypic. Any differences in size and darkness of plumage, sometimes grounds for claims of more subspecies than one, seem slight and variable.

1stW, Britain, Oct: note combination of diagnostic grey-brown wash with whitish scalloping below (flanks often more uniformly greyish olive-brown, and undertail-coverts near-white), and uniform olive-brown upperparts, though wing-feathers and uppertail rufous-fringed. Aged by retained buff-tipped juv greater coverts. (R. Nason)

1stW, Denmark, Oct: variation in degree of scalloping below produces birds with less of it, here almost confined to breast. Note pale eye-ring, large-looking black eye, and pale pinkish legs. Beware confusion with Thrush Nightingale, but note shorter tail, rustier wings and dusky-grey underparts with at least some whitish spots on breast. At least some outer greater coverts retained juv. (M. Bentzon Hansen)

SIBERIAN BLUE ROBIN *Larvivora cyane* (Pallas, 1776)

Channel Islands one (Oct 1975), Britain three (Oct 2000, Oct 2001, Oct 2011), Spain one (Oct 2000). A widely distributed and often abundant taiga species, frequently noted by its loud and characteristic song delivered from an exposed perch, often a horizontal branch high up in a tree. Found in C & E Siberia from Ob east to Okhotsk, south to N Altai, Tuva, Baikal, N & E Manchuria, Japan and North Korea; winters South-East Asia. **IDENTIFICATION** A small (**L** 12–13 cm) compact 'chat' with *strong, pointed bill, short neck and tail*, and strong pale feet. Spends much time on ground, hopping quickly in brief bursts, then remains very still for seconds. Sometimes quivers tail like Redstart. Carriage crouching with head and tail held up, creating rather horizontal outline to back. Ad ♂ *dark blue from crown to uppertail*, whereas *underparts are pure white*. Between the blue and white, from lores to lower flanks, runs a diffuse streak of black and grey, *lores, ear-coverts and sides of neck being blackest*. Bill black, *legs a rather light greyish-pink*. ♀ has crown to back olive grey-brown, while rump and uppertail-coverts are bluish-tinged, at least when fresh. Note that many outer, rarely all, tail-feathers in worn ♀ plumage may appear just dark brown-grey not showing any blue colour (perhaps most commonly, or only, in 1stS). Underparts dusky-white with a faint buff-brown wash on breast and flanks and *dark vermiculations or mottling on lower cheeks, lower throat and breast*. Belly and undertail-coverts purer white. 1stW ♂ like ♀ but usually distinguished by a variably extensive bluish wash on upperparts, and a few median and greater coverts moulted to adult type, being tinged bluish already in autumn. Call a deep, hard clicking *tuck*, or several more complex subdued short clicking notes. Alarm a scolding *brscheeh*. **AGEING & SEXING** Ages of ♂ separable during 1stY, of ♀ at least in 1stW. Sexes readily separable from 1stS, often also in 1stW. – SPRING **Ad** ♂ Deep blue upperparts. Lores, sides of head and neck black, flanks greyish. Rest of underparts pure white. All wing-feathers uniformly dark, edges blue. **1stS** ♂ Much as ad ♂, but flight-feathers, primary-coverts and variable number of outer greater coverts unmoulted juv type, brownish-olive without any blue tinge, but with buff-white tips, and most wing-feathers now abraded. Tertials either same (rarely with pale tips) or moulted to bluish ad type. ♀ Crown to back olive brown-grey. Rump and uppertail-coverts bluish. (Very rarely also a faint bluish cast on mantle and back but not on wing-coverts.) A hint of a paler short supercilium. Tail-feathers either dark olive-grey or same with faint bluish edges. Underparts off-white with buff-brown tinge on sides of throat, lower throat, breast and flanks. Lower throat and breast vermiculated or mottled darker grey. Wings dark grey-brown, feathers edged slightly paler olive-brown. – SUMMER **Ad** Worn. Plumages as in spring. **Juv** Fresh. Brown upperparts with buff spots at many feather-tips. Cream-buff underparts with dark-tipped feathers, giving spotted appearance. – AUTUMN **Ad** As in spring (thus sexes readily separable) but freshly moulted, all tertials and greater coverts uniformly brown without paler tips. **1stW** ♂ Frequently like ad ♀ but generally separated by some moulted greater and median coverts and scapulars being bluish, and usually a bluish wash over much of upperparts. Rarely extensively bluish above on body and wings, but head remains more like ♀, and primary-coverts and outer greater coverts never moulted. **1stW** ♀ Usually like ad ♀ but a few separated by some unmoulted juv outer greater coverts being buff-tipped, and tips of tail-feathers rather more worn from Sep. Tail-feathers on average with fainter blue than ad ♀, and sometimes without any. **GEOGRAPHICAL VARIATION & RANGE** Monotypic. – In Far East (Japan, Sakhalin, Ussuri, lower Amur) darker birds claimed to be more frequent (*bochaiensis* Shulpin, 1928), but this was questioned in BWP (1988) as being perhaps too variable, and we are unable to discern any difference at all. – A further race has been described in Far East from S Sakhalin, Kurile Is and Japan (*nechaevi* Redkin, 2006) being slightly larger and having a subtly more pointed wing (p2 invariably > 6), but since no material for evaluation has been seen, it has not been further dealt with here.

Ad ♂, Thailand, Jan: compact shape with shortish tail and quite long, pale legs. Upperparts of ♂ become purer and deeper blue through winter and in spring. Striking black sides of head and neck and pure white underparts make it unmistakable in ♂ plumage. Compared to young ♂, fringes of ad remiges and primary-coverts are bluish, not olive-brown.

1stW ♂, Thailand, Feb: foreparts still resemble ad ♀, but some bluish feathers on mantle, scapulars and smaller wing-coverts. Much of wing, including greater coverts, is juv (mostly olive-brown and already slightly abraded, and being finely buffish-tipped).

1stS ♀, Mongolia, Jun: faint buff-brown or ochraceous wash on breast and flanks with dark mottling on breast provides some contrast with purer white belly. Large head, large dark eye, squat body and quite short tail, but notably long and pale pink legs. Aged by juv wing-feathers that are worn and bleached brownish. (H. Shirihai)

DAURIAN REDSTART *Phoenicurus auroreus* (Pallas, 1776)

Russia one (Sep 2006). An E Asian redstart, breeding in SC Siberia, Mongolia, Korea, China and NE India; northern populations migratory, southern ones make short altitudinal movements only. Breeds from forested river valleys and open forests in lowlands up to the tree-line, with sparse scrub and bushes; also in clearings, plantations and near habitation. Winters in NE India, N South-East Asia, S & E China, and S Japan. **IDENTIFICATION** Very slightly larger than Common Redstart (**L** 14–15 cm) with stronger build and proportionately longer tail. Both sexes have conspicuous *white wing patch*. ♂ has diagnostic *pale grey-white crown and nape* (to some degree extending onto upper mantle), with wear becoming variably whiter, and a *black bib* extending to upper breast. Most of *underparts* below bib *rufous-orange*. Young ♂ develops adult-looking appearance from first autumn. ♀ is buff-brown or tinged yellow-buff below (more greyish on breast), close to ♀ Common Redstart or eastern races of Black Redstart, but with diagnostic white wing-mark of typical triangular shape (or, when worn or in young, a narrower rectangular patch on bases of inner primaries; any patch in Common or Black Redstarts is formed by pale edges that create a less solid and more longitudinal panel, almost along entire length of the edges of secondaries and tertials). Song a short

Variation in ♂♂, with ad (left: Mongolia, Jun) and 1stW (right: China, Oct): note diagnostic large triangular-shaped white wing patch, largest on ad ♂ (left). In summer, the light ash-grey cap and black parts of plumage become purer. In first winter and spring, young ♂ develops quite ad-like plumage, but the retained juv wing has smaller white wing patch. Also, when still fresh, grey crown and black mantle are extensively concealed by broad brownish tips. (Left: H. Shirihai; right: D. Occhiato)

and rhythmic strophe with clear whistled notes but also trilling elements, less soft and 'poetic' as usually Common Redstart song, more akin to song of Whinchat. Main calls (contact, alarm) recall Northern Wheatear or western Black Redstart, a sharp *wheet* followed by a few quickly repeated clicking notes, *tac-tac-tac-...*, the latter being somewhat 'fatter' than the corresponding dry notes of Wheatear. **AGEING & SEXING Ad** Sexes differ clearly (see above), but some older ad ♀♀ may show partly developed ♂-like plumage, especially underparts with extensive rufous-orange tinge. In ad ♂, primary-coverts, alula and all greater coverts are evenly black. **1stW** In ♂, primary-coverts, some or all alula-feathers and a varying number of outer greater coverts are retained juv feathers, contrastingly browner and paler than moulted blacker feathers. Note also rather prominent brown tips to scapulars and back (not thin brown or grey as in ad). Due to absence of colour contrast and individual variation in wear usually not possible to age ♀♀ unless tail-feather shape and wear help. **GEOGRAPHICAL VARIATION & RANGE** Two subspecies generally recognised, but the slight differences in plumage and size can only be evaluated when comparing series in a museum. Northern *auroreus* (SC Siberia and Mongolia, in the east to Korea and NE China) has a longer migration and is the race most likely to have reached the treated region. The other race is *leucopterus* (C & E China and NE India).

♀, presumed 1stW, Hong Kong, Dec: unlike ♀ Common Redstart or eastern races of Black Redstart, note diagnostic triangular-shaped white wing patch (more of long, narrow panel along secondaries and longest tertial in the other two species). Outer greater coverts seem to be juv but difficult to be sure. (M. & P. Wong)

BLUE WHISTLING THRUSH *Myophonus caeruleus* (Scopoli, 1786)

Iran one (Nov 1900). Breeds widely in Central, E & SE Asia, and through the Himalayas reaches west to Afghanistan, where frequents rivers and torrents in forest and wooded banks in lower mountain valleys, shaded by canopy and with patches of scrub and herbs mixed with boulders and cliffs; sedentary or partly migratory. **IDENTIFICATION** A large dark thrush (**L** 30–33 cm) having the general appearance of a giant Blackbird, with a *longish slender bill* and a *long, broad, rounded tail*. Entire plumage *dark purplish-blue* with Common Starling-like *silvery-white iridescent spots and streaks*, and a noticeably *bright silvery blue-grey forehead*. Whitish marks on crown and underparts in form of streaks, but on mantle and scapulars and (most noticeable) on median and lesser wing-coverts in form of round spots. Some birds appear to have a *dotty white wing-bar* on the median coverts. Edges of remiges, tertials, lesser wing-coverts and tail-feathers brightest blue. Bill usually largely *yellow* (but in some E Asian races black), legs dark. Song can be heard year-round and is a loud whistling strophe mixing high and low notes, slowly given in irregular pattern, often ending with a couple of very thin, high-pitched notes. Commonly uttered call a harsh sharp piercing whistle, *dzhee*, and a loud screeching note in alarm, e.g. *eer-ee-ee*, *kreee* or *scree*. **AGEING & SEXING Ad** Sexes very similar or the same, both in size and plumage. **1stY** Juv develops quite ad-like flight-feathers with blue colour very similar to ad, but in summer told by dull sooty grey-brown head and underparts (some blue only on lateral crown). Ad-like head- and body-feathers attained in late summer–autumn, after which ageing becomes difficult. Tail-feathers average narrower than in ad, but some overlap, and not easy to use even when birds are handled. **GEOGRAPHICAL VARIATION & RANGE** Polytypic, with usually six subspecies recognised. Western race *temminckii* (W Pakistan and Afghanistan north through Tajikistan and E Uzbekistan to W & N Tien Shan), most likely the one to have straggled to Iran, is among the races having yellow bill, the bill also being comparatively short. This race has well-developed white tips to median coverts, and the blue plumage is quite glossy with prominent white shaft-marks on head, neck and breast.

M. c. temminckii, presumed ad, Kazakhstan, May: combination of large size, dark purplish-blue plumage (striking in bright light) with silvery white iridescent spots and streaks, including silvery blue-grey forehead, and largely yellow bill makes this thrush unmistakable in the treated region. Sexes alike. Primary-coverts appear blue-edged and ad-like. (A. Katuncev)

VARIED THRUSH *Ixoreus naevius* (J. F. Gmelin, 1789)

Britain one (Nov 1982), Iceland one (May 2004). These two records are remarkable for such an improbable transatlantic vagrant considering its mostly W North American breeding range (from Alaska to N California), although it is migratory. Mostly found in dense, moist woodland, especially coniferous forests. **IDENTIFICATION** Rather small to medium-sized thrush (**L** 23–25.5 cm) with remarkable ♂ plumage comprising greyish nape and back, *orange supercilium, bright orange underparts, an attractive black mask* and *a black breast-band*, but also noted for its highly *variegated wing pattern with buffy-orange bars and patches* (notable in flight on both wing surfaces). *Supercilium starts above the eye, broadening backwards*. The ♀, however, is much duller and more weakly patterned, but distinguished by *pale orange supercilium and wing-bars*, and *dusky breast-band*. Throat pale buff-orange, and whitish-buff chest and belly marbled dark. Usual call a soft, low *chuk* or *tchook*, and it also utters a trilling *whirr*. **AGEING & SEXING Ad** Sexes strongly differentiated as described above, and both are evenly fresh after complete post-nuptial moult. **1stY** As ad, but after post-juv moult shows moult limits in wing with retained outer greater coverts worn and edged brownish, contrasting with fresher, darker and grey-edged renewed inner ones. Young ♂♂ further distinctive by their brownish tinge on grey upperparts, with head pattern sometimes visibly less boldly patterned in autumn. **GEOGRAPHICAL VARIATION** Four races recognised, but it has not been established which has reached Europe; on range, all four seem equally improbable. In a very rare colour morph, all orange colours are replaced by white.

Above: ♂, presumed ad, Iceland, May: in ad plumage unmistakable, especially given the rich orange throat, supercilium and belly, throat sharply delimited by blackish breast-band and mask. Upperparts blue-grey, while blackish wings sport interesting orange-yellow pattern with panels and bars. Bright plumage and being overall fresh infer age and sex. (S. G. Þórisson)

Left: ♀, Canada, early spring: a rather small but stocky thrush (often looking fat-bellied) with a relatively short tail. ♀ generally mirrors ♂'s plumage, but is usually clearly duller with poorly marked dark breast-band and less pure grey mantle. (T. Zurowski)

WOOD THRUSH *Hylocichla mustelina* (J. F. Gmelin, 1789)

Azores two (around or prior to 1902, Oct 2012), Iceland one (Oct 1967), Britain one (Oct 1987). Breeds in E USA and extreme SE Canada; winters in Mexico and Central America. Prefers moist deciduous or mixed woods with thickets, and also found in plantations. **IDENTIFICATION** Fairly small but large-headed, sturdy thrush (**L** 18–20.5 cm) with *reddish-brown fore upperparts*, brightest on crown and nape (rufous to russet, shading to cinnamon), rump and tail brownish-olive. Most characteristic, however, are the conspicuous *broad white eye-ring on dark-streaked face*, and the *white underparts*, which are *more boldly and extensively black-spotted* than on most other vagrant thrushes. These characteristics should readily eliminate the only superficially similar native Song Thrush. Upper mandible and tip of bill dark, lower pale. *Legs pale flesh.* Buffish underwing-coverts (forming central band) visible in flight. Tends to be rather secretive, like most other North American thrushes. Calls include a rapid light *uit-uit-uit-uit* or *pwit-pwit-pwit-pwit*, and also a lower clucking *whe-whe-whe-wheh*. **AGEING & SEXING Ad** Sexes similar, and ad evenly fresh after complete post-nuptial moult. Only subtly paler rufous tips to greater coverts blend smoothly into slightly darker centres. **1stY** Similar to ad but normally shows at least some retained juv outer greater coverts, with buffish or pale rufous spots or fringes at tips, and tips more contrasting pale than in ad-type coverts. **GEOGRAPHICAL VARIATION** Monotypic.

1stW, Azores, Oct: pot-bellied and short-tailed thrush, with big head and eye (latter emphasised by broad white eye-ring). Distinctive marks include warm reddish-brown upperparts and contrastingly pure white underparts with bold black spots. For ageing, note pale rufous tips to retained greater coverts. (V. Legrand)

Presumed ad, Costa Rica, Dec: apart from rufous-brown upperparts and boldly black-spotted underparts, note pale and pink legs. Face is often densely dark-streaked. All wing-feathers and tail-feathers seem ad. (H. Shirihai)

TICKELL'S THRUSH *Turdus unicolor* Tickell, 1833

Germany one (Oct 1932). A Himalayan specialty, breeding in N Pakistan and N India east to Bhutan; winters in foothills and south to C & E India and Bangladesh, thus a short-range migrant and an extremely unlikely vagrant to Europe. Lives in montane deciduous forests. **IDENTIFICATION** Rather small to medium-sized (**L** 21–25 cm) *plain-coloured* thrush. ♂ is characteristically uniform *medium lead-grey on head, upperparts*, tail and wings, *paler grey* (often a bit marbled) *on underparts*, with *dusky lores*, brownish-tinged ear-coverts and whitish belly and wing panel. *Underwing-coverts diagnostically rufous-cinnamon.* Bill, orbital-ring and legs *bright to dull yellow.* ♀ plumage variable, but is distinctly more *brownish-olive on head and upperparts*, with rather blotchy-patterned underparts and has *dusky lateral throat-stripe*, with rows of dusky triangular spots on lower throat and upper breast. Rest of underparts whitish to pale grey with *deep olive-buff breast-sides and flanks.* Bare parts rather greyish-brown with limited dull yellow. Song has desolate ring, a little like Eyebrowed Thrush or fragments of Mistle Thrush song. Call in alarm a chattering *chat-chat-chat-chat*, hard almost like Ring Ouzel. **AGEING & SEXING Ad** Sexes differ clearly in ad plumage as described above. **1stY ♂** as ad ♂ but more olive-grey, and has traces of ♀-like marks on sides of head, throat and breast. Young ♀ can only be aged if a slight contrast in wear and colour can be detected between retained juv flight-feathers and newer coverts. Juv outer greater coverts have paler and more contrasting tips than inner moulted ad type. Other seasonal plumage variation slight and only due to effect of wear and bleaching. **GEOGRAPHICAL VARIATION** Monotypic.

Ad ♂, India, spring: unmistakable given almost uniform pale lead-grey plumage with only somewhat pale-mottled underparts, this in combination with yellow orbital ring and bill. Wing-feathers appear uniform and of ad type, inferring age. (M. McDonald)

♀, presumed ad, India, Jun: more brownish-olive above than ♂. Note also weak dusky lateral throat-stripe, and some irregular dusky mottling and streaks on lower throat and upper breast. Yellowish bill and orbital ring tinged brownish. Limited dark marks to fore parts in combination with apparent rather freshly feathered wing suggest ad. (P. J Saikia)

1stW, ♀-like, Nepal, Jan: distinctly more brownish-olive above than ad ♂, and has blotchier underparts, while moulted inner greater coverts rather pure grey. Again, note diffuse dusky lateral throat-stripe, and some irregular dusky mottling and streaks on lower throat and upper breast. Yellowish bill tinged brownish, orbital ring thin. Still, safe sexing of imm birds difficult and best refrained from. (N. Bowman)

GRAY'S WARBLER (GRAY'S GRASSHOPPER WARBLER) *Locustella fasciolata* (G. R. Gray, 1861)

France one (Sep 1913), Denmark one (Sep 1955); a third record, of a bird claimed to have been found on Ile d'Ouessant, France in Sep 1933 (Meinertzhagen 1948) is now believed to be fraudulent. Confined as a summer visitor to SC & SE Siberia, E Altai, to Sakhalin and Japan; possibly also N Korea. Winters in the Philippines, E Indonesia and New Guinea, and migration to there seems to pass through SE Siberia and Japan, but not through China. Habitat often less aquatic than congeners, often found well away from water in thickets at forest edges, or in brushy taiga clearings. **IDENTIFICATION** The *largest Locustella* warbler (**L** 16.5–18 cm), grey-brown with a *strong bill* and therefore possible to confuse with the larger *Acrocephalus* species, but with slightly smaller head and *fuller tail* (rather long, broad and strongly rounded) characteristic of a *Locustella*. Plumage somewhat variable, but *essentially dark and unstreaked*, thereby resembling most its smaller relatives Savi's and River Warblers. Apart from size, it is usually readily separated from the other two by the better developed head pattern, with more distinct *grey-white supercilium* and eye-ring, and there is often a grey tinge to crown, ear-coverts, sides of neck and breast. Upperparts are dark brown with varying *rufous tinge*, especially noticeable on rump and uppertail-coverts. Breast and flanks are buffish-olive or grey, *breast and lower throat often diffusely mottled*, and contrast somewhat with the more whitish upper throat and centre of belly. The *undertail-coverts are plain cinnamon-buff and virtually lack the pale tips* of the other two species. The straight and relatively strong and heavy *bill is dark* with variably paler lower mandible. Legs appear stout and robust being greyish-pink to pale pinkish-brown. Primary projection exceeds 2/3 of the exposed tertial length, and p3 forms the wingtip, p3 also being emarginated. Song is short, very loud and repetitive, stuttering in bulbul fashion and accelerating at the end, almost in a flourish, e.g. *chuk, chek chuk-chek-chuk-chekeruchetteritriah.* Main calls include a strong *tek-tek* and a sharp, deep *chuck*; also utters a guttural series of harsh, clicking notes, *che-che-che-che-...* **AGEING & SEXING Ad** Sexes similar. Seasonal plumage variation restricted to effects of wear and bleaching. Upperparts usually dull dark rufous-brown to earth-brown. Lower throat, breast and neck-sides greyish, upper belly and flanks dull olive-ochre (lacking lemon tinge of young birds). Undergoes partial post-nuptial moult on breeding grounds in Jul–Aug, including head and body, occasionally some or all of the rectrices, and rarely also replaces 1–4 outer primaries, leaving complete moult to be concluded or occur in the winter quarters (Jan–Feb). Autumn migrants thus usually have worn tips to primaries, or contrast between old and new. **Juv** Quite distinct with more rufous-tinged upperparts, a sulphur-yellow or lemon-buff suffusion below, and often some fine dusky specks and arcs on ear-coverts, throat, and upper breast. 1stY retains juv plumage till at least early winter, after which it moults completely, too. Thus, during first autumn, remiges and rectrices of young birds are of same generation and rather fresh. There are no reliable ageing differences after complete pre-nuptial moult in winter. **GEOGRAPHICAL VARIATION & RANGE** Two subspecies often recognised, *fasciolata* (SC & SE Siberia, Russian Far East, NE China) and *amnicola* (Kurile Is, Sakhalin, N Japan). Both European records refer to the former. Differences between the two described subspecies are often very slight, and the distinctness of *amnicola* and its exact range should be further studied. Two studies (Drovetski et al. 2004, Alström et al. 2011) even suggested the two could merit separate species status based on genetic distance, though both studies concluded that a more comprehensive study was required.

Amur, Siberia, Jun: Great Reed Warbler-sized and strong-billed *Locustella* warbler. Note plain brown upperparts and dusky brown-white underparts. Unlike the smaller Savi's and River Warblers, has prominent pale supercilium and plain undertail-coverts (no pale tips; right). Head and breast are slightly more grey-tinged than brownish wings (best visible on left image). Breast and lower throat often diffusely mottled. (I. Ukolov)

ORIENTAL REED WARBLER *Acrocephalus orientalis* (Temminck & Schlegel, 1847)

Israel two (Feb–Apr 1988, May 1990). A summer visitor to E Mongolia, Transbaikalia, Russian Far East, Sakhalin, W & N China and south-east to Japan; winters in NE India and south and east through SE Asia. Breeds in reedbeds and damp tall grasslands in temperate and subtropical regions. **IDENTIFICATION** Slightly smaller than Great Reed Warbler (**L** 16–18.5 cm) with *moderately long, somewhat rounded wings and intermediate primary projection* (about 2/3 to 3/4 of exposed tertial length). *Tail proportionately long and rounded* with quite long undertail-coverts. Bill as Great Reed but appears somewhat slenderer. In spring, *throat often has distinct streaking extending onto upper and even lower breast or sides* (more prominent and extensive streaking on average than in ad Great Reed). In autumn, Oriental Reed is warmer coloured both above and below, and streaking of throat/breast is obscured to a varying degree. Face pattern rather pronounced: *supercilium cream-coloured or white, quite long and sharply-defined*. Most, especially when fresh, have *broader and whiter tips to rectrices* (buffier and narrower on Great Reed). Bare parts much as Great Reed, as are gait and general behaviour, but has proportionately longer tail which may afford a somewhat Clamorous Reed Warbler-like impression. Song resembles Great Reed Warbler but is not quite as loud, slightly lower-pitched and partly slower, at times creating frog-like impression; sometimes contains Thrush Nightingale-like rapid clicking series. Calls include both sharp clicking *chuck* and rolling *trrrr*. **AGEING & SEXING** Sexes alike. Both ad and juv generally undergo complete moult in late summer–early autumn. However, judging by some extremely heavily worn spring birds not all 1stY moult flight-feathers and tail. Pre-nuptial moult apparently often absent or at most partial. **Ad** Iris warm brown. Plumage freshly moulted or worn but lacking moult contrasts or faintly darker subterminal marks on tail-feathers. **1stW** As ad but iris averages duller and greyer brown. **Juv** In summer before complete post-juv moult recognised on paler rufous and ochre-buff plumage with some pale-tipped feathers with faintly darker subterminal marks, notably on some outer tail-feathers. Plumage soft and fluffy. **GEOGRAPHICAL VARIATION & RANGE** Monotypic. – Has been described as being polytypic, with breeders of Japan north to Hokkaido (*orientalis* s.s.) claimed to differ in being larger and having longer wing and bill than breeders of continental E Asia ('*magnirostris* Swinhoe, 1860') (Malykh & Redkin 2011). It has been difficult to examine long enough series of confirmed breeders from Japan, but based on material in NHM, AMNH, MNHN, ZMB, NRM and YIO (n = 16 against 104 *magnirostris*) the difference appears rather slight and hardly grounds for separation; average difference in wing length was *c.* 4 mm (5.2%) and bill length 0.6 mm (2.7%), with extensive overlap in all measurements.

Mongolia, Jun: a recent split from the similar Great Reed Warbler, breeding in E Asia. In spring both 1stS and older birds have similarly fresh plumage, and upperparts are usually tinged olive-brown. (H. Shirihai)

China, Apr: compared to Great Reed Warbler, wings are slightly shorter making tail look a little longer, while bill is subtly smaller. When worn, tinged olive-grey above, and throat-sides and breast frequently diffusely flecked. Quite long whitish supercilium. At this time of year, ages are indistinguishable by plumage or moult. (J. Martinez)

EASTERN CROWNED WARBLER *Phylloscopus coronatus* (Temminck & Schlegel, 1847)

Germany two (Oct 1843, Oct 2012), Norway one (Sep 2002), Finland one (Oct 2004), Netherlands two (Oct 2007, Oct 2016), Britain five (Oct 2009, Oct 2010, Oct 2011, Oct–Nov 2014, Oct 2016), Belgium one (Oct 2016), France one (Oct 2016). Breeds in NE Asia, from C & NE China to Amurland and Ussuriland, Sakhalin, Japan and Korea; winters in SE Asia to W Indonesia and in NE India. Breeds in mixed and broadleaf forests in lowlands and mountain foothills, but in non-breeding season found in all types of wooded habitats. **IDENTIFICATION** A fairly large leaf warbler (**L** 11–12 cm) with relatively strongly marked head pattern, *broad dark olive lateral crown-stripes* and *paler olive median crown-stripe* (from centre of crown to nape; hard to see unless viewed from above or behind). Supercilium long and sharply defined but rather narrow and pale yellow in front of and above eye, becoming wider and white at rear (can appear all white when worn), often slightly up-turned at rear, bordered below by a broad dark olive-grey eye-stripe. The rest of the upperparts are rather pale olive-green, the *underparts mainly white with* nearly always *yellowish vent*; cheeks, breast and flanks sometimes sullied grey. *Wing-bend* and fore underwing-coverts *lemon-yellow*, often visible at edge of folded wing. There is a *single narrow, yellowish*

England, Nov: note strongly marked broad dark lateral crown-stripes and prominent but narrow pale median crown-stripe. Supercilium long, yellow-tinged and quite narrow in its forepart, widening and becoming more whitish behind eye, often ending up-turned. Note also narrow yellowish greater coverts wing-bar and shorter median coverts wing-bar (variable; can even be absent). (G. Sellors)

wing-bar (white when worn). This together with the long supercilium and dark olive crown make it resemble Arctic Warbler, but that species lacks a paler median crown-stripe and lacks contrast between whitish belly and yellowish vent. Confusion with Greenish Warbler also possible, but note in Eastern Crowned e.g. *dark olive crown-sides strongly contrasting with paler green mantle*. Longish *bill rather thick and broad-based*, dark grey or brown above, *pale orange or bright yellow below, lacking dark tip*. Legs dark pink to grey or greyish-brown. (Very similar to several other leaf warblers, but these so far unrecorded within the treated region, including e.g. Western Crowned Warbler *P. occipitalis* and Blyth's Leaf Warbler *P. reguloides*; not treated.) Very unobtrusive in winter, keeping to canopy of large trees. The simple song is usually rhythmic and consists of varied combinations of loud, sweet, liquid or slurred notes, e.g. *pitschu-**pitsch**-pitschu-**pitsch**-pitsch* (a little like Coal Tit), *chewee-chewee-chewee-chewee* (almost like Marsh Tit!) or *chiiyo-chiiyo-chiiyo-bshiiiih*, latter example of typical variation with final note very distinctively drawn-out, strained and nasal. Not so vocal outside breeding season, only irregularly gives strong whistling *chiu*, a soft *phit-phit*, or harsher, buzzy nasal *dwee*. **AGEING & SEXING Ad** Sexes similar. As in other *Phylloscopus*, shows rather limited age-related and seasonal plumage variation, apart from the effect of wear (when becoming duller and greyer above, whiter on the underparts and supercilium, and the wing-bar reduced). Performs complete post-nuptial moult in late summer, after which ageing becomes difficult or impossible, since young birds are very similar. **GEOGRAPHICAL VARIATION** Monotypic.

England, Oct: seen from below difficult to appreciate typically rather light olive-green upperparts (especially rear parts), while underparts are mainly white (though vent nearly always yellowish-tinged). That cheeks, breast and flanks are faintly sullied grey not always easy to see. Note longish bill with pale orange-yellow lower mandible, lacking a dark tip. (G. Thoburn)

PALE-LEGGED LEAF WARBLER *Phylloscopus tenellipes* Swinhoe, 1860

Britain one (Oct 2016); a further record exists of either this species or Sakhalin Leaf Warbler *P. borealoides* (Oct 2012; not treated). Breeds in NE China, N Korea and Russian Far East (Ussuri, Amur); winters in SE Asia south to W Malaysia. Breeds in scrubland thickets or open broadleaf or mixed forests in lowlands and mountain foothills, but in non-breeding season found in all types of lowland wooded habitats. **IDENTIFICATION** A fairly large leaf warbler (**L** 11–12 cm) with *well-marked, long and narrow cream-white supercilium, long broad and dark grey eye-stripe* and a *weak pale wing-bar* (often a second bar visible on median coverts). Upperparts olive-green with *clearly darker and greyer crown, underparts mainly dusky-white*, cheeks, throat and breast being irregularly smudged or spotted greyish. Can recall Greenish Warbler, but differs by *voice, darker crown than mantle, much paler pink legs* and a faint rufous or *warm brown tinge to greenish rump and uppertail*. Wing-bars usually buffish-yellow when fresh in autumn, not as white as in Greenish (but bleaches to more yellowish-white in summer). Dark bill small with hint of paler pinkish base to lower mandible. Often dips tail downwards in Chiffchaff fashion. Song a rapid, short, straight, high-pitched, insect-like trill, *zezezezezezeze*, rather like a grasshopper or recalling song of Lanceolated Warbler. Less vocal outside breeding season; call a high-pitched, short and metallic *zreet*. **AGEING & SEXING Ad** Sexes similar. Limited age-related and seasonal plumage variation, apart from the effect of wear and bleaching. Performs complete post-nuptial moult in late summer, after which ageing becomes difficult or impossible, since young birds are very similar. **GEOGRAPHICAL VARIATION** Monotypic.

Thailand, Mar: note well-marked, long whitish supercilium, bordered below by dark eye-stripe widening at rear. Also pale pink legs, and shortish but strong bill. Wing-bars variable, prominent or more subdued, either single or double, commonly more obvious on greater coverts. Tail-sides basally quite brightly ochraceous-green. (Thailand Wildlife)

Thailand, spring: note olive-grey tinge to crown and mantle in contrast to bright bronzy-green upperwing and tail-sides. This bird has rather broad and prominent double wing-bars. Ear-coverts and breast typically irregularly smudged greyish. Cutting-edges and lower mandible of bill pinkish.

LARGE-BILLED LEAF WARBLER *Phylloscopus magnirostris* Blyth, 1843

United Arab Emirates one (Oct 2014). Breeds in NE Afghanistan, N Pakistan and Himalayas east to Myanmar and C China; winters in SE Asia including S India. Breeds in mixed or evergreen forests in mountain foothills or forested alpine valleys, preferably near running water; in non-breeding season found in all types of wooded habitats. **IDENTIFICATION** A fairly large leaf warbler (**L** 12–13 cm) characterised by relatively *long, dark bill with yellowish base to lower mandible*. Size and *well-marked whitish supercilium* (often slightly yellow-tinged over forepart) and *bold dark eye-stripe*, plus hint of weak pale wing-bars, makes it rather similar to Arctic Warbler. However, note longer bill, usually *darker legs* and *different call*. Song a short (five-syllable), initially faltering, then descending whistled phrase, notes straight and uttered rather mechanically. Call a disyllabic strongly rising, sharp *chu-zees* like a creaking gate or rusty pump, or a longer *chu-zee-see*, again rising in pitch. **AGEING & SEXING Ad** Sexes similar. Limited age-related and seasonal plumage variation, apart from the effects of wear and bleaching. Performs complete post-nuptial moult in late summer, after which ageing becomes difficult or impossible, since young birds are very similar. **GEOGRAPHICAL VARIATION** Monotypic.

Sri Lanka, Nov: note strong and long, dark bill with yellowish base to lower mandible. Also, well-marked long whitish supercilium, bordered below by very bold dark eye-stripe. Thin but distinct pale double wing-bars, although by Nov the median coverts bar usually gets very thin through abrasion. Seen at this angle, the prominent olive-grey suffusion to breast throat and lower cheeks is clearly visible. (H. Shirihai)

VAGRANTS TO THE REGION

SULPHUR-BELLIED WARBLER *Phylloscopus griseolus* Blyth, 1847

Denmark one (May–Jun 2016). Breeds in Central Asia, from NE Afghanistan and Tajikistan south-east to N Pakistan and NW India, north-east to Altai, Tuva and C Mongolia; winters in N & C India. Breeds in mountains and foothills with boulders and mixed broadleaf forests; in non-breeding season found in all types of wooded habitats. **IDENTIFICATION** A fairly large leaf warbler (**L** 11–12 cm) with rather plain and dusky plumage, being *dull greyish-brown above* without olive tinge, and *pale greyish-drab below with yellowish tinge in most* (in some, yellow restricted to centre of belly, difficult to see). *Supercilium long, narrow and well-marked, usually bright lemon-yellow in front of and above eye*, slightly paler yellow or whitish behind. Bill narrow and dark, *lower mandible yellowish-pink* with darker tip. *Legs dull reddish-brown*. Perhaps closest to Dusky Warbler with similarly well-defined supercilium and brownish upperparts lacking olive tinge or yellow-green primary edges (though underparts generally obviously yellowish, not ochre-buff and white, and lemon fore supercilium stands out). Can resemble Radde's Warbler, but has less strong and blunt-tipped bill, supercilium is not blurred in outline near forehead, and lacks obvious olive tinge above. The simple song is surprisingly similar to that of Dusky Warbler, i.e. a short rhythmic strophe with 4–6 repeated liquid notes (mono- or disyllabic), all notes the same but often a few variations inserted. Usually gives a single call note at start of each strophe. Call a short clicking *chip* or more liquid *chlit*. Can be repeated in rapid rattle when agitated. **AGEING & SEXING Ad** Sexes similar. Limited age-related and seasonal plumage variation, apart from the effect of wear and bleaching. Performs complete post-nuptial moult in late summer, after which ageing becomes difficult or impossible, since young birds are very similar. **GEOGRAPHICAL VARIATION** Monotypic.

Denmark (above) and Mongolia, Jun: readily identified by combination of dull greyish-brown upperparts (no olive tinge), warmer greyish-drab underparts (in some with yellow tinge evident), and well-marked long supercilium, usually bright lemon-yellow in front of and above eye, more washed-out at rear. Also typical is narrow and dark bill with yellowish-pink lower mandible. (Above: T. Varto Nielsen; below: H. Shirihai).

RUBY-CROWNED KINGLET *Regulus calendula* (L., 1766)

Iceland three (Nov 1987, Oct 1998, Oct 2013), Azores one (Oct–Nov 2015). Breeds widely in Canada except in extreme northernmost areas, also in NE & W USA; winters from British Columbia and Maryland south to Guatemala. Found in woodland with thickets. **IDENTIFICATION** Resembles Goldcrest due to *tiny size* (**L** 9.5–10.5 cm), but also by having rather similar bold wing pattern with prominent *double white wing-bars*, emphasized by *thick blackish bar across base of secondaries*; *yellowish edgings to flight-feathers* often very evident, too. Unlike Goldcrest *lacks dark lateral crown-stripes* and has diagnostic *broad, broken white eye-ring* (white oddly confined to in front of and behind eye). ♂'s *red crown patch* diagnostic, too (grenadine-red), but often concealed (mainly visible if crown-feathers raised during song or in excitement). ♀ differs from Goldcrest by the *lack of any colourful crown patch*. Underparts greyish-cream, washed buffish olive (can appear rather dusky or yellowish-tinged), especially on chest and flanks. Confusion risk could involve juv Goldcrest, which has a plain crown. Active, flicks wings rapidly. Song lively, varied and loud, with high, clear notes, typically repeated a few times; can recall sections of Pallas's Leaf Warbler song. Call a thin *see see* or a dry sparrow-like chattering *dr-dr-drt* or *che-che-chut*, often prolonged when scolding. **AGEING & SEXING Ad** Sexes differ as described above. **1stY** Sometimes distinguishable from ad by on average narrower and more pointed tail-feathers, but overlap and variation make it an unreliable clue.

♂, presumed 1stS, E Canada, spring: note largely plain greyish head almost recalling juvenile Goldcrest. Red crown-patch small but diagnostic for sex as long as not concealed. Sharply pointed tail-feathers fit best with 1stS.

Note that detecting moult limits among the greater and median coverts is very difficult in this species, even in autumn. **GEOGRAPHICAL VARIATION & RANGE** Three subspecies recognised (one of which, *obscurus*, may be extinct). Considering range, ssp. *calendula* (C & E Canada and C & E portion of the range in USA) is the likeliest race to have reached Europe.

Presumed ♀, Azores, Nov: lack of any colourful crown patch or black lateral crown-stripes of Goldcrest is diagnostic discrimination (but beware of rarely seen juv Goldcrest with plain crown). Age difficult to judge, but rather broad and seemingly obtusely tipped tail-feathers could indicate ad. (D. Monticelli)

CHECKLIST OF THE BIRDS OF THE WESTERN PALEARCTIC – PASSERINES

This checklist is a complete list of all the species covered in Volumes 1 and 2, including vagrants. The sequence adopted is, for practical reasons, a traditional one, being very similar to the one used in both *The Birds of the Western Palearctic* vol. V–IX (Cramp et al. 1988–94) and *Collins Bird Guide* (Svensson, Mullarney & Zetterström 2009). Please see a brief overview on page 27 explaining modern trends in systematics and how handbooks in the future probably will be arranged to better mirror evolution and true relationships.

The delimitation of the Western Palearctic as defined here is explained on page 8. The main difference compared to most previously published works is the inclusion of the whole of Arabia and Iran, resulting in many additional species. The recently published checklist, *Birds of Europe, North Africa and the Middle East* (Mitchell 2017) also includes Arabia and Iran but, unlike that work, we prefer to follow international country borders. This has resulted in the omission of a few African species claimed to have been recorded in the north of several countries bordering the region in the south (in Mauritania, Niger and Sudan). We have also omitted some localised or recently established introductions in several central and southern European countries.

Species limits and nomenclature follow our preferences based on our own studies, as explained in 'Species taxonomy' in the Introduction on pages 9–10.

In order to make the list more useful to readers, volume and page numbers are given for each species, together with a coded general status:

B = breeder, resident or summer visitor

B (I) = established introduced breeder

B (E) = endemic breeder (91 species breed only within the covered area)

M = migrant (or winter visitor) within the treated region

S = scarce visitor, in some parts of the region almost annual, with a total of > 10 records up to the end of 2016

V = vagrant, usually < 10 records ever up to the end of 2016; these are presented in a separate section at the end of each volume

TYRANNIDAE
Eastern Kingbird *Tyrannus tyrannus* **V** – Vol 1: 620
Eastern Phoebe *Sayonis phoebe* **V** – Vol 1: 620
Acadian Flycatcher *Empidonax virescens* **V** – Vol 1: 621
Alder Flycatcher *Empidonax alnorum* **V** – Vol 1: 621
Least Flycatcher *Empidonax minimus* **V** – Vol 1: 621
Eastern Wood Pewee *Contopus virens* **V** – Vol 1: 622

PITTIDAE
Indian Pitta *Pitta brachyura* **V** – Vol 1: 622

ALAUDIDAE
Singing Bush Lark *Mirafra cantillans* **B** – Vol 1: 31
Black-crowned Sparrow Lark *Eremopterix nigriceps* **B** – Vol 1: 33
Chestnut-headed Sparrow Lark *Eremopterix signatus* **V** – Vol 1: 622
Dunn's Lark *Eremalauda dunni* **B** – Vol 1: 36
Arabian Lark *Eremalauda eremodites* **B** (E) – Vol 1: 38
Bar-tailed Lark *Ammomanes cinctura* **B** – Vol 1: 40
Desert Lark *Ammomanes deserti* **B** – Vol 1: 42
(Greater) Hoopoe Lark *Alaemon alaudipes* **B** – Vol 1: 46
Dupont's Lark *Chersophilus duponti* **B** (E) – Vol 1: 49
Thick-billed Lark *Ramphocoris clotbey* **B** – Vol 1: 51
Calandra Lark *Melanocorypha calandra* **B** – Vol 1: 54
Bimaculated Lark *Melanocorypha bimaculata* **B** – Vol 1: 57
Black Lark *Melanocorypha yeltoniensis* **S** – Vol 1: 59
(Greater) Short-toed Lark *Calandrella brachydactyla* **B** – Vol 1: 62
Rufous-capped Lark *Calandrella eremica* **B** – Vol 1: 66
Hume's Short-toed Lark *Calandrella acutirostris* **B** – Vol 1: 68
Lesser Short-toed Lark *Calandrella rufescens* **B** – Vol 1: 70
(Indian) Sand Lark *Calandrella raytal* **B** – Vol 1: 75
Crested Lark *Galerida cristata* **B** – Vol 1: 77
Thekla's Lark *Galerida theklae* **B** – Vol 1: 81
Woodlark *Lullula arborea* **B** (E) – Vol 1: 84
Oriental Skylark *Alauda gulgula* **B** – Vol 1: 86
(Common) Skylark *Alauda arvensis* **B** – Vol 1: 88
White-winged Lark *Alauda leucoptera* **S** – Vol 1: 91
Raso Lark *Alauda razae* **B** (E) – Vol 1: 94
Horned Lark *Eremophila alpestris* **B** – Vol 1: 96
Temminck's Lark *Eremophila bilopha* **B** (E) – Vol 1: 101

HIRUNDINIDAE
Banded Martin *Neophedina cincta* **V** – Vol 1: 623
Brown-throated Martin *Riparia paludicola* **B** – Vol 1: 103
Chinese Martin *Riparia chinensis* **S** – Vol 1: 105
Sand Martin *Riparia riparia* **B** – Vol 1: 107
Pale Martin *Riparia diluta* **S** – Vol 1: 110
Tree Swallow *Tachycineta bicolor* **V** – Vol 1: 623
Purple Martin *Progne subis* **V** – Vol 1: 623
Rock Martin *Ptyonoprogne fuligula* **B** – Vol 1: 113
Crag Martin *Ptyonoprogne rupestris* **B** – Vol 1: 116
Barn Swallow *Hirundo rustica* **B** – Vol 1: 118
Wire-tailed Swallow *Hirundo smithii* **S** – Vol 1: 123
Ethiopian Swallow *Hirundo aethiopica* **V** – Vol 1: 624
Red-rumped Swallow *Cecropis daurica* **B** – Vol 1: 125
Lesser Striped Swallow *Cecropis abyssinica* **V** – Vol 1: 624
Streak-throated Swallow *Petrochelidon fluvicola* **S** – Vol 1: 128
(American) Cliff Swallow *Petrochelidon pyrrhonota* **S** – Vol 1: 129
Asian House Martin *Delichon dasypus* **V** – Vol 1: 624
(Common) House Martin *Delichon urbicum* **B** – Vol 1: 130

MOTACILLIDAE
Richard's Pipit *Anthus richardi* **S** – Vol 1: 132
African Pipit *Anthus cinnamomeus* **B** – Vol 1: 135
Blyth's Pipit *Anthus godlewskii* **S** – Vol 1: 136
Paddyfield Pipit *Anthus rufulus* **V** – Vol 1: 625
Tawny Pipit *Anthus campestris* **B** – Vol 1: 139
Long-billed Pipit *Anthus similis* **B** – Vol 1: 142
Berthelot's Pipit *Anthus berthelotii* **B** (E) – Vol 1: 145
Olive-backed Pipit *Anthus hodgsoni* **S** – Vol 1: 147
Tree Pipit *Anthus trivialis* **B** – Vol 1: 149
Pechora Pipit *Anthus gustavi* **B** – Vol 1: 152
Meadow Pipit *Anthus pratensis* **B** – Vol 1: 154
Red-throated Pipit *Anthus cervinus* **B** – Vol 1: 156
Rock Pipit *Anthus petrosus* **B** (E) – Vol 1: 159
Water Pipit *Anthus spinoletta* **B** – Vol 1: 163
Buff-bellied Pipit *Anthus rubescens* **S** – Vol 1: 166
Golden Pipit *Tmetothylacus tenellus* **V** – Vol 1: 625
Forest Wagtail *Dendronanthus indicus* **S** – Vol 1: 169

Yellow Wagtail *Motacilla flava* **B** – Vol 1: 170
Citrine Wagtail *Motacilla citreola* **B** – Vol 1: 180
Grey Wagtail *Motacilla cinerea* **B** – Vol 1: 185
White / Pied Wagtail *Motacilla alba* **B** – Vol 1: 188
African Pied Wagtail *Motacilla aguimp* **B** – Vol 1: 196

PYCNONOTIDAE
White-eared Bulbul *Pycnonotus leucotis* **B** – Vol 1: 198
White-spectacled Bulbul *Pycnonotus xanthopygos* **B** (E) – Vol 1: 199
Common Bulbul *Pycnonotus barbatus* **B** – Vol 1: 200
Red-vented Bulbul *Pycnonotus cafer* **B** (I) – Vol 1: 201

BOMBYCILLIDAE
Cedar Waxwing *Bombycilla cedrorum* **V** – Vol 1: 626
(Bohemian) Waxwing *Bombycilla garrulus* **B** – Vol 1: 202
(Grey) Hypocolius *Hypocolius ampelinus* **B** – Vol 1: 205

CINCLIDAE
(White-throated) Dipper *Cinclus cinclus* **B** – Vol 1: 207

TROGLODYTIDAE
(Eurasian) Wren *Troglodytes troglodytes* **B** – Vol 1: 210

MIMIDAE
Northern Mockingbird *Mimus polyglottos* **V** – Vol 1: 626
Brown Thrasher *Toxostoma rufum* **V** – Vol 1: 627
Grey Catbird *Dumetella carolinensis* **V** – Vol 1: 627

PRUNELLIDAE
Dunnock *Prunella modularis* **B** – Vol 1: 213
Siberian Accentor *Prunella montanella* **B** – Vol 1: 216
Black-throated Accentor *Prunella atrogularis* **B** – Vol 1: 219
Radde's Accentor *Prunella ocularis* **B** (E) – Vol 1: 221
Alpine Accentor *Prunella collaris* **B** – Vol 1: 224

TURDIDAE
Rufous-tailed Scrub Robin *Cercotrichas galactotes* **B** – Vol 1: 227
Black Scrub Robin *Cercotrichas podobe* **B** – Vol 1: 230
(European) Robin *Erithacus rubecula* **B** – Vol 1: 232
Rufous-tailed Robin *Larvivora sibilans* **V** – Vol 1: 627
Siberian Blue Robin *Larvivora cyane* **V** – Vol 1: 628
White-throated Robin *Irania gutturalis* **B** – Vol 1: 235
Thrush Nightingale *Luscinia luscinia* **B** – Vol 1: 237
(Common) Nightingale *Luscinia megarhynchos* **B** – Vol 1: 240
Bluethroat *Luscinia svecica* **B** – Vol 1: 243
Siberian Rubythroat *Calliope calliope* **B** – Vol 1: 248
Red-flanked Bluetail *Tarsiger cyanurus* **B** – Vol 1: 250
Eversmann's Redstart *Phoenicurus erythronotus* **S** – Vol 1: 253
Black Redstart *Phoenicurus ochruros* **B** – Vol 1: 255
(Common) Redstart *Phoenicurus phoenicurus* **B** – Vol 1: 262
Daurian Redstart *Phoenicurus auroreus* **V** – Vol 1: 628
Moussier's Redstart *Phoenicurus moussieri* **B** (E) – Vol 1: 267
Güldenstädt's Redstart *Phoenicurus erythrogastrus* **B** – Vol 1: 269
Little Rock Thrush *Monticola rufocinereus* **B** – Vol 1: 272
(Common) Rock Thrush *Monticola saxatilis* **B** – Vol 1: 273
Blue Rock Thrush *Monticola solitarius* **B** – Vol 1: 276
Blue Whistling Thrush *Myophonus caeruleus* **V** – Vol 1: 629
Whinchat *Saxicola rubetra* **B** – Vol 1: 279
Fuerteventura Stonechat *Saxicola dacotiae* **B** (E) – Vol 1: 282
(European) Stonechat *Saxicola rubicola* **B** (E) – Vol 1: 284
Eastern Stonechat *Saxicola maurus* **B** – Vol 1: 287
African Stonechat *Saxicola torquatus* **B** – Vol 1: 294
Pied Stonechat *Saxicola caprata* **B** – Vol 1: 295
Isabelline Wheatear *Oenanthe isabellina* **B** – Vol 1: 297
Red-breasted Wheatear *Oenanthe bottae* **B** – Vol 1: 300
(Northern) Wheatear *Oenanthe oenanthe* **B** – Vol 1: 302
Seebohm's Wheatear *Oenanthe seebohmi* **B** (E) – Vol 1: 306
Pied Wheatear *Oenanthe pleschanka* **B** – Vol 1: 309
Cyprus Wheatear *Oenanthe cypriaca* **B** (E) – Vol 1: 315
Black-eared Wheatear *Oenanthe hispanica* **B** (E) – Vol 1: 319
Desert Wheatear *Oenanthe deserti* **B** – Vol 1: 325
Finsch's Wheatear *Oenanthe finschii* **B** – Vol 1: 328
Red-rumped Wheatear *Oenanthe moesta* **B** (E) – Vol 1: 332
Kurdish Wheatear *Oenanthe xanthoprymna* **B** (E) – Vol 1: 334
Persian Wheatear *Oenanthe chrysopygia* **B** – Vol 1: 337
Blyth's Wheatear *Oenanthe picata* **B** – Vol 1: 339
Mourning Wheatear *Oenanthe lugens* **B** – Vol 1: 343
Maghreb Wheatear *Oenanthe halophila* **B** (E) – Vol 1: 346
Basalt Wheatear *Oenanthe warriae* **B** (E) – Vol 1: 348
Arabian Wheatear *Oenanthe lugentoides* **B** (E) – Vol 1: 350
Hooded Wheatear *Oenanthe monacha* **B** – Vol 1: 353
Hume's Wheatear *Oenanthe albonigra* **B** – Vol 1: 355
White-crowned Wheatear *Oenanthe leucopyga* **B** – Vol 1: 356
Black Wheatear *Oenanthe lucura* **B** (E) – Vol 1: 358
Blackstart *Oenanthe melanura* **B** – Vol 1: 360
White's Thrush *Zoothera aurea* **B** – Vol 1: 362
Siberian Thrush *Geokichla sibirica* **S** – Vol 1: 364
Varied Thrush *Ixoreus naevius* **V** – Vol 1: 629
Wood Thrush *Hylocichla mustelina* **V** – Vol 1: 630
Hermit Thrush *Catharus guttatus* **S** – Vol 1: 366
Swainson's Thrush *Catharus ustulatus* **S** – Vol 1: 367
Grey-cheeked Thrush *Catharus minimus* **S** – Vol 1: 368
Veery *Catharus fuscescens* **S** – Vol 1: 369
Yemen Thrush *Turdus menachensis* **B** (E) – Vol 1: 370
Tickell's Thrush *Turdus unicolor* **V** – Vol 1: 630
Ring Ouzel *Turdus torquatus* **B** – Vol 1: 371
(Common) Blackbird *Turdus merula* **B** – Vol 1: 375
Eyebrowed Thrush *Turdus obscurus* **S** – Vol 1: 378
American Robin *Turdus migratorius* **S** – Vol 1: 380
Naumann's Thrush *Turdus naumanni* **S** – Vol 1: 382
Dusky Thrush *Turdus eunomus* **S** – Vol 1: 385
Red-throated Thrush *Turdus ruficollis* **S** – Vol 1: 388
Black-throated Thrush *Turdus atrogularis* **B** – Vol 1: 391
Fieldfare *Turdus pilaris* **B** – Vol 1: 394
Song Thrush *Turdus philomelos* **B** – Vol 1: 397
Redwing *Turdus iliacus* **B** – Vol 1: 400
Mistle Thrush *Turdus viscivorus* **B** – Vol 1: 402

SYLVIIDAE
Cetti's Warbler *Cettia cetti* **B** – Vol 1: 405
Zitting Cisticola *Cisticola juncidis* **B** – Vol 1: 408
Cricket Warbler *Spiloptila clamans* **B** – Vol 1: 411
Graceful Prinia *Prinia gracilis* **B** – Vol 1: 414
Levant Scrub Warbler *Scotocerca inquieta* **B** – Vol 1: 418
Saharan Scrub Warbler *Scotocerca saharae* **B** – Vol 1: 421
Pallas's Grasshopper Warbler *Locustella certhiola* **S** – Vol 1: 423
Lanceolated Warbler *Locustella lanceolata* **B** – Vol 1: 425
(Common) Grasshopper Warbler *Locustella naevia* **B** – Vol 1: 427
River Warbler *Locustella fluviatilis* **B** – Vol 1: 430
Savi's Warbler *Locustella luscinioides* **B** – Vol 1: 433
Gray's Warbler *Locustella fasciolata* **V** – Vol 1: 630
Moustached Warbler *Acrocephalus melanopogon* **B** – Vol 1: 436
Aquatic Warbler *Acrocephalus paludicola* **B** (E) – Vol 1: 439
Sedge Warbler *Acrocephalus schoenobaenus* **B** – Vol 1: 441
Paddyfield Warbler *Acrocephalus agricola* **B** – Vol 1: 443

Blyth's Reed Warbler *Acrocephalus dumetorum* **B** – Vol 1: 445
Marsh Warbler *Acrocephalus palustris* **B** – Vol 1: 448
(Common) Reed Warbler *Acrocephalus scirpaceus* **B** – Vol 1: 452
Cape Verde Warbler *Acrocephalus brevipennis* **B** (E) – Vol 1: 457
Clamorous Reed Warbler *Acrocephalus stentoreus* **B** – Vol 1: 459
Oriental Reed Warbler *Acrocephalus orientalis* **V** – Vol 1: 631
Great Reed Warbler *Acrocephalus arundinaceus* **B** – Vol 1: 463
Basra Reed Warbler *Acrocephalus griseldis* **B** (E) – Vol 1: 465
Thick-billed Warbler *Iduna aedon* **S** – Vol 1: 467
Isabelline Warbler *Iduna opaca* **B** (E) – Vol 1: 469
Olivaceous Warbler *Iduna pallida* **B** – Vol 1: 471
Booted Warbler *Iduna caligata* **B** – Vol 1: 475
Sykes's Warbler *Iduna rama* **B** – Vol 1: 477
Upcher's Warbler *Hippolais languida* **B** (E) – Vol 1: 480
Olive-tree Warbler *Hippolais olivetorum* **B** (E) – Vol 1: 483
Icterine Warbler *Hippolais icterina* **B** – Vol 1: 485
Melodious Warbler *Hippolais polyglotta* **B** (E) – Vol 1: 487
Marmora's Warbler *Sylvia sarda* **B** (E) – Vol 1: 490
Balearic Warbler *Sylvia balearica* **B** (E) – Vol 1: 493
Dartford Warbler *Sylvia undata* **B** (E) – Vol 1: 495
Tristram's Warbler *Sylvia deserticola* **B** (E) – Vol 1: 498
Spectacled Warbler *Sylvia conspicillata* **B** (E) – Vol 1: 501
Western Subalpine Warbler *Sylvia inornata* **B** (E) – Vol 1: 505
Eastern Subalpine Warbler *Sylvia cantillans* **B** (E) – Vol 1: 509
Moltoni's Warbler *Sylvia subalpina* **B** (E) – Vol 1: 513
Ménétries's Warbler *Sylvia mystacea* **B** – Vol 1: 516
Sardinian Warbler *Sylvia melanocephala* **B** (E) – Vol 1: 520
Cyprus Warbler *Sylvia melanothorax* **B** (E) – Vol 1: 524
Rüppell's Warbler *Sylvia ruppeli* **B** (E) – Vol 1: 528
Asian Desert Warbler *Sylvia nana* **B** – Vol 1: 531
African Desert Warbler *Sylvia deserti* **B** (E) – Vol 1: 533
Arabian Warbler *Sylvia leucomelaena* **B** – Vol 1: 535
Yemen Warbler *Sylvia buryi* **B** (E) – Vol 1: 537
Western Orphean Warbler *Sylvia hortensis* **B** (E) – Vol 1: 539
Eastern Orphean Warbler *Sylvia crassirostris* **B** – Vol 1: 542
Barred Warbler *Sylvia nisoria* **B** – Vol 1: 546
Lesser Whitethroat *Sylvia curruca* **B** – Vol 1: 549
Common Whitethroat *Sylvia communis* **B** – Vol 1: 554
Garden Warbler *Sylvia borin* **B** – Vol 1: 558
Blackcap *Sylvia atricapilla* **B** – Vol 1: 560
Eastern Crowned Warbler *Phylloscopus coronatus* **V** – Vol 1: 631
Pale-legged Leaf Warbler *Phylloscopus tenellipes* **V** – Vol 1: 632
Brown Woodland Warbler *Phylloscopus umbrovirens* **B** – Vol 1: 564
Arctic Warbler *Phylloscopus borealis* **B** – Vol 1: 565
Large-billed Leaf Warbler *Phylloscopus magnirostris* **V** – Vol 1: 632
Green Warbler *Phylloscopus nitidus* **B** (E) – Vol 1: 567
Greenish Warbler *Phylloscopus trochiloides* **B** – Vol 1: 569
Two-barred Warbler *Phylloscopus plumbeitarsus* **S** – Vol 1: 572
Pallas's Leaf Warbler *Phylloscopus proregulus* **S** – Vol 1: 574
Yellow-browed Warbler *Phylloscopus inornatus* **B** – Vol 1: 576
Hume's Leaf Warbler *Phylloscopus humei* **S** – Vol 1: 578
Radde's Warbler *Phylloscopus schwarzi* **S** – Vol 1: 581
Dusky Warbler *Phylloscopus fuscatus* **S** – Vol 1: 583
Sulphur-bellied Warbler *Phylloscopus griseolus* **V** – Vol 1: 633
Western Bonelli's Warbler *Phylloscopus bonelli* **B** (E) – Vol 1: 585
Eastern Bonelli's Warbler *Phylloscopus orientalis* **B** (E) – Vol 1: 588
Wood Warbler *Phylloscopus sibilatrix* **B** – Vol 1: 590
Plain Leaf Warbler *Phylloscopus neglectus* **B** – Vol 1: 592
Common Chiffchaff *Phylloscopus collybita* **B** – Vol 1: 594

Mountain Chiffchaff *Phylloscopus lorenzii* **B** (E) – Vol 1: 600
Iberian Chiffchaff *Phylloscopus iberiae* **B** (E) – Vol 1: 603
Canary Islands Chiffchaff *Phylloscopus canariensis* **B** (E) – Vol 1: 606
Willow Warbler *Phylloscopus trochilus* **B** – Vol 1: 608
Ruby-crowned Kinglet *Regulus calendula* **V** – Vol 1: 633
Goldcrest *Regulus regulus* **B** – Vol 1: 611
(Common) Firecrest *Regulus ignicapilla* **B** (E) – Vol 1: 616
Madeira Firecrest *Regulus madeirensis* **B** (E) – Vol 1: 619

MUSCICAPIDAE
Blue-and-white Flycatcher *Cyanoptila cyanomelana* **V** – Vol 2: 551
(Asian) Verditer Flycatcher *Eumyias thalassinus* **V** – Vol 2: 552
Dark-sided Flycatcher *Muscicapa sibirica* **V** – Vol 2: 552
(Asian) Brown Flycatcher *Muscicapa dauurica* **S** – Vol 2: 31
Spotted Flycatcher *Muscicapa striata* **B** – Vol 2: 33
Gambaga Flycatcher *Muscicapa gambagae* **B** – Vol 2: 37
Ultramarine Flycatcher *Ficedula superciliaris* **V** – Vol 2: 552
Mugimaki Flycatcher *Ficedula mugimaki* **V** – Vol 2: 553
Red-breasted Flycatcher *Ficedula parva* **B** – Vol 2: 39
Taiga Flycatcher *Ficedula albicilla* **B** – Vol 2: 42
Semicollared Flycatcher *Ficedula semitorquata* **B** (E) – Vol 2: 45
Pied Flycatcher *Ficedula hypoleuca* **B** – Vol 2: 48
Atlas Flycatcher *Ficedula speculigera* **B** (E) – Vol 2: 52
Collared Flycatcher *Ficedula albicollis* **B** (E) – Vol 2: 55

MONARCHIDAE
African Paradise Flycatcher *Terpsiphone viridis* **B** – Vol 2: 58
Indian Paradise Flycatcher *Terpsiphone paradisi* **V** – Vol 2: 554
Black-naped Monarch *Hypothymis azurea* **V** – Vol 2: 555

TIMALIIDAE
Bearded Reedling *Panurus biarmicus* **B** – Vol 2: 61
Iraq Babbler *Turdoides altirostris* **B** (E) – Vol 2: 64
Common Babbler *Turdoides caudata* **B** – Vol 2: 66
Arabian Babbler *Turdoides squamiceps* **B** (E) – Vol 2: 68
Fulvous Babbler *Turdoides fulva* **B** – Vol 2: 70

AEGITHALIDAE
Long-Tailed Tit *Aegithalos caudatus* **B** – Vol 2: 72

PARIDAE
Marsh Tit *Poecile palustris* **B** – Vol 2: 77
Willow Tit *Poecile montanus* **B** – Vol 2: 80
Caspian Tit *Poecile hyrcanus* **B** (E) – Vol 2: 84
Siberian Tit *Poecile cinctus* **B** – Vol 2: 86
Sombre Tit *Poecile lugubris* **B** (E) – Vol 2: 88
Crested Tit *Lophophanes cristatus* **B** (E) – Vol 2: 91
Coal Tit *Periparus ater* **B** – Vol 2: 93
(European) Blue Tit *Cyanistes caeruleus* **B** – Vol 2: 97
African Blue Tit *Cyanistes teneriffae* **B** (E) – Vol 2: 100
Azure Tit *Cyanistes cyanus* **B** – Vol 2: 103
Great Tit *Parus major* **B** – Vol 2: 106
Cinereous Tit *Parus cinereus* **B** – Vol 2: 111
Turkestan Tit *Parus bokharensis* **B** – Vol 2: 113

SITTIDAE
Krüper's Nuthatch *Sitta krueperi* **B** (E) – Vol 2: 115
Corsican Nuthatch *Sitta whiteheadi* **B** (E) – Vol 2: 117
Algerian Nuthatch *Sitta ledanti* **B** (E) – Vol 2: 119
Red-breasted Nuthatch *Sitta canadensis* **V** – Vol 2: 555
(Eurasian) Nuthatch *Sitta europaea* **B** – Vol 2: 121
Eastern Rock Nuthatch *Sitta tephronota* **B** – Vol 2: 125
(Western) Rock Nuthatch *Sitta neumayer* **B** (E) – Vol 2: 128

TICHODROMIDAE
Wallcreeper *Tichodroma muraria* **B** – Vol 2: 131

CERTHIIDAE
(Eurasian) Treecreeper *Certhia familiaris* **B** – Vol 2: 134
Short-toed Treecreeper *Certhia brachydactyla* **B** (E) – Vol 2: 137

REMIZIDAE
(Eurasian) Penduline Tit *Remiz pendulinus* **B** – Vol 2: 140
Black-headed Penduline Tit *Remiz macronyx* **B** – Vol 2: 145
White-crowned Penduline Tit *Remiz coronatus* **S** – Vol 2: 147

NECTARINIIDAE
Nile Valley Sunbird *Hedydipna metallica* **B** – Vol 2: 149
Purple Sunbird *Cinnyris asiaticus* **B** – Vol 2: 152
Shining Sunbird *Cinnyris habessinicus* **B** – Vol 2: 155
Palestine Sunbird *Cinnyris osea* **B** – Vol 2: 157

ZOSTEROPIDAE
Abyssinian White-Eye *Zosterops abyssinicus* **B** – Vol 2: 160
Oriental White-Eye *Zosterops palpebrosus* **B** – Vol 2: 162

ORIOLIDAE
Black-naped Oriole *Oriolus chinensis* **V** – Vol 2: 555
(Eurasian) Golden Oriole *Oriolus oriolus* **B** – Vol 2: 163

LANIIDAE
Rosy-patched Bush-Shrike *Rhodophoneus cruentus* **B** – Vol 2: 167
Black-crowned Tchagra *Tchagra senegalus* **B** – Vol 2: 169
Brown Shrike *Lanius cristatus* **S** – Vol 2: 171
Turkestan Shrike *Lanius phoenicuroides* **B** – Vol 2: 174
Isabelline Shrike *Lanius isabellinus* M – Vol 2: 178
Red-backed Shrike *Lanius collurio* **B** – Vol 2: 185
Bay-backed Shrike *Lanius vittatus* **B** – Vol 2: 188
Long-tailed Shrike *Lanius schach* **S** – Vol 2: 190
Lesser Grey Shrike *Lanius minor* **B** – Vol 2: 192
Great Grey Shrike *Lanius excubitor* **B** – Vol 2: 195
Northern Shrike *Lanius borealis* **B** – Vol 2: 207
Iberian Grey Shrike *Lanius meridionalis* **B** (E) – Vol 2: 210
Woodchat Shrike *Lanius senator* **B** (E) – Vol 2: 213
Masked Shrike *Lanius nubicus* **B** (E) – Vol 2: 218

DICRURIDAE
Black Drongo *Dicrurus macrocercus* **V** – Vol 2: 556
Ashy Drongo *Dicrurus leucophaeus* **V** – Vol 2: 556

CORVIDAE
(Eurasian) Jay *Garrulus glandarius* **B** – Vol 2: 221
Siberian Jay *Perisoreus infaustus* **B** – Vol 2: 227
Azure-winged Magpie *Cyanopica cyanus* **B** – Vol 2: 229
(Common) Magpie *Pica pica* **B** – Vol 2: 231
Pleske's Ground Jay *Podoces pleskei* **B** (E) – Vol 2: 236
(Spotted) Nutcracker *Nucifraga caryocatactes* **B** – Vol 2: 238
Alpine Chough *Pyrrhocorax graculus* **B** – Vol 2: 241
(Red-billed) Chough *Pyrrhocorax pyrrhocorax* **B** (E) – Vol 2: 244
(Western) Jackdaw *Corvus monedula* **B** – Vol 2: 247
Daurian Jackdaw *Corvus dauuricus* **S** – Vol 2: 250
House Crow *Corvus splendens* **B** (I) – Vol 2: 252
Rook *Corvus frugilegus* **B** – Vol 2: 254
Carrion Crow *Corvus corone* **B** – Vol 2: 257
Hooded Crow *Corvus cornix* **B** – Vol 2: 260
Large-billed Crow *Corvus macrorhynchos* **V** – Vol 2: 557
Pied Crow *Corvus albus* **B** – Vol 2: 264
Brown-necked Raven *Corvus ruficollis* **B** – Vol 2: 265
(Common) Raven *Corvus corax* **B** – Vol 2: 268
Fan-tailed Raven *Corvus rhipidurus* **B** – Vol 2: 272

STURNIDAE
Tristram's Starling *Onychognathus tristramii* **B** (E) – Vol 2: 274
(Common) Starling *Sturnus vulgaris* **B** – Vol 2: 276
Spotless Starling *Sturnus unicolor* **B** (E) – Vol 2: 280
Pied Myna *Gracupica contra* **B** (I) – Vol 2: 283
Rose-coloured Starling *Pastor roseus* **B** – Vol 2: 285
Wattled Starling *Creatophora cinerea* **S** – Vol 2: 289
Brahminy Starling *Sturnia pagodarum* **S** – Vol 2: 291
Daurian Starling *Agropsar sturninus* **V** – Vol 2: 557
Chestnut-tailed Starling *Sturnia malabarica* **V** – Vol 2: 558
Common Myna *Acridotheres tristis* **B** (I) – Vol 2: 292
Bank Myna *Acridotheres ginginianus* **B** (I) – Vol 2: 294
Amethyst Starling *Cinnyricinclus leucogaster* **B** – Vol 2: 296
Red-billed Oxpecker *Buphagus erythrorhynchus* **V** – Vol 2: 558

PASSERIDAE
Saxaul Sparrow *Passer ammodendri* **B** – Vol 2: 298
House Sparrow *Passer domesticus* **B** – Vol 2: 300
Italian Sparrow *Passer italiae* **B** (E) – Vol 2: 305
Spanish Sparrow *Passer hispaniolensis* **B** – Vol 2: 307
Sind Sparrow *Passer pyrrhonotus* **B** – Vol 2: 310
Dead Sea Sparrow *moabiticus* **B** (E) – Vol 2: 311
Iago Sparrow *Passer iagoensis* **B** (E) – Vol 2: 314
Desert Sparrow *Passer simplex* **B** – Vol 2: 316
Zarudny's Sparrow *Passer zarudnyi* **B** – Vol 2: 318
(Eurasian) Tree Sparrow *Passer montanus* **B** – Vol 2: 319
Arabian Golden Sparrow *Passer euchlorus* **B** – Vol 2: 321
Sudan Golden Sparrow *Passer luteus* **B** – Vol 2: 323
Pale Rock Sparrow *Carpospiza brachydactyla* **B** – Vol 2: 325
Yellow-throated Sparrow *Gymnoris xanthocollis* **B** – Vol 2: 327
Bush Sparrow *Gymnoris dentata* **B** – Vol 2: 329
(Common) Rock Sparrow *Petronia petronia* **B** – Vol 2: 330
(White-winged) Snowfinch *Montifringilla nivalis* **B** – Vol 2: 333

PLOCEIDAE
Streaked Weaver *Ploceus manyar* **B** (I) – Vol 2: 336
Rüppell's Weaver *Ploceus galbula* **B** – Vol 2: 338
Village Weaver *Ploceus cucullatus* **V** – Vol 2: 558

ESTRILDIDAE
Common Waxbill *Estrilda astrild* **B** – Vol 2: 340
Arabian Waxbill *Estrilda rufibarba* **B** (E) – Vol 2: 342
Red Avadavat *Amandava amandava* **B** – Vol 2: 343
Zebra Waxbill *Amandava subflava* **B** – Vol 2: 345
Indian Silverbill *Euodice malabarica* **B** – Vol 2: 346
African Silverbill *Euodice cantans* **B** – Vol 2: 348
Red-billed Firefinch *Lagonosticta senegala* **B** – Vol 2: 350

VIREONIDAE
White-eyed Vireo *Vireo griseus* **V** – Vol 2: 589
Yellow-throated Vireo *Vireo flavifrons* **V** – Vol 2: 589
Philadelphia Vireo *Vireo philadelphicus* **V** – Vol 2: 589
Red-eyed Vireo *Vireo olivaceus* **S** – Vol 2: 351

FRINGILLIDAE
(Common) Chaffinch *Fringilla coelebs* **B** – Vol 2: 353
Blue Chaffinch *Fringilla teydea* **B** (E) – Vol 2: 359
Brambling *Fringilla montifringilla* **B** – Vol 2: 361
Red-fronted Serin *Serinus pusillus* **B** – Vol 2: 364
(European) Serin *Serinus serinus* **B** (E) – Vol 2: 367
Syrian Serin *Serinus syriacus* **B** (E) – Vol 2: 370
(Atlantic) Canary *Serinus canaria* **B** (E) – Vol 2: 372
Arabian Serin *Crithagra rothschildi* **B** (E) – Vol 2: 375

Yemen Serin *Crithagra menachensis* **B** (E) – Vol 2: 377
Golden-winged Grosbeak *Rhynchostruthus socotranus* **B** – Vol 2: 379
(European) Greenfinch *Chloris chloris* **B** – Vol 2: 381
(European) Goldfinch *Carduelis carduelis* **B** – Vol 2: 385
Citril Finch *Carduelis citrinella* **B** (E) – Vol 2: 389
Corsican Finch *Carduelis corsicana* **B** (E) – Vol 2: 392
(Eurasian) Siskin *Spinus spinus* **B** – Vol 2: 394
(Common) Linnet *Linaria cannabina* **B** – Vol 2: 397
Yemen Linnet *Linaria yemenensis* **B** (E) – Vol 2: 402
Twite *Linaria flavirostris* **B** – Vol 2: 404
Common Redpoll *Acanthis flammea* **B** – Vol 2: 407
Arctic Redpoll *Acanthis hornemanni* **B** – Vol 2: 412
Two-barred Crossbill *Loxia leucoptera* **B** – Vol 2: 417
Common Crossbill *Loxia curvirostra* **B** – Vol 2: 420
Parrot Crossbill *Loxia pytyopsittacus* **B** (E) – Vol 2: 425
Asian Crimson-winged Finch *Rhodopechys sanguineus* **B** – Vol 2: 427
African Crimson-winged Finch *Rhodopechys alienus* **B** (E) – Vol 2: 430
Desert Finch *Rhodospiza obsoleta* **B** – Vol 2: 432
Mongolian Finch *Bucanetes mongolicus* **B** – Vol 2: 435
Trumpeter Finch *Bucanetes githagineus* **B** – Vol 2: 438
Common Rosefinch *Carpodacus erythrinus* **B** – Vol 2: 441
Sinai Rosefinch *Carpodacus synoicus* **B** (E) – Vol 2: 444
Pallas's Rosefinch *Carpodacus roseus* **V** – Vol 2: 560
Great Rosefinch *Carpodacus rubicilla* **B** – Vol 2: 446
Long-tailed Rosefinch *Carpodacus sibiricus* **B** – Vol 2: 448
Pine Grosbeak *Pinicola enucleator* **B** – Vol 2: 450
(Common) Bullfinch *Pyrrhula pyrrhula* **B** – Vol 2: 452
Azores Bullfinch *Pyrrhula murina* **B** (E) – Vol 2: 455
White-winged Grosbeak *Mycerobas carnipes* **B** – Vol 2: 457
Hawfinch *Coccothraustes coccothraustes* **B** – Vol 2: 459
Evening Grosbeak *Hesperiphona vespertina* **V** – Vol 2: 560

PARULIDAE
Ovenbird *Seiurus aurocapilla* **S** – Vol 2: 462
Northern Waterthrush *Parkesia noveboracensis* **S** – Vol 2: 464
Black-and-white Warbler *Mniotilta varia* **S** – Vol 2: 466
Golden-winged Warbler *Vermivora chrysoptera* **V** – Vol 2: 591
Blue-winged Warbler *Vermivora cyanoptera* **V** – Vol 2: 591
Tennessee Warbler *Oreothlypis peregrina* **S** – Vol 2: 468
(Common) Yellowthroat *Geothlypis trichas* **S** – Vol 2: 471
American Redstart *Setophaga ruticilla* **S** – Vol 2: 473
Northern Parula *Setophaga americana* **S** – Vol 2: 475
(American) Yellow Warbler *Setophaga petechia* **S** – Vol 2: 477
Chestnut-sided Warbler *Setophaga pensylvanica* **V** – Vol 2: 595
Cerulean Warbler *Setophaga cerulea* **V** – Vol 2: 592
Black-throated Blue Warbler *Setophaga caerulescens* **V** – Vol 2: 595
Yellow-throated Warbler *Setophaga dominica* **V** – Vol 2: 596
Black-throated Green Warbler *Setophaga virens* **V** – Vol 2: 597
Blackburnian Warbler *Setophaga fusca* **V** – Vol 2: 594
Prairie Warbler *Setophaga discolor* **V** – Vol 2: 597
Cape May Warbler *Setophaga tigrina* **V** – Vol 2: 592
Magnolia Warbler *Setophaga magnolia* **V** – Vol 2: 593
Yellow-rumped Warbler *Setophaga coronata* **S** – Vol 2: 481
Palm Warbler *Setophaga palmarum* **V** – Vol 2: 596
Blackpoll Warbler *Setophaga striata* **S** – Vol 2: 479
Bay-breasted Warbler *Setophaga castanea* **V** – Vol 2: 594
Hooded Warbler *Setophaga citrina* **V** – Vol 2: 591
Wilson's Warbler *Cardellina pusilla* **V** – Vol 2: 598
Canada Warbler *Cardellina canadensis* **V** – Vol 2: 598

THRAUPIDAE
Summer Tanager *Piranga rubra* **V** – Vol 2: 599
Scarlet Tanager *Piranga olivacea* **S** – Vol 2: 484

EMBERIZIDAE
Eastern Towhee *Pipilo erythrophthalmus* **V** – Vol 2: 599
Lark Sparrow *Chondestes grammacus* **V** – Vol 2: 599
Savannah Sparrow *Passerculus sandwichensis* **V** – Vol 2: 600
Fox Sparrow *Passerella iliaca* **V** – Vol 2: 600
American Tree Sparrow *Passerella arborea* **V** – Vol 2: 601
Song Sparrow *Melospiza melodia* **S** – Vol 2: 486
Lincoln's Sparrow *Melospiza lincolnii* **V** – Vol 2: 571
White-crowned Sparrow *Zonotrichia leucophrys* **S** – Vol 2: 488
White-throated Sparrow *Zonotrichia albicollis* **S** – Vol 2: 490
Dark-eyed Junco *Junco hyemalis* **S** – Vol 2: 492
Snow Bunting *Plectrophenax nivalis* **B** – Vol 2: 494
Lapland Bunting *Calcarius lapponicus* **B** – Vol 2: 499
Black-faced Bunting *Emberiza spodocephala* **S** – Vol 2: 502
Pine Bunting *Emberiza leucocephalos* **B** – Vol 2: 505
Yellowhammer *Emberiza citrinella* **B** – Vol 2: 509
Cirl Bunting *Emberiza cirlus* **B** – Vol 2: 512
White-capped Bunting *Emberiza stewartii* **V** – Vol 2: 601
Rock Bunting *Emberiza cia* **B** – Vol 2: 516
Meadow Bunting *Emberiza cioides* **V** – Vol 2: 602
Striolated Bunting *Emberiza striolata* **B** – Vol 2: 520
House Bunting *Emberiza sahari* **B** – Vol 2: 523
Cinnamon-breasted Bunting *Emberiza tahapisi* **B** – Vol 2: 525
Cinereous Bunting *Emberiza cineracea* **B** (E) – Vol 2: 528
Ortolan Bunting *Emberiza hortulana* **B** – Vol 2: 531
Grey-necked Bunting *Emberiza buchanani* **B** – Vol 2: 534
Cretzschmar's Bunting *Emberiza caesia* **B** (E) – Vol 2: 537
Chestnut-eared Bunting *Emberiza fucata* **V** – Vol 2: 603
Yellow-browed Bunting *Emberiza chrysophrys* **S** – Vol 2: 540
Rustic Bunting *Emberiza rustica* **B** – Vol 2: 542
Little Bunting *Emberiza pusilla* **B** – Vol 2: 545
Chestnut Bunting *Emberiza rutila* **S** – Vol 2: 548
Yellow-breasted Bunting *Emberiza aureola* **B** – Vol 2: 551
(Common) Reed Bunting *Emberiza schoeniclus* **B** – Vol 2: 554
Pallas's Reed Bunting *Emberiza pallasi* **S** – Vol 2: 560
Red-headed Bunting *Emberiza bruniceps* **B** – Vol 2: 564
Black-headed Bunting *Emberiza melanocephala* **B** – Vol 2: 567
Corn Bunting *Emberiza calandra* **B** – Vol 2: 570
Dickcissel *Spiza americana* **V** – Vol 2: 603
Rose-breasted Grosbeak *Pheucticus ludovicianus* **S** – Vol 2: 572
Indigo Bunting *Passerina cyanea* **V** – Vol 2: 574

ICTERIDAE
Bobolink *Dolichonyx oryzivorus* **S** – Vol 2: 576
Brown-headed Cowbird *Molothrus ater* **V** – Vol 2: 604
Common Grackle *Quiscalus quiscula* **V** – Vol 2: 604
Yellow-headed Blackbird *Xanthocephalus xanthocephalus* **V** – Vol 2: 605
Baltimore Oriole *Icterus galbula* **S** – Vol 2: 578

PHOTOGRAPHIC CREDITS

Front cover, main image: Basalt Wheatear by Amir Ben Dov. Small images clockwise from top left: Thick-billed Lark (Amir Ben Dov), Melodious Warbler (Daniele Occhiato), White-spectacled Bulbul (Hadoram Shirihai), European Robin (Daniele Occhiato), Bohemian Waxwing (Martin Mecnarowski/Shutterstock), Siberian Stonechat (Hadoram Shirihai), Western Subalpine Warbler (Carlos N. G. Bocos), Siberian Accentor (James Lowen/jameslowen.com), Lanceolated Warbler (Daniele Occhiato), Firecrest (Kasperczyk Bogdan/Shutterstock), Eurasian Wren (Hadoram Shirihai), Bluethroat (Markus Varesvuo).

Spine image: Rock Thrush by Daniele Occhiato.

Back cover, central trio (l-r): Common Chiffchaff *tristis* (Aurélien Audevard), Mountain Chiffchaff (Rami Mizrachi), Common Chiffchaff *collybita* (Aurélien Audevard). Small images clockwise from top left: Olive-backed Pipit (Daniele Occhiato), Red-rumped Swallow (Hadoram Shirihai), Black-throated Thrush (Daniele Occhiato), Western Bonelli's Warbler (Carlos N. G. Bocos), Yellow Wagtail (Rafal Szozda/Shutterstock), White-throated Dipper (Frank Fichtmueller/Shutterstock), Güldenstädt's Redstart (Hadoram Shirihai), Grey Hypocolius (Hanne & Jens Eriksen), Clamorous Reed Warbler (Hadoram Shirihai), Clamorous Reed Warbler (Hadoram Shirihai), Eastern Mourning Wheatear (Hadoram Shirihai).

Photographic credits for images that appear in this book. l = left; r = right; t = top; b = bottom; c = centre.

AbdulRahman Al-Sirhan Alenezi 247cl; 292tl; 292cl; 352tl; 352tr; 462tl; 462tc; 462cl; 462cr; 465br; 466br; 473tr; 504br; 518tl; 531br; 537cl; 544br; 548tr; 557tr; 589cr: **Abel Fernández** 604bl; 605cr: **Abhishek Gulshan** 105cl: **Adrian Drummond-Hill** 24bl; 177cl; 416cr; 480bl: **Ahmed Abdallah Al Ali** 323tr: **Ahmet Karataş** 71bl: **Alain Fossé** 82tl: **Alan Murphy** / AlanMurphyPhotography.com 626bct; 626bcb: **Alan Tate** / www.aabirdpix.com 323tl: **Alastair Wilson** www.amwphotos.co.uk 479tr: **Albert de Jong** 482tl: **Albert R. Salemgareyev (ACBK)** 92bl: **Aleix Comas** 491tl: **Alejandro Torés Sánchez** 85cl; 151cl; 268bl; 284bl; 331tc; 346br; 410tr; 497tr: **Alex Ash** 486cr: **Alexander C. Lees** 411br; 412br: **Alexander Hellquist** 553tl; 553tc; 553tr; 602bc: Alexander Katuncev / http://birds.kz 629cr: **Alexander Varlamov** 291tl; 291tr: **Alexandr Belyaev** 454tr: **Alexey Levashkin** 475bl: **Ali Atahan** - Birdwatcher and SubaşıKuş 222cr: **Ali Atıcı** 324tl: **Ali Murtaza Doğan** 238bl: **Ali Sadr** / http://birdsofiran.com 338bl: **Alison McArthur** 278tc; 315tr; 316cl; 329cr; 437cr; 522cr; 525cl; 525tr; 526br; 526tr; 526cl: **Amir Ben Dov** 10cl; 22tr; 51cl; 52tr; 69tr; 85bl; 86br; 90br; 100cl; 100bl; 120bl; 120br; 125tr; 125cr; 194cr; 134tl; 134tr; 140tl; 140tr; 157bl; 157br; 158cr; 165bl; 165bc; 165br; 168tl; 171tl; 176bl; 177tl; 189cr; 259tl; 259tr; 259bl; 275bl; 304cl; 312tc; 312cc; 323bl; 329bl; 329br; 330tl; 331tl; 336cr; 343br; 354br; 350br; 415bl; 435bl; 460tl; 472cr; 473bl; 479bl; 481cr; 482br; 483bl; 483br; 499br; 547c; 561tl: **Amir Talebui Gol** 339cl: **Anders Blomdahl** 47br; 79bl; 127tl: **Andreas Gygax** 373cl: **Andreas Uppstu** 548br: **Andrés M. Dominguez** 78tl; 278tl; 278bl; 286br; 303cl; 319bl; 320tl; 373br; 403tr; 489tr: **Andrew M. Allport**/Shutterstock 548cl: **Andrew W. Clarke** 162cr: **Andrey Vilyayev** 177tr: **Annika Forsten** 91br: **António Antunes Gonçalves** 141br: **António Guerra** 500br; 503br: **Antonio Manglano** 453bl: **Antonio Tamayo Guerrero** 540cr: **Antti Below** 35cr; 143tr; 171cr; 325tr; 431bl: **Arend Wassink** 261tr; 313br; 451br: **Arie Ouwerkerk** 48cl; 75bl; 76bl; 113cr; 144br; 179tl; 193tr; 202bl; 228tr; 338tl; 372br; 391cl; 397cr; 398cr; 471cl; 577tl; 577cr: **Arijit Banerjee** 106cl: **Armando Caldas** 604tr: **Arnoud B van den Berg/The Sound Approach** 53tl; 60tl; 61bl; 82c; 95bl; 103cl; 194tr; 306cr; 377bl; 377tl; 380bl; 474tr: **Aron Edman & Simon S. Christiansen** 24cl: **Arpit Deomurari** 64cr; 545cr: **Arto Juvonen** 108bl; 108bc; 108br; 282tr; 283cr; 431tr: **Artur Stankiewicz**/www.arturstankiewicz.com 239b: **Ashley Banwell** / www.ashleybanwellsbirding.com 424bl; 488br: **Askar Isabekov** / birds.kz 60cr; 92tr; 218br; 219bl; 219br; 247tr; 258tr; 271tl; 393tl: **Astrid Kant** 89cr: **Augusto Faustino** 102bl; 236br; 333br; 469bl: **Aurélien AUDEVARD** 44tl; 129bl; 129bc; 192cr; 137tl; 148tl; 163tl; 166tl; 174tl; 176tl; 178bl; 239tr; 242tl; 275tr; 287br; 337tl; 351tl; 355cl; 383tl; 390tl; 416tl; 426tl; 453tl; 453bl; 487br; 496tl; 532tl; 592bl; 595br; 598cl; 609cl; 610cr; 613tr: **Avner Rinot** 335br: **Axel Halley** 64tr; 71cr; 79tl; 137cl; 205cl; 205tr; 280cl; 286bl; 297tr; 299cl; 322cr; 335tl; 338cl; 444cr: **Ayhan Öztürk- Dörtceker Wildlife Photography** (www.dortceker.org) 246br; 526br: **Ayuwat Jearwattanakanok** 131cr; 378tl: **Bas van den Boogaard** 153br; 184tr; 289br; 317cr; 468tl; 581tl; 597cr: **Bayram Gocmen** / http://www.bayramgocmen.com/ 290br: **Ben Porter** / www.benporterwildlife.co.uk 486br: **Beneharo Rodríguez** - www.gohnic.org 502tl: **Benoit Maire** 307tr; 307tr; 308c; 308tr; 308bl; 473br; 534bl: **Bernat Ramis** 493bl; 493br; 494cl: **Bernhard Herren** 518cl: **Bert de Bruin** 393bl: **Bill Baston** (www.billbaston.com) 54cl; 116bl; 283br; 495br: **Björn Johansson** / bjornjohansson.dinstudio.se 261bl: **blickwinkel** / Alamy Stock Photo 522br; 617tl: **Bo Tureby** 151bl: **Bob Gibbons/FLPA** 396br: **Bonnie Taylor Barrie/Shutterstock** 620br: **Boris Nikolov/ Biota Films** 55bl; 165tl; 243tr; 298br; 310cr; 344cr; 443cr: **Bosse Carlsson** 563bl; 563br: **Brenda Veen** 224br: **Brett Spencer** 551cr: **Brian Lasenby/Shutterstock** 620bl: **Brian Liggins** 130bc: **Callan Cohen** / www.birdingafrica.com 455cl; 455bl; 455br; 474cr: **Carl Bovis** / carlbovisnaturephotography.blogspot.co.uk 214bl; 286tl: **Carl-Johan Svensson** / www.pbase.com/carljohansvensson/birds 249cr: **Carlos M. Martín** 49tr: **Carlos N. G. Bocos** 9cr; 36cl; 46br; 62br; 75tl; 75tr; 78tc; 81bl; 81br; 173cr; 173tl; 210bl; 215cr; 228tl; 238br; 241tl; 241tr; 241cr; 244bl; 256tr; 295tr; 295cl; 298cr; 331br; 341tl; 341cl; 349cr; 350cl; 358bl; 358br; 361tl; 405br; 412c; 429br; 454tl: **Cem Doğut** 542br: **chakapong/shutterstock** 628tl: **Choi Soonkyoo** 178tl: **Choi Wai Mun** 169cl: **Chris Batty** 48tr; 53br; 268br; 366br: **Chris Galvin** 363tl; 363tr: **Chris Turner** 598br: **Chris van Rijswijk** / birdshooting.nl 80tr; 285tr; 475br: **Christian Tiller** / www.fuglebilder.no 460br; 518br; 575cl; 575tl; 621cl: **Claus Halkjaer** 167tl: **Claus Johansson** / www.digitalinside.eu 606cr: **Clement Francis** 68bl; 184tl; 220tr: **Colin Bradshaw** 57cr; 70tl; 72tr; 160tr; 192bl; 242cr; 254tr; 297cr; 323tr; 552tr; 598tr: **Colin G. Manville** 366br: **Craig Robson** 123bl: **Cristian Zea** 586cr: **Dan Brown** (www.NaturalWorldConsultants.com) 413tl: **Daniele Occhiato** 22cr; 24br; 34br; 35tl; 47br; 52tl; 54tr; 56tr; 56tl; 57bl; 58tr; 63tl; 63tr; 79br; 82bl; 88bl; 89tr; 97cr; 98br; 117tl; 117br; 119cr; 120tr; 120cl; 126tr; 193tl; 195tl; 195cr; 133bl; 133br; 139bl; 143bl; 150br; 151tl; 155cl; 155cr; 156bl; 157tr; 158bl; 162bl; 163tr; 164bl; 167cr; 167bl; 171tr; 173br; 173bl; 174tr; 175tr; 179tr; 182cl; 182bl; 182bl; 183tl; 187br; 188br; 189br; 190cl; 190bl; 196bl; 198tl; 199tr; 200cl; 200bl; 200br; 210tr; 211tl; 211br; 213tr; 213bl; 214cr; 225cl; 225br; 229tl; 229tr; 233tr; 233br; 235cl; 235br; 245tr; 245cl; 245br; 256cl; 257cc; 257tr; 263br; 263br; 273br; 274br; 277br; 279bl; 285tr; 285br; 285bc; 287br; 288cr; 288cr; 298bc; 299tr; 299br; 299bc; 305bl; 315cl; 321bl; 322tr; 322bl; 322br; 325cl; 327tl; 327bl; 332br; 333bl; 334br; 335cl; 335br; 347cc; 349bc; 351tl; 351b; 353bl; 355br; 356br; 358tl; 359tl; 359br; 361cl; 377cc; 377br; 404tr; 405bl; 406tl; 406cr; 406bl; 408br; 415br; 435cr; 437tl; 437bl; 438cl; 438cr; 440br; 445bl; 451br; 453cl; 457cr; 459cc; 464tl; 464cl; 471br; 478tl; 478cl; 480bl; 481tl; 483tr; 487tr; 487cl; 488br; 497cr; 497bl; 498br; 499tr; 501bl; 502cr; 509tl; 513bl; 513br; 514tl; 514tr; 514br; 515tl; 515tr; 515bl; 516tl; 517bl; 517br; 520bl; 524cl; 528bl; 532tl; 543bl; 543br; 549tl; 551br; 554bl; 554br; 557cl; 559br; 559bl; 561bl; 561tr; 567bl; 585br; 588br; 589bl; 594bl; 596tr; 599tl; 599bc; 601tr; 605br; 605tr; 612br; 614cl; 614tr; 617cl; 617bl; 618cl; 618bl; 628br: **Dave Barnes** / www.pbase.com/davebarnes 78bl; 78bc; 102tc: **Dave Clark** 106br; 112br; 323bc; 327cr; 480br; 516bl; 527cr; 547br: **Dave Curtis** / www.flickr.com/photos/davethebird 479cl: **Dave Hutton** 218tl: **Dave Stewart** 368cl; 491cr: **David Cottridge** 492bl: **David Erterius** 132cl; 580cl: **David Jirovsky** 373tl; 501br; 525tr; 589br: **David Monticelli** pbase.com/david_monticelli 34tl; 48br; 51tr; 72cl; 83cl; 197cl; 198br; 219cr; 228bl; 267tr; 267br; 326cr; 345tl; 357br; 411tl; 422cr; 473cr; 620cr; 497br; 540cr; 621tr; 621tl; 623cr; 626tl; 626cl; 633br: **David Pérez** - ornitoaddiction.blogspot.com.es 534tl: **David Tipling** 408bl: **David Tipling Photo Library** / Alamy Stock Photo 258tr; 374tr; 374cr: **David Tipling/FLPA** 131cl: **David Tromans** 428cl: **David Verdonck** 278br: **Dennis F. Asselbergs** 451br; 451bl: **Dennis Morrison** 369cl: **Derviş KÖKENEK, Bird and Nature Photographer, Adana /TURKEY** 417bl; 417tr: **Devaram Thirunavukkarasu** / www.gamebirds.me 384bl; 384tr: **Dick Forsman** / www.dickforsman.com 536tl: **Dolly Laishram** 270br: **Dominic Mitchell** (www.birdingetc.com) 286tr; 307cl; 307bl; 307tr: **Doug Gochfeld** 231bl: **Dougie Preston** 479tl: **Dr. R.B.Balar, Ahmedabad** 69br: **Dubi Kalay** 519tr; 523cr; 523cr; 557tl: **Duha a alhashimi** 231br: **Eduardo Alba Padilla** 507br; 604br: **Edwin Winkel** 33cl; 45br; 80bl; 89br; 95cl; 99bl; 99br; 100tr; 121c; 126br; 158br; 191tl; 194br; 274bc; 275bc; 324cr; 401tl; 407cl; 473bl;

519br: Eigil Ødegaard 445cl: Emin Yoğurtcouğlu 59cl; 59br; 91bl; 169bl; 238cr; 349cl; 443br; 476cl; 476cr; 478tl; 479cr; 510bc: Espen Bergersen / www.Natur-Galleriet.no 162bl: Eva Foss Henriksen 302bl: Eyal Bartov 544tr; 560bl: Fabio Balanti 209br; 241bl: Fatih Izler 266br: Fedor Karpov 432br: Felix Heintzenberg 259cr; 578bl: Ferran López 37tr; 412tr; 493cl; 603br: Fikret Yorgancıoğlu 529bl; 530br, 555cl: Fran Trabalon 103br: Frank Joisten 184bl: Frédéric Jiguet 492br; 494cc; 515br: Gabriel Norevik 572br: Gabriel Schuler 152bl; 208bl; 252cr; 481cl; 511cr; Gal Shon 237bl: Gallinago_media/shutterstock 1: Garry Bakker 289br: Gary Brown, Sultan Qaboos University, Muscat, Oman 38bl; 417br; 464cr: Gary Jenkins 393cr; 492tr; 566bl: Gary Thoburn (GaryTsPhotos) 164tr; 172tr; 172cr; 172bl; 256cr; 273cl; 274br; 321tl; 380br; 398br; 442bl; 471br; 488bl; 489cl; 548cr; 550tl; 582br; 609tr; 612tl; 632tr: Gavin Thomas 478br: Gediminas Gražulevičius 257tr: Gennadiy Dyakin 407bl: George Petrie 591br: George Reszeter 99tr; 312cr; 328cl; 329tl; 329cr; 381bl; 503br; 527bl; 527bc; 584br; 621cr: Gerhard Loidolt, Austria 591cl: Ghulam Mughal 261cr; 261br; 269cl: Gianni Conca pbase.com/birdclick 53bl 101cl; 102cl; 304bl; 404cl; 502bl; 506tr: Giorgi Darchiashvili 270tl; 271tr; 271bl; Giorgio Di Liddo 500tr: Giuseppe Falcone 521br: Glyn Sellors - www.glyn-sellorsphotography.com 631br: Gökhan Coral 530bl: Göran Ekström 41tl; 69tl; 342tr; 425bl; 425br; 528tr: Graham P Catley 172tl; 318br: Graham R. Lobley 201bl; 272cl: Greg Baker 517cr: Gunnar Gundersen / www.listafuglestasjon.no 256br: Gustavo Peña-Tejera 282bl; 282br; 283tr: Hadoram Shirihai 9cl; 12cr; 13tr; 15bl; 23cl; 23bl; 24tl; 39tl; 39cr; 40bl; 41br; 45tc; 45tr; 50tr; 56bl; 57cl; 58cl; 65bl; 65br; 67br; 70bl; 70br; 74tl; 74tc; 74tr; 78br; 81tr; 81cr; 85cr; 85br; 86cl; 86cr; 90tl; 90cl; 90cr; 94tr; 94br; 98tl; 98cr; 101br; 104bl; 104br; 109bl; 109br; 110cl; 111tl; 111br; 112bc; 114tc; 114tr; 117cl; 119br; 122cl; 122cr; 122bl; 122br; 124bl; 125bl; 132cr; 136cl; 136bl; 136br; 139tr; 141tr; 142bl; 142bc; 142br; 145cl; 147cr; 148bl; 151tr; 168tr; 177bl; 178tr; 180bc; 180br; 183cr; 185tl; 187tr; 196br; 196cl; 198cl; 199cl; 200bc; 201tr; 201bl; 211cr; 212br; 214tr; 215tr; 225tr; 228cl; 230cr; 232tr; 232bl; 233bl; 234tl; 234tc; 234tr; 234bc; 247cr; 248bl; 248br; 251tl; 253br; 255bl; 260tl; 260tr; 260bl; 265tr; 265cr; 269cr; 269br; 271cr; 272tr; 272br; 276bl; 283bc; 288tr; 295br; 298tl; 298cl; 301br; 310tl; 310tc; 313br; 313cl; 316tl; 316tr; 316cc; 316bl; 316bc; 316br; 317tl; 317cl; 317cr; 318tl; 318bl; 318bc; 320cr; 323cl; 326cc; 327cl; 328bl; 328br; 330br; 335bl; 336tl; 343cl; 343cr; 344tr; 344cl; 345cr; 348cl; 349tl; 349cr; 349cc; 349bl; 353cr; 354cl; 354cr; 357cr; 360cl; 366bc 367br; 378br; 389tl; 407tl; 414tl; 414bl; 414br; 415bl; 418bl; 419bl; 423tr; 433bl; 435tl; 447cl; 448br; 452bl; 455tr; 457tl; 457bl; 458cl; 458br; 468tr; 472tr; 481br; 483cl; 484br; 497tl; 507tl; 509bl; 510br; 511tl; 511br; 512br; 521tr; 523tr; 523cl; 525cr; 526tl; 526bl; 528br; 529tr; 529cl; 529cr; 530tr; 530cr; 531cl; 531br; 532br; 535tl; 535bl; 536br; 542cl; 543cr; 543br; 551cl; 552bl; 532bl; 563tl; 568cl; 568br; 568cc; 573bl; 573br; 574cl; 576br; 579tl; 579tr; 581br; 583bl; 583br; 588bl; 591tr; 600cl; 606br; 610tl; 614bl; 614br; 615cr; 619cl; 620cl; 621br; 622tr; 622cr; 622bl; 622br; 623tr; 623tl; 623ccl; 624cr; 624bl; 624br; 625bl; 625br; 625cl; 625br; 626tr; 627tb; 628cl; 628cr; 628bl; 630cr; 631cl; 632br; 633bc: Hakan Kahraman 226bl; 226br: Hanna & Janne Aalto 40cl; 324tl: Hanne & Jens Eriksen 24tr; 31tr; 31cr; 31bl; 32 br; 35tc; 43tr; 44bl; 66cl; 66bl; 73br; 87cr; 90; 115br; 128tr; 128cl; 143tl; 143br; 147cl; 165bl; 165cr; 167bl; 170bl; 187bl; 205br; 206br; 206bl; 223br; 223br; 230cl; 253tr; 254bl; 254cl; 254cr; 260cr; 294tr; 294cl; 294br; 300cl; 301br; 330cl; 331br; 338bl; 339bl; 340tl; 340tr; 351br; 355br; 370tl; 370cr; 419br; 420b; 462tr; 462br; 472cl; 477tc; 518cr; 538br; 557cr; 564tr; 564bl; 564bc: Hannes Nussbaumer 146bl; 203tl; 607tr; 613cl; 615bl: Hans Gebuis 244tr; 434tr; 561bl; 595tl: Hans Henrik Larsen 502br: Hans Overduin 194tl: Harri Taavetti 126cl; 126cr; 257tl: Harry J. Lehto 369bl; 369bl; 547bl: Harvey van Diek 197tr; 206tl: Henrik Brandt 275tl; 463br: Henrik F. Nielsen 284tr: Himaru Iozawa 385br: Hugh Harrop 152bl; 153tl; 156bl; 167cl; 181bl; 183bl; 203br; 212bl; 212bl; 246tl; 256bl; 362cl; 367tr; 369bl; 426bl; 429bl; 444tr; 444cl; 446tr; 467tr; 476tr; 476bl; 484tr; 485tr; 489cl; 489br; 566tr; 566br; 570cr; 570bl; 570br; 575tr; 579cr; 582bl; 587br; 589cl; 591bl; 609bl: Hüseyin Meşe 222cl; Hüseyin Yılmaz 258tr; 550tr: Huw Roberts 24cl; 124tr; 128bl; 128br; 137br; 481tr; 482br; 482cl; 482cl; 517cl; 518bl; 518br; 545br; 592br: Iain H. Leach 246tr; 452tr; 582cl: Ian Boustead 65tl; 144tr; 236tr; 242cl; 261cl; 386bl; 486cl: Ian Dickey / Irishbirdphotography.com 130bl: Ian Fisher/Cahow Photography 148tr; 249cl; 259br; 381tr; 572bl: Ian Fulton 127cl; 127br: Ibrahim Hangül 281bl: Ignacio Yúfera, www.iyufera.com 71tl; 193br: Ilya Ukolov 428bl; 602tl; 631tl; 631tr: Imran Shah 404br: Ingo Waschkies 248tr; 327cc; 327cr; 379bl: Ingo Weiß 121bl; 122tl: J. A. van den Bosch 401cl: Jahan Shah 519tl: Jainy Kuriakose 68tl: Jake Gearty 257cl: James Eaton / Birdtour Asia 105bl: Jan Ravnborg 424br: Jan van Holten 311tr: Jan-Michael Breider 93br; 246cr: Jari Peltomäki 96tr; 96bl; 107tr; 137tr; 184br; 190br; 425tr: Jaysukh Parekh Suman, India 123cl: Jeffrey Wang 106tr: Jens Hering 456bl; 456br: Jens Morin 220cr: Jens Schmitz/Getty Images 532cl: Jens Søgaard Hansen 114cr; 200tr; 461bl: Jeremy Babbington 13tl; 135tr; 135br; 301tr; 538cr: Jerry Ting 105br: Johan Buckens 527tl: Johan Stenlund 392cr; 403bl: Jóhann Óli Tilmarsson 621bl: Johan Tufvesson 89cl; 311br: John & Jemi Holmes 625tl; 625tr: John Anderson / www.pbase.com/crail_birder 283tc; 403br: John Coveney Photography www.johncoveney.ie 491bl: John East 322tl; 524br: John Hawkins/FLPA 410bl: John Larsen 401bc: John Tymon 255tr: Johnny Salomonsson 392tr: Jonas Rosquist, Åkarp, Sweden 218tr: Jonathan Martinez 386tr; 467bl; 467br; 631cr: Jörgen Dam 162tr: Jorma Tenovuo/www.jtenovuo.com 22tl; 64br; 263bl; 464br; 466tr; 552br; 565br; 590tr: Juan José Bazán Hiraldo 116cr; 587br: Juan Luis Muñoz 320bl; 504cr; 505tl; 506bl; 508tl; 540br: Juan Matute / Vultour Naturaleza / Vultour.es 544cl; 544cr: Juan Sagardía 440bl; 508br; 510bl; 541tr; 603cl: Juha Niemi 535bl; 536br: Jukka J. Nurmi 121br; 371bl; 619bl: Jules Fouarge, Aves, Belgium 319br; 387cl; 586br; 590bl; 613bl; 616cl; 618br: Julia Knorr Alonso 228cr: Julian Bell / www.naturalbornbirder.com 447tr: Julien DAUBIGNARD 49cl: Jussi Vakkala 145tr; 257cr: Jyrki Normaja 22cr; 92cc; 110tr; 110br; 175br; 186cr; 186bl; 186br; 189cl; 189tr; 245tr; 249bl; 309tr; 388bl; 423cl; 434tl; 446tr; 475tr; 569bl; 583cl; 613br; 623br: Kaajal Dasgupta 106bl; 112br: Kadir Dabak 512bl: Kai Gauger 72tl; 236cl; 265tl; 296tr: Karel Mauer 107tl; 108tr; 508tr: Kari Haataja 175cl; 175br; 461cr; 481bl; 601br: Kazunori Kimura 382tl: Khalifa Al Dhaheri 545tr: Kim Aaen 362br: Kim Hyun-Tae 168br; 217tr: Kjell Johansson 578bl: Klaas Felix Jachmann 584bl; 584cr: Klaus Dichmann 442tl: Klaus Drissner 470cl: Klaus Malling Olsen 97tr; 341br; 342bl; 406br; 472cl; 525tl; 597tl: Knud Pedersen 167tr; Kris De Rouck 42bl; 50tr; 102tr; 198cr; 229cl; 303cc; 306cl; 306br; 440tl; 440tr: Lars Jensen 584tr: Lars Lundmark, Sweden / larslundmark.se/ 440cl: Lars Svensson 15cr; 15br; 16tl; 16cl: Lars Svensson/Natural History Museum 11tl: Lars Svensson/NRM 20tl; 20tr; 20cl; 20cr: Lasse Olsson / birding.se 190cr; 216b; 494bc; 617tr: Lefteris Stavrakis 548bl: Lennart Waara 263tr: Leo J. R. Boon/Cursorius 87cl: Lesley van Loo - www.LTDphoto.com 372cr; 449bl; 605bl: Lieven De Temmerman (www.batumiraptorcount.org) 292tr: Lior Kislev 58bl; 65tr; 76tr; 87bl; 90bl; 120cr; 121tl; 134cl; 134cr; 141cr; 142tr; 171cl; 174br; 226cr; 264cl; 274br; 281tl; 292cl; 292br; 311bc; 312br; 324bl; 329cl; 331br; 356br; 407tr; 415cr; 437bl; 438bl; 442br; 451br; 454cr; 454bl; 460tr; 460cl; 461tl; 504cl; 516br; 556c; 556br: Lisle Gwynn - tropicalbirding.com 366bl: Lubos Mráz / Naturfoto.cz 431cr; 432cr: Luigi Sebastiani/www.birds.it 287bc; 490cl; 491tr; 595tr: Luis Barrón (FOTOFONTALBA) / www.facebook.com/foto.fontalba/ 470br: Lutz Dürselen, www.duerselen.eu 124cl: M. Refik Kaleli 304tr: Machiel Valkenburg 254br: Magnus Hellström 131bl; 131br; 293bl; 293br; 446tc; 573bl; 573tr: Magnus Johansson 255br: Magnus Ullman 567br; 571cr; 592cl; 593bl; 593cr: Mansur Al Fahan 538tr: Manuel Estébanez Ruiz 289br: Marek Walford/www.marekwalford.co.uk 391br: Marjin Prins / www.marijnprins.nl 321cl: Mark Andrews 384cl: Mark Breaks 239cl; 239cr; 424cr; 426cr: Mark L. Stanley/Getty Images 441bl: Markku Huhta-Koivisto 354tr: Markku Saarinen 49br: Markus Lagerqvist (pbase.com/lagerqvist) 257bl: Markus Römhild 532cl; 619bl: Markus Tallroth 593br: Markus Varesvuo 21cr; 22cl; 22br; 55tr; 57br; 58br; 84tr; 79tr; 104tr; 108cl; 113cl; 113cr; 114tl; 118bl; 119tl; 119tr; 119bl; 120tl; 131tl; 133tl; 133cl; 134bl; 140br; 149tr; 150tl; 150tr; 154bl; 158bl; 161tl; 161bl; 170br; 180tr; 180bl; 188bl; 189tl; 202tl; 202br; 203tl; 203bl; 204cl; 204bl; 204br; 207tr; 207bl; 208tl; 208tr; 233cr; 237bc; 238tl; 238tr; 240tr; 243bl; 250bl; 251bl; 262tr; 262bl; 297cc; 298tr; 323tc; 326tr; 326bl; 347bl; 372tr; 375tr; 375bl; 376tl; 376br; 379cl; 383br; 394bl; 395cl; 395cr; 395tr; 400tr; 400bl; 401cr; 402bl; 406bl; 409tr; 427cl; 441tr; 485tr; 486tr; 488cr; 520br; 522tr; 546tr; 546bl; 547tr; 558bl; 558br; 559tr; 565cl; 566tl; 569cr; 571tr; 571bl; 596cl; 596br; 609cr; 611tr; 612cl; 612cr: Martin Bennett 496tc; 496tr: Martin Hale/FLPA 386tr: Martin Lofgren / Wildbirdgallery.com 21br: Martin P Goodey 160bl; 312tr: Mary McDonald/NaturePL 630ctl: Mateusz Matysiak www.fotomatysiak.pl 439bl; 439br; 439tr: Mathias Putze 103tr; 571cr; 582tr: Mathias Schäf / living-nature.eu 35tr; 43tl; 46tr; 52bl; 52br; 62tr;

— 640 —

PHOTOGRAPHIC CREDITS

62cl; 83cr; 89bl; 97br; 101tr; 114br; 115tl; 115tr; 155bc; 170br; 174tl; 174cl; 181bl; 199cr; 279tr; 332cl; 332bl; 333tl; 333tr; 361tr; 376br; 409cr; 461cl; 499bl; 533tl; 577tr; 595cr; 608br: **Mats Waern** 391bl; 610bl; 610br: **Matt Fidler** 496cl: **Matthieu VASLIN** 55br; 159tr; 176br; 186tr; 253bl; 309cl; 403tl; 424tr; 578br: **Mattias Ullman / pbase.com/mull** 32tl; 32tr; 473cl: **Maxim Koshkin / kazakh-steppe.com** 91cl: **Mehmet Goren** 544tl: **Mehmet ÜNLÜ - Bird Watcher** 226 tr: **Menno Hornman** 73tl; 73tr; 341tl: **Michael Heiß** 40tr; 92cr; 191tr; 211bl: **Michael Pope** 175bl; 229bl; 236cr; 236bl; 264br; 265br; 266t; 310tl; 310bl; 317br; 435br; 450bl; 451tr; 461br; 464bl; 466bl; 517cl; 589br: **Michael Rose/FLPA** 160br: **Michael Sammut** 277tl: **Michael Southcott** 47cl; 47cc; 47cr: **Michael Westerbjerg Andersen** 270bl; 460cc; 477tl; 478bl: **Michele Viganò** 117cr: **Michelle & Peter Wong** 386cl; 390br; 629tr: **Mika Bruun** 567cl; 580tl: **Mikael Nord** 136tl; 161cl: **Mike Barth** 112tr; 148cr; 169cr; 169br; 242tl; 242tr; 416bl; 416br; 419tr; 419cl; 432tl; 435tr; 450tr; 517tl: **Mike Buckland** 50tl: **Mike Danzenbaker / avesphoto.com** 619cr: **Mike Grimes/www.oretani.com** 192cl: **Mike Lane/FLPA** 401tr: **Mike Lawrence** 159br; 179cl: **Mike Parker** 217cr; 251tr; 364tr; 365tl; 365cr; 379tr; 387bl; 388br: **Mikkel Høegh Post** 145br; 524tr; 607cr; 607bl: **Mohamed Almazrouei** 112tl; 123tr: **moose henderson / Alamy Stock Photo** 468bl: **Morten Bentzon Hansen / Netfugl.dk** 627bb: **Naki Tez** 550cr: **Natalino Fenech** 276tr; 277br; 280tl; 504tr; **Neil Bowman/FLPA** 192br; 290bl; 630cbl: **Nial Moores** 387br: **Nic Hallam** 160br: **Nicholas Galea** 533b: **Nicholas R. Brown** 264cr: **Nick Moran** 111br; 168cr; 254bl; 374bl: **Nicolas Martinez** 345bl; 345br; 348bl; 348br: **Nigel Blake** 23cr; 59tr; 172br; 192tr: **Nikhil Devasar** 183br; 404cr; 475tr; 545cl: **Nikolay Loginov, Komsomolsk-on-Amur, Russia / fotki.yandex.ru/users/loghinov15** 127cl: **Nitin Srinivasa Murthy** 76bl; 106tl; 106cr: **non15/Shutterstock** 632cb: **Norman Deans van Swelm** 63bl; 87bl; 97tl; 108tl; 155br; 164tl; 164cr; 192tr; 261bc; 281br; 281tl; 286br; 302br; 304br; 305cl; 305cr; 376cr; 395cr; 396cl; 398cl; 442tr; 448bl; 560tr: **Ohad Sherer** 218bl: **OKYAY BULUT the wild life photographer** 486tr; 512tl: **Ola Elleström** 266bl: **Old Apple/Shutterstock** 384br: **Ole Amstrup** 477br: **Ole Krogh** 486bl: **Oleg Khromushin** 430bl; 434br; 449cl: **Olof Jönsson** 470tr: **Ömer Necipoğlu** 484tr; 484cl; 484cr; 512tr: **Oscar Campbell** 313cr; 314tr; 314cr; 450tl: **Ottenby Bird Observatory** 23tl: **Otto Pfister** 270tr: **Otto Plantema (www.pbase.com/otto1)** 43br; 63cr; 210br: **Otto Samwald** 367bc: **Oz Horine/ www.facebook.com/BirdsFamiliesOfTheWorld** 336cl: **panda3800/shutterstock** 628tl: **Paolo Casali / http://www.pbase.com/lep** 492cl: **Park Jong-Gil** 178br: **Patrick Palmen (www.patrickpalmen.nl)** 479br; 583tr: **Paul Baxter** 426cr: **Paul Cools / pbase.com/paulcoolsphotography** 84bl; 85tr; 224bl; 252cl: **Pauli Dernjatin** 511br: **Paul Hobson / naturepl.com** 590br: **Paul Smith BPE3* - www.pdsdigital.co.uk** 396bl: **Pavel Parkhaev** 171cl; 247bl; 252bc; 281cr; 423br; 428tr; 475tr; 546bl: **Pekka Fågel** 177tl; 222tr; 236bc; 429t; 465cl: **Pekka Komi** 372cl; 521cl: **Per Alström** 580bl: **Per-Göran Bentz** 556tr: **Per Poulson** 312bc: **Per Schans Christensen** 119cl; 130br; 598bl: **Pete Blanchard / www.flickr.com/photos/flyingfast/** 549br: **Pete Morris** 153tr: **Peter Adriaens** 37tr; 346tr: **Peter Arras** 229cr; 449cr: **Peter Dam / www.pdfoto.dk** 141tl: **Peter de Knijff** 244cr: **Peter Kaestner** 50br: **Peter Leigh of Firecrest Wildlife Photography** 440br: **Peter Lundgren** 216tr: **Peter van Rij** 289cr; 441bl: **Petri Kuhno** 421tl: **Petros Petrou** 519tc; 605br: **Petteri Hytönen** 120cr; 577br: **Phil Roberts** 456tr: **Philip Round** 447bl: **Philippe Dubois** 181cr: **Pia Öberg flickr.com/photos/hummingbirder/** 193cr: **Piero Alberti** 72bl; 158bc; 179cr; 268tr; 273tr; 346tr; 355bl; 356tr; 360tr; 500bl: **Pieter Bison** 370br: **Pranjal J. Saikia** 630ctr: **Prasad Ganpule** 76cl: **Quique Marcelo / www.micuaderno-decampo.com** 173tr; 422bl; 436bl; 436br; 438bl; 470tl; 541br: **Radosław Gwóźdź / www.radoslawgwozdz.pl** 413bl: **Rafael Armada (www.rafaelarmada.net)** 33b; 34tr; 34cr; 34bl; 36br; 36bl; 36tl; 37tr; 41tr; 37tl; 71br; 82TL; 146tr; 197tl; 245tr; 304tl; 328tr; 357tc; 359tc; 359bl; 416tl; 463br; 476tl: **Rafael Cediel-Algovia** 190tr: **Rafael Herrero Viturtia** 453br: **Rajneesh Suvarna / naturechronicles.com** 342tl: **Ralph Martin / www.visual-nature.de** 304cr; 305tr; 1491br; 494tl: **Rami Mizrachi** 38br; 231br; 432tr; 601bl: **Ran Schols** 41tc; 41bl; 47cl; 237br; 249tr; 267bl; 330tr; 364br; 488tl; 534tr; 576tr; 576cl **Rashed Al-Hajji** 174cc; 229bl; 291bl; 314tl; 314br; 337tl; 344br: **Raül Aymi** 321cr: **Raymond Wilson** 56tr; 299br; 309br: **Rebecca Nason** 147bl; 368bl; 426cl; 627bt: **René Lortie (http://rlortie.ca)** 381br: **René Pop** 44cr; 64bl; 66br; 83tr; 77tl; 77bl; 82tc; 94cl; 95tr; 98bl; 99cr; 191bl; 126bl; 129cr; 195bl; 146br; 155cl; 155tr; 155bl; 156br; 157cl; 182tl; 182tr; 183cl; 185br; 204tl; 209cl; 227tr; 240bl; 280cc; 288cc; 289cl; 303cr; 312tl; 312cl; 320cl; 325tr; 326cl; 368br; 374br; 393tr; 396c; 421br; 463bl; 474tl; 549tr: **René van Rossum** 98tl; 485bl: **Reto Burri** 498tl: **Richard Bonser** 206tr; 336br: **Richard Short** 308tr: **Richard Smith** 357bl: **Richard Somers Cocks** 365cr: **Richard Steel / wildlifephotographic.blogspot.com** 494tr; 495tr: **Rick van der Weijde** 70cl; 150bl; 418br: **Rob Robinson / www.robrob.photo** 399br: **Robert Newlin** 21bl: **Roberto Parmiggiani** 510br: **Robin Chittenden (www.robinchittenden.co.uk)** 23tr; 60bl; 61tr; 62cl; 62cr; 93t; 171br; 390bl; 495bl: **Roger Tidman** 444bl: **Roger Tidman/FLPA** 587br: **Roland Jansen** 209tr: **Rolf Kunz** 187cr: **Roman T. Brewka** 627cr: **Ron McIntyre** 494bl: **Ronald Messemaker** 280tr; 433br; 499cl: **Ronny Svensson, Allerum, Sweden** 320br: **Ruben Martinez Fraga** 275cr: **Rudi Debruyne** 181tr; 245cr; 277bl; 398tr; 545tl; 615cr; 615bl: **Ruedi Aeschlimann, Switzerland / www.vogelwarte.ch/de/voegel/voegel-der-schweiz** 161cr; 264tr; 373cr; 585br: **Rune Bisp Christensen** 527tr: **Ruud GM Altenburg** 347cl: **Ruud Wielinga** 276br: S. **Gatto / Arco Images GmbH / Alamy Stock Photo** 458bl: **S. J. M. Gantlett** 109tr; 374cl; 600cr: **S. N. Mostafawi/WCS Afghanistan** 447br: **Sami Tuomela** 91cl: **Sampo Kunttu** 401br: **Samuel Peregrina** 193cl: **Sean Cronin** 367tl: **Selim Imamoğlu** 321tr: **Sergey Cherenkov** 216cl: **Serhat Tigrel** 221cl: **Serkan Mutan** 258bc: **Sharad Sridhar** 138tl; 138br: **Sheila Mary Castelino** 69bl: **Sigmundur Ásgeirsson / www.flickr.com/photos/simmi25** 623cl: **Silas K. K. Olofson/www.birdingfaroes.wordpress.com** 401bl; 424bc: **Simon Lloyd** 128cr: **Simon Rosenkilde Waagner** 291br: **Sjaak Schilperoort** 96br: **Skarphéðinn G. Þórisson - hreindyr.com** 629btl; 629btr: **Soili Leveelahti** 383bl; 503cr; 596cl: **Soner Bekir** 217bl: **Søren Kristoffersen /www. pbase.com/nature_pix** 311bl; 450cr: **Søren Skov** 212tr: **Steen E. Jensen** 612tr: **Stefan Hage, www.birds.se** 54br; 311tr; 340br: **Stefan McElwee** 444tl; 444tc: **Stefan Persson** 449tl: **Stefan Pfützke / Green-Lens.de** 137cl; 213bl; 249tl; 324br; 490br; 555tr: **Steve Arlow (www.birdersplayground.co.uk)** 363cr; 503cl: **Steve Ashton / www.steveashtonwildlifephotography.co.uk** 496cr: **Steve Fletcher** 227cl; 410cr; 496br; 507tr; 521cr; 559br; 586cl: **Steve Garvie** 371br; 409br; 472br: **Steve N. G. Howell** 129tr: **Steve Rooke / www.sunbirdtours.co.uk** 385tl: **Steve Round (stevenround-birdphotography.com)** 131cc; 208tr; 215cl; 399tr; 427bl: **Stuart Elsom LRPS** 252tr; 609tl: **Stuart Fisher** 212cr; 575b; 616br: **Stuart Piner** 368tr; 458cl: **Stuart Price/ blog.hakodatebirding.com** 625t: **Subhadeep Ghosh** 602cr: **Subhoranjan Sen/Indian Forest Service** 128 cl: **Sumit K Sen** 105tr; 447bl: **Sunil Singhal** 128cr; 330cr; 340bl: **T J Tams** 572cl: **Theo Bakker** 311br: **Tadao Shimba** 379cr; 382br; 383br; 386br; 387tr: **tahirsphotography/shutterstock** 2-3: **Tamer Yilmaz** 555bl: **Terje Kolaas / www.terjekolaas.com** 117bl; 500cr: **Terry Townshend/birdingbeijing.com** 365cl: **Thailand Wildlife/Alamy** 632ct: **Theodoros Gaitanakis** 543br: **Thierry Quelennec** 317tr; 337cl: **Thomas Grüner** 41tl; 42tr; 50tl; 226tl; 303cr; 347tr; 373cr; 608bl: **Thomas Krumenacker** 231br; 258bl; 315br; 323br; 334cl; 334bl; 353br: **Thomas Kuppel** 619tr: **Thomas Langenberg** 310cc; 388tl; 350br; 389cl: **Thomas Luiten / www.pbase.com/thomasluiten** 263cr; 432cl: **Thomas Varto Nielsen** 461tr; 633tc: **Thorsten Krüger / thorsten-krueger.com** 597tr; 606cl; 618cl: **Tim Zurowski/Getty** 623br; 629bb: **tntphototravis/Shutterstock** 633bl: **Tom Bedford www.tombedford.co.uk** 469tr; 506cl: **Tom Beeke** 132tr: **Tom Lindroos** 60tr; 92cl; 115tr; 115tr; 116cl; 332tr; 350br; 389tl; 574br: **Tomas Svensson** 422br; 540tl: **Tomi Muukonen** 38cl; 104tr; 310br; 312bl; 438cl; 449bl; 561cr; 570tr: **Tommy Holmgren** 220bl; 282cl; 522bl: **Tommy P. Pedersen** 111bl; 151cl; 241br; 296tr; 482bl; 482bc; 593bl; 593tr: **Tonny Ravn Kristiansen** 403cr: **Tony Peral** 346cl: **Tony Small** 459cl: **Tuncer Tozsin** 226cr: **Uku Paal** 444cr: **Ulf Liedén** 361br: **Ulf Ståhle** 9bl; 67tl; 67tc; 67tr; 223tl; 223tr; 300br; 370bl; 537br: **Ümit Özgür** 550br; 611br: **Vadim Ivushkin / flickr.com/photos/91544025@N05/** 389bl; 389tr; 393br: **Vasil Ananian (Armenia)** 191cl; 245tr; 290tl; 290tr; 602tr: **Vasilly Fedorenko** 92br: **Vaughan & Svetlana Ashby/Birdfinders** 197cl; 313bl; 363cr; 322tc; 568br: **Veysel Karakullukçu** 303bl; 322tc; 568br: **Vincent Legrand** 80cl; 367br; 372bc; 392br; 412bl; 413br; 415cr; 422t; 443cl; 459br; 469cl; 469br; 474cl; 474bc; 474br; 478cl; 519cl; 532cr; 568tr; 605cl; 622tl; 623ccr; 624tl; 624br; 626cl; 627tt; 630tl: **Vitantonio Dell'Orto/exuviaphoto.com** 244tl: **Vivek Tiwari / flickr.com/spiderhunters** 390cr: **Volkan Pek** 417tl: **Vytautas Jusys** 446tl: **Werner Müller** 9br; 68br; 221tl; 221br; 269bl; 272cl; 274bc; 300tr; 352cr; 420tl; 466cl; 564cr: **Yann Kolbeinsson / Birding Iceland** 321cr; 431br: **Yathin Krishnappa** 568bl: **Yevgeny Belousov** 176tl; 176tl; 176cr; 191bl; 195br; 287bl; 288tl; 289cl; 552bl; 553bl; 579br: **Yoav Perlman** 39br; 54bl; 199bc; 230br; 296tl: **Zafer KURNUÇ-TURKEY** 164br; 248tr; 373bl.

INDEX OF SCIENTIFIC NAMES

A

abbotti, Luscinia svecica 247
abietinus, Phylloscopus collybita 597
abyssinica, Cecropis 624
abyssinica, Cecropis abyssinica 624
acredula, Phylloscopus trochilus 610
ACROCEPHALUS 436, 631
acutirostris, Calandrella 68
acutirostris, Calandrella acutirostris 69
adamsi, Calandrella raytal 76
aedon, Iduna 467
aethiopica, Hirundo 624
aethiopica, Hirundo aethiopica 624
africana, Luscinia megarhynchos 241
agricola, Acrocephalus 443
aguimp, Motacilla 196
aharonii, Calandrella rufescens 73
airensis, Oenanthe melanura 361
akyildizi, Prinia gracilis 416
ALAEMON 46
ALAUDA 86
alaudipes, Alaemon 46
alaudipes, Alaemon alaudipes 48
alba, Motacilla 188
alba, Motacilla alba 194
albigula, Eremophila alpestris 100
albistriata, Sylvia cantillans 512
albiventris, Acrocephalus melanopogon 437
albiventris, Cettia cetti 407
albocoeruleus, Tarsiger cyanurus 251
albonigra, Oenanthe 355
alexanderi, Galerida cristata 80
algeriensis, Ammomanes deserti 43
alnorum, Empidonax 621
alpestris, Eremophila 96
alpestris, Eremophila alpestris 98
alpestris, Turdus torquatus 374
althaea, Sylvia curruca 552
altirostris, Galerida cristata 80
alulensis, Iduna pallida 474
amadoni, Hirundo aethiopica 624
ambiguus, Acrocephalus scirpaceus 455
amicorum, Turdus torquatus 374
AMMOMANES 40
amnicola, Locustella fasciolata 631
ampelinus, Hypocolius 205
annae, Ammomanes deserti 44
annae, Anthus cinnamomeus 135
ANTHUS 132, 625
apetzii, Calandrella rufescens 73
aquaticus, Cinclus cinclus 209
arabica, Ptyonoprogne fuligula 115
arabicus, Anthus similis 144
arborea, Lullula 84
arborea, Lullula arborea 85
arenicola Galerida cristata 80
arenicolor, Ammomanes cinctura 41
armenica Alauda arvensis 89
arsinoe, Pycnonotus barbatus 200
arundinaceus, Acrocephalus 463
arundinaceus, Acrocephalus arundinaceus 464
arvensis, Alauda 88
arvensis, Alauda arvensis 89
atlas, Eremophila alpestris 97
atricapilla, Sylvia 560

atricapilla, Sylvia atricapilla 562
atrogularis, Prunella 219
atrogularis, Prunella atrogularis 220
atrogularis, Turdus 391
aurea, Zoothera 362
aurita (var.), Oenanthe hispanica 319
auroreus, Phoenicurus 628
auroreus, Phoenicurus auroreus 629
avicenniae, Acrocephalus scirpaceus 456
axillaris (var.), Phylloscopus collybita 599
azorensis, Turdus merula 377
azoricus, Regulus regulus 614
azizi, Ammomanes deserti 45
azuricollis, Luscinia svecica 246

B

badia, Prunella montanella 217
balcanica, Eremophila alpestris 99
balearica, Sylvia 493
balearicus, Regulus ignicapilla 618
barbatus, Pycnonotus 200
barbatus, Pycnonotus barbatus 200
barnesi, Oenanthe finschii 331
beema, Motacilla flava 177
berthelotii, Anthus 145
berthelotii, Anthus berthelotii 146
bicolor, Saxicola caprata 296
bicolor, Tachycineta 623
bicornis, Eremophila alpestris 100
bilopha, Eremophila 101
bimaculata, Melanocorypha 57
bimaculata, Melanocorypha bimaculata 58
blanfordi, Sylvia leucomelaena 536
blythi, Sylvia curruca 552
boavistae, Alaemon alaudipes 47
BOMBYCILLA 202, 626
bonapartei, Turdus viscivorus 404
bonelli, Phylloscopus 585
borealis, Phylloscopus 565
borealis, Troglodytes troglodytes 211
borin, Sylvia 558
boscaweni, Oenanthe lugentoides 352
bottae, Oenanthe 300
bottae, Oenanthe bottae 301
brachydactyla, Calandrella 62
brachydactyla, Calandrella brachydactyla 63
brachyura, Pitta 622
brandti, Eremophila alpestris 98
brevipennis, Acrocephalus 457
brunnescens, Acrocephalus stentoreus 462
buryi, Scotocerca inquieta 420
buryi, Sylvia 537

C

cabrerae, Turdus merula 377
caeruleus, Myophonus 629
cafer, Pycnonotus 201
cafer, Pycnonotus cafer 201
calandra, Melanocorypha 54
calandra, Melanocorypha calandra 55
CALANDRELLA 66
calcarata, Motacilla citreola 184
calendula, Regulus 633
calendula, Regulus calendula 633
caligata, Iduna 475

CALLIOPE 248
calliope, Calliope 248
campestris, Anthus 139
canariensis, Phylloscopus 606
canariensis, Phylloscopus canariensis 607
cantillans, Mirafra 31
cantillans, Sylvia 509
cantillans, Sylvia cantillans 511
capistrata, Oenanthe 342
caprata, Saxicola 295
captus, Anthus similis 144
carolinae, Galerida theklae 83
carolinensis, Dumetella 627
carthaginis, Galerida cristata 78
cashmiriense, Delichon dasypus 625
CATHARUS 366
caucasica, Sylvia curruca 551
caucasicus, Cinclus cinclus 209
caucasicus, Erithacus rubecula 234
CECROPIS 125, 624
cedrorum, Bombycilla 626
centralasiae, Locustella certhiola 424
CERCOTRICHAS 227
certhiola, Locustella 423
certhiola, Locustella certhiola 424
cervinus, Anthus 156
CETTIA 405
cetti, Cettia 405
cetti, Cettia cetti 407
cheleensis, Calandrella rufescens 74
CHERSOPHILUS 49
chinensis, Riparia 105
chrysopygia, Oenanthe 337
CINCLUS 207
cinclus, Cinclus 207
cinclus, Cinclus cinclus 208
cincta, Neophedina 623
cinctura, Ammomanes 40
cinctura, Ammomanes cinctura 41
cinerea, Motacilla 185
cinerea, Motacilla cinerea 187
cinereocapilla, Motacilla flava 175
cinnamomeus, Anthus 135
CISTICOLA 408
cisticola, Cisticola juncidis 410
citreola, Motacilla 180
citreola, Motacilla citreola 184
clamans, Spiloptila 411
clarkei, Turdus philomelos 399
clotbey, Ramphocoris 51
coatsi, Regulus regulus 615
coburni, Turdus iliacus 401
collaris, Prunella 224
collaris, Prunella collaris 226
collybita, Phylloscopus 594
collybita, Phylloscopus collybita 597
communis, Sylvia 554
communis, Sylvia communis 556
conspicillata, Sylvia 501
coronatus, Phylloscopus 631
CONTOPUS 622
coutellii, Anthus spinoletta 165
crassirostris, Sylvia 542
crassirostris, Sylvia crassirostris 545
cristata, Galerida 77

INDEX OF SCIENTIFIC NAMES

cristata, Galerida cristata 78
curruca, Sylvia 549
curruca, Sylvia curruca 551
cyanecula, Luscinia svecica 245
cyane, Larvivora 628
cyanurus, Tarsiger 250
cyanurus, Tarsiger cyanurus 251
cypriaca, Oenanthe 315
cypriotes, Troglodytes troglodytes 212
cyrenaicae, Sylvia hortensis 541

D

daaroodensis, Calandrella eremica 67
dacotiae, Saxicola 282
dacotiae, Saxicola dacotiae 283
dammholzi, Sylvia atricapilla 562
dartfordiensis, Sylvia undata 497
dasypus, Delichon 624
dasypus, Delichon dasypus 625
daurica, Cecropis 125
daurica, Cecropis daurica 127
davisoni, Geokichla sibirica 365
decaptus, Anthus similis 144
deichleri, Turdus viscivorus 404
DELICHON 130, 624
deltae, Prinia gracilis 417
DENDRONANTHUS 169
deserti, Ammomanes 42
deserti, Ammomanes deserti 44
deserti, Oenanthe 325
deserti, Oenanthe deserti 327
deserti, Sylvia 533
deserticola, Sylvia 498
desertorum, Alaemon alaudipes 48
diluta, Riparia 110
diluta, Riparia diluta 112
doriae, Alaemon alaudipes 48
dukhunensis, Calandrella 65
dukhunensis, Motacilla alba 194
dulcivox, Alauda arvensis 89
DUMETELLA 627
dumetorum, Acrocephalus 445
dunni, Eremalauda 36
duponti, Chersophilus 49
duponti, Chersophilus duponti 50

E

elaeica, Iduna pallida 473
ellenthalerae, Regulus regulus 614
EMPIDONAX 621
EREMALAUDA 36
eremica, Calandrella 66
eremica, Calandrella eremica 67
eremodites, Eremalauda 38
EREMOPHILA 96
EREMOPTERIX 33, 622
ERITHACUS 232
erlangeri, Galerida theklae 83
erlangeri, Neophedina cincta 623
ernesti, Oenanthe leucopyga 357
erythrogaster, Hirundo rustica 120
erythrogastrus, Phoenicurus 269
erythronotus, Phoenicurus 253
eunomus, Turdus 385
eximius, Anthus cinnamomeus 135
exsul, Phylloscopus canariensis 607

F

fagani, Prunella ocularis 223
familiaris, Cercotrichas galactotes 228
fasciolata, Locustella 630
fasciolata, Locustella fasciolata 631
faxoni, Catharus guttatus 366
feldegg, Motacilla flava 176
felix, Saxicola torquatus 294
filifera, Hirundo smithii 124
finschii, Oenanthe 328
finschii, Oenanthe finschii 331
flava, Eremophila alpestris 97
flava, Motacilla 170
flava, Motacilla flava 174
flavissima, Motacilla flava 175
fluviatilis, Locustella 430
fluvicola, Petrochelidon 128
fohkienensis, Riparia diluta 112
frenata, Oenanthe bottae 301
fuligula, Ptyonoprogne 113
'fulvescens', Phylloscopus collybita 599
fusca, Locustella luscinioides 434
fuscatus, Phylloscopus 583
fuscescens, Catharus 369
fuscescens, Catharus fuscescens 369
fuscus, Acrocephalus scirpaceus 455

G

galactotes, Cercotrichas 227
galactotes, Cercotrichas galactotes 228
GALERIDA 81
garrulus, Bombycilla 202
garrulus, Bombycilla garrulus 204
GEOKICHLA 364
gibraltariensis, Phoenicurus ochruros 259
godlewskii, Anthus 136
golzii, Luscinia megarhynchos 242
gracilis, Prinia 414
gracilis, Prinia gracilis 417
griseldis, Acrocephalus 465
griseolus, Phylloscopus 632
gularis, Cinclus cinclus 208
gularis, Sylvia atricapilla 563
gulgula, Alauda 86
gustavi, Anthus 152
gustavi, Anthus gustavi 153
guttatus, Catharus 366
gutturalis, Irania 235

H

haemorrhousus, Pycnonotus cafer 201
halimodendri, Sylvia curruca 552
halophila, Oenanthe 346
harringtoni, Anthus trivialis 151
harterti, Alauda arvensis 89
hebridensis, Turdus philomelos 399
heinei, Calandrella rufescens 73
heineken, Sylvia atricapilla 563
hemprichii, Saxicola maurus 291
hermonensis, Calandrella brachydactyla 64
hibernans, Saxicola rubicola 286
HIPPOLAIS 480
hirtensis, Troglodytes troglodytes 211
HIRUNDO 118, 624
hispanica, Oenanthe 319
hispanica, Oenanthe hispanica 324

hodgsoni, Anthus 147
hodgsoni, Anthus hodgsoni 148
homochroa, Oenanthe deserti 326
hortensis, Sylvia 539
hortensis, Sylvia hortensis 541
hufufae, Prinia gracilis 417
humei, Phylloscopus 578
humei, Phylloscopus humei 579
huttoni, Prunella atrogularis 220
HYLOCICHLA 630
HYPOCOLIUS 205
hyrcanus, Erithacus rubecula 234

I

iberiae, Motacilla flava 175
iberiae, Sylvia inornata 508
ibericus, Phylloscopus 603
icterina, Hippolais 485
icterops, Sylvia communis 556
IDUNA 467
ignicapilla, Regulus 616
ignicapilla, Regulus ignicapilla 618
ijimae, Riparia riparia 109
iliacus, Turdus 400
iliacus, Turdus iliacus 401
inconspicua, Alauda gulgula 87
indica, Riparia diluta 112
indicus, Dendronanthus 169
inermis, Regulus regulus 614
inornata, Sylvia 505
inornata, Sylvia inornata 508
inornatus, Phylloscopus 576
inquieta, Scotocerca 418
inquieta, Scotocerca inquieta 419
insularis, Ammomanes deserti 45
intermedius, Pycnonotus cafer 201
intermedius, Turdus merula 377
interni, Regulus regulus 614
IRANIA 235
isabellina, Oenanthe 297
islandicus, Troglodytes troglodytes 211
IXOREUS 629

J

japonica, Cecropis daurica 127
japonicus, Anthus rubescens 167
jerdoni, Sylvia crassirostris 545
jordansi, Galerida cristata 80
juncidis, Cisticola 408
juncidis, Cisticola juncidi 409
juniperi, Troglodytes troglodytes 212

K

katharinae, Ammomanes deserti 44
kleinschmidti, Galerida cristata 78
koenigi, Troglodytes troglodytes 212

L

lagopodum, Delichon urbicum 131
lanceolata, Locustella 425
languida, Hippolais 480
LARVIVORA 627
lepida, Prinia gracilis 416
leucocephala, Motacilla flava 178
leucomelaena, Sylvia 535
leucomelaena, Sylvia leucomelaena 536

leucophaea, Calandrella rufescens 74
leucopsis, Motacilla alba 195
leucoptera, Alauda 91
leucopyga, Oenanthe 356
leucopyga, Oenanthe leucopyga 357
leucorhoa, Oenanthe oenanthe 305
leucotis, Pycnonotus 198
leucotis, Pycnonotus leucotis 198
leucura, Oenanthe 358
leucura, Oenanthe leucura 359
levantinus, Acrocephalus stentoreus 460
libanotica, Oenanthe oenanthe 304
littoralis, Anthus petrosus 162
LOCUSTELLA 423, 630
longipennis, Calandrella brachydactyla 64
lorenzii, Phylloscopus 600
ludlowi, Phylloscopus trochiloides 571
lugens, Oenanthe 343
lugens, Oenanthe lugens 344
lugentoides, Oenanthe 350
lugentoides, Oenanthe lugentoides 352
LULLULA 84
luristanica, Luscinia svecica 246
LUSCINIA 237
luscinia, Luscinia 237
luscinioides, Locustella 433
luscinioides, Locustella luscinioides 434
lutea, Motacilla flava 178
lypura, Oenanthe melanura 361

M

macronyx, Motacilla flava 179
macrorhyncha, Galerida cristata 80
maculata, Galerida cristata 79
madeirensis, Anthus berthelotii 146
madeirensis, Regulus 619
madoci, Monticola solitarius 278
magna, Galerida cristata 80
magnirostris, Phylloscopus 632
mandellii, Phylloscopus humei 579
margaritae, Chersophilus duponti 50
margelanica, Sylvia curruca 553
mauritanica, Riparia paludicola 104
mauritanicus, Turdus merula 377
maurus, Saxicola 287
maurus, Saxicola maurus 293
megarhynchos, Luscinia 240
megarhynchos, Luscinia megarhynchos 241
melanauchen, Eremopterix nigriceps 35
melanocephala, Sylvia 520
melanocephala, Sylvia melanocephala 522
MELANOCORYPHA 57
melanoleuca, Oenanthe hispanica 324
melanopogon, Acrocephalus 436
melanopogon, Acrocephalus melanopogon 437
melanoptera, Cercotrichas podobe 231
melanothorax, Sylvia 524
melanura, Oenanthe 360
melanura, Oenanthe melanura 361
melopnilus, Erithacus rubecula 234
menachensis, Turdus 370
menzbieri, Anthus gustavi 153
menzbieri, Phylloscopus collybita 599
merula, Turdus 375
merula, Turdus merula 377
mesopotamia, Pycnonotus leucotis 198
migratorius, Turdus 380

migratorius, Turdus migratorius 381
mimicus, Acrocephalus melanopogon 437
MIMUS 626
minimus, Catharus 368
minimus, Empidonax 621
minor, Calandrella rufescens 73
minor, Cercotrichas galactotes 229
minula, Sylvia curruca 553
MIRAFRA 31
modularis, Prunella 213
modularis, Prunella modularis 215
moesta, Oenanthe 332
momus, Sylvia melanocephala 522
monacha, Oenanthe 353
montana, Prunella collaris 226
montanella, Prunella 216
montanella, Prunella montanella 217
MONTICOLA 272
MOTACILLA 170
moussieri, Phoenicurus 267
murielae, Saxicola dacotiae 283
mustelina, Hylocichla 630
mya, Ammomanes deserti 44
MYOPHONUS 629
mystacea, Sylvia 516
mystacea, Sylvia mystacea 519

N

naevia, Locustella 427
naevia, Locustella naevia 429
naevius, Ixoreus 629
namnetum, Luscinia svecica 246
nana, Sylvia 531
natronensis, Prinia gracilis 417
naumanni, Turdus 382
negevensis, Sylvia leucomelaena 536
neglectus, Phylloscopus 592
NEOPHEDINA 623
neumanni, Oenanthe melanura 361
nicolli, Calandrella rufescens 73
nigricans, Galerida cristata 79
nigriceps, Eremopterix 33
nigriceps, Eremopterix nigriceps 35
nigrideus, Turdus migratorius 381
nigrimentale, Delichon dasypus 625
nisoria, Sylvia 546
nitidus, Phylloscopus 567
nivescens, Anthus similis 144
norissae, Sylvia melanocephala 523

O

obscura, Prunella modularis 215
obscuratus, Phylloscopus trochiloides 571
obscurus, Turdus 378
obsoleta, Ptyonoprogne fuligula 114
occidentalis, Prunella modularis 215
ochruros, Phoenicurus 255
ochruros, Phoenicurus ochruros 259
ocularis, Prunella 221
ocularis, Prunella ocularis 223
OENANTHE 360
oenanthe, Oenanthe 302
oenanthe, Oenanthe oenanthe 304
olivetorum, Hippolais 483
opaca, Iduna 469
opistholeuca, Oenanthe 342
oreophila, Oenanthe deserti 327

orientalis, Acrocephalus 631
orientalis, Cettia cetti 407
orientalis, Phylloscopus 588
orinus, Acrocephalus 446

P

palaestinae, Prinia gracilis 417
pallida, Iduna 471
pallida, Iduna pallida 473
pallida, Lullula arborea 85
pallidior, Tarsiger rufilatus 252
pallidogularis, Luscinia svecica 247
paludicola, Acrocephalus 439
paludicola, Riparia 103
paludicola, Riparia paludicola 104
palustris, Acrocephalus 448
pandoo, Monticola solitarius 278
parvirostris, Ammomanes deserti 45
patriciae, Motacilla cinerea 187
pauluccii, Sylvia atricapilla 562
payni, Ammomanes deserti 43
penicillata, Eremophila alpestris 99
perpallida, Ptyonoprogne fuligula 115
persica, Calandrella rufescens 74
persica, Oenanthe lugens 344
persicus, Cinclus cinclus 209
personata, Motacilla alba 194
PETROCHELIDON 128
petrosus, Anthus 159
petrosus, Anthus petrosus 161
philippensis, Monticola solitarius 278
philomelos, Turdus 397
philomelos, Turdus philomelos 399
phoebe, Sayornis 620
phoenicuroides, Ammomanes deserti 45
phoenicuroides, Phoenicurus ochruros 260
PHOENICURUS 253, 628
phoenicurus, Phoenicurus 262
phoenicurus, Phoenicurus phoenicurus 265
PHYLLOSCOPUS 564, 631
picata, Oenanthe 339
pilari, Turdus 394
PITTA 622
platyura, Scotocerca inquieta 420
pleschanka, Oenanthe 309
plumbeitarsus, Phylloscopus 572
podobe, Cercotrichas 230
podobe, Cercotrichas podobe 231
polatzeki, Calandrella rufescens 73
polyglotta, Hippolais 487
polyglottos, Mimus 626
polyglottos, Mimus polyglottos 626
pratensis, Anthus 154
PRINIA 414
PROGNE 623
proregulus, Phylloscopus 574
PRUNELLA 213
psammochroa, Melanocorypha calandra 56
pseudobaetica, Calandrella rufescens 73
PTYONOPROGNE 113
PYCNONOTUS 198
pygmaea, Motacilla flava 177
pyrrhonota, Petrochelidon 129
pyrrhonota, Petrochelidon pyrrhonota 129

R

rama, Iduna 477

RAMPHOCORIS 51
randonii, Galerida cristata 79
raytal, Calandrella 75
razae, Alauda 94
REGULUS 611, 633
regulus, Regulus 611
regulus, Regulus regulus 613
reiseri, Iduna pallida 474
relicta (var.), *Turdus atrogularis* 393
richardi, Anthus 132
riggenbachi, Galerida cristata 78
riggenbachi, Oenanthe leucura 359
RIPARIA 103
riparia, Riparia 107
riparia, Riparia riparia 109
rossorum, Saxicola caprata 296
rubecula, Erithacus 232
rubecula, Erithacus rubecula 234
rubescens, Anthus 166
rubescens, Anthus rubescens 167
rubescens, Locustella certhiola 424
rubescens, Sylvia mystacea 519
rubetra, Saxicola 279
rubicola, Saxicola 284
rubicola, Saxicola rubicola 286
rubicola, Sylvia communis 556
rufescens, Calandrella 70
rufescens, Calandrella rufescens 72
rufescens, Melanocorypha bimaculata 58
ruficollis, Turdus 388
ruficolor, Galerida theklae 83
rufilatus, Tarsiger rufilatus 252
rufiventris, Cinclus cinclus 209
rufiventris, Phoenicurus ochruros 260
rufocinereus, Monticola 272
rufocinereus, Monticola rufocinereus 272
rufula, Cecropis daurica 126
rufulus, Anthus 625
rufulus, Anthus rufulus 626
rufum, Toxostoma 627
rufum, Toxostoma rufum 627
rupestris, Ptyonoprogne 116
ruppeli, Sylvia 528
rustica, Hirundo 118
rustica, Hirundo rustica 119

S

saharae, Scotocerca 421
saharae, Scotocerca saharae 421
samamisicus, Phoenicurus phoenicurus 265
samharensis, Ammomanes deserti 45
sanctaemariae, Regulus regulus 614
sarda, Sylvia 490
saturata, Ammomanes deserti 44
savignii, Hirundo rustica 120
saxatili, Monticola 273
SAXICOLA 279
SAYORNIS 620
schmitzi, Motacilla cinerea 187
schoenobaenus, Acrocephalus 441
schwarzi, Phylloscopus 581
scirpaceus, Acrocephalus 452
scirpaceus, Acrocephalus scirpaceus 455
sclateri, Monticola rufocinereus 272

SCOTOCERCA 418
seebohmi, Oenanthe 306
semirufus, Phoenicurus ochruros 259
shelleyi, Riparia riparia 109
sibilans, Larvivora 627
sibilatrix, Phylloscopus 590
sibirica, Geokichla 364
sibirica, Geokichla sibirica 365
signatus, Eremopterix 622
similis, Anthus 142
simplex, Mirafra cantillans 32
sindianus, Phylloscopus 602
smithii, Hirundo 123
smithii, Hirundo smithii 124
solitarius, Monticola 276
solitarius, Monticola solitarius 278
spatzi, Ptyonoprogne fuligula 114
SPILOPTILA 411
spinoletta, Anthus 163
spinoletta, Anthus spinoletta 165
stapazina (var.), *Oenanthe hispanica* 319
stejnegeri, Saxicola maurus 293
stentoreus, Acrocephalus 459
stentoreus, Acrocephalus stentoreus 460
straminea, Locustella naevia 429
striata, Scotocerca inquieta 419
subalpina, Prunella collaris 226
subalpina, Sylvia 513
subis, Progne 623
subpersonata, Motacilla alba 194
subtaurica, Galerida cristata 80
superbus, Erithacus rubecula 234
superflua, Galerida theklae 83
svecica, Luscinia 243
svecica, Luscinia svecica 245
swainsoni, Catharus ustulatus 367
SYLVIA 490
Sylvia atricapilla 560
syriaca, Cercotrichas galactotes 228
syriacus, Turdus merula 377

T

TACHYCINETA 623
taivana, Motacilla flava 179
TARSIGER 250
tataricus, Erithacus rubecula 234
telengitica, Sylvia curruca 553
temminckii, Myophonus caeruleus 629
tenellipes, Phylloscopus 632
tenellus, Tmetothylacus 625
teneriffae, Regulus regulus 614
theklae, Galerida 81
theklae, Galerida theklae 83
theresae, Scotocerca saharae 421
thunbergi, Motacilla flava 174
tianschanicus, Troglodytes troglodytes 212
tibetana, Riparia diluta 112
toni, Sylvia undata 497
torquatus, Saxicola 294
torquatus, Turdus 371
torquatus, Turdus torquatus 373
TOXOSTOMA 627
TMETOTHYLACUS 625
transitiva, Hirundo rustica 119

tristis, Phylloscopus collybita 598
tristis, Regulus regulus 615
trivialis, Anthus 149
trivialis, Anthus trivialis 150
trochiloides, Phylloscopus 569
trochiloides, Phylloscopus trochiloides 571
trochilus, Phylloscopus 608
trochilus, Phylloscopus trochilus 610
TROGLODYTES 210
troglodytes, Troglodytes 210
troglodytes, Troglodytes troglodytes 211
tschutschensis, Motacilla flava 179
turcmenica, Sylvia mystacea 519
TURDUS 370, 630
TYRANNUS 620
tyrannus, Tyrannus 620
tytleri, Hirundo rustica 120

U

umbrovirens, Phylloscopus 564
undata, Sylvia 495
undata, Sylvia undata 497
unicolor, Turdus 630
uralensis, Cinclus cinclus 209
urbicum, Delichon 130
urbicum, Delichon urbicum 131
uropygialis, Cisticola juncidis 410
ustulatus, Catharus 367

V

variegatus, Saxicola maurus 292
vidua, Motacilla aguimp 197
virens, Contopus 622
virescens, Empidonax 621
viridanus, Phylloscopus trochiloides 570
viscivorus, Turdus 402
viscivorus, Turdus viscivorus 404
vittata (var.), *Oenanthe pleschanka* 313

W

warriae, Oenanthe 348
werae, Motacilla citreola 183
whitakeri, Ammomanes deserti 44
wolfi (var.), *Luscinia svecica* 246

X

xanthoprymna, Oenanthe 334
xanthopygos, Pycnonotus 199

Y

yakutensis, Phylloscopus trochilus 610
yarrellii, Motacilla alba 191
yeltoniensis, Melanocorypha 59
yemenensis, Phylloscopus umbrovirens 564
yemenensis, Prinia gracilis 417
yunnanensis, Anthus hodgsoni 148

Z

zarudnyi, Acrocephalus arundinaceus 464
zarudnyi, Ammomanes cinctura 41
ZOOTHERA 362

INDEX OF ENGLISH NAMES

A

Accentor, Alpine 224
    Black-throated 219
    Hedge 213
    Radde's 221
    Siberian 216
    'Yemen' 223

B

Blackbird, Common 375
Blackcap 560
Blackstart 360
Bluetail, Himalayan 252
    Red-flanked 250
Bluethroat 243
    'Red-spotted' 245
    'White-spotted' 245
Bulbul, Common 200
    Red-vented 201
    White-eared 198
    White-spectacled 199
    Yellow-vented 199
Bushchat, Pied 295
    Rufous 227

C

Catbird, Grey 627
Chiffchaff, Canary Islands 606
    Caucasian 600
    Caucasian Mountain 600
    Common 594
    Iberian 603
    Kashmir 602
    Mountain 600
    'Siberian' 598
Cisticola, Fan-tailed 408
    Zitting 408

D

Dipper 207
    White-throated 207
Dunnock 213

F

Fieldfare 394
Firecrest, Common 616
    Madeira 619
Flycatcher, Acadian 621
    Alder 621
    Least 621

G

Goldcrest 611
    'Caspian' 615
    'Tenerife' 614

H

Hypocolius 205
    Grey 205

K

Kingbird, Eastern 620

Kinglet, Ruby-crowned 633

L

Lark, Arabian 38
    'Asian' Short-toed 74
    Bar-tailed 40
    Bimaculated 57
    Black 59
    Black-crowned Sparrow 33
    Calandra 54
    Chestnut-headed Sparrow 622
    Crested 77
    Desert 42
    Dunn's 36
    Dupont's 49
    Greater Hoopoe 46
    Greater Short-toed 62
    Hoopoe 46
    Horned 96
    Hume's Short-toed 68
    Indian Sand 75
    Lesser Short-toed 70
    'Levant' Horned 99
    Mongolian Short-toed 65
    Raso 94
    Rufous-capped 66
    Sand 75
    Shore 96
    Short-toed 62
    Singing Bush 31
    Temminck's 101
    Temminck's Horned 101
    Thekla's 81
    Thick-billed 51
    White-winged 91
Longtail, Cricket 411
    Scaly 411
    Scaly-fronted 411

M

Martin, Asian House 624
    Banded 623
    Brown-throated 103
    Chinese 105
    Collared Sand 107
    Common House 130
    Common Sand 107
    Crag 116
    Grey-throated 105
    House 130
    Pale 110
    Pale Crag 113
    Pale Sand 110
    Plain 103
    Purple 623
    Rock 113
    Sand 107
Mockingbird, Northern 626

N

Nightingale, Common 240
    Thrush 237

O

Ouzel, Ring 371

P

Parisoma, Yemen 537
Peewee, Eastern Wood 622
Phoebe, Eastern 620
Pipit, African 135
    American 166
    Berthelot's 145
    Blyth's 136
    Buff-bellied 166
    Golden 625
    Grassland 135
    Long-billed 142
    Meadow 154
    Olive-backed 147
    Paddyfield 625
    Pechora 152
    Red-throated 156
    Richard's 132
    Rock 159
    Tawny 139
    Tree 149
    Water 163
Pitta, Indian 622
Prinia, Cricket 411
    Graceful 414
    Scaly 411

R

Redstart, Black 255
    Common 262
    Daurian 628
    'Ehrenberg's' 265
    Eversmann's 253
    Güldenstädt's 269
    'Levant Black' 259
    Moussier's 267
    'Red-bellied' Black 260
Redwing 400
Robin, American 380
    Black Bush 230
    Black Scrub 230
    Eurasian 232
    European 232
    Rufous Bush 227
    Rufous Scrub 227
    Rufous-tailed 627
    Rufous-tailed Scrub 227
    Siberian Blue 628
    'Tenerife' 234
    White-throated 235
Rubythroat, Siberian 248

S

Skylark 88
    Common 88
    Oriental 86
    Small 86
Stonechat, African 294
    'Armenian' 292
    Asian 287
    Canary Islands 282

INDEX OF ENGLISH NAMES

'Caspian' 291
Common 284
Eastern 287
European 284
Fuerteventura 282
Pied 295
Siberian 287, 293
'Stejneger's' 293
White-rumped 287
Swallow, American Cliff 129
 Bank 107
 Barn 118
 Cliff 129
 Ethiopian 624
 Indian Cliff 128
 Lesser Striped 624
 Red-rumped 125
 Streak-throated 128
 Tree 623
 Wire-tailed 123

T

Thrasher, Brown 627
Thrush, Black-throated 391
 Blue Rock 276
 Blue Whistling 629
 Common Rock 273
 Dusky 385
 Eyebrowed 378
 Gray-cheeked 368
 Grey-cheeked 368
 Hermit 366
 Little Rock 272
 Mistle 402
 Mountain Ground 362
 Naumann's 382
 Northern Scaly 362
 Red-throated 388
 Rock 273
 Siberian 364
 Song 397
 Swainson's 367
 Tickell's 630
 Varied 629
 White's 362
 Wood 630
 Yemen 370

V

Veery 369

W

Wagtail, African Pied 196
 'Ashy-headed' 175
 'Black-headed' 176
 'Blue-headed' 174
 'British' Yellow 175
 Citrine 180
 'Eastern' Yellow 179
 'Egyptian' Yellow 177
 Forest 169
 'Green-headed' 179
 Grey 185
 'Grey-headed' 174
 'Iberian' 175
 'Manchurian' 179
 'Masked' 194
 'Moroccan' 194
 'Mountain' Citrine 184
 Pied 188
 'Sykes's' 177
 Tree 169
 White 188
 'White-headed' 178
 Yellow 170
 'Yellow-headed' 178
Warbler, African Desert 533
 African Reed 456
 Aquatic 439
 Arabian 535
 Arctic 565
 Asian Desert 531
 Balearic 493
 Barred 546
 Basra Reed 465
 Blyth's Reed 445
 Booted 475
 Bright-green 567
 Brown Woodland 564
 Cape Verde 457
 Cape Verde Cane 457
 Cape Verde Swamp 457
 'Caspian' Reed 455
 Cetti's 405
 Clamorous Reed 459
 Common Grasshopper 427
 Common Reed 452
 Cricket 411
 Cyprus 524
 Dartford 495
 Dusky 583
 Eastern Bonelli's 588
 Eastern Crowned 631
 Eastern Olivaceous 471
 Eastern Orphean 542
 Eastern Subalpine 509
 Eurasian Reed 452
 European Reed 452
 Fan-tailed 408
 Garden 558
 Grasshopper 427
 Gray's 630
 Gray's Grasshopper 630
 Great Reed 463
 Green 567
 Green Leaf 567
 Greenish 569
 Hume's 578
 Hume's Leaf 578
 Icterine 485
 Isabelline 469
 Lanceolated 425
 Large-billed Leaf 632
 Large-billed Reed 446
 Levant Scrub 418
 'Mandelli's Leaf' 579
 Marmora's 490
 Marsh 448
 Melodious 487
 Ménétries's 516
 Moltoni's 513
 Moustached 436
 Olivaceous 471
 Olive-tree 483
 Oriental Reed 631
 Paddyfield 443
 Pale-legged Leaf 632
 Pallas's 574
 Pallas's Leaf 574
 Pallas's Grasshopper 423
 Plain Leaf 592
 Radde's 581
 Red Sea 535
 'Red Sea' Reed 456
 Reed 452
 River 430
 Rüppell's 528
 Rusty-rumped 423
 Saharan Scrub 421
 Sardinian 520
 Savi's 433
 Scaly 411
 Scaly-fronted 411
 Sedge 441
 Spectacled 501
 Sulphur-bellied 633
 Sykes's 477
 Thick-billed 467
 Tristram's 498
 Two-barred 572
 Two-barred Greenish 572
 Upcher's 480
 Western Bonelli's 585
 Western Olivaceous 469
 Western Orphean 539
 Western Subalpine 505
 Willow 608
 Wood 590
 Yellow-browed 576
 Yemen 537
Waxwing 202
 Bohemian 202
 Cedar 626
Wheatear, Arabian 350
 Basalt 348
 Black 358
 Black-eared 319
 Blyth's 339
 Botta's 300
 Buff-breasted 300
 Cyprus 315
 Cyprus Pied 315
 Desert 325
 'Eastern' Black-eared 324
 Eastern Pied 339
 Finsch's 328
 Gould's 342
 'Greenland' 305
 Hooded 353
 Hume's 355
 Isabelline 297
 Kurdish 334
 Maghreb 346
 Mourning 343
 Northern 302
 Persian 337
 Pied 309
 Red-breasted 300
 Red-rumped 332
 Red-tailed 337
 Seebohm's 306
 South Arabian 350
 Strickland's 342

    'Western' Black-eared 324
    White-bellied 339
    White-crowned 356
Whinchat 279
Whitethroat, 'Caucasian' Lesser 551
    'Chuia' Lesser 553
    Common 554
    'Desert' Lesser 553
    'Gansu' Lesser 553
    'Hume's' Lesser 552
    Lesser 549
    'Siberian' Lesser 552
    'Turkestan' Lesser 552
Woodlark 84
Wren 210
    Eurasian 210